Handbook of
Analytic-Computational Methods in Applied Mathematics

edited by

George Anastassiou

Department of Mathematical Sciences
The University of Memphis
Memphis, Tennessee

CHAPMAN & HALL/CRC

Boca Raton London New York Washington, D.C.

Library of Congress Cataloging-in-Publication Data

Handbook on analytic-computational methods in applied mathematics / edited by George Anastassiou.
p. cm.
Includes bibliographical references and index.
ISBN 1-58488-135-6 (alk. paper)
1. Mathematics—Handbooks, manuals, etc. I. Anastassiou, George A., 1952–.
QA40 .H358 2000
510—dc21 00-036089
CIP

Direct all inquiries to CRC Press LLC, 2000 N.W. Corporate Blvd., Boca Raton, Florida 33431.

International Standard Book Number 1-58488-135-6
Library of Congress Card Number 00-036089
Printed in the United States of America 1 2 3 4 5 6 7 8 9 0
Printed on acid-free paper

Dedicated to my lovely wife,
Koula

Contents

List of Contributors

George A. Anastassiou
Department of Mathematical Sciences
The University of Memphis
Memphis, TN 38152
E-Mail: anastasg@msci.memphis.edu; ganastss@memphis.edu

Ioannis K. Argyros
Department of Mathematics
Cameron University
Lawton, OK 73505
E-mail: ioannisa@cameron.edu

J. Marshall Ash
Mathematics Department
DePaul University
Chicago, IL 60614
E-mail: mash@math.depaul.edu

Kai Bittner
Institute of Biomathematics and Biometrics
GSF-National Research Center for Environment and Health
D-85764 Neuherberg, Germany
E-mail: bittner@gsf.de

Jesus de la Cal
Departamento de Matematica Aplicada
Estadistica e Investigacion Operativa
Facultad de Ciencias, Universidad del Pais Vasco
48080 Bilbao, Spain
E-mail: mepcaagj@lg.ehu.es

Michele Campiti
Department of Mathematics
Polytechnic of Bari
University Degli Studi di Bari
70125 Bari, Italy
E-mail: campiti@dm.uniba.it

Pietro Cerone
School of Communications and Informatics
Victoria University of Technology
Melbourne City, MC Victoria 8001, Australia
E-mail: pietro.cerone@vu.edu.au

Marco Dall'Aglio
Dipartimento di Scienze
Universita "G. d'Annunzio"
Viale Pindaro, 42, 65127-Pescara, Italy
E-mail: maglio@sci.unich.it

Sever S. Dragomir
School of Communications and Informatics
Victoria University of Technology
Melbourne City, MC Victoria 8001, Australia
E-mail: sever.dragomir@vu.edu.au

Frank Filbir
Institute of Biomathematics and Biometrics
GSF-National Research Center for Environment and Health
D-85764 Neuherberg, Germany
E-mail: filbir@gsf.de

S. Gal
Department of Mathematics
University of Oradea
Str. Armatei Romane 5
3700 Oradea, Romania
E-mail: galso@math.uoradea.ro

M. Ganzburg
Department of Mathematics
Hampton University
Hampton, VA 23668
E-mail: michael.ganzburg@hamptonu.edu

C. W. Groetsch
Department of Mathematical Sciences
University of Cincinnati
Cincinnati, OH 45221-0025
E-mail: groetsch@email.uc.edu

Tian-Xiao He
Department of Mathematics and Computer Science
Illinois Wesleyan University
Bloomington, IL 61702-2900
E-mail: the@sun.iwu.edu

Don Hong
Department of Mathematics
East Tennessee State University
Johnson City, TN 37614-0663
E-mail: hong@etsu.edu

Irina N. Khindanova
Department of Economics
University of California at Santa Barbara
Santa Barbara, CA 93106-3110
E-mail: khindan@econ.ucsb.edu

Rupert Lasser
Institute of Biomathematics and Biometrics
GSF-National Research Center for Environment and Health
D-85764 Neuherberg, Germany
E-mail: lasser@gsf.de

Josef Obermaier
Institute of Biomathematics and Biometrics
GSF-National Research Center for Environment and Health
D-85764 Neuherberg, Germany
E-mail: josef.obermaier@gsf.de

Stanley Osher
Department of Mathematics
University of California at Los Angeles
Los Angeles, CA 90095
E-mail: sjo@math.ucla.edu

Charles E. M. Pearce
Applied Mathematics Department
The University of Adelaide
Adelaide, SA 5005, Australia
E-mail: cpearce@maths.adelaide.edu.au

Josip Pecaric
Applied Mathematics Department
The University of Adelaide
Adelaide, SA 5005, Australia
E-mail: jpecaric@maths.adelaide.edu.au

Svetlozar T. Rachev
Institute of Statistics and Mathematical Economics
School of Economics
University of Karlsruhe, Kollegium and Schloss, Bau II
20.12, R210, Postfach 6980
D-76128, Karlsruhe, Germany
E-mail: rachev@lsoe.uni-karlsruhe.de

Jianhong Shen
Department of Mathematics
University of California at Los Angeles
Los Angeles, CA 90095
E-mail: jhshen@math.ucla.edu

Manfred Tasche
Medical University of Luebeck
Institute of Mathematics
Wallstr. 40, D-23560 Luebeck, Germany
E-mail: tasche@math.mu-luebeck.de

Yesim Tokat
Institute of Statistics and Mathematical Economics
School of Economics
University of Karlsruhe, Kollegium and Schloss, Bau II
20.12, R210, Postfach 6980, D-76128, Karlsruhe, Germany
E-mail: yesim@econ.ucsb.edu

Ana M. Valle
Departamento de Matematica Aplicada
Estadistica e Investigacion Operativa
Facultad de Ciencias
Universidad de Pais Vasco
Apartado 644, 48080 Bilbao, Spain
E-mail: mepvamaa@lg.ehu.es

Sanja Varosanec
Department of Mathematics
University of Zagreb
Bijenicka 30, 10000 Zagreb, Croatia
E-mail: varosans@math.hr

Hansmartin Zeuner
Medical University of Luebeck
Institute of Mathematics
Wallstr. 40, D-23560 Luebeck, Germany
E-mail: zeuner@math.mu-luebeck.de

Preface

Analytic-Computation Methods in Applied Mathematics are engaged in solving specific mathematical problems related to the real world within the framework of various sciences. Working in the above spirit has become one of the main trends in Mathematical Sciences the last thirty or so years and many scientists around the world use Analytic-Computational Methods in their everyday research.

This handbook contains long survey/expository articles, including original results, as chapters on the main core of contemporary Analytic-Computational Methods in Applied Mathematics and other related sciences. They are written by top experts of the field from important universities from around the world.

The important topics that are discussed in the volume include approximation related to order statistics, inverse problems, inequalities for trapezoidal and midpoint rules, numerical methods for solving equations in Banach spaces, computational harmonic analysis, fast trigonometric transform methods, wavelet methods, classical inequalities, optimality in approximation theory, approximation methods in random and fuzzy environments, global smoothness methods in approximation theory, orthogonal polynomials, digitized PDE methods, quadrature formulas applications, mathematical finance methods, optimization methods, connections of Bernstein approximations to real life sciences, and finally, applications of splines.

This handbook is appropriate for libraries in the mathematical sciences, engineering and economics, and is a good text for individual researchers and graduate students (classroom and research) of the related fields.

The editor would like to thank Mrs. Kate MacDougall of Warren, RI, for an excellent typing job, on time, preparing this volume.

The Editor
George A. Anastassiou
University of Memphis, Memphis, TN
February 1, 2000

Chapter 1

Degree of Approximation of Order Statistics Functionals, Dependent Case

George A. Anastassiou

Department of Mathematical Sciences, The University of Memphis, Memphis, TN 38152
E-mail: anastasg@msci.memphis.edu or ganastss@memphis.edu

Abstract Using approximation theory methods, we investigate the rate of convergence of expected L-estimates to a non-trivial associated integral limit in the d.i.d. case. We produce quantitative results in many different settings; their related inequalities are sharp and give the degree of approximation of the associated non-positive linear functionals to the above fixed limit. The involved weight functions carry minimal smoothness assumptions.

Contents

1.1 Introduction

Throughout this work we assume that $X_1, \ldots, X_n$, $(n \in \mathbb{N})$, are possibly dependent identically distributed random variables with common distribution function F. Also $X_{1:n}, \ldots, X_{n:n}$ denote the same random variables, ordered in increasing order and called *order statistics* (1st, ..., nth one, resp.). Their linear combinations are called L-estimates or L-statistics; $\sum_{i=1}^{n} c_i X_{i:n}$, $c_i \in \mathbb{R}$ are frequently used in statistical inference (see, e.g., Balakrishnan and Cohen (1991), [2], and David (1981), [3]).

When we analyze their asymptotics, the weights c_i are typically generated by a regular (smooth) function, say $D : [0,1] \to \mathbb{R}$ such that

$$c_i = \int_{(i-1)/n}^{i/n} D(u)\, du.$$

If D is nondecreasing, then

$$\sup_{\substack{\text{over all dependent } X_i \\ \text{with common marginal c.d.f. } F}} E\left(\sum_{i=1}^{n} c_i X_{i:n}\right)$$

$$= \sum_{i=1}^{n} \int_{(i-1)/n}^{i/n} n c_i F^{-1}(y)\, dy, \tag{1.1}$$

where $F^{-1}(y) := \sup\{t : F(t) \le y\}$, the quantile function, and E stands for the expectation. In general, if D is not nondecreasing, and $c_i \in \mathbb{R}$, $i = 1, \ldots, n$ we have the inequality

$$E\left(\sum_{i=1}^{n} c_i X_{i:n}\right) \le \int_0^1 F^{-1}(y) \sum_{i=1}^{n} d_i \delta_i(y)\, dy, \tag{1.2}$$

where

$$d_i := n\left(c\left(\frac{i}{n}\right) - c\left(\frac{i-1}{n}\right)\right), \quad i = 1, \ldots, n$$

with c being the greatest convex function on $[0,1]$ that satisfies

$$c\left(\frac{j}{n}\right) \le \sum_{i=1}^{j} c_i, \quad j = 0, 1, \ldots, n,$$

and δ_i denoting the indicator function on $\left[\left(\frac{i-1}{n}\right), \frac{i}{n}\right)$. Furthermore inequality (1.2) is sharp, i.e., it becomes equality for some joint distribution with given common marginals F. All the above were established by T. Rychlik (1993), see [5].

In particular the above-mentioned results were used for determining sharp bounds for values of L-estimates, and their expectations of dependent samples in terms of various location and dispersion parameters for the sample, general F, and F from restricted families of c.d.f.'s in the works of T. Rychlik (1993), [6], L. Gajek and T. Rychlik (1996), [4], and of T. Rychlik (1998), [7]. In this work the author investigates *the rate of convergence of expected L-estimates to the limit*

$$\int_0^1 F^{-1}(y) D(y)\, dy, \quad D : [0,1] \to \mathbb{R}$$

for many different cases. Here we are specific, namely, we have either

$$c_i = \int_{(i-1)/n}^{i/n} D(u)\, du,$$

or its approximation

$$c_i = \frac{1}{n} D\left(\frac{i}{n}\right), \quad i = 1, \ldots, n.$$

More precisely here the author studies the rate of convergence (as $n \to +\infty$) to $\int_0^1 F^{-1}(y) D(y)\, dy$, of the following sequences of order-statistics motivated/induced linear functionals (not necessarily positive)

$$\mathcal{R}_{1n}(D) := \int_0^1 F^{-1}(y) \left(\sum_{i=1}^{n} n d_i(D) \chi_{\left[\frac{i-1}{n}, \frac{i}{n}\right]}(y) \right) dy, \tag{1.3}$$

where

$$d_i(D) := \int_{(i-1)/n}^{i/n} D(u)\, du, \quad i = 1, \ldots, n \tag{1.4}$$

and χ is the indicator function on $\left[\frac{i-1}{n}, \frac{i}{n}\right]$, and

$$\mathcal{R}_{2n}(D) := \int_0^1 F^{-1}(y) \left(\sum_{i=1}^{n} D\left(\frac{i}{n}\right) \chi_{\left[\frac{i-1}{n}, \frac{i}{n}\right]}(y) \right) dy, \quad n \in \mathbb{N}. \tag{1.5}$$

D is typically called the weight function.

Linear functionals $\mathcal{R}_{1n}$ and $\mathcal{R}_{2n}$ were recommended for study to the author by T. Rychlik during his visit at the University of Memphis (1998–99). The linear functionals $\mathcal{R}_{1n}$, $\mathcal{R}_{2n}$ are acting on the spaces $C([0,1])$, $C^1([0,1])$, $\mathcal{L}_1(\mathbb{R})$, and $\mathcal{L}_1^1(\mathbb{R})$.

Notice that

$$E(X_1 D(F(X_1))) = \int_0^1 F^{-1}(y) D(y)\, dy. \tag{1.6}$$

We need throughout this work that

$$\int_0^1 |F^{-1}(y) D(y)|\, dy < +\infty. \tag{1.7}$$

Clearly we have that

$$\mathcal{R}_{1n}(D) = \sum_{i=1}^{n} n d_i(D) E\left(X_1 \chi_{[\frac{i-1}{n}, \frac{i}{n}]}(F(X_1))\right),$$

and

$$\mathcal{R}_{2n}(D) = \sum_{i=1}^{n} D\left(\frac{i}{n}\right) E\left(X_1 \chi_{[\frac{i-1}{n}, \frac{i}{n}]}(F(X_1))\right). \tag{1.8}$$

One also has

$$E(|X_1|) = \int_0^1 |F^{-1}(y)|\, dy. \tag{1.9}$$

Assuming that

$$E(|X_1|) < +\infty \tag{1.10}$$

we easily get

$$|\mathcal{R}_{1n}(D)|, |\mathcal{R}_{2n}(D)| < +\infty. \tag{1.11}$$

The presented results are quantitative and appear for the first time in this subject.

1.2 Main Results

We present our first related result

THEOREM 1.1
Let $D \in C([0,1])$, $n \in \mathbb{N}$,

$$d_i(D) := \int_{(i-1)/n}^{i/n} D(u)\, du, \quad i = 1, \ldots, n. \tag{1.4}$$

Consider the sequence of linear functionals on $C([0,1])$:

$$\mathcal{R}_{1n}(D) := \int_0^1 F^{-1}(y) \left(\sum_{i=1}^{n} n d_i(D) \chi_{[\frac{i-1}{n}, \frac{i}{n}]}(y) \right) dy, \tag{1.3}$$

where F stands for the common probability distribution function of the involved sequence of possibly dependent random variables $X_1, \ldots, X_n$, while $F^{-1}(y) := \sup\{t : F(t) \leq y\}$ is the quantile function, and $\chi_{[\frac{i-1}{n}, \frac{i}{n}]}$ stands for the indicator function on $\left[\frac{i-1}{n}, \frac{i}{n}\right]$.

Assume that

$$\int_0^1 |F^{-1}(y)| dy < +\infty \tag{1.12}$$

(which is the same as (1.10)). Then

$$\left| \mathcal{R}_{1n}(D) - \int_0^1 F^{-1}(y) D(y)\, dy \right| \leq \left(\int_0^1 |F^{-1}(y)|\, dy \right) \omega_1 \left(D, \frac{1}{n} \right), \tag{1.13}$$

where ω_1 stands for the first modulus of continuity of D, i.e.,

$$\omega_1(D, \delta) := \sup_{\substack{x,y \in [0,1]: \\ |x-y| \leq \delta}} |D(x) - D(y)|, \quad 0 < \delta \leq 1.$$

Inequality (1.13) is attained when D is a constant function.

PROOF We notice that

$$\left| \mathcal{R}_{1n}(D) - \int_0^1 F^{-1}(y) D(y)\, dy \right|$$

$$= \left| \sum_{i=1}^{n} \int_{(i-1)/n}^{i/n} F^{-1}(y) n d_i(D)\, dy - \sum_{i=1}^{n} \int_{(i-1)/n}^{i/n} F^{-1}(y) D(y)\, dy \right|$$

$$\leq \sum_{i=1}^{n} \int_{(i-1)/n}^{i/n} |F^{-1}(y)|\, |nd_i(D) - D(y)|\, dy$$

$$= \int_0^1 |F^{-1}(y)| \left(\sum_{i=1}^{n} |nd_i(D) - D(y)| \chi_{[\frac{i-1}{n}, \frac{i}{n}]}(y) \right) dy$$

$$= (*).$$

For any $u, y \in \left[\frac{i-1}{n}, \frac{i}{n}\right]$ we have $|u - y| \leq \frac{1}{n}$.

And

$$|nd_i(D) - D(y)| = \left| \int_{(i-1)/n}^{i/n} nD(u)\, du - \int_{(i-1)/n}^{i/n} nD(y)\, du \right|$$

$$\leq n \int_{(i-1)/n}^{i/n} |D(u) - D(y)|\, du \leq n \int_{(i-1)/n}^{i/n} \omega_1(D, |u - y|)\, du$$

$$\leq n \int_{(i-1)/n}^{i/n} \omega_1 \left(D, \frac{1}{n} \right) du = \omega_1 \left(D, \frac{1}{n} \right);$$

i.e.

$$|nd_i(D) - D(y)| \leq \omega_1 \left(D, \frac{1}{n} \right), \quad \forall y \in \left[\frac{(i-1)}{n}, \frac{i}{n} \right].$$

That is,

$$|nd_i(D) - D(y)| \chi_{[\frac{(i-1)}{n}, \frac{i}{n}]}(y) \leq \omega_1 \left(D, \frac{1}{n} \right) \chi_{[\frac{(i-1)}{n}, \frac{i}{n}]}(y), \quad \forall y \in [0, 1].$$

Hence

$$\sum_{i=1}^{n} |nd_i(D) - D(y)| \chi_{[\frac{(i-1)}{n}, \frac{i}{n}]}(y) \leq \omega_1 \left(D, \frac{1}{n} \right) \left(\sum_{i=1}^{n} \chi_{[\frac{(i-1)}{n}, \frac{i}{n}]}(y) \right)$$

$$= \omega_1 \left(D, \frac{1}{n} \right) \cdot 1 = \omega_1 \left(D, \frac{1}{n} \right), \quad \forall y \in [0, 1].$$

Therefore

$$(*) \leq \int_0^1 |F^{-1}(y)|\omega_1\left(D, \frac{1}{n}\right) dy = \left(\int_0^1 |F^{-1}(y)|\, dy\right) \omega_1\left(D, \frac{1}{n}\right).$$

So that inequality (1.13) has been established. ∎

COROLLARY 1.1

All terms and assumptions are as in Theorem 1.1. Additionally, assume that

$$|D(x) - D(y)| \leq K|x-y|^{\alpha}, \quad 0 < \alpha \leq 1,\ x, y \in [0,1],\ K > 0. \tag{1.14}$$

Then

$$\left|\mathcal{R}_{1n}(D) - \int_0^1 F^{-1}(y)D(y)\, dy\right| \leq \left(\int_0^1 |F^{-1}(y)|\, dy\right) \frac{K}{n^{\alpha}}, \quad n \in \mathbb{N}. \tag{1.15}$$

PROOF By inequality (1.13) and noticing that (1.14) is equivalent to $\omega_1(D, \delta) \leq K\delta^{\alpha}$, $0 < \delta \leq 1$. ∎

The counterpart of Theorem 1.1 follows.

THEOREM 1.2

Let $D \in C([0,1])$, $n \in \mathbb{N}$. Consider the linear functional on $C([0,1])$:

$$\mathcal{R}_{2n}(D) := \int_0^1 F^{-1}(y) \left(\sum_{i=1}^{n} D\left(\frac{i}{n}\right) \chi_{[\frac{(i-1)}{n}, \frac{i}{n}]}(y)\right) dy, \tag{1.16}$$

where $F, \chi_{[\frac{(i-1)}{n}, \frac{i}{n}]}$ as in Theorem 1.1. Then

$$\left|\mathcal{R}_{2n}(D) - \int_0^1 F^{-1}(y)D(y)\, dy\right| \leq \left(\int_0^1 |F^{-1}(y)|\, dy\right) \omega_1\left(D, \frac{1}{n}\right). \tag{1.17}$$

Inequality (1.17) is attained when D is a constant function.

PROOF Let $y \in \left[\frac{(i-1)}{n}, \frac{i}{n}\right]$, then

$$\left|D\left(\frac{i}{n}\right) - D(y)\right| \leq \omega_1\left(D, \left|\frac{i}{n} - y\right|\right) \leq \omega_1\left(D, \frac{1}{n}\right).$$

Hence

$$\left|D\left(\frac{i}{n}\right)-D(y)\right|\chi_{[\frac{(i-1)}{n},\frac{i}{n}]}(y)\le \omega_1\left(D,\frac{1}{n}\right)\chi_{[\frac{(i-1)}{n},\frac{i}{n}]}(y),\quad \forall y\in[0,1].$$

Therefore

$$\sum_{i=1}^{n}\left|D\left(\frac{i}{n}\right)-D(y)\right|\chi_{[\frac{(i-1)}{n},\frac{i}{n}]}(y)$$

$$\le \sum_{i=1}^{n}\omega_1\left(D,\frac{1}{n}\right)\chi_{[\frac{(i-1)}{n},\frac{i}{n}]}(y)=\omega_1\left(D,\frac{1}{n}\right),\quad \forall y\in[0,1].$$

So that

$$\left|\mathcal{R}_{2n}(D)-\int_0^1 F^{-1}(y)D(y)\,dy\right|$$

$$=\left|\sum_{i=1}^{n}\left(\int_{(i-1)/n}^{i/n}F^{-1}(y)D\left(\frac{i}{n}\right)dy-\int_{(i-1)/n}^{i/n}F^{-1}(y)D(y)\,dy\right)\right|$$

$$\le \sum_{i=1}^{n}\left(\int_{(i-1)/n}^{i/n}|F^{-1}(y)|\left|D\left(\frac{i}{n}\right)-D(y)\right|dy\right)$$

$$=\int_0^1|F^{-1}(y)|\left(\sum_{i=1}^{n}\left|D\left(\frac{i}{n}\right)-D(y)\right|\chi_{[\frac{i-1}{n},\frac{i}{n}]}(y)\right)dy$$

$$\le\left(\int_0^1|F^{-1}(y)|\,dy\right)\omega_1\left(D,\frac{1}{n}\right).$$

That establishes inequality (1.17). ∎

COROLLARY 1.2

All terms and assumptions are as in Theorem 1.2. Additionally, assume that

$$|D(x)-D(y)|\le K|x-y|^{\alpha},\quad 0<\alpha\le 1,\ x,y\in[0,1],\ K>0. \tag{1.14}$$

Then

$$\left|\mathcal{R}_{2n}(D)-\int_0^1 F^{-1}(y)D(y)\,dy\right| \le \left(\int_0^1 |F^{-1}(y)|\,dy\right)\frac{K}{n^{\alpha}}, \quad n\in\mathbb{N}. \tag{1.18}$$

PROOF From inequality (1.17) and seeing that (1.14) is equivalent to

$$\omega_1(D,\delta)\le K\delta^{\alpha}, \quad 0<\delta\le 1.$$

∎

Next we prove related theorems when D is continuously differentiable.

THEOREM 1.3
Let $D\in C^1([0,1])$, $n\in\mathbb{N}$. Let $\mathcal{R}_{1n}$ be acting on $C^1([0,1])$. The rest of the terms and assumptions are as in Theorem 1.1. Then

$$\left|\mathcal{R}_{1n}(D)-\int_0^1 F^{-1}(y)D(y)\,dy\right|$$

$$\le \left(\int_0^1 |F^{-1}(y)|\left\{\sum_{i=1}^{n}\left(n\int_{(i-1)/n}^{i/n}|D'(u)|\,|y-u|\,du\right)\cdot \chi_{[\frac{(i-1)}{n},\frac{i}{n}]}(y)\right\}dy\right)$$

$$+\frac{1}{n}\left(\int_0^1 |F^{-1}(y)|\,dy\right)\omega_1\left(D',\frac{1}{n}\right). \tag{1.19}$$

Inequality (1.19) is attained when D is a constant function.

PROOF Let $y,u\in\left[\frac{(i-1)}{n},\frac{i}{n}\right]$, then

$$D(y)-D(u)=D'(u)(y-u)+\int_u^y (D'(t)-D'(u))\,dt.$$

That is

$$|D(y)-D(u)|\le |D'(u)|\,|y-u|+\left|\int_u^y |D'(t)-D'(u)|\,dt\right|.$$

Let $t, u \in \left[\frac{(i-1)}{n}, \frac{i}{n}\right]$, then

$$|D'(t) - D'(u)| \leq \omega_1\left(D', \frac{1}{n}\right),$$

and

$$\left|\int_u^y |D'(t) - D'(u)|\, dt\right| \leq \frac{1}{n}\omega_1\left(D' \frac{1}{n}\right).$$

Therefore

$$|D(y) - D(u)| \leq |D'(u)|\,|y-u| + \frac{1}{n}\omega_1\left(D', \frac{1}{n}\right), \forall y, u \in \left[\frac{(i-1)}{n}, \frac{i}{n}\right].$$

Furthermore we see that

$$|nd_i(D) - D(y)| = \left|\int_{(i-1)/n}^{i/n} nD(u)\, du - \int_{(i-1)/n}^{i/n} nD(y)\, du\right|$$

$$\leq n\int_{(i-1)/n}^{i/n} |D(u) - D(y)|\, du$$

$$\leq n\int_{(i-1)/n}^{i/n} |D'(u)|\,|y-u|\, du + \frac{1}{n}\omega_1\left(D', \frac{1}{n}\right).$$

Finally we obtain

$$|nd_i(D) - D(y)|\chi_{[\frac{(i-1)}{n}, \frac{i}{n}]}(y)$$

$$\leq \left(n\int_{(i-1)/n}^{i/n} |D'(u)|\,|y-u|\, du\right)\chi_{[\frac{(i-1)}{n}, \frac{i}{n}]}(y)$$

$$+ \frac{1}{n}\omega_1\left(D', \frac{1}{n}\right)\chi_{[\frac{i-1}{n}, \frac{i}{n}]}(y), \quad \forall y \in [0,1].$$

And

$$\sum_{i=1}^{n} |nd_i(D) - D(y)|\chi_{[\frac{(i-1)}{n}, \frac{i}{n}]}(y)$$

$$\leq \sum_{i=1}^{n} \left(n \int_{(i-1)/n}^{i/n} |D'(u)|\,|y-u|\,du \right) \chi_{[\frac{(i-1)}{n},\frac{i}{n}]}(y)$$

$$+\frac{1}{n}\omega_1\left(D',\frac{1}{n}\right). \tag{1.20}$$

Now as in the proof of Theorem 1.1 we have

$$\left| \mathcal{R}_{1n}(D) - \int_0^1 F^{-1}(y)D(y)\,dy \right| \leq (*) \tag{1.21}$$

there. Multiplying (1.20) by $|F^{-1}(y)|$, integrating against y over $[0,1]$, and using (1.21) we obtain inequality (1.19). ∎

COROLLARY 1.3

All terms and assumptions are as in Theorem 1.3. Additionally, assume that

$$|D'(x) - D'(y)| \leq K|x-y|^{\alpha}, \quad 0 < \alpha \leq 1, \; x, y \in [0,1], \; K > 0.$$

Then

$$\left| \mathcal{R}_{1n}(D) - \int_0^1 F^{-1}(y)D(y)\,dy \right|$$

$$\leq \left(\int_0^1 |F^{-1}(y)| \left\{ \sum_{i=1}^{n} \left(n \int_{(i-1)/n}^{i/n} |D'(u)|\,|y-u|\,du \right) \cdot \chi_{[\frac{(i-1)}{n},\frac{i}{n}]}(y) \right\} dy \right)$$

$$+ \left(\int_0^1 |F^{-1}(y)|\,dy \right) \frac{K}{n^{1+\alpha}}, \quad n \in \mathbb{N}. \tag{1.22}$$

The counterpart of Theorem 1.3 follows.

THEOREM 1.4

Let $D \in C^1([0,1])$, $n \in \mathbb{N}$. Consider the linear functional $\mathcal{R}_{2n}$ acting on $C^1([0,1])$, where $F, \chi_{[\frac{(i-1)}{n},\frac{i}{n}]}$ are as in Theorem 1.1. Then

$$\left| \mathcal{R}_{2n}(D) - \int_0^1 F^{-1}(y)D(y)\,dy \right|$$

$$\leq \int_0^1 |F^{-1}(y)| \left(\sum_{i=1}^n \left| D'\left(\frac{i}{n}\right) \right| \left(\frac{i}{n} - y\right) \chi_{[\frac{(i-1)}{n}, \frac{i}{n}]}(y) \right) dy$$

$$+ \frac{1}{n} \left(\int_0^1 |F^{-1}(y)| \, dy \right) \omega_1 \left(D', \frac{1}{n} \right). \tag{1.23}$$

Inequality (1.23) is attained when D is a constant function.

PROOF Let $y \in \left[\frac{(i-1)}{n}, \frac{i}{n}\right]$, then

$$D(y) - D\left(\frac{i}{n}\right) = D'\left(\frac{i}{n}\right)\left(y - \frac{i}{n}\right) + \int_{i/n}^{y} \left(D'(t) - D'\left(\frac{i}{n}\right) \right) dt.$$

Hence

$$\left| D(y) - D\left(\frac{i}{n}\right) \right| \leq \left| D'\left(\frac{i}{n}\right) \right| \left| y - \frac{i}{n} \right| + \left| \int_{i/n}^{y} \left| D'(t) - D'\left(\frac{i}{n}\right) \right| dt \right|. \tag{1.24}$$

Since

$$\left| D'(t) - D'\left(\frac{i}{n}\right) \right| \leq \omega_1 \left(D', \frac{1}{n} \right),$$

we get

$$\left| \int_{i/n}^{y} \left| D'(t) - D'\left(\frac{i}{n}\right) \right| dt \right| \leq \frac{1}{n} \omega_1 \left(D', \frac{1}{n} \right). \tag{1.25}$$

From (1.24) and (1.25) we obtain

$$\left| D(y) - D\left(\frac{i}{n}\right) \right| \chi_{[\frac{(i-1)}{n}, \frac{i}{n}]}(y) \leq \left| D'\left(\frac{i}{n}\right) \right| \left| y - \frac{i}{n} \right| \chi_{[\frac{(i-1)}{n}, \frac{i}{n}]}(y)$$

$$+ \frac{1}{n} \omega_1 \left(D', \frac{1}{n} \right) \chi_{[\frac{(i-1)}{n}, \frac{i}{n}]}(y), \quad \forall y \in [0, 1]. \tag{1.26}$$

And

$$\sum_{i=1}^n \left| D(y) - D\left(\frac{i}{n}\right) \right| \chi_{[\frac{(i-1)}{n}, \frac{i}{n}]}(y)$$

$$\leq \sum_{i=1}^n \left| D'\left(\frac{i}{n}\right) \right| \left| y - \frac{i}{n} \right| \chi_{[\frac{(i-1)}{n}, \frac{i}{n}]}(y) + \frac{1}{n} \omega_1 \left(D', \frac{1}{n} \right). \tag{1.27}$$

Next we see that

$$\left|\mathcal{R}_{2n}(D) - \int_0^1 F^{-1}(y)D(y)\,dy\right|$$

$$= \left|\sum_{i=1}^{n}\left(\int_{(i-1)/n}^{i/n} F^{-1}(y)D\left(\frac{i}{n}\right)dy - \int_{(i-1)/n}^{i/n} F^{-1}(y)D(y)\,dy\right)\right|$$

$$\leq \sum_{i=1}^{n}\int_{(i-1)/n}^{i/n} |F^{-1}(y)|\left|D\left(\frac{i}{n}\right) - D(y)\right|dy$$

$$= \int_0^1 |F^{-1}(y)|\left(\sum_{i=1}^{n}\left|D\left(\frac{i}{n}\right) - D(y)\right|\chi_{[\frac{(i-1)}{n},\frac{i}{n}]}(y)\right)dy$$

$$\overset{(1.27)}{\leq} \int_0^1 |F^{-1}(y)|\left(\sum_{i=1}^{n}\left|D'\left(\frac{i}{n}\right)\right|\left(\frac{i}{n} - y\right)\chi_{[\frac{(i-1)}{n},\frac{i}{n}]}(y)\right)dy$$

$$+ \frac{1}{n}\left(\int_0^1 |F^{-1}(y)|\,dy\right)\omega_1\left(D', \frac{1}{n}\right).$$

This proves (1.23). ∎

COROLLARY 1.4

All terms and assumptions are as in Theorem 1.4. Additionally, assume that

$$|D'(x) - D'(y)| \leq K|x - y|^{\alpha}, \quad 0 < \alpha \leq 1, \ x, y \in [0,1], \ K > 0.$$

Then

$$\left|\mathcal{R}_{2n}(D) - \int_0^1 F^{-1}(y)D(y)\,dy\right|$$

$$\leq \int_0^1 |F^{-1}(y)|\left(\sum_{i=1}^{n}\left|D'\left(\frac{i}{n}\right)\right|\left(\frac{i}{n} - y\right)\chi_{[\frac{(i-1)}{n},\frac{i}{n}]}(y)\right)dy$$

$$+ \left(\int_0^1 |F^{-1}(y)|\,dy\right)\frac{K}{n^{1+\alpha}}, \quad n \in \mathbb{N}. \tag{1.28}$$

Next follows related $\mathcal{L}_1$ results.

THEOREM 1.5

Let $D \in \mathcal{L}_1(\mathbb{R})$, $n \in \mathbb{N}$. Assume that the common probability distribution function F has finite support. Then

$$\left| \mathcal{R}_{1n}(D) - \int_0^1 F^{-1}(y)D(y)\,dy \right|$$

$$\leq 2\|F^{-1}\|_{ess.sup\,[0,1]} \cdot \omega_1\left(D, \frac{1}{n}\right)_{1,\mathbb{R}}. \tag{1.29}$$

Here

$$\omega_1(f,\ell)_{1,\mathbb{R}} := \sup_{0\leq h\leq \ell} \|f(t+h) - f(t)\|_{\mathcal{L}_1(\mathbb{R})}$$

$$= \sup_{0\leq h\leq \ell} \left(\int_{\mathbb{R}} |f(t+h) - f(t)|\,dt \right). \tag{1.30}$$

PROOF First we treat

$$\Gamma := \sum_{i=1}^n \left\| (nd_i(D) - D(y))\chi_{[\frac{(i-1)}{n},\frac{i}{n}]}(y) \right\|_{1,[0,1]}$$

$$= \sum_{i=1}^n \left(\int_0^1 |nd_i(D) - D(y)|\,\chi_{[\frac{(i-1)}{n},\frac{i}{n}]}(y)\,dy \right)$$

$$= \sum_{i=1}^n \left(\int_{(i-1)/n}^{i/n} |nd_i(D) - D(y)|\,dy \right);$$

i.e.

$$\Gamma = \sum_{i=1}^n \|nd_i(D) - D(y)\|_{1,[\frac{(i-1)}{n},\frac{i}{n}]}. \tag{1.31}$$

Let $u, y \in \left[\frac{(i-1)}{n}, \frac{i}{n}\right]$, then $|u-y| \leq \frac{1}{n}$.

Consider

$$\|nd_i(D) - D(y)\|_{1,[\frac{(i-1)}{n},\frac{i}{n}]}$$

$$= \left\| \int_{(i-1)/n}^{i/n} nD(u)\,du - \int_{(i-1)/n}^{i/n} nD(y)\,du \right\|_{1,[\frac{(i-1)}{n},\frac{i}{n}]}$$

$$= n \left\| \int_{(i-1)/n}^{i/n} (D(u) - D(y))\,du \right\|_{1,[\frac{(i-1)}{n},\frac{i}{n}]}$$

$$\leq n \left\| \int_{(i-1)/n}^{i/n} |D(u) - D(y)|\,du \right\|_{1,[\frac{(i-1)}{n},\frac{i}{n}]}$$

$$= n \left\| \int_{(i-1)/n}^{y} |D(u) - D(y)|\,du + \int_{y}^{i/n} |D(u) - D(y)|\,du \right\|_{1,[\frac{(i-1)}{n},\frac{i}{n}]}$$

$$\leq n \left\{ \left\| \int_{(i-1)/n}^{y} |D(u) - D(y)|\,du \right\|_{1,[\frac{(i-1)}{n},\frac{i}{n}]} + \left\| \int_{y}^{i/n} |D(u) - D(y)|\,du \right\|_{1,[\frac{(i-1)}{n},\frac{i}{n}]} \right\}$$

$$\leq n \left\{ \left\| \int_{y-\frac{1}{n}}^{y} |D(u) - D(y)|\,du \right\|_{1,[\frac{(i-1)}{n},\frac{i}{n}]} + \left\| \int_{y}^{y+\frac{1}{n}} |D(u) - D(y)|\,du \right\|_{1,[\frac{(i-1)}{n},\frac{i}{n}]} \right\}$$

$$= n \left\{ \left\| \int_{0}^{1/n} |D(y) - D(y-u)|\,du \right\|_{1,[\frac{(i-1)}{n},\frac{i}{n}]} + \left\| \int_{0}^{1/n} |D(u+y) - D(y)|\,du \right\|_{1,[\frac{(i-1)}{n},\frac{i}{n}]} \right\} \text{ (by Fubini's theorem)}$$

$$= n \left\{ \int_{0}^{1/n} \|D(y) - D(y-u)\|_{1,[\frac{(i-1)}{n},\frac{i}{n}]} \cdot du \right.$$

$$+\int_0^{1/n} \|D(u+y)-D(y)\|_{1,[\frac{(i-1)}{n},\frac{i}{n}]}\cdot du\Big\};$$

i.e.

$$\|nd_i(D)-D(y)\|_{1,[\frac{(i-1)}{n},\frac{i}{n}]} \le n\Big\{\int_0^{1/n} \|D(y)-D(y-u)\|_{1,[\frac{(i-1)}{n},\frac{i}{n}]}\cdot du$$

$$+\int_0^{1/n} \|D(u+y)-D(y)\|_{1,[\frac{(i-1)}{n},\frac{i}{n}]}\cdot du\Big\}. \tag{1.32}$$

Therefore by (1.31) and (1.32) we get

$$\Gamma \le n\sum_{i=1}^{n}\Big\{\int_0^{1/n} \|D(y)-D(y-u)\|_{1,[\frac{(i-1)}{n},\frac{i}{n}]}du$$

$$+\int_0^{1/n} \|D(u+y)-D(y)\|_{1,[\frac{(i-1)}{n},\frac{i}{n}]}du\Big\}$$

$$= n\Big\{\int_0^{1/n}\left(\sum_{i=1}^{n}\|D(y)-D(y-u)\|_{1,[\frac{(i-1)}{n},\frac{i}{n}]}\right)du$$

$$+\int_0^{1/n}\left(\sum_{i=1}^{n}\|D(u+y)-D(y)\|_{1,[\frac{(i-1)}{n},\frac{i}{n}]}\right)du\Big\}$$

$$= n\Big\{\int_0^{1/n} \|D(y)-D(y-u)\|_{1,[0,1]}du$$

$$+\int_0^{1/n} \|D(u+y)-D(y)\|_{1,[0,1]}\,du\Big\}$$

$$\le n\left\{\int_0^{1/n} \|D(y)-D(y-u)\|_{1,\mathbb{R}}\,du+\int_0^{1/n} \|D(u+y)-D(y)\|_{1,\mathbb{R}}\,du\right\}$$

$$= 2n\left(\int_0^{1/n} \|D(u+y)-D(y)\|_{1,\mathbb{R}}\,du\right)$$

$$\leq 2n \int_0^{1/n} \omega_1(D,u)_{1,\mathbb{R}}\, du \leq 2\omega_1\left(D,\frac{1}{n}\right)_{1,\mathbb{R}}.$$

That is

$$\Gamma \leq 2\omega_1\left(D,\frac{1}{n}\right)_{1,\mathbb{R}} \tag{1.33}$$

Here we mention again

$$\mathcal{R}_{1n}(D) := \int_0^1 F^{-1}(y)\left(\sum_{i=1}^n n d_i(D)\chi_{[\frac{i-1}{n},\frac{i}{n}]}(y)\right) dy. \tag{1.3}$$

Thus

$$\left|\mathcal{R}_{1n}(D) - \int_0^1 F^{-1}(y)D(y)\,dy\right|$$

$$= \left|\sum_{i=1}^n \int_{(i-1)/n}^{i/n} F^{-1}(y) n d_i(D)\,dy - \sum_{i=1}^n \int_{(i-1)/n}^{i/n} F^{-1}(y)D(y)\,dy\right|$$

$$\leq \sum_{i=1}^n \int_{(i-1)/n}^{i/n} |F^{-1}(y)|\,|n d_i(D) - D(y)|\,dy$$

$$= \int_0^1 |F^{-1}(y)| \left(\sum_{i=1}^n |n d_i(D) - D(y)|\chi_{[\frac{(i-1)}{n},\frac{i}{n}]}(y)\right) dy$$

$$\leq \|F^{-1}\|_{ess.sup\,[0,1]} \left\|\sum_{i=1}^n |n d_i(D) - D(y)|\chi_{[\frac{(i-1)}{n},\frac{i}{n}]}(y)\right\|_{1,[0,1]}$$

$$\leq \|F^{-1}\|_{ess.sup\,[0,1]} \cdot \left\{\sum_{i=1}^n \left\|(n d_i(D) - D(y))\chi_{[\frac{(i-1)}{n},\frac{i}{n}]}(y)\right\|_{1,[0,1]}\right\}$$

$$= \|F^{-1}\|_{ess.sup\,[0,1]}\Gamma \overset{(1.33)}{\leq} \|F^{-1}\|_{ess.sup\,[0,1]} \cdot 2\omega_1\left(D,\frac{1}{n}\right)_{1,\mathbb{R}}$$

This proves (1.29). ∎

COROLLARY 1.5

Let $D \in \mathcal{L}_1(\mathbb{R})$ of compact support, such that

$$\omega_1(D,\delta)_{1,\mathbb{R}} = O(\delta^\alpha), \quad \forall \delta > 0,\ 0 < \alpha \le 1. \tag{1.34}$$

All the rest is as in Theorem 1.5. Then

$$\left|\mathcal{R}_{1n}(D) - \int_0^1 F^{-1}(y)D(y)\,dy\right| = O\left(\frac{1}{n^\alpha}\right), \quad n \in \mathbb{N}. \tag{1.35}$$

Next we treat the differentiable analog of Theorem 1.5.

THEOREM 1.6

Let $\mathcal{L}_1^1(\mathbb{R}) := \{D : D \in AC(\mathbb{R})\ \&\ D' \in \mathcal{L}_1(\mathbb{R})\}$. Let $D \in \mathcal{L}_1^1(\mathbb{R})$, such that D' is nondecreasing over $[0,1]$, $n \in \mathbb{N}$. Assume that the common probability distribution function F has finite support. Then

$$\left|\mathcal{R}_{1n}(D) - \int_0^1 F^{-1}(y)D(y)\,dy\right| \le \|F^{-1}\|_{ess.sup\,[0,1]}\left\{\frac{1}{n}\int_0^1 |D'(u)|\,du + \frac{2}{n}\omega_1\left(D', \frac{1}{n}\right)_{1,\mathbb{R}}\right\}. \tag{1.36}$$

Here $\omega_1\left(D', \frac{1}{n}\right)_{1,\mathbb{R}}$ is defined as in (1.30). Inequality (1.36) is attained when D is a constant function.

PROOF Let $y, u \in \left[\frac{(i-1)}{n}, \frac{i}{n}\right]$, then

$$D(y) - D(u) = D'(u)(y-u) + \int_u^y (D'(t) - D'(u))\,dt,$$

and

$$\begin{aligned} |D(y) - D(u)| &\le |D'(u)|\,|y-u| + \left|\int_u^y |D'(t) - D'(u)|\,dt\right| \\ &\le |D'(u)|\,|y-u| + \frac{1}{n}|D'(y) - D'(u)|. \end{aligned} \tag{1.37}$$

Next we treat

$$\begin{aligned}\Gamma &:= \sum_{i=1}^{n} \left\| (nd_i(D) - D(y))\chi_{[\frac{(i-1)}{n},\frac{i}{n}]}(y) \right\|_{1,[0,1]} \\ &= \sum_{i=1}^{n} \left(\int_0^1 |nd_i(D) - D(y)| \chi_{[\frac{(i-1)}{n},\frac{i}{n}]}(y)\, dy \right) \\ &= \sum_{i=1}^{n} \left(\int_{(i-1)/n}^{i/n} |nd_i(D) - D(y)|\, dy \right);\end{aligned}$$

i.e.

$$\Gamma = \sum_{i=1}^{n} \|nd_i(D) - D(y)\|_{1,[\frac{(i-1)}{n},\frac{i}{n}]}. \tag{1.38}$$

Consider

$$\begin{aligned}\gamma_i &:= \|nd_i(D) - D(y)\|_{1,[\frac{(i-1)}{n},\frac{i}{n}]} \\ &= \left\| \int_{(i-1)/n}^{i/n} nD(u)\, du - \int_{(i-1)/n}^{i/n} nD(y)\, du \right\|_{1,[\frac{(i-1)}{n},\frac{i}{n}]} \\ &= n \left\| \int_{(i-1)/n}^{i/n} (D(u) - D(y))\, du \right\|_{1,[\frac{(i-1)}{n},\frac{i}{n}]} \\ &\le n \left\| \int_{(i-1)/n}^{i/n} |D(u) - D(y)|\, du \right\|_{1,[\frac{(i-1)}{n},\frac{i}{n}]} \\ &\overset{(1.37)}{\le} n \Bigg\{ \left\| \int_{(i-1)/n}^{i/n} |D'(u)|\, |y-u|\, du \right\|_{1,[\frac{(i-1)}{n},\frac{i}{n}]} \\ &\quad + \frac{1}{n} \left\| \int_{(i-1)/n}^{i/n} |D'(y) - D'(u)|\, du \right\|_{1,[\frac{(i-1)}{n},\frac{i}{n}]} \Bigg\};\end{aligned}$$

i.e.

$$\gamma_i \leq n \left\| \int_{(i-1)/n}^{i/n} |D'(u)|\,|u-y|\,du \right\|_{1,[\frac{(i-1)}{n},\frac{i}{n}]} + \left\| \int_{(i-1)/n}^{i/n} |D'(u)-D'(y)|\,du \right\|_{1,[\frac{(i-1)}{n},\frac{i}{n}]} \tag{1.39}$$

We observe that

$$\left\| \int_{(i-1)/n}^{i/n} |D'(u)-D'(y)|\,du \right\|_{1,[\frac{(i-1)}{n},\frac{i}{n}]}$$

$$= \left\| \int_{(i-1)/n}^{y} |D'(u)-D'(y)|\,du + \int_{y}^{i/n} |D'(u)-D'(y)|\,du \right\|_{1,[\frac{(i-1)}{n},\frac{i}{n}]}$$

$$\leq \left\| \int_{(i-1)/n}^{y} |D'(u)-D'(y)|\,du \right\|_{1,[\frac{(i-1)}{n},\frac{i}{n}]} + \left\| \int_{y}^{i/n} |D'(u)-D'(y)|\,du \right\|_{1,[\frac{(i-1)}{n},\frac{i}{n}]}$$

$$\leq \left\| \int_{y-\frac{1}{n}}^{y} |D'(u)-D'(y)|\,du \right\|_{1,[\frac{(i-1)}{n},\frac{i}{n}]} + \left\| \int_{y}^{y+\frac{1}{n}} |D'(u)-D'(y)|\,du \right\|_{1,[\frac{(i-1)}{n},\frac{i}{n}]}$$

$$= \left\| \int_{0}^{1/n} |D'(y)-D'(y-u)|\,du \right\|_{1,[\frac{(i-1)}{n},\frac{i}{n}]} + \left\| \int_{0}^{1/n} |D'(u+y)-D'(y)|\,du \right\|_{1,[\frac{(i-1)}{n},\frac{i}{n}]}$$

$$= \int_0^{1/n} \|D'(y) - D'(y-u)\|_{1,[\frac{(i-1)}{n},\frac{i}{n}]} \cdot du$$

$$+ \int_0^{1/n} \|D'(u+y) - D'(y)\|_{1,[\frac{(i-1)}{n},\frac{i}{n}]} \cdot du. \tag{1.40}$$

Therefore from (1.39) and (1.40) we get

$$\gamma_i \leq n \left\| \int_{(i-1)/n}^{i/n} |D'(u)|\,|u-y|\,du \right\|_{1,[\frac{(i-1)}{n},\frac{i}{n}]}$$

$$+ \int_0^{1/n} \|D'(y) - D'(y-u)\|_{1,[\frac{(i-1)}{n},\frac{i}{n}]} du$$

$$+ \int_0^{1/n} \|D'(u+y) - D'(y)\|_{1,[\frac{(i-1)}{n},\frac{i}{n}]} du.$$

So that from (1.38) and the last inequality we obtain

$$\Gamma \leq n \sum_{i=1}^{n} \left\| \int_{(i-1)/n}^{i/n} |D'(u)|\,|u-y|\,du \right\|_{1,[\frac{(i-1)}{n},\frac{i}{n}]}$$

$$+ \int_0^{1/n} \left(\sum_{i=1}^{n} \|D'(y) - D'(y-u)\|_{1,[\frac{(i-1)}{n},\frac{i}{n}]} \right) du$$

$$+ \int_0^{1/n} \left(\sum_{i=1}^{n} \|D'(u+y) - D'(y)\|_{1,[\frac{(i-1)}{n},\frac{i}{n}]} \right) du$$

$$= n \sum_{i=1}^{n} \left\| \int_{(i-1)/n}^{i/n} |D'(u)|\,|u-y|\,du \right\|_{1,[\frac{(i-1)}{n},\frac{i}{n}]}$$

$$+ \int_0^{1/n} \|D'(y) - D'(y-u)\|_{1,[0,1]} du$$

$$+ \int_0^{1/n} \|D'(u+y) - D'(y)\|_{1,[0,1]} du$$

$$\leq n\sum_{i=1}^{n}\left\|\int_{(i-1)/n}^{i/n}|D'(u)|\,|u-y|\,du\right\|_{1,[\frac{(i-1)}{n},\frac{i}{n}]}$$

$$+\int_0^{1/n}\|D'(y)-D'(y-u)\|_{1,\mathbb{R}}\,du+\int_0^{1/n}\|D'(u+y)-D'(y)\|_{1,\mathbb{R}}\,du$$

$$=n\sum_{i=1}^{n}\left\|\int_{(i-1)/n}^{i/n}|D'(u)|\,|u-y|\,du\right\|_{1,[\frac{(i-1)}{n},\frac{i}{n}]}$$

$$+\,2\int_0^{1/n}\|D'(u+y)-D'(y)\|_{1,\mathbb{R}}\,du$$

$$\leq n\sum_{i=1}^{n}\left\|\int_{(i-1)/n}^{i/n}|D'(u)|\,|u-y|\,du\right\|_{1,[\frac{(i-1)}{n},\frac{i}{n}]}+2\int_0^{1/n}\omega_1(D',u)_{1,\mathbb{R}}\,du;$$

i.e.

$$\Gamma\leq n\sum_{i=1}^{n}\left\|\int_{(i-1)/n}^{i/n}|D'(u)|\,|u-y|\,du\right\|_{1,[\frac{(i-1)}{n},\frac{i}{n}]}+\frac{2}{n}\omega_1\left(D',\frac{1}{n}\right)_{1,\mathbb{R}}. \quad (1.41)$$

Since $u,y\in\left[\frac{(i-1)}{n},\frac{i}{n}\right]$ we get $|u-y|\leq\frac{1}{n}$. And

$$\int_{(i-1)/n}^{i/n}|D'(u)|\,|u-y|\,du\leq\frac{1}{n}\int_{(i-1)/n}^{i/n}|D'(u)|\,du.$$

Consequently

$$\left\|\int_{(i-1)/n}^{i/n}|D'(u)|\,|u-y|\,du\right\|_{1,[\frac{i-1}{n},\frac{i}{n}]}\leq\frac{1}{n^2}\int_{(i-1)/n}^{i/n}|D'(u)|\,du.$$

That is

$$n\sum_{i=1}^{n}\left\|\int_{(i-1)/n}^{i/n}|D'(u)|\,|u-y|\,du\right\|_{1,[\frac{(i-1)}{n},\frac{i}{n}]}$$

$$\leq n \sum_{i=1}^{n} \frac{1}{n^2} \int_{(i-1)/n}^{i/n} |D'(u)|\, du = \frac{1}{n} \int_0^1 |D'(u)|\, du.$$

From (1.41) and the last estimate we obtain

$$\Gamma \leq \frac{1}{n} \int_0^1 |D'(u)|\, du + \frac{2}{n} \omega_1 \left(D', \frac{1}{n}\right)_{1,\mathbb{R}} \tag{1.42}$$

Remember that

$$\mathcal{R}_{1n}(D) := \int_0^1 F^{-1}(y) \left(\sum_{i=1}^{n} n d_i(D) \chi_{[\frac{(i-1)}{n}, \frac{i}{n}]}(y) \right) dy. \tag{1.3}$$

Then

$$\left| \mathcal{R}_{1n}(D) - \int_0^1 F^{-1}(y) D(y)\, dy \right|$$

$$= \left| \sum_{i=1}^{n} \int_{(i-1)/n}^{i/n} F^{-1}(y) n d_i(D)\, dy - \sum_{i=1}^{n} \int_{(i-1)/n}^{i/n} F^{-1}(y) D(y)\, dy \right|$$

$$\leq \sum_{i=1}^{n} \int_{(i-1)/n}^{i/n} |F^{-1}(y)|\, |n d_i(D) - D(y)|\, dy$$

$$= \int_0^1 |F^{-1}(y)| \left(\sum_{i=1}^{n} |n d_i(D) - D(y)| \chi_{[\frac{(i-1)}{n}, \frac{i}{n}]}(y) \right) dy$$

$$\leq \|F^{-1}\|_{ess.sup\,[0,1]} \left\| \sum_{i=1}^{n} |n d_i(D) - D(y)| \chi_{[\frac{(i-1)}{n}, \frac{i}{n}]}(y) \right\|_{1,[0,1]}$$

$$\leq \|F^{-1}\|_{ess.sup\,[0,1]} \cdot \left\{ \sum_{i=1}^{n} \|(n d_i(D) - D(y)) \cdot \chi_{[\frac{(i-1)}{n}, \frac{i}{n}]}(y)\|_{1,[0,1]} \right\}$$

$$= \|F^{-1}\|_{ess.sup\,[0,1]} \cdot \Gamma$$

$$\overset{(1.42)}{\leq} \|F^{-1}\|_{ess.sup\,[0,1]} \left\{ \frac{1}{n} \int_0^1 |D'(u)|\,du + \frac{2}{n}\omega_1\left(D', \frac{1}{n}\right)_{1,\mathbb{R}} \right\}.$$

That is proving (1.36). ∎

COROLLARY 1.6

Let $D \in \mathcal{L}_1^1(\mathbb{R})$ with D' of compact support, such that

$$\omega_1(D', \delta)_{1,\mathbb{R}} \leq K \cdot \delta^{\alpha}, \quad K > 0,\ \forall \delta > 0,\ 0 < \alpha \leq 1. \tag{1.43}$$

All the rest is as in Theorem 1.6. Then

$$\left| \mathcal{R}_{1n}(D) - \int_0^1 F^{-1}(y) D(y)\,dy \right| \leq \|F^{-1}\|_{ess.sup\,[0,1]}$$

$$\cdot \left\{ \frac{1}{n} \int_0^1 |D'(u)|\,du + \frac{2K}{n^{1+\alpha}} \right\}, \quad n \in \mathbb{N}. \tag{1.44}$$

In the following we establish $\mathcal{L}_1$ results for $\mathcal{R}_{2n}$.

THEOREM 1.7

Let $D \in \mathcal{L}_1(\mathbb{R})$, D nondecreasing on $[0,2]$, $n \in \mathbb{N}$. Assume that the common probability distribution function F has finite support.

Here

$$\mathcal{R}_{2n}(D) := \int_0^1 F^{-1}(y) \left(\sum_{i=1}^{n} D\left(\frac{i}{n}\right) \chi_{[\frac{(i-1)}{n}, \frac{i}{n}]}(y) \right) dy. \tag{1.5}$$

Then

$$\left| \mathcal{R}_{2n}(D) - \int_0^1 F^{-1}(y) D(y)\,dy \right| \leq \|F^{-1}\|_{ess.sup\,[0,1]} \cdot \omega_1\left(D, \frac{1}{n}\right)_{1,\mathbb{R}}. \tag{1.45}$$

PROOF For $i = 1, \ldots, n$, and $\frac{(i-1)}{n} \leq y \leq \frac{i}{n}$ we get

$$0 \leq D\left(\frac{i}{n}\right) - D(y) \leq D\left(\frac{1}{n} + y\right) - D(y).$$

Next we treat

$$\theta := \sum_{i=1}^{n} \left\| \left(D\left(\frac{i}{n}\right) - D(y) \right) \chi_{[\frac{(i-1)}{n}, \frac{i}{n}]}(y) \right\|_{1,[0,1]} \tag{1.46}$$

$$= \sum_{i=1}^{n} \int_0^1 \left| D\left(\frac{i}{n}\right) - D(y) \right| \chi_{[\frac{(i-1)}{n}, \frac{i}{n}]}(y)\, dy$$

$$= \sum_{i=1}^{n} \int_{(i-1)/n}^{i/n} \left(D\left(\frac{i}{n}\right) - D(y) \right) dy$$

$$\leq \sum_{i=1}^{n} \int_{(i-1)/n}^{i/n} \left(D\left(\frac{1}{n} + y\right) - D(y) \right) dy$$

$$= \int_0^1 \left(D\left(\frac{1}{n} + y\right) - D(y) \right) dy$$

$$= \left\| D\left(\frac{1}{n} + y\right) - D(y) \right\|_{1,[0,1]}$$

$$\leq \left\| D\left(\frac{1}{n} + y\right) - D(y) \right\|_{1,\mathbb{R}} \leq \omega_1\left(D, \frac{1}{n}\right)_{1,\mathbb{R}}.$$

Thus

$$\theta \leq \omega_1\left(D, \frac{1}{n}\right)_{1,\mathbb{R}} \tag{1.47}$$

We observe that

$$\left| \mathcal{R}_{2n}(D) - \int_0^1 F^{-1}(y) D(y)\, dy \right|$$

$$= \left| \sum_{i=1}^{n} \left(\int_{(i-1)/n}^{i/n} F^{-1}(y) D\left(\frac{i}{n}\right) dy - \int_{(i-1)/n}^{i/n} F^{-1}(y) D(y)\, dy \right) \right|$$

$$\leq \sum_{i=1}^{n} \int_{(i-1)/n}^{i/n} |F^{-1}(y)| \left| D\left(\frac{i}{n}\right) - D(y) \right| dy$$

$$= \int_0^1 |F^{-1}(y)| \left(\sum_{i=1}^{n} \left| D\left(\frac{i}{n}\right) - D(y) \right| \chi_{[\frac{(i-1)}{n}, \frac{i}{n}]}(y) \right) dy$$

$$\leq \|F^{-1}\|_{ess.sup\,[0,1]} \left\| \sum_{i=1}^{n} \left| D\left(\frac{i}{n}\right) - D(y) \right| \chi_{[\frac{(i-1)}{n}, \frac{i}{n}]}(y) \right\|_{1,[0,1]}$$

$$\leq \|F^{-1}\|_{ess.sup\,[0,1]} \left\{ \sum_{i=1}^{n} \left\| \left(D\left(\frac{i}{n}\right) - D(y) \right) \chi_{[\frac{(i-1)}{n}, \frac{i}{n}]}(y) \right\|_{1,[0,1]} \right\}$$

$$\overset{(1.46)}{=} \|F^{-1}\|_{ess.sup\,[0,1]} \cdot \theta \overset{(1.47)}{\leq} \|F^{-1}\|_{ess.sup\,[0,1]} \cdot \omega_1 \left(D, \frac{1}{n} \right)_{1,\mathbb{R}};$$

i.e. establishing (1.45). ∎

COROLLARY 1.7
Let $D \in \mathcal{L}_1(\mathbb{R})$ of compact support, such that

$$\omega_1(D, \delta)_{1,\mathbb{R}} = O(\delta^{\alpha}), \quad \forall \delta > 0, \ 0 < \alpha \leq 1. \tag{1.34}$$

All the rest is as in Theorem 1.7. Then

$$\left| \mathcal{R}_{2n}(D) - \int_0^1 F^{-1}(y) D(y)\, dy \right| = O\left(\frac{1}{n^{\alpha}} \right), \quad n \in \mathbb{N}. \tag{1.48}$$

Next follows the differentiable analog of Theorem 1.7.

THEOREM 1.8
Let $D \in \mathcal{L}_1^1(\mathbb{R}) := \{D : D \in AC(\mathbb{R})\ \&\ D' \in \mathcal{L}_1(\mathbb{R})\}$ such that D' is nondecreasing over $[0,2]$ and bounded on $[0,1]$, $n \in \mathbb{N}$. Assume that the common probability distribution function F has finite support. Then

$$\left| \mathcal{R}_{2n}(D) - \int_0^1 F^{-1}(y) D(y)\, dy \right| \leq \|F^{-1}\|_{ess.sup\,[0,1]}$$

$$\cdot \left\{ \frac{1}{2n^2} \left(\sum_{i=1}^{n} \left| D'\left(\frac{i}{n}\right) \right| \right) + \frac{1}{n} \omega_1 \left(D', \frac{1}{n} \right)_{1,\mathbb{R}} \right\}. \tag{1.49}$$

Inequality (1.49) is attained when D is a constant function.

PROOF Let $y \in \left[\frac{(i-1)}{n}, \frac{i}{n}\right]$, $i \in \{1, \ldots, n\}$, then

$$D(y) - D\left(\frac{i}{n}\right) = D'\left(\frac{i}{n}\right)\left(y - \frac{i}{n}\right) + \int_{i/n}^{y}\left(D'(t) - D'\left(\frac{i}{n}\right)\right) dt.$$

Hence

$$\left|D(y) - D\left(\frac{i}{n}\right)\right| \leq \left|D'\left(\frac{i}{n}\right)\right|\left|y - \frac{i}{n}\right|$$

$$+ \frac{1}{n}\left|D'\left(\frac{i}{n}\right) - D'(y)\right|, \quad \forall y \in \left[\frac{(i-1)}{n}, \frac{i}{n}\right].$$

That is,

$$\left|D(y) - D\left(\frac{i}{n}\right)\right| \chi_{[\frac{(i-1)}{n}, \frac{i}{n}]}(y) \leq \left|D'\left(\frac{i}{n}\right)\right|\left|y - \frac{i}{n}\right| \chi_{[\frac{(i-1)}{n}, \frac{i}{n}]}(y)$$

$$+ \frac{1}{n}\left|D'\left(\frac{i}{n}\right) - D'(y)\right| \chi_{[\frac{(i-1)}{n}, \frac{i}{n}]}(y), \quad \forall y \in [0,1].$$

Furthermore we have

$$\left\|\left(D\left(\frac{i}{n}\right) - D(y)\right) \chi_{[\frac{(i-1)}{n}, \frac{i}{n}]}(y)\right\|_{1,[0,1]}$$

$$\leq \left|D'\left(\frac{i}{n}\right)\right| \left\|\left|y - \frac{i}{n}\right| \chi_{[\frac{(i-1)}{n}, \frac{i}{n}]}(y)\right\|_{1,[0,1]}$$

$$+ \frac{1}{n}\left\|\left(D'\left(\frac{i}{n}\right) - D'(y)\right) \chi_{[\frac{(i-1)}{n}, \frac{i}{n}]}(y)\right\|_{1,[0,1]}$$

Notice that

$$\Omega := \sum_{i=1}^{n}\left\|\left(D\left(\frac{i}{n}\right) - D(y)\right) \chi_{[\frac{(i-1)}{n}, \frac{i}{n}]}(y)\right\|_{1,[0,1]} \tag{1.50}$$

$$\leq \sum_{i=1}^{n}\left|D'\left(\frac{i}{n}\right)\right| \left\|\left(y - \frac{i}{n}\right) \chi_{[\frac{(i-1)}{n}, \frac{i}{n}]}(y)\right\|_{1,[0,1]}$$

$$+\frac{1}{n}\sum_{i=1}^{n}\left\|\left(D'\left(\frac{i}{n}\right)-D'(y)\right)\chi_{[\frac{(i-1)}{n},\frac{i}{n}]}(y)\right\|_{1,[0,1]}$$

$$=\sum_{i=1}^{n}\left|D'\left(\frac{i}{n}\right)\right|\left(\int_{(i-1)/n}^{i/n}\left(\frac{i}{n}-y\right)dy\right)$$

$$+\frac{1}{n}\sum_{i=1}^{n}\left(\int_{(i-1)/n}^{i/n}\left(D'\left(\frac{i}{n}\right)-D'(y)\right)dy\right)$$

$$\leq\frac{1}{2n^2}\left(\sum_{i=1}^{n}\left|D'\left(\frac{i}{n}\right)\right|\right)+\frac{1}{n}\sum_{i=1}^{n}\left(\int_{(i-1)/n}^{i/n}\left(D'\left(\frac{1}{n}+y\right)-D'(y)\right)dy\right)$$

$$=\frac{1}{2n^2}\left(\sum_{i=1}^{n}\left|D'\left(\frac{i}{n}\right)\right|\right)+\frac{1}{n}\int_0^1\left(D'\left(\frac{1}{n}+y\right)-D'(y)\right)dy$$

$$=\frac{1}{2n^2}\left(\sum_{i=1}^{n}\left|D'\left(\frac{i}{n}\right)\right|\right)+\frac{1}{n}\left\|D'\left(\frac{1}{n}+y\right)-D'(y)\right\|_{1,[0,1]}$$

$$\leq\frac{1}{2n^2}\left(\sum_{i=1}^{n}\left|D'\left(\frac{i}{n}\right)\right|\right)+\frac{1}{n}\left\|D'\left(\frac{1}{n}+y\right)-D'(y)\right\|_{1,\mathbb{R}}$$

$$\leq\frac{1}{2n^2}\left(\sum_{i=1}^{n}\left|D'\left(\frac{i}{n}\right)\right|\right)+\frac{1}{n}\omega_1\left(D',\frac{1}{n}\right)_{1,\mathbb{R}};$$

i.e.

$$\Omega\leq\frac{1}{2n^2}\left(\sum_{i=1}^{n}\left|D'\left(\frac{i}{n}\right)\right|\right)+\frac{1}{n}\omega_1\left(D',\frac{1}{n}\right)_{1,\mathbb{R}} \tag{1.51}$$

Finally we see that

$$\left|\mathcal{R}_{2n}(D)-\int_0^1 F^{-1}(y)D(y)\,dy\right|$$

$$=\left|\sum_{i=1}^{n}\left(\int_{(i-1)/n}^{i/n}F^{-1}(y)D\left(\frac{i}{n}\right)dy-\int_{(i-1)/n}^{i/n}F^{-1}(y)D(y)\,dy\right)\right|$$

$$\leq \sum_{i=1}^{n} \int_{(i-1)/n}^{i/n} |F^{-1}(y)| \left| D\left(\frac{i}{n}\right) - D(y) \right| dy$$

$$= \int_0^1 |F^{-1}(y)| \left(\sum_{i=1}^{n} \left| D\left(\frac{i}{n}\right) - D(y) \right| \chi_{[\frac{(i-1)}{n}, \frac{i}{n}]}(y) \right) dy$$

$$\leq \|F^{-1}\|_{ess.sup\,[0,1]} \left\| \sum_{i=1}^{n} \left| D\left(\frac{i}{n}\right) - D(y) \right| \chi_{[\frac{(i-1)}{n}, \frac{i}{n}]}(y) \right\|_{1,[0,1]}$$

$$\leq \|F^{-1}\|_{ess.sup\,[0,1]} \left\{ \sum_{i=1}^{n} \left\| \left(D\left(\frac{i}{n}\right) - D(y) \right) \chi_{[\frac{(i-1)}{n}, \frac{i}{n}]}(y) \right\|_{1,[0,1]} \right\}$$

$$\overset{(1.50)}{=} \|F^{-1}\|_{ess.sup\,[0,1]} \cdot \Omega$$

$$\overset{(1.51)}{\leq} \|F^{-1}\|_{ess.sup\,[0,1]} \cdot \left\{ \frac{1}{2n^2} \left(\sum_{i=1}^{n} \left| D'\left(\frac{i}{n}\right) \right| \right) + \frac{1}{n} \omega_1 \left(D', \frac{1}{n} \right)_{1,\mathbb{R}} \right\};$$

i.e. proving inequality (1.49). ∎

COROLLARY 1.8
Let $D \in \mathcal{L}_1^1(\mathbb{R})$ with D' of compact support, such that

$$\omega_1(D', \delta)_{1,\mathbb{R}} \leq K \cdot \delta^{\alpha}, \quad K > 0,\ \forall \delta > 0,\ 0 < \alpha \leq 1. \tag{1.43}$$

All the rest is as in Theorem 1.8.
Then

$$\left| \mathcal{R}_{2n}(D) - \int_0^1 F^{-1}(y) D(y)\, dy \right| \leq \|F^{-1}\|_{ess.sup\,[0,1]}$$

$$\cdot \left\{ \frac{1}{2n^2} \left(\sum_{i=1}^{n} \left| D'\left(\frac{i}{n}\right) \right| \right) + \frac{K}{n^{1+\alpha}} \right\}, \quad n \in \mathbb{N}. \tag{1.52}$$

REMARK 1.1 The right-hand sides of all inequalities appearing in the statements of Theorems 1.1–1.8 and of Corollaries 1.1–1.8, are finite for any $n \in \mathbb{N}$. Furthermore as $n \to +\infty$, all these right-hand sides converge to zero, giving quantitatively the rate of convergence of $\mathcal{R}_{1n}(D)$, $\mathcal{R}_{2n}(D)$ to $\int_0^1 F^{-1}(y)D(y)\,dy$ in the various cases. ∎

REMARK 1.2 The proofs of the presented results are non-probabilistic and plain mathematical. Therefore in these results one can replace the quantile function F^{-1} by other more general functions, and then the results can stand as general approximation theory results. However, since the motivation and significance of them came from and lie in order statistics theory, we stay there. ∎

References

1. G.A. Anastassiou and T. Rychlik, Refined rates of bias convergence for generalized L-statistics in the I.I.D. case, submitted.
2. N. Balakrishnan and A.G. Cohen, *Order Statistics and Inference*, Academic Press, Boston, 1991.
3. H.A. David, *Order Statistics*, 2nd ed., Wiley, New York, 1981.
4. L. Gajek and T. Rychlik, Projection method for moment bounds on order statistics from restricted families. I. Dependent case, *J. Multivariate Anal.* **57**, (1996), 156–174.
5. T. Rychlik, Bounds for expectation of L-estimates for dependent samples, *Statistics* **24**, (1993), 1–7.
6. T. Rychlik, Sharp bounds on L-estimates and their expectations for dependent samples, *Commun. Statistics-Theor. Meth.* **22**, (1993), 1053–1068.
7. T. Rychlik, Bounds on expectation of L-estimates. In: *Order Statistics: Theory and Methods* (N. Balakrishnan, C.R. Rao, eds.), *Handbook of Statistics*, Vol. 16, North-Holland, Amsterdam, pp. 105–145, 1998.

Chapter 2

Regularization and Stabilization of Inverse Problems

Charles W. Groetsch

Department of Mathematical Sciences, University of Cincinnati, Cincinnati, OH 45221-0025. E-mail: groetsch@email.uc.edu

Abstract A variety of inverse problems in the natural sciences, including identification of coefficients, reconstruction of initial data, estimation of source functions, and discovery of boundary conditions, demand the solution of ill-posed operator equations or the evaluation of unbounded operators. Instability is a common feature of these inverse problems and special means, known as regularization or stabilization procedures, are required for their numerical solution. We survey some of the basic theories of Tikhonov's regularization method and related stabilization techniques in a general Hilbert space context.

Contents

2.1 Introduction

There seems to be general agreement on what is meant by a *direct problem* for a partial differential equation. A differential operator is viewed as a transformation whose complete description requires elaboration of its domain in terms of smoothness requirements and initial/boundary conditions, as well as specification of structural parameters, in the form of coefficients. In addition, a target function, in the guise of a "right-hand side" or source function, is specified. The direct problem then consists of finding the solution of the partial differential equation. The process of solving the direct problem can be represented schematically by

$$\left\{\begin{array}{c}\text{InitialValues}\\ \text{BoundaryValues}\\ \text{Sources}\\ \text{Coefficients}\end{array}\right\} \quad \rightarrow \quad \text{PDE} \quad \rightarrow \quad \text{Solution}$$

For us, an *inverse problem* for a partial differential equation will be loosely defined as recovering either initial values, boundary values, coefficients, or source functions, given the PDE and (some profile of) the solution. In this chapter we survey, in fairly general terms, a methodology for such inverse problems. This methodology — regularization theory — was conceived in the early sixties (Phillips (1962); Tikhonov (1963)) as a means of mitigating the chief numerical hazard encountered when resolving inverse problems: instability. Before describing regularization theory, we illustrate the main issues on a simple partial differential equation: the lowly one-dimensional heat equation.

2.2 A Matrix of Linear Inverse Problems

We exemplify the intrinsic instability of linear inverse problems with a matrix $[EX_{i,j}]$ of examples for the heat equation. Our aim is to show that solutions of such inverse problems, whether expressed in implicit or explicit form, are unstable with respect to data perturbations in natural norms. The examples are coded as follows: the "row" index $i \in \{\mathrm{I}, \mathrm{B}, \mathrm{S}, \mathrm{C}\}$ indicates initial values, boundary values, sources, or coefficients, respectively, while the "column" index $j \in \{\mathrm{I}, \mathrm{E}\}$ indicates an implicit or explicit, respectively, representation of the solution.

2.2.1 $\text{EX}_{\text{I,I}}$ The Backward-in-Time Heat Equation

Here we wish to determine the initial temperature distribution $f(x) = u(x, 0)$ for $x \in [0, \pi]$, given a later temperature distribution $g(x) = u(x, 1)$, where the evolution of the temperatures is governed by the partial differential equation:

$$\frac{\partial u}{\partial t} = \frac{\partial^2 u}{\partial x^2}, \quad 0 < x < \pi, \quad t > 0$$

and satisfies the boundary conditions $u(0, t) = u(\pi, t) = 0$. Separation of variables leads to the relationship

$$g(x) = \int_0^\pi k(x, y) f(y) dy$$

where

$$k(x, y) = \frac{2}{\pi} \sum_{n=1}^{\infty} e^{-n^2} \sin nx \sin ny.$$

We may express the inverse problem abstractly by the implicit equation $Kf = g$, where K is the compact linear operator on $L^2[0, \pi]$ with eigenvalues $\sigma(K) = \{e^{-n^2} : n = 1, 2, 3, ...\}$. The most notable feature of K is its highly smoothing effect, accomplished by the damping action of the eigenvalues. This smoothing can mask very large variations in high frequency components of f, making the reconstruction of f from the data g a daunting task, to say the least. □

2.2.2 $\text{EX}_{\text{I,E}}$ The Same

Formally, one can also write the solution of the inverse problem of the previous example as $f = Lg$ where

$$Lg(x) = \int_0^\pi l(x, y) g(y) dy$$

$$l(x, y) = \frac{2}{\pi} \sum_{n=1}^{\infty} e^{n^2} \sin nx \sin ny.$$

In this explicit formulation of the inverse problem, the solution f is obtained from the data g via the *unbounded* operator L defined on

$$\mathcal{D}(L) = \left\{ g \in L^2[0, \pi] : \sum_{m=1}^{\infty} e^{2m^2} a_m^2 < \infty, \quad a_m = \int_0^\pi g(y) \sin my dy \right\}$$

Here the instability is apparent: small (in L^2-norm) perturbations of the data g can, thanks to the factors e^{n^2} in the kernel, translate into very large changes in the solution f. More precisely, L is a closed linear operator defined on the dense subspace $\mathcal{D}(L)$ of $L^2[0,\pi]$ which is unbounded with respect to the L^2-norm (see Section 2.3.2). □

2.2.3 $\mathrm{EX}_{\mathrm{B,I}}$ The Semi-infinite Rod

Here one is given a temperature history $g(t) = u(a,t)$ at some point $a > 0$ and the challenge is to reconstruct the boundary value $f(t) = u(0,t)$, where

$$\frac{\partial u}{\partial t} = \frac{\partial^2 u}{\partial x^2}, \quad 0 < x < \infty, \quad t > 0$$

and $u(x,0) = 0$. Formal Laplace transform techniques lead to the representation

$$g(t) = \frac{a}{2\sqrt{\pi}} \int_0^t \frac{\exp(-\frac{a^2}{4(t-\tau)})}{(t-\tau)^{3/2}} f(\tau) d\tau$$

or $g = Kf$ where K is again a highly smoothing integral operator. □

2.2.4 $\mathrm{EX}_{\mathrm{B,E}}$ The Finite Rod For All Time

The problem of determining the boundary condition $f(t) = u(1,t)$ given $g(t) = u(0,t)$, where $u(x,t)$ satisfies

$$\frac{\partial u}{\partial t} = \frac{\partial^2 u}{\partial x^2}, \quad 0 < x < 1, \quad -\infty < t < \infty$$

$$\frac{\partial u}{\partial x}(0,t) = 0$$

has, under appropriate conditions, the explicit solution

$$f(t) = \frac{1}{\sqrt{2\pi}} \int_{-\infty}^{\infty} e^{\imath\omega t} \cosh(\sqrt{i\omega}) \hat{g}(\omega) d\omega$$

where $\hat{g}$ is the Fourier transform of g. The explicit solution has the form $f = Lg$ where the operator L is unbounded in the L^2-norm (see, e.g., Engl, Hanke, and Neubauer (1996)). □

2.2.5 $\text{EX}_{\text{S,I}}$ A General Source

Consider the heat problem with sources

$$\frac{\partial u}{\partial t} = \frac{\partial^2 u}{\partial x^2} + f(x,t), \quad 0 < x < \pi, \quad 0 < t < \infty$$

subject to the boundary and initial conditions

$$u(0,t) = u(\pi,t) = 0, \quad u(x,0) = 0.$$

If we wish to reconstruct the sources from the temperature history $g(t) = u(a,t)$ at some interior point $a \in (0,\pi)$, then formal separation of variables leads to the representation

$$g(t) = \int_0^t \int_0^\pi k(s, t-\tau) f(s,\tau) ds d\tau$$

where

$$k(s,z) = \frac{2}{\pi} \sum_{n=1}^{\infty} e^{-n^2 z} \sin na \sin ns.$$

That is, $g = Kf$, where K is a compact operator on $L^2([0,\pi] \times [0,\infty))$. The instability of the mapping from the data g to the solution f is easy to see in this problem. For example, if $g_n(t) = n^{-3/2}(2 - e^{-n^2 t}) \sin na$, then the solution f is independent of t and is given by

$$f(x) = 2\sqrt{n} \sin nx.$$

It is then apparent that $g_n \to 0$ uniformly in t, while

$$\max_{x \in [0,\pi]} |f(x)| = 2\sqrt{n} \to \infty$$

as $n \to \infty$. $\square$

2.2.6 $\text{EX}_{\text{S,E}}$ Sources in Space

If in the model

$$\frac{\partial u}{\partial t} = \frac{\partial^2 u}{\partial x^2} + f(x), \quad 0 < x < \pi, \quad 0 < t < 1$$

where $u(x,t)$ is subject to the boundary and initial conditions

$$u(0,t) = u(\pi,t) = 0, \quad u(x,0) = 0$$

one wishes to reconstruct the source distribution $f(x)$ from the spatial temperature distribution $g(x) = u(x,1)$, one is led to the explicit representation

$$f(x) = \int_0^\pi l(x,s)g(s)ds$$

where

$$l(x,s) = \frac{2}{\pi}\sum_{n=1}^{\infty}\frac{n^2}{1-e^{-n^2}}\sin nx \sin ns.$$

That is, $f = Lg$, where L is the linear operator on $L^2[0,\pi]$ with domain

$$\mathcal{D}(L) = \left\{ g : \sum_{m=1}^{\infty} m^4 a_m^2 < \infty, \quad a_m = \frac{2}{\pi}\int_0^\pi g(s)\sin msds \right\}$$

defined by the kernel $l(x,s)$. We note that $\mathcal{D}(L)$ is dense in $L^2[0,\pi]$ since all finite linear combinations of the eigenfunctions $g_n(s) = \sin ns$ are contained in $\mathcal{D}(L)$. Moreover, L is unbounded since $\|g_n\|^2 = \pi/2$ while

$$\|Lg_n\|^2 = \frac{\pi n^2}{2(1-e^{-n^2})} \to \infty$$

as $n \to \infty$. Finally, L is a *closed* operator, that is, its graph is a closed subspace of $L^2[0,\pi] \times L^2[0,\pi]$ (see Section 2.3.2). □

2.2.7 EX$_{\mathrm{C,I}}$ A Coefficient in Space

Here we consider the problem of reconstructing the space-dependent coefficient $\rho(x)$ in the model

$$\rho(x)\frac{\partial u}{\partial t} = \frac{\partial^2 u}{\partial x^2}, \quad 0 < x < 1, \quad 0 < t$$

$$u(0,t) = u(1,t) = 0$$

given the two distributions $g(x) = u(x,0)$ and $u_1(x) = \frac{\partial u}{\partial t}(x,0)$. The relationship between ρ, g and u_1 is then given by

$$g(x) = \int_0^1 k(x,s)\rho(s)ds, \quad k(x,s) = [(x-s)_+ - x(1-s)]u_1(s)$$

or, $g = K\rho$, for a compact operator K. $\square$

2.2.8 $\text{EX}_{\text{C,I}}$ A Coefficient in Time

In the model (see, Cannon (1984))

$$\frac{\partial u}{\partial t} = a(t)\frac{\partial^2 u}{\partial x^2} \quad 0 < x < 1, \quad 0 < t$$

$$u(0,t) = u(1,t) = 0, \quad u(x,0) = \sin \pi x$$

The coefficient $a(t)$ is related to the profile $h(t) = u(.5, t)$ by

$$a(t) = -\frac{h^{'}(t)}{\pi^2 h(t)}$$

That is, the coefficient is related to the processed profile

$$g(t) = (-\ln|h(t)|)/\pi^2$$

by the unbounded operator $L = \frac{d}{dt}$, namely, $a = Lg$. $\square$

For more examples of inverse problems formulated as operator equations see Ang et al. (1995), Baumeister (1986), Bertero and Boccaci (1998), Cannon and Hornung (1986), Colton et al. (1990), Colton and Kress (1992), Engl and Groetsch (1987), Engl, Hanke and Neubauer (1996), Glasko (1988), Groetsch (1990b, 1993), Hofmann (1986), Isakov (1990), Kirsch (1996), Kubo (1992), Lavrent'ev et al. (1986), Lavrent'ev and Savel'ev (1995), Lebedev, Vorovich and Gladwell (1996), Louis (1989), Payne (1975), Tikhonov and Arsenin (1977), Tikhonov and Goncharsky (1987), Wing (1991), Yamaguti (1991), and Zakhariev and Suzko (1990).

2.3 A Framework for Linear Inverse Problems

The examples of inverse problems in the previous section fall into two classes: implicit problems of the form $Kf = g$, where g is given data and K is a highly smoothing operator, or explicit problems of the form $f = Lg$, where L is an unbounded operator. In this section a general Hilbert space framework for the study of such problems is suggested.

2.3.1 Implicit Problem: Generalized Inverses

Suppose $K : H_1 \to H_2$ is a compact linear operator (for much of what follows, it suffices that K be merely bounded, or even closed) from a Hilbert space H_1 into a Hilbert space H_2. We neither assume that K is one-to-one, nor that it is surjective, and we seek a solution of $Kf = g$ for a given $g \in H_2$. This equation has no solution if g is not in the range of K, and hence we seek least squares solutions, that is, solutions of the minimization problem

$$\|Kf - g\|_2 = \inf_{u \in H_1} \|Ku - g\|_2$$

or, equivalently, solutions of the equation

$$K^*Kf = K^*g$$

where $K^* : H_2 \to H_1$ is the adjoint of $K : H_1 \to H_2$. Such a least squares solution exists if and only if the orthogonal projection of g onto the closure of the range of K actually lies in $R(K)$, the range of K. That is, a least squares solution exists if and only if g lies in the dense subspace

$$\mathcal{D}(K^\dagger) := R(K) + R(K)^\perp$$

of H_2. For a given $g \in \mathcal{D}(K^\dagger)$ the set of least squares solutions forms an affine space parallel to $N(K)$, the nullspace of K. Therefore, there is always a unique least squares solution of minimum norm, $K^\dagger g$. The operator $K^\dagger : \mathcal{D}(K^\dagger) \to H_1$, which associates with $g \in \mathcal{D}(K^\dagger)$ the unique least squares solution of smallest norm (equivalently, $K^\dagger g$ is the unique least squares solution lying $N(K)^\perp$), is called the Moore-Penrose inverse of K (see, e.g., Nashed (1976), Groetsch (1977)); it provides a mechanism for producing a unique generalized solution of the implicit inverse problem.

The operator $K^\dagger$ is linear, and it is *closed*; that is, its graph is a closed subspace of the product space $H_1 \times H_2$. Indeed, if

$$g_n \subseteq \mathcal{D}(K^\dagger), \quad g_n \to g, \quad \text{and} \quad K^\dagger g_n \to f,$$

then $f \in N(K)^\perp$, since $N(K)^\perp$ is closed. Also, since K^* and K^*K are continuous,

$$K^*g = \lim_{n\to\infty} K^*g_n = \lim_{n\to\infty} K^*KK^\dagger g_n = K^*Kf$$

and hence f is a least squares solution lying in $N(K)^\perp$. Therefore, $g \in \mathcal{D}(K^\dagger)$ and $K^\dagger g = f$, and hence the graph of $K^\dagger$ is closed.

Unfortunately, the closed densely defined linear operator $K^\dagger$ is bounded if and only if $R(K)$ is closed in H_2. Indeed, if $R(K)$ is closed, then $\mathcal{D}(K^\dagger) = H_2$ and therefore, by the *closed graph theorem* (see, e.g., Groetsch (1980)), $K^\dagger$ is continuous. On the other hand, if $K^\dagger$ is continuous and $Kf_n \to g$, where we may assume that $f_n \subset N(K)^\perp$, then

$$f_n = K^\dagger K f_n \to K^\dagger g \quad \text{and} \quad K f_n \to K K^\dagger g$$

and hence $g = KK^\dagger g \in R(K)$; that is, $R(K)$ is closed. For compact operators the range is closed if and only if it is finite dimensional; that is, in the case of an integral operator, if and only if the kernel of the integral operator is degenerate. This is seldom the case in interesting applications (it never happened in any of the examples of the previous section). Therefore, in applications $K^\dagger$ is typically unbounded and hence small perturbations in the data g can lead to very large changes in the generalized solution $K^\dagger g$. The method of Tikhonov regularization was developed to try to smooth out these instabilities by requiring an extra norm constraint on the Tikhonov approximations to the minimal norm least squares solution.

2.3.2 Explicit Problem: Unbounded Operators

In the case when an explicit solution $f = Lg$ of the inverse problem is known, it often happens (as in the examples above) that the operator $L : \mathcal{D}(L) :\to H_1$ is unbounded and only densely defined. Therefore, if g^δ is a small perturbation of "proper" data $g \in \mathcal{D}(L)$, it may happen that $g^\delta \notin \mathcal{D}(L)$ and hence the explicit solution procedure, i.e., L, cannot be applied to the actual data g^δ. Even if $g^\delta \in \mathcal{D}(L)$ and $g^\delta \to g$ as $\delta \to 0$, it may happen that $Lg^\delta \not\to Lg$, as the operator L is not continuous. Methods which seek to stabilize an explicit solution procedure for inverse problems are sometimes called mollification methods (see, e.g., Murio (1993)).

A general class of unbounded, closed densely defined linear operators can be developed in any separable Hilbert space. This class of operators includes the operators previously discussed in Sections 2.2.2 and 2.2.6. Suppose H is a real separable Hilbert space with complete orthonormal system $\{\phi_n\}_{n=1}^\infty$, then

$$f = \sum_{n=1}^\infty a_n \phi_n \in H \quad \Longleftrightarrow \quad \sum_{n=1}^\infty |a_n|^2 < \infty.$$

In this way the space H is seen to be isomorphic to the space l^2. Suppose $\{w_n\}_{n=1}^\infty$ is a sequence of positive weights with $w_n \to \infty$. Let

$$\mathcal{D}(L) = \left\{ f \in H : \sum_{n=1}^\infty w_n^2 | < f, \phi_n > |^2 < \infty \right\},$$

where $< \cdot, \cdot >$ is the inner product in H. Then $\mathcal{D}(L)$ is a linear subspace which is dense in H since it contains all finite linear combinations of the basis vectors $\{\phi_n\}$. Define $L : \mathcal{D}(L) \longrightarrow H$ by

$$Lf = \sum_{n=1}^{\infty} w_n < f, \phi_n > \phi_n .$$

Then L is unbounded since $\|L\phi_k\| = w_k \to \infty$, while $\|\phi_k\| = 1$. Moreover, L is closed. Indeed, if $\{f_m\} \subset \mathcal{D}(L)$ and $f_m \to f \in H$ as $m \to \infty$, then for each n, $< f_m, \phi_n > \to < f, \phi_n >$ as $m \to \infty$. If, in addition, $Lf_m \to g \in H$, then $< g - Lf_m, \phi_n > \to 0$ as $m \to \infty$, for each n. But,

$$< g - Lf_m, \phi_n > = < g, \phi_n > - w_n < f_m, \phi_n > \to < g, \phi_n > - w_n < f, \phi_n > .$$

Therefore,

$$\sum_{n=1}^{\infty} w_n^2 | < f, \phi > |^2 = \|g\|^2 < \infty$$

and hence $f \in \mathcal{D}(L)$ and $Lf = g$.

Another concrete example of closed unbounded operator is provided by the Dirichlet to Neumann map. Consider for simplicity the unit disk D and unit circle ∂D. For a given function g on ∂D we denote by u the function which is harmonic in D and takes boundary values g. The operator L is then defined by the normal (i.e., radial) derivative: $Lg = \frac{\partial u}{\partial n}$. To be more specific, L is the closed operator defined on the dense subspace

$$\mathcal{D}(L) = \left\{ g \in L^2(\partial D) : \sum_{n \in \mathbf{Z}} |n|^2 |\hat{g}(n)|^2 < \infty \right\}$$

of $L^2(\partial D)$ by

$$(Lg)(e^{\imath\theta}) = \sum_{n \in \mathbf{Z}} |n| \hat{g}(n) \exp(in\theta)$$

where $\hat{g}(n)$ is the Fourier coefficient

$$\hat{g}(n) = \frac{1}{2\pi} \int_0^{2\pi} g(t) e^{-\imath nt} \, dt .$$

2.4 Tikhonov Regularization

Tikhonov regularization (Tikhonov (1963), see also Phillips (1962)) strives to stabilize the computation of the minimum norm least squares solution by minimizing a weighted least squares functional of the form

$$F_\alpha(u;g) = \|Ku - g\|_2^2 + \alpha\|u\|_1^2$$

where $\alpha > 0$ is a *regularization parameter.* The unique minimizer, f_α, of this functional satisfies the Euler equation

$$K^*Kf_\alpha + \alpha f_\alpha = K^*g$$

where K^* is the adjoint of the operator K. The important point is that this equation is *well-posed*; that is, the solution f_α always exists, is unique, and depends continuously on the data g. This is so because the self-adjoint operator $K^*K + \alpha I$ has spectrum bounded away from zero, and hence the inverse operator $(K^*K + \alpha I)^{-1}$ is bounded. Furthermore, it is not difficult to show that $f_\alpha \to K^\dagger g$ as $\alpha \to 0^+$ if $g \in \mathcal{D}(K^\dagger)$, while $\|f_\alpha\|_1 \to \infty$ if $g \notin \mathcal{D}(K^\dagger)$. Therefore, the Tikhonov method with exact data $g \in \mathcal{D}(K^\dagger)$ converges to the minimal norm least squares solution, while if $g \notin \mathcal{D}(K^\dagger)$, no subsequence of the approximations converges even in the weak sense.

Analysis of the Tikhonov method is complicated by the practical circumstance that the data g are seldom available in a "clean" form. Rather only some measured estimate g^δ of the data, which satisfies $\|g - g^\delta\|_2 < \delta$, where δ is a bound for the measurement error, is accessible. Therefore, the successful practice of regularization requires that proper attention be paid to the regularization parameter, the data error, the details of the discretization procedure used to transfer the operator from function space to the realm of numerical computation, and the interrelationships of these entities. This will be a recurring theme in the sequel.

2.4.1 Convergence in Function Space

Given more information about $K^\dagger g$ one can obtain a rate of convergence. The extra information can be viewed as an abstract smoothness assumption, and is often called a "source" condition. Specifically, if $K^\dagger g \in R(K^*K)$ (or, equivalently, $g \in R(KK^*K)$), then

$$\|f_\alpha - K^\dagger g\|_1 = O(\alpha) \quad \text{as} \quad \alpha \to 0^+.$$

Indeed, if $K^\dagger g = K^*Kw$, then

$$f_\alpha - K^\dagger g = (K^*K + \alpha I)^{-1}K^*KK^\dagger g - K^\dagger g$$

$$= -\alpha K^*K(K^*K + \alpha I)^{-1}w$$

and, since $\|K^*K(K^*K + \alpha I)^{-1}\| \leq 1$, we have

$$\|f_\alpha - K^\dagger g\|_1 \leq \alpha \|w\|_1.$$

In practical circumstances the data function g is subject to error and hence only an estimate is available, say a vector g^δ satisfying

$$\|g - g^\delta\|_2 < \delta.$$

One then computes the regularized approximation f_α^δ by

$$(K^*K + \alpha I)f_\alpha^\delta = K^*g^\delta$$

using the available data g^δ. To gauge the accuracy of this approximation a bound is needed for the stability error $\|f_\alpha^\delta - f_\alpha\|_1^2$. Such a bound is easy to find, for

$$\|f_\alpha^\delta - f_\alpha\|_1^2$$

$$= < K^*(KK^* + \alpha I)^{-1}(g^\delta - g), K^*(KK^* + \alpha I)^{-1}(g^\delta - g) >$$

$$= < (KK^* + \alpha I)^{-1}(g^\delta - g), KK^*(KK^* + \alpha I)^{-1}(g^\delta - g) >$$

while

$$\|KK^*(KK^* + \alpha I)^{-1}\| \leq 1 \quad \text{and} \quad \|(KK^* + \alpha I)^{-1}\| \leq 1/\alpha.$$

Therefore,

$$\|f_\alpha^\delta - f_\alpha\|_1 < \delta/\sqrt{\alpha}$$

Combining this stability bound with the error estimate for the method with "clean" data $g \in R(KK^*K)$ we have

$$\|K^\dagger g - f_\alpha^\alpha\| = O(\alpha) + \delta/\sqrt{\alpha}$$

and hence

$$\|K^\dagger g - f_\alpha^\delta\|_1 = O(\delta^{\frac{2}{3}})$$

if $K^\dagger g \in R(K^*K)$ and α is chosen according to the a priori rule $\alpha = C\delta^{\frac{2}{3}}$. It turns out that this two-thirds power for the order of convergence is a true "brick wall" for ordinary Tikhonov regularization (Groetsch (1983a)). In fact, it can be shown that for compact operators K, if a strictly stronger rate of the form $o(\delta^{2/3})$ was attainable by some choice $\alpha = \alpha(\delta)$ of the regularization parameter, then the operator K has finite rank, and hence the original problem is well-posed.

2.4.2 A Posteriori Parameter Choice

For computational purposes, a priori choice strategies for the regularization parameter are generally unsuitable. It is preferable to have a parameter choice strategy that works in "real time" depending on the current state of the computations. In the mid-sixties Morozov (1966) (see also Phillips (1962)) proposed an a posteriori parameter choice strategy known as the *discrepancy principle.* According to this principle, if the signal-to-noise ratio in the data is strictly greater than one, i.e., if $\|g^\delta\|_2 > \delta \geq \|g - g^\delta\|_2$, then the discrepancy equation

$$\|Kf_\alpha^\delta - g^\delta\|_2 = \delta$$

has a unique solution $\alpha = \alpha(\delta)$ and $f_\alpha^\delta \to K^\dagger g$ as $\delta \to 0^+$. If, in addition, $K^\dagger g \in R(K^*)$, then it can be shown (Groetsch (1983b)) that when the regularization parameter is chosen according to the discrepancy principle, then $\|f_{\alpha(\delta)}^\delta - K^\dagger\|_1 = O(\sqrt{\delta})$, but for compact operators with infinite rank (i.e., those which lead to ill-posed problems) this order is the best possible.

Raus (1985), Gfrerer (1987), and Engl, Hanke, Neubauer (1996) have developed a modified discrepancy principle which achieves the optimal two-thirds order of convergence for Tikhonov regularization. The method relies on monitoring the functional

$$\Phi_\alpha(z) = \alpha^3 < (\alpha I + K^*K)^{-3}Qz, Qz >$$

where Q is the orthogonal projector of H_2 onto the closure of the range of K. For a given constant $C \geq 1$, it can be shown that under suitable conditions the equation

$$\Phi_\alpha(g^\delta) = C\delta^2$$

has a unique solution $\alpha = \alpha(\delta)$ and that if $K^\dagger g \in R(K^*K)$, then

$$\|f_{\alpha(\delta)}^\delta - K^\dagger g\|_1 = O(\delta^{2/3}).$$

2.4.3 Discretization

Some sort of discretization scheme is necessary to implement Tikhonov regularization as a computational method. This could be done by discretizing the original operator equation (for example, by a quadrature method if the operator is expressed as an integral operator) and then regularizing the resulting discrete problem (see, e.g., Hanke and Hansen (1993) and Hansen (1998)). An alternative to this would be to apply Tikhonov regularization to the infinite-dimensional problem and then discretize the resulting infinite-dimensional well-posed problem. That is, one has the option of discretizing and then regularizing, or regularizing and then discretizing (see, e.g., Engl and Groetsch (1988)). In the latter case, if the operator is an integral operator acting on the space $L^2[a,b]$, one could apply a quadrature rule to the iterated kernel to produce a degenerate kernel method for the second kind equation which results from regularization (see Groetsch (1990a)). That is, the operator K^*K is generated by the kernel

$$\tilde{k}(s,t) = \int_a^b k(u,s)k(u,t)du,$$

where $k(\cdot,\cdot)$ is the kernel generating the operator K. Applying a quadrature rule with weights $\{w_j^{(n)}\}$ and nodes $\{u_j^{(n)}\}$ to this kernel converts the Tikhonov equation to an operator equation of the form

$$\alpha x_n(\alpha) + \tilde{K}_n x_n(\alpha) = K^* g$$

where $\tilde{K}_n$ is the finite rank operator defined by

$$\tilde{K}_n y = \sum_{j=1}^{n} w_j^{(n)} < k_j, y > k_j$$

where $< \cdot,\cdot >$ is the L^2-inner product, and

$$k_j(s) = k(u_j^{(n)}, s).$$

The convergence of this finite-dimensional version of the Tikhonov method relies on the interplay between the regularization parameter α, the data error δ, and the parameter $\gamma_n = \|\tilde{K}_n - K^*K\|$ which gauges the quality of the finite-dimensional approximating operator. For example, if the source condition $K^\dagger g \in R(K^*K)$ is satisfied, and if $\delta_n \to 0$ while $\alpha_n = C\delta_n^{2/3}$ and $\gamma_n = O(\alpha_n^2)$, then $x_n^{\delta_n}(\alpha_n) \to K^\dagger g$ as $n \to \infty$. The behavior of γ_n

depends on the details of the quadrature rule used and on the nature of the kernel. If the quadrature rule is convergent and the kernel, and its partial derivative with respect to the second variable, are continuous, then $\gamma_n \to 0$ as $n \to \infty$. Rates of convergence for γ_n can be obtained for specific quadrature rules, assuming appropriate smoothness of the kernel. For example, under suitable conditions on the kernel, the trapezoidal rule yields the rate $\gamma_n = O(n^{-2})$.

Another general approach to discretization that has a certain appeal is to use the variational characterization of the regularized solution to construct finite element methods. Suppose $V_1 \subset V_2 \subset \cdots$ is an increasing sequence of finite-dimensional subspaces of H_1 whose union is dense in H_1. Let $f^{\delta}_{\alpha,m} \in V_m$ be the minimizer of the Tikhonov functional $F_\alpha(\cdot, g^\delta)$ over the finite-dimensional subspace V_m. The success of the approximation then relies not only on the relationship between the regularization parameter α and the error level δ in the data, but also on the extent to which the subspaces $\{V_m\}$ faithfully support the operator K. The quality of the latter approximation may be gauged by the numbers

$$\gamma_m = \|K(I - P_m)\|$$

where P_m is the orthogonal projector of H_1 onto the subspace V_m. If K is compact, then it is easy to see that $\gamma_m \to 0$ as $m \to \infty$. In asymptotic regularization theory, the trick is to balance all the error terms (regularization error, approximation error, and stability error) so as to obtain convergent approximations. For this to succeed, the parameters α, γ_m, and δ must link hands and dance to the same tune. Specifically, if $g \in \mathcal{D}(K^\dagger)$ and $m = m_\delta \to \infty$, and if $\alpha = \alpha(m_\delta)$ and $\gamma = \gamma(m_\delta)$ satisfy $\gamma = O(\sqrt{\alpha})$ and $\delta = o(\sqrt{\alpha})$, then $f^{\delta}_{\alpha,m} \to K^\dagger g$ as $\delta \to 0$. Furthermore, the optimal order of convergence $O(\delta^{2/3})$ is achievable with the finite element method. In fact, if $K^\dagger g \in R(K^*K)$, $\alpha = C\delta^{2/3}$, and $\gamma = \gamma(m_\delta) = O(\delta^{2/3})$, then

$$\|f^{\delta}_{\alpha,m} - K^\dagger g\|_1 = O(\delta^{2/3})$$

(see, Groetsch, King and Murio (1982)). Finally, one can achieve the same order of convergence as the discrepancy method with a finite element version of the Tikhonov regularization method (Groetsch (1983b)). Namely, suppose for some constant $\tau > 1$ that $\|Q_m g^\delta\| > \tau\delta$ for all $m \geq M$, where Q_m is the orthogonal projector of H_2 onto $K(V_m)$. This condition says that the signal-to-noise ratio for the *projected* data is strictly greater than one. It may then be shown that the equation

$$\|K f_{\alpha,m} - Q_m g^\delta\|_2 = \tau\delta$$

has a unique solution $\alpha = \alpha(m)$. Furthermore, if $K^\dagger g \in R(K^*)$, then

$$\|f^\delta_{m,\alpha} - K^\dagger\|_1 = O(\sqrt{\delta})$$

where $\alpha = \alpha(m(\delta))$.

2.4.4 Iterated Regularization

The stark fact remains that the ordinary Tikhonov method discussed above has an optimal order of approximation $O(\delta^{2/3})$ for data errors of order $O(\delta)$. This order can be surpassed by use of iterative methods. For example, in nonstationary iterated Tikhonov regularization one sequentially minimizes the functionals

$$F_n(u; g^\delta) = \|Ku - g^\delta\|_2^2 + \alpha_n \|f^\delta_{n-1} - u\|_1^2$$

where f^δ_{n-1} is the minimizer of $F_{n-1}(\cdot; g^\delta)$ and $\{\alpha_n\}$ is a suitable sequence of regularization parameters. This is equivalent to defining the sequence of regularized approximations $\{f^\delta_n\}$ by

$$(K^*K + \alpha_n I) f^\delta_n = \alpha_n f^\delta_{n-1} + K^* g^\delta.$$

A typical choice for the sequence of regularization parameters is the geometric scheme; that is, $\alpha_n = \alpha q^{n-1}$ where $0 < q < 1$ represents the geometric decay rate of the parameters. In Hanke and Groetsch (1998) it is shown that the condition

$$0 < c \leq \sum_{j=1}^{n-1} \frac{\alpha_n}{\alpha_j}$$

for all large enough n is a sufficient (and essentially necessary) condition for convergence of this nonstationary iterated Tikhonov method. It is also possible to formulate a discrepancy principle for the nonstationary iterated Tikhonov method that can achieve a $O(\delta^\nu)$ order of convergence for any $\nu \in [0, 1)$. In fact, if for a some $\tau > 1$, the signal-to-noise ratio satisfies $\|g^\delta\|_2/\delta > \tau$, then there is a first iteration number, say $n = n(\delta)$ for which

$$\|g^\delta - K f^\delta_{n(\delta)}\|_2 \leq \delta\tau.$$

For a given $\nu \in [0, 1)$ one then has, if $K^\dagger g$ is sufficiently smooth (a source condition depending on ν),

$$\|f^\delta_{n(\delta)} - K^\dagger g\|_1 = O(\delta^\nu).$$

For an approach to regularization using a class of nonlinear nonstationary iterative methods, see Hanke (1995).

2.5 Stabilized Evaluation

In this section we discuss the explicit inverse problem $f = Lg$ where $L : \mathcal{D}(L) \subset H_1 \to H_2$ is a closed, unbounded linear operator defined on the dense subspace $\mathcal{D}(L)$ of a Hilbert space H_1. Recall that the main issue here is that the "true" data g is not directly available, and we have no guarantee that the measured data g^δ is in the domain of the operator L. And even if $g^\delta \in \mathcal{D}(L)$, we cannot be sure that $Lg^\delta \to Lg$ as $\delta \to 0$ since the operator is unbounded. So some sort of stable, convergent, smoothing technique is needed to map the data into $\mathcal{D}(L)$.

2.5.1 A General Scheme

We now sketch an abstract scheme (Groetsch (1992)) for "smoothing" or mollifying the data g^δ so as to stablize the application of the operator L to the data g^δ. The scheme relies on an old theorem of von Neumann (see, Riesz and Sz.-Nagy (1955)), namely, that the operators

$$\check{L} := (I + L^*L)^{-1} \quad \text{and} \quad L\check{L}$$

are defined *everywhere* on H_1 and are *bounded.* Von Neumann's proof makes elegant use of the graph of L

$$\mathcal{G}(L) = \{(x, Lx) : x \in \mathcal{D}(L)\}$$

considered as a closed subspace of the Hilbert space $H_1 \times H_2$ with inner product

$$< (x, y), (w, z) >=< x, w > + < y, z >$$

(we use the symbol $< \cdot, \cdot >$ indiscriminately to denote inner products). Since $\mathcal{G}(L)$ is closed, the product Hilbert space may be decomposed as

$$H_1 \times H_2 = \mathcal{G}(L) + \mathcal{G}(L)^\perp.$$

However, it is easy to see that

$$\mathcal{G}(L)^\perp = \{(-L^*y, y) : y \in \mathcal{D}(L^*) \subset H_2\}.$$

Therefore, for any $h \in H_1$, we have the representation

$$(h, 0) = (x, Lx) + (-L^*y, y)$$

for some $x \in \mathcal{D}(L)$ and $y \in \mathcal{D}(L^*)$. But then $y = -Lx$ and $h = x - L^*y$, that is, $x \in \mathcal{D}(L^*L)$ and $h = (I + L^*L)x$. Hence $(I + L^*L)^{-1}$ is defined on all of H_1 and, since $\|h\|_1^2 \geq \|x\|_1^2$,

$$\|\check{L}\| = \|(I + L^*L)^{-1}\| \leq 1.$$

We also have

$$\|L\check{L}x\|_2^2 =< L(I + L^*L)^{-1}x, L(I + L^*L)^{-1}x >=< \check{L}x, x - \check{L}x >\leq \|x\|_1^2$$

and hence $\|L\check{L}\| \leq 1$.

The mollified version of the data g^δ is the function

$$g_\alpha^\delta := \check{L}T_\alpha(\check{L})g^\delta$$

where $\{T_\alpha\}$ is a family of continuous functions on $[0, 1]$ satisfying $T_\alpha(t) \to 1/t$ as $\alpha \to 0^+$ for each $t \in (0, 1]$ and $|tT_\alpha(t)|$ is uniformly bounded. Now the mapping

$$g^\delta \longmapsto L\check{L}T_\alpha(\check{L})g^\delta$$

is everywhere defined and continuous, since both $L\check{L}$ and $T_\alpha(\check{L})$ are bounded operators, i.e., the evaluation of the mollified data is *stable*. Furthermore, if $\delta\sqrt{r(\alpha)} \to 0$ as $\delta \to 0$, where $r(\alpha) := \max|tT_\alpha(t)|$, then $Lg_\alpha^\delta \to f$ as $\alpha \to 0^+$. If greater regularity is assumed of g, then rates of convergence for the general method can be had. For example, if $g \in \mathcal{D}(LL^*L)$, then

$$\|Lg - Lg_\alpha^\delta\|_2 = O(\omega(\alpha)) + \delta\sqrt{r(\alpha(\delta))}$$

where $\omega(\alpha) := \max|(1 - tT_\alpha(t))t|$.

2.5.2 Tikhonov-Morozov Stabilization

We will be concerned with the special case $T_\alpha(t) = [\alpha + (1 - \alpha)t]^{-1}$ which was studied extensively by Morozov (see, Morozov (1969), (1993)). This leads to

$$f_\alpha^\delta := L(\alpha L^*L + I)^{-1}g^\delta.$$

We now summarize some results in Groetsch and Scherzer (1994). The first involves convergence rates. If $g \in \mathcal{D}(LL^*L)$ and $\alpha = \alpha(\delta)$ satisfies $\alpha^3/\delta^2 \to C > 0$ as $\delta \to 0^+$, then

$$\|f_\alpha^\delta - Lg\|_2 = O(\delta^{2/3})$$

and it can be shown that this order is optimal in the sense that if the stronger rate $o(\delta^{2/3})$ is achievable for all data g^δ, then g is in the nullspace of L and hence the evaluation of Lg is trivial. To see this, we first get an estimate on the regularization parameter. In so doing, we will use the notation $\hat{L} = (I + LL^*)^{-1}$.

We now show that if $g \notin N(L)$ and $\|Lg_\alpha^\delta - Lg\|_2 = o(\delta^{\frac{2}{3}})$ for all g^δ satisfying $\|g - g^\delta\|_1 \leq \delta$, then $\alpha = o(\delta^{\frac{2}{3}})$. To see this, let $g^\delta = g - \delta u$, where u is a unit vector and let $e_\alpha^\delta = Lg_\alpha^\delta - Lg$. Then

$$\begin{aligned}[\alpha I + (1-\alpha)\hat{L}]e_\alpha^\delta &= [\alpha I + (1-\alpha)\hat{L}]\left(L\check{L}(\alpha I + (1-\alpha)\check{L})^{-1}g - Lg\right) \\ &\quad - \delta[\alpha I + (1-\alpha)\hat{L}]L\check{L}(\alpha I + (1-\alpha)\check{L})^{-1}u \\ &= \alpha(\hat{L} - I)Lg - \delta L\check{L}u.\end{aligned}$$

Since $\|e_\alpha^\delta\|_2 = o(\delta^{\frac{2}{3}})$, by assumption, and since

$$\|\delta L\check{L}u\| \leq \delta\|L\check{L}\| = o(\delta^{\frac{2}{3}}),$$

we find that

$$\frac{\alpha}{\delta^{\frac{2}{3}}}\|(\hat{L} - I)Lg\| \to 0, \text{ as } \delta \to 0.$$

But, by assumption, $g \notin N(L) = N((\hat{L} - I)L)$ and hence $\alpha = o\left(\delta^{\frac{2}{3}}\right)$.

We now show that for a wide class of operators the order of convergence $O\left(\delta^{\frac{2}{3}}\right)$ cannot be improved. We only consider the important class of operators L^*L which have a divergent sequence of eigenvalues. Such is the case if L is the derivative operator, when $-L^*L$ is the Laplacian operator, or more generally whenever L is a differential operator for which $\check{L}$ is compact. If L^*L has eigenvalues $\mu_n \to \infty$ and $\|Lg_\alpha^\delta - Lg\|_2 = o\left(\delta^{\frac{2}{3}}\right)$ for all g^δ with $\|g - g^\delta\|_1 \leq \delta$, then we will show that $g \in N(L)$. Indeed, if $g \notin N(L)$, then, as we have just shown, $\alpha = o(\delta^{\frac{2}{3}})$. Let $e_\alpha^\delta = Lg_\alpha^\delta - Lg$, then

$$\|e_\alpha^\delta\|_2^2 = \|Lg_\alpha - Lg\|_2^2 + 2 < Lg_\alpha - Lg, Lg_\alpha^\delta - Lg_\alpha > + \|Lg_\alpha^\delta - Lg_\alpha\|_2^2$$

and by hypothesis

$$\|Lg_\alpha - Lg\|_2^2/\delta^{\frac{4}{3}} \to 0$$

as $\delta \to 0$ (since $g^\delta = g$ satisfies $\|g - g^\delta\|_1 \leq \delta$). Therefore, we must have

$$\frac{2 < Lg_\alpha - Lg, Lg_\alpha^\delta - Lg_\alpha > + \|Lg_\alpha^\delta - Lg_\alpha\|_2^2}{\delta^{\frac{4}{3}}} \to 0 \text{ as } \delta \to 0.$$

Suppose that $\{u_n\}$ are orthonormal eigenvectors of L^*L associated with $\{\mu_n\}$. Then $\{u_n\}$ are eigenvectors of $\check{L}$ associated with the eigenvalues $\lambda_n = 1/1 + \mu_n$ and $\lambda_n \to 0$ as $n \to \infty$. Now let $g^\delta = g + \delta u_n$. Then

$$\|Lg_\alpha^\delta - Lg_\alpha\|_2^2 = \delta^2 \langle \check{L}(\alpha I + (1-\alpha)\check{L})^{-1} u_n, L^*L\check{L}(\alpha I + (1-\alpha)\check{L})^{-1} u_n \rangle$$

$$= \delta^2 \lambda_n^2 \mu_n (\alpha + (1-\alpha)\lambda_n)^{-2}$$

$$= \delta^2 \lambda_n (1 - \lambda_n)(\alpha + (1-\alpha)\lambda_n)^{-2}.$$

Therefore, if $\delta = \delta_n = \lambda_n^{\frac{3}{2}}$, then $\delta_n \to 0$ as $n \to \infty$ and

$$\frac{\|Lg_\alpha^{\delta_n} - Lg_\alpha\|_2^2}{\delta_n^{\frac{4}{3}}} = (1 - \lambda_n)\left(\frac{\alpha}{\delta_n^{\frac{2}{3}}} + 1 - \alpha\right)^{-2} \to 1 \text{ as } n \to \infty.$$

Finally, we have

$$\frac{| < Lg_\alpha - Lg, Lg_\alpha^{\delta_n} - Lg_\alpha > |}{\delta_n^{\frac{4}{3}}} \leq \frac{\|Lg_\alpha - Lg\|_2}{\delta_n^{\frac{2}{3}}} \frac{\|Lg_\alpha^{\delta_n} - Lg_\alpha\|_2}{\delta_n^{\frac{2}{3}}} \to 0.$$

This contradiction establishes that $g \in N(L)$ and hence that the order $O(\delta^{2/3})$ is essentially the best possible.

A choice of the parameter by the discrepancy principle

$$\|(\alpha L^*L + I)^{-1} g^\delta - g^\delta\|_1 = \delta$$

leads, under appropriate conditions, to a suboptimal convergence rate $O(\sqrt{\delta})$, which generally cannot be improved. The optimal order can be recovered, however, by using a modified discrepancy principle. Namely, if the regularization parameter is chosen according to the equation

$$\|(\alpha L^*L + I)^{-1} g^\delta - g^\delta\|_1 = \delta^2/\alpha^2$$

then it can be shown that under suitable conditions, the optimal rate $O(\delta^{2/3})$ is attained.

2.5.3 An Illustration

We illustrate the use of the general stabilization technique in identifying a nonlinear boundary condition in a simple heat transfer problem (see Groetsch and Hanke (1994)). Consider a spherical shell $b \le r \le a$, which is initially at temperature zero, whose inner surface $r = b$ is constantly at temperature zero, and whose outer surface $r = a$ is heated under the influence of a spatially homogeneous ambient temperature. In suitable units the temperature $u(r,t)$ satisfies

$$u_t = \Delta u = u_{rr} + \frac{2}{r}u_r, \quad b < r < a, \quad t > 0$$

$$u(r,0) = 0, \quad b \le r \le a$$

$$u(b,t) = 0, \quad t > 0$$

We assume that the flux at the active surface is given by

$$u_r(a,t) = h(u(a,t)) - \phi(t)$$

where $\phi(t)$ is a given source term and h is an unknown function to be determined. To simplify the notation somewhat we let $g(t) := u(a,t)$ be the temperature history at the outer surface. We assume that g is differentiable and that all functions of time are sufficiently regular to allow application of standard Laplace transform techniques. If we denote Laplace transforms (with respect to time) by corresponding upper case letters, e.g., $U = U(r,s) = \mathcal{L}\{u(r,t)\}$, then the equations above transform to

$$r^2U'' + 2rU' - sr^2U = 0$$

$$U(b) = 0, U'(a) = H$$

where H is the Laplace transform of $h(u(a,t)) - \phi(t)$ and primes signify differentiation with respect to r. It is a routine, though tedious, matter to show that

$$U = \frac{a^2H\left[e^{\sqrt{s}r}/r - e^{-\sqrt{s}(r-2b)}/r\right]}{\left[(a\sqrt{s}-1)e^{a\sqrt{s}} + (a\sqrt{s}+1)e^{-(a-2b)\sqrt{s}}\right]}$$

and in the limit as $r \to a^-$, we obtain

$$G = H \frac{a(e^{a\sqrt{s}} - e^{-(a-2b)\sqrt{s}})}{(a\sqrt{s}-1)e^{a\sqrt{s}} + (a\sqrt{s}+1)e^{-(a-2b)\sqrt{s}}},$$

where G is the Laplace transform of the surface temperature history $g(t) = u(a,t)$. Therefore

$$H(s) = \{sG(s)\} \times$$

$$\left\{ \frac{(a\sqrt{s}-1)e^{a\sqrt{s}} + (a\sqrt{s}+1)e^{-(a-2b)\sqrt{s}}}{as(e^{a\sqrt{s}} - e^{-(a-2b)\sqrt{s}})} \right\}$$

and hence, by the convolution theorem,

$$h(g(t)) - \phi(t) = \int_0^t g'(\tau)k(t-\tau)d\tau$$

where

$$k(t) = \frac{1}{a}\mathcal{L}^{-1}\left\{ \frac{(a\sqrt{s}-1)e^{a\sqrt{s}} + (a\sqrt{s}+1)e^{-(a-2b)\sqrt{s}}}{s(e^{a\sqrt{s}} - e^{-(a-2b)\sqrt{s}})} \right\}$$

$$= -\frac{1}{a} + \frac{1}{\sqrt{\pi t}} \sum_{n=-\infty}^{\infty} \exp(-(a-b)^2 n^2/t).$$

We see that identifying the unknown function h which controls the heat transfer at the surface involves computing values of the function

$$\gamma(t) = \int_0^t g'(\tau)k(t-\tau)d\tau.$$

Note that this entails evaluating the derivative of the surface temperature history — an unstable procedure. We now present a simple technique to accomplish this.

Suppose $T > 0$ and let n be a positive integer. Lay out a uniform grid on $[0,T]$ by setting $t_i = i\Delta t$, $i = 0, 1, \ldots, n$, where $\Delta t = T/n$. Let ℓ_j be the piecewise linear function on $[0,T]$ satisfying $\ell_j(t_j) = \delta_{ij}$. We approximate

$\gamma(t_i)$ by replacing g with its piecewise linear interpolant,

$$S_h g := \sum_{j=1}^{n} g(t_j)\ell_j$$

(note that the initial condition $S_h g(0) = 0$ is then satisfied). The approximation is then

$$\gamma(t_i) \approx \int_0^{t_i} (S_h g)'(\tau)k(t_i - \tau)d\tau$$

$$= \sum_{j=1}^{i} a_{ij} g(t_j)$$

where,

$$a_{ij} = \sigma_{i-j} - \sigma_{i-j-1}, \quad \sigma_{-1} = 0$$

and

$$\sigma_m := \int_0^1 k((m+v)\Delta t)dv.$$

The numbers σ_m may be approximated effectively by use of Gaussian integration, summing the series to machine accuracy (see Groetsch (1995)).

The approximation procedure for h is simply this: for a selection of points $z_k = g(t_k)$, approximate $g(z_k)$ by $\gamma(t_k) + \phi(t_k)$ and plot.

The general stabilization technique of the previous section can now be applied to the special case when L is the differentiation operator acting on

$$\mathcal{D}(L) = \{x \in AC[0,T] : x' \in L^2[0,T], x(0) = 0\},$$

where $AC[0,T]$ stands for the space of absolutely continuous functions on $[0,T]$. The adjoint, L^*, of L is then defined on the space

$$\mathcal{D}(L^*) = \{y \in AC[0,T] : y' \in L^2[0,T], y(T) = 0\}$$

and $L^*y = -y'$. In this case the approximations y_α converge *uniformly* to $Lx = x'$ on closed subintervals of $[0,T)$. To see this, note that $y_\alpha \in \mathcal{D}(L^*)$ and hence $y_\alpha(T) = 0$. Also, if $x \in \mathcal{D}(L^*L)$, then $x'(T) = 0$ and hence for any t with $0 \le t < T$, we have

$$|y_\alpha(t) - x'(t)| = \left| \int_t^T (x'' - y'_\alpha)(\tau)d\tau \right|$$

$$\leq \|L^* y_\alpha - L^* Lx\|_2 \sqrt{T}$$

$$\leq \|I - \hat{L}T_\alpha(\hat{L})\| \|L^* Lx\|_2 \sqrt{T} \to 0, \text{ as } \alpha \to 0,$$

by (14).

These theoretical results suggest an approach to regularizing the computation of

$$\gamma(t) = \int_0^t g'(\tau)k(t-\tau)d\tau, \quad 0 \leq t < T$$

in the case where the surface temperature history is corrupted by errors.

We assume that the surface temperature history g belongs to $\mathcal{D}(L)$. Then $g' = Lg$ may be approximated by the stable approximation Lw_α where

$$w_\alpha = \check{L}T_\alpha(\check{L})g.$$

In general, these stable approximations converge in the L^2-sense, but under appropriate conditions, as shown above, the convergence is uniform on proper subintervals of $[0, T)$. As the kernel k belongs to $L^1[0, T]$, one obtains uniform convergence of the approximations to $\gamma(t)$ on proper subintervals of $[0, T)$ in this case.

As a specific example, we consider the Tikhonov-Morozov method, where $T_\alpha(t) = [\alpha + (1-\alpha)t]^{-1}$, $\alpha > 0$. In this case

$$w_\alpha = (I + \alpha L^* L)^{-1} g$$

and hence w_α satisfies the boundary value problem

$$w_\alpha - \alpha w_\alpha'' = g$$

$$w_\alpha(0) = w_\alpha'(T) = 0.$$

Denoting the inner product on $L^2[0, T]$ by $< \cdot, \cdot >$, this implies the weak form

$$< w_\alpha, \psi > + \alpha < w_\alpha', \psi' > = < g, \psi >$$

for $\psi \in \mathcal{D}(L)$. This in turn suggests a way to regularize the approximations of $\gamma(t)$ considered above. Namely, we replace g' by Lw_α. As an approximation to w_α we use

$$w_\alpha \approx \sum_{j=1}^{n-1} c_j \ell_j + c_{n-1} \ell_n,$$

where ℓ_j are the piecewise linear spline functions introduced above (note these approximations to w_α satisfy the required boundary conditions). The coefficients c_j are determined by

$$(A + \alpha B)\mathbf{c} = \mathbf{g}$$

where the matrices A and B are given by

$$A = [< \ell_i, \ell_j >], \qquad B = [< \ell_i', \ell_j' >]$$

and $\mathbf{g} = A[g(t_j)]$. The regularized values c_j are then used instead of the raw data $g(t_j)$. A numerical illustration of this technique is found in Groetsch and Hanke (1994).

2.5.4 Iterated Stabilization

Recently, a nonstationary iterated version of this method has been developed which can, under suitable conditions, attain the order $O(\delta^\nu)$ for any given $\nu \in [0, 1)$ (Groetsch and Scherzer (1999)). The method constructs smooth approximations $\{g_n^\delta\} \subset \mathcal{D}(L)$ for which $Lg_n^\delta \to Lg$ as $\delta \to 0$ and $n \to \infty$ in a suitable fashion. The approximation g_n^δ is the minimizer of the functional

$$G_n(z) = \|g^\delta - z\|_1^2 + \alpha_n \|Lz - Lg_{n-1}^\delta\|_2^2$$

over $\mathcal{D}(L)$ (this minimizer will necessarily be contained in the subspace $\mathcal{D}(L^*L)$ of smoother functions). This is equivalent to the iteratively defined method

$$(I + \alpha_n L^*L)g_n^\delta = g^\delta + \alpha_n L^* L g_{n-1}^\delta, \quad g_0^\delta = 0,$$

where $\{\alpha_n\}$ is a suitable sequence of positive regularization parameters.

Before presenting convergence results for this nonstationary iterated Tikhonov-Morozov method, we provide some motivation. The basic idea of mollification methods is to produce smooth versions g_n^δ of the data g^δ. One can imagine such smoothing is accomplished by the evolution problem

$$\frac{dw}{dt} = -L^*Lw, \quad w(0) = g^\delta.$$

For example, if $L = \nabla$, the gradient operator, then this is the classical heat equation, the smoothing properties of which are legendary. Since L^*L has unbounded spectrum, $w(t)$ will be a smoothed version of $w(0)$, even for small values of t. The smoothing effect may be approximated by a single step of the implicit Euler method (a method known for its stability) with step size α_1. Setting $g_0^\delta = w(0) = g^\delta$ and $g_1^\delta \approx w(\alpha_1)$, we have

$$\frac{g_1^\delta - g_0^\delta}{\alpha_1} = -L^*Lg_1^\delta \quad \text{or} \quad (I + \alpha_1 L^*L)g_1^\delta = g_0^\delta.$$

In the approximation g_1^δ, certain noise components are damped by the factors $(1+\alpha_1\mu_n)^{-1}$, where μ_n are eigenvalues of L^*L. One is naturally led to the possibility of subtracting out this smoothed approximation and continuing the filtration process on the remaining noise. That is, one takes g_2^δ to be the approximation to $w(\alpha_2)$ obtained by applying the implicit Euler method, with step size α_2, to the problem

$$\frac{dw}{dt} = -L^*Lw, \quad w(0) = g^\delta - g_1^\delta.$$

This leads to the next approximation g_2^δ, where

$$(I + \alpha_2 L^*L)(g_2^\delta - g_1^\delta) = g^\delta - g_1^\delta.$$

Continuing in this way we obtain a sequence of filtered approximations $\{g_n^\delta\}$ defined by

$$(I + \alpha_n L^*L)g_n^\delta = g^\delta + \alpha_n L^*Lg_{n-1}^\delta, \quad g_0^\delta = 0.$$

It can be shown that if the iteration number $n = n(\delta)$ is chosen so that $\delta\sigma_{n(\delta)} \to 0$ as $\delta \to 0$, where $\sigma_n = \sum_{j=1}^n \alpha_j^{-1}$, then $f_{n(\delta)}^\delta \to g$ in the graph norm of L, i.e., $f_{n(\delta)}^\delta \to g$ and $Lf_{n(\delta)}^\delta \to Lg$. Moreover, an a posteriori parameter strategy may be used which, under certain conditions, achieves an order of convergence $O(\delta^\nu)$ for any $\nu \in [0,1)$. In fact, if the signal-to-noise ratio is strictly greater than one, i.e., $\|g^\delta\|_1 > \tau\delta$, for some $\tau > 1$, and if the stopping index of the iteration is chosen to be the first index $n = n(\delta)$ satisfying

$$\|g_{n(\delta)}^\delta - g^\delta\|_1 \le \tau\delta,$$

then

$$\|Lg_{n(\delta)}^\delta - Lg^\delta\|_2 = O(\delta^{1-\frac{1}{\nu}})$$

assuming that the true data satisfies the source condition $g \in R((I+\tilde{L})^{-\nu})$ for some $\nu \in (0,1)$. This shows that the nonstationary iterated Tikhonov-Morozov method does not suffer the saturation order $O(\delta^{2/3})$ of the ordinary Tikhonov-Morozov method, but in fact can attain an essentially optimal order of convergence.

2.6 Regularization of Nonlinear Problems

We close this survey with a brief sketch of the convergence theory of Tikhonov regularization for nonlinear ill-posed problems. For other approaches to nonlinear ill-posed problems, see Vogel (1987), Engl, Hanke and Neubauer (1996), Hanke (1997), and Tikhonov et al. (1998). Our treatment of Tikhonov regularization for nonlinear operators is based, as was much of the development above, on the notion of the graph of an operator. The analysis we present (see Binder et al. (1995)) has a distinctly geometric flavor and is based on the fact that the distance of a point from a weakly closed set is a strongly continuous functional of the point.

Suppose that $F : \mathcal{D}(F) \subseteq H_1 \to H_2$ is a weakly closed operator; that is, its graph is weakly closed in the product space $H_1 \times H_2$. We are concerned with least squares solutions of the equation $F(f) = g$, meaning with vectors $x_0 \in \mathcal{D}(F)$ satisfying

$$\|F(x_0) - g\|_2 = \inf\{\|F(f) - g\|_2 : f \in \mathcal{D}(F)\}.$$

Note that the set of least squares solutions is weakly closed in H_1, although it is not necessarily nonempty, nor is it necessarily convex. The Tikhonov functional on $\mathcal{D}(F)$ is defined by

$$T_\alpha(z; g) = \|F(z) - g\|_2^2 + \alpha\|z\|_1^2$$

where α is a given regularization parameter. If one defines the inner product $[\cdot,\cdot]_\alpha$ on the product space $H_1 \times H_2$ by

$$[(u, v), (x, w)]_\alpha = \alpha < u, x > + < v, w >,$$

then it is immediately seen that

$$\inf\{T_\alpha(z, g) : z \in \mathcal{D}(F)\} = \text{dist}^2\left(\mathcal{G}(F), (0, g)\right)$$

where $\mathcal{G}(F)$ is the graph of F and $\text{dist}(S, (u, v))$ measures the distance from the point (u, v) to the set $S \subseteq H_1 \times H_2$ in the norm induced on $H_1 \times H_2$ by the inner product $[\cdot,\cdot]_\alpha$. Since $\mathcal{G}(F)$ is weakly closed the Tikhonov functional $T_\alpha(\cdot; g)$ possesses (generally nonunique) minima and we denote the set of all such minima by $M_\alpha(g)$; that is,

$$M_\alpha(g) = \{x \in H_1 : T_\alpha(x; g) = \min\{T_\alpha(y; g) : y \in \mathcal{D}(F)\}\}.$$

We note that since $\mathcal{G}(F)$ is assumed to be weakly closed, $x \in M_\alpha(g)$ if and only if $x \in \mathcal{D}(F)$ and

$$\text{dist}((x, F(x)), (0, g)) = \text{dist}(\mathcal{G}(F), (0, g)).$$

Moreover, the set of minimizers of the Tikhonov functional are *stable* in the following sense: if $g_k \to g$, then each sequence $\{x_k\}$ with $x_k \in M_\alpha(g_k)$ has a strongly convergent subsequence whose limit lies in $M_\alpha(g)$.

We provide a single example of a convergence result. For more general results and extensions see Binder et al. (1995). Suppose that the equation $F(z) = g$ has a unique least squares solution $f^\dagger$ and that $\{g^{\delta_n}\}$ is a sequence of approximate data satisfying $\|g - g^{\delta_n}\| \leq \delta_n \to 0$ as $n \to \infty$. Let $\{\alpha_n\}$ be a sequence of positive regularization parameters satisfying $\alpha_n \to 0$, while δ_n/α_n is bounded. Then $f_{\alpha_n}^{\delta_n} \to f^\dagger$ for each sequence $f_{\alpha_n}^{\delta_n} \in M_{\alpha_n}(g^{\delta_n})$. We establish this by showing that any subsequence of $\{f_{\alpha_n}^{\delta_n}\}$, which we continue to denote by $\{f_{\alpha_n}^{\delta_n}\}$, in turn has a subsequence $\{f_{\alpha_k}^{\delta_k}\}$ with $f_{\alpha_k}^{\delta_k} \to f^\dagger$.

First note that

$$T_{\alpha_n}(f_{\alpha_n}^{\delta_n}; g^{\delta_n}) \leq T_{\alpha_n}(f^\dagger; g^{\delta_n}) \leq \|F(f^\dagger) - g^{\delta_n}\|_2^2$$

and hence

$$\limsup_{n\to\infty} \|F(f_{\alpha_n}^{\delta_n}) - g^{\delta_n}\|_2 \leq \|F(f^\dagger) - g\|_2.$$

Also,

$$(\|F(f_{\alpha_n}^{\delta_n}) - g\|_2^2 - \|F(f^\dagger) - g\|_2^2) + \alpha_n \|f_{\alpha_n}^{\delta_n}\|_1^2$$

$$\leq 2\langle F(f_{\alpha_n}^{\delta_n}) - F(f^\dagger), g^{\delta_n} - g\rangle + \alpha_n \|f^\dagger\|_1^2$$

and hence

$$\|f_{\alpha_n}^{\delta_n}\|_1^2 \leq 2\|F(f_{\alpha_n}^{\delta_n}) - F(f^\dagger)\|_2 \times \frac{\delta_n}{\alpha_n} + \|f^\dagger\|_1^2.$$

Since F is weakly closed, there is an $\hat{f} \in \mathcal{D}(F)$ and a subsequence $\{f_{\alpha_k}^{\delta_k}\}$ with

$$F(f_{\alpha_k}^{\delta_k}) \rightharpoonup F(\hat{f}) \quad \text{and} \quad f_{\alpha_k}^{\delta_k} \rightharpoonup \hat{f}$$

($\rightharpoonup$ indicates weak convergence) and we then have

$$\|F(\hat{f}) - g\|_2^2 = \lim_{k\to\infty} < F(f_{\alpha_k}^{\delta_k}) - g^{\delta_k}, F(\hat{f}) - g >$$

$$\leq \|F(f^\dagger) - g\|_2 \cdot \|F(\hat{f}) - g\|_2$$

and hence $\hat{f} = f^\dagger$, since $f^\dagger$ is the unique least squares solution, and $F(f_{\alpha_k}^{\delta_k}) \to F(f^\dagger)$. It now follows that $\limsup_{k\to\infty} \|f_{\alpha_k}^{\delta_k}\|_1 \leq \|f^\dagger\|_1$ and hence $f_{\alpha_k}^{\delta_k} \to f^\dagger$, as promised.

Acknowledgement This survey is an expanded version of an address presented at the SIAM Great Lakes Section meeting in Spring, 1999, and a series of lectures given at the National University of Colombia - Medellin in Fall, 1998. The work was supported in part by the National Science Foundation.

References

[1] R.S. Anderssen and G.S. Newsom (Eds.), *Special Program on Inverse Problems*, Centre for Mathematical Analysis, Australian National University, Canberra, 1986.

[2] D.D. Ang et al. (Eds.), *Inverse Problems and Applications in Geophysics, Industry, Medicine and Technology*, Ho Chi Minh City Mathematical Society, Ho Chi Minh City, 1995.

[3] J. Baumeister, *Stable Solution of Inverse Problems*, Vieweg Verlag, Braunschweig, 1986.

[4] M. Bertero and P. Boccaci, *Introduction to Inverse Problems in Imaging*, Institute of Physics, Philadelphia, 1998.

[5] A. Binder et al., Weakly closed nonlinear operators and parameter identification in parabolic equations by Tikhonov regularization, *Applicable Analysis* **55**(1995), 215-234.

[6] J.R. Cannon, *The One-dimensional Heat Equation*, Addison-Wesley, Menlo Park, CA, 1984.

[7] J.R. Cannon and U. Hornung (Eds.), *Inverse Problems*, Birkhäuser, Basel, 1986.

[8] K. Chadan, D. Colton, L. Päivärinta, and W. Rundell, *An Introduction to Inverse Scattering and Inverse Spectral Theory*, SIAM, Philadelphia, 1997.

[9] D. Colton, R. Ewing, and W. Rundell (Eds.), *Inverse Problems in Partial Differential Equations*, SIAM, Philadelphia, 1990.

[10] D. Colton and R. Kress, *Inverse Acoustic and Electromagnetic Scattering Theory*, Springer Verlag, New York, 1992.

[11] J.P. Corones et al. (Eds.), *Invariant Imbedding and Inverse Problems*, SIAM, Philadelphia, 1992.

[12] I. Craig and J. Brown, *Inverse Problems in Astronomy*, Adam Hilger, Bristol, 1986.

[13] H.W. Engl, *Inverse Problems*, Sociedad Matematica Mexicana, Mexico City, 1995.

[14] H.W. Engl and H. Gfrerer, A posteriori parameter choice for general regularization methods for solving linear ill-posed problems, *Applied Numerical Mathematics* **4**(1988), 395-417.

[15] H.W. Engl and C.W. Groetsch (Eds.), *Inverse and Ill-posed Problems*, Academic Press, Orlando, 1987.

[16] H.W. Engl and C.W. Groetsch, Projection-regularization methods for operator equations of the first kind, in Anderssen and Newsom (1988), pp. 17-30.

[17] H.W. Engl, M. Hanke, and A. Neubauer, *Regularization of Inverse Problems*, Kluwer, Dordrecht, 1996.

[18] H.W. Engl and A. Neubauer, A parameter choice strategy for (iterated) Tikhonov regularization of ill-posed problems leading to superconvergence with optimal rates, *Applicable Analysis* **27**(1988), 5-18.

[19] H. Gfrerer, An a-posteriori parameter choice for ordinary and iterated Tikhonov regularization of ill-posed problems leading to optimal convergence rates, *Mathematics of Computation* **49**(1987), 507-522.

[20] V.B. Glasko, *Inverse Problems in Mathematical Physics*, American Institute of Physics, New York, 1988.

[21] C.W. Groetsch, *Generalized Inverses of Linear Operators: Representation and Approximation*, Marcel Dekker, New York, 1977.

[22] C.W. Groetsch, *Elements of Applicable Functional Analysis*, Marcel Dekker, New York, 1980.

[23] C.W. Groetsch, On the asymptotic order of accuracy of Tikhonov regularization, *Journal of Optimization Theory and Applications*

41(1983), 293-298.

[24] C.W. Groetsch, Comments on Morozov's discrepancy principle, in *Improperly Posed Problems and Their Numerical Treatment* (G. Hämmerlin and K.-H. Hoffmann, Eds.), Birkhäuser, Basel, 1983, pp. 97-104.

[25] C.W. Groetsch, *The Theory of Tikhonov Regularization for Fredholm Equations of the First Kind*, Pitman, Boston, 1984.

[26] C.W. Groetsch, Convergence analysis of a regularized degenerate kernel method for Fredholm integral equations of the first kind, *Integral Equations and Operator Theory* **13**(1990), 69-75.

[27] C.W. Groetsch, Some integral equations of the first kind in mechanics and regularization of ill-posed operator equations, in *Inverse Problems in Engineering Sciences* (M. Yamaguti, et al., Eds.), Springer Verlag, Tokyo, 1990, pp. 11-29.

[28] C.W. Groetsch, Spectral methods for linear inverse problems with unbounded operators, *Journal of Approximation Theory* **70**(1992), 16-28.

[29] C.W. Groetsch, *Inverse Problems in the Mathematical Sciences*, Vieweg Verlag, Braunschweig, 1993.

[30] C.W. Groetsch, Numerical methods for some inverse problems in nonlinear heat transfer, *Computer Mathematics and Its Applications* **1**(1995), 49-60.

[31] C.W. Groetsch and M. Hanke, Regularization of some one-dimensional inverse problems for identification of nonlinear surface phenomena, *Proceedings of the 1994 Design Engineering Technical Conferences*, Vol. 3C, American Society of Mechanical Engineers, New York, 1994, pp. 903-908.

[32] C.W. Groetsch and M. Hanke, A general framework for regularized evaluation of unstable operators, *Journal of Mathematical Analysis and Applications* **203**(1996), 451-463.

[33] C.W. Groetsch, J.T. King, and D. Murio, Asymptotic analysis of a finite element method for Fredholm equations of the first kind, in *Treatment of Integral Equations by Numerical Methods* (C.T.H. Baker and G.F. Miller, Eds.), Academic Press, London, 1982.

[34] C.W. Groetsch and O. Scherzer, The optimal order of convergence for stable evaluation of differential operators, *Electronic Journal of*

Differential Equations, **4** (1994), 1-10.

[35] C.W. Groetsch and O. Scherzer, Nonstationary iterated Tikhonov-Morozov method and third order differential equations for the evaluation of unbounded operators, preprint, 1999.

[36] M. Hanke, *Conjugate Gradient Type Methods for Ill-posed Problems*, Longman, Harlow, U.K., 1995.

[37] M. Hanke, A regularizing Levenberg-Marquardt scheme, with applications to inverse groundwater filtration problems, *Inverse Problems*, **13**(1997), 79-95.

[38] M. Hanke and C.W. Groetsch, Nonstationary iterated Tikhonov regularization, *Journal of Optimization Theory and Applications* **98**(1998), 37-53.

[39] M. Hanke and P.C. Hansen, Regularization methods for large-scale problems, *Surveys of Mathematics for Industry* **3**(1993), 253-315.

[40] P.C. Hansen, *Rank-deficient and Discrete Ill-posed Problems*, SIAM, Philadelphia, 1998.

[41] B. Hofmann, *Regularization for Applied Inverse and Ill-posed Problems*, Teubner, Leipzig, 1986.

[42] V. Isakov, *Inverse Source Problems*, American Mathematical Society, Providence, 1990.

[43] A. Kirsch, *An Introduction to the Mathematical Theory of Inverse Problems*, Springer Verlag, New York, 1996.

[44] S. Kubo (Ed.), *Inverse Problems*, Atlanta Technical Publications, Atlanta, 1992.

[45] M.M. Lavrent'ev, *Some Improperly Posed Problems of Mathematical Physics*, Springer Verlag, New York, 1967.

[46] M.M. Lavrent'ev, V. Romanov, and S. Shishatskii, *Ill-posed Problems of Mathematical Physics and Analysis*, American Mathematical Society, Providence, 1986.

[47] M.M. Lavrent'ev and L. Ya. Savel'ev, *Linear Operators and Ill-posed Problems*, Consultants Bureau, New York, 1995.

[48] L.P. Lebedev, I.I. Vorovich, and G.M.L. Gladwell, *Functional Analysis: Applications in Mechanics and Inverse Problems*, Kluwer, Dordrecht, 1996.

[49] A.K. Louis, *Inverse und Schlect Gestellte Probleme*, Teubner, Stuttgart, 1989.

[50] V.A. Morozov, On the solution of functional equations by the method of regularization, *Soviet Mathematics Doklady* **7**(1966), 414-417.

[51] V.A. Morozov, A stable method for computation of values of unbounded operators, *Soviet Mathematics Doklady* **10**(1969), 339-342.

[52] V.A. Morozov, *Methods for Solving Incorrectly Posed Problems*, Springer Verlag, New York, 1984.

[53] V.A. Morozov, *Regularization Methods for Ill-posed Problems*, CRC Press, Boca Raton, 1993.

[54] D.A. Murio, *The Mollification Method for the Numerical Solution of Ill-posed Problems*, Wiley, New York, 1993.

[55] M.Z. Nashed (Ed.), *Generalized Inverses and Applications*, Academic Press, New York, 1976.

[56] L.E. Payne, *Improperly Posed Problems in Partial Differential Equations*, SIAM, Philadelphia, 1975.

[57] V. Petkov and R. Lazarov (Eds.), *Integral Equations and Inverse Problems*, Longman, Harlow, UK, 1991.

[58] D.L. Phillips, A technique for the numerical solution of certain integral equations of the first kind, *Journal of the Association for Computing Machinery* **9**(1962), 84-97.

[59] J. Pöschel and E. Trubowitz, *Inverse Spectral Theory*, Academic Press, New York, 1987.

[60] A.G. Ramm, *Multidimensional inverse scattering problems*, Longman, Harlow, UK, 1992.

[61] T. Raus, The principle of residual in the solution of ill-posed problems with non-self-adjoint operator, *Tartu Riikl. Ül. Toimetised* **715**(1985), 12-20.(in Russian)

[62] F. Riesz and B. Sz.-Nagy, *Functional Analysis*, Ungar, New York, 1955.

[63] P. Sabatier (Ed.), *Inverse Problems in Action*, Springer Verlag, New York, 1990.

[64] F. Santosa et al. (Eds.), *Inverse Problems of Acoustic and Electromagnetic Waves*, SIAM, Philadelphia, 1984.

[65] G. Talenti (Ed.), *Inverse Problems*, Springer Verlag, Berlin, 1986.

[66] A.N. Tikhonov, Solution of incorrectly formulated problems and the regularization method, *Soviet Mathematics Doklady* **4**(1963), 1035-1038.

[67] A.N. Tikhonov and V.Y. Arsenin, *Solutions of Ill-posed Problems*, Wiley, New York, 1977.

[68] A.N. Tikhonov and A.V. Goncharsky, *Ill-posed Problems in the Natural Sciences*, MIR Publishers, Moscow, 1987.

[69] A.N. Tikhonv, A.S. Leonov and A.G. Yagola, *Nonlinear Ill-posed Problems*, Chapman and Hall, London, 1998.

[70] C.R. Vogel, An overview of numerical methods for nonlinear ill-posed problems, in Engl and Groetsch (1987), pp. 231-246.

[71] G. Wahba, *Spline Models of Observational Data*, SIAM, Philadelphia, 1990.

[72] G.M. Wing, *A Primer on Integral Equations of the First Kind*, SIAM, Philadelphia, 1991.

[73] M. Yamaguti et al. (Eds.), *Inverse Problems in Engineering Sciences*, Springer-Verlag, Tokyo, 1991.

[74] B.N. Zakhariev and A.A. Suzko, *Direct and Inverse Problems*, Springer-Verlag, Berlin, 1990.

Chapter 3

Trapezoidal-Type Rules from an Inequalities Point of View

Pietro Cerone and Sever S. Dragomir

School of Communications and Informatics, Victoria University of Technology,
P.O. Box 14428, Melbourne City MC, Victoria 8001, Australia
E-mail: Pietro.Cerone@vu.edu.au, Sever.Dragomir@vu.edu.au
Contact Author: S. Dragomir

Abstract This chapter investigates trapezoidal-type rules and obtains explicit bounds through the use of a Peano kernel approach and the modern theory of inequalities. Both Riemann-Stieltjes and Riemann integrals are evaluated with a variety of assumptions about the integrand enabling the characterisation of the bound in terms of a variety of norms. Perturbed quadrature rules are obtained through the use of Grüss, Chebychev and Lupaş inequalities, producing a variety of tighter bounds. The implementation is demonstrated through the investigation of a variety of composite rules based on inequalities developed. The analysis allows the determination of the partition required that would assure that the accuracy of the result would be within a prescribed error tolerance.

Contents

3.1 Introduction

The following inequality is well known in the literature as the *trapezoid inequality*:

$$\left| \int_a^b f(x)\,dx - \frac{b-a}{2}\left[f(a)+f(b)\right] \right| \leq \frac{(b-a)^3}{12}\left\|f''\right\|_\infty, \tag{3.1}$$

where the mapping $f:[a,b]\subset\mathbb{R}\to\mathbb{R}$ is assumed to be twice differentiable on the interval (a,b), with the second derivative bounded on (a,b). That is, $\|f''\|_\infty := \sup_{x\in(a,b)}|f''(x)| < \infty$.

Now if we assume that $I_n : a = x_0 < x_1 < \ldots < x_{n-1} < x_n = b$ is a partition of the interval $[a,b]$ and f is as above, then we can approximate the integral $\int_a^b f(x)\,dx$ by the *trapezoidal quadrature formula* $A_T(f,I_n)$, having an error given by $R_T(f,I_n)$, where

$$A_T(f,I_n) := \frac{1}{2}\sum_{i=0}^{n-1}\left[f(x_i)+f(x_{i+1})\right]h_i, \tag{3.2}$$

and the remainder satisfies the estimation

$$|R_T(f,I_n)| \leq \frac{1}{12}\left\|f''\right\|_\infty \sum_{i=0}^{n-1} h_i^3, \tag{3.3}$$

with $h_i := x_{i+1} - x_i$ for $i = 0,\ldots,n-1$.

Expression (3.2) is known as the trapezoidal rule, if $n = 1$, and as the composite trapezoidal rule for $n > 1$. The trapezoidal rule is the simplest closed Newton-Cotes quadrature rule in which function evaluation is restricted at the ends of equispaced intervals.

The trapezoidal rule is widely used in practice since it is easy to implement in an efficient fashion, especially if the partitioning is done in a uniform manner. It is also very accurate for periodic functions. It forms the basic building block for intricate closed Newton-Cotes formulae (Press et al. [33])

The current work investigates trapezoidal-type rules and obtains explicit bounds through the use of a Peano kernel approach and the modern theory of inequalities. This approach allows for the investigation of quadrature rules that place fewer restrictions on the behaviour of the integrand and thus allow us to cope with larger classes of functions. Expression (3.1) relies on the behaviour of the second derivative whereas bounds for the trapezoidal rules are obtained in terms of Riemann-Stieltjes integrals in Sections 3.2, 3.3 and 3.4 for functions that are of bounded variation, monotonic and Lipschitzian, respectively. In Section 3.5, trapezoidal-type rules are obtained for $f^{(n)} \in L_p[a,b]$, implying that

$$\left\|f^{(n)}\right\|_p := \left(\int_a^b \left|f^{(n)}(x)\right|^p dx\right)^{\frac{1}{p}} < \infty \quad \text{for } p \geq 1$$

and $\left\|f^{(n)}\right\|_\infty := \sup_{x\in[a,b]} \left|f^{(n)}(x)\right|$.

Perturbed trapezoidal-type rules are obtained in Section 3.3 using what are termed **premature** variants of Grüss, Chebychev and Lupaş inequalities. Atkinson [3] uses an asymptotic error estimate technique to obtain what he defines as a corrected trapezoidal rule. His approach, however, does not readily produce a bound on the error.

In Section 3.6, non-symmetric bounds are obtained for a trapezoidal-type rule for functions whose derivative is bounded above and below. Section 3.7 utilises a Grüss-type inequality to obtain trapezoidal rules whose bound relies on $f'(x) - S$ where $S = \frac{f(b)-f(a)}{b-a}$ is the secant slope. Finally, in Section 3.8, trapezoidal rules whose error bound involves the second derivative belonging to a variety of norms are investigated. This allows for greater flexibility since any of them may be best for different functions.

The current work brings together results for trapezoidal-type rules giving explicit error bounds, using Peano type kernels and results from the modern theory of inequalities. Although bounds through the use of Peano kernels have been obtained in some classical review books on numerical integration such as Stroud [34], Engels [26] and Davis and Rabinowitz [8]. These do not seem to be utilised to the extent that perhaps they should be. So much so that even in the more recent comprehensive monograph by Krommer and Ueberhuber [28], a constructive approach is taken via Taylor or interpolating polynomials to obtain quadrature results. This approach does not readily provide explicit error bounds but rather gives the order of the approximation.

3.2 Estimates of the Remainder for Mappings of Bounded Variation

In this section we develop Trapezoidal-type quadrature rules for functions that are of bounded variation. This covers a very large class of functions, unlike the traditional Trapezoidal rule which relies on the second derivative of the function for its error approximation.

3.2.1 Some Integral Inequalities

Let us start with the following integral inequality for mappings of bounded variation [18]:

THEOREM 3.1
Let $f : [a, b] \to \mathbb{R}$ be a mapping of bounded variation.
We then have the inequality:

$$\left| \int_a^b f(t)\, dt - [(x-a) f(a) + (b-x) f(b)] \right| \\ \leq \left[\frac{1}{2}(b-a) + \left| x - \frac{a+b}{2} \right| \right] \bigvee_a^b (f) \tag{3.4}$$

holding for all $x \in [a, b]$, where $\bigvee_a^b (f)$ denotes the total variation of f on the interval $[a, b]$.
The constant $\frac{1}{2}$ is the best possible one.

PROOF Using the integration by parts formula for a Riemann-Stieltjes integral, we have

$$\int_a^b (x-t)\, df(t) = (x-t) f(t) \Big|_a^b + \int_a^b f(t)\, dt$$

from which we get the identity

$$\int_a^b f(t)\,dt = (x-a)\,f(a) + (b-x)\,f(b) + \int_a^b (x-t)\,df(t) \tag{3.5}$$

for all $x \in [a,b]$.

It is well known [2, p. 159] that if $g, v : [a,b] \to \mathbb{R}$ are such that g is continuous on $[a,b]$ and v is of bounded variation on $[a,b]$, then $\int_a^b g(t)\,dv(t)$ exists and [2, p. 177]

$$\left| \int_a^b g(t)\,dv(t) \right| \le \sup_{t\in[a,b]} |g(t)| \bigvee_a^b (v). \tag{3.6}$$

Applying inequality (3.6) we can state that

$$\left| \int_a^b (x-t)\,df(t) \right| \le \sup_{t\in[a,b]} |x-t| \bigvee_a^b (f). \tag{3.7}$$

As

$$\sup_{t\in[a,b]} |x-t| = \max\{x-a, b-x\} = \frac{1}{2}(b-a) + \left| x - \frac{a+b}{2} \right|$$

then by (3.7) and (3.5) we get (3.4).

Now to prove that $\frac{1}{2}$ is the best possible. Suppose that (3.4) holds with a constant $c > 0$. That is,

$$\left| \int_a^b f(t)\,dt - [f(b)(b-x) + f(a)(x-a)] \right|$$

$$\le \left[c(b-a) + \left| x - \frac{a+b}{2} \right| \right] \bigvee_a^b (f) \tag{3.8}$$

for all $x \in [a,b]$. For $x = \frac{a+b}{2}$, we get

$$\left| \int_a^b f(t)\,dt - \frac{f(a)+f(b)}{2}(b-a) \right| \le c(b-a) \bigvee_a^b (f). \tag{3.9}$$

Consider the mapping $f : [a, b] \to \mathbb{R}$ given by

$$f(x) = \begin{cases} 0 & \text{if } x = a \\ 1 & \text{if } x \in (a, b) \\ 0 & \text{if } x = b, \end{cases}$$

then f is of bounded variation on $[a, b]$ and

$$\int_a^b f(x)\, dx = b - a, \qquad \bigvee_a^b (f) = 2.$$

Hence, from inequality (3.9) applied for this particular mapping we have

$$(b - a) \le 2c\,(b - a)$$

from which we get $c \ge \frac{1}{2}$ and thus showing that $\frac{1}{2}$ is the best constant in (3.4). ∎

REMARK 3.1 If we choose $x = \frac{a+b}{2}$, then we get (see also [13]):

$$\left| \int_a^b f(t)\, dt - \frac{f(a) + f(b)}{2}(b - a) \right| \le \frac{1}{2}(b - a) \bigvee_a^b (f) \tag{3.10}$$

which is the "trapezoid" inequality. Note that the trapezoid inequality (3.10) is in a sense the best possible inequality we can get from (3.4). Also, the constant $\frac{1}{2}$ is the best possible, as shown earlier. ∎

If we assume that the mapping f is continuous and differentiable on $[a, b]$, then we get the following corollary.

COROLLARY 3.1
Let $f \in C^{(1)}[a, b]$. Then we have the inequality

$$\left| \int_a^b f(t)\, dt - [(x - a) f(a) + (b - x) f(b)] \right|$$

$$\le \left[\frac{1}{2}(b - a) + \left| x - \frac{a + b}{2} \right| \right] \|f'\|_1 \tag{3.11}$$

for all $x \in [a,b]$, *where* $\|\cdot\|_1$ *is the* L_1*-norm, namely,*

$$\|f'\|_1 := \int_a^b |f'(t)|\,dt.$$

The following corollaries are also interesting.

COROLLARY 3.2
Let $f : [a,b] \to \mathbb{R}$ *be a Lipschitzian mapping with the constant* $L > 0$. *Then we have the inequality:*

$$\left|\int_a^b f(t)\,dt - [(x-a)f(a) + (b-x)f(b)]\right|$$

$$\leq \left[\frac{1}{2}(b-a) + \left|x - \frac{a+b}{2}\right|\right](b-a)L \tag{3.12}$$

for all $x \in [a,b]$.

PROOF The mapping f being L-Lipschitzian on $[a,b]$ implies it is also of bounded variation on $[a,b]$, since

$$\bigvee_a^b(f) = \sup_{I_n \in Div[a,b]} \sum_{i=0}^{n-1} |f(x_{i+1}) - f(x_i)|$$

$$\leq L \sup_{I_n \in Div[a,b]} |x_{i+1} - x_i| = (b-a)L,$$

and the inequality (3.12) is proved, upon using (3.4). ∎

The case of monotonic mappings is embodied in the following corollary which is a special case of Theorem 3.1.

COROLLARY 3.3
Let $f : [a,b] \to \mathbb{R}$ *be a monotonic mapping on* $[a,b]$. *Then we have the inequality:*

$$\left|\int_a^b f(t)\,dt - [(x-a)f(a) + (b-x)f(b)]\right|$$

$$\leq \left[\frac{1}{2}(b-a)+\left|x-\frac{a+b}{2}\right|\right]|f(b)-f(a)| \tag{3.13}$$

for all $x \in [a,b]$.

REMARK 3.2 The following inequality is well known in the literature as the Hermite-Hadamard inequality (see, for example, [32, p. 137]):

$$f\left(\frac{a+b}{2}\right) \leq \frac{1}{b-a}\int_a^b f(x)\,dx \leq \frac{f(a)+f(b)}{2} \tag{3.14}$$

where $f:[a,b] \to \mathbb{R}$ is a convex mapping on $[a,b]$. ▮

Using the above results, we are able to point out the following counterparts of the second Hermite-Hadamard inequality, namely, from (3.10),

$$0 \leq \frac{f(a)+f(b)}{2} - \frac{1}{b-a}\int_a^b f(t)\,dt \leq \frac{1}{2}\bigvee_a^b(f) \tag{3.15}$$

provided that f is convex and of bounded variation on $[a,b]$.

If f is convex and Lipschitzian with the constant L on $[a,b]$, then we get from (3.12)

$$0 \leq \frac{f(a)+f(b)}{2} - \frac{1}{b-a}\int_a^b f(t)\,dt \leq \frac{1}{2}(b-a)L. \tag{3.16}$$

If f is convex and monotonic on $[a,b]$, then we have from (3.13)

$$0 \leq \frac{f(a)+f(b)}{2} - \frac{1}{b-a}\int_a^b f(t)\,dt \leq \frac{1}{2}|f(b)-f(a)|. \tag{3.17}$$

Finally, if f is continuous, differentiable and convex and $f' \in L_1(a,b)$, then, from (3.11),

$$0 \leq \frac{f(a)+f(b)}{2} - \frac{1}{b-a}\int_a^b f(t)\,dt \leq \frac{1}{2}\|f'\|_1. \tag{3.18}$$

3.2.2 Applications for Quadrature Formulae

Let $I_n : a = x_0 < x_1 < \ldots < x_{n-1} < x_n = b$ be an arbitrary division of $[a,b]$ and $\xi_i \in [x_i, x_{i+1}]$, $i = 0, \ldots, n-1$; be intermediate points. Put

$h_i := x_{i+1} - x_i$ and define the sum:

$$T_P(f, I_n, \boldsymbol{\xi}) := \sum_{i=0}^{n-1} [(\xi_i - x_i) f(x_i) + (x_{i+1} - \xi_i) f(x_{i+1})].$$

We have the following result concerning the approximation of the integral $\int_a^b f(x)\,dx$ in terms of T_P [18].

THEOREM 3.2

Let $f : [a,b] \to \mathbb{R}$ be a mapping of bounded variation on $[a,b]$, then we have

$$\int_a^b f(x)\,dx = T_P(f, I_n, \boldsymbol{\xi}) + R_P(f, I_n, \boldsymbol{\xi}). \tag{3.19}$$

The remainder term $R_P(f, I_n, \boldsymbol{\xi})$ satisfies the estimate

$$|R_P(f, I_n, \boldsymbol{\xi})| \tag{3.20}$$

$$\leq \left[\frac{1}{2}\nu(h) + \max_{i=\overline{0,n-1}} \left|\xi_i - \frac{x_i + x_{i+1}}{2}\right|\right] \bigvee_a^b (f) \leq \nu(h) \bigvee_a^b (f)$$

where $\nu(h) := \max\{h_i \mid i = \overline{0, n-1}\}$.

The constant $\frac{1}{2}$ is the best possible.

PROOF Apply Theorem 3.1 on the intervals $[x_i, x_{i+1}]$ $(i = 0, ..., n-1)$ to get

$$\left|\int_{x_i}^{x_{i+1}} f(t)\,dt - [f(x_i)(\xi_i - x_i) + f(x_{i+1})(x_{i+1} - \xi_i)]\right|$$

$$\leq \left[\frac{1}{2}h_i + \left|\xi_i - \frac{x_i + x_{i+1}}{2}\right|\right] \bigvee_{x_i}^{x_{i+1}} (f),$$

for all $i \in \{0, ..., n-1\}$.

Using this and the generalized triangle inequality, we have successively

$$|R_P(f, I_n, \boldsymbol{\xi})|$$

$$\leq \sum_{i=0}^{n-1} \left|\int_{x_i}^{x_{i+1}} f(t)\,dt - [f(x_i)(\xi_i - x_i) + f(x_{i+1})(x_{i+1} - \xi_i)]\right|$$

$$\leq \sum_{i=0}^{n-1}\left[\frac{1}{2}h_i+\left|\xi_i-\frac{x_i+x_{i+1}}{2}\right|\right]\bigvee_{x_i}^{x_{i+1}}(f)$$

$$\leq \max_{i=\overline{0,n-1}}\left[\frac{1}{2}h_i+\left|\xi_i-\frac{x_i+x_{i+1}}{2}\right|\right]\sum_{i=0}^{n-1}\bigvee_{x_i}^{x_{i+1}}(f)$$

$$\leq \left[\frac{1}{2}\nu(h)+\max_{i=\overline{0,n-1}}\left|\xi_i-\frac{x_i+x_{i+1}}{2}\right|\right]\bigvee_a^b(f)$$

and the first inequality in (3.20) is proved.

For the second inequality, we observe that

$$\left|\xi_i-\frac{x_i+x_{i+1}}{2}\right|\leq\frac{1}{2}h_i,\qquad i=0,...,n-1;$$

and then

$$\max_{i=\overline{0,n-1}}\left|\xi_i-\frac{x_i+x_{i+1}}{2}\right|\leq\frac{1}{2}\nu(h).$$

Thus the theorem is proved. ∎

REMARK 3.3 If we choose $\xi_i=\frac{x_i+x_{i+1}}{2}$, then we get (see also [13]):

$$\int_a^b f(x)\,dx=T(f,I_n)+R_T(f,I_n),$$

where $T(f,I_n)$ is the "trapezoid rule", namely,

$$T(f,I_n):=\frac{1}{2}\sum_{i=0}^{n-1}\left[f(x_i)+f(x_{i+1})\right]$$

and the remainder satisfies the estimate

$$|R_T(f,I_n)|\leq\frac{1}{2}\nu(h)\bigvee_a^b(f).$$

Note that the trapezoid inequality is in a certain sense the best possible one we can get from Theorem 3.2. ∎

The following corollaries can be useful in practice.

COROLLARY 3.4
Let $f : [a,b] \to \mathbb{R}$ be a Lipschitzian mapping with the constant $L > 0$, and I_n is as above. Then we have the formula (3.19) *and the remainder satisfies the estimate*

$$|R_T(f, I_n, \xi)| \leq L \left[\frac{1}{2} \nu(h) + \max_{i=\overline{0,n-1}} \left| \xi_i - \frac{x_i + x_{i+1}}{2} \right| \right] \leq L \nu(h) . \quad (3.21)$$

COROLLARY 3.5
Let $f : [a,b] \to \mathbb{R}$ be a monotonic mapping on $[a,b]$. Then we have the quadrature formula (3.19) *and the remainder satisfies the inequality*

$$|R_T(f, I_n, \xi)| \leq \left[\frac{1}{2} \nu(h) + \max_{i=\overline{0,n-1}} \left| \xi_i - \frac{x_i + x_{i+1}}{2} \right| \right] |f(b) - f(a)|$$

$$\leq \nu(h) |f(b) - f(a)| . \quad (3.22)$$

3.3 An Estimate of the Remainder for Monotonic Mappings

Some bounds were obtained in Corollary 3.2 for monotonic mappings as a particular instance in the development for functions of bounded variation. This section treats specifically monotonic mappings, enabling tighter bounds to be determined.

3.3.1 Some Integral Inequalities

We know that, from Corollary 3.3, for a monotonic mapping $f : [a,b] \to \mathbb{R}$, the following inequality holds

$$\left| \int_a^b f(t)\,dt - [(x-a) f(a) + (b-x) f(b)] \right|$$

$$\leq \left[\frac{1}{2}(b-a) + \left| x - \frac{a+b}{2} \right| \right] |f(b) - f(a)| \quad (3.23)$$

for all $x \in [a,b]$.

Using a firmer argument, we can improve this result by obtaining tighter bounds as follows:

THEOREM 3.3
Let $f : [a, b] \to \mathbb{R}$ be a monotonic nondecreasing mapping on $[a, b]$. Then we have the inequality

$$\begin{aligned}
&\left|\int_a^b f(t)\,dt - [(x-a) f(a) + (b-x) f(b)]\right| \\
&\qquad \leq (b-x) f(b) - (x-a) f(a) + \int_a^b \operatorname{sgn}(x-t) f(t)\,dt \\
&\qquad \leq (x-a)[f(x) - f(a)] + (b-x)[f(b) - f(x)] \\
&\qquad \leq \left[\frac{1}{2}(b-a) + \left|x - \frac{a+b}{2}\right|\right][f(b) - f(a)], \qquad (3.24)
\end{aligned}$$

for all $x \in [a, b]$.
The above inequalities are sharp.

PROOF Using the integration by parts formula for a Riemann-Stieltjes integral, we have the identity as given by (3.5).

Now, assume that $\Delta_n : a = x_0^{(n)} < x_1^{(n)} < \ldots < x_{n-1}^{(n)} < x_n^{(n)} = b$ is a sequence of divisions with $\nu(\Delta_n) \to 0$ as $n \to \infty$, where $\nu(\Delta_n) := \max_{i=\overline{0,n-1}}\left(x_{i+1}^{(n)} - x_i^{(n)}\right)$ and $\xi_i^{(n)} \in \left[x_i^{(n)}, x_{i+1}^{(n)}\right]$. If $p : [a, b] \to \mathbb{R}$ is a continuous mapping on $[a, b]$ and v is monotonic nondecreasing on $[a, b]$, then

$$\begin{aligned}
\left|\int_a^b g(t)\,dv(t)\right| &= \left|\lim_{\nu(\Delta_n)\to 0} \sum_{i=0}^{n-1} p\left(\xi_i^{(n)}\right)\left[\nu\left(x_{i+1}^{(n)}\right) - \nu\left(x_i^{(n)}\right)\right]\right| \\
&\leq \lim_{\nu(\Delta_n)\to 0} \sum_{i=0}^{n-1} \left|p\left(\xi_i^{(n)}\right)\right| \left|\nu\left(x_{i+1}^{(n)}\right) - \nu\left(x_i^{(n)}\right)\right| \\
&= \lim_{\nu(\Delta_n)\to 0} \sum_{i=0}^{n-1} \left|p\left(\xi_i^{(n)}\right)\right| \left(\nu\left(x_{i+1}^{(n)}\right) - \nu\left(x_i^{(n)}\right)\right)
\end{aligned}$$

$$= \int_a^b |g(t)|\, dv(t). \tag{3.25}$$

Applying the inequality (3.25), we can state that

$$\begin{aligned}
&\left|\int_a^b (x-t)\, df(t)\right| \\
&\leq \int_a^b |x-t|\, df(t) = \int_a^x (x-t)\, df(t) + \int_x^b (t-x)\, df(t) \\
&= (x-t) f(t) \big|_a^x + \int_a^x f(t)\, dt + (t-x) f(t) \big|_t^b + \int_x^b f(t)\, dt \\
&= -(x-a) f(a) + \int_a^x f(t)\, dt + (b-x) f(b) - \int_x^b f(t)\, dt \\
&= (b-x) f(b) - (x-a) f(a) + \int_a^b \operatorname{sgn}(x-t) f(t)\, dt
\end{aligned} \tag{3.26}$$

and the first inequality in (3.24) is proved on utilising identity (3.2).

As f is monotonic nondecreasing on $[a, b]$, we can state that

$$\int_a^x f(t)\, dt \leq (x-a) f(x)$$

and

$$\int_x^b f(t)\, dt \geq (b-x) f(x)$$

and then

$$\begin{aligned}
\int_a^b \operatorname{sgn}(x-t) f(t)\, dt &= \int_a^x f(t)\, dt - \int_x^b f(t)\, dt \\
&\leq (x-a) f(x) + (x-b) f(x).
\end{aligned}$$

Therefore,

$$(b-x) f(b) - (x-a) f(a) + \int_a^b \operatorname{sgn}(x-t) f(t)\, dt$$

$$\leq (b-x) f(b) - (x-a) f(a) + (x-a) f(x) + (x-b) f(x)$$

$$= (x-a)[f(x) - f(a)] + (b-x)[f(b) - f(x)], \tag{3.27}$$

which proves the second inequality in (3.24).

As f is monotonic nondecreasing on $[a,b]$, we have $f(a) \leq f(x) \leq f(b)$ for all $x \in [a,b]$ and so

$$(x-a)\big(f(x) - f(a)\big) + (b-x)\big(f(b) - f(x)\big)$$

$$\leq \max\{x-a, b-x\}[f(x) - f(a) + f(b) - f(x)]$$

$$= \left[\frac{1}{2}(b-a) + \left|x - \frac{a+b}{2}\right|\right][f(b) - f(a)] \tag{3.28}$$

and the inequality (3.24) is completely proved.

Now to demonstrate the sharpness of the inequalities in (3.24), let $f_0 : [a,b] \to \mathbb{R}$ be given by

$$f_0(x) = \begin{cases} 0, & \text{if } x \in [a,b), \\ 1 & \text{if } x = b. \end{cases}$$

Then f_0 is monotonic nondecreasing on $[a,b]$ and realizes the equality in (3.24) for $x = \frac{a+b}{2}$, as a simple calculation shows that

$$\left|\int_a^b f(t)\,dt - \frac{(b-a)}{2}(f(a) + f(b))\right|$$

$$= \frac{(b-a)}{2}(f(b) - f(a)) + \int_a^b \operatorname{sgn}\left(\frac{a+b}{2} - t\right) f(t)\,dt$$

$$= \frac{1}{2}(b-a)(f(b) - f(a)) = \frac{b-a}{2}. \tag{3.29}$$

The theorem is thus completely proved. ∎

REMARK 3.4 For a more general result containing both the Ostrowski inequality and Simpson's inequality, see the recent paper [24]. ∎

REMARK 3.5 If we choose $x = \frac{a+b}{2}$ in (3.24), we have

$$\left| \int_a^b f(t)\,dt - \frac{(b-a)}{2}[f(a)+f(b)] \right|$$

$$\leq \frac{1}{2}(b-a)[f(b)-f(a)] + \int_a^b \operatorname{sgn}\left(\frac{a+b}{2}-t\right) f(t)\,dt$$

$$\leq \frac{1}{2}(b-a)[f(b)-f(a)], \tag{3.30}$$

which is the "trapezoid inequality". ∎

Note that the trapezoid inequality (3.30) is, in a sense, the best possible inequality we can obtain from (3.24). Moreover, the constant $\frac{1}{2}$ is the best possible for both inequalities.

REMARK 3.6 The following inequality is well known in the literature as the Hermite-Hadamard inequality (see also (3.22))

$$f\left(\frac{a+b}{2}\right) \leq \frac{1}{b-a}\int_a^b f(x)\,dx \leq \frac{f(a)+f(b)}{2}, \tag{3.31}$$

provided that $f : [a,b] \to \mathbb{R}$ is a convex mapping on $[a,b]$. ∎

Using the above inequality (3.30), we can state that

$$0 \leq \frac{f(a)+f(b)}{2} - \frac{1}{b-a}\int_a^b f(t)\,dt$$

$$\leq \frac{1}{2}[f(b)-f(a)] + \frac{1}{b-a}\int_a^b \left(\frac{a+b}{2}-t\right) f(t)\,dt$$

$$\leq \frac{1}{2}[f(b)-f(a)], \tag{3.32}$$

provided that f is monotonic nondecreasing and convex on $[a,b]$.

3.3.2 Applications for Quadrature Formulae

Let $I_n : a = x_0 < x_1 < ... < x_{n-1} < x_n = b$ be an arbitrary division of $[a, b]$ and $\xi_i \in [x_i, x_{i+1}]$ $(i = 0, ..., n-1)$ be intermediate points. Put $h_i := x_{i+1} - x_i$ and consider the sum

$$T_P(f, I_n, \boldsymbol{\xi}) := \sum_{i=0}^{n-1} \left[(\xi_i - x_i) f(x_i) + (x_{i+1} - \xi_i) f(x_{i+1}) \right].$$

In Section 3.2, Corollary 3.5, we proved that for a monotonic mapping $f : [a, b] \to \mathbb{R}$ we have

$$\int_a^b f(t)\, dt = T_P(f, I_n, \boldsymbol{\xi}) + R_P(f, I_n, \boldsymbol{\xi}) \tag{3.33}$$

and the remainder $R_P(f, I_n, \boldsymbol{\xi})$ satisfies the bound as given by (3.22) where $\nu(h)$ is the norm of the division I_n, that is, $\nu(h) = \max_{i=\overline{0,n-1}} h_i$.

We can improve this result as follows.

THEOREM 3.4

Let $f : [a, b] \to \mathbb{R}$ be a monotonic nondecreasing mapping on $[a, b]$ and $I_n, \boldsymbol{\xi}$ as above. Then we have the formula (3.33) and the remainder $R_P(f, I_n, \boldsymbol{\xi})$ satisfies the estimate

$$\begin{aligned}
&|R_P(f, I_n, \boldsymbol{\xi})| \\
&\quad \le \sum_{i=0}^{n-1} \left[(x_{i+1} - \xi_i) f(x_{i+1}) - (\xi_i - x_i) f(x_i) \right] \\
&\qquad + \sum_{i=0}^{n-1} \int_{x_i}^{x_{i+1}} \operatorname{sgn}(\xi_i - t) f(t)\, dt \\
&\quad \le \sum_{i=0}^{n-1} (\xi_i - x_i) \left[f(\xi_i) - f(x_i) \right] + \sum_{i=0}^{n-1} (x_{i+1} - \xi_i) \left[f(x_{i+1}) - f(\xi_i) \right] \\
&\quad \le \sum_{i=0}^{n-1} \left[\frac{1}{2} h_i + \left| \xi_i - \frac{x_{i+1} - x_i}{2} \right| \right] (f(x_{i+1}) - f(x_i))
\end{aligned}$$

$$\leq \left[\frac{1}{2}\nu(h) + \max_{i=\overline{0,n-1}} \left| \xi_i - \frac{x_{i+1}-x_i}{2} \right| \right] (f(b) - f(a))$$

$$\leq \nu(h)(f(b) - f(a)). \tag{3.34}$$

The proof is obvious by Theorem 3.3 applied on the intervals $[x_i, x_{i+1}]$ $(i = 0, ..., n)$. We omit the details.

Now, if we consider the classical trapezoidal formula

$$T(f, I_n) := \frac{1}{2}\sum_{i=1}^{n-1} [f(x_i) + f(x_{i+1})] h_i,$$

then we can state the following corollary.

COROLLARY 3.6

Let $f : [a, b] \to \mathbb{R}$ be a monotonic nondecreasing mapping on $[a, b]$. Then we have

$$\int_a^b f(t)\,dt = T(f, I_n) + R_T(f, I_n), \tag{3.35}$$

where the remainder satisfies the estimate

$$|R_T(f, I_n)|$$

$$\leq \frac{1}{2}\sum_{i=0}^{n-1} h_i [f(x_{i+1}) - f(x_i)] + \sum_{i=0}^{n-1}\int_{x_i}^{x_{i+1}} \operatorname{sgn}\left(\frac{x_i + x_{i+1}}{2} - t\right) f(t)\,dt$$

$$\leq \frac{1}{2}\sum_{i=0}^{n-1} h_i [f(x_{i+1}) - f(x_i)]$$

$$\leq \frac{1}{2}\nu(h)(f(b) - f(a)). \tag{3.36}$$

3.4 An Estimate of the Remainder for Lipschitzian Mappings

We know that, from Corollary 3.2, for a mapping $f : [a, b] \to \mathbb{R}$, which is L-Lipschitzian, so that $f(\cdot)$ satisfies

$$|f(x) - f(y)| \leq L|x - y| \text{ for all } x, y \in [a, b], \tag{3.37}$$

where $L > 0$ is given, we have the inequality

$$\left| \int_a^b f(t)\, dt - [f(a)(x-a) + f(b)(b-x)] \right| \leq \left[\frac{1}{2}(b-a) + \left| x - \frac{a+b}{2} \right| \right] (b-a) L \tag{3.38}$$

for all $x \in [a, b]$.

Using a firmer argument, we are able now to improve this result as follows.

3.4.1 Inequalities for Lipschitzian Mappings

The following theorem holds [16].

THEOREM 3.5
Let $f : [a, b] \to \mathbb{R}$ be an L-Lipschitzian mapping on $[a, b]$. Then we have the inequality

$$\left| \int_a^b f(t)\, dt - [(x-a) f(a) + (b-x) f(b)] \right| \leq \left[\frac{1}{4}(b-a)^2 + \left(x - \frac{a+b}{2} \right)^2 \right] L \tag{3.39}$$

for all $x \in [a, b]$.

The constant $\frac{1}{4}$ is the best in (3.39).

PROOF As f is L-Lipschitzian and thus of bounded variation, the following Riemann-Stieltjes integral $\int_a^b (x-t)\,df(t)$ exists and (see (3.5))

$$\int_a^b f(t)\,dt = (x-a)\,f(a) + (b-x)\,f(b) + \int_a^b (x-t)\,df(t).$$

Now, assume that $\Delta_n := a = x_0^{(n)} < x_1^{(n)} < \ldots < x_{n-1}^{(n)} < x_n^{(n)} = b$ is a sequence of divisions with $\nu(\Delta_n) \to 0$ as $n \to \infty$, where $\nu(\Delta_n) := \max_{i=\overline{0,n-1}} \left(x_{i+1}^{(n)} - x_i^{(n)}\right)$ and $\xi_i^{(n)} \in \left[x_i^{(n)}, x_{i+1}^{(n)}\right]$.

If $p : [a,b] \to \mathbb{R}$ is continuous on $[a,b]$ and $\nu : [a,b] \to \mathbb{R}$ is L-Lipschitzian, then the Riemann-Stieltjes integral $\int_a^b p(x)\,dv(x)$ exists and

$$\begin{aligned}
&\left|\int_a^b p(x)\,dv(x)\right| \\
&= \left|\lim_{\nu(\Delta_n)\to 0} \sum_{i=0}^{n-1} p\left(\xi_i^{(n)}\right)\left[\nu\left(x_{i+1}^{(n)}\right) - \nu\left(x_i^{(n)}\right)\right]\right| \\
&\leq \lim_{\nu(\Delta_n)\to 0} \sum_{i=0}^{n-1} \left|p\left(\xi_i^{(n)}\right)\right| \left(x_{i+1}^{(n)} - x_i^{(n)}\right) \left|\frac{\nu\left(x_{i+1}^{(n)}\right) - \nu\left(x_i^{(n)}\right)}{x_{i+1}^{(n)} - x_i^{(n)}}\right| \\
&\leq L \lim_{\nu(\Delta_n)\to 0} \sum_{i=0}^{n-1} \left|p\left(\xi_i^{(n)}\right)\right| \left(x_{i+1}^{(n)} - x_i^{(n)}\right) \\
&= L \int_a^b |p(x)|\,dx. \qquad (3.40)
\end{aligned}$$

Applying the inequality (3.40) for $p(t) = x - t$ and $v(t) = f(t)$, $t \in [a,b]$, we obtain

$$\begin{aligned}
\left|\int_a^b (x-t)\,df(t)\right| &\leq L \int_a^b |x-t|\,dt \\
&= L\left[\int_a^x (x-t)\,dt + \int_x^b (t-x)\,dt\right]
\end{aligned}$$

$$= L\left[\frac{(x-a)^2+(b-x)^2}{2}\right]$$

$$= L\left[\frac{1}{4}(b-a)^2+\left(x-\frac{a+b}{2}\right)^2\right],$$

and the inequality (3.39) is proved.

To prove the sharpness of the constant $\frac{1}{4}$, assume that (3.39) holds with a constant $C>0$ instead. That is,

$$\left|\int_a^b f(t)\,dt-[(b-x)f(b)+(x-a)f(a)]\right|$$

$$\leq\left[C(b-a)^2+\left(x-\frac{a+b}{2}\right)^2\right], \tag{3.41}$$

for any L-Lipschitzian mapping and any $x\in[a,b]$.

Consider the mapping $f_0:[a,b]\to M$, $f_0(x)=\left|x-\frac{a+b}{2}\right|$. Then

$$|f(x)-f(y)|=\left|\left|x-\frac{a+b}{2}\right|-\left|y-\frac{a+b}{2}\right|\right|\leq|x-y|$$

for all $x,y\in[a,b]$, which shows that f_0 is L-Lipschitzian with the constant $L=1$.

We have

$$\int_a^b f_0(x)\,dx-\frac{f_0(a)+f_0(b)}{2}(b-a)$$

$$=\frac{(b-a)^2}{4}-\frac{(b-a)^2}{2}=-\frac{(b-a)^2}{4}$$

and

$$L(b-a)^2=(b-a)^2,$$

which shows that, for $x=\frac{a+b}{2}$, the inequality (3.41) becomes

$$\left|\int_a^b f(x)\,dx-\frac{f(a)+f(b)}{2}(b-a)\right|\leq CL(b-a)^2,$$

which is equivalent to

$$\frac{(b-a)^2}{4} \leq C(b-a)^2,$$

thus implying that $C \geq \frac{1}{4}$, and the theorem is thus proved. ∎

REMARK 3.7 If we choose $x = \frac{a+b}{2}$, then we have

$$\left| \int_a^b f(t)\,dt - \frac{f(a)+f(b)}{2}(b-a) \right| \leq \frac{1}{4}(b-a)^2 L, \tag{3.42}$$

which is the "trapezoid inequality". Note that the trapezoid inequality (3.42) is, in a sense, the best possible inequality we can obtain from (3.39). In addition, the constant $\frac{1}{4}$ is the best possible one, providing the sharpest bound in the class. ∎

COROLLARY 3.7

Let $f \in C^{(1)}[a,b]$. That is, f is differentiable on (a,b) and the derivative is continuous on (a,b), and put $\|f'\|_\infty = \sup_{t\in(a,b)} |f'(t)| < \infty$. Then we have the inequality

$$\left| \int_a^b f(t)\,dt - [(x-a)f(a) + (b-x)f(b)] \right|$$

$$\leq \left[\frac{1}{4}(b-a)^2 + \left(x - \frac{a+b}{2}\right)^2 \right] \|f'\|_\infty \tag{3.43}$$

for all $x \in [a,b]$.

REMARK 3.8 Now, if we assume that $f:[a,b] \to \mathbb{R}$ is differentiable convex on (a,b) and the derivative $f':(a,b)\to\mathbb{R}$ is bounded on (a,b), i.e., $\|f'\|_\infty < \infty$, then we have the following converse of the second Hermite-Hadamard inequality

$$0 \leq \frac{f(a)+f(b)}{2} - \frac{1}{b-a}\int_a^b f(t)\,dt \leq \frac{1}{4}(b-a)\|f'\|_\infty. \tag{3.44}$$

∎

3.4.2 Applications for Quadrature Formulae

Let us reconsider the generalised trapezoid quadrature formula

$$T_P(f, I_n, \xi) := \sum_{i=0}^{n-1} [(\xi_i - x_i) f(x_i) + (x_{i+1} - \xi_i) f(x_{i+1})],$$

provided that $I_n : a = x_0 < x_1 < ... < x_{n-1} < x_n = b$ is an arbitrary division of $[a, b]$, $\xi_i \in [x_i, x_{i+1}]$ $(i = 0, ..., n-1)$ are intermediate points, $h_i := x_{i+1} - x_i$ are the step sizes and $\nu(h) := \max_{i=\overline{0,n}} \{h_i\}$ is the norm of the division.

We can improve Corollary 3.4 in the following manner.

THEOREM 3.6
Let $f : [a, b] \to \mathbb{R}$ be a L-Lipschitzian mapping and I_n, $\boldsymbol{\xi}$ be as above. Then we have the formula

$$\int_a^b f(t)\,dt = T_P(f, I_n, \boldsymbol{\xi}) + R_P(f, I_n, \boldsymbol{\xi}), \tag{3.45}$$

where the remainder $R_P(f, I_n, \boldsymbol{\xi})$ is such that it satisfies the estimate

$$|R_P(f, I_n, \boldsymbol{\xi})| \leq \frac{1}{4} L \sum_{i=0}^{n-1} h_i^2 + L \sum_{i=0}^{n-1} \left(\xi_i - \frac{x_{i+1} + x_i}{2}\right)^2$$

$$\leq \frac{1}{2} L \sum_{i=0}^{n-1} h_i^2 \leq \frac{1}{2} L (b - a) \nu(h). \tag{3.46}$$

The proof follows by Theorem 3.5 applied on the interval $[x_i, x_{i+1}]$ $(i = 0, ..., n-1)$.

REMARK 3.9 If we choose $\xi_i = \frac{x_i + x_{i+1}}{2}$, then we obtain the trapezoid formula where the remainder $R_T(f, I_n)$ satisfies the estimate

$$|R_T(f, I_n)| \leq \frac{1}{4} L \sum_{i=0}^{n-1} h_i^2 \leq \frac{1}{4} L (b - a) \nu(h), \tag{3.47}$$

where $\nu(h) = \max\{h_i | i = 0, 1, ..., n-1\}$. ∎

3.5 A Generalization for Derivatives Which are Absolutely Continuous

3.5.1 Integral Identities

We start with the following result [7]. For work on related matters see [11, 12, 15, 20, 21, 23, 27]

THEOREM 3.7
Let $f : [a,b] \to \mathbb{R}$ be a mapping such that the derivative $f^{(n-1)}$ $(n \geq 1)$ is absolutely continuous on $[a,b]$. Then

$$\int_a^b f(t)\,dt = \sum_{k=0}^{n-1} \frac{1}{(k+1)!} \left[(x-a)^{k+1} f^{(k)}(a) + (-1)^k (b-x)^{k+1} f^{(k)}(b) \right]$$

$$+ \frac{1}{n!} \int_a^b (x-t)^n f^{(n)}(t)\,dt, \tag{3.48}$$

for all $x \in [a,b]$.

PROOF The proof is by mathematical induction.

For $n = 1$, we have to prove that

$$\int_a^b f(t)\,dt = (x-a) f(a) + (b-x) f(b) + \int_a^b (x-t) f^{(1)}(t)\,dt, \tag{3.49}$$

which is straightforward as may be seen by the integration by parts formula applied for the integral

$$\int_a^b (x-t) f^{(1)}(t)\,dt.$$

Assume that (3.48) holds for "n" and let us prove it for "$n+1$". That is, we wish to show that

$$\int_a^b f(t)\,dt = \sum_{k=0}^{n} \frac{1}{(k+1)} \left[(x-a)^{k+1} f^{(k)}(a) + (-1)^k (b-x)^{k+1} f^{(k)}(b) \right]$$

$$+\frac{1}{(n+1)!}\int_a^b (x-t)^{n+1} f^{(n+1)}(t)\,dt. \tag{3.50}$$

For this purpose, we apply formula (3.49) for the mapping $g(t) := (x-t)^n f^{(n)}(t)$, which is absolutely continuous on $[a,b]$, and then, we can write

$$\begin{aligned}
&\int_a^b (x-t)^n f^{(n)}(t)\,dt \\
&= (x-a)(x-a)^n f^{(n)}(a) + (b-x)(x-b)^n f^{(n)}(b) \\
&+ \int_a^b (x-t)\frac{d}{dt}\left[(x-t)^n f^{(n)}(t)\right] dt \\
&= \int_a^b (x-t)\left[-n(x-t)^{n-1} f^{(n)}(t) + (x-t)^n f^{(n+1)}(t)\right] dt \\
&+ (x-a)^{n+1} f^{(n)}(a) + (-1)^n (b-x)^{n+1} f^{(n)}(b) \\
&= -n\int_a^b (x-t)^n f^{(n)}(t)\,dt + \int_a^b (x-t)^{n+1} f^{(n+1)}(t)\,dt \\
&+ (x-a)^{n+1} f^{(n)}(a) + (-1)^n (b-x)^{n+1} f^{(n)}(b).
\end{aligned} \tag{3.51}$$

From identity (3.51) we can get

$$\begin{aligned}
&\int_a^b (x-t)^n f^{(n)}(t)\,dt \\
&= \frac{1}{n+1}\int_a^b (x-t)^{n+1} f^{(n+1)}(t)\,dt \\
&+ \frac{1}{n+1}\left[(x-a)^{n+1} f^{(n)}(a) + (-1)^n (b-x)^{n+1} f^{(n)}(b)\right].
\end{aligned} \tag{3.52}$$

Now, using the induction hypothesis, we have

$$\int_a^b f(t)\,dt = \sum_{k=0}^{n-1} \frac{1}{(k+1)!}\left[(x-a)^{k+1} f^{(k)}(a) + (-1)^k (b-x)^{k+1} f^{(k)}(b)\right]$$

$$+\frac{1}{n!}\left[\frac{1}{n+1}\int_a^b (x-t)^{n+1} f^{(n+1)}(t)\,dt\right.$$

$$\left.+\frac{1}{n+1}\left[(x-a)^{n+1} f^{(n)}(a)+(b-x)^{n+1} f^{(n)}(b)\right]\right]$$

$$=\sum_{k=0}^{n}\frac{1}{(k+1)!}\left[(x-a)^{k+1} f^{(k)}(a)+(-1)^k (b-x)^{k+1} f^{(k)}(b)\right]$$

$$+\frac{1}{(n+1)!}\int_a^b (x-t)^{n+1} f^{(n+1)}(t)\,dt$$

and the identity (3.50) is obtained. This completes the proof. ∎

The following corollary is useful in practice.

COROLLARY 3.8
With the above assumptions for f and R, we have the particular identities (which can also be obtained by using Taylor's formula with the integral remainder)

$$\int_a^b f(t)\,dt=\sum_{k=0}^{n-1}\frac{(-1)^k}{(k+1)!}(b-a)^{k+1} f^{(k)}(b)$$

$$+\frac{(-1)^n}{n!}\int_a^b (t-a)^n f^{(n)}(t)\,dt, \tag{3.53}$$

$$\int_a^b f(t)\,dt=\sum_{k=0}^{n-1}\frac{1}{(k+1)!}(b-a)^{k+1} f^{(k)}(a)+\frac{1}{n!}\int_a^b (b-t)^n f^{(n)}(t)\,dt, \tag{3.54}$$

and the identity (see also [10])

$$\int_a^b f(t)\,dt=\sum_{k=0}^{n-1}\frac{1}{(k+1)!}\left(\frac{b-a}{2}\right)^{k+1}\left[f^{(k)}(a)+(-1)^k f^{(k)}(b)\right]$$

$$+\frac{(-1)^n}{n!}\int_a^b\left(t-\frac{a+b}{2}\right)^n f^{(n)}(t)\,dt. \tag{3.55}$$

Here (3.53), (3.54) and (3.55) are obtained from (3.48) with $t = b,\ a,\ \frac{a+b}{2}$, *respectively.*

REMARK 3.10

a) For $n = 1$, we get the identity (3.49) which is a generalization of the trapezoid rule. Further, with $x = \frac{a+b}{2}$, we get ([14]), (3.55) with $n = 1$. Namely,

$$\int_a^b f(t)\,dt = \frac{b-a}{2}(f(a)+f(b)) - \int_a^b \left(t-\frac{a+b}{2}\right) f'(t)\,dt \tag{3.56}$$

which is the "trapezoid rule".

b) For $n = 2$, we get the identity:

$$\int_a^b f(t)\,dt = (x-a) f(a) + (b-x) f(b)$$
$$+\frac{1}{2}\left[(x-a)^2 f'(a) + (b-x)^2 f'(b)\right]$$
$$+\frac{1}{2}\int_a^b (x-t)^2 f''(t)\,dt. \tag{3.57}$$

Further, with $x = \frac{a+b}{2}$, we capture the "perturbed trapezoid rule" [10], (or, equivalently, $n = 2$ in (3.55))

$$\int_a^b f(t)\,dt = \frac{b-a}{2}[f(a)+f(b)] + \frac{(b-a)^2}{8}(f'(a)-f'(b))$$
$$+\frac{1}{2}\int_a^b \left(t-\frac{a+b}{2}\right)^2 f''(t)\,dt. \tag{3.58}$$

∎

3.5.2 Integral Inequalities

Using the integral representation by Theorem 3.7, we can prove the following inequality [7].

THEOREM 3.8
Let $f : [a,b] \to \mathbb{R}$ be a mapping so that the derivative $f^{(n-1)}$ $(n \geq 1)$ is absolutely continuous on $[a,b]$. Then

$$\left| \int_a^b f(t)\,dt - \sum_{k=0}^{n-1} \frac{1}{(k+1)!} \left[(x-a)^{k+1} f^{(k)}(a) + (-1)^k (b-x)^{k+1} f^{(k)}(b) \right] \right|$$

$$\leq \begin{cases} \frac{\|f^{(n)}\|_\infty}{(n+1)!} \left[(x-a)^{n+1} + (b-x)^{n+1} \right] & \text{if } f^{(n)} \in L_\infty[a,b], \\ \frac{\|f^{(n)}\|_p}{n!} \left[\frac{(x-a)^{nq+1} + (b-x)^{nq+1}}{nq+1} \right]^{\frac{1}{q}} & \text{if } p > 1, \frac{1}{p} + \frac{1}{q} = 1 \\ & \text{and } f^{(n)} \in L_p[a,b], \\ \frac{\|f^{(n)}\|_1}{n!} \left[\frac{1}{2}(b-a) + \left| x - \frac{a+b}{2} \right| \right]^n & \text{if } f^{(n)} \in L_1[a,b] \end{cases} \tag{3.59}$$

for all $x \in [a,b]$.

PROOF From equation (3.48) and the properties of the modulus, we have

$$\left| \int_a^b f(t)\,dt - \sum_{k=0}^{n-1} \frac{1}{(k+1)!} \left[(x-a)^{k+1} f^{(k)}(a) + (-1)^k (b-x)^{k+1} f^{(k)}(b) \right] \right|$$

$$\leq \frac{1}{n!} \int_a^b |x-t|^n \left| f^{(n)}(t) \right| dt =: R(x).$$

Observe that

$$R(x) \leq \left[\frac{1}{n!} \int_a^b |x-t|^n\,dt \right] \left\| f^{(n)} \right\|_\infty$$

$$= \frac{\left\| f^{(n)} \right\|_\infty}{n!} \left[\int_a^x (x-t)^n\,dt + \int_x^b (t-x)^n\,dt \right]$$

$$= \frac{\left\| f^{(n)} \right\|_\infty}{n!} \left[\frac{(x-a)^{n+1} + (b-x)^{n+1}}{n+1} \right]$$

and the first inequality in (3.59) is proved.

Using Hölder's integral inequality, we also have

$$R(x) \leq \frac{1}{n!}\left(\int_a^b \left|f^{(n)}(t)\right|^p dt\right)^{\frac{1}{p}} \left(\int_a^b |x-t|^{nq}\, dt\right)^{\frac{1}{q}}$$

$$= \frac{1}{n!}\left\|f^{(n)}\right\|_p \left[\frac{(x-a)^{nq+1} + (b-x)^{nq+1}}{nq+1}\right]^{\frac{1}{q}},$$

which proves the second inequality in (3.59).

Finally, let us observe that

$$R(x) \leq \frac{1}{n!} \sup_{t\in[a,b]} |x-t|^n \int_a^b \left|f^{(n)}(t)\right| dt$$

$$= \frac{1}{n!}\left[\sup_{t\in[a,b]} |x-t|\right]^n \left\|f^{(n)}\right\|_1$$

$$= \frac{1}{n!}\left[\max(x-a, b-x)\right]^n \left\|f^{(n)}\right\|_1$$

$$= \frac{1}{n!}\left[\frac{1}{2}(b-a) + \left|x - \frac{a+b}{2}\right|\right]^n \left\|f^{(n)}\right\|_1$$

and the theorem is completely proved. ∎

The following corollary is useful in practice.

COROLLARY 3.9

With the above assumptions for f and n, we have the particular inequalities (see also [10])

$$\left|\int_a^b f(t)\, dt - \sum_{k=0}^{n-1} \frac{1}{(k+1)!}\left(\frac{b-a}{2}\right)^{k+1} \left[f^{(k)}(a) + (-1)^k f^{(k)}(b)\right]\right|$$

$$\leq \begin{cases} \frac{\left\|f^{(n)}\right\|_\infty}{2^n(n+1)!}(b-a)^{n+1} & \textit{if } f^{(n)} \in L_\infty[a,b], \\ \frac{\left\|f^{(n)}\right\|_p}{2^n n!(nq+1)^{1/q}}(b-a)^{n+\frac{1}{q}} & \textit{if } p>1, \frac{1}{p}+\frac{1}{q}=1 \\ & \textit{and } f^{(n)} \in L_p[a,b], \\ \frac{\left\|f^{(n)}\right\|_1}{2^n n!}(b-a)^n; & \textit{if } f^{(n)} \in L_1[a,b], \end{cases} \tag{3.60}$$

respectively, which are the sharpest possible from (3.59) with $x = \frac{a+b}{2}$.

REMARK 3.11 If we put $n = 1$ in (3.59), we capture the inequality

$$\left| \int_a^b f(t)\,dt - [(x-a)f(a) + (b-x)f(b)] \right|$$

$$\leq \begin{cases} \left[\frac{1}{4}(b-a)^2 + \left(x - \frac{a+b}{2}\right)^2\right] \|f'\|_\infty & \text{if} \quad f' \in L_\infty[a,b], \\ \|f'\|_p \left[\frac{(x-a)^{q+1}+(b-x)^{q+1}}{q+1}\right]^{\frac{1}{q}} & \text{if} \quad p>1, \frac{1}{p}+\frac{1}{q}=1 \\ & \text{and} \quad f' \in L_p[a,b], \\ \left[\frac{1}{2}(b-a) + \left|x - \frac{a+b}{2}\right|\right] \|f'\|_1; & \text{if} \quad f' \in L_1[a,b] \end{cases} \tag{3.61}$$

for all $x \in [a,b]$, and, in particular, the "trapezoid" inequality

$$\left| \int_a^b f(t)\,dt - \frac{b-a}{2}[f(a) + f(b)] \right|$$

$$\leq \begin{cases} \frac{\|f'\|_\infty}{4}(b-a)^2 & \text{if } f' \in L_\infty[a,b], \\ \frac{\|f'\|_p}{2(q+1)^{1/q}}(b-a)^{1+\frac{1}{q}} & \text{if } f' \in L_p[a,b], \\ \frac{\|f'\|_1}{2}(b-a) & \text{if } f' \in L_1[a,b] \end{cases} \tag{3.62}$$

is obtained by taking $x = \frac{a+b}{2}$. ∎

REMARK 3.12 If we put $n = 2$ in (3.59), we get the inequality

$$\left| \int_a^b f(t)\,dt - [(x-a) f(a) + (b-x) f(b)] \right.$$

$$\left. -\frac{1}{2}\left[(x-a)^2 f'(a) - (b-x)^2 f'(b)\right] \right|$$

$$\leq \begin{cases} \frac{\|f''\|_\infty}{6}\left[(b-a)^3 + (b-x)^3\right] & \text{if} \quad f'' \in L_\infty[a,b]; \\ \frac{\|f''\|_p}{2}\left[\frac{(x-a)^{2q+1}+(b-x)^{2q+1}}{2q+1}\right]^{\frac{1}{q}} & \text{if} \quad p>1, \frac{1}{p}+\frac{1}{q}=1 \\ & \text{and } f'' \in L_p[a,b]; \\ \frac{\|f''\|_1}{2}\left[\frac{1}{2}(b-a) + \left|x - \frac{a+b}{2}\right|\right]^2 & \text{if} \quad f'' \in L_1[a,b] \end{cases} \tag{3.63}$$

for all $x \in [a,b]$, and, in particular, the "perturbed trapezoid" inequality

$$\left| \int_a^b f(t)\,dt - \frac{b-a}{2}[f(a)+f(b)] + \frac{(b-a)^2}{8}(f'(b)-f'(a)) \right|$$

$$\leq \begin{cases} \frac{\|f''\|_\infty}{24}(b-a)^3 & \text{if } f'' \in L_\infty[a,b]; \\ \frac{\|f''\|_p}{8(2q+1)^{1/q}}(b-a)^{2+\frac{1}{q}} & \text{if } f'' \in L_p[a,b]; \\ \frac{\|f''\|_1}{8}(b-a)^2 & \text{if } f'' \in L_1[a,b] \end{cases} \tag{3.64}$$

is obtained on taking $x = \frac{a+b}{2}$. ∎

In practice the perturbed trapezoid inequality only involves the evaluation of the derivatives at the boundary points for a uniform partition of the interval.

3.5.3 A Perturbed Version

A *premature* Grüss inequality is embodied in the following theorem which was considered and applied for the first time in the paper [29] by Matić, Pečarić and Ujević. Dragomir [12] obtains an estimate of the remainder for a Taylor series using the Grüss inequality (see also [9] and [17]).

THEOREM 3.9

Let h, g be integrable functions defined on $[a,b]$ and let $d \leq g(t) \leq D$. Then

$$|T(h,g)| \leq \frac{D-d}{2}\left[T(h,h)\right]^{\frac{1}{2}}, \tag{3.65}$$

where

$$T(h,g) = \frac{1}{b-a}\int_a^b h(t)\,g(t)\,dt - \frac{1}{b-a}\int_a^b h(t)\,dt \cdot \frac{1}{b-a}\int_a^b g(t)\,dt.$$

REMARK 3.13 For some applications of this result for three-point quadrature formulae see [5]. ∎

Using the above theorem, the following result may be stated [7].

THEOREM 3.10

Let $f : [a,b] \to \mathbb{R}$ so that the derivative $f^{(n-1)}$, $n \geq 1$ is absolutely continuous on $[a,b]$. Assume that there exist constants $\gamma, \Gamma \in \mathbb{R}$ such that $\gamma \leq f^{(n)}(t) \leq \Gamma$ a.e on $[a,b]$. Then, the following inequality holds

$$\begin{aligned}|P_T(x)| :=& \left| \int_a^b f(t)\,dt - \sum_{k=0}^{n-1}\left(\frac{1}{(k+1)!}\times \right.\right.\\ &\left.\left[(x-a)^{k+1} f^{(k)}(a) + (-1)^k (b-x)^{k+1} f^{(k)}(b)\right]\right)\\ &\left. - \frac{(x-a)^{n+1} + (-1)^n (b-x)^{n+1}}{(n+1)!}\left[\frac{f^{(n-1)}(b) - f^{(n-1)}(a)}{b-a}\right]\right|\\ \leq& \frac{\Gamma-\gamma}{2}\cdot\frac{1}{n!} I(x,n)\\ \leq& \frac{\Gamma-\gamma}{2}\cdot\frac{n}{n+1}\cdot\frac{(b-a)^{n+1}}{\sqrt{2n+1}},\end{aligned} \tag{3.66}$$

where

$$I(x,n) = \frac{1}{(n+1)\sqrt{2n+1}}\left\{n^2 (b-a)\left[(x-a)^{2n+1} + (b-x)^{2n+1}\right]\right.$$

$$+ (2n+1)(x-a)(b-x)\left[(x-a)^n - (x-b)^n\right]^2 \Big\}^{\frac{1}{2}}. \quad (3.67)$$

PROOF Applying the premature Grüss result (3.65) on $(x-t)^n$ and $f^{(n)}(t)$, we have

$$\left| \frac{1}{b-a} \int_a^b (x-t)^n f^{(n)}(t)\,dt - \frac{1}{b-a} \int_a^b (x-t)^n \,dt \cdot \frac{1}{b-a} \int_a^b f^{(n)}(t)\,dt \right|$$

$$\leq \frac{\Gamma-\gamma}{2} \left\{ \frac{1}{b-a} \int_a^b (x-t)^{2n}\,dt - \left[\frac{1}{b-a} \int_a^b (x-t)^n\,dt \right]^2 \right\}^{\frac{1}{2}}.$$

Therefore,

$$\left| \frac{1}{b-a} \int_a^b (x-t)^n f^{(n)}(t)\,dt \right.$$

$$\left. - \frac{(x-a)^{n+1} + (-1)^n (b-x)^{n+1}}{(n+1)(b-a)} \cdot \frac{f^{(n-1)}(b) - f^{(n-1)}(a)}{b-a} \right|$$

$$\leq \frac{\Gamma-\gamma}{2} \left\{ \frac{(x-a)^{2n+1} + (b-x)^{2n+1}}{(2n+1)(b-a)} \right.$$

$$\left. - \left[\frac{(x-a)^{n+1} + (-1)^n (b-x)^{n+1}}{(b-a)(n+1)} \right]^2 \right\}^{\frac{1}{2}}.$$

Further, simplification of the above result by multiplying throughout by $\frac{b-a}{n!}$ gives

$$\left| \frac{1}{n!} \int_a^b (x-t)^n f^{(n)}(t)\,dt \right.$$

$$\left. - \frac{(x-a)^{n+1} + (-1)^n (b-x)^{n+1}}{(n+1)!} \cdot \left[\frac{f^{(n-1)}(b) - f^{(n-1)}(a)}{b-a} \right] \right|$$

$$\leq \frac{\Gamma - \gamma}{2} \cdot \frac{1}{n!} J(x, n), \tag{3.68}$$

where

$$J^2(x, n) = \frac{1}{(2n+1)(n+1)^2} \left\{ (n+1)^2 (A+B) \left(A^{2n+1} + B^{2n+1} \right) \right.$$

$$\left. - (2n+1) \left(A^{n+1} + (-1)^n B^{n+1} \right)^2 \right\} \tag{3.69}$$

with $A = x - a$, $B = b - x$.

Now, from (3.69),

$$(2n+1)(n+1)^2 J^2(x, n)$$

$$= n^2 (A+B) \left(A^{2n+1} + B^{2n+1} \right)$$

$$+ (2n+1) \left[(A+B) \left(A^{2n+1} + B^{2n+1} \right) - \left(A^{n+1} + (-1)^n B^{n+1} \right)^2 \right]$$

$$= n^2 (A+B) \left(A^{2n+1} + B^{2n+1} \right)$$

$$+ (2n+1) \left[AB \left(A^{2n} + B^{2n} \right) - 2A^{n+1} \cdot (-1)^n B^{n+1} \right]$$

$$= n^2 (A+B) \left[A^{2n+1} + B^{2n+1} \right] + (2n+1) AB \left[A^n - (-B)^n \right]^2.$$

Now, substitution of $A = x - a$, $B = b - x$ and the fact that $A + B = b - a$ gives $I(x, n) = \frac{J(x,n)}{(n+1)\sqrt{2n+1}}$, as presented in (3.67). Substitution of identity (3.48) into (3.68) gives (3.66) and the first part of the theorem is thus proved.

The upper bound is obtained by taking either $I(a, n)$ or $I(b, n)$ since $I(x, n)$ is convex and symmetric. Hence the theorem is completely proved. ∎

COROLLARY 3.10

Let the conditions of Theorem 3.10 hold. Then the following result holds:

$$\left| \int_a^b f(t)\, dt - \sum_{k=0}^{n-1} \frac{1}{(k+1)!} \left(\frac{b-a}{2} \right)^{k+1} \left[f^{(k)}(a) + (-1)^k f^{(k)}(b) \right] \right.$$

$$- \left(\frac{b-a}{2}\right)^{n+1} \frac{[1+(-1)^n]}{(n+1)!} \left[\frac{f^{(n-1)}(b) - f^{(n-1)}(a)}{b-a}\right]\Bigg|$$

$$\leq \frac{\Gamma-\gamma}{2} \cdot \frac{1}{n!} \left(\frac{b-a}{2}\right)^{n+1} \cdot \frac{1}{\sqrt{2n+1}} \cdot \begin{cases} \frac{2n}{n+1}, & n \;\; even \\ 2, & n \;\; odd \end{cases}. \tag{3.70}$$

PROOF Taking $x = \frac{a+b}{2}$ in (3.66) gives (3.70), where

$$I\left(\frac{a+b}{2}, n\right) = \frac{1}{(n+1)\sqrt{2n+1}} \left(\frac{b-a}{2}\right)^{n+1} \{4n^2 + (2n+1)[1+(-1)^n]^2\}^{\frac{1}{2}}.$$

Examining the above expression for n even or n odd readily gives the result (3.70). ∎

REMARK 3.14 For n odd, the third term in the modulus sign vanishes and thus there is no perturbation to the trapezoidal rule (3.70). ∎

THEOREM 3.11

Let the conditions of Theorem 3.10 be satisfied. Further, suppose that $f^{(n)}$ is differentiable and is such that

$$\left\|f^{(n+1)}\right\|_{\infty} := \sup_{t \in [a,b]} \left|f^{n+1}(t)\right| < \infty.$$

Then

$$|P_T(x)| \leq \frac{b-a}{\sqrt{12}} \left\|f^{(n+1)}\right\|_{\infty} \cdot \frac{1}{n!} I(x,n), \tag{3.71}$$

where $P_T(x)$ is the perturbed trapezoidal-type rule given by the left-hand side of (3.66) and $I(x,n)$ is as given by (3.67).

PROOF Let $h, g : [a,b] \to \mathbb{R}$ be absolutely continuous and h', g' be bounded. Then Chebychev's inequality holds (see [32, p. 207])

$$|T(h,g)| \leq \frac{(b-a)^2}{12} \sup_{t \in [a,b]} |h'(t)| \cdot \sup_{t \in [a,b]} |g'(t)|.$$

Matić, Pečarić and Ujević [29] using a *premature* Grüss-type argument proved that

$$|T(h,g)| \leq \frac{(b-a)}{\sqrt{12}} \sup_{t\in[a,b]} |g'(t)| \sqrt{T(h,h)}. \tag{3.72}$$

Thus, associating $f^{(n)}(\cdot)$ with $g(\cdot)$ and $(x-t)^n$ with h in (3.72) readily produces (3.71) where $I(x,n)$ is as given by (3.67). ∎

THEOREM 3.12

Let the conditions of Theorem 3.10 be satisfied. Further, suppose that $f^{(n)}$ is locally absolutely continuous on (a,b) and let $f^{(n+1)} \in L_2(a,b)$. Then

$$|P_T(x)| \leq \frac{b-a}{\pi} \left\| f^{(n+1)} \right\|_2 \cdot \frac{1}{n!} I(x,n), \tag{3.73}$$

where $P_T(x)$ is the perturbed trapezoidal-type rule given by the left-hand side of (3.66) and $I(x,n)$ is as given in (3.67).

PROOF The following result was obtained by Lupaş (see [32, p. 210]). For $h,g:(a,b)\to\mathbb{R}$ locally absolutely continuous on (a,b) and $h',g' \in L_2(a,b)$, then

$$|T(h,g)| \leq \frac{(b-a)^2}{\pi^2} \|h'\|_2 \|g'\|_2,$$

where

$$\|h\|_2 := \left(\frac{1}{b-a}\int_a^b |h(t)|^2\right)^{\frac{1}{2}} \quad \text{for } h \in L_2(a,b).$$

Matić, Pečarić and Ujević [29] further show that

$$|T(h,g)| \leq \frac{(b-a)}{\pi} \|g'\|_2 \sqrt{T(h,h)}. \tag{3.74}$$

Now, associating $f^{(n)}(\cdot)$ with $g(\cdot)$ and $(x-t)^n$ with h in (3.74) gives (3.73), where $I(x,n)$ is found in (3.67). ∎

REMARK 3.15 Results (3.71) and (3.73) are not readily comparable to that obtained in Theorem 3.10 since the bound now involves the behaviour of $f^{(n+1)}(\cdot)$ rather than $f^{(n)}(\cdot)$. ∎

3.5.3.1 Application in Numerical Integration

Consider the partition $I_m : a = x_0 < x_1 < ... < x_{m-1} < x_m = b$ of the interval $[a, b]$ and the intermediate points $\xi = (\xi_0, ..., \xi_{m-1})$, where $\xi_j \in [x_j, x_{j+1}]$ $(j = 0, ..., m-1)$. Put $h_j := x_{j+1} - x_j$ and $\nu(h) = \max\{h_j | j = 0, ..., m-1\}$.

In [7], the authors considered the following generalization of the trapezoid formula

$$T_{m,n}(f, I_m) := \sum_{j=0}^{m-1} \sum_{k=0}^{n-1} \frac{h_j^{k+1}}{(k+1)!} \left[\frac{f^{(k)}(x_j) + (-1)^k f^{(k)}(x_{j+1})}{2} \right] \tag{3.75}$$

and proved the following theorem:

THEOREM 3.13

Let $f : [a, b] \to \mathbb{R}$ be such that its derivative $f^{(n-1)}$ is absolutely continuous on $[a, b]$. Then we have

$$\int_a^b f(t)\, dt = T_{m,n}(f, I_m) + R_{m,n}(f, I_m), \tag{3.76}$$

where the reminder $R_{m,n}(f, I_m)$ satisfies the estimate

$$|R_{m,n}(f, I_m)| \leq \frac{C_n}{(n+1)!} \left\| f^{(n)} \right\|_\infty \sum_{j=0}^{m-1} h_j^{n+1}, \tag{3.77}$$

and

$$C_n := \begin{cases} 1 & \text{if } n = 2r \\ \frac{2^{2r+1}-1}{2^{2r}} & \text{if } n = 2r+1. \end{cases}$$

Now, let us define the even more generalized quadrature formula

$$\tilde{T}_{m,n}(f, \boldsymbol{\xi}, I_m) := \sum_{j=0}^{m-1} \sum_{k=0}^{n-1} \frac{1}{(k+1)!} \left[(\xi_j - x_j)^{k+1} f^{(k)}(x_j) \right.$$

$$\left. + (-1)^k (x_{j+1} - \xi_j)^{k+1} f^{(k)}(x_{j+1}) \right],$$

where x_j, ξ_j $(j = 0, ..., m-1)$ are as above.

The following theorem holds [7].

THEOREM 3.14

Let f be as in Theorem 3.13. Then we have the formula

$$\int_a^b f(t)\,dt = \tilde{T}_{m,n}(f,\boldsymbol{\xi},I_m) + \tilde{R}_{m,n}(f,\boldsymbol{\xi},I_m), \tag{3.78}$$

where the remainder satisfies the estimate

$$\left|\tilde{R}_{m,n}(f,\boldsymbol{\xi},I_m)\right| \tag{3.79}$$

$$\leq \begin{cases} \frac{1}{(n+1)!}\left\|f^{(n)}\right\|_\infty \sum\limits_{j=0}^{m-1}\left[(\xi_j - x_j)^{n+1} + (x_{j+1}-\xi_j)^{n+1}\right], \\ \frac{1}{n!(nq+1)^{1/q}}\left\|f^{(n)}\right\|_p \left[\sum\limits_{j=0}^{m-1}(\xi_j - x_j)^{nq+1} + \sum\limits_{j=0}^{m-1}(x_{j+1}-\xi_j)^{nq+1}\right]^{\frac{1}{q}}, \\ \frac{1}{n!}\left\|f^{(n)}\right\|_1\left[\frac{1}{2}\nu(h) + \max\limits_{j=0,\ldots,m-1}\left|\xi_j - \frac{x_j+x_{j+1}}{2}\right|\right]^n. \end{cases}$$

PROOF Apply the inequality (3.59) on the subinterval $[x_j, x_{j+1}]$ to get

$$\left|\int_{x_j}^{x_{j+1}} f(t)\,dt - \sum_{k=0}^{n-1}\frac{1}{(k+1)!}\right.$$

$$\left.\times\left[(\xi_j - x_j)^{k+1} f^{(k)}(x_j) + (-1)^k (x_{j+1}-\xi_j)^{k+1} f^{(k)}(x_{j+1})\right]\right|$$

$$\leq \begin{cases} \frac{1}{(n+1)!}\sup\limits_{t\in[x_j,x_{j+1}]}\left|f^{(n)}(t)\right|\left[(\xi_j - x_j)^{n+1} + (x_{j+1}-\xi_j)^{n+1}\right], \\ \frac{1}{n!}\left(\int_{x_j}^{x_{j+1}}\left|f^{(n)}(s)\right|^p ds\right)^{\frac{1}{p}}\left[\frac{(\xi_j - x_j)^{nq+1} + (x_{j+1}-\xi_j)^{nq+1}}{nq+1}\right]^{\frac{1}{q}}, \\ \frac{1}{n!}\left(\int_{x_j}^{x_{j+1}}\left|f^{(n)}(s)\right| ds\right)\left[\frac{1}{2}h_j + \left|\xi_j - \frac{x_j+x_{j+1}}{2}\right|\right]^n. \end{cases}$$

Summing over j from 0 to $m-1$ and using the generalized triangle inequality, we have

$$\left|\tilde{R}_{m,n}(f,\boldsymbol{\xi},I_m)\right| \leq \left|\sum_{j=0}^{m-1}\int_{x_j}^{x_{j+1}} f(t)\,dt - \sum_{k=0}^{n-1}\frac{1}{(k+1)!}\right.$$

$$\left.\times\left[(\xi_j - x_j)^{k+1} f^{(k)}(x_j) + (-1)^k (x_{j+1}-\xi_j)^{k+1} f^{(k)}(x_{j+1})\right]\right|$$

$$:= \begin{cases} \frac{1}{(n+1)!}\sum\limits_{j=0}^{m-1}\sup\limits_{t\in[x_j,x_{j+1}]}\left|f^{(n)}(t)\right|\left[(\xi_j - x_j)^{n+1} + (x_{j+1}-\xi_j)^{n+1}\right], \\ \frac{1}{n!}\sum\limits_{j=0}^{m-1}\left(\int_{x_j}^{x_{j+1}}\left|f^{(n)}(s)\right|^p ds\right)^{\frac{1}{p}}\left[\frac{(\xi_j - x_j)^{nq+1}+(x_{j+1}-\xi_j)^{nq+1}}{nq+1}\right]^{\frac{1}{q}}, \\ \frac{1}{n!}\sum\limits_{j=0}^{m-1}\left(\int_{x_j}^{x_{j+1}}\left|f^{(n)}(s)\right| ds\right)\left[\frac{1}{2}h_j + \left|\xi_j - \frac{x_j - x_{j+1}}{2}\right|\right]^n. \end{cases}$$

As $\sup\limits_{t\in[x_j,x_{j+1}]}\left|f^{(n)}(t)\right| \leq \left\|f^{(n)}\right\|_\infty$, the first inequality in (3.79) readily follows.

Now, using the discrete Hölder inequality, we have

$$\frac{1}{(nq+1)^{1/q}}\sum_{j=0}^{m-1}\left(\int_{x_j}^{x_{j+1}}|f^{(n)}(s)|^p ds\right)^{\frac{1}{p}}\left[(\xi_j - x_j)^{nq+1} + (x_{j+1}-\xi_j)^{nq+1}\right]^{\frac{1}{q}}$$

$$\leq \frac{1}{(nq+1)^{1/q}}\left[\sum_{j=0}^{m-1}\left[\left(\int_{x_j}^{x_{j+1}}\left|f^{(n)}(s)\right|^p ds\right)^{\frac{1}{p}}\right]^p\right]^{\frac{1}{p}}$$

$$\times\left[\sum_{j=0}^{m-1}\left[\left[(\xi_j - x_j)^{nq+1} + (x_{j+1}-\xi_j)^{nq+1}\right]^{\frac{1}{q}}\right]^q\right]^{\frac{1}{q}}$$

$$= \frac{1}{(nq+1)^{1/q}}\left\|f^{(n)}\right\|_p\left[\sum_{j=0}^{m-1}(\xi_j - x_j)^{nq+1} + \sum_{j=0}^{m-1}(x_{j+1}-\xi_j)^{nq+1}\right]^{\frac{1}{q}}$$

and thus the second inequality in (3.79) is proved.

Finally, let us observe that

$$\frac{1}{n!}\sum_{j=0}^{m-1}\left(\int_{x_j}^{x_{j+1}}\left|f^{(n)}(s)\right|ds\right)\left[\frac{1}{2}h_j+\left|\xi_j-\frac{x_j+x_{j+1}}{2}\right|\right]^n$$

$$\leq \max_{j=0,\ldots,m-1}\left[\frac{1}{2}h_j+\left|\xi_j-\frac{x_j+x_{j+1}}{2}\right|\right]^n\sum_{j=0}^{m-1}\left(\int_{x_j}^{x_{j+1}}\left|f^{(n)}(s)\right|ds\right)$$

$$\leq \left[\frac{1}{2}h_j+\max_{j=0,\ldots,m-1}\left|\xi_j-\frac{x_j+x_{j+1}}{2}\right|\right]^n\left\|f^{(n)}\right\|_1$$

and the last part of (3.79) is proved. ∎

REMARK 3.16 As $(x-a)^\alpha+(b-x)^\alpha\leq(b-a)^\alpha$ for $\alpha\geq 1$, $x\in[a,b]$, then we remark that the first branch of (3.79) can be bounded by

$$\frac{1}{(n+1)!}\left\|f^{(n)}\right\|_\infty\sum_{j=0}^{m-1}h_j^{n+1} \tag{3.80}$$

The second branch can be bounded by

$$\frac{1}{n!(nq+1)^{1/q}}\left\|f^{(n)}\right\|_p\left[\sum_{j=0}^{m-1}h_j^{nq+1}\right]^{\frac{1}{q}} \tag{3.81}$$

and finally, the last branch in (3.79) can be bounded by

$$\frac{1}{n!}\left[\nu(h)\right]^n\left\|f^{(n)}\right\|_1 \tag{3.82}$$

Note that all the bounds provided by (3.80)-(3.82) are uniform bounds for $\tilde{R}_{m,n}(f,\xi,I_m)$ in terms of the intermediate points ξ. ∎

A further inequality that we can obtain from (3.79) is the one that results from taking $\xi_j=\frac{x_j+x_{j+1}}{2}$. Consequently, we can state the following corollary (see also [10]):

COROLLARY 3.11
Let f be as in Theorem 3.14. Then we have the formula

$$\int_a^b f(t)\,dt = \tilde{T}_{m,n}(f, I_m) + \tilde{R}_{m,n}(f, I_m), \tag{3.83}$$

where

$$\tilde{T}_{m,n}(f, I_m) = \sum_{j=0}^{m-1}\sum_{k=0}^{n-1} \frac{1}{2^{k+1}(k+1)!}\left[f^{(k)}(x_j) + (-1)^k f^{(k)}(x_{j+1})\right] h_j^{n+1} \tag{3.84}$$

and the remainder $\tilde{R}$ satisfies the estimate

$$\left|\tilde{R}_{m,n}(f, I_m)\right| \le \begin{cases} \frac{1}{2^n(n+1)!}\left\|f^{(n)}\right\|_\infty \sum_{j=0}^{m-1} h_j^{n+1}, \\ \\ \frac{1}{2^n n!(nq+1)^{1/q}}\left\|f^{(n)}\right\|_p \left[\sum_{j=0}^{m-1} h_j^{n+1}\right]^{\frac{1}{q}}, \\ \\ \frac{1}{2^n n!}\left[\nu(h)\right]^n \left\|f^{(n)}\right\|_1. \end{cases}$$

REMARK 3.17 Similar results can be stated by using the "perturbed" versions embodied in Theorems 3.10, 3.11 and 3.12, but we omit the details. ∎

3.6 Trapezoidal-Type Rules for Functions whose Derivative is Bounded Above and Below

3.6.1 Introduction

In 1938, Iyengar proved the following theorem obtaining bounds for a trapezoidal quadrature rule for functions whose derivative is bounded (see, for example, [31, p. 471]).

THEOREM 3.15
Let f be a differentiable function on (a, b) and assume that there is a

constant $M > 0$ *such that* $|f'(x)| \leq M$, *for all* $x \in (a,b)$. *Then we have*

$$\left| \int_a^b f(x)\,dx - \frac{b-a}{2}[f(a)+f(b)] \right|$$

$$\leq \frac{M(b-a)^2}{4} - \frac{1}{4M}(f(a)-f(b))^2. \tag{3.85}$$

Using the classical inequality due to Hayashi (see, for example, [30, pp. 311–312]), Agarwal and Dragomir proved in [1] the following generalization of Theorem 3.15 involving the Trapezoidal rule, while Dragomir and Wang [25] apply it to numerical quadrature.

THEOREM 3.16

Let $f : I \subseteq \mathbb{R} \to \mathbb{R}$ *be a differentiable mapping in* $\overset{\circ}{I}$, *the interior of* I, *and let* $a, b \in \overset{\circ}{I}$ *with* $a < b$. *Let* $M = \sup_{x\in[a,b]} f'(x) < \infty$ *and* $m = \inf_{x\in[a,b]} f'(x) > -\infty$. *If* $m < M$ *and* f' *is integrable on* $[a,b]$, *then we have*

$$\left| \int_a^b f(t)\,dt - \frac{b-a}{2}[f(a)+f(b)] \right|$$

$$\leq \frac{[f(b)-f(a)-m(b-a)][M(b-a)-f(b)+f(a)]}{2(M-m)}. \tag{3.86}$$

Thus, by placing $m = -M$ in (3.86) then Iyengar's result (3.85) is recovered.

In this section we point out further results in connection with the trapezoid inequality.

3.6.2 Integral Inequalities

The following theorem due to Hayashi [30, pp. 311–312] will be required and thus it is stated for convenience.

THEOREM 3.17

Let $h : [a,b] \longrightarrow \mathbb{R}$ *be a nonincreasing mapping on* $[a,b]$ *and* $g : [a,b] \longrightarrow \mathbb{R}$ *an integrable mapping on* $[a,b]$ *with*

$$0 \leq g(x) \leq A, \quad \textit{for all } x \in [a,b],$$

then

$$A\int_{b-\lambda}^{b} h(x)\,dx \leq \int_{a}^{b} h(x)\,g(x)\,dx \leq A\int_{a}^{a+\lambda} h(x)\,dx \tag{3.87}$$

where

$$\lambda = \frac{1}{A}\int_{a}^{b} g(x)\,dx.$$

Using this result we can state the following trapezoid inequality. See [6] for further details.

THEOREM 3.18
Let $f : I \subseteq \mathbb{R} \longrightarrow \mathbb{R}$ be a differentiable mapping on $\mathring{I}$ ($\mathring{I}$ is the interior of I) and $[a,b] \subset \mathring{I}$ with $M = \sup_{x\in[a,b]} f'(x) < \infty$, $m = \inf_{x\in[a,b]} f'(x) > -\infty$ and $M > m$. If f' is integrable on $[a,b]$, then the following inequalities hold:

$$\left|\int_{a}^{b} f(x)\,dx - \frac{b-a}{2}\left[f(a)+f(b)\right]\right|$$

$$\leq \frac{(b-a)^2}{2(M-m)}(S-m)(M-S) \tag{3.88}$$

$$\leq \frac{M-m}{2}\left(\frac{b-a}{2}\right)^2 \tag{3.89}$$

where $S = \frac{f(b)-f(a)}{b-a}$.

PROOF Let $h(x) = \theta - x$, $\theta \in [a,b]$ and $g(x) = f'(x) - m$. Then, from Hayashi's inequality (3.87)

$$L \leq I \leq U \tag{3.90}$$

where

$$I = \int_{a}^{b} (\theta - x)\left(f'(x) - m\right) dx,$$

$$\lambda = \frac{1}{M-m}\int_{a}^{b} \left(f'(x) - m\right) dx,$$

and

$$L = (M - m) \int_{b-\lambda}^{b} (\theta - x)\, dx,$$

$$U = (M - m) \int_{a}^{a+\lambda} (\theta - x)\, dx.$$

It is now a straightforward matter to evaluate and simplify the above expansions to give

$$I = \int_a^b f(u)\, du \tag{3.91}$$

$$- \left[m(b-a)\left(\theta - \frac{b+a}{2}\right) + (b-\theta) f(b) + (\theta - a) f(a) \right],$$

$$\lambda = \frac{1}{M-m}\left[f(b) - f(a) - m(b-a)\right] = \frac{b-a}{M-m}(S-m), \tag{3.92}$$

$$L = \frac{(M-m)}{2}\lambda\left[\lambda + 2(\theta - b)\right], \tag{3.93}$$

and

$$U = \frac{(M-m)}{2}\lambda\left[2(\theta - a) - \lambda\right]. \tag{3.94}$$

In addition, it may be noticed from (3.90) that

$$\left| I - \frac{U+L}{2} \right| \leq \frac{U-L}{2}, \tag{3.95}$$

where, upon using (3.93) and (3.94),

$$\frac{U+L}{2} = (M-m)\lambda\left(\theta - \frac{b+a}{2}\right) \tag{3.96}$$

and

$$\frac{U-L}{2} = \frac{(M-m)}{2}\lambda(b - a - \lambda). \tag{3.97}$$

Equation (3.95) is then (3.88) upon using (3.91), (3.92), (3.96) and (3.97) together with some routine simplification.

Now, for inequality (3.89). Consider the right-hand side of (3.88). Completing the square gives

$$\frac{(b-a)^2}{2(M-m)}(S-m)(M-S)$$

$$= \frac{2}{M-m}\left(\frac{b-a}{2}\right)^2 \times \left[\left(\frac{M-m}{2}\right)^2 - \left(S - \frac{M+m}{2}\right)^2\right] \quad (3.98)$$

and (3.89) is readily determined by neglecting the negative term. ∎

REMARK 3.18 The above theorem was proved independently of the value of θ. Agarwal and Dragomir [1] proved an equivalent result with effectively $\theta = a$. It may be noticed from the above development, however, that if $\theta = \frac{a+b}{2}$, then there is some simplification for I and $\frac{U+L}{2} = 0$. ∎

REMARK 3.19 For $\|f'\|_\infty = \sup_{x\in[a,b]} |f'(x)| < \infty$ and let $m = -\|f'\|_\infty$, $M = \|f'\|_\infty$ in (3.88). Then the result obtained by Iyengar [31, p. 471] using geometrical means, is recovered. It should also be noted that if either both m and M are positive or both negative, then the bound obtained here is tighter than that of Iyengar as given by (3.85). ∎

Bounds for the generalized trapezoidal rule will now be developed in the following theorem.

THEOREM 3.19

Let f satisfy the conditions of Theorem 3.18, then the following result holds:

$$\beta_L \le \int_a^b f(x)\,dx - (b-a)\left[\left(\frac{\theta-a}{b-a}\right)f(a) + \left(\frac{b-\theta}{b-a}\right)f(b)\right] \le \beta_U \quad (3.99)$$

where

$$\beta_U = \frac{(b-a)^2}{2(M-m)}\left[S(2\gamma_U - S) - mM\right], \quad (3.100)$$

$$\beta_L = \frac{(b-a)^2}{2(M-m)}\left[S(S - 2\gamma_L) + mM\right], \quad (3.101)$$

$$\gamma_U = \left(\frac{\theta - a}{b - a}\right) M + \left(\frac{b - \theta}{b - a}\right) m, \ \gamma_L = M + m - \gamma_U, \tag{3.102}$$

and

$$S = \frac{f(b) - f(a)}{b - a}. \tag{3.103}$$

PROOF From (3.90) and (3.91) it may be readily seen that

$$\beta_U = U + m(b - a)\left(\theta - \frac{a + b}{2}\right) \tag{3.104}$$

and

$$\beta_L = L + m(b - a)\left(\theta - \frac{a + b}{2}\right). \tag{3.105}$$

Now, from (3.104) and using (3.94), (3.92) gives

$$\begin{aligned}\beta_U = &\frac{1}{2(M - m)}\Big\{(b - a)(S - m)\left[2(M - m)(\theta - a) - (b - a)(S - m)\right] \\ &+ 2m(b - a)(M - m)\left(\theta - \frac{a + b}{2}\right)\Big\} \\ = &\frac{(b - a)^2}{2(M - m)}\Big\{(S - m)\left[S - m + 2(M - m)\left(\frac{\theta - a}{b - a}\right)\right] \\ &+ 2m\left(\frac{M - m}{b - a}\right)\left(\theta - \frac{a + b}{2}\right)\Big\}.\end{aligned}$$

Expanding in powers of S and after simplification we produce the expression (3.100). In a similar fashion, (3.101) may be derived from (3.105) and using (3.93), (3.92) gives

$$\begin{aligned}\beta_L = &\frac{1}{2(M - m)}\{(b - a)(S - m)\left[(b - a)(S - m) + 2(M - m)(\theta - b)\right] \\ &+ 2m(b - a)(M - m)\left(\theta - \frac{a + b}{2}\right)\Big\} \\ = &\frac{(b - a)^2}{2(M - m)}\Big\{(S - m)\left[S - m + 2(M - m)\left(\frac{\theta - b}{b - a}\right)\right]\end{aligned}$$

$$+ 2m\left(\frac{M-m}{b-a}\right)\left(\theta - \frac{a+b}{2}\right)\bigg\}.$$

Again, expanding in powers of S produces (3.101) after some algebra and thus the proof of the theorem is complete. ∎

REMARK 3.20 Allowing $\theta = \frac{a+b}{2}$ gives

$$\beta_L = -\beta_U = \frac{(b-a)^2}{2(M-m)}(S-m)(M-S),$$

thus reproducing the result of Theorem 3.18. ∎

REMARK 3.21 It may be shown from (3.100) and (3.101) that for any $\theta \in [a,b]$, the size of the bound interval for the generalized trapezoidal rule is

$$\beta_U - \beta_L = \frac{(b-a)^2}{(M-m)}\left[\left(\frac{M-m}{2}\right)^2 - \left(S - \frac{M+m}{2}\right)^2\right].$$

This is the same size as that for the symmetric bounds for the trapezoidal rule of Theorem 3.18 which seems, at first, surprising though on observing (3.104) and (3.105) may be less so. ∎

REMARK 3.22 The difference between the upper and lower bounds is always positive since

$$\beta_U - \beta_L = \frac{(b-a)^2}{M-m}(S-m)(M-S) \geq 0$$

where S, from (3.103), is the slope of the secant and $m \leq S \leq M$. ∎

REMARK 3.23 For $\|f'\|_\infty = \sup_{x\in[a,b]}|f'(x)| < \infty$, let $m = -\|f'\|_\infty$ and $M = \|f'\|$ in (3.99)–(3.103), then an Iyengar-type result for the generalized trapezoidal rule will be obtained. ∎

COROLLARY 3.12

Let f satisfy the conditions of Theorems 3.18 and 3.19. Then

$$\frac{(b-a)^2}{2(M-m)}\left(mM-\gamma_L^2\right)$$

$$\leq \int_a^b f(u)\,du-(b-a)\left[\left(\frac{\theta-a}{b-a}\right)f(a)+\left(\frac{b-\theta}{b-a}\right)f(b)\right]$$

$$\leq \frac{(b-a)^2}{2(M-m)}\left[\gamma_U^2-mM\right] \tag{3.106}$$

where γ_U and γ_L are as defined in (3.102).

PROOF From (3.99) and (3.100) it may be shown by completing the square that

$$\beta_U=\frac{(b-a)^2}{2(M-m)}\left[\gamma_U^2-mM-(S-\gamma_U)^2\right]$$

and

$$\beta_L=\frac{(b-a)^2}{2(M-m)}\left[(S-\gamma_L)^2+mM-\gamma_L^2\right].$$

The result (3.106) follows from neglecting the negative term from β_U and the positive term from β_L. ∎

REMARK 3.24 The results obtained in this section could also be implemented by constructing composite quadrature rules as previously. This, however, will not be pursued further here. ∎

3.7 Grüss-Type Bounds

In 1935, G. Grüss (see, for example, [30, p. 296]) proved the following integral inequality which gives an approximation for the integral of a product in terms of the product of integrals:

THEOREM 3.20

Let $f,g:[a,b]\to\mathbb{R}$ be two integrable mappings so that $\varphi\leq h(x)\leq\Phi$ and $\gamma\leq g(x)\leq\Gamma$ for all $x\in[a,b]$, where $\varphi,\Phi,\gamma,\Gamma$ are real numbers. Then we *have*

$$
\begin{aligned}
|T(h,g)| :&= \left| \frac{1}{b-a} \int_a^b h(x)\, g(x)\, dx \right. \\
&\quad \left. - \frac{1}{b-a} \int_a^b h(x)\, dx \cdot \frac{1}{b-a} \int_a^b g(x)\, dx \right| \\
&\leq \frac{1}{4} (\Phi - \varphi)(\Gamma - \gamma) \qquad (3.107)
\end{aligned}
$$

and the inequality is sharp, in the sense that the constant $\frac{1}{4}$ *cannot be replaced by a smaller one.*

For a simple proof of this fact as well as for extensions, generalizations, discrete variants and other associated material, see [30, p. 296], and the papers [17]-[9] where further references are given.

In this section, we point out a different Grüss-type inequality and apply it for trapezoid formula.

3.7.1 A Grüss-Type Result and Applications for the Trapezoid Inequality

We start with the following result of Grüss type [22].

THEOREM 3.21

Let $h, g : [a,b] \to \mathbb{R}$ *be two integrable mappings. Then we have the following Grüss-type inequality:*

$$
\begin{aligned}
&\left| \frac{1}{b-a} \int_a^b h(x)\, g(x)\, dx - \frac{1}{b-a} \int_a^b h(x)\, dx \cdot \frac{1}{b-a} \int_a^b g(x)\, dx \right| \\
&\leq \frac{1}{b-a} \int_a^b \left| \left(h(x) - \frac{1}{b-a} \int_a^b h(y)\, dy \right) \right. \\
&\quad \left. \cdot \left(g(x) - \frac{1}{b-a} \int_a^b g(y)\, dy \right) \right| dx. \qquad (3.108)
\end{aligned}
$$

The inequality (3.108) *is sharp.*

PROOF First of all, let us observe that

$$I := \frac{1}{b-a}\int_a^b \left(h(x) - \frac{1}{b-a}\int_a^b h(y)\,dy\right)\cdot\left(g(x) - \frac{1}{b-a}\int_a^b g(y)\,dy\right)dx$$

$$= \frac{1}{b-a}\int_a^b \left(h(x)g(x) - g(x)\cdot\frac{1}{b-a}\int_a^b h(y)\,dy - h(x)\cdot\frac{1}{b-a}\int_a^b g(y)\,dy\right.$$

$$\left. + \frac{1}{b-a}\int_a^b h(y)\,dy\cdot\frac{1}{b-a}\int_a^b g(y)\,dy\right)dx$$

$$= \frac{1}{b-a}\int_a^b h(x)g(x)\,dx - \frac{1}{b-a}\int_a^b g(x)\,dx\cdot\frac{1}{b-a}\int_a^b h(y)\,dy$$

$$- \frac{1}{b-a}\int_a^b h(x)\,dx\cdot\frac{1}{b-a}\int_a^b g(y)\,dy + \frac{1}{b-a}\int_a^b h(y)\,dy\cdot\frac{1}{b-a}\int_a^b g(y)\,dy$$

$$= \frac{1}{b-a}\int_a^b h(x)g(x)\,dx - \frac{1}{b-a}\int_a^b g(x)\,dx\cdot\frac{1}{b-a}\int_a^b h(x)\,dx.$$

On the other hand, by the use of modulus properties, we have

$$|I| \le \frac{1}{b-a}\int_a^b \left|\left(h(x) - \frac{1}{b-a}\int_a^b h(y)\,dy\right)\cdot\left(g(x) - \frac{1}{b-a}\int_a^b g(y)\,dy\right)\right|dx$$

and the inequality (3.108) is proved.
Choosing $h(x) = g(x) = \operatorname{sgn}\left(x - \frac{a+b}{2}\right)$, equality is satisfied in (3.108). ∎

The following corollaries follow immediately.

COROLLARY 3.13

Let $f : [a,b] \to \mathbb{R}$ be a differentiable mapping on (a,b) having the first derivative $f' : (a,b) \to \mathbb{R}$ bounded on (a,b). Then we have the inequality:

$$\left|\frac{b-a}{2}[f(a)+f(b)] - \int_a^b f(x)\,dx\right|$$

$$\leq \frac{(b-a)^2}{4} \max_{x\in(a,b)} \left| f'(x) - \frac{f(b)-f(a)}{b-a} \right| \tag{3.109}$$

PROOF A simple integration by parts gives that

$$\frac{b-a}{2}[f(a)+f(b)] - \int_a^b f(x)\,dx = \int_a^b \left(x - \frac{a+b}{2}\right) f'(x)\,dx. \tag{3.110}$$

Applying the inequality (3.108) we find that

$$\left| \int_a^b \frac{1}{b-a}\left(x - \frac{a+b}{2}\right) f'(x)\,dx \right.$$

$$\left. - \frac{1}{b-a}\int_a^b \left(x - \frac{a+b}{2}\right) dx \cdot \frac{1}{b-a}\int_a^b f'(x)\,dx \right|$$

$$\leq \frac{1}{b-a}\int_a^b \left| \left(x - \frac{a+b}{2} - \frac{1}{b-a}\int_a^b \left(y - \frac{a+b}{2}\right) dy\right) \right.$$

$$\left. \cdot \left(f'(x) - \frac{1}{b-a}\int_a^b f'(y)\,dy\right) \right| dx.$$

As

$$\int_a^b \left(x - \frac{a+b}{2}\right) dx = 0,$$

we obtain

$$\left| \int_a^b \left(x - \frac{a+b}{2}\right) f'(x)\,dx \right|$$

$$\leq \int_a^b \left| \left(x - \frac{a+b}{2}\right) \left(f'(x) - \frac{f(b)-f(a)}{b-a}\right) \right| dx$$

$$\leq \max_{x\in(a,b)} \left| f'(x) - \frac{f(b)-f(a)}{b-a} \right| \int_a^b \left| \left(x - \frac{a+b}{2}\right) \right| dx$$

$$= \frac{(b-a)^2}{4} \max_{x\in(a,b)} \left| f'(x) - \frac{f(b)-f(a)}{b-a} \right|. \tag{3.111}$$

Now, using the identity (3.110), the inequality(3.111) becomes the desired result (3.109). ∎

COROLLARY 3.14

Suppose $p,q>1$ satisfy $\frac{1}{p}+\frac{1}{q}=1$. Let $f:[a,b]\to\mathbb{R}$ be a differentiable mapping on (a,b) having the first derivative $f':(a,b)\to\mathbb{R}$ being p-integrable on (a,b). Then we have the inequality

$$\left|\frac{b-a}{2}[f(a)+f(b)]-\int_a^b f(x)\,dx\right|$$

$$\leq\frac{1}{2}\left(\frac{b-a}{q+1}\right)^{\frac{1}{q}+1}\left(\int_a^b\left|f'(x)-\frac{f(b)-f(a)}{b-a}\right|^p dx\right)^{\frac{1}{p}}. \quad (3.112)$$

PROOF Using Hölder's inequality, we have that

$$\int_a^b\left|\left(x-\frac{a+b}{2}\right)\left(f'(x)-\frac{f(b)-f(a)}{b-a}\right)\right|dx$$

$$\leq\left(\int_a^b\left|x-\frac{a+b}{2}\right|^q dx\right)^{\frac{1}{q}}\left(\int_a^b\left|f'(x)-\frac{f(b)-f(a)}{b-a}\right|^p dx\right)^{\frac{1}{p}}.$$

A simple computation shows that

$$\int_a^b\left|x-\frac{a+b}{2}\right|^q dx=\int_a^{\frac{a+b}{2}}\left(\frac{a+b}{2}-x\right)^q dx+\int_{\frac{a+b}{2}}^b\left(x-\frac{a+b}{2}\right)^q dx$$

$$=\frac{(b-a)^{q+1}}{(q+1)\,2^q}$$

and so

$$\int_a^b\left|\left(x-\frac{a+b}{2}\right)\left(f'(x)-\frac{f(b)-f(a)}{b-a}\right)\right|dx$$

$$\leq\frac{(b-a)^{\frac{1}{q}+1}}{2(q+1)^{\frac{1}{q}}}\left(\int_a^b\left|f'(x)-\frac{f(b)-f(a)}{b-a}\right|^p dx\right)^{\frac{1}{p}}.$$

Now, using the first part of (3.111) and the identity (3.110), we obtain the desired result (3.112). ∎

The following result also holds.

COROLLARY 3.15
Let $f : [a,b] \to \mathbb{R}$ be a differentiable mapping on (a,b) and suppose that $f' : (a,b) \to \mathbb{R}$ is integrable on (a,b). Then we have the inequality

$$\left| \frac{b-a}{2} [f(a) + f(b)] - \int_a^b f(x)\,dx \right| \leq \frac{b-a}{2} \int_a^b \left| f'(x) - \frac{f(b)-f(a)}{b-a} \right| dx \tag{3.113}$$

PROOF We have

$$\begin{aligned} &\int_a^b \left| \left(x - \frac{a+b}{2} \right) \left(f'(x) - \frac{f(b)-f(a)}{b-a} \right) \right| dx \\ &\leq \max_{x \in (a,b)} \left| x - \frac{a+b}{2} \right| \times \int_a^b \left| f'(x) - \frac{f(b)-f(a)}{b-a} \right| dx \\ &= \frac{b-a}{2} \int_a^b \left| f'(x) - \frac{f(b)-f(a)}{b-a} \right| dx. \end{aligned}$$

Using the first part of (3.111) and the identity (3.110), we obtain the desired result (3.113). ∎

REMARK 3.25 The results of this section may be compared with those as presented in equation (3.62). Either may be tighter depending on the specific function $f(\cdot)$. ∎

REMARK 3.26 Theorem 3.21 may be utilised with $h(t) = \frac{(x-t)^n}{n!}$ and $g(t) = f^{(n)}(t)$ to obtain perturbed generalised trapezoidal-type rules. However, this will not be pursued further here.

Taking $x = \frac{a+b}{2}$ and $n = 1$ would produce the results of this section. ∎

3.8 Trapezoidal Rules with an Error Bound Involving the Second Derivative

In this section, via the use of some classical results from the Theory of Inequalities (Hölder's inequality, Grüss inequality and the Hermite-Hadamard inequality), we produce some *quasi-trapezoid quadrature formulae* for which the remainder term is smaller than the classical one.

This section focuses on the trapezoidal rule in which the error bound involves the behaviour of the second derivative in terms of a variety of norms. Section 3.5, on the other hand, examined the generalised trapezoidal rule in which the bound on the error involved the $\left\|f^{(n)}\right\|_{\infty}$ norm.

For other results in connection with trapezoid inequalities, see Chapter XV of the recent book by Mitrinović et al. [31].

3.8.1 Some Integral Inequalities

We shall start with the following theorem which is also interesting in its own right [19].

THEOREM 3.22

Let $f:[a,b]\to\mathbb{R}$ be a twice differentiable mapping on (a,b). Then we have the estimation

$$\left|\int_a^b f(x)\,dx-\frac{b-a}{2}\left[f(a)+f(b)\right]\right|$$

$$\leq\begin{cases}\frac{\|f''\|_{\infty}}{12}(b-a)^3 & \text{if } f''\in L_{\infty}[a,b]\\[2ex] \frac{1}{2}\|f''\|_p\left[B(q,q)\right]^{\frac{1}{q}}(b-a)^{2+\frac{1}{q}},\ \frac{1}{p}+\frac{1}{q}=1, p>1 & \\ \qquad\text{if } f''\in L_p[a,b] & \\[2ex] \frac{\|f''\|_1}{8}(b-a)^2 & \text{if } f''\in L_1[a,b]\end{cases}, \quad (3.114)$$

where

$$\|f''\|_{\infty}:=\sup_{t\in[a,b]}\left|f''(t)\right|,$$

$$\|f''\|_1 := \int_a^b |f''(t)|\, dt,$$

$$\|f''\|_p := \left(\int_a^b |f''(t)|^p\, dt\right)^{\frac{1}{p}}, \ p > 1$$

and B is the Beta function of Euler; that is,

$$B(r, s) := \int_0^1 t^{r-1} (1-t)^{s-1}\, dt, \ r, s > 0.$$

PROOF Integrating by parts we can state that

$$\begin{aligned}
&\int_a^b (x-a)(b-x) f''(x)\, dx \\
&= \left[(x-a)(b-x) f'(x)\right]_a^b - \int_a^b \left[(a+b) - 2x\right] f'(x)\, dx \\
&= \int_a^b \left[2x - (a+b)\right] f'(x)\, dx \\
&= f(x)\left[2x - (a+b)\right]_a^b - 2\int_a^b f(x)\, dx \\
&= (b-a)(f(a) + f(b)) - 2\int_a^b f(x)\, dx,
\end{aligned}$$

from which we get the well-known identity

$$\int_a^b f(x)\, dx = \frac{b-a}{2}\left[f(a) + f(b)\right] - \frac{1}{2}\int_a^b (x-a)(b-x) f''(x) dx. \quad (3.115)$$

Thus, using properties of the modulus gives

$$\left|\int_a^b f(x) dx - \frac{b-a}{2}\left[f(a) + f(b)\right]\right| \le \frac{1}{2}\int_a^b (x-a)(b-x)\left|f''(x)\right| dx. \quad (3.116)$$

Now, first, let us observe that

$$\int_a^b (x-a)(b-x)|f''(x)|\,dx \le \|f''\|_\infty \int_a^b (x-a)(b-x)\,dx$$

$$= \frac{\|f''\|_\infty}{6}(b-a)^3.$$

Thus, by (3.116), we get the first inequality in (3.114).

Further, by Hölder's integral inequality we obtain

$$\int_a^b (x-a)(b-x)|f''(x)|\,dx$$

$$\le \left(\int_a^b (x-a)^q (b-x)^q\,dx\right)^{\frac{1}{q}} \|f''\|_p,$$

where $\frac{1}{p}+\frac{1}{q}=1$, $p>1$ and $\|f''\|_p$ is as given above. Now, using the transformation $x=(1-t)a+tb$, $t\in[0,1]$, we get

$$(x-a)^q (b-x)^q = (b-a)^{2q} t^q (1-t)^q,$$

$$dx = (b-a)\,dt$$

and thus

$$\int_a^b (x-a)^q (b-x)^q\,dx = (b-a)^{2q+1}\int_0^1 t^q (1-t)^q\,dt$$

$$= (b-a)^{2q+1} B(q+1,q+1), \qquad (3.117)$$

where B is the Beta function of Euler; and the second inequality in (3.114) is proved.

Finally, we have that

$$\int_a^b (x-a)(b-x)|f''(x)|\,dx \le \max_{x\in[a,b]}[(x-a)(b-x)]\|f''\|_1$$

and, since

$$\max_{x\in[a,b]}[(x-a)(b-x)]=\frac{(b-a)^2}{4},$$

at $x=\frac{a+b}{2}$, we deduce the last part of (3.114). ∎

REMARK 3.27 The inequalities in (3.114) that provided a variety of norms involving the second derivative give flexibility since any of them may be tighter depending on the function which we wish to approximate (see [19]) for further details). ∎

The following theorem is of interest since it provides another integral inequality in connection with the trapezoid formula, giving a perturbed rule.

THEOREM 3.23

Let $f:[a,b]\to\mathbb{R}$ be a twice differentiable mapping on (a,b) and assume that

$$\gamma:=\inf_{x\in(a,b)}f''(x)>-\infty \text{ and } \Gamma:=\sup_{x\in(a,b)}f''(x)<\infty. \tag{3.118}$$

Then, we have the estimation

$$\left|\int_a^b f(x)\,dx-\frac{b-a}{2}[f(a)+f(b)]+\frac{(b-a)^2}{12}(f'(b)-f'(a))\right|$$

$$\leq\frac{(b-a)^3(\Gamma-\gamma)}{32}. \tag{3.119}$$

PROOF We shall apply the celebrated Grüss inequality as given by (3.107).

Now, if we choose in (3.107), $h(x)=(x-a)(b-x)$, $g(x)=f''(x)$, $x\in[a,b]$, we get

$$\phi=0,\ \Phi=\frac{(b-a)^2}{4},$$

and we can state that

$$\left|\frac{1}{b-a}\int_a^b(x-a)(b-x)f''(x)\,dx\right.$$

$$- \frac{1}{b-a}\int_a^b (x-a)(b-x)\,dx \cdot \frac{1}{b-a}\int_a^b f''(x)\,dx\Bigg|$$

$$\leq \frac{(b-a)^2(\Gamma-\gamma)}{16}. \tag{3.120}$$

A simple calculation gives us that

$$\int_a^b (x-a)(b-x)\,dx = \frac{(b-a)^3}{6} \text{ and } \int_a^b f''(x)\,dx = f'(b) - f'(a),$$

then, from (3.120),

$$\left|\int_a^b (x-a)(b-x) f''(x)\,dx - \frac{(b-a)^2}{6}(f'(b)-f'(a))\right|$$

$$\leq \frac{(b-a)^3(\Gamma-\gamma)}{16}.$$

Finally, using the identity (3.115) gives

$$\left|\frac{f(a)+f(b)}{2}(b-a) - \int_a^b f(x)\,dx - \frac{(b-a)^2}{12}(f'(b)-f'(a))\right|$$

$$\leq \frac{(b-a)^3(\Gamma-\gamma)}{32}$$

and the theorem is proved. ∎

THEOREM 3.24
Let f have the properties of Theorem 3.23. Then the estimation

$$\left|\int_a^b f(x)\,dx - \frac{b-a}{2}[f(a)+f(b)] + \frac{(b-a)^2}{12}(f'(b)-f'(a))\right|$$

$$\leq \frac{1}{24\sqrt{5}} \cdot (b-a)^3(\Gamma-\gamma) \tag{3.121}$$

holds with γ, Γ as given by (3.118).

PROOF The proof utilises the premature Grüss inequality as given by Theorem 3.9, equation (3.65) rather than the Grüss inequality given by Theorem 3.20, equation (3.107). Only the bound varies from that of Theorem 3.23 and so taking $h(x) = (x-a)(b-x)$, $g(x) = f''(x)$, $x \in [a,b]$, we have that the bound as given by (3.65)

$$\frac{b-a}{2} \cdot \frac{\Gamma - \gamma}{2} \left[T(h,h)\right]^{\frac{1}{2}}, \tag{3.122}$$

where

$$T(h,h) = \frac{1}{b-a}\int_a^b h^2(x)\,dx - \left(\frac{1}{b-a}\int_a^b h(x)\,dx\right)^2.$$

Utilising (3.117), we have that

$$\begin{aligned} T(h,h) &= (b-a)^4 B(3,3) - \left[(b-a)^2 B(2,2)\right]^2 \\ &= \frac{(b-a)^4}{180}, \end{aligned} \tag{3.123}$$

since B is Euler's Beta function.

Thus, using (3.122), (3.123) gives from (3.65), (3.121) and the theorem is proved. ∎

REMARK 3.28 A comparison of the bounds in (3.119) and (3.121) shows that the premature Grüss inequality is 1.26% better than that obtained using the Grüss inequality. ∎

REMARK 3.29 Atkinson [3] terms the quadrature rule in (3.119) or (3.121) as a corrected trapezoidal rule and obtains it using an asymptotic error estimate approach which does not provide an expression for the error bound. He does state that the corrected trapezoidal rule is $O(h^4)$ compared with $O(h^2)$ for the trapezoidal rule. Atkinson does subsequently find an explicit bound using the Euler-Maclaurin summation formula. The bound given in (3.119) depends only on the bounds for $f'(\cdot)$ and not the fourth derivative. ∎

REMARK 3.30 From equation (3.70) of Corollary 3.10, an alternate premature Grüss inequality may be obtained with $n = 2$, giving, after some simplification,

$$\left| \int_a^b f(t)\,dt - \frac{b-a}{2}[f(a)+f(b)] + (f'(b)-f'(a)) \right|$$

$$\leq \frac{1}{40\sqrt{5}} \cdot (b-a)^3 (\Gamma - \gamma). \tag{3.124}$$

This is superior to (3.121) obtained from a different kernel. A premature Grüss inequality was used to obtain a generalised trapezoidal-type rule containing an unspecified $x \in [a,b]$. Equation (3.70) is obtained from taking $x = \frac{a+b}{2}$ to give the tightest bound. Equation (3.121) was obtained directly without the extra degree of freedom, which may explain its inferiority.

In comparing results (3.119), (3.121) and (3.124), the natural question that may be asked, which is to our best knowledge an open problem, is what should the best constant C be that satisfies

$$\left| \int_a^b f(x)\,dx - \frac{b-a}{2}[f(a)+f(b)] + \frac{(b-a)^2}{12}[f'(b)-f'(a)] \right|$$

$$\leq C(b-a)^3(\Gamma-\gamma),$$

where γ, Γ are as given by (3.118).

It may further be observed that the first inequality in (3.64) gives a different perturbed trapezoidal-type rule with a bound involving $\|f''\|_\infty$. Now, since $0 \leq \Gamma - \gamma \leq 2\|f''\|_\infty$, then (3.121) and (3.124) are all tighter bounds, while (3.119) is not by this coarse bound; however, it may be in practice. The fact that the constant of the perturbation is different has little consequence in practice since the perturbation only affects the end points in a composite rule. ∎

Finally, using a classical result on convex functions due to Hermite and Hadamard we have the following theorem concerning a double integral inequality [19].

THEOREM 3.25

Let $f : [a,b] \to \mathbb{R}$ be a twice differentiable mapping on (a,b) and suppose that $-\infty < \gamma \leq f''(x) \leq \Gamma < \infty$ for all $x \in (a,b)$.

Then we have the double inequality

$$\frac{\gamma}{12}(b-a)^2 \leq \frac{f(a)+f(b)}{2} - \frac{1}{b-a}\int_a^b f(x)\,dx$$

$$\leq \frac{\Gamma}{12}(b-a)^2, \tag{3.125}$$

and the estimation

$$\left|\frac{f(a)+f(b)}{2}(b-a) - \frac{\Gamma+\gamma}{24}(b-a)^2 - \int_a^b f(x)\,dx\right|$$

$$\leq \frac{(\Gamma-\gamma)(b-a)^3}{24}. \tag{3.126}$$

PROOF We shall use the following inequality for convex mappings $g:[a,b]\to\mathbb{R}$:

$$\frac{1}{b-a}\int_a^b g(x)\,dx \leq \frac{g(a)+g(b)}{2}, \tag{3.127}$$

which is well known in the literature as the Hermite-Hadamard inequality. Let us choose first $g:[a,b]\to\mathbb{R}$, $g(x)=f(x)-\frac{\gamma}{2}x^2$. Then g is twice differentiable on $[a,b]$ and

$$g'(x)=f'(x)-\gamma x,\ g''(x)=f''(x)-\gamma\geq 0 \text{ on } (a,b),$$

hence, g is convex on $[a,b]$. Thus, we can apply (3.127) for g to get

$$\frac{1}{b-a}\int_a^b \left(f(x)-\frac{\gamma}{2}x^2\right)dx \leq \frac{f(a)+f(b)}{2} - \frac{\gamma}{4}\left(a^2+b^2\right),$$

giving on rearrangement

$$\frac{1}{b-a}\int_a^b f(x)\,dx \leq \frac{f(a)+f(b)}{2} - \frac{\gamma}{12}(b-a)^2,$$

which is ostensibly identical to the first inequality in (3.125).

The second part in (3.125) follows by (3.127) applied for the convex (and twice differentiable mapping) $h:[a,b]\to\mathbb{R}$, $h(x)=\frac{\Gamma}{2}x^2-f(x)$. Now, it is straightforward to see that for $\alpha\leq t\leq\beta$, $\left|t-\frac{\alpha+\beta}{2}\right|\leq\frac{\beta-\alpha}{2}$. On taking

$\alpha = \frac{\gamma}{12}(b-a)^2$ and $\beta = \frac{\Gamma}{12}(b-a)^2$ we get the desired estimation (3.126). ∎

3.8.2 Some Trapezoid Quadrature Rules

We now consider applications of the integral inequalities developed in 3.8.1 [19].

THEOREM 3.26
Let $f : [a,b] \to \mathbb{R}$ be as in Theorem 3.22.
If $I_n : a = x_0 < x_1 < \ldots < x_{n-1} < x_n = b$ is a partition of the interval $[a,b]$, then we have

$$\int_a^b f(x)\,dx = A_T(f, I_n) + R_T(f, I_n), \tag{3.128}$$

where

$$A_T(f, I_n) := \frac{1}{2}\sum_{i=0}^{n-1}\left[f(x_i) + f(x_{i+1})\right] h_i$$

is the trapezoid quadrature rule and the remainder $R_T(f, I_n)$ satisfies the relation:

$$|R_T(f, I_n)| \tag{3.129}$$

$$\leq \begin{cases} \frac{1}{2}\|f''\|_\infty \sum_{i=0}^{n-1} h_i^3 \\ \\ \frac{1}{2}\|f''\|_p \left[B(q+1, q+1)\right]^{\frac{1}{q}} \left(\sum_{i=0}^{n-1} h_i^{2q+1}\right)^{\frac{1}{q}}, & \frac{1}{p}+\frac{1}{q} = 1,\ q > 1, \\ \\ \frac{1}{8}\|f''\|_1 \nu^2(I_n) \end{cases}$$

where $h_i := x_{i+1} - x_i$, $i = 0, \ldots, n-1$ and $\nu(I_n) = \max\{h_i|,\ i = 0, \ldots, n-1\}$.

PROOF Applying the first inequality, (3.114), we get

$$\left|\int_{x_i}^{x_{i+1}} f(x)\,dx - \frac{f(x_i) + f(x_{i+1})}{2}(x_{i+1} - x_i)\right| \leq \frac{\|f''\|_\infty}{12} h_i^3$$

for all $i \in \{0, \ldots, n-1\}$.

Summing over i from 0 to $n-1$ we get the first part of (3.129). The second inequality in (3.114) gives us

$$\left|\int_{x_i}^{x_{i+1}} f(x)\,dx - \frac{f(x_i)+f(x_{i+1})}{2}(x_{i+1}-x_i)\right|$$

$$\leq \frac{1}{2}h_i^{2+\frac{1}{q}}\left[B(q+1,q+1)\right]^{\frac{1}{p}}\left(\int_{x_i}^{x_{i+1}}|f''(t)|^p\,dt\right)^{\frac{1}{p}},$$

for all $i=0,...,n-1$.

Summing and using Hölder's discrete inequality, we get

$$\left|\int_a^b f(x)\,dx - A_T(f,I_n)\right|$$

$$\leq \frac{1}{2}\left[B(q+1,q+1)\right]^{\frac{1}{q}}\sum_{i=0}^{n-1}h_i^{\frac{2q+1}{q}}\left(\int_{x_i}^{x_{i+1}}|f''(t)|^p\,dt\right)^{\frac{1}{p}}$$

$$\leq \frac{1}{2}\left[B(q+1,q+1)\right]^{\frac{1}{q}}\left[\sum_{i=0}^{n-1}\left(h_i^{\frac{2q+1}{q}}\right)^q\right]^{\frac{1}{q}}\left[\sum_{i=0}^{n-1}\left[\left(\int_{x_i}^{x_{i+1}}|f''(t)|^p\,dt\right)^{\frac{1}{p}}\right]^p\right]^{\frac{1}{p}}$$

$$= \frac{1}{2}\left[B(q+1,q+1)\right]^{\frac{1}{q}}\left(\sum_{i=0}^{n-1}h^{2q+1}\right)^{\frac{1}{q}}\sum_{i=0}^{n-1}\left(\int_{x_i}^{x_{i+1}}|f''(t)|^p\,dt\right)^{\frac{1}{p}}$$

$$= \frac{1}{2}\left[B(q+1,q+1)\right]^{\frac{1}{q}}\|f''\|_p\left(\sum_{i=0}^{n-1}h^{2q+1}\right)^{\frac{1}{q}},$$

and the second inequality in (3.129) is proved. In the last part, we have by (3.114), that

$$|R_T(f,I_n)| \leq \frac{1}{8}\sum_{i=0}^{n-1}\left(\int_{x_i}^{x_{i+1}}|f''(t)|\,dt\right)h_i^2$$

$$\leq \frac{1}{8}\max_{i=\overline{0,n-1}}h_i^2\sum_{i=0}^{n-1}\left(\int_{x_i}^{x_{i+1}}|f''(t)|\,dt\right)$$

$$= \frac{1}{8}\nu^2 (I_n) \|f''\|_1 ,$$

and the theorem is proved. ∎

REMARK 3.31 We would like to note that in every book on numerical integration encountered by the authors only the first estimation in (3.129) is used. Sometimes, when $\|f''\|_p$ $(p > 1)$ or $\|f''\|_1$ are easier to compute, it would perhaps be more appropriate to use the second or the third estimation. ∎

We shall now investigate the case where we have an equidistant partitioning of $[a, b]$ given by

$$I_n : x_i = a + \frac{b-a}{n} \cdot i,\ i = 0, 1, ..., .n.$$

The following result is a consequence of Theorem 3.26.

COROLLARY 3.16

Let $f : [a, b] \to \mathbb{R}$ be a twice differentiable mapping and $\|f''\|_\infty < \infty$. Then we have

$$\int_a^b f(x)\, dx = A_{T,n}(f) + R_{T,n}(f),$$

where

$$A_{T,n}(f) = \frac{b-a}{2n} \sum_{i=0}^{n-1} \left[f\left(a + \frac{b-a}{n} i\right) + f\left(a + \frac{b-a}{n}(i+1)\right)\right]$$

and the remainder $R_{T,n}(f)$ satisfies the estimation

$$|R_{T,n}(f)| \le \begin{cases} \frac{(b-a)^3 \|f''\|_\infty}{12n^2} \\ \frac{(b-a)^2 [B(q+1,q+1)]^{2+\frac{1}{q}} \|f''\|_p}{2n^2}, & \frac{1}{p} + \frac{1}{q} = 1,\ p > 1, \\ \frac{(b-a)^2 \|f''\|_1}{8n^2} \end{cases}$$

for all $n \ge 1$.

The following theorem gives a perturbed trapezoid formula using Theorem 3.24, which is sometimes more appropriate.

THEOREM 3.27
Let $f : [a,b] \to \mathbb{R}$ be as in Theorem 3.24 and I_n be an arbitrary partition of the interval $[a,b]$. Then we have

$$\int_a^b f(x)\,dx = A_T(f, f', I_n) + \tilde{R}_T(f, I_n), \tag{3.130}$$

where

$$A_T(f, f', I_n) := \frac{1}{2}\sum_{i=0}^{n-1}\left[f(x_i) + f(x_{i+1})\right]h_i + \frac{1}{12}\sum_{i=0}^{n-1}\left(f'(x_{i+1}) - f'(x_i)\right)h_i^2$$

is a perturbed trapezoidal rule and the remainder term $\tilde{R}_T(f, I_k)$ satisfies the estimation:

$$\left|\tilde{R}_T(f, I_k)\right| \le \frac{1}{24\sqrt{5}}(\Gamma - \gamma)\sum_{i=0}^{n-1} h_i^3, \tag{3.131}$$

where the h_i are as above.

PROOF Writing the inequality (3.119) on the intervals $[x_i, x_{i+1}]$ $(i = 0, ..., n-1)$ we get

$$\left|\int_{x_i}^{x_{i+1}} f(x)\,dx - \frac{f(x_i) + f(x_{i+1})}{2}\cdot h_i + \frac{1}{12}\left(f'(x_{i+1}) - f'(x_i)\right)h_i^2\right|$$

$$\le \frac{1}{24\sqrt{5}}(\Gamma - \gamma)\cdot h_i^3$$

for all $i = 0, ..., n-1$.
Summing over i from 0 to $n-1$, we deduce the desired estimation (3.131).
∎

It should be noted that similar results could be obtained for (3.119) or (3.124). All that would change would be the coefficient for the bound.

REMARK 3.32 As

$$0 \leq \Gamma - \gamma \leq 2 \left\| f'' \right\|_{\infty},$$

then

$$\frac{1}{24\sqrt{5}} (\Gamma - \gamma) \leq \frac{\left\| f'' \right\|_{\infty}}{12\sqrt{5}} < \frac{\left\| f'' \right\|_{\infty}}{12},$$

and so the approximation of the integral $\int_a^b f(x)dx$ by the use of $A_T(f, f', I_n)$ is better than that provided by the classical trapezoidal formulae $A_T(f, I_n)$ for every partition I_n of the interval $[a, b]$. Atkinson [3] calls this the corrected trapezoidal rule. However, only the classical $\left\| f'' \right\|_{\infty}$ norm is used as the bound on the error. Atkinson [3] uses the idea of an asymptotic error estimate rather than the inequality by Grüss. ∎

The following corollary of Theorem 3.27 holds:

COROLLARY 3.17
Let $f : [a, b] \to \mathbb{R}$ be as in Theorem 3.23. Thus we have

$$\int_a^b f(x)\, dx = A_{T,n}(f, f') + \tilde{R}_{T,n}(f),$$

where

$$A_{T,n}(f, f') = \frac{b-a}{2n} \sum_{i=0}^{n-1} \left[f\left(a + \frac{b-a}{n} \cdot i \right) + f\left(a + \frac{b-a}{n} \cdot (i+1) \right) \right]$$

$$+ \frac{(b-a)^2}{12n^2} (f'(b) - f'(a)),$$

and the remainder $\tilde{R}_T(f)$ satisfies the estimation:

$$\left| \tilde{R}_{T,n}(f) \right| \leq \frac{(\Gamma - \gamma)(b-a)^3}{24\sqrt{5}n^2},$$

for all $n \geq 1$.

Now, if we apply Theorem 3.25, we can state the following quadrature formulae which is a quasi-trapezoid formula or a perturbed trapezoid formula.

THEOREM 3.28

Let f be a in Theorem 3.25. If I_n is a partition of the interval $[a,b]$, then we have

$$\int_a^b f(x)\,dx = A_{T,\gamma,\Gamma}(f, I_n) + R_{T,\gamma,\Gamma}(f, I_n), \tag{3.132}$$

where

$$A_{T,\gamma,\Gamma}(f, I_n) = \sum_{i=0}^{n-1} \frac{f(x_i) + f(x_{i+1})}{2} \cdot h_i - \frac{\Gamma + \gamma}{24} \sum_{i=0}^{n-1} h_i^3$$

and

$$|R_{T,\gamma,\Gamma}(f, I_n)| \le \frac{\Gamma + \gamma}{24} \sum_{i=0}^{n-1} h_i^3. \tag{3.133}$$

PROOF Applying the inequality (3.126) on $[x_i, x_{i+1}]$, we get

$$\left| \int_{x_i}^{x_{i+1}} f(x)\,dx - \frac{f(x_i) + f(x_{i+1})}{2} \cdot h_i + \frac{\Gamma + \gamma}{24} \cdot h_i^2 \right| \le \frac{\Gamma + \gamma}{24} \cdot h_i^3$$

for all $i \in \{0, ..., n-1\}$.
Summing over i from 0 to $n-1$ we get the representation (3.132) over the estimation (3.133). ∎

COROLLARY 3.18

Let f be as above. Then we have

$$\int_a^b f(x)\,dx = A_{T,\gamma,\Gamma,n}(f) + R_{T,\gamma,\Gamma,n}(f),$$

where

$$A_{T,\gamma,\Gamma,n}(f) = \frac{1}{2} \sum_{i=0}^{n-1} \left[f\left(a + i \cdot \frac{b-a}{n}\right) - f\left(a + (i+1) \cdot \frac{b-a}{n}\right) \right]$$

$$+ \frac{\Gamma + \gamma}{12} \cdot \frac{(b-a)^2}{n}$$

and the remainder term $R_{T,\gamma,\Gamma,n}(f)$ satisfies

$$|R_{T,\gamma,\Gamma,n}(f)| \le \frac{(\Gamma - \gamma)(b-a)^3}{24n^2}.$$

REMARK 3.33 As $0 \leq \Gamma - \gamma \leq 2\|f''\|_{\infty}$, then the approximation given by $A_{T,\gamma,\Gamma,n}(f)$ to the integral $\int_a^b f(x)\,dx$ is better than the classical trapezoidal rule. ∎

3.9 Concluding Remarks

The current work has demonstrated the development of trapezoidal-type rules. Identities are obtained by using a Peano kernel approach which enables us, through the use of the modern theory of inequalities, to obtain bounds in terms of a variety of norms. This is useful in practice as the behaviour of the function would necessitate the use of one norm over another. Although not all inequalities have been developed into composite quadrature rules, we believe that enough demonstrations have been given that would enable the reader to proceed further.

Rules have been developed that do not necessarily require the second derivative to be well behaved or indeed, exist, thus allowing the treatment of a much larger class of functions. Rules have been developed by examining the Riemann-Stieltjes integral. Additionally, the rules also allow for a non-uniform partition, thus giving the user the option of choosing a partition that minimises the bound or enabling the calculation of the bound given a particular partition.

If we wish to approximate the integral $\int_a^b f(x)\,dx$ using a quadrature rule $Q(f, I_n)$ with bound $B(n)$, where I_n is a uniform partition, for example, with an accuracy of $\varepsilon > 0$, then we will need $n_\varepsilon \in \mathbb{N}$ where

$$n_\varepsilon \geq \left[B^{-1}(\varepsilon)\right] + 1$$

with $[x]$ denoting the integer part of x.

This approach enables the user to predetermine the partition required to *assure* that the result is within a certain tolerance rather than utilizing the commonly used method of halving the mesh size and comparing the resulting estimation.

Acknowledgment The authors would like to express their sincere gratitude to Ms. Pui Ling Pang for the preparation of the manuscript. Further, the selfless contribution to the typesetting of the work by their colleague John Roumeliotis is gratefully acknowledged. This work would not have been possible without their dedication well beyond the call!

References

[1] R.P. Agarwal and S.S. Dragomir, An application of Hayashi's inequality for differentiable functions, *Computers Math. Appl.*, **32**(6), 95–99, 1996.

[2] T. M. Apostol, *Mathematical Analysis,* 2nd Ed., Addison-Wesley Publishing Company, 1975.

[3] K.E. Atkinson, *An Introduction to Numerical Analysis*, 2nd Ed., Wiley and Sons, 1989.

[4] N.S. Barnett, S.S. Dragomir and C.E.M. Pearce, A quasi-trapezoid inequality for double integrals, (submitted, 1999).

[5] P. Cerone and S.S. Dragomir, Three-point quadrature rules involving, at most, a first derivative, (submitted, 1999). [ONLINE] *RGMIA Res. Rep. Coll.*, **2**(4) Article 8, 1999. `http://rgmia.vu.edu.au/v2n4.html`

[6] P. Cerone and S.S. Dragomir, Lobatto type quadrature rules for functions with bounded derivative, *Mathematical Inequalities and Applications*, accepted, 1999. [ONLINE] *RGMIA Res. Rep. Coll.*, **2**(4) Article 8, 1999. `http://rgmia.vu.edu.au/v2n2.html`

[7] P. Cerone, S.S. Dragomir, J. Roumeliotis and J. Sunde, A new generalization of the trapezoid formula for n-time differentiable mappings and applications, *Preprint: RGMIA Res. Rep. Coll.*, **2**(5), Article 7, 1999. [ONLINE] `http://rgmia.vu.edu.au/v2n5.html`

[8] P.J. Davis and P. Rabinowitz, *Methods of Numerical Integration*, 2nd Ed., Academic Press, New York, 1994.

[9] S.S. Dragomir, A Grüss type integral inequality for mappings of r-Hölder's type and applications for trapezoid formula, *Tamkang Journal of Mathematics*, accepted, 1999.

[10] S.S. Dragomir, A Taylor like formula and application in numerical integration,[1] (submitted, 1999).

[1]A great number of these papers are available from the *RGMIA* at `http://rgmia.vu.edu.au/`

[11] S.S. Dragomir, Grüss inequality in inner product spaces, *The Australian Math. Gazette* **26**(2), 66-70, 1999.

[12] S.S. Dragomir, New estimation of the remainder in Taylor's formula using Grüss' type inequalities and applications, *Mathematical Inequalities and Applications*, **2**(2), 183-194, 1999.

[13] S.S. Dragomir, On the trapezoid formula for mappings of bounded variation and applications, (submitted, 1999). [ONLINE] `http://rgmia.vu.edu.au/severdvi/TRAPB.dvi`

[14] S.S. Dragomir, On the trapezoid inequality for absolutely continuous mappings, (submitted, 1999).

[15] S.S. Dragomir, On the trapezoid quadrature formula and applications, (submitted, 1999).

[16] S.S. Dragomir, On the trapezoid quadrature formula for Lipschitzian mappings and applications, *Tamkang J. of Math.*, (in press), 1999. [ONLINE] `http://rgmia.vu.edu.au/severdvi/TRAPL.dvi`

[17] S.S. Dragomir, Some integral inequalities of Grüss type, *Italian J. of Pure and Appl. Math.*, accepted, 1999.

[18] S.S. Dragomir, P. Cerone and C. E. M. Pearce, Generalizations of the trapezoid inequality for mappings of bounded variation and applications, (submitted, 1999).

[19] S.S. Dragomir, P. Cerone and A. Sofo, Some remarks on the trapezoid rule in numerical integration, *Indian J. of Pure and Appl. Math.*, (in press), 1999. [ONLINE] *RGMIA Res. Rep. Coll.*, **2**(5), Article 1, 1999.

[20] S.S. Dragomir, Y.J. Cho and S.S. Kim, Some remarks on the Milovanović-Pečarić inequality and applications for special means and numerical integration, *Tamkang Journal of Mathematics*, accepted, 1999.

[21] S.S. Dragomir and I. Fedotov, An inequality of Grüss type for Riemann-Stieltjes integral and applications for special means, *Tamkang J. of Math.*, **29**(4), 287-292, 1998.

[22] S.S. Dragomir and A. McAndrew, On trapezoid inequality via a Grüss type result and applications, *Preprint: RGMIA Res. Rep. Coll.*, **2**(2), Article 6, 1999. [ONLINE] `http://rgmia.vu.edu.au/v2n2.html`

[23] S.S. Dragomir and T.C. Peachey, New estimation of the remainder in the trapezoidal formula with applications, (submitted, 1998). [ONLINE] `http://rgmia.vu.edu.au/v1n2.html`

[24] S.S. Dragomir, J.E. Pečarić and S. Wang, The unified treatment of trapezoid, Simpson and Ostrowski type inequality for monotonic mappings and applications, *Preprint: RGMIA Res. Rep. Coll.*, **2**(4), Article 3, 1999. [ONLINE] `http://rgmia.vu.edu.au/v2n4.html`

[25] S.S. Dragomir and S. Wang, Applications of Iyengar's type inequalities to the estimation of error bounds for the trapezoidal quadrature rule, *Tamkang Journal of Mathematics*, **29**(1), 55-58, 1998.

[26] H. Engels, *Numerical Quadrature and Cubature*, Academic Press, New York, 1980.

[27] I. Fedotov and S.S. Dragomir, Another approach to quadrature methods, *Indian Journal of Pure and Applied Mathematics*, (in press), 1999.

[28] A.R. Krommer and C.W. Ueberhuber, *Numerical Integration on Advanced Computer Systems*, Lecture Notes in Computer Science, 848, Springer-Verlag, Berlin, 1994.

[29] M. Matić, J.E. Pečarić and N. Ujević, On new estimation of the remainder in Generalised Taylor's Formula, *Mathematical Inequalities and Applications*, **2**(3), 343-361, 1999.

[30] D.S. Mitrinović, J.E. Pečarić and A.M. Fink, *Classical and New Inequalities in Analysis*, Kluwer Academic Publishers, 1993.

[31] D.S. Mitrinović, J.E. Pečarić and A.M. Fink, *Inequalities for Functions and Their Integrals and Derivatives*, Kluwer Academic Publishers, 1994.

[32] J.E. Pečarić, F. Proschan and Y.L. Tong, *Convex Functions, Partial Orderings, and Statistical Applications*, Academic Press, 1992.

[33] W.H. Press, B.P. Flannery, S.A. Telkolsky and W.T. Verling, *Numerical Recipes: The Art of Scientific Computing*, Cambridge University Press, Cambridge, 1986.

[34] A.H. Stroud, *Numerical Quadrature and Solution of Ordinary Differential Equations*, Springer-Verlag, New York, 1974.

Chapter 4

Midpoint-Type Rules from an Inequalities Point of View

Pietro Cerone and Sever S. Dragomir

School of Communications and Informatics, Victoria University of Technology, P.O. Box 14428, Melbourne City MC, Victoria 8001, Australia
E-mail: Pietro.Cerone@vu.edu.au, Sever.Dragomir@vu.edu.au
Contact Author: S. Dragomir

Abstract This paper investigates interior point rules which contain the midpoint as a special case, and obtains explicit bounds through the use of a Peano kernel approach and the modern theory of inequalities. Thus the simplest open Newton-Cotes rules are examined. Both Riemann-Stieltjes and Riemann integrals are evaluated with a variety of assumptions about the integrand enabling the characterisation of the bound in terms of a variety of norms. Perturbed quadrature rules are obtained through the use of Grüss, Chebychev and Lupaş inequalities, producing a variety of tighter bounds. The implementation is demonstrated through the investigation of a variety of composite rules based on inequalities developed. The analysis allows the determination of the partition required that would assure that the accuracy of the result would be within a prescribed error tolerance. It is demonstrated that the bounds of the approximations are equivalent to those obtained from a Peano kernel that produces trapezoidal-type rules.

Contents

4.1 Introduction

The following inequality is well known in the literature as the midpoint inequality:

$$\left| \int_a^b f(x)\,dx - (b-a) f\left(\frac{a+b}{2}\right) \right| \leq \frac{(b-a)^3}{24} \left\| f'' \right\|_\infty , \tag{4.1}$$

where the mapping $f : [a,b] \subset \mathbb{R} \to \mathbb{R}$ is assumed to be twice differentiable on the interval (a,b) and having the second derivative bounded on (a,b). That is, $\|f''\|_\infty := \sup_{x \in (a,b)} |f''(x)| < \infty$.

Now, if we assume that $I_n : a = x_0 < x_1 < \ldots < x_{n-1} < x_n = b$ is a partition of the interval $[a,b]$ and f is as above, then we can approximate the integral $\int_a^b f(x)\,dx$ by the *midpoint quadrature* formula $A_M(f, I_n)$ having an error given by $R_M(f, I_n)$, where

$$A_M(f, I_n) = \sum_{i=0}^{n-1} f\left(\frac{x_{i+1} + x_i}{2}\right) h_i \tag{4.2}$$

and the remainder $R_M(f, I_n)$ satisfies the estimation

$$|R_M(f, I_n)| \leq \frac{\|f''\|_\infty}{24} \sum_{i=0}^{n-1} h_i^3, \tag{4.3}$$

where $h_i = x_{i+1} - x_i$ for $i = 0, 1, 2, ..., n-1$.

Equation (4.2) is known as the midpoint rule for $n = 1$ and as the composite midpoint rule for $n > 1$. The midpoint rule is the most basic open Newton-Cotes quadrature in which function evaluations occur at the midpoints of equispaced intervals (Press et al. [36]).

The current work investigates an interior point (which contains the midpoint as a special case) and obtains explicit bounds through the use of a Peano kernel approach and the modern theory of inequalities. This approach allows for the investigation of quadrature rules that place fewer restrictions on the behaviour of the integrand and thus allow us to cope with larger classes of functions. Expression (4.1) relies on the behaviour of the second derivative whereas bounds for the interior point are obtained in terms of Riemann-Stieltjes integrals in Sections 4.2, 4.3 and 4.4 for functions that are of bounded variation, Lipschitzian and monotonic, respectively. In Section 4.5, interior point rules are obtained for $f^{(n)} \in L_p[a, b]$, implying that

$$\left\|f^{(n)}\right\|_p := \left(\int_a^b \left|f^{(n)}(x)\right|^p dx\right)^{\frac{1}{p}} < \infty \text{ for } p \geq 1$$

and $\left\|f^{(n)}\right\|_\infty := \sup_{x\in[a,b]} \left|f^{(n)}(x)\right|$.
Further, a generalised Taylor series representation is presented that enables an expansion about any point on an interval.

In Section 4.6, perturbed interior point rules are obtained using what are termed as **premature** variants of Grüss, Chebychev and Lupaş inequalities. Atkinson [2] uses an asymptotic error estimate technique to obtain what he defines as a corrected rule. His approach, however, does not readily produce a bound on the error.

Further, in 4.6.2, alternate Grüss-type results are obtained to produce perturbed interior point rules with bounds given in terms of norms associated with $f'(x) - S$, where $S = \frac{f(b)-f(a)}{b-a}$ is the secant slope.

Finally, in Section 4.7, a perturbed interior point rule is obtained whose perturbation involves S and not $\frac{f'(b)-f'(a)}{b-a}$. The bound relies on the behaviour of $f''(\cdot)$.

The current work brings together results for interior point-type rules giving explicit error bounds, using Peano type kernels and results from the modern theory of inequalities. Although bounds through the use of Peano kernels have been obtained in some classical review books on numerical integration such as Stroud [38], Engels [29], and Davis and Rabinowitz [9], these do not seem to be utilised to perhaps the extent that they should be. So much so that even in the more recent comprehensive monograph by Krommer and Ueberhuber [31], a constructive approach is taken via Taylor or interpolating polynomials to obtain quadrature results. This approach does not readily provide explicit error bounds but rather gives the order of the approximation.

4.2 The Ostrowski Inequality for Mappings of Bounded Variation

In this section we develop interior point-type quadrature rules for functions that are of bounded variation. It includes the midpoint rule as a special case. Functions of bounded variation include a very large class in contrast to traditional interior or specifically midpoint rules which rely on the second derivative of the function for its error approximation.

4.2.1 Some Integral Inequalities

The following inequality for mappings of bounded variation holds [15]:

THEOREM 4.1

Let $u : [a, b] \to \mathbb{R}$ be a mapping of bounded variation on $[a, b]$. Then for all $x \in [a, b]$, we have the inequality

$$\left| \int_a^b u(t)\,dt - (b-a)\,u(x) \right| \leq \left[\frac{1}{2}(b-a) + \left| x - \frac{a+b}{2} \right| \right] \bigvee_a^b (u), \tag{4.4}$$

where $\bigvee_a^b (u)$ denotes the total variation of u.
The constant $\frac{1}{2}$ is the best possible one.

PROOF Using the integration by parts formula for Riemann-Stieltjes integrals we have

$$\int_a^x (t-a)\,du(t) = u(x)(x-a) - \int_a^x u(t)\,dt$$

and

$$\int_x^b (t-b)du(t) = u(x)(b-x) - \int_x^b u(t)\,dt.$$

If we add the above two equalities, we obtain

$$(b-a)\,u(x) - \int_a^b u(t)\,dt = \int_a^b p(x,t)\,du(t)\ , \tag{4.5}$$

where

$$p(x,t) := \begin{cases} t-a, & \text{if } t \in [a,x) \\ t-b, & \text{if } t \in [x,b], \end{cases}$$

for all $x, t \in [a,b]$.
It is well known [1] that if $p : [a,b] \to \mathbb{R}$ is continuous on $[a,b]$ and $v : [a,b] \to \mathbb{R}$ is of bounded variation on $[a,b]$, then

$$\left| \int_a^b p(x)\,dv(x) \right| \le \sup_{x\in[a,b]} |p(x)| \bigvee_a^b (v). \tag{4.6}$$

Applying the inequality (4.6) for $p(x,t)$ as above and $v(x) = u(x)$, $x \in [a,b]$, we get

$$\left| \int_a^b p(x,t)\,du(t) \right| \le \sup_{t\in[a,b]} |p(x,t)| \bigvee_a^b (u)$$

$$= \max\{x-a, b-x\} \bigvee_a^b (u)$$

$$= \left[\frac{b-a}{2} + \left| x - \frac{a+b}{2} \right| \right] \bigvee_a^b (u) \tag{4.7}$$

and then by (4.7), via the identity (4.5), we deduce the desired inequality (4.4).
Now to prove that $\frac{1}{2}$ is the best possible constant assume that the inequality (4.4) holds with a constant $C > 0$. That is,

$$\left| \int_a^b u(t)\,dt - u(x)(b-a) \right| \le \left[C(b-a) + \left| x - \frac{a+b}{2} \right| \right] \bigvee_a^b (u) \tag{4.8}$$

for all $x \in [a, b]$.
Consider the mapping $u : [a, b] \to \mathbb{R}$, given by

$$u(x) = \begin{cases} 0 & \text{if } x \in [a, b] \setminus \{\frac{a+b}{2}\} \\ 1 & \text{if } x = \frac{a+b}{2} \end{cases}$$

in (4.8). Then u is of bounded variation on $[a, b]$, and

$$\bigvee_a^b (u) = 2, \int_a^b u(t)\,dt = 0.$$

For $x = \frac{a+b}{2}$, we get in (4.8)

$$1 \leq 2C,$$

which implies that $C \geq \frac{1}{2}$ and the theorem is completely proved. ∎

The following corollary holds for monotonic mappings.

COROLLARY 4.1

Let $u : [a, b] \to \mathbb{R}$ be a monotonic mapping on $[a, b]$. Then we have the inequality

$$\left| \int_a^b u(t)\,dt - (b-a)\,u(x) \right| \leq \left[\frac{1}{2}(b-a) + \left| x - \frac{a+b}{2} \right| \right] |u(b) - u(a)|.$$

The case of Lipschitzian mappings is embodied in the following corollary.

COROLLARY 4.2

Let $u : [a, b] \to \mathbb{R}$ be an L-Lipschitzian mapping on $[a, b]$. That is, we recall

$$|u(x) - u(y)| \leq L\,|x - y| \quad \textit{for all } x, y \in [a, b].$$

Then, for all $x \in [a, b]$ we have the inequality

$$\left| \int_a^b u(t)\,dt - (b-a)\,u(x) \right| \leq L \left[\frac{1}{2}(b-a) + \left| x - \frac{a+b}{2} \right| \right] (b-a),$$

giving a midpoint-type rule.

COROLLARY 4.3
Let $u : [a,b] \to \mathbb{R}$ be as above. Then we have the inequality

$$\left| \int_a^b u(t)\, dx - (b-a)\, u\left(\frac{a+b}{2}\right) \right| \le \frac{1}{2}(b-a) \bigvee_a^b (u). \tag{4.9}$$

Similar inequalities can be found if we assume that u is monotonic or Lipschitzian on $[a,b]$ by taking $x = \frac{a+b}{2}$ in Corollaries 4.1 and 4.2, respectively.

REMARK 4.1 If we assume that u is continuous differentiable on (a,b) and u' is integrable on (a,b), then, by (4.4), we get

$$\left| \int_a^b u(t)\, dx - (b-a)\, u(x) \right| \le \left[\frac{1}{2}(b-a) + \left| x - \frac{a+b}{2} \right| \right] \|u'\|_1 ,$$

which is the inequality obtained by Dragomir and Wang in the recent paper [26]. ∎

REMARK 4.2 It is well known that if $f : [a,b] \to \mathbb{R}$ is a convex mapping on $[a,b]$, then the **Hermite-Hadamard** inequality (see, for example, [35, p. 137])

$$f\left(\frac{a+b}{2}\right) \le \frac{1}{b-a} \int_a^b f(x)\, dx \le \frac{f(a)+f(b)}{2} \tag{4.10}$$

holds [34].
Now, if we assume that $f : I \subset \mathbb{R} \to \mathbb{R}$ is convex on I and $a, b \in Int(I)$, $a < b$; then f'_+ is monotonic nondecreasing on $[a,b]$ and, by Corollary 4.3, we obtain

$$0 \le \frac{1}{b-a} \int_a^b f(x)\, dx - f\left(\frac{a+b}{2}\right) \le \frac{1}{2} \|f'_+\|_1 , \tag{4.11}$$

which gives a counterpart for the first membership of Hadamard's inequality. ∎

Similar results can be obtained if we assume that f is convex and monotonic or convex and Lipschitzian on $[a,b]$.

4.2.2 A Quadrature Formula of Riemann Type

Let $I_n : a = x_0 < x_1 < ... < x_{n-1} < x_n = b$ be a division of the interval $[a,b]$ and $\xi_i \in [x_i, x_{i+1}]$ $(i = 0, ..., n-1)$ a sequence of intermediate points for I_n. Construct the Riemann sums

$$R_n(f, I_n, \boldsymbol{\xi}) = \sum_{i=0}^{n-1} f(\xi_i) h_i ,$$

where $h_i := x_{i+1} - x_i$.

We have the following quadrature formula [15].

THEOREM 4.2

Let $f : [a,b] \to \mathbb{R}$ be a mapping of bounded variation on $[a,b]$ and I_n, ξ_i $(i = 0, ..., n-1)$ be as above. Then we have the Riemann quadrature formula

$$\int_a^b f(x)\, dx = R_n(f, I_n, \boldsymbol{\xi}) + W_n(f, I_n, \boldsymbol{\xi}\ ), \tag{4.12}$$

where the remainder satisfies the estimation

$$\begin{aligned} |W_n(f, I_n, \boldsymbol{\xi})| &\leq \sup_{i=0,...,n} \left[\frac{1}{2}h_i + \left|\xi_i - \frac{x_i + x_{i+1}}{2}\right|\right] \bigvee_a^b (f) \\ &\leq \left[\frac{1}{2}\nu(h) + \sup_{i=0,...,n}\left|\xi_i - \frac{x_i + x_{i+1}}{2}\right|\right] \bigvee_a^b (f) \\ &\leq \nu(h) \bigvee_a^b (f) \end{aligned} \tag{4.13}$$

for all ξ_i $(i = 0, ..., n-1)$ as above, where $\nu(h) := \max\{h_i | i = 0, 1, ..., n\}$. *The constant $\frac{1}{2}$ is sharp in (4.13).*

PROOF Apply Theorem 4.1 on the interval $[x_i, x_{i+1}]$ to get

$$\left|\int_{x_i}^{x_{i+1}} f(x)\, dx - f(\xi_i) h_i\right| \leq \left[\frac{1}{2}h_i + \left|\xi_i - \frac{x_i + x_{i+1}}{2}\right|\right] \bigvee_{x_i}^{x_{i+1}} (f). \tag{4.14}$$

Summing over i from 0 to $n-1$ and using the generalized triangle inequality we get

$$
\begin{aligned}
|W_n(f, I_n, \boldsymbol{\xi})| &\leq \sum_{i=0}^{n-1} \left| \int_{x_i}^{x_{i+1}} f(x)\,dx - f(\xi_i)\,h_i \right| \\
&\leq \sum_{i=0}^{n-1} \left[\frac{1}{2} h_i + \left| \xi_i - \frac{x_i + x_{i+1}}{2} \right| \right] \bigvee_{x_i}^{x_{i+1}} (f) \\
&\leq \sup_{i=0,\dots,n} \left[\frac{1}{2} h_i + \left| \xi_i - \frac{x_i + x_{i+1}}{2} \right| \right] \sum_{i=0}^{n-1} \bigvee_{x_i}^{x_{i+1}} (f) \\
&= \sup_{i=0,\dots,n} \left[\frac{1}{2} h_i + \left| \xi_i - \frac{x_i + x_{i+1}}{2} \right| \right] \bigvee_a^b (f).
\end{aligned}
$$

The second inequality follows by the properties of $\sup(\cdot)$.
Now, as

$$
\left| \xi_i - \frac{x_i + x_{i+1}}{2} \right| \leq \frac{1}{2} h_i
$$

for all $\xi_i \in [x_i, x_{i+1}]$ $(i = 0, \dots, n-1)$ the last part of (4.13) is also proved. ▮

COROLLARY 4.4

Let $f : [a,b] \to \mathbb{R}$ be a monotonic mapping on $[a,b]$ and I_n, ξ_i $(i = 0, \dots, n-1)$ be as above. Then we have the Riemann quadrature formula (4.12) where the remainder satisfies the estimation

$$
\begin{aligned}
|W_n(f, I_n, \boldsymbol{\xi})| &\leq \sup_{i=0,\dots,n} \left[\frac{1}{2} h_i + \left| \xi_i - \frac{x_i + x_{i+1}}{2} \right| \right] |f(b) - f(a)| \\
&\leq \left[\frac{1}{2} \nu(h) + \sup_{i=0,\dots,n} \left| \xi_i - \frac{x_i + x_{i+1}}{2} \right| \right] |f(b) - f(a)| \\
&\leq \nu(h) |f(b) - f(a)|
\end{aligned}
$$

for all ξ_i $(i = 0, \dots, n-1)$ as above.

The case of Lipschitzian mappings is embodied within the following corollary.

COROLLARY 4.5
Let $f : [a,b] \to \mathbb{R}$ be an L-Lipschitzian mapping on $[a,b]$ and I_n, ξ_i $(i = 0,...,n-1)$ be as above. Then we have the Riemann quadrature formula (4.12) where the remainder satisfies the estimation

$$|W_n(f, I_n, \boldsymbol{\xi})| \leq L \sum_{i=0}^{n-1} \left[\frac{1}{2}h_i + \left|\xi_i - \frac{x_i + x_{i+1}}{2}\right|\right] h_i$$

$$\leq L \sum_{i=0}^{n-1} h_i^2.$$

The proof is obvious by Corollary 4.2 applied on the intervals $[x_i, x_{i+1}]$ and summing the resulting inequalities.

We shall omit the details.

Note that the best estimation we can get from (4.13) is the one for which $\xi_i = \frac{x_i + x_{i+1}}{2}$, obtaining the following midpoint formula for functions of bounded variation.

COROLLARY 4.6
Let f, I_n be as Theorem 4.2. Then we have the midpoint rule

$$\int_a^b f(x)\,dx = M_n(f, I_n) + S_n(f, I_n)\,,$$

where

$$M_n(f, I_n) = \sum_{i=0}^{n-1} f\left(\frac{x_i + x_{i+1}}{2}\right) h_i$$

and the remainder $S_n(f, I_n)$ satisfies the estimation

$$|S_n(f, I_n)| \leq \frac{1}{2}\nu(h) \bigvee_a^b (f)\,.$$

Similar results can be obtained from Corollaries 4.4 and 4.5.

REMARK 4.3 If we assume that $f : [a,b] \to \mathbb{R}$ is differentiable on (a,b) and whose derivative f' is integrable on (a,b), we can put instead of $\bigvee_a^b (f)$ the L_1-norm $\|f'\|_1$ obtaining the estimation due to Dragomir and Wang from the paper [26]. ∎

4.3 An Inequality for Monotonic Mappings

Bounds were obtained for monotonic mappings in Corollary 4.2 as a particular case in the development of functions of bounded variation. This section treats specifically monotonic functions and obtains tighter bounds.

4.3.1 Integral Inequalities

The following results of the Ostrowski type hold [16].

THEOREM 4.3
Let $u : [a,b] \to \mathbb{R}$ be a monotonic nondecreasing mapping on $[a,b]$. Then for all $x \in [a,b]$, we have the inequality

$$\left|(b-a)\,u(x) - \int_a^b u(t)\,dt\right|$$

$$\leq [2x-(a+b)]u(x) + \int_a^b \operatorname{sgn}(t-x)u(t)\,dt$$

$$\leq (x-a)(u(x)-u(a)) + (b-x)(u(b)-u(x))$$

$$\leq \left[\frac{b-a}{2} + \left|x - \frac{a+b}{2}\right|\right](u(b)-u(a)). \tag{4.15}$$

All the inequalities in (4.15) are sharp and the constant $\frac{1}{2}$ is the best possible one.

PROOF Using the integration by parts formula for Riemann-Stieltjes integrals, we have the identity as given by (4.5).

Now, assume that $\Delta_n : a = x_0^{(n)} < x_1^{(n)} < \ldots < x_{n-1}^{(n)} < x_n^{(n)} = b$ is a sequence of divisions with $\nu(\Delta_n) \to 0$ as $n \to \infty$, where $\nu(\Delta_n) := \max_{i\in\{0,\ldots,n-1\}}(x_{i+1}^{(n)} - x_i^{(n)})$ and $\xi_i^{(n)} \in \left[x_i^{(n)}, x_{i+1}^{(n)}\right]$. If $p : [a,b] \to \mathbb{R}$ is continuous on $[a,b]$ and $v:[a,b]\to\mathbb{R}$ is monotonic nondecreasing on $[a,b]$, then

$$\begin{aligned}\left|\int_a^b p(x)\,dv(x)\right| &= \left|\lim_{\nu(\Delta_n)\to 0}\sum_{i=0}^{n-1} p\left(\xi_i^{(n)}\right)\left[v\left(x_{i+1}^{(n)}\right)-v\left(x_i^{(n)}\right)\right]\right| \\ &\le \lim_{\nu(\Delta_n)\to 0}\sum_{i=0}^{n-1}\left|p\left(\xi_i^{(n)}\right)\right|\left|v\left(x_{i+1}^{(n)}\right)-v\left(x_i^{(n)}\right)\right| \\ &\le \lim_{\nu(\Delta_n)\to 0}\sum_{i=0}^{n-1}\left|p\left(\xi_i^{(n)}\right)\right|\left(v\left(x_{i+1}^{(n)}\right)-v\left(x_i^{(n)}\right)\right) \\ &= \int_a^b |p(x)|\,dv(x).\end{aligned}$$

As u is monotonic nondecreasing on $[a,b]$, and $p(x,\cdot)$ is continuous on the intervals, then using the above inequality we can state that

$$\left|\int_a^b p(x,t)\,du(t)\right| \le \int_a^b |p(x,t)|\,du(t). \tag{4.16}$$

Now, let us observe that

$$\begin{aligned}&\int_a^b |p(x,t)|\,du(t) \\ &= \int_a^x |t-a|\,du(t)+\int_x^b |t-b|\,du(t) \\ &= \int_a^x (t-a)du(t)+\int_x^b (b-t)du(t) \\ &= (t-a)u(t)\,]_a^x-\int_a^x u(t)\,dt-(b-t)u(t)\,]_x^b+\int_x^b u(t)\,dt \\ &= [2x-(a+b)]u(x)-\int_a^x u(t)\,dt+\int_x^b u(t)\,dt\end{aligned}$$

$$= [2x - (a+b)]u(x) + \int_a^b \operatorname{sgn}(t-x)u(t)\,dt.$$

Using the inequality (4.16) and the identity (4.5) we get the first part of (4.15).

Now let us observe that

$$\int_a^b \operatorname{sgn}(t-x)u(t)\,dt = -\int_a^x u(t)\,dt + \int_x^b u(t)\,dt.$$

As u is monotonic nondecreasing on $[a,b]$, we can state that

$$\int_a^x u(t)\,dt \geq (x-a)u(a) \text{ and } \int_x^b u(t)\,dt \leq (b-x)u(b)$$

so that

$$\int_a^b \operatorname{sgn}(t-x)u(t)\,dt \leq (b-x)u(b) - (x-a)u(a).$$

Consequently

$$\begin{aligned}
&[2x - (a+b)]u(x) + \int_a^b \operatorname{sgn}(t-x)u(t)\,dt \\
&\quad \leq [2x-(a+b)]u(x) + (b-x)u(b) - (x-a)u(a) \\
&\quad = (b-x)(u(b)-u(x)) + (x-a)(u(x)-u(a))
\end{aligned}$$

and the second part of (4.15) is proved.

Finally, let us observe that

$$\begin{aligned}
&(b-x)(u(b)-u(x)) + (x-a)(u(x)-u(a)) \\
&\quad \leq \max\{b-x, x-a\}[u(b)-u(x)+u(x)-u(a)] \\
&\quad = \left[\frac{b-a}{2} + \left|x - \frac{a+b}{2}\right|\right](u(b)-u(a))
\end{aligned}$$

and the inequality (4.15) is thus proved.

Now for the sharpness of the inequalities, assume that (4.15) holds with a constant $C > 0$ instead of $\frac{1}{2}$. That is,

$$\left|(b-a)\,u(x) - \int_a^b u(t)\,dt\right|$$

$$\leq [2x-(a+b)]\,u(x) + \int_a^b \operatorname{sgn}(t-x)u(t)\,dt$$

$$\leq\ (x-a)(u(x)-u(a)) + (b-x)(u(b)-u(x))$$

$$\leq \left[C(b-a) + \left|x - \frac{a+b}{2}\right|\right](u(b)-u(a)). \tag{4.17}$$

Consider the mapping $u_0 : [a,b] \to \mathbb{R}$ given by

$$u_0(x) := \begin{cases} -1 & \text{if } x = a \\ 0 & \text{if } x \in (a,b] \end{cases}.$$

Putting in (4.17) $u = u_0$ and $x = a$, we have

$$\left|u(x) - \int_a^b u(t)\,dt\right|$$

$$= [2x-(a+b)]u(x) + \int_a^b \operatorname{sgn}(t-x)u(t)\,dt$$

$$= (x-a)(u(x)-u(a)) + (b-x)(u(b)-u(x)) = 1$$

$$\leq \left[C(b-a) + \left|x - \frac{a+b}{2}\right|\right](u(b)-u(a))$$

$$= \left(C + \frac{1}{2}\right)(b-a),$$

which proves the sharpness of the first two inequalities and the fact that C should not be less than $\frac{1}{2}$. ∎

The following corollaries are interesting.

COROLLARY 4.7
Let u be as above. Then we have the midpoint inequality

$$\left|(b-a)\,u\left(\frac{a+b}{2}\right)-\int_a^b u(t)\,dt\right| \le \int_a^b \operatorname{sgn}\left(t-\frac{a+b}{2}\right)u(t)\,dt$$

$$\le \frac{b-a}{2}\left[u(b)-u(a)\right]. \qquad (4.18)$$

Also, we have the following "*trapezoid inequality*" for monotonic nondecreasing mappings.

COROLLARY 4.8
Under the above assumptions, we have

$$\left|\frac{b-a}{2}\left[u(a)+u(b)\right]-\int_a^b u(t)\,dt\right| \le \frac{b-a}{2}\left[u(b)-u(a)\right]. \qquad (4.19)$$

PROOF Taking $x - a$ and $x = b$ in Theorem 4.3, summing, using the triangle inequality and dividing by 2, we get the desired inequality (4.19). ∎

4.3.2 A Quadrature Formula

Let $I_n : a = x_0 < x_1 < ... < x_{n-1} < x_n = b$ be a division of the interval $[a, b]$ and $\xi_i \in [x_i, x_{i+1}]$ $(i = 0, ..., n-1)$ a sequence of intermediate points for I_n. Construct the Riemann sums

$$R_n(f, I_n, \boldsymbol{\xi}) = \sum_{i=0}^{n-1} f(\xi_i)\,h_i\,,$$

where $h_i := x_{i+1} - x_i$.

We have the following quadrature formula.

THEOREM 4.4
Let $f : [a, b] \to \mathbb{R}$ be a monotonic nondecreasing mapping on $[a, b]$ and I_n, ξ_i $(i = 0, ..., n-1)$ be as above. Then we have the Riemann quadrature formula

$$\int_a^b f(x)\,dx = R_n(f, I_n, \boldsymbol{\xi}) + W_n(f, I_n, \boldsymbol{\xi})\,, \qquad (4.20)$$

where the remainder satisfies the estimation

$$
\begin{aligned}
&|W_n(f, I_n, \boldsymbol{\xi})| \\
&\leq 2\sum_{i=0}^{n-1}\left(\xi_i - \frac{x_i + x_{i+1}}{2}\right) f(\xi_i) + \int_a^b S(t, I_n, \boldsymbol{\xi}) f(t)\, dt \\
&\leq \sum_{i=0}^{n-1}[(\xi_i - x_i)(f(\xi_i) - f(x_i)) + (x_{i+1} - \xi_i)(f(x_{i+1}) - f(\xi_i))] \\
&\leq \sup_{i=0,\ldots,n}\left[\frac{1}{2}h_i + \left|\xi_i - \frac{x_i + x_{i+1}}{2}\right|\right](f(b) - f(a)) \\
&\leq \left[\frac{1}{2}\nu(h) + \sup_{i=0,\ldots,n}\left|\xi_i - \frac{x_i + x_{i+1}}{2}\right|\right](f(b) - f(a)) \\
&\leq \nu(h)(f(b) - f(a))
\end{aligned}
\tag{4.21}
$$

for all ξ_i $(i = 0, ..., n-1)$ *as above, where* $\nu(h) := \max_{i=0,\ldots,n}\{h_i\}$ *and*

$$S(t, I_n, \boldsymbol{\xi} = \operatorname{sgn}(t - \zeta_i) \text{ if } t \in [x_i, x_{i+1})(i = 0, ..., n-1).$$

PROOF Apply Theorem 4.3 on the interval $[x_i, x_{i+1}]$ to get

$$
\begin{aligned}
&\left|\int_{x_i}^{x_{i+1}} f(x)\, dx - f(\xi_i) h_i\right| \\
&\leq 2(\xi_i - \frac{x_i + x_{i+1}}{2}) f(\xi_i) + \int_{x_i}^{x_{i+1}} S(t, I_n, \boldsymbol{\xi}) f(t)\, dt \\
&\leq (\xi_i - x_i)(f(\xi_i) - f(x_i)) + (x_{i+1} - \xi_i)(f(x_{i+1}) - f(\xi_i)) \\
&\leq \left[\frac{1}{2}h_i + \left|\xi_i - \frac{x_i + x_{i+1}}{2}\right|\right](f(x_{i+1}) - f(x_i)).
\end{aligned}
$$

Summing over i from 0 to $n-1$ and using the generalized triangle inequality we get

$$|W_n(f, I_n, \boldsymbol{\xi})|$$

$$\leq \sum_{i=0}^{n-1} \left| \int_{x_i}^{x_{i+1}} f(x)\,dx - f(\xi_i)\,h_i \right|$$

$$\leq \quad 2\sum_{i=0}^{n-1} \left[\left(\xi_i - \frac{x_i + x_{i+1}}{2} \right) f(\xi_i) + \int_{x_i}^{x_{i+1}} S(t, I_n, \boldsymbol{\xi})f(t)\,dt \right]$$

$$\leq \sum_{i=0}^{n-1} [(\xi_i - x_i)(f(\xi_i) - f(x_i)) + (x_{i+1} - \xi_i)(f(x_{i+1}) - f(\xi_i))]$$

$$\leq \sum_{i=0}^{n-1} \left[\frac{1}{2} h_i + \left| \xi_i - \frac{x_i + x_{i+1}}{2} \right| \right] (f(x_{i+1}) - f(x_i))$$

$$\leq \sup_{i=0,\ldots,n} \left[\frac{1}{2} h_i + \left| \xi_i - \frac{x_i + x_{i+1}}{2} \right| \right] \sum_{i=0}^{n-1} (f(x_{i+1}) - f(x_i))$$

$$= \sup_{i=0,\ldots,n} \left[\frac{1}{2} h_i + \left| \xi_i - \frac{x_i + x_{i+1}}{2} \right| \right] (f(b) - f(a)).$$

The fourth inequality follows by the properties of $\sup(\cdot)$.
Now, as

$$\left| \xi_i - \frac{x_i + x_{i+1}}{2} \right| \leq \frac{1}{2} h_i$$

for all $\xi_i \in [x_i, x_{i+1}] (i = 0, ..., n-1)$ the last part of (4.21) is also proved. ∎

COROLLARY 4.9

Let f, I_n be as in Theorem 4.4. Then we have the midpoint rule

$$\int_a^b f(x)\,dx = M_n(f, I_n) + S_n(f, I_n),$$

where

$$M_n(f, I_n) = \sum_{i=0}^{n-1} f\left(\frac{x_i + x_{i+1}}{2} \right) h_i$$

and the remainder $S_n(f, I_n)$ *satisfies the estimation*

$$|S_n(f, I_n)| \leq \int_a^b \mu(I_n) f(t)\, dt \leq \frac{1}{2}\nu(h)(f(b) - f(a)),$$

where

$$\mu(I_n) = \operatorname{sgn}\left(t - \frac{x_i + x_{i+1}}{2}\right) \text{ if } t \in [x_i, x_{i+1}) \quad (i = 0, ..., n-1).$$

4.4 Ostrowski Inequality for Lipschitzian Mappings

In Corollary 4.2, bounds were obtained for an interior point rule for Lipschitzian mappings as a special instance of functions of bounded variation. Specifically treating Lipschitzian mappings, tighter bounds are now obtained.

4.4.1 Integral Inequalities

The following inequality for Lipschitzian mappings holds [14].

THEOREM 4.5
Let $u : [a, b] \to \mathbb{R}$ *be an* $L-$*Lipschitzian mapping on* $[a, b]$. *That is,*

$$|u(x) - u(y)| \,|\leq L\,|x - y| \text{ for all } x, y \in [a, b].$$

Then we have the inequality

$$\left|\int_a^b u(t)dt - (b-a)u(x)\right| \leq L\left[\frac{(b-a)^2}{4} + (x - \frac{a+b}{2})^2\right], \tag{4.22}$$

for all $x \in [a, b]$.
The constant $\frac{1}{4}$ *is the best possible one.*

PROOF Using the integration by parts formula for Riemann-Stieltjes integrals we have the identity as given in (4.5).
Now, assume that $\Delta_n : a = x_0^{(n)} < x_1^{(n)} < ... < x_{n-1}^{(n)} < x_n^{(n)} = b$ is a sequence of divisions with $\nu(\Delta_n) \to 0$ as $n \to \infty$, where $\nu(\Delta_n) :=$

$\max_{i\in\{0,\ldots,n-1\}}\left(x_{i+1}^{(n)}-x_i^{(n)}\right)$ and $\xi_i^{(n)}\in\left[x_i^{(n)},x_{i+1}^{(n)}\right]$. If $p:[a,b]\to\mathbb{R}$ is Riemann integrable on $[a,b]$ and $v:[a,b]\to\mathbb{R}$ is L-Lipschitzian on $[a,b]$, then

$$\begin{aligned}
&\left|\int_a^b p(x)dv(x)\right| \\
&= \left|\lim_{\nu(\Delta_n)\to 0}\sum_{i=0}^{n-1} p\left(\xi_i^{(n)}\right)\left[v\left(x_{i+1}^{(n)}\right)-v\left(x_i^{(n)}\right)\right]\right| \\
&\le \lim_{\nu(\Delta_n)\to 0}\sum_{i=0}^{n-1}\left|p\left(\xi_i^{(n)}\right)\right|\left(x_{i+1}^{(n)}-x_i^{(n)}\right)\left|\frac{v\left(x_{i+1}^{(n)}\right)-v\left(x_i^{(n)}\right)}{x_{i+1}^{(n)}-x_i^{(n)}}\right| \\
&\le L\lim_{\nu(\Delta_n)\to 0}\sum_{i=0}^{n-1}\left|p\left(\xi_i^{(n)}\right)\right|\left(x_{i+1}^{(n)}-x_i^{(n)}\right)
\end{aligned}$$

and so

$$\left|\int_a^b p(x)dv(x)\right|\le L\int_a^b |p(x)|\,dx. \tag{4.23}$$

Applying the inequality (4.23) for $p(x,t)$ as given in (4.5) and $v(x)=u(x)$, $x\in[a,b]$, we get

$$\begin{aligned}
\left|\int_a^b p(x,t)du(t)\right| &\le L\left[\int_a^x |t-a|\,dt+\int_x^b |t-b|dt\right] \\
&= \frac{L}{2}\left[(x-a)^2+(b-x)^2\right] \\
&= L\left[\frac{(b-a)^2}{4}+\left(x-\frac{a+b}{2}\right)^2\right] \qquad (4.24)
\end{aligned}$$

and so by (4.24), via the identity (4.5), we deduce the desired inequality (4.22).

Now to determine the best constant, assume that the inequality (4.22) holds with a constant $C > 0$. That is,

$$\left| \int_a^b u(t)dt - (b-a)u(x) \right| \le L \left[C(b-a)^2 + \left(x - \frac{a+b}{2} \right)^2 \right] \tag{4.25}$$

for all $x \in [a, b]$.
Consider the mapping $f : [a, b] \to \mathbb{R}$, $f(x) = x$ in (4.25). Then

$$\left| x - \frac{a+b}{2} \right| \le C(b-a)^2 + \left(x - \frac{a+b}{2} \right)^2$$

for all $x \in [a, b]$; and then for $x = a$, we get

$$\frac{b-a}{2} \le \left(C + \frac{1}{4} \right) (b-a)$$

which implies that $C \ge \frac{1}{4}$ and the theorem is completely proved. ∎

The following corollary holds, giving a midpoint rule for Lipschitzian functions.

COROLLARY 4.10
Let $u : [a, b] \to \mathbb{R}$ be as above. Then we have the inequality

$$\left| \int_a^b u(t)dx - (b-a)u\left(\frac{a+b}{2} \right) \right| \le \frac{1}{4} L(b-a)^2. \tag{4.26}$$

REMARK 4.4 It is well known that if $f : [a, b] \to \mathbb{R}$ is a convex mapping on $[a, b]$, then the **Hermite-Hadamard** inequality holds

$$f\left(\frac{a+b}{2} \right) \le \frac{1}{b-a} \int_a^b f(x)dx \le \frac{f(a)+f(b)}{2}. \tag{4.27}$$

Now, if we assume that $f : I \subset \mathbb{R} \to \mathbb{R}$ is convex on I and $a, b \in Int(I)$, $a < b$; then f'_+ is monotonic nondecreasing on $[a, b]$ and by Theorem 4.5 we obtain

$$0 \le \frac{1}{b-a} \int_a^b f(x)dx - f\left(\frac{a+b}{2} \right) \le \frac{1}{4} f'_+(b)(b-a), \tag{4.28}$$

which gives a counterpart for the first membership of Hadamard's inequality. ∎

4.4.2 A Quadrature Formula of Riemann Type

Let $I_n : a = x_0 < x_1 < ... < x_{n-1} < x_n = b$ be a division of the interval $[a, b]$ and $\xi_i \in [x_i, x_{i+1}]$ $(i = 0, ..., n-1)$ a sequence of intermediate points for I_n. Construct the Riemann sums

$$R_n(f, I_n, \xi) = \sum_{i=0}^{n-1} f(\xi_i) h_i ,$$

where $h_i := x_{i+1} - x_i$.

We have the following quadrature formula [14].

THEOREM 4.6

Let $f : [a, b] \to \mathbb{R}$ be an L-Lipschitzian mapping on $[a, b]$ and I_n, ξ_i $(i = 0, ..., n-1)$ be as above. Then we have the Riemann quadrature formula

$$\int_a^b f(x)dx = R_n(f, I_n, \xi) + W_n(f, I_n, \xi) , \tag{4.29}$$

where the remainder satisfies the estimation

$$|W_n(f, I_n, \xi)| \le \frac{1}{4} L \sum_{i=0}^{n-1} h_i^2 + L \sum_{i=0}^{n-1} \left(\xi_i - \frac{x_i + x_{i+1}}{2} \right)^2$$

$$\le \frac{1}{2} L \sum_{i=0}^{n-1} h_i^2 \tag{4.30}$$

for all ξ_i $(i = 0, ..., n-1)$ as above. The constant $\frac{1}{4}$ is sharp in (4.30).

PROOF Apply Theorem 4.5 on the interval $[x_i, x_{i+1}]$ to get

$$\left| \int_{x_i}^{x_{i+1}} f(x)dx - f(\xi_i) h_i \right| \le L \left[\frac{1}{4} h_i^2 + \left(\xi_i - \frac{x_i + x_{i+1}}{2} \right)^2 \right] . \tag{4.31}$$

Summing over i from 0 to $n-1$ and using the generalized triangle inequality we get

$$|W_n(f, I_n, \xi)| \le \sum_{i=0}^{n-1} \left| \int_{x_i}^{x_{i+1}} f(x)dx - f(\xi_i)h_i \right|$$

$$\le L \sum_{i=0}^{n-1} \left[\frac{1}{4}h_i^2 + (\xi_i - \frac{x_i + x_{i+1}}{2})^2 \right].$$

Now, as

$$\left(\xi_i - \frac{x_i + x_{i+1}}{2} \right)^2 \le \frac{1}{4}h_i^2$$

for all $\xi_i \in [x_i, x_{i+1}] (i = 0, ..., n-1)$ the second part of (4.30) is also proved. ∎

Note that the best estimation we can get from (4.30) is the one for which $\xi_i = \frac{x_i + x_{i+1}}{2}$ obtaining the following midpoint formula.

COROLLARY 4.11

Let f, I_n be as above. Then we have the midpoint rule

$$\int_a^b f(x)dx = M_n(f, I_n) + S_n(f, I_n)$$

where

$$M_n(f, I_n) = \sum_{i=0}^{n-1} f\left(\frac{x_i + x_{i+1}}{2} \right) h_i$$

and the remainder $S_n(f, I_n)$ satisfies the estimation

$$|S_n(f, I_n)| \le \frac{1}{4} L \sum_{i=0}^{n-1} h_i^2.$$

REMARK 4.5 If we assume that $f : [a, b] \to \mathbb{R}$ is differentiable on (a, b) and whose derivative f' is bounded on (a, b) we can put instead of L the infinity norm $\|f'\|_\infty$, obtaining the estimation due to Dragomir and Wang from the paper [27]. ∎

4.5 A Generalization for Derivatives that are Absolutely Continuous

In 1938, Ostrowski (see, for example, [34, p. 468]) proved the following integral inequality.

Let $f : I \subseteq \mathbb{R} \to \mathbb{R}$ be a differentiable mapping on I° (I° is the interior of I), and let $a, b \in I^\circ$ with $a < b$. If $f' : (a,b) \to \mathbb{R}$ is bounded on (a,b), i.e., $\|f'\|_\infty := \sup_{t\in(a,b)} |f'(t)| < \infty$, then we have the inequality

$$\left| f(x) - \frac{1}{b-a}\int_a^b f(t)\,dt \right| \leq \left[\frac{1}{4} + \frac{\left(x - \frac{a+b}{2}\right)^2}{(b-a)^2} \right] (b-a)\|f'\|_\infty$$

for all $x \in [a,b]$.

The constant $\frac{1}{4}$ is sharp in the sense that it cannot be replaced by a smaller one.

For applications of Ostrowski's inequality to some special means and some numerical quadrature rules, we refer the reader to the recent paper [27] by S.S. Dragomir and S. Wang.

In 1976, G.V. Milovanović and J.E. Pečarić (see, for example, [34, p. 468]), proved the following generalization of Ostrowski's result.

Let $f : [a,b] \to \mathbb{R}$ be an n-times differentiable function, $n \geq 1$, such that $\left\|f^{(n)}\right\|_\infty := \sup_{t\in(a,b)} \sup \left|f^{(n)}(t)\right| < \infty$. Then

$$\left| \frac{1}{n}\left(f(x) + \sum_{k=1}^{n-1} \frac{n-k}{k!} \cdot \frac{f^{(k-1)}(a)(x-a)^k - f^{(k-1)}(b)(x-b)^k}{b-a} \right) \right.$$

$$\left. - \frac{1}{b-a}\int_a^b f(t)\,dt \right| \leq \frac{\left\|f^{(n)}\right\|_\infty}{n(n+1)!} \cdot \frac{(x-a)^{n+1} + (b-x)^{n+1}}{b-a}$$

for all $x \in [a,b]$.

In [4], P. Cerone, S.S. Dragomir and J. Roumeliotis proved the following Ostrowski-type inequality for twice differentiable mappings:

Let $f : [a,b] \to \mathbb{R}$ be a twice differentiable mapping on (a,b) and $f'' : (a,b) \to \mathbb{R}$ is bounded, i.e., $\|f''\|_\infty = \sup_{t\in(a,b)} |f''(t)| < \infty$. Then we have the inequality

$$\left| f(x) - \frac{1}{b-a}\int_a^b f(t)\,dt - \left(x - \frac{a+b}{2}\right) f'(x) \right|$$

$$\leq \left[\frac{1}{24} (b-a)^2 + \frac{1}{2} \left(x - \frac{a+b}{2} \right)^2 \right] \|f''\|_\infty \leq \frac{(b-a)^2}{6} \|f''\|_\infty$$

for all $x \in [a, b]$.

In this section we establish another generalization of the Ostrowski inequality for n-time differentiable mappings which naturally generalizes the result from [4]. Further, work on representation in terms of power series expansion of functions on an interval is presented, giving a generalisation of Taylor series which produces an expansion about a point.

4.5.1 Integral identities

The following theorem holds [8] (see also [37]).

THEOREM 4.7

Let $f : [a, b] \to \mathbb{R}$ be a mapping such that $f^{(n-1)}$ is absolutely continuous on $[a, b]$. Then for all $x \in [a, b]$ we have the identity:

$$\int_a^b f(t)\,dt = \sum_{k=0}^{n-1} \left[\frac{(b-x)^{k+1} + (-1)^k (x-a)^{k+1}}{(k+1)!} \right] f^{(k)}(x)$$

$$+ (-1)^n \int_a^b K_n(x,t) f^{(n)}(t)\,dt \tag{4.32}$$

where the kernel $K_n : [a, b]^2 \to \mathbb{R}$ is given by

$$K_n(x,t) := \begin{cases} \frac{(t-a)^n}{n!} & \text{if } t \in [a, x] \\ \frac{(t-b)^n}{n!} & \text{if } t \in (x, b] \end{cases}, \quad x \in [a, b] \tag{4.33}$$

and n is a natural number, $n \geq 1$.

PROOF The proof is by mathematical induction [8]. For a different argument, see [37].

For $n = 1$, we have to prove the equality

$$\int_a^b f(t)\,dt = (b-a) f(x) - \int_a^b K_1(x,t) f^{(1)}(t)\,dt \tag{4.34}$$

where

$$K_1(x,t) := \begin{cases} t-a & \text{if } t \in [a,x] \\ t-b & \text{if } t \in (x,b] \end{cases}.$$

Integrating by parts, we have

$$\begin{aligned}
&\int_a^b K_1(x,t) f^{(1)}(t)\,dt \\
&= \int_a^x (t-a) f'(t)\,dt + \int_x^b (t-b) f'(t)\,dt \\
&= (t-a) f(t)\big|_a^x - \int_a^x f(t)\,dt + (t-b) f(t)\big|_x^b - \int_x^b f(t)\,dt \\
&= (x-a) f(x) + (b-x) f(x) - \int_a^b f(t)\,dt \\
&= (b-a) f(x) - \int_a^b f(t)\,dt
\end{aligned}$$

and the identity (4.34) is proved.

Assume that (4.32) holds for "n" and let us prove it for "$n+1$". That is, we have to prove the equality

$$\int_a^b f(t)\,dt = \sum_{k=0}^{n} \left[\frac{(b-x)^{k+1} + (-1)^k (x-a)^{k+1}}{(k+1)!}\right] f^{(k)}(x) + (-1)^{n+1} \int_a^b K_{n+1}(x,t) f^{(n+1)}(t)\,dt. \tag{4.35}$$

We have, using (4.33),

$$\begin{aligned}
&\int_a^b K_{n+1}(x,t) f^{(n+1)}(t)\,dt \\
&= \int_a^x \frac{(t-a)^{n+1}}{(n+1)!} f^{(n+1)}(t)\,dt + \int_x^b \frac{(t-b)^{n+1}}{(n+1)!} f^{(n+1)}(t)\,dt
\end{aligned}$$

and integrating by parts gives

$$\int_a^b K_{n+1}(x,t) f^{(n+1)}(t)\,dt$$

$$= \frac{(t-a)^{n+1}}{(n+1)!} f^{(n)}(t)\Big|_a^x - \frac{1}{n!}\int_a^x (t-a)^n f^{(n)}(t)\,dt$$

$$+ \frac{(t-b)^{n+1}}{(n+1)!} f^{(n)}(t)\Big|_x^b - \frac{1}{n!}\int_x^b (t-b)^n f^{(n)}(t)\,dt$$

$$= \frac{(x-a)^{n+1} + (-1)^{n+2}(b-x)^{n+1}}{(n+1)!} f^{(n)}(x) - \int_a^b K_n(x,t) f^{(n)}(t)\,dt.$$

That is,

$$\int_a^b K_n(x,t) f^{(n)}(t)\,dt = \frac{(x-a)^{n+1} + (-1)^{n+2}(b-x)^{n+1}}{(n+1)!} f^{(n)}(x)$$

$$- \int_a^b K_{n+1}(x,t) f^{(n+1)}(t)\,dt.$$

Now, using the mathematical induction hypothesis, we get

$$\int_a^b f(t)\,dt = \sum_{k=0}^{n-1} \left[\frac{(b-x)^{k+1} + (-1)^k (x-a)^{k+1}}{(k+1)!}\right] f^{(k)}(x)$$

$$+ \frac{(b-x)^{n+1} + (-1)^n (x-a)^{n+1}}{(n+1)!} f^{(n)}(x)$$

$$-(-1)^n \int_a^b K_{n+1}(x,t) f^{(n+1)}(t)\,dt$$

$$= \sum_{k=0}^{n} \left[\frac{(b-x)^{k+1} + (-1)^k (x-a)^{k+1}}{(k+1)!}\right] f^{(k)}(x)$$

$$+(-1)^{n+1}\int_a^b K_{n+1}(x,t)\,f^{(n+1)}(t)\,dt.$$

That is, identity (4.35) and the theorem is thus proved. ∎

COROLLARY 4.12

With the above assumptions, we have the representation

$$\int_a^b f(t)\,dt=\sum_{k=0}^{n-1}\left[\frac{1+(-1)^k}{(k+1)!}\right]\frac{(b-a)^{k+1}}{2^{k+1}}f^{(k)}\left(\frac{a+b}{2}\right)$$

$$+(-1)^n\int_a^b M_n(t)\,f^{(n)}(t)\,dt \tag{4.36}$$

where

$$M_n(t):=\begin{cases}\frac{(t-a)^n}{n!} & \text{if } t\in\left[a,\frac{a+b}{2}\right]\\ \frac{(t-b)^n}{n!} & \text{if } t\in\left(\frac{a+b}{2},b\right]\end{cases}.$$

The proof follows by Theorem 4.7 by choosing $x=\frac{a+b}{2}$ so that $M_n(t)=K_n\left(\frac{a+b}{2},t\right)$.

COROLLARY 4.13

With the above assumptions, we have the representation

$$\int_a^b f(t)\,dt=\sum_{k=0}^{n-1}\frac{(b-a)^{k+1}}{(k+1)!}\left[\frac{f^{(k)}(a)+(-1)^k f^{(k)}(b)}{2}\right]$$

$$+\int_a^b T_n(t)\,f^{(n)}(t)\,dt \tag{4.37}$$

where

$$T_n(t):=\frac{1}{n!}\left[\frac{(b-t)^n+(-1)^n(t-a)^n}{2}\right],\qquad t\in[a,b]. \tag{4.38}$$

PROOF Choose $x=a$ and $x=b$ in (4.32), then summing the resulting identities and dividing by 2 gives

$$\int_a^b f(t)\,dt = \sum_{k=0}^{n} \frac{(b-a)^{k+1}}{(k+1)!}\left[\frac{f^{(k)}(a)+(-1)^k f^{(k)}(b)}{2}\right]$$

$$+\int_a^b T_n(t)\,f^{(n)}(t)\,dt$$

and the corollary is proved. ∎

The following Taylor-like formula with integral remainder also holds.

COROLLARY 4.14
Let $g:[a,y]\to\mathbb{R}$ be a mapping such that $g^{(n)}$ is absolutely continuous on $[a,y]$. Then for all $x\in[a,y]$, we have the identity

$$g(y) = g(a) + \sum_{k=0}^{n-1} \frac{\left[(y-x)^{k+1}+(-1)^k(x-a)^{k+1}\right]}{(k+1)!} g^{(k+1)}(x)$$

$$+(-1)^n \int_a^y \widetilde{K}_n(x,t)\,g^{(n+1)}(t)\,dt. \tag{4.39}$$

The proof is obvious by Theorem 4.7, choosing $f=g'$ and $b=y$ where $\widetilde{K}\equiv K$ with $b=y$.

REMARK 4.6 If we choose $n=1$ in (4.32), we get the identity (4.34) which is the identity employed by S.S. Dragomir and S. Wang to prove an Ostrowski-type inequality in paper [27]. ∎

If in (4.36) we choose $n=1$, then we get

$$\int_a^b f(t)\,dt = (b-a)\,f\left(\frac{a+b}{2}\right) - \int_a^b M_1(t)\,f'(t)\,dt \tag{4.40}$$

where

$$M_1(t) = \begin{cases} t-a & \text{if } t\in\left[a,\frac{a+b}{2}\right] \\ t-b & \text{if } t\in\left(\frac{a+b}{2},b\right] \end{cases}$$

which gives the midpoint-type identity useful in numerical analysis, although here only the first derivative is involved.

Also, if we put $n = 1$ in (4.37), we get the trapezoid identity

$$\int_a^b f(t)\,dt = \frac{b-a}{2}(f(a) + f(b)) + \int_a^b T_1(t)\,f'(t)\,dt \tag{4.41}$$

where

$$T_1(t) = \frac{a+b}{2} - t, \qquad t \in [a,b].$$

Finally, if in the Taylor-like formula (4.39) we put $n = 1$, we get

$$g(y) = g(a) + (y-a)\,g'(x) - \int_a^y K_1(y,t)\,g^{(2)}(t)\,dt$$

where $x \in [a,y]$.

REMARK 4.7 If we choose $n = 2$ in (4.32), we get the identity

$$\int_a^b f(t)\,dt = (b-a)\,f(x) - \left(x - \frac{a+b}{2}\right) f'(x)$$

$$+ \int_a^b K_2(x,t)\,f''(t)\,dt \tag{4.42}$$

where $K_2(x,t)$ is as given in (4.33), which is the identity employed by P. Cerone, S.S. Dragomir and J. Roumeliotis to prove some Ostrowski-type inequalities for twice differentiable mappings in the paper [4]. ∎

If in (4.36) we choose $n = 2$, then we get

$$\int_a^b f(t)\,dt = (b-a)\,f\left(\frac{a+b}{2}\right) + \int_a^b M_2(t)\,f''(t)\,dt \tag{4.43}$$

where

$$M_2(t) = \begin{cases} \frac{(t-a)^2}{2} & \text{if } t \in \left[a, \frac{a+b}{2}\right] \\ \frac{(t-b)^2}{2} & \text{if } t \in \left(\frac{a+b}{2}, b\right] \end{cases}$$

which is the classical midpoint identity.

Also, if we put $n = 2$ in (4.37), we get the identity

$$\int_a^b f(t)\,dt = \frac{b-a}{2}(f(a)+f(b)) + \frac{(b-a)^2}{2}\cdot\frac{f'(a)-f'(b)}{2} + \int_a^b T_2(t)\,f''(t)\,dt \tag{4.44}$$

where $T_2(t)$ is as given in (4.38).

Finally, if we put $n = 2$ in (4.39), we get

$$g(y) = g(a) + (y-a)\,g'(x) - (y-a)\left(x - \frac{a+y}{2}\right) g''(x) + \int_a^y \widetilde{K}_2(x,t)\,g^{(3)}(t)\,dt, \tag{4.45}$$

where $\widetilde{K}_2 \equiv K_2$ as given in (4.33) with $a \le x \le y$ and $b = y$.

4.5.2 Some Integral Inequalities

The following theorem holds (see also [8] for the case of the $\|\cdot\|_\infty$ norm).

THEOREM 4.8

Let $f : [a,b] \to \mathbb{R}$ be a mapping such that $f^{(n-1)}$ is absolutely continuous on $[a,b]$. Then for all $x \in [a,b]$, we have the inequalities

$$\left|\int_a^b f(t)\,dt - \sum_{k=0}^{n-1}\left[\frac{(b-x)^{k+1}+(-1)^k(x-a)^{k+1}}{(k+1)!}\right] f^{(k)}(x)\right|$$

$$\le \begin{cases} \frac{\|f^{(n)}\|_\infty}{(n+1)!}\left[(x-a)^{n+1}+(b-x)^{n+1}\right] & \text{if } f^{(n)} \in L_\infty[a,b], \\ \frac{\|f^{(n)}\|_p}{n!}\left[\frac{(x-a)^{nq+1}+(b-x)^{nq+1}}{nq+1}\right]^{\frac{1}{q}} & \text{if } p>1,\ \frac{1}{p}+\frac{1}{q}=1 \\ & \text{and } f^{(n)} \in L_p[a,b], \\ \frac{\|f^{(n)}\|_1}{n!}\left[\frac{1}{2}(b-a)+\left|x-\frac{a+b}{2}\right|\right]^n & \text{if } f^{(n)} \in L_1[a,b], \end{cases} \tag{4.46}$$

where

$$\left\|f^{(n)}\right\|_{\infty}:=\sup_{t\in[a,b]}\left|f^{(n)}(t)\right|<\infty \quad and \quad \left\|f^{(n)}\right\|_{p}:=\left(\int_a^b\left|f^{(n)}(t)\right|^p dt\right)^{\frac{1}{p}}.$$

PROOF Using the identity (4.32), we have

$$\left|\int_a^b f(t)\,dt-\sum_{k=0}^{n-1}\left[\frac{(b-x)^{k+1}+(-1)^k(x-a)^{k+1}}{(k+1)!}\right]f^{(k)}(x)\right|$$

$$=\left|\int_a^b K_n(x,t)\,f^{(n)}(t)\,dt\right|:=Q(x).$$

Now, observe that

$$Q(x)\le\left\|f^{(n)}\right\|_{\infty}\|K_n(x,\cdot)\|_1=\left\|f^{(n)}\right\|_{\infty}\int_a^b|K_n(x,t)|\,dt$$

and so using (4.33),

$$Q(x)\le\left\|f^{(n)}\right\|_{\infty}\left[\int_a^x\frac{(t-a)^n}{n!}dt+\int_x^b\frac{(b-t)^n}{n!}dt\right]$$

$$=\frac{\left\|f^{(n)}\right\|_{\infty}}{(n+1)!}\left[(x-a)^{n+1}+(b-x)^{n+1}\right],$$

and the first part of inequality (4.46) is proved.

Further, using Hölder's integral inequality, we have

$$Q(x)\le\left\|f^{(n)}\right\|_{p}\left(\int_a^b|K_n(x,t)|^q\,dt\right)^{\frac{1}{q}}$$

$$=\frac{\left\|f^{(n)}\right\|_{p}}{n!}\left[\int_a^x(t-a)^{nq}\,dt+\int_x^b(b-t)^{nq}\,dt\right]^{\frac{1}{q}},$$

and so, on evaluation of the above integrals, the second inequality in (4.46) is proven.

Finally, let us observe that

$$\begin{aligned} Q(x) &\leq \|K_n(x,\cdot)\|_\infty \left\|f^{(n)}\right\|_1 \\ &= \left\|f^{(n)}\right\|_1 \sup_{t\in[a,b]} |K_n(x,t)| \\ &= \frac{\|f^{(n)}\|_1}{n!} \left[\max\{x-a, b-x\}\right]^n \\ &= \frac{\|f^{(n)}\|_1}{n!} \left[\frac{b-a}{2} + \left|x - \frac{a+b}{2}\right|\right]^n \end{aligned}$$

and the theorem is completely proved. ∎

REMARK 4.8 It may be noticed that the expressions for the bounds of a generalised interior point rule such as that given by (4.46) are upper bounded by taking $x = a$ or $x = b$ in the bound while keeping a general $x \in [a,b]$ for the rule. The sharpest bound is obtained by taking $x = \frac{a+b}{2}$, giving the result expressed in the following corollary. ∎

COROLLARY 4.15
Let the conditions of Theorem 4.8 hold. Then

$$\left| \int_a^b f(t)\,dt - \sum_{k=0}^{n-1} \left(\frac{b-a}{2}\right)^{k+1} \left[1 + (-1)^k\right] \frac{f^{(k)}\left(\frac{a+b}{2}\right)}{(k+1)!} \right|$$

$$\leq \begin{cases} \frac{(b-a)^{n+1}}{2^n(n+1)!} \left\|f^{(n)}\right\|_\infty & \text{if } \ f^{(n)} \in L_\infty[a,b], \\ \frac{(b-a)^{n+\frac{1}{q}}}{2^n n!(nq+1)^{\frac{1}{q}}} \left\|f^{(n)}\right\|_p & \text{if } \ p > 1,\ \frac{1}{p}+\frac{1}{q} = 1 \\ & \text{and } f^{(n)} \in L_p[a,b], \\ \frac{(b-a)^n}{2^n n!} \left\|f^{(n)}\right\|_1 & \text{if } \ f^{(n)} \in L_1[a,b]. \end{cases} \tag{4.47}$$

REMARK 4.9 Taking $n = 1$ in Theorem 4.8 and Corollary 4.14 reproduces some of the results obtained by Dragomir and Wang ([25]-[28]) while $n = 2$ reproduces the results of Cerone, Dragomir and Roumeliotis ([4]-[7]). It is important to note that assuming that the behaviour of the first derivatives determines the bound on the rule allows greater flexibility than assumptions about the second derivative. Taking $n = 2$ allows a comparison with traditional midpoint $\left(x = \frac{a+b}{2}\right)$ or interior point rules. It should further be noted that only even derivatives occur in the rule given in (4.47). ∎

REMARK 4.10 It is most interesting to observe that the bounds given in (4.46) for a generalised interior point method obtained from investigating various norms of $K_n(x,t)$ as given by (4.33) are the same as the bounds obtained from the generalised trapezoidal-type rule resulting from various norms of the Peano kernel given by $\frac{(x-t)^n}{n!}$. ∎

The following corollary is a generalisation of the trapezoidal inequality [8].

COROLLARY 4.16

With the above assumptions, we have the inequality

$$\left|\int_a^b f(t)\,dt - \sum_{k=0}^{n-1} \frac{(b-a)^{k+1}}{(k+1)!}\left[\frac{f^{(k)}(a) + (-1)^k f^{(k)}(b)}{2}\right]\right|$$

$$\leq \frac{(b-a)^{n+1}}{(n+1)!}\left\|f^{(n)}\right\|_\infty \times \begin{cases} 1 & \text{if } n = 2r \\ \frac{2^{2r+1}-1}{2^{2r}} & \text{if } n = 2r+1 \end{cases}. \tag{4.48}$$

PROOF Using the identity (4.37), we get

$$\left|\int_a^b f(t)\,dt - \sum_{k=0}^{n-1} \frac{(b-a)^{k+1}}{(k+1)!}\left[\frac{f^{(k)}(a) + (-1)^k f^{(k)}(b)}{2}\right]\right|$$

$$= \left|\int_a^b T_n(t)\, f^{(n)}(t)\,dt\right| \leq \left\|f^{(n)}\right\|_\infty \int_a^b |T_n(t)|\,dt. \tag{4.49}$$

If $n = 2r$, then

$$\int_a^b |T_n(t)|\,dt$$

$$= \frac{1}{(2r)!} \int_a^b \left[\frac{(b-t)^{2r} + (t-a)^{2r}}{2} \right] dt$$

$$= \frac{1}{(2r)!} \cdot \frac{1}{2} \left[\frac{(b-a)^{2r+1}}{(2r+1)} + \frac{(b-a)^{2r+1}}{(2r+1)} \right] = \frac{(b-a)^{2r+1}}{(2r+1)!}. \quad (4.50)$$

For $n = 2r+1$, put $h_{2r+1}(t) := (b-t)^{2r+1} - (t-a)^{2r+1}, \quad t \in [a,b]$. Observe that $h_{2r+1}(t) = 0$ iff $t = \frac{a+b}{2}$ and $h_{2r+1}(t) > 0$ if $t \in [a, \frac{a+b}{2})$ and $h_{2r+1}(t) < 0$ if $t \in (\frac{a+b}{2}, b]$.

Then

$$\int_a^b |T_{2r+1}(t)|\, dt$$

$$= \int_a^{\frac{a+b}{2}} \left[(b-t)^{2r+1} - (t-a)^{2r+1} \right] dt + \int_{\frac{a+b}{2}}^b \left[(t-a)^{2r+1} - (b-t)^{2r+1} \right] dt$$

$$= 2\frac{(b-a)^{2r+2}}{2r+2} - \frac{4\left(\frac{b-a}{2}\right)^{2r+2}}{2r+2}$$

$$= \frac{1}{2r+2} \left[2(b-a)^{2r+2} - \frac{(b-a)^{2r+2}}{2^{2r}} \right]$$

$$= \frac{(b-a)^{2r+2}}{2r+2} \left(2 - \frac{1}{2^{2r}} \right) = \frac{(b-a)^{2r+2}}{2r+2} \cdot \frac{2^{2r+1}-1}{2^{2r}}.$$

Using (4.48) we get the desired inequality (4.49). ∎

The following inequalities for the Taylor-like expansion (4.39) also hold.

COROLLARY 4.17

Let g be as in Corollary 4.14. Then we have the inequality

$$\left| g(y) - g(a) - \sum_{k=0}^{n-1} \frac{\left[(y-x)^{k+1} + (-1)^k (x-a)^{k+1} \right]}{(k+1)!} g^{(k+1)}(x) \right|$$

$$\leq \begin{cases} \frac{\|g^{(n+1)}\|_\infty}{(n+1)!}\left[(x-a)^{n+1}+(y-x)^{n+1}\right], & g^{(n+1)}\in L_\infty[a,b], \\ \frac{\|g^{(n+1)}\|_p}{n!}\left[\frac{(x-a)^{nq+1}+(y-x)^{nq+1}}{nq+1}\right]^{\frac{1}{q}}, & g^{(n+1)}\in L_p[a,b] \\ & \frac{1}{p}+\frac{1}{q}=1,\ p>1, \\ \frac{\|g^{(n+1)}\|_1}{n!}\left[\frac{1}{2}(y-a)+\left|x-\frac{a+y}{3}\right|\right]^n, & g^{(n+1)}\in L_1[a,b] \end{cases} \tag{4.51}$$

for all $a \leq x \leq y$.

PROOF From equation (4.39) and using norms, or else from (4.46) on choosing $f = g'$ and $b = y$, readily produces the above result. ∎

REMARK 4.11 Since the right-hand side of (4.51) is a convex function, then upper bounds may be found by taking either $x = a$ or b on the right. It is well-known that for the classical Taylor expansion around a we have the inequality

$$\left|g(y)-\sum_{k=0}^{n}\frac{(y-a)^k}{k!}g^{(k)}(a)\right| \leq \frac{(y-a)^{n+1}}{(n+1)!}\left\|g^{(n+1)}\right\|_\infty \tag{4.52}$$

for all $y \geq a$. It is clear now that the above approximation (4.51) around the arbitrary point $x \in [a, y]$ provides a better approximation for the mapping g at the point y than the classical Taylor expansion around the point a. ∎

If in (4.51) we choose $x = \frac{a+y}{2}$, then we get

$$\left|g(y)-g(a)-\sum_{k=1}^{n}\frac{\left[1+(-1)^{k-1}\right]}{k!}\frac{(y-a)^k}{2^k}g^{(k)}\left(\frac{a+y}{2}\right)\right|$$

$$\leq \begin{cases} \frac{(y-a)^{n+1}}{(n+1)!2^n}\left\|g^{(n+1)}\right\|_\infty, & g^{(n+1)}\in L_\infty[a,b], \\ \frac{(y-a)^{n+\frac{1}{q}}}{2^n n!(nq+1)^{\frac{1}{q}}}\left\|g^{(n+1)}\right\|_p, & g^{(n+1)}\in L_p[a,b] \\ & \text{with } \frac{1}{p}+\frac{1}{q}=1,\ p>1, \\ \frac{(y-a)^n}{2^n n!}\left\|g^{(n+1)}\right\|_1, & g^{(n+1)}\in L_1[a,b]. \end{cases} \tag{4.53}$$

The above inequality (4.53) shows that for $g \in C^{\infty}[a,b]$ the series

$$g(a) + \sum_{k=0}^{\infty} \frac{\left[1+(-1)^k\right]}{(k+1)!} \frac{(y-a)^{k+1}}{2^{k+1}} g^{(k+1)}\left(\frac{a+y}{2}\right)$$

converges more rapidly to $g(y)$ than the usual one

$$\sum_{k=0}^{\infty} \frac{(y-a)^k g^{(k)}(a)}{k!}$$

which comes from Taylor's expansion. Further, it should be noted that the Taylor-like expansion in (4.53) only involves odd derivatives of g.

REMARK 4.12 If in the inequality (4.46) we choose $n = 1$, we get

$$\left| \int_a^b f(t)\,dt - (b-a) f(x) \right| \le \frac{(x-a)^2 + (b-x)^2}{2} \|f'\|_{\infty}.$$

As a simple calculation shows that

$$\frac{1}{2}\left[(x-a)^2 + (b-x)^2\right] = \frac{1}{4}(b-a)^2 + \left(x - \frac{a+b}{2}\right)^2,$$

consequently we obtain the Ostrowski inequality

$$\left| \int_a^b f(t)\,dt - (b-a) f(x) \right| \le \left[\frac{1}{4} + \frac{\left(x - \frac{a+b}{2}\right)^2}{(b-a)^2}\right] (b-a)^2 \|f'\|_{\infty} \tag{4.54}$$

for all $x \in [a,b]$. ∎

If in (4.49) we put $n = 1$, we get the midpoint inequality

$$\left| \int_a^b f(t)\,dt - (b-a) f\left(\frac{a+b}{2}\right) \right| \le \frac{1}{4}(b-a)^2 \|f'\|_{\infty}. \tag{4.55}$$

From the inequality (4.48), for $n = 1$, we get the trapezoid inequality

$$\left| \int_a^b f(t)\,dt - \frac{b-a}{2}[f(a) + f(b)] \right| \le \frac{1}{2}(b-a)^2 \|f'\|_{\infty}. \tag{4.56}$$

Also, from (4.51) we deduce

$$|g(y)-g(a)-(y-a)g'(x)| \leq \left[\frac{(y-a)^2}{4}+\left(x-\frac{a+y}{2}\right)^2\right]\|g''\|_\infty \tag{4.57}$$

for all $a \leq x \leq y$.

REMARK 4.13 If in the inequality (4.46) we choose $n=2$, then we get

$$\left|\int_a^b f(t)\,dt-(b-a)f(x)+(b-a)\left(x-\frac{a+b}{2}\right)f'(x)\right|$$

$$\leq \frac{1}{6}\left[(x-a)^3+(b-x)^3\right]\|f''\|_\infty$$

for all $x \in [a,b]$. ∎

From which, on noting that

$$(x-a)^3+(b-x)^3=(b-a)\left[\left(\frac{b-a}{2}\right)^2+3\left(x-\frac{a+b}{2}\right)^2\right],$$

we recapture the result obtained in [4], namely,

$$\left|\int_a^b f(t)\,dt-(b-a)f(x)+(b-a)\left(x-\frac{a+b}{2}\right)f'(x)\right|$$

$$\leq \left[\frac{1}{24}+\frac{1}{2}\frac{\left(x-\frac{a+b}{2}\right)^2}{(b-a)^2}\right](b-a)^3\|f''\|_\infty . \tag{4.58}$$

For $f'' \in L_\infty[a,b]$ with $n=2$ in (4.47), we get the classical midpoint inequality

$$\left|\int_a^b f(t)\,dt-(b-a)f\left(\frac{a+b}{2}\right)\right| \leq \frac{1}{24}(b-a)^3\|f''\|_\infty , \tag{4.59}$$

while from (4.48) we get a perturbed trapezoid inequality,

$$\left|\int_a^b f(t)\,dt - \frac{b-a}{2}[f(a)+f(b)] - \frac{(b-a)^2}{4}[f'(b)-f'(a)]\right|$$

$$\leq \frac{(b-a)^3}{6}\left\|f''\right\|_\infty. \tag{4.60}$$

Finally, if we put $n = 2$ in (4.51) for $g''' \in L_\infty[a,b]$, then we get the inequality

$$\left|g(y) - g(a) - (y-a)g'(x) + (y-a)\left(x - \frac{a+y}{2}\right)g''(x)\right|$$

$$\leq \left[\frac{1}{24} + \frac{1}{2}\cdot\frac{\left(x-\frac{a+y}{2}\right)^2}{(y-a)^2}\right](y-a)^3\left\|g'''\right\|_\infty,$$

valid for any $x \in [a,y]$.

Equivalent results may also be produced which involve the $L_p[a,b]$ spaces for $p \geq 1$.

The following particular case for euclidean norms is of particular interest.

COROLLARY 4.18

Let $f : [a,b] \to \mathbb{R}$ be twice differentiable on (a,b) and $f'' \in L_2(a,b)$. Then we have the following inequality

$$\left|\int_a^b f(t)\,dt - (b-a)\left[f(x) - \left(x-\frac{a+b}{2}\right)f'(x)\right]\right|$$

$$\leq \frac{(b-a)^{\frac{1}{2}}}{2\sqrt{5}}\left[\left(\frac{b-a}{2}\right)^4 + 10\left(\frac{b-a}{2}\right)^2\left(x-\frac{a+b}{2}\right)^2 + 5\left(x-\frac{a+b}{2}\right)^4\right]^{\frac{1}{2}}$$

$$\leq \frac{(b-a)^{\frac{5}{2}}}{2\sqrt{5}}. \tag{4.61}$$

PROOF Let $p = q = 2$ and $n = 2$ in (4.46) to give the left-hand side of

(4.61) with the right-hand side given by

$$\frac{\|f^{(2)}\|_2}{2!\sqrt{5}}\left[(x-a)^5+(b-x)^5\right]^{\frac{1}{2}}.$$

Now, expansion of $(x-a)^5+(b-x)^5$ in a Taylor series about $x=\frac{a+b}{2}$ readily produces the first bound in (4.61). The coarser bound is obtained from evaluating at either of the end points. Hence, the corollary is proved. ∎

REMARK 4.14 The optimal rule from (4.61) in terms of obtaining the tightest bound occurs on taking $x=\frac{a+b}{2}$ since the bound is symmetric and convex. ∎

Now, using the celebrated Hermite-Hadamard integral inequality for convex functions, $g:[a,b]\to\mathbb{R}$, which may be written as

$$g\left(\frac{a+b}{2}\right)\leq\frac{1}{b-a}\int_a^b g(x)\,dx\leq\frac{g(a)+g(b)}{2}, \tag{4.62}$$

we obtain the following theorem [19].

THEOREM 4.9
Let $f:[a,b]\to\mathbb{R}$ be twice differentiable and $\gamma\leq f(x)\leq\Gamma$ for all $x\in(a,b)$. Then we have the following double inequality

$$\frac{\gamma(b-a)^2}{24}\leq\frac{1}{b-a}\int_a^b f(x)\,dx-f\left(\frac{a+b}{2}\right)\leq\frac{\Gamma(b-a)^2}{24} \tag{4.63}$$

and the estimation

$$\left|\int_a^b f(x)\,dx-(b-a)f\left(\frac{a+b}{2}\right)-\frac{(\gamma+\Gamma)(b-a)^3}{48}\right|\leq\frac{(\Gamma-\gamma)(b-a)^3}{48}. \tag{4.64}$$

PROOF Let us choose in (4.62) $g(x)=f(x)-\frac{\gamma x^2}{2}$, then $g(x)$ is a convex function in x, since $g''(x)\geq 0$, and hence

$$f\left(\frac{a+b}{2}\right)-\frac{\gamma(a+b)^2}{8}\leq\frac{1}{b-a}\left(\int_a^b f(x)\,dx-\frac{\gamma\left(b^3-a^3\right)}{6}\right),$$

which is equivalent to

$$\frac{1}{b-a}\int_a^b f(x)\,dx - f\left(\frac{a+b}{2}\right) \geq \frac{\gamma}{2}\left(\frac{b^3-a^3}{3(b-a)} - \left(\frac{a+b}{2}\right)^2\right)$$

$$= \frac{\gamma(b-a)^2}{24},$$

and the first part of (4.63) is therefore obtained. For the second part, let $g(x) = \frac{x^2\Gamma}{2} - f(x)$, and similar manipulations, as previously lead to the second part of (4.63). The inequality (4.64) is now obvious by (4.63). The details have been omitted. ∎

4.5.3 Applications for Numerical Integration

Consider the partition $I_m : a = x_0 < x_1 < \ldots < x_{m-1} < x_m = b$ of the interval $[a,b]$ and the intermediate points $\boldsymbol{\xi} = (\xi_0, \ldots, \xi_{m-1})$ where $\xi_j \in [x_j, x_{j+1}]$, $j = 0, \ldots, m-1$. Define the formula

$$F_{m,n}(f, I_m, \boldsymbol{\xi}) := \sum_{j=0}^{m-1}\sum_{k=0}^{n-1} \frac{\left[(x_{j+1}-\xi_j)^{k+1} + (-1)^k(\xi_j - x_j)^{k+1}\right]}{(k+1)!} f^{(k)}(\xi_j)$$

which can be regarded as a perturbation of Riemann's sum

$$\Gamma(f, I_m, \boldsymbol{\xi}) = \sum_{j=0}^{m-1} f(\xi_j)\,h_j$$

where $h_j := x_{j+1} - x_j, \quad j = 0, \ldots, m-1$.

The following theorem holds.

THEOREM 4.10

Let $f : [a,b] \to \mathbb{R}$ be a mapping such that $f^{(n-1)}$ is absolutely continuous on $[a,b]$ and I_m a partitioning of $[a,b]$ as above. Then we have the quadrature formula

$$\int_a^b f(x)\,dx = F_{m,n}(f, I_m, \boldsymbol{\xi}) + R_{m,n}(f, I_m, \boldsymbol{\xi}) \tag{4.65}$$

where $F_{m,n}$ is defined above and the remainder $R_{m,n}$ satisfies the estimation

$$|R_{m,n}(f, I_m, \boldsymbol{\xi})|$$

$$\leq \begin{cases} \frac{\|f^{(n)}\|_\infty}{(n+1)!} \sum_{j=0}^{m-1} \left[(\xi_j - x_j)^{n+1} + (x_{j+1} - \xi_j)^{n+1} \right] \\ \text{for } f^{(n)} \in L_\infty [a,b], \\ \frac{\|f^{(n)}\|_p}{n!(nq+1)^{\frac{1}{q}}} \left[\sum_{j=0}^{m-1} (\xi_j - x_j)^{nq+1} + \sum_{j=0}^{m-1} (x_{j+1} - \xi_j)^{nq+1} \right]^{\frac{1}{q}} \\ \text{for } f^{(n)} \in L_p [a,b], \text{ where } \frac{1}{p} + \frac{1}{q} = 1, \ p > 1, \\ \frac{\|f^{(n)}\|_1}{n!} \left[\frac{1}{2}\nu(h) + \max_{j=0,\dots,m-1} \left| \xi_j - \frac{x_j + x_{j+1}}{2} \right| \right]^n \\ \text{for } f^{(n)} \in L_1 [a,b], \end{cases} \tag{4.66}$$

for all $\boldsymbol{\xi}$ *as above and* $\nu(h) = \max\{h_j | j = 0, ..., m-1\}$.

PROOF Apply Theorem 4.8 on the interval $[x_j, x_{j+1}]$ to get

$$\left| \int_{x_j}^{x_{j+1}} f(t)\,dt - \sum_{k=0}^{n-1} \left[\frac{\left[(x_{j+1} - \xi_j)^{k+1} + (-1)^k (\xi_j - x_j)^{k+1} \right]}{(k+1)!} \right] f^{(k)}(\xi_j) \right|$$

$$\leq \begin{cases} \frac{1}{(n+1)!} \sup_{t \in [x_j, x_{j+1}]} \left| f^{(n)}(t) \right| \left[(\xi_j - x_j)^{n+1} + (x_{j+1} - \xi_j)^{n+1} \right], \\ \frac{1}{n!} \left(\int_{x_j}^{x_{j+1}} \left| f^{(n)}(s) \right|^p ds \right)^{\frac{1}{p}} \left[\frac{(\xi_j - x_j)^{nq+1} + (x_{j+1} - \xi_j)^{nq+1}}{nq+1} \right]^{\frac{1}{q}}, \\ \frac{1}{n!} \left(\int_{x_j}^{x_{j+1}} \left| f^{(n)}(s) \right| ds \right) \left[\frac{1}{2} h_j + \left| \xi_j - \frac{x_j + x_{j+1}}{2} \right| \right]^n. \end{cases}$$

Summing over j from 0 to $m-1$ and using the generalized triangle inequality, we have

$$|R_{m,n}(f, I_n, \boldsymbol{\xi})|$$

$$\leq \begin{cases} \frac{1}{(n+1)!} \sum_{j=0}^{m-1} \sup_{t\in[x_j,x_{j+1}]} \left|f^{(n)}(t)\right| \left[(\xi_j - x_j)^{n+1} + (x_{j+1}-\xi_j)^{n+1}\right], \\ \frac{1}{n!} \sum_{j=0}^{m-1} \left(\int_{x_j}^{x_{j+1}} \left|f^{(n)}(s)\right|^p ds\right)^{\frac{1}{p}} \left[\frac{(\xi_j - x_j)^{nq+1} + (x_{j+1}-\xi_j)^{nq+1}}{nq+1}\right]^{\frac{1}{q}}, \\ \frac{1}{n!} \sum_{j=0}^{m-1} \left(\int_{x_j}^{x_{j+1}} \left|f^{(n)}(s)\right| ds\right) \left[\frac{1}{2}h_j + \left|\xi_j - \frac{x_j - x_{j+1}}{2}\right|\right]^n. \end{cases}$$

As $\sup_{t\in[x_j,x_{j+1}]} \left|f^{(n)}(t)\right| \leq \left\|f^{(n)}\right\|_\infty$, the first inequality in (4.66) readily follows.

Now, using the discrete Hölder inequality, we have

$$\frac{1}{(nq+1)^{1/q}} \sum_{j=0}^{m-1} \left(\int_{x_j}^{x_{j+1}} |f^{(n)}(s)|^p ds\right)^{\frac{1}{p}} \left[(\xi_j - x_j)^{nq+1} + (x_{j+1}-\xi_j)^{nq+1}\right]^{\frac{1}{q}}$$

$$\leq \frac{1}{(nq+1)^{1/q}} \left[\sum_{j=0}^{m-1} \left[\left(\int_{x_j}^{x_{j+1}} \left|f^{(n)}(s)\right|^p ds\right)^{\frac{1}{p}}\right]^p\right]^{\frac{1}{p}}$$

$$\times \left[\sum_{j=0}^{m-1} \left[\left[(\xi_j - x_j)^{nq+1} + (x_{j+1}-\xi_j)^{nq+1}\right]^{\frac{1}{q}}\right]^q\right]^{\frac{1}{q}}$$

$$= \frac{1}{(nq+1)^{1/q}} \left\|f^{(n)}\right\|_p \left[\sum_{j=0}^{m-1} (\xi_j - x_j)^{nq+1} + \sum_{j=0}^{m-1} (x_{j+1}-\xi_j)^{nq+1}\right]^{\frac{1}{q}}$$

and thus the second inequality in (4.66) is proved.

Finally, let us observe that

$$\frac{1}{n!} \sum_{j=0}^{m-1} \left(\int_{x_j}^{x_{j+1}} \left|f^{(n)}(s)\right| ds\right) \left[\frac{1}{2}h_j + \left|\xi_j - \frac{x_j + x_{j+1}}{2}\right|\right]^n$$

$$\leq \max_{j=0,\dots,m-1} \left[\frac{1}{2}h_j + \left|\xi_j - \frac{x_j + x_{j+1}}{2}\right|\right]^n \sum_{j=0}^{m-1} \left(\int_{x_j}^{x_{j+1}} \left|f^{(n)}(s)\right| ds\right)$$

$$\leq \left[\frac{1}{2}h_j + \max_{j=0,\ldots,m-1}\left|\xi_j - \frac{x_j + x_{j+1}}{2}\right|\right]^n \left\|f^{(n)}\right\|_1$$

and the last part of (4.66) is proved. ∎

REMARK 4.15 As $(x-a)^\alpha + (b-x)^\alpha \leq (b-a)^\alpha$ for $\alpha \geq 1$, $x \in [a,b]$, then we remark that the first branch of (4.66) can be bounded by

$$\frac{1}{(n+1)!}\left\|f^{(n)}\right\|_\infty \sum_{j=0}^{m-1} h_j^{n+1}. \tag{4.67}$$

The second branch can be bounded by

$$\frac{1}{n!(nq+1)^{1/q}}\left\|f^{(n)}\right\|_p \left[\sum_{j=0}^{m-1} h_j^{nq+1}\right]^{\frac{1}{q}} \tag{4.68}$$

and finally, the last branch in (4.66) can be bounded by

$$\frac{1}{n!}\left[\nu(h)\right]^n \left\|f^{(n)}\right\|_1. \tag{4.69}$$

Note that all the bounds provided by (4.67)-(4.69) are uniform bounds for $R_{m,n}(f, I_m, \boldsymbol{\xi})$ in terms of the intermediate points $\boldsymbol{\xi}$. ∎

As an interesting particular case, we can consider the following perturbed midpoint formula

$$M_{m,n}(f, I_m) := \sum_{j=0}^{m-1}\sum_{k=0}^{n-1}\left[\frac{1+(-1)^k}{(k+1)!}\right]\frac{h_j^{k+1}}{2^{k+1}} f^{(k)}\left(\frac{x_j + x_{j+1}}{2}\right),$$

which in effect involves only even k.

We state the following result concerning the estimation of the remainder term. The traditional midpoint rule (4.2), (4.3) is obtained if we take $n = 2$ and $f'' \in L_2[a,b]$.

COROLLARY 4.19
Let f and I_m be as in Theorem 4.10. Then we have

$$\int_a^b f(t)\,dt = M_{m,n}(f, I_m) + R_{m,n}(f, I_m) \tag{4.70}$$

and the remainder term $R_{m,n}$ satisfies the estimation

$$|R_{m,n}(f, I_m)| \le \begin{cases} \frac{\left\|f^{(n)}\right\|_\infty}{2^n (n+1)!} \sum_{j=0}^{m-1} h_j^{n+1}, & f^{(n)} \in L_\infty[a,b] \\ \frac{\left\|f^{(n)}\right\|_p}{2^n n!(nq+1)^{1/q}} \left[\sum_{j=0}^{m-1} h_j^{n+1}\right]^{\frac{1}{q}}, & f^{(n)} \in L_p[a,b] \\ & \frac{1}{p}+\frac{1}{q}=1,\ p>1 \\ \frac{\left\|f^{(n)}\right\|_1}{2^n n!} [\nu(h)]^n, & f^{(n)} \in L_1[a,b]. \end{cases} \tag{4.71}$$

We can consider the following perturbed version of the trapezoid formula:

$$T_{m,n}(f, I_m) := \sum_{j=0}^{m-1} \sum_{k=0}^{n-1} \frac{h_j^{k+1}}{(k+1)!} \left[\frac{f^{(k)}(x_j) + (-1)^k f^{(k)}(x_{j+1})}{2}\right].$$

By the use of Corollary 4.16, we have the following approximation of the integral $\int_a^b f(t)\,dt$ in terms of $T_{m,n}(f, I_m)$:

COROLLARY 4.20
Let f and I_m be as in Theorem 4.10. Then we have

$$\int_a^b f(t)\,dt = T_{m,n}(f, I_m) + \tilde{R}_{m,n}(f, I_m) \tag{4.72}$$

and the remainder $\tilde{R}_{m,n}(f, I_m)$ satisfies the inequality

$$\left|\tilde{R}_{m,n}(f, I_m)\right| \le \frac{C_n}{(n+1)!} \left\|f^{(n)}\right\|_\infty \sum_{j=0}^{m-1} h_j^{n+1}$$

where

$$C_n := \begin{cases} 1 & \text{if } n = 2r \\ \frac{2^{2r+1}-1}{2^{2r}} & \text{if } n = 2r+1 \end{cases}.$$

REMARK 4.16

a) If we choose $n = 1$ in the above quadrature formulae (4.65) and (4.70), we recapture some results from the paper [27].

b) If we put $n = 2$, then by the above Theorem 4.10 and Corollary 4.19, we recover some results from the paper [4].

∎

We omit the details.

4.5.4 Application of Taylor-Like Expansions for some particular mappings

a) Consider $g : \mathbb{R} \to \mathbb{R}$, $g(x) = e^x$. Then $g^{(n)}(x) = e^x$, $n \in \mathbb{N}$ and

$$\left\| g^{(n+1)} \right\|_\infty = \sup_{t \in [a,y]} \left| g^{(n+1)}(t) \right| = e^y.$$

Using inequality (4.51), we have

$$\left| e^y - e^a - e^x \sum_{k=0}^{n-1} \frac{\left[(y-x)^{k+1} + (-1)^k (x-a)^{k+1} \right]}{(k+1)!} \right|$$

$$\leq \frac{e^y}{(n+1)!} \left[(y-x)^{n+1} + (x-a)^{n+1} \right]$$

$$\leq \frac{e^y}{(n+1)!} (y-a)^{n+1} \tag{4.73}$$

for all $a \leq x \leq y$.

In particular, if we choose $a = 0$, then we get

$$\left| e^y - 1 - e^x \sum_{k=0}^{n-1} \frac{\left[(y-x)^{k+1} + (-1)^k x^{k+1} \right]}{(k+1)!} \right|$$

$$\leq \frac{e^y}{(n+1)!}\left[(y-x)^{n+1}+x^{n+1}\right] \leq \frac{e^y}{(n+1)!}y^{n+1}. \quad (4.74)$$

Moreover, if we choose $x = \frac{y}{2}$, then we get

$$\left|e^y - 1 - e^{\frac{y}{2}}\sum_{k=0}^{n-1}\frac{1+(-1)^k}{(k+1)!}\cdot\frac{y^{k+1}}{2^{k+1}}\right| \leq \frac{e^y y^{n+1}}{2^n (n+1)!} \quad (4.75)$$

for all $y \geq 0$.

b) Consider $g:(0,\infty)\to\mathbb{R}$, $g(x)=\ln x$. Then

$$g^{(n)}(x) = \frac{(-1)^{n-1}(n-1)!}{x^n}, \qquad n \geq 1, x > 0$$

and

$$\left\|g^{(n+1)}\right\|_\infty = \sup_{t\in[a,y]}\left|\frac{(-1)^n n!}{t^{n+1}}\right| = \frac{n!}{a^{n+1}}, \qquad a > 0.$$

Using the inequality (4.51) we can state:

$$\left|\ln y - \ln a - \sum_{k=0}^{n-1}\frac{(y-x)^{k+1}+(-1)^k(x-a)^{k+1}}{(k+1)!}\cdot\frac{(-1)^k k!}{x^{k+1}}\right|$$

$$\leq \frac{n!}{(n+1)!a^{n+1}}\left[(y-x)^{n+1}+(x-a)^{n+1}\right]$$

$$\leq \frac{n!}{(n+1)!a^{n+1}}(y-a)^{n+1}$$

which is equivalent to

$$\left|\ln\left(\frac{y}{a}\right) - \sum_{k=0}^{n-1}\frac{1}{k+1}\cdot\frac{(x-a)^{k+1}+(-1)^k(y-x)^{k+1}}{x^{k+1}}\right|$$

$$\leq \frac{(y-x)^{n+1}+(x-a)^{n+1}}{(n+1)a^{n+1}}$$

$$\leq \frac{1}{(n+1)a^{n+1}}(y-a)^{n+1}. \quad (4.76)$$

Now, if we choose in (4.76) $y = z+1, x = w+1, a = 1, z \geq w \geq 0$, then we get

$$\left| \ln(z+1) - \sum_{k=0}^{n-1} \frac{1}{k+1} \cdot \frac{w^{k+1} + (-1)^k (z-w)^{k+1}}{(w+1)^{k+1}} \right|$$

$$\leq \frac{(z-w)^{n+1} + w^{n+1}}{n+1} \leq \frac{1}{(n+1)} z^{n+1}. \tag{4.77}$$

Finally, if we choose in (4.76), $y = ua, x = wa$ with $u \geq w > 1$, then we have

$$\left| \ln u - \sum_{k=0}^{n-1} \frac{1}{k+1} \frac{(w-1)^{k+1} + (-1)^k (u-w)^{k+1}}{w^{k+1}} \right|$$

$$\leq \frac{(u-w)^{n+1} + (w-1)^{m+1}}{n+1} \leq \frac{(u-1)^{n+1}}{(n+1)}.$$

4.6 Perturbed Interior Point Rules Through Grüss-Type Inequalities

In 1935, G. Grüss (see, for example, [33]), proved the following integral inequality which gives an approximation for the integral of a product in terms of the product of integrals.

THEOREM 4.11

Let $f, g : [a,b] \to \mathbb{R}$ be two integrable mappings so that $\phi \leq h(x) \leq \Phi$ and $\gamma \leq g(x) \leq \Gamma$ for all $x \in [a,b]$, where $\phi, \Phi, \gamma, \Gamma$ are real numbers. Then we have

$$|T(h,g)| \leq \frac{1}{4} (\Phi - \phi)(\Gamma - \gamma) \tag{4.78}$$

where

$$T(h,g) = \frac{1}{b-a} \int_a^b h(x) g(x)\, dx - \frac{1}{b-a} \int_a^b h(x)\, dx \cdot \frac{1}{b-a} \int_a^b g(x)\, dx \tag{4.79}$$

and the inequality is sharp, in the sense that the constant $\frac{1}{4}$ cannot be replaced by a smaller one.

For a simple proof of this fact as well as for extensions, generalisations, discrete variants and other associated material, see [33], and the papers [10]-[13], [17], [21] and [30] where further references are given.

A *premature* Grüss inequality is embodied in the following theorem which was proved in the paper [32]. It provides a sharper bound than the above Grüss inequality. The term *premature* is used to denote the fact that the result is obtained from not completing the proof of the Grüss inequality if one of the functions is known explicitly.

THEOREM 4.12

Let h, g be integrable functions defined on $[a,b]$ and let $d \leq g(t) \leq D$. Then

$$|T(h,g)| \leq \frac{D-d}{2}[T(h,h)]^{\frac{1}{2}}, \tag{4.80}$$

where $T(h,g)$ is as defined in (4.79).

The above Theorem 4.12 will now be used to provide a perturbed generalised interior point rule.

4.6.1 Perturbed Rules From Premature Inequalities

We start with the following result.

THEOREM 4.13

Let $f:[a,b] \to \mathbb{R}$ be such that the derivative $f^{(n-1)}$, $n \geq 1$ is absolutely continuous on $[a,b]$. Assume that there exist constants γ, $\Gamma \in \mathbb{R}$ such that $\gamma \leq f^{(n)}(t) \leq \Gamma$ a.e on $[a,b]$. Then the following inequality holds:

$$\begin{aligned}
|P_M(x)| :&= \Bigg| \int_a^b f(t)\,dt - \sum_{k=0}^{n-1} \frac{\left[(b-x)^{k+1} + (-1)^k (x-a)^{k+1}\right]}{(k+1)!} f^{(k)}(x) \\
&\quad - \frac{(x-a)^{n+1} + (-1)^n (b-x)^{n+1}}{(n+1)!} \left[\frac{f^{(n-1)}(b) - f^{(n-1)}(a)}{b-a}\right] \Bigg| \\
&\leq \frac{\Gamma-\gamma}{2} \cdot \frac{1}{n!} I(x,n) \\
&\leq \frac{\Gamma-\gamma}{2} \cdot \frac{n}{n+1} \cdot \frac{(b-a)^{n+1}}{\sqrt{2n+1}},
\end{aligned} \tag{4.81}$$

where

$$I(x,n) = \frac{1}{(n+1)\sqrt{2n+1}} \left\{ n^2 (b-a) \left[(x-a)^{2n+1} + (b-x)^{2n+1} \right] \right.$$

$$\left. + (2n+1)(x-a)(b-x)\left[(x-a)^n - (x-b)^n\right]^2 \right\}^{\frac{1}{2}}. \qquad (4.82)$$

PROOF Applying the premature Grüss result (4.80) by associating $f^{(n)}(t)$ with $g(t)$ and $h(t)$ with $K_n(x,t)$, from (4.33), gives

$$\left| \int_a^b K_n(x,t) f^{(n)}(t)\, dt - \left(\int_a^b K_n(x,t)\, dt \right) \frac{f^{(n-1)}(b) - f^{(n-1)}(a)}{b-a} \right|$$

$$\leq (b-a) \frac{\Gamma - \gamma}{2} \left[T(K_n, K_n) \right]^{\frac{1}{2}}, \qquad (4.83)$$

where from (4.79)

$$T(K_n, K_n) = \frac{1}{b-a} \int_a^b K_n^2(x,t)\, dt - \left(\frac{1}{b-a} \int_a^b K_n(x,t)\, dt \right)^2.$$

Now, from (4.33),

$$\frac{1}{b-a} \int_a^b K_n(x,t)\, dt = \frac{1}{b-a} \left[\int_a^x \frac{(t-a)^n}{n!} dt + \int_x^b \frac{(t-b)^n}{n!} dt \right]$$

$$= \frac{1}{(b-a)(n+1)!} \left[(x-a)^{n+1} + (-1)^n (b-x)^{n+1} \right]$$

and

$$\frac{1}{b-a} \int_a^b K_n^2(x,t)\, dt = \frac{1}{(b-a)(n!)^2} \left[\int_a^x (t-a)^{2n}\, dt + \int_x^b (b-t)^{2n}\, dt \right]$$

$$= \frac{1}{(b-a)(n!)^2} \left[\frac{(x-a)^{2n+1} + (b-x)^{2n+1}}{(2n+1)} \right].$$

Hence, substitution into (4.83) gives

$$\left| \int_a^b K_n(x,t) f^{(n)}(t)\,dt - \frac{(x-a)^{n+1} + (-1)^n (b-x)^{n+1}}{(n+1)!} \cdot \left[\frac{f^{(n-1)}(b) - f^{(n-1)}(a)}{b-a} \right] \right| \le \frac{\Gamma - \gamma}{2} \frac{1}{n!} J(x,n) \tag{4.84}$$

where

$$J^2(x,n) = \frac{1}{(2n+1)(n+1)^2} \left\{ (n+1)^2 (A+B)\left(A^{2n+1} + B^{2n+1}\right) - (2n+1)\left(A^{n+1} + (-1)^n B^{n+1}\right)^2 \right\}$$

with $A = x - a$, $B = b - x$.

Now, in the proof of Theorem 31 in the previous chapter relating to the trapezoid-type inequalities it was shown that $J(x,n) = (n+1)\sqrt{2n+1} I(x, n)$, where $I(x,n)$ is as given by (4.82). Thus, using identity (4.32) into (4.84) readily produces the result (4.81) and the first part of the theorem is proved. The upper bound results from noticing that $I(x,n)$ is convex and symmetric so that the maximum occurs at either of the end points. Thus, the theorem is now completely proven. ∎

COROLLARY 4.21

Let the conditions of Theorem 31 hold. Then the following result holds:

$$\left| \int_a^b f(t)\,dt - \sum_{k=0}^{n-1} \frac{1}{(k+1)!} \left(\frac{b-a}{2}\right)^{k+1} \left[1 + (-1)^k\right] f^{(k)}\left(\frac{a+b}{2}\right) - \left(\frac{b-a}{2}\right)^{n+1} \frac{\left[1 + (-1)^n\right]}{(n+1)!} \left[\frac{f^{(n-1)}(b) - f^{(n-1)}(a)}{b-a} \right] \right|$$
$$\le \frac{\Gamma - \gamma}{2} \cdot \frac{1}{n!} \left(\frac{b-a}{2}\right)^{n+1} \cdot \frac{1}{\sqrt{2n+1}} \cdot \begin{cases} \frac{2n}{n+1}, & n \text{ even} \\ 2, & n \text{ odd} \end{cases}. \tag{4.85}$$

PROOF Taking $x = \frac{a+b}{2}$ in (4.81) gives (4.85), where

$$I\left(\frac{a+b}{2}, n\right) = \frac{1}{(n+1)\sqrt{2n+1}}\left(\frac{b-a}{2}\right)^{n+1}\left\{4n^2 + (2n+1)\left[1+(-1)^n\right]^2\right\}^{\frac{1}{2}}.$$

Examining the above expression for n even or n odd readily gives the result (4.85). ∎

REMARK 4.17 For n odd, then the third term in the modulus sign vanished and thus there is no perturbation to the generalised midpoint rule (4.85). Further, it may be noticed that only even derivatives are present. For $n = 1$ then there is no perturbation term, giving

$$\left|\int_a^b f(t)\,dt - (b-a) f\left(\frac{a+b}{2}\right)\right| \le \frac{\Gamma - \gamma}{\sqrt{3}}\left(\frac{b-a}{2}\right)^2,$$

where $\gamma \le f'(t) \le \Gamma$.
The above result may be compared with (4.55).
If $n = 2$ is taken, then there is a perturbation term giving

$$\left|\int_a^b f(t)\,dt - (b-a) f\left(\frac{a+b}{2}\right) - \frac{(b-a)^2}{24}\left(f'(b) - f'(a)\right)\right|$$

$$\le \frac{\Gamma - \gamma}{5\sqrt{5}}\left(\frac{b-a}{2}\right)^3,$$

where $\gamma \le f''(t) \le \Gamma$. ∎

THEOREM 4.14
Let the condition of Theorem 4.13 be satisfied. Further, suppose that $f^{(n)}$ is differentiable and is such that

$$\left\|f^{(n+1)}\right\|_\infty := \sup_{t\in[a,b]}\left|f^{n+1}(t)\right| < \infty.$$

Then

$$|P_M(x)| \le \frac{b-a}{\sqrt{12}}\left\|f^{(n+1)}\right\|_\infty \cdot \frac{1}{n!} I(x,n), \tag{4.86}$$

where $P_M(x)$ is the perturbed interior point rule given by the left-hand side of (4.81) and $I(x,n)$ is as given by (4.82).

PROOF Let $h, g : [a,b] \to \mathbb{R}$ be absolutely continuous and h', g' be bounded. Then Chebychev's inequality holds (see [33]):

$$|T(h,g)| \leq \frac{(b-a)^2}{\sqrt{12}} \sup_{t\in[a,b]} |h'(t)| \cdot \sup_{t\in[a,b]} |g'(t)| .$$

Matić, Pečarić and Ujević [32] using a *premature* Grüss-type argument proved that

$$|T(h,g)| \leq \frac{(b-a)}{\sqrt{12}} \sup_{t\in[a,b]} |g'(t)| \sqrt{T(h,h)}. \tag{4.87}$$

Thus, associating $f^{(n)}(\cdot)$ with $g(\cdot)$ and $K(x,\cdot)$, from (4.33), with $h(\cdot)$ in (4.87) produces (4.86) where $I(x,n)$ is as given by (4.82). ∎

THEOREM 4.15

Let the conditions of Theorem 4.13 be satisfied. Further, suppose that $f^{(n)}$ is locally absolutely continuous on (a,b) and let $f^{(n+1)} \in L_2(a,b)$. Then

$$|P_M(x)| \leq \frac{b-a}{\pi} \left\| f^{(n+1)} \right\|_2 \cdot \frac{1}{n!} I(x,n), \tag{4.88}$$

where $P_M(x)$ is the perturbed generalised interior point rule given by the left-hand side of (4.81) and $I(x,n)$ is as given in (4.82).

PROOF The following result was obtained by Lupaş (see [33]). for $h, g : (a,b) \to \mathbb{R}$ locally absolutely continuous on (a,b) and $h', g' \in L_2(a,b)$, then

$$|T(h,g)| \leq \frac{(b-a)^2}{\pi^2} \|h'\|_2 \|g'\|_2 ,$$

where

$$\|k\|_2 := \left(\frac{1}{b-a} \int_a^b |k(t)|^2 \right)^{\frac{1}{2}} \text{ for } k \in L_2(a,b).$$

Matić, Pečarić and Ujević [32] further show that

$$|T(h,g)| \leq \frac{b-a}{\pi} \|g'\|_2 \sqrt{T(h,h)}. \tag{4.89}$$

Now, associating $f^{(n)}(\cdot)$ with $g(\cdot)$ and $K(x,\cdot)$, from (4.33) with $h(\cdot)$ in (4.89) gives (4.88), where $I(x,n)$ is as found in (4.82). ∎

REMARK 4.18 Results (4.86) and (4.88) are not readily comparable to that obtained in Theorem 4.13 since the bound now involves the behaviour of $f^{(n+1)}(\cdot)$ rather than $f^{(n)}(\cdot)$. ∎

4.6.2 Alternate Grüss-Type Results for Perturbed Interior Point Rules

Let

$$\sigma(h(x)) = h(x) - \mathcal{M}(h) \tag{4.90}$$

where

$$\mathcal{M}(h) = \frac{1}{b-a}\int_a^b h(u)\,du. \tag{4.91}$$

Then from (4.79),

$$\mathcal{T}(h,g) = \mathcal{M}(hg) - \mathcal{M}(h)\mathcal{M}(g). \tag{4.92}$$

Dragomir and McAndrew [22] showed effectively, that

$$\mathcal{T}(h,g) = \mathcal{T}(\sigma(h),\sigma(g)) \tag{4.93}$$

and proceeded to obtain bounds for the trapezoidal rule. We now apply identity (4.93) to obtain interior point rules.

THEOREM 4.16
Let $f:[a,b]\to\mathbb{R}$ be a mapping such that $f(\cdot)$ is absolutely continuous on $[a,b]$. Then for all $x\in[a,b]$

$$\left|\int_a^b f(t)\,dt - (b-a)f(x) + (b-a)\left(x-\frac{a+b}{2}\right)S\right|$$

$$\leq \begin{cases} \left(\frac{b-a}{2}\right)^2 \|\sigma(f')\|_\infty, & \text{if } f'\in L_\infty[a,b] \\ \frac{b-a}{2}\left(\frac{b-a}{q+1}\right)^{\frac{1}{q}} \|\sigma(f')\|_p, & \text{if } f'\in L_p[a,b] \\ & \text{with } \frac{1}{p}+\frac{1}{q}=1,\ p>1, \\ \frac{b-a}{2}\|\sigma(f')\|_1, & \text{if } f'\in L_1[a,b], \end{cases} \tag{4.94}$$

where $S=\frac{f(b)-f(a)}{b-a}$, the secant slope.

PROOF Using identity (4.93), associate with $h(t)$, $K_1(x,t)$ from (4.33) and $f'(t)$ for $g(t)$ then

$$\int_a^b K_1(x,t) f'(t)\,dt - (b-a)\mathcal{M}(K_1(x,\cdot))S$$

$$= \int_a^b [K_1(x,t) - \mathcal{M}(K_1(x,\cdot))][f'(t) - S]\,dt, \tag{4.95}$$

where, from (4.33),

$$\mathcal{M}(K_1(x,t)) = \frac{1}{b-a}\left[\int_a^x (t-a)\,dt + \int_x^b (t-b)\,dt\right]$$

$$= x - \frac{a+b}{2} \tag{4.96}$$

and from (4.90),

$$\sigma(K_1(x,\cdot)) = \begin{cases} t - a + \frac{a+b}{2} - x, & t \in [a,x] \\ t - b + \frac{a+b}{2} - x, & t \in (x,b] \end{cases}. \tag{4.97}$$

Thus, (4.95), on utilising (4.96), (4.97) and (4.34) may be written, on taking the modulus, as

$$\left|\int_a^b f(t)\,dt - (b-a)f(x) + (b-a)\left(x - \frac{a+b}{2}\right)S\right|$$

$$= \left|\int_a^b \sigma(K_1(x,t))(f'(t) - S)\,dt\right| := B(x). \tag{4.98}$$

Now, observe that

$$B(x) \le \|f'(\cdot) - S\|_\infty \int_a^b |\sigma(K_1(x,t))|\,dt$$

$$= \|f'(\cdot) - S\|_\infty \left[\int_{\frac{a+b}{2}-x}^{\frac{b-a}{2}} |u|\,du + \int_{-\left(\frac{b-a}{2}\right)}^{\frac{a+b}{2}-x} |v|\,dv\right]$$

and so after some algebra, the first bound in (4.94) is obtained.

Now, from (4.98) using Hölder's inequality we have that

$$B(x) \leq \|\sigma(f')\|_p \left(\int_a^b |\sigma(K_1(x,t))|^q \, dt \right)^{\frac{1}{q}}. \tag{4.99}$$

Now,

$$\int_a^b |\sigma(K_1(x,t))|^q \, dt = \int_{\frac{a+b}{2}-x}^{\frac{b-a}{2}} |u|^q \, du + \int_{-\left(\frac{b-a}{2}\right)}^{\frac{a+b}{2}-x} |v|^q \, dv.$$

That is,

$$L = L_1 + L_2, \quad \text{say.}$$

For $x < \frac{a+b}{2}$,

$$(q+1) L_1 = \int_{\frac{a+b}{2}-x}^{\frac{b-a}{2}} u^q du = \left(\frac{b-a}{2}\right)^{q+1} - \left(\frac{a+b}{2} - x\right)^{q+1}$$

and

$$(q+1) L_2 = \int_{-\frac{b-a}{2}}^{0} |v|^q \, dv + \int_0^{\frac{a+b}{2}-x} v^q dv$$

$$= \left(\frac{b-a}{2}\right)^{q+1} - \left(\frac{a+b}{2} - x\right)^{q+1}.$$

Hence,

$$L = \frac{b-a}{q+1} \cdot \left(\frac{b-a}{2}\right)^q.$$

A similar argument holds for $x > \frac{a+b}{2}$ and so from (4.99) and (4.98), the second inequality in (4.94) is obtained.

For the third inequality we note from (4.98) that

$$B(x) \leq \|\sigma(f')\|_1 \|\sigma(K(x,\cdot))\|_\infty$$

$$= \sup_{t \in [a,b]} |\sigma(K(x,t))| \cdot \|\sigma(f')\|_1 \, . \tag{4.100}$$

Now,

$$\sup_{t\in[a,b]} |\sigma (K (x,t))| = \max \left\{ \frac{b-a}{2}, \left| x - \frac{b-a}{2} \right| \right\}$$

and using the result

$$\max \{X,Y\} = \frac{X+Y}{2} + \frac{1}{2} |Y-X|$$

gives, on treating the cases $x > \frac{a+b}{2}$ and $x < \frac{a+b}{2}$ separately,

$$\max \left\{ \frac{b-a}{2}, \left| x - \frac{b-a}{2} \right| \right\} = \frac{b-a}{2}.$$

Thus, on substitution into (4.100), we obtain the third result in (4.94), and the theorem is completely proved. ▮

REMARK 4.19 The results of this section allow the consideration of $f' \in L_p [a,b]$, $p \in [1,\infty)$ whereas the results of Section 4.6.1 to produce the perturbed rules are valid using $\|\cdot\|_\infty$. The working, however, for explicit results with $K_n (x,t)$ as given by (4.33), is somewhat more difficult with the methodology of the current section. ▮

This section considers results involving $\sigma (f')$ which may be more useful when information is known about deviations of the slope from its mean.

4.7 An Ostrowski-Type Inequality for Mappings Whose Second Derivatives Are Bounded

In [27], S.S. Dragomir and S. Wang obtained the Ostrowski-type inequality using for the proof essentially, the identity

$$f(x) = \frac{1}{b-a} \int_a^b f(t)\,dt + \frac{1}{b-a} \int_a^b p(x,t) f'(t)\,dt \tag{4.101}$$

for all $x \in [a,b]$, where f is as above and the kernel, $p(\cdot,\cdot) : [a,b]^2 \to \mathbb{R}$, is given by

$$p(x,t) := \begin{cases} t-a & \text{if } t \in [a,x] \\ t-b & \text{if } t \in (x,b] \end{cases}. \tag{4.102}$$

Identity (4.101) is easily proven from considering $\int_a^b p(x,t) f'(t)\,dt$ and integrating by parts.

The main aim of this section is to obtain a perturbed interior point rule in which the perturbation does not involve derivative evaluations.

4.7.1 A New Integral Inequality

The following results hold (see, for example, [18]).

THEOREM 4.17
Let $f : [a,b] \to \mathbb{R}$ be a continuous on $[a,b]$ and twice differentiable function on (a,b), whose second derivative $f'' : (a,b) \to \mathbb{R}$ is bounded on (a,b). Then we have the inequality

$$\left| f(x) - \frac{1}{b-a}\int_a^b f(t)\,dt - \frac{f(b)-f(a)}{b-a}\left(x - \frac{a+b}{2}\right)\right|$$

$$\leq \frac{1}{2}\left\{\left[\frac{\left(x-\frac{a+b}{2}\right)^2}{(b-a)^2} + \frac{1}{4}\right]^2 + \frac{1}{12}\right\}(b-a)^2 \|f''\|_\infty$$

$$\leq \frac{\|f''\|_\infty}{6}(b-a)^2 \tag{4.103}$$

for all $x \in [a,b]$.

PROOF Applying the identity (4.101) for $f'(\cdot)$ we can state

$$f'(t) = \frac{1}{b-a}\int_a^b f'(s)\,ds + \frac{1}{b-a}\int_a^b p(t,s) f''(s)\,ds,$$

which is equivalent to

$$f'(t) = \frac{f(b)-f(a)}{b-a} + \frac{1}{b-a}\int_a^b p(t,s) f''(s)\,ds.$$

Substituting $f'(t)$ in the right-hand side of (4.101) we get

$$f(x) = \frac{1}{b-a}\int_a^b f(t)\,dt$$

$$+\frac{1}{b-a}\int_a^b p(x,t)\left[\frac{f(b)-f(a)}{b-a}+\frac{1}{b-a}\int_a^b p(t,s)\,f''(s)\,ds\right]dt$$

$$=\frac{1}{b-a}\int_a^b f(t)\,dt+\frac{f(b)-f(a)}{(b-a)^2}\int_a^b p(x,t)\,dt$$

$$+\frac{1}{(b-a)^2}\int_a^b\int_a^b p(x,t)\,p(t,s)\,f''(s)\,dsdt$$

and as

$$\int_a^b p(x,t)\,dt=\int_a^x (t-a)\,dt+\int_x^b (t-b)\,dt$$

$$=(b-a)\left(x-\frac{a+b}{2}\right),$$

the integral identity

$$f(x)=\frac{1}{b-a}\int_a^b f(t)\,dt+\frac{f(b)-f(a)}{b-a}\left(x-\frac{a+b}{2}\right)$$

$$+\frac{1}{(b-a)^2}\int_a^b\int_a^b p(x,t)\,p(t,s)\,f''(s)\,dsdt \tag{4.104}$$

results for all $x\in[a,b]$.

Now, using the identity (4.104), we get

$$\left|f(x)-\frac{1}{b-a}\int_a^b f(t)\,dt-\frac{f(b)-f(a)}{b-a}\left(x-\frac{a+b}{2}\right)\right|$$

$$\leq\frac{1}{(b-a)^2}\int_a^b\int_a^b |p(x,t)\,p(t,s)|\,|f''(s)|\,dsdt$$

$$\leq\frac{\|f''\|_\infty}{(b-a)^2}\int_a^b\int_a^b |p(x,t)|\,|p(t,s)|\,dsdt$$

$$:= \frac{\|f''\|_\infty}{(b-a)^2} A(x). \tag{4.105}$$

We have

$$\int_a^b |p(t,s)|\, ds = \frac{(t-a)^2 + (b-t)^2}{2}$$

and so, from (4.105),

$$A(x) = \int_a^b |p(x,t)| \left[\frac{(t-a)^2 + (b-t)^2}{2} \right] dt$$

$$= \frac{1}{2} \left[\int_a^x (t-a)\left[(t-a)^2 + (b-t)^2\right] dt + \int_x^b (b-t)\left[(t-a)^2 + (b-t)^2\right] dt \right]$$

$$= \frac{1}{2} \left[\int_a^x \left[(t-a)^3 + (t-a)(b-t)^2\right] dt + \int_x^b \left[(t-a)^2(b-t) + (b-t)^3\right] dt \right].$$

Note that

$$\int_a^x (t-a)^3\, dt = \frac{(x-a)^4}{4}, \quad \int_x^b (t-b)^3\, dt = \frac{(x-b)^4}{4},$$

$$\int_a^x (t-a)(b-t)^2\, dt = -\frac{1}{3}(b-x)^3(x-a) - \frac{1}{12}(b-x)^4 + \frac{1}{12}(b-x)^4$$

and

$$\int_x^b (t-b)(t-a)^2\, dt = \frac{1}{3}(x-a)^3(b-x) - \frac{1}{12}(b-a)^4 + \frac{1}{12}(x-a)^4.$$

Consequently, we have

$$A(x) = \frac{1}{12}\Big[(x-a)^4 - 2(b-x)^3(x-a) - 2(x-a)^3(b-x)$$

$$+ (b-x)^4 + (b-a)^4\Big],$$

which may be simplified in a variety of ways to give

$$A(x)=\frac{1}{12}\left[6\left(x-\frac{a+b}{2}\right)^4+3(b-a)^2\left(x-\frac{a+b}{2}\right)^2+\frac{7}{8}(b-a)^4\right].$$

Now, using the inequality (4.105) and simple algebraic manipulations, we get the first result in (4.103).

The second part is obvious from the fact that

$$\left|x-\frac{a+b}{2}\right|\le\frac{b-a}{2}$$

for all $x\in[a,b]$. ∎

4.7.2 Applications in Numerical Integration

Let $I_n : a=x_0<x_1<...<x_{n-1}<x_n=b$ be a division of the interval $[a,b]$, $\xi_i\in[x_i,x_{i+1}]$ $(i=0,1,...,n-1)$ a sequence of intermediate points and $h_i:=x_{i+1}-x_i$ $(i=0,1,...,n-1)$. As in [18], consider the perturbed Riemann's sum defined by

$$A_G(f,I_n,\boldsymbol{\xi}):=\sum_{i=0}^{n-1}f(\xi_i)h_i-\sum_{i=0}^{n-1}\left(\xi_i-\frac{x_i+x_{i+1}}{2}\right)(f(x_{i+1})-f(x_i)).\tag{4.106}$$

In that paper Dragomir and Wang [25] proved the following result:

THEOREM 4.18

Let $f:[a,b]\to\mathbb{R}$ be continuous on $[a,b]$ and differentiable on (a,b), whose derivative $f':(a,b)\to\mathbb{R}$ is bounded on (a,b) and assume that

$$\gamma\le f'(x)\le\Gamma\quad\textit{for all}\quad x\in(a,b).\tag{4.107}$$

Then we have the quadrature formula:

$$\int_a^b f(x)\,dx=A_G(f,I_n,\boldsymbol{\xi})+R_G(f,I_n,\boldsymbol{\xi}),\tag{4.108}$$

where the remainder $R_G(f,I_n,\boldsymbol{\xi})$ satisfies the estimation

$$|R_G(f,I_n,\boldsymbol{\xi})|\le\frac{1}{4}(\Gamma-\gamma)\sum_{i=0}^{n-1}h_i^2,\tag{4.109}$$

for all $\boldsymbol{\xi} = (\xi_0, ..., \xi_{n-1})$ *as above.*

Here, we prove another type of estimation for the remainder $R_G(f, I_n, \boldsymbol{\xi})$ in the case when f is twice differentiable [18].

THEOREM 4.19
Let $f : [a, b] \to \mathbb{R}$ be continuous on $[a, b]$ and a twice differentiable function on (a, b), whose second derivative, $f'' : (a, b) \to \mathbb{R}$, is bounded on (a, b). Denote $\|f''\|_\infty := \sup_{t\in(a,b)} |f''(t)| < \infty$. Then we have the quadrature formula (4.108), where the remainder $R_G(f, I_n, \boldsymbol{\xi})$ satisfies the estimation

$$\begin{aligned}
&|R_G(f, I_n, \boldsymbol{\xi})| \\
&\leq \frac{\|f''\|_\infty}{2} \sum_{i=0}^{n-1} \left\{ \left[\frac{\left(\xi_i - \frac{x_i + x_{i+1}}{2}\right)^2}{h_i^2} + \frac{1}{4} \right]^2 + \frac{1}{12} \right\} h_i^3 \\
&\leq \frac{\|f''\|_\infty}{6} \sum_{i=0}^{n-1} h_i^3,
\end{aligned} \tag{4.110}$$

for all ξ_i as above.

PROOF Apply Theorem 4.17 on the interval $[x_i, x_{i+1}]$ $(i = 0, ..., n-1)$ to obtain

$$\begin{aligned}
&\left| f(\xi_i) h_i - \int_{x_i}^{x_{i+1}} f(t)\,dt - \left(\xi_i - \frac{x_i + x_{i+1}}{2}\right)(f(x_{i+1}) - f(x_i)) \right| \\
&\leq \frac{\|f''\|_\infty}{2} \left[\left[\frac{\left(\xi_i - \frac{x_i + x_{i+1}}{2}\right)^2}{h_i^2} + \frac{1}{4} \right]^2 + \frac{1}{12} \right] h_i^3 \leq \frac{\|f''\|_\infty}{6} h_i^3
\end{aligned}$$

for all $\xi_i \in [x_i, x_{i+1}]$ and $i \in \{0, ..., n-1\}$.

Summing over i from 0 to $n-1$ and using the generalized triangle inequality, we get the desired inequality (4.110).

We omit the details. ∎

4.8 Concluding Remarks

The current work has demonstrated the development of interior point rules which contain the midpoint rule as a special case. Identities are obtained by using a Peano kernel approach which enables us, through the use of the modern theory of inequalities, to obtain bounds in terms of a variety of norms. This is useful in practice as the behaviour of the function would necessitate the use of one norm over another. Although not all inequalities have been developed into composite quadrature rules, we believe that enough demonstrations have been given that would enable the reader to proceed further.

It has been shown that the bounds for interior point rules are the same as those obtained for the trapezoidal rules of the previous chapter as highlighted in Remark 4.10.

Rules have also been developed that do not necessarily require the second derivative to be well behaved or indeed, exist, thus allowing the treatment of a much larger class of functions. Rules have been developed by examining the Riemann-Stieltjes integral. Additionally, the results also allow for a non-uniform partition, thus giving the user the option of choosing a partition that minimises the bound or enabling the calculation of the bound given a particular partition.

If we wish to approximate the integral $\int_a^b f(x)\,dx$ using a quadrature rule $Q(f, I_n)$ with bound $B(n)$, where I_n is a uniform partition, for example, with an accuracy of $\varepsilon > 0$, then we will need $n_\varepsilon \in \mathbb{N}$ where

$$n_\varepsilon \geq \left[B^{-1}(\varepsilon)\right] + 1$$

with $[x]$ denoting the integer part of x.

This approach enables the user to predetermine the partition required to *assure* that the result is within a certain tolerance rather than utilizing the commonly used method of halving the mesh size and comparing the resulting estimation.

We conclude the work by bringing to the attention of the reader that three-point rules may be obtained by taking a convex combination of trapezoidal-type identities, $I_T(x)$ of the previous chapter and interior-point identities of the current chapter, $I_M(x)$. That is,

$$\lambda I_T(x) + (1-\lambda) I_M(x).$$

Simpson-type rules would result from taking $\lambda = \frac{1}{3}$ and $x = \frac{a+b}{2}$.

This will not be presented here.

For a three-point quadrature rule involving at most the first derivative, see Cerone and Dragomir [3]. For other results that involve three-point rules, the reader is referred to [20, 23, 24].

Acknowledgement The authors would like to express their sincere gratitude to Ms. Pui Ling Pang for the preparation of the manuscript. Further, the selfless contribution to the typesetting of the work by their colleague John Roumeliotis is gratefully acknowledged. This work would not have been possible without their dedication well beyond the call!

References

[1] T. M. Apostol, *Mathematical Analysis*, 2nd Ed., Addison-Wesley Publishing Company, 1975.

[2] K.E. Atkinson, *An Introduction to Numerical Analysis*, Wiley and Sons, 2nd Ed., 1989.

[3] P. Cerone and S. S. Dragomir, Three point quadrature rules involving, at most, a first derivative, (submitted, 1999). [ONLINE] *RGMIA Res. Rep Coll.*, **2**(4), Article 8, 1999, `http://rgmia.vu.edu.au/v2n4.html`

[4] P. Cerone, S.S. Dragomir and J. Roumeliotis,[1] An inequality of Ostrowski type for mappings whose second derivatives are bounded and applications, *Preprint: RGMIA Res. Rep. Coll.*, **1**(1), Article 4, 1998. [ONLINE] `http://rgmia.vu.edu.au/v1n1.html`

[5] P. Cerone, S.S. Dragomir and J. Roumeliotis, An inequality of Ostrowski-Grüss type for twice differentiable mappings and applications, *RGMIA Res. Rep. Coll.*, 1(2), Article 8, 1998. [ONLINE] `http://rgmia.vu.edu.au/v1n2.html`

[6] P. Cerone, S.S. Dragomir and J. Roumeliotis, An Ostrowski type inequality for mappings whose second derivatives belong to $L_p(a, b)$ and applications, *Preprint: RGMIA Res. Rep. Coll.*, **1**(1), Article 5, 1998. [ONLINE] `http://rgmia.vu.edu.au/v1n1.html`

[1]All the papers from RGMIA Res. Rep. Coll. are available at the address: `http://rgmia.vu.edu.au/reports.html`

[7] P. Cerone, S.S. Dragomir and J. Roumeliotis, On Ostrowski type for mappings whose second derivatives belong to $L_1(a,b)$ and applications, *Preprint: RGMIA Res. Rep. Coll.*, **1**(2), Article 7, 1998. [ONLINE] `http://rgmia.vu.edu.au/v1n2.html`

[8] P. Cerone, S.S. Dragomir and J. Roumeliotis, Some Ostrowski type inequalities for n-time differentiable mappings and applications, *Demonstratio Math.*, in press, 1999. [ONLINE] *RGMIA Res. Rep Coll.*, **1**(2), Article 6, 1998, `http://rgmia.vu.edu.au/v1n2.html`

[9] P.J. Davis and P. Rabinowitz, *Methods of Numerical Integration*, 2nd Ed., Academic Press, New York, 1994.

[10] S.S. Dragomir, A Grüss type integral inequality for mappings of r-Hölder's type and applications for trapezoid formula, *Tamkang Journal of Mathematics,* accepted, 1999.

[11] S.S. Dragomir, A Taylor like formula and application in numerical integration, (submitted, 1999).

[12] S.S. Dragomir, Grüss inequality in inner product spaces, *The Australian Math. Gazette*, **26**(2), 66-70, 1999.

[13] S.S. Dragomir, New estimation of the remainder in Taylor's formula using Grüss' type inequalities and applications, *Mathematical Inequalities and Applications*, **2**(2), 183-194, 1999.

[14] S.S. Dragomir, On the Ostrowski's integral inequality for Lipschitzian mappings and applications, *Computers and Mathematics with Appl.*, in press, 1999.

[15] S.S. Dragomir, On the Ostrowski's integral inequality for mappings with bounded variation and applications, (submitted, 1999).

[16] S.S. Dragomir, Ostrowski's Inequality for monotonic mappings and applications, *J. KSIAM*, in press, 1999.

[17] S.S. Dragomir, Some integral inequalities of Grüss type, *Indian J. of Pure and Appl. Math.*, (accepted, 1999).

[18] S.S. Dragomir and N.S. Barnett, An Ostrowski type inequality for mappings whose second derivatives are bounded and applications, *Preprint: RGMIA Res. Rep. Coll.*, **1**(2), Article 9, 1998. [ONLINE] `http://rgmia.vu.edu.au/v1n2.html`

[19] S.S. Dragomir, P. Cerone and A. Sofo, Some remarks on the midpoint rule in numerical integration, (submitted, 1999).

[20] S.S. Dragomir, Y.J. Cho and S.S. Kim, Some remarks on the Milovanović-Pečarić inequality and in applications for special means and numerical integration, *Tamkang J. of Math.*, (accepted, 1999).

[21] S.S. Dragomir and I. Fedotov, An inequality of Grüss type for Riemann-Stieltjes integral and applications for special means, *Tamkang J. of Math.*, **29**(4), 287-292, 1998.

[22] S.S. Dragomir and A. McAndrew, On trapezoid inequality via Grüss type results and applications, *RGMIA Res. Rep. Coll.*, **2**(2) (1999), Article 6.

[23] S.S. Dragomir, J.E. Pečarić and S. Wang, The Unified treatment of trapezoid, Simpson and Ostrowski type inequality for monotonic mappings and applications, *Preprint: RGMIA Res. Rep. Coll.*, **2**(4), Article 3, 1999. [ONLINE] `http://rgmia.vu.edu.au/v2n4.html`

[24] S.S. Dragomir and A. Sofo, An integral inequality for twice differentiable mappings and applications, *Preprint: RGMIA Research Report Collection*, **2**(2), Article 9, 1999. [ONLINE] `http://rgmia.vu.edu.au/v2n2.html`

[25] S.S. Dragomir and S. Wang, An inequality of Ostrowski-Grüss type and its applications to the estimation of error bounds for some special means and for some numerical quadrature rules, *Computers Math. Applic.*, **33**, 15-20, 1997.

[26] S.S. Dragomir and S. Wang, A new inequality of Ostrowski's type in L_1 norm and applications to some special means and to some numerical quadrature rules, *Tamkang J. of Math.*, **28**, 239-244, 1997.

[27] S.S. Dragomir and S. Wang, Applications of Ostrowski's inequality to the estimation of error bounds for some special means and some numerical quadrature rules, *Appl. Math. Lett.*, **11**, 105-109, 1998.

[28] S.S. Dragomir and S. Wang, A new inequality of Ostrowski's type in L_p- norm, *Indian Journal of Mathematics*, (in press), 1999.

[29] H. Engels, *Numerical Quadrature and Cubature*, Academic Press, New York, 1980.

[30] I. Fedotov and S.S. Dragomir, Another approach to quadrature methods, *Indian Journal of Pure and Applied Mathematics*, (in press), 1999.

[31] A.R. Krommer and C.W. Ueberhuber, *Numerical Integration on Advanced Computer Systems*, Lecture Notes in Computer Science, 848,

Springer-Verlag, Berlin, 1994.

[32] M. Matić, J.E. Pečarić and N. Ujević, On new estimation of the remainder in generalised Taylor's formula, *M.I.A.*, Vol. **2**(3), 343-361, 1999.

[33] D.S. Mitrinović, J.E. Pečarić and A.M. Fink, *Classical and New Inequalities in Analysis*, Kluwer Academic Publishers, Dordrecht, 1993.

[34] D.S. Mitrinović, J.E. Pečarić and A.M. Fink, *Inequalities for Functions and their Integrals and Derivatives*, Kluwer Academic Publishers, 1994.

[35] J.E. Pečarić, F. Proschan and Y.L. Tong, *Convex Functions, Partial Orderings, and Statistical Applications*, Academic Press, 1992.

[36] W.H. Press, B.P. Flannery, S.A. Telkolsky and W.T. Verling, *Numerical Recipes: The Art of Scientific Computing*, Cambridge University Press, Cambridge, 1986.

[37] J. Sándor, On the Jensen-Hadamard inequality, *Studia Univ. Babes-Bolyai, Math.*, **36**(1), 9-15, 1991.

[38] A.H. Stroud, *Numerical Quadrature and Solution of Ordinary Differential Equations*, Springer-Verlag, New York, 1974.

Chapter 5

A Unified Approach for Solving Equations, Part I: On Infinite-Dimensional Spaces

Ioannis K. Argyros

Department of Mathematics, Cameron University, Lawton, OK 73505
E-mail: ioannisa@cameron.edu

Abstract We introduce some new very general ways of constructing fast two-step methods to approximate a locally unique solution of a nonlinear operator equation in a Banach space setting. We provide existence-uniqueness theorems as well as an error analysis for the iterations involved using Newton–Kantorovich-type hypotheses and the majorant method. Our results depend on the existence of a Lipschitz function defined on a closed ball centered at a certain point and of a fixed radius and with values into the positive real axis. Special choices of this function lead to favorable comparisons with results already in the literature. The monotone convergence is also examined in a partially ordered topological space setting. Some applications to the solution of nonlinear integral equations appearing in radiative transfer as well as to the solution of integral equations of Uryson-type are also provided. Finally, Newton methods using outer or generalized inverses as well as rates of convergence are being investigated.

Contents

5.1 Two-Step Iterative Procedures

5.1.1 Introduction

This study has been divided into Part I and II. Part I deals with equations on infinite-dimensional spaces whereas Part II examines equations on finite-dimensional spaces. In Part I we introduce some new very general ways of constructing fast two-step methods to approximate a locally unique solution of a nonlinear operator equation in a Banach space setting. We provide existence-uniqueness theorems as well as an error analysis for the iterations involved using Newton–Kantorovich-type hypotheses and the majorant method. Our results depend on the existence of a Lipschitz function defined on a closed ball centered at a certain point and of a fixed radius and with values into the positive real axis. Special choices of this function lead to favorable comparisons with results already in the literature. The monotone convergence is also examined in a partially ordered topological space setting. Some applications to the solution of nonlinear integral equations appearing in radiative transfer as well as to the solution of integral equations of Uryson-type are also provided.

In this study we are concerned with the problem of approximating a locally unique solution x^* of the equation

$$F(x) = 0, \tag{5.1}$$

where F is a nonlinear operator defined on some convex subset D of a Banach space E_1 with values in a Banach space E_2.

Let $x_0 \in D$ and define the two-step Newton method for all $n \geq 0$ by

$$y_n = x_n - F'(x_n)^{-1}F(x_n), \tag{5.2}$$

$$x_{n+1} = y_n - z_n. \tag{5.3}$$

Here $F'(x_n)$ denotes the Fréchet derivative of F evaluated at $x = x_n$, and $z_n \in E_1$ are points to be determined for all $n \geq 0$. Some choices for z_n can be given by $z_n = 0$ or $z_n = F'(x_n)^{-1}F(y_n)$ or $z_n = F'(y_n)^{-1}F(y_n)$ for all $n \geq 0$. The first choice gives the ordinary Newton method (single step). The second choice leads to a two-step Newton method, where two function evaluations and one inverse are required at each step. The third choice leads to another two-step method, where two function evaluations and two inverses are required at each step. Here we will give general conditions for the selection of the z_n's $(n \geq 0)$ that will guarantee that the sequence $\{x_n\}$ $(n \geq 0)$ generated by the approximations (5.2)–(5.3) converges to a solution x^* of the equation (5.1).

Let $x_0 \in D$ and $R > 0$ and assume that with the notion $U(x_0, R) = \{x \in E_1 \mid \|x - x_0\| \leq R\}$, there exists a function $\bar{\alpha} : U^3(x_0, R) \to [0, +\infty)$ such that

$$\|F'(x_0)^{-1}[F(y) - F(x) - F'(x)(y - x) + F'(y)(z - y)]\| \leq \bar{\alpha}(x, y, z) \tag{5.4}$$

for all $x, y, z \in U(x_0, R) \subseteq D$.

Using the majorant theory and the above condition (5.4) we will show that under certain hypotheses the iteration $\{x_n\}$ $(n \geq 0)$ generated by (5.2)–(5.3) converges to a locally unique solution x^* of equation (5.1). The order of convergence of the iteration $\{x_n\}$ $(n \geq 0)$ is also examined here. Under special choices of the function $\bar{\alpha}$ and the points z_n $(n \geq 0)$ our results can be reduced to the ones in [1]–[22]. The monotone convergence of this method is also examined here.

Some applications of our results to the solution of nonlinear integral equations appearing in radiative transfer [1], [2], [8], [9], [17] are provided.

Finally, our results are extended to include two-step Newton-like methods (see Subsection 5.1.5).

5.1.2 Convergence Analysis for Two-Step Newton Methods

We will need the definition of a divided difference of order one [8], [9], [14], [17], [19].

DEFINITION 5.1 *Let F be a nonlinear operator defined on a subset D_0 of a linear space E_3 with values in a linear space E_4, and let x, y be two points of D_0. A linear operator from E_3 into E_4, denoted by $[x, y]$ which satisfies the condition*

$$[x, y](x - y) = F(x) - F(y) \tag{5.5}$$

is called a divided difference of order one of F at the points x and y.

The condition (5.5) does not uniquely determine the divided difference, with the exception of the case when E_3 is one dimensional. An operator $[\cdot, \cdot] : D_0 \times D_0 \to L(E_3, E_4)$ satisfying condition (5.5) is called a divided difference of order one of F on D_0. If we fix the first variable, we get an operator

$$[x_0, \cdot] : D_0 \to L(E_3, E_4).$$

From now on we assume $D_0 = D$, $E_3 = E_1$ and $E_4 = E_2$.

We will need to introduce the constants:

$$t_0 = 0, \quad s_0 \geq \|y_0 - x_0\|, \quad I_o \geq \|z_0\|, \tag{5.6}$$

the sequences for all $n \geq 0$

$$s_{n+1} = t_{n+1} + \frac{1}{1 - Lt_{n+1}} h_{n+1}, \quad L = 2L_0, \tag{5.7}$$

$$t_{n+1} = s_n + I_n, \tag{5.8}$$

$$h_{n+1} = \frac{L}{2}(t_{n+1} - s_n)^2 + \alpha_n \tag{5.9}$$

for some fixed $L_0 > 0$ and some given real sequences $\{\alpha_n\}$ and $\{I_n\}$ with $\alpha_n \geq 0$, $I_n \geq 0$,

$$\sum_{i=0}^{k+1} \alpha_i \leq \gamma_0 r_0 + \gamma_1 \quad \text{and} \quad \sum_{i=0}^{k+1} I_i \leq \gamma_2 r_0 \tag{5.10}$$

for all nonnegative integers k, some fixed real constants γ_0, γ_1, γ_2 and some fixed $r_0 \in [0, R]$.

Moreover, we define the sequences for all $n \geq 0$

$$\bar{h}_{n+1} = \frac{L}{2}\|x_{n+1} - y_n\|^2 + \bar{\alpha}_n, \tag{5.11}$$

where $\bar{\alpha}_n$ denotes $\bar{\alpha}(x_n, y_n, x_{n+1})$ with

$$\bar{\alpha}_n \leq \alpha_n \quad \text{for all } n \geq 0, \tag{5.12}$$

$$\bar{\theta}_n = \left[1 - \frac{L}{2}(\|x^* - x_0\| + \|x_{n+1} - x_0\|)\right]^{-1}, \tag{5.13}$$

$$\theta_{n+1} = \left[1 - \frac{L}{2}(r_0 + t_{n+1})\right]^{-1}, \tag{5.14}$$

$$p_n = \frac{L}{2}\|x^* - x_n\|^2, \tag{5.15}$$

and the function

$$T(r) = s_0 + \frac{1}{1 - Lr}\left[\frac{L}{2}(r - s_0)^2 + \gamma_0 r + \gamma_1\right] + \gamma_2 r \tag{5.16}$$

on $[0, R]$.

We can now state the main result:

THEOREM 5.1

Let $F : D \subseteq E_1 \to E_2$ be a nonlinear operator whose divided difference $[x, y]$ satisfies

$$\|F'(x_0)^{-1}([x, y] - [u, v])\| \leq L_0(\|x - u\| + \|y - v\|) \tag{5.17}$$

for some $L_0 > 0$ and all $x, y, u, v \in U(x_0, R)$. Moreover, we assume:

(i) condition (5.4) is satisfied;

(ii) sequences $\{\alpha_n\}$, $\{I_n\}$ and $\{\bar{\alpha}_n\}$ $(n \geq 0)$ satisfy conditions (5.10), (5.12) and $\|z_n\| \leq I_n$ for all $n \geq 0$;

(iii) there exists a minimum nonnegative number r_0 such that

$$T(r_0) \leq r_0; \tag{5.18}$$

(iv) the following estimates are also true for the numbers r_0, R:

$$r_0 \leq R, \quad L_0(r_0 + R) < 1, \quad U(x_0, R) \subseteq D. \tag{5.19}$$

Then

(a) scalar sequence $\{t_n\}$ $(n \geq 0)$ generated by relations (5.17)–(5.8) is monotonically increasing and bounded above by its limit, which is number r_0;

(b) sequence $\{x_n\}$ $(n \geq 0)$ generated by relations (5.2)–(5.3) is well defined, remains in $U(x_0, r_0)$ for all $n \geq 0$, and convergences to a solution x^* of the equation $F(x) = 0$, which is unique in $U(x_0, R)$.

Moreover, the following estimates are true for all $n \geq 0$:

$$\|y_n - x_n\| \leq s_n - t_n, \tag{5.20}$$

$$\|x_{n+1} - y_n\| \leq t_{n+1} - s_n, \tag{5.21}$$

$$\|x_n - x^*\| \leq r_0 - t_n, \quad \|y_n - x^*\| \leq r_0 - s_n, \tag{5.22}$$

$$\|F'(x_0)^{-1}F(x_{n+1})\| \leq \bar{h}_{n+1}, \tag{5.23}$$

$$\|x^* - x_{n+1}\| \leq \bar{e}_{n+1}\bar{h}_{n+1} \leq e_{n+1}h_{n+1} \leq r_0 - t_{n+1} \tag{5.24}$$

and

$$\|y_n - x_n\| \leq \|x^* - x_n\| + \frac{p_n}{1 - L\|x_n - x_0\|}. \tag{5.25}$$

(We will be concerned only with the case $r_0 > 0$, since when $r_0 = 0$, $x_0 = x$.)

PROOF (a) Using relations (5.6), (5.7), (5.8), (5.10) and (5.18) we deduce that the scalar sequence $\{t_n\}$ $(n \geq 0)$ is monotonically increasing, nonnegative, and $t_0 \leq s_0 \leq t_1 \leq s_1 \leq r_0$. Let us assume that $t_k \leq s_k \leq t_{k+1} \leq s_{k+1} \leq r_0$ for $k = 0, 1, 2, \ldots, n$. Then, by relations (5.7), (5.8) and (5.10) we can have in turn

$$t_{k+2} = s_{k+1} + I_{k+1} = t_{k+1} + \frac{1}{1 - Lt_{k+1}}\left[\frac{L}{2}(t_{k+1} - s_k)^2 + \alpha_k\right] + I_{k+1}$$

$$\leq t_{k+1} + \frac{1}{1 - Lr_0}\left[\frac{L}{2}(r_0 - s_0)(t_{k+1} - s_k) + \alpha_k\right] + I_{k+1}$$

$$\leq \cdots \leq s_0 + \frac{1}{1 - Lr_0}\left[\frac{L}{2}(r_0 - s_0)\sum_{i=0}^{k+1}(t_{i+1} - s_i) + \sum_{i=0}^{k}\alpha_i\right] + \sum_{i=0}^{k+1} I_i$$

$$\leq s_0 + \frac{1}{1 - Lr_0}\left[\frac{L}{2}(r_0 - s_0)^2 + \gamma_0 r_0 + \gamma_1\right] + \gamma_2 r_0 = T(r_0) \leq r_0,$$

by hypothesis (5.18).

Hence, the scalar sequence $\{t_n\}$ $(n \geq 0)$ is bounded above by r_0. By hypothesis (5.18) the number r_0 is the minimum nonnegative zero of the equation $T(r) - r = 0$ on $[0, r_0]$ and from the above, $r_0 = \lim_{n\to\infty} t_n$.

(b) Using relations (5.2), (5.3), (5.6), (5.7) and (5.8) we deduce that $x_1, y_0 \in U(x_0, r_0)$ and that estimates (5.20) and (5.21) are true for $n = 0$. Let us assume that they are true for $k = 0, 1, 2, \ldots, n-1$. Using the induction hypothesis we can have in turn

$$\begin{aligned}\|x_{k+1} - x_0\| \leq \|x_{k+1} - y_0\| + \|y_0 - x_0\| &\leq \|x_{k+1} - y_k\| \\ &+ \|y_k - y_0\| + \|y_0 - x_0\| \\ &\leq \cdots \leq (t_{k+1} - s_k) + (s_k - s_0) + s_0 \leq t_{k+1} \leq r_0,\end{aligned}$$

and

$$\begin{aligned}\|y_{k+1} - x_0\| \leq \|y_{k+1} - y_0\| + \|y_0 - x_0\| &\leq \|y_{k+1} - x_{k+1}\| + \|x_{k+1} - y_k\| \\ &+ \|y_k - y_0\| + \|y_0 - x_0\| \\ &\leq \cdots \leq (s_{k+1} - t_{k+1}) + (t_{k+1} - s_k) + (s_k - s_0) + s_0 \\ &\leq s_{k+1} \leq r_0.\end{aligned}$$

That is, $x_n, y_n \in U(x_0, r_0)$ for all $n \geq 0$.

Note that by hypothesis (5.17) $F'(x) = [x, x]$ for all $x \in D$, and for all $x_k \in U(x_0, r_0)$

$$\begin{aligned}\|F'(x_0)^{-1}(F'(x_k) - F'(x_0))\| &\leq L_0(\|x_k - x_0\| + \|x_k - x_0\|) \\ &= L\|x_k - x_0\| \leq L(t_k - t_0) \\ &= Lt_k \leq Lr_0 < 1,\end{aligned}$$

by hypothesis (5.19). It now follows from the Banach Lemma on invertible operators [14] that the linear operator $F'(x_k)$ is invertible, and

$$\|F'(x_k)^{-1}F'(x_0)\| \leq \frac{1}{1 - L\|x_k - x_0\|} \leq \frac{1}{1 - Lt_k} \tag{5.26}$$

for all k.

We can now have

$$F(x_{k+1}) = [F(x_{k+1}) - F(y_k) - F'(y_k)(x_{k+1} - y_k)]$$
$$+ [F(y_k) + F'(y_k)(x_{k+1} - y_k)]$$
$$= \int_0^1 [F'(y_k + t(x_{k+1} - y_k)) - F'(y_k)](x_{k+1} - y_k)dt$$
$$+ [F(y_k) + F'(y_k)(x_{k+1} - y_k)].$$

Using hypotheses (5.4) and (5.17) we obtain

$$\|F'(x_0)^{-1}F(x_{k+1})\|$$
$$\leq \int_0^1 \|F'(x_0)^{-1}(F'(y_k + t(x_{k+1} - y_k)) - F'(y_k))\| \, \|x_{k+1} - y_k\|dt$$
$$+ \|F'(x_0)^{-1}[F(y_k) - F(x_k) - F'(x_k)(y_k - x_k)$$
$$+ F'(y_k)(x_{k+1} - y_k)]\|$$
$$\leq \frac{L}{2}\|x_{k+1} - y_k\|^2 + \bar{\alpha}(x_k, y_k, x_{k+1})$$
$$= \frac{L}{2}\|x_{k+1} - y_k\|^2 + \bar{\alpha}_k = \bar{h}_{k+1}$$
$$\leq \frac{L}{2}(t_{k+1} - s_k)^2 + \alpha_k = h_{k+1},$$

by hypotheses (5.12), (5.21) and relations (5.11) and (5.9). Hence, we showed estimate (5.23) for all $k \geq 0$.

Using relations (5.2), (5.7), (5.23) and (5.26) we obtain

$$\|y_{k+1} - x_{k+1}\| \leq \|F'(x_{k+1})^{-1}F'(x_0)\| \cdot \|F'(x_0)^{-1}(F(x_{k+1})\|$$
$$\leq \frac{1}{1 - L\|x_{k+1} - x_0\|}\bar{h}_{k+1} \leq \frac{1}{1 - Lt_{k+1}}h_{k+1} = s_{k+1} - t_{k+1},$$

which shows estimate (5.20) for all $n \geq 0$.

Similarly, from relations (5.3), (5.8) and hypothesis (ii) we obtain

$$\|x_{k+1} - y_k\| = \| - z_k\| = \|z_k\| \leq I_k = t_{k+1} - s_k,$$

from which it follows that estimate (5.21) is true for all $n \geq 0$.

It now follows from the estimates (5.20) and (5.21) that the sequence $\{x_k\}$ is Cauchy in a Banach space E_1 and as such it converges to some $x^* \in U(x_0, r_0)$, which by taking the limit as $k \to \infty$ in (5.2), we obtain $F(x^*) = 0$.

To show uniqueness, we assume that there exists another solution y^* of equation (5.1) in $U(x_0, R)$. From hypothesis (5.17) we get

$$\int_0^1 \|F'(x_0)^{-1}[F'(x^* + t(y^* - x^*)) - F'(x_0)]\|dt$$

$$\leq L \int_0^1 \|x^* + t(y^* - x^*) - x_0\|dt$$

$$\leq L \int_0^1 [(1-t)\|y^* - x_0\| + t\|x^* - x_0\|]dt$$

$$\leq L_0(r_0 + R) < 1, \quad \text{by hypothesis (5.19).}$$

It now follows that the linear operator

$$\int_0^1 F'(x^* + t(y^* - x^*))dt \text{ is invertible.}$$

and from the approximation

$$F(y^*) - F(x^*) = \int_0^1 F'(x^* + t(y^* - x^*))dt(y^* - x^*),$$

it follows that $x^* = y^*$.

Estimates (5.22) follow easily from estimates (5.20) and (5.21), respectively, for all $n \geq 0$.

Finally, using the approximations

$$x_{k+1} - x^* = (D_{k+1}^{-1} F'(x_0))(F'(x_0)^{-1} F(x_{k+1})),$$

$$D_{k+1} = \int_0^1 F'(x^* + t(x_{k+1} - x^*))dt,$$

$$y_k - x_k = x^* - x_k + (F'(x_k)^{-1}F'(x_0)) \int_0^1$$

$$\cdot\{F'(x_0)^{-1}[F'(x_k + t(x^* - x_k)) - F'(x_k)](x^* - x_k)\}dt,$$

the estimate

$$\int_0^1 \|F'(x_0)^{-1}[F'(x^* + t(x_{k+1} - x^*)) - F'(x_0)]\|dt$$

$$\leq L\int_0^1 \|x^* + t(x_{k+1} - x^*) - x_0\|dt$$

$$\leq L\int_0^1 [(1-t)\|x^* - x_0\| + t\|x_{k+1} - x_0\|]dt$$

$$\leq Lr_0 < 1, \quad \text{by hypothesis (5.19)}$$

and

$$\|D_{k+1}^{-1}F'(x_0)\| \leq \bar{e}_{k+1},$$

we can immediately obtain estimates (5.24) and (5.25), where we have also used relations (5.13), (5.14) and (5.15).

That completes the proof of the theorem. ∎

REMARK 5.1 (a) Theorem 5.1 was proved by using only the weaker condition

$$\|F'(x_0)^{-1}([x,x] - [u,u])\| \leq L\|x - u\|, \quad L > 0, \ x, y \in U(x_0, R) \quad (5.27)$$

instead of (5.17).

(b) Theorem 5.1 can be further generalized if we assume instead of (5.17) that

$$\|F'(x_0)^{-1}([x,x] - [u,u])\| \leq q(r)\|x - y\| \text{ for all } x, u \in U(x_0, r), 0 \leq r \leq R \quad (5.28)$$

where $q(r)$ is a nondecreasing function on the interval $[0, R]$ with $q(0) = 0$.

Let us define the functions

$$w(r) = \int_0^r q(t)dt, \quad \varphi(r) = \int_0^r w(t)dt - r + s_0, \tag{5.29}$$

$$T(r) = s_0 - \frac{1}{\varphi'(r)}\left[\int_{s_0}^r w(t)dt - w(s_0)(r - s_0)\right.$$

$$\left. + \gamma_0 r + \gamma_1\right] + \gamma_2 r, \tag{5.30}$$

iterations for all $n \geq 0$

$$\bar{h}_{n+1} = \int_0^1 [w(\|y_n - x_0\| + t\|x_{n+1} - y_n\|)$$

$$- w(\|y_n - x_0\|)]\|x_{n+1} - y_n\|dt + \bar{\alpha}_n,$$

$$h_{n+1} = \varphi(t_{n+1}) - \varphi(s_n) - \varphi'(s_n)(t_{n+1} - s_n) + \alpha_n,$$

$$\bar{e}_{n+1} = (1 - w((1-t)\|x_0 - x^*\| + t\|x_{n+1} - x_0\|))^{-1},$$

$$e_{n+1} = (1 - w((1-t)r_0 + tt_{n+1}))^{-1},$$

and

$$p_n = \int_0^1 (w(\|x_n - x_0\| + t\|x_n - x^*\|) - w(\|x_n - x_0\|))\|x_n - x^*\|dt.$$

Replace in relations (5.7) and (5.25) the iterations $\frac{1}{1-Lt_{n+1}}$, $\frac{1}{1-L\|x_n-x_0\|}$ by $-\frac{1}{\varphi'(t_{n+1})}$ and $-\frac{1}{\varphi'(\|x_n-x_0\|)}$, respectively. Moreover, replace condition (5.17) with condition (5.28) and condition (5.19) with $w((1-t)r_0+tR) \leq 1$. Then following the proof of Theorem 5.1 step by step, we can produce a more general theorem under exactly the same hypotheses (with the modifications introduced above) and the same conditions. See also the proofs in [4] and [9].

In Subsection 5.1.3, Part B we show how to choose the function $q(r)$, when it comes to solving nonlinear integral equations of the Uryson type [9], [14].

(c) We will now find the order of convergence for the iterations $\{x_n\}$, $\{t_n\}$ $(n \geq 0)$, by first showing that conditions (5.18) and (5.19) can somehow correspond to standard Newton-Kantorovich-type hypotheses.

Let us assume that there exist nonnegative constants η, h, σ and $K \geq 0$ such that

$$\beta K \leq L, \quad \eta \leq s_0, \quad \sigma \leq 2h = 2K\eta \leq 1, \tag{5.31}$$

$$r_0 \leq 2\eta \text{ and } \sigma = 1 - \left(\frac{2\eta}{r_0} - 1\right)^2 \text{ provided that } r_0 \neq 0. \tag{5.32}$$

If $r_0 = 0$, we choose $s_0 = \eta = h = 0$, and our conditions reduce to $K \leq L$ only. Moreover, we define a scalar function

$$g(t) = \frac{K}{2}t^2 - t + \eta, \tag{5.33}$$

and iterations

$$v_n = w_n - \frac{g(w_n)}{g'(w_n)}, \quad w_0 = 0 \tag{5.34}$$

$$w_{n+1} = v_n - \delta_n, \quad \delta_n \leq 0 \tag{5.35}$$

$$q_n = \frac{2}{K}[g(v_n) - \delta_n g'(v_n)] \tag{5.36}$$

with

$$L_0 q_n \leq \alpha_n \quad \text{and} \quad -\delta_n \leq I_n \text{ for all } n \geq 0. \tag{5.37}$$

Furthermore, let us define

$$r_1 = \frac{1 - \sqrt{1-2h}}{h}\eta, \tag{5.38}$$

$$r_2 = \frac{1 + \sqrt{1-2h}}{h}\eta \tag{5.39}$$

and

$$\theta = \frac{r_1}{r_2}. \tag{5.40}$$

Note that the constants given by relations (5.38) and (5.39) are the real solutions of the equation $g(t) = 0$, where g is given by (5.33) (provided that $2h \leq 1$). ∎

We can now prove the following proposition:

PROPOSITION 5.1
Let us assume:
(i) hypotheses of Theorem 5.1 are true;
(ii) hypotheses (5.31), (5.32) and (5.37) are also true.
Then the following estimates hold: $w_n \leq t_n$, $v_n \leq s_n$, $w_{n+1} - v_n \leq t_{n+1} - s_n$, $v_n - w_n \leq s_n - t_n$, $r_0 = r_1$ *and* $r_0 - t_n \leq r_1 - w_n$ *for all* $n \geq 0$.

PROOF We will use induction on n. Using relations (5.7), (5.8), (5.34) and (5.35) and the hypotheses for $n = 0$, we get in turn

$$w_1 = v_0 - \delta_0 = \eta - \delta_0 \leq s_0 + I_0 - t_1 \Rightarrow w_1 \leq t_1,$$

$$w_1 - v_0 = -\delta_0 \leq I_0 = t_1 - s_0 \Rightarrow w_1 - v_0 \leq t_1 - s_0,$$

$$v_1 = w_1 + \frac{K}{2(1 - Kw_1)}[(w_1 - v_0)^2 + q_0]$$

$$\leq t_1 + \frac{1}{1 - Lt_1}\left[\frac{L}{2}(w_1 - v_0)^2 + L_0 q_0\right]$$

$$\leq t_1 + \frac{1}{1 - Lt_1}\left[\frac{L}{2}(t_1 - s_0)^2 + \alpha_0\right] = s_1 \Rightarrow v_1 \leq s_1,$$

and from the same arguments $w_1 - s_1 \leq s_1 - t_1$.

That is, we have shown all the inequalities for $n = 0$. We now assume that they are true for $k = 0, 1, 2, \ldots, n - 1$. The induction can now easily be completed if we repeat the proof we gave for $n = 0$, by observing that we can replace the subscripts 1 and 0 with $k + 1$ and k, respectively.

From hypothesis $\sigma < 2h$ it can easily follow that $r_0 \leq r_1$, and by taking the limit as $n \to \infty$ in the estimate $w_n \leq v_n$, we obtain $r_1 \leq r_0$. Hence, we get $r_0 = r_1$.

The proof of the proposition is now complete. ∎

From the proof of the proposition and the theorem it can now easily be seen that the uniqueness of the solution x^* can be extended in the ball $U(x_0, r)$ with $r_1 \leq r \leq r_2$, provided that $U(x_0, r) \subseteq D$ and $L_0(r_1 + r_2) < 1$.

It also follows from the above proposition that the order of convergence of iterations $\{x_n\}$ and $\{t_n\}$ is comparable to the order of convergence of the iteration $\{w_n\}$ $(n \geq 0)$.

The conditions $\sigma \leq 2h$ and $r_0 \leq 2\eta$ are used only to show that $r_0 \leq r_1$. If these conditions are violated, we can still show that the order of convergence of iterations $\{x_n\}$ and $\{t_n\}$ $(n \geq 0)$ is "at least asymptotically" equal to the order of convergence of the iteration $\{w_n\}$ $(n \geq 0)$. See Cases 1, 2 and 3 that follow.

REMARK 5.2 Let us examine some special choices for the z_n's. Set

$$z_n = A_n^{-1} B_n (y_n - x_n) \tag{5.41}$$

for all $n \geq 0$, some linear operators B_n and A_n with A_n being invertible for all $n \geq 0$. Here the linear operators A_n and B_n may depend on x_n or y_n or both or neither. Condition (5.4) can now be replaced by one of the following three sets of conditions for all $x_n, x_{n+1}, y_n \in U(x_0, R)$ $(n \geq 0)$

$$\left\| F'(x_0)^{-1} \left[\int_0^1 [F'(x_n + t(y_n - x_n)) - (F'(x_n) + B_n)](y_n - x_n) dt \right.\right.$$

$$\left.\left. + (F'(y_n) - A_n)(x_{n+1} - y_n) \right] \right\| \leq \bar{a}_n \|x_n - y_n\| + \bar{n}_n \|x_{n+1} - y_n\|$$

or by

$$\left\| F'(x_0)^{-1} \left[\int_0^1 [F'(x_n + t(y_n - x_n)) - (F'(x_n) + B_n)](y_n - x_n) dt \right] \right\|$$

$$\leq \bar{a}_n \|x_n - y_n\|,$$

and

$$\|F'(x_0)^{-1}(F'(y_n) - A_n)(x_{n+1} - y_n)\| \leq \bar{b}_n \|x_{n+1} - y_n\|,$$

or by

$$\left\| F'(x_0)^{-1} \left[\int_0^1 [F'(x_n + t(y_n - x_n)) - (F'(x_n) + B_n)] \right] dt \right\| \leq \bar{a}_n,$$

and

$$\|F'(x_0)^{-1}(F'(y_n) - A_n)\| \leq \bar{b}_n.$$

The sequences $\{\bar{a}_n\}$, $\{b_n\}$ $(n \geq 0)$ may differ in each part of the above conditions, but we use the same letters. The above conditions can be written in the more inclusive form (5.4) if x_n, y_n, x_{n+1}, $\bar{a}_n$, b_n, B_n, A_n are replaced by x, y, z, $a(x,y,z)$, $b(x,y,z)$, $B(x,y)$, $A(x,y)$, respectively, for all $x,y,z \in U(x_0,R)$. Here $a(x,y,z)$ and $b(x,y,z)$ denote functions from $U^3(x_0,R)$ into $[0,+\infty)$, and $B(x,y)$, $A(x,y)$ denote linear operators for fixed points $x,y \in U(x_0,R)$.

However, we will use the discrete conditions for our purposes from now on. ∎

Let us now assume that there exist sequences $\{a_n\}$, $\{b_n\}$, $(n \geq 0)$ and positive numbers a and b which may depend on r_0 such that

$$\bar{a}_n \leq a_n \leq a \quad \text{and} \quad \bar{b}_n \leq b_n \leq b.$$

Then we can set

$$\bar{\alpha}_n = \bar{a}_n \|y_n - x_n\| + \bar{b}_n \|x_{n+1} - y_n\|,$$

$$\alpha_n = a_n(s_n - t_n) + b_n(t_{n+1} - s_n)$$

to obtain that

$$\sum_{i=0}^{k} \alpha_i \leq ar_0 + b(r_0 - s_0) = \gamma_0 r_0 + \gamma_1$$

where $\gamma_0 = a + b$ and $\gamma_1 = -bs_0$ (see also relation (5.10)).

Moreover, let us assume that there exist sequences $\{\bar{c}_n\}$, $\{c_n\}$, $(n \geq 0)$ and a number c which may depend on r_0 such that

$$\|A_n^{-1} B_n\| \leq c_n \leq c_n \leq c \quad \text{for all } n \geq 0.$$

Then we can set

$$I_n = c_n(s_n - t_n),$$

and therefore

$$\sum_{i=0}^{k+1} I_i \leq c \sum_{i=0}^{k+1} (s_n - t_n) \leq cr_0.$$

That is, by relation (5.10) $\gamma_2 = c$.

From now on we will assume that the function $\bar{\alpha}$ appearing in condition (5.4) can be chosen by any of the three sets of conditions mentioned above.

We can now show how to choose the operators A_n and B_n for all $n \geq 0$.

Case 1. Let $A_n = F'(x_n)$ and $B_n = [x_n, y_n] - [x_n, x_n]$ for all $n \geq 0$. The iteration $\{x_n\}$ generated by (5.2)–(5.3) with the above choices has been examined in [9]. It is an Euler-Chebysheff-type method [9], [15], [17], [18], [20], [21]. We can now set for all $n \geq 0$

$$\bar{b}_n = L\|y_n - x_n\| \leq L(s_n - t_n) = b_n \leq Lr_0 = b,$$

$$\bar{a}_n = L_0\|y_n - x_n\| \leq L_0(s_n - t_n) = a_n \leq L_0 r_0 = a,$$

$$\bar{c}_n = \frac{L_0\|y_n - x_n\|}{1 - L\|x_n - x_0\|} \leq \frac{L_0(s_n - t_n)}{1 - L(t_n - t_0)} = c_n \leq \frac{L_0 r_0}{1 - Lr_0} = c,$$

$$\delta_n = \frac{K}{2}\frac{(v_n - w_n)^2}{g'(w_n)}, \quad q_n = -\frac{K(v_n - w_n)^2}{g'(w_n)}$$

and for $h \leq .48528137\ldots$

$$r_1 - w_n = \frac{(1 - \theta^2)\eta}{1 - \frac{1}{\sqrt{2}}(\sqrt{2}\theta)^{3^n}}(\sqrt{2}\theta)^{3^n - 1}.$$

The rest of the conclusions of Theorem 5.1 hold for this method.

Hence, the order of convergence for this method is almost three.

Asymptotic Case 1. As mentioned after the proposition, the conditions $c \leq 2h$ and $r_0 \leq 2\eta$ are used only to show that $r_0 \leq r_1$. If these conditions are violated we can reason as follows. For sufficiently large n and since $[x_n - y_n] - [x_n, x_n] = [x_n, y_n, x_n](y_n - x_n)$, the operator $[x_n, y_n, x_n]$ can be "approximated" by $\frac{1}{2}F''(x_n)$. Here $[x_n, y_n, x_n]$ denotes a divided difference of order two for F and $F''(x_n)$ the second Fréchet derivative of F evaluated at $x = x_n$. Replace now the difference operator in the approximation by $\frac{1}{2}F''(x_n)$ and use the same letters for the new iteration $\{x_n\}$ $(n \geq 0)$. Moreover, let us assume that $\|F'(x_0)^{-1}F''(x)\| \leq M$,

$$\|F'(x_0)^{-1}(F''(x) - F''(y))\| \leq N\|x - y\| \quad \text{for all } x, y \in D,$$

$$\|y_0 - x_0\| \leq \eta, \quad \text{and} \quad \left(M^2 + \frac{N}{3}\right)^{1/2} \leq K.$$

Then the sequence $\{w_n\}$ $(n \geq 0)$ majorizes the sequence $\{x_n\}$ also and in particular $\|x_n - x^*\| \leq r_1 - w_1$ for all $n \geq 0$. For the proof of this result see [9].

Case 2. Let $A_n = F'(x_n)$ and $B_n = (I - E_n)^{-1}([x_n, y_n] - [x_n, x_n])$ with $E_n = -F'(x_n)^{-1}([x_n, y_n] - [x_n, x_n])$ for all $n \geq 0$. The iteration $\{x_n\}$ ($n \geq 0$) generated by (5.2)–(5.3) with the above choices has been examined in [9]. It is a Chebysheff-Halley-type method [9], [12], [15], [17], [18], [20], [21]. We can now set

$$\bar{b}_n = L\|y_n - x_n\| \leq L(s_n - t_n) = b_n \leq Lr_0 = b,$$

$$\bar{a}_n = \frac{L_0\|y_n - x_n\|^2 + (1 - 2L_0\|x_n - x_0\|)\|y_n - x_n\|}{1 - L_0(2\|x_n - x_0\| + \|y_n - x_n\|)}$$

$$\leq \frac{L_0(s_n - t_n)^2 + (s_n - t_n)}{1 - L_0(s_n + t_n)} = a_n \leq \frac{L_0 r_0^2 + r_0}{1 - 2L_0 r_0} = a,$$

$$\bar{c}_n = \frac{L_0\|y_n - x_n\|}{1 - L_0(2\|x_n - x_0\| + \|y_n - x_n\|} \leq \frac{L_0(s_n - t_n)}{1 - L_0(s_n + t_n)}$$

$$= c_n \leq \frac{L_0 r_0}{1 - 2L_0 r_0} = c,$$

$$q_n = -\frac{K}{2}\frac{(v_n - w_n)^3 g'(w_n)^{-1}}{1 + \frac{K}{2}(v_n - w_n)g'(w_n)^{-1}},$$

$$\delta_n = \frac{K}{2}\frac{g'(w_n)^{-1}(v_n - w_n)^2}{1 + \frac{K}{2}g'(w_n)^{-1}(v_n - w_n)}$$

$$r_1 - w_n = \frac{(1 - \theta^2)\eta}{1 - \theta^{3^n}}\theta^{3^n - 1} \quad \text{for all } n \geq 0.$$

The rest of the conclusions of Theorem 5.1 hold for this method.

Hence, the order of convergence for this method is almost three.

Asymptotic Case 2. For the asymptotic case we reason exactly as in Case 1. The condition on K becomes

$$\left(3M^2 + \frac{2N}{3}\right)^{1/2} \leq K,$$

but the rest of the hypotheses remain as in Case 1. For the proof of this result see [9].

Case 3. Let $A_n = F'(x_n)$ and

$$B_n = \frac{3}{4}\left(F'\left(x_n + \frac{2}{3}(y_n - x_n)\right) - F'(x_n)\right)\left(I - \frac{3}{2}H_n\right),$$

$$H_n = F'(x_n)^{-1}\left(F'\left(x_n + \frac{2}{3}(y_n - x_n)\right) - F'(x_n)\right) \quad \text{for all } n \geq 0.$$

The iteration $\{x_n\}$ $(n \geq 0)$ generated by (5.2)–(5.3) with the above choices has been examined in [9].

We can now set

$$\bar{b}_n = L\|y_n - x_n\| \leq L(s_n - t_n) = b_n \leq Lr_0 = b,$$

$$\bar{a}_n = L_0\left(2 + \frac{L\|y_n - x_n\|}{1 - L\|x_n - x_0\|}\right)\|y_n - x_n\| \leq L_0\left(2 + \frac{L(s_n - t_n)}{1 - L(s_n - t_0)}\right) = a_n$$

$$\bar{c}_n = \frac{L_0\|y_n - x_n\|}{1 - L\|x_n - x_0\|}\left(1 + \frac{L\|y_n - x_n\|}{1 - L\|x_n - x_0\|}\right)$$

$$\leq \frac{L_0}{1 - L(t_n - t_0)}\left(1 + \frac{L(s_n - t_n)}{1 - L(t_n - t_0)}\right)(s_n - t_n)$$

$$= c_n \leq \frac{L_0}{1 - Lr_0}\left(1 + \frac{Lr_0}{1 - Lr_0}\right) = \frac{L_0 r_0}{(1 - Lr_0)^2} = c,$$

$$\delta_n = \frac{3}{4}g'(w_n)^{-1}\left(g'\left(v_n + \frac{2}{3}(v_n - w_n)\right) - g'(v_n)\right)$$

$$\cdot\left[1 - \frac{3}{2}g'(w_n)^{-1}\left(g'\left(v_n + \frac{2}{3}(v_n - w_n)\right) - g'(v_n)\right)\right],$$

and for $h \leq .46568\ldots$,

$$r_1 - w_n = \frac{(1 - \theta^2)\eta}{1 - \frac{1}{\sqrt[3]{5}}[\sqrt[3]{5}\theta]^{4^n}}[\sqrt[3]{5}\theta]^{4^n - 1}.$$

The rest of the conclusions of Theorem 5.1 hold for this method.

Hence, the order of convergence for this method is almost four.

Asymptotic Case 3. For the asymptotic case we reason exactly as in Case 1. The condition on K becomes

$$\left(M^2 + \frac{N}{6}\right)^{1/2} \leq K,$$

and the rest of the hypotheses remain as in Case 2. (See also [9] for the proof.)

For the computational details in the remaining cases, see [9].

Case 4. Choose $z_n = 0$ for all $n \geq 0$ in approximation (5.3). Then by approximations (5.2)–(5.3) we obtain the single-step Newton-Kantorovich method which is of order two. Note that we can set $\bar{\alpha}_n = 0$, $\alpha_n = I_n = \delta_n = q_n = 0$, $t_n = v_n$ and $s_n = w_n$ for all $n \geq 0$.

Case 5. Choose $z_n = M^{-1}\varepsilon_n$, $\varepsilon_n \in E_2$ and set $M_n = F'(x_n)$, $\varepsilon_n = F(y_n)$ for all $n \geq 0$. That is, we obtain the two-step Newton-Kantorovich method that requires two function evaluations and are inverse at every step. We can then have that

$$r_1 - w_n = \frac{(1-\theta^2)\eta}{1-(d\theta)^{3^n}}(d\theta)^{3^n-1}, \quad (\bar{\alpha}_n = L\|y_n - x_n\| \leq L(s_n - t_n) = \alpha_n \text{ etc.}),$$

provided that $h \leq \frac{2d}{(d+1)^2}$, where $d = \sqrt[6]{2}$.

Case 6. Choose $z_n = M_n^{-1}\varepsilon_n$, $\varepsilon_n \in E_1$ and set $M_n = F'(y_n)$, $\varepsilon_n = F(y_n)$ for all $n \geq 0$. That is we obtain the two-step Newton-Kantorovich method that requires two function evaluations and two inverses at every step. We can then have that $\bar{a}_n = 0$, $a_n = 0$, $\bar{a}_n = a_n = \bar{b}_n = b_n = 0$ and

$$r_1 - w_n = \frac{(1-\theta^2)\eta}{1-\theta^{4^n}}\theta^{4^n-1} \quad \text{for all } n \geq 0.$$

Other choices are also possible.

REMARK 5.3 (a) The error estimates (5.21)–(5.27) can be improved even further, if as we did for Newton-like methods, Ptak-type or Zabrejko-Nguen or generalized Zabrejko-Nguen-type conditions are assumed [5], [9], [11], [16], [22].

(b) As noted earlier the full strength of condition (5.17) is not used in the proof of Theorem 5.1, since condition (5.27) is sufficient. However if the z_n's are chosen as in special cases 1–3, then the construction of the sequence $\bar{\alpha}_n$ requires condition (5.17). Similarly, condition (5.28) can be replaced by condition

$$\|F'(x_0)^{-1}([x,y]-[u,v])\| \le q_1(r)\|x-u\| + q_2(r)\|y-v\| \tag{5.42}$$

for all $x, y, u, v \in U(x_0, r)$, $r \in [0, R]$, where $q_1(r)$ and $q_2(r)$ are nondecreasing functions on the interval $[0, R]$ with $q_1(0) = q_2(0) = 0$. We then just need to replace the function q by $q_1 + q_2$ in all definitions that followed Remark 5.1(b) after Theorem 5.1.

(c) Due to the estimate $[x,y]-[u,v] = ([x,y]-[u,u]) - ([u,v]-[u,u])$ all the above mentioned results remain valid if conditions (5.17), (5.42) are replaced by the weaker conditions where $u = v$. ∎

5.1.3 Applications for Two-Step Newton Methods

(A) We will give an example for the Asymptotic Case 1.

Example 5.1
Our method becomes

$$y_n = x_n - F'(x_n)^{-1}F(x_n), \tag{5.43}$$

$$x_{n+1} = y_n - \frac{1}{2}F'(x_n)^{-1}F''(x_n)(y_n - x_n)^2 \quad (n \ge 0). \tag{5.44}$$

Note that by eliminating y_n between the above approximations we can obtain the method of tangent parabolas (or Euler-Chebysheff method) which has been extensively studied in [9], [12], [15], [17], [18], [20], [21]. In all but our references it is assumed that $N > 0$, which means that their results cannot apply to solve quadratic operator equations of the form

$$P(x) = Qx^2 + Lx + z,$$

where Q, L are bounded quadratic and linear operators, respectively, with z fixed in E_1 [2], [3], [9], [10].

We have that $P'(x) = 2Qx + L$ and $P''(x) = 2Q$. Hence, we get $M = 2\|Q\|$ and $N = 0$. Integral equations of the form $P(x) = 0$ have very important applications in radiative transfer [3], [4], [9], [10].

As a specific example, let us consider the solutions of quadratic integral equations of the form

$$x(s) = y(x) + \lambda x(s)\int_0^1 q(s,t)x(t)dt \tag{5.45}$$

in the space $E_1 = C[0,1]$ of all functions continuous on the interval $[0,1]$, with norm

$$\|x\| = \max_{0 \le s \le 1} |x(s)|.$$

Here we assume that λ is a real number called the "albedo" for scattering and the kernel $q(s,t)$ is a continuous function of two variables s, t with $0 < s$, $t < 1$ satisfying

(i) $0 < q(s,t) < 1$, $q(0,0) = 1$, $0 < s, t \le 1$;

(ii) $q(s,t) + q(t,s) = 1$, $0 \le s, t \le 1$.

The function $y(s)$ is a given continuous function defined on $[0,1]$, and finally $x(s)$ is the unknown function sought in $[0,1]$.

Equations of this type are closely related with the work of S. Chandrasekhar [10] (Nobel prize for physics, 1983), and arise in the theories of radiative transfer, neutron transport and in the kinetic theory of gasses [2], [3], [9], [10].

There exists an extensive literature on equations like (5.45) under various assumptions on the kernel $q(s,t)$ and λ is a real or complex number. One can refer to the recent work [3], [4], [9] and the references there. Here we demonstrate that the theorem via the iteration procedures (5.43)–(5.44) provides existence results for (5.45). Note that the computational cost is no higher than the corresponding one of previous methods [1], [11]–[22].

For simplicity (without loss of generality) we will assume that

$$q(s,t) = \frac{s}{s+t} \quad \text{for all } \ 0 < s, t \le 1 \text{ and } q(0,0) = 1.$$

Note that $q(s,t)$ so defined satisfies (i) and (ii) above.

Let us now choose $\lambda = .25$, $y(s) = 1$ for all $s \in [0,1]$; and define the operator P on E_1 by

$$P(x) = \lambda x(s) \int_0^1 \frac{s}{s+t} x(t)dt - x(s) + 1.$$

Note that every zero of the equation $P(x) = 0$ satisfies equation (5.45).

Set $x_0(s) = 1$, use the definition of the first and second Fréchet derivatives of the operator P to obtain using the theorem, $\|P'(1)^{-1}\| = 1.53039421$,

$$N = 0, \ K = M = 2|\lambda| \max_{0 \le s \le 1} \left| \int_0^1 \frac{s}{s+t} dt \right| \|P'(1)^{-1}\|$$

$$= 2|\lambda| \ln 2 \|P'(1)^{-1}\| = .530394215,$$

$$\eta \geq \|P'(1)^{-1}P(1)\| \geq \beta\lambda \ln 2 = .265197107,$$

$$h = .140659011 < .48528137, \ldots,$$

$$r_1 = .28704852, \quad r_2 = 3.4837317$$

and

$$\theta = .08239685.$$

(For detailed computations, see also [3], [4] and [9].)

Therefore equation (5.45) has a solution x^* and the two-point method (5.43)–(5.44) converges to x^*.

Note that the results obtained in [12], [15], [17], [18], [20], [21] cannot apply here since $N = 0$. The rest of the cases can similarly be used to solve equation (5.45). $\square$

(B) We will provide an example under Case 6 that shows how to choose the constants s_0, γ_0, γ_1, γ_2 and the functions $q(r)$, $w(r)$, $\varphi(r)$, $T(r)$ (see relations (5.29) and (5.30)).

Let us assume that $E_1 = E_2 = C = C[0,1]$ the space of continuous functions on $[0,1]$ equipped with the usual supremum norm. We consider Uryson-type nonlinear integral equations of the form

$$F(x)(t) = x(t) - \int_0^1 K(t, sx(s))ds. \tag{5.46}$$

We assume for simplicity that $x_0 = 0$, and make use of the following standard result whose proof can be found, for example, in [1], [9] or [14], and the references there.

THEOREM 5.2

The Lipschitz condition (5.28) for the Fréchet derivative F' of the operator (5.46) holds if and only if the second derivative $K^*_{uu}(t,s,u)$ exists for all t and almost all s and u, and

$$\sup_{t\in[0,1]} \int_0^1 \sup_{|u|\leq r} |K^*_{uu}(t, su)|ds < \infty. \tag{5.47}$$

Moreover, the left-hand side in relation (5.47) is then the minimal Lipschitz constant $\frac{q(r)}{\beta}$, $\beta = \|F'(x_0)^{-1}\|$ in (5.28).

Furthermore, the constants s_0, β *are given by*

$$s_0 = \sup_{t\in[0,1]} \left| \int_0^1 K(t,s,0)ds + \int_0^1 r(t,s) \int_0^1 K(s,p,0)dp\, ds \right| \tag{5.48}$$

and

$$\beta = 1 + \sup_{t\in[0,1]} \int_0^1 |rt,s)|ds, \tag{5.49}$$

where $r(t,s)$ *is the resolvent kernel of the equation*

$$h(t) - \int_0^1 K_u^*(t,s,0)h(s)ds = -\int_0^1 K(t,s0)ds. \tag{5.50}$$

Example 5.2

Let us consider a simple example. Suppose that $K(t,s,u) = c_1(t)c_2(s)c_3(u)$ with two continuous functions c_1 and c_2, and $c_3 \in C^2$. We set

$$d_1 = \int_0^1 c_2(s)ds, \quad d_2 = \int_0^1 c_1(s)c_2(s)ds.$$

Then relation (5.50) becomes

$$h(t) = [c_4'c_3'(0) - d_1c_3(0)]c_1(t), \tag{5.51}$$

where

$$c_4' = \int_0^1 c_2(s)h(s)ds. \tag{5.52}$$

Substituting relation (5.51) into (5.52), one may conclude c_4' and hence find the resolvent kernel $r(t,s)$ in case $d_2c_3'(0) < 1$, to get

$$r(t,s) = \frac{c_1(t)c_2(t)c_3'(0)}{1 - d_2c_3'(0)}.$$

Using relations (5.47)–(5.49), we obtain

$$q(t) = \|c_1\|d_1 \sup_{\|u\|\leq r} |c_3'(u)|,$$

$$\eta = \frac{d_1c_3(0)}{1 - d_2c_3'(0)}\|c_1\|$$

and

$$\beta = 1 + \frac{d_1 c_3(0)}{1 - d_2 c_3'(0)} \|c_1\|.$$

Thus, in this case a complete and explicit computation of the function T given by relation (5.30) is possible. As an example, let us choose

$$c_1(t) = \frac{3}{10}t, \; c_2(s) = \frac{2}{10}s \quad \text{and} \quad c_3(u) = \frac{1}{3}u^3 + \frac{1}{10}u + 1$$

on $[0, 1]$. Then using the above relations (5.30), (5.29) and (5.10), we get

$$d_1 = \frac{1}{10}, \; d_2 = \frac{2}{100}, \; d_2 c_3'(0) = \frac{2}{1000} < 1,$$

$$r(t, s) = \frac{30}{499}ts, \; q(r) = \frac{6\beta}{100}r, \; w(r) = \frac{3\beta}{100}r^2, \; \varphi(r) = \frac{\beta}{100}r^3 - r + s_0,$$

$$\int_0^1 w(t)dt = \frac{\beta}{100}r^3, \; s_0 = \frac{15}{499}, \; \beta = \frac{514}{499},$$

$$\alpha = \bar{\alpha}_n = \alpha_n = \gamma_0 = \gamma_1 = 0, \; \gamma_2 = \frac{\beta(r^2 - 3s_0^2)}{100 - 3\beta r^2}.$$

Conditions (5.18) and (5.19) of Theorem 5.1 become, respectively,

$$r^3 - .09018036r^2 - 97.0762904r + 2.91823361 \leq 0,$$

and

$$R \leq 5.688635222 = R_0.$$

That is, the hypotheses of Theorem 5.1 are satisfied for $r_0 = .030061239$ and $R = R_0$. The conclusions of Theorem 5.1 for the iteration under Case 6 can now follow. Similar work can immediately follow for the rest of the six cases mentioned earlier. $\square$

5.1.4 Monotone Convergence of Two-Step Newton Methods

We will assume that the reader is familiar with the notion of a partially ordered topological space, and that E_1 and E_2 are POTL-spaces [8], [9], [17], [19], [20].

We can now prove the main result:

THEOREM 5.3
Let F be a nonlinear operator defined on a convex subset D of a regular POTL-space E_1 with values in a POTL-space E_2. Let $\bar{x}_0$ and x_0 be two points of D such that

$$\bar{x}_0 \le x_0 \quad \text{and} \quad F(\bar{x}_0) \le 0 \le F(x_0). \tag{5.53}$$

Suppose that F has a divided difference of order one on

$$D_0 = \langle \bar{x}_0, x_0 \rangle = \{x \in E_1 \mid \bar{x}_0 \le x \le x_0\} \subseteq D \text{ satisfying} \tag{5.54}$$

$$P_0 = [x_0, x_0] \text{ has a continuous nonnegative left subinverse } Q_0 \tag{5.55}$$

$$[x, v] - [x, y] \le 0 \quad \text{if } v \le y. \tag{5.56}$$

Consider also the iterations

$$F(x_n) + P_n(y_n - x_n) = 0, \quad P_n = [x_n, x_n], \tag{5.57}$$

$$z_n + x_{n+1} - y_n = 0, \tag{5.58}$$

$$F(\bar{x}_n) + P_n(\bar{y}_n - \bar{x}_n) = 0, \tag{5.59}$$

$$\bar{z}_n + \bar{x}_{n+1} - \bar{y}_n = 0 \tag{5.60}$$

for some sequences $\{z_n\}$, $\{\bar{z}_n\}$, $(n \ge 0)$ selected so that

$$\bar{z}_n - P_n(x_n - \bar{x}_n) \ge 0, \tag{5.61}$$

$$z_n - P_n(x_n - \bar{x}_n) \le 0, \tag{5.62}$$

$$(I - B_n[x_n, \bar{x}_n])(x_n - \bar{x}_n) - B_n(z_n - \bar{z}_n) \ge 0 \tag{5.63}$$

$$z_n \ge 0 \tag{5.64}$$

and

$$\bar{z}_n \ge 0 \quad \text{for all } n \ge 0, \tag{5.65}$$

where B_n denote the continuous nonnegative left subinverses of A_n $(n \ge 0)$.

Then there exist two sequences $\{\bar{x}_n\}$, $\{x_n\}$ $(n \geq 0)$ *satisfying the approximations (5.57)–(5.60),*

$$\bar{x}_0 \leq \bar{y}_0 \leq \bar{x}_1 \leq \cdots \leq \bar{y}_n \leq \bar{x}_{n+1}$$

$$\leq x_{n+1} \leq y_n \leq \cdots \leq x_1 \leq y_0 \leq x_0, \tag{5.66}$$

$$\lim_{n\to\infty} \bar{x}_n = \bar{x}, \ \lim_{n\to\infty} x_n = x \quad \textit{and} \quad \bar{x}, x \in D_0 \ \textit{with} \ \bar{x} \leq x. \tag{5.67}$$

Moreover, if the operators $P_n = [x_n, x_n]$ *are inverse nonnegative, then any solution* u *of the equation* $F(x) = 0$ *in* $\langle \bar{x}_0, x_0 \rangle$ *belongs to* $\langle \bar{x}, x \rangle$.

PROOF Let us define the operator

$$G_1 : \langle 0, x_0 - \bar{x}_0 \rangle \to E_1, \ G_1(x) = x - Q_0(F(\bar{x}_0) + P_0(x)).$$

This operator is isotone and continuous. We can have in turn

$$G_1(0) = -Q_0F(\bar{x}_0) \leq 0, \quad \text{(by (5.53))},$$

$$G_1(x_0 - \bar{x}_0) = x_0 - \bar{x}_0 - Q_0F(x_0) + Q_0(F(x_0) - F(\bar{x}_0) - P_0(x_0 - \bar{x}_0))$$

$$\leq x_0 - \bar{x}_0 + Q_0([x_0, \bar{x}_0] - [x_0, x_0])(x_0 - \bar{x}_0)$$

$$\leq x_0 - \bar{x}_0 \quad \text{(by (5.56))}.$$

By Kantorovich's Theorem [8], [9], [14], [17], [19] the operator G_1 has a fixed point $u_1 \in \langle 0, x_0 - \bar{x}_0 \rangle : G_1(u_1) = u_1$. Set $\bar{y}_0 = \bar{x}_0 + u_1$, then we have the estimates

$$F(\bar{x}_0) + P_0(\bar{y}_0 - \bar{x}_0) = 0,$$

$$F(\bar{y}_0) = F(\bar{y}_0) - F(\bar{x}_0) - P_0(\bar{y}_0 - \bar{x}_0) \leq 0,$$

$$F(\bar{x}_1) \leq ([\bar{x}_1, \bar{x}_0] - P_0)(\bar{x}_1 - \bar{x}_0) \leq 0 \quad \text{(by (5.62))}$$

and $\bar{x}_0 \leq \bar{y}_0 \leq x_0$.

Let us now define the operator

$$G_2 : \langle 0, x_0 - \bar{y}_0 \rangle \to E_1, \ G_2(x) = x + Q_0(F(x_0) - P_0(x)).$$

This operator is isotone and continuous. We have in turn $G_2(0) = Q_0F(x_0) \geq 0$ (by (5.53)),

$$G_2(x_0 - \bar{y}_0) = x_0 - \bar{y}_0 + Q_0F(\bar{y}_0) + Q_0(F(x_0) - F(\bar{y}_0) - P_0(x_0 - \bar{y}_0))$$

$$\leq x_0 - \bar{y}_0 + Q_0([x_0, \bar{y}_0] - [x_0, x_0])(x_0 - \bar{y}_0)$$

$$\leq x_0 - \bar{y}_0 \quad \text{(by (5.56))}.$$

By Kantorovich's theorem there exists $u_2 \in \langle 0, x_0 - \bar{y}_0 \rangle$ such that $G_2(u_2) = u_2$. Set $y_0 = x_0 - u_2$. Then we have the estimates

$$F(x_0) + P_0(y_0 - x_0) = 0,$$

$$F(y_0) = F(y_0) - F(x_0) - P_0(y_0 - x_0) \geq 0,$$

$$F(x_1) \geq ([x_1, x_0] - P_0)(x_1 - x_0) \geq 0 \quad \text{(by (5.61))}$$

and

$$\bar{x}_0 \leq \bar{y}_0 \leq y_0 \leq x_0.$$

From the approximations (5.58), (5.60) and estimates (5.61)–(5.65) for $n = 0$ we obtain, respectively, that

$$\bar{x}_1 - \bar{y}_0 = -\bar{z}_0 \geq 0 \Rightarrow \bar{x}_1 \geq \bar{y}_0,$$

$$x_1 - y_0 = -z_0 \leq 0 \Rightarrow x_1 \leq y_0$$

and $\bar{x}_1 \leq x_1$. Hence, we get $\bar{x}_0 \leq \bar{y}_0 \leq \bar{x}_1 \leq x_1 \leq y_0 \leq x_0$.

By hypothesis (5.56) it follows that the operator P_n has a continuous nonnegative left subinverse Q_n for all $n \geq 0$. Proceeding by induction we can show that there exist two sequences $\{\bar{x}_n\}$, $\{x_n\}$ $(n \geq 0)$ satisfying relations (5.66) and (5.67) in a regular space E_1 and as such they converge to some $\bar{x}, x \in D_0$, respectively. That is, we have

$$\lim_{n \to \infty} \bar{x}_n = \bar{x} \leq x = \lim_{n \to \infty} x_n.$$

If $\bar{x}_0 \leq u \leq x_0$ and $F(u) = 0$, then we can obtain

$$P_0(y_0 - u) = P_0(x_0 - Q_0F(x_0)) - P_0u$$

$$= P_0(x_0 - u) - P_0Q_0(F(x_0) - F(u))$$

$$= P_0(I - Q_0[x_0, u])(x_0 - u) \geq 0,$$

since $Q_0[x_0, u] \leq Q_0P_0 \leq I$ by (5.56).

Similarly, we show $P_0(\bar{y}_0 - u) \leq 0$.

If the operator P_0 is inverse nonnegative, then it follows from the above that $\bar{y}_0 \leq u \leq y_0$. Proceeding by induction we deduce that $\bar{y}_n \leq u \leq y_n$, from which it follows that $\bar{y}_n \leq \bar{x}_n \leq \bar{y}_{n+1} \leq u \leq y_{n+1} \leq x_n \leq y_n$ for all $n \geq 0$.

That is, we have $\bar{x}_n \leq u \leq x_n$ for all $n \geq 0$. Hence, we get $\bar{x} \leq u \leq x$.

That completes the proof of the theorem. ∎

Let us now consider the following sets of conditions:

(C_1) Let $\bar{x}_0$ and x_0 be two points of D such that $\bar{x}_0 \leq x_0$ and $F(\bar{x}_0) \leq 0 \leq F(x_0)$;

(C_2) F has a divided difference of order one on $D_0 = \langle \bar{x}_0, x_0 \rangle \subseteq D$;

(C_3) $P_0 = [x_0, y_0]$ has a continuous nonnegative subinverse Q_0;

(C_4) $P_n = [x_n, x_n]$ are inverse nonnegative for all $n \geq 0$;

(C_5) $N_n = [x_n, y_n]$ are inverse nonnegative for all $n \geq 0$.

The following conditions are satisfied:

(C_6) $[x_0, y] \geq 0$ for all $\bar{x}_0 \leq y \leq x_0$;

(C_7) $[u, v] \leq [x, y]$ if $u \leq x$ and $v \leq y$;

(C_8) $[z, w] + [w, q] - [z, z] - [v, z] \geq 0$ if $v \leq w \leq z$ for some $q \in \langle v, z \rangle$;

(C_9) $[x, y] + [y, x] + 2[y, y] - 2[x, x] \geq 0$ if $y \leq x$;

(C_{10}) There exists a positive number c such that

$$[x, y] + [y, x] + 2[y, y] - (c + 2)[x, x] \leq 0,$$

and

$$\frac{c}{2}\left[[x', y] + [y, x] + 2[y, y]\right] + [z, x] \leq [p, q]$$

for all $v \leq y \leq p \leq q \leq x$.

Case 7. Under Case 1, we choose $A_n = P_n = [x_n, x_n]$, $B_n = [x_n, y_n] - [x_n, x_n]$, $z_n = -B_n(y_n - x_n)$ and $z_n = -B_n(y_n - x_n)$ for all $n \geq 0$. Let us assume that conditions (C_1)–(C_4), (C_6)–(C_8) are satisfied. Then the conditions (5.61)–(5.65) are satisfied with the above choices of the sequences $\{z_n\}$, $\{\bar{z}_n\}$ $(n \geq 0)$. Therefore the conclusions of Theorem 5.3 follow.

Case 8. Under Case 2 we choose $A_n = P_n = [x_n, x_n]$, $B_n = \bar{E}_n([x_n, y_n] - [x_n, x_n])$, $E_n = -Q_n([x_n, y_n] - [x_n, x_n])$, $z_n = -B_n(y_n - x_n)$, $\bar{z}_n = -B_n(y_n - x_n)$ and $\bar{E}_n$ denote the continuous nonnegative left subinverses of $I - E_n$ for all $n \geq 0$. Let us assume that conditions (C_1)–(C_8) are satisfied. Then the conditions (5.61)–(5.65) are satisfied. Therefore the conclusions of Theorem 5.3 follow.

Case 9. Under Case 3 we choose $A_n = P_n = [x_n, x_n]$,

$$B_n = \frac{3}{4}\left[\left(x_n + \frac{2}{3}(y_n - x_n), x_n + \frac{2}{3}(y_n - x_n)\right) - A_n\right]\left(I - \frac{3}{2}H_n\right),$$

$$H_n = Q_n\left[\left(x_n + \frac{2}{3}(y_n - x_n), x_n + \frac{2}{3}(y_n - x_n)\right) - A_n\right],$$

$$z_n = -B_n(y_n - x_n), \quad \bar{z}_n = -B_n(\bar{y}_n - \bar{x}_n)$$

for all $n \geq 0$. Let us assume that conditions (C_1)–(C_4), (C_6), (C_7), (C_9) and (C_{10}) are satisfied. Then the conditions (5.61)–(5.65) are satisfied. Therefore the conclusions of Theorem 3 follow.

The choices for the sequences $\{z_n\}$, $\{z_n\}$ $(n \geq 0)$ under Cases 4–6 are well documented in the literature and are omitted (see, e.g., [8], [9], [17], [20] and the references there).

REMARK 5.4 (a) Conditions for $x^* = y^*$ and $F(x^*) = F(y^*) = 0$ can be found in [8], [9], [17], [20].

(b) Conditions (C_7)–(C_8) correspond to conditions (5.49) and (5.50) in [17, p. 98]. Note that the results in that paper are proved only for the single step Euler-Chebysheff method of order 1.839..., where hypotheses divided differences of order two have been used.

(c) Similar results can immediately follow if the divided difference $[x_0, x_0]$ is replaced by $[x_0, q_0]$ for some $q_0 \in \langle \bar{x}_0, x_0 \rangle$, and $[x_n, x_n]$ is replaced by $[x_n, y_{n-1}]$ $(n \geq 1)$.

(d) In cases $E_1 = E_2 = R$, conditions (C_6)–(C_8) are satisfied if and only if F is differentiable on D_0, and F, F' are convex on D_0. ∎

5.1.5 Convergence Analysis for Newton-Like Methods

Let $x_0 \in D$ and define the two-step Newton-like methods for all $n \geq 0$ by

$$y_n = x_n - A(x_n)^{-1}F(x_n) \tag{5.68}$$

$$x_{n+1} = y_n - z_n. \tag{5.69}$$

Here $A(x_n)$ denotes a linear operator which is an approximation to the Fréchet derivative $F'(x_n)$ of F evaluated at $x = x_n$ for all $n \geq 0$.

Some choices for the operator A and the points z_n are given by $A(x_n) = F'(x_n)$, $z_n = 0$ or $A(x_n) = F'(x_n)$, $z_n = F'(x_n)^{-1}F(y_n)$ or $A(x_n) = F'(x_n)$, $z_n = F'(y_n)^{-1}F(y_n)$ for all $n \geq 0$.

Let $R > 0$, and assume that there exists a function $\bar{\alpha} : U^3(x_0, R) \to [0, +\infty)$ such that

$$\|A(x_0)^{-1}[F(y) - F(x) - A(x)(y - x) + F'(y)(z - y)]\| \leq \bar{\alpha}(x, y, z) \quad (5.70)$$

for all $x, y, z \in U(x_0, R) \subseteq D$. Note that for $A(x) = F'(x)$ for all $x \in U(x_0, R)$, condition (5.70) reduces to (5.4).

Using the majorant theory we will show that under certain choices of the points z_n $(n \geq 0)$ the iteration $\{x_n\}$ $(n \geq 0)$ generated by relations (5.68)–(5.69) converges to a locally unique solution x^*. The order of convergence of our iteration is also examined.

We will need to introduce the constants

$$t_0 = 0, \quad s_0 \geq \|y_0 - x_0\|, \quad I_0 \geq \|z_0\|, \quad (5.71)$$

the iterations for all $n \geq 0$

$$s_{n+1} = t_{n+1} + (1 - C_0(0, t_{n+1}))^{-1} h_{n+1}, \quad (5.72)$$

$$t_{n+1} = s_n + I_n, \quad (5.73)$$

$$h_{n+1} = \int_{s_n}^{t_{n+1}} C(s_n t)dt + C(s_n, s_n)(t_{n+1} - s_n) + \alpha_n \quad (5.74)$$

for some given real functions C_0, C_1, C_2, $C = C_1 + C_2$ of two variables and some given sequences $\{\alpha_n\}$ and $\{I_n\}$ $(n \geq 0)$ with

$$\alpha_n \geq 0, \ I_n \geq 0, \ \sum_{i=0}^{k+1} \alpha_i \leq \gamma_0 r_0 + \gamma_1, \text{ and } \sum_{i=0}^{k+1} I_i \leq \gamma_2 r_0 \quad (5.75)$$

for all positive integers k and some fixed real constants (which may depend on r_0) γ_0, γ_1, γ_2 and some fixed $r_0 \in [0, R]$.

Moreover, we define the iterations for all $n \geq 0$

$$\bar{h}_{n+1} = \int_0^1 C(\|y_n - x_0\|, \|y_n - x_0\| + t\|x_{n+1} - y_n\|)\|x_{n+1} - y_n\|dt$$

$$+ C(\|y_n - x_0\|, \|y_n - x_0\|)\|x_{n+1} - y_n\| + \bar{\alpha}_n, \tag{5.76}$$

where $\bar{\alpha}_n = \bar{\alpha}(x_n, y_n, x_{n+1})$ with

$$\bar{\alpha}_n \leq \alpha_n \quad \text{for all } n \geq 0, \tag{5.77}$$

$$\bar{e}_{n+1} = \left[1 - \int_0^1 C(0, (1-t)\|x_{n+1} - x_0\| + t\|x^* - x_0\|)dt\right]^{-1}, \tag{5.78}$$

$$e_{n+1} = \left[1 - \int_0^1 C(0, (1-t)t_{n+1} + tr_0)dt\right]^{-1}, \tag{5.79}$$

$$p_n = \int_0^1 C(\|x_n - x_0\|, \|x_n - x_0\| + t\|x^* - x_n\|)\|x_n - x^*\|dt, \tag{5.80}$$

and the function

$$T(r) = s_0 + \frac{1}{1 - C_0(0,r)}\left[\int_{s_0}^{r} C(r,t)dt + C(r,r)(r - s_0) + \gamma_0 r + \gamma_1\right] + \gamma_2 r \tag{5.81}$$

on $[0, R]$. The functions C_0, C_1, C_2 are assumed to vanish at the origin, be continuous two variables, and such that if one variable is fixed, then they are nondecreasing functions of the other on $[0, R]$.

We can now state the main result:

THEOREM 5.4

Let $F : D \subseteq E_1 \to E_2$ be a nonlinear Fréchet-differentiable oeprator whose divided difference $[x, y]$ satisfies

$$\|A(x_0)^{-1}([x+h_1, x+h_2] - A(x))\| \leq C_1(r, r+\|h_1\|) + C_2(r, r+\|h_2\|), \tag{5.82}$$

and

$$\|A(x_0)^{-1}(A(x) - A(x_0))\| \leq C_0(r, r + \|h\|) \tag{5.83}$$

for all $x \in U(x_0, r) \subseteq U(x_0, R)$ and $0 \leq \|h\|$, $\|h_1\|$, $\|h_2\| \leq R - r$.

Moreover we assume:

(i) condition (5.70) is satisfied;

(ii) iterations $\{\alpha_n\}$, $\{I_n\}$ and $\{\bar{\alpha}_n\}$ ($n \geq 0$ satisfy conditions (5.75)-(5.77), and

$$\|z_n\| \leq I_n \quad \text{for all } n \geq 0;$$

(iii) there exists a minimum nonnegative number r_0 such that

$$T(r_0) \leq r_0; \tag{5.84}$$

(iv) the following estimates are also true for r_0, R:

$$r_0 \leq R, \int_0^1 C(0,(1-t)R + tr_0)dt < 1, \; C_0(0, r_0) < 1, \; U(x_0, R) \subseteq D. \tag{5.85}$$

Then

(a) scalar sequence $\{t_n\}$ $(n \geq 0)$ generated by relations (5.72)–(5.74) is monotonically increasing and bounded above by its limit, which is number r_0;

(b) sequence $\{x_n\}$ $(n \geq 0)$ generated by relations (5.68)–(5.69) is well defined, remains in $U(x_0, r_0)$ for all $n \geq 0$, and converges to a solution x^* of the equation $F(x) = 0$, which is unique in $U(x_0, R)$.

Moreover, the following estimates are true for all $n \geq 0$

$$\|y_n - x_n\| \leq s_n - t_n, \tag{5.86}$$

$$\|x_{n+1} - y_n\| \leq t_{n+1} - s_n, \tag{5.87}$$

$$\|x_n - x^*\| \leq r_0 - t_n, \tag{5.88}$$

$$\|y_n - x^*\| \leq r_0 - s_n, \tag{5.89}$$

$$\|A(x_0)^{-1}F(x_{n+1})\| \leq \bar{h}_{n+1}, \tag{5.90}$$

$$\|x^* - x_{n+1}\| \leq \bar{e}_{n+1}\bar{h}_{n+1} \leq e_{n+1}h_{n+1} \leq r_0 - t_{n+1} \tag{5.91}$$

and

$$\|y_n - x_n\| \leq \|x^* - x_n\| + \tag{5.92}$$

$$\frac{\int_0^1 C(\|x_n - x_0\|, \|x_n - x_0\| + t\|x^* - x_n\|)}{1 - C_0(0, \|x_-x_0\|}\|x_n - x^*\|dt.$$

(We will be concerned only with the case $r_0 > 0$, since when $r_0 = 0$, $x_0 = x^*$.)

PROOF (a) Using relations (5.71), (5.72), (5.73), (5.75), and (5.84), we deduce that the scalar sequence $\{t_n\}$ $(n \geq 0)$ is monotonically increasing, nonnegative, and $t_0 \leq s_0 \leq t_1 \leq s_1 \leq r_0$. Let us assume that $t_k \leq s_k \leq t_{k+1} \leq s_{k+1} \leq r_0$ for $k = 0, 1, 2, \ldots, n$. Then by relations (5.72), (5.73) and (5.75) we can have in turn

$$
\begin{aligned}
t_{k+2} &= s_{k+1} + I_k \leq t_{k+1} \\
&\quad + \frac{1}{1 - C_0(0, r_0)} \left[\int_{s_k}^{t_{k+1}} C(r_0, t)dt + C(r_0, r_0)(t_{k+1} - s_k) \right. \\
&\quad \left. + \alpha_k \right] + I_{k+1} \leq \cdots \leq s_0 + \frac{1}{1 - C_0(0, r_0)} \left[\int_{s_0}^{t_{k+1}} C(r_0, t)dt \right. \\
&\quad \left. + C(r_0, r_0)(t_{k+1} - s_0) + \sum_{i=0}^{k} \alpha_i \right] + \sum_{i=0}^{k+1} I_i \qquad (5.93) \\
&\leq s_0 + \frac{1}{1 - C_0(0, r_0)} \left[\int_{s_0}^{r_0} C(r_0, t)dt + C(r_0, r_0)(r_0 - s_0) \right. \\
&\quad \left. + \gamma_0 r_0 + \gamma_1 \right] + \gamma_2 r_0 = T(r_0) \leq r_0. \qquad (5.94)
\end{aligned}
$$

Hence, the scalar sequence $\{t_n\}$ $(n \geq 0)$ is bounded above by r_0. By hypothesis (5.84) the number r_0 is the minimum nonnegative zero of the equation $T(r) - r = 0$ on $[0, r_0]$, and from the above $r_0 = \lim_{n \to \infty} t_n$.

(b) Using relations (5.68), (5.69), (5.71), (5.72) and (5.73) we deduce that $x_1, y_0 \in U(x_0, r_0)$ and that estimates (5.86) and (5.87) are true for $n = 0$. Let us assume that they are true for $k = 0, 1, 2, \ldots, n-1$. Using the induction hypothesis, we can have in turn

$$
\begin{aligned}
\|x_{k+1} - x_0\| &\leq \|x_{k+1} - y_0\| + \|y_0 - x_0\| \\
&\leq \|x_{k+1} - y_k\| + \|y_k - y_0\| + \|y_0 - x_0\| \\
&\leq \cdots \leq (t_{k+1} - s_k) + (s_k - s_0) + s_0 \leq t_{k+1} \leq r_0,
\end{aligned}
$$

and

$$
\|y_{k+1} - x_0\| \leq \|y_{k+1} - y_0\| + \|y_0 - x_0\|
$$

$$\leq \|y_{k+1} - x_{k+1}\| + \|x_{k+1} - y_k\| + \|y_k - y_0\| + \|y_0 - x_0\|$$

$$\leq \cdots \leq (s_{k+1} - t_{k+1}) + (t_{k+1} - s_k) + (s_k - s_0) + s_0$$

$$\leq s_{k+1} \leq r_0.$$

Tht is $x_n, y_n \in U(x_0, r_0)$ for all $n \geq 0$.

Using hypothesis (5.83) we now have

$$\begin{aligned}\|A(x_0)^{-1}(A(x_k) - A(x_0))\| &\leq C_0(0, 0 + \|x_k - x_0\|) \\ &= C_0(0, \|x_k - x_0\|) \\ &\leq C_0(0, t_k - t_0) \leq C_0(0, r_0) \leq 1, \quad (5.95)\end{aligned}$$

by condition (5.85). It now follows from the Banach lemma on invertible operators [13] that the linear operator $A(x_k)$ is invertible, and

$$\|A(x_k)^{-1}A(x_0)\| \leq (1 - C_0(0, \|x_k - x_0\|))^{-1} \leq (1 - C_0(0, t_k))^{-1} \quad (5.96)$$

for all $k \geq 0$.

We can now have from relations (5.68) and (5.69)

$$\begin{aligned}F(x_{k+1}) &= F(x_{k+1}) - F(y_k) - A(y_k)(x_{k+1} - y_k) \\ &\quad + A(y_k)(x_{k+1} - y_k) + F(y_k) \\ &= \int_0^1 F'(y_k + t(x_{k+1} - y_k)) - A(y_k)](x_{k+1} - y_k)dt \\ &\quad + (A(y_k) - F'(y_k))(x_{k+1} - y_k) \\ &\quad + (F(y_k) - F(x_k) - A(x_k)(y_k - x_k) + F'(y_k)(x_{k+1} - y_k)).\end{aligned}$$

Hence, by using hypotheses (5.70) and (5.82) we obtain in turn

$$\begin{aligned}&\|A(x_0)^{-1}F(x_{k+1})\| \\ &\leq \int_0^1 \|A(x_0)^{-1}[F'(y_k + t(x_{k+1} - y_k)) - A(y_k)]\| \, \|x_{k+1} - y_k\|dt\end{aligned}$$

$$+ \|A(x_0)^{-1}(A(y_k) - F'(y_k))\| \, \|x_{k+1} - y_k\|$$

$$+ \|A(x_0)^{-1}[F(y_k) - F(x_k) - A(x_k)(y_k - x_k) + F'(y_k)(x_{k+1} - y_k)]\|$$

$$= \bar{h}_{k+1} \leq h_{k+1},$$

by hypotheses (5.77), relations (5.74) and (5.76). Hence, we showed estimate (5.93) for all $n \geq 0$.

Using relations (5.68), (5.72), (5.90) and (5.96), we obtain

$$\|y_{k+1} - x_{k+1}\| \leq \|A(x_{k+1})^{-1}A(x_0)\| \cdot \|A(x_0)^{-1}F(x_{k+1})\|$$

$$\leq (1 - C_0, \|x_{k+1} - x_0\|))^{-1}\bar{h}_{k+1}$$

$$\leq (1 - C_0(0, t_{k+1}))^{-1}h_{k+1} = s_{k+1} - t_{k+1},$$

which shows estimate (5.86) for all $n \geq 0$.

Similarly, from relations (5.69), (5.73) and hypotheses (ii) we obtain

$$\|x_{k+1} - y_k\| = \| - z_k\| \leq I_k = t_{k+1} - s_k,$$

from which it follows that estimate (5.87) is true for all $n \geq 0$.

It now follows from the estimates (5.86) and (5.87) that the sequence $\{x_n\}$ $(n \geq 0)$ is Cauchy in a Banach space E_1 and as such it converges to some $x^* \in U(x_0, r_0)$, which by taking the limit as $n \to \infty$ in (5.68), we obtain $F(x^*) = 0$.

To show uniqueness, we assume that there exists another solution y^* of equation (5.1) in $U(x_0, R)$. From hypothesis (5.82) we get

$$\int_0^1 \|A(x_0)^{-1}[F'(x^* + t(y^* - x^*)) - A(x_0)]\|dt$$

$$\leq \int_0^1 C(0, \|x^* + t(y^* - x^*) - x_0\|)dt$$

$$\leq \int_0^1 C(0, (1-t)\|y^* - x_0\| + t\|x^* - x_0\|)dt$$

$$\leq \int_0^1 C(0, (1-t)R + tr_0)dt < 1 \quad \text{by hypothesis (5.85).}$$

It now follows that the linear operator $\int_0^1 F'(x^*+t(y^*-x^*))dt$ is invertible, and from the approximation

$$F(y^*)-F(x^*)=\int_0^1 F'(x^*+t(y^*-x^*))dt(y^*-x^*)$$

it follows that $x^*=y^*$.

Estimates (5.88) and (5.89) follow from (5.86) and (5.87), respectively, by using the standard arguments in majorant theory.

Finally, using the approximations for all $n\geq 0$

$$x_{n+1}-x^*=(D_{n+1}^{-1}A(x_0))(A(x_0)^{-1}F(x_{n+1})),$$

$$D_{n+1}=\int_0^1 F'(x^*+t(x_{n+1}-x^*))dt,$$

$$y_n-y_n=x^*-x_n+(A(x_n)^{-1}A(x_0))\int_0^1 A(x_0)^{-1}[F'(x_n+t(x^*-x_n))$$
$$-A(x_n)](x^*-x_n)dt,$$

and the estimate

$$\int_0^1 \|A(x_0)^{-1}[F'(x^*+t(x_{n+1}-x^*))-A(x_0)]\|dt$$
$$\leq \int_0^1 C(0,(1-t)\|x_{n+1}-x_0\|+t\|x^*-x_0\|)dt$$
$$\leq \int_0^1 C(0,(1-t)r_0+tr_0)dt=C(0,r_0)<1$$

(by (5.85)), we deduce that

$$\|D_{n+1}^{-1}A(x_0)\|\leq \bar{e}_{n+1}.$$

Hence, we can immediately obtain estimates (5.91) and (5.92) for all $n\geq 0$, where we have also used relations (5.78), (5.79) and (5.80).

That completes the proof of the theorem. ∎

REMARK 5.5 Let us examine some special choices for the points z_n $(n \geq 0)$ as in (5.41). Condition (5.70) can now be replaced by one of the following three sets of conditions for all $x_n, x_{n+1} \in U(y_n, r(x_n, x_{n+1}))$

$$\left\| A(x_0)^{-1} \left[\int_0^1 (F'(x_n + t(y_n - x_n)) - A(x_n) + B_n))(y_n - x_n) dt \right.\right.$$

$$\left.\left. + (F'(y_n) - L_n)(x_{n+1} - y_n) \right] \right\|$$

$$\leq \bar{a}_n \|x_n - y_n\| + \bar{b}_n \|x_{n+1} - y_n\|,$$

or by

$$\left\| A(x_0)^{-1} \left[\int_0^1 (F'(x_n + t(y_n - x_n)) - (A(x_n) + B_n)) \right] (y_n - x_n) dt \right\|$$

$$\leq \bar{a}_n \|y_n - x_n\|,$$

$$\|A(x_0)^{-1}(F'(y_n) - L_n)(x_{n+1} - y_n)\| \leq \bar{b}_n \|x_{n+1} - y_n\|$$

or by

$$\left\| A(x_0)^{-1} \left[\int_0^1 (F'(x_n + t(y_n - x_n)) - (A(x_n) + B_n) \right] dt \right\| \leq \bar{a}_n,$$

$$\|A(x_0)^{-1}(F'(y_n) - L_n)\| \leq \bar{b}_n \quad \text{for all } n \geq 0.$$

We can easily see that the conditions above reduce to the corresponding ones in Remark 5.4 if we set $A(x) = F'(x)$ $(x \in D)$. Hence similar comments can follow. ∎

We will need the lemma.

LEMMA 5.1
Let $f : U^2(x_0, R) \to E_2$ be a nonlinear operator satisfying:

$$\|f(x, y) - f(z, z)\| \leq k_1(r)\|x - z\| + k_2(r)\|y - z\| \tag{5.97}$$

for all $x, y, z \in U(x_0, r)$, $r \in [0, R]$, and for some nondecreasing functions k_1 and k_2 on the interval $[0, R]$ with $k_1(0) = k_2(0) = 0$.

Then we have

$$\|f(x+h_1,y+h_2)-f(x,y)\| \leq v_1(w_1+\|h_1\|)-v_1(w_1)$$

$$+ v_2(w_2+\|h_2\|)-v_2(w_2) \quad (5.98)$$

for all $x \in U(x_0,w_1)$, $y \in U(x_0,w_2)$, $\|h_1\| \leq R-w_1$, $\|h_2\| \leq R-w_2$ *with* $v_1(r)=\int_0^1 k_1(t)dt$ *and* $v_2(r)=\int_0^r k_2(t)dt$.

PROOF Let $x \in U(x_0,w_1)$, $y \in U(x_0,w_2)$, $\|h_1\| \leq R-w_1$ and $\|h_2\| \leq R-w_2$. Using relation (5.97) for $m \in N$, we obtain in turn

$$\|f(x+h_1,y+h_2)-f(x,y)\| \leq \sum_{j=1}^{m} \|f(x+m^{-1}jh_1,y+m^{-1}jh_2)$$

$$+ f(x+m^{-1}(j-1)h_1,y+m^{-1}(j-1)h_2)\|$$

$$\leq \sum_{j=1}^{m} k_1(w_1+m^{-1}j\|h_1\|)m^{-1}\|h_1\| + \sum_{j=1}^{m} k_2(w_2+m^{-1}j\|h_2\|)m^{-1}\|h_2\|$$

$$\leq v_1(w_1+\|h_1\|)-v_1(w_1)+v_2(w_2+\|h_2\|)-v_2(w_2) \text{ as } m \to \infty,$$

by the monotonicity of the functions k_1, k_2 and the definition of the Riemann integral. ∎

REMARK 5.6 We will now show how to choose the functions C_1, C_2 and C_0 appearing in conditions (5.82) and (5.83). The most popular choice (but not the only one) for the nonlinear operator A is given by $A(x)=F'(x)$. Then conditions (5.82) and (5.83) become

$$\|F'(x_0)^{-1}([x+h_1,x+h_2]-F'(x))\| \leq C_1(r,r+\|h_1\|)+C_2(r,r+\|h_2\|)$$

and

$$\|F'(x_0)^{-1}(F'(x)-F'(x_0))\| \leq C_0(r,r+\|h\|).$$

Therefore if we assume that

$$\|F'(x_0)^{-1}([x,y]-[z,z])\| \leq k_1(r)\|x-z\|+k_2(r)\|y-z\|, \quad (5.99)$$

then by the lemma, we can choose

$$C_1(r, r + \|h_1\|) = \int_r^{r+\|h_1\|} k_1(t)dt, \quad C_2(r, r + \|h_2\|) = \int_r^{r+\|h_2\|} k_2(t)dt,$$

and

$$C_0(r, r + \|h\|) = C_0(\|h\|) = C_1(0, \|h\|) + C_2(0, \|h\|).$$

Note that the function C_0 in (5.83) can be simply denoted by $C_0(\|h\|)$.

For simplicity, let us assume that condition (5.99) is true with $k_1(r) = k_2(r) = q_0$, for some fixed nonnegative constant q_0 and all $r \in [0, R]$. Choose also the points z_n $(n \geq 0)$ as in (5.41) with $L_n = F'(x_n)$ and $B_n = [x_n, y_n] - [x_n, x_n]$ for all $n \geq 0$. The iteration $\{x_n\}$ $(n \geq 0)$ generated by (5.68)–(5.69) with the above choices has already been examined in [5], [6], [9]. It is an Euler-Chebysheff-type method, whose order of convergence is almost three. This method has been compared with the method of tangent hyperbolas (or Chebysheff-Halley method). We refer the reader to [9], [12], [15], [16], [17], [18], [20], [21].

We can now set

$$\bar{b}_n = q\|y_n - x_n\| \leq q(s_n - t_n) = b_n \leq qr_0 = b, \; q = 2q_0,$$

$$\bar{a}_n = q_0\|y_n - x_n\| \leq q_0(s_n - t_n) = a_n \leq q_0 r_0 = a,$$

and

$$\bar{c}_n = \frac{q_0\|y_n - x_n\|}{1 - q\|x_n - x_0\|} \leq \frac{q_0(s_n - t_n)}{1 - q(t_n - t_0)} = c_n \leq \frac{q_0 r_0}{1 - qr_0} = c \text{ for all } n \geq 0.$$

With the above choices, under the hypotheses of the theorem the conclusions will follow for the Euler-Chebysheff-type method given by

$$y_n = x_n - F(x_n)^{-1}F(x_n)$$

$$x_{n+1} = y_n - F'(x_n)^{-1}([x_n, y_n] - [x_n, x_n])(y_n - x_n)$$

for all $n \geq 0$, in this case. ∎

REMARK 5.7 We note that in the proof of the theorem we really used the weaker condition

$$\|A(x_0)^{-1}([x + h, x + h] - A(x))\| \leq C_1(r, r + \|h\|) + C_2(r, r + \|h\|)$$

$$= C(r, r + \|h\|) \tag{5.100}$$

for all $x \in U(x_0, r) \subseteq U(x_0, R)$ and $0 \le \|h\| \le R - r$ instead of (5.82). But as it turns out condition (5.82) wil be used for the explicit computation of the function $\bar{\alpha}$ (see also [3]–[9]). ∎

5.1.6 Applications for Two-Step Newton-Like Methods

Let us now consider the method of tangent parabolas (5.43)–(5.44) studied also in Subsection 5.1.3, Example 5.1.

Let us now assume instead of hypotheses (5.82) and (5.83) that hypothesis

$$\|F'(x_0)^{-1}([x, x] - [u, u]\| \le q(r)\|x - u\| \tag{5.101}$$

for all $x, y \in U(x_0, r) \subseteq U(x_0, R)$ with $q(r) = L$ for all $r \in [0, R]$ and some nonnegative number L and $C(r, r + \|h\|) = \int_r^{r+\|h\|} q(t)dt$.

Furthermore, let us assume that

$$\|F'(x_0)^{-1}F''(x)\| \le M \tag{5.102}$$

and

$$\|F'(x_0)^{-1}(F''(x) - F''(y))\| \le N\|x - y\| \tag{5.103}$$

for all $x, y \in U(x_0, R)$.

Then using relations (5.41), as in Remark 5.2 we can obtain in turn

$$\bar{a}_n = \frac{N}{6}\|y_n - x_n\|^2, \quad \bar{b}_n = M\|y_n - x_n\|,$$

$$\bar{c}_n = \frac{M}{2(1 - L\|x_n - x_0\|)}\|y_n - x_n\| \quad (n \ge 0).$$

Hence, we have

$$\bar{\alpha}_n = \frac{N}{6}\|y_n - x_n\|^3 + M\|y_n - x_n\| \, \|x_{n+1} - y_n\|$$

$$\le \frac{N}{6}(s_n - t_n)^3 + M(s_n - t_n)(t_{n+1} - s_n) = \alpha_n,$$

$$\sum_{i=0}^{k} \alpha_i = \frac{N}{6}\sum_{i=0}^{k}(s_i - t_i)^3 + M\sum_{i=0}^{k}(s_i - t_i)(t_{i+1} - s_i)$$

$$\leq \frac{N}{6}r_0^3 + Mr_0(r_0 - s_0) = \gamma_0 r_0 + \gamma_1,$$

where

$$\gamma_0 = \left(\frac{N}{6}r_0 + M\right) r_0, \quad \gamma = -Ms_0,$$

$$\bar{c}_n \leq \frac{M}{2(1-Lt_n)}(s_n - t_n) = c_n \leq \frac{Mr_0}{2(1-Lr_0)} = c = \gamma_2.$$

Hence, using (5.81) with the above values, we obtain that

$$T(r) = s_0 + \frac{1}{6(1-Lr)}\left[Nr^3 + 3L(r-s_0)^2 + 6Mr^2 + 3Mr^2 - 6Ms_0 r\right].$$

Therefore the hypotheses of Theorem 5.4 are satisfied provided that there exists a minimum nonnegative number r_0 such that

$$Nr_0^3 + 9(M+L)r_0^2 - 6[1 + (2L+M)s_0]r_0 + 3s_0(2 + Ls_0) \leq 0$$

and $1 - Lr_0 > 0$, with $r_0 \leq R$.

Another approach is to define the function

$$g(t) = \frac{K}{2}t^2 - t + s_0$$

for some number K satisfying

$$\left(M^2 + \frac{N}{3}\right)^{1/2} \leq K,$$

numbers

$$r_1 = \frac{1-\sqrt{1-2h}}{h}s_0, \quad r_2 = \frac{1+\sqrt{1-2h}}{h}s_0,$$

$$h = Ks_0,$$

$$\theta = \frac{1-\sqrt{1-2h}}{1+\sqrt{1-2h}},$$

and iteration

$$s_n = t_n - \frac{g(t_n)}{g'(t_n)}, \quad t_0 = 0$$

$$t_{n+1} = s_n - \frac{1}{2}(s_n - t_n)^2 \frac{g''(t_n)}{g'(t_n)} \quad \text{for all } n \geq 0.$$

Then as in the proof of Theorem 5.4 we can obtain (see also [9])

$$\begin{aligned}\|F'(x_0)^{-1}F(x_{n+1})\| &\leq \frac{M}{2}\|x_{n+1} - y_n\|^2 + \frac{M}{2}\frac{\|y_n - x_n\|^3}{1 - M\|x_n - x_0\|} \\ &\quad + \frac{N}{6}\|y_n - x_n\|^3 \\ &\leq \frac{M}{2}\|x_{n+1} - y_n\|^2 + \left(\frac{M^2}{2} + \frac{N}{6}\right)\frac{\|y_n - x_n\|^3}{1 - M\|x_n - x_0\|} \\ &\leq \frac{K}{2}(t_{n+1} - s_n)^2 + \frac{K^2}{2}\frac{(s_n - t_n)^3}{1 - Kt_n} = g(t_{n+1}).\end{aligned}$$

Hence, provided that

$$h \leq .485281374\ldots, \tag{5.104}$$

we will have

$$\|x_n - y_n\| \leq s_n - t_n,$$

$$\|x_{n+1} - y_n\| \leq t_{n+1} - s_n,$$

$$\|x_n - x^*\| \leq r_1 - t_n,$$

$$\|y_n - x^*\| \leq r_1 - s_n,$$

$$a_n = r_1 - t_n, \; b_n = r_2 - t_n, \; g(t_n) = \frac{K}{2}a_n b_n, \; g(r_1) = g(r_2) = 0,$$

$$\delta_n = \delta_{n-1}^3 \frac{\delta_{n-1} + 2}{1 + 2\delta_{n-1}}, \quad \delta_n = \frac{a_n}{b_n},$$

and

$$a_n = \frac{(1-\theta^2)s_0}{1-\frac{1}{\sqrt{2}}(\sqrt{2}\theta)^{3n}}(\sqrt{2}\theta)^{3^n-1} \quad \text{for all } n \geq 0 \tag{5.105}$$

(see also Case 1).

REMARK 5.8 We note that hypotheses (5.101), (5.102), (5.103), and (5.104) constitute the usual Newton-Kantorovich-type hypotheses [9], [16]. By (5.105) we deduce that the order of convergence of the iteration $\{x_n\}$ $(n \geq 0)$ generated by (5.43)–(5.44) is three. ∎

Example 5.3

With the notation of Example 5.1 in Subsection 5.1.3 for the iteration (5.68)–(5.69), we obtain

$$\beta = \|P'(1)^{-1}\| = 1.53039421,$$

$$N = 0,\ L = K = M = 2|\lambda| \max_{0\leq s\leq 1}\left|\int_0^1 \frac{s}{s+t}dt\right| \beta = 2|\lambda| \ln 2\beta = .530394215$$

$$s_0 \geq \|P'(1)^{-1}P(1)\| \geq .485281374\ldots$$

$$h = .140659011 < .485281374\ldots$$

$$r_1 = .28704852, \quad r_2 = 3.4837317,$$

$$\theta = .08239685.$$

(For detailed computations, see also [3], [9] and [10].) Moreover relations (5.84), (5.85) become, respectively,

$$9.54709587r^2 - 8.531862205r + 1.703089731 \leq 0$$

and

$$1 - .530394215R > 0.$$

Hence, we deduce that

$$r_0 = .30099125 \quad \text{and} \quad R = 1.885390096.$$

Furthermore, we note that the hypotheses of the theorem for the method of tangent parabolas hold in the ball $U(x_0, r_3)$ with the uniqueness of the solution x^* of equation (5.45) extended in the ball $U(x_0, R_0)$, where r_3 denotes r_0 or r_1. Furthermore we note that the results in [12], [15], [17], [18], [20], [21] cannot apply here since $N = 0$. □

REMARK 5.9 The monotone convergence of iteration (5.68)–(5.69) can be examined similarly as iteration (5.2)–(5.3) in Subsection 5.1.4. The details are left to the motivated reader (see also [9], Chapter 6). ∎

5.1.7 A Result on a Generalized Iteration

Let $x_0, w_0, z_0 \in D$ be fixed and define the two-step method for all $n \geq 0$ by

$$y_n = x_n - w_n \tag{5.106}$$

$$x_{n+1} = y_n - z_n. \tag{5.107}$$

We will choose the sequences $\{w_n\}$ and $\{z_n\}$ $(n \geq 0)$ in such a way that the iteration $\{x_n\}$ $(n \geq 0)$ generated by relations (5.106)–(5.107) converges to a solution x^* of equation (5.1). The convergence of the iteration $\{x_n\}$ $(n \geq 0)$ has already been examined in some special cases. For example, by setting $z_n = 0$ and $w_n = F'(x_n)^{-1}F(x_n)$ for all $n \geq 0$, we obtain the well-known Newton-Kantorovich method for $z_n = 0$ and $w_n = A(x_n)^{-1}F(x_n)$ $(n \geq 0)$, we obtain Newton-like methods. For $w_n = F'(x_n)^{-1}F(x_n)$ and $z_n = F'(y_n)^{-1}F(y_n)$ $(n \geq 0)$, we obtain a two-step Newton method which is of order four [9].

We will need to introduce the constants

$$t_0 = 0, \quad m_0 \geq \|y_0 - x_0\|, \quad I_0 \geq \|z_0\|, \tag{5.108}$$

the scalar iterations for all $n \geq 0$

$$s_n = t_n + m_n \tag{5.109}$$

$$t_{n+1} = s_n + I_n \tag{5.110}$$

and the scalar function

$$T_1(r) = s_0 + T_2(r), \quad \text{on } [0, R] \text{ for some } R \tag{5.111}$$

where the function T_2 is such that

$$\sum_{i=0}^{k+1} I_k + \sum_{i=1}^{k+1} m_k \leq T_2(r_0) \tag{5.112}$$

for all $k \geq 0$ and some fixed $r_0 \in [0, R]$.

We can now state and prove the main result of this section.

PROPOSITION 5.2

Let $F : D \subseteq E_1 \to E_2$ be a nonlinear operator. Assume:

(i) sequences $\{w_n\}$, $\{z_n\}$ $(n \geq 0)$ satisfy

$$\|w_n\| \leq m_n, \quad \|z_n\| \leq I_n \quad \textit{for all } n \geq 0; \tag{5.113}$$

(ii) the sequences $\{m_n\}$ and $\{I_n\}$ $(n \geq 0)$ satisfy condition (5.112);

(iii) there exists a minimum number $r_0 \in [0, R]$ for some fixed R such that

$$T_1(r_0) \leq r_0 \quad \textit{and the ball} \quad U(x_0, R) \subseteq D. \tag{5.114}$$

Then

(a) scalar sequence $\{t_n\}$ $(n \geq 0)$, generated by relations (5.109)-(5.110) is monotonically increasing and bounded above by its limit, which is number r_0;

(b) sequence $\{x_n\}$ $(n \geq 0)$, generated by relations (5.106)-(5.107) remains in $U(x_0, r_0)$ for all $n \geq 0$ and converges to some $x^* \in U(x_0, r_0)$.

Moreover, the following estimates are true for all $n \geq 0$;

$$\|y_n - x_n\| \leq s_n - t_n \tag{5.115}$$

$$\|x_{n+1} - y_n\| \leq t_{n+1} - s_n \tag{5.116}$$

$$\|x_n - x^*\| \leq r_0 - t_n \tag{5.117}$$

and

$$\|y_n - x^*\| \leq r_0 - s_n. \tag{5.118}$$

PROOF (a) Using relations (5.108), (5.109), (5.110) and (5.113), we deduce that the scalar sequence $\{t_n\}$ $(n \geq 0)$ generated by relations (5.109)–(5.110) is monotonically increasing, nonnegative, and $t_0 \leq s_0 \leq t_1 \leq s_1 \leq r_0$. Let us assume that $t_k \leq s_k \leq t_{k+1} \leq s_{k+1} \leq r_0$ for $k = 1, 2, 3 \ldots, n$.

Then by relations (5.109), (5.110), (5.111), (5.112) and (5.114) we can have in turn

$$t_{k+2} = s_{k+1} + I_{k+1} = t_{k+1} + m_{k+1} + I_{k+1} = t_k + (m_k + m_{k+1}) + I_k + I_{k+1}).$$

Hence, the scalar sequence $\{t_n\}$ $(n \geq 0)$ is bounded above by r_0. By hypotheses (5.114), the number r_0 is the minimum nonnegative zero of the equation $T_1(r) - r = 0$ on $[0, r_0]$ and from the above $r_0 = \lim_{n\to\infty} t_n$.

(b) Using relations (5.106)–(5.110), (5.113) and (5.114) we deduce that $x_1, y_0 \in U(x_0, r_0)$ and that estimates (5.115) and (5.116) are true for $n = 0$. Let us assume that they are true for $k = 0, 1, 2, \ldots, n-1$. Using the induction hypotheses we can have in turn

$$\begin{aligned}\|x_{k+1} - x_0\| &\leq \|x_{k+1} - y_0\| + \|y_0 - x_0\| \\ &\leq \|x_{k+1} - y_k\| + \|y_k - y_0\| + \|y_0 - x_0\| \\ &\leq \cdots \leq (t_{k+1} - s_k) + (s_k - s_0) + s_0 \leq t_{k+1} \leq r_0,\end{aligned}$$

and

$$\begin{aligned}\|y_{k+1} - x_0\| &\leq \|y_{k+1} - y_0\| + \|y_0 - x_0\| \\ &\leq \|y_{k+1} - x_{k+1}\| + \|x_{k+1} - y_k\| + \|y_k - y_0\| + \|y_0 - x_0\| \\ &\leq \cdots \leq (s_{k+1} - t_{k+1}) + (t_{k+1} - s_k) + (s_k - s_0) + s_0 \\ &\leq s_{k+1} \leq r_0.\end{aligned}$$

That is, $x_n, y_n \in U(x_0, r_0)$ for all $n \geq 0$.

Moreover, from the induction hypotheses and relations (5.113) we obtain in turn

$$\|y_k - x_k\| = \| - w_k\| = \|w_k\| \leq m_k = s_k - t_k$$

and

$$\|x_{k+1} - y_k\| = \| - z_k\| = \|z_k\| \leq I_k = t_{k+1} - s_k$$

which show relations (5.115) and (5.116) for all $n \geq 0$.

Relations (5.115) and (5.116) show that the iteration $\{x_n\}$ $(n \geq 0)$ is Cauchy in a Banach space E_1 and as such it converges to some $x^* \in U(x_0, r_0)$.

Finally relations (5.117) and (5.118) follow from relations (5.115) and (5.116) using standard techniques in the majorant method.

That completes the proof of the proposition. ∎

REMARK 5.10 (a) By setting $w_n = A(x_n)^{-1}F(x_n)$ for all $n \geq 0$, we can select

$$m_n = (1 - C_0(0, t_n))^{-1}h_n, \quad n \geq 1, \quad m_0 \geq \|y_0 - x_0\|$$

and

$$T_2(r) = T(r) - s_0$$

where $T(r)$ is given by relation (5.16). Then under the rest of the hypotheses of Theorem 5.4, the conclusions of that theorem will hold for the iteration $\{x_n\}$ $(n \geq 0)$ generated by (5.106)–(5.107).

(b) Similar remarks can be made by setting $w_n = F'(x_n)^{-1}F(x_n)$ for all $n \geq 0$ and relating Theorem 5.5 and the proposition. ∎

5.1.8 A Result on the Monotone Convergence of a Generalized Iteration

Concerning the monotone convergence of the generalized iteration (5.106)–(5.107) we can show:

THEOREM 5.5

Let F be a nonlinear operator defined on a convex subset D of a regular POTL-space E_1 with values in a POTL-space E_2. Let $\bar{x}_0$ and x_0 be two points of D such that

$$\bar{x}_0 \leq x_0.$$

Consider the iterations for all $n \geq 0$

$$y_n = x_n - w_n, \tag{5.119}$$

$$x_{n+1} = y_n - z_n, \tag{5.120}$$

and

$$\bar{y}_n = \bar{x} - \bar{w}_n, \tag{5.121}$$

$$\bar{x}_{n+1} = \bar{y}_n - \bar{z}_n \tag{5.122}$$

for some sequences $\{w_n\}$, $\{z_n\}$, $\{\bar{w}_n\}$, $\{\bar{z}_n\}$ $(n \geq 0)$ *satisfying*

$$w_n \geq 0, \; z_n \geq 0, \; \bar{w}_n \leq 0, \; \bar{z}_n \leq 0 \tag{5.123}$$

and

$$w_n \leq \bar{w}_n, \; z_n \leq \bar{z}_n \quad \textit{for all } n \geq 0. \tag{5.124}$$

Then there exist two sequences $\{\bar{x}_n\}$, $\{x_n\}$ $(n \geq 0)$ *satisfying the approximations (5.119)–(5.122)*

$$\bar{x}_0 \leq \bar{y}_0 \leq \bar{x}_1 \leq \cdots \leq \bar{y}_n \leq \bar{x}_{n+1} \leq y_n \leq \cdots \leq x_1 \leq y_0 \leq x_0, \tag{5.125}$$

$$\lim_{n\to\infty} \bar{x}_n = \bar{x}, \; \lim_{n\to\infty} x_n - x \; \textit{and} \; \bar{x}, x \in D_0 \; \textit{with} \; \bar{x} \leq x.$$

PROOF From hypothesis (5.123) we have in turn

$$\bar{y}_n = \bar{x}_n - \bar{w}_n \geq \bar{x}_n$$

$$\bar{x}_{n+1} = \bar{y}_n - \bar{z}_n \geq \bar{y}_n,$$

$$y_n = x_n - w_n \leq x_n,$$

and

$$x_{n+1} = y_n - z_n \leq y_n \quad \text{for all } n \geq 0.$$

Hence, we get $\bar{x}_n \leq \bar{y}_n \leq \bar{x}_{n+1}$ and $x_{n+1} \leq y_n \leq x_n$ for all $n \geq 0$. Also, from (5.123) and (5.124) we get

$$\bar{y}_n = \bar{x}_n - \bar{w}_n \leq x_n - w_n = y_n$$

and

$$\bar{x}_{n+1} = \bar{y}_n - \bar{z}_n \leq y_n - z_n = x_{n+1} \quad \text{for all } n \geq 0.$$

∎

5.1.9 Related Lipschitz Conditions and Single-Step Newton Methods

In this subsection we assume $z_n = 0$ $(n \geq 0)$ in (5.3). We present several theorems without proofs. The proofs can be found in [2], [9] or can be carried out along the lines of the preceding Theorems 5.1 and 5.2.

We assume that

$$\|F'(x) - F'(y)\| \leq w(\|x - y\|) \quad \text{for all } x, y \in U(x_0, R) \tag{5.126}$$

with some monotonically increasing function w satisfying $\lim_{t\to 0} w(t) = 0$ and some fixed $R \geq 0$, or, more generally, that

$$\|F'(x)-F'(y)\| \leq w(r, \|x-y\|) \quad \text{for all } x, y \in \bar{U}(x_0, r),\ 0 < r < R \tag{5.127}$$

with some monotonically increasing (in both arguments) positive function w satisfying

$$\lim_{t\to 0} w(r,t) = 0 \quad (0 \leq r \leq R). \tag{5.128}$$

Define parameters

$$a = \|F'(x_0)^{-1}F(x_0)\|, \quad b = \|F'(x_0)^{-1}\| \quad \text{for } x_0 \text{ fixed}, \tag{5.129}$$

function

$$\psi(r) = a + b\int_0^r w(t)dt - r \quad (0 \leq r \leq R) \tag{5.130}$$

and equation

$$\psi(r) = 0. \tag{5.131}$$

THEOREM 5.6

Suppose that equation (5.131) has a unique solution $r^* \in [0, R]$, and that $\psi(R) \leq 0$. Then equation (5.1) has a solution $x^* \in \bar{U}(x_0, r^*)$; this solution is unique in $U(x_0, R)$.

It is convenient to pass from the function w to function

$$\bar{w}(r) = \sup\{w(u) + w(v) : u + v = r\}. \tag{5.132}$$

For example, in the Hölder case $w(r) = cr^p$ $(0 < p < 1)$ we have $\bar{w}(r) = 2^{1-p}cr^p$. More generally, if w is concave on $[0, R]$, we have $\bar{w}(r) = 2w(\frac{r}{2})$. It is clear that $\bar{w}$ is always increasing and convex, and that $\bar{w}(r) \geq w(r)$ for $0 \leq r \leq R$.

In an analogy to (5.130), we set

$$\bar{\psi}(r) = a + b\int_0^r \bar{w}(t)dt - r \quad (0 \leq r \leq R) \tag{5.133}$$

and define the scalar equation

$$\bar{\psi}(r) = 0. \tag{5.134}$$

THEOREM 5.7

Suppose that equation (5.134) has a unique solution $q^* \in [0, R]$. Then iteration (5.2) is well defined for all n and belongs in $U(x_0, q^*)$. Moreover define the sequence $\{q_n\}$ $(n \geq 0)$ by

$$q_{n+1} = q_n - \frac{\bar{\psi}(q_n)}{\bar{\psi}'(q_n)} \quad (n \geq 0). \tag{5.135}$$

Then, $\lim_{n\to\infty} q_n = q^*$, *and the error estimates*

$$\|x_{n+1} - x_n\| \leq q_{n+1} - q_n \tag{5.136}$$

$$\|x^* - x_n\| \leq x^* - q_n \tag{5.137}$$

hold for all $n \geq 0$. It is useful to introduce numerical sequences

$$r_0 = 0, \quad r_1 = a,$$

$$r_{n+1} = r_n + \frac{b}{1 - bw(r_n)} \int_0^{r_n - r_{n-1}} w(t)dt \ (n \geq 1) \tag{5.138}$$

$$\delta_0 = \|x^* - x_0\|, \ \delta_{n+1} = \frac{b}{1 - bw(r_n)} \int_0^{\delta_n} w(t)dt \ (n \geq 0). \tag{5.139}$$

THEOREM 5.8

Suppose that the sequence $\{r_n\}$ $(n \geq 0)$ given by (5.138) converges for given a to some limit $r_\infty(a)$. Then iteration (5.2) is well defined and belongs to $U(x_0, r_\infty(a))$ for all $n \geq 0$. Moreover, the error estimates

$$\|x_{n+1} - x_n\| \leq r_{n+1} - r_n \tag{5.140}$$

and

$$\|x^* - x_n\| \leq \delta_n \tag{5.141}$$

hold for all $n \geq 0$.

Note also that under the hypotheses of Theorem 5.7

$$r_{n+1} - r_n \leq q_{n+1} - q_n \quad (n \geq 0) \tag{5.142}$$

$$r_{n+1} - r_n \leq \frac{b}{1 - bw(r^*)} \int_0^{r_n - r_{n-1}} w(t)dt \tag{5.143}$$

and

$$\delta_{n+1} \leq \frac{b}{1 - bw(r^*)} \int_0^{\delta_n} w(t)dt. \tag{5.144}$$

In the remainder of this subsection, we consider Kantorovich quasi-majorants. Given a bounded operator G between two Banach spaces, we set

$$w_G(u,v) = \sup\{\|G(x+h) - G(x)\| : \|x - x_0\| \leq u, \|h\| \leq v - u\}. \tag{5.145}$$

Function w_G is called the Kantorovich quasi-majorant for the operator G. Comparing with (5.127) we get

$$w(u,v) \leq w(v, v-u), \quad w(r,t) \leq w(r, r+l). \tag{5.146}$$

If $w(r,t) = c(r)t$, then

$$w(u,v) = \int_u^v c(t)dt \tag{5.147}$$

and in this case the function $n(r) = w(0,r)$ is the classical Kantorovich majorant.

It is convenient for us to define for $0 \leq r \leq R$ functions

$$n(r) = w(0,r), \tag{5.148}$$

$$\bar{n}(r) = \sup\{w(0,s) + w(s,r) : 0 \leq s \leq r\} \tag{5.149}$$

where $w = w_{F'}$,

$$\phi(r) = a + b\int_0^r n(t)dt - r, \tag{5.150}$$

$$\tilde{\phi}(r) = a + b\int_0^r \bar{n}(t)dt - r, \tag{5.151}$$

and equations

$$\phi(r) = 0 \tag{5.152}$$

$$\bar{\phi}(r) = 0. \tag{5.153}$$

Then along the lines of Theorems 5.6, 5.7 and 5.8 we can prove Theorems 5.9, 5.10 and 5.11, respectively.

THEOREM 5.9

Suppose that equation (5.152) has a unique solution $r^* \in [0, R]$, and that $\phi(R) \leq 0$. Then equation (5.1) has a solution $x^* \in \bar{U}(x_0, r^*)$; this solution is unique in $U(x_0, R)$.

THEOREM 5.10

Suppose that equation (5.153) has a unique solution $q^* \in [0, R]$. Then iteration (5.2) is well defined and belongs in $U(x_0, q^*)$. Define sequence $\{q_n\}$ $(n \geq 0)$ by

$$q_{n+1} = q_n - \frac{\bar{\phi}(q_n)}{\bar{\phi}'(q_n)} \quad (n \geq 0). \tag{5.154}$$

Then $\lim_{n\to\infty} q_n = q^*$ and estimates (5.136) and (5.137) hold for all $n \geq 0$.

In order to formulate a result parallel to Theorem 5.8, we define sequences

$$r_0 = 0, \ r_1 = a,$$

$$r_{n+1} = r_n + \frac{b}{1 - bn(r_n)} \int_{r_{n-1}}^{r_n} w(r_{n-1}, t)dt \ (n \geq 1), \tag{5.155}$$

$$\delta_0 = \|x^* - x_0\|,$$

$$\delta_{n+1} = \frac{b}{1 - bn(r_n)} \int_{r_n}^{r_n + \delta_n} w(r_n, t)dt \ (n \geq 0). \tag{5.156}$$

THEOREM 5.11

Suppose that sequence $\{r_n\}$ $(n \geq 0)$ given by (5.155) converges for a given a to some limit $r_\infty(a)$. Then iteration (5.2) is well defined and belongs to $U(x_0, r_\infty(a))$, for all $n \geq 0$. Moreover, error estimates (5.140) and (5.141) hold for all $n \geq 0$.

Various aspects of the preceding theorems by means of Uryson integral operators have been discussed in [2].

References

1. J. Appel, E. DePascale, and P.P. Zabrejko, On the application of the Newton-Kantorovich method to nonlinear integral equations of Uryson type, *Numer. Funct. Anal. Optimiz.* **12**, 3 (1991), 271–283.

2. J. Appel, E. DePascale, J. Lysenko, and P.P. Zabrejko, New results on Newton-Kantorovich approximations with applications to nonlinear integral equations, *Numer. Funct. Anal. Optimiz.* **18**, 1, 2 (1997), 1–17.

3. I.K. Argyros, Quadratic equations and applications to Chandrasekhar's and related equations, *Bull. Austral. Math. Soc.* **32** (1985), 275–292.

4. I.K. Argyros, On a class of nonlinear integral equations arising in neutron transport, *Aequationes Mathematicae* **32** (1988), 99–111.

5. I.K. Argyros, On the solution of equations with nondifferentiable operators and the Ptăk error estimates, *BIT* **90** (1990), 752–754.

6. I.K. Argyros, On the convergence of a certain class of iterative procedures under relaxed conditions with applications, *J. Comp. Appl. Math.* **94** (1998), 13–21.

7. I.K. Argyros, *Polynomial Operator Equations in Abstract Spaces and Applications*, Chapman Hall/CRC Press, Boca Raton, Florida, 1998.

8. I.K. Argyros and F. Szidarovszky, On the monotone convergence of general Newton-like methods, *Bull. Austral. Math. Soc.* **45** (1992), 489–502.

9. I.K. Argyros and F. Szidarovszky, *The Theory and Applications of Iteration Methods*, CRC Press, Inc., Boca Raton, Florida, 1993.

10. S. Chandrasekhar, *Radiative Transfer*, Dover Publ., New York, 1960.

11. X. Chen and T. Yamamoto, Convergence domains of certain iterative methods for solving nonlinear equations, *Numer. Funct. Anal. and Optimiz.* **10** (1 and 2), (1989), 37–48.

12. X. Chen and M. Z. Nashed, Convergence of Newton-like methods

for singular operator equations using outer inverses, *Numer. Math.* **66** (1993), 235–257.

13. R.S. Dembo, S.C. Eisenstat, and T. Steihaug, Inexact Newton methods, *SIAM J. Numer. Anal.* **19**, 2 (1982), 400–408.

14. L.V. Kantorovich and G.P. Akilov, *Functional Analysis in Normed Spaces*, Pergamon Press, New York, 1964.

15. M.T. Necepurenko, On Chebysheff's method for functional equations (Russian), *Usephi Mat. Nauk.* **9** (1954), 163–170.

16. F.A. Potra and V. Ptăk, Sharp error bounds for Newton's process, *Numer. Math.* **34** (1980), 63–72.

17. F.A. Potra, On an iterative algorithm of order 1.839... for solving nonlinear operator equations, *Numer. Funct. Anal. and Optimiz.* **7**, (1) (1984–85), 75–106.

18. R.A. Safiev, The method of tangent hyperbolas, *Sov. Math. Dokl.* **4** (1963), 482–485.

19. A.E. Taylor, *Introduction to Functional Analysis*, Wiley Publ., New York, 1957.

20. S. Ul'm, Iteration methods with divided differences of the second order (Russian), *Dokl. Akad. Nauk. SSSR* **158** (1964), 55–58. *Soviet Math. Dokl.* **5** 1187–1190.

21. T. Yamamoto, On the method of tangent hyperbolas in Banach spaces, *J. Comput. Appl. Math.* **21** (1988), 75–86.

22. P.P. Zabrejko and D.F. Nguen, The majorant method in the theory of Newton-Kantorovich approximations and the Ptăk error estimates, *Numer. Funct. Anal. and Optimiz.* **9**, (5 and 6) (1987), 671–684.

Chapter 6

A Unified Approach for Solving Equations, Part II: On Finite-Dimensional Spaces

Ioannis K. Argyros

Department of Mathematics, Cameron University, Lawton, OK 73505
E-mail: ioannisa@cameron.edu

Abstract We study inexact Newton-like methods for the solution of nonlinear operator equations in a Banach space involving a nondifferentiable term, and their discretized versions in connection with the mesh independence principle. This principle asserts that the behavior of the discretized process is asymptotically the same as that for the original iteration and consequently, the number of steps required by the two processes to converge to within a given tolerance is essentially the same. So far this result has been proved by others using Newton's method for certain classes of boundary value problems and even more generally by considering a Lipschitz uniform (or not) discretization. In some of our earlier papers we extended these results to include Newton-like methods under more general conditions. However, all previous results assume that the iterates can be computed exactly. This is not true in general. That is why we use inexact Newton-like methods and even more general conditions. Our results, on the one hand, extend, and on the other hand, make more practical and applicable previous results. Moreover, we can solve a wider range of problems and find sharper error bounds. Several applications of our results to two-point boundary value problems are illustrated. This study has been divided into two sections, followed by corresponding applications. This way the reader does not have to study Section 6.1 if interested only in Section 6.2.

Contents

6.1 Nondifferential Equations

6.1.1 Introduction to Nonlinear Equations Containing a Nondifferentiable Term

The goal of this study is to extend the validity of the mesh independence principle to include inexact Newton-like methods. Let us consider the problem of approximating a locally unique solution x^* of the equation

$$F(x) + Q(x) = 0 \tag{6.1}$$

where F, Q are nonlinear operators defined on some closed convex subset D of a Banach space E_1 with values in a Banach space E_2.

Let $x_0 \in D$ be fixed and define the inexact Newton-like method for all $n \geq 0$ by

$$y_n = x_n - A(x_n)^{-1}(F(x_n) + Q(x_n)) \tag{6.2}$$

and

$$x_{n+1} = y_n - z_n. \tag{6.3}$$

The operator F is Fréchet differentiable on D, whereas Q is only continuous there. $A(x_n)$ denotes a linear operator which is an approximation to the Fréchet derivative $F'(x_n)$ of F evaluated at $x = x_n$ for all $n \geq 0$. The points $z_n \in D$ for all $n \geq 0$, and are determined in such a way that the iteration $\{x_n\}$ $n \geq 0$ converges to a solution x^* of equation (6.1). By setting $z_n = 0$ for all $n \geq 0$, $Q(x) = 0$ for all $x \in D$, we obtain the Newton-like method. Moreover, setting $A(x) = F'(x)$ and $Q(x) = 0$ for all $x \in D$, we obtain Newton's method. The convergence of both methods have been examined extensively by others and us, [1], [2], [4], [15], in connection with the mesh independence principle. Since the iterates of the inexact Newton-like method (6.2)–(6.3) (for $z_n = 0$, $n \geq 0$ or not) can rarely be computed in infinite-dimensional spaces, (6.2)–(6.3) can be replaced in practice by a family of discretized equations

$$P(a) + P_1(a) = 0 \tag{6.4}$$

indexed by some real number $h > 0$, where P, P_1 are nonlinear operators between finite-dimensional spaces E^1 and E^2. Let the discretization on E_1 be defined by the bounded linear operators $L : E_1 \to E^1$. Consider also the iteration $\{a_n\}$ $n \geq 0$ given for all $n \geq 0$ by

$$b_n = a_n - S(a_n)^{-1}(P(a_n) + P_1(a_n)), \quad a_0 = L(x_0) \tag{6.5}$$

$$a_{n+1} = b_n - d_n. \tag{6.6}$$

$S(a_n)$ denotes a linear operator which is an approximation to the Fréchet derivative $P'(a_n)$ of P evaluated at $a = a_n$ for all $n \geq 0$. The points $d_n \in E^1$ for all $n \geq 0$, and are determined in such a way that the iteration $\{a_n\}$ $n \geq 0$ converges to a solution a^* of equation (6.4). Note that all symbols introduced in (6.4)–(6.6) really depend on h. That is, $P = P_h$, $L = L_h$, $S = S_h$, etc. But we do not use the latter to simplify the notation.

In practice, the iterates y_n or even b_n can rarely be computed exactly. That is why we need to "correct" at every step by introducing z_n or d_n, respectively, for the iterations under consideration. This is the factor that the results in the studies mentioned above have not taken into account when proving the mesh independence principle. The mesh independence principle (proved in the studies listed above) asserts that the number of steps required by the two processes to converge to within a given tolerance is essentially the same.

Here we show that this is true for our perturbed Newton-like method (6.2)–(6.3). Our results can be reduced to the ones obtained earlier for appropriate choices of the factors involved.

The importance of the formulation of an efficient mesh size strategy based upon the mesh independence has been extensively discussed in [1], [2], [4], [15] and the references there. Some applications to two-point boundary value problems are also illustrated.

6.1.2 Convergence Analysis for Newton-Like Methods and the Mesh-Independence Principle

The norms in all spaces will be denoted by the same symbol $\| \ \|$. For any bounded linear operator from E_1 to E_2 or from E^1 to E^2, the induced norm will be used.

We find it convenient to introduce the following:

(G_1) Let $R > 0$ and using the notation $U(x_0, R) = \{x \in E_1 \mid \|x - x_0\| \leq R\}$, assume there exist $x_0 \in D$ and a function $\bar{\alpha} : U^3(x_0, R) \to [0, +\infty)$ such that

$$\|A(x_0)^{-1}[F(y) - F(x) - A(x)(y - x) + F'(y)(z - y)]\| \leq \bar{\alpha}(x, y, z) \quad (6.7)$$

for all $x, y, z \in U(x_0, R) \subseteq D$.

(G_2) There exist continuous, nondecreasing functions w, w_1 and w_2 such that

$$w : D \to [0, +\infty), \quad w_1, w_2 : [0, R] \to [0, +\infty)$$

with $w_1(0) = w_2(0) = 0$ and a sequence $\{z_n\}$ $(n \geq 0)$ of points from D with

$$\|z_i\| \leq w(z_i) \leq w_1(r) \quad \text{for all } i \geq 0 \quad (6.8)$$

and for all $k \in N$

$$\sum_{i=0}^{k+1} \|z_i\| \leq \sum_{i=0}^{k+1} w(z_i) \leq w_2(r) \quad r \in [0, R] \quad (6.9)$$

for all $z_i \in D$. We note that (6.9) implies that the sequence $\{z_n\}$ $(n \geq 0)$ is null.

(G_3) There exist continuous, nondecreasing functions $w_3, w_4 : [0, R] \to [0, +\infty)$ with $w_3(0) = w_4(0) = 0$ such that

$$\bar{\alpha}_i = \bar{\alpha}(x_i, y_i, x_{i+1}) \leq w_3(r) \quad \text{for all } i \geq 0 \quad (6.10)$$

and for all $k \in N$

$$\sum_{i=0}^{k+1} \bar{\alpha}_i \leq w_4(r) \quad (6.11)$$

for all $x_i, y_i, x_{i+1} \in U(x_0, r) \subseteq U(x_0, R)$.

(G_4) Let $F, Q : D \subseteq E_1 \to E_2$ be nonlinear operators satisfying

$$\|A(x_0)^{-1}(F'(x + t(y - x)) - A(x))\| \leq C_1(r, r + t\|\ell\|)$$

$$+ C_2(r, r + t\|\ell\|) + b_1 \text{ with } b_1 \in [0, 1), \quad (6.12)$$

$$\|A(x_0)^{-1}(Q(x+\ell)-Q(x))\| \leq C_3(r,\|\ell\|) \leq C_4(r,\|\ell\|)\|\ell\| \tag{6.13}$$

and

$$\|A(x_0)^{-1}(A(x)-A(x_0))\| \leq C_0(\|x-x_0\|)+b_2 \quad \text{with } b_2 \in [0,1) \tag{6.14}$$

for all $t \in [0,1)$, $x, y \in U(x_0, r) \subseteq U(x_0, R)$ and $\ell = y - z$. The functions C_1, C_2, C_3 and C_4 are assumed to be continuous in both variables on $[0, R]\times[0, R]$ and such that if one variable is fixed, then they are nondecreasing functions of the other on $[0, R]$ with $C_1(0,0) = C_2(0,0) = C_3(0,0) = C_4(0,0) = 0$. Moreover, C_3 is linear in the right variable. We also set $C = C_1 + C_2 + b_1$. The function C_0 is continuous and nondecreasing on $[0, R]$ with $C_0(0) = 0$.

We will need to introduce the constants:

$$t_0 = 0, \quad s_0 \geq \|y_0 - x_0\|, \quad \ell_0 \geq \|z_0\|,$$

the iterations for all $n \geq 0$

$$s_{n+1} = t_{n+1} + [1-(C_0(t_{n+1})+b_2)]^{-1}v_{n+1}, \tag{6.15}$$

$$t_{n+1} = s_n + \ell_n, \quad \ell_n = w(z_n), \tag{6.16}$$

$$v_{n+1} = \int_{s_n}^{t_{n+1}} C(s_n, t)dt + C(s_n, s_n)(t_{n+1} - s_n)$$

$$+ C_3(t_n, t_{n+1} - t_n) + \alpha_n \tag{6.17}$$

for some given sequences $\{\alpha_n\}$ and $\{\ell_n\}$ $n \geq 0$ with

$$\alpha_n \geq 0,\ \ell_n \geq 0,\ \sum_{i=0}^{k+1} \alpha_i \leq w_4(r_0), \text{ and } \sum_{i=0}^{k+1} \ell_i \leq w_2(r_0) \tag{6.18}$$

for all positive integers k, and some fixed $r_0 \in [0, R]$.

Moreover, we define the iterations for all $n \geq 0$

$$\bar{v}_{n+1} = \int_0^1 C(\|y_n - x_0\|, \|y_n - x_0\| + t\|x_{n+1} - y_n\|)\|x_{n+1} - y_n\|dt$$

$$+ C(\|y_n - x_0\|, \|y_n - x_0\|)\|x_{n+1} - y_n\|$$

$$+ C_3(\|x_n - x_0\|, \|x_{n+1} - x_n\|) + \bar{\alpha}_n \tag{6.19}$$

where $\bar{\alpha}_n = \bar{\alpha}(x_n, y_n, x_{n+1})$ with

$$\bar{\alpha}_n \leq \alpha_n \quad \text{for all } n \geq 0 \tag{6.20}$$

$$\bar{e}_{n+1} = \left[1 - \int_0^1 C(0, (1-t)\|x_{n+1} - x_0\| + t\|x^* - x_0\|)dt\right]^{-1}, \tag{6.21}$$

$$e_{n+1} = \left[1 - \int_0^1 C(0, (1-t)t_{n+1} + tr_0)dt\right]^{-1} \tag{6.22}$$

$$p_n = \int_0^1 C(\|x_n - x_0\|, \|x_n - x_0\| + t\|x^* - x_n\|)\|x_n - x^*\|dt \tag{6.23}$$

and the function

$$T(r) = s_0 + \frac{1}{1 - C_0(r)}\left[\int_{s_0}^r C(r,t)dt + C(r,r)(r - s_0)\right.$$

$$\left. + C_3(r,r) + w_4(r)\right] + w_2(r) \tag{6.24}$$

on $[0, R]$.

We can now state the main result on semilocal convergence:

THEOREM 6.1

Let F, Q be operators defined on a closed convex subset D of a Banach space E_1 with values in a Banach space E_2.

Assume:

(i) F is Fréchet-differentiable on D whereas Q is only continuous there;

(ii) condition (G_4) is satisfied;

(iii) condition (G_1) is satisfied;

(iv) iterations $\{\alpha_n\}$, $\{\ell_n\}$ and $\{\bar{\alpha}_n\}$ $(n \geq 0)$ satisfy conditions (6.18) and (6.20), and

$$\|z_n\| \leq \ell_n \quad \textit{for all } n \geq 0; \tag{6.25}$$

(v) there exists a minimum nonnegative number r_0 such that

$$T(r_0) \leq r_0; \tag{6.26}$$

(vi) the following estimates are true:

$$r_0 \leq R, \tag{6.27}$$

$$\theta(r_0) = \frac{\int_0^1 C(R, R + t(r_0 + R))dt + C_0(r_0) + C_0(R) + C_3(r_0) + 2b_2}{1 - (C_0(r_0) + b_2)} < 1$$

and

$$C_0(r_0) + b_2 < 1, \tag{6.28}$$

and

(vii) the ball

$$U(x_0, R) \subseteq D, \tag{6.29}$$

and the points z_n $(n \geq 0)$ are chosen to satisfy the inclusion in (G_1). Then

(a) the scalar sequence $\{t_n\}$ $(n \geq 0)$ generated by relations (6.15)-(6.17) is monotonically increasing and bounded above by its limit, which is number r_0;

(b) the sequences $\{x_n\}$, $\{y_n\}$ $(n \geq 0)$ generated by relations (6.2)-(6.3) are well defined, remain in $U(x_0, r_0)$ for all $n \geq 0$, and converge to a solution x^* of the equation $F(x) = 0$, which is unique in $U(x_0, R)$.

Moreover, the following estimates are true for all $n \geq 0$

$$\|y_n - x_n\| \leq s_n - t_n, \tag{6.30}$$

$$\|x_{n+1} - y_n\| \leq t_{n+1} - s_n, \tag{6.31}$$

$$\|x_n - x^*\| \leq r_0 - t_n, \tag{6.32}$$

$$\|y_n - x^*\| \leq r_0 - s_n, \tag{6.33}$$

$$\|A(x_0)^{-1}(F(x_{n+1}) + Q(x_{n+1}))\| \leq \bar{v}_{n+1}, \tag{6.34}$$

$$\begin{aligned} \|x^* - x_{n+1}\| &\leq \bar{e}_{n+1}(\bar{v}_{n+1} + C_3(\|x_{n+1} - x_0\|, \|x_{n+1} - x^*\|)) \\ &\leq e_{n+1}(v_{n+1} + C_3(t_{n+1}, \|x_{n+1} - x^*\|)) \leq r_0 - t_{n+1} \end{aligned} \tag{6.35}$$

and

$$\|y_n - x_n\| \leq \|x^* - x_n\| + \tag{6.36}$$

$$\frac{\int_0^1 C(\|x_n - x_0\|, \|x_n - x_0\| + t\|x^* - x_n\|)\|x_n - x^*\|dt + C_3(\|x_n - x_0\|, \|x^* - x_n\|)}{1 - C_0(\|x_n - x_0\|)}.$$

(We will be concerned only with the case $r_0 > 0$, since when $r_0 = 0$, $x_0 = x^*$).

PROOF (a) Using relations (6.14), (6.15), (6.16), (6.18), (6.26) and (6.28), we deduce that the scalar sequence $\{t_n\}$ $(n \geq 0)$ is monotonically increasing, nonnegative, and $t_0 \leq s_0 \leq t_1 \leq s_1 \leq r_0$. Let us assume that $t_k \leq s_k \leq t_{k+1} \leq s_{k+1} \leq r_0$ for $k = 0, 1, 2, \ldots, n$. Then by relations (6.15), (6.16) and (6.18), we can have in turn

$$\begin{aligned}
t_{k+2} &= s_{k+1} + \ell_{k+1} \\
&\leq t_{k+1} + \frac{1}{1 - C_0(r_0)}\left[\int_{s_k}^{t_{k+1}} C(r_0, t)dt + C(r_0, r_0)(t_{k+1} - s_k)\right. \\
&\quad \left. + C_3(r_0, t_{k+1} - t_k) + \alpha_k\right] + \ell_{k+1} \\
&\leq \cdots \leq s_0 + \frac{1}{1 - C_0(r_0)}\left[\int_{s_0}^{t_{k+1}} C(r_0, t)dt + C(r_0, r_0)(t_{k+1} - s_0)\right. \\
&\quad \left. + C_3(r_0, t_{k+1} - t_0) + \sum_{i=0}^{k} \alpha_i\right] + \sum_{i=0}^{k+1} \ell_i \\
&\leq s_0 + \frac{1}{1 - C_0(r_0)}\left[\int_{s_0}^{r_0} C(r_0, t)dt + C(r_0, r_0)(r_0 - s_0)\right. \\
&\quad \left. + C_3(r_0, r_0) + w_4(r_0)\right] + w_2(r_0) \\
&= T(r_0) \leq r_0.
\end{aligned}$$

Hence, the scalar sequence $\{t_n\}$ $n \geq 0$ is bounded above by r_0. By hypothesis (6.26), the number r_0 is the minimum nonnegative zero of the equation $T(r) - r = 0$ on $[0, r_0]$, and from the above $r_0 = \lim_{n\to\infty} t_n$.

(b) Using relations (6.2), (6.3), (6.15), (6.16), (6.24) and (6.26), we deduce that $x_1, y_0 \in U(x_0, r_0)$ and that estimates (6.30) and (6.31) are true for $n = 0$. Let us assume that they are true for $k = 0, 1, 2, \ldots, n-1$. Using the induction hypothesis, we can have in turn

$$\begin{aligned}\|x_{k+1}-x_0\| &\leq \|x_{k+1}-y_0\|+\|y_0-x_0\| \\ &\leq \|x_{k+1}-y_k\|+\|y_k-y_0\|+\|y_0-x_0\| \\ &\leq \cdots \leq (t_{k+1}-s_k)+(s_k-s_0)+s_0 \leq t_{k+1} \leq r_0,\end{aligned}$$

and

$$\begin{aligned}\|y_{k+1}-x_0\| &\leq \|y_{k+1}-y_0\|+\|y_0-x_0\| \\ &\leq \|y_{k+1}-x_{k+1}\|+\|x_{k+1}-y_k\|+\|y_k-y_0\|+\|y_0-x_0\| \\ &\leq \cdots \leq (s_{k+1}-t_{k+1})+(t_{k+1}-s_k)+(s_k-s_0)+s_0 \\ &\leq s_{k+1} \leq r_0.\end{aligned}$$

That is, $x_n, y_n \in U(x_0, r_0)$ for all $n \geq 0$.

Using hypothesis (6.14) we now have

$$\begin{aligned}\|A(x_0)^{-1}(A(x_k)-A(x_0))\| &\leq C_0(\|x_k-x_0\|)+b_2 \\ &\leq C_0(t_k-t_0)+b_2 \leq C_0(r_0)+b_2 < 1\end{aligned}$$

by condition (6.28). It now follows from the Banach lemma on invertible operators [16] that the linear operator $A(x_k)$ is invertible, and

$$\|A(x_k)^{-1}A(x_0)\| \leq [1-(C_0(\|x_k-x_0\|)+b_2)]^{-1} \leq [1-(C_0(t_k)+b_2)]^{-1} \tag{6.37}$$

for all $k \geq 0$.

We can now have from relations (6.2) and (6.3)

$$\begin{aligned}F(x_{k+1})+Q(x_{k+1}) &= F(x_{k+1})-F(y_k)-A(y_k)(x_{k+1}-y_k) \\ &\quad + A(y_k)(x_{k+1}-y_k)+F(y_k)+Q(x_{k+1}) \\ &= \int_0^1 [F'(y_k+t(x_{k+1}-y_k))-A(y_k)](x_{k+1}-y_k)dt\end{aligned}$$

$$+ (A(y_k) - F'(y_k))(x_{k+1} - y_k) + (Q(x_{k+1}) - Q(x_k))$$

$$+ (F(y_k) - F(x_k) - A(x_k)(y_k - x_k) + F'(y_k)(x_{k+1} - y_k)).$$

Hence, by using hypotheses (6.7) and (6.12) we obtain in turn

$$\begin{aligned}
&\|A(x_0)^{-1}(F(x_{k+1}) + Q(x_{k+1}))\| \\
&\leq \int_0^1 \|A(x_0)^{-1}[F'(y_k + t(x_{k+1} - y_k)) - A(y_k)]\| \, \|x_{k+1} - y_k\| dt \\
&+ \|A(x_0)^{-1}(A(y_k) - F'(y_k))\| \, \|x_{k+1} - y_k\| \\
&+ \|A(x_0)^{-1}(Q(x_{k+1}) - Q(x_k))\| \\
&+ \|A(x_0)^{-1}[F(y_k) - F(x_k) - A(x_k)(y_k - x_k) + F'(y_k)(x_{k+1} - y_k)]\| \\
&= \bar{v}_{k+1} \leq v_{k+1}
\end{aligned}$$

by hypotheses (6.20) and relations (6.17) and (6.19). Hence, we showed estimate (6.34) for all $n \geq 0$.

Using relations (6.2), (6.15), (6.34) and (6.37), we obtain

$$\begin{aligned}
\|y_{k+1} - x_{k+1}\| &\leq \|A(x_{k+1})^{-1}A(x_0)\| \cdot \|A(x_0)^{-1}(F(x_{k+1}) + Q(x_{k+1}))\| \\
&\leq [1 - (C_0(\|x_{k+1} - x_0\|) + b_2)]^{-1}\bar{v}_{k+1} \\
&\leq [1 - (C_0(t_{k+1}) + b_2)]^{-1}v_{k+1} = s_{k+1} - t_{k+1},
\end{aligned}$$

which show estimate (6.30) for all $n \geq 0$.

Similarly, from relations (6.3), (6.16) and hypotheses (iv), we obtain

$$\|x_{k+1} - y_k\| = \| - z_k\| \leq \ell_k = t_{k+1} - s_k,$$

from which it follows that estimate (6.31) is true for all $n \geq 0$.

It now follows from the estimates (6.30) and (6.31) that the sequences $\{x_n\}$, $\{y_n\}$ $(n \geq 0)$ are Cauchy in a Banach space E_1 and as such they both converge to some $x^* \in U(x_0, r_0)$, which by taking the limit as $n \to \infty$ in (6.2), we obtain $F(x^*) + Q(x^*) = 0$.

To show uniqueness, we assume that there exists another solution $y^* \neq x^*$ of equation (6.1) in $U(x_0, R)$. Then using (6.2) we can obtain the approximation

$$\begin{aligned} y_{n+1} - y^* &= x_{n+1} - y^* - A(x_{n+1})^{-1}[(F(x_{n+1}) - F(y^*)) \\ &\quad + (Q(x_{n+1}) - Q(y^*))] \\ &= (A(x_{n+1})^{-1}A(x_0))\{A(x_0)^{-1}[(A(x_{n+1})(x_{n+1} - y^*)) \\ &\quad - (F(x_{n+1}) - F(y^*)) - (Q(x_{n+1}) - Q(y^*))]\} \\ &= (A(x_n)^{-1}A(x_0))A(x_0)^{-1}\bigg\{\int_0^1 [F'(y^* + t(x_n - y^*)) \\ &\quad - A(y^*)](x_n - y^*) + (A(y^*) - A(x_0))(x_n - y^*) \\ &\quad + (A(x_0) - A(x_n))(x_n - y^*) - (Q(x_{n+1}) - Q(y^*))\bigg\}. \end{aligned}$$

Hence, using (6.12), (6.13) and (6.37) as above, we obtain that

$$\|y_{n+1} - y^*\| \leq \theta(r_0)\|x_{n+1} - y^*\|.$$

Using (6.3) we get

$$\|x_{n+1} - y^*\| \leq \|y_n - y^*\| + \|z_n\|.$$

Therefore, we get

$$\|y_{n+1} - y^*\| \leq \theta(r_0)^2\|y_n - y^*\| + \theta(r_0)\|z_n\|,$$

and by letting $n \to \infty$ in the above estimate we obtain

$$\|x^* - y^*\| \leq \theta(r_0)^2\|x^* - y^*\| < \|x^* - y^*\|.$$

(Since $\lim_{n\to\infty} \|z_n\| = 0$ by (6.9).)

That shows that $x^* = y^*$.

Estimates (6.32) and (6.33) follow from (6.30) and (6.31), respectively, by using the standard arguments in majorant theory.

Finally, using the approximations

$$x_{n+1} - x^* = H_{n+1}^{-1} A(x_0)[A(x_0)^{-1}(F(x_{n+1}) + Q(x_{n+1}))$$
$$+ A(x_0)^{-1}(Q(x^*) - Q(x_{n+1}))],$$

$$H_{n+1} = \int_0^1 F'(x^* + t(x_{n+1} - x^*))dt,$$

$$y_n - x_n = x^* - x_n + (A(x_n)^{-1}A(x_0))\Bigg[\int_0^1 A(x_0)^{-1}[F'(x_n + t(x^* - x_n))$$
$$- A(x_n)](x^* - x_n)dt + A(x_0)^{-1}(Q(x^*) - Q(x_n))\Bigg]$$

and the estimate

$$\int_0^1 \|A(x_0)^{-1}[F'(x^* + t(x_{n+1} - x^*)) - A(x_0)]\|dt$$
$$\leq \int_0^1 C(0, (1-t)\|x_{n+1} - x_0\| + t\|x^* - x_0\|)dt$$
$$\leq \int_0^1 C(0, (1-t)r_0 + tr_0)dt = C(0, r_0) < 1$$

(by (6.27)), and consequently

$$\|H_{n+1}^{-1} A(x_0)\| \leq \bar{e}_{n+1},$$

we can immediately obtain estimates (6.35) and (6.36) for all $n \geq 0$, where we have also used relations (6.21), (6.22) and (6.23).

That completes the proof of the theorem. ∎

Let us define the functions $\varphi, \varphi_1, \varphi_2 : [0, R] \to [0, +\infty)$ by

$$\varphi(r) = r - w_2(r) - \frac{1}{1 - [C_0(r) + b_2]}$$

$$\cdot \left[\int_0^r C(r,t)dt + C(r,r)r + C_3(r,r) + w_4(r) \right], \tag{6.38}$$

$$\varphi_1(r) = r - \varphi(r) \tag{6.39}$$

and

$$\varphi_2(r) = \frac{1}{1 - [C_0^*(r) + b_2^*]} \left[\int_0^1 C^*(0, rt)dt + C_3^*(0,r) \right]$$

$$+ w_5(r) \quad \text{for } r \in [0, R]. \tag{6.40}$$

Here $C^* = C_1^* + C_2^* + b_1^*$ and C_1^*, C_2^*, C_0^*, C_3^*, C_4^*, b_1^*, b_2^* are as the corresponding ones above without the "stars," and $w_5 : [0, R] \to [0, +\infty)$ is continuous and nondecreasing with $w_5(0) = 0$.

By the hypotheses on the C and w functions above, there exist constants p, h_1, δ_0, δ_1, δ_2, δ_3, c_6, c_5, c_4, c_3, with $c_4 > \delta_3$ such that for

$$0 < \delta_0 \le c_4 - c_5 h^p \le r_0(h) = c_3 h^p \le \delta_1 \le r_0 \le R$$

$$(\text{with } r(h) = 0 \text{ when } x_h^0 = x_h^* \text{ or } h = 0) \tag{6.41}$$

the following are true for a sufficiently small $R > 0$ and $h \in (0, h_1]$

$$0 < \delta_2 \le \varphi_1(r(h)) \le \delta_3, \quad C_0(r(h)) + b_2 \text{ and } \theta(r(h)) < 1. \tag{6.42}$$

We can now show that for all $h \in (0, h_2]$, where

$$h_2 = \min\left\{ h_1, \left(\frac{\delta_2}{c_3 - c_6}\right)^{\frac{1}{p}}, \left(\frac{c_4 - \delta_3}{c_2 + c_5}\right)^{\frac{1}{p}} \right\} \text{ with } c_3 > c_6,\ c_6 \le c_2 \tag{6.43}$$

the following is true

$$0 < c_2 h^p \le \varphi(r(h)) \le c_6 h^p. \tag{6.44}$$

Indeed from (6.42), we get $r(h) - \delta_3 \le \varphi(r(h)) \le r(h) - \delta_2$. It is enough to show that $r(h) - \delta_3 \ge c_2 h^p$ and $r(h) - \delta_2 \le c_6 h^p$, which will be true if $c_4 - c_5 h^p - \delta_3 \ge c_2 h^p$ and $c_3 h^p - \delta_2 \le c_6 h^p$, respectively. The last inequalities are true by the choice of h and (6.43).

Similar arguments can show that for sufficiently small $h \in (0, h_2]$ there exist δ_4, δ_5, δ_6, δ_7, c_7, c_8, c_9 such that for

$$0 < \delta_4 \leq c_7 - c_8 h^p \leq \|L(x^*) - a^*\| = r^*(h)$$

$$\leq c_9 h^p \leq \delta_5 \leq r^* = \|x_0 - x^*\|, \ \delta_5 \leq \delta_1 \tag{6.45}$$

(with $r^*(h) = 0$ when $L(x^*) = a^*$ or $h = 0$) the following are true for a sufficiently small $r^* \leq r_0$

$$0 < \delta_6 \leq \varphi_2(r^*(h)) \leq \delta_7 < 1 \ \text{ and } \ C_0^*(r^*(h)) + b_2^*) < 1. \tag{6.46}$$

We will need the following result on local convergence.

THEOREM 6.2

Let $F, Q : D \subset E_1 \to E_2$ be nonlinear operators as in Theorem 6.1 and assume:

(i) There exists a solution $x^* \in D$ of the equation $F(x) + Q(x) = 0$ such that the linear operator $A(x^*)$ has a bounded inverse;

(ii) F is a Fréchet-differentiable operator satisfying

$$\|A(x^*)^{-1}(F'(x + t\ell) - A(x))\| \leq C_1^*(r, r + t\|\ell\|)$$

$$+ C_2^*(r, r + t\|\ell\|) + b_1^* \ \textit{with} \ b_1^* \in [0, 1),$$

$$\|A(x^*)^{-1}(Q(x + \ell) - Q(x))\| \leq C_3^*(r, \|\ell\|) \leq c_4^*(r, \|\ell\|)\|\ell\| \tag{6.47}$$

and

$$\|A(x^*)^{-1}(A(x) - A(x^*))\| \leq C_0^*(\|\ell\|) + b_2^* \ \ \textit{with} \ b_2^* \in [0, 1), \tag{6.48}$$

for all $t \in [0, 1]$, $x, y \in U(x^*, r) \subseteq U(x^*, r^*)$, $x_0 \in D$, $\ell = y - x$ with $U(x^*, r^*) \subseteq D$. Moreover, C_0^*, C_1^*, C_2^*, C_3^*, C_4^*, b_1^*, b_2^* are as C_0, C_1, C_2, C_3, C_4, b_1 and b_2, respectively, with $C_0^*(0) = C_1^*(0, 0) = C_2^*(0, 0) = C_3^*(0, 0) = C_4^*(0, 0) = 0$; and

(iii) There exists a sequence $\{z_n\}$ $(n \geq 0)$ of points from D satisfying the inclusion in (G_1) *and*

$$\|z_n\| \leq g_n = g(z_n) \leq w_5(r) \tag{6.49}$$

where $g : U(x^*, r^*) \to [0, +\infty)$ is continuous; and

(iv) The constants b_1^*, b_2^* are such that $b_1^* + b_2^* \in [0, 1)$.
Then the following are true:
(a) For a sufficiently small $r^* \in (0, R]$

$$C_0^*(r^*) + b_2^* < 1 \tag{6.50}$$

and

$$0 < \varphi_2(r^*) < 1 \tag{6.51}$$

where $C^* = C_1^* + C_2^* + b_1^*$.

(b) The sequences $\{y_n\}$, $\{x_n\}$ $(n \geq 0)$ are well defined, remain in $U(x^*, r^*)$ for all $n \geq 0$ and $\lim_{n\to\infty} x_n = \lim_{n\to\infty} y_n = x^*$, provided that $x_0 \in U(x^*, r^*)$. Moreover, the solution x^* of equation (1) is unique in $U(x^*, r^*)$.

Furthermore, the following estimates are true for all $n \geq 0$:

$$\|x_{n+1} - x^*\| \leq \gamma_n \|x_n - x^*\| \leq \gamma \|x_n - x^*\| \tag{6.52}$$

and

$$\|y_n - x^*\| \leq \delta_n \|x_n - x^*\| \leq \delta \|x_n - x^*\|, \tag{6.53}$$

where

$$\delta_n = \frac{\int_0^1 C^*(0, \|x_n - x^*\|t)dt + C_3^*(0, \|x_n, -x^*\|)}{1 - C_0^*(\|x_n - x^*\|) + b_2^*} \tag{6.54}$$

$$\delta = \frac{\int_0^1 C^*(0, r^*t)dt + C_3^*(0, r^*)}{1 - C_0^*(r^*) + b_2^*} \tag{6.55}$$

$$\gamma_n = \delta_n + g_n \tag{6.56}$$

and

$$\gamma = \delta + w_5(r^*). \tag{6.57}$$

PROOF (a) By hypotheses (ii)–(iv) we have $b_1^* + b_2^* \in [0, 1)$, $C_0^*(0) = C_1^*(0,0) = C_2^*(0,0) = C_3^*(0,0) = C_4^*(0,0) = C^*(0,0) = w_5(0) = 0$, and that all these functions are continuous and nondecreasing on $[0, r^*]$. Hence, we can find a sufficiently large r^* such that estimates (6.50) and (6.51) are satisfied.

(b) Using the Banach lemma on invertible operators [16] and condition (6.48), we deduce

$$\|A(x^*)^{-1}(A(x) - A(x^*))\| \leq C_0^*(\|x - x^*\|) + b_2^* \leq C_0^*(r^*) + b_2^* < 1 \tag{6.58}$$

(by (6.50)) for all $x \in U(x^*, r^*)$.

Hence we deduce that the linear operator $A(x)$ is invertible on $U(x^*, r^*)$ and

$$\|A(x)^{-1}A(x^*)\| \leq \frac{1}{1-(C_0^*(\|x-x^*\|)+b_2^*)} \leq \frac{1}{1-(C_0^*(r^*)+b_2^*)}. \tag{6.59}$$

In particular, (6.58), (6.59) are true for $x = x_0$, since $x_0 \in U(x^*, r^*)$. Let us assume that $x_m \in U(x^*, r^*)$ for $m = 0, 1, 2, \ldots, k$. Using (6.2) we get

$$y_k - x^* = -A(x_k)^{-1}[F(x_k) - F(x^*) - A(x_k)(x_k - x^*) + Q(x_k) - Q(x^*)]. \tag{6.60}$$

We also introduce the approximation

$$\begin{aligned} &Q(x_k) - Q(x^*) + F(x_k) - F(x^*) - A(x_k)(x_k - x^*) \\ &\quad = \int_0^1 [F'(x^* + t(x_k - x^*)) - A(x_k)](x_k - x^*)dt + Q(x_k) - Q(x^*) \\ &\quad = \int_0^1 [F'(x^* + t(x_k - x^*)) - A(x_k)](x_k - x^*)dt \\ &\quad + Q(x_k) - Q(x^*). \end{aligned} \tag{6.61}$$

We now compose both sides of (6.61) by $A(x^*)^{-1}$ and then by taking norms and using (6.26), we obtain that the left-hand side of (6.61) is bounded above by

$$\begin{aligned} &\int_0^1 [C_1^*(0, 0 + t\|x_k - x^*\|) + C_2^*(0, 0 + t\|x_k - x^*\|)] \\ &\qquad \cdot \|x_k - x^*\|dt + C_3^*(0, \|x_k - x^*\|) \\ &\quad = \int_0^1 C^*(0, \|x_k - x^*\|t)\|x_k - x^*\|dt + C_3^*(0, \|x_k - x^*\|). \end{aligned} \tag{6.62}$$

From (6.59), (6.60) and (6.62) we now have

$$\|y_k - x^*\| \leq \|A(x_k)^{-1}A(x^*)\| \Big\| A(x^*)^{-1}$$

$$\cdot \int_0^1 [F'(x^* + t(x_k - x^*)) - A(x_k)(x_k - x^*)dt$$

$$+ (Q(x_k) - Q(x^*))] \Bigg\| + C_3^*(0, \|x_k, x^*\|)$$

$$\leq \delta_k \|x_k - x^*\| \leq \delta \|x_k - x^*\|. \tag{6.63}$$

The above estimate shows that (6.53) is true and that $y_k \in U(x^*, r^*)$ since $\delta_k \leq \delta < 1$ (by (6.51)).

Moreover, from (6.3), (6.63) and (6.49), we get

$$\|x_{k+1} - x^*\| \leq \|y_k - x^*\| + \|z_k\| \leq \gamma_k \|x_k - x^*\| \leq \gamma \|x_k - x^*\| \tag{6.64}$$

which shows (6.52) and that $x_{k+1} \in U(x^*, r^*)$. Hence the sequences $\{x_n\}$, $\{y_n\}$ are well defined, remain in $U(x^*, r^*)$ and satisfy (6.52) and (6.53) for all $n \geq 0$.

Let $m \geq 0$. Then by (6.52) we get

$$\begin{aligned} \|x_{n+m} - x^*\| &\leq \gamma_{n+m-1} \|x_{n+m-1} - x^*\| \\ &\leq \gamma_{n+m-1} \gamma_{n+m-2} \|x_{n+m-2} - x^*\| \\ &\leq \cdots \leq \gamma^m \|x_n - x^*\|. \end{aligned} \tag{6.65}$$

Similarly by (6.53) and (6.52) we get

$$\|y_{n+m} - x^*\| \leq \delta \cdot \gamma^m \|x_n - x^*\|. \tag{6.66}$$

Finally by letting $m \to \infty$ in (6.65) and (6.66) we obtain $\lim_{n\to\infty} x_n = \lim_{n\to\infty} y_n = x^*$ (since $0 < \gamma < 1$). The proof of the uniqueness of the solution x^* in $U(x^*, r^*)$ is omitted as identical to the one in Theorem 6.1.

The points z_n $(n \geq 0)$ appearing in (6.8) depend on x_n $(n \geq 0)$, whereas the ones appearing in (6.49) depend on x_n $(n \geq 0)$ and (maybe) the point x^*. That is why we can choose the functions g, w_5 (see (6.44)) to be the same or different from the functions w, w_1, respectively (see (6.8)).

That completes the proof of the theorem. ∎

In many applications it turns out that the solution x^* of equation (6.1) as well as the iterates x_n, y_n have "better smoothness" properties than the elements of E_1. This is a motivation for considering a subset $E_3 \subseteq E_1$ such that

$$x^* \in E_3,\ x_n, y_n \in E_3,\ x_n - x^*, y_n - x^* \in E_3,$$

$$x_{n+1} - x_n, y_{n+1} - y_n \in E_3 \quad (n \geq 0). \tag{6.67}$$

We consider a family

$$\{P, P_1, L, L_0\}, \quad h > 0 \tag{6.68}$$

where

$$P, P_1 : E_4 \subseteq E^1 \to E^2, \quad h > 0$$

are nonlinear operators and

$$L : E_1 \to E^1,\ L_0 : E_2 \to E^2, \quad h > 0$$

are bounded linear discretization operators such that

$$L(E_3 \cap U(x^*, \delta_5)) \subseteq E_4, \quad h > 0. \tag{6.69}$$

The discretization (6.68) is called uniform if there exist a number δ_1 such that

$$U(L(x^*), \delta_1) \subseteq E_4, \quad h > 0 \tag{6.70}$$

and $(P, P_1, S, L(x^*))$ satisfies the "G" conditions that (F, Q, A, x_0) satisfy for all $h > 0$ in the ball $U(L(x^*), \delta_1)$.

The operators P, P_1, L, L_0 depend on h. That is, $P = P_h$, etc. To simplify the notation we assume that this is understood and hence we avoid the subscript h.

Moreover, the discretization family (6.68) is called bounded if there is a constant $q > 0$ such that

$$\|L(x)\| \leq q\|x\|, \quad x \in E_3, \quad h > 0, \tag{6.71}$$

stable if there is a constant $\sigma > 0$ such that

$$\|S(L(x))^{-1}\| \leq \sigma,\ x \in E_3 \cap U(L(x^*), \delta_5), \quad h > 0, \tag{6.72}$$

consistent of order $p > 0$ if there are two constants $c_0, c_1 > 0$ such that

$$\|S(L(x^*))^{-1}(L_0(F + Q)(x) - (P + P_1)(L(x)))\| \leq c_0 h^p,$$

$$x \in E_3 \cap U(L(x^*), \delta_5), \quad h > 0 \text{ with } \sigma c_0 \leq c_2 \tag{6.73}$$

and

$$\|S(L(x^*))^{-1}(L_0(A(x)(y) - S(L(x))L(y)))\| \leq c_1 h^p,$$
$$x \in E_3 \cap U(L(x^*), \delta_5), \; y \in E_3, \quad h > 0. \tag{6.74}$$

The C^*, b^* appearing in Theorem 6.2 are in general different from C, b appearing in condition (G_4). Under the hypotheses of Theorem 6.1, the C^*, b^* can be expressed in terms of the corresponding C, b. Let $x_0, x^* \in D$ be fixed and assume that

$$U(x^*, r^*) \subseteq U(x_0, r_0).$$

Then using the Banach lemma on invertible operators, (6.14) and (6.48), we can easily see that

$$C_0^*(r) = \frac{1}{1 - C_0(r^*)}(C_0(r) + C_0(r^*)) \quad \text{and} \quad b_2^* = \frac{b_2}{1 - C_0(r^*)} \quad \text{on } [0, R].$$

The inclusion above will certainly be true if

$$\|x^* - x_0\| + r^* = 2r^* \leq r_0,$$

which can certainly be satisfied for x_0 sufficiently close to x^* (or equivalently r^* sufficiently small). Similarly we can define the rest of the C^*, b^*. However, other choices are also possible. Indeed, let s_0 be an upper bound of the function C_0 on $[0, s_1]$ for some $s_1 \leq R$. Then for sufficiently small s_1, r^* the number s_0 will be an upper bound for the function C_0^* in $[0, s_1]$. For example, s_1, r^* can be chosen so that

$$C_0(r) \leq \frac{s_0}{2} \quad \text{and} \quad C_0(r^*) \leq \frac{s_0}{1 + s_0}.$$

Hence the C, b can be used (identified) instead of the C^*, b^* and vice versa in cases where upper bounds of these expressions are needed.

Similarly, the C, C^*, b, b^* associated with the family $(P, P_1, S, L(x^*))$ can be identified with each other in the sense that we described above for sufficiently small r^*, R and h.

Consider the condition:

(C_5) The C^h, w^h, $\bar{\alpha}^h$, $b(h)$, $b^*(h)$ are identified with the corresponding C, w, $\bar{\alpha}$, b, b^*, respectively, for all $h \in (0, h_0]$ for some fixed $h_0 > 0$, $r(h) \in (0, \delta_1]$ and $r^*(h) \in (0, \delta_5]$ provided that

$$U(L(x^*), \delta_5) \subseteq U(x_0, R) \subseteq D.$$

The left-hand side inclusion will certainly be true if

$$L(x_0) \in U\left(x_0, \frac{R}{3}\right), \; r^* \leq \frac{R}{3q} \quad \text{and} \quad \delta_5 \leq \frac{R}{3}.$$

We also note that if $0 \in U(L(x^*), \delta_5) \subseteq U(x_0, R)$ for all h sufficiently small, then the null sequences $\{z_n\}$, $\{d_n\}$ $(n \geq 0)$ can be chosen from $U(L(x^*), \delta_5)$ instead of the set D.

Finally, we can easily see from (6.9) and (6.49) that the function w_2 can be identified with the function w_5. Choose for example $z_n = d_n$ for all $n \geq 0$, $h > 0$. However, we do not need this to prove our discretization results.

With the notation introduced above we can now formulate our main result.

THEOREM 6.3

Let $F, Q : D \subseteq E_1 \to E_2$ be nonlinear operators as in Theorem 1. *Assume:*

(i) The hypotheses of Theorem 6.2 are satisfied;

(ii) The discretization (6.68) is bounded, stable, and consistent of order p;

(iii) Condition (G_5) *is satisfied.*

Then

(a) Equation (6.4) has a locally unique solution

$$a^*(h) = a^* = L(x^*) + O(h^p) \tag{6.75}$$

for all $h \in (0, \bar{h}_0]$ with $\bar{h}_0$ being a fixed constant.

(b) There exist constants $\bar{h}_1 \in (0, \bar{h}_0]$, $r_1 \in (0, \delta_5]$ such that the discrete iteration (6.5)–(6.6) converges to a^*, and

(c) if there exist constants b_1, b_2, c_0, c_1, c_{10}, c_{11} with $5(b_1c_{11} + 4qc_3b_2 + c_0 + c_1 + c_{10}b_2) \leq c_{10} \leq c_{11}$ and for all $n \geq 0$, the following estimate is true;

$$\|d_n - L(z_n)\| \leq (c_{11} - c_{10})h^p, \; h \in (0, \bar{h}_0], \; r \in (0, \delta_5]. \tag{6.76}$$

Then there exist constants $\bar{h}_1 \in (0, \bar{h}_0]$, $r_3 \in (0, r_1]$ such that the following estimates are true for all $n \geq 0$

$$b_n = L(y_n) + O(h^p), \tag{6.77}$$

$$a_n = L(x_n) + O(h^p), \tag{6.78}$$

$$S(b_n)^{-1}(P + P_1)(b_n) = S(b_n)^{-1}L_0((F + Q)(y_n)) + O(h^p), \tag{6.79}$$

(provided that $P + P_1$ is b-Lipschitz continuous on $U(L(x^*), \delta_5)$)

$$S(a_n)^{-1}(P + P_1)(a_n) = S(a_n)^{-1}L_0((F + Q)(x_n)) + O(h^p), \tag{6.80}$$

(provided that $P + P_1$ is b-Lipschitz continuous on $U(L(x^*), \delta_5)$)

$$b_n - a^* = L(y_n - x^*) + O(h^p) \tag{6.81}$$

and

$$a_n - a^* = L(x_n - x^*) + O(h^p). \tag{6.82}$$

PROOF (a) The C, w functions are continuous, vanish at the origin and $b_1 + b_2 \in [0, 1)$. Hence we can find intervals $(0, h_0)$ and $(0, \delta_1]$ such that conditions (6.27) and (6.28) are satisfied for all $h \in (0, h_0]$ and $r(h) \in (0, \delta_1]$. Set $h_3 = \min\{h_0, h_2, \left(\frac{\delta_1}{c_3}\right)^{1/p}\}$. Then using (6.41), (6.72) and (6.73) we obtain in turn

$$s_0(h) = \|S(L(x^*))^{-1}(P + P_1)(L(x^*))\|$$

$$\leq \sigma\|(P + P_1)(L(x^*)) - L_0(F + Q)(x^*)\|$$

$$\leq \sigma c_0 h^p \leq c_2 h^p \leq \varphi(r(h)), \tag{6.83}$$

and

$$r(h) \leq c_3 h^p \leq \delta_1, \quad \text{for all } h \in (0, h_3], \tag{6.84}$$

which shows that (6.26), (6.27) and (6.28) hold for all $h \in (0, h_3]$. Since all hypotheses of Theorem 6.1 are satisfied, equation (6.4) has a solution $a^*(h) = a^* \in U(L(x^*), r(h))$ which is a unique solution of (6.4) in $U(L(x^*), \delta_1)$. Thus (6.75) follows from

$$\|a^* - L(x^*)\| \leq r(h) \leq c_3 h^p \leq \delta_1 \tag{6.85}$$

by setting $\bar{h}_0 = h_3$.

(b) As in part (a) we can find intervals $(0, h_3]$ and $(0, r^*]$ such that conditions (6.45), (6.50) and (6.51) are satisfied for all $h \in (0, h_3]$ and $r^*(h) \in (0, \delta_5]$. By applying Theorem 6.2 to (6.4) we see that the sequence (6.5)–(6.6) converges to a^* if

$$\|L(x_0) - a^*\| < r^*(h) \tag{6.86}$$

and

$$U(a^*, \|L(x_0) - a^*\|) \subseteq U(L(x^*), \delta_1). \tag{6.87}$$

But (6.87) holds if

$$\|a^* - L(x^*)\| + \|L(x_0) - a^*\| \leq \delta_1. \tag{6.88}$$

By (6.71) and (6.84) we obtain

$$\begin{aligned}\|L(x_0) - a^*\| &\leq \|L(x_0) - L(x^*)\| + \|L(x^*) - a^*\| \\ &\leq q\|x_0 - x^*\| + c_3 h^p. \end{aligned}\tag{6.89}$$

Thus (6.86), (6.87) are satisfied if

$$q\|x_0 - x^*\| + 2c_3 h^p \leq \delta_1 \tag{6.90}$$

and

$$q\|x_0 - x^*\| + c_3 h^p \leq c_7 - c_8 h^p \tag{6.91}$$

hold, respectively. Conditions (6.90) and (6.91) will certainly hold if

$$q\|x_0 - x^*\| \leq \frac{\delta_1}{2}, \quad 2c_3 h^p \leq \frac{\delta_1}{2}, \quad q\|x_0 - x^*\| \leq \frac{c_7}{2}$$

and

$$c_3 h^p \leq \frac{c_7}{2} - c_8 h^p.$$

We choose

$$\|x_0 - x^*\| \leq r_1 = \min\left\{\frac{\delta_1}{2q}, \frac{c_7}{2q}\right\},$$

and

$$h_4 = \min\left\{h_2, h_3, \left(\frac{\delta_1}{4c_3}\right)^{\frac{1}{p}}, \left[\frac{c_7}{2(c_3 + c_8)}\right]^{\frac{1}{p}}\right\}.$$

It is now easily verified that (6.86) and (6.87) are satisfied for all $h \in (0, h_4]$ and $x_0 \in U(x^*, r_1)$. Therefore, for these h and x_0, the iteration (6.5)–(6.6) converges to a^*.

(c) We will now show that there exists $\bar{h}_1 \in (0, \bar{h}_0]$, $r_3 \in (0, r_1]$ such that

$$\|a_n - L(x_n)\| \leq c_{11} h^p \tag{6.92}$$

for $n = 0$ (6.92) is true since $a_0 = L(x_0)$. Suppose that (6.92) holds for $n = 0, 1, \ldots, i$. We note that if we show that

$$\|b_n - L(y_n)\| \le c_{10}h^p, \tag{6.93}$$

then from (6.5), (6.6), (6.93), (6.76) and the estimate

$$\|a_{i+1} - L(x_{i+1})\| = \|b_i - d_i - L(y_i - z_i)\| \le \|b_i - L(y_i)\| + \|d_i - L(z_i)\|$$

$$\le c_{10}h^p + (c_{11} - c_{10})h^p = c_{11}h^p, \tag{6.94}$$

we can complete the induction for (6.92). But (6.93) is true for $n = 0$ by (6.76). We now suppose that (6.93) is true for $n = 0, 1, \ldots, i$. Using (6.2), (6.3), (6.5) and (6.6) we can obtain the approximation

$$b_i - L(y_i) = S(a_i)^{-1}\{[S(a_i)(a_i - L(x_i)) - (P + P_1)(a_i) + (P + P_1)(L(x_i))]$$

$$+ [(S(a_i) - S(L(x_i)))L(A(x_i)^{-1}(F + Q)(x_i))]$$

$$+ [S(L(x_i))L(A(x_i)^{-1}(F + Q)(x_i)) - L_0((F + Q)(x_i))]$$

$$+ [L_0((F + Q)(x_i)) - (P + P_1)(L(x_i))]\}. \tag{6.95}$$

From (6.71) and (6.92) we obtain

$$\|a_i - L(x^*)\| \le \|a_i - L(x_i)\| + \|L(x_i) - L(x^*)\| \le c_{11}h^p + qr_1. \tag{6.96}$$

As in (6.58), (6.59) and using (6.14) and (6.96) we obtain

$$\|S(a_i)^{-1}S(L(x^*))\| \le \frac{1}{1 - [C_0(\|a_i - L(x^*)\|) + b_2]}$$

$$\le \frac{1}{1 - [C_0(c_{11}h^p + qr_1) + b_2]}. \tag{6.97}$$

By composing the first bracket in (6.95) by $S(L(x^*))^{-1}$ and by taking norms, using (6.12) and (6.96), we obtain that this term is bounded above by

$$\left\| \int_0^1 S(L(x^*))^{-1}\{[P'(a_i + t(L(x_i) - a_i)) - S(a_i)]\right.$$

$$\cdot (a_i - L(x_i))dt + (P_1(L(x_i)) - P_1(a_i))\}\Bigg\|$$

$$\leq \int_0^1 C(\|a_i - L(x^*)\|, \|a_i - L(x^*)\| + t\|a_i - L(x_i)\|)\|a_i - L(x_i)\|dt$$

$$+ C_3(\|a_i - L(x^*)\|, \|a_i - L(x_i)\|)$$

$$\leq \int_0^1 C(c_{11}h^p + qr_1, qr_1 + (t+1)c_{11}h^p))c_{11}h^p dt$$

$$+ C_4(c_{11}h^p + qr_1, c_{11}h^p)c_{11}h^p \text{ (by (6.13))}. \tag{6.98}$$

Moreover, by adding and subtracting $S(L(x^*))$ inside the parenthesis of the second bracket, composing by $S(L(x^*))^{-1}$ and using (6.14), (6.96), and (6.71), we obtain that this term is bounded above by

$$[C_0(\|a_i - L(x^*)\|) + C_0(\|L(x^*) - L(x_i)\|) + 2b_2]q(\|x_i - x^*\| + \|y_i - x^*\|)$$

$$\leq 2[C_0(c_{11}h^p + qr_1) + C_0(r_1) + 2b_2]qc_3h^p \quad \text{(by (6.20))}. \tag{6.99}$$

Furthermore, using (6.74) and (6.73), we obtain that the third and fourth bracket in (6.95), after being composed by $S(L(x^*))^{-1}$, are bounded above by c_1h^p and c_0h^p, respectively.

Finally, by collecting all the above majorizations, we obtain that estimate (6.93) will be true if

$$\int_0^1 (C_1 + C_2)(c_{11}h^p + qr_1, qr_1 + (t+1)c_{11}h^p)c_{11}dt$$

$$+ C_4(c_{11}h^p + qr_1, c_{11}h^p)c_{11} + 2qc_3C_0(c_{11}h^p + qr_1)$$

$$+ c_{10}C_0(c_{11}h^p + qr_1) + (4qc_3b_2 + c_0 + c_1$$

$$+ c_{10}b_2 + b_1c_{11}) \leq c_{10}. \tag{6.100}$$

Inequality (6.100) will certainly be true if each term at the left-hand side of it, divided by c_{10}, is bounded above by $\frac{1}{5}$. Since the C, w functions vanish at the origin, we can find h_5, $r_2 > 0$ such that this will happen for the first four terms. By hypothesis (c), the last term is also bounded above by $\frac{1}{5}$.

Finally, set $\bar{h}_1 = \min\{h_4, h_5\}$ and $r_3 = \min\{r_1, r_2\}$. With the above choices of $\bar{h}_1$ and r_3, estimates (6.77) and (6.78) follow.

Using (6.73), (6.93), we obtain

$$\begin{aligned}
&\|S(b_n)^{-1}((P+P_1)(b_n) - L_0((F+Q)(y_n)))\| \\
&\leq \|S(b_n)^{-1}((P+P_1)(b_n) - (P+P_1)(L(y_n)))\| \\
&\quad + \|S(b_n)^{-1}((P+P_1)(L(y_n)) - L_0((F+Q)(y_n)))\| \\
&\leq \frac{1}{1-[C_0(c_{10}h^p + qr_3) + b_2]}(b\|b_n - L(y_n)\| + c_0 h^p) \\
&\leq \frac{bc_{10} + c_0}{1-[C_0(c_{10}h^p + qr_3) + b_2]} h^p, \qquad (6.101)
\end{aligned}$$

where we have used the hypothesis that $P+P_1$ is b-Lipschitz continuous on $U(L(x^*), \delta_1)$.

Estimate (6.101) shows (6.79). Estimate (6.80) is obtained similarly by replacing b_n, y_n and c_{10} by a_n, x_n and c_{11}.

Moreover, from (6.93) and (6.94) we obtain

$$\begin{aligned}
\|(b_n - a^*) - L(y_n - x^*)\| &\leq \|b_n - L(y_n)\| + \|a^* - L(x^*)\| \\
&\leq c_{10}h^p + c_3 h^p = (c_{10} + c_3)h^p \qquad (6.102)
\end{aligned}$$

which shows (6.81).

Furthermore, from (6.92) and (6.84) we finally obtain

$$\begin{aligned}
\|(a_n - a^*) - L(x_n - x^*)\| &\leq \|a_n - L(x_n)\| + \|a^* - L(x^*)\| \\
&\leq c_{11}h^p + c_3 h^p = (c_{11} + c_3)h^p \qquad (6.103)
\end{aligned}$$

from which (6.82) follows.

That completes the proof of the theorem. ∎

We note that conditions $\delta_1 \leq r_0$ and $\delta_2 \leq r^*$ (see (6.41) and (6.45)) are not used in the proof. However, in many practical applications we may want these conditions to be true.

We can now prove the mesh-independence principle for perturbed Newton-like methods.

THEOREM 6.4
Assume:
(a) the hypotheses of Theorem 6.3 are true;
(b) there exists a constant $\delta > 0$ such that

$$\liminf_{h\to 0} \|L(u)\| \geq \delta\|u\| \quad \text{for each } u \in E_3. \tag{6.104}$$

Then for some $r_6 \in (0, r_3]$, and for any fixed $\varepsilon > 0$ and $x_0 \in U(x^*, r_6)$ there exists a constant $\bar{h} = \bar{h}(\varepsilon, x_0) \in (0, \bar{h}_1]$ such that

$$|\min\{n \geq 0, \|x_n - x^*\| < \varepsilon\} - \min\{n \geq 0, \|a_n - a^*\| < \varepsilon\} \leq 1 \tag{6.105}$$

for all $h \in (0, \bar{h}]$.

PROOF By hypotheses there exists a unique integer $i > 0$ such that

$$\|x_{i+1} - x^*\| < \varepsilon \leq \|y_i - x^*\| \tag{6.106}$$

and $h_6 = h_6(x_0)$ such that

$$\|L(y_i - x^*)\| \geq \delta\|y_i - x^*\| \quad \text{for all } h \in (0, h_6]. \tag{6.107}$$

We will prove that the theorem holds for

$$r_6 = \min\left\{r_3, \frac{r_4}{q}\right\}, \; \beta = \min\left\{\delta, 2q, \frac{1 - [C_0(r_5) + b_2]}{2q[\int_0^1 C(0, r_5) + C_4(0, r_5)]}\right\}, \tag{6.108}$$

$$\bar{h} = \min\left\{\bar{h}_1, h_6, \left(\frac{\beta\varepsilon}{2c_{12}}\right)^{\frac{1}{p}}, \left(\frac{\delta\varepsilon}{2c_{13}}\right)^{\frac{1}{p}}\right\}, \tag{6.109}$$

where $c_{12} = c_3 + c_{11}$, $c_{13} = c_{10} + c_3$ and $r_5 = \delta_1 + r_4$.
From (6.103) and (6.109) it follows that

$$\|a_{i+1} - a^*\| \leq \|L(x_{i+1} - x^*)\| + c_{12}h^p \leq q\varepsilon + \frac{\beta\varepsilon}{2} < 2q\varepsilon. \tag{6.110}$$

Using (6.108), (6.101) and Theorem 6.2 we obtain in turn that

$$\|b_{i+1} - a^*\| \leq \frac{\left[\int_0^1 C(0, \|a_{i+1} - a^*\|t)dt + C_4(0, \|a_{i+1} - a^*\|)\right]}{1 - [C_0(\|a_{i+1} - a^*\|) + b_2]} \|a_{i+1} - a^*\|$$

$$\leq \frac{\left[\int_0^1 C(0,r_5)dt + C_4(0,r_5)\right]}{1-[C_0(r_5)+b_2]} 2\beta q\varepsilon < \varepsilon \tag{6.111}$$

(since $\|a_{i+1}-a^*\| \leq \|a_0-a^*\| = \|L(x_0)-a^*\| \leq \|L(x_0)-L(x^*)\|+\|L(x^*)-a^*\| \leq qr_3 + c_3h^p \leq r_4+\delta_1 = r_5$).

Moreover from (6.107) and (6.102) we obtain

$$\varepsilon \leq \|y_i - x^*\| \leq \frac{1}{\delta}\|L(y_i-x^*)\| \leq \frac{1}{\delta}(\|b_i-a^*\|+c_{13}h^p)$$

or

$$\|b_i - a^*\| \geq \delta\varepsilon - c_{13}h^p \geq \delta\varepsilon - \frac{\delta\varepsilon}{2} = \frac{\delta\varepsilon}{2}. \tag{6.112}$$

Furthermore, if $\|a_i - a^*\| < \varepsilon$, then as in (6.111) we get

$$\|b_i - a^*\| < \frac{1}{2}\beta\varepsilon \leq \frac{\delta\varepsilon}{2}$$

which contradicts (6.112).

Hence we must have

$$\|a_i - a^*\| \geq \varepsilon. \tag{6.113}$$

The result now follows from (6.106), (6.111) and (6.113).

That completes the proof of the theorem. ∎

As it was observed in [1], [2], [4], [15], condition (6.104) follows from the condition

$$\lim_{h\to 0}\|L(u)\| = \|u\| \quad \text{for each } u \in E_3, \tag{6.114}$$

which is standard in most discretization studies. In fact, for some discretization studies, we have

$$\lim_{h\to 0}\|L(u)\| = \|u\| \quad \text{uniformly for } u \in E_3. \tag{6.115}$$

If this is the case, we can have a stronger version of the mesh-independence principle.

COROLLARY 6.1

Assume:

(a) the hypotheses of Theorem 6.3 are true;

(b) condition (6.115) holds uniformly for $u \in E_3$.

Then there exists a constant $r_7 \in (0, r_3]$ and, for any fixed $\varepsilon > 0$, some $\bar{h}_2 = \bar{h}_2(\varepsilon) \in (0, \bar{h}_1]$ such that (6.105) holds for all $h \in (0, \bar{h}_2]$ and all starting points $x_0 \in (x^*, r_7)$.

6.1.3 Applications

REMARK 6.1 (1) In [5], [6], [10], [11] (for $t = 1$, $Q(x) = 0$ for all $x \in D$ and $b_1 = b_2 = 0$) we show how to choose the functions C_0, C_1, C_2, C_3, C_4, $\bar{\alpha}$, w, w_1, w_2, w_3, w_4, w_5 and the sequence $\{z_n\}$ $(n \geq 0)$. We also show that special choices of the above can reduce our results to earlier ones involving single step methods (Newton's method, Secant method, the method of tangent parabolas, the method of tangent hyperbolas and others) as well as two-step methods [5], [6], [10], [11] and by making use of Theorem 6.1. In particular for $Q(x) = 0$, $A(x) = F'(x)$ for all $x \in D$, $z_n = 0$ for all $n \geq 0$, $b_1 = b_2 = 0$, and the choices of the α, w, C functions as given in [5] for this case our results reduce to the ones in [1]. With the notation used in [1], we can define the crucial "c" constants appearing in (6.41), (6.45) and (6.76) as follows:

$$c_3 = 2\sigma c_0, \quad c_4 = \frac{1}{\sigma L}, \quad c_5 = \sqrt{\mathrm{e}}\ (0 < \mathrm{e} < 1) \ \text{ for } h \leq \left(\frac{1-\mathrm{e}}{2\sigma^2 L c_0}\right)^{\frac{1}{p}},$$

$$c_7 = \frac{2}{3L\sigma}, \quad c_8 = \frac{4\sigma c_0}{3}, \quad c_{11} = c_{12} = 8\sigma \max\{c_0, c_1\} \ \text{ and } \ \delta_1 = p$$

(see the proof of Theorem 2 in [1, pp. 163–164]). Moreover, if the C function is bounded above by a Lipschitz constant (using the notation in [1]) and $\|x_n - x^*\| < \varepsilon$ under the hypotheses in [1], then we can find $n_1 \leq n$ so that $\|x_{n_1} - x^*\| < \varepsilon$ also. That improves the error bounds and finds smaller n satisfying the desired error tolerance. Several examples involving the solution of nonlinear integral equations were also given there.

(2) As an application of Theorem 1 (for $Q = 0$), we note that this theorem can be realized for operators F which satisfy an autonomous differential equation of the form

$$F'(x) = B(F(x)), \quad \text{for some given operator } B \text{ [13]}.$$

Assume for simplicity that $A(x) = F'(x)$ for all $x \in D$. As $F'(x^*) = B(0)$, the inverse $F'(x^*)^{-1}$ can be evaluated without knowing the actual solution x^*. Consider, for example, the scalar equation

$$F(x) = 0 \tag{6.116}$$

where F is given by $F(x) = e^x - s$, $s > 0$. Note that $F'(x) = F(x) + s$. That is $F'(x^*) = s$. Under the hypotheses of Theorem 6.1 and provided that $x_0 \in U(x^*, r^*)$, the iteration (6.2)–(6.3) converges to the solution $x^* = \ln(s)$ of equation (6.116).

(3) As we mentioned in Remark 1, our results can be reduced to the ones in [1]. Moreover, they can be reduced to the ones in [2], [4] for $z_n = 0$ for all $n \geq 0$, $Q = 0$ and $t = 1$. Furthermore, our condition (6.105) and the corresponding ones in [1], [2], [4], [15] state that if

$$\min\{n \geq 0, \|x_n - x^*\| < \varepsilon\} = i + 1, \quad i > 0, \tag{6.117}$$

then

$$\min\{n \geq 0, \|a_n - a^*\| < \varepsilon\} = i + 1, \quad \text{or } i, \text{ or } i + 2. \tag{6.118}$$

However we can actually show that if (6.117) is true, then

$$\min\{n \geq 0, \|a_n - a^*\| < \varepsilon\} = i + 1 \text{ or } i, \tag{6.119}$$

which improves (6.118).

Let us assume that $q \in (0, \gamma^*)$, for some $\gamma^* \in (0,1)$, and under the hypotheses of Theorem 6.4, set

$$\bar{h}_3 = \min\left\{\bar{h}, \left(\frac{(\gamma^* - q)\varepsilon}{c_{12}}\right)^{\frac{1}{p}}\right\}. \tag{6.120}$$

The estimate (6.110) can also be written as

$$\|a_{i+1} - a^*\| \leq q\varepsilon + c_{12}h^p \leq \gamma^*\varepsilon < \varepsilon, \quad \text{for } h \in (0, \bar{h}_3], \tag{6.121}$$

which shows (6.119).

If $q \geq \gamma^*$ in (6.71), we can consider the linear operators M, M_0 instead of L, L_0 given by $M = \lambda L$ and $M_0 = \lambda L_0$, where λ is such that $|\lambda| q < \gamma^*$ and $\lambda \neq 0$. Then conditions (6.71), (6.73) and (6.74) will still be true with $\bar{q}$, $\bar{\sigma}$, $\bar{c_0}$, $\bar{c_1}$ replacing q, σ, c_0, c_1 and given by $\bar{q} = |\lambda| q$, $\bar{\sigma} = \frac{\sigma}{|\lambda|}$, $\bar{c_0} = |\lambda| c_0$ and $\bar{c_1} = |\lambda| c_1$.

(4) Comparing the choices of the "corrector" sequences $\{z_n\}$ and $\{d_n\}$ $n \geq 0$ appearing in (6.3) and (6.6), respectively, we state the following. Once the z_n's $n \geq 0$ are chosen (see Remark 1), then the d_n's will be chosen in such a way that condition (6.76) is satisfied. Note that condition

(6.76) will certainly be satisfied if we simply set $d_n = L(z_n)$ for all $n \geq 0$, which is a logical choice but not the only one.

(5) Our conditions (6.12), (6.13) and (6.14) are more general than the recent ones of Chen-Yamamoto [12], which in turn are more general than Potra-Ptak [19], [20], and Zajrejko-Nguen [23]. In particular, our conditions can be reduced to the ones given by Chen-Yamamoto, if by using their notation, we choose special cases of our functions given by

$$C(r, r + t\|\ell\|) = w(r + t\|\ell\|) - w_0(r),$$

$$C_0(r) = w_0(r), \quad C_3(r, q) = e(r)q, \quad b = b_1, \quad c = b_2,$$
$$z_n = 0 \ (n \geq 0), \quad \alpha = 0 \quad \text{and} \quad w_1 = w_2 = w_3 = w_4 = w_5 = w = 0,$$

for all $t \in [0, 1]$ and all $r, \ell \in [0, R]$. Hence they can be used to solve a wider range of problems.

(6) We can easily see from the proof of Theorem 6.1 that all the results obtained here remain true if (6.12) is replaced by the weaker condition

$$\left\| \int_0^1 A(x_0)^{-1}(F'(x + t\ell) - A(x))\ell \, dt \right\| \leq C_4(r, \|\ell\|)\|\ell\|.$$

This condition is useful. It can obviously be used to improve further the error bounds obtained in this study and under special choices of the functions involved (see also Remarks 1 and 5 above). The corresponding error bounds can be found in [5], [6] and [12]. Since we have that

$$\left\| \int_0^1 A(x_0)^{-1}(F'(x + t\ell) - A(x))\ell \, dt \right\|$$
$$\leq \int_0^1 \|A(x_0)^{-1}(F'(x + t\ell) - A(x))\ell\| \, dt$$
$$\leq \int_0^1 \|A(x_0)^{-1}(F'(x + t\ell) = A(x))\| \, \|\ell\| \, dt,$$

it is hoped that the function C_4 can be chosen to be smaller or equal to the function $\int_0^1 C(r, r + t\|\ell\|)dt$ which is the one used in the proof of Theorem 6.1.

(7) Our results can be extended to include more general iterations of the form

$$y_n = x_n - A(x_n)^{-1}(F(x_n) + Q(x_n)), \quad x_{n+1} = e_n y_n - z_n \ (n \geq 0) \text{ for } e_n \in R.$$

We note that for $e_n = 1$ $(n \geq 0)$ the above iteration reduces to iteration (6.2)–(6.3). Moreover it can easily be seen from the proof of Theorem 6.1 that if we just replace z_i by $(e_i - 1)y_i - z_i$ $(i \geq 0)$ in (6.9), all the results obtained here will hold for this more general iteration.

(8) The constants q, σ, c_0, c_1 and b may be chosen to be the upper bounds of functions C_5, C_6, C_7, C_8, and C_9, respectively, having the form and properties of the C function. It can easily be seen that all the proofs can still be carried out. For example, C_7 can be such that $C_7(\|L(x)-L(x^*)\|)+b_3 \leq C_7(q, r^*) + b_3 = c_0$. We will also need to assume that the left-hand side of (6.73) is bounded above by the left-hand side of the above inequality.

(9) As we showed in [2], [4] (see also [1], [11], [15]) the discretization method $\{P, P_1, L, L_0\}$ can be used to solve boundary value problems involving operators F of the form

$$F(y) = \{y' - f(x, y, y'), 0 \leq x \leq 1, y(0) - v, y(1) - w\}$$

or

$$F(y) = \{y' - f(x, y), 0 \leq x \leq 1, sy(0) + ty(1) - v\},$$

or integral operators of the form

$$(F(y))(x) = y(x) - \int_0^1 f(x, b, y(t))dt + g(x), \quad 0 \leq x \leq 1$$

or operators of the form

$$F(y) = \{-y_{x_1x_1} - y_{x_2x_2} + f(x_1, x_2, y, y_{x_1}, y_{x_2}) \text{ in } \Omega, \quad y = 0 \text{ on } \partial\Omega\}$$

involving partial differential equation boundary value problems. ∎

6.2 Inexact Method and Mysovskii Hypotheses

6.2.1 Introduction to Mysovskii-Type Hypotheses

The goal of this section is to extend further the validity of the mesh-independence principle to include inexact Newton-like methods. Let us consider the problem of approximating a locally unique solution x^* of the equation

$$F(x) + Q(x) = 0 \tag{6.122}$$

where F, Q are nonlinear operators defined on some closed convex subset D of a Banach space E_1 with values in a Banach space E_2.

Let $x_0 \in D$ be fixed and define the inexact Newton-like method for all $n \geq 0$ by

$$y_n = x_n - A(x_n)^{-1}(F(x_n) + Q(x_n)) \tag{6.123}$$

and

$$x_{n+1} = y_n - z_n. \tag{6.124}$$

P. Deuflhard and G. Heindl in [14] and P. Deuflhard and F. Potra in [15] used affine invariant versions of the so-called Newton-Mysovskii-type hypotheses [10], [11] of the form

$$\|F'(y)^{-1}(F'(x + t(y-x)) - F'(x))\| \leq wt\|y - x\| \tag{6.125}$$

for all $x, y \in D$, some $w \in \mathbf{R}$ for all $t \in [0,1]$, and

$$\|F'(z)^{-1}(F'(u) - F'(x))(u - x)\| \leq w\|u - x\|^2 \tag{6.126}$$

for all $u, x, z \in D$, for some $w \in \mathbf{R}$. They provided sufficient conditions for the convergence of iteration (6.123) to a solution x^* of equation (6.122), when $Q(x) = 0$, $A(x) = F'(x)$ for all $x \in D$ and $z_n = 0$ $(n \geq 0)$. Error bounds on the distances $\|x_{n+1} - x_n\|$ and $\|x_n - x^*\|$ $(n \geq 0)$ were also given. Here under weaker and more general conditions (see (6.145)–(6.148)) we provide sharper error bounds on the same distances. This is important since if the error tolerance on the distances is ε and n_1 is the smallest nonnegative integer for which $\|x_{n_1} - x^*\| < \varepsilon$, then under (6.145)–(6.148) $\|x_{n_2} - x^*\| < \varepsilon$ and $n_2 \leq n_1$.

Since the iterates of the inexact Newton-like method (6.123)–(6.124) (whether $z_n = 0$, $n \geq 0$ or not) can rarely be computed in infinite-dimensional spaces, (6.123)–(6.124) can be replaced in practice by a family of discretized equations (as in Section 1.1)

$$P(a) + P_1(a) = 0. \tag{6.127}$$

Consider also the iteration $\{a_n\}$ $(n \geq 0)$ given for all $n \geq 0$ by

$$b_n = a_n - S(a_n)^{-1}P(a_n), \quad a_0 = L(x_0) \tag{6.128}$$

and

$$a_{n+1} = b_n - d_n. \tag{6.129}$$

6.2.2 Convergence Analysis For Inexact Newton-Like Methods

We find it convenient to introduce the following:

(G_1) Let $R > 0$ and using the notation $U(x_0, R) = \{x \in E_1 \mid \|x - x_0\| \leq R\}$, assume there exist $x_0 \in D$ and a function $\bar{\alpha}: U^3(x_0, R) \to [0, +\infty)$ such that

$$\|A(y)^{-1}[F(y) - F(x) - A(x)(y - x) + F'(y)(z - y)]\| \leq \bar{\alpha}(x, y, z) \quad (6.130)$$

for all $x, y, z \in U(x_0, R) \subseteq D$.

(G_2) There exist continuous, nondecreasing functions w, w_1 and w_2 such that

$$w: D \to [0, +\infty), \quad w_1, w_2: [0, R] \to [0, +\infty)$$

with $w_1(0) = w_2(0) = 0$ and a sequence $\{z_n\}$ $(n \geq 0)$ of points from D with

$$\|z_i\| \leq w(z_i) \leq w_1(r) \quad \text{for all } i \geq 0 \quad (6.131)$$

and for all $k \in N$

$$\sum_{i=0}^{k+1} \|z_i\| \leq \sum_{i=0}^{k+1} w(z_i) \leq w_2(r), \quad r \in [0, R] \quad (6.132)$$

for all $z_i \in U(x_0, r) \subseteq U(x_0, R)$. We note that (6.10) implies that the sequence $\{z_n\}$ $(n \geq 0)$ is null.

(G_3) Let $F, Q: D \subseteq E_1 \to E_2$ be nonlinear operators satisfying

$$\left\| \int_0^1 A(y)^{-1}(F'(x + t(y - x)) - A(x))(y - x)dt \right\|$$

$$\leq [C(\|x - x_0\|, \|y - x_0\|, \|y - x\|) + b_1]\|y - x\|, \quad (6.133)$$

$$\|A(y)^{-1}(F'(x) - A(x))(y - x)\|$$

$$\leq [C_1(\|x - x_0\|, \|y - x_0\|) + b_2]\|y - x\|, \quad (6.134)$$

$$\|A(y)^{-1}(Q(y) - Q(x))\|$$

$$\leq [C_2(\|x - x_0\|, \|y - x_0\|) + b_3]\|y - x\|, \quad (6.135)$$

and

$$\|A(y)^{-1}(A(x) - A(y))(y - x)\|$$

$$\leq [C_3(\|x - x_0\|, \|y - x_0\|, \|y - x\|) + b_4]\|y - x\| \tag{6.136}$$

for all $x, y \in U(x_0, r) \subseteq U(x_0, R) \subseteq D$. The real functions C, C_3, C_1, and C_2 are assumed to be continuous and nondecreasing on $[0, R]^3$, $[0, R]^3$, $[0, R]^2$, and $[0, R]^2$, respectively. Moreover $C(0,0,0) = C_1(0,0) = C_2(0,0) = C_3(0,0) = 0$ and b_1, b_2 b_3, b_4 are fixed nonnegative real numbers.

(G_4) There exist continuous, nondecreasing functions $w_3, w_4\colon [0, R] \to [0, +\infty)$ with $w_3(0) = w_4(0) = 0$ such that

$$\overline{\alpha_i} = \bar{\alpha}(x_i, y_i, x_{i+1}) \leq w_3(r) \quad \text{for all } i \geq 0 \tag{6.137}$$

and for all $k \in N$

$$\sum_{i=0}^{k+1} \overline{\alpha_i} \leq w_4(r) \tag{6.138}$$

for all $x_i, y_i, x_{i+1} \in U(x_0, r) \subseteq U(x_0, R)$.

Let us define the functions $\varphi, \varphi_1, \varphi_2\colon [0, R] \to [0, +\infty)$ by

$$\varphi(r) = r - [w_2(r) + (C(r, r, r) + C_1(r, r)$$

$$+ C_2(r, r) + b_1 + b_2 + b_3)r + w_4(r)], \tag{6.139}$$

$$\varphi_1(r) = r - \varphi(r), \tag{6.140}$$

and

$$\varphi_2(r) = C(0, r, r) + C_2(0, r) + C_3(0, r, r) + b_1 + b_4 + w_2(r), \tag{6.141}$$

$r \in [0, R]$.

By the hypotheses on the C and w functions above, there exist constants p, h_1, δ_0, δ_1, δ_2, δ_3, c_6, c_5, c_4, c_3 with $c_4 > \delta_3$ such that for

$$0 < \delta_0 \leq c_4 - c_5 h^p \leq r_0(h) = r(h) \leq c_3 h^p \leq \delta_1 \leq r_0 \leq R \tag{6.142}$$

and $r(h) = 0$ when $a_0 = a^*$ or $h = 0$, the following are true for sufficiently small r_0, $R > 0$ and $h \in (0, h_1]$

$$0 < \delta_2 \leq \varphi_1(r(h)) \leq \delta_3 \quad \text{and} \quad \theta(r(h)) < 1, \tag{6.143}$$

where

$$\theta(r) = C(R, r, r + R) + C_2(R, r) + C_3(R, r, r + R) + b_1 + b_4 \qquad (6.144)$$

provided that $b_1 + b_4 \in [0, 1)$.

We can now show that for all $h \in (0, h_2]$, where

$$h_2 = \min\left\{h_1, \left(\frac{\delta_2}{c_3 - c_6}\right)^{1/p}, \left(\frac{c_4 - \delta_3}{c_2 + c_5}\right)^{1/p}\right\} \text{ with } c_3 > c_6,\ c_6 \geq c_2 \qquad (6.145)$$

the following is true:

$$0 < c_2 h^p \leq \varphi(r(h)) \leq c_6 h^p. \qquad (6.146)$$

Indeed from (6.143), (6.146), we get $r(h) - \delta_3 \leq \varphi(r(h)) \leq r(h) - \delta_2$. It is enough to show that $r(h) - \delta_3 \geq c_2 h^p$ and $r(h) - \delta_2 \leq c_6 h^p$, which will be true if $c_4 - c_5 h^p - \delta_3 \geq c_2 h^p$ and $c_3 h^p - \delta_2 \leq c_6 h^p$, respectively. The last inequalities are true by the choice of h and (6.145).

Similar arguments can show that for sufficiently small $h \in (0, h_2]$ there exist δ_4, δ_5, δ_6, δ_7, c_7, c_8, c_9 such that for

$$0 < \delta_4 \leq c_7 - c_8 h^p \leq r^*(h) = \|L(x^*) - a^*\|$$

$$\leq c_9 h^p \leq \delta_5 \leq r^* = \|x_0 - x^*\|, \quad \delta_5 \leq \delta_1 \qquad (6.147)$$

and $r^*(h) = 0$ when $L(x^*) = a^*$ or $h = 0$, the following are true for a sufficiently small $r^* \leq r_0$

$$0 < \delta_6 \leq \varphi_2(r^*(h)) \leq \delta_7 < 1. \qquad (6.148)$$

We will need to introduce constants s_0, ℓ_0, such that

$$t_0 = 0, \quad s_0 \geq \|y_0 - x_0\|, \quad \ell_0 \geq \|z_0\|, \qquad (6.149)$$

iterations for all $n \geq 0$

$$s_{n+1} = t_{n+1} + h_{n+1}, \qquad (6.150)$$

$$t_{n+1} = s_n + \ell_n, \quad \ell_n = w(z_n), \qquad (6.151)$$

$$h_{n+1} = [C(s_n, t_{n+1}, t_{n+1} - s_n) + b_1](t_{n+1} - s_n)$$

$$+ [C_1(s_n, t_{n+1}) + b_2](t_{n+1} - s_n)$$

$$+ [C_2(t_n, t_{n+1}) + b_3](t_{n+1} - t_n) \tag{6.152}$$

for some given sequences $\{\alpha_n\}$ and $\{\ell_n\}$ $(n \geq 0)$ with

$$\alpha_n \geq 0, \quad \sum_{i=0}^{k+1} \alpha_i \leq w_4(r), \quad \text{and} \quad \sum_{i=0}^{k+1} \ell_i \leq w_2(r) \tag{6.153}$$

for all positive integers k, some fixed real constants (which may depend on r_0) γ_0, γ_1, γ_2 and some fixed $r_0 \in [0, R]$.

Moreover we define the iterations for all $n \geq 0$

$$\bar{v}_{n+1} = [C(\|y_n - x_0\|, \|x_{n+1} - x_0\|, \|x_{n+1} - y_n\|) + b_1]\|x_{n+1} - y_n\|$$

$$+ [C_1(\|y_n - x_0\|, \|x_{n+1} - x_0\|) + b_2]\|x_{n+1} - y_n\|$$

$$+ [C_2(\|x_n - x_0\|, \|x_{n+1} - x_0\|) + b_3]\|x_{n+1} - x_n\| + \bar{\alpha}_n \tag{6.154}$$

where $\bar{\alpha}_n = \bar{\alpha}(x_n, y_n, x_{n+1})$ with

$$\bar{\alpha}_n \leq \alpha_n \quad \text{for all } n \geq 0 \tag{6.155}$$

and the function

$$T(r) = s_0 + \varphi_1(r) \tag{6.156}$$

on $[0, R]$.

We can now state the main result on semilocal convergence.

THEOREM 6.5

Let F, Q be operators defined on a closed convex subset D of a Banach space E_1 with values in a Banach space E_2.

Assume:

(i) F is Fréchet-differentiable on D whereas Q is only continuous there;

(ii) conditions (G_1) and (G_3) are satisfied;

(iii) iterations $\{\alpha_n\}$, $\{\ell_n\}$, $\{\bar{\alpha}_n\}$, $\{z_n\}$ $(n \geq 0)$ satisfy conditions (6.153), (6.155), and

$$\|z_n\| \leq \ell_n \quad \textit{for all } n \geq 0; \tag{6.157}$$

(iv) there exists a minimum nonnegative number r_0 such that

$$T(r_0) \leq r_0 \quad \textit{and} \quad r_0 \leq R; \tag{6.158}$$

(v) moreover r_0, R *satisfy the inequality*

$$\theta(r_0) < 1 \quad \text{for } r_0 \leq R, \tag{6.159}$$

where the function θ *is given by (6.144); and*

(vi) the ball $U(x_0, R) \subseteq D$.

Then

(a) scalar sequences $\{t_n\}$ $\{s_n\}$ $(n \geq 0)$ *generated by relations (6.149)–(6.152) are monotonically increasing and bounded above by their limit, which is* r_0;

(b) sequences $\{x_n\}$, $\{y_n\}$ $(n \geq 0)$ *generated by relations (6.123)–(6.124) are well defined, remain in* $U(x_0, r_0)$ *for all* $n \geq 0$, *and converge to a solution* x^* *of the equation* $F(x) + Q(x) = 0$, *which is unique in* $U(x_0, R)$.

Moreover, the following estimates hold for all $n \geq 0$:

$$\|y_n - x_n\| \leq s_n - t_n, \tag{6.160}$$

$$\|x_{n+1} - y_n\| \leq t_{n+1} - s_n, \tag{6.161}$$

$$\|x_n - x^*\| \leq r_0 - t_n, \tag{6.162}$$

$$\|y_n - x^*\| \leq r_0 - s_n, \tag{6.163}$$

$$\|y_{n+1} - x_{n+1}\| = \|A(x_{n+1})^{-1}(F(x_{n+1}) + Q(x_{n+1}))\| \leq \bar{v}_{n+1}, \tag{6.164}$$

and

$$\|y_n - x_n\| \leq \|x^* - x_n\| + [C(\|x_n - x_0\|, \|x^* - x_0\|, \|x^* - x_n\|)$$

$$+ C_2(\|x_n - x_0\|, \|x^* - x_0\|) + b_1 + b_3]\|x_n - x^*\| \tag{6.165}$$

(We will be concerned only with the case $r_0 > 0$, *since when* $r_0 = 0$, $x_0 = x^*$).

PROOF (a) Using relations (6.149), (6.150), (6.151), (6.153), (6.158), we deduce that the scalar sequence $\{t_n\}$ $n \geq 0$ is monotonically increasing, nonnegative, and $t_0 \leq s_0 \leq t_1 \leq s_1 \leq r_0$. Let us assume that $t_k \leq s_k \leq t_{k+1} \leq s_{k+1} \leq r_0$ for $k = 0, 1, 2, \ldots, n$. Then by relations (6.150), (6.151) and (6.153) we can have in turn

$$t_{k+2} = s_{k+1} + \ell_{k+1} \leq t_{k+1} + C(s_k, t_{k+1}, t_{k+1} - s_k)(t_{k+1} - s_k)$$

$$+ C_1(s_k, t_{k+1})(t_{k+1} - s_k) + (b_1 + b_2)(t_{k+1} - s_k)$$

$$+ [C_2(t_k, t_{k+1}) + b_3](t_{k+1} - t_k) + \alpha_k + \ell_{k+1}$$

$$\leq \cdots s_0 + [C(r_0, r_0, r_0) + C_1(r_0, r_0) + C_2(r_0, r_0)$$

$$+ b_1 + b_2 + b_3]t_{k+1} + \sum_{i=0}^{k} \alpha_i + \sum_{i=0}^{k+1} \ell_i \leq T(r_0) \leq r_0.$$

Hence, the scalar sequence $\{t_n\}$ $(n \geq 0)$ is bounded above by r_0. By hypothesis (6.158) the number r_0 is the minimum nonnegative zero of the equation $T(r) - r = 0$ on $[0, r_0]$, and from the above $r_0 = \lim_{n\to\infty} t_n$.

(b) Using relations (6.123), (6.124), (6.149), (6.150), and (6.151) we deduce that $x_1, x_0 \in U(x_0, r_0)$ and that estimates (6.160) and (6.161) are true for $n = 0$. Let us assume that they are true for $k = 0, 1, 2, \ldots, n-1$. Using the induction hypothesis, we have

$$\|x_{k+1} - x_0\| \leq \|x_{k+1} - y_0\| + \|y_0 - x_0\|$$

$$\leq \|x_{k+1} - y_k\| + \|y_k - y_0\| + \|y_0 - x_0\|$$

$$\leq \cdots \leq (t_{k+1} - s_k) + (s_k - s_0) + s_0 \leq t_{k+1} \leq r_0,$$

and

$$\|y_{k+1} - x_0\| \leq \|y_{k+1} - y_0\| + \|y_0 - x_0\|$$

$$\leq \|y_{k+1} - x_{k+1}\| + \|x_{k+1} - y_k\| + \|y_k - y_0\| + \|y_0 - x_0\|$$

$$\leq \cdots \leq (s_{k+1} - t_{k+1}) + (t_{k+1} - s_k) + (s_k - s_0) + s_0$$

$$\leq s_{k+1} \leq r_0.$$

That is, $x_n, y_n \in U(x_0, r_0)$ for all $n \geq 0$.

We can now have from approximations (6.123) and (6.124)

$$F(x_{k+1}) + Q(x_{k+1}) = F(x_{k+1}) - F(y_k) - A(y_k)(x_{k+1} - y_k)$$

$$+ A(y_k)(x_{k+1} - y_k) + F(y_k) + Q(x_{k+1})$$

$$= \int_0^1 [F'(y_k + t(x_{k+1} - y_k)) - A(y_k)](x_{k+1} - y_k)dt$$

$$+ (A(y_k) - F'(y_k))(x_{k+1} - y_k) + (Q(x_{k+1}) - Q(x_k))$$

$$+ (F(y_k) - F(x_k) - A(x_k)(y_k - x_k) + F'(y_k)(x_{k+1} - y_k)).$$

Hence, by using hypotheses (6.133)–(6.136) we obtain in turn

$$\|A(x_{k+1})^{-1}(F(x_{k+1}) + Q(x_{k+1}))\|$$

$$\leq \left\| \int_0^1 A(x_{k+1})^{-1}[F'(y_k + t(x_{k+1} - y_k)) - A(y_k)](x_{k+1} - y_k) \right\| dt$$

$$+ \|A(x_{k+1})^{-1}(A(y_k) - F'(y_k))(x_{k+1} - y_k)\|$$

$$+ \|A(x_{k+1})^{-1}(Q(x_{k+1}) - Q(x_k))\|$$

$$+ \|A(x_{k+1})^{-1}[F(y_k) - F(x_k) - A(x_k)(y_k - x_k)$$

$$+ F'(y_k)(x_{k+1} - y_k)]\| = \bar{v}_{k+1} \leq v_{k+1} \tag{6.166}$$

by hypotheses (6.155) and relations (6.152) and (6.154). Hence, we have shown estimate (6.164) for all $n \geq 0$.

Using relations (6.123), (6.150), and (6.164) we obtain

$$\|y_{k+1} - x_{k+1}\| = \|A(x_{k+1})^{-1}(F(x_{k+1}) + Q(x_{k+1}))\|$$

$$\leq \bar{v}_{k+1} \leq v_{k+1} = s_{k+1} - t_{k+1},$$

which shows estimate (6.160) for all $n \geq 0$.

Similarly, from relations (6.124), (6.151) and (6.157) we obtain

$$\|x_{k+1} - y_k\| = \| - z_k\| \leq w(z_k) = t_{k+1} - s_k,$$

from which it follows that estimate (6.161) is true for all $n \geq 0$.

It now follows from the estimates (6.160) and (6.161) that the sequence $\{x_n\}$ $(n \geq 0)$ is Cauchy in a Banach space E_1 and as such it converges to some $x^* \in U(x_0, r_0)$, which by taking the limit as $n \to \infty$ in (6.123), we obtain $F(x^*) + Q(x^*) = 0$.

Using (6.131) we deduce that $\lim_{n\to\infty} z_n = 0$. Moreover from (6.124) we can get

$$\lim_{n\to\infty} y_n = \lim_{n\to\infty} (x_{n+1} + z_n) = \lim_{n\to\infty} x_{n+1} + \lim_{n\to\infty} z_n = x^*.$$

Estimates (6.162) and (6.163) follow from (6.155) and (6.161), respectively, by using standard arguments in majorant theory [6], [11].

To show uniqueness, we assume that there exists another solution y^* of Equation (6.122) in $U(x_0, R)$ with $x^* \neq y^*$. From the approximation

$$y_n - y^* = -A(x_n)^{-1}\left\{\int_0^1 [F'(y^* + t(x_n - y^*)) - A(y^*)](x_n - y^*)dt\right.$$

$$\left. + (A(y^*) - A(x_n))(x_n - y^*) + (Q(x_n) - Q(y^*))\right\},$$

we can obtain

$$\begin{aligned}\|y_n - y^*\| &\leq [C(\|y^* - x_0\|, \|x_n - x_0\|, \|x_n - y^*\|) + b_1]\|x_n - y^*\| \\ &\quad + [C_3(\|y^* - x_0\|, \|x_n - x_0\|, \|x_n - y^*\|) + b_4]\|x_n - y^*\| \\ &\quad + C_2(\|y^* - x_0\|, \|x_n - x_0\|)\|x_n - y^*\| \\ &\leq \theta(r_0)\|x_n - y^*\|. \end{aligned} \tag{6.167}$$

By letting $n \to \infty$ in (6.167) we get

$$\|x^* - y^*\| \leq \theta(r_0)\|x^* - y^*\| < \|x^* - y^*\|. \tag{6.168}$$

Hence, we conclude from (6.168) that $x^* = y^*$. Finally using the approximation

$$y_n - x_n = x^* - x_n + A(x_n)^{-1}\left[\int_0^1 (F'(x_n + t(x^* - x_n))\right.$$

$$\left. - A(x_n))(x^* - x_n)dt + Q(x^*) - Q(x_n))\right]$$

and from conditions (6.133)–(6.136), (6.166) we obtain (6.165).

That completes the proof of the theorem. ∎

We will need the following result on local convergence.

THEOREM 6.6
Let $F, Q: D \subset E_1 \to E_2$ be nonlinear operators as in Theorem 6.5, and assume:

(i) there exists a regular solution $x^* \in D$ of the equation $F(x)+Q(x)=0$;

(ii) condition (G_3) is satisfied on $U(x^*, r^*)$ (for $x_0 = x^*$) and $U(x^*, r^*) \subseteq D$;

(iii) there exists a sequence $\{z_n\}$ $(n \geq 0)$ of points from D satisfying

$$\|z_n\| \leq g_n = g(z_n) \leq w_5(r) \tag{6.169}$$

where $g: U(x^*, r^*) \to [0, +\infty)$ is continuous; and

(iv) constants b_1, b_4 are such that $b_1 + b_4 \in [0, 1)$.

Then the following hold:

(a) For sufficiently small $r^* \in (0, R]$

$$0 < \varphi_2(r^*) < 1. \tag{6.170}$$

(b) Sequences $\{y_n\}$, $\{x_n\}$ $(n \geq 0)$ are well defined, remain in $U(x^*, r^*)$ for all $n \geq 0$ and $\lim_{n\to\infty} x_n = \lim_{n\to\infty} y_n = x^*$. Moreover, the solution x^* of equation (6.122) is unique in $U(x^*, r^*)$.

Furthermore, the following estimates hold for all $n \geq 0$:

$$\|x_{n+1} - x^*\| \leq \gamma_n \|x_n - x^*\| \leq \gamma \|x_n - x^*\| \tag{6.171}$$

and

$$\|y_n - x^*\| \leq \delta_n \|x_n - x^*\| \leq \delta \|x_n - x^*\|, \tag{6.172}$$

where

$$\delta_n = C(0, \|x_n - x^*\|, \|x_n - x^*\|) + C_2(0, \|x_n - x^*\|)$$

$$+ C_3(0, \|x_n - x^*\|, \|x_n - x^*\|) + b_1 + b_4, \tag{6.173}$$

$$\delta = C(0, r^*, r^*) + C_2(0, r^*) + C_3(0, r^*, r^*) + b_1 + b_4, \tag{6.174}$$

$$\gamma_n = \delta_n + g_n \tag{6.175}$$

and

$$\gamma = \delta + w_5(r^*). \tag{6.176}$$

PROOF (a) By hypotheses (ii), (iii), and (iv) we have $C_0(0,0,0) = C_1(0,0) = C_2(0,0) = w_5(0) = 0$, $b_1 + b_2 \in [0,1)$ and that all these functions are continuous and nondecreasing on $[0, \|x^* - x_0\|]$. Hence, we can find a number $r^* \in (0, \|x^* - x_0\|$ such that the estimate (6.170) holds.

(b) Let us assume that

$$x_m \in U(x^*, r^*) \quad \text{for } m = 0, 1, 2, \ldots, k \tag{6.177}$$

since $x_0 \in U(x^*, r^*)$. Using (6.123) we get

$$y_k - x^* = -A(x_k)^{-1}[F(x_k) - F(x^*) - A(x_k)(x_k - x^*) + Q(x_k) - Q(x^*)]. \tag{6.178}$$

We also introduce the approximation

$$\begin{aligned} &F(x_k) - F(x^*) - A(x_k)(x_k - x^*) + Q(x_k) - Q(x^*) \\ &= \int_0^1 [F'(x^* + t(x_k - x^*)) - A(x_k)](x_k - x^*)dt + Q(x_k) - Q(x^*) \\ &= \int_0^1 [F'(x^* + t(x_k - x^*)) - A(x_k)](x_k - x^*)dt \\ &\quad + Q(x_k) - Q(x^*). \end{aligned} \tag{6.179}$$

We now compose both sides of (6.179) by $A(x_k)^{-1}$, and then by taking norms and using (6.145)–(6.148), we obtain that the left-hand side of (6.179) is bounded above by

$$\begin{aligned} &[C(0, \|x_k - x^*\|, \|x_k - x^*\|) + b_1 + C_3(0, \|x_k - x^*\|, \|x_k - x^*\|) \\ &\quad + b_4 + C_2(0, \|x_k - x^*\|)]\|x_k - x^*\|. \end{aligned} \tag{6.180}$$

From (6.177), (6.178), and (6.180) we now have

$$\begin{aligned} \|y_k - x^*\| &\leq \|A(x_k)^{-1} \int_0^1 [F'(x^* + t(x_k - x^*)) - A(x_k)] \\ &\qquad \cdot (x_k - x^*)dt + (Q(x_k) - Q(x^*))]\| \\ &\leq \delta_k\|x_k - x^*\| \leq \delta\|x_k - x^*\|. \end{aligned} \tag{6.181}$$

The above estimate shows that (6.172) is true and that $y_k \in U(x^*, r^*)$ since $\delta_k \leq \delta < 1$ (by (6.170)).

Moreover, from (6.124), (6.181), and (6.169), we get

$$\|x_{k+1} - x^*\| \leq \|y_k - x^*\| + \|z_k\| \leq \gamma_k \|x_k - x^*\| \leq \gamma \|x_k - x^*\|, \tag{6.182}$$

which shows (6.171) and that $x_{k+1} \in U(x^*, r^*)$. Hence the sequences $\{x_n\}$, $\{y_n\}$ $(n \geq 0)$ are well defined, remain in $U(x^*, r^*)$ and satisfy (6.171) and (6.172) for all $n \geq 0$.

Let $m \geq 0$. Then by (6.171) we get

$$\begin{aligned} \|x_{n+m} - x^*\| &\leq \gamma_{n+m-1} \|x_{n+m-1} - x^*\| \\ &\leq \gamma_{n+m-1}\gamma_{n+m-2} \|x_{n+m-2} - x^*\| \\ &\leq \cdots \leq \gamma^m \|x_n - x^*\|. \end{aligned} \tag{6.183}$$

Similarly by (6.171) and (6.172) we get

$$\|y_{n+m} - x^*\| \leq \delta \cdot \gamma^m \|x_n - x^*\|. \tag{6.184}$$

Finally by letting $m \to \infty$ in (6.183) and (6.184) we obtain $\lim_{n\to\infty} x_n = \lim_{n\to\infty} y_n = x^*$ (since $0 < \gamma < 1$). The proof of the uniqueness of the solution x^* of Equation (6.122) in $U(x^*, r^*)$ is omitted as it is identical to the corresponding one in Theorem 6.5.

That completes the proof of the theorem. ∎

The points z_n $(n \geq 0)$ appearing in (6.131) depend on x_n $(n \geq 0)$. The points $\{z_n\}$ $(n \geq 0)$ in (6.169) depend on x_n $(n \geq 0)$ and may be the point x^*. That is why we can choose the functions g, w_5 (see (6.169)) to be the same or different from the functions w, w_1 (see (6.131)), respectively.

In many applications it turns out that the solution x^* of Equation (6.122) as well as iterates x_n, y_n have "better smoothness" properties than the elements of E_1. This is a motivation for considering a subset $E_3 \subseteq E_1$ such that

$$\begin{aligned} &x^* \in E_3, x_n, y_n \in E_3, \; x_n - x^*, y_n - x^* \in E_3, \\ &x_{n+1} - x_n, y_{n+1} - y_n \in E_3 \; (n \geq 0). \end{aligned} \tag{6.185}$$

We consider a family

$$\{P, P_1, L, L_0\}, \quad h > 0 \tag{6.186}$$

where

$$P, P_1\colon E_4 \subseteq E^1 \to E^2, \quad h > 0$$

are nonlinear operators and

$$L\colon E_1 \to E^1, \quad L_0\colon E_2 \to E^2, \quad h > 0$$

are bounded linear discretization operators such that

$$L(E_3 \cap U(x^*, \delta_5)) \subseteq E_4, \quad h > 0. \tag{6.187}$$

The operators P, P_1, L, L_0, S depend on h. That is, $P = P_h$, etc. To simplify the notation we assume that this is understood and hence we avoid the use of the subscript h.

The discretization (6.171) is called uniform if there exist a number δ_1 such that

$$U(L(x^*), \delta_1) \subseteq E_4, \quad h > 0 \tag{6.188}$$

and the triplet $(P, Q, L(x^*))$ satisfies the "G" conditions that the triplet (F, A, x_0) satisfies for all $h > 0$ in the ball $U(L(x^*), \delta_1)$.

Moreover, the discretization family (6.186) is called *bounded* if there is a constant $q > 0$ such that

$$\|L(x)\| \leq q\|x\|, \quad x \in E_3, \quad h > 0, \tag{6.189}$$

consistent of order $p > 0$ if there are two constants $c_0, c_1 > 0$ such that

$$\|S(L(x))^{-1}(L_0(F + Q) - (P + P_1)(L(x)))\| \leq c_0 h^p,$$

$$x \in E_3 \cap U(L(x^*), \delta_5), \quad h > 0 \tag{6.190}$$

and

$$\|S(L(x))^{-1}(L_0(A(x)(y) - S(L(x))L(y)))\| \leq c_1 h^p,$$

$$x \in E_3 \cap U(L(x^*), \delta_5), y \in E_3, h > 0. \tag{6.191}$$

Concerning the function w_5, we can easily see from (6.132) and (6.169) that it can be identified with the function w_2. Choose, for example, $z_n = d_n$ for all $n \geq 0$, $h > 0$. However we do not need this to prove our discretization results. By choosing $L(x_0) \in U(x_0, \frac{R}{3})$, $\delta_5 \leq \frac{R}{3}$ and $r^* \leq \frac{R}{3q}$, one can easily show that $U(L(x^*), \delta_5) \subseteq U(x_0, R)$. Hence the C^h, w^h, $\bar{\alpha}^h$, b^h are just the C, w, $\bar{\alpha}$, b, respectively, restricted on $U(L(x^*), \delta_5)$.

With the notation introduced above we can now formulate our main result.

THEOREM 6.7

Let $F, Q\colon D \subseteq E_1 \to E_2$ be nonlinear operators as in Theorem 6.5. Assume:

(i)] hypotheses of Theorem 6.6 are satisfied;

(ii) discretization (6.186) is bounded, stable, and consistent of order p and $U(L(x^*), \delta_5) \subseteq U(x_0, R)$ for all $h \in (0, h_2]$;

(iii) the following estimate holds:

$$\|S(v)^{-1}(S(w) - S(v))\| \leq C_4(\|v - L(x^*)\|, \|w - L(x^*)\|, \|v - w\|) + b_5, \tag{6.192}$$

for all $v, w \in U(L(x^*), \delta_5)$, some $b_5 \in [0, 1)$ and a function C_4 satisfying the same properties as the C function.

Then

(a) Equation (6.127) has a locally unique solution

$$a^*(h) = a^* = L(x^*) + O(h^p) \tag{6.193}$$

for all $h \in (0, \overline{h_0}]$ with $\overline{h_0}$ being a constant.

(b) There exist constants $\overline{h_1} \in (0, \overline{h_0}]$, $r_1 \in (0, \delta_5]$ such that the discrete iteration (6.128)-(6.129) converges to a^*, and

(c) If there exist constants c_{10}, c_{11} with $4(b_1c_{11} + b_3c_{11} + 2qc_3b_5 + c_0 + c_1) \leq c_{10} \leq c_{11}$, such that for all $n \geq 0$,

$$\|d_n - L(z_n)\| \leq (c_{11} - c_{10})h^p, \quad h \in (0, \overline{h_0}], \quad r \in (0, \delta_5], \tag{6.194}$$

then there exist constants $\overline{h_1} \in (0, \overline{h_0}]$, $r_3 \in (0, r_1]$ such that the following estimates hold for all $n \geq 0$

$$b_n = L(y_n) + O(h^p), \tag{6.195}$$

$$a_n = L(x_n) + O(h^p), \tag{6.196}$$

$$S(b_n)^{-1}(P + P_1)(b_n) = S(b_n)^{-1}L_0((F + Q)(y_n)) + O(h^p), \tag{6.197}$$

(provided that $(P+P_1)$ *is b-Lipschitz continuous on* $U(L(x^*),\delta_5)$

$$S(a_n)^{-1}(P+P_1)(a_n)=S(a_n)^{-1}L_0((F+Q)(x_n))+O(h^p), \quad (6.198)$$

$$b_n-a^*=L(y_n-x^*)+O(h^p) \quad (6.199)$$

and

$$a_n-a^*=L(x_n-x^*)+O(h^p). \quad (6.200)$$

PROOF (a) The C,w functions are continuous, vanish at the origin and $b_1+b_4\in[0,1)$. Hence we can find intervals $(0,h_0]$ and $(0,1]$ such that conditions (6.158) and (6.159) are satisfied for all $h\in(0,h_0]$ and $r(h)\in(0,\delta_1]$. Set $h_3=\min\{h_0,h_2,(\frac{\delta_1}{c_3})^{1/p}\}$. Then using (6.142) and (6.190) we obtain in turn

$$\begin{aligned} s_0(h) &= \|S(L(x))^{-1}(P+P_1)(L(x^*))\| \\ &\le \|(P+P_1)(L(x^*))-L_0(F+Q)(x^*))\| \\ &\le c_0h^p\le c_2h^p\le\varphi(r(h)), \end{aligned} \quad (6.201)$$

and

$$r(h)\le c_3h^p\le\delta_1, \quad \text{for all } h\in(0,h_3], \quad (6.202)$$

which shows that (6.158) and (6.159) hold for all $h\in(0,h_3]$. Since all hypotheses of Theorem 6.122 are satisfied, Equation (6.127) has a solution $a^*(h)=a^*\in U(L(x^*),r(h))$ which is a unique solution in $U(L(x^*),\delta_1)$. Thus (6.193) follows from

$$\|a^*-L(x^*)\|\le r(h)\le c_3h^p\le\delta_1 \quad (6.203)$$

by setting $\overline{h_0}=h_3$.

(b) As in part (a) we can find intervals $(0,h_3]$ and $(0,\delta_5]$ such that conditions (6.139) and (6.170) are satisfied for all $h\in(0,h_3]$ and $r^*(h)\in(0,\delta_5]$. By applying equation 6.122 to (6.127) we see that the sequence (6.128)–(6.129) converges to a^* if

$$\|L(x_0)-a^*\|<r^*(h) \quad (6.204)$$

and

$$U(a^*,\|L(x_0)-a^*\|)\subseteq U(L(x^*),\delta_1). \quad (6.205)$$

But (6.205) certainly holds if

$$\|a^* - L(x^*)\| + \|L(x_0) - a^*\| \leq \delta_1. \tag{6.206}$$

By (6.189) and (6.202) we obtain

$$\|L(x_0)-a^*\| \leq \|L(x_0)-L(x^*)\|+\|L(x^*)-a^*\| \leq q\|x_0-x^*\|+c_3h^p. \tag{6.207}$$

Thus (6.204), (6.205) hold if

$$q\|x_0 - x^*\| + 2c_3h^p \leq \delta_1, \tag{6.208}$$

and

$$q\|x_0 - x^*\| + c_3h^p \leq c_7 - c_8h^p \tag{6.209}$$

hold, respectively. Conditions (6.208) and (6.209) will certainly hold if

$$q\|x_0 - x^*\| \leq \frac{\delta_1}{2}, \quad 2c_3h^p \leq \frac{\delta_1}{2}, \quad q\|x_0 - x^*\| \leq \frac{c_7}{2}$$

and

$$c_3h^p \leq \frac{c_7}{2} - c_8h^p.$$

We choose

$$\|x_0 - x^*\| \leq r_1 = \min\left\{\frac{\delta_1}{2q}, \frac{c_7}{2q}\right\},$$

and

$$h_4 = \min\left\{h_2, h_3, \left(\frac{\delta_1}{4c_3}\right)^{1/p}, \left[\frac{c_7}{2(c_3+c_8)}\right]^{1/p}\right\}.$$

It is now easily verified that (6.204) and (6.205) are satisfied for all $h \in (0, h_4]$ and $x_0 \in U(x^*, r_1)$. Therefore, for these h and x_0, the iteration (6.128)–(6.129) converges to a^*.

(c) We will now show that there exist $\overline{h_1} \in (0, \overline{h_0}]$, $r_3 \in (0, r_1]$ such that

$$\|a_n - L(x_n)\| \leq c_{11}h^p. \tag{6.210}$$

For $n = 0$ (6.210) is true since $a_0 = L(x_0)$. Suppose that (6.210) holds for $n = 0, 1, \ldots, i$. We note that if we show that

$$\|b_n - L(y_n)\| \leq c_{10}h^p, \tag{6.211}$$

then from (6.128), (6.129), (6.211), (6.194) and the estimate

$$\begin{aligned}\|a_{i+1} - L(x_{i+1})\| &= \|b_i - d_i - L(y_i - z_i)\| \\ &\leq \|b_i - L(y_i)\| + \|d_i - L(z_i)\| \\ &\leq c_{10}h^p + (c_{11} - c_{10})h^p = c_{11}h^p, \end{aligned} \tag{6.212}$$

we can complete the induction for (6.210). But (6.211) is true for $n = 0$ by (6.194). We now suppose that (6.196) is true for $n = 0, 1, \ldots, i$. Using (6.123), (6.124), (6.128) and (6.129) we can obtain the approximation

$$\begin{aligned}b_i - L(y_i) = S(a_i)^{-1}\{&[S(a_i)(a_i - L(x_i)) - (P + P_1)(a_i) \\ &+ (P + P_1)(L(x_i))] + [(S(a_i) - S(L(x_i)))L(A(x_i)^{-1}(F + Q)(x_i))] \\ &+ [S(L(x_i))L(A(x_i)^{-1}(F + Q)(x_i)) - L_0((F + Q)(x_i))] \\ &+ [L_0((F + Q)(x_i)) - (P + P_1)(x_i))]\}. \end{aligned} \tag{6.213}$$

From (6.189) and (6.210) we obtain

$$\|a_i - L(x^*)\| \leq \|a_i - L(x_i)\| + \|L(x_i) - L(x^*)\| \leq c_{11}h^p + qr_1. \tag{6.214}$$

By composing the first bracket in (6.213) by $S(a_i)^{-1}$ and taking norms, using (6.133), (6.136), and (6.214), we obtain that this term is bounded above by

$$\begin{aligned}&\left\| \int_0^1 S(a_i)^{-1}\{[P'(L(x_i) + t(a_i - L(x_i)) - S(a_i)] \right. \\ &\qquad \left. \cdot(a_i - L(x_i))dt + (P_1(L(x_i)) - P_1(a_i))\} \right\| \\ &\leq [C(\|L(x_i) - L(x^*)\|, \|a_i - L(x^*)\|, \|a_i - L(x_i)\| + b_1]\|a_i - L(x_i)\| \\ &\quad + [C_2(\|L(x_i) - L(x^*)\|, \|a_i - L(x^*)\|) + b_3]\|a_i - L(x_i)\| \qquad (6.215) \\ &\leq [C(qr_1, c_{11}h^p + qr_1, c_{11}h^p) + b_1]c_{11}h^p \end{aligned}$$

$$+ [C_2(qr_1, c_{11}h^p + qr_1) + b_3]c_{11}h^p. \quad (6.216)$$

Moreover, by adding and subtracting $S(a_i)^{-1}$ inside the parenthesis of the second bracket, composing by $S(L(a_i)^{-1}$ and using (6.214) and (6.189), we obtain that this term is bounded above by

$$[C_4(\|L(x_i) - L(x^*)\|, \|a_i - L(x^*)\|, \|a_i$$

$$- L(x_i)\| + b_5]q(\|y_i - x^*\| + \|x_i - x^*\|)$$

$$\leq 2qc_3[C_4(qr_1, c_{11}h^p + qr_1, c_{11}h^p) + b_5]h^p \quad \text{(by (6.138))} \quad (6.217)$$

Furthermore using (6.191) and (6.193), we obtain that the third and fourth bracket in (6.213), after being composed by $S(a_i)^{-1}$, are bounded above by c_1h^p and c_0h^p, respectively.

Finally, by collecting all the above majorizations, we obtain that estimate (6.211) will be true if

$$C(qr_1, c_{11}h^p + qr_1, c_{11}h^p)c_{11} + C_2(qr_1, c_{11}h^p + qr_1)c_{11}$$

$$+ 2qC_4(qr_1, c_{11}h^p + qr_1, c_{11}h^p)c_3$$

$$+ (b_1c_{11} + b_3c_{11} + 2qc_3b_5 + c_1 + c_1) \leq c_{10}. \quad (6.218)$$

Inequality (6.218) will certainly be true if each term on the left is bounded above by $\frac{1}{4}$. Since the functions C and C_0 vanish at the origin, we can find h_5, $r_2 > 0$ such that this will happen for the first three terms. By the hypothesis of (c), the last term is also bounded above by $\frac{1}{4}$. Finally, set $\overline{h_1} = \min\{h_4, h_5, (\frac{1}{4c_{11}})^{1/p}\}$ and $r_3 = \min\{r_1, r_2\}$. With the above choices of $\overline{h_1}$ and r_3, estimates (6.195) and (6.196) follow.

Using (6.190) and (6.211) we obtain

$$\|S(b_n)^{-1}((P + P_1)(b_n) - L_0((F + Q)(y_n)))\|$$

$$\leq \|S(b_n)^{-1}((P + P_1)(b_n) - (P + P_1)(L(y_n)))\|$$

$$+ \|S(b_n)^{-1}((P + P_1)(L(y_n)) - L_0((F + Q)(y_n)))\|$$

$$\leq b\|b_n - L(y_n)\| + c_0h^p \leq (bc_{10} + c_0)h^p. \quad (6.219)$$

Estimate (6.219) shows (6.197). Estimate (6.198) is obtained similarly by replacing b_n, y_n and c_{10} by a_n, x_n and c_{11}.

Moreover, from (6.211) and (6.202) we obtain

$$\|(b_n - a^*) - L(y_n - x^*)\| \leq \|b_n - L(y_n)\| + \|a^* - L(x^*)\|$$

$$\leq c_{10}h^p + c_3h^p = (c_{10} + c_3)h^p, \quad (6.220)$$

which shows (6.199).

Furthermore, from (6.210) and (6.202), we finally obtain

$$\|(a_n - a^*) - L(x_n - x^*)\| \leq \|a_n - L(x_n)\| + \|a^* - L(x^*)\|$$

$$\leq c_{11}h^p + c_3h^p = (c_{11} + c_3)h^p, \quad (6.221)$$

from which (6.200) follows.

That completes the proof of the theorem. ∎

We note that conditions $\delta_1 \leq r_0$ and $\delta_5 \leq r^*$ (see (6.142) and (6.147)) are not used in the proof. However, in many practical applications we may want these conditions to be true.

We can now state the mesh-independence principle for perturbed Newton-like methods. The proof is similar to the one in Section 1.1 and is omitted.

THEOREM 6.8

Assume:

(i) hypotheses of Theorem 6.7 hold;

(ii) there exists a constant $\delta > 0$ such that

$$\liminf_{h\to 0} \|L(u)\| \geq \delta\|u\| \quad \textit{for each } u \in E_3. \quad (6.222)$$

Then for some $r_6 \in (0, r_3]$, and for any fixed $\varepsilon > 0$ and $x_0 \in U(x^*, r_6)$ there exists a constant $\bar{h} = \bar{h}(\varepsilon, x_0) \in (0, \overline{h_1}]$ such that

$$|\min\{n \geq 0, \|x_n - x^*\| < \varepsilon\} - \min\{n \geq 0, \|a_n - a^*\| < \varepsilon\} \leq 1 \quad (6.223)$$

for all $h \in (0, \bar{h}]$.

As in Section 1.1, we can have a stronger version of the mesh-independence principle.

COROLLARY 6.2

Assume:

(i) hypotheses of Theorem 6.7 hold;

(ii) condition $\lim_{h\to 0}\|L(u)\| = \|u\|$ *holds uniformly for* $u \in E_3$.

Then there exists a constant $r_7 \in (0, r_3]$ *and, for any fixed* $\varepsilon > 0$*, some* $\overline{h_2} = \overline{h_2}(\varepsilon) \in (0, \overline{h_1}]$ *such that (6.223) holds for all* $h \in (0, \overline{h_2}]$ *and all starting points* $x_0 \in (x^*, r_7)$.

6.2.3 Applications

REMARK 6.2 (1) In all our previous results we assumed that $A(y)$ is invertible for all $y \in D$. It turns out that our results hold under the weaker condition that $A(x_0)$ is invertible only. For Theorem 6.5 replace $A(y)^{-1}$ by $A(x_0)^{-1}$ in (6.133), (6.134), (6.135) and (6.136), and add the hypotheses

$$\|A(x_0)^{-1}(A(x) - A(x_0))\| \leq C_0(\|x - x_0\|) + b_0, \quad b_0 \in [0, 1) \tag{6.224}$$

and

$$C_0(r_0) + b_0 < 1 \tag{6.225}$$

for all $x \in U(x_0, r) \subseteq U(x_0, R)$, $r \in [0, R]$, where the function C_0 is continuous and nondecreasing on $[0, R]$ with $C_0(0) = 0$. Moreover define the function C_5 on $[0, R]$ by

$$C_5(r) = [1 - (C_0(r) + b_0)]^{-1}. \tag{6.226}$$

Then using (6.224), (6.226) and the Banach lemma on invertible operators we get that $A(x_n)$ is invertible and

$$\|A(x_n)^{-1}A(x_0)\| \leq C_5(\|x_n - x_0\|). \tag{6.227}$$

Furthermore multiply the "C", w_4, $\bar{\alpha}$ functions by $C_5(r)$ (or $C_5(\|x_n - x_0\|)$) and the "b" constants by $C_5(r)$ (or $C_5(\|x_n - x_0\|)$.

With the above modifications one can easily see that the conclusions of Theorem 6.5 can now follow.

(2) Similarly, for Theorem 6.6, we can argue as in Remark (1) but with the following modifications: x_0, C_0, b_0, C_5, w_4 are x^*, C_0^*, b_0^*, C_5^* and w_5, respectively. The "C^*" functions and the point b_0^* have properties similar to the ones without the stars. We note that they can even be taken to be equal to each other, and if this is true the rest of the results in this study can follow.

(3) The results obtained in this study will also hold if the left-hand side of (6.133) is replaced by the conditions

$$\|A(y)^{-1}(F'(x+t(y-x))-A(x))(y-x)\| \quad \text{for all } t \in [0,1],$$

or

$$\int_0^1 \|A(y)^{-1}(F'(x+t(y-x))-A(x))(y-x)\|dt$$

or

$$\|A(y)^{-1}(F'(x+t(y-x))-A(x))\|\,\|y-x\| \quad \text{for all } t \in [0,1],$$

or

$$\int_0^1 \|A(y)^{-1}(F'(x+t(y-x))-A(x))\|\,\|y-x\|dt$$

or any combination of the above conditions in the nonaffine form whether $A(y) = A(x_0)$ (or not). (See the proof of Theorem 6.5.) In particular see relation (6.166).) In all cases the function C and the point b_1 appearing at the right-hand side of (6.133) may become larger which will result to larger upper bounds on the distances

$$\|y_n - x_n\|, \|x_{n+1} - y_n\|, \|x_n - x^*\| \text{ and } \|y_n - x^*\| \quad (n \geq 0).$$

Similar remarks can be made for the left-hand sides of conditions (6.134), (6.135) and (6.136).

(4) Theorem 6.5 can be reduced to Theorem 6.5 in [14] (see also [17], [18]). Indeed, set $Q(x) = 0$, $A(x) = F'(x)$ $(x \in D)$, $z_n = 0$ $(n \geq 0)$, $C_1 = C_2 = 0$, $b_2 = b_3 = 0$. Assume that condition (6.125) is satisfied. Then (6.133) is also satisfied if we set

$$2C_3 = C(\|x-x_0\|, \|y-x_0\|, \|y-x\|) = \frac{1}{2}\omega\|y-x\| \text{ and } b_1 = b_4 = 0.$$

(5) Our conditions (6.133)–(6.136) are more general than the ones by Chen-Yamamoto [12], which in turn are more general than Potra-Ptak [19], [20], Vainikko [22], Yamamoto [22], and Zabrejko-Nguen [23]. Hence they can be used to solve a wider range of problems. In particular our conditions can be reduced to the ones obtained by Chen-Yamamoto if by using their versions and the last choice in Remark 3, we choose special cases of our functions given by $C = w(\|x-x_0\| + t\|y-x\|) - w_0(\|x-x_0\|)$, $(A(y) = A(x_0))$, $C_2 = e(\|x-x_0\|)\|x-y\|$, $C_3 = w_0(\|x-x_0\|)$, $C_1 = C_3 = 0$, $b_2 = b_4 = 0$, $b_1 = c$, $b_4 = b$ and $A(y)^{-1} = A(x_0)^{-1}$ for all $t \in [0,1]$, $x, y \in U(x_0, R)$. ∎

References

1. E.L. Allgower, K. Böhmer, F.A. Potra, and W.C. Rheinboldt, A mesh-independence principle for operator equations and their discretizations, *SIAM J. Numer. Anal.* **23**, 1 (1986), 160–169.

2. I.K. Argyros, A mesh independence principle for operator equations and their discretizations under mild differentiability conditions, *Computing* **45** (1990), 265–268.

3. I.K. Argyros, On the solution of equations with nondifferentiable operators, *Tamkang J. Math.* **24**, No. 3 (1993), 237–249.

4. I.K. Argyros, On the discretization of Newton-like methods, *Internat. J. Computer. Math.* **52** (1994), 161–170.

5. I.K. Argyros, A unified approach for constructing fast two-step Newton-like methods, *Monatshefte für Mathematik* **119** (1995), 1–22.

6. I.K. Argyros, Results on Newton Methods; Part I: A unified approach for constructing perturbed Newton-like methods in Banach space and their applications, *Appl. Math. and Comp.* **74** (1996), 119–141.

7. I.K. Argyros, Inexact Newton methods and nondifferentiable operator equations on Banach spaces with a convergence structure, *Approx. Th. Applic.* **13**, 3 (1997), 91–104.

8. I.K. Argyros, On a new Newton-Mysovskii-type theorem with applications to inexact Newton-like methods and their discretizations, *IMAJ Num. Anal. Journal of the Inverse of Mathematics and Applications* **18** (1997), 37–56.

9. I.K. Argyros, *Polynomial Operator Equations in Abstract Spaces and Applications*, Chapman Hall/CRC Press LLC, Boca Raton, Florida, 1998.

10. I.K. Argyros, A unified approach for solving equations and applications, Part I: On infinite-dimensional spaces, this handbook.

11. I.K. Argyros and F. Szidarovszky, *The Theory and Applications of Iteration Methods*, CRC Press, Inc. Boca Raton, Florida, 1993.

12. X. Chen and T. Yamamoto, Convergence domains of certain iterative methods for solving nonlinear equations, *Numer. Funct. Anal. and Optimiz.* **10** (1 and 2), (1989), 37–48.

13. L. Collatz, *The Numerical Treatment of Differential Equations*, Grundlehren der Mathematik, Bd 8, 2nd edition, Springer-Verlag, Heidelberg, (1959).

14. P. Deuflhard and G. Heindl, Affine invariant convergence theorems for Newton's method and extensions to related methods, *SIAM J. Numer. Anal.* **16**, No. 1 (1979), 1–10.

15. P. Deuflhard and F. Potra, Asymptotic mesh independence of Newton-Galerkin methods via a refined Mysovskii theorem, *SIAM J. Numer. Anal.* **29**, No. 5, (1992), 1395–1412.

16. L.V. Kantorovich and E.P. Akilov, *Functional Analysis in Normed Spaces*, Pergamon Press, New York, 1964.

17. I. Mysovskii, On the convergence of L.V. Kantorovich's method of solution of functional equations and its applications, *Dokl. Akad. Nauk SSSR* **70** (1950), 565–568 (in Russian).

18. I. Mysovskii, On the convergence of Newton's method, *Trudy Mat. Inst. Steklov* **28** (1949), 145–147 (in Russian).

19. F.A. Potra and V. Ptak, Sharp error bounds for Newton's process, *Numer. Math.* **34** (1980), 63–72.

20. F.A. Potra, On an iterative algorithm of order 1,839... for solving nonlinear operator equations, *Numer. Funct. Anal. and Optimiz.* **7**, (1) (1984–85), 75–106.

21. G.M. Vainikko, Galerkin's perturbation method and the general theory of approximate methods for non-linear equations, *USSR Comp. Math. and Math. Phys.* **4** (1967), 723–751.

22. T. Yamamoto, On the method of tangent hyperbolas in Banach spaces, *J. Comput. Appl. Math.* **21** (1988), 75–86.

23. P.P. Zabrejko and D.F. Nguen, The majorant method in the theory of Newton-Kantorovich approximations and the Ptak error estimates, *Numer. Funct. Anal. Optimiz.* **9**, (5 and 6), (1987), 671–684.

Chapter 7

Uniqueness for Spherically Convergent Multiple Trigonometric Series

J. Marshall Ash

Mathematics Department, DePaul University, Chicago, IL 60614
E-mail: mash@math.depaul.edu

Abstract In 1870 Cantor proved that representation of a function of one variable by a trigonometric series can be done in only one way. In 1996 Bourgain proved the same thing for spherical convergence and multiple trigonometric series. His proof involves injecting a lot of new ideas into the theory of uniqueness. We give here an exposition of Bourgain's proof, specialized to the case of dimension 2. Our exposition includes a fairly general method for finding maximal elements without resorting to the Axiom of Choice.

Contents

7.1 Background

The first major question that arose in the history of Fourier series was this: Determine which functions mapping the interval $[0, 2\pi) = \mathbb{T}$ into the complex numbers can be represented in the form

$$S(x) = \frac{a_0}{2} + \sum_{n=1}^{\infty} a_n \cos nx + b_n \sin nx. \tag{7.1}$$

Fourier himself thought that all functions could. Considering the fact that there was not even consensus at the time about just what a function was, this was not a bad guess. (For example, if f is continuous and of bounded variation, then its Fourier series has the form (7.1) and converges to f both pointwise and in L^2 norm.)

The second major question is the question of uniqueness. How many functions can share the same representation? Depending on the way this question is made precise, its answer can be easy, or quite difficult. Perhaps the simplest form of this question is the following. Consider only $L^2(\mathbb{T})$, the set of functions f satisfying

$$\left(\frac{1}{2\pi}\int_0^{2\pi} |f(x)|^2 dx\right)^{1/2} = \|f\|_{L^2} < \infty.$$

Say that a series of the form (7.1) represents f in the L^2 sense if its partial sums

$$s_N(x) = \frac{a_0}{2} + \sum_{n=1}^{N} a_n \cos nx + b_n \sin nx$$

tend to $f(x)$ in L^2, that is, if $\|s_N(x) - f(x)\|_{L^2}$ tends to zero as N tends to infinity. Then it is easy to show that for each L^2 function f, the Fourier series of f, given by choosing the coefficients in the series (7.1) to be

$$a_n = \frac{1}{\pi}\int_0^{2\pi} f(x) \cos nx dx \text{ and } b_n = \frac{1}{\pi}\int_0^{2\pi} f(x) \sin nx dx, \; n = 0, 1, 2, ...,$$

is the one and only series that represents f.

However, there is a more naive notion of "representation" which is perhaps more natural: say that the series (7.1) represents f if it converges to it at each point x, i.e., if for every point x, we have

$$\lim_{N\to\infty} s_N(x) = f(x).$$

Here the question of uniqueness becomes the following.

$$(*) \quad \left\{ \begin{array}{l} \text{Suppose that a function } f(x) \text{ and a series } S(x) \text{ of the form (7.1)} \\ \text{have been found so that } \lim\limits_{N\to\infty} s_N(x) = f(x) \text{ for every } x \text{ and sup-} \\ \text{pose that another series } S'(x) = \frac{a_0'}{2} + \sum\limits_{n=1}^{\infty} a_n' \cos nx + b_n' \sin nx \\ \text{also converges at every } x \text{ to } f(x). \text{ Must we have } S \text{ identical to} \\ S' \text{ (written } S \equiv S') \text{ in the sense that for every } n, a_n = a_n' \\ \text{and } b_n = b_n'? \end{array} \right.$$

This question can be rephrased in a tidy way by considering the difference series $\frac{a_0 - a_0'}{2} + \sum_{n=1}^{\infty}(a_n - a_n') \cos nx + (b_n - b_n') \sin nx$ which has partial sums converging to $f(x) - f(x) = 0$. The system of equations $\left\{ \begin{array}{ll} a_n = a_n', & n = 0, 1, 2, ... \\ b_n = b_n', & n = 1, 2, ... \end{array} \right.$ is, of course, identical with the system $\left\{ \begin{array}{ll} a_n - a_n' = 0, & n = 0, 1, 2, ... \\ b_n - b_n' = 0, & n = 1, 2, ... \end{array} \right.$. Thus $S \equiv S'$ if and only if $S - S' \equiv 0$. This allows us to rephrase our uniqueness question $(*)$ as follows.

$$(**) \quad \left\{ \begin{array}{l} \text{Suppose that a series } S(x) = \frac{a_0}{2} + \sum_{n=1}^{\infty} a_n \cos nx + b_n \sin nx \\ \text{satisfies } \lim\limits_{N\to\infty} s_N(x) = 0 \text{ for every } x. \text{ Must we have } S \equiv 0, \\ \text{i.e., for every } n, a_n = 0 \text{ and } b_n = 0? \end{array} \right.$$

The following theorem asserts that the answer to this question is yes.

THEOREM 7.1

(Cantor) *Suppose that* $S(x) = \frac{a_0}{2} + \sum_{n=1}^{\infty} a_n \cos nx + b_n \sin nx$ *converges to* 0 *at every point of* $\mathbb{T}$. *Then* $S \equiv 0$.

This answer lies quite a bit deeper than does the corresponding answer to the L^2 question. It was given in 1870 by Georg Cantor, who used Bernard Riemann's brilliant idea of associating to the trigonometric series $S(x)$ its second formal integral $F(x) = \frac{a_0}{2}\frac{x^2}{2} - \sum_{n=1}^{\infty} \frac{a_n}{n^2} \cos nx + \frac{b_n}{n^2} \sin nx$. (Incidentally, this idea of passing from the original series to what is now called the Riemann function lies at the foundation of the modern theory of distributions.)

Here are the steps of Cantor's proof.

1. Form the Riemann function F and use the Cantor-Lebesgue Theorem to see that it is a continuous function.
2. Observe that the second symmetric derivative of F is identically 0.
3. By a theorem of Schwarz, F must be a linear function.
4. Apply the L^2 theory of uniqueness to F to see that all the coefficients of F are 0. Therefore all the coefficients of S are 0 also.

For the details of Cantor's proof, see the survey paper [A].

7.2 Multiple Trigonometric Series

In order to discuss multiple trigonometric series, it will be helpful to develop a more concise notation. So instead of writing $a_n \cos nx + b_n \sin nx$, we will make use of Euler's formula $e^{ix} = \cos x + i \sin x$ and write $c_n e^{inx} + c_{-n} e^{-inx}$. To assure that this expression be real, we demand that c_{-n} be the complex conjugate of c_n. For simplicity, we will restrict ourselves to dimension 2, so by a multiple trigonometric series we mean

$$S(x) = \sum_{n \in \mathbb{Z} \times \mathbb{Z}} c_n e^{inx}, c_{-n} = \overline{c_n}$$

where $n = (n_1, n_2), x = (x_1, x_2)$, and $nx = n_1 x_1 + n_2 x_2$. It is not as clear as in the one-dimensional case what one should mean by "converges." We will make four definitions, discuss the first three briefly, and then concentrate all our efforts on the fourth. Let $s^N = \sum_{|n_1| \leq N_1, |n_2| \leq N_2} c_n$ be a rectangular partial sum of the series $S = \sum_{n \in \mathbb{Z} \times \mathbb{Z}} c_n$. Say that S converges unrestrictedly rectangularly to s if $\lim_{\min\{N_1, N_2\} \to \infty} s^N = s$; say that S is square convergent to s if $\lim_{k \to \infty} s^{(k,k)} = s$; and say that S is restrictedly rectangularly convergent to s if, for every (very big) $M > 1$, $\lim_{\min\{(N_1, N_2): M^{-1} < N_1/N_2 < M\} \to \infty} s^N = s$. Finally, let $|n| = \sqrt{n_1^2 + n_2^2}$, let $s_R = \sum_{|n| \leq R} c_n$ be a circular partial sum of S, and say that S converges spherically (circularly) to s if $\lim_{R \to \infty} s_R = s$. There are, correspondingly, four potential generalizations of Theorem 7.1. Two of these are theorems, while the other two are open questions. Here is the list.

THEOREM 7.2
Suppose that $S(x) = \sum_{n \in \mathbb{Z}^d} c_n e^{inx}$ converges unrestrictedly rectangularly to 0 at every point of $\mathbb{T}^d$. Then $S \equiv 0$.

CONJECTURE 7.1 *Suppose that $S(x) = \sum_{n \in \mathbb{Z}^d} c_n e^{inx}$ converges squarely to 0 at every point of $\mathbb{T}^d$. Then $S \equiv 0$.*

CONJECTURE 7.2 *Suppose that $S(x) = \sum_{n \in \mathbb{Z}^d} c_n e^{inx}$ converges restrictedly rectangularly to 0 at every point of $\mathbb{T}^d$. Then $S \equiv 0$.*

THEOREM 7.3
Suppose that $S(x) = \sum_{n \in \mathbb{Z}^d} c_n e^{inx}$ converges spherically to 0 at every point of $\mathbb{T}^d$. Then $S \equiv 0$.

The names associated with Theorem 7.2 are Ash, Freiling, Rinne, Shapiro, Tetunashvili, and Welland. The names associated with Theorem 7.3 are Bourgain, Connes, Cooke, Shapiro and Zygmund. For details as to who proved what and when, see the survey paper [AWa]. For extensions of Theorem 7.3, see [AWa1] and [AWa2]. The conjectures seem to be much more difficult to prove, even when $d = 2$. They should more properly be called open questions since there is a good amount of empirical evidence against them.

The remainder of this chapter will be devoted to the proof of Theorem 7.3. One element of the proof that the reader may find of special interest is Lemma 7.1, which is a fairly general method for finding maximal elements of partially ordered sets without resorting to the Axiom of Choice. This lemma appears in Section 7.4.

7.3 Proof of Theorem 7.3

We will give the proof only for the case when the dimension d is 2. The first proof when $d = 2$ was the work of Shapiro and Cooke. However we will sketch here the proof of Bourgain, which incorporates much of Shapiro's earlier work. Although Bourgain's work is very delicate, difficult to understand, and not really necessary unless $d \geq 3$, we have made this choice in order to bring some of his interesting and useful techniques to the attention of the mathematical community. We will also assume that

$$c_0 = c_{(0,0)} = 0.$$

This is simply for notational convenience and has no effect on the method of proof. (It can even be shown to not involve any loss of generality. See page 49 of [AWa].)

First notice that the possible values of R for which $A_R(x) = \sum_{|n|=R} c_n e^{inx}$ may be non-zero are square roots of integers. So if $R = \sqrt{k}$, in the sequence of partial sums the predecessor of s_R is $s_{\sqrt{k-1}} = s_{\sqrt{R^2-1}}$. In particular, as R tends to ∞, since $s_R(x)$ tends to 0 everywhere on $\mathbb{T}^2$, $s_R(x) - s_{\sqrt{R^2-1}}(x) = A_R(x)$ tends to 0 for all $x \in \mathbb{T}^2$. A generalization of the Cantor-Lebesgue Theorem, due to Cooke [Coo], [Z] when $d = 2$ and Connes [Con] when $d \geq 3$ asserts that $A_R(x)$ does not converge to zero by virtue of cancellation, but rather because

$$\varepsilon_R = \sqrt{\sum_{|n|=R} |c_n|^2} \text{ tends to 0 as } R \text{ tends to } \infty. \tag{7.2}$$

Following the original proof of Cantor, consider the series

$$T(x) = -\sum_{n \neq 0} \frac{c_n}{|n|^2} e^{inx}. \tag{7.3}$$

The series T converge spherically everywhere, because summing by parts gives

$$-\left\{ \sum_{k=1}^{\infty} \frac{1}{k} \sum_{|n|^2=k} c_n e^{inx} \right\}$$

$$= -\left\{ \lim_{K\to\infty} \sum_{k=1}^{K-1} \left(\sum_{|n|^2=1}^{k} c_n e^{inx} \right) \left(\frac{1}{k} - \frac{1}{k+1} \right) + \left(\sum_{|n|^2=1}^{K} c_n e^{inx} \right) \frac{1}{K} \right\}$$

$$= -\left\{ \lim_{K\to\infty} \sum_{k=1}^{K-1} \left(\sum_{|n|^2=1}^{k} c_n e^{inx} \right) \frac{1}{k(k+1)} \right\},$$

which converges, since for each fixed x, the sequence $s_{\sqrt{k}}(x) = \sum_{|n|^2=1}^{k} c_n e^{inx}$ is bounded and converges to 0 as $k \to \infty$. Define the Riemann function $F(x)$ at each x to be the value of this series. On the other hand, the coefficients of the series T are in $\ell^2(\mathbb{T}^2)$ since by condition (7.2)

$$\sum_{n\neq 0}\left|\frac{c_n}{|n|^2}\right|^2 = \sum_{k=1}^{\infty}\frac{1}{k^2}\sum_{|n|^2=k}|c_n|^2 = \sum_{k=1}^{\infty}\frac{(\varepsilon_{\sqrt{k}})^2}{k^2} < \infty.$$

By the Riesz-Fischer theorem, there is a function in $L^2(\mathbb{T}^2)$ so that T is the Fourier series of that function. Now the Fourier series of a function is almost everywhere spherically Abel summable to that function (Theorem 2 of [Sh1]), and spherical Abel summability is consistent with spherical convergence (sum by parts to see this), so that T is spherically Abel summable to $F(x)$ everywhere. It follows that $T(x)$ is the Fourier series of the Riemann function $F(x)$.

This completes the analogue of the first step of Cantor's proof of Theorem 7.1 mentioned above. But things are more difficult than in the one-dimensional case. The problem is that when $d \geq 2$ the condition $\varepsilon_R = o(1)$ does not imply the convergence of $\sum \frac{|c_n|}{|n|^2}$, so it is not clear that $F(x)$ is continuous. This will turn out to be the main obstacle to extending Cantor's proof.

Turning to the second step of Cantor's proof, we need an analogue for F having an everywhere 0 second symmetric derivative. Notice that the Laplacian of e^{inx} is $\Delta e^{inx} = (\partial^2_{x_1} + \partial^2_{x_2})e^{in_1x_1}e^{in_2x_2} = -|n|^2e^{inx}$, so that if term by term differentiation were allowed, we would have $\Delta F(x) = S(x) = 0$ and F would be harmonic. This motivates considering the second generalized Laplacian operator for a function $f \in L^1(\mathbb{T}^2)$ which is defined as

$$\tilde{\Delta} f(x) = \lim_{\rho\to 0} 8\frac{A_\rho f(x) - f(x)}{\rho^2},$$

where $A_\rho f(x) = \frac{1}{\pi\rho^2}\int_{B(x,\rho)} f(y)dy$ is the average of f over $B(x,\rho)$, the open disk of radius ρ centered at x, provided the limit exists. To see that this is a good definition, suppose that f is C^2 and write the Taylor expansion about x, $f(x+y) = f(x) + f_1(x)y_1 + f_2(x)y_2 + \frac{y_1^2}{2}f_{11}(x) + y_1y_2f_{12}(x) + \frac{y_2^2}{2}f_{22}(x) + o(|y|^2)$. Integrating gives $\int_{|y|\leq\rho} f(x+y)dy = f(x)\pi\rho^2 + 0 + 0 + \frac{f_{11}(x)}{2}\int_0^\rho\int_0^{2\pi}(r\cos\theta)^2 rdrd\theta + 0 + \frac{f_{22}(x)}{2}\int_0^\rho\int_0^{2\pi}(r\sin\theta)^2 rdrd\theta + o(\rho^4)$ so that $A_\rho f(x) = f(x) + \frac{1}{\pi\rho^2}(\frac{\pi f_{11}(x)}{8}\rho^4 + \frac{\pi f_{22}(x)}{8}\rho^4) + o(\rho^2) = f(x) + \frac{\Delta f(x)}{8}\rho^2 + o(\rho^2)$. Thus the generalized Laplacian agrees with the ordinary Laplacian for C^2 functions.

To compute $\tilde{\Delta}F(x)$, let $\chi(x) = \frac{1}{\pi}\chi_{B(0,1)}(x)$, where $\chi_{B(0,1)}$ is the characteristic function of the unit disk, and calculate

$$8\frac{A_\rho F(x) - F(x)}{\rho^2} = 8\sum_{n\neq 0} \frac{c_n}{|n|^2} \frac{\frac{1}{\pi\rho^2}\int_{B(0,\rho)} e^{in(x+y)}dy - e^{inx}}{\rho^2} \tag{7.4}$$

$$= 8\sum_{n\neq 0} \frac{c_n}{|n|^2} \frac{\frac{1}{\rho^2}\int_{R^2} \chi(y)e^{iny}dy - 1}{\rho^2} e^{inx}$$

$$= \sum_{n\neq 0} \left(8\frac{\hat{\chi}(\rho n) - 1}{|n|^2\rho^2}\right) c_n e^{inx}.$$

Here $\hat{\chi}$ denotes the Fourier transform of χ. To continue we will need two properties of $\hat{\chi}$. First choose the first coordinate axis to be in the direction of $n = (n_1, n_2)$ to get

$$\hat{\chi}(\rho n) - 1 = \frac{1}{\pi}\int_{|x|\leq 1} (e^{ix\pi n} - 1)dx$$

$$= \frac{1}{\pi}\int_{|x|\leq 1} (e^{ix_1\rho|n|} - 1)dx$$

$$= \frac{1}{\pi}\int_{-1}^{1} (\cos\rho|n|x_1 - 1)\left(\int_{-\sqrt{1-x_1^2}}^{\sqrt{1-x_1^2}} dx_2\right) dx_1$$

$$= \frac{2}{\pi}\int_{-1}^{1} (\cos\rho|n|x_1 - 1)\sqrt{1 - x_1^2}dx_1.$$

Using this and the fact that $\lim_{t\to 0} \frac{\cos t - 1}{t^2} = -\frac{1}{2}$, we have

$$\lim_{\rho\to 0} \frac{\hat{\chi}(\rho n) - 1}{(\rho|n|)^2} = \lim_{\rho\to 0} \frac{2}{\pi}\int_{-1}^{1} \left(\frac{\cos\rho|n|x_1 - 1}{(\rho|n|)^2}\right)\sqrt{1 - x_1^2}dx_1$$

$$= -\frac{1}{\pi}\int_{-1}^{1} \sqrt{1 - x_1^2}dx_1 \tag{7.5}$$

$$= -\frac{1}{2}.$$

If $|n| = 1$, then

$$\frac{\partial}{\partial r}\left[\frac{\hat{\chi}(\rho r n) - 1}{\rho^2 r^2}\right] = \frac{2}{\pi}\int_{-1}^{1}\frac{\partial}{\partial r}\left(\frac{e^{i\rho r x_1} - 1}{(\rho r)^2}\right)\sqrt{1 - x_1^2}dx_1$$

$$= -\frac{4}{\pi}\int_{-1}^{1}\frac{(1 - \frac{1}{2}ir\rho x_1)e^{i\rho r x_1} - 1}{\rho^2 r^3}\sqrt{1 - x_1^2}dx_1.$$

Using this and making the change of variable $u = \rho r$, we find that if $|n| = 1$,

$$\int_0^\infty \left|\frac{\partial}{\partial r}\left[\frac{\hat{\chi}(\rho r n) - 1}{\rho^2 r^2}\right]\right| dr$$

$$= \frac{4}{\pi}\int_0^\infty \left|\int_{-1}^{1}\frac{\left(1 - \frac{1}{2}ir\rho x_1\right)e^{i\rho r x_1} - 1}{\rho^2 r^3}\sqrt{1 - x_1^2}dx_1\right| dr$$

$$= \frac{4}{\pi}\int_0^\infty \left|\int_{-1}^{1}\frac{\left(1 - \frac{1}{2}iux_1\right)e^{iux_1} - 1}{u^3}\sqrt{1 - x_1^2}dx_1\right| du$$

$$= c_\chi < \infty. \tag{7.6}$$

To see that c_χ is finite, note that

$$\text{for } u \geq 1, \left|\int_{-1}^{1}\frac{(1 - \frac{1}{2}iux_1)e^{iux_1} - 1}{u^3}\sqrt{1 - x_1^2}\,dx_1\right| = O\left(\frac{1}{u^2}\right);$$

while for $0 < u < 1$,

$$(1 - \frac{1}{2}iux_1)e^{iux_1} - 1 = (1 - \frac{1}{2}iux_1)\left(1 + iux_1 + \frac{1}{2}(iux_1)^2 + O(u^3)\right) - 1$$

$$= \frac{1}{2}iux_1 + O(u^3),$$

so taking into account $\frac{i}{2u}\int_{-1}^{1} x_1\sqrt{1 - x_1^2}\,dx_1 = 0$,

$$\text{for } 0 < u < 1, \left|\int_{-1}^{1}\frac{(1 - \frac{1}{2}iux_1)e^{iux_1} - 1}{u^3}\sqrt{1 - x_1^2}\,dx_1\right| = O(1).$$

We can now finish the calculation of $\tilde{\Delta}F$ begun with equations (7.4) above. The computation leading to relation (7.5) shows that $\hat{\chi}(n_1) = \hat{\chi}(n_1)$ if $|n_1| = |n_2|$. Thus, we may abuse our notation and write $\hat{\chi}(n) = \hat{\chi}(|n|)$. We thus have

$$
\begin{aligned}
8\frac{A_\rho F(x) - F(x)}{\rho^2} &= \sum_{n \neq 0} \left(8\frac{\hat{\chi}(\rho n) - 1}{|n|^2 \rho^2} \right) c_n e^{inx} \\
&= \lim_{K\to\infty} \sum_{k=1}^{K} \left(8\frac{\hat{\chi}(\rho\sqrt{k}) - 1}{\rho k} \right) \sum_{|n|^2 = k} c_n e^{inx} \\
&= \lim_{K\to\infty} \sum_{k=1}^{K} b(\rho, k) A_k(x) \\
&= \lim_{K\to\infty} \sum_{k=1}^{K-1} \left(\begin{array}{c} s_{\sqrt{k}}(x)\left(b(\rho,k) - b(\rho,k+1)\right) \\ + s_{\sqrt{K}}(x) b(\rho, K) \end{array} \right) \\
&= \sum_{k=1}^{\infty} s_{\sqrt{k}}(x)\left(b(\rho,k) - b(\rho,k+1)\right), \qquad (7.7)
\end{aligned}
$$

where the last equality holds since $|\hat{\chi}(n)| \leq \frac{1}{\pi} \int_{|x|\leq 1} |e^{inx}| dx = 1$ implies that $\lim_{K\to\infty} b(\rho, K) = 0$. Now for fixed $x, s_{\sqrt{k}}(x) = \sum_{\nu=1}^{k} A_\nu(x) = \sum_{|n|^2 \leq k} c_n e^{inx}$ is a partial sum of a series tending to zero, and the calculations (7.5) and (7.6) performed above show that

$$
\lim_{\rho\to 0} b(\rho,k) - b(\rho,k+1) = 0 \text{ for each } k, \text{ and} \qquad (7.8)
$$

$$
\begin{aligned}
\sum_{k=1}^{\infty} \left| b(\rho,k+1) - b(\rho,k) \right| &= \sum_{k=1}^{\infty} \left| \int_k^{k+1} \frac{\partial}{\partial r} b(\rho, r) dr \right| \\
&\leq \int_0^\infty \left| \frac{\partial}{\partial r} b(\rho, r) \right| dr \leq c_\chi \text{ for each } \rho. \quad (7.9)
\end{aligned}
$$

Because of equation (7.7) we have that

$$\tilde{\Delta}F(x) = \lim_{\rho\to 0}\sum_{k=1}^{\infty} s_{\sqrt{k}}(x)\big(b(\rho,k) - b(\rho,k+1)\big) \text{ for every } x.$$

It is now easy to see that

$$\tilde{\Delta}F(x) = 0 \text{ for every } x,$$

for, given any $\varepsilon > 0$, find k_0 so large that $k \geq k_0$ implies $\left|s_{\sqrt{k}}(x)\right| < \frac{\varepsilon}{c_x}$. Decompose the last sum into $\sum_{k=1}^{k_0-1} + \sum_{k=k_0}^{\infty}$. By relation (7.9) the second term is bounded by ε for all ρ, while, as ρ tends to zero, the first term tends to zero by relation (7.8).

REMARK 7.1 The calculation that $\tilde{\Delta}F(x) = 0$ shows that just as in Cantor's one-dimensional proof the Schwarz derivative of the Riemann function is a linear mean of the original series, also here the generalized Laplacian of the Riemann function is a linear mean of the original series. There was so much detail given because a certain uniformity observation will be required later in the proof. ▮

We are "almost" ready to carry out the last two steps of Cantor's program. *Assume* that we know that the Riemann function F is continuous. Then a classical theorem of Rado asserts that continuous functions with identically 0 generalized Laplacian are harmonic, so F is harmonic. (A generalization of Rado's theorem will be proved below.) Also, since F is continuous, it is bounded on the compact set $[0, 2\pi]^2$, and so, being 2π periodic in each variable, it is bounded on all of $\mathbb{R}^2$. But a well-known generalization of Liouville's theorem asserts that a bounded harmonic function must be constant ([PW], p. 140). So $F(x)$ is a constant function and hence all the non-zero indexed coefficients of its Fourier series are zero, that is to say $-c_n/|n|^2 = 0$ for all $n \neq 0$. But we already know that c_0 is 0, so all c_n are 0 and our theorem is proved. Thus we need only show that the Riemann function F is continuous. Since everything we have done to this point is classical, we see that the heart of the matter is establishing the continuity of F.

Let $Z = \{x \in \mathbb{T}^2 : F \text{ is not continuous at } x\}$. Our goal is to prove that Z is empty. Assume that $Z \neq \emptyset$.

By assumption, $\sum_{|m|\leq r} c_m e^{imx}$ is uniformly bounded in r for each fixed

x; thus

$$\bigcup_{n=1}^{\infty}\bigcap_{k=1}^{\infty}\left\{x\in\bar{Z}:\left|\sum_{|m|\le k}c_m e^{imx}\right|\le n\right\}=\bar{Z}.$$

Baire's category theorem applied to the complete metric space $\bar{Z}$ concludes that for some positive integer n,

$$\bigcap_{k=1}^{\infty}\left\{x\in\bar{Z}:\left|\sum_{|m|\le k}c_m e^{imx}\right|\le n\right\}$$

contains a portion of $\bar{Z}$. Explicitly, for some $p\in Z$, there exists an open disk $B(p,\rho_0)$ such that

$$\sup_k \max_{z\in B(p,\rho_0)\cap\bar{Z}}\left|\sum_{|m|\le k}c_m e^{imx}\right|\le n. \tag{7.10}$$

Consequently, by the Weierstrass M-test,

$$F(x)=-\sum_{k=1}^{\infty}\left(\sum_{|n|^2=1}^{k}c_n e^{inx}\right)\frac{1}{k(k+1)}$$

converges uniformly on $B(p,\rho_0)\cap\bar{Z}$. Thus, when restricted to $B(p,\rho_0)\cap\bar{Z}$, $F(x)$ is continuous. In addition, the argument that was used above to prove that $\tilde{\Delta}F(x)=0$, when combined with inequality (7.10) shows that there is a constant C so that

$$\left|F(x)-A_\rho F(x)\right|\le C\rho^2 \tag{7.11}$$

uniformly on $B(p,\rho_0)\cap\bar{Z}$.

A major tool for the remaining argument is harmonic measure. Fix an open set $G\subset\mathbb{T}^2$ and fix a point $x\in G$. Then the harmonic measure $\omega(G,\cdot,x)$ is a probability measure on ∂G, the boundary of G. If F is a subset of ∂G, then its harmonic measure $\omega(G,F,x)$ is the probability that a Brownian motion starting at x will first intersect the boundary of G at some point of F. The reason for the name is that if we switch perspectives by fixing F and G and then letting x vary, the resulting function of x is harmonic. In fact it can be thought of as the solution to Dirichlet's problem, the problem of finding a harmonic function in G which tends to 1 on any sequence of points tending to a point of F and which tends to 0 on any sequence of points tending to a point of $\partial G \setminus F$. It is this sort of frequent change of point of view that motives using the democratic notation that we do, instead of something like $\omega_{G,x}(F)$. By the maximum principle, for every $x \in G$, $0 < \omega(G, F, x) < 1$, unless $F = \partial G$ so that $\omega \equiv 1$ or $F = \emptyset$ so that $\omega \equiv 0$.

Formally, let $\{X_t\}_{t \geq 0}$ be the standard Brownian motion in R^2 and let $(\Omega, \mathfrak{F}, P^x)$ be the corresponding probability space such that $P^x(X_0 = x) = 1$. The stopping time of X_t hitting ∂G is defined as $T = \inf\{t \geq 0 : X_t \in \partial G\}$, with the convention that $\inf \emptyset = 0$. Then the harmonic measure on ∂G at $x \in \partial G$ is the distribution of X_T under P^x. In other words, for $F \subset \partial G, \omega(G, F, x) = P^x(X_T \in F) = E^x \chi_F(X_T)$, where E^x is the expectation operator under P^x. However, I will not deal with harmonic measure in a formal way here; instead I will intuitively motivate a small number of its properties and then apply those properties in an axiomatic way.

We are ready to state three major theorems which we will need to show that the assumption that $Z \neq \emptyset$ is untenable. The first one is the promised generalization the theorem of Rado mentioned above. Let

$$Z_\varepsilon = \{x \in \mathbb{T}^2 : \sup_{|x - x_i| < \delta, i=1,2} \left|F(x_1) - F(x_2)\right| > \varepsilon \text{ for all } \delta > 0\}.$$

Even though Z_ε is defined by strict inequalities, it is easy to see that it contains its limit points and is therefore a closed set.

THEOREM 7.4

Let F be a bounded function defined on the ball $\bar{B}(p, r)$ and for each $\varepsilon \in I = \left\{1, \frac{1}{2}, \frac{1}{3}, \ldots\right\}$ let

$$\begin{aligned} W_\varepsilon &= Z_\varepsilon \cap \bar{B}(p, r) \\ &= \{x \in \bar{B}(p, r) : \sup_{|x - x_i| < \delta, i=1,2} \left|F(x_1) - F(x_2)\right| > \varepsilon \textit{ for all } \delta > 0\} \end{aligned}$$

so that $W = \cup_{\varepsilon \in I} W_\varepsilon$ is the set of discontinuities of F in $\bar{B}(p, r)$. Suppose that F is the Riemann function for a series which is spherically convergent to 0 at every point of $\bar{B}(p, r)$. Further suppose that for all $x \in B(p, r) \setminus \bar{W}(p, r)$ and all $\varepsilon \in I$, the harmonic measure $\omega\big(B(p, r) \setminus W_\varepsilon, \partial W_\varepsilon, x\big)$ is equal to 0, then F is harmonic on $B(p, r)$ provided that F is continuous when restricted to $\bar{W}$.

THEOREM 7.5

Let F be the Riemann function, $B(p,\rho_0)$ the disk constructed above so that conditions (7.10) and (7.11) hold on $B(p,\rho_0)\cap\bar{Z}$, where Z is the set of discontinuities of F. If $p_1\in Z$, $B(p_1,\rho_1)\subset B\left(p,\frac{1}{2}\rho_0\right)$ and $p_2\in B\left(p_1,\frac{1}{2}\rho_1\right)$, then

$$\left|F(p_1)-F(p_2)\right| \tag{7.12}$$

$$\leq c\left(\begin{array}{c}\left[F(p_1)+\frac{1}{\rho_1^{\frac{3}{4}}}\right]\left[1-\omega\left(B(p_1,\rho_1)\setminus\bar{Z},\partial\left(Z\cap B(p_1,\rho_1)\right),p_2\right)\right]^{\frac{1}{4}}\\ +\sup_{q\in B(p_1,2\rho_1)\cap\bar{Z}}|F(p_1)-F(q)|\end{array}\right).$$

THEOREM 7.6

Let $B(p_0,r)$ be a disk in $\mathbb{T}^2$ and F a closed set such that $B(p_0,r)\cap F\neq\emptyset$. Suppose that for some $x\in B(p_0,r)\cap F$,

$$\omega\left(B(p_0,r)\setminus\bar{F},\partial\big(B(p_0,r)\cap F\big),x\right)>0.$$

Then there exists a point $p_1\in B(p_0,r)\cap F$ such that

$$\inf_{\delta_1>0}\liminf_{\delta_2\to 0}\inf_{x\in B(p_1,\delta_2)}\omega\left(B(p_1,\delta_1)\setminus\bar{F},\partial\big(B(p_1,\delta_1)\cap F\big),x\right)=1.$$

Leaving the proofs of these three theorems to later sections, we resume our study of the assumption $Z\neq\emptyset$.

Consider the following dichotomy. Either

1. for every $\varepsilon\in I=\left\{1,\frac{1}{2},\frac{1}{3},\ldots\right\}$ and all $x\in B(p,r)\setminus Z_\varepsilon$, the harmonic measure

$$\omega\left(B(p,\rho_0)\setminus Z_\varepsilon,\partial(B(p,\rho_0)\cap Z_\varepsilon),x\right)\quad\text{is equal to }0,$$

or

2. for some $\varepsilon\in I$,

$$\omega\left(B(p,\rho_0)\setminus Z_\varepsilon,\partial(B(p,\rho_0)\cap Z_\varepsilon),x\right)>0.$$

We will focus on the portion of Z centered about the point p of Z such that the Riemann function F is well behaved when restricted to the points of Z near p. Here is a rough sketch of why both cases lead to a contradiction.

In case 1, Theorem 7.4 (with a small assist from Theorem 7.5) will imply that Z is so thin that there is none of it at all in a small neighborhood of p, contrary to $p \in Z$. In case 2, Theorem 7.6 will produce a point p_1 of Z_ε which has "a lot" of Z_ε near it. At p_1, on the one hand, by the definition of Z_ε there will be points x arbitrarily close to p_1 where $bigl|F(x) - F(p_1)\big|$ is big, while on the other hand, the good behavior of F on Z and Theorem 7.5 will show that for all x near p_1, $\big|F(x) - F(p_1)\big|$ will be small. This also will be a contradiction, and Theorem 7.3 will follow.

Here are the details of the proof sketched in the last paragraph. First suppose that case 1 holds. Use Theorem 7.5 with $B(p_1, \rho_1) = B\left(p, \frac{1}{2}\rho_0\right)$: inequality (7.12) implies that F is bounded on $B\left(p, \frac{1}{4}\rho_0\right)$. Then because of Theorem 7.4 with $B(p,r) = B\left(p, \frac{1}{4}\rho_0\right)$, F the Riemann function, and $W = \bar{B}(p,r) \cap Z$, it follows that F is harmonic, and hence continuous on all of $B\left(p, \frac{1}{4}\rho_0\right)$. This contradicts the fact that p is a point of discontinuity of F. Now suppose that case 2 holds. By Theorem 7.6, there exists $p_1 \in B\left(p, \frac{1}{8}\rho_0\right) \cap Z_\varepsilon$ such that

$$\inf_{\delta_1 > 0} \liminf_{\delta_2 \to 0} \inf_{x \in B(p_1, \delta_2)} \omega\left(B(p_1, \delta_1) \setminus Z_\varepsilon, \partial\big(B(p_1, \delta_1) \cap Z_\varepsilon\big), x\right) = 1. \quad (7.13)$$

Because F restricted to $B\left(p, \frac{1}{8}\rho_0\right) \cap \bar{Z}$ is continuous, there exists a positive δ_1 such that

$$\big|F(x) - F(p_1)\big| \leq \frac{\varepsilon}{10} \text{ for all } x \in B(p_1, 2\delta_1) \cap \bar{Z}. \quad (7.14)$$

Let $\eta > 0$ be any positive number. By relation (7.13), there exists a positive $\delta_2 = \delta_2(\eta)$ such that

$$\omega\left(B(p_1, \delta_1) \setminus Z_\varepsilon, \partial\big(B(p_1, \delta_1) \cap Z_\varepsilon\big), x\right) > 1 - \eta \text{ for all } x \in B(p_1, \delta_2). \quad (7.15)$$

Since $p_1 \in Z_\varepsilon$, by the definition of Z_ε there exists $p_2 \in B(p_1, \delta_2)$ such that

$$\frac{\varepsilon}{2} \leq |F(p_1) - F(p_2)|.$$

Combine this inequality with Theorem 7.5 with $\rho_1 = \delta_1$ and take inequalities (7.14) and (7.15) into account to get

$$\frac{\varepsilon}{2} \leq \big|F(p_1) - F(p_2)\big| \leq \left[F(p_1) + \delta_1^{-\frac{3}{4}(d-1)}\right] \eta^{1/4} + \frac{\varepsilon}{10}.$$

But the choice of η was arbitrary, so choosing η sufficiently small leads to a contradiction. Thus Theorem 7.3 will be proved as soon as Theorems 7.4, 7.5, and 7.6 are.

7.4 Proof of Theorem 7.4

The circle average of F over the edge of the disk $B(x,\rho)$ is given by

$$S_\rho F(x) = \frac{1}{2\pi}\int_0^{2\pi} F\left(x + (\rho\cos\theta, \rho\sin\theta)\right) d\theta.$$

An alternative definition of generalized Laplacian could have been made using circle averages rather than area averages. Let

$$\Delta^a F(x) = \lim_{\rho\to 0} 4\frac{S_\rho F(x) - F(x)}{\rho^2}$$

be the alternate generalized Laplacian. Then the Riemann function F satisfies

$$\Delta^a F(x) = 0 \text{ for all } x. \tag{7.16}$$

The proof of this is almost identical to the proof given above for $\tilde{\Delta}$. As before, testing on a Taylor polynomial of degree 2 shows that Δ^a agrees with the usual Laplacian when operating on C^2 functions. As before, the proof begins by calculating $4\frac{S_\rho F(x)-F(x)}{\rho^2}$ by means of termwise integration. All the calculations are the same, except that the function $\hat{\chi}(n)$ is replaced by $\hat{\sigma}(n) = \frac{1}{2\pi}\int_0^{2\pi} e^{i(n_1\cos\theta + n_2\sin\theta)}d\theta$, so the only estimates required are 1° $|\hat{\sigma}(n)| \le 1$, 2° $\lim_{\rho\to 0}\frac{\hat{\sigma}(\rho n)-1}{(\rho|n|)^2} = -\frac{1}{4}$, and 3° $\int_0^\infty \left|\frac{\partial}{\partial r}\left[\frac{\hat{\sigma}(\rho r n)-1}{\rho^2 r^2}\right]\right| dr \le c < \infty$ for every $\rho > 0$. The integration involves the dot product of $n = (n_1, n_2)$ with a unit vector $(\cos\theta, \sin\theta)$ which rotates though one full 360° rotation, so without loss of generality we can take $n = (|n|, 0)$ and write $\hat{\sigma}(n) = \hat{\sigma}(|n|) = \frac{1}{2\pi}\int_0^{2\pi} e^{i|n|\cos\theta}d\theta$. Since $\left|e^{i|n|\cos\theta}\right| = 1$, we have 1°. Replacing the integrand $e^{i\rho|n|\cos\theta} - 1$ in $\hat{\sigma}(\rho n) - 1$ by $i\rho|n|\cos\theta + \frac{(i\rho|n|\cos\theta)^2}{2} + o(\rho^2)$ leads to 2°. For 3°, we may assume $n = (1,0)$ and calculate

$$\frac{\partial}{\partial r}\left[\frac{\hat{\sigma}(\rho r n) - 1}{\rho^2 r^2}\right] = \frac{\partial}{\partial r}\left[\frac{1}{2\pi\rho^2 r^2}\int_0^{2\pi}(e^{i\rho r\cos\theta} - 1)d\theta\right]$$

$$= -\frac{1}{\pi\rho^2 r^3}\int_0^{2\pi}\left[\left(1 - \frac{1}{2}i\rho r\cos\theta\right)e^{i\rho r\cos\theta} - 1\right] d\theta,$$

so that after making the substitution $u = \rho r$, we find

$$\int_0^\infty \left|\frac{\partial}{\partial r}\left[\frac{\hat{\sigma}(\rho r n) - 1}{\rho^2 r^2}\right]\right| dr$$

$$= \frac{1}{\pi}\int_0^\infty \left|\int_0^{2\pi}\left[\left(1-\frac{1}{2}iu\cos\theta\right)e^{iu\cos\theta}-1\right]d\theta\right|\frac{du}{u^3} = c < \infty.$$

The proof that c is finite is similar to the proof that c_χ is finite in equation (7.6) above: if $u > 1$, the inner integral is $O(u)$ as $u \to \infty$, while if $0 < u < 1$, Taylor expanding the inner integrand in u and taking $\int_0^{2\pi}\cos\theta d\theta = 0$ into account shows that the inner integral is $O(u^3)$ as $u \to 0$.

Harmonic measure, like any other nonnegative measure, naturally gives rise to an integral. The idea is that if χ is the characteristic function of a subset F of ∂G, define $\int_{\partial G}\chi(q)\omega(G,dq,x)$ to be $\omega(G,F,x)$, then extend the integral to simple functions by linearity, and then pass to limits. If you fix a continuous function f defined on ∂G, then varying x produces a function $H_f(x) = \int_{\partial G} f(q)\omega(G,dq,x)$ defined on G. If the boundary of G is fairly smooth, for example, if at every point q of ∂G there is a triangle having one corner at q whose interior is disjoint from G, then H_f is continuous on $\bar{G}$, harmonic on G, and $H_f(x)$ tend to $f(q)$ as x tends to q from inside G. See [[He]], for example, for details. We may use another terminology here and say that H_f is a solution to Dirichlet's problem: Given a continuous function f defined on the boundary of a region, find a harmonic function $H(x)$ defined on the region whose boundary values agree with f. A very important special case occurs when G is a disk $B(q,r_1)$. Here the solution to Dirichlet's problem is known to be unique and is given by

$$H(x) = \frac{1}{2\pi}\int_0^{2\pi}\frac{r_1^2-|q-x|^2}{|q-x+r_1e^{i\theta}|^2}f\big(q+(r_1\cos\theta, r_1\sin\theta)\big)d\theta.$$

But since H_f also solves Dirichlet's problem, the fact that there is only one solution tells us that

$$\omega\big(B(q,r_1),dz,x\big) = \frac{1}{2\pi r_1}\frac{r_1^2-|q-x|^2}{|q-x+r_1e^{i\theta}|^2}r_1 d\theta$$

and that we may also write

$$H(x) = \int_{\partial B(q,r_1)} f(z)\omega\big(B(q,r_1),dz,x\big).$$

Again, [He] can be used as a reference for all these facts.

Motivated by the formula in the last paragraph, we are going to show below that if $B(q,r_1) \subset B(p,r)$ and $y \in B(q,r_1)\setminus\bar{Z}$, then

$$F(y) = \int_{\partial B(q,r_1)} F(z)\omega\big(B(q,r_1),dz,y\big) \tag{7.17}$$

$$= \frac{1}{2\pi} \int_0^{2\pi} \frac{r_1^2 - |q-y|^2}{\left|q - y + (r_1 \cos\theta, r_1 \sin\theta)\right|^2} F(q + r_1 e^{i\theta}) d\theta.$$

Temporarily assuming this formula, it is easy to finish the proof of Theorem 7.4. Assume that $Z \cap B(p,r) \neq \emptyset$ and let $q \in Z \cap B(p,r)$. Pick r_1 very small and suppose that $|q-y| < r_1$. We have

$$\begin{aligned} \left|F(q) - F(y)\right| &\leq \left|F(q) - S_{r_1}F(q)\right| \\ &\quad + \left|S_{r_1}F(q) - \int_{\partial B(q,r_1)} F(z)\omega(B(q,r_1), dz, y)\right| \\ &\leq I + II. \end{aligned} \tag{7.18}$$

Because the circular generalized Laplacian of F is 0 at q, if r_1 is sufficiently small, we have that

$$I = \left|F(q) - S_{r_1}F(q)\right| \leq r_1^2. \tag{7.19}$$

By equation (7.17) we also have that

$$II \leq \frac{1}{2\pi} \int_0^{2\pi} |F(q + (r_1 \cos\theta, r_1 \sin\theta))| \left|1 - \frac{r_1^2 - |q-y|^2}{|q - y + (r_1 \cos\theta, r_1 \sin\theta)|^2}\right| d\theta. \tag{7.20}$$

If $|q-y| < \frac{r_1}{2}$, then choosing the first coordinate axis in the direction of $q - y$, we have

$$\begin{aligned} \left|1 - \frac{r_1^2 - |q-y|^2}{|q - y + (r_1 \cos\theta, r_1 \sin\theta)|^2}\right| &= \frac{2(|q-y| + r_1 \cos\theta)|q-y|}{(|q-y| + r_1\cos\theta)^2 + (r_1 \sin\theta)^2} \\ &= \frac{2(r_1 \cos\theta + |q-y|)|q-y|}{(r_1 + |q-y|\cos\theta)^2 + (|q-y|\sin\theta)^2} \\ &\leq \frac{2(r_1 + \frac{1}{2}r_1)|q-y|}{\left(r_1 - \frac{1}{2}r_1\right)^2} \\ &= 12\frac{|q-y|}{r_1}. \end{aligned}$$

But F is bounded by hypothesis, so combining this with inequalities (7.18)–(7.20) yields

$$\left|F(q) - F(y)\right| \le r_1^2 + C\frac{|q-y|}{r_1}.$$

Fixing r_1 and letting y tend to q yields

$$\limsup_{\substack{y \to q \\ y \notin \bar{Z}}} |F(q) - F(y)| \le r_1^2.$$

But since r_1 can be chosen arbitrarily small, this implies that

$$\lim_{\substack{y \to q \\ y \notin \bar{Z}}} |F(q) - F(y)| = 0.$$

But since F was assumed to be continuous with respect to $\bar{Z}$, this implies that F is continuous at q, contrary to the assumption that $q \in Z$. Thus $B(p,r) \cap Z = \emptyset$. Hence equation (7.17) holds at every point y of $B(p,r)$, so that F is harmonic.

It only remains to prove relation (7.17). It is enough to prove that for fixed $y \in B(q,r_1) \setminus \bar{Z} \subset B(q,r_1) \subset B(p,r)$,

$$F(y) \le \frac{1}{2\pi}\int_0^{2\pi} \frac{r_1^2 - |q-y|^2}{\left|q - y + (r_1\cos\theta, r_1\sin\theta)\right|^2} F(q + r_1e^{i\theta})d\theta, \tag{7.21}$$

because this relation would then also hold with F replaced by $-F$, and equation (7.17) would follow.

Consider Brownian motions starting at y. Let T be the exit time of a Brownian motion from the disk $B(q,r_1)$. This means that for each Brownian motion starting at y, X_T, the position at time T belongs to the boundary of $B(q,r_1)$. Denote by $E[F(X_T)]$ the value of $F(X_T)$ averaged over the positions of all Brownian motions starting at y, each motion proceeding for the time T. The connection between harmonic measure and Brownian motion is that

$$E[F(X_T)] = \int_{\partial B(q,r_1)} F(z)\omega\big(B(q,r_1), dz, y\big).$$

Combining this with equation (7.17) and inequality (7.21) reduces our goal to proving

$$F(y) \le E[F(X_T)]. \tag{7.22}$$

Our derivation of inequality (7.22) will require the following lemma.

LEMMA 7.1

Let $(\mathfrak{S}, \prec)$ be a nonempty set equipped with a partial order $\prec$. Suppose that every increasing sequence $\{S_i\} \subset \mathfrak{S}$ has an upper bound $S_\infty \in \mathfrak{S}$ such that $S_i \prec S_\infty$ for all $i \in \mathbb{N}$. If in addition, there is a mapping $T : \mathfrak{S} \to [0,1]$ such that for $S_1, S_2 \in \mathfrak{S}$,

$$S_1 \prec S_2 \text{ implies that } TS_1 \leq TS_2, \text{ and}$$

$$S_1 \precneqq S_2 \text{ implies that } TS_1 < TS_2,$$

then there is a maximal element $S^* \in \mathfrak{S}$.

PROOF Let

$$\mathfrak{S}_0 = \mathfrak{S} \text{ and } R_0 = \{TS : S \in \mathfrak{S}_0\}.$$

Select $S_1 \in \mathfrak{S}$ such that

$$TS_1 > \sup R_0 - 1.$$

Let

$$\mathfrak{S}_1 = \{S \in \mathfrak{S} : S_1 \prec S\} \text{ and } R_1 = \{TS : S \in \mathfrak{S}_1\}.$$

If $\mathfrak{S}_1 = \{S_1\}$, then S_1 is a maximal element in $\mathfrak{S}$. Otherwise, select $S_2 \in \mathfrak{S}_1$ such that

$$TS_2 > \sup R_1 - \frac{1}{2}.$$

Note that $S_1 \prec S_2$ and $TS_1 \leq TS_2$. After selecting $S_1 \prec \cdots \prec S_n$ let

$$\mathfrak{S}_{n+1} = \{S \in S : S_n \prec S\} \text{ and } R_{n+1} = \{TS : S \in \mathfrak{S}_{n+1}\}.$$

Note $\mathfrak{S}_{n+1} \subset \mathfrak{S}_n \subset \cdots \subset \mathfrak{S}_1 \subset \mathfrak{S}_0$. Then either $\mathfrak{S}_{n+1} = \{S_n\}$ in which case S_n is a maximal element of $\mathfrak{S}$ or $\mathfrak{S}_{n+1} \neq \{S_n\}$ in which case we can select $S_{n+1} \in \mathfrak{S}_{n+1}$ such that

$$TS_{n+1} > \sup R_{n+1} - \frac{1}{n+1}.$$

Repeat this process. Either we will get a maximal element $S_m \in \mathfrak{S}$ if for some $m \in \mathbb{N}$, $\mathfrak{S}_{m+1} = \{S_m\}$ or we get a sequence of elements $S_1 \prec S_2 \prec \cdots$ such that $S_i \in \mathfrak{S}_{i+1} \subset \mathfrak{S}_i$, and $TS_{i+1} > \sup\{TS : S \in \mathfrak{S}_{i+1}\} - \frac{1}{i+1}$, $i \in \mathbb{N}$.

By assumption, there exists $S_\infty \in \mathfrak{S}$ such that $S_i \prec S_\infty$ for all $i \in \mathbb{N}$. We claim that S_∞ is a maximal element of $\mathfrak{S}$. In fact, if there is an element $S^* \in \mathfrak{S}$ such that $S_\infty \precneqq S^*$, then first $S^* \in \mathfrak{S}_n$ as $S_n \prec S_\infty \prec S^*$ for all $n \in \mathbb{N}$. Second, by the order-preserving property of T, we have $0 < \epsilon = TS^* - TS_\infty$. Thus,

$$\epsilon + TS_\infty = TS^* \leq \sup R_n < TS_n + \frac{1}{n} \leq TS_\infty + \frac{1}{n}.$$

We get a contradiction by letting $n \to \infty$. ∎

Example 7.1
Let $\mathfrak{S}$ be a set of stopping times for a stochastic process. Define a partial order $\prec$ by $\tau_1 \prec \tau_2$ if $\tau_1 \leq \tau_2$ a.e. Define a mapping T to be

$$T\tau = \frac{1}{\pi} E\left[\arctan \tau\right] + \frac{1}{2}.$$

(Actually, $\tau \geq 0$, so $T\tau = \frac{2}{\pi} E[\arctan \tau]$ would also work.) Clearly, T satisfies the order-preserving property. So $\mathfrak{S}$ has a maximal element if it is nonempty and if its increasing sequences have upper bounds in $\mathfrak{S}$. □

Example 7.2
Let $\mathfrak{S}$ be a set of integrable functions defined in $D \subset \mathbb{R}^n$. Define a partial order $\prec$ by $f_1 \prec f_2$ if $f_1 \leq f_2$ a.e. in D. Define a mapping T to be

$$T\tau = \frac{1}{\pi |D|} \int_D \arctan f(x) dx + \frac{1}{2}.$$

Clearly, T satisfies the order-preserving property. So $\mathfrak{S}$ has a maximal element if it is nonempty and if its increasing sequences have upper bounds in $\mathfrak{S}$. □

COROLLARY 7.1
Let $(\mathfrak{S}, \prec)$ be a nonempty set equipped with a partial order $\prec$. Suppose the assumption of Zorn's Lemma is satisfied by $(\mathfrak{S}, \prec)$. Namely, if $\{S_\alpha\} \subset \mathfrak{S}$ is a chain (linearly ordered set), then there is $S_\infty \in \mathfrak{S}$ such that every $S_\alpha \prec S_\infty$. If in addition, there is a mapping $T : \mathfrak{S} \to [0,1]$ such that for

$S_1, S_2 \in \mathfrak{S}$,

$$S_1 \prec S_2 \textit{ implies that } TS_1 \le TS_2, \textit{ and}$$

$$S_1 \precneqq S_2 \textit{ implies that } TS_1 < TS_2,$$

then there is a maximal element $S^* \in \mathfrak{S}$.

PROOF If all chains have upper bounds in $\mathfrak{S}$, in particular all countable chains (increasing sequences) have upper bounds in $\mathfrak{S}$. ∎

REMARK 7.2 Lemma 7.1 provides an alternative to Zorn's lemma which, when it can be applied, allows us to avoid the axiom of choice. As Corollary 7.1 makes explicit, an additional advantage of Lemma 7.1 is that it requires a slightly weaker upper bound condition than does Zorn's Lemma itself. The idea of the proof was pointed out to me by Robert Kaufman. It appears on page 121 of [Ha]. ∎

We are now ready to prove inequality (7.22). We first show that if S is a stopping time satisfying $S \le T$, then the probability that the associated motion X_S is in Z at time S is zero. In other words, the set Z is "invisible" to Brownian motions starting at y. Since $Z \subset \cup_{\varepsilon \in I} Z_\varepsilon$, it is enough to see that for each $\varepsilon \in I$, the probability that a motion is in $\bar{Z}_\varepsilon$ at time S is zero. We denote this as

$$P\big(X_S \in \bar{Z}_\varepsilon\big) = 0. \tag{7.23}$$

Fix $\varepsilon \in I$ and let R be the time it takes the motion to hit $\partial(B(q, r_1) \setminus \bar{Z}_\varepsilon)$. Then $R \le T$. Our assumption that $\omega\big(B(q,r_1) \setminus \bar{Z}_\varepsilon, \partial\big(B(q,r_1) \setminus \bar{Z}_\varepsilon\big), y\big) = 0$ translates into the statement that the probability that the motion is in $\bar{Z}_\varepsilon$ at time R is 0. But $R < T$ only if the motion is in ∂Z_ε at time R so the probability that $R < T$ is 0. The probability that the motion is in $\bar{Z}_\varepsilon$ at time T is the sum of two probabilities: the probability that the motion is in $\bar{Z}_\varepsilon$, conditional on $R = T$ and the probability that the motion is in $\bar{Z}_\varepsilon$, conditional on $R < T$. Both of these are 0 since the former is $\le$ the probability that the motion is in $\bar{Z}_\varepsilon$ at time R, which is 0 and the latter is $\le$ the probability that $R < T$, which is also 0. Hence equation (7.23) is proved when $S = T$.

Returning to the stopping time S, for any $\tau > 0$ there exists an open set G such that $\bar{Z}_\varepsilon \subset G$ and $\omega\big(B(q,r_1) \backslash G, \partial\big(B(q,r_1) \cap \bar{G}\big), y\big) < \tau$. Define a function u on $\bar{B}(q, r_1)$ as follows:

$$u(x) = \begin{cases} \omega\big(B(q,r_1)\backslash G, \partial\big(B(q,r_1) \cap \bar{G}\big), x\big) & x \in B(q,r_1)\backslash G \\ 1 & x \in \overline{B(q,r_1) \cap G} \\ 0 & x \in B(q,r_1) \setminus \partial\big(B(q,r_1) \cap G\big). \end{cases}$$

Then u is superharmonic on $B(q, r_1)$. Let $\tilde{r}_n \nearrow r_1$ where $\tilde{r}_1$ is so large that $y \in B(q, \tilde{r}_1)$. Let T_n be the time when the Brownian motion leaves $B(q, \tilde{r}_n)$. Clearly $T_n \nearrow T$. Since Brownian motion is continuous, we have

$$\{X_S \in \bar{Z}_\varepsilon \text{ for some } S < T\} \subset \bigcup_{n=1}^{\infty} \{X_{S \wedge T_n} \in \bar{Z}_\varepsilon\},$$

where $S \wedge T_n = \min\{S, T_n\}$. But

$$\begin{aligned} P(X_S \in \bar{Z}_\varepsilon) &= P(X_S \in \bar{Z}_\varepsilon \text{ conditional on } S = T) \\ &\qquad + P(X_S \in \bar{Z}_\varepsilon \text{ conditional on } S < T) \\ &\leq P(X_T \in \bar{Z}_\varepsilon) + P(X_S \in \bar{Z}_\varepsilon \text{ conditional on } S < T) \\ &= 0 + P(X_S \in \bar{Z}_\varepsilon \text{ conditional on } S < T) \\ &\leq \lim_{n \to \infty} P(X_{S \wedge T_n} \in \bar{Z}_\varepsilon \text{ conditional on } S < T), \end{aligned}$$

so equation (7.23) will follow if we can prove

$$\text{for each positive } n, P\big(X_{S \wedge T_n} \in \bar{Z}_\varepsilon\big) = 0. \tag{7.24}$$

Fix n. The superharmonic function u can be approximated on the domain $B(q, \tilde{r}_n)$ by an increasing sequence of C^2 superharmonic functions u_j. Itô's formula asserts that

$$E[u_j(X_{S \wedge T_n})] \leq u_j(y).$$

Let $j \to \infty$ to see that

$$\tau \geq u(y) \geq E[u(X_{S \wedge T_n})] \geq P\left(X_{S \wedge T_n} \in \overline{B(q,r_1) \cap G}\right) \geq P\left(X_{S \wedge T_n} \in \bar{Z}_\varepsilon\right).$$

Letting $\tau \to 0$ establishes equations (7.24), hence equations (7.23), and so the probability that a motion is in Z at time S is zero. Because of this and the continuity of F on $B(q, r_1)$, if $S_n \nearrow S_\infty \leq T$, almost surely a Brownian motion starting at y satisfies

$$\lim_{n\to\infty} F(X_{S_n}) \leq F(X_{S_\infty}). \tag{7.25}$$

Let $\eta > 0$. By (7.16) there is a $\beta_{y,\eta} \in (0, r_1 - |y-q|)$ such that for all $\beta \leq \beta_{y,\eta}$

$$F(y) - S_\beta F(y) \leq \eta\beta^2. \tag{7.26}$$

Consider a family of stopping times

$$\mathfrak{S} = \left\{S \leq T : F(y) - E[F(X_S)] \leq \eta E[|y - X_S|^2]\right\}.$$

We will show that

$$\mathfrak{S} \text{ is nonempty and the upper bound of any chain in } \mathfrak{S} \text{ is also in } \mathfrak{S}. \tag{7.27}$$

Define

$$S_0 = \inf_{t\geq 0} \left\{|X_t - y| \geq \beta_{y,\eta}\right\}.$$

Then X_{S_0} is uniformly distributed on the circle $\partial B(y, \beta_{y,\eta})$. So by inequality (7.26), we have

$$F(y) - E[F(X_{S_0})] \leq \eta E[|y - X_S|^2].$$

Thus $S_0 \in \mathfrak{S}$ and $\mathfrak{S}$ is therefore nonempty. Also, for an increasing sequence $\{S_n\}$ of stopping times in $\mathcal{G}$, let $S_\infty = \lim_{n\to\infty} S_n$. Then by inequality (7.25) and Fatou's lemma, we have

$$F(y) - E[F(X_{S_\infty})] \leq \eta E[|y - X_{S_\infty}|^2].$$

So $S_\infty \in \mathfrak{S}$ and statement (7.27) holds. It follows from Lemma 7.1 that there exists $S^* \in \mathfrak{S}$ such that it is a maximal element of $\mathfrak{S}$. We show that $S^* = T$ for almost every Brownian motion starting at y. In fact, if $S^* < T$ with positive probability, then define

$$S_1^* = \inf\left\{t \in [S^*, T] : |X_t - X_{S^*}| \geq \beta_{X_{S^*},\eta}\right\},$$

where for $x \in \partial B(q, r_1)$ we define $\beta_{x,\eta}$ to be 0. Then $S_1^* \geq S^*$, with strict inequality whenever $S^* < T$. On the other hand, conditional on X_{S^*}, $X_{S_1^*}$ is uniformly distributed on the circle $\partial B(X_{S^*}, \beta_{X_{S^*},\eta})$ whenever $S^* < S_1^*$.

Above we have been writing $E[F(X_t)]$ for the value of the function F averaged over the positions of all Brownian motions starting at y and moving for a time of t. We now make the dependence on the starting point explicit by writing $E^y[F(X_t)]$. From (7.16) we have

$$F(X_{S^*}) - E^{X_{S^*}}[F(X_{S^*})] = F(X_{S^*}) - S_{\beta_{X_{S^*}},\eta}F(X_{S^*})$$

$$\leq \eta(\beta_{X_{S^*}})^2$$

$$= \eta E^{X_{S^*}}[|X_{S^*} - X_{S_1^*}|^2].$$

Using this, the strong Markovian property, orthogonality between $X_{S^*} - X_{S_1^*}$ and $X_{S^*} - y$, and $S^* \in \mathfrak{S}$, we have

$$F(y) - E^y[F(X_{S_1^*})] = F(y) - E^y[F(X_{S^*})] + E^y[F(X_{S^*}) - F(X_{S_1^*})]$$

$$\leq \eta E^y[|y - X_{S^*}|^2]$$

$$+ E^y[F(X_{S^*}) - F(X_{S_1^*}), S^* < S_1^*]$$

$$= \eta E^y[|y - X_{S^*}|^2]$$

$$+ E^y[F(X_{S^*}) - E^{X_{S^*}}F(X_{S_1^*}), S^* < S_1^*]$$

$$\leq \eta E^y[|y - X_{S^*}|^2]$$

$$+ \eta E^y[E^{X_{S^*}}[|X_{S^*} - X_{S_1^*}|^2], S^* < S_1^*]$$

$$= \eta E^y[|y - X_{S^*}|^2] + \eta E^y[|X_{S^*} - X_{S_1^*}|^2]$$

$$= \eta E^y[|y - X_{S_1^*}|^2].$$

Thus $S_1^* \in \mathfrak{S}$. This contradicts the maximality of S^* in $\mathfrak{S}$ since $S_1^* \geq S^*$ and $S_1^* \neq S^*$. Thus we have shown that $T \in G$. It follows that

$$F(y) - E^y[F(X_T)] \leq \eta E^y[|X_T - y|^2] \leq \eta(2r_1^2).$$

Letting η tend to 0 leads to inequality (7.22). This completes the proof of Theorem 7.4.

7.5 Proof of Theorem 7.5

The first Lemma that we will prove is well known.

LEMMA 7.2
Let $x \in \mathbb{T}^2$, let $\mathfrak{R}$ be the family of all rectangles containing x and having sides parallel to the coordinate axes, and let

$$Mh(x) = \sup_{R \in \mathfrak{R}} \frac{1}{|R|} \int_R |h(z)| dz$$

be the maximal function of the nth degree two variable trigonometric polynomial $h(x) = \sum_{|m| \leq n} c_m e^{imx}$. Then if $|x - y| \leq \frac{c}{n}$, there exists a constant $C = C(c)$ such that

$$|h(x)| \leq CMh(y).$$

REMARK 7.3 The operator M differs from the standard strong maximal operator M_S in that we do not demand that the chosen rectangles be centered at x. However, since any rectangle R containing x can be contained in a rectangle centered at x and having volume no more than $4|R|$, it is easy to see that $Mh(x) \leq 4M_S h(x)$. ∎

PROOF (Proof of Lemma) The ideas used in this proof can be found in Chapter III of [Z1].

Let $K_n(t) = \frac{2}{n+1} \left(\frac{\sin \frac{1}{2}(n+1)t}{2 \sin \frac{1}{2}t} \right)^2$ be the Fejer kernel and

$$\tau_n(t) = 2\big(2K_{2n-1}(t) - K_{n-1}(t)\big) = \frac{1}{n} \frac{\sin^2(nt) - \sin^2(nt/2)}{\sin^2(t/2)}$$

be the delayed arithmetic means. Then the Fourier coefficients of τ_n satisfy $\hat{\tau}_n(m) = 1$ whenever $|m| \leq n$ and it is easy to show that $|\tau_n(t)| \leq \psi_n(t) = \frac{100n}{1+(nt)^2}$. Let $m \in \mathbb{Z}^2$. If $|m| \leq n$, then $|m_1| \leq n$ and $|m_2| \leq n$, so that convolution with $\tau_n(x)\tau_n(y)$ multiplies $e^{im_1x}e^{im_2y}$ by 1 and hence maps $h(x, y)$ into itself. We have

$$|h(x)| = \left| (2\pi)^{-2} \int_{-\pi}^{\pi} \int_{-\pi}^{\pi} h(z)\tau_n(x_1 - z_1)\tau_n(x_2 - z_2) dz_1 dz_2 \right|$$

$$\leq (2\pi)^{-2} \int_{-\pi}^{\pi} \int_{-\pi}^{\pi} |h(z)| \psi_n(x_1 - z_1) \psi_n(x_2 - z_2) dz_1 dz_2.$$

Fix y and make the dyadic decomposition

$$[-\pi, \pi)^2 = \left(\bigcup_{j=0}^{\infty} I_j(y_1) \right) \times \left(\bigcup_{k=0}^{\infty} I_k(y_2) \right),$$

where $I_j(t) = \{s \in [-\pi, \pi) : \frac{c}{n} a_j \leq |s - t| < \frac{c}{n} 2^j\}$, where $a_0 = 0$ and $a_j = 2^{j-1}$ if $j \geq 1$. Then make the estimates

$$\int_{I_j(y_1) \times I_k(y_2)} |h(z)| \psi_n(x_1 - z_1) \psi_n(x_2 - z_2) dz_1 dz_2$$

$$\leq \psi_n \left((2^{j-1} - 1) \frac{c}{n} \right) \psi_n \left((2^{k-1} - 1) \frac{c}{n} \right) |I_j(y_1)| |I_k(y_2)| M h(y).$$

Add everything up to complete the proof. ∎

LEMMA 7.3

Let $B(p_1, \rho_1)$ be any open ball contained in $\left(-\frac{\pi}{2}, \frac{\pi}{2}\right)^2$. Let $F_k(x) = -\sum_{2^{k-1} \leq |m| < 2^k} \frac{c_m}{|m|^2} e^{imx}$ be a trigonometric polynomial in two variables. If

$$\sup_r \varepsilon_r^2 = \sup_r \sum_{|m|=r} |c_m|^2 \leq M < \infty,$$

then

$$\int_0^{2\pi} |F_k(p_1 + (\rho_1 \cos\theta, \rho_1 \sin\theta))|^{4/3} d\theta \leq c(M+1) 2^{-2k/3},$$

where the constant c is independent of k, M, p_1, and ρ_1.

PROOF Let $E = E_{k,\gamma}$ be a set of 2^{-k}-separated points x of $\partial B(p_1, \rho_1)$ (i.e., if r and s are distinct points of E, then $|r - s| > 2^{-k}$) satisfying $|F_k(x)| > \gamma$. Then

$$|E| = \sum_{x \in E} 1 \leq \sum_{x \in E} \left(\frac{|F_k(x)|}{\gamma} \right)^2 = \frac{1}{\gamma^2} \sum_{x \in E} |F_k(x)|^2. \tag{7.28}$$

By Lemma 7.2, whenever $y \in B(x, 2^{-k})$, $|F_k(x)| \le CMF_k(y)$. So

$$|F_k(x)|^2 = \frac{1}{\pi(2^{-k})^2} \int_{B(x,2^{-k})} |F_k(x)|^2 dy$$

$$\le \frac{1}{\pi 2^{-2k}} \int_{B(x,2^{-k})} CMF_k(y)^2 dy.$$

Combining this with inequality (7.28) and observing that 2^{-k}-disks about distinct points of E cannot overlap more than 5-fold (for if a and b are in E and if x is within 2^{-k} of both a and b, then the angle axb is greater than $60°$), we get

$$\gamma^2|E| \le \frac{C}{\pi 2^{-2k}} \sum_{x\in E} \int_{B(x,2^{-k})} MF_k(y)^2 dy$$

$$\le \frac{5C}{\pi 2^{-2k}} \int_{\mathbb{T}^2} MF_k(y)^2 dy.$$

Since the strong maximal operator is bounded on $L^2(\mathbb{T}^2)$, so M is also by Remark 7.3, i.e., there is a constant c_M such that for every $f \in L^2(\mathbb{T}^2)$, $\int_{\mathbb{T}^2} Mf(y)^2 dy \le c_M \int_{\mathbb{T}^2} |f|^2 dy$. Using the last inequality, the fact that M is bounded on $L^2(\mathbb{T}^2)$, and Plancherel's equation, we get

$$\gamma^2|E| \le \frac{5Cc_M}{\pi 2^{-2k}} \int_{\mathbb{T}^2} |F_k(y)|^2 dy$$

$$= \frac{5Cc_M}{\pi 2^{-2k}} (2\pi)^2 \sum_{2^{k-1} \le |m| < 2^k} \left| \frac{c_n}{|n|^2} \right|^2$$

$$= 20Cc_M\pi 2^{2k} \sum_{2^{2k-2} \le j < 2^{2k}} \frac{1}{j^2} \varepsilon^2_{\sqrt{j}}$$

$$\le \frac{20Cc_M\pi 2^{2k}}{(2^{2k-2})^2} (2^{2k} - 2^{2k-2}) \left\{ \frac{1}{2^{2k} - 2^{2k-2}} \sum_{2^{2k-2} \le j < 2^{2k}} \varepsilon^2_{\sqrt{j}} \right\}$$

$$= 120Cc_M\pi\{(\delta_k)^2\}$$

$$\leq 120 C c_M \pi M,$$

where $(\delta_k)^2$, the quantity in curly brackets is the average of a number of $\varepsilon^2_{\sqrt{j}}$'s and hence is also bounded by M. Thus,

$$|E| \leq c\gamma^{-2} M. \tag{7.29}$$

It is geometrically evident that if E is chosen to be a maximal set of 2^{-k}-separated points of $S = \{x \in \partial B(p_1, \rho_1) : |F_k(x)| > \gamma\}$, then S is covered by disks of radius 2^{-k} centered about the points of E. Combine this with inequality (7.29) to get

$$\sigma\{|F_k(x)| > \gamma\} \leq c\gamma^{-2} 2^{-k} M, \tag{7.30}$$

where σ denotes arclength along the edge of the disk $B(p_1, \rho_1)$. We also have the trivial estimate

$$\sigma\{|F_k(x)| > \gamma\} \leq \sigma\{\partial B(p_1, \rho_1)\} = 2\pi\rho_1 < c. \tag{7.31}$$

We now use the representation of the integral in terms of the distribution function to write

$$\frac{1}{2\pi}\int_0^{2\pi} |F(p_1 + (\rho_1 \cos\theta, \rho_1 \sin\theta))|^{4/3} d\theta = \frac{4}{3}\int_0^{\infty} \gamma^{1/3}\sigma\{|F_k(x)| > \gamma\} d\gamma$$

$$= \frac{4}{3}\int_0^{2^{-k/2}} + \frac{4}{3}\int_{2^{-k/2}}^{\infty}.$$

By inequality (7.31), the first integral is bounded by $\int_0^{2^{-k/2}} c\gamma^{1/3} d\gamma = c\frac{3}{4}2^{-(2/3)k}$, and by inequality (7.30), the second integral is bounded by $\int_{2^{-k/2}}^{\infty} c\gamma^{-5/3} 2^{-k} M d\gamma = c\frac{3}{2} 2^{(1/3)k} 2^{-k} M = c\frac{3}{2} 2^{-(2/3)k} M$, so that the lemma is established. ∎

COROLLARY 7.2

Let $B(p_1, \rho_1)$ be any open ball contained in $\left(-\frac{\pi}{2}, \frac{\pi}{2}\right)^2$ and let $\{c_n\}$ be a double sequence of complex numbers satisfying

$$\sup_r \varepsilon_r^2 = \sup_r \sum_{|m|=r} |c_m|^2 \leq M < \infty.$$

Then the $L^2(\mathbb{T}^2)$ function defined by

$$F(x) = -\sum_{n \neq 0} \frac{c_n}{|n|^2} e^{inx}$$

satisfies

$$\frac{1}{2\pi} \int_0^{2\pi} \left| F\left(p_1 + (\rho_1 \cos\theta, \rho_1 \sin\theta)\right) \right|^{4/3} d\theta < c(M+1) < \infty,$$

where c is independent of M, p_1, and ρ_1.

PROOF Decompose the function F into dyadic blocks. For $k = 1, 2, \ldots$, we set

$$F_k(x) = -\sum_{2^{k-1} \leq |m| < 2^k} \frac{c_m}{|m|^2} e^{imx} = -\sum_{|m| \sim 2^k} \frac{c_m}{|m|^2} e^{imx},$$

where

$$|m| \sim 2^k$$

means that $2^{k-1} \leq |m| < 2^k$ or $2^{2k-2} \leq |m|^2 < 2^{2k}$, so that m lies on one of $2^{2k} - 2^{2k-2} = \frac{3}{4} 2^{2k}$ possible circles, each of whose radii is the square root of an integer. Then

$$F(x) = \sum_{k=1}^{\infty} F_k(x).$$

By Minkowski's inequality and Lemma 7.3,

$$\left(\frac{1}{2\pi} \int_0^{2\pi} |F(p_1 + (\rho_1 \cos\theta, \rho_1 \sin\theta))|^{4/3} d\theta \right)^{3/4}$$

$$\leq \sum_{k=1}^{\infty} \left(\frac{1}{2\pi} \int_0^{2\pi} |F_k(p_1 + (\rho_1 \cos\theta, \rho_1 \sin\theta))|^{4/3} d\theta \right)^{3/4}$$

$$\leq c(M+1) \sum_{k=1}^{\infty} \frac{1}{\left(\sqrt{2}\right)^k}$$

$$< \infty.$$

∎

Before going further, we will gather some of the proof of Lemma 7.3 into a separate lemma for later use.

LEMMA 7.4

Let k be a positive integer, $\gamma > 0$, and $\eta \in (0, 2^{-k}]$. Let $E_{k,\gamma,\eta}$ be a set of η-separated points x satisfying

$$\left| \sum_{|m| \sim 2^k} \frac{b_m}{|m|^2} e^{ixm} \right| > \gamma.$$

Then the cardinality of $E_{k,\gamma,\eta}$ satisfies

$$|E_{k,\gamma,\eta}| \leq c\gamma^{-2}\eta^{-2}2^{-2k}\nu_k^2,$$

where c is an absolute constant and

$$\nu_k^2 = \frac{4}{3} 2^{-2k} \sum_{|m| \sim 2^k} |b_m|^2.$$

We will not write out the proof because this lemma is proved in exactly the same way that the first part of the proof of Lemma 7.3 led to inequality (7.29). In that case η was exactly 2^{-k}, so that the factor $\eta^{-2}2^{-2k}$ was equal to 1.

We will also need three more properties enjoyed by harmonic measure.

LEMMA 7.5

Let $F_0 \subset F_1 \subset F_2$ be closed subsets of a bounded open set G. Then for $x \in G \setminus F_2$,

$$\omega\big(G \setminus F_2, \partial F_2, x\big) \geq \omega\big(G \setminus F_1, \partial F_1, x\big), \tag{7.32}$$

$$\omega\big(G \setminus F_1, \partial F_1, x\big) \geq \omega\big(G \setminus F_1, \partial F_0, x\big), \tag{7.33}$$

and

$$\omega\big(G \setminus F_1, \partial F_0, x\big) \geq \omega\big(G \setminus F_2, \partial F_0, x\big). \tag{7.34}$$

PROOF The proofs of the first and third inequalities are quite geometrical, so the reader is encouraged to draw pictures. For the first inequality, recall that $\omega\big(G \setminus F_1, \partial F_1, x\big)$ is the measure of the set of all Brownian motions starting out at x and first exiting $G \setminus F_1$ at some point of ∂F_1. One such path must exit $G \setminus F_1$ at a point p, where either $p \in \partial F_1 \cap \partial F_2$ or $p \in \partial F_1 \setminus \partial F_2$. In the former case, the path hits ∂F_2 before it hits $G \setminus F_1$ and so all the more before it hits the smaller $G \setminus F_2$. In the latter case, since $F_1 \subset F_2$, p must be an interior point of F_2, which means that the path hits points of ∂F_2 even before reaching p, hence before hitting $G \setminus F_1$ and hence, before hitting $G \setminus F_2$. So in either case, the Brownian motion hits ∂F_2 before hitting $G \setminus F_2$. Thus every Brownian motion that contributes to the calculation of $\omega(G \setminus F_1, \partial F_1, x)$ also contributes to the calculation of $\omega(G \setminus F_2, \partial F_2, x)$, which proves inequality (7.32). The second inequality holds because $\omega(G \setminus F_1, \cdot, x)$ is a measure and $\partial(G \setminus F_1) \cap \partial F_1$ contains $\partial(G \setminus F_1) \cap \partial F_0$ and hence has measure at least as large. For the third inequality, consider a Brownian motion which starts at x and first leaves $G \setminus F_2$ at a point $p \in \partial F_0$. As in the case of inequality (7.32), it suffices to show that this motion exits $G \setminus F_1$ at p. Since $G \setminus F_2 \subset G \setminus F_1$, all the points of the Brownian motion that precede p are also in the larger set $G \setminus F_1$, so the Brownian motion also exits $G \setminus F_1$ at p. ∎

It is possible to prove directly that the Riemann Function's average function $A_\rho F(x)$ is a continuous function of x, but the proof would be special to dimension 2. The following lemma discovers a weaker, but useful property of $A_\rho Ft(x)$.

LEMMA 7.6

Let F be the Riemann function, $B(p, \rho_0)$ the disk constructed above so that conditions (7.10) and (7.11) hold on $B(p, \rho_0) \cap \bar{Z}$, where Z is the set of discontinuities of F. Let $p_1 \in Z, B(p_1, \rho_1) \subset B\left(p, \frac{1}{2}\rho_0\right)$, let $p_2 \in B\left(p_1, \frac{1}{2}\rho_1\right)$, and let τ be the distance from p_2 to $\bar{Z}$. Let $\kappa \ll \tau$ and define

$$G_\kappa = \left\{x \in B(p_1, \rho_1) : dist(x, \bar{Z}) < \kappa\right\}$$

and to each point $x \in \partial G_\kappa$, associate a closest point $\tilde{x} = \tilde{x}(x)$ in $\bar{Z}$ such that $|x - \tilde{x} = |dist(x, \bar{Z})|$. Then

$$\lim_{\kappa \to 0} \int_{\partial(G_\kappa \cap B(p_1, \rho_1))} |A_\kappa F(x) - A_\kappa F(\tilde{x})| \, \omega\left(B(p_1, \rho_1) \setminus \bar{G}_\kappa, dx, p_2\right) = 0.$$

PROOF In this proof we will use c to denote a generic constant whose value may be different from line to line. Using once more the formula connecting the integral with the integral of the distribution function, we see that it suffices to prove

$$\lim_{\kappa \to 0} \int_0^\infty \omega\left(B(p_1, \rho_1) \setminus \bar{G}_\kappa, S_\alpha, p_2\right) d\alpha = 0,$$

where

$$S_\alpha = \{x \in \partial(G_\kappa \cap B(p_1, \rho_1)) : |A_\kappa F(x) - A_\kappa F(\tilde{x})| > \alpha\}.$$

For each positive integer k, define

$$\alpha_k = \begin{cases} d\alpha \left[\log(1 + 2^k \kappa)\right]^{-2}, & \text{for } 2^k \geq \kappa^{-1}, \\ d\alpha \left[\log(1 + 2^{-k}\kappa^{-1})\right]^{-2}, & \text{for } 2^k < \kappa^{-1}, \end{cases}$$

where the absolute constant d is chosen such that $\alpha = \sum_{k=1}^{\infty} \alpha_k$. We denote

$$S_{\kappa,k} = \{x \in \partial(G_\kappa \cap B(p_1, \rho_1)) : |A_\kappa F_k(x) - A_\kappa F_k(\tilde{x})| > \alpha_k\},$$

where F_k is the part of F with frequencies n satisfying $2^{k-1} \leq |n| < 2^k$. It is clear that

$$S_\alpha \subset \bigcup_{k=1}^{\infty} S_{\kappa,k}. \tag{7.35}$$

If $x \in S_{\kappa,k}$, by the mean value theorem there is an $\bar{x}$ on the line segment joining x and $\tilde{x}$ such that $|\nabla A_\kappa F_k(\bar{x})| > \kappa^{-1}\alpha_k$. Therefore it will be useful to also consider the sets

$$S'_{\kappa,k} = \left\{\bar{x} \in B(p_1, 2\rho_1) : |\nabla A_\kappa F_k(\bar{x})| > \frac{\alpha_k}{\kappa}\right\}.$$

Recall the easy estimate $|\hat{\chi}(m)| \leq 1$, where $\chi(x) = \pi^{-1}\chi_{B(0,1)}(x)$. Using contour integration (see page 158 of [SW]), we also have

$$\hat{\chi}(m) = O(|m|^{-3/2}) \text{ as } |m| \to \infty.$$

Since $\nabla e^{im\bar{x}} = \left(im_1 e^{im\bar{x}}, im_2 e^{im\bar{x}}\right)$,

$$|\nabla A_\kappa F_k| = \sqrt{\left|\sum_{|m|\sim 2^k} \frac{im_1\hat{\chi}(\kappa m)c_m e^{im\bar{x}}}{|m^2|}\right|^2 + \left|\sum_{|m|\sim 2^k} \frac{im_2\hat{\chi}(\kappa m)c_m e^{im\bar{x}}}{|m^2|}\right|^2},$$

so $S'_{\kappa,k} \subset \left\{\bar{x} \in B(p_1, 2\rho_1)\text{: } \left|\sum_{|m|\sim 2^k} \frac{im_1\hat{\chi}(\kappa m)c_m e^{im\bar{x}}}{|m^2|}\right|^2 > \frac{1}{2}\left(\frac{\alpha_k}{\kappa}\right)^2\right\} \cup \left\{\bar{x} \in B(p_1, 2\rho_1)\text{: } \left|\sum_{|m|\sim 2^k} \frac{im_2\hat{\chi}(\kappa m)c_m e^{im\bar{x}}}{|m^2|}\right|^2 > \frac{1}{2}\left(\frac{\alpha_k}{\kappa}\right)^2\right\} = A_{k,\kappa} \cup B_{k,\kappa}$. Now

$\sum_{|m|\sim 2^k} |im_1\hat{\chi}(\kappa m)c_m|^2$ is dominated by

$$\sum_{|m|\sim 2^k} |m|^2|\hat{\chi}(\kappa m)|^2|c_m|^2 \leq 2^{2k} \sup_{|m|\sim 2^k} |\hat{\chi}(\kappa m)|^2 \sum_{|m|\sim 2^k} |c_m|^2$$

$$\leq c2^{4k} \sup_{|m|\sim 2^k} |\hat{\chi}(\kappa m)|^2 \frac{4}{3} 2^{-2k} \sum_{R\sim 2^k} \varepsilon_R^2 \quad (7.36)$$

$$\leq \begin{cases} c2^{4k}(\kappa 2^k)^{-3}\delta_k^2 & \text{if } \kappa 2^k \geq 1 \\ c2^{4k}\delta_k^2 & \text{if } \kappa 2^k < 1, \end{cases}$$

where

$$\delta_k^2 = \frac{1}{2^{2k} - 2^{2k-2}} \sum_{2^{2k-2}\leq j<2^{2k}} \varepsilon_{\sqrt{j}}^2 = \frac{4}{3} 2^{-2k} \sum_{R\sim 2^k} \varepsilon_R^2.$$

Notice that since δ_k^2 is an average of values of ε_R^2, from relation (7.2) it follows that

$$\delta_k^2 \to 0 \text{ as } k \to \infty.$$

We may apply Lemma 7.4 with $\gamma = \frac{1}{\sqrt{2}}\frac{\alpha_k}{\kappa}$ and ν_k^2 dominated by 2^{-2k} times the quantity in inequality (7.36) to see that if $F_{k,\kappa}$ is a set of η-separated points of $A_{k,\kappa}$, then

$$|F_{k,\kappa}| \leq \begin{cases} c\left(\frac{\kappa}{\alpha_k}\right)^2 \eta^{-2}(\kappa 2^k)^{-3}\delta_k^2, & \text{if } \kappa 2^k \geq 1 \\ c\left(\frac{\kappa}{\alpha_k}\right)^2 \eta^{-2}\delta_k^2, & \text{if } \kappa 2^k < 1. \end{cases}$$

Since the estimates for $B_{k,\kappa}$ are identical, the same holds for $S_{k,\kappa}$. Thus if $E_{k,\kappa}$ is a maximal set of η-separated points of $S'_{k,\kappa}$, then

$$|E_{k,\kappa}| \leq \begin{cases} c\alpha_k^{-2}\kappa^{-1}\eta^{-2}2^{-3k}\delta_k^2, & \text{if } \kappa 2^k \geq 1 \\ c\alpha_k^{-2}\kappa^2\eta^{-2}\delta_k^2, & \text{if } \kappa 2^k < 1. \end{cases} \quad (7.37)$$

Pick κ so small that $\kappa < \sqrt{\kappa} \ll \tau$. We distinguish three cases.

Case (i) $\kappa^{-1} \leq 2^k$

Set $\eta = 2^{-k}$ and construct the set $E_{k,\kappa}$. By the mean value theorem, every point of $S_{\kappa,k}$ is within κ of a point of $S'_{\kappa,k}$ and, since $\kappa \geq 2^{-k}$, every point of $S'_{\kappa,k}$ is within κ of a point of $E_{\kappa,k}$. It follows that the union of the disks of radius 2κ centered at the points of $E_{\kappa,k}$

contains $S_{\kappa,k}$. Call these disks $\{B_i\}$, $i = 1, 2, ..., N$. From inequality (7.37) we have that $N \leq c\alpha_k^{-2}\kappa^{-1}2^{-k}\delta_k^2$. Then

$$\begin{aligned}
\omega\left(B(p_1,\rho_1)\setminus\bar{G}_\kappa, S_{\kappa,k}, p_2\right) &\leq \omega\left(B(p_1,\rho_1)\setminus S_{\kappa,k}, S_{\kappa,k}, p_2\right)\\
&\leq \omega\left(B(p_1,\rho_1)\setminus\overline{\cup B_i}, \partial(\cup B_i), p_2\right)\\
&\leq \sum_{i=1}^{N}\omega\left(B(p_1,\rho_1)\setminus\overline{\cup B_i}, \partial B_i, p_2\right)\\
&\leq N \leq c\alpha_k^{-2}(\kappa 2^k)^{-1}\delta_k^2. \qquad (7.38)
\end{aligned}$$

The first line follows from inequality (7.34); the second line follows from inequality (7.32); the third line follows from subadditivity of harmonic measure in the second coordinate; and the last line follows from combining the trivial estimate $\omega(\cdot,\cdot,\cdot) \leq 1$ and inequality (7.37) with $\eta = 2^{-k}$.

Case (ii) $\kappa^{-1/2} \leq 2^k < \kappa^{-1}$

Again set $\eta = 2^{-k}$ and construct the set $E_{k,\kappa}$. Now disks of radius $2 \cdot 2^{-k}$ centered at the points of $E_{\kappa,k}$ cover $S_{\kappa,k}$ and by inequality (7.37) we have that $N \leq c\alpha_k^{-2}(\kappa 2^k)^2\delta_k^2$, so that

$$\omega\left(B(p_1,\rho_1)\setminus\bar{G}_\kappa, S_{\kappa,k}, p_2\right) \leq N \leq c\alpha_k^{-2}(\kappa 2^k)^2\delta_k^2. \qquad (7.39)$$

Case (iii) $2^k < \kappa^{-1/2}$

This time set $\eta = \sqrt{\kappa}$ and construct the set $E_{\kappa,k}$. Now disks of radius $2\sqrt{\kappa}$ centered at the points of $E_{\kappa,k}$ cover $S_{\kappa,k}$ and by inequality (7.37) we have that $N \leq c\alpha_k^{-2}\kappa\delta_k^2$, so that

$$\omega\left(B(p_1,\rho_1)\setminus\bar{G}_\kappa, S_{\kappa,k}, p_2\right) \leq N \leq c\alpha_k^{-2}\kappa\delta_k^2. \qquad (7.40)$$

Combining inequalities (7.38), (7.39), and (7.40), it follows from (7.35) and the definitions of α_k that

$$\omega\big(B(p_1,\rho_1)\setminus\bar{G}_\kappa, S_\alpha, p_2\big)$$

$$\leq c\left\{\begin{array}{c}\sum\limits_{2^k\geq\kappa^{-1}}\alpha_k^{-2}(\kappa 2^k)^{-1}\delta_k^2\\ +\sum\limits_{\kappa^{-1}>2^k\geq\kappa^{-1/2}}\alpha_k^{-2}(\kappa 2^k)^2\delta_k^2+\sum\limits_{\kappa^{-1/2}>2^k}\alpha_k^{-2}\kappa\delta_k^2\end{array}\right\}$$

$$= c\{I + II + III\}. \tag{7.41}$$

Thinking of $2^k\kappa$ as a variable with nth value in $[2^n, 2^{n+1})$, we have

$$I \le c\alpha^{-2} \max_{2^k > \kappa^{-1/2}} \delta_k^2 \sum_{2^k\kappa \ge 1} \frac{\log^4(1 + 2^k\kappa)}{2^k\kappa}$$

$$\le c\alpha^{-2} \max_{2^k > \kappa^{-1/2}} \delta_k^2 \sum_{n=0}^{\infty} \frac{\log^4(1 + 2^{n+1})}{2^n}$$

$$\le c\alpha^{-2} \max_{2^k > \kappa^{-1/2}} \delta_k^2.$$

Similarly

$$II \le c\alpha^{-2} \sup_{2^k > \kappa^{-1/2}} \delta_k^2 \sum_{\kappa^{-1} > 2^k \ge \kappa^{-1/2}} \log^4\big(1 + (2^k\kappa)^{-1}\big)(\kappa 2^k)^2$$

$$\le c\alpha^{-2} \sup_{2^k > \kappa^{-1/2}} \delta_k^2 \sum_{n=0}^{\infty} \frac{\log^4(1 + 2^{n+1})}{4^n}$$

$$\le c\alpha^{-2} \sup_{2^k > \kappa^{-1/2}} \delta_k^2.$$

Also

$$III \le c\alpha^{-2} \sup_k \delta_k^2 \kappa \sum_{\kappa^{-1/2} > 2^k} \log^4\big(1 + (\kappa 2^k)^{-1}\big)$$

$$\le c\alpha^{-2}\kappa \log^5(1/\kappa),$$

since the last sum consists of $O\big(\log(1/\kappa)\big)$ terms, each of size $O\big(\log^4(1/\kappa)\big)$. Using our estimates for I, II, and III, we find that

$$\omega\big(B(p_1, \rho_1) \setminus \bar{G}_\kappa, S_\alpha, p_2\big) \le c\alpha^{-2}\beta_k^2,$$

where

$$\beta_\kappa = \left\{ \sup_{2^k > \kappa^{-1/2}} \delta_k^2 + \kappa \log^5(1/\kappa) \right\}^{1/2}.$$

Use the trivial estimate $\omega \leq 1$ on $[0, \beta_\kappa]$ and this estimate on $[\beta_\kappa, \infty)$ to get

$$\int_0^\infty \omega\left(B(p_1,\rho_1) \setminus \bar{G}_\kappa, S_\alpha, p_2\right) d\alpha \leq \int_0^{\beta_\kappa} d\alpha + c\beta_k^2 \int_{\beta_\kappa}^\infty \alpha^{-2} d\alpha$$

$$= \beta_\kappa(1+c).$$

Since β_κ tends to zero as κ tends to 0, Lemma 7.6 is proved. ∎

We need one more lemma before we can complete the proof of Theorem 7.5.

LEMMA 7.7
Let $F(x)$ be bounded and $\tilde{\Delta} F(x) = 0$ on $\overline{B(p,\rho_0)}$. Let

$$Z = \left\{ x \in \overline{B(p,\rho_0)} : F \text{ is discontinuous at } x \right\}.$$

Let $p_1 \in Z$ and $\rho_1 > 0$ be chosen so that $B(p_1,\rho_1) \subset B\left(p, \frac{1}{2}\rho_0\right)$. Let $p_2 \notin \bar{Z}$ be chosen so close to p_1 that $\tau = |p_2 - p_1| < \frac{1}{2}\rho_1$. Let $\kappa \ll \tau$ and define G_κ as above. Then

$$\int_{\partial B(p_1,\rho_1)} |F(p_1) - F(x)|\, \omega\left(B(p_1,\rho_1) \setminus \bar{G}_\kappa, dx, p_2\right)$$

$$\leq c\left[|F(p_1)| + \rho_1^{-3/4}\right]\left[1 - \omega(B(p_1,\rho_1) \setminus \bar{Z}, \partial\left(Z \cap B(p_1,\rho_1)\right), p_2)\right]^{1/4}$$

PROOF Let S be a subset of the circumference of $B(p_1,\rho_1)$. Then

$$\omega\left(B(p_1,\rho_1) \setminus \bar{G}_\kappa, S, p_2\right) \leq \omega\left(B(p_1,\rho_1), S, p_2\right). \tag{7.42}$$

The right-hand side is the probability that a Brownian motion starting at p_2 will leave $B(p_1,\rho_1)$ at a point belonging to S. Intuitively, this inequality asserts that a Brownian motion will have less trouble reaching S if G_κ is not in the way. For a formal proof, see the proof of Lemma 4.5 in [AWa].

We will compare three different measures defined on the circumference C of $B(p_1,\rho_1)$, $\omega(\cdot) = \omega\big(B(p_1,\rho_1) \setminus \bar{G}_\kappa, \cdot, p_2\big)$, $\omega_1(\cdot) = \omega\big(B(p_1,\rho_1), \cdot, p_2\big)$, and ordinary arclength $\sigma(\cdot)$. Inequality (7.42) may be restated as asserting that ω is absolutely continuous with respect to ω_1 and that the Radon-Nikodym derivative $\frac{d\omega}{d\omega_1}$ is bounded by 1. But the harmonic measure ω_1 is absolutely continuous with respect to σ and the Radon-Nikodym derivative is the Poisson kernel $\frac{1}{2\pi\rho_1}\frac{\rho_1^2-|p_2-p_1|^2}{|x-p_2|^2}$. In other words,

$$\omega_1(S)=\omega\left(B(p_1,\rho_1),S,p_2\right)=\frac{1}{2\pi\rho_1}\int_S\frac{\rho_1^2-|p_2-p_1|^2}{|x-p_2|^2}d\sigma.$$

Since $|p_2-p_1|\leq\frac{1}{2}\rho_1$, we have the estimate

$$\frac{d\omega_1}{d\sigma}=\frac{1}{2\pi\rho_1}\frac{\rho_1^2-|p_2-p_1|^2}{|x-p_2|^2}\leq\frac{c}{\rho_1}\text{ if }x\in\partial B(p_1,\rho_1),$$

so, when restricted to $\partial B(p_1,\rho_1)$, the harmonic measure $\omega(\cdot)$ is absolutely continuous with respect to the arclength σ. Moreover the Radon-Nikodym derivative $\frac{d\omega}{d\sigma}$ satisfies

$$\frac{d\omega}{d\sigma}=\frac{d\omega}{d\omega_1}\frac{d\omega_1}{d\sigma}\leq\frac{c}{\rho_1}. \tag{7.43}$$

By Holder's inequality with $(p,p')=(4/3,4)$, Corollary 7.2, and inequality (7.43), we may estimate $\int_{\partial B(p_1,\rho_1)}|F(p_1)-F(x)|\omega(B(p_1,\rho_1)\setminus\bar{G}_\kappa,dx,p_2)$ by

$$|F(p_1)|\omega\big(B(p_1,\rho_1)\setminus\bar{G}_\kappa,\partial B(p_1,\rho_1),p_2\big)+\|F\|_{L^{\frac{4}{3}}(d\sigma)}\left\|\frac{d\omega}{d\sigma}\right\|_{L^4(d\sigma)}$$

$$\leq|F(p_1)|\omega\left(B(p_1,\rho_1)\setminus\bar{G}_\kappa,\partial B(p_1,\rho_1),p_2\right)+c\left\|\frac{d\omega}{d\sigma}\right\|_{L^4(d\sigma)}$$

$$\leq|F(p_1)|\omega\left(B(p_1,\rho_1)\setminus\bar{G}_\kappa,\partial B(p_1,\rho_1),p_2\right)^{\frac{1}{4}}+c\left(\int_{\partial B(p_1,\rho_1)}\left|\frac{d\omega}{d\sigma}\right|^3d\omega\right)^{\frac{1}{4}}$$

$$\leq\left\{\begin{array}{l}|F(p_1)|\omega\left(B(p_1,\rho_1)\setminus\bar{G}_\kappa,\partial B(p_1,\rho_1),p_2\right)^{\frac{1}{4}}\\+c\rho_1^{-3/4}\omega\left(B(p_1,\rho_1)\setminus\bar{G}_\kappa,\partial B(p_1,\rho_1),p_2\right)^{\frac{1}{4}}\end{array}\right\}$$

$$\leq c\left[|F(p_1)|+c\rho_1^{-3/4}\right]\left[1-\omega\left(B(p_1,\rho_1)\setminus\bar{Z},\partial\left(Z\cap B(p_1,\rho_1)\right),p_2\right)\right]^{1/4},$$

where the last inequality requires yet two more facts. The first is

$$\omega\left(B(p_1,\rho_1)\setminus\bar{G}_\kappa,\partial B(p_1,\rho_1),p_2\right)$$

$$= 1 - \omega\left(B(p_1, \rho_1) \setminus \bar{G}_\kappa, \partial\left(G_\kappa \cap B(p_1, \rho_1)\right), p_2\right)$$

and the second, which follows in a straightforward manner from the continuity of Brownian motion is that as κ tends to zero,

$$\omega\left(B(p_1, \rho_1) \setminus \bar{G}_\kappa, \partial\left(B(p_1, \rho_1) \cap G_\kappa\right), p_2\right)$$

$$\searrow \omega\left(B(p_1, \rho_1) \setminus \bar{Z}, \partial\left(Z \cap B(p_1, \rho_1)\right), p_2\right). \tag{7.44}$$

This completes the proof of Lemma 7.6. ∎

We are now ready to prove Theorem 7.5.

PROOF Let p_1, ρ_1, τ, and G_κ be as in Lemma 7.6. By Theorem 7.4, F is harmonic on $B(p_1, \rho_1) \setminus \bar{Z}$. Since $\bar{Z} \cap B(p_1, \rho_1) \subset G_\kappa$ $F(p_1) - F(x)$ is harmonic on $B(p_1, \rho_1) \setminus \bar{G}_\kappa$. Consequently, we have

$$F(p_1) - F(p_2) = \int_{\partial\left(B(p_1,\rho_1)\setminus\bar{G}_\kappa\right)} (F(p_1) - F(x))\, \omega\left(B(p_1, \rho_1) \setminus \bar{G}_\kappa, dx, p_2\right).$$

Therefore

$$|F(p_1) - F(p_2)| \le \int_{\partial B(p_1,\rho_1)} (F(p_1) - F(x))\, \omega\left(B(p_1, \rho_1) \setminus \bar{G}_\kappa, dx, p_2\right)$$

$$+ \int_{\partial(B(p_1,\rho_1)\cap G_\kappa)} (F(p_1) - F(x))\, \omega\left(B(p_1, \rho_1) \setminus \bar{G}_\kappa, dx, p_2\right)$$

$$= I_1 + I_2.$$

Applying Lemma 7.7 to I_1, we get the first half of inequality (7.12). To estimate I_2, for each $x \in \partial G_\kappa$, let $\tilde{x}$ be a point of $\bar{Z}$ such that $|x - \tilde{x}| = \kappa$. So if x is a point of $\partial\left(B(p_1, \rho_1) \cap G_\kappa\right)$, $|\tilde{x} - p_1| \le |\tilde{x} - x| + |x - p_1| \le \kappa + \rho_1 < 2\rho_1$ and

$$|F(p_1) - F(x)| \le \sup_{q \in \bar{Z} \cap B(p_1, 2\rho_1)} |F(p_1) - F(q)| + |F(x) - F(\tilde{x})|. \tag{7.45}$$

Since $x \in \partial G_\kappa$, $B(x, \kappa) \cap \bar{Z} = \emptyset$, so by harmonicity,

$$F(x) = A_\kappa F(x).$$

But from inequality (7.11),

$$|F(\tilde{x}) - A_\kappa F(\tilde{x})| \leq C\kappa^2$$

uniformly for all $\tilde{x} \in B(p_1, 2\rho_1) \cap \bar{Z}$. These two relations imply that

$$|F(x) - F(\tilde{x})| \leq |A_\kappa F(x) - A_\kappa F(\tilde{x})| + C\kappa^2.$$

Apply Lemma 7.6 and inequality (7.45) and then let $\kappa \to 0$ to see that

$$I_2 \leq \sup_{q \in \bar{Z} \cap B(p_1, 2\rho_1)} |F(p_1) - F(q)|.$$

This completes the proof. ∎

7.6 Proof of Theorem 7.6

We start with the following Lemma.

LEMMA 7.8

Let B be an open disk and let F be a closed set such that $F \subset B$. If $\omega(B \setminus F, \partial F, x) > 0$ for some $x \in B \setminus F$, then for any $\eta > 0$, there exists $p \in F$ and $\tilde{\delta} > 0$ such that

$$\omega(B \setminus F, \partial F, z) > 1 - \eta \textit{ for all } z \in B(p, \tilde{\delta}).$$

PROOF We will omit several details of the proof; for these, see pages 65–66 of [AWa]. As in the proof of Theorem 7.5 cover F with open sets G_κ. If κ is sufficiently small, G_κ approximates F closely and there is a point $y \in \partial G_\kappa$ such that $\omega(B \setminus F, \partial F, y)$ is very close to 1, say $\omega(B \setminus F, \partial F, y) > 1 - \eta_1$, where η_1 will be chosen later. To see this, think about a Brownian motion moving from x to F as moving from x to ∂G_κ and then from ∂G_κ to ∂F. This point of view makes it intutively clear that

$$\omega(B \setminus F, \partial F, x) \leq \omega\left(B \setminus \overline{G_\kappa}, \partial G_\kappa, x\right) \cdot \sup_{z \in \partial G_\kappa} \omega(B \setminus F, \partial F, z),$$

an inequality that follows rigorously from the strong Markovian property. But from relation (7.44), it follows that $\omega(B \setminus F, \partial F, x)$ and $\omega\left(B \setminus \overline{G_\kappa}, \partial G_\kappa, x\right)$

are almost equal, so that the supremum is almost 1. Choose y to be a point of ∂G_κ where the supremum is almost achieved.

Let $\tau = dist(y, F)$. Next we will pick δ, γ, and Γ, both very close to 1 such that $\delta < 1 < \gamma < \Gamma$ and consider the 4 concentric disks $B_{\delta\tau}$, $B_\tau, B_{\gamma\tau}$, $B_{\Gamma\tau}$ centered at y of radii $\delta\tau, \tau, \gamma\tau$ and $\Gamma\tau$. The idea is to let p be a point of F so that $|p - y| = \tau$ and then to pick $\tilde{\delta}$ so small that $B(p, \tilde{\delta})$ lies in the annulus A bounded by the circumferences of $B_{\delta\tau}$ and $B_{\gamma\tau}$. So the goal is to make the four choices $\eta_1, \delta, \gamma, \Gamma$ in such a way that $\omega(B \setminus F, \partial F, z)$ exceeds $1 - \eta$ in the annulus A. Apply Harnack's inequality to the disk B_τ and the positive harmonic function $1 - \omega(B \setminus F, \partial F, y)$ to get that for all z in $B_{\delta\tau}$,

$$\frac{1 - \omega(B \setminus F, \partial F, z)}{1 - \omega(B \setminus F, \partial F, y)} \leq \frac{1+\delta}{1-\delta} = c_\delta.$$

Thus, we have

$$\omega(B \setminus F, \partial F, z) > 1 - c_\delta \eta_1 \text{ if } z \in B_{\delta\tau}. \tag{7.46}$$

Pick Γ so small that $B_{\Gamma\tau} \subset B$. By the maximal principle for harmonic functions

$$\omega(B \setminus \overline{B_{\delta\tau}}, \partial B_{\delta\tau}, x) \geq \omega(B_{\Gamma\tau} \setminus \overline{B_{\delta\tau}}, \partial B_{\delta\tau}, x),$$

since both functions are 1 on $\partial B_{\delta\tau}$, while on $\partial B_{\Gamma\tau}$ the left function is positive and the right function is 0. By direct calculation, $\left(\partial^2_{x_1} + \partial^2_{x_2}\right) \log\left(x_1^2 + x_2^2\right) = 0$, so that $a \log |x - y| + b$ is a harmonic function of x. Since there is only one harmonic function that is 1 on $\partial B_{\delta\tau}$ and 0 on $\partial B_{\Gamma\tau}$ we see that

$$\omega\left(B_{\Gamma\tau} \setminus \overline{B_{\delta\tau}}, \partial B_{\delta\tau}, x\right) = \frac{\log \frac{\Gamma\tau}{|x-y|}}{\log \frac{\Gamma}{\delta}}.$$

Now since $\lim_{\varepsilon \to 0} \log\left(\frac{\Gamma}{1+\varepsilon}\right) / \log\left(\frac{\Gamma}{1-\varepsilon}\right) = 1$, we may pick $\delta < 1$ so big and $\gamma \in (1, \Gamma)$ so small that

$$\frac{\log \frac{\Gamma\tau}{\gamma\tau}}{\log \frac{\Gamma}{\delta}} > 1 - \frac{\eta}{2}.$$

From the last three relations we conclude that

$$\omega(B \setminus \overline{B_{\delta\tau}}, \partial B_{\delta\tau}, x) > 1 - \frac{\eta}{2} \text{ for all } x \in A. \tag{7.47}$$

Having picked δ, we now pick η_1 so small that $c_\delta \eta_1 < \eta/2$. Then from inequality (7.46) it follows that

$$\omega(B \setminus F, \partial F, z) > 1 - \frac{\eta}{2} \text{ if } z \in B_{\delta\tau}. \tag{7.48}$$

Let F_1 be the disjoint union of F and $B_{\delta\tau}$. Then ∂F_1 is the disjoint union of ∂F and $\partial B_{\delta\tau}$. Let $x \in A$ and consider $\omega(B \setminus F, \partial F, x)$ as the probability of a Brownian motion starting at x hitting ∂F. Such a motion either hits $\partial B_{\delta\tau}$ or not on its way to ∂F. Intuitively, the probability of the former is bounded below by $\omega(B \setminus F_1, \partial B_{\delta\tau}, x) \cdot \inf_{z\in B_{\delta\tau}} \omega(B \setminus F, \partial B_{\delta\tau}, z)$ since the particle first goes to $B_{\delta\tau}$ and then goes on to ∂F, while the probability of the latter is $\omega(B \setminus F_1, \partial F, x)$. So taking inequality (7.47) into account, we have

$$\begin{aligned} \omega(B \setminus F, \partial F, x) &\geq \omega\big(B \setminus F_1, \partial B_{\delta\tau}, x\big) \cdot \left(1 - \frac{\eta}{2}\right) + \omega\big(B \setminus F_1, \partial F, x\big) \\ &\geq \left(1 - \frac{\eta}{2}\right) \left(\omega(B \setminus F_1, \partial B_{\delta\tau}, x) + \omega(B \setminus F_1, \partial F, x)\right) \\ &= \left(1 - \frac{\eta}{2}\right) \omega\big(B \setminus F_1, \partial F_1, x\big). \end{aligned} \tag{7.49}$$

But $F_1 \supset B_{\delta\tau}$, so inequalities (7.32) and (7.48) imply

$$\omega\big(B \setminus F_1, \partial F_1, x\big) \geq \omega\big(B \setminus B_{\delta\tau}, \partial B_{\delta\tau}, x\big) \geq 1 - \frac{\eta}{2}.$$

Combine this with inequality (7.49) to finish the lemma,

$$\omega\big(B \setminus F, \partial F, x\big) \geq \left(1 - \frac{\eta}{2}\right)^2 > 1 - \eta.$$

∎

We now prove Theorem 7.6.

We need to find $p_1 \in B(p_0, r) \cap F$, such that for any $\eta \in (0, 1)$ and any $\delta_1 > 0$, there exists $\delta_2 = \delta_2(\eta) \in (0, \delta_1)$ such that

$$\omega\left(B(p_1, \delta_1) \setminus F, \partial(B(p_1, \delta_1) \cap F), x\right) > 1 - \eta \tag{7.50}$$

for all $x \in B(p_1, \delta_2)$.

By the maximum principle for harmonic functions, for fixed p_1, δ_2, and x, the left-hand side of inequality (7.50) is an increasing function of δ_1.

Thus inequality (7.50) would be implied by the following: There exists $p_1 \in B(p_0, r) \cap F$ and a sequence of positive numbers $\delta_1 > \tilde{\delta}_1 > \delta_2 > \ldots > \delta_n > \tilde{\delta}_n > \ldots$ decreasing to zero, such that

$$\omega\left(B(p_1, \delta_n) \setminus F, \partial\left(B(p_1, \delta_n) \cap F\right), x\right) > 1 - \frac{1}{n} \tag{7.51}$$

for all $x \in B(p_1, \tilde{\delta}_n)$.

To prove this, denote $B(p_0, r)$ as B_0 and note that $\omega\big(B_0 \setminus F, \partial B_0, x\big) > 0$ implies that for any $\varepsilon > 0$ there exists a disk $B_\varepsilon = B\big(q, \varepsilon r'\big)$, where $r' = r'(q)$ is such that $B(q, r') \subset B_0$, and such that $\omega\big(B_0 \setminus F, \partial(B_\varepsilon \cap F), x\big) > 0$. Choose $\varepsilon_1 \in (0, \frac{1}{9})$ such that

$$\frac{\log \frac{D}{2/9r'}}{\log \frac{D}{2\varepsilon_1 r'}} < \frac{1}{4}, \tag{7.52}$$

where D is the diameter of B_0. Let $F_1 = \overline{B_{\varepsilon_1}} \cap F$. Notice that if $q_1 \in F_1$ and $x \in B\left(q_1, \frac{2}{9}r'\right)$, $|x - q| \leq |x - q_1| + |q_1 - q| < \frac{2}{9}r' + \varepsilon_1 r' < r'$, so $B\left(q_1, \frac{2}{9}r'\right) \subset B_0$. Since by inequality (7.34),

$$\omega\big(B_0 \setminus F_1, \partial F_1, x\big) \geq \omega\big(B_0 \setminus F, \partial F_1, x\big) > 0,$$

we may apply Lemma 7.8 to see that there exists $q_1 \in F_1$, and $\tilde{\delta}_1 > 0$ such that $B(q_1, \tilde{\delta}_1) \subset \overline{B(q_1, 2\varepsilon_1 r')} \subset B\left(q_1, \frac{2}{9}r'\right) \subset B_0$, and such that

$$\omega\big(B_0 \setminus F_1, \partial F_1, z\big) > 3/4 \tag{7.53}$$

for all $z \in B(q_1, \tilde{\delta}_1)$. Choose such a q_1 and set $B = B\big(q_1, \frac{2}{9}r'\big)$.

Let $z \in B(q_1, \tilde{\delta}_1)$. Then $\omega\big(B_0 \setminus F_1, \partial F_1, z\big)$, the probability that a Brownian path starting at z hits ∂F_1 before ∂B_0, is dominated by

$$\sup_{x \in \partial B} \omega(B_0 \setminus F_1, \partial F_1, x) + \omega(B \setminus F_1, \partial F_1, z),$$

since such a path must either hit ∂B or hit F_1 before hitting ∂B. Thus

$$\omega\big(B \setminus F_1, \partial F_1, z\big) \geq \omega\big(B_0 \setminus F_1, \partial F_1, z\big) - \sup_{x \in \partial B} \omega\big(B_0 \setminus F_1, \partial F_1, x\big). \tag{7.54}$$

But if $x \in \partial B$, by inequality (7.32), we have

$$\omega\big(B_0 \setminus F_1, \partial F_1, x\big) \leq \omega\left(B_0 \setminus \overline{B(q_1, 2\varepsilon_1 r')}, \partial B(q_1, 2\varepsilon_1 r'), x\right). \tag{7.55}$$

By the maximum principle we have

$$\omega\left(B_0 \setminus \overline{B(q_1, 2\varepsilon_1 r')}, \partial B(q_1, 2\varepsilon_1 r'), x\right) \le \frac{\log \frac{D}{|x-q_1|}}{\log \frac{D}{2\varepsilon_1 r'}}, \tag{7.56}$$

since both harmonic functions equal 1 when $|x - q_1| = 2\varepsilon_1 r'$, while on ∂B_0 the left-hand function is 0 and the right-hand function is nonnegative. Chain together inequalities (7.55), (7.56), and (7.52); and then substitute this and inequality (7.53) into inequality (7.54) to obtain

$$\omega\left(B \setminus F_1, \partial F_1, z\right) > \frac{3}{4} - \frac{1}{4} = \frac{1}{2} \text{ for } z \in B\left(q_1, \tilde{\delta}_1\right). \tag{7.57}$$

Apply inequality (7.32) one more time and then use inequality (7.57) to finally arrive at

$$\omega\left(B \setminus \left(\overline{B(q_1, 2\varepsilon_1 r')} \cap F_1\right), \partial(B(q_1, 2\varepsilon_1 r') \cap F), z\right) \ge \omega\left(B \setminus F_1, \partial F_1, z\right) > \frac{1}{2}.$$

We have shown that there exists $q_1 \in B(q, r') \cap F, \gamma_1 = \frac{2}{9}r' > \eta_1 = 2\varepsilon_1 r' > \tilde{\eta}_1 = \tilde{\delta}_1 > 0$ such that

$$\omega\left(B(q_1, \gamma_1) \setminus (\overline{B(q_1, \eta_1)} \cap F), \partial\left(B(q_1, \eta_1) \cap F\right), z\right) > \frac{1}{2} \text{ for all } z \in B(q_1, \tilde{\eta}_1).$$

Adjoin the additional constrain to $\tilde{\eta}_1$ that it be so small that

$$\frac{\log \frac{D}{\gamma_1 - \tilde{\eta}_1}}{\log \frac{D}{\eta_1 + \tilde{\eta}_1}} < \frac{1}{2},$$

which can be done in view of inequality (7.52).

Repeating the above process with $B(p_0, r)$ replaced by $B(q_1, \tilde{\eta}_1)$, we can find q_2 in $B(q_1, \tilde{\eta}_1) \cap F$, $\frac{2}{9}\tilde{\eta}_1 > \gamma_2 > \eta_2 > \tilde{\eta}_2 > 0$ such that

$$\omega\left(B(q_2, \gamma_2) \setminus (\overline{B(q_2, \eta_2)} \cap F), \partial(B(q_2, \eta_2) \cap F), z\right) > \frac{3}{4} \text{ for all } z \in B(q_2, \tilde{\eta}_2),$$

and such that

$$\frac{\log \frac{D}{\gamma_2 - \tilde{\eta}_2}}{\log \frac{D}{\eta_2 + \tilde{\eta}_2}} < \frac{1}{4}.$$

In general, for any $n \geq 2$, there exists q_n in $B(q_{n-1}, \tilde{\eta}_{n-1}) \cap F$, $\frac{2}{9}\tilde{\eta}_{n-1} > \gamma_n > \eta_n > \tilde{\eta}_n > 0$ such that

$$\omega\left(B(q_n, \gamma_n) \setminus (\overline{B(q_n, \eta_n)} \cap F), \partial\left(B(q_n, \eta_n) \cap F\right), z\right) > 1 - \frac{1}{2n} \tag{7.58}$$

for all $z \in B(q_n, \tilde{\eta}_n)$ and such that

$$\frac{\log \frac{D}{\gamma_n - \tilde{\eta}_n}}{\log \frac{D}{\eta_n + \tilde{\eta}_n}} < \frac{1}{2n}. \tag{7.59}$$

It is clear that $\{\eta_n\}$ and $\{\tilde{\eta}_n\}$ are decreasing to 0.

Let $\{p_1\} = \bigcap\limits_{n \geq 1} B(q_n, \eta_n) \cap F = \bigcap\limits_{n \geq 1} B(q_n, \tilde{\eta}_n) \cap F$. For any $n \geq 1$, since $p_1 \in B(q_n, \tilde{\eta}_n) \subset B(q_n, \eta_n)$, there exists $\tilde{\delta}_n < \tilde{\eta}_n$ such that $B(p_1, \tilde{\delta}_n) \subset B(q_n, \tilde{\eta}_n)$. Let $\tilde{\gamma}_n = \tilde{\eta}_n + \eta_n$. From inequality (7.59) there follows $\eta_n + \tilde{\eta}_n < \gamma_n - \tilde{\eta}_n$, which is equivalent to $\tilde{\gamma}_n < \gamma_n - \tilde{\eta}_n$. So $B(q_n, \eta_n) \subset B(p_1, \tilde{\gamma}_n) \subset B(q_n, \gamma_n)$. Thus by inequality (7.32) and inequality (7.58), we have

$$\begin{aligned} &\omega\left(B(q_n, \gamma_n) \setminus (\overline{B(p_1, \tilde{\gamma}_n)} \cap F), \partial\left(B(p_1, \tilde{\gamma}_n) \cap F\right), z\right) \\ &\geq \omega\left(B(q_n, \gamma_n) \setminus (\overline{B(q_n, \eta_n)} \cap F), \partial\left(B(q_n, \eta_n) \cap F\right), z\right) \\ &\geq 1 - \frac{1}{2n} \text{ for all } z \in B(p_1, \tilde{\delta}_n). \end{aligned} \tag{7.60}$$

We can now finish the proof by repeating the part of the proof that led from inequality (7.52) to inequality (7.57) keeping everything centered at p_1. The correspondences are as follows. Replace B_0 by $B(q_n, \gamma_n)$, q_1 with p_1, $\frac{2}{9}r'$ with $\delta_n = \gamma_n - \tilde{\eta}_n$ and F_1 with $F_2 = B(p_1, \tilde{\gamma}_n) \cap F$. Then inequality (7.53) is replaced by inequality (7.60), and inequality (7.54) becomes

$$\begin{aligned} &\omega\left(B(p_1, \delta_n) \setminus F_2, \partial F_2, z\right) \\ &\geq \omega\left(B(q_n, \gamma_n) \setminus F_2, \partial F_2, z\right) - \sup_{x \in \partial B(p_1, \delta_n)} \omega\left(B(q_n, \gamma_n) \setminus F_2, \partial F_2, x\right), \end{aligned}$$

for all $z \in B(p_1, \tilde{\delta}_n)$. Now substituting inequalities (7.59) and (7.60) into this we get

$$\omega\left(B(p_1, \delta_n) \setminus F_2, F_2, z\right) > \left(1 - \frac{1}{2n}\right) - \frac{1}{2n} = 1 - \frac{1}{n}$$

for all $z \in B(p_1, \tilde{\delta}_n)$. Since $\delta_n > \tilde{\delta}_n > \delta_{n+1}$ decreases to 0, and since $B(p_1, \delta_n) \cap F \supset B(p_1, \tilde{\gamma}_n) \cap F = F_2$, one final application of inequality (7.32) yields

$$\omega\left(B(p_1, \delta_n) \setminus F, \partial\left(B(p_1, \delta_n) \cap F\right), z\right) \geq \omega\left(B(p_1, \delta_n) \setminus F_2, F_2, z\right)$$

$$> 1 - \frac{1}{n} \text{ for all } z \in B(p_1, \tilde{\delta}_n),$$

which establishes inequality (7.51) and hence Theorem 7.6.

Acknowledgment The writing of this article was partially supported by NSF grant DMS 9707011.

References

[A] J. M. Ash, Uniqueness of representation by trigonometric series, *Amer. Math. Monthly* **96** (1989), 873–885.

[AWa] J. M. Ash and G. Wang, *A survey of uniqueness questions in multiple trigonometric series*, A Conference in Harmonic Analysis and Nonlinear Differential Equations in Honor of Victor L. Shapiro, *Contemporary Mathematics* **208** (1997), 35–71.

[AWa1] J. M. Ash and G. Wang, Some spherical uniqueness theorems for multiple trigonometric series, *Annals of Math.* (2000), to appear.

[AWa2] J. M. Ash and G. Wang, Sets of uniqueness for spherically convergent multiple trigonometric series, Preprint.

[AWe] J. M. Ash and G. V. Welland, Convergence, uniqueness, and summability of multiple trigonometric series, *Trans. Amer. Math. Soc.* **163** (1972), 401–436.

[B] J. Bourgain, Spherical summation and uniqueness of multiple trigonometric series, *Internat. Math. Res. Notices* (1996), 93–107.

[Con] B. Connes, Sur les coefficients des séries trigonometriques convergents sphériquement, *CRA Sc Paris t. Ser. A* **283** (1976), 159–161.

[Coo] R. Cooke, A Cantor-Lebesgue theorem in two dimensions, *Proc. Amer. Math. Soc.* **30** (1971), 547–550.

[Ha] P. R. Halmos, *Measure Theory*, D. Van Nostrand, Princeton, 1956.

[He] L. Helms, *Introduction to Potential Theory*, John Wiley & Sons, New York, 1969.

[PW] M. H. Protter and H. F. Weinberger, *Maximum Principles in Differential Equations*, Prentice-Hall, Englewood Cliffs, NJ, 1967.

[Sh] V. L. Shapiro, Uniqueness of multiple trigonometric series, *Ann. Math.* **66** (1957), 467–480.

[Sh1] V. L. Shapiro, Fourier series in several variables, *Bull. Amer. Math. Soc.* **70** (1964), 48–93.

[SW] E. Stein and G. Weiss, *Introduction to Fourier Analysis on Euclidean Spaces*, Princeton Univ. Press, Princeton, 1971.

[Z] A. Zygmund, A Cantor-Lebesgue theorem for double trigonometric series, *Studia Math.* **43** (1972), 173–178.

[Z1] A. Zygmund, *Trigonometric Series*, Vol. I, 2nd rev. ed., Cambridge Univ. Press, New York, 1959.

Chapter 8

Roundoff Error Analysis for Fast Trigonometric Transforms

Manfred Tasche[1] and Hansmartin Zeuner[2]

[1] *University of Rostock, Department of Mathematics, D–18055 Rostock, Germany.*
Email: manfred.tasche@mathematik.uni-rostock.de

[2] *Medical University of Lübeck, Institute of Mathematics, Wallstr. 40, D–23560 Lübeck, Germany.*
Email: zeuner@math.mu-luebeck.de

Contact Author: M. Tasche.

Abstract The fast Fourier transform (FFT) and related fast trigonometric transforms are nowadays essential tools for practical computations. It is very important that fast algorithms work stable in a floating point arithmetic. This survey paper presents both worst case and average case analysis of roundoff errors occuring in floating point computation of FFTs with precomputed twiddle factors and show the strong influence of precomputation errors on the numerical stability of FFT. The results are mainly based on a factorization of the Fourier matrix into a product of almost unitary and sparse matrices. Finally we consider roundoff errors of fast convolutions and fast algorithms for discrete cosine/sine transforms.

Contents

8.1 Introduction

Traditionally, a fast algorithm of the discrete Fourier transform (DFT) is called a fast Fourier transform (FFT). Discrete trigonometric transforms, such as the discrete cosine transform (DCT) or the discrete sine transform (DST), are closely related to the DFT. Nowadays, FFTs and fast trigonometric transforms are essential tools for numerous practical computations. It is very important that the fast algorithms work stable in a floating point arithmetic with unit roundoff u. In this survey paper, it is shown that FFT and fast trigonometric transforms are very sensitive with respect to the accuracy of the precomputation and that under certain conditions these algorithms can be remarkably stable.

Let $\boldsymbol{F}_n \in \mathbb{C}^{n \times n}$ be the unitary Fourier matrix

$$\boldsymbol{F}_n := n^{-1/2}\left(\omega_n^{jk}\right)_{j,k=0}^{n-1}$$

with $\omega_n := \exp(-2\pi\mathrm{i}/n)$. In a worst case study, for every input vector $\boldsymbol{x} \in \mathbb{C}^n$ let $\boldsymbol{y} := \boldsymbol{F}_n\boldsymbol{x}$ be the exact output vector. By $\hat{\boldsymbol{y}} \in \mathbb{C}^n$ we denote the computed vector of $\boldsymbol{F}_n\boldsymbol{x}$, where the twiddle factors ω_n^k have been precomputed using a suitable method. Then $\hat{\boldsymbol{y}}$ can be represented in the form $\hat{\boldsymbol{y}} = \boldsymbol{F}_n(\boldsymbol{x} + \Delta\boldsymbol{x})$ $(\Delta\boldsymbol{x} \in \mathbb{C}^n)$. An algorithm for computing $\boldsymbol{F}_n\boldsymbol{x}$ is called *normwise backward stable* (see [6], p. 142), if there exists a constant $k_n > 0$ such that for all $\boldsymbol{x} \in \mathbb{C}^n$

$$\|\Delta\boldsymbol{x}\|_2 \leq (k_n u + \mathcal{O}(u^2))\|\boldsymbol{x}\|_2 \tag{8.1}$$

and $k_n u \ll 1$. Here $\|\boldsymbol{x}\|_2 := \left(\sum_{k=0}^{n-1} |x_k|^2\right)^{1/2}$ denotes the Euclidean norm of $\boldsymbol{x} = (x_k)_{k=0}^{n-1} \in \mathbb{C}^n$. Since $\boldsymbol{F}_n$ is unitary and the Euclidean norm is unitary invariant, we also have *normwise forward stability* by

$$\|\hat{\boldsymbol{y}} - \boldsymbol{y}\|_2 \leq (k_n u + \mathcal{O}(u^2))\|\boldsymbol{x}\|_2.$$

In an average case study, we assume that all components of an input vector and the resulting roundoff errors are random variables. We are interested in the distribution of the vector $\Delta\boldsymbol{X} := \boldsymbol{F}_n^{-1}(\hat{\boldsymbol{Y}} - \boldsymbol{F}_n\boldsymbol{X})$, where $\hat{\boldsymbol{Y}}$ denotes the computed vector $\boldsymbol{F}_n\boldsymbol{X}$. Then we measure the average case backward numerical stability in terms of the expected value

$$\mathbb{E}(\|\Delta\boldsymbol{X}\|_2^2) = \left(\bar{k}_n^2 u^2 + \mathcal{O}(u^3)\right)\mathbb{E}((\|\boldsymbol{X}\|_2^2) \tag{8.2}$$

with some constant $\bar{k}_n > 0$ such that $\bar{k}_n u \ll 1$. Under the central limit theorem, the standard deviation of the backward error is asymptotically

$$\sigma(\|\Delta\boldsymbol{X}\|_2) \sim \frac{1}{\sqrt{2n}}\bar{k}_n\mathbb{E}(\|\boldsymbol{X}\|_2^2)^{1/2}u\,,$$

thus confirming a property which has been observed in many simulations.

An FFT can be interpreted as the application of a factorization of $\boldsymbol{F}_n$ into a product of sparse, almost unitary matrices. Almost all considerations on numerically stable FFTs are based on this fact. G. U. Ramos [10] and P. Y. Yalamov [18] have studied the worst case roundoff errors of the Cooley–Tukey FFT with the twiddle factors precomputed accurately. The interesting thesis [3] of M. Y. Chu contains a comprehensive worst case analysis of the roundoff errors of the Cooley–Tukey and Gentleman–Sande FFT, based on the analysis of the roundoff errors in the precomputation of the twiddle factors. An average case analysis of the roundoff errors for direct DFT and Cooley–Tukey FFT was given by D. Calvetti [2] for accurately precomputed twiddle factors and non-random input data. J. C. Schatzman [12] has presented an interesting report on the roundoff error behaviour of direct DFT and FFTs in a series of numerical tests. He proposed that all twiddle factors should be precomputed by the most accurate method, direct call. M. Arioli, H. Munte–Kaas and L. Valdettaro [1] have studied both worst case and average case analysis of roundoff errors for the Gentleman–Sande FFT with accurate twiddle factors. Recently, the authors [16] have studied both worst case and average case analysis of roundoff errors for FFTs paying special attention to the influence of the precomputation of the twiddle factors.

This survey paper is organized as follows: In Section 8.2, we introduce Wilkinson's model for the worst case study as well as a related model for the average case study of roundoff errors. In Section 8.3 we compare several methods for the precomputation of the twiddle factors in the light of the worst and average case behaviour of the roundoff errors. In Section 8.4, we develop the factorization of the Fourier matrix $\boldsymbol{F}_n$ for $n = p^t$ $(p, t \geq 2)$ into a product of sparse, almost unitary matrices, using divide–and–conquer techniques and Kronecker products. From the two principle types of factorization, decimation–in–time and decimation–in–frequency, methods particularly suitable for the use on parallel and on vector computers are derived. These factorizations are closely related to stride permutation matrices and digit–reversal permutation matrices. The important special cases $p = 2$ and $p = 4$ are considered in Section 8.5. Further we describe an FFT for real data based on complex FFT of half the length. In Section 8.6, we present a new factorization of $\boldsymbol{F}_n$ with $n = 2^t$ which leads to the split–radix FFT. This FFT has the lowest arithmetic complexity and it will be shown later that it also has the lowest average size of the roundoff error.

The key point of this paper is Section 8.7. Let $\boldsymbol{A}$ denote a unitary matrix with a known factorization into a product of sparse, almost unitary matrices. Using this factorization, we can compute step by step the product $\boldsymbol{Ax}$ with arbitrary $\boldsymbol{x} \in \mathbb{C}^n$. In Theorems 8.3 and 8.4, we present a new unified approach to the stability constants k_n and $\bar{k}_n$ in the worst case and average case, respectively (see (8.1)–(8.2)). This general roundoff error analysis is applied to Fourier matrix and cosine/sine matrices in Sections 8.8 and 8.9. Using the factorizations of $\boldsymbol{F}_n$ from Sections 8.4–8.6, we now immediately obtain the stability constants for various FFTs, also for FFTs on parallel or vector computers. The best algorithm with respect to arithmetical complexity and numerical stability is the split–radix FFT, where the twiddle factors are precomputed by direct call (Algorithm 8.1) or repeated subvector scaling (Algorithm 8.3). Then in Section 8.9, we consider orthogonal cosine and sine matrices of type II–IV which are closely related to the Fourier matrix. A cosine matrix is factorized either into a product of sparse, almost orthogonal matrices or into a product of sparse, almost unitary *complex* matrices (using a factorization of a Fourier matrix). Applying the general roundoff error analysis of Section 8.7, we obtain the corresponding stability constants for real fast DCTs and complex fast DCTs via FFT. We see that complex arithmetic is very useful for the numerical stability of fast DCTs and DSTs. Finally in Section 8.10, we consider the numerical stability of fast convolutions and deconvolutions.

8.2 Models of roundoff error analysis

For the *worst case analysis* of an algorithm we use Wilkinson's standard model for the binary floating point arithmetic of real numbers (see [6], p. 44). For arbitrary $x, y \in \mathbb{R}$ and any arithmetical operation $\circ \in \{+, -, \times, \div\}$

the exact value $x \circ y$ and the computed value $\mathrm{fl}(x \circ y)$ are related by

$$\mathrm{fl}(x \circ y) = (x \circ y)(1 + \varepsilon^{\circ}) \qquad (|\varepsilon^{\circ}| \le u), \tag{8.3}$$

where u denotes the *unit roundoff* (or *machine precision*), as long as we disregard underflow and overflow. By $\hat{x}$ we denote the computed value $\mathrm{fl}(x)$ of a variable x. In the IEEE arithmetic of single precision (24 bits for the mantissa with 1 sign bit, 8 bits for the exponent), we have

$$u = 2^{-24} \approx 5.96 \times 10^{-8}$$

and for double precision (53 bits for the mantissa with 1 sign bit, 11 bits for the exponent), we have

$$u = 2^{-53} \approx 1.11 \times 10^{-16}$$

(see [6], p. 45).

Usually the total roundoff error in the result of an algorithm is composed of a number of such errors. To make their origin clear by our notation, we will use superscripts in general for the operation and subscripts for the step of the algorithm. Usually the symbol ε designates the relative error and δ the absolute error.

Since complex arithmetic is implemented using real arithmetic, we can derive the following bounds for the roundoff error of complex floating point operations:

LEMMA 8.1

Let $z, w \in \mathbb{C}$. Then we have

$$\mathrm{fl}(z + w) = (z + w)(1 + \varepsilon^{+}) \qquad (|\varepsilon^{+}| \le u), \tag{8.4}$$

$$\mathrm{fl}(z \times w) = (z \times w)(1 + \varepsilon^{\times}) \qquad \left(|\varepsilon^{\times}| \le \mu_{\mathbb{C}} u + \mathcal{O}(u^2)\right) \tag{8.5}$$

where $\mu_{\mathbb{C}} := \frac{4\sqrt{3}}{3}$ is the best possible constant under the Wilkinson model (8.3). In the case $z \in \mathbb{R} \cup \mathrm{i}\mathbb{R}$ and $w \in \mathbb{C}$ we have

$$\mathrm{fl}(z \times w) = (z \times w)(1 + \varepsilon^{\times}) \qquad (|\varepsilon^{\times}| \le u). \tag{8.6}$$

If $w \neq 0$, then

$$\mathrm{fl}\left(\tfrac{z}{w}\right) = \tfrac{z}{w}(1 + \varepsilon^{\div}) \qquad (|\varepsilon^{\div}| \le (\mu_{\mathbb{C}} + 3)u + \mathcal{O}(u^2)). \tag{8.7}$$

PROOF The first and third assertions are obviously true. The second was proved in [16]. Other proofs can be found in [6], pp. 79–80 (with $\mu_{\mathbb{C}} = 2\sqrt{2}$) and in [3], pp. 53–56 (with $\mu_{\mathbb{C}} = 1 + \sqrt{2}$).

Let $w = a + ib \neq 0$. By (8.3) we get

$$\mathrm{fl}(w \times \bar{w}) = (a^2(1+\varepsilon_1^{\times}) + b^2(1+\varepsilon_2^{\times}))(1+\varepsilon^{+})$$

with $|\varepsilon_1^{\times}|, |\varepsilon_2^{\times}|, |\varepsilon^{+}| \leq u$ such that

$$\mathrm{fl}(w \times \bar{w}) = (w \times \bar{w})(1+\varepsilon^{\times}) \qquad (|\varepsilon^{\times}| \leq 2u + u^2). \tag{8.8}$$

Further, from (8.3) it follows that

$$\mathrm{fl}\Big(\frac{1}{w}\Big) = \frac{a}{\mathrm{fl}(w \times \bar{w})}(1+\varepsilon_1^{\div}) - \mathrm{i}\,\frac{b}{\mathrm{fl}(w \times \bar{w})}(1+\varepsilon_2^{\div}) = \frac{\bar{w}}{\mathrm{fl}(w \times \bar{w})}(1+\varepsilon_3^{\div})$$

with $|\varepsilon_1^{\div}|, |\varepsilon_2^{\div}|, |\varepsilon_3^{\div}| \leq u$ such that by (8.8)

$$\mathrm{fl}\Big(\frac{1}{w}\Big) = \frac{\bar{w}}{w \times \bar{w}}\,\frac{1+\varepsilon_3^{\div}}{1+\varepsilon^{\times}} = \frac{1}{w}(1+\varepsilon_4^{\div}) \qquad (|\varepsilon_4^{\div}| \leq 3u + \mathcal{O}(u^2)). \tag{8.9}$$

Finally, we consider $\mathrm{fl}(\frac{z}{w})$. By (8.5) and (8.8)–(8.9) we obtain that

$$\mathrm{fl}\Big(\frac{z}{w}\Big) = \mathrm{fl}\Big(z \times \mathrm{fl}\Big(\frac{1}{w}\Big)\Big) = \frac{z}{w}\,(1+\varepsilon^{\times})(1+\varepsilon_4^{\div}) = \frac{z}{w}\,(1+\varepsilon^{\div})$$

with $|\varepsilon^{\div}| \leq (\mu_{\mathbb{C}} + 3)u + \mathcal{O}(u^2)$. Note that in [6], pp. 79–80 the larger upper bound $4\sqrt{2}u + \mathcal{O}(u^2)$ is derived. For the direct method of calculating z/w it can be shown that the roundoff error satisfies the smaller bound $\frac{8}{7}\sqrt{14}\,u + \mathcal{O}(u^2)$. ∎

In the analysis of some of the algorithms we will benefit from the fact that multiplications (or divisions) with the trivial factors ± 1, $\pm\mathrm{i}$ or powers of 2 do not produce any roundoff errors.

For the *average case study* of the roundoff errors we also start from Wilkinson's model (8.3) for the relative error. But now we assume that the input data X and Y are real or complex random variables and so is the total error $\delta^{\circ} := \mathrm{fl}(X \circ Y) - X \circ Y$ of the arithmetical operation $\circ \in \{+, -, \times, \div\}$. From this, a simple, but nevertheless realistic model is derived in [19]. In it, the size of the error depends mainly on the L^2–norm $\|X \circ Y\|$ of the result of the operation, defined by $\|Z\|^2 := \mathbb{E}(|Z|^2)$ for every random variable with finite second moment. We make the following assumptions of our average case analysis:

(M1) The roundoff error δ° of an arithmetic operation $\circ$ in the number field $\mathbb{K} = \mathbb{R}$ or $\mathbb{C}$ is independent of the operands X and Y and has expectation and variance

$$\mathbb{E}(\delta^{\circ}) = 0, \qquad \mathbb{V}(\delta^{\circ}) = \|X \circ Y\|^2\,\sigma_{\mathbb{K}\circ}^2 u^2. \tag{8.10}$$

The constant $\sigma_{\mathbb{K},\circ}$ depends on the operation $\circ$ and the number field $\mathbb{K}$, but in our model it does not depend on the distribution of the data.

(M2) In the case $\mathbb{K} = \mathbb{C}$ we assume that $\Re(\delta^\circ)$ and $\Im(\delta^\circ)$ are uncorrelated and $\mathbb{V}(\Re(\delta^\circ)) = \mathbb{V}(\Im(\delta^\circ))$.

(M3) All the input variables are uncorrelated, each of them having expectation $\mathbb{E}(X) = 0$ and known variance $\mathbb{V}(X) = \|X\|^2 < \infty$. Without loss of generality we assume that $\|X\| > 0$, i.e. $X \neq 0$ with positive probability. If the data are complex, the real and imaginary parts have to be uncorrelated with equal variances.

(M4) Roundoff errors occuring in one step of an algorithm are propagated linearly through the later steps (i.e., we disregard error terms of order $\mathcal{O}(u^2)$). Further, all roundoff errors are independent.

In numerical simulations it has been observed that the variance of the error is practically independent of the data distribution. The following values for the constants have been found for standard normal data: $\sigma_{\mathbb{R},+} \approx \sigma_{\mathbb{C},+} \approx 0.406$, $\sigma_{\mathbb{R},\times} \approx \sigma_{\mathbb{R},\div} \approx 0.425$ and $\sigma_{\mathbb{C},\times} \approx 0.616$ (see [19]). As in (8.6), the constant $\sigma_{\mathbb{R},\times}$ has to be used for products of a real with a complex number. Note that all these constants are relative to the machine precision u, slightly different from the notation in [16] and [19].

In order to determine the constant for complex division, it has been shown in [19] that the relative error $\varepsilon^{(-1)}$ from inversion, defined by $\mathrm{fl}(Z^{-1}) = Z^{-1} + \varepsilon^{(-1)}|Z|^{-1}$, has variance $\kappa\sigma^2_{\mathbb{R},\times} + \sigma^2_{\mathbb{R},\div} + \sigma^2_{\mathbb{R},+}$ where

$$\kappa := 2\mathbb{E}\Big(\frac{\Re(Z)^2(\Re(Z)^4 + \Im(Z)^4)}{|Z|^6}\Big)$$

which is ≈ 0.75 for a wide range of distributions of Z. Therefore $\frac{W}{Z} = \frac{1}{Z}W$ is calculated with a roundoff error of variance $\sigma^2_{\mathbb{C},\div}\|\frac{W}{Z}\|^2u^2$, where $\sigma^2_{\mathbb{C},\div} = \kappa\sigma^2_{\mathbb{R},\times} + \sigma^2_{\mathbb{R},\div} + \sigma^2_{\mathbb{R},+} + \sigma^2_{\mathbb{C},\times}$, and hence $\sigma_{\mathbb{C},\div} \approx 0.928$ in accordance with our simulations.

8.3 Precomputation of twiddle factors

All of the algorithms used for the discrete trigonometric transforms require the precomputation of a large number of trigonometric function values. If this is done by calls to library functions, the numerical accuracy is usually good. But since they usually take much computing time, it can be necessary to use faster methods relying mainly on arithmetic operations and only few function calls. However, one should try avoid to do this at the expense of a much worse numerical accuracy.

Since we usually need sine and cosine values of the same angles at the same time, we will study methods to compute $\omega_n^k = \cos\frac{2\pi k}{n} - \mathrm{i}\sin\frac{2\pi k}{n}$ $(k = 1, \ldots, n-1)$, where $\omega_n := \exp(-\frac{2\pi \mathrm{i}}{n})$. By using symmetries of the trigonometric functions, it is possible to construct a large portion of these values without actual calculations (see [7], p. 28 and [16]): In general it is sufficient to compute ω_n^k for $k = 1, \ldots, \lfloor\frac{n-1}{2}\rfloor$, and if n is a multiple of 4, only the values for $k \le \lfloor n/8 \rfloor$ have to be calculated. Since this last condition is fulfilled in almost all of the applications, the bounds below are usually valid in the range $k \le \lfloor n/8 \rfloor$.

In most examples we have $n = 2^t$ $(t \ge 3)$. Let us introduce the notation $n_s := 2^{t-s} = n/2^s$ $(s = 1, \ldots, t)$. In the following examples, we will study the worst case and average case behaviour of the roundoff error

$$\Delta_{s,k} := \hat{\omega}_n^{kn_s} - \omega_n^{kn_s} \qquad (s = 3, \ldots, t,\ k = 0, \ldots, 2^s - 1).$$

It is necessary to consider powers where the exponent is a multiple of n_s since in the fast algorithms for FFT and DCT most of the matrices only contain such powers. So the accuracy of the results can be improved. For $s = t$ we get back $\Delta_{t,k} = \hat{\omega}_n^k - \omega_n^k$. The results in this section have been proved in [16].

ALGORITHM 8.1 (Direct call)
Input: $n = 2^t$ $(t \ge 3)$, $\theta = 2\pi/n$.
For $k = 1 : 1 : n_3$ form
$\hat{\omega}_n^k := \mathrm{fl}(\cos(k\theta)) - \mathrm{i}\ \mathrm{fl}(\sin(k\theta))$.
Output: precomputed values $\hat{\omega}_n^k$ of ω_n^k $(k = 1, \ldots, n_3)$.

This algorithm involves almost n_2 trigonometric function calls. If the library routines for sine and cosine are of high quality, then very accurate roots of unity are precomputed and we have

$$|\Delta_{s,k}| \le \frac{\sqrt{2}}{2} u \qquad (k = 1, \ldots, n_3),$$

if we assume that the trigonometric functions are internally computed to higher precision and then rounded towards the next machine number. In the average case study we obtain

$$\mathbb{V}(\Delta_{s,k}) = \frac{u^2}{6}.$$

Since the direct call of n_2 trigonometric functions is time consuming, we have to consider other methods, in particular for the on–line computation of ω_n^k.

The following algorithm uses only two trigonometric calls and is based on repeated multiplications $\omega_n^k = \omega_n \times \omega_n^{k-1}$.

ALGORITHM 8.2 (Repeated multiplication)
Input: $n = 2^t$ $(t \geq 4)$, $\theta = 2\pi/n$.
1. By direct call form
$\hat{\omega}_n := \text{fl}(\cos\theta) - \mathrm{i}\ \text{fl}(\sin\theta)$.
2. For $k = 2 : 1 : n_3$ form
$\hat{\omega}_n^k := \text{fl}(\hat{\omega}_n \times \hat{\omega}_n^{k-1})$.
Output: precomputed values $\hat{\omega}_n^k$ of ω_n^k $(k = 1, \ldots, n_3)$.

In this algorithm, $3n_2$ flops and 2 direct calls are used. As a *flop* we understand a real multiplication or real addition in floating point arithmetic. For the roundoff errors we have the upper bound

$$|\Delta_{s,k}| \leq \left(\mu_{\mathrm{C}} + \frac{\sqrt{2}}{2}\right) kn_s u \leq \left(\mu_{\mathrm{C}} + \frac{\sqrt{2}}{2}\right) n_3 u \qquad (k = 1, \ldots, 2^{s-3}). \tag{8.11}$$

In the average case analysis we obtain

$$\mathbb{V}(\Delta_{s,k}) = \left(\frac{1}{12}k^2 n_s^2 + \mathcal{O}(kn_s)\right) u^2.$$

The roundoff error from the direct call dominates all the arithmetical roundoff errors and the average error is of the same order of magnitude as in the worst case study.

Note that by using the system of two–term recursions

$$\begin{aligned} \gamma_k &= \rho\omega_n^{k-1} + \gamma_{k-1}\,, \\ \omega_n^k &= \omega_n^{k-1} - \gamma_k \qquad\qquad (k \geq 2) \end{aligned}$$

with $\gamma_1 := \frac{1}{2}\rho + \mathrm{i}\sin\theta$ and $\rho := 4\sin^2\frac{\theta}{2}$ one can obtain the better precomputation method of the *stabilized forward recursion* (see [16]) which has the same complexity, memory requirement and worst case roundoff error bound, but an average case error of only $\mathbb{V}(\Delta_{s,k}) = \mathcal{O}(kn_s)u^2$ (see Figure 1).

The third method combines the accuracy of Algorithm 8.1 with the arithmetical simplicity of Algorithm 8.2:

ALGORITHM 8.3 (Repeated subvector scaling)
Input: $n = 2^t$ $(t \geq 4)$, $\theta = 2\pi/n$.
1. For $s = 0 : 1 : t - 3$ form
$\hat{\omega}_n^{2^s} := \text{fl}(\cos\theta_s) - \mathrm{i}\ \text{fl}(\sin\theta_s)$
with $\theta_s := \theta 2^s$ by direct calls.
2. For $s = 1 : 1 : t - 4$ form
$\left(\hat{\omega}_n^k\right)_{k=2^s+1}^{2^{s+1}-1} := \text{fl}\left(\hat{\omega}_n^{2^s} \times \left(\hat{\omega}_n^k\right)_{k=1}^{2^s-1}\right)$.
Output: precomputed values $\hat{\omega}_n^k$ of ω_n^k $(k = 1, \ldots, n_3)$.

Algorithm 8.3 requires about $3n_2$ flops and $2\log_2 n$ direct calls. The roundoff error is bounded by

$$|\Delta_{s,k}| \le \begin{cases} \frac{1}{2}\sqrt{2}\,u & \text{if } k = 2^r, \quad r = 0, \ldots, s-3, \\ (\mu_{\mathbb{C}} + \frac{\sqrt{2}}{2})(r+1)u & \text{if } 2^r < k < 2^{r+1}, \quad r = 1, \ldots, s-4. \end{cases}$$

Hence we obtain

$$|\Delta_{s,k}| \le \left(\mu_{\mathbb{C}} + \frac{\sqrt{2}}{2}\right)(s-3)\,u \qquad (k = 1, \ldots, 2^{s-3}). \tag{8.12}$$

The variance of the error is

$$\mathbb{V}(\Delta_{s,k}) = (\jmath(k) - 1)\,\sigma^2_{\mathbb{C}\times} u^2 + \frac{\jmath(k)}{12} u^2 \qquad (k \ge 1). \tag{8.13}$$

where $\jmath(k) := \sum_{l=0}^{t-1} k_l$ and $k = (k_{t-1}, \ldots, k_0)_2$ is the binary representation of k.

A variant of Algorithm 8.3 can be used for $n = p^t$. Since usually p is an odd number, we calculate the powers ω_n^k for all $k \le \lfloor p^t/2 \rfloor$ unlike in the rest of this section.

ALGORITHM 8.4 (Radix–p repeated subvector scaling)
Input: $n = p^t$ $(p \ge 3,\ t \ge 2)$, $\theta = 2\pi/n$.
1. For $s = 0 : 1 : t-1$ and $j = 1 : 1 : p-1$ form
$\hat{\omega}_n^{jp^s} := \mathrm{fl}(\cos(j\hat{\theta}_s)) - \mathrm{i}\,\mathrm{fl}(\sin(j\hat{\theta}_s))$
with $\hat{\theta}_s := \theta p^s$ by direct calls.
2. For $s = 1 : 1 : t-1$ and $j = 1 : 1 : p-1$ form
$\left(\hat{\omega}_n^k\right)_{k=jp^s+1}^{(j+1)p^s-1} := \mathrm{fl}\left(\hat{\omega}_n^{jp^s} \times \left(\hat{\omega}_n^k\right)_{k=1}^{p^s-1}\right).$

Output: Precomputed values $\hat{\omega}_n^k$ of ω_n^k $(0 \le k \le \lfloor n/2 \rfloor)$.

If only the powers ω_n^k for $k \le \lfloor n/2 \rfloor$ are calculated, this algorithm uses $2(p-1)t$ direct calls and $3n$ flops. Let $n_s := p^{t-s}$. In order to analyze the roundoff error of $\omega_n^{kn_s}$ for $s = 1, \ldots, t$, $k = 1, \ldots, p^s - 1$, we use the p–adic representation $k = \sum_{r=0}^{s-1} k_r p^r$. If $k_r \ge 1$, then the roundoff error in $\omega_n^{k_r p^r n_s}$ is bounded by $\frac{\sqrt{2}}{2}u$. One more multiplication occurs in this step of the algorithm except for the smallest r with $k_r \ge 1$. Hence the total roundoff error is

$$|\Delta_{k,s}| \le (\jmath(k) - 1)\mu_{\mathbb{C}} u + \jmath(k)\frac{\sqrt{2}}{2}u + \mathcal{O}(u^2),$$

where $\jmath(k) := \#\{r \in \{0, \ldots, s-1\} : k_r > 0\}$. By the same argument, in the average case study we obtain the variance

$$\mathbb{V}(\Delta_{s,k}) = (\jmath(k) - 1)\sigma^2_{\mathbb{C}\times} u^2 + \jmath(k)\frac{u^2}{12}.$$

Other methods for the precomputation of twiddle factors can be found in [3], [7], pp. 22–28, and [16]. For corresponding estimates of the roundoff error see [3] and [16].

8.4 Factorizations of the Fourier matrix

It is well known that FFTs are based on a factorization of the Fourier matrix

$$\boldsymbol{F}_n = n^{-1/2} \left(\omega_n^{jk}\right)_{j,k=0}^{n-1}$$

with $\omega_n = \exp(-\frac{2\pi \mathrm{i}}{n})$ into a product of sparse matrices (see Algorithm 8.5 in Section 8.5). These matrices are Kronecker products, diagonal matrices or permutation matrices. The special structure of these matrices can be used in the design of algorithms for parallel and vector computers. First we summarize some known results concerning Kronecker products and special permutation matrices (see [17], pp. 28–40 or [7], pp. 7–8).

The *Kronecker product* or *tensor product* $\boldsymbol{x} \otimes \boldsymbol{y}$ of two vectors $\boldsymbol{x} := (x_j)_{j=0}^{m-1} \in \mathbb{C}^m$ and $\boldsymbol{y} \in \mathbb{C}^n$ is defined as the vector

$$\boldsymbol{x} \otimes \boldsymbol{y} := (x_j \boldsymbol{y})_{j=0}^{m-1} \in \mathbb{C}^{mn}.$$

The *Kronecker product* or *tensor product* of two matrices $\boldsymbol{A} := (a_{jk})_{j,k=0}^{m-1} \in \mathbb{C}^{m \times m}$ and $\boldsymbol{B} \in \mathbb{C}^{n \times n}$ is by definition the block matrix

$$\boldsymbol{A} \otimes \boldsymbol{B} := (a_{jk} \boldsymbol{B})_{j,k=0}^{m-1} \in \mathbb{C}^{mn \times mn}.$$

Obviously we have

$$(\boldsymbol{A} \otimes \boldsymbol{B})(\boldsymbol{x} \otimes \boldsymbol{y}) = (\boldsymbol{A}\boldsymbol{x}) \otimes (\boldsymbol{B}\boldsymbol{y}),$$
$$\boldsymbol{A} \otimes \boldsymbol{B} = (\boldsymbol{A} \otimes \boldsymbol{I}_n)(\boldsymbol{I}_m \otimes \boldsymbol{B}) = (\boldsymbol{I}_m \otimes \boldsymbol{B})(\boldsymbol{A} \otimes \boldsymbol{I}_n)$$

with the *identity matrix* $\boldsymbol{I}_n := (\delta(j-k))_{j,k=0}^{n-1}$ and $\delta(0) := 1$, $\delta(j) := 0$ $(j \in \mathbb{Z} \setminus \{0\})$.

The *direct sum* of $\boldsymbol{A} \in \mathbb{C}^{m \times m}$ and $\boldsymbol{B} \in \mathbb{C}^{n \times n}$ is defined by

$$\boldsymbol{A} \oplus \boldsymbol{B} := \mathrm{diag}(\boldsymbol{A}, \boldsymbol{B}) := \begin{pmatrix} \boldsymbol{A} & \boldsymbol{O} \\ \boldsymbol{O} & \boldsymbol{B} \end{pmatrix} \in \mathbb{C}^{(m+n)\times(m+n)}.$$

REMARK 8.1 Let $\boldsymbol{z} \in \mathbb{C}^{mn}$ and $\boldsymbol{B} \in \mathbb{C}^{n \times n}$ be given. We split $\boldsymbol{z}$ into m vectors $\boldsymbol{z}_k := (z_{kn+j})_{j=0}^{n-1} \in \mathbb{C}^n$ $(k = 0, \ldots, m-1)$. Then the matrix–vector product $(\boldsymbol{I}_m \otimes \boldsymbol{B})\boldsymbol{z}$ can be implemented as a *parallel operation,* since

$$(\boldsymbol{I}_m \otimes \boldsymbol{B})\boldsymbol{z} = (\boldsymbol{B}\boldsymbol{z}_k)_{k=0}^{m-1}$$

can be computed in parallel by m independent processors.

For the matrix–vector product $(\boldsymbol{A} \otimes \boldsymbol{I}_n)\boldsymbol{z}$ with $\boldsymbol{A} = (a_{j,k})_{j,k=0}^{m-1} \in \mathbb{C}^{m \times m}$, the situation is quite different. By

$$(\boldsymbol{A} \otimes \boldsymbol{I}_n)\boldsymbol{z} = (a_{j,0}\boldsymbol{z}_0 + \ldots + a_{j,m-1}\boldsymbol{z}_{m-1})_{j=0}^{m-1}$$

we obtain an operation suitable for implementation on a *vector computer*. ∎

Let $n = pq$ $(p, q \geq 2)$. The *p–stride permutation matrix* $\boldsymbol{P}_n(p) \in \mathbb{R}^{n \times n}$ is defined by the property

$$\boldsymbol{P}_n(p)\boldsymbol{x} = \left((x_{pk+l})_{k=0}^{q-1}\right)_{l=0}^{p-1},$$

i.e. $\boldsymbol{P}_n(p) = (\delta(\langle pk - l\rangle_{n-1}))_{k,l=0}^{n-1}$, where $\langle m\rangle_{n-1}$ denotes the non–negative residuum of m modulo $n - 1$. For $\boldsymbol{x} \in \mathbb{C}^p$, $\boldsymbol{y} \in \mathbb{C}^q$ and $n = pq$ we have

$$\boldsymbol{P}_n(p)(\boldsymbol{y} \otimes \boldsymbol{x}) = \boldsymbol{x} \otimes \boldsymbol{z}.$$

If n is a multiple of pq, then we have the following properties

$$\boldsymbol{P}_n(p)^{-1} = \boldsymbol{P}_n(p)^{\mathrm{T}} = \boldsymbol{P}_n(\tfrac{n}{p}), \tag{8.14}$$

$$\boldsymbol{P}_n(pq) = \boldsymbol{P}_n(p)\boldsymbol{P}_n(q) = \boldsymbol{P}_n(q)\boldsymbol{P}_n(p). \tag{8.15}$$

The Kronecker product possesses the following properties:

THEOREM 8.1

Let $\boldsymbol{A}, \boldsymbol{A}' \in \mathbb{C}^{m \times m}$, $\boldsymbol{B}, \boldsymbol{B}' \in \mathbb{C}^{n \times n}$, *and* $\boldsymbol{C} \in \mathbb{C}^{p \times p}$ *be given. Then we have*

(K1) $(\boldsymbol{A} \otimes \boldsymbol{B})(\boldsymbol{A}' \otimes \boldsymbol{B}') = (\boldsymbol{A}\boldsymbol{A}') \otimes (\boldsymbol{B}\boldsymbol{B}')$,

(K2) $(\boldsymbol{A} \otimes \boldsymbol{B})^{\mathrm{T}} = \boldsymbol{A}^{\mathrm{T}} \otimes \boldsymbol{B}^{\mathrm{T}}$,

(K3) $(\boldsymbol{A} \otimes \boldsymbol{B}) \otimes \boldsymbol{C} = \boldsymbol{A} \otimes (\boldsymbol{B} \otimes \boldsymbol{C})$,

(K4) $(\boldsymbol{A} \oplus \boldsymbol{B}) \otimes \boldsymbol{C} = (\boldsymbol{A} \otimes \boldsymbol{C}) \oplus (\boldsymbol{B} \otimes \boldsymbol{C})$,

(K5) $\boldsymbol{P}_{mn}(n)\,(\boldsymbol{A} \otimes \boldsymbol{B})\,\boldsymbol{P}_{mn}(m) = \boldsymbol{B} \otimes \boldsymbol{A}$.

From Remark 8.1 and (K5) it follows that stride permutation matrices interchange parallel and vector operations. For a proof of Theorem 8.1 see [17], pp. 28–40 or [7], pp. 7–8.

Many FFTs are based on the following result:

LEMMA 8.2

Let $n = pq$ *and* $\boldsymbol{W}_q(n) := \mathrm{diag}\,(\omega_n^j)_{j=0}^{q-1}$. *Then we have*

$$\boldsymbol{P}_n(p)\boldsymbol{F}_n = (\boldsymbol{I}_p \otimes \boldsymbol{F}_q)\left(\boldsymbol{I}_q \oplus \boldsymbol{W}_q(n) \oplus \ldots \oplus \boldsymbol{W}_q(n)^{p-1}\right)(\boldsymbol{F}_p \otimes \boldsymbol{I}_q).$$

PROOF Let $\boldsymbol{y} := \boldsymbol{F}_n \boldsymbol{x}$ ($\boldsymbol{x} \in \mathbb{C}^n$). By the divide–and–conquer method, we split each sum

$$y_k = n^{-1/2} \sum_{j=0}^{n-1} x_j \omega_n^{jk} = n^{-1/2} \sum_{s=0}^{q-1} \left(\sum_{r=0}^{p-1} x_{rq+s} \omega_p^{rk} \right) \omega_n^{sk} \qquad (k = 0, \ldots, n-1)$$

into q parts by substituting $j = rq+s$. For $k = pm+l$ (with $m = 0, \ldots, q-1$; $l = 0, \ldots, p-1$) this reads as follows:

$$y_{pm+l} = q^{-1/2} \sum_{s=0}^{q-1} \left(p^{-1/2} \sum_{r=0}^{p-1} x_{rq+s} \omega_p^{rl} \right) \omega_n^{ls} \omega_q^{sm}. \tag{8.16}$$

Since

$$\boldsymbol{P}_n(p)\boldsymbol{y} = \left((y_{pm+l})_{m=0}^{q-1} \right)_{l=0}^{p-1},$$

$$\left(\left(p^{-1/2} \sum_{r=0}^{p-1} x_{rq+s} \omega_p^{rl} \right)_{s=0}^{q-1} \right)_{l=0}^{p-1} = (\boldsymbol{F}_p \otimes \boldsymbol{I}_q)\boldsymbol{x},$$

the result is a consequence of (8.16). ∎

COROLLARY 8.1
Let $n = pq$ ($p, q \geq 2$). Then $\boldsymbol{F}_n$ can be factorized as follows:

(i) *Decimation–in–frequency*

$$\boldsymbol{F}_n = \boldsymbol{P}_n(q)(\boldsymbol{I}_p \otimes \boldsymbol{F}_q) \left(\boldsymbol{I}_q \oplus \boldsymbol{W}_q(n) \oplus \ldots \oplus \boldsymbol{W}_q(n)^{p-1} \right) (\boldsymbol{F}_p \otimes \boldsymbol{I}_q),$$

(ii) *Decimation–in–time*

$$\boldsymbol{F}_n = (\boldsymbol{F}_p \otimes \boldsymbol{I}_q) \left(\boldsymbol{I}_q \oplus \boldsymbol{W}_q(n) \oplus \ldots \oplus \boldsymbol{W}_q(n)^{p-1} \right) (\boldsymbol{I}_p \otimes \boldsymbol{F}_q)\boldsymbol{P}_n(p),$$

(iii) *Parallel form*

$$\boldsymbol{F}_n = \boldsymbol{P}_n(p)(\boldsymbol{I}_q \otimes \boldsymbol{F}_p)\boldsymbol{P}_n(q) \left(\boldsymbol{I}_q \oplus \boldsymbol{W}_q(n) \oplus \ldots \oplus \boldsymbol{W}_q(n)^{p-1} \right) (\boldsymbol{I}_p \otimes \boldsymbol{F}_q)\boldsymbol{P}_n(p),$$

(iv) *Vector form*

$$\boldsymbol{F}_n = (\boldsymbol{F}_p \otimes \boldsymbol{I}_q) \left(\boldsymbol{I}_q \oplus \boldsymbol{W}_q(n) \oplus \ldots \oplus \boldsymbol{W}_q(n)^{p-1} \right) \boldsymbol{P}_n(p)(\boldsymbol{F}_q \otimes \boldsymbol{I}_p).$$

PROOF The first factorization follows immediately from Lemma 8.2 and (8.14). Transposing (i), we obtain (ii) by (8.14). Using (K5) and (8.14), (iii) and (iv) follow from (ii). ∎

These results can be recursively extended to the radix–p case $n = p^t$ $(p, t \geq 2)$. Setting $n_s := p^{t-s}$ $(s = 1, \ldots, t)$, we introduce the *radix–p butterfly matrices*

$$\boldsymbol{B}_n^{(s)} := \boldsymbol{I}_{n_s} \otimes \left(\sqrt{p}\boldsymbol{F}_p\right) \otimes \boldsymbol{I}_{p^{s-1}} \qquad (s = 1, \ldots, t)$$

and the *radix–p twiddle matrices*

$$\boldsymbol{T}_n^{(1)} := \boldsymbol{I}_n \,,$$
$$\boldsymbol{T}_n^{(s)} := \boldsymbol{I}_{n_s} \otimes \left(\boldsymbol{I}_{p^{s-1}} \oplus \boldsymbol{W}_{p^{s-1}}(p^s) \oplus \ldots \oplus \boldsymbol{W}_{p^{s-1}}(p^s)^{p-1}\right) \qquad (s = 2, \ldots, t)$$

with $\boldsymbol{W}_{p^{s-1}}(p^s) := \operatorname{diag}(\omega_{p^s}^k)_{k=0}^{p^{s-1}-1}$. Note that $\boldsymbol{B}_n^{(s)}$ and $\boldsymbol{T}_n^{(s)}$ are sparse, almost unitary matrices. A matrix $\boldsymbol{A} \in \mathbb{C}^{n \times n}$ is called *almost unitary,* if there exists an $\alpha := \alpha(\boldsymbol{A}) > 0$ such that $\alpha^{-1}\boldsymbol{A}$ is unitary. An almost unitary, real matrix is called *almost orthogonal.* It is a consequence of Theorems 8.3 and 8.4 below that FFTs based on a factorization of $\boldsymbol{F}_n$ into a product of almost unitary matrices possess a good numerical stability. All our factorizations in Sections 8.4–8.6 have almost unitary factors.

For $p = 2$ or 4 the multiplication with $\boldsymbol{B}_n^{(s)}$ only uses summations. Our definition of butterfly matrices is slightly different from [7], p. 18, because the separation into a multiplication part and a trivial butterfly operation will improve the roundoff error bounds in the sequel. For $p = 2$ by [7], p. 18 and pp. 29–29, the matrices $\boldsymbol{B}_n^{(s)}\boldsymbol{T}_n^{(s)}$ $(s = 1, \ldots, t)$ generate so–called Cooley–Tukey butterflies, whereas by [7], p. 66 the matrices $\boldsymbol{T}_n^{(s)}\boldsymbol{B}_n^{(s)} = \left(\boldsymbol{B}_n^{(s)}\boldsymbol{T}_n^{(s)}\right)^{\mathrm{T}}$ generate so–called Gentleman–Sande butterflies.

In the case $n = p^t$ it is very useful to introduce a special permutation matrix. Each integer $k \in \{0, \ldots, n-1\}$ has the *p–adic representation*

$$k = (k_{t-1}, \ldots, k_1, k_0)_p := \sum_{s=0}^{t-1} k_s p^s$$

with digits $k_s \in \{0, \ldots, p-1\}$ $(s = 0, \ldots, t-1)$. Then the *radix–p digit–reversal number* of k is defined by

$$\operatorname{rev}_n(k) := (k_0, \ldots, k_{t-1})_p = \sum_{s=0}^{t-1} k_s p^{t-s-1}.$$

Obviously we have

$$\operatorname{rev}_n(\operatorname{rev}_n(k)) = k \qquad (k = 0, \ldots, n-1). \tag{8.17}$$

Then

$$\boldsymbol{R}_n(p) := (\delta(\operatorname{rev}_n(k) - l))_{k,l=0}^{n-1}$$

is called *radix-p digit-reversal permutation matrix*. By (8.17), we see that

$$\boldsymbol{R}_n(p) = \boldsymbol{R}_n(p)^{\mathrm{T}} = \boldsymbol{R}_n(p)^{-1}. \tag{8.18}$$

LEMMA 8.3
Let $n = p^t$ $(p, t \geq 2)$. Then we have

$$\boldsymbol{R}_n(p) = \left(\boldsymbol{I}_{n_2} \otimes \boldsymbol{P}_{p^2}(p)\right) \ldots \left(\boldsymbol{I}_p \otimes \boldsymbol{P}_{n_1}(p)\right) \boldsymbol{P}_n(p), \tag{8.19}$$

$$\boldsymbol{R}_n(p)\boldsymbol{F}_n = n^{-1/2}\boldsymbol{B}_n^{(1)}\boldsymbol{T}_n^{(2)}\boldsymbol{B}_n^{(2)} \ldots \boldsymbol{T}_n^{(t)}\boldsymbol{B}_n^{(t)}. \tag{8.20}$$

PROOF 1. Define

$$\boldsymbol{R}'_n(p) := \left(\boldsymbol{I}_{n_2} \otimes \boldsymbol{P}_{p^2}(p)\right) \ldots \left(\boldsymbol{I}_p \otimes \boldsymbol{P}_{n_1}(p)\right) \boldsymbol{P}_n(p). \tag{8.21}$$

We show by induction in t that $\boldsymbol{R}_n(p) = \boldsymbol{R}'_n(p)$ for $n = p^t$. In the case $t = 2$, we have $\boldsymbol{R}_n(p) = \boldsymbol{P}_n(p) = \boldsymbol{R}'_n(p)$ with $n = p^2$.

Assume that the assertion $\boldsymbol{R}'_{n_1}(p) = \boldsymbol{R}_{n_1}(p)$ is true for some $t \geq 3$. Then it follows by (K1) and (K3) that

$$\boldsymbol{I}_p \otimes \boldsymbol{R}'_{n_1}(p) = \left(\boldsymbol{I}_{n_2} \otimes \boldsymbol{P}_{p^2}(p)\right) \ldots \left(\boldsymbol{I}_{p^2} \otimes \boldsymbol{P}_{n_2}(p)\right)\left(\boldsymbol{I}_p \otimes \boldsymbol{P}_{n_1}(p)\right).$$

Since for $\boldsymbol{x} = (x(k))_{k=0}^{n-1} \in \mathbb{C}^n$

$$\boldsymbol{P}_n(p)\boldsymbol{x} = (x(p(k_{t-2}, \ldots, k_0)_p + k_{t-1}))_{k=0}^{n-1} = (x((k_{t-2}, \ldots, k_0, k_{t-1})_p))_{k=0}^{n-1},$$

we obtain by (8.21) and induction

$$\begin{aligned}\boldsymbol{R}'_n(p)\boldsymbol{x} &= \left(\boldsymbol{I}_p \otimes \boldsymbol{R}_{n_1}(p)\right)\boldsymbol{P}_n(p)\boldsymbol{x} = (x((k_0, \ldots, k_{t-2}, k_{t-1})_p))_{k=0}^{n-1} \\ &= (x(\mathrm{rev}_n(k)))_{k=0}^{n-1} = \boldsymbol{R}_n(p)\boldsymbol{x}.\end{aligned}$$

2. The factorization (8.20) can be shown by induction in t, too. From Lemma 8.2 the result follows immediately for $t = 2$. Assume that assertion (8.20) is true for $t - 1$ $(t \geq 3)$, i.e.

$$\boldsymbol{R}_{n_1}(p)\boldsymbol{F}_{n_1} = n_1^{-1/2}\boldsymbol{B}_{n_1}^{(1)}\boldsymbol{T}_{n_1}^{(2)}\boldsymbol{B}_{n_1}^{(2)} \ldots \boldsymbol{T}_{n_1}^{(t-1)}\boldsymbol{B}_{n_1}^{(t-1)}. \tag{8.22}$$

Using Lemma 8.2 with $q = n_1$ and the definitions of $\boldsymbol{T}_n^{(t)}$ and $\boldsymbol{B}_n^{(t)}$, we obtain

$$\boldsymbol{P}_n(p)\boldsymbol{F}_n = p^{-1/2}\left(\boldsymbol{I}_p \otimes \boldsymbol{F}_{n_1}\right)\boldsymbol{T}_n^{(t)}\boldsymbol{B}_n^{(t)}.$$

Hence, by (K1) and the first part of this proof we conclude that

$$\boldsymbol{R}_n(p)\boldsymbol{F}_n = \left(\boldsymbol{I}_p \otimes \boldsymbol{R}_{n_1}(p)\right)\boldsymbol{P}_n(p)\boldsymbol{F}_n = p^{-1/2}\left(\boldsymbol{I}_p \otimes \boldsymbol{R}_{n_1}(p)\boldsymbol{F}_{n_1}\right)\boldsymbol{T}_n^{(t)}\boldsymbol{B}_n^{(t)}$$

such that the induction hypothesis (8.22) and (K1) together with

$$\boldsymbol{B}_n^{(s)} = \boldsymbol{I}_p \otimes \boldsymbol{B}_{n_1}^{(s)}, \qquad \boldsymbol{T}_n^{(s)} = \boldsymbol{I}_p \otimes \boldsymbol{T}_{n_1}^{(s)} \qquad (s = 1, \ldots, t-1)$$

and $n^{-1/2} = n_1^{-1/2} p^{-1/2}$ yield the result. ∎

REMARK 8.2 The only butterfly matrix in *parallel form* is $\boldsymbol{B}_n^{(1)} = \boldsymbol{I}_{n_1} \otimes \sqrt{p}\boldsymbol{F}_p$. Using (K3) and (K5), the other butterfly matrices $\boldsymbol{B}_n^{(s)}$ $(s = 2, \ldots, t)$ can also be represented in parallel form

$$\boldsymbol{B}_n^{(s)} = (\boldsymbol{I}_{n_s} \otimes \sqrt{p}\boldsymbol{F}_p) \otimes \boldsymbol{I}_{p^{s-1}} = \boldsymbol{P}_n(n_{s-1}) \boldsymbol{B}_n^{(1)} \boldsymbol{P}_n(p^{s-1}).$$

Using (K1), (K3) and (K5), we obtain another parallel form

$$\boldsymbol{B}_n^{(s)} = \boldsymbol{I}_{n_s} \otimes \left(\sqrt{p}\boldsymbol{F}_p \otimes \boldsymbol{I}_{p^{s-1}}\right) = \left(\boldsymbol{I}_{n_s} \otimes \boldsymbol{P}_{p^s}(p)\right) \boldsymbol{B}_n^{(1)} \left(\boldsymbol{I}_{n_s} \otimes \boldsymbol{P}_{p^s}(p^{s-1})\right).$$

The only butterfly matrix in *vector form* is $\boldsymbol{B}_n^{(t)} = \sqrt{p}\boldsymbol{F}_p \otimes \boldsymbol{I}_{n_1}$. Similar as before, using the identities

$$\boldsymbol{B}_n^{(s)} = \boldsymbol{P}_n(n_s) \boldsymbol{B}_n^{(t)} \boldsymbol{P}_n(p^s), \tag{8.23}$$

$$\boldsymbol{B}_n^{(s)} = \left(\boldsymbol{P}_{n_{s-1}}(n_s) \otimes \boldsymbol{I}_{p^{s-1}}\right) \boldsymbol{B}_n^{(t)} \left(\boldsymbol{P}_{n_{s-1}}(p) \otimes \boldsymbol{I}_{p^{s-1}}\right) \tag{8.24}$$

the other butterfly matrices $\boldsymbol{B}_n^{(s)}$ $(s = 1, \ldots, t-1)$ can be transferred into vector form. ∎

Thus we obtain the main result of this section:

THEOREM 8.2
Let $n = p^t$ $(p, t \geq 2)$. Then $\boldsymbol{F}_n$ can be factorized into products of sparse, almost unitary matrices as follows:
(i) *Decimation–in–frequency*

$$\boldsymbol{F}_n = n^{-1/2} \boldsymbol{R}_n(p) \boldsymbol{B}_n^{(1)} \boldsymbol{T}_n^{(2)} \boldsymbol{B}_n^{(2)} \ldots \boldsymbol{T}_n^{(t)} \boldsymbol{B}_n^{(t)}, \tag{8.25}$$

(ii) *Decimation–in–time*

$$\boldsymbol{F}_n = n^{-1/2} \boldsymbol{B}_n^{(t)} \boldsymbol{T}_n^{(t)} \ldots \boldsymbol{B}_n^{(2)} \boldsymbol{T}_n^{(2)} \boldsymbol{B}_n^{(1)} \boldsymbol{R}_n(p). \tag{8.26}$$

By representing the butterfly matrices $\boldsymbol{B}_n^{(s)}$ in parallel or vector form (see Remark 8.2), we obtain a corresponding factorization of $\boldsymbol{F}_n$ in parallel or vector form.

PROOF The factorization (8.25) follows directly from (8.20) and (8.18). The transpose of (8.25) yields (8.26). ∎

Theorem 8.2 can be extended to the mixed–radix case $n = p_1 \ldots p_t$ $(p_s \geq 2,\ t \geq 2)$; see [7], pp. 95–101 and [17], pp. 81–87.

8.5 Radix–2 and radix–4 FFT

Stepwise matrix–vector multiplications corresponding to the factorizations of $\boldsymbol{F}_n$ (see Theorem 8.2) yield various FFTs for computing $\boldsymbol{F}_n\boldsymbol{x}$ ($\boldsymbol{x} \in \mathbb{C}^n$). In the case $p = 2$, the factorization (8.25) provides the *radix–2 decimation–in–frequency algorithm* or *Gentleman–Sande* FFT:

ALGORITHM 8.5 (Gentleman–Sande FFT)
Input: $n = 2^t$ $(t \geq 2)$, $\boldsymbol{x}^{(0)} := \boldsymbol{x} \in \mathbb{C}^n$.

0. Precompute ω_n^k $(k = 1, \ldots, n_1 - 1)$ by a method of Section 8.3.
1. For $s = 1 : 1 : t-1$ form
   $\boldsymbol{z}^{(s)} := \boldsymbol{B}_n^{(t-s+1)}\boldsymbol{x}^{(s-1)}$,
   $\boldsymbol{x}^{(s)} := \boldsymbol{T}_n^{(t-s+1)}\boldsymbol{z}^{(s)}$.
2. Compute
   $\boldsymbol{x}^{(t)} := \boldsymbol{B}_n^{(1)}\boldsymbol{x}^{(t-1)}$.
3. Permute
   $\boldsymbol{x}^{(t+1)} := \boldsymbol{R}_n(2)\boldsymbol{x}^{(t)}$.
4. Multiply $\boldsymbol{y} := n^{-1/2}\boldsymbol{x}^{(t+1)}$.

Output: $\boldsymbol{y} = \boldsymbol{F}_n\boldsymbol{x}$.

Similarly, the factorization (8.26) provides the *decimation–in–time algorithm* or *Cooley–Tukey* FFT. A corresponding factorization of $\boldsymbol{F}_n$ in parallel or vector form yields an FFT *for a parallel* or *vector computer* (see [17], pp. 76–79).

Now let $n = 4^t = 2^{2t}$ $(t \geq 2)$. In order to further reduce the arithmetical complexity, we consider a *modified radix–4* FFT, where the standard radix–2 butterfly operations are used (see [7], pp. 101–107; [14], pp. 139–140). In contrast to Theorem 8.2 with $p = 4$, this modified radix–4 FFT is based on the following factorization of $\boldsymbol{F}_n$. By Lemma 8.2 we have

$$\boldsymbol{P}_n(4)\boldsymbol{F}_n = \left(\boldsymbol{I}_4 \otimes \boldsymbol{F}_{n/4}\right)\Big(\boldsymbol{I}_{n/4} \oplus \boldsymbol{W}_{n/4}(n) \oplus \oplus \boldsymbol{W}_{n/4}(n)^2 \oplus \boldsymbol{W}_{n/4}(n)^3\Big)\left(\boldsymbol{F}_4 \otimes \boldsymbol{I}_{n/4}\right) \tag{8.27}$$

with $\boldsymbol{W}_{n/4}(n) = \operatorname{diag}\left(\omega_n^j\right)_{j=0}^{n/4-1}$.

By the same lemma and $\boldsymbol{P}_4(2) = \boldsymbol{P}_4(2)^{-1}$ (see (8.14)), it follows that

$$\boldsymbol{F}_4 = \boldsymbol{P}_4(2)\left(\boldsymbol{I}_2 \otimes \boldsymbol{F}_2\right)\operatorname{diag}(1, 1, 1, -\mathrm{i})\left(\boldsymbol{F}_2 \otimes \boldsymbol{I}_2\right)$$

and hence by Theorem 8.1

$$\boldsymbol{F}_4 \otimes \boldsymbol{I}_{n/4} = \tfrac{1}{2}\left(\boldsymbol{P}_4(2) \otimes \boldsymbol{I}_{n/4}\right)\boldsymbol{B}_n^{(2t-1)}\boldsymbol{T}_n^{(2t)}\boldsymbol{B}_n^{(2t)} \tag{8.28}$$

with the radix–2 butterfly matrices

$$B_n^{(s)} = I_{2^{2t-s}} \otimes \left(\sqrt{2}F_2\right) \otimes I_{2^{s-1}}$$

and the trivial twiddle matrix

$$T_n^{(2t)} := \operatorname{diag}(1,1,1,-\mathrm{i}) \otimes I_{n/4}.$$

Substituting (8.28) into (8.27), we obtain

$$\begin{aligned} P_n(4)F_n &= \tfrac{1}{2}\left(I_4 \otimes F_{n/4}\right)\left(P_4(2) \otimes I_{n/4}\right) T_n^{(2t-1)} B_n^{(2t-1)} T_n^{(2t)} B_n^{(2t)} \\ &= \tfrac{1}{2}\left(P_4(2) \otimes I_{n/4}\right)\left(I_4 \otimes F_{n/4}\right) T_n^{(2t-1)} B_n^{(2t-1)} T_n^{(2t)} B_n^{(2t)} \end{aligned} \quad (8.29)$$

with the nontrivial twiddle matrix

$$\begin{aligned} T_n^{(2t-1)} &:= \left(P_4(2) \otimes I_{n/4}\right)\Big(I_{n/4} \oplus W_{n/4}(n) \oplus \\ &\qquad \oplus W_{n/4}(n)^2 \oplus W_{n/4}(n)^3\Big)\left(P_4(2) \otimes I_{n/4}\right) \\ &= I_{n/4} \oplus W_{n/4}(n)^2 \oplus W_{n/4}(n) \oplus W_{n/4}(n)^3. \end{aligned}$$

It is easy to see that $P_n(4) = \left(P_4(2) \otimes I_{n/4}\right)\left(I_4 \otimes P_{n/4}(2)\right) P_n(2)$, and so by (8.29)

$$F_n = \tfrac{1}{2} P_n(2)^{\mathrm{T}}\left(I_2 \otimes P_{n/2}(2)^{\mathrm{T}}\right)\left(I_4 \otimes F_{n/4}\right) T_n^{(2t-1)} B_n^{(2t-1)} T_n^{(2t)} B_n^{(2t)}. \quad (8.30)$$

Using $R_n(2) = P_n(2)^{\mathrm{T}}\left(I_2 \otimes P_{n/2}(2)^{\mathrm{T}}\right) \ldots \left(I_{n/4} \otimes P_4(2)^{\mathrm{T}}\right)$ (see (8.18)-(8.19)), we get by repeated application of (8.30) the following factorization

$$F_n = 2^{-t} R_n(2) B_n^{(1)} T_n^{(2)} B_n^{(2)} \ldots T_n^{(2t-1)} B_n^{(2t-1)} T_n^{(2t)} B_n^{(2t)} \quad (8.31)$$

with the twiddle matrices

$$\begin{aligned} T_n^{(2s)} &:= I_{2^{2t-2s}} \otimes \operatorname{diag}(1,1,1,-\mathrm{i}) \otimes I_{2^{2s-2}}, \\ T_n^{(2s-1)} &:= I_{2^{2t-2s}} \otimes \left(I_{2^{2s-2}} \oplus W_{2^{2s-2}}(2^{2s})^2 \oplus W_{2^{2s-2}}(2^{2s}) \oplus W_{2^{2s-2}}(2^{2s})^3\right) \end{aligned}$$

for $s = 1, \ldots, t$. Note that $T_n^{(1)} = I_n$.

REMARK 8.3 If $p \geq 3$, the FFT corresponding to the factorization of F_n in Theorem 8.2 uses $\leq t(p-1)p^t$ complex multiplications and $t(p-1)p^t$ complex summations, i.e. $\leq 4t(p-1)p^t$ real products and the same number of additions. In total, this amounts to $8t(p-1)p^t = \frac{8(p-1)}{\log_2 p} n \log_2 n$ flops. For $p = 2$, $B_n^{(s)}$ has only entries $0, \pm 1$, and so $5t\,2^t = 5n \log_2 n$ flops are sufficient. The radix–4 FFT based on the factorization (8.31) takes $\frac{17}{2} tn = 4.25\, n \log_2 n$ flops. ∎

We conclude this section with an FFT which cuts down the number of operations by about a half, when the data are *real.* Here and in the following, we use the notation $\boldsymbol{J}_n := (\delta(j+k+1-n))_{j,k=0}^{n-1}$ for the *counteridentity matrix.*

If the data vector $\boldsymbol{x} \in \mathbb{R}^n$ is *real,* then $\boldsymbol{y} := \boldsymbol{F}_n\boldsymbol{x}$ satisfies $y_0 \in \mathbb{R}$ and $y_j = \overline{y_{n-j}}$ $(j = 1, \ldots, n-1)$, i.e., $\boldsymbol{y}$ is *conjugate even.* If $n = 2^t$ $(t \geq 4)$ then this means that $\boldsymbol{U}_n\boldsymbol{y}$ is again a real vector, where

$$\boldsymbol{U}_n := \begin{pmatrix} \sqrt{2} & \mathbf{O} & 0 & \mathbf{O} \\ \mathbf{O} & \boldsymbol{I}_{n_1-1} & \mathbf{O} & \boldsymbol{J}_{n_1-1} \\ 0 & \mathbf{O} & \sqrt{2} & \mathbf{O} \\ \mathbf{O} & -\mathrm{i}\boldsymbol{I}_{n_1-1} & \mathbf{O} & \mathrm{i}\boldsymbol{J}_{n_1-1} \end{pmatrix}$$

is almost unitary. Note that $\boldsymbol{U}_n^{-1} = \frac{1}{2}\boldsymbol{U}_n^*$ and so $\frac{1}{2}\boldsymbol{U}_n\boldsymbol{y}$ contains the real part of $\boldsymbol{y}$ in the first half and the imaginary part in the second. Therefore, we can easily reconstruct $\boldsymbol{y}$ from $\boldsymbol{U}_n\boldsymbol{y}$. It follows from Corollary 8.1, (ii) that we have

$$\boldsymbol{F}_n = 2^{-1/2}\boldsymbol{B}_n^{(t)}\left(\boldsymbol{I}_{n_1} \oplus \boldsymbol{W}_{n_1}(n)\right)\left(\boldsymbol{F}_{n_1} \oplus \boldsymbol{F}_{n_1}\right)\boldsymbol{P}_n(2).$$

Therefore we can calculate $\boldsymbol{U}_n\boldsymbol{y}$ in two steps: Let $\boldsymbol{x}^{(0)} := (x_{2j})_{j=0}^{n_1-1}$ be the first half of $\boldsymbol{P}_n(2)\boldsymbol{x}$ and $\boldsymbol{x}^{(1)} := (x_{2j+1})_{j=0}^{n_1-1}$ the second. Then the first half of $\boldsymbol{v} := (\boldsymbol{U}_{n_1} \oplus \boldsymbol{U}_{n_1})(\boldsymbol{F}_{n_1} \oplus \boldsymbol{F}_{n_1})\boldsymbol{P}_n(2)\boldsymbol{x}$ is the real part of $\boldsymbol{w} := \boldsymbol{U}_{n_1}\boldsymbol{F}_{n_1}(\boldsymbol{x}^{(0)} + \mathrm{i}\boldsymbol{x}^{(1)})$, and the second half is the imaginary part of $\boldsymbol{w}$ (using again that $\boldsymbol{U}_{n_1}\boldsymbol{F}_{n_1}$ is a real matrix). Here we apply some FFT of *half the length.* In the second step, we compute $\boldsymbol{U}_n\boldsymbol{y} = 2^{-1/2}\boldsymbol{V}_n\boldsymbol{v}$ where $\boldsymbol{V}_n := \boldsymbol{U}_n\boldsymbol{B}_n^{(t)}\left(\boldsymbol{I}_{n_1} \oplus \boldsymbol{W}_{n_1}(n)\right)\left(\boldsymbol{U}_{n_1} \oplus \boldsymbol{U}_{n_1}\right)^{-1}$, i.e.

$$\boldsymbol{V}_n = \begin{pmatrix} 1 & \mathbf{O} & 0 & \mathbf{O} & 1 & \mathbf{O} & 0 & \mathbf{O} \\ \mathbf{O} & \boldsymbol{I}_{n_2-1} & \mathbf{O} & \mathbf{O} & \mathbf{O} & \boldsymbol{C}_{n_2-1} & \mathbf{O} & \boldsymbol{S}_{n_2-1} \\ 0 & \mathbf{O} & \sqrt{2} & \mathbf{O} & 0 & \mathbf{O} & 0 & \mathbf{O} \\ \mathbf{O} & \boldsymbol{J}_{n_2-1} & \mathbf{O} & \mathbf{O} & \mathbf{O} & -\boldsymbol{J}_{n_2-1}\boldsymbol{C}_{n_2-1} & \mathbf{O} & -\boldsymbol{J}_{n_2-1}\boldsymbol{S}_{n_2-1} \\ 1 & \mathbf{O} & 0 & \mathbf{O} & -1 & \mathbf{O} & 0 & \mathbf{O} \\ \mathbf{O} & \mathbf{O} & \mathbf{O} & \boldsymbol{I}_{n_2-1} & \mathbf{O} & -\boldsymbol{S}_{n_2-1} & \mathbf{O} & \boldsymbol{C}_{n_2-1} \\ 0 & \mathbf{O} & 0 & \mathbf{O} & 0 & \mathbf{O} & -\sqrt{2} & \mathbf{O} \\ \mathbf{O} & \mathbf{O} & \mathbf{O} & -\boldsymbol{J}_{n_2-1} & \mathbf{O} & -\boldsymbol{J}_{n_2-1}\boldsymbol{S}_{n_2-1} & \mathbf{O} & \boldsymbol{J}_{n_2-1}\boldsymbol{C}_{n_2-1} \end{pmatrix}$$

and $\boldsymbol{C}_{n_2-1} := \operatorname{diag}\left((\cos\frac{2\pi j}{n})_{j=1}^{n_2-1}\right)$, $\boldsymbol{S}_{n_2-1} := \operatorname{diag}\left((\sin\frac{2\pi j}{n})_{j=1}^{n_2-1}\right)$. Note that $\boldsymbol{V}_n$ is sparse and almost orthogonal with $\boldsymbol{V}_n\boldsymbol{V}_n^T = 2\boldsymbol{I}_n$. We summarize (see [7], pp. 215–223):

ALGORITHM 8.6 (FFT with real data)
Input: $n = 2^t$ $(t \geq 4)$, $\boldsymbol{x} \in \mathbb{R}^n$.

0. Precompute $\cos\frac{2\pi j}{n}, \sin\frac{2\pi j}{n}$ $(j = 1, \ldots, n_2 - 1)$ by a method of Section 8.3.

1. Set $\boldsymbol{x}^{(0)} := (x_{2j})_{j=0}^{n_1-1}$ and $\boldsymbol{x}^{(1)} := (x_{2j+1})_{j=0}^{n_1-1}$.
2. Compute by a radix–2 FFT
   $\boldsymbol{x}^{(2)} := \boldsymbol{F}_{n_1}(\boldsymbol{x}^{(0)} + \mathrm{i}\boldsymbol{x}^{(1)})$.
3. Form $\boldsymbol{w} := \boldsymbol{U}_{n_1}\boldsymbol{x}^{(2)}$.
4. Set $\boldsymbol{v} := \begin{pmatrix} \Re \boldsymbol{w} \\ \Im \boldsymbol{w} \end{pmatrix}$.
5. Compute $\boldsymbol{z} := 2^{-1/2}\boldsymbol{V}_n\boldsymbol{v}$.

Output: $\boldsymbol{z} = \boldsymbol{U}_n\boldsymbol{F}_n\boldsymbol{x}$.

If the Cooley–Tukey or Gentleman–Sande FFT is used in step 2, then $\boldsymbol{z} = \boldsymbol{U}_n\boldsymbol{F}_n\boldsymbol{x}$ can be computed in $\frac{1}{2}(5t + 7)n$ flops. Note that this method can also be applied, if we only assume that n is a multiply of 4. Then we have to use a suitable mixed–radix FFT in step 2.

8.6 Split–radix FFT

A standard FFT of length $n = 2^t$ $(t \geq 2)$ is based upon the fast synthesis of two DFTs of half length n_1. The split–radix FFT is based upon a clever synthesis of radix–2 and radix–4 FFTs. The idea of this algorithm, due to [4], can be described shortly as follows (compare with [7], pp. 111–119):

$$y_{2j} = \frac{1}{\sqrt{n}} \sum_{k=0}^{n_1-1} (x_k + x_{n_1+k})\omega_{n_1}^{jk}$$

for $j = 0, \ldots, n_1 - 1$ and

$$y_{4j+1} = \frac{1}{\sqrt{n}} \sum_{k=0}^{n_2-1} \left((x_k - x_{n_1+k}) - \mathrm{i}(x_{n_2+k} - x_{3n_2+k})\right) \omega_n^k \omega_{n_2}^{jk},$$

$$y_{4j+3} = \frac{1}{\sqrt{n}} \sum_{k=0}^{n_2-1} \left((x_k - x_{n_1+k}) + \mathrm{i}(x_{n_2+k} - x_{3n_2+k})\right) \omega_n^{3k} \omega_{n_2}^{jk}$$

for $j = 0, \ldots, n_2 - 1$. We are interested in the corresponding factorization of $\boldsymbol{F}_n$ in order to study the numerical stability of the split–radix FFT.

The *odd Fourier matrix* $\boldsymbol{F}_n^{(1)} \in \mathbb{C}^{n \times n}$ is defined by

$$\boldsymbol{F}_n^{(1)} := n^{-1/2} \left(\omega_{2n}^{(2j+1)k}\right)_{j,k=0}^{n-1}.$$

This matrix is unitary, too. Sometimes, $\boldsymbol{F}_n^{(0)} := \boldsymbol{F}_n$ is called *even Fourier matrix*. In the following we show that the split–radix FFT is based upon a synthesis of even and odd DFTs. In other words, the split–radix algorithm splits recursively $\boldsymbol{F}_n$ into $\boldsymbol{F}_{n_1} \oplus \boldsymbol{F}^{(1)}_{n_1}$ and then $\boldsymbol{F}^{(1)}_{n_1}$ into $\boldsymbol{F}_{n_2} \oplus \boldsymbol{F}_{n_2}$.

We start with step 1 of the split–radix FFT of computing $\boldsymbol{y} := \boldsymbol{F}_n \boldsymbol{x}$ ($\boldsymbol{x} \in \mathbb{C}^n$). By the divide–and–conquer method we get

$$y_{2j} = n^{-1/2} \sum_{k=0}^{n-1} x_k \omega_n^{2jk} = n_1^{-1/2} \sum_{k=0}^{n_1-1} x_k^{(0)} \omega_{n_1}^{jk}, \tag{8.32}$$

$$y_{2j+1} = n^{-1/2} \sum_{k=0}^{n-1} x_k \omega_n^{(2j+1)k} = n_1^{-1/2} \sum_{k=0}^{n_1-1} x_k^{(1)} \omega_n^{(2j+1)k} \tag{8.33}$$

for $j = 0, \ldots, n_1 - 1$ with

$$x_k^{(l)} := 2^{-1/2}(x_k + (-1)^l x_{n_1+k}) \qquad (k = 0, \ldots, n_1 - 1;\ l = 0, 1).$$

In matrix form, (8.32)–(8.33) can be written as

$$\boldsymbol{P}_n(2)\boldsymbol{y} = \left(\boldsymbol{F}_{n_1} \oplus \boldsymbol{F}^{(1)}_{n_1}\right) \begin{pmatrix} \boldsymbol{x}^{(0)} \\ \boldsymbol{x}^{(1)} \end{pmatrix}, \qquad \begin{pmatrix} \boldsymbol{x}^{(0)} \\ \boldsymbol{x}^{(1)} \end{pmatrix} = (\boldsymbol{F}_2 \otimes \boldsymbol{I}_{n_1})\, \boldsymbol{x}$$

with $\boldsymbol{x}^{(l)} := \left(x_k^{(l)}\right)_{k=0}^{n_1-1}$ ($l = 0, 1$).

The same technique can be applied in the next steps. Then step 2 yields

$$y_{4j} = n_2^{-1/2} \sum_{k=0}^{n_2-1} x_k^{(0,0)} \omega_{n_2}^{jk} \qquad (j = 0, \ldots, n_2 - 1), \tag{8.34}$$

$$y_{4j+2} = n_2^{-1/2} \sum_{k=0}^{n_2-1} x_k^{(0,1)} \omega_{n_2}^{(2j+1)k} \qquad (j = 0, \ldots, n_2 - 1) \tag{8.35}$$

with

$$x_k^{(0,l)} := 2^{-1/2}(x_k^{(0)} + (-1)^l x_{n_2+k}^{(0)}) \qquad (k = 0, \ldots, n_2 - 1;\ l = 0, 1).$$

This means that by (8.34)–(8.35) we have

$$\boldsymbol{P}_{n_1}(2)(y_{2j})_{j=0}^{n_1-1} = \left(\boldsymbol{F}_{n_2} \oplus \boldsymbol{F}^{(1)}_{n_2}\right) \begin{pmatrix} \boldsymbol{x}^{(0,0)} \\ \boldsymbol{x}^{(0,1)} \end{pmatrix}, \qquad \begin{pmatrix} \boldsymbol{x}^{(0,0)} \\ \boldsymbol{x}^{(0,1)} \end{pmatrix} = (\boldsymbol{F}_2 \otimes \boldsymbol{I}_{n_2})\, \boldsymbol{x}^{(0)}$$

with $\boldsymbol{x}^{(0,j)} := \left(x_k^{(0,j)}\right)_{k=0}^{n_2-1}$ ($j = 0, 1$).

In step 2 we have now to consider the odd indices. Using the same technique and $\omega_n^{(2j+1)n_2} = -\mathrm{i}(-1)^j$, we obtain

$$y_{2j+1} = n_1^{-1/2} \sum_{k=0}^{n_2-1} \left(x_k^{(1)} - \mathrm{i}(-1)^j x_{n_2+k}^{(1)}\right) \omega_n^{(2j+1)k} \qquad (j = 0, \ldots, n_1 - 1),$$

such that

$$y_{4j+1} = n_2^{-1/2} \sum_{k=0}^{n_2-1} x_k^{(1,0)} \omega_{n_2}^{jk} \qquad (j = 0, \ldots, n_2 - 1), \tag{8.36}$$

$$y_{4j+3} = n_2^{-1/2} \sum_{k=0}^{n_2-1} x_k^{(1,1)} \omega_{n_2}^{jk} \qquad (j = 0, \ldots, n_2 - 1) \tag{8.37}$$

with

$$x_k^{(1,l)} := 2^{-1/2} \left(x_k^{(1)} - \mathrm{i}(-1)^l x_{n_2+k}^{(1)} \right) \omega_n^{(2j+1)k} \quad (k = 0, \ldots, n_2 - 1;\ l = 0, 1).$$

In matrix–vector form, (8.36)–(8.37) read as follows:

$$\boldsymbol{P}_{n_1}(2)(y_{2j+1})_{j=0}^{n_1-1} = (\boldsymbol{F}_{n_2} \oplus \boldsymbol{F}_{n_2}) \begin{pmatrix} \boldsymbol{x}^{(1,0)} \\ \boldsymbol{x}^{(1,1)} \end{pmatrix},$$

$$\begin{pmatrix} \boldsymbol{x}^{(1,0)} \\ \boldsymbol{x}^{(1,1)} \end{pmatrix} = \left(\boldsymbol{V}_{n_2} \oplus \boldsymbol{V}_{n_2}^3 \right) \left(\boldsymbol{F}_2 \otimes \boldsymbol{I}_{n_2} \right) \left(\mathrm{diag}(1, -\mathrm{i}) \otimes \boldsymbol{I}_{n_2} \right) \boldsymbol{x}^{(1)}$$

with $\boldsymbol{x}^{(1,j)} := \left(x_k^{(1,j)} \right)_{k=0}^{n_2-1}$ and $\boldsymbol{V}_{n_2} := \mathrm{diag}(\omega_n^k)_{k=0}^{n_2-1}$. In summary, the whole second step reads as follows:

$$(\boldsymbol{I}_2 \otimes \boldsymbol{P}_{n_1}(2)) \, \boldsymbol{P}_n(2) \boldsymbol{y} = \left(\boldsymbol{F}_{n_2} \oplus \boldsymbol{F}_{n_2}^{(1)} \oplus \boldsymbol{F}_{n_2} \oplus \boldsymbol{F}_{n_2} \right) \begin{pmatrix} \boldsymbol{x}^{(0,0)} \\ \boldsymbol{x}^{(0,1)} \\ \boldsymbol{x}^{(1,0)} \\ \boldsymbol{x}^{(1,1)} \end{pmatrix},$$

$$\begin{pmatrix} \boldsymbol{x}^{(0,0)} \\ \boldsymbol{x}^{(0,1)} \\ \boldsymbol{x}^{(1,0)} \\ \boldsymbol{x}^{(1,1)} \end{pmatrix} = \frac{1}{2} \tilde{\boldsymbol{T}}_n^{(t-1)} \boldsymbol{B}_n^{(t-1)} \boldsymbol{T}_n^{(t)} \boldsymbol{B}_n^{(t)} \boldsymbol{x},$$

with the radix–2 butterfly matrices $\boldsymbol{B}_n^{(s)} = \boldsymbol{I}_{n_s} \otimes (\sqrt{2}\, \boldsymbol{F}_2) \otimes \boldsymbol{I}_{2^{s-1}}$ and the diagonal matrices

$$\boldsymbol{T}_n^{(t)} := \mathrm{diag}(1, 1, 1, -\mathrm{i}) \otimes \boldsymbol{I}_{n_2}, \qquad \tilde{\boldsymbol{T}}_n^{(t-1)} := \boldsymbol{I}_{n_1} \oplus \boldsymbol{V}_{n_2} \oplus \boldsymbol{V}_{n_2}^3.$$

Likewise in step 3, we can see that

$$\boldsymbol{T}_n^{(t-1)} := \left(\mathrm{diag}(1, 1, 1, -\mathrm{i}) \otimes \boldsymbol{I}_{n_3} \right) \oplus \boldsymbol{V}_{n_2} \oplus \boldsymbol{V}_{n_2}^3$$

is the twiddle matrix.

As shown, the split–radix FFT is based on recursive splitting of $\boldsymbol{F}_n$ into $\boldsymbol{F}_{n_1} \oplus \boldsymbol{F}_{n_1}^{(1)}$ and $\boldsymbol{F}_{n_1}^{(1)}$ into $\boldsymbol{F}_{n_2} \oplus \boldsymbol{F}_{n_2}$. This is illustrated by the following diagram.

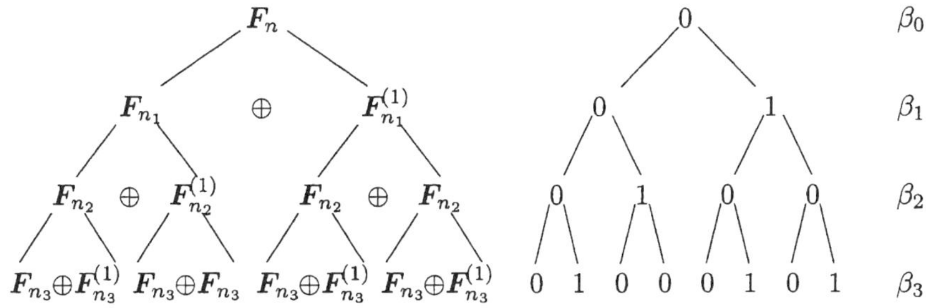

To specify whether even or odd DFT of length n_s is used in step s, we define a binary vector $\beta_s = (\beta_s(1), \ldots, \beta_s(2^s))$ $(s = 0, \ldots, t)$ by

$$\beta_0 = (0), \quad \beta_1 = (0,1), \quad \beta_2 = (0,1,0,0), \quad \beta_3 = (0,1,0,0,0,1,0,1)$$

and generally

$$\beta_{s+1} := (\beta_s, \tilde{\beta}_s),$$

where $\tilde{\beta}_s$ equals β_s with the exception that the last bit position is reversed (see [7], pp. 115–119). Using this notation, it can be shown that

$$\boldsymbol{T}_n^{(s)} := \boldsymbol{T}_{2^s}^{(s)}(1) \oplus \ldots \oplus \boldsymbol{T}_{2^s}^{(s)}(2^{t-s}-1) \oplus \boldsymbol{T}_{2^s}^{(s)}(2^{t-s}) \qquad (s = 2, \ldots, t)$$

with

$$\boldsymbol{T}_{2^s}^{(s)}(l) := \begin{cases} \operatorname{diag}(1,1,1,-\mathrm{i}) \otimes \boldsymbol{I}_{2^{s-1}} & \text{if } \beta_{t-s}(l) = 0, \\ \boldsymbol{V}_{2^{s-1}} \oplus \boldsymbol{V}_{2^{s-1}}^3 & \text{if } \beta_{t-s}(l) = 1, \end{cases}$$

$$\boldsymbol{V}_{2^{s-1}} := \operatorname{diag}(\omega_{2^{s+1}}^k)_{k=0}^{2^{s-1}-1}.$$

Using (8.19), the corresponding factorization of the split–radix algorithm reads as follows

$$\boldsymbol{F}_n = n^{-1/2}\boldsymbol{R}_n(2)\boldsymbol{B}_n^{(1)}\boldsymbol{T}_n^{(2)}\boldsymbol{B}_n^{(2)} \ldots \boldsymbol{T}_n^{(t)}\boldsymbol{B}_n^{(t)}. \tag{8.38}$$

The split–radix FFT based on factorization (8.38) takes $4tn = 4n\log_2 n$ flops. Compared with other radix–2 and radix–4 FFTs described in this article, the split–radix FFT possesses the lowest arithmetical complexity.

8.7 Error analysis for matrix–vector products

Now we consider a matrix–vector product $\boldsymbol{A}\boldsymbol{x}$, where $\boldsymbol{x} \in \mathbb{C}^n$ and $\boldsymbol{A} \in \mathbb{C}^{n\times n}$. Assume that a factorization of $\boldsymbol{A}$ into a product of sparse matrices is known:

$$\boldsymbol{A} = \prod_{s=1}^{t} \boldsymbol{A}^{(s)} := \boldsymbol{A}^{(t)} \ldots \boldsymbol{A}^{(2)}\boldsymbol{A}^{(1)}. \tag{8.39}$$

Note that the matrix product is formed from right to left with the first factor $\boldsymbol{A}^{(1)}$. Using (8.39), the matrix–vector product can be computed stepwise by the following

ALGORITHM 8.7
Input: $\boldsymbol{x}^{(0)} := \boldsymbol{x} \in \mathbb{C}^n$; $\boldsymbol{A}^{(s)} \in \mathbb{C}^{n\times n}$ $(s = 1, \dots t)$ sparse.
For $s = 1 : 1 : t$ compute $\boldsymbol{x}^{(s)} := \boldsymbol{A}^{(s)}\boldsymbol{x}^{(s-1)}$.
Output: $\boldsymbol{A}\boldsymbol{x} = \boldsymbol{x}^{(t)}$.

Special cases of Algorithm 8.7 are the FFTs described in Sections 8.5–8.6. Since the factors $\boldsymbol{A}^{(s)}$ are sparse, the complexity of Algorithm 8.7 will be low. In many applications, the entries $a_{j,k}^{(s)}$ of $\boldsymbol{A}^{(s)}$ must be precomputed. If the precomputed value of $a_{j,k}^{(s)}$ is denoted by $\hat{a}_{j,k}^{(s)}$, then $\hat{\boldsymbol{A}}^{(s)} := \left(\hat{a}_{j,k}^{(s)}\right)_{j,k=0}^{n-1}$. The computed intermediary results of Algorithm 8.7 are $\hat{\boldsymbol{x}}^{(s)} := \mathrm{fl}\left(\hat{\boldsymbol{A}}^{(s)}\hat{\boldsymbol{x}}^{(s-1)}\right)$ $(s = 1, \dots, t)$ with $\hat{\boldsymbol{x}}^{(0)} := \boldsymbol{x}$. In this section we are interested in normwise roundoff error estimates for Algorithm 8.7.

In the following, we introduce the notations $|\boldsymbol{A}| := (|a_{j,k}|)_{j,k=0}^{n-1}$ for $\boldsymbol{A} = (a_{j,k})_{j,k=0}^{n-1} \in \mathbb{C}^{n\times n}$ and $|\boldsymbol{x}| := (|x_j|)_{j=0}^{n-1}$ for $\boldsymbol{x} = (x_j)_{j=0}^{n-1} \in \mathbb{C}^n$.

LEMMA 8.4
Let $\boldsymbol{A} = (a_{j,k})_{j,k=0}^{n-1} \in \mathbb{C}^{n\times n}$ and $\boldsymbol{x} \in \mathbb{C}^n$. Assume that each row of $\boldsymbol{A}$ contains at most λ non-zero entries and that the precomputed values $\hat{a}_{j,k}$ satisfy

$$|\hat{a}_{j,k} - a_{j,k}| \le \eta u \qquad (j, k = 0, \dots, n-1). \tag{8.40}$$

with $\eta > 0$. Define $\tilde{\boldsymbol{A}} = (\tilde{a}_{j,k})_{j,k=0}^{n-1}$ with

$$\tilde{a}_{j,k} := \begin{cases} 1 & \text{if } a_{j,k} \notin \{0, \pm 1, \pm \mathrm{i}\}, \\ 0 & \text{otherwise.} \end{cases} \tag{8.41}$$

Then the error vector $\Delta\boldsymbol{y} := \mathrm{fl}(\hat{\boldsymbol{A}}\boldsymbol{x}) - \boldsymbol{A}\boldsymbol{x}$ satisfies the componentwise inequality

$$|\Delta\boldsymbol{y}| \le \left((\lambda - 1 + \mu_{\mathbb{C}})u + \mathcal{O}(u^2)\right)|\boldsymbol{A}|\,|\boldsymbol{x}| + \eta\tilde{\boldsymbol{A}}|\boldsymbol{x}|(u + \mathcal{O}(u^2))$$

for sequential summation,

$$|\Delta\boldsymbol{y}| \le \left((\lceil \log_2 \lambda \rceil + \mu_{\mathbb{C}})u + \mathcal{O}(u^2)\right)|\boldsymbol{A}|\,|\boldsymbol{x}| + \eta\tilde{\boldsymbol{A}}|\boldsymbol{x}|(u + \mathcal{O}(u^2))$$

for cascade summation, and

$$|\Delta\boldsymbol{y}| \le |\boldsymbol{A}\boldsymbol{x}|u + (\mu_{\mathbb{C}}u + \mathcal{O}(u^2)|\boldsymbol{A}|\,|\boldsymbol{x}| + \eta\tilde{\boldsymbol{A}}|\boldsymbol{x}|(u + \mathcal{O}(u^2))$$

for $\lambda = 2$.

PROOF First we observe that by (8.40)–(8.41)

$$|\Delta \boldsymbol{y}| \leq |\operatorname{fl}(\hat{\boldsymbol{A}}\boldsymbol{x}) - \hat{\boldsymbol{A}}\boldsymbol{x}| + |\hat{\boldsymbol{A}}\boldsymbol{x} - \boldsymbol{A}\boldsymbol{x}| \leq |\operatorname{fl}(\hat{\boldsymbol{A}}\boldsymbol{x}) - \hat{\boldsymbol{A}}\boldsymbol{x}| + |\hat{\boldsymbol{A}} - \boldsymbol{A}|\,|\boldsymbol{x}| \\ \leq |\operatorname{fl}(\hat{\boldsymbol{A}}\boldsymbol{x}) - \hat{\boldsymbol{A}}\boldsymbol{x}| + \eta\tilde{\boldsymbol{A}}u|\boldsymbol{x}|. \tag{8.42}$$

We will now prove only the case of cascade summation. Consider the j–th component of $\operatorname{fl}(\hat{\boldsymbol{A}}\boldsymbol{x})$. At the first stage of the cascade summation, terms

$$\operatorname{fl}(\operatorname{fl}(\hat{a}_{j,2k}x_{2k}) + \operatorname{fl}(\hat{a}_{j,2k+1}x_{2k+1})) \\ = \big(\hat{a}_{j,2k}x_{2k}(1+\varepsilon_{2k}^{\times}) + \hat{a}_{j,2k+1}x_{2k+1}(1+\varepsilon_{2k+1}^{\times})\big)(1+\varepsilon_k^{+}) \tag{8.43}$$

with $|\varepsilon_k^{\times}| \leq \mu_{\mathbb{C}}u$ and $|\varepsilon_k^{+}| \leq u$ (see Lemma 8.1) are computed. In total, the roundoff error of stage 1 is bounded by

$$\big((\mu_{\mathbb{C}}+1)u + \mathcal{O}(u^2)\big)|\hat{\boldsymbol{a}}_j^{\mathrm{T}}|\,|\boldsymbol{x}|,$$

where $\hat{\boldsymbol{a}}_j$ denotes the j–th row of $\hat{\boldsymbol{A}}$. At the subsequent stages of cascade summation, only additions occur. Therefore the roundoff error at each such stage is bounded by

$$(u + \mathcal{O}(u^2))|\hat{\boldsymbol{a}}_j^{\mathrm{T}}|\,|\boldsymbol{x}|.$$

Since the number of stages is at most $\lceil \log_2 \lambda \rceil$, the arithmetic roundoff error of the j–th component of $\operatorname{fl}(\hat{\boldsymbol{A}}\boldsymbol{x}) - \boldsymbol{A}\boldsymbol{x}$ is bounded by

$$\big((\mu_{\mathbb{C}} + \lceil \log_2 \lambda \rceil)u + \mathcal{O}(u^2)\big)|\hat{\boldsymbol{a}}_j^{\mathrm{T}}|\,|\boldsymbol{x}|.$$

Consequently,

$$|\operatorname{fl}(\hat{\boldsymbol{A}}\boldsymbol{x}) - \hat{\boldsymbol{A}}\boldsymbol{x}| \leq \big((\mu_{\mathbb{C}} + \lceil \log_2 \lambda \rceil)u + \mathcal{O}(u^2)\big)|\hat{\boldsymbol{A}}|\,|\boldsymbol{x}|.$$

By (8.40)–(8.41), we conclude that $|\hat{\boldsymbol{A}}| \leq |\boldsymbol{A}| + \eta u\tilde{\boldsymbol{A}}$ and hence

$$|\operatorname{fl}(\hat{\boldsymbol{A}}\boldsymbol{x}) - \hat{\boldsymbol{A}}\boldsymbol{x}| \leq \big((\mu_{\mathbb{C}} + \lceil \log_2 \lambda \rceil)u + \mathcal{O}(u^2)\big)|\boldsymbol{A}|\,|\boldsymbol{x}| + \eta\tilde{\boldsymbol{A}}|\boldsymbol{x}|\mathcal{O}(u^2). \tag{8.44}$$

The assertion now follows from (8.40) and (8.44). Note that by (8.40) and (8.43), $\operatorname{fl}(\hat{\boldsymbol{A}}\boldsymbol{x})$ can be represented in the form (see [6], p. 76)

$$\operatorname{fl}(\hat{\boldsymbol{A}}\boldsymbol{x}) = (\boldsymbol{A} + \Delta\boldsymbol{A})\boldsymbol{x},$$

where $\Delta\boldsymbol{A} \in \mathbb{C}^{n\times n}$ satisfies the componentwise inequality

$$|\Delta\boldsymbol{A}| \leq \big((\mu_{\mathbb{C}} + \lceil \log_2 \lambda \rceil)u + \mathcal{O}(u^2)\big)|\boldsymbol{A}| + \eta\tilde{\boldsymbol{A}}(u + \mathcal{O}(u^2))$$

in the case of cascade summation. ∎

THEOREM 8.3 (see [13])

Let $\boldsymbol{A} \in \mathbb{C}^{n\times n}$ *be a unitary matrix with a factorization* (8.39), *where* $\boldsymbol{A}^{(s)} = (a_{j,k}^{(s)})_{j,k=0}^{n-1} \in \mathbb{C}^{n\times n}$ $(s = 1, \ldots, t)$ *are almost unitary (with* $\alpha_s := \alpha(\boldsymbol{A}^{(s)})$*) and block–diagonal (up to some permutation) with blocks of size* $\leq \lambda_s$. *Assume that in each row and in each column at least* κ_s *of the nonzero entries are in* $\{\pm 1, \pm \mathrm{i}\}$ *and that all entries* $a_{j,k}^{(s)} \notin \{0, \pm 1, \pm \mathrm{i}\}$ *are precomputed with* $|\hat{a}_{j,k}^{(s)} - a_{j,k}^{(s)}| \leq \eta_s u$ $(\eta_s > 0)$.

Then for arbitrary $\boldsymbol{x} \in \mathbb{C}^n$, *the backward error vector*

$$\Delta \boldsymbol{x} := \boldsymbol{A}^{-1}\left(\mathrm{fl}(\widehat{\boldsymbol{A}}^{(t)}\,\mathrm{fl}(\ldots\widehat{\boldsymbol{A}}^{(2)}\,\mathrm{fl}(\widehat{\boldsymbol{A}}^{(1)}\boldsymbol{x}))) - \boldsymbol{A}\boldsymbol{x}\right)$$

satisfies

$$\|\Delta \boldsymbol{x}\|_2 \leq \left(k_n u + \mathcal{O}(u^2)\right)\|\boldsymbol{x}\|_2$$

with

$$k_n := \sum_{s=1}^{t}\left(\zeta_s + \sqrt{\lambda_s}\,\mu_{\mathbb{C}} + \frac{\eta_s}{\alpha_s}(\lambda_s - \kappa_s)\right), \tag{8.45}$$

$$\zeta_s := \begin{cases} (\lambda_s - 1)\sqrt{\lambda_s} & \textit{for sequential summation,} \\ \lceil \log_2 \lambda_s \rceil \sqrt{\lambda_s} & \textit{for cascade summation,} \\ 1 & \textit{for } \lambda_s = 2. \end{cases}$$

PROOF The assertion will be proved in the case of cascade summation. Set $\hat{\boldsymbol{x}}^{(0)} := \boldsymbol{x}$ and $\hat{\boldsymbol{x}}^{(s)} := \mathrm{fl}(\hat{\boldsymbol{A}}^{(s)}\hat{\boldsymbol{x}}^{(s-1)})$ $(s = 1, \ldots, t)$ according to Algorithm 8.7. Let $\Delta \boldsymbol{x}^{(s)} := \mathrm{fl}(\hat{\boldsymbol{A}}^{(s)}\hat{\boldsymbol{x}}^{(s-1)}) - \boldsymbol{A}^{(s)}\hat{\boldsymbol{x}}^{(s-1)}$ denote the error vector of step s. Then it follows from Lemma 8.4 that

$$\begin{aligned}|\Delta \boldsymbol{x}^{(s)}| \leq \left((\lceil \log_2 \lambda_s \rceil + \mu_{\mathbb{C}})u + \mathcal{O}(u^2)\right)|\boldsymbol{A}^{(s)}|\,|\hat{\boldsymbol{x}}^{(s-1)}| \\ + \eta_s \tilde{\boldsymbol{A}}^{(s)}|\hat{\boldsymbol{x}}^{(s-1)}|\,(u + \mathcal{O}(u^2)).\end{aligned} \tag{8.46}$$

For every block $\boldsymbol{B}$ of the block–diagonal matrix $|\boldsymbol{A}^{(s)}|$, the spectral norm $\|\boldsymbol{B}\|_2$ is bounded by the Frobenius norm $\|\boldsymbol{B}\|_F$. Hence we get

$$\|\boldsymbol{B}\|_2 \leq \|\boldsymbol{B}\|_F \leq \alpha_s\sqrt{\lambda_s},$$

since $\boldsymbol{B}$ has at most λ_s rows and since for every row we have

$$\sum_k b_{j,k}^2 = \sum_{k=0}^{n-1} |a_{j,k}^{(s)}|^2 = \alpha_s^2.$$

Then the spectral norm of $|\boldsymbol{A}^{(s)}|$ is the maximum over the norms of all blocks $\boldsymbol{B}$ of $|\boldsymbol{A}^{(s)}|$ and such that $\|\,|\boldsymbol{A}^{(s)}|\,\|_2 \leq \alpha_s\sqrt{\lambda_s}$ and $\|\,|\boldsymbol{A}^{(s)}|\,|\hat{\boldsymbol{x}}^{(s-1)}|\,\|_2 \leq \alpha_s\sqrt{\lambda_s}\,\|\hat{\boldsymbol{x}}^{(s-1)}\|_2$.

By our assumption, we know that $\|\tilde{\boldsymbol{A}}^{(s)}\|_1, \|\tilde{\boldsymbol{A}}^{(s)}\|_\infty \le \lambda_s - \kappa_s$. Using the known estimate (see [6], p. 125)

$$\|\tilde{\boldsymbol{A}}^{(s)}\|_2 \le \sqrt{\|\tilde{\boldsymbol{A}}^{(s)}\|_1 \|\tilde{\boldsymbol{A}}^{(s)}\|_\infty},$$

we obtain by (8.46) that

$$\begin{aligned} \|\Delta \boldsymbol{x}^{(s)}\|_2 &\le \big((\lceil \log_2 \lambda_s \rceil + \mu_c)u + \mathcal{O}(u^2)\big) \|\,|\boldsymbol{A}^{(s)}|\,\|_2 \|\hat{\boldsymbol{x}}^{(s-1)}\|_2 \\ &\qquad + \eta_s \|\tilde{\boldsymbol{A}}^{(s)}\|_2 \|\hat{\boldsymbol{x}}^{(s-1)}\|_2 \\ &\le \big(\zeta_s \alpha_s + \mu_c \alpha_s \sqrt{\lambda_s} + \eta_s(\lambda_s - \kappa_s) + \mathcal{O}(u)\big) \|\hat{\boldsymbol{x}}^{(s-1)}\|_2 u. \end{aligned} \tag{8.47}$$

From $\|\boldsymbol{x}^{(s-1)}\|_2 = \alpha_1 \dots \alpha_{s-1} \|\boldsymbol{x}\|_2$ and

$$\|\hat{\boldsymbol{x}}^{(s-1)}\|_2 \le \|\boldsymbol{x}^{(s-1)}\|_2 + \|\hat{\boldsymbol{x}}^{(s-1)} - \boldsymbol{x}^{(s-1)}\|_2,$$

it follows by Lemma 8.4 that

$$\|\hat{\boldsymbol{x}}^{(s-1)}\|_2 \le (\alpha_1 \dots \alpha_{s-1} + \mathcal{O}(u)) \|\boldsymbol{x}\|_2 \qquad (s = 2, \dots, t). \tag{8.48}$$

Using telescope summation, we get

$$\begin{aligned} \Delta \boldsymbol{x} &= \boldsymbol{A}^{-1} \sum_{s=1}^{t-1} \boldsymbol{A}^{(t)} \dots \boldsymbol{A}^{(s+1)} \Delta \hat{\boldsymbol{x}}^{(s)} + \boldsymbol{A}^{-1} \Delta \hat{\boldsymbol{x}}^{(t)} \\ &= \sum_{s=1}^{t} (\boldsymbol{A}^{(1)})^{-1} \dots (\boldsymbol{A}^{(s)})^{-1} \Delta \hat{\boldsymbol{x}}^{(s)} \end{aligned}$$

and hence by (8.47)–(8.48)

$$\begin{aligned} \|\Delta \boldsymbol{x}\|_2 &\le \sum_{s=1}^{t} \alpha_1^{-1} \dots \alpha_s^{-1} \|\Delta \hat{\boldsymbol{x}}^{(s)}\|_2 \\ &\le \sum_{s=1}^{t} \alpha_1^{-1} \dots \alpha_s^{-1} \big(\zeta_s \alpha_s + \mu_c \alpha_s \sqrt{\lambda_s} + \eta_s(\lambda_s - \kappa_s) + \mathcal{O}(u)\big) \times \\ &\qquad\qquad \times\ (\alpha_1 \dots \alpha_{s-1} + \mathcal{O}(u)) \|\boldsymbol{x}\|_2 u \\ &\le \sum_{s=1}^{t} \big(\zeta_s + \mu_c \sqrt{\lambda_s} + \frac{\eta_s}{\alpha_s}(\lambda_s - \kappa_s) + \mathcal{O}(u)\big) \|\boldsymbol{x}\|_2 u. \end{aligned}$$

This completes the proof. ∎

In many applications, another representation of the error vector $\Delta \boldsymbol{x}$ is useful:

COROLLARY 8.2
Under the assumptions of Theorem 8.3, the computed vector

$$\hat{\boldsymbol{y}} := \mathrm{fl}(\hat{\boldsymbol{A}}^{(t)}\,\mathrm{fl}(\ldots\hat{\boldsymbol{A}}^{(2)}\,\mathrm{fl}(\hat{\boldsymbol{A}}^{(1)}\boldsymbol{x})))$$

can be represented in the form $\hat{\boldsymbol{y}} = (\boldsymbol{A} + \Delta\boldsymbol{A})\boldsymbol{x}$ *with* $\Delta\boldsymbol{A} \in \mathbb{C}^{n\times n}$ *and*

$$\|\Delta\boldsymbol{A}\|_2 \le k_n u + \mathcal{O}(u^2).$$

PROOF By Lemma 8.4 or [6], p. 76, we know that for $\boldsymbol{z} \in \mathbb{C}^n$,

$$\mathrm{fl}(\hat{\boldsymbol{A}}^{(s)}\boldsymbol{z}) = (\boldsymbol{A}^{(s)} + \boldsymbol{E}^{(s)})\boldsymbol{z} \qquad (s = 1, \ldots, t)$$

with some "error matrix" $\boldsymbol{E}^{(s)} \in \mathbb{C}^{n\times n}$. Hence we obtain

$$\hat{\boldsymbol{y}} = (\boldsymbol{A}^{(t)} + \boldsymbol{E}^{(t)})\ldots(\boldsymbol{A}^{(1)} + \boldsymbol{E}^{(1)})\boldsymbol{x} = (\boldsymbol{A} + \Delta\boldsymbol{A})\boldsymbol{x}$$

with $\Delta\boldsymbol{A} \in \mathbb{C}^{n\times n}$. Since $\boldsymbol{A}$ is unitary, we can estimate by Theorem 8.3

$$\begin{aligned}\|\Delta\boldsymbol{A}\boldsymbol{x}\|_2 &= \|\hat{\boldsymbol{y}} - \boldsymbol{A}\boldsymbol{x}\|_2 = \|\boldsymbol{A}\Delta\boldsymbol{x}\|_2 \le \|\boldsymbol{A}\|_2\|\Delta\boldsymbol{x}\|_2 = \|\Delta\boldsymbol{x}\|_2\\ &\le (k_n u + \mathcal{O}(u^2))\|\boldsymbol{x}\|_2.\end{aligned}$$

This yields the assertion. ∎

REMARK 8.4 For any matrix $\boldsymbol{A}^{(s)}$ in the above factorization with only real entries we may replace the constant $\mu_{\mathbb{C}}$ (introduced in Lemma 8.1) by 1. If all entries of $\boldsymbol{A}^{(s)}$ are in $\{0, \pm 1, \pm \mathrm{i},\}$ (i.e. $\kappa_s = \lambda_s$), then we may replace $\mu_{\mathbb{C}}$ by 0. ∎

Example 8.1
For a diagonal almost unitary matrix $\boldsymbol{A}^{(s)}$, the corresponding term in the definition of k_n is $\mu_{\mathbb{C}} + \frac{\eta_s}{\alpha_s}$. If $\boldsymbol{A}^{(s)}$ is a radix–2 butterfly matrix, then its contribution to k_n is 1. For a block–diagonal matrix, where each block is a real rotation or reflection 2×2–matrix, it is $1 + \sqrt{2} + 2\eta_s$. □

We will now present the corresponding result for the average case analysis.

THEOREM 8.4
Let $\boldsymbol{A} \in \mathbb{C}^{n\times n}$ *be a unitary matrix with a factorization* (8.39), *where* $\boldsymbol{A}^{(s)}$ $(s = 1, \ldots, t)$ *are almost unitary (with* $\alpha_s := \alpha(\boldsymbol{A}^{(s)})$*) and have* $2^{\nu_{s,j}}$ *non–zero entries in the* j*–th row, of which* $\kappa_{s,j}$ *are in* $\{\pm 1, \pm \mathrm{i}\}$. *Assume that the*

non-trivial entries in row j of $A^{(s)}$ have precomputation errors of variance $\sigma^2_{s,j,k}u^2$ $(k = 1, \ldots, 2^{\nu_{s,j}} - \kappa_{s,j})$. Furthermore, let the data vector X have uncorrelated components of equal variances. If the summation is done by the cascade method, then the backward error satisfies

$$\mathbb{E}(\|\Delta X\|_2^2) = (\bar{k}_n^2 u^2 + \mathcal{O}(u^3))\,\mathbb{E}(\|X\|_2^2)$$

with

$$\bar{k}_n^2 := t\sigma^2_{\mathbb{C},\times} + \sum_{s=1}^{t}\left(\bar{\nu}_s\sigma^2_{\mathbb{C},+} + \alpha_s^{-2}(\bar{\sigma}_s^2 - \bar{\kappa}_s\sigma^2_{\mathbb{C},\times})\right), \tag{8.49}$$

where

$$\bar{\nu}_s := \frac{1}{n}\sum_j \nu_{s,j}, \qquad \bar{\kappa}_s := \frac{1}{n}\sum_j \kappa_{s,j}, \qquad \bar{\sigma}_s^2 := \frac{1}{n}\sum_{j,k}\sigma^2_{s,j,k}.$$

PROOF If ρ^2 denotes the variance of the components of X, then $\mathbb{E}(\|X\|_2^2) = n\rho^2$ and the intermediary result $X^{(s)} := A^{(s)}X^{(s-1)}$ after the s-th step of Algorithm 8.7 has uncorrelated components of variance $\rho_s^2 := \rho^2\prod_{r=1}^{s}\alpha_r^2 = \frac{1}{n}\mathbb{E}(\|X\|_2^2)\prod_{r=1}^{s}\alpha_r^2$.

Let $\Delta X^{(s)}$ be the error vector from step s. Then the j-th component of $\Delta X^{(s)}$ can be decomposed into three parts. From the precomputations we obtain an error of variance $\rho_{s-1}^2\sum_k\sigma^2_{s,j,k}$. Since

$$\sum_{\substack{k \\ a^{(s)}_{j,k}\notin\{\pm 1,\pm \mathrm{i}\}}} |a^{(s)}_{j,k}|^2 = \alpha_s^2 - \kappa_{s,j},$$

the multiplications with the non-trivial entries generate a roundoff error of total variance $\sigma^2_{\mathbb{C},\times}(\alpha_s^2 - \kappa_{s,j})\rho_{s-1}^2$. Finally the summation is the origin of an error of variance $\nu_{s,j}\sigma^2_{\mathbb{C},+}\alpha_s^2\rho_s^2$. By our model assumptions, these are all independent. We obtain

$$\begin{aligned}\mathbb{E}(\|\Delta X^{(s)}\|_2^2) &= \sum_{j=0}^{n-1}\Big(\sum_k\sigma^2_{s,j,k} + \sigma^2_{\mathbb{C},\times}(\alpha_s^2 - \kappa_{s,j}) + \nu_{s,j}\sigma^2_{\mathbb{C},+}\alpha_s^2 + \mathcal{O}(u^3)\Big)\rho_{s-1}^2 \\ &= \Big(\bar{\sigma}_s^2 + \sigma^2_{\mathbb{C},\times}(\alpha_s^2 - \bar{\kappa}_s) + \bar{\nu}_s\sigma^2_{\mathbb{C},+}\alpha_s^2 + \mathcal{O}(u^3)\Big)\,\mathbb{E}(\|X\|_2^2)\prod_{r=1}^{s-1}\alpha_r^2.\end{aligned}$$

Since $\mathbb{E}(\|(A^{(1)})^{-1}\ldots(A^{(s)})^{-1}\Delta X^{(s)}\|_2^2) = \alpha_1^{-2}\ldots\alpha_s^{-2}\,\mathbb{E}(\|\Delta X^{(s)}\|_2^2)$, the proof now proceeds like the end of the proof of Theorem 8.3. ∎

The above theorem gives no information on single components of the total roundoff error vector ΔX. In general, the variances of the components will not be equal. Although each $\Delta \boldsymbol{X}^{(s)}$ has uncorrelated components, this need not be true for the total roundoff error $\Delta \boldsymbol{X}$.

REMARK 8.5 If in Theorem 8.4 the number $\lambda_{s,j}$ of non–zero entries in row j of $\boldsymbol{A}^{(s)}$ is *not* a power of 2, we need the additional assumption that they all have equal modulus. Then it has be shown in [19] that for the optimal selection of the unmatched terms in the cascade summation the above result holds with $\bar{\nu}_s$ replaced by

$$\bar{\nu}_s := \frac{1}{n} \sum_j \left(\lceil \log_2 \lambda_{s,j} \rceil + 1 - 2^{\lceil \log_2 \lambda_{s,j} \rceil - \log_2 \lambda_{s,j}} \right).$$

Clearly, Theorem 8.5 and its generalization also hold for orthogonal matrices, if we replace $\sigma_{\mathbb{C},+}$ and $\sigma_{\mathbb{C},\times}$ by $\sigma_{\mathbb{R},+}$ and $\sigma_{\mathbb{R},\times}$. ∎

REMARK 8.6 In many practical situations, the matrices $\boldsymbol{A}^{(s)}$ of Theorems 8.3 and 8.4 are constructed using smaller building blocks, either as the direct sum $\boldsymbol{A} \oplus \boldsymbol{B}$ or the Kronecker product $\boldsymbol{A} \otimes \boldsymbol{B}$ of almost unitary matrices $\boldsymbol{A} \in \mathbb{C}^{l \times l}$ and $\boldsymbol{B} \in \mathbb{C}^{m \times m}$. For the corresponding constants λ, κ, η, $\bar{\nu}$, $\bar{\kappa}$ and $\bar{\sigma}^2$ used in these results, we obtain the formulas

$$\begin{aligned}
\lambda(\boldsymbol{A} \oplus \boldsymbol{B}) &= \max(\lambda(\boldsymbol{A}), \lambda(\boldsymbol{B})), & \bar{\nu}(\boldsymbol{A} \oplus \boldsymbol{B}) &= \frac{l\bar{\nu}(\boldsymbol{A}) + m\bar{\nu}(\boldsymbol{B})}{m+l}, \\
\kappa(\boldsymbol{A} \oplus \boldsymbol{B}) &= \min(\kappa(\boldsymbol{A}), \kappa(\boldsymbol{B})), & \bar{\kappa}(\boldsymbol{A} \oplus \boldsymbol{B}) &= \frac{l\bar{\kappa}(\boldsymbol{A}) + m\bar{\kappa}(\boldsymbol{B})}{m+l}, \\
\eta(\boldsymbol{A} \oplus \boldsymbol{B}) &= \max(\eta(\boldsymbol{A}), \eta(\boldsymbol{B})), & \bar{\sigma}^2(\boldsymbol{A} \oplus \boldsymbol{B}) &= \frac{l\bar{\sigma}^2(\boldsymbol{A}) + m\bar{\sigma}^2(\boldsymbol{B})}{m+l}
\end{aligned}$$

if $\alpha(\boldsymbol{A}) = \alpha(\boldsymbol{B})$, as well as

$$\begin{aligned}
\lambda(\boldsymbol{A} \otimes \boldsymbol{P}) = \lambda(\boldsymbol{P} \otimes \boldsymbol{A}) &= \lambda(\boldsymbol{A}), & \bar{\nu}(\boldsymbol{A} \otimes \boldsymbol{P}) = \bar{\nu}(\boldsymbol{P} \otimes \boldsymbol{A}) &= \bar{\nu}(\boldsymbol{A}), \\
\kappa(\boldsymbol{A} \otimes \boldsymbol{P}) = \kappa(\boldsymbol{P} \otimes \boldsymbol{A}) &= \kappa(\boldsymbol{A}), & \bar{\kappa}(\boldsymbol{A} \otimes \boldsymbol{P}) = \bar{\kappa}(\boldsymbol{P} \otimes \boldsymbol{A}) &= \bar{\kappa}(\boldsymbol{A}), \\
\eta(\boldsymbol{A} \otimes \boldsymbol{P}) = \eta(\boldsymbol{P} \otimes \boldsymbol{A}) &= \eta(\boldsymbol{A}), & \bar{\sigma}^2(\boldsymbol{A} \otimes \boldsymbol{P}) = \bar{\sigma}^2(\boldsymbol{P} \otimes \boldsymbol{A}) &= \bar{\sigma}^2(\boldsymbol{A})
\end{aligned}$$

if $\boldsymbol{P}$ is a permutation matrix. ∎

REMARK 8.7 For a factorization (8.39) of a unitary matrix $\boldsymbol{A} \in \mathbb{C}^{n \times n}$, the computation of $\boldsymbol{A}^* \boldsymbol{x}$ ($\boldsymbol{x} \in \mathbb{C}^n$) can be based on

$$\boldsymbol{A}^* = \left(\boldsymbol{A}^{(1)}\right)^* \left(\boldsymbol{A}^{(2)}\right)^* \ldots \left(\boldsymbol{A}^{(t)}\right)^*.$$

Then the assertion of Theorems 8.3 holds for the computation of $\boldsymbol{A}^* \boldsymbol{x}$ with the *same* constants k_n. Theorem 8.4 also holds with the same $\bar{k}_n$ if the matrices $\boldsymbol{A}^{(s)}$ are block–diagonal (up to permutation). ∎

8.8 Error analysis for FFT

Let $\boldsymbol{A} \in \mathbb{C}^{n\times n}$ be a unitary matrix with a factorization (8.39), where $\boldsymbol{A}^{(s)} \in \mathbb{C}^{n\times n}$ $(s = 1, \ldots t)$ are almost unitary with $\alpha_s = \alpha(\boldsymbol{A}^{(s)})$ and block-diagonal (up to some permutation) with blocks of size $\leq \lambda_s = \lambda(\boldsymbol{A}^{(s)})$. Assume that in each row and in each column, at least $\kappa_s = \kappa(\boldsymbol{A}^{(s)})$ of the non-zero entries are in $\{\pm 1, \pm \mathrm{i}\}$ and that all entries $a_{j,k}^{(s)} \notin \{0, \pm 1, \pm \mathrm{i}\}$ are precomputed with $|\hat{a}_{j,k}^{(s)} - a_{j,k}^{(s)}| \leq \eta_s u$ and $\eta_s = \eta(\boldsymbol{A}^{(s)}) > 0$. By $\bar{\nu}_s = \bar{\nu}(\boldsymbol{A}^{(s)})$, $\bar{\kappa}_s = \bar{\kappa}(\boldsymbol{A}^{(s)})$ and $\bar{\sigma}_s^2 = \bar{\sigma}^2(\boldsymbol{A}^{(s)})$ we denote the average case constants introduced in Theorem 8.4.

In this section, we choose $\boldsymbol{A} = \boldsymbol{F}_n$ with $n = p^t$ $(p, t \geq 2)$. In Section 8.9, $\boldsymbol{A}$ will be an orthogonal cosine/sine matrix. Now we derive the corresponding stability constants k_n and $\bar{k}_n$ by means of a factorization of $\boldsymbol{F}_n$ (see Sections 8.4–8.6) and Theorems 8.3–8.4. First we summarize the bounds of the precomputation errors $\delta_l = \hat{\omega}_n^l - \omega_n^l$ $(l = 0, \ldots, n)$ by one constant. Let $n = p^t$ and $n_s := n/p^s$ $(s = 1, \ldots, t)$. Note that with the notations of Section 8.3 we have $\Delta_{s,k} = \delta_{kn_s}$. Then

$$c_{n,s} := \max\left\{\frac{|\delta_{kn_s}|}{u} : 0 < kn_s < n\right\} \qquad (1 \leq s \leq t) \tag{8.50}$$

(or an upper bound thereof) contains the information needed to determine the constant η_s in Theorem 8.3. All these constants are relative to the machine accuracy u. Using the symmetry properties of the roots of unity, it is sufficient to take the maximum for $kn_s \leq \lfloor \frac{n-1}{2} \rfloor$ only. If n is a multiple of 4, even $kn_s \leq \lfloor \frac{n}{8} \rfloor$ are enough.

Likewise, the average of the non-zero variances

$$\bar{c}_{n,s}^2 := \frac{1}{p^s - 1} \sum_{k=1}^{p^s-1} \frac{\mathbb{V}(\delta_{kn_s})}{u^2} \qquad (1 \leq s \leq t) \tag{8.51}$$

is essential to determine the constant $\bar{\sigma}_s^2$ in Theorem 8.4. The same simplification by the use of symmetries applies here. In some applications, the more general numbers

$$\bar{c}_{n,s}^2(m) := \frac{1}{\lceil m/n_s \rceil u^2} \sum_{k=1}^{\lceil m/n_s \rceil} \mathbb{V}(\delta_{kn_s}) \qquad (1 \leq m < p^t, s = 1, \ldots, t)$$

are needed. Clearly, all these constants reflect the method used for the precomputation of ω_n^k.

Example 8.2 (Precomputation of twiddle factors)
For the method of direct call (Algorithm 8.1) we obviously have $c_{n,s} = \frac{\sqrt{2}}{2}$ and $\bar{c}_{n,s}^2 = \frac{1}{6}$ ($s = 3, \ldots, t$ for $p = 2$, resp. $s = 1, \ldots, t$ for p odd). Repeated multiplication (Algorithm 8.2) is described by the constants

$$c_{n,s} = \frac{1}{8}\left(\mu_{\mathbb{C}} + \frac{\sqrt{2}}{2}\right) n, \qquad \bar{c}_{n,s}^2 = \frac{1}{2^{s-3}} \sum_{k=1}^{2^{s-3}} \frac{\mathbb{V}(\delta_{kn_s})}{u^2} = \frac{1}{2304} n^2 + \mathcal{O}(n)$$

($s = 1, \ldots, t,\ p = 2$). For repeated subvector scaling we conclude from Algorithm 8.3 that

$$c_{n,3} = \frac{1}{2}\sqrt{2}, \qquad c_{n,s} = \left(\mu_{\mathbb{C}} + \frac{1}{2}\sqrt{2}\right)(s-3),$$

$$\bar{c}_{n,3}^2 = \frac{1}{6}, \qquad \bar{c}_{n,s}^2 = \frac{s-2}{24} + \frac{s-4}{2}\sigma_{\mathbb{C}\times}^2 \qquad (s = 4, \ldots, t,\ p = 2).$$

For the radix–p variant, Algorithm 8.4, we study the slightly more general situation that in step 1 an *arbitrary* precomputation method may be used for $\omega_n^{jp^s}$. If c_p resp. $\bar{c}_p^2$ denote the maximum error resp. the average error variance for this precomputation method, then $c_{n,s} \le s\left((p-1)\mu_{\mathbb{C}} + c_p\right)$ and asymptotically $\bar{c}_{n,s}^2 \sim s\left(\frac{p-1}{2}\sigma_{\mathbb{C}\times}^2 + \bar{c}_p^2\right)$ ($s = 1, \ldots, t$). In particular, if direct call is used in step 1, then we have $c_p = \frac{\sqrt{2}}{2}$ and $\bar{c}_p^2 = \frac{1}{6}$. □

Example 8.3 (Direct computation of $\boldsymbol{F}_n\boldsymbol{x}$)
For small n, the DFT $\boldsymbol{y} = \boldsymbol{F}_n\,\boldsymbol{x}$ can be computed directly by one matrix–vector multiplication and one scaling. Since we know that in each row of $\boldsymbol{A}^{(1)} := \left(\omega_n^{jk}\right)_{j,k=0}^{n-1}$ the entries with $k = 0$ are 1, we have $\kappa_1 = 1$, $\lambda_1 = n$ and $\alpha_1 = \sqrt{n}$. By (8.50) we can choose $\eta_1 = c_{n,t}$. Furthermore, for $\boldsymbol{A}^{(2)} := n^{-1/2}\boldsymbol{I}_n$ the precomputation error is $\le \eta_2 u = n^{-1/2}u$. It is now easy to determine the worst case constant

$$\begin{aligned} k_n &= \sum_{s=1}^{2}\left((\lambda_s - 1 + \mu_{\mathbb{C}})\sqrt{\lambda_s} + (\lambda_s - \kappa_s)\frac{\eta_s}{\alpha_s}\right) \\ &= (n - 1 + \mu_{\mathbb{C}})\sqrt{n} + \frac{n-1}{\sqrt{n}}c_{n,t} + 2 \\ &= n^{3/2} + c_{n,t}\sqrt{n} + \mathcal{O}(\sqrt{n}) \end{aligned}$$

in the case of sequential summation. For a cascade summation we have $k_n = \sqrt{n}\log_2 n + c_{n,t}\sqrt{n} + \mathcal{O}(\sqrt{n})$.

In the average case analysis for cascade summation, we obtain $\bar{\kappa}_1 = 2 - \frac{1}{n}$, since the first row and column of $\boldsymbol{A}^{(1)}$ are trivial, and $\bar{\nu}_1 = \lceil \log_2 n \rceil +$

$1 - 2^{\lceil \log_2 n \rceil - \log_2 n}$. Since there are $(n-1)^2$ non-trivial entries, we have $\bar{\sigma}_1^2 = \frac{(n-1)^2}{n}\bar{c}_{n,t}^2$, and since the precomputation error for $n^{-1/2}$ has variance $\frac{u^2}{3n}$, we get $\bar{\sigma}_2^2 = \frac{1}{3n}$. The scaling factor $n^{-1/2}$ is a real number and hence from Theorem 8.4 and Remark 8.5 we obtain

$$\begin{aligned}\bar{k}_n^2 &= \sigma_{\mathbb{C},\times}^2 + \sigma_{\mathbb{R},\times}^2 + \bar{\nu}_1\sigma_{\mathbb{C},+}^2 + \frac{1}{n}\left(\bar{\sigma}_1^2 - \bar{\kappa}_1\sigma_{\mathbb{C},\times}^2\right) + n\,\frac{1}{3n}\\ &= \sigma_{\mathbb{C},+}^2\log_2 n + \bar{c}_{n,t}^2 + \mathcal{O}(1)\end{aligned}$$

as long as $\bar{c}_{n,t}^2 = \mathcal{O}(n)$, which is true in all of our precomputation methods except repeated multiplication. $\square$

In the subsequent examples we will neglect the final scaling by $n^{-1/2}$, since in many applications it combines with a similar factor in the inverse transform.

Example 8.4 (Decimation–in–time FFT)
This algorithm is based on the decimation–in–time factorization (8.26). Since $\boldsymbol{B}_n^{(s)} = \boldsymbol{I}_{n_s} \otimes (\sqrt{p}\,\boldsymbol{F}_p) \otimes \boldsymbol{I}_{p^{s-1}}$, by Remark 8.6 we can use the results in Example 8.3 to obtain

$$\begin{aligned}&\lambda(\boldsymbol{B}_n^{(s)}) = \lambda(\sqrt{p}\boldsymbol{F}_p) = p, &&\kappa(\boldsymbol{B}_n^{(s)}) = \kappa(\sqrt{p}\boldsymbol{F}_p) = 1,\\ &\alpha(\boldsymbol{B}_n^{(s)}) = \sqrt{p}, &&\eta(\boldsymbol{B}_n^{(s)}) = c_{n,1} \qquad (s = 1,\dots,t),\end{aligned}$$

since the entries of $\sqrt{p}\boldsymbol{F}_p$ are the roots of unity $\omega_p^k = \omega_n^{kn_1}$. The diagonal matrix $\boldsymbol{T}_n^{(s)}$ satisfies

$$\begin{aligned}&\lambda(\boldsymbol{T}_n^{(s)}) = 1, &&\kappa(\boldsymbol{T}_n^{(s)}) = 0,\\ &\alpha(\boldsymbol{T}_n^{(s)}) = 1, &&\eta(\boldsymbol{T}_n^{(s)}) = c_{n,s} \qquad (s = 2,\dots,t).\end{aligned}$$

The permutation matrix $\boldsymbol{R}_n(p)$ does not contribute to the roundoff error.

Using Theorem 8.3, the worst case analysis for cascade summation and $p > 2$ is now a simple calculation. We obtain the following constant for normwise backward stability:

$$\begin{aligned}k_n &= \sum_{s=1}^{t}\left(\left(\log_2(\lambda(\boldsymbol{B}_n^{(s)})) + \mu_{\mathbb{C}}\right)\sqrt{\lambda(\boldsymbol{B}_n^{(s)})} + \left(\lambda(\boldsymbol{B}_n^{(s)}) - \kappa(\boldsymbol{B}_n^{(s)})\right)\frac{\eta(\boldsymbol{B}_n^{(s)})}{\alpha(\boldsymbol{B}_n^{(s)})}\right)\\ &+ \sum_{s=2}^{t}\left(\left(\lambda(\boldsymbol{T}_n^{(s)}) - 1 + \mu_{\mathbb{C}}\right)\sqrt{\lambda(\boldsymbol{T}_n^{(s)})} + \left(\lambda(\boldsymbol{T}_n^{(s)}) - \kappa(\boldsymbol{T}_n^{(s)})\right)\frac{\eta(\boldsymbol{T}_n^{(s)})}{\alpha(\boldsymbol{T}_n^{(s)})}\right)\\ &= \sum_{s=1}^{t}\left((\lceil\log_2 p\rceil + \mu_{\mathbb{C}})\sqrt{p} + (p-1)p^{-1/2}c_{n,1}\right) + \sum_{s=2}^{t}(\mu_{\mathbb{C}} + c_{n,s})\\ &\le t\sqrt{p}\left(\lceil\log_2 p\rceil + c_{n,1} + \mu_{\mathbb{C}}\right) + (t-1)\mu_{\mathbb{C}} + \sum_{s=2}^{t} c_{n,s}.\end{aligned}$$

For the average case analysis, we have to compute a different set of constants: As in Example 8.3, for the radix–p butterfly matrices we obtain

$$\bar\nu(\boldsymbol{B}_n^{(s)}) = \lceil \log_2 p \rceil + 1 - 2^{\lceil \log_2 p\rceil - \log_2 p} =: \gamma(p),$$
$$\bar\kappa(\boldsymbol{B}_n^{(s)}) = 2 - \tfrac{1}{p}, \qquad \bar\sigma^2(\boldsymbol{B}_n^{(s)}) = \frac{(p-1)^2}{p}\bar c_{n,1}^2.$$

For the radix–p twiddle matrices, $\bar\nu(\boldsymbol{T}_n^{(s)}) = 0$, $\bar\kappa(\boldsymbol{T}_n^{(s)}) = \frac{1}{p} + p^{1-s} - p^{-s}$, and

$$\bar\sigma^2(\boldsymbol{T}_n^{(s)}) = \bar\sigma^2\Bigl(\bigoplus_{j=0}^{p-1} \operatorname{diag}(\omega_n^{jkn_s})_{k=0}^{p^s-1}\Bigr)$$
$$= \frac{1}{p}\Bigl(\sum_{j=1}^{(p-1)/2} \bar c_{n,s}^2\Bigl(\frac{jn}{p}\Bigr) + \frac{p-1}{2}\bar c_{n,s}^2\Bigr) =: \tilde c_{n,s}^2$$

for $s = 2, \ldots, t$. Therefore, as long as p is an odd number,

$$\bar k_n^2 = (2t-1)\sigma_{\mathbb{C},\times}^2 + \sum_{s=1}^{t}\Bigl(\gamma(p)\sigma_{\mathbb{C},+}^2 + \frac{1}{p}\Bigl(\frac{(p-1)^2}{p}\bar c_{n,1}^2 - \Bigl(2-\frac{1}{p}\Bigr)\sigma_{\mathbb{C},\times}^2\Bigr)\Bigr)$$
$$+ \sum_{s=2}^{t}\Bigl(\tilde c_{n,s}^2 - \Bigl(\frac{1}{p} + p^{1-s} - p^{-s}\Bigr)\sigma_{\mathbb{C},\times}^2\Bigr)$$
$$= t\left(\gamma(p)\sigma_{\mathbb{C},+}^2 + \Bigl(2 - \frac{3}{p} + \frac{1}{p^2}\Bigr)\sigma_{\mathbb{C},\times}^2 + \Bigl(1-\frac{1}{p}\Bigr)^2\bar c_{n,1}^2\right) + \sum_{s=2}^{t}\tilde c_{n,s}^2 + \mathcal{O}(1)$$

with $\gamma(p) = \lceil \log_2 p\rceil + 1 - 2^{\lceil \log_2 p\rceil - \log_2 p} = \log_2 p + \mathcal{O}(1)$ by Remark 8.5.

By Remark 8.7, the decimation–in–frequency FFT, based on (8.25), has the same constants k_n resp. $\bar k_n^2$ as the decimation–in–time FFT. ▯

Example 8.5 (Cooley–Tukey FFT)
In the special case $p = 2$, we obtain the Cooley–Tukey FFT. Compared to the calculations in the general case above, some simplifications occur, since $\boldsymbol{B}_n^{(s)}$ has only entries 0 or ±1:

$$\kappa(\boldsymbol{B}_n^{(s)}) = \bar\kappa(\boldsymbol{B}_n^{(s)}) = 2,\ \eta(\boldsymbol{B}_n^{(s)}) = \bar\sigma^2(\boldsymbol{B}_n^{(s)}) = 0 \qquad (s = 1, \ldots, t)$$

for the radix–2 butterfly matrices, and $\kappa(\boldsymbol{T}_n^{(2)}) = 1$, $\eta(\boldsymbol{T}_n^{(2)}) = 0$,

$$\bar\kappa(\boldsymbol{T}_n^{(2)}) = 1, \qquad \bar\kappa(\boldsymbol{T}_n^{(s)}) = \frac{1}{2} + 2^{1-s} \qquad (s = 3, \ldots, t),$$
$$\bar\sigma^2(\boldsymbol{T}_n^{(2)}) = 0, \qquad \bar\sigma^2(\boldsymbol{T}_n^{(s)}) = \Bigl(\frac{1}{2} - 2^{1-s}\Bigr)\bar c_{n,s}^2 \qquad (s = 3, \ldots, t)$$

for the twiddle matrices. The remaining constants are unchanged.

Therefore by Example 8.1 we obtain the worst case constant

$$k_n = t + (t-2)\mu_{\mathbb{C}} + \sum_{s=3}^{t} c_{n,s} \tag{8.52}$$

and for the average case analysis

$$\begin{aligned}\bar{k}_n^2 &= (2t-1)\sigma_{\mathbb{C},\times}^2 + \sum_{s=1}^{t}\left(\bar{\nu}(\boldsymbol{B}_n^{(s)})\sigma_{\mathbb{C},+}^2 + \frac{\bar{\sigma}^2(\boldsymbol{B}_n^{(s)}) - \bar{\kappa}(\boldsymbol{B}_n^{(s)})\sigma_{\mathbb{C},\times}^2}{\alpha(\boldsymbol{B}_n^{(s)})^2}\right) \\ &\quad + \sum_{s=2}^{t}\left(\bar{\nu}(\boldsymbol{T}_n^{(s)})\sigma_{\mathbb{C},+}^2 + \frac{\bar{\sigma}^2(\boldsymbol{T}_n^{(s)}) - \bar{\kappa}(\boldsymbol{T}_n^{(s)})\sigma_{\mathbb{C},\times}^2}{\alpha(\boldsymbol{T}_n^{(s)})^2}\right) \\ &= t\sigma_{\mathbb{C},+}^2 + \left(\frac{t}{2} - 2 + \frac{2}{n}\right)\sigma_{\mathbb{C},\times}^2 + \frac{1}{2}\sum_{s=3}^{t}(1-2^{2-s})\bar{c}_{n,s}^2 \end{aligned} \tag{8.53}$$

by Theorem 8.4. □

Example 8.6 (Gentleman–Sande FFT)
By Remark 8.7 we obtain the same stability constants as in Example 8.5:

$$k_n = t + (t-2)\mu_{\mathbb{C}} + \sum_{s=3}^{t} c_{n,s},$$

$$\bar{k}_n^2 = t\sigma_{\mathbb{C},+}^2 + \left(\frac{t}{2} - 2 + \frac{2}{n}\right)\sigma_{\mathbb{C},\times}^2 + \frac{1}{2}\sum_{s=3}^{t}(1-2^{2-s})\bar{c}_{n,s}^2.$$

The parallel form of these algorithms (see Remark 8.2), as well as the vector form, have the same constants, since they only differ in permutation matrices. □

Example 8.7 (Radix–4 FFT)
For this algorithm we use the factorization (8.31). The butterfly matrices have trivial entries, and so

$$\begin{aligned} &\lambda(\boldsymbol{B}_n^{(s)}) = 2, \quad && \kappa(\boldsymbol{B}_n^{(s)}) = 2, \quad && \eta(\boldsymbol{B}_n^{(s)}) = 0, \\ &\bar{\nu}(\boldsymbol{B}_n^{(s)}) = 1, && \bar{\kappa}(\boldsymbol{B}_n^{(s)}) = 2, && \bar{\sigma}^2(\boldsymbol{B}_n^{(s)}) = 0 \end{aligned}$$

and $\alpha(\boldsymbol{B}_n^{(s)}) = \sqrt{2}$ for $s = 1, \ldots, 2t$. The twiddle matrices $\boldsymbol{T}_n^{(2s)}$ ($s = 1, \ldots, t$) do not produce roundoff errors, whereas

$$\begin{aligned} &\lambda(\boldsymbol{T}_n^{(2s-1)}) = 1, \quad && \kappa(\boldsymbol{T}_n^{(2s-1)}) = 0, \quad && \eta(\boldsymbol{T}_n^{(2s-1)}) = c_{n,s}, \\ &\bar{\nu}(\boldsymbol{T}_n^{(2s-1)}) = 0, && \bar{\kappa}(\boldsymbol{T}_n^{(2s-1)}) = \tfrac{1}{4} + 4^{1-s}, && \bar{\sigma}^2(\boldsymbol{T}_n^{(2s-1)}) = \left(\frac{3}{4} - 4^{1-s}\right)\bar{c}_{n,s}^2 \end{aligned}$$

for $s = 2, \ldots, t$. Hence we obtain $k_n = 2t + (t-1)\mu_{\mathbb{C}} + \sum_{s=2}^{t} c_{n,s}$ and

$$\bar{k}_n^2 = 2t\sigma_{\mathbb{C}+}^2 + \left(\tfrac{3}{4}t - \tfrac{13}{12} + \tfrac{4}{3n}\right)\sigma_{\mathbb{C}\times}^2 + \sum_{s=2}^{t}\left(\tfrac{3}{4} - 4^{1-s}\right)\bar{c}_{n,s}^2.$$

Note that because of $p = 4$ the meaning of n_s in the definitions (8.50)–(8.51) of $c_{n,s}$ and $\bar{c}_{n,s}^2$ is $n/4^s$. We can express these numbers by the usual *binary* definition simply by replacing s by $2s$ (and $c_{n,s}$ by $c_{n,2s}$, now with $n_{2s} = n/2^{2s}$). With the new meaning of the $c_{n,s}$ and with $n = 2^{t'}$ ($t' = 2t$ even) we can rewrite

$$k_n = t' + \frac{t'-2}{2}\mu_{\mathbb{C}} + \sum_{s=2}^{t'/2} c_{n,2s},$$

$$\bar{k}_n^2 = t'\sigma_{\mathbb{C}+}^2 + \left(\frac{3}{8}t' - \frac{13}{12} + \frac{4}{3n}\right)\sigma_{\mathbb{C}\times}^2 + \sum_{s=2}^{t'/2}\left(\frac{3}{4} - 4^{1-s}\right)\bar{c}_{n,2s}^2.$$

With the result represented in this form, we see that the radix–4 FFT compares favourably with radix–2 FFTs. By Remark 8.7, it holds both for the radix–2 decimation–in–time and the radix–2 decimation–in–frequency FFT. The same constants for backward stability arise for FFTs based directly on one of the factorizations in Theorem 8.2. □

Example 8.8 (Split–radix FFT)
The split–radix FFT is based on the factorization (8.38). We use the notations from Section 8.6. As in Example 8.5, for the butterfly matrices $\boldsymbol{B}_n^{(s)}$ we have $\lambda(\boldsymbol{B}_n^{(s)}) = 2 = \kappa(\boldsymbol{B}_n^{(s)}) = 2$, $\alpha(\boldsymbol{B}_n^{(s)}) = \sqrt{2}$ and $\eta(\boldsymbol{B}_n^{(s)}) = 0$. The twiddle matrices $\boldsymbol{T}_n^{(s)}$ are again diagonal and so $\lambda(\boldsymbol{T}_n^{(s)}) = 1$, $\kappa(\boldsymbol{T}_n^{(t)}) = 1$, $\kappa(\boldsymbol{T}_n^{(s)}) = 0$ ($s = 2, \ldots, t-1$) as well as $\eta(\boldsymbol{T}_n^{(t)}) = 0$ and $\eta(\boldsymbol{T}_n^{(s)}) = c_{n,s+1}$, the latter because of $\eta(\boldsymbol{T}_{2^s}^{(s)}(l)) = c_{n,s+1}$ if $\beta_{t-s}(l) = 1$ ($s = 2, \ldots, t-1$). Hence we obtain the worst case constant

$$k_n = t + (t-2)\mu_{\mathbb{C}} + \sum_{s=3}^{t} c_{n,s},$$

which is the same as for the Cooley–Tukey FFT (see Example 8.5).

For the average case analysis we observe that

$$\bar{\sigma}^2(\boldsymbol{T}_{2^s}^{(s)}(l)) = \begin{cases} 0 & \text{if } \beta_{t-s}(l) = 0, \\ (1 - 2^{1-s})\bar{c}_{n,s+1}^2 & \text{if } \beta_{t-s}(l) = 1. \end{cases}$$

Since it can be shown by induction that the number of ones in β_s equals $\frac{1}{3}(2^s - (-1)^s)$, we have

$$\bar{\sigma}^2(\boldsymbol{T}_n^{(s)}) = \frac{1}{3}\left(1 - \left(-\frac{1}{2}\right)^{t-s}\right)(1 - 2^{1-s})\bar{c}_{n,s+1}^2 \qquad (s = 2, \ldots, t).$$

Furthermore, $\bar{\nu}(\boldsymbol{T}_n^{(s)}) = 0$ and $\bar{\kappa}(\boldsymbol{T}_n^{(s)}) = \frac{1}{3}\big(2+\big(-\frac{1}{2}\big)^{t-s}+2^{1-s}-(-1)^{t-s}2^{1-t}\big)$. Therefore we obtain

$$\bar{k}_n^2 = t\sigma_{\mathbb{C},+}^2+\Big(\frac{t}{3}-\frac{8}{9}+\frac{9-(-1)^t}{9n}\Big)\sigma_{\mathbb{C},\times}^2+\frac{1}{3}\sum_{s=3}^{t}\Big(1-\Big(-\frac{1}{2}\Big)^{t+1-s}\Big)(1-2^{2-s})\,\bar{c}_{n,s}^2.$$

This is slightly smaller than for the radix–4 FFT. ▯

Example 8.9 (FFT with real data)
With the notations of Algorithm 8.6, in the calculation of $\boldsymbol{v}$ a backward roundoff error $\Delta\boldsymbol{x}^{(v)}$ arises, which is bounded by

$$\|\Delta\boldsymbol{x}^{(v)}\|_2 \le \left(t-1+(t-3)\mu_{\mathbb{C}}+\sum_{s=3}^{t-1}c_{n_1,s}+(1+\sqrt{2}\,c_{n,3})+\mathcal{O}(u)\right)\|\boldsymbol{x}\|_2 u$$

if the Cooley–Tukey or Gentleman–Sande FFT is used. The conversions between real vectors of length n and complex vectors of length n_1 do not contribute to the error and the norm remains unchanged.

The computation of the result $\boldsymbol{z} = \boldsymbol{U}_n\boldsymbol{y} = 2^{-1/2}\boldsymbol{V}_n\boldsymbol{v}$ produces a backward error $\Delta\boldsymbol{x}^{(z)}$ with $\|\Delta\boldsymbol{x}^{(z)}\|_2 \le \big(3+\sqrt{2}\,c_{n,t}\big)\|\boldsymbol{x}\|_2 u$, since the precomputation errors for the sine and cosine values are bounded by the same constant $c_{n,t}$ as for the complex exponential. Therefore, in total

$$k_n \le t+3+(t-3)\mu_{\mathbb{C}}+\sum_{s=3}^{t-1}c_{n_1,s}+\sqrt{2}(c_{n,3}+c_{n,t}).$$

In the average case analysis,

$$\begin{aligned}\mathbb{E}(\|\Delta\boldsymbol{X}^{(v)}\|_2^2) &= \Big((t-1)\sigma_{\mathbb{C},+}^2+\tfrac{t-5}{2}\sigma_{\mathbb{C},\times}^2+\tfrac{1}{2}\sum_{s=3}^{t-1}\bar{c}_{n_1,s}^2\\ &\qquad +\big(1-\tfrac{2}{n}\big)\sigma_{\mathbb{R},+}^2+\tfrac{2}{n}\sigma_{\mathbb{R},\times}^2+\tfrac{1}{3n}\Big)\mathbb{E}(\|\boldsymbol{X}\|_2^2)u^2,\\ \mathbb{E}(\|\Delta\boldsymbol{X}^{(z)}\|_2^2) &= \Big(\big(\tfrac{3}{2}-\tfrac{4}{n}\big)\sigma_{\mathbb{R},+}^2+\tfrac{1}{2}\sigma_{\mathbb{R},\times}^2+\big(\tfrac{1}{2}-\tfrac{2}{n}\big)\bar{c}_{n,t}^2+\tfrac{1}{3n}\Big)\mathbb{E}(\|\boldsymbol{X}\|_2^2)u^2.\end{aligned}$$

From this

$$\bar{k}_n^2 = (t-1)\sigma_{\mathbb{C},+}^2+\tfrac{5}{2}\sigma_{\mathbb{R},+}^2+\tfrac{t-5}{2}\sigma_{\mathbb{C},\times}^2+\tfrac{1}{2}\sigma_{\mathbb{R},\times}^2+\tfrac{1}{2}\sum_{s=3}^{t-1}\bar{c}_{n_1,s}+\big(\tfrac{1}{2}-\tfrac{2}{n}\big)\bar{c}_{n,t}^2+\mathcal{O}\big(\tfrac{1}{n}\big)$$

follows. ▯

To illustrate the results above, in the following diagrams we compare the upper bounds obtained in Example 8.5 to the maximum of the error norms in 200 simulations for $n = 2^5, \ldots, 2^{18}$ and four precomputation methods.

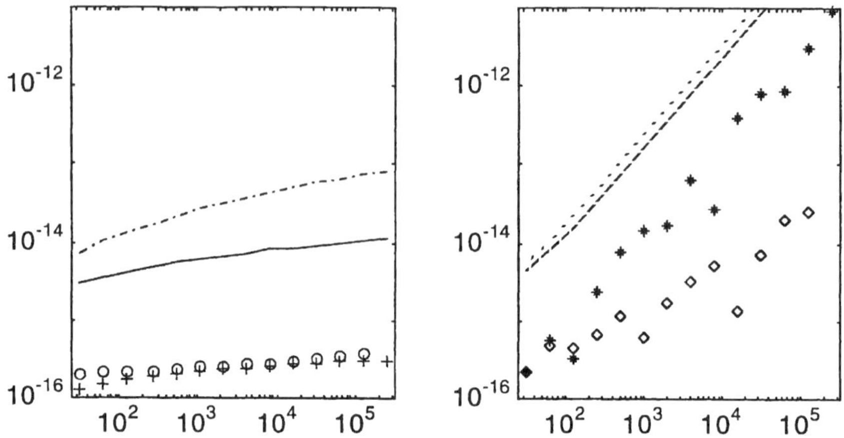

Figure 1: Maximum of the norm of 200 simulated roundoff errors with the Cooley–Tukey FFT using four precomputation methods. Left diagram: Repeated subvector scaling (∘) and direct call (+); dashed resp. solid lines: Upper bound (8.52). Right diagram: Repeated multiplication (∗) and stabilized forward recursion (◊); dotted and dashed lines: upper bound.

The following diagrams show a comparison between the estimates from Example 8.5 and the average error norm resp. the standard deviation obtained in 200 simulations.

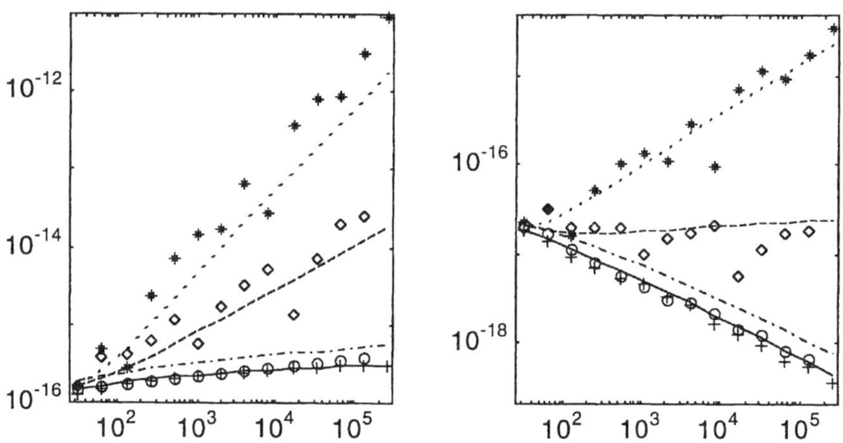

Figure 2: Left diagram: Average of 200 simulated roundoff error norms using the Cooley–Tukey FFT and four precomputation methods, compared with the estimation in (8.53). Right diagram: Standard deviation of the simulated error norms. Markers and line styles as in Figure 1.

It becomes clear from our results and simulations that the precomputation method has a dramatic influence on the roundoff errors for all FFTs studied in this article. Therefore only high quality methods for precomputation should be used if n is large.

8.9 Error analysis for DCT

In this section we will study fast algorithms for DCTs which are defined by the *cosine matrices of type* II–IV

$$\begin{aligned} C_n^{\mathrm{II}} &:= \sqrt{\frac{2}{n}}\left(\eta_j^{(n)} \cos\frac{j(2k+1)\pi}{2n}\right)_{j,k=0}^{n-1}, \\ C_n^{\mathrm{III}} &:= (C^{\mathrm{II}})^{\mathrm{T}}, \\ C_n^{\mathrm{IV}} &:= \sqrt{\frac{2}{n}}\left(\cos\frac{(2j+1)(2k+1)\pi}{4n}\right)_{j,k=0}^{n-1} \end{aligned}$$

with $\eta_0^{(n)} = \eta_n^{(n)} := 2^{-1/2}$, $\eta_j^{(n)} := 1$ $(j = 1, \ldots, n-1)$. It is known that these cosine matrices are orthogonal (see [11], pp. 12–16, [15], [13]).

The precomputed constants in this section are cosine and sine values separately, and are not combined into the complex exponential as in Section 8.3. For the roundoff errors $\delta_j^{\cos} = \mathrm{fl}(\cos\frac{j\pi}{n}) - \cos\frac{j\pi}{n}$ and $\delta_j^{\sin} = \mathrm{fl}(\sin\frac{j\pi}{n}) - \sin\frac{j\pi}{n}$, we assume the worst case bounds

$$|\delta_{jn2^{-s}}^{\cos}| \leq c_{2n,s+1}u, \quad |\delta_{jn2^{-s}}^{\sin}| \leq c_{2n,s+1}u \quad (j = 0, \ldots, 2^{s+1}-1), \tag{8.54}$$

where $n = 2^t$ and $s = 0, \ldots t$. Obviously, the values $c_{2n,s+1}$ from (8.50) can be used as an upper bound, but for particular precomputation methods a smaller bound may be possible. For example for the direct call, we have $c_{n,s} = \frac{1}{2}$, a smaller bound than for the complex exponential in Algorithm 8.1. Note that in the algorithms for data of length n studied below, it is necessary to calculate trigonometric functions in points $\frac{2\pi j}{n'}$, where n' is a multiple of $2n$.

For the average case situation, in all examples (with the exception of diagonal entries $\sqrt{2}$) only rotation and reflection 2×2–matrices occur as building blocks and hence each row with a sine value also contains the corresponding cosine value and vice versa. Therefore their error variances σ_{s,j,k_1}^2 and σ_{s,j,k_2}^2 in Theorem 8.4 satisfy $\sigma_{s,j,k_1}^2 + \sigma_{s,j,k_2}^2 = \mathbb{V}(\delta_k^{\exp})$ with $\delta_k^{\exp} = \mathrm{fl}(\exp(-\frac{2k\pi \mathrm{i}}{2n})) - \exp(-\frac{2k\pi \mathrm{i}}{2n})$, and the constant $\bar{c}_{2n,s+1}^2$ (which has the same meaning as in Section 8.8) will again be useful to express the term $\bar{\sigma}_s^2$ occuring in the assertion of Theorem 8.4.

Example 8.10 (Fast DCT-IV)

This algorithm relies on the following factorization of the cosine matrix C_n^{IV} for $n = 2^t$ ($t \geq 3$) into a product of sparse, almost orthogonal matrices (see [13]):

$$C_n^{IV} = 2^{-\frac{t-1}{2}} P_n^{(t)} \left(\prod_{s=1}^{t-1} \left[(I_{2^{s-1}} \otimes (I_{n_s} \oplus G_{n_s}))\, B_n^{(t-s+1)} P_n^{(s)} \right] \right) D_n P_n^{(0)},$$

where

$$G_m := \bigoplus_{j=0}^{m/2-1} \begin{pmatrix} \cos\frac{(2j+1)\pi}{2m} & (-1)^j \sin\frac{(2j+1)\pi}{2m} \\ (-1)^j \sin\frac{(2j+1)\pi}{2m} & -\cos\frac{(2j+1)\pi}{2m} \end{pmatrix},$$

$$D_m := \bigoplus_{j=0}^{m/2-1} \begin{pmatrix} \cos\frac{(2j+1)\pi}{4m} & \sin\frac{(2j+1)\pi}{4m} \\ (-1)^j \sin\frac{(2j+1)\pi}{4m} & -(-1)^j \cos\frac{(2j+1)\pi}{4m} \end{pmatrix},$$

$B_n^{(s)}$ is the radix–2 butterfly matrix, and $P_n^{(s)}$ a permutation matrix ($s = 0, \ldots, t$). Not counting the final multiplication by $2^{-(t-1)/2}$, which often can be avoided, the corresponding fast algorithm for computing $C_n^{IV} x$ ($x \in \mathbb{R}^n$) takes $\frac{5}{2}nt + \frac{t}{2}$ flops.

Obviously, the permutation matrices do not cause any roundoff errors. Using Example 8.1, we see that the contribution to the backward constant k_n by the radix–2 butterfly matrix $B_n^{(t-s+1)}$ is 1. For the rotation matrix $I_{2^{s-1}} \otimes (I_{n_s} \oplus G_{n_s})$, trigonometric functions at points $j\frac{2\pi}{4n_s}$ ($j \in \mathbb{N}$ odd) have to be precomputed and hence its contribution is $1 + \sqrt{2} + 2c_{4n,t+2-s}$ ($s = 1, \ldots, t-1$). Similarly, D_n is responsible for a term $1 + \sqrt{2} + 2c_{8n,t+3}$. Hence from Theorem 8.3 we obtain the following worst case constant

$$\begin{aligned} k_n &= \sum_{s=1}^{t-1} \left((1 + \sqrt{2} + 2c_{8n,t+2-s}) + 1 \right) + (1 + \sqrt{2} + 2c_{8n,t+3}) \\ &= (2 + \sqrt{2})t - 1 + 2 \sum_{s=3}^{t+1} c_{8n,s} + 2c_{8n,t+3}\,. \end{aligned}$$

Likewise, the constants needed to apply Theorem 8.4 are as follows:

$$\begin{aligned} \bar{\nu}(I_{2^{s-1}} \otimes (I_{n_s} \oplus G_{n_s})) &= \frac{1}{2} (\bar{\nu}(I_{n_s}) + \bar{\nu}(G_{n_s})) &= \frac{1}{2}, \\ \bar{\kappa}(I_{2^{s-1}} \otimes (I_{n_s} \oplus G_{n_s})) &= \frac{1}{2} (\bar{\kappa}(I_{n_s}) + \bar{\kappa}(G_{n_s})) &= \frac{1}{2}, \\ \bar{\sigma}^2(I_{2^{s-1}} \otimes (I_{n_s} \oplus G_{n_s})) &= \frac{1}{2} (\bar{\sigma}^2(I_{n_s}) + \bar{\sigma}^2(G_{n_s})) &= \frac{1}{2} \bar{c}_{8n,t+2-s}^2 \end{aligned}$$

($s = 1, \ldots, t-1$) and $\bar{\nu}(D_n) = 1$, $\bar{\kappa}(D_n) = 0$, $\bar{\sigma}^2(D_n) = \bar{c}_{8n,t+3}^2$ for the rotation matrices. As in Example 8.5, the contribution of $B_n^{(t-s+1)}$ to $\bar{k}_n^2$

is $\sigma^2_{\mathbb{R},+}$. We obtain the average case constant

$$\begin{aligned}\bar{k}_n^2 &= t\sigma^2_{\mathbb{R},\times} + \sum_{s=1}^{t-1}\left(\left(\frac{1}{2}\sigma^2_{\mathbb{R},+} + \frac{1}{2}\bar{c}^2_{8n,t+2-s} - \frac{1}{2}\sigma^2_{\mathbb{R},\times}\right) + \sigma^2_{\mathbb{R},+}\right) + \sigma^2_{\mathbb{R},+} + \bar{c}^2_{8n,t+3} \\ &= \frac{3t+1}{2}\sigma^2_{\mathbb{R},+} + \frac{t+1}{2}\sigma^2_{\mathbb{R},\times} + \frac{1}{2}\sum_{s=3}^{t+1}\bar{c}^2_{8n,s} + \bar{c}^2_{8n,t+3}\end{aligned}$$

by inserting these $\bar{\nu}_s$, $\bar{\kappa}_s$, $\bar{\sigma}_s^2$ into (8.49). □

Example 8.11 (Fast DCT-III)
This algorithm relies on the following factorization of the cosine matrix $\boldsymbol{C}_n^{\mathrm{III}}$ for $n = 2^t$ $(t \geq 3)$ into a product of sparse, almost orthogonal matrices (see [13]):

$$\boldsymbol{C}_n^{\mathrm{III}} = 2^{-\frac{t-2}{2}}\left(\prod_{s=0}^{t-1}\tilde{\boldsymbol{R}}_n^{(s)}\right)\tilde{\boldsymbol{P}}_n^{(t-1)}\left(\prod_{s=1}^{t-2}\left[\tilde{\boldsymbol{G}}_n^{(s)}\tilde{\boldsymbol{B}}_n^{(s)}\tilde{\boldsymbol{P}}_n^{(s)}\right]\right)\tilde{\boldsymbol{D}}_n\tilde{\boldsymbol{P}}_n^{(0)},$$

where, using the notations from Example 8.10,

$$\begin{aligned}\tilde{\boldsymbol{R}}_n^{(s)} &= \tfrac{1}{\sqrt{2}}\begin{pmatrix}\boldsymbol{I}_{2^s} & \boldsymbol{I}_{2^s}\\ \boldsymbol{J}_{2^s} & -\boldsymbol{J}_{2^s}\end{pmatrix}\oplus\boldsymbol{I}_{n-2^{s+1}} && (s = 0,\ldots,t-1),\\ \tilde{\boldsymbol{G}}_n^{(s)} &= \boldsymbol{I}_{n_s}\oplus(\boldsymbol{I}_{2^s-1}\otimes\boldsymbol{G}_{n_s}) && (s = 1,\ldots,t-2),\\ \tilde{\boldsymbol{B}}_n^{(s)} &= \sqrt{2}\boldsymbol{I}_{n_s}\oplus\left(\boldsymbol{I}_{2^s-1}\otimes\begin{pmatrix}1 & 1\\ 1 & -1\end{pmatrix}\otimes\boldsymbol{I}_{n_{s+1}}\right) && (s = 1,\ldots,t-2),\\ \tilde{\boldsymbol{D}}_n &= \boldsymbol{I}_2\oplus\bigoplus_{s=1}^{t-1}\boldsymbol{D}_{2^s},\end{aligned}$$

and $\tilde{\boldsymbol{P}}_n^{(s)}$ are permutation matrices. Using Remark 8.6, the computation of the constants is now straightforward:

$$\begin{aligned}\lambda(\tilde{\boldsymbol{R}}_n^{(s)}) &= 2, & \kappa(\tilde{\boldsymbol{R}}_n^{(s)}) &= 0, & \eta(\tilde{\boldsymbol{R}}_n^{(s)}) &= c_{4n,3} && (s = 0,\ldots,t-1),\\ \lambda(\tilde{\boldsymbol{B}}_n^{(s)}) &= 2, & \kappa(\tilde{\boldsymbol{B}}_n^{(s)}) &= 0, & \eta(\tilde{\boldsymbol{B}}_n^{(s)}) &= 2c_{4n,3} && (s = 1,\ldots,t-2),\end{aligned}$$

since $\sqrt{2} = 2\cos\frac{\pi}{4}$, and $\alpha(\tilde{\boldsymbol{B}}_n^{(s)}) = \sqrt{2}$. Again, by Example 8.1 the contribution of the rotation matrices in (8.45) is $1 + \sqrt{2} + 2c_{4n,t+2-s}$ for $\tilde{\boldsymbol{G}}_n^{(s)}$ and $1 + \sqrt{2} + 2c_{4n,t+2}$ for $\tilde{\boldsymbol{D}}_n$.

Therefore it follows from Theorem 8.3 that the worst case constant is

$$k_n = t(1+\sqrt{2})(3+2c_{4n,3}) + 2\sum_{s=4}^{t+2}c_{4n,s} - (1+\sqrt{2})(3+4c_{4n,3}).$$

In order to study the average case behaviour of the norm of the error, we have to compute

$$\begin{aligned}
\bar{\nu}(\tilde{R}_n^{(s)}) &= 2^{s+1-t}, & \bar{\kappa}(\tilde{R}_n^{(s)}) &= 1-2^{s+1-t}, & \bar{\sigma}^2(\tilde{R}_n^{(s)}) &= 2^{s+1-t}\,\bar{c}_{4n,3}^2,\\
\bar{\nu}(\tilde{G}_n^{(s)}) &= 1-2^{s-t}, & \bar{\kappa}(\tilde{G}_n^{(s)}) &= 2^{s-t}, & \bar{\sigma}^2(\tilde{G}_n^{(s)}) &= (1-2^{s-t})\bar{c}_{4n,t+2-s}^2,\\
\bar{\nu}(\tilde{B}_n^{(s)}) &= 1-2^{s-t}, & \bar{\kappa}(\tilde{B}_n^{(s)}) &= 1-2^{s-t}, & \bar{\sigma}^2(\tilde{B}_n^{(s)}) &= 2^{s+1-t}\,\bar{c}_{4n,3}^2,\\
\bar{\nu}(\tilde{D}_n) &= 1-2^{1-t}, & \bar{\kappa}(\tilde{D}_n) &= 2^{1-t}, & \bar{\sigma}^2(\tilde{D}_n) &= (1-2^{1-t})\bar{c}_{4n,t+2}^2
\end{aligned}$$

and conclude from Theorem 8.4 (after some computation) that

$$\bar{k}_n^2 = 2t\sigma_{\mathbb{R},+}^2 + \frac{3}{2}t\sigma_{\mathbb{R},\times}^2 + \sum_{s=3}^{t+2}(1-2^{2-s})\bar{c}_{4n,s}^2 + \mathcal{O}(1)$$

as long as $\bar{c}_{4n,3}^2 = \mathcal{O}(1)$.

Since C_n^{II} is the transpose of C_n^{III}, it follows from Remark 8.7 that the constants k_n and $\bar{k}_n$ for the corresponding fast DCT–II take the same values as above. $\square$

The following algorithm (see [8], pp. 347–349, [5], and [9]) calculates the DCT-IV using an FFT of half the length:

ALGORITHM 8.8 (Fast DCT-IV via FFT)
Input: $n = 2^t$ $(t \geq 3)$, $\boldsymbol{x} \in \mathbb{R}^n$.
1. Form $\boldsymbol{u} := (x_{2j})_{j=0}^{n_1-1} + \mathrm{i}(x_{n-1-2j})_{j=0}^{n_1-1} \in \mathbb{C}^{n_1}$ with $n_1 = n/2$.
2. Compute $\boldsymbol{v} := \left(\mathrm{diag}\left(\exp(-\frac{\mathrm{i}\pi(4j+1)}{4n})\right)_{j=0}^{n_1-1}\right)\boldsymbol{v}$.
3. Form $\hat{\boldsymbol{v}} := \boldsymbol{F}_{n_1}\boldsymbol{v}$ via FFT.
4. Calculate $\boldsymbol{w} := \left(\mathrm{diag}\left(\exp(-\frac{\mathrm{i}\pi j}{n})\right)_{j=0}^{n_1-1}\right)\hat{\boldsymbol{v}}$.
5. Set $y_{2j} := \Re(w_j)$, $y_{2j+1} := -\Im(w_{n_1-1-j})$ $(j = 0, \ldots, n_1 - 1)$.
Output: $\boldsymbol{y} = C_n^{\mathrm{IV}}\boldsymbol{x}$.

When the Cooley–Tukey or the Gentleman–Sande FFT is used for step 3, the DCT-IV is computed in $\frac{5}{2}nt + \frac{t}{2}$ flops.

Example 8.12 (Fast DCT-IV via FFT)
Steps 1 and 5 of Algorithm 8.8 obviously do not contribute a roundoff error and also preserve the norm. Using the results from Section 8.8 for step 3, we obtain the worst case constant

$$k_n = (\mu_{\mathrm{c}} + c_{8n,t+3}) + (t-1) + (t-3)\mu_{\mathrm{c}} + \sum_{s=3}^{t-1} c_{n_1,s} + (\mu_{\mathrm{c}} + c_{2n,t+1})$$

$$= (t-1)(\mu_{\mathrm{C}}+1) + \sum_{s=3}^{t-1} c_{n_1,s} + c_{2n,t+1} + c_{8n,t+3}$$

and the average case constant

$$\bar{k}_n^2 = (t-1)\sigma_{\mathrm{C},+}^2 + \left(\tfrac{t-1}{2} - \tfrac{9}{n}\right) + \tfrac{1}{2}\sum_{s=3}^{t-1} \bar{c}_{n_1,s}^2 + \left(1-\tfrac{1}{n}\right)\bar{c}_{2n,t+1}^2 + \bar{c}_{8n,t+3}^2.$$

Comparing with Example 8.10, we see that both versions of fast DCT-IV have almost the same arithmetical complexity and almost the same stability constants under the assumption of accurately precomputed cosine/sine values. However, if we apply the split–radix FFT in step 3 of Algorithm 8.8, then it is useful to compute a DCT-IV via FFT in complex arithmetic. □

Similarly to the cosine matrices, the *sine matrices of type* II–IV are defined as

$$\boldsymbol{S}_n^{\mathrm{II}} := \sqrt{\frac{2}{n}}\left(\eta_{j+1}^{(n)} \sin\frac{(j+1)(2k+1)\pi}{2n}\right)_{j,k=0}^{n-1},$$
$$\boldsymbol{S}_n^{\mathrm{III}} := \left(\boldsymbol{S}^{\mathrm{II}}\right)^{\mathrm{T}},$$
$$\boldsymbol{S}_n^{\mathrm{IV}} := \sqrt{\frac{2}{n}}\left(\sin\frac{(2j+1)(2k+1)\pi}{4n}\right)_{j,k=0}^{n-1}.$$

They are closely related to the cosine matrices of the same type (see [13]):

$$\boldsymbol{S}_n^{\mathrm{II}} = \boldsymbol{J}_n \boldsymbol{C}_n^{\mathrm{II}} \boldsymbol{\Sigma}_n, \qquad \boldsymbol{S}_n^{\mathrm{III}} = \boldsymbol{\Sigma}_n \boldsymbol{C}_n^{\mathrm{III}} \boldsymbol{J}_n, \qquad \boldsymbol{S}_n^{\mathrm{IV}} = \boldsymbol{\Sigma}_n \boldsymbol{C}_n^{\mathrm{IV}} \boldsymbol{J}_n$$

with $\boldsymbol{\Sigma}_n := \mathrm{diag}\big((-1)^j\big)_{j=0}^{n-1}$. In particular, the sine matrices of type II–IV are orthogonal. Any fast DST of type II–IV can be realized by a fast DCT of the same type. Hence the stability constants for a fast DST of type II–IV coincide with the stability constants for a fast DCT of the same type.

8.10 Fast convolution and deconvolution

For given $\boldsymbol{a} \in \mathbb{C}^n$, the *circulant matrix* is defined by

$$\mathrm{circ}\,\boldsymbol{a} := \left(a_{\langle j-k\rangle_n}\right)_{j,k=0}^{n-1} \in \mathbb{C}^{n\times n}.$$

This is a special Toeplitz matrix in which the diagonals wrap around. Then the *cyclic convolution* of $\boldsymbol{x}, \boldsymbol{y} \in \mathbb{C}^n$ is by definition the vector $\boldsymbol{z} = \boldsymbol{x} * \boldsymbol{y} :=$

$(\operatorname{circ} \boldsymbol{x})\boldsymbol{y}$, i.e.

$$z_j = \sum_{k=0}^{n-1} x_{\langle j-k \rangle_n} y_k = \sum_{k=0}^{j} x_{j-k} y_k + \sum_{k=j+1}^{n-1} x_{n+j-k} y_k \qquad (j = 0, \ldots, n-1).$$

Cyclic convolution is a commutative, associative, and distributive multiplication in $\mathbb{C}^n$. First we consider the roundoff error for the straightforward computation by matrix–vector product. It takes about $7n^2$ flops for sequential summation and $6n^2 + 2n \log_2 n$ flops for cascade summation.

LEMMA 8.5

(i) *Let $\boldsymbol{x}, \boldsymbol{y} \in \mathbb{C}^n \setminus \{o\}$ be given. For the computed value $\hat{\boldsymbol{z}} := \mathrm{fl}(\boldsymbol{x} * \boldsymbol{y})$ of $\boldsymbol{z} = \boldsymbol{x} * \boldsymbol{y}$ we have*

$$\frac{\|\hat{\boldsymbol{z}} - \boldsymbol{z}\|_2}{\|\boldsymbol{x}\|_2 \|\boldsymbol{y}\|_2} \le \begin{cases} \sqrt{n}\big(n + \mu_{\mathbb{C}} - 1 + \mathcal{O}(u)\big)u & \text{for sequential summation,} \\ \sqrt{n}\big(\lceil \log_2 n \rceil + \mu_{\mathbb{C}} + \mathcal{O}(u)\big)u & \text{for cascade summation.} \end{cases}$$

(ii) *Let $\boldsymbol{X}, \boldsymbol{Y} \in \mathbb{C}^n$ be independent random vectors satisfying the conditions in Section* 8.2. *For the computed value $\hat{\boldsymbol{Z}} := \mathrm{fl}(\boldsymbol{X} * \boldsymbol{Y})$ of $\boldsymbol{Z} = \boldsymbol{X} * \boldsymbol{Y}$ we have*

$$\frac{\|\hat{\boldsymbol{Z}} - \boldsymbol{Z}\|^2}{\|\boldsymbol{X}\|^2 \|\boldsymbol{Y}\|^2} = \begin{cases} \big((\frac{n+1}{2} - \frac{1}{n})\sigma^2_{\mathbb{C},+} + \sigma^2_{\mathbb{C},+}\big)u^2 & \text{for sequential summation,} \\ \big(\lceil \log_2 n \rceil \sigma^2_{\mathbb{C},+} + \sigma^2_{\mathbb{C},\times}\big)u^2 & \text{for cascade summation.} \end{cases}$$

The proof is straightforward and therefore omitted.

For real valued data the above results hold with $\mu_{\mathbb{C}}$ replaced by 1 and $\sigma^2_{\mathbb{C},+}$ and $\sigma^2_{\mathbb{C},\times}$ replaced by $\sigma^2_{\mathbb{R},+}$ and $\sigma^2_{\mathbb{R},\times}$.

The DFT has the important *cyclic convolution property,* i.e. for $\boldsymbol{x}, \boldsymbol{y} \in \mathbb{C}^n$,

$$\boldsymbol{F}_n(\boldsymbol{x} * \boldsymbol{y}) = \sqrt{n} \operatorname{diag}(\boldsymbol{F}_n \boldsymbol{x}) \boldsymbol{F}_n \boldsymbol{y}. \tag{8.55}$$

Using $\boldsymbol{F}_n^{-1} = \boldsymbol{F}_n^* = \boldsymbol{J}'_n \boldsymbol{F}_n$ with the *flip matrix* $\boldsymbol{J}'_n := \big(\delta(\langle j + k \rangle_n)\big)_{j,k=0}^{n-1}$, we obtain the formula

$$\boldsymbol{x} * \boldsymbol{y} = \sqrt{n} \boldsymbol{J}'_n \boldsymbol{F}_n \operatorname{diag}(\boldsymbol{F}_n \boldsymbol{x}) \boldsymbol{F}_n \boldsymbol{y}, \tag{8.56}$$

which yields the idea of *fast convolution.*

ALGORITHM 8.9 (Fast convolution)

Input: $n = 2^t$ $(t \ge 3)$; $\boldsymbol{x}, \boldsymbol{y} \in \mathbb{C}^n$.

1. Compute $\boldsymbol{a} := \boldsymbol{F}_n \boldsymbol{x}$ via FFT.
2. Compute $\boldsymbol{b} := \boldsymbol{F}_n \boldsymbol{y}$ via FFT.
3. For $j = 0 : 1 : n-1$ multiply $c_j := a_j b_j$.

4. Form $z := \sqrt{n} J_n' F_n (c_j)_{j=0}^{n-1}$ via FFT.
Output: $z = x * y \in \mathbb{C}^n$.

The fast convolution involves mainly three FFTs and a diagonal scaling. Depending on the used FFT, for $n = 2^t$ it needs between $12n \log_2 n + \mathcal{O}(n)$ (split–radix FFT) and $15n \log_2 n + \mathcal{O}(n)$ flops (Cooley–Tukey or Gentleman–Sande FFT). Now we analyze the roundoff errors. Let k_n resp. $\bar{k}_n^2$ be the constants for backward stability of the FFT used. For the worst case analysis we have from Corollary 8.2

$$\hat{a} = (F_n + \Delta_1)x, \qquad \hat{b} = (F_n + \Delta_2)y, \qquad \|\Delta_j\|_2 \le k_n u \ \ (j = 1, 2).$$

Step 3 yields by (8.5)

$$\hat{c} = \big(\operatorname{diag} \hat{a} + \operatorname{diag}(\varepsilon_j^{\times} \hat{a}_j)_{j=0}^{n-1}\big)\hat{b} = (\operatorname{diag} a + \Delta)(F_n + \Delta_2)y$$

with $|\varepsilon_j^{\times}| \le \mu_{\mathbb{C}}$ and a diagonal matrix

$$\Delta := \operatorname{diag}(\Delta_1 x) + \operatorname{diag}(\varepsilon_j^{\times} \hat{a}_j)_{j=0}^{n-1}.$$

The spectral norm of Δ can be estimated as follows

$$\begin{aligned}\|\Delta\|_2 &\le \|\Delta_1 x\|_\infty + \mu_{\mathbb{C}} u \|\hat{a}\|_\infty \le \|\Delta_1 x\|_2 + (\mu_{\mathbb{C}} u + \mathcal{O}(u^2))\|x\|_2 \\ &\le \|\Delta_1\|_2 \|x\|_2 + (\mu_{\mathbb{C}} u + \mathcal{O}(u^2))\|x\|_2 \le (k_n u + \mu_{\mathbb{C}} u + \mathcal{O}(u^2))\|x\|_2.\end{aligned}$$

Step 4 yields

$$\begin{aligned}\hat{z} &= \sqrt{n} J_n' (F_n + \Delta_3)\hat{c} = \sqrt{n} J_n' (F_n + \Delta_3)(\operatorname{diag} a + \Delta)(F_n + \Delta_2)y \\ &= z + \sqrt{n} J_n' \Big(F_n (\operatorname{diag} a)\Delta_2 + F_n \Delta\, F_n + \\ &\qquad + \Delta_3 (\operatorname{diag} a) F_n + \mathcal{O}(u^2)\|x\|_2\Big) y \qquad (8.57)\end{aligned}$$

with $\|\Delta_3\|_2 \le k_n u$. Thus we can estimate the roundoff error

$$\begin{aligned}\|\hat{z} - z\|_2 &\le \sqrt{n}\Big(\|(\operatorname{diag} a)\Delta_2\|_2 + \|\Delta\|_2 + \|\Delta_3 (\operatorname{diag} a)\|_2 + \|x\|_2 \mathcal{O}(u^2)\Big)\|y\|_2 \\ &\le \sqrt{n}\Big(3k_n + \mu_{\mathbb{C}} + \mathcal{O}(u)\Big)\|x\|_2 \|y\|_2 u\end{aligned}$$

For the average case analysis we observe that in (8.57) we have

$$\begin{aligned}|\!|\!| F_n (\operatorname{diag} F_n X)\Delta_2 Y |\!|\!|^2 &= \sum_{j=0}^{n-1} \mathbb{E}(|(F_n X)_j|^2)\mathbb{E}(|(\Delta_2 Y)_j|^2) = |\!|\!| X_0 |\!|\!|^2 |\!|\!| \Delta_2 Y |\!|\!|^2 \\ &= \frac{1}{n}\bar{k}_n^2 |\!|\!| X |\!|\!|^2 |\!|\!| Y |\!|\!|^2 u^2,\end{aligned}$$

since $\|X_j\|^2 = \frac{1}{n}\|\boldsymbol{X}\|^2$ $(j = 0, \ldots, n-1)$. Similarly, $\|\Delta_3(\operatorname{diag} \boldsymbol{F}_n\boldsymbol{X})\boldsymbol{F}_n\boldsymbol{Y}\|^2 = \frac{\bar{k}_n^2}{n}\|\boldsymbol{X}\|^2\|\boldsymbol{Y}\|^2u^2$ and $\|\boldsymbol{F}_n\boldsymbol{\Delta}\,\boldsymbol{F}_n\boldsymbol{Y}\|^2 = \frac{1}{n}\big(\bar{k}_n^2\|\boldsymbol{X}\|^2\|\boldsymbol{Y}\|^2 + \sigma_{\mathbb{C},\times}^2\|\boldsymbol{X}\|^2\|\boldsymbol{Y}\|^2\big)u^2$.
Hence we can summarize:

THEOREM 8.5
(i) *Let $\boldsymbol{x}, \boldsymbol{y} \in \mathbb{C}^n \setminus \{\boldsymbol{o}\}$ be given and $\boldsymbol{z} := \boldsymbol{x} * \boldsymbol{y}$. For the Algorithm* 8.9 *of fast convolution we have*

$$\frac{\|\hat{\boldsymbol{z}} - \boldsymbol{z}\|_2}{\|\boldsymbol{x}\|_2\|\boldsymbol{y}\|_2} \leq \sqrt{n}\big((3k_n + \mu_{\mathbb{C}})u + \mathcal{O}(u^2)\big).$$

assuming that the backward error of the FFT *has the stability constant k_n.*
(ii) *Let $\boldsymbol{X}, \boldsymbol{Y}$ be complex random vectors satisfying the conditions in Section* 8.2 *and $\boldsymbol{Z} = \boldsymbol{X} * \boldsymbol{Y}$. Then we have*

$$\frac{\|\hat{\boldsymbol{Z}} - \boldsymbol{Z}\|^2}{\|\boldsymbol{X}\|^2\|\boldsymbol{Y}\|^2} = (3\bar{k}_n^2 + \sigma_{\mathbb{C},\times}^2)u^2.$$

The convolution via FFT is faster than ordinary convolution as soon as $t \geq 3$ (resp. $t \geq 6$ for real data). If sequential summation is used, the roundoff errors of ordinary convolution are of order $\mathcal{O}(n^2)$ and larger than those obtained via FFT with every accurate precomputation method (i.e. direct call or repeated subvector multiplication). For cascade summation the roundoff errors are of the same order of magnitude (though smaller for the ordinary computation), if the twiddle factors are precomputed by direct call.

Circulant matrices have the important property that they can be diagonalized by $\boldsymbol{F}_n$, namely

$$\boldsymbol{A}_n := \operatorname{circ} \boldsymbol{a} = \sqrt{n}\boldsymbol{J}'_n\boldsymbol{F}_n(\operatorname{diag} \boldsymbol{F}_n\boldsymbol{a})\boldsymbol{F}_n. \tag{8.58}$$

This follows immediately from (8.56). Hence a linear system

$$\boldsymbol{A}_n\boldsymbol{x} = \boldsymbol{a} * \boldsymbol{x} = \boldsymbol{f} \qquad (\boldsymbol{f} \in \mathbb{C}^n) \tag{8.59}$$

can be transformed into

$$\sqrt{n}(\operatorname{diag} \boldsymbol{F}_n\boldsymbol{a})\boldsymbol{F}_n\boldsymbol{x} = \boldsymbol{F}_n\boldsymbol{f}.$$

Let $\boldsymbol{b} := \boldsymbol{F}_n\boldsymbol{a}$. If $\operatorname{diag} \boldsymbol{b}$ is invertible, i.e. $b_j \neq 0$ for all $j = 0, \ldots, n-1$, then the linear system (8.59) is uniquely solvable and the solution reads as follows

$$\boldsymbol{x} = n^{-1/2}\boldsymbol{J}'_n\boldsymbol{F}_n\big(\operatorname{diag} \boldsymbol{b}\big)^{-1}\boldsymbol{F}_n\boldsymbol{f}.$$

For $n = 2^t$ $(t \geq 3)$, this solution can be computed in $\mathcal{O}(n \log n)$ operations with the aid of FFT.

ALGORITHM 8.10 (Fast deconvolution)
Input: $n = 2^t$ $(t \geq 3)$; $\boldsymbol{a}, \boldsymbol{f} \in \mathbb{C}^n$.
1. Precompute $\boldsymbol{b} := \boldsymbol{F}_n \boldsymbol{a}$ via FFT.
2. Compute $\boldsymbol{g} := \boldsymbol{F}_n \boldsymbol{f}$ via FFT.
3. For $j = 0 : 1 : n-1$ form $h_j := g_j / b_j$.
4. Form $\boldsymbol{x} := \frac{1}{\sqrt{n}} \boldsymbol{J}'_n \boldsymbol{F}_n (h_j)_{j=0}^{n-1}$ via FFT.

Output: $\boldsymbol{x} \in \mathbb{C}^n$ with $\boldsymbol{a} * \boldsymbol{x} = \boldsymbol{f}$.

The fast deconvolution involves mainly three FFTs and a diagonal scaling and therefore needs $\mathcal{O}(n \log n)$ operations. Now we analyze the effect of roundoff errors. By Corollary 8.2, steps 1 and 2 yield

$$\hat{\boldsymbol{b}} = (\boldsymbol{F}_n + \boldsymbol{\Delta}_1)\boldsymbol{a} \qquad (\|\boldsymbol{\Delta}_1\|_2 \leq k_n u), \tag{8.60}$$

$$\hat{\boldsymbol{g}} = (\boldsymbol{F}_n + \boldsymbol{\Delta}_2)\boldsymbol{f} \qquad (\|\boldsymbol{\Delta}_2\|_2 \leq k_n u). \tag{8.61}$$

Then (8.60) implies that $\operatorname{diag} \hat{\boldsymbol{b}} = \operatorname{diag} \boldsymbol{b} + \operatorname{diag} \boldsymbol{\Delta}_1 \boldsymbol{a}$. Using Lemma 8.1, step 3 yields the result

$$\hat{\boldsymbol{h}} = (\operatorname{diag} \boldsymbol{b} + \boldsymbol{D})^{-1} \hat{\boldsymbol{g}} \tag{8.62}$$

with the diagonal matrix $\boldsymbol{D} := \operatorname{diag} \boldsymbol{\Delta}_1 \boldsymbol{a} + \operatorname{diag}(\varepsilon_j^{\div} \hat{b}_j)_{j=0}^{n-1}$ and $|\varepsilon_j^{\div}| \leq (\mu_{\mathbb{C}} + 3)u$. Note that we have

$$\begin{aligned}
\|\operatorname{diag} \boldsymbol{\Delta}_1 \boldsymbol{a}\|_2 &= \|\boldsymbol{\Delta}_1 \boldsymbol{a}\|_\infty \leq \|\boldsymbol{\Delta}_1\|_2 \leq k_n u \|\boldsymbol{a}\|_2, \\
\|\operatorname{diag}(\varepsilon_j^{\div} \hat{b}_j)_{j=0}^{n-1}\|_2 &\leq (\mu_{\mathbb{C}} + 3)u \|\hat{\boldsymbol{b}}\|_\infty \leq (\mu_{\mathbb{C}} + 3)u \|\hat{\boldsymbol{b}}\|_2 \\
&\leq (\mu_{\mathbb{C}} + 3 + \mathcal{O}(u)) \|\boldsymbol{a}\|_2 u
\end{aligned}$$

and so $\|\boldsymbol{D}\|_2 \leq (k_n + \mu_{\mathbb{C}} + 3 + \mathcal{O}(u)) \|\boldsymbol{a}\|_2 u$. Finally step 4 yields

$$\hat{\boldsymbol{x}} = \frac{1}{\sqrt{n}} \boldsymbol{J}'_n (\boldsymbol{F}_n + \boldsymbol{\Delta}_3) \hat{\boldsymbol{h}} \qquad (\|\boldsymbol{\Delta}_3\|_2 \leq k_n u). \tag{8.63}$$

Putting (8.61)–(8.63) together, we obtain

$$\boldsymbol{x} = \frac{1}{\sqrt{n}} \boldsymbol{J}'_n (\boldsymbol{F}_n + \boldsymbol{\Delta}_3)(\operatorname{diag} \boldsymbol{b} + \boldsymbol{D})^{-1} (\boldsymbol{F}_n + \boldsymbol{\Delta}_2) \boldsymbol{f}.$$

Assume that $k_n u \ll 1$. Then we have by working to first order in u that

$$\begin{aligned}
\boldsymbol{f} &= \sqrt{n} (\boldsymbol{F}_n + \boldsymbol{\Delta}_2)^{-1} (\operatorname{diag} \boldsymbol{b} + \boldsymbol{D}) (\boldsymbol{F}_n + \boldsymbol{\Delta}_3)^{-1} \boldsymbol{J}'_n \hat{\boldsymbol{x}} \\
&\approx \sqrt{n} (\boldsymbol{F}_n^{-1} - \boldsymbol{F}_n^{-1} \boldsymbol{\Delta}_2 \boldsymbol{F}_n^{-1}) (\operatorname{diag} \boldsymbol{b} + \boldsymbol{D}) (\boldsymbol{F}_n^{-1} - \boldsymbol{F}_n^{-1} \boldsymbol{\Delta}_3 \boldsymbol{F}_n^{-1}) \boldsymbol{J}'_n \hat{\boldsymbol{x}} \\
&= (\boldsymbol{A}_n + \Delta \boldsymbol{A}) \hat{\boldsymbol{x}}
\end{aligned}$$

with

$$\begin{aligned}
\Delta \boldsymbol{A} := \sqrt{n} \Big(&-\boldsymbol{F}_n^{-1} (\operatorname{diag} \boldsymbol{b}) \boldsymbol{F}_n^{-1} \boldsymbol{\Delta}_3 \boldsymbol{F}_n^{-1} + \boldsymbol{F}_n^{-1} \boldsymbol{D} \boldsymbol{F}_n^{-1} \\
&- \boldsymbol{F}_n^{-1} \boldsymbol{\Delta}_2 \boldsymbol{F}_n^{-1} (\operatorname{diag} \boldsymbol{b}) \boldsymbol{F}_n^{-1} + \mathcal{O}(u^2) \|\boldsymbol{a}\|_2 \Big).
\end{aligned}$$

Estimating the spectral norm, we obtain

$$\begin{aligned}\|\Delta A\|_2 &\le \sqrt{n}\big(\|\operatorname{diag} b\|_2\|\Delta_3\|_2 + \|D\|_2 + \|\Delta_2\|_2\|\operatorname{diag} b\|_2 + \mathcal{O}(u^2)\|a\|_2\big)\\ &\le \sqrt{n}\big(3k_n + \mu_c + 3 + \mathcal{O}(u)\big)\|a\|_2 u.\end{aligned}$$

Using the Frobenius norm, we have $\|a\|_2 = n^{-1/2}\|\operatorname{circ} a\|_F \le \|\operatorname{circ} a\|_2 = \|A_n\|_2$ and hence

$$\|f - A_n\hat{x}\|_2 \le \|\Delta A\|_2\|\hat{x}\|_2 \le \sqrt{n}\big(3k_n + \mu_c + 3 + \mathcal{O}(u)\big)\|A_n\|_2\|\hat{x}\|_2 u.$$

Thus we obtain the following stability result:

THEOREM 8.6
Let $n = 2^t$ $(t \ge 3)$ with $k_n u \ll 1$ and $a, f \in \mathbb{C}^n \setminus \{o\}$ be given. Let $A_n = \operatorname{circ} a \in \mathbb{C}^{n\times n}$ be an invertible, circulant matrix. Then the Algorithm 8.10 *of fast deconvolution is normwise backward stable (see* [6], *p.* 142) *by*

$$\frac{\|f - A_n\hat{x}\|_2}{\|A_n\|_2\|\hat{x}\|_2 + \|f\|_2} \le \sqrt{n}(3k_n + \mu_c + 3)u + \mathcal{O}(u^2).$$

A similar theorem can be found in [6], pp. 468–470. The normwise forward stability of Algorithm 8.10 can be obtained by applying standard perturbation theory (see [6], p. 133). Note that

$$\operatorname{cond}_2 A_n = \|A_n\|_2\|A_n^{-1}\|_2 = \frac{\max\{|b_j| : j = 0,\ldots,n-1\}}{\min\{|b_j| : j = 0,\ldots,n-1\}}.$$

In the following we set

$$\varepsilon_n := \sqrt{n}(3k_n + \mu_c + 3)u.$$

COROLLARY 8.3
Let $n = 2^t$ $(t \ge 3)$ with $a, f \in \mathbb{C}^n \setminus \{o\}$ be given. Let $A_n = \operatorname{circ} a \in \mathbb{C}^{n\times n}$ be an invertible, circulant matrix. Assume that

$$\varepsilon_n \operatorname{cond}_2 A_n < 1$$

Then the Algorithm 8.10 *of fast deconvolution is normwise forward stable (see* [6], *p.* 142) *by*

$$\frac{\|\hat{x} - x\|_2}{\|x\|_2} \le \frac{\varepsilon_n \operatorname{cond}_2 A_n}{1 - \varepsilon_n \operatorname{cond}_2 A_n}.$$

If $A_n = \operatorname{circ} a$ $(a \in \mathbb{C}^n)$ is not invertible, then we can consider the *pseudo–inverse* of A_n defined by

$$A_n^+ := \frac{1}{\sqrt{n}} F_n^*(\operatorname{diag} b)^+ F_n \qquad (b = F_n a)$$

with $(\operatorname{diag} \boldsymbol{b})^+ = \operatorname{diag}(b_j^+)_{j=0}^{n-1}$ and

$$b_j^+ := \begin{cases} 0 & \text{if } b_j = 0, \\ 1/b_j & \text{if } b_j \neq 0. \end{cases}$$

Then $\boldsymbol{x} = \boldsymbol{A}_n^+ \boldsymbol{f}$ is the *least squares solution of minimal norm* of the system $\boldsymbol{A}_n\boldsymbol{x} = \boldsymbol{f}$, i.e., $\boldsymbol{x} \in \mathbb{C}^n$ is the solution of the least squares problem

$$\|\boldsymbol{A}_n\boldsymbol{x} - \boldsymbol{f}\|_2 = \min\{\|\boldsymbol{A}_n\boldsymbol{z} - \boldsymbol{f}\|_2 : \boldsymbol{z} \in \mathbb{C}^n\} \tag{8.64}$$

and among all the solutions of (8.64), $\boldsymbol{x}$ is the solution of minimal Euclidean norm. For the proof we observe that (8.58) implies

$$\|\boldsymbol{A}_n\boldsymbol{z} - \boldsymbol{f}\|_2 = \|\boldsymbol{F}_n^*\big(\sqrt{n}(\operatorname{diag} \boldsymbol{b})\boldsymbol{F}_n\boldsymbol{z} - \boldsymbol{F}_n\boldsymbol{f}\big)\|_2 = \|\sqrt{n}(\operatorname{diag} \boldsymbol{b})\boldsymbol{F}_n\boldsymbol{z} - \boldsymbol{F}_n\boldsymbol{f}\|_2$$

and hence the minimum is attained when $\boldsymbol{F}_n\boldsymbol{z} = \frac{1}{\sqrt{n}}(\operatorname{diag} \boldsymbol{b})^+\boldsymbol{F}_n\boldsymbol{f}$. Therefore

$$\boldsymbol{x} = \frac{1}{\sqrt{n}}\boldsymbol{F}_n^*(\operatorname{diag} \boldsymbol{b})^+\boldsymbol{F}_n\boldsymbol{f}$$

is the least squares solution of minimal norm of the system $\boldsymbol{A}_n\boldsymbol{x} = \boldsymbol{f}$.

References

[1] M. Arioli, H. Munthe-Kaas, and L. Valdettaro. Componentwise error analysis for FFTs with applications to fast Helmholtz solvers. *Numerical Algorithms*, 12:65–88, 1996.

[2] D. Calvetti. A stochastic roundoff error analysis for the fast fourier transform. *Math. Comp.*, 56:755 – 774, 1991.

[3] C. Y. Chu. *The fast Fourier transform on hypercube parallel computers.* PhD thesis, Cornell University, 1987.

[4] P. Duhamel and H. Hollmann. Split radix FFT algorithms. *Electron. Lett.*, 20:14 – 16, 1984.

[5] P. Duhamel, Y. Maheux, and J. Petit. A fast algorithm for the implementation of filter banks based on time domain aliasing cancellation. In *Proc. IEEE Int. Conf. Acoust. Speech Signal Proc.*, volume 20, pages 2209 – 2212, Toronto, 1991.

[6] N. J. Higham. *Accuracy and Stability of Numerical Algorithms.* SIAM, Philadelphia, 1996.

[7] C. Van Loan. *Computational Frameworks for the Fast Fourier Transform.* SIAM, Philadelphia, 1992.

[8] S. Mallat. *A Wavelet Tour of Signal Processing.* Academic Press, San Diego, 1998.

[9] H. S. Malvar. *Signal Processing with Lapped Transforms.* Artech House, Norwood, 1992.

[10] G. U. Ramos. Roundoff error analysis of the fast Fourier transform. *Math. Comp.*, 25:757 – 768, 1971.

[11] K. R. Rao and P. Yip. *Discrete Cosine Transform: Algorithm, Advantages and Applications.* Academic Press, San Diego, 1990.

[12] J. C. Schatzman. Accuracy of the discrete Fourier transform and the fast Fourier transform. *SIAM J. Sci. Comput.*, 17:1150 – 1166, 1996.

[13] U. Schreiber. *Schnelle und numerisch stabile trigonometrische Transformationen.* PhD thesis, University of Rostock, 1999.

[14] G. Steidl and M. Tasche. *Schnelle Fouriertransformationen: Theorie und Anwendungen.* Fernuniversität Hagen, 1998.

[15] G. Strang. The discrete cosine transform. *SIAM Rev.*, 41:135 – 147, 1999.

[16] M. Tasche and Hm. Zeuner. Roundoff error analysis for the fast Fourier transform with precomputed twiddle factors. Technical report, Medical University of Luebeck, 1999.

[17] R. Tolimieri, M. An, and C. Lu. *Algorithms for Discrete Fourier Transforms and Convolution.* Springer–Verlag, New York, 1997.

[18] P. Y. Yalamov. Improvements of some bounds on the stability of fast Helmholtz solvers. Technical report, University of Rousse, 1998.

[19] Hm. Zeuner. Stochastic roundoff error analysis with applications to DFT and DCT. Technical report, Medical University of Luebeck, 1999.

Chapter 9

Biorthogonal Local Trigonometric Bases

Kai Bittner

Institute of Biomathematics and Biometrics, GSF – National Research Center for Environment and Health, D-85764 Neuherberg, Germany
E-mail: bittner@gsf.de

Abstract Local trigonometric bases consist of cosines and sines multiplied by smooth, well-localized window functions in order to have basis functions with good time-frequency localization. On the one hand, bases in the two-overlapping setting are considered. In particular, the development of such bases from the orthonormal bases of Coifman and Meyer to the general approach for the construction of biorthogonal bases introduced by Chui and Shi is reviewed. On the other hand, a new generalized theory for biorthogonal Wilson bases is presented which includes former approaches. Connections between the two-overlapping bases and the Wilson bases are pointed out. Numerous examples illustrate the theoretical results.

Contents

9.1 Introduction

Fourier methods have many applications in signal and image processing. The Fourier transform

$$\hat{f}(\xi) = \int_{\mathbb{R}} f(t)e^{-2\pi i t\xi}\, dt, \quad \xi \in \mathbb{R}, \tag{9.1}$$

of a function $f \in L^2(\mathbb{R})$ yields a frequency representation of f, i.e., $\hat{f}(\xi)$ represents the amplitude of the frequency term $e^{-2\pi i x\xi}$. The main disadvantage of the Fourier transform is that it does not represent local information in the time domain. A local perturbation of f may result in a perturbation of $\hat{f}$ in the whole frequency domain. To preserve local information in time so-called block transforms are used in signal processing. Here, the real axis is divided into disjoint intervals and Fourier series expansions are considered on each interval. This method, however, has several disadvantages. Fourier series converge rapidly when the function is smooth and periodic. Evidently the restriction of a smooth function to an interval is in general not smooth and periodic. The convergence will thus be slow. Furthermore, since each interval is handled separately, data processing, as e.g., thresholding or quantization, may cause discontinuities which appear as "blocking effects" in the reconstructed signal. Another disadvantage of block transforms is the poor frequency localization caused by the discontinuities of the basis functions used in this transform.

To obtain good localization in time and frequency Gabor [31] proposed to represent a signal as a superposition of elementary signals

$$g_{nm}(x) := g(x - an)e^{2\pi i bmx}, \quad m, n \in \mathbb{Z},$$

where a and b are some positive real numbers and $g \in L^2(\mathbb{R})$ is well localized around the origin of the time-frequency plane. By this definition the elementary signals g_{nm} are translations and modulations of g such that they are essentially localized at the point (na, mb) of the time-frequency plane. In particular, if g is chosen as the Gaussian e^{-x^2} the time-frequency localization is optimal, in the sense of Heisenberg. To obtain a unique, numerical stable representation for each function $f \in L^2(\mathbb{R})$ by a linear combination of the g_{nm} we need that the Gabor system $\{g_{nm}\}$ is a Riesz basis (see Section 9.2). But from the *Balian-Low theorem* [5, 8, 25, 24, 27, 38] it follows that if the functions g_{nm} constitute a Riesz basis of $L^2(\mathbb{R})$, which is possible only for $ab = 1$, then

$$\int_{\mathbb{R}} x^2 |g(x)|\, dx \int_{\mathbb{R}} \xi^2 |\hat{g}(\xi)|\, d\xi = \infty,$$

i.e., g has poor time-frequency localization. Only if one gives up uniqueness can one obtain Gabor frames with excellent time-frequency localization for $ab < 1$. For a detailed description of such Gabor frames we refer to [30].

To construct orthonormal bases with good time-frequency localization properties Sullivan et al. [43] and Wilson [49] suggested considering functions which are localized around the positive and negative frequencies of the same order. This idea was used by Daubechies [25] and Daubechies, Jaffard, and Journé [26] to construct orthonormal "Wilson bases" which consist of windowed sines and cosines (see Section 9.6) instead of the exponentials $e^{2\pi i b m \cdot}$. For such bases the disadvantage described in the Balian-Low theorem is completely removed.

Independently from these results Malvar [39, 40] introduced an efficient algorithm based on a lapped orthogonal discrete cosine transform (DCT) to remove "block artifacts" induced by block transforms. This algorithm is based on orthonormal local trigonometric bases consisting of the functions $w(\cdot - j)\cos(k + \frac{1}{2})\pi(\cdot - j)$, $j \in \mathbb{Z}$, $k \in \mathbb{N}_0$, in a so-called two-overlapping setting. Here, two-overlapping means that the window function w is compactly supported such that only consecutive translates of w are allowed to have overlapping supports. Based on the results from [26, 37, 40] Coifman and Meyer [21] investigated more general orthonormal local trigonometric bases, where the translates $w(\cdot - j)$ are replaced by a sequence of window functions w_j with variable support length which satisfy a two-overlapping condition (see Section 9.3). A detailed description of these Coifman-Meyer bases with further generalizations can be found in [4]. In [3], a connection between the bases of Coifman-Meyer and the orthonormal Wilson bases from [26] was pointed out.

The conditions of Coifman and Meyer for the orthonormality of local trigonometric bases allow only a small class of window functions. To obtain more freedom biorthogonal bases and Riesz bases are usually considered instead of orthonormal bases. Therefore, Matviyenko [41] as well as Jawerth and Sweldens [36] investigated particular biorthogonal local trigonometric bases in the two-overlapping setting in order to obtain a more effective representation of constant functions. A general theory of biorthogonal local trigonometric bases in the two-overlapping setting was introduced by Chui and Shi [17, 18].

Biorthogonal Wilson bases were studied by Coifman and Meyer in [22] for the particular case of a Gaussian window $w(x) := e^{-\zeta x^2}$. A detailed investigation of biorthogonal Wilson bases with a symmetric window function w of arbitrary shape was carried out by Chui and Shi in [18, 19]. A more general approach can be found in [13, 15].

The folding operators introduced by Wickerhauser in [47, 48] proved to be an important tool for the investigation of local trigonometric bases. Such folding operators, with numerous generalizations, were successfully used by many authors [11, 13, 14, 15, 16, 17, 28, 36, 41] in order to gain new results on local trigonometric bases. The main property of such folding operators is that they map the local trigonometric functions onto an orthonormal basis of trigonometric functions. Thanks to this property results from classical harmonical analysis can be applied.

Furthermore, we want to mention the application of best basis algorithms for the adaptive choice of the window size in [20, 23, 45, 48] and the construction of bivariate bases in [11, 16, 50, 51]. Results on the approximation with local trigonometric bases can be found in [14, 15, 29, 33]. For the design of optimized window functions we refer to [12, 14, 15, 36, 41]. To reduce blocking effects local trigonometric bases have been applied in speech processing [45] and image compression [1, 2, 35, 40]. Furthermore, efficient algorithms based on the fast Fourier transform (FFT) have been described using the folding operator [35, 41].

This chapter gives a detailed overview of the construction of local trigonometric bases. For the two-overlapping bases as well as for the Wilson bases we will present a general theory of biorthogonal Riesz bases. Therefore, the main assertions of the chapter deal with Riesz stability and biorthogonality. Folding and unfolding operators play a central role for our studies. Numerous examples are included to illustrate the theoretical results.

The chapter is organized as follows. In Section 9.2, Riesz bases are introduced. We will establish the results which we will need for our studies in the following sections. In particular, the application of folding operators is based on these results. In Sections 9.3–9.5, we consider the two-overlapping setting. We start with the orthonormal bases of Coifman and Meyer [21] in Section 9.3. Furthermore, the folding operators of Wickerhauser [47] are described. Section 9.4 gives a short overview of the biorthogonal bases of Jawerth and Sweldens [36]. The general theory of Chui and Shi [17] is presented in Section 9.5. Here, the theorems about Riesz stability and dual bases are proved using an unfolding operator. Finally, two particular examples of such local trigonometric bases are presented.

In Section 9.6 a general theory of biorthogonal Wilson bases developed in [13, 15] is introduced. In particular, the generalization of ideas from Chui and Shi [17] for the two-overlapping setting leads to new results for biorthogonal Wilson bases. Therefore, the main results point out the affinity between the bases of Chui and Shi and the biorthogonal Wilson bases. In several examples, former results [19, 22, 26] are reviewed with respect to the new theory. Furthermore, it turns out that biorthogonal Wilson bases with two-overlapping window functions can be seen as a particular case of the bases of Chui and Shi in [17].

9.2 Riesz Bases

To describe and investigate a complex signal it is often useful to express it as a superposition of simple well-understood objects. For many applications it is important that such representations are unique and numerical stable. These conditions are satisfied if the elementary signals form a Riesz basis.

DEFINITION 9.1 *Let H be a separable complex Hilbert space with inner product $\langle \cdot, \cdot \rangle$ and $\mathcal{I}$ a countable index set. A sequence $\{x_n : n \in \mathcal{I}\} \subset H$ is a* Riesz basis *of H, if $\{x_n\}$ is complete in H and there exist constants $A, B > 0$ such that for each finite subset $\mathcal{J} \subset \mathcal{I}$ it holds*

$$A \sum_{n \in \mathcal{J}} |a_n|^2 \leq \left\| \sum_{n \in \mathcal{J}} a_n x_n \right\|_H^2 \leq B \sum_{n \in \mathcal{J}} |a_n|^2 \tag{9.2}$$

with arbitrary coefficients $a_n \in \mathbb{C}$. The constants A and B are called Riesz bounds.

The notion of Riesz bases were introduced by Barry [6, 7]. Riesz bases can be described in different ways (cf. [52, p. 32–36 and p. 188–189]). In particular, a Riesz basis can be defined as the image of an orthonormal basis under a topological isomorphism. A linear operator $T : H \to H$ is a *topological isomorphism* if it is bounded and bijective and if the inverse operator T^{-1} is bounded, too.

THEOREM 9.1
Let $\{x_n : n \in \mathcal{I}\}$ be a sequence in a separable Hilbert space H. Then, the following assertions are equivalent.

(i) *The sequence $\{x_n : n \in \mathcal{I}\}$ is a Riesz basis with Riesz bounds A and B.*

(ii) *For each orthonormal basis $\{e_n\}$ of H there exists a topological* *isomorphism* $T : H \to H$ *with* $\|T\|^2 \leq B$, $\|T^{-1}\|^2 \leq A^{-1}$ *and* $Te_n = x_n$, $n \in \mathcal{I}$.

PROOF (i)⇒(ii). Assume $\{x_n\}$ is a Riesz basis. First, we will show that for each $y \in H$ there exist uniquely determined coefficients $a_n(y)$ such that $y = \sum_{n\in\mathcal{I}} a_n(y)x_n$. Since $\{x_n\}$ is complete a sequence $\{y_N : N \in \mathbb{N}\} \subset \operatorname{span}\{x_n\}$ exists for each $y \in H$ so that

$$y_N = \sum_{n\in\mathcal{J}_N} a_{N,n}x_n \to y, \quad N \to \infty,$$

where the index sets $\mathcal{J}_N \subset \mathcal{I}$ are finite with $|\mathcal{J}_N| = N$. We set $a_{N,n} := 0$ for $n \notin \mathcal{J}_N$ such that $\boldsymbol{a}_N := \{a_{N,n} : n \in \mathcal{I}\} \in \ell^2(\mathcal{I})$. Because $\{y_N\}$ is a Cauchy sequence in H we obtain from the lower Riesz inequality in (9.2) that

$$\sum_{n\in\mathcal{J}_N} |a_{N,n} - a_{M,n}|^2 \leq A^{-1}\|y_N - y_M\|_H^2 \to 0, \quad N > M, \ M \to \infty.$$

Hence, $\{\boldsymbol{a}_N\}$ is a Cauchy sequence in $\ell^2(\mathcal{I})$. From the completeness of $\ell^2(\mathcal{I})$ we deduce the existence of a sequence $\boldsymbol{a}(y) := \{a_n(y)\} \in \ell^2(\mathcal{I})$ with $\boldsymbol{a}_N \to \boldsymbol{a}(y)$. Analogously, the upper Riesz inequality in (9.2) implies that for any given sequence $\{a_n\} \in \ell^2(\mathcal{I})$ a Cauchy sequence in H is given by $\{\sum_{n\in\mathcal{J}_N} a_n x_n : N \in \mathbb{N}\}$. Hence, $\sum_{n\in\mathcal{I}} a_n x_n$ is convergent in the Hilbert space H and from (9.2) we conclude

$$A\sum_{n\in\mathcal{I}} |a_n|^2 \leq \Big\|\sum_{n\in\mathcal{I}} a_n x_n\Big\|_H^2 \leq B\sum_{n\in\mathcal{I}} |a_n|^2. \tag{9.3}$$

In particular, for the sequence $\{a_n(y)\} \in \ell^2(\mathcal{I})$ we obtain

$$\Big\|\sum_{n\in\mathcal{J}_N} a_{N,n}x_n - \sum_{n\in\mathcal{I}} a_n(y)x_n\Big\|_H^2 \leq B\sum_{n\in\mathcal{I}} |a_{N,n} - a_n(y)|^2 \to 0, \quad N \to \infty.$$

Now, by triangle inequality it follows immediately that $y = \sum_{n\in\mathcal{I}} a_n(y)x_n$.

Let us assume that $y = \sum_{n\in\mathcal{I}} b_n x_n$ for some coefficients b_n. The lower inequality in (9.3) implies

$$\sum_{n\in\mathcal{I}} |a_n(y) - b_n|^2 \leq A^{-1}\Big\|\sum_{n\in\mathcal{I}} a_n(y)x_n - \sum_{n\in\mathcal{J}_n} b_n x_n\Big\|_H^2 = 0,$$

i.e., $a_n(y) = b_n$. Hence, the coefficients $a_n(y)$ are uniquely determined by $y = \sum_{n\in\mathcal{I}} a_n(y)x_n$ for each $y \in H$.

Now, for an arbitrary orthonormal basis $\{e_n\}$ of H we can find the topological isomorphism T as follows. For $y \in H$ we set $Ty := \sum_{n\in\mathcal{I}}\langle y, e_n\rangle x_n$, such that $Te_n = x_n$. From the upper Riesz inequality in (9.3) it follows that

$$\|Ty\|_H^2 \le B\sum_{n\in\mathcal{I}} |\langle y, e_n\rangle|^2 = B\|y\|_H^2,$$

i.e., T is bounded. Furthermore, it is easy to check that the inverse operator is given by $T^{-1}y = \sum_{n\in\mathcal{I}} a_n(y)e_n$. The lower Riesz inequality in (9.3) yields

$$\|T^{-1}y\|_H^2 = \sum_{n\in\mathcal{I}} |a_n(y)|^2 \le A^{-1}\|y\|_H^2.$$

Hence, T is a topological isomorphism which satisfies the conditions in (ii).

(ii)⇒(i). Let T be a topological isomorphism with $\|T\|^2 \le B$, $\|T^{-1}\|^2 \le A^{-1}$ and $Te_n = x_n$, $n \in \mathcal{I}$, where $\{e_n\}$ is an orthonormal basis of H. Since T^{-1} is a bounded operator $T^{-1}y$ is contained in H for every $y \in H$. Therefore, $T^{-1}y$ has the unique representation $T^{-1}y = \sum_{n\in\mathcal{I}}\langle T^{-1}y, e_n\rangle e_n$. Applying the continuous operator T we obtain

$$y = \sum_{n\in\mathcal{I}}\langle T^{-1}y, e_n\rangle Te_n = \sum_{n\in\mathcal{I}} a_n(y)x_n$$

with $a_n(y) := \langle T^{-1}y, e_n\rangle$. Hence, the sequence $\{x_n\}$ is complete.

Since $\{e_n\}$ is an orthonormal basis we have for any finite index set $\mathcal{J} \subset \mathcal{I}$

$$\Big\|\sum_{n\in\mathcal{J}} a_n x_n\Big\|_H^2 = \Big\|T\Big(\sum_{n\in\mathcal{J}} a_n e_n\Big)\Big\|_H^2 \le B\Big\|\sum_{n\in\mathcal{J}} a_n e_n\Big\|_H^2 = B\sum_{n\in\mathcal{J}} |a_n|^2$$

and

$$A\sum_{n\in\mathcal{I}} |a_n|^2 = A\Big\|\sum_{n\in\mathcal{J}} a_n e_n\Big\|_H^2 = A\Big\|T^{-1}\Big(\sum_{n\in\mathcal{J}} a_n x_n\Big)\Big\|_H^2 \le \Big\|\sum_{n\in\mathcal{J}} a_n x_n\Big\|_H^2,$$

i.e., the Riesz inequality (9.2) is shown. ∎

The equivalence in Theorem 9.1 was already proved by Young [52, p. 32–36], without considering the relationship between Riesz bounds and the norms of T and its inverse. The proof of Theorem 9.1 gives us some further assertions on Riesz bases. Obviously, every $y \in H$ has the representation $y = \sum_{n\in\mathcal{I}} a_n(y)x_n$ with the unique coefficients $a_n(y) := \langle T^{-1}y, e_n\rangle$, i.e., we have a unique representation of $y \in H$ by the coefficients $a_n(y)$. Furthermore, inequality (9.3) states that this representation is stable so that a small change of y with respect to the H-norm causes only a small change of the sequence $\{a_n(y)\}$ with respect to the norm in $\ell^2(\mathcal{I})$ and vice versa. In particular, the stability is better if A and B are closer to each other. Moreover, we know that $\sum_{n\in\mathcal{I}} a_n x_n$ converges absolutely for any sequence $\{a_n\} \in \ell^2(\mathcal{I})$. Now, we want to give an explicit description of the functionals $a_n(y)$. Since H is a Hilbert space we know from the Riesz's representation theorem that these functionals can be identified with elements of H.

THEOREM 9.2

Let $\{x_n\}$ be a Riesz basis of the Hilbert space H with Riesz bounds A and B. Then, there exists a uniquely determined dual basis $\{\tilde{x}_n\}$ which is biorthogonal to $\{x_n\}$, i.e.,

$$\langle x_n, \tilde{x}_m\rangle = \delta_{n,m}, \quad n, m \in \mathcal{I}. \tag{9.4}$$

The sequence $\{\tilde{x}_n\}$ is a Riesz basis with Riesz bounds B^{-1} and A^{-1}. The basis representations of $y \in H$ are given by

$$y = \sum_{n\in\mathcal{I}} \langle y, \tilde{x}_n\rangle x_n \qquad \text{and} \qquad y = \sum_{n\in\mathcal{I}} \langle y, x_n\rangle \tilde{x}_n. \tag{9.5}$$

If an orthonormal basis $\{e_n\}$ and a topological isomorphism T are given as in Theorem 9.1(ii) *the elements of the dual basis can be determined by $\tilde{x}_n = T^{-\star}e_n$, where $T^{-\star} := (T^{-1})^\star = (T^\star)^{-1}$ is the adjoint operator of T^{-1}.*

PROOF Let $\{e_n\}$ be an orthonormal basis. Since $\{x_n\}$ is a Riesz basis there exists a topological isomorphism T such that $Te_n = x_n$. Since

$$\langle x_n, T^{-\star}e_m\rangle = \langle T^{-1}Te_n, e_m\rangle = \langle e_n, e_m\rangle = \delta_{n,m}$$

the existence of a biorthogonal sequence is shown. On the other hand for any sequence $\{\tilde{x}_n\}$ biorthogonal to $\{x_n\}$ it holds that

$$\delta_{n,m} = \langle x_n, \tilde{x}_m\rangle = \langle e_n, T^\star\tilde{x}_m\rangle.$$

Hence, $T^\star\tilde{x}_n = e_n$ which proves the uniqueness of $\{\tilde{x}_n\} = \{T^{-\star}e_n\}$. Since T is a topological isomorphism $T^{-\star}$ is a topological isomorphism with

$\|T^{-\star}\| = \|T^{-1}\| \leq A^{-1}$ and $\|(T^{-\star})^{-1}\| = \|T\| \leq B$. Thus, $\{\tilde{x}_n\}$ is a Riesz basis with Riesz bounds B^{-1} and A^{-1}. The first identity in (9.5) follows immediately from the proof of Theorem 9.1 using that $a_n(y) = \langle T^{-1}y, e_n\rangle = \langle y, T^{-\star}e_n\rangle$. Finally, the second identity in (9.5) is implied by the fact that $\{x_n\}$ is the dual basis with respect to $\{\tilde{x}_n\}$. ∎

In the following theorem we will see that orthonormal bases can be described as particular Riesz bases. Furthermore, orthonormal bases can be characterized by a unitary isomorphism, which is a linear map $T : H \to H$ with $TT^\star = T^\star T = \text{id}$, where id is the identity operator.

THEOREM 9.3

Let $\{x_n : n \in \mathcal{I}\}$ be a sequence in a separable Hilbert space H. Then, the following assertions are equivalent.

(i) *The sequence $\{x_n : n \in \mathcal{I}\}$ is an orthonormal basis of H.*

(ii) *The sequence $\{x_n : n \in \mathcal{I}\}$ is a Riesz basis of H with Riesz bounds $A = B = 1$.*

(iii) *For each orthonormal basis $\{e_n\}$ of H there exists a unitary isomorphism $T : H \to H$ such that $Te_n = x_n$, $n \in \mathcal{I}$.*

PROOF **(i)⇒(ii).** Every orthonormal basis is complete. The orthonormality implies

$$\Big\| \sum_{n\in\mathcal{J}} a_n x_n \Big\|_H^2 = \sum_{n\in\mathcal{J}} |a_n|^2$$

for each finite subset $\mathcal{J} \subset \mathcal{I}$. Hence, the Riesz inequality (9.2) holds with $A = B = 1$.

(ii)⇒(iii). From Theorem 9.1 we know that there exists a topological isomorphism with $\|T\| \leq 1$ and $\|T^{-1}\| \leq 1$ which maps e_n onto x_n. Therefore, we conclude for all $y \in H$ that $\|Ty\|_H \leq \|y\|_H$ and $\|y\|_H = \|T^{-1}Ty\|_H \leq \|Ty\|_H$. Summing up, we obtain $\|Ty\|_H = \|y\|_H$ and analogously $\|T^\star y\|_H = \|y\|_H$. Then, we have

$$\|y\|_H^2 = \langle T^\star Ty, y\rangle \leq \|T^\star Ty\|_H \, \|y\|_H = \|Ty\|_H \, \|y\|_H = \|y\|_H^2,$$

i.e., equality is attained in the Cauchy-Schwartz inequality above. Hence, $T^\star Ty = \lambda y$, $\lambda \in \mathbb{C}$. From $\|y\|_H^2 = \langle T^\star Ty, y\rangle = \lambda\langle y, y\rangle$ we conclude $\lambda = 1$. Now, it follows immediately that $T^\star T = TT^\star = \text{id}$ and we have shown that T is a unitary isomorphism.

(iii)⇒(i). Obviously, T is a topological isomorphism. Then, Theorem 9.1 implies that $\{x_n\}$ is a Riesz basis and therefore complete. Furthermore, by the unitarity of T it follows that

$$\langle x_n, x_m \rangle = \langle Te_n, Te_m \rangle = \langle e_n, e_m \rangle = \delta_{n,m}.$$

Hence, $\{x_n\}$ is a complete orthonormal system, i.e., $\{x_n\}$ is an orthonormal basis. ∎

9.3 Coifman-Meyer Bases and Folding

Let the real axis be divided into intervals $[a_j, a_{j+1})$, where $\{a_j : j \in \mathbb{Z}\}$ is an increasing sequence with $\lim_{j\to\pm\infty} a_j = \pm\infty$. We denote the length of the interval $[a_j, a_{j+1})$ by $h_j := a_{j+1} - a_j$. Furthermore, let positive values $\varepsilon_j > 0$ with $\varepsilon_j + \varepsilon_{j+1} \leq h_j$ be given and set $a_j^{\pm} := a_j \pm \varepsilon_j$. Then, the intervals (a_j^-, a_j^+) are disjoint neighbourhoods of the points a_j. For each $j \in \mathbb{Z}$ we introduce a window function $w_j \in L^2(\mathbb{R})$ with $\operatorname{supp} w_j \subset [a_j^-, a_{j+1}^+]$. By this construction we obtain a two-overlapping setting, i.e., only the immediate neighbouring windows w_{j-1} and w_j, $j \in \mathbb{Z}$, have overlapping support (see Fig. 9.1). In detail, the window functions w_j shall fulfill the following conditions:

$$w_j(x) = 0, \qquad x \notin (a_j^-, a_{j+1}^+), \tag{9.6}$$

$$0 \leq w_j(x) \leq 1, \qquad x \in \mathbb{R}, \tag{9.7}$$

$$w_j(x) = 1, \qquad x \in [a_j^+, a_{j+1}^-], \tag{9.8}$$

$$w_j(a_j + x) = w_{j-1}(a_j - x), \qquad |x| < \varepsilon_j, \tag{9.9}$$

$$w_j^2(a_j + x) + w_{j-1}^2(a_j + x) = 1, \qquad |x| < \varepsilon_j. \tag{9.10}$$

Furthermore, we consider (trigonometric) functions t_j^k, $j \in \mathbb{Z}$, $k \in \mathbb{N}_0$

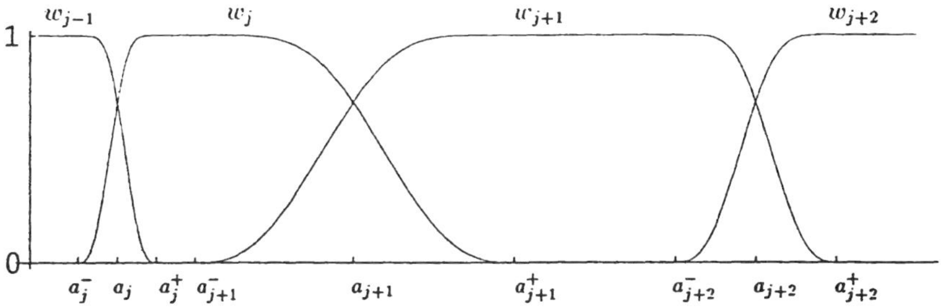

FIGURE 9.1
Two-overlapping window functions of Coifman and Meyer.

which for all $j \in \mathbb{Z}$ satisfy:

$$\{t_j^k : k \in \mathbb{N}_0\} \text{ is an orthonormal basis of } L^2([a_j, a_{j+1}]), \tag{9.11}$$

$$t_j^k(a_j + x) = \varsigma_j \, t_j^k(a_j - x), \quad |x| < \varepsilon_j, \tag{9.12}$$

$$t_{j-1}^k(a_j + x) = -\varsigma_j \, t_{j-1}^k(a_j - x), \quad |x| < \varepsilon_j, \tag{9.13}$$

where ς_j is either 1 or -1. The condition (9.12) means that all the functions t_j^k, $k \in \mathbb{N}_0$, have the same parity in a_j, i.e., are even resp. odd in a_j if $\varsigma_j = 1$ resp. $\varsigma_j = -1$. We say a function f is even resp. odd in a, if $f(a+x) = f(a-x)$ resp. $f(a+x) = -f(a-x)$. Furthermore, from (9.13) we see that the t_{j-1}^k, $k \in \mathbb{N}_0$, are even resp. odd in a_j, if the t_j^k, $k \in \mathbb{N}_0$, are odd resp. even in a_j. This setting implies that the product $t_j^k \, t_{j-1}^\ell$ is odd in a_j for any choice of $k, \ell \in \mathbb{N}_0$. For any admissible choice of the ς_j there exist trigonometric functions t_j^k which satisfy (9.11)-(9.13).

LEMMA 9.1
Let the functions t_j^k, $j \in \mathbb{Z}$, $k \in \mathbb{N}_0$, be given by

$$t_j^k(x) = \begin{cases} \sqrt{\frac{2}{h_j}} \cos\left(\left(k + \frac{1}{2}\right)\pi\left(\frac{x-a_j}{h_j}\right)\right), & \text{if } \varsigma_j = \varsigma_{j+1} = 1, \\ \sqrt{\frac{2}{h_j}} \sin\left(\left(k + \frac{1}{2}\right)\pi\left(\frac{x-a_j}{h_j}\right)\right), & \text{if } \varsigma_j = \varsigma_{j+1} = -1, \\ \sqrt{\frac{\epsilon_k}{h_j}} \cos\left(k\pi\left(\frac{x-a_j}{h_j}\right)\right), & \text{if } \varsigma_j = -\varsigma_{j+1} = 1, \\ \sqrt{\frac{2}{h_j}} \sin\left((k+1)\pi\left(\frac{x-a_j}{h_j}\right)\right), & \text{if } -\varsigma_j = \varsigma_{j+1} = 1, \end{cases}$$

where

$$\epsilon_k := \begin{cases} 1, & \text{if } k = 0, \\ 2, & \text{otherwise.} \end{cases}$$

Then, the conditions (9.11)–(9.13) *are fulfilled.*

PROOF Obviously, the functions t_j^k satisfy (9.12) and (9.13). We still have to show that for a given $j \in \mathbb{Z}$ the sequence is an orthonormal basis of $L^2([a_j, a_{j+1}])$. Without loss of generality we assume $a_j = 0$ and $a_{j+1} = 1$. Every function $f \in L^2([0,1])$ can be extended uniquely to a function $g \in L^2([-2,2])$ with

$$f(x) = g(x) = \varsigma_j g(-x) = \varsigma_{j+1} g(2-x) = \varsigma_{j+1}\varsigma_j g(x-2), \quad x \in (0,1).$$

With this parity condition we obtain by Fourier series expansion

$$f(x) = g(x) = a_0 + \sum_{k=1}^{\infty} \left(a_k \cos\left(\tfrac{k\pi x}{2}\right) + b_k \sin\left(\tfrac{k\pi x}{2}\right)\right)$$

$$= \sum_{k\in\mathbb{N}_0} \tfrac{1}{4} \int_{-2}^{2} g(\xi)\, t_j^k(\xi)\, d\xi\; t_j^k(x) = \sum_{k\in\mathbb{N}_0} \int_0^1 f(\xi)\, t_j^k(\xi)\, d\xi\; t_j^k(x), \quad x \in [0,1].$$

Note, that the Fourier coefficient a_k resp. b_k vanishes if $g(x)$ and $\cos\left(\frac{k\pi x}{2}\right)$ resp. $\sin\left(\frac{k\pi x}{2}\right)$ have different parities in $x = 0$ or $x = 1$. Hence, each $f \in L^2([0,1])$ can be expanded in terms of the orthonormal system $\{t_j^k : k \in \mathbb{N}_0\}$. ∎

We can now construct smooth orthonormal bases of localized trigonometric functions.

THEOREM 9.4

Coifman and Meyer [21] *Let the window functions w_j fulfill the conditions* (9.6)–(9.10). *Further, the functions t_j^k shall satisfy* (9.11)–(9.13). *Then, the functions $\psi_j^k := w_j t_j^k$, $j \in \mathbb{Z}$, $k \in \mathbb{N}_0$, (see* Fig. 9.3*) constitute an orthonormal basis of $L^2(\mathbb{R})$.*

Coifman and Meyer have proved this theorem for the case that $\varsigma_j = \varsigma_{j+1}$ for all $j \in \mathbb{Z}$ by showing orthonormality and completeness of $\{\psi_j^k\}$. In [4], the proof was generalized to arbitrary functions t_j^k which satisfy (9.11)–(9.13). A particularly elegant proof was given by Wickerhauser in [47], where so-called folding operators are used. Such folding operators are very useful in theory and applications because they lead to a better understanding of local trigonometric bases. In this section, we will describe the

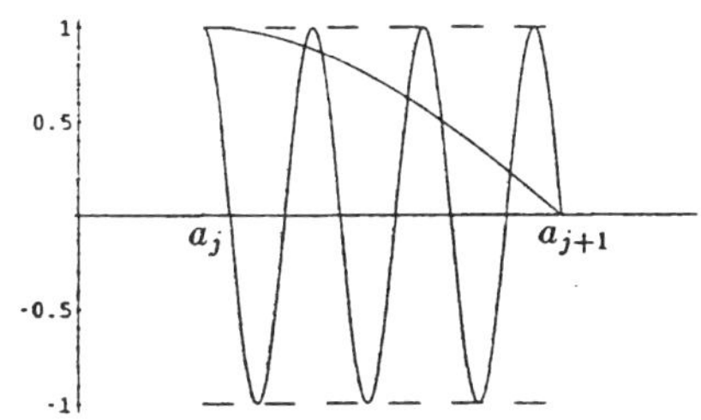

FIGURE 9.2
The functions $\chi_{[j,j+1]}t_j^0$ and $\chi_{[j,j+1]}t_j^6$.

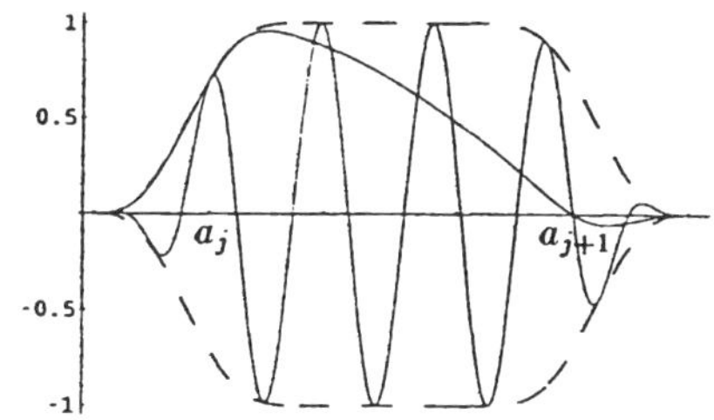

FIGURE 9.3
The functions ψ_j^0 and ψ_j^6 with the envelopes w_j and $-w_j$.

folding operators of Wickerhauser. Furthermore we will use these operators to prove the theorem of Coifman and Meyer.

DEFINITION 9.2 Wickerhauser [47]
Let $a \in \mathbb{R}$, $\varepsilon > 0$, and $m \in \mathbb{N}_0$ be given and let r be a real-valued function in $C^m(\mathbb{R})$ with

$$r^2(a+x) + r^2(a-x) = 1, \quad x \in \mathbb{R} \quad \text{and} \quad r(x) = \begin{cases} 0, & \text{if } x \le a - \varepsilon, \\ 1, & \text{if } x \ge a + \varepsilon. \end{cases}$$

Then, the folding operator $U := U(r,a) : L^2(\mathbb{R}) \to L^2(\mathbb{R})$ is defined by

$$Uf(a+x) := \begin{cases} r(a+x)f(a+x) + r(a-x)f(a-x), & \text{if } x \ge 0, \\ r(a-x)f(a+x) - r(a+x)f(a-x), & \text{if } x \le 0. \end{cases}$$

The application of the folding operator means that one multiplies $f \in L^2(\mathbb{R})$ by a smooth "cut-off function" r (see Fig. 9.4), "folds" the interval $(-\infty, a)$ by reflection onto (a, ∞) and obtains $Uf(x)$ for $x > a$ as the sum of both parts. Analogously, one obtains $Uf(x)$ for $x < a$ by multiplication of $f(x)$ and $r(2a - x)$, "folding" of (a, ∞) onto $(-\infty, a)$ and subtraction (see Fig. 9.5).

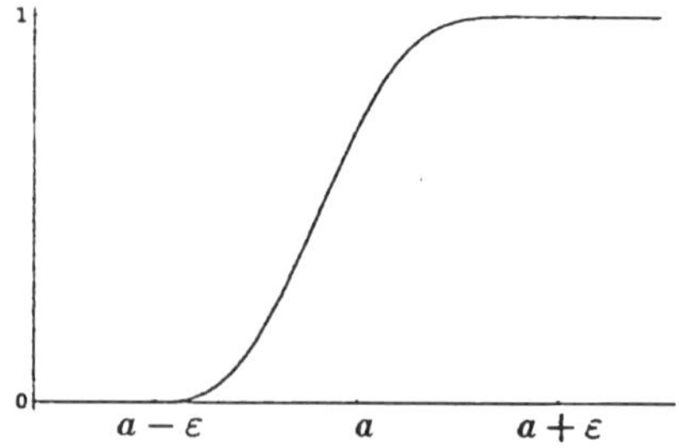

FIGURE 9.4
A cut-off function r.

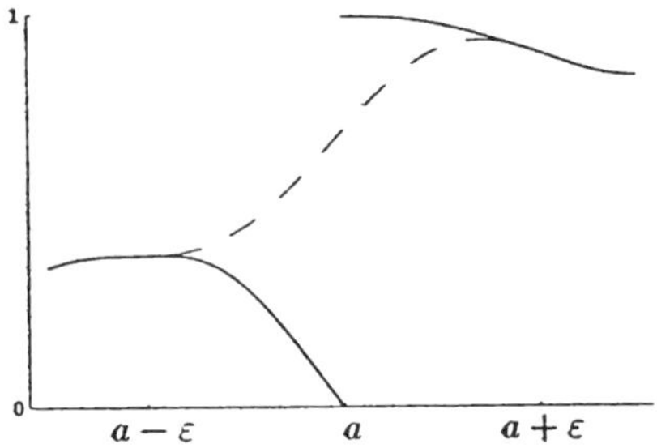

FIGURE 9.5
Folding Uf of a function f (dashed).

Obviously, the adjoint operator $U^\star = U^\star(r,a)$, which will be called *unfolding operator* in the sequel, is given by

$$U^\star f(a+x) := \begin{cases} r(a+x)f(a+x) - r(a-x)f(a-x), & \text{if } x \geq 0, \\ r(a-x)f(a+x) + r(a+x)f(a-x), & \text{if } x \leq 0. \end{cases}$$

For $|a-x| \geq \varepsilon$, it is easy to see $Uf(x) = U^\star f(x) = f(x)$. Furthermore, if $x \in (0,\varepsilon)$ we have

$$\begin{aligned} U^\star Uf(a+x) &= r(a+x)Uf(a+x) - r(a-x)Uf(a-x) \\ &= r(a+x)\big(r(a+x)f(a+x) + r(a-x)f(a-x)\big) \\ &\qquad -r(a-x)\big(r(a+x)f(a-x) - r(a-x)f(a+x)\big) \\ &= \left(r^2(a+x) + r^2(a-x)\right) f(a+x) = f(a+x). \end{aligned} \tag{9.14}$$

By analogous computations one shows $U^\star Uf(a+x) = f(a+x)$, $x \in (-\varepsilon, 0)$, as well as $UU^\star f(a+x) = f(a+x)$, $x \in (-\varepsilon, \varepsilon)$. Summing up, it holds that

$$UU^\star f(x) = U^\star U f(x) = f(x), \quad x \in \mathbb{R} \setminus \{a\}.$$

Hence, U is a unitary isomorphism on $L^2(\mathbb{R})$.

Note that for $f \in L^2(\mathbb{R})$ it does not matter how we define $Uf(a)$ or $U^\star f(a)$. But for f satisfying certain smoothness and boundary limit conditions there exists a smooth extension of $U^\star f$ across $x = a$. Furthermore, the folding operator preserves the smoothness of f for $x \neq a$ and Uf can be extended by smooth functions, even or odd at $x = a$ (see Fig. 9.5).

LEMMA 9.2

Let $m \in \mathbb{N}_0$ and $r \in C^m(\mathbb{R})$ be given. If $f \in C^m(\mathbb{R})$, then $Uf \in C^m(\mathbb{R} \setminus \{a\})$, the limits $(Uf)^{(\mu)}(a+0)$ and $(Uf)^{(\mu)}(a-0)$ exist for all $\mu = 0, \ldots, m$, and we have

$$(Uf)^{(\mu)}(a-0) = 0 \text{ if } \mu \text{ is even} \quad \text{and} \quad (Uf)^{(\mu)}(a+0) = 0 \text{ if } \mu \text{ is odd.}$$

Conversely, if $f \in C^m(\mathbb{R} \setminus \{a\})$ has limits $f^{(\mu)}(a+0)$ and $f^{(\mu)}(a-0)$ for all $\mu = 0, \ldots, m$ which satisfy the equations

$$f^{(\mu)}(a-0) = 0 \text{ if } \mu \text{ is even} \quad \text{and} \quad f^{(\mu)}(a+0) = 0 \text{ if } \mu \text{ is odd,}$$

then $U^\star f$ has a unique continuous extension across $x = a$ which belongs to $C^m(\mathbb{R})$.

This lemma was found by Wickerhauser [47] and the proof follows immediately from the definition of U.

Assume a family of window functions w_j, $j \in \mathbb{Z}$, satisfies the conditions (9.6)–(9.10). Then, we can write $w_j(x) := r_j(x) r_{j+1}(-x)$ with cut-off functions

$$r_j(x) := \begin{cases} 0, & \text{if } x \leq a_j^-, \\ w_j(x), & \text{if } a_j^- < x < a_j^+, \\ 1, & \text{if } x \geq a_j^+. \end{cases}$$

From (9.9) and (9.10), it follows that $|r_j(a_j + x)|^2 + |r_j(a_j - x)|^2 = 1$, $x \in \mathbb{R}$. Now, we define $U_j := U(r_j, a_j)$. Since $U_j f(x) = U_j^\star f(x) = f(x)$ for $x \notin (a_j^-, a_j^+)$ and the intervals (a_j^-, a_j^+) are disjoint, the operators $U_j^\star$ and U_j, $j \in \mathbb{Z}$, all commute. Using the parity conditions (9.12) and (9.13) on

t_j^k one obtains for ψ_j^k the representation

$$\psi_j^k = \begin{cases} U_j^\star U_{j+1}^\star\big(\chi_{[a_j,a_{j+1}]}t_j^k\big), & \text{if } 1 = \varsigma_j = \varsigma_{j+1}, \\ U_j^\star U_{j+1}\big(\chi_{[a_j,a_{j+1}]}t_j^k\big), & \text{if } 1 = \varsigma_j = -\varsigma_{j+1}, \\ U_j U_{j+1}^\star\big(\chi_{[a_j,a_{j+1}]}t_j^k\big), & \text{if } 1 = -\varsigma_j = \varsigma_{j+1}, \\ U_j U_{j+1}\big(\chi_{[a_j,a_{j+1}]}t_j^k\big), & \text{if } 1 = -\varsigma_j = -\varsigma_{j+1}, \end{cases} \tag{9.15}$$

with ς_j as in (9.12) and (9.13). Here,

$$\chi_I(x) := \begin{cases} 1, & \text{if } x \in I, \\ 0, & \text{otherwise}, \end{cases}$$

denotes the characteristic function of the interval I. With these results we can prove the theorem of Coifman and Meyer.

PROOF of Theorem 9.4. We define an operator $\mathcal{T}_w : L^2(\mathbb{R}) \to L^2(\mathbb{R})$ by

$$\mathcal{T}_w := \Big(\prod_{\substack{j\in\mathbb{Z}\\ \varsigma_j=1}} U_j\Big)\Big(\prod_{\substack{j\in\mathbb{Z}\\ \varsigma_j=-1}} U_j^\star\Big). \tag{9.16}$$

Obviously, $\mathcal{T}_w$ can be written as

$$\mathcal{T}_w f(x) := \begin{cases} U_j f(x), & \text{if } x \in (a_j^-, a_j^+),\ j \in \mathbb{Z},\ \varsigma_j = 1, \\ U_j^\star f(x), & \text{if } x \in (a_j^-, a_j^+),\ j \in \mathbb{Z},\ \varsigma_j = -1, \\ f(x), & \text{if } x \in [a_j^+, a_{j+1}^-],\ j \in \mathbb{Z}. \end{cases}$$

Since the operators $U_j^\star$ and U_j, $j \in \mathbb{Z}$, all commute it follows immediately that the adjoint operator, which will be called the total unfolding operator, is given by

$$\mathcal{T}_w^\star = \mathcal{U}_w := \Big(\prod_{\substack{j\in\mathbb{Z}\\ \varsigma_j=1}} U_j^\star\Big)\Big(\prod_{\substack{j\in\mathbb{Z}\\ \varsigma_j=-1}} U_j\Big). \tag{9.17}$$

Furthermore, one obtains by the commutativity of the folding and unfolding operators that $\mathcal{T}_w\mathcal{T}_w^\star = \mathcal{T}_w^\star\mathcal{T}_w = \mathrm{id}$. Thus, $\mathcal{T}_w$ and $\mathcal{U}_w$ are unitary isomorphisms on $L^2(\mathbb{R})$, too. If $\operatorname{supp} f$ and (a_j^-, a_j^+) are disjoint we have $U_j f = U_j^\star f = f$. Therefore, from (9.15) it follows that

$$\psi_j^k = \mathcal{U}_w\left(\chi_{[a_j,a_{j+1}]}t_j^k\right). \tag{9.18}$$

With the adjoint operator $\mathcal{T}_w$ the identity (9.18) can be written equivalently as

$$\langle f, \psi_j^k \rangle = \int_{a_j}^{a_{j+1}} \mathcal{T}_w f(x)\, t_j^k(x)\, dx. \tag{9.19}$$

Since $\mathcal{U}_w$ is a unitary isomorphism and $\{\chi_{[a_j,a_{j+1}]} t_j^k : j \in \mathbb{Z},\ k \in \mathbb{N}_0\}$ is an orthonormal basis, condition (iii) in Theorem 9.3 is fulfilled. Hence, $\{\psi_j^k\}$ is an orthonormal basis. ∎

The operator $\mathcal{T}_w$ was introduced by Jawerth and Sweldens [36] and is called total folding operator. On the interval (a_j^-, a_{j+1}^-), $j \in \mathbb{Z}$, it is given by

$$\mathcal{T}_w f(x) = \begin{cases} r_j(x) f(x) + \varsigma_j r_j(2a_j - x) f(2a_j - x), & \text{if } x \in (a_j, a_j^+), \\ r_j(2a_j - x) f(x) - \varsigma_j r_j(x) f(2a_j - x), & \text{if } x \in (a_j^-, a_j), \\ f(x), & \text{if } x \in [a_j^+, a_{j+1}^-]. \end{cases} \tag{9.20}$$

Analogously to Lemma 9.2 it follows that $\mathcal{T}_w f$ is smooth in (a_j, a_{j+1}) if f and w_j are smooth functions on the whole real line. Furthermore, $\mathcal{T}_w f$ has smooth extensions across the boundaries of the interval (a_j, a_{j+1}) which have the same parity as the functions t_j^k. In particular, for the trigonometric bases from Lemma 9.1 one can show by partial integration that the basis coefficients $\langle f, \psi_j^k \rangle$ decay fast for $k \to \infty$.

The reason for the importance of the total folding operator is that it describes a relation between the functions ψ_j^k and the simple structured orthonormal basis $\{\chi_{[a_j,a_{j+1}]} t_j^k\}$ (cf. equalities (9.18) and (9.19)). Therefore, we can derive many properties of $\{\psi_j^k\}$ from properties of the total folding operator. In the sequel, we will use various generalizations of the (total) folding operator given in (9.20). All these operators will satisfy the property (9.18) and (9.19) which we will need for our proofs.

9.4 Biorthogonal Local Trigonometric Bases of Jawerth and Sweldens

To construct local trigonometric bases for which constant functions have an economical representation Jawerth and Sweldens [36] investigated biorthogonal local trigonometric bases in a two-overlapping setting. Similar bases were discovered independently by Matviyenko [41]. To obtain more freedom for the choice of the window functions they removed the condition (9.10) of Coifman and Meyer which is very restrictive. Then, for

$x \in (a_j^-, a_j^+)$ one can choose $w_j(x)$, $j \in \mathbb{Z}$, as an arbitrary function with $0 \le w_j(x) \le 1$. With these window functions we cannot expect to obtain an orthonormal basis. But if we replace (9.10) by a weaker condition the local trigonometric functions still form a Riesz basis.

THEOREM 9.5

Jawerth and Sweldens [36] *Let the window functions w_j fulfill the conditions (9.6)–(9.9) and let*

$$A \le w_j^2(a_j + x) + w_{j-1}^2(a_j + x) \le B, \quad |x| \le \varepsilon_j. \tag{9.21}$$

Further, the t_j^k shall satisfy (9.11) – (9.13). Then, the functions $\psi_j^k := w_j t_j^k$, $j \in \mathbb{Z}$, $k \in \mathbb{N}_0$, constitute a Riesz basis of $L^2(\mathbb{R})$ with Riesz bounds A and B.

PROOF As in the previous section we introduce cut-off functions

$$r_j(x) := \begin{cases} 0, & \text{if } x \le a_j^-, \\ w_j(x), & \text{if } a_j^- < x < a_j^+, \\ 1, & \text{if } x \ge a_j^+. \end{cases}$$

Let a folding operator $\mathcal{T}_w : L^2(\mathbb{R}) \to L^2(\mathbb{R})$ defined by (9.20). Obviously, the adjoint operator $\mathcal{T}_w^\star = \mathcal{U}_w$ is given by

$$\mathcal{U}_w f(x) = \begin{cases} r_j(x) f(x) - \varsigma_j r_j(2a_j - x) f(2a_j - x), & \text{if } x \in (a_j, a_j^+), \\ r_j(2a_j - x) f(x) + \varsigma_j r_j(x) f(2a_j - x), & \text{if } x \in (a_j^-, a_j), \\ f(x), & \text{if } x \in [a_j^+, a_{j+1}^-]. \end{cases} \tag{9.22}$$

If $x \in [a_j^+, a_{j+1}^-]$ we have $\mathcal{T}_w f(x) = \mathcal{U}_w f(x) = f(x)$. Analogous to (9.14) we obtain for $x \in (a_j^-, a_j^+)$ the equality

$$\begin{aligned} \mathcal{U}_w \mathcal{U}_w^\star f(x) = \mathcal{U}_w^\star \mathcal{U}_w f(x) &= \left(r_j^2(x) + r_j^2(2a_j - x)\right) f(x) \\ &= \left(w_j^2(x) + w_{j-1}^2(x)\right) f(x). \end{aligned}$$

Computing the norm of $\mathcal{U}_w f$ we conclude

$$\|\mathcal{U}_w f\|_{L^2(\mathbb{R})} = \langle \mathcal{U}_w^\star \mathcal{U}_w f, f\rangle$$

$$= \sum_{j\in\mathbb{Z}} \left(\int_{a_j^-}^{a_j^+} \left(w_j^2(x) + w_{j-1}^2(x)\right) |f(x)|^2\, dx + \int_{a_j^+}^{a_{j+1}^-} |f(x)|^2\, dx \right).$$

Then, by lower and upper estimates one obtains the inequality

$$A\|f\|_{L^2(\mathbb{R})} \le \|\mathcal{U}_w f\|_{L^2(\mathbb{R})} \le B\|f\|_{L^2(\mathbb{R})}.$$

Therefore, $\mathcal{U}_w : L^2(\mathbb{R}) \to L^2(\mathbb{R})$ is a topological isomorphism with $\|\mathcal{U}_w\| \le B$ and $\|\mathcal{U}_w^{-1}\| \le A^{-1}$. From the parity conditions (9.12) and (9.13) on t_j^k one concludes again that $\mathcal{U}_w$ satisfies (9.18). Hence, $\{\psi_j^k\}$ is the image of an orthonormal basis under a topological isomorphism. Then, from Theorem 9.1 it follows immediately that $\{\psi_j^k\}$ is a Riesz basis of $L^2(\mathbb{R})$ with Riesz bounds A and B. ∎

From the two-overlapping condition (9.6) it follows immediately that $\psi_j^k \perp \psi_r^l$ for $|j-r| \ge 2$ because the supports of these functions do not overlap, but also for immediately neighbouring functions one has orthonormality. Indeed, using the symmetry condition (9.9) for the windows and the parity conditions (9.12) and (9.13) for the functions t_j^k we obtain

$$\begin{aligned}\langle \psi_j^k, \psi_{j-1}^\ell \rangle &= \int_{-\varepsilon_j}^{\varepsilon_j} w_j(a_j+x)w_j(a_j-x)\, t_j^k(a_j+x)t_{j-1}^\ell(a_j+x)\, dx \\ &= \int_0^{\varepsilon_j} \Big(w_j(a_j+x)w_j(a_j-x) - w_j(a_j-x)w_j(a_j+x) \Big) \\ &\qquad\qquad \times t_j^k(a_j+x)t_{j-1}^\ell(a_j+x)\, dx \\ &= 0.\end{aligned}$$

Obviously, window functions which satisfy the assumptions of Theorem 9.5 are easy to find. Furthermore, if one has such window functions w_j one can construct an orthonormal basis as follows. We define new window functions

$\breve{w}_j$ by

$$\breve{w}_j(x) = \begin{cases} \frac{w_j(x)}{\sqrt{w_j^2(x)+w_{j-1}^2(x)}}, & \text{if } x \in (a_j^-, a_j^+), \\ \frac{w_j(x)}{\sqrt{w_j^2(x)+w_{j+1}^2(x)}}, & \text{if } x \in (a_{j+1}^-, a_{j+1}^+), \\ 1, & \text{if } x \in [a_j^+, a_{j+1}^-], \\ 0, & \text{otherwise.} \end{cases}$$

These window functions satisfy (9.6)–(9.10). Therefore, the functions $\psi_j^k := \breve{w}_j t_j^k$ form an orthonormal basis. Since this method preserves the smoothness of the window functions, we can find orthonormal bases of arbitrary smoothness.

On the other hand, it is often useful to give up orthonormality to obtain other important features. Therefore, we will remove all constraints on the window functions which are not necessary for Riesz stability. In the following sections we will consider biorthogonal local trigonometric bases in a generalized setting.

9.5 Biorthogonal Bases of Chui and Shi

In this section we introduce a generalized approach for local trigonometric bases in the two-overlapping setting given by Chui and Shi in [17, 18]. Let the sequences $(a_j)_{j\in\mathbb{Z}}$, $(a_j^+)_{j\in\mathbb{Z}}$, $(a_j^-)_{j\in\mathbb{Z}}$ be given as in Section 9.3. Furthermore, let the window functions w_j satisfy the two-overlapping condition (9.6), i.e.,

$$\operatorname{supp} w_j \subset [a_j^-, a_{j+1}^+]. \tag{9.23}$$

In particular, the window functions need not satisfy the conditions (9.7)–(9.10) of Coifman and Meyer.

9.5.1 Generalization of folding operators

Now, we want to investigate the basis properties of $\{\psi_j^k\} = \{w_j t_j^k\}$, where the t_j^k are (trigonometric) functions with the properties (9.11)–(9.13). Again, we will use folding and unfolding operators. For given functions w_j and t_j^k we define the unfolding operator $\mathcal{U}_w$ by

$$\mathcal{U}_w f(x) := \begin{cases} w_j(x)f(x) - \varsigma_j w_{j-1}(x) f(2a_j - x), & \text{if } x \in (a_j, a_j^+), \\ w_{j-1}(x)f(x) + \varsigma_j w_j(x) f(2a_j - x), & \text{if } x \in (a_j^-, a_j), \\ w_j(x)f(x), & \text{if } x \in [a_j^+, a_{j+1}^-], \end{cases} \tag{9.24}$$

(cf. [14]). Using the parity properties (9.12) and (9.13) one obtains

$$\mathcal{U}_w\left(\chi_{[a_j,a_{j+1}]}t_j^k\right)(x)$$

$$= \begin{cases} w_j(x)t_j^k(x), & \text{if } x \in (a_j, a_{j+1}), \\ \varsigma_j w_j(x)t_j^k(2a_j - x) = w_j(x)t_j^k(x), & \text{if } x \in (a_j^-, a_j), \\ -\varsigma_{j+1} w_j(x)t_j^k(2a_{j+1} - x) = w_j(x)t_j^k(x), & \text{if } x \in (a_{j+1}, a_{j+1}^+), \\ 0, & \text{otherwise.} \end{cases}$$

Summing up, we have

$$\mathcal{U}_w\left(\chi_{[a_j,a_{j+1}]}t_j^k\right) = \psi_j^k, \tag{9.25}$$

i.e., $\mathcal{U}_w$ maps the elements of the orthonormal basis $\{\chi_{[a_j,a_{j+1}]}t_j^k\}$ onto the functions ψ_j^k. For the interval (a_j^-, a_j^+) a particularly useful representation of $\mathcal{U}_w$ is given by

$$\begin{pmatrix} \mathcal{U}_w f(x) \\ \mathcal{U}_w f(2a_j - x) \end{pmatrix} = \boldsymbol{M}_j(x) \begin{pmatrix} f(x) \\ f(2a_j - x) \end{pmatrix}, \quad x \in (a_j, a_j^+), \tag{9.26}$$

with the matrices

$$\boldsymbol{M}_j(x) := \begin{pmatrix} w_j(x) & -\varsigma_j\, w_{j-1}(x) \\ \varsigma_j\, w_j(2a_j - x) & w_{j-1}(2a_j - x) \end{pmatrix}$$

(cf. [17, 18]). By easy computations one obtains that the adjoint operator $\mathcal{U}_w^\star$ is given by

$$\begin{aligned} \mathcal{U}_w^\star f(x) &:= w_j(x)\, f(x), & \text{for } x \in [a_j^+, a_{j+1}^-], \\ \begin{pmatrix} \mathcal{U}_w^\star f(x) \\ \mathcal{U}_w^\star f(2a_j - x) \end{pmatrix} &:= \boldsymbol{M}_j^H(x) \begin{pmatrix} f(x) \\ f(2a_j - x) \end{pmatrix}, & \text{for } x \in (a_j, a_j^+). \end{aligned}$$

If the window functions satisfy the conditions (9.6)–(9.9) the adjoint operator $\mathcal{U}_w^\star$ coincides with the total folding operator of Jawerth and Sweldens defined in (9.20). Therefore, we denote $\mathcal{T}_w := \mathcal{U}_w^\star$ as the folding operator for the generalized bases of Chui and Shi. In particular, from (9.25) it follows immediately that the generalized operator $\mathcal{T}_w$ still satisfies the condition (9.19).

9.5.2 Riesz stability and the dual basis

Applying the folding operator $\mathcal{U}_w$, we can give a criterion for the Riesz stability of $\{\psi_j^k\}$. For the proof of the following theorem we will need the Euclidean norm $|\boldsymbol{x}| := \sqrt{|x_1|^2 + \ldots + |x_n|^2}$ of a vector $\boldsymbol{x} \in \mathbb{R}^n$ and the spectral norm $\|\boldsymbol{A}\|_2 := \sup_{|x|=1} |\boldsymbol{A}\boldsymbol{x}| = \sqrt{\rho(\boldsymbol{A}^H \boldsymbol{A})}$ of a matrix $\boldsymbol{A}$.

THEOREM 9.6

Chui and Shi [17] *Let the measurable functions w_j satisfy the two-overlapping condition* (9.23). *Further, the functions t_j^k shall satisfy* (9.11)–(9.13). *The functions $\psi_j^k := w_j t_j^k$, $j \in \mathbb{Z}$, $k \in \mathbb{N}_0$, constitute a Riesz basis with Riesz bounds A and B if and only if $0 < A \le A_0 \le B_0 \le B < \infty$, where*

$$\begin{aligned} A_0 := \inf_{j\in\mathbb{Z}} \min\Bigg(&\operatorname*{ess\,inf}_{x\in(a_j^+, a_{j+1}^-)} |w_j(x)|^2, \\ &\operatorname*{ess\,inf}_{x\in(a_j, a_j^+)} \left(\frac{\Delta_j(x)}{2} - \sqrt{\frac{\Delta_j^2(x)}{4} - \left|\det \boldsymbol{M}_j(x)\right|^2} \right)\Bigg), \\ B_0 := \sup_{j\in\mathbb{Z}} \max\Bigg(&\operatorname*{ess\,sup}_{x\in(a_j^+, a_{j+1}^-)} |w_j(x)|^2, \\ &\operatorname*{ess\,sup}_{x\in(a_j, a_j^+)} \left(\frac{\Delta_j(x)}{2} + \sqrt{\frac{\Delta_j^2(x)}{4} - \left|\det \boldsymbol{M}_j(x)\right|^2} \right)\Bigg) \end{aligned} \tag{9.27}$$

with $\Delta_j(x) = |w_j(x)|^2 + |w_j(2a_j - x)|^2 + |w_{j-1}(x)|^2 + |w_{j-1}(2a_j - x)|^2$.

PROOF Assume $0 < A \le A_0 \le B_0 \le B < \infty$. Let the unfolding operator $\mathcal{U}_w$ be defined by (9.24). By Hölder's inequality it follows that for $f \in L^2(\mathbb{R})$

$$\begin{aligned} \int_{a_j^+}^{a_{j+1}^-} |\mathcal{U}_w f(x)|^2\,dx &= \int_{a_j^+}^{a_{j+1}^-} |w_j(x) f(x)|^2\,dx \\ &\le \operatorname*{ess\,sup}_{x\in(a_j^+, a_{j+1}^-)} |w_j(x)|^2 \int_{a_j^+}^{a_{j+1}^-} |f(x)|^2\,dx \qquad (9.28) \\ &\le B_0 \int_{a_j^+}^{a_{j+1}^-} |f(x)|^2\,dx. \end{aligned}$$

Again, with Hölder's inequality we conclude

$$\int_{a_j^-}^{a_j^+} |\mathcal{U}_w f(x)|^2\, dx = \int_{a_j}^{a_j^+} \left| \begin{pmatrix} \mathcal{U}_w f(x) \\ \mathcal{U}_w f(2a_j - x) \end{pmatrix} \right|^2 dx$$

$$\leq \int_{a_j}^{a_j^+} \|\boldsymbol{M}_j(x)\|_2^2 \left| \begin{pmatrix} f(x) \\ f(2a_j - x) \end{pmatrix} \right|^2 dx$$

$$\leq \operatorname*{ess\,sup}_{x\in(a_j,a_j^+)} \|\boldsymbol{M}_j(x)\|_2^2 \int_{a_j}^{a_j^+} \left| \begin{pmatrix} f(x) \\ f(2a_j - x) \end{pmatrix} \right|^2 dx.$$

Computing the eigenvalues of $\boldsymbol{M}_j^H(x)\boldsymbol{M}_j(x)$ we obtain

$$\|\boldsymbol{M}_j(x)\|_2^2 = \tfrac{\Delta_j(x)}{2} + \sqrt{\tfrac{\Delta_j^2(x)}{4} - \left|\det \boldsymbol{M}_j(x)\right|^2}$$

such that

$$\int_{a_j}^{a_j^+} |\mathcal{U}_w f(x)|^2\, dx \leq B_0 \int_{a_j^-}^{a_j^+} |f(x)|^2\, dx.$$

Summing up, it follows for $f \in L^2(\mathbb{R})$ that

$$\begin{aligned} \|\mathcal{U}_w f\|_{L^2(\mathbb{R})}^2 &= \sum_{j\in\mathbb{Z}} \left(\int_{a_j^-}^{a_j^+} |\mathcal{U}_w f(x)|^2\, dx + \int_{a_j^+}^{a_{j+1}^-} |\mathcal{U}_w f(x)|^2\, dx \right) \\ &\leq B_0 \sum_{j\in\mathbb{Z}} \left(\int_{a_j^-}^{a_j^+} |f(x)|^2\, dx + \int_{a_j^+}^{a_{j+1}^-} |f(x)|^2\, dx \right) \\ &\leq B_0\, \|f\|_{L^2(\mathbb{R})}^2. \end{aligned} \tag{9.29}$$

Hence, it is proved that $\|\mathcal{U}_w\|_{L^2(\mathbb{R})\to L^2(\mathbb{R})}^2 \leq B_0 \leq B$.

Obviously,

$$\|\boldsymbol{M}_j(x)^{-1}\|_2^{-2} = \tfrac{\Delta_j(x)}{2} - \sqrt{\tfrac{\Delta_j^2(x)}{4} - \left|\det \boldsymbol{M}_j(x)\right|^2} > 0$$

if and only if $\det \boldsymbol{M}_j(x) \neq 0$. Therefore, $A_0 > 0$ implies that $\boldsymbol{M}_j^{-1}(x)$ resp. $\frac{1}{w_j(x)}$ are well defined a.e. on (a_j, a_j^+) resp. $[a_j^+, a_{j+1}^-]$. Hence, for a given

$f \in L^2(\mathbb{R})$ the function $\mathcal{V}_w f$ given by

$$\begin{aligned} \mathcal{V}_w f(x) &:= \frac{f(x)}{w_j(x)}, & \text{for } x \in [a_j^+, a_{j+1}^-], \\ \begin{pmatrix} \mathcal{V}_w f(x) \\ \mathcal{V}_w f(2a_j - x) \end{pmatrix} &:= \boldsymbol{M}_j^{-1}(x) \begin{pmatrix} f(x) \\ f(2a_j - x) \end{pmatrix}, & \text{for } x \in (a_j, a_j^+), \end{aligned} \tag{9.30}$$

is well defined a.e. on $\mathbb{R}$. Analogous to (9.29) one shows that

$$\|\mathcal{V}_w f\|_{L^2(\mathbb{R})}^2 \le A_0^{-1} \, \|f\|_{L^2(\mathbb{R})}^2.$$

Furthermore, we obtain by simple inspection that

$$\mathcal{V}_w \mathcal{U}_w f(x) = \mathcal{U}_w \mathcal{V}_w f(x) = f(x) \quad \text{a.e. on } \mathbb{R}.$$

Hence, $\mathcal{U}_w$ is invertible and has the bounded inverse $\mathcal{U}_w^{-1} = \mathcal{V}_w$ with $\|\mathcal{U}_w^{-1}\|_{L^2(\mathbb{R}) \to L^2(\mathbb{R})}^2 \le A_0^{-1} \le A^{-1}$. Thus, $\mathcal{U}_w$ is an topological isomorphism with $\|\mathcal{U}_w^{-1}\|_{L^2(\mathbb{R}) \to L^2(\mathbb{R})}^2 \le A^{-1}$ and $\|\mathcal{U}_w\|_{L^2(\mathbb{R}) \to L^2(\mathbb{R})}^2 \le B$. Since $\mathcal{U}_w$ furthermore maps an orthonormal basis onto $\{\psi_j^k\}$ (cf. (9.25)) condition (ii) in Theorem 9.6 is satisfied and therefore $\{\psi_j^k\}$ is a Riesz basis with Riesz bounds A and B.

Conversely, let $\{\psi_j^k\}$ be a Riesz basis with Riesz bounds A and B. Then, there exists a topological isomorphism T with $\|T\|_{L^2(\mathbb{R}) \to L^2(\mathbb{R})}^2 \le B$, $\|T^{-1}\|_{L^2(\mathbb{R}) \to L^2(\mathbb{R})}^2 \le A^{-1}$, and $T\left(\chi_{[a_j, a_{j+1}]} t_j^k\right) = \psi_j^k$. From (9.25) we see that $\mathcal{U}_w$ coincides with T on $\operatorname{span}\{\chi_{[a_j, a_{j+1}]} t_j^k\}$ which is a dense subset of $L^2(\mathbb{R})$ and it follows that $\mathcal{U}_w = T$ on $L^2(\mathbb{R})$.

Assume now that for the upper Riesz bound it holds that $B < B_0$. Then, there exists an integer $j \in \mathbb{Z}$ such that

$$B < \max\left(\operatorname*{ess\,sup}_{x \in (a_j^+, a_{j+1}^-)} |w_j(x)|^2, \operatorname*{ess\,sup}_{x \in (a_j, a_j^+)} \left(\frac{\Delta_j(x)}{2} + \sqrt{\frac{\Delta_j^2(x)}{4} - \left|\det \boldsymbol{M}_j(x)\right|^2} \right) \right).$$

Let on the one hand $B < \operatorname{ess\,sup}_{x \in (a_j^+, a_{j+1}^-)} |w_j(x)|^2$. Then, we can choose a function $f \subset L^2(\mathbb{R})$ with $\operatorname{supp} f \in [a_j^+, a_{j+1}^-]$ such that

$$\int_{a_j^+}^{a_{j+1}^-} |w_j(x) f(x)|^2 \, dx > B \int_{a_j^+}^{a_{j+1}^-} |f(x)|^2 \, dx$$

because

$$\operatorname*{ess\,sup}_{x \in (a_j^+, a_{j+1}^-)} |w_j(x)|^2$$

is the smallest possible constant in the Hölder's inequality (9.28). From the support property of f it follows $\operatorname{supp}\mathcal{U}_w f \subset [a_j^+, a_{j+1}^-]$. Summing up we obtain $\|\mathcal{U}_w f\|_{L^2(\mathbb{R})} > B\, \|f\|_{L^2(\mathbb{R})}$ which is a contradiction.

If on the other hand

$$B < \operatorname*{ess\,sup}_{x \in (a_j, a_j^+)} \left(\frac{\Delta_j(x)}{2} + \sqrt{\frac{\Delta_j^2(x)}{4} - \left|\det \boldsymbol{M}_j(x)\right|^2} \right) \tag{9.31}$$

we conclude by analogous arguments that there exists a non-negative function $g \in L^2([a_j, a_j^+])$ such that

$$\int_{a_j}^{a_j^+} \|\boldsymbol{M}_j(x)\|_2^2\, |g(x)|^2\, dx > B \int_{a_j}^{a_j^+} |g(x)|^2\, dx.$$

Now, we choose $f \in L^2(\mathbb{R})$ with $\operatorname{supp} f \subset [a_j^-, a_j^+]$ such that for $x \in [a_j, a_j^+]$ the vector $(f(x),\ f(2a_j - x))^T$ is an eigenvector of $\boldsymbol{M}_j^H(x)\boldsymbol{M}_j(x)$ corresponding to the maximal eigenvalue $\rho\big(\boldsymbol{M}_j^H(x)\boldsymbol{M}_j(x)\big) = \|\boldsymbol{M}_j(x)\|_2^2$ and $\left|(f(x),\ f(2a_j - x))^T\right| = g(x)$. Since $\operatorname{supp}\mathcal{U}_w f \subset [a_j^-, a_j^+]$ we obtain for this f that

$$\begin{aligned} \|\mathcal{U}_w f\|_{L^2(\mathbb{R})} &= \int_{a_j}^{a_j^+} \left| \boldsymbol{M}_j(x) \begin{pmatrix} f(x) \\ f(2a_j - x) \end{pmatrix} \right|^2 dx \\ &= \int_{a_j}^{a_j^+} \|\boldsymbol{M}_j(x)\|_2^2\, |g(x)|^2\, dx \;>\; B\, \|f\|_{L^2(\mathbb{R})} \end{aligned}$$

which is again a contradiction. Hence, $B_0 \le B < \infty$. Analogously, one shows $0 < A \le A_0$. Since $A_0 \le B_0$ follows immediately from the definitions of A_0 and B_0 the theorem is proved. ∎

Theorem 9.6 states that the functions ψ_j^k form a Riesz basis of $L^2(\mathbb{R})$ if and only if the unfolding operator $\mathcal{U}_w$ is a topological isomorphism. In particular, the best possible Riesz bounds are given by $A_0 = \|\mathcal{U}_w^{-1}\|^{-2}$ and $B_0 = \|\mathcal{U}_w\|$. The application of Theorem 9.6 gives us a sufficient and necessary condition for the Riesz stability of $\{\psi_j^k\}$ which is easy to check.

COROLLARY 9.1

Let the functions ψ_j^k satisfy the assumptions of Theorem 9.11. *Then, $\{\psi_j^k\}$ is a Riesz basis of $L^2(\mathbb{R})$, if and only if there exist constants $c, C > 0$ such* *that for all $j \in \mathbb{Z}$ it holds that $|w_j(x)| \geq c$ a.e. on (a_j^+, a_{j+1}^-), $|\det \boldsymbol{M}_j(x)| \geq c$ a.e. on (a_j, a_j^+), and $\|w_j\|_{L^\infty(\mathbb{R})} \leq C$. In particular, the Riesz bounds can be estimated by*

$$\inf_{j\in\mathbb{Z}} \min\left(\operatorname*{ess\,inf}_{x\in(a_j^+,a_{j+1}^-)} |w_j(x)|^2, \operatorname*{ess\,inf}_{x\in(a_j,a_j^+)} \frac{|\det \boldsymbol{M}_j(x)|^2}{B_0} \right) \tag{9.32}$$

$$\leq A_0 \leq \inf_{j\in\mathbb{Z}} \min\left(\operatorname*{ess\,inf}_{x\in(a_j^+,a_{j+1}^-)} |w_j(x)|^2, \operatorname*{ess\,inf}_{x\in(a_j,a_j^+)} |\det \boldsymbol{M}_j(x)| \right),$$

$$\sup_{j\in\mathbb{Z}} \max\left(\operatorname*{ess\,sup}_{x\in(a_j^+,a_{j+1}^-)} |w_j(x)|^2, \operatorname*{ess\,sup}_{x\in(a_j,a_j^+)} \frac{\Delta_j(x)}{2} \right) \tag{9.33}$$

$$\leq B_0 \leq \sup_{j\in\mathbb{Z}} \max\left(\operatorname*{ess\,sup}_{x\in(a_j^+,a_{j+1}^-)} |w_j(x)|^2, \operatorname*{ess\,sup}_{x\in(a_j,a_j^+)} \Delta_j(x) \right).$$

PROOF From

$$\Delta_j(x) - 2\,|\det \boldsymbol{M}_j(x)|$$

$$= \left(|w_j(x)| - |w_{j-1}(2a_j - x)|\right)^2 + \left(|w_{j-1}(x)| - |w_j(2a_j - x)|\right)^2 \geq 0$$

it follows that $0 \leq |\det \boldsymbol{M}_j(x)| \leq \frac{\Delta_j(x)}{2}$. Applying these inequalities to the definition of B_0 in (9.27) we obtain the desired estimates in (9.33). Then, it is obvious that the existence of an upper Riesz bound is equivalent to $w_j \in L^\infty(\mathbb{R})$, where $\|w_j\|_{L^\infty(\mathbb{R})}$ is uniformly bounded for all $j \in \mathbb{Z}$.

Now, we denote the two eigenvalues of $\boldsymbol{M}_j^H(x)\boldsymbol{M}_j(x)$ by $\lambda_j(x)$ and $\mu_j(x)$, where $\lambda_j(x) \leq \mu_j(x)$. Then,

$$\lambda_j(x)\mu_j(x) = \det\left(\boldsymbol{M}_j^H(x)\boldsymbol{M}_j(x)\right) = |\det \boldsymbol{M}_j(x)|^2.$$

From

$$|\det \boldsymbol{M}_j(x)| = \sqrt{\lambda_j(x)\mu_j(x)} \geq \lambda_j(x) = \|\boldsymbol{M}_j^{-1}(x)\|^{-2} \geq A_0$$

we obtain the upper estimate in (9.32). On the other hand

$$\operatorname*{ess\,inf}_{x\in(a_j^-,a_j^+)} |\det \boldsymbol{M}_j(x)|^2 \leq \left(\operatorname*{ess\,sup}_{x\in(a_j^-,a_j^+)} \mu_j(x)\right)\left(\operatorname*{ess\,inf}_{x\in(a_j^-,a_j^+)} \lambda_j(x)\right)$$

$$\leq \left(\operatorname*{ess\,sup}_{x\in(a_j^-,a_j^+)} \|\boldsymbol{M}_j(x)\|_2^2\right)\left(\operatorname*{ess\,inf}_{x\in(a_j^-,a_j^+)} \|\boldsymbol{M}_j^{-1}(x)\|_2^{-2}\right)$$

$$\leq B_0 \operatorname*{ess\,inf}_{x\in(a_j^-,a_j^+)} \|\boldsymbol{M}_j^{-1}(x)\|_2^{-2}$$

yields the lower estimate in (9.32). Finally, from (9.32) it follows immediately that the conditions $|w_j(x)| \geq c$ a.e. in (a_j^+, a_{j+1}^-) and $|\det \boldsymbol{M}_j(x)| \geq c$ a.e. in (a_j, a_j^+), with $c > 0$ are sufficient and necessary for $A_0 > 0$. ∎

In particular, Corollary 9.1 provides an assertion on Riesz bases generated by non-negative window functions. Let $w_j(x) \geq 0$, $x \in \mathbb{R}$, and $w_j(x) \geq c$, $x \in (a_j, a_{j+1})$, where $c > 0$ is independent of j. Then, one checks easily that $\det M_j(x) \geq c^2$ if $x \in [a_j^-, a_j^+]$. Hence, $\{\psi_j^k\}$ is a Riesz basis if the windows w_j are uniformly bounded, non-negative and every window w_j is sufficiently large at $[a_j, a_{j+1}]$. That is, we obtain Riesz bases for a wide class of window functions.

The dual basis functions can be determined using the inverse folding operator.

THEOREM 9.7

Chui and Shi [17] *Let the functions ψ_j^k constitute a Riesz basis. Then, the dual basis is constituted by the functions $\tilde{\psi}_j^k := \tilde{w}_j t_j^k$, $j \in \mathbb{Z}$, $k \in \mathbb{N}_0$, with the dual window functions*

$$\tilde{w}_j(x) := \begin{cases} \dfrac{1}{\overline{w_j(x)}}, & \text{if } x \in [a_j^+, a_{j+1}^-], \\[2ex] \dfrac{\overline{w_{j-1}(2a_j - x)}}{\overline{\det \boldsymbol{M}_j(x)}}, & \text{if } x \in (a_j^-, a_j^+), \\[2ex] \dfrac{\overline{w_{j+1}(2a_{j+1} - x)}}{\overline{\det \boldsymbol{M}_{j+1}(x)}}, & \text{if } x \in (a_{j+1}^-, a_{j+1}^+), \\ 0, & \text{otherwise}. \end{cases} \tag{9.34}$$

PROOF Following Theorem 9.2 the dual basis is given by

$\tilde{\psi}_j^k = (\mathcal{U}_w^{-1})^\star(\chi_{[a_j,a_{j+1}]}t_j^k)$. From the proof of Theorem 9.6 we know that $(\mathcal{U}_w^{-1})^\star = \mathcal{V}_w^\star$ with $\mathcal{V}_w$ defined in (9.30). For $x \in [a_j^+, a_{j+1}^-]$ we conclude

$$\tilde{\psi}_j^k(x) = \mathcal{V}_w^\star(\chi_{[a_j,a_{j+1}]}t_j^k)(x) = \frac{1}{w_j(x)}\, t_j^k(x) = \tilde{w}_j(x)\, t_j^k(x).$$

By Cramer's rule it follows that

$$\boldsymbol{M}_j^{-1}(x) = \frac{1}{\det \boldsymbol{M}_j(x)} \begin{pmatrix} w_{j-1}(2aj - x) & \varsigma_j w_{j-1}(x) \\ -\varsigma_j w_j(2a_j - x) & w_j(x) \end{pmatrix}.$$

Hence, we obtain for $x \in (a_j, a_j^+)$ that

$$\begin{aligned} \begin{pmatrix} \tilde{\psi}_j^k(x) \\ \tilde{\psi}_j^k(2a_j - x) \end{pmatrix} &= \begin{pmatrix} \mathcal{V}_w^*\left(\chi_{[a_j,a_{j+1}]}t_j^k\right)(x) \\ \mathcal{V}_w^*\left(\chi_{[a_j,a_{j+1}]}t_j^k\right)(2a_j - x) \end{pmatrix} \\ &= \left(\boldsymbol{M}_j^{-1}(x)\right)^H \begin{pmatrix} t_j^k(x) \\ 0 \end{pmatrix} \\ &= \begin{pmatrix} \tilde{w}_j(x)\, t_j^k(x) \\ \tilde{w}_j(2a_j - x)\, \varsigma_j t_j^k(x) \end{pmatrix}. \end{aligned}$$

Using the parity condition (9.12) for t_j^k it follows now that $\tilde{\psi}_j^k(x) = \tilde{w}_j(x)t_j^k(x)$ for $x \in (a_j^-, a_j^+)$. Analogously, one shows this equality if $x \in (a_{j+1}^-, a_{j+1}^+)$. Obviously, $\mathcal{V}_w^\star(\chi_{[a_j,a_{j+1}]}t_j^k) = 0$ if $x \notin (a_j^-, a_{j+1}^+)$. ∎

An important question is how the smoothness of the window functions is connected with the smoothness of the dual windows.

THEOREM 9.8

Let $m \in \mathbb{N}_0$ be given and let the functions $\psi_{j,k} = w_j(x)t_j^k(x)$ form a Riesz basis, where the window functions w_j satisfy the two-overlapping condition (9.23). Then the dual window functions $\tilde{w}_j$, $j \in \mathbb{Z}$, are contained in $C^m(\mathbb{R})$ if and only if $w_j \in C^m(\mathbb{R})$ for all $j \in \mathbb{Z}$.

PROOF Assume $w_j \in C^m(\mathbb{R})$ for all $j \in \mathbb{Z}$. Since $\{\psi_{j,k}\}$ is a Riesz basis and the windows are continuous we know that $w_j(x)$ resp. $\det \boldsymbol{M}_j(x)$ do not vanish for $x \in [a_j^+, a_{j+1}^-]$ resp. $x \in [a_j^-, a_j^+]$. Therefore, in every interval (a_j^-, a_j^+), $[a_j^+, a_{j+1}^-]$ and (a_{j+1}^-, a_{j+1}^+) the dual window $\tilde{w}_j$ is a fraction of functions from $C^m(\mathbb{R})$ with non-vanishing denominator. Hence, it is shown that $\tilde{w}_j$ is m-times continuous differentiable in $(-\infty, a_j^-)$, (a_j^-, a_j^+), (a_j^+, a_{j+1}^-), (a_{j+1}^-, a_{j+1}^+) and (a_{j+1}^+, ∞).

From the smoothness of w_{j-1} and the support property (9.23) we know that $w_{j-1}^{(\mu)}(a_j^+) = 0$, $\mu = 0, \ldots, m$ which implies

$$\tilde{w}_j^{(\mu)}(a_j^- + 0) = 0 = \tilde{w}_j^{(\mu)}(a_j^- - 0) = \tilde{w}_j^{(\mu)}(a_j^-), \quad \mu = 0, \ldots, m,$$

i.e., $\tilde{w}_j$ is m-times continuously differentiable in $x = a_j^-$. Furthermore, we have

$$\frac{1}{w_j(x)} - \frac{w_{j-1}(2a_j - x)}{\det \boldsymbol{M}_j(x)} = \frac{w_j(2a_j - x)\, w_{j-1}(x)}{w_j(x) \det \boldsymbol{M}_j(x)}.$$

The derivatives up to m-th order of the numerator $w_j(2a_j - x)\, w_{j-1}(x)$ vanish for $x = a_j^+$ and hence $\tilde{w}_j^{(\mu)}(a_j^+ + 0) - \tilde{w}_j^{(\mu)}(a_j^+ - 0) = 0$, $\mu = 0, \ldots, m$. From the definition of $\tilde{w}_j$ in (9.34) we know $\tilde{w}_j(a_j^+) = \tilde{w}_j(a_j^+ + 0)$. Summing up, $\tilde{w}_j$ is m-times continuously differentiable in $x = a_j^+$, too. For $x = a_{j+1}^-$ and $x = a_{j+1}^+$ the proof is analogous. Therefore, we have shown that $\tilde{w}_j$, $j \in \mathbb{Z}$, is contained in $C^m(\mathbb{R})$ if $w_j \in C_m(\mathbb{R})$, $j \in \mathbb{Z}$.

Because the functions w_j are the dual windows with respect to $\tilde{w}_j$ it immediately follows now that $\tilde{w}_j \in C^m(\mathbb{R})$, $j \in \mathbb{Z}$, implies $w_j \in C^m(\mathbb{R})$, $j \in \mathbb{Z}$. ▮

9.5.3 Examples of local trigonometric bases

Orthonormal bases. In [21], Coifman and Meyer have given sufficient conditions for the orthonormality of local trigonometric bases (cf. Section 9.3). The result of Chui and Shi also allows finding necessary conditions for orthonormality in the two-overlapping setting.

THEOREM 9.9

Chui and Shi [17] *Let the functions ψ_j^k satisfy the assumptions of* Theorem 9.6. *Then, the system $\{\psi_j^k\}$ is an orthonormal basis of $L^2(\mathbb{R})$ if and only if for all $j \in \mathbb{Z}$ it holds*

$$|w_j(x)| = 1, \qquad a.e.\ on\ (a_j^+, a_{j+1}^-), \tag{9.35}$$

$$|w_j(x)|^2 + |w_{j-1}(x)|^2 = 1, \qquad a.e.\ on\ (a_j^-, a_j^+), \tag{9.36}$$

$$w_{j-1}(x) = w_j(2a_j - x)e^{i\beta_j(|x-a_j|)}, \qquad a.e.\ on\ (a_j^-, a_j^+), \tag{9.37}$$

with an arbitrary measurable function $\beta_j : [0, \varepsilon_j) \to [0, 2\pi)$.

PROOF Assume the functions ψ_j^k constitute an orthonormal basis. Analogous to the proof of Theorem 9.6 it follows from Theorem 9.3 that $\mathcal{U}_w$ is a unitary isomorphism. For $x \in [a_j^+, a_{j+1}^-]$ we conclude

$$f(x) = \mathcal{U}_w \mathcal{U}_w^\star f(x) = \overline{w_j(x)} w_j(x) f(x)$$

which implies (9.35). Analogously, we obtain that $\boldsymbol{M}_j(x)\boldsymbol{M}_j^H(x) = \boldsymbol{I}$ for $x \in (a_j, a_j^+)$. The computation of the entries of $\boldsymbol{M}_j(x)\boldsymbol{M}_j^H(x)$ leads on the one hand directly to (9.36) and on the other hand to

$$w_j(x)\overline{w_j(2a_j - x)} = w_{j-1}(x)\overline{w_{j-1}(2a_j - x)}, \quad x \in (a_j^-, a_j^+). \tag{9.38}$$

We use (9.38) to eliminate $w_j(x)$ in (9.36) and obtain

$$1 = \frac{|w_{j-1}(x)|^2}{|w_j(2a_j - x)|^2}\left(\left|w_{j-1}(2a_j - x)\right|^2 + \left|w_j(2a_j - x)\right|^2\right) = \frac{|w_{j-1}(x)|^2}{|w_j(2a_j - x)|^2}.$$

Hence, it holds that $|w_{j-1}(x)| = |w_j(2a_j - x)|$ which is equivalent to $w_{j-1}(x) = w_j(2a_j - x)e^{\imath\beta_j(x - a_j)}$. Applying this identity to (9.38) we conclude that β_j is an even function and the identity (9.37) is shown.

On the other hand, if (9.36) and (9.37) are valid one shows by simple computations that $\boldsymbol{M}_j(x)\boldsymbol{M}_j^H(x) = \boldsymbol{I}$. Hence, $\mathcal{U}_w \mathcal{U}_w^\star f(x) = f(x)$ for $x \in (a_j, a_j^+)$. For $x \in (a_j^+, a_{j+1}^-)$ the same equality follows from (9.35), immediately. Thus, $\mathcal{U}_w$ is a unitary isomorphism and from Theorem 9.3 we know that $\{\psi_j^k\}$ is an orthonormal basis. ∎

If the window functions are real-valued and non-negative the conditions (9.35)–(9.37) coincide with the Coifman-Meyer conditions (9.8)–(9.10). Thus, for the case of non-negative, two-overlapping windows the equations (9.8)–(9.10) are not only sufficient but also necessary conditions for the orthonormality of $\{\psi_j^k\}$.

Biorthogonal bases of Jawerth and Sweldens. Jawerth and Sweldens [36] have considered real-valued, two-overlapping window functions, which are non-negative and satisfy the conditions

$$w_j(x) = 1, \qquad x \in [a_j^+, a_{j+1}^-], \tag{9.39}$$

$$w_j(a_j + x) = w_{j-1}(a_j - x), \qquad |x| < \varepsilon_j. \tag{9.40}$$

Equality (9.40) implies that for $x \in (a_j^-, a_j^+)$

$$\det \boldsymbol{M}_j(x) = \tfrac{\Delta_j(x)}{2} = w_j^2(x) + w_{j-1}^2(x). \tag{9.41}$$

Then, from Theorem 9.6 we obtain the best possible Riesz constants

$$A_0 = \inf_{j\in\mathbb{Z}} \min\Big(1, \operatorname*{ess\,inf}_{x\in(a_j,a_j^+)} \big(w_j^2(x) + w_{j-1}^2(x)\big)\Big),$$

$$B_0 = \sup_{j\in\mathbb{Z}} \max\Big(1, \operatorname*{ess\,sup}_{x\in(a_j,a_j^+)} \big(w_j^2(x) + w_{j-1}^2(x)\big)\Big).$$

In particular, the result of Jawerth and Sweldens in Theorem 9.5 can be derived from the Riesz constants given above.

Applying (9.39), (9.40) and (9.41) to Theorem 9.7 we obtain the dual window functions

$$\tilde{w}_j(x) = \begin{cases} 1, & \text{if } x \in [a_j^+, a_{j+1}^-], \\ \dfrac{w_j(x)}{w_j^2(x) + w_{j-1}^2(x)}, & \text{if } x \in (a_j^-, a_j^+), \\ \dfrac{w_j(x)}{w_j^2(x) + w_{j+1}^2(x)}, & \text{if } x \in (a_{j+1}^{,} a_{j+1}^+), \\ 0, & \text{otherwise.} \end{cases}$$

9.6 Wilson Bases

Now we want to investigate bases of windowed trigonometric functions without restriction on the support of the window functions. Here we consider the trigonometric functions

$$d_j^k(x) := \begin{cases} \epsilon_k \cos(2k\pi x), & \text{if } j \text{ is even}, \\ 2\sin(2(k+1)\pi x), & \text{if } j \text{ is odd}, \end{cases} \qquad k \in \mathbb{N}_0,\ j \in \mathbb{Z},$$

where

$$\epsilon_k := \begin{cases} \sqrt{2}, & \text{if } k = 0, \\ 2, & \text{otherwise.} \end{cases}$$

In particular, for each $j \in \mathbb{Z}$ the functions d_j^k, $k \in \mathbb{N}_0$, form an orthonormal basis of $L^2([\frac{j}{2}, \frac{j+1}{2}])$ (cf. Lemma 9.1). Furthermore, we introduce window functions $w_j \in L^2(\mathbb{R})$ which are integer translates of w_0 resp. w_{-1} for even resp. odd $j \in \mathbb{Z}$. Namely, we have $w_{2j+s}(x) = w_s(x-j)$, $j \in \mathbb{Z}$, $s \in \{-1, 0\}$.

DEFINITION 9.3 *The Wilson system for the window functions* $w_0, w_{-1} \in L^2(\mathbb{R})$ *is given by* $\{\psi_j^k : j \in \mathbb{Z}, k \in \mathbb{N}_0\}$ *with*

$$\psi_j^k(x) := w_j(x)\, d_j^k(x) = \begin{cases} w_0\big(x - \frac{j}{2}\big)\, \epsilon_k \cos(2k\pi x), & \text{if } j \text{ is even}, \\ w_{-1}\big(x - \frac{j+1}{2}\big)\, 2\sin(2(k+1)\pi x), & \text{if } j \text{ is odd}. \end{cases}$$

In particular, if the Wilson system $\{\psi_j^k\}$ *is a Riesz basis we call it a Wilson basis.*

The notation of Wilson bases was introduced by Daubechies, Jaffard and Journé in [26]. To construct orthonormal bases with fast decay in time and frequency domain they followed an approach of Sullivan et al. [43] and Wilson [49]. He had observed that for many applications it is often not necessary to distinguish between positive and negative frequencies of the same order. In particular, the functions ψ_j^k of the Wilson system in Definition 9.3 are localized around k and $-k$ in the frequency domain. It turns out that for these functions the restriction from the Balian-Low is completely removed.

9.6.1 Folding operators for Wilson Bases

To investigate the basis properties of Wilson systems we will again use an unfolding operator, which maps the orthonormal basis $\{\chi_{[j,j+\frac{1}{2}]}\, t_j^k\}$ onto the Wilson system $\{\psi_j^k\}$. For this purpose we will apply the Zak transform. The Zak transform has been introduced by Gelfand [32], and was rediscovered by Weil [44] and independently by Zak [53]. A review of the Zak transform and its use in signal analysis can be found in [34].

DEFINITION 9.4 *The Zak transform of* $f \in L^2(\mathbb{R})$ *is defined by*

$$Zf(x,\xi) := \sum_{n\in\mathbb{Z}} f(x+n)\, e^{2\pi i n\xi}, \quad x, \xi \in \mathbb{R}.$$

For a fixed $x \in \mathbb{R}$ the Zak transform can be interpreted as a Fourier series with coefficients $f(x+n)$. This point of view results in some features (cf. [10]) which turn out to be very useful for our purposes.

LEMMA 9.3

a) *The Zak transform is quasi-periodic in the sense that*

$$Zf(x+n,\xi+m) = e^{-2\pi in\xi} Zf(x,\xi), \;\; x,\xi \in \mathbb{R};\; n,m \in \mathbb{Z}. \tag{9.42}$$

Thus, Zf is completely determined on $\mathbb{R}^2$ by its values on the square $Q := [-\frac{1}{2}, \frac{1}{2})^2$.

b) *The Zak transform is a unitary map, i.e., $Z : L^2(\mathbb{R}) \to L^2(Q)$ is bijective and*

$$\|f\|_{L^2(\mathbb{R})} = \|Zf\|_{L^2(Q)}. \tag{9.43}$$

c) *For $f, g \in L^2(\mathbb{R})$ it holds that*

$$\langle f, g\rangle = \iint_Q Zf(x,\xi)\, \overline{Zg(x,\xi)}\, dx\, d\xi. \tag{9.44}$$

d) *Every function $f \in L^2(\mathbb{R})$ can be determined from Zf by*

$$f(x) = \int_{-\frac{1}{2}}^{\frac{1}{2}} Zf(x,\xi)\, d\xi. \tag{9.45}$$

e) *The relationship between the Zak transform of $f \in L^2(\mathbb{R})$ and its Fourier transform (see* (9.1)*) is given by*

$$Zf(x,\xi) = e^{-2\pi ix\xi}\, Z\hat{f}(-\xi, x). \tag{9.46}$$

PROOF One obtains assertion a) immediately by applying Definition 9.4 of the Zak transform.

Now, we will show b). Obviously, $\{f(x+j)e^{2\pi ij\xi} : j \in \mathbb{Z}\}$ is an orthogonal system in $L^2(Q)$. Thus, for $f \in L^2(\mathbb{R})$ it holds that

$$\|Zf\|^2_{L^2(Q)} = \sum_{j\in\mathbb{Z}} \iint_Q \left|f(x+j)e^{2\pi ij\xi}\right|^2 dx\, d\xi = \|f\|^2_{L^2(\mathbb{R})}$$

and (9.43) is shown. From (9.43) it follows immediately that Z is injective. Furthermore, for each $G \in L^2(Q)$ we can find a function $f \in L^2(\mathbb{R})$ with

$Zf = G$ as follows. We define $f(t)$ for each $t \in \mathbb{R}$ by

$$f(x+j) := \int_{-\frac{1}{2}}^{\frac{1}{2}} G(x,\xi)\, e^{-2\pi i j\xi}\, d\xi, \quad x \in [-\tfrac{1}{2}, \tfrac{1}{2}),\ j \in \mathbb{Z}.$$

By Parseval's identity we conclude

$$\begin{aligned} \|G\|^2_{L^2(Q)} &= \sum_{j,k\in\mathbb{Z}} \left| \iint_Q G(x,\xi)\, e^{-2\pi i(kx+j\xi)}\, d\xi\, dx \right|^2 \\ &= \sum_{j,k\in\mathbb{Z}} \left| \int_{-\frac{1}{2}}^{\frac{1}{2}} f(x+j)\, e^{-2\pi i kx}\, dx \right|^2 \\ &= \sum_{j\in\mathbb{Z}} \int_{-\frac{1}{2}}^{\frac{1}{2}} |f(x+j)|^2\, dx = \|f\|_{L^2(\mathbb{R})}, \end{aligned}$$

i.e., $f \in L^2(\mathbb{R})$. Since the Zak transform of f is given by

$$Zf(x,\xi) = \sum_{j\in\mathbb{Z}} \int_{-\frac{1}{2}}^{\frac{1}{2}} G(x,\eta)\, e^{-2\pi i j\eta}\, d\eta\; e^{2\pi i j\xi} = G(x,\xi)$$

it follows that Z is surjective. Hence, b) is proved.

Now, from the quasi-periodicity of the Zak transform one obtains

$$f(t+j) = \int_{-\frac{1}{2}}^{\frac{1}{2}} G(t,\xi)\, e^{-2\pi i j\xi}\, d\xi = \int_{-\frac{1}{2}}^{\frac{1}{2}} Zf(t+j,\xi)\, d\xi, \quad t \in [-\tfrac{1}{2}, \tfrac{1}{2}),\ j \in \mathbb{Z}.$$

By the substitution $t + j = x \in \mathbb{R}$ assertion d) follows. The equality (9.44) can be shown to be analogous to (9.43) using that $f(x+j)e^{2\pi i j\xi}$, $j \in \mathbb{Z}$, and $g(x+r)e^{2\pi i r\xi}$, $r \in \mathbb{Z}$, are orthogonal if $r \neq j$.

To prove e) we assume f to be in the Schwartz space $S(\mathbb{R})$. Applying the Poisson summation formula on $f\, e^{2\pi i\xi\cdot}$, it follows

$$\begin{aligned} Zf(x,\xi) &= e^{-2\pi i x\xi} \sum_{j\in\mathbb{Z}} f(x+j)\, e^{2\pi i(x+j)\xi} \\ &= e^{-2\pi i x\xi} \sum_{k\in\mathbb{Z}} \hat{f}(k-\xi)\, e^{2\pi i kx} = e^{-2\pi i x\xi}\, Z\hat{f}(-\xi, x). \end{aligned}$$

Since $S(\mathbb{R})$ is dense in $L^2(\mathbb{R})$ the equality holds for all $f \in L^2(\mathbb{R})$. ∎

Now we can define an unfolding operator for Wilson bases (cf. [15]).

THEOREM 9.10
Let w_0 and w_{-1} satisfy the inequality

$$|w_s(x)| \leq C(1+|x|)^{-1-\varepsilon}, \quad s \in \{-1, 0\}, \tag{9.47}$$

for some positive constant C and some $\varepsilon > 0$. Furthermore, let the matrix

$$\boldsymbol{M}(x,\xi) := \begin{pmatrix} Zw_0(x,\xi) & -Zw_{-1}(x,\xi) \\ Zw_0(-x,\xi) & Zw_{-1}(-x,\xi) \end{pmatrix}$$

be given. Then, a bounded operator $\mathcal{U}_w : L^2(\mathbb{R}) \to L^2(\mathbb{R})$ is defined by

$$\begin{pmatrix} Z\mathcal{U}_w f(x,\xi) \\ Z\mathcal{U}_w f(-x,\xi) \end{pmatrix} = \boldsymbol{M}(x,\xi) \begin{pmatrix} Zf(x,\xi) \\ Zf(-x,\xi) \end{pmatrix}, \tag{9.48}$$

$$(x,\xi) \in Q^+ := (0, \tfrac{1}{2}) \times [-\tfrac{1}{2}, \tfrac{1}{2}).$$

This operator satisfies the equality

$$\mathcal{U}_w \left(\chi_{[\frac{j}{2}, \frac{j+1}{2}]} d_j^k \right) = \psi_j^k. \tag{9.49}$$

PROOF From (9.47) it follows that $w_0, w_{-1} \in L^2(\mathbb{R})$ and we obtain

$$|Zw_s(x,\xi)| \leq \sum_{n\in\mathbb{Z}} |w_s(x+n)| \leq C \sum_{n\in\mathbb{Z}} (1+|x+n|)^{-1-\varepsilon} < \infty.$$

Hence, Zw_0 and Zw_{-1} are contained in $L^\infty(Q)$ which implies $Z\mathcal{U}_w f \in L^2(Q)$ if $Zf \in L^2(Q)$. By the unitarity of the Zak transform (9.43) it follows that $\mathcal{U}_w$ is a bounded linear mapping from $L^2(\mathbb{R})$ into $L^2(\mathbb{R})$.

Using the 1-periodicity of d_j^k we obtain

$$\begin{aligned} Z\psi_{2j+s}^k(x,\xi) &= \sum_{n\in\mathbb{Z}} w_s(x+n-j)\, e^{2\pi i n\xi}\, d_s^k(x) \\ &= Zw_s(x,\xi)\, e^{2\pi i j\xi}\, d_s^k(x). \end{aligned} \tag{9.50}$$

In particular, we have for $w_0 = \chi_{[0,\frac{1}{2}]}$ that

$$Z\left(\chi_{[j,j+\frac{1}{2}]}d_{2j}^k\right)(x,\xi) = \begin{cases} e^{2\pi ij\xi}\,\epsilon_k\cos(2k\pi x), & \text{if } x\in[0,\frac{1}{2}],\\ 0, & \text{if } x\in(-\frac{1}{2},0),\end{cases} \tag{9.51}$$

and for $w_{-1} = \chi_{[-\frac{1}{2},0]}$ that

$$Z\left(\chi_{[j-\frac{1}{2},j]}d_{2j-1}^k\right)(x,\xi) = \begin{cases} 0, & \text{if } x\in(0,\frac{1}{2}),\\ e^{2\pi ij\xi}\,2\sin(2(k+1)\pi x), & \text{if } x\in[-\frac{1}{2},0].\end{cases} \tag{9.52}$$

Since $\cos(2k\pi\cdot)$ is an even function it follows for $(x,\xi)\in Q^+$

$$\begin{aligned}\begin{pmatrix} Z\mathcal{U}_w\big(\chi_{[j,j+\frac{1}{2}]}d_{2j}^k\big)(x,\xi)\\ Z\mathcal{U}_w\big(\chi_{[j,j+\frac{1}{2}]}d_{2j}^k\big)(-x,\xi)\end{pmatrix} &= \boldsymbol{M}(x,\xi)\begin{pmatrix} e^{2\pi ij\xi}\,\epsilon_k\cos(2k\pi x)\\ 0\end{pmatrix}\\ &= \begin{pmatrix} Zw_0(x,\xi)\,e^{2\pi ij\xi}\,\epsilon_k\cos(2k\pi x)\\ Zw_0(-x,\xi)\,e^{2\pi ij\xi}\,\epsilon_k\cos(-2k\pi x)\end{pmatrix}\\ &= \begin{pmatrix} Z\psi_{2j}^k(x,\xi)\\ Z\psi_{2j}^k(-x,\xi)\end{pmatrix}.\end{aligned}$$

That is, $Z\mathcal{U}_w\big(\chi_{[j,j+\frac{1}{2}]}d_{2j}^k\big)(x,\xi) = Z\psi_{2j}^k(x,\xi)$, $(x,\xi)\in Q$, and by the unitarity of the Zak transform we conclude $\mathcal{U}_w\big(\chi_{[j,j+\frac{1}{2}]}d_{2j}^k\big) = \psi_{2j}^k$. Analogously, it follows that $\mathcal{U}_w\big(\chi_{[j-\frac{1}{2},j]}d_{2j-1}^k\big) = \psi_{2j-1}^k$ and (9.49) is shown. ∎

Note that the definition of $\mathcal{U}_w$ in (9.48) has a similar structure as equality (9.26), which describes the unfolding operator for two-overlapping bases in the overlapping interval (a_j^-, a_j^+). Therefore, we will obtain similar results as for the bases of Chui and Shi and the proofs will be based on analogous ideas. Furthermore, the property (9.49) of $\mathcal{U}_w$ corresponds to the property (9.25) of the unfolding operator for local trigonometric bases in the two-overlapping setting. Therefore, we will denote the operator $\mathcal{U}_w$ defined in (9.48) as **unfolding operator** for the Wilson system $\{\psi_j^k\}$. Analogously, we denote the dual operator $\mathcal{T}_w = \mathcal{U}_w^\star$ as **folding operator** for the Wilson system $\{\psi_j^k\}$. By simple calculations it follows that $\mathcal{T}_w$ is given by

$$\begin{pmatrix} Z\mathcal{T}_w f(x,\xi)\\ Z\mathcal{T}_w f(-x,\xi)\end{pmatrix} = \boldsymbol{M}^H(x,\xi)\begin{pmatrix} Zf(x,\xi)\\ Zf(-x,\xi)\end{pmatrix}, \quad (x,\xi)\in Q^+. \tag{9.53}$$

From equality (9.49) one concludes immediately

$$\langle f, \psi_j^k \rangle = \int_{\frac{j}{2}}^{\frac{j+1}{2}} \mathcal{T}_w f(x)\, d_j^k(x)\, dx. \tag{9.54}$$

The operators $\mathcal{U}_w$ and $\mathcal{T}_w$ can also be described without using the Zak transform.

COROLLARY 9.2

Let the window functions w_0 and w_{-1} be given as in Theorem 9.10. *For $f \in L^2(\mathbb{R})$ and $w_{2j+s} = w_s(\cdot - j)$ the unfolding operator $\mathcal{U}_w$ defined in (9.48) satisfies*

$$\mathcal{U}_w f(j \pm x) = \sum_{r\in\mathbb{Z}} \Big(w_0(r \pm x) f(j + r + x) \mp w_{-1}(r \pm x) f(j + r - x) \Big),$$

where $x \in (0, \frac{1}{2})$, $j \in \mathbb{Z}$. Furthermore the folding operator $\mathcal{T}_w$ defined in (9.53) satisfies

$$\mathcal{T}_w f(x) = \sum_{r\in\mathbb{Z}} \Big(\overline{w_j(r + x)} f(r + x) + (-1)^j\, \overline{w_j(r - x)} f(r - x) \Big), \tag{9.55}$$

where $x \in [\frac{j}{2}, \frac{j+1}{2})$, $j \in \mathbb{Z}$.

PROOF Since the Zak transform is defined as a Fourier series we have

$$\begin{aligned} Zf(x,\xi)\overline{Zw_s(x,\xi)} &= \sum_{j\in\mathbb{Z}} \Big(\sum_{r\in\mathbb{Z}} f(x + j)\overline{w_s(x + r)} \Big) e^{2\pi i (j-r)\xi} \\ &= \sum_{j\in\mathbb{Z}} \Big(\sum_{r\in\mathbb{Z}} f(x + j + r)\overline{w_s(x + r)} \Big) e^{2\pi i j\xi}. \end{aligned}$$

From (9.53) it follows now by comparison of coefficients that for $x \in [0, \frac{1}{2})$

$$\mathcal{T}_w f(x + j) = \sum_{r\in\mathbb{Z}} \Big(f(x + r + j)\overline{w_0(x + r)} + f(-x + r + j)\overline{w_0(-x + r)} \Big).$$

By the substitution $x \to x - j$, $j \in \mathbb{Z}$, we obtain (9.55) for $x \in [j, j + \frac{1}{2})$. Analogously, one shows (9.55) for $x \in [j - \frac{1}{2}, j)$. Therefore, identity (9.55) is proved. The assertion for the unfolding operator $\mathcal{U}_w$ follows by analogous arguments. ∎

Corollary 9.2 states that one obtains the folding operator $\mathcal{T}_w f$ in $[\frac{j}{2}, \frac{j+1}{2})$ by "folding" the function $w_j f$ repeatedly at the points $\frac{j}{2}$ and adding or subtracting the folded parts. Similar to the classical folding operator of Wickerhauser we have for smooth window functions w_s and smooth f that $\mathcal{T}_w f$ is smooth in each interval $(\frac{j}{2}, \frac{j+1}{2})$. Furthermore, $\mathcal{T}_w f$ has smooth even resp. odd extensions over the interval boundaries for even resp. odd $j \in \mathbb{Z}$ (see Fig. 9.6).

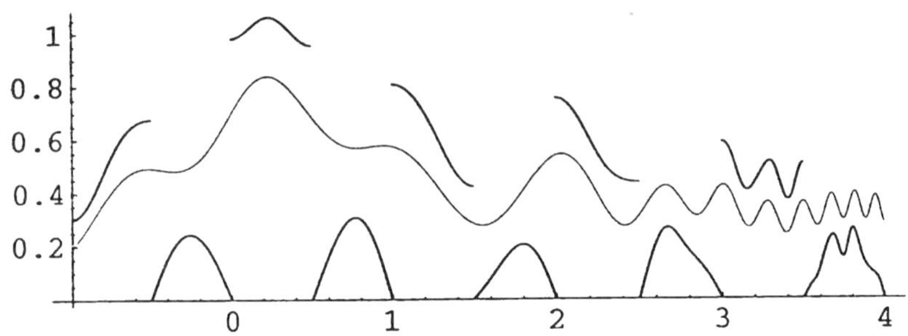

FIGURE 9.6
Folding of a smooth function.

REMARK 9.1 It is also possible to construct Wilson bases using the anti-periodic functions $c_j^k(x) := \cos((2k+1)\pi(x - \frac{j}{2}))$ instead of the periodic functions d_j^k (see, e.g., [15, 19, 22]). An unfolding operator which satisfies (9.54) with these functions c_j^k is given by

$$\begin{pmatrix} Z_2\mathcal{U}_w f(x,\xi) \\ Z_2\mathcal{U}_w f(-x,\xi) \end{pmatrix} = M(x, \xi + \tfrac{1}{2}) \begin{pmatrix} Z_2 f(x,\xi) \\ Z_2 f(-x,\xi) \end{pmatrix}, \quad (x,\xi) \in Q^+.$$

Since we obtain similar statements for these bases we will consider only the functions d_j^k in the sequel. ∎

9.6.2 Biorthogonal Wilson bases and Riesz bounds

With the unfolding operator $\mathcal{U}_w$ and the folding operator $\mathcal{T}_w$ introduced in (9.48) and (9.24) we can investigate the basis properties of Wilson systems. In particular, we are interested in the question for which windows w_0 and w_{-1} the functions ψ_j^k constitute a Riesz basis.

THEOREM 9.11

Let the functions $\psi^k_{2j+s}(x) := w_s(x-2j)d^k_{2j+s}(x)$, $j \in \mathbb{Z}$, $s \in \{-1,0\}$, $k \in \mathbb{N}_0$, *be given, where the window functions* w_0 *and* w_{-1} *satisfy the decay property* (9.47). *Furthermore, we define*

$$A_0 \quad := \quad \operatorname*{ess\,inf}_{(x,\xi)\in Q^+} \tfrac{\Delta(x,\xi)}{2} - \sqrt{\tfrac{\Delta^2(x,\xi)}{4} - |\det \boldsymbol{M}(x,\xi)|^2}, \tag{9.56}$$

$$B_0 \quad := \quad \operatorname*{ess\,sup}_{(x,\xi)\in Q^+} \tfrac{\Delta(x,\xi)}{2} + \sqrt{\tfrac{\Delta^2(x,\xi)}{4} - |\det \boldsymbol{M}(x,\xi)|^2} \tag{9.57}$$

with

$$\Delta(x,\xi) := |Zw_0(x,\xi)|^2 + |Zw_0(-x,\xi)|^2 + |Zw_{-1}(x,\xi)|^2 + |Zw_{-1}(-x,\xi)|^2.$$

The Wilson system $\{\psi^k_j\}$ *is a Riesz basis of* $L^2(\mathbb{R})$ *with the Riesz bounds* A *and* B *if and only if* $0 < A \le A_0 \le B_0 \le B < \infty$.

PROOF Assume $0 < A \le A_0 \le B_0 \le B < \infty$. Let the unfolding operator $\mathcal{U}_w$ be defined by (9.48). For $f \in L^2(\mathbb{R})$ we conclude from unitarity of the Zak transform

$$\|\mathcal{U}_w f\|^2_{L^2(\mathbb{R})} = \|Z\,\mathcal{U}_w f\|_{L^2(Q)} = \iint_{Q^+} \left| \begin{pmatrix} Z\,\mathcal{U}_w f(x,\xi) \\ Z\,\mathcal{U}_w f(-x,\xi) \end{pmatrix} \right|^2 dx\, d\xi$$

$$= \quad \iint_{Q^+} \left| \boldsymbol{M}(x,\xi) \begin{pmatrix} Zf(x,\xi) \\ Zf(-x,\xi) \end{pmatrix} \right|^2 dx\, d\xi$$

$$\le \quad \iint_{Q^+} \|\boldsymbol{M}(x,\xi)\|_2^2 \left(|Zf(x,\xi)|^2 + |Zf(-x,\xi)|^2\right) dx\, d\xi$$

$$\le \quad \operatorname*{ess\,sup}_{(x,\xi)\in Q^+} \|\boldsymbol{M}(x,\xi)\|_2^2 \iint_{Q} |Zf(x,\xi)|^2\, dx\, d\xi$$

$$= \quad \operatorname*{ess\,sup}_{(x,\xi)\in Q^+} \|\boldsymbol{M}(x,\xi)\|_2^2\, \|f\|^2_{L^2(\mathbb{R})}.$$

Since

$$\|\boldsymbol{M}(x,\xi)\|_2^2 = \tfrac{\Delta(x,\xi)}{2} + \sqrt{\tfrac{\Delta^2(x,\xi)}{4} - |\det \boldsymbol{M}(x,\xi)|^2}$$

it is proved that $\|\mathcal{U}_w\|^2_{L^2(\mathbb{R})\to L^2(\mathbb{R})} \le B_0 \le B$.

Analogous to the proof of Theorem 9.6 one shows that from $A_0 > 0$ it follows that the matrix $\boldsymbol{M}(x,\xi)$ is invertible a.e. on Q^+ and

$$\|\boldsymbol{M}^{-1}(x,\xi)\|_2^2 = \tfrac{\Delta(x,\xi)}{2} - \sqrt{\tfrac{\Delta^2(x,\xi)}{4} - |\det \boldsymbol{M}(x,\xi)|^2}.$$

Now, it follows immediately that the inverse unfolding operator $\mathcal{U}_w^{-1} : L^2(\mathbb{R}) \to L^2(\mathbb{R})$ exists and is completely described by

$$\begin{pmatrix} Z\mathcal{U}_w^{-1} f(x,\xi) \\ Z\mathcal{U}_w^{-1} f(-x,\xi) \end{pmatrix} = \boldsymbol{M}^{-1}(x,\xi) \begin{pmatrix} Zf(x,\xi) \\ Zf(-x,\xi) \end{pmatrix}, \quad (x,\xi) \in Q^+. \tag{9.58}$$

Analogous to the boundedness of $\mathcal{U}_w$ we show

$$\|\mathcal{U}_w^{-1}\|^2_{L^2(\mathbb{R})\to L^2(\mathbb{R})} \le \operatorname*{ess\,sup}_{(x,\xi)\in Q^+} \|\boldsymbol{M}^{-1}(x,\xi)\|_2^2 = A_0^{-1} \le A^{-1}.$$

Thus, $\mathcal{U}_w$ is an topological isomorphism with $\|\mathcal{U}_w^{-1}\|^2_{L^2(\mathbb{R})\to L^2(\mathbb{R})} \le A^{-1}$ and $\|\mathcal{U}_w\|^2_{L^2(\mathbb{R})\to L^2(\mathbb{R})} \le B$. Since $\mathcal{U}_w$ furthermore maps an orthonormal basis onto $\{\psi_j^k\}$ (see (9.49)) the assumptions of Theorem 9.1(ii) are satisfied and therefore $\{\psi_j^k\}$ is a Riesz basis with Riesz bounds A and B.

Conversely, let $\{\psi_j^k\}$ be a Riesz basis with Riesz bounds A and B. Then, it follows as in the proof of Theorem 9.6 that $0 < A \le A_0 \le B_0 \le B < \infty$. In particular, one can show that $B < B_0$ leads to a contradiction by the same arguments as for inequality (9.31). The inequality $A \le A_0$ is shown analogously. ▮

As for the local trigonometric bases of Chui and Shi we can derive a simple sufficient and necessary condition for Riesz stability.

COROLLARY 9.3

Let the functions ψ_j^k satisfy the assumptions of Theorem 9.11. *Then, the Wilson system $\{\psi_j^k\}$ is a Riesz basis of $L^2(\mathbb{R})$, if and only if*

$$|\det \boldsymbol{M}(x,\xi)| \ge c > 0. \tag{9.59}$$

In particular, the Riesz bounds can be estimated by

$$\frac{1}{B_0} \operatorname*{ess\,inf}_{(x,\xi)\in Q^+} |\det \boldsymbol{M}(x,\xi)|^2 \;\le\; A_0 \;\le\; \operatorname*{ess\,inf}_{(x,\xi)\in Q^+} |\det \boldsymbol{M}(x,\xi)|, \tag{9.60}$$

$$\operatorname*{ess\,sup}_{(x,\xi)\in Q^+} \frac{\Delta(x,\xi)}{2} \;\le\; B_0 \;\le\; \operatorname*{ess\,sup}_{(x,\xi)\in Q^+} \Delta(x,\xi). \tag{9.61}$$

PROOF Since the assumptions of Theorem 9.11 are satisfied it follows from Theorem 9.10 that the entries of $\boldsymbol{M}(x,\xi)$ are elements of $L^\infty(Q)$. Then, Theorem 9.11 implies that B_0 is finite. Analogous to the proof of Corollary 9.1 we prove the inequalities (9.60) and (9.61) as well as the equivalence of (9.59) and $A_0 > 0$. ∎

If $\{\psi_j^k\}$ is a Riesz basis with Riesz bounds A and B, then the dual basis $\{\tilde{\psi}_j^k\}$ exists. To describe this dual basis we introduce the dual window functions $\tilde{w}_0$ and $\tilde{w}_{-1}$ which are defined by

$$\tilde{w}_0(x) := \int_{-\frac{1}{2}}^{\frac{1}{2}} \frac{\overline{Zw_{-1}(-x,\xi)}}{\overline{\det \boldsymbol{M}(x,\xi)}}\, d\xi \quad \text{and} \quad \tilde{w}_{-1}(x) := \int_{-\frac{1}{2}}^{\frac{1}{2}} \frac{\overline{Zw_0(-x,\xi)}}{\overline{\det \boldsymbol{M}(x,\xi)}}\, d\xi.$$

It turns out that the dual basis is a Wilson basis, too.

THEOREM 9.12

Let the functions ψ_j^k *satisfy the assumptions of* Theorem 9.11. *Then, the dual basis is constituted by the functions*

$$\tilde{\psi}_j^k(x) := \tilde{w}_j(x) d_j^k(x)$$

$$= \begin{cases} \tilde{w}_0\left(x - \frac{j}{2}\right) \epsilon_k \cos(2k\pi x), & \textit{if } j \textit{ is even}, \\ \tilde{w}_{-1}\left(x - \frac{j+1}{2}\right) 2\sin(2(k+1)\pi x), & \textit{if } j \textit{ is odd}, \end{cases} \quad j \in \mathbb{Z},\ k \in \mathbb{N}_0.$$

PROOF By Theorem 9.2 we know that the dual basis is given by $\tilde{\psi}_j^k = (\mathcal{U}_w^{-1})^\star(\chi_{[\frac{j}{2},\frac{j+1}{2}]} d_j^k)$ with the inverse of the unfolding operator $\mathcal{U}_w$ defined in (9.48). By Cramer's rule it follows that

$$\boldsymbol{M}^{-1}(x,\xi) = \frac{1}{\det \boldsymbol{M}(x,\xi)} \begin{pmatrix} Zw_{-1}(-x,\xi) & Zw_{-1}(x,\xi) \\ -Zw_0(-x,\xi) & Zw_0(x,\xi) \end{pmatrix}.$$

Using the assertions on the inverse unfolding operator in (9.58) and on the Zak transform of $\chi_{[j,j+\frac{1}{2}]} d_0^k$ in (9.51) we obtain

$$\begin{pmatrix} Z\tilde{\psi}_{2j}^k(x,\xi) \\ Z\tilde{\psi}_{2j}^k(-x,\xi) \end{pmatrix} = \begin{pmatrix} Z(\mathcal{U}_w^{-1})^\star(\chi_{[j,j+\frac{1}{2}]} d_{2j}^k)(x,\xi) \\ Z(\mathcal{U}_w^{-1})^\star(\chi_{[j,j+\frac{1}{2}]} d_{2j}^k)(-x,\xi) \end{pmatrix}$$

$$= \left(\boldsymbol{M}^{-1}(x)\right)^H \begin{pmatrix} e^{2\pi i j\xi}\ \epsilon_k \cos(2k\pi x) \\ 0 \end{pmatrix}$$

$$= \frac{e^{2\pi i j\xi}\ \epsilon_k \cos(2k\pi x)}{\det \boldsymbol{M}(x,\xi)} \begin{pmatrix} \overline{Zw_{-1}(-x,\xi)} \\ \overline{Zw_{-1}(x,\xi)} \end{pmatrix}.$$

From $\det \boldsymbol{M}(x,\xi) = \det \boldsymbol{M}(-x,\xi)$ and $\cos(x) = \cos(-x)$ it follows that

$$Z\tilde{\psi}_{2j}^k(x,\xi) = \frac{\overline{Zw_{-1}(-x,\xi)}}{\det \boldsymbol{M}(x,\xi)}\ e^{2\pi i j\xi}\ \epsilon_k \cos(2k\pi x).$$

With the quasi-periodicity (9.42) of the Zak transform we conclude

$$Z\tilde{\psi}_{2j}^k(x+j,\xi) = \frac{\overline{Zw_{-1}(-x,\xi)}}{\det \boldsymbol{M}(x,\xi)}\ \epsilon_k \cos(2k\pi x).$$

Now, the application of the inverse Zak transform (9.45) leads to

$$\tilde{\psi}_{2j}^k(x+j) = \int_{-\frac{1}{2}}^{\frac{1}{2}} \frac{\overline{Zw_{-1}(-x,\xi)}}{\det \boldsymbol{M}(x,\xi)}\ d\xi\ \epsilon_k \cos(2k\pi x)$$

and thus $\tilde{\psi}_{2j}^k(x) = \tilde{w}_0(x-j)\ \epsilon_k \cos(2k\pi x)$. Analogously, one shows $\tilde{\psi}_{2j-1}^k(x) = \tilde{w}_{-1}(x-j)\ 2\sin(2(k+1)\pi x)$. ∎

9.6.3 Examples of Wilson bases

In the sequel, we consider various examples of Wilson bases to illustrate the application of our results. Up to now, Wilson bases with one real-valued, symmetric window have been investigated by several authors [19, 22, 26]. We will derive general results for this setting. As particular cases of Wilson bases with a real-valued, symmetric window we present the orthonormal bases of Daubechies, Jaffard and Journé [26] as well as bases introduced by Coifman and Meyer [22] with the Gaussian bell as the window function.

As an example with two non-symmetric windows we consider Wilson bases in a two-overlapping setting, which can also be seen as a particular case of the biorthogonal bases of Chui and Shi [17] in the two-overlapping setting.

Wilson bases with one real-valued, symmetric window. We consider the functions $\psi_j^k := w(\cdot - j)\, d_j^k$ with a *real-valued, symmetric window function* w. Obviously, $\{\psi_j^k : j \in \mathbb{Z},\ k \in \mathbb{N}_0\}$ is a Wilson system as introduced in Definition 9.3, where $w_0(x) = w_{-1}(x - \frac{1}{2}) = w(x) = w(\frac{1}{2} - x)$. Applying the Zak transform we obtain

$$\begin{aligned} Zw_{-1}(-x,\xi) &= \sum_{n\in\mathbb{Z}} w(-x+n+\tfrac{1}{2})e^{2\pi i n\xi} = \sum_{n\in\mathbb{Z}} w(x-n)e^{2\pi i n\xi} \\ &= \sum_{n\in\mathbb{Z}} w(x+n)e^{-2\pi i n\xi} = \overline{Zw(x,\xi)}. \end{aligned} \tag{9.62}$$

Hence,

$$\det \boldsymbol{M}(x,\xi) = |Zw(x,\xi)|^2 + |Zw(-x,\xi)|^2 = \tfrac{\Delta(x,\xi)}{2} \tag{9.63}$$

and from Theorem 9.6.2 it follows that the best possible Riesz bounds are given by

$$\begin{aligned} A_0 &= \operatorname*{ess\,inf}_{(x,\xi)\in Q^+} \Big(|Zw(x,\xi)|^2 + |Zw(-x,\xi)|^2\Big), \\ B_0 &= \operatorname*{ess\,sup}_{(x,\xi)\in Q^+} \Big(|Zw(x,\xi)|^2 + |Zw(-x,\xi)|^2\Big). \end{aligned} \tag{9.64}$$

Furthermore, from Theorem 9.12 we conclude that the dual basis is constituted by the functions $\tilde{\psi}_j^k = \tilde{w}(\cdot - j)\, d_j^k$, $j \in \mathbb{Z}$, $k \in \mathbb{N}_0$ with the dual window

$$\tilde{w}(x) := \int_{-\frac{1}{2}}^{\frac{1}{2}} \frac{Zw(x,\xi)}{|Zw(x,\xi)|^2 + |Zw(-x,\xi)|^2}\, d\xi. \tag{9.65}$$

Orthonormal Wilson bases. In [26], Daubechies, Jaffard, and Journé have investigated orthonormal Wilson bases constituted by the functions $\psi_j^k := w(\cdot - j)\, d_j^k$ with a real-valued, symmetric window function. Here, we will describe under which conditions the more general Wilson system from Definition 9.3 is an orthonormal basis.

From Theorem 9.3 we know that $\{\psi_j^k\}$ is an orthonormal basis if the unfolding operator $\mathcal{U}_w$ is a unitary isomorphism. According to the definition of $\mathcal{U}_w$ in (9.48) the unfolding operator is unitary if and only if the matrix $\boldsymbol{M}(x,\xi)$ is unitary for all $(x,\xi) \in Q^+$, i.e.,

$$\boldsymbol{M}^H(x,\xi)\boldsymbol{M}(x,\xi) = \boldsymbol{M}(x,\xi)\boldsymbol{M}^H(x,\xi) = \boldsymbol{I}, \quad (x,\xi) \in Q^+.$$

Computing the entries of $\boldsymbol{M}(x,\xi)\boldsymbol{M}^H(x,\xi)$ we conclude that $\{\psi_j^k\}$ is an orthonormal basis if and only if

$$\left|Zw_0(x,\xi)\right|^2 + \left|Zw_{-1}(x,\xi)\right|^2 = 1, \tag{9.66}$$

$$Zw_0(x,\xi)\overline{Zw_0(-x,\xi)} = Zw_{-1}(x,\xi)\overline{Zw_{-1}(-x,\xi)} \tag{9.67}$$

for $(x,\xi) \in Q$.

In particular, if we restrict ourselves to the case of one real-valued, symmetric window we have the setting of Daubechies, Jaffard, and Journé [26]. Here, we will shortly review their results.

THEOREM 9.13

Daubechies, Jaffard, Journé [26] *Let w be a measurable, real-valued function such that $w(x) = w(\frac{1}{2} - x)$ and $|w(x)| \leq C(1+|x|)^{-1-\varepsilon}$ for some positive constant C and some $\varepsilon > 0$. Then, the following assertions are equivalent.*

(i) *The functions $\psi_j^k := w(\cdot - j)\, d_j^k$, $j \in \mathbb{Z}$, $k \in \mathbb{N}_0$, constitute an orthonormal basis.*

(ii) *For the Zak transform of w it holds that*

$$|Zw(x,\xi)|^2 + |Zw(-x,\xi)|^2 = 1, \quad (x,\xi) \in Q^+.$$

(iii) *For all $j \in \mathbb{Z}$ the window function w satisfies the equality*

$$\sum_{r\in\mathbb{Z}} \Big(w(j+r+x)\, w(r+x) + w(j+r-x)\, w(r-x)\Big) = \delta_{j,0}, \quad x \in (0, \tfrac{1}{2}).$$

(iv) *For the Fourier transform of w it holds that*

$$2\sum_{r\in\mathbb{Z}} \hat{w}(2j+r+\xi)\, \overline{\hat{w}(r+\xi)} = \delta_{j,0}, \quad j \in \mathbb{Z},\ \xi \in [0,1).$$

PROOF As in the previous example we have $w_0(x) = w_{-1}(x - \frac{1}{2}) = w(x) = w(\frac{1}{2} - x)$ and thus $Zw_{-1}(-x,\xi) = \overline{Zw(x,\xi)} = \overline{Zw_0(x,\xi)}$ (cf. (9.62)). Therefore, (9.67) is satisfied and $\{\psi_j^k\}$ is an orthonormal basis if and only (9.66) holds. Replacing $Zw_0(x,\xi)$ and $\overline{Zw_{-1}(-x,\xi)}$ by $Zw(x,\xi)$ in (9.66) the equivalence of (i) and (ii) follows immediately.

Applying Definition 9.4 of the Zak transform the equality in (ii) reads as

$$\sum_{j,r\in\mathbb{Z}} \Big(w(j+r+x)\,w(r+x) + w(j+r-x)\,w(r-x)\Big)e^{2\pi i j\xi} = 1.$$

Now, by comparison of the Fourier coefficients we obtain the equivalence of (ii) and (iii).

To show that (ii) and (iv) are equivalent we use that $Zw(-x,\xi) = \overline{Zw(x+\frac{1}{2},\xi)}$ (cf. (9.62)), i.e., (ii) can be written as

$$1 = |Zw(x,\xi)|^2 + |Zw(x+\tfrac{1}{2},\xi)|^2.$$

This equality is in turn equivalent to

$$1 = |Z\hat{w}(\xi,x)|^2 + |Z\hat{w}(\xi,x+\tfrac{1}{2})|^2, \quad (x,\xi)\in Q^+, \tag{9.68}$$

as one can see from the assertion on the Zak transform of $\hat{w}$ in (9.46). Again, a comparison of coefficients proves the equivalence of (9.68) and (iv). ∎

In particular, condition (ii) leads to a possible construction of orthonormal Wilson bases (cf. [26]).

THEOREM 9.14
Let v be a measurable, real-valued function such that $v(x) = v(\frac{1}{2}-x)$ and $|v(x)| \le c_1(1+|x|)^{-1-\varepsilon}$ for some positive constant c_1 and some $\varepsilon > 0$. Furthermore, let $|Zv(x,\xi)|^2 + |Zv(-x,\xi)|^2 \ge c_2 > 0$. Then,

$$w(x) := \int_{-\frac{1}{2}}^{\frac{1}{2}} \frac{Zv(x,\xi)}{\sqrt{|Zv(x,\xi)|^2 + |Zv(-x,\xi)|^2}}\, d\xi \tag{9.69}$$

is a real-valued function such that $w(x) = w(\frac{1}{2}-x)$ and the functions $\psi_j^k := w(\cdot - j)\, d_j^k$, $j\in\mathbb{Z}$, $k\in\mathbb{N}_0$, constitute an orthonormal basis.

PROOF Since v is real-valued it holds that $Zv(x,-\xi) = \overline{Zv(x,\xi)}$. Hence,

$$\overline{w(x)} = \int_{-\frac{1}{2}}^{\frac{1}{2}} \frac{Zv(x,-\xi)}{\sqrt{|Zv(x,-\xi)|^2 + |Zv(-x,-\xi)|^2}}\, d\xi = w(x),$$

i.e., w is real-valued, too. Because v is also symmetric we know that $Zv(\frac{1}{2}-x,\xi)=\overline{Zv(x,\xi)}$ (cf. (9.62)). Then, we see from

$$w(\tfrac{1}{2}-x)=\int_{-\frac{1}{2}}^{\frac{1}{2}}\frac{\overline{Zv(x,\xi)}}{\sqrt{|Zv(x,\xi)|^2+|Zv(-x,\xi)|^2}}\,d\xi=\overline{w(x)}=w(x)$$

that w is a symmetric window function, too. Since v and w are real-valued we have

$$\begin{aligned} w(x) &= \operatorname{Re} w(x)=\int_{-\frac{1}{2}}^{\frac{1}{2}}\frac{\operatorname{Re}\, Zv(x,\xi)}{\sqrt{|Zv(x,-\xi)|^2+|Zv(-x,-\xi)|^2}}\,d\xi \\ &\le \frac{1}{c_2}\operatorname{Re}\int_{-\frac{1}{2}}^{\frac{1}{2}} Zv(x,\xi)\,d\xi = \frac{v(x)}{c_2} \le \frac{c_1}{c_2}(1+|x|)^{-1-\varepsilon}. \end{aligned}$$

The inequality $w(x)\ge -\frac{c_1}{c_2}(1+|x|)^{-1-\varepsilon}$ follows analogously. Hence, it is shown that $|w(x)|\le\frac{c_1}{c_2}(1+|x|)^{-1-\varepsilon}$. Summing up, the window function w satisfies the assumptions of Theorem 9.13.

From the definition of w it follows immediately that

$$|Zw(x,\xi)|^2+|Zw(-x,\xi)|^2=\frac{|Zv(x,\xi)|^2+|Zv(-x,\xi)|^2}{|Zv(x,\xi)|^2+|Zv(-x,\xi)|^2}=1,$$

i.e., condition (ii) in Theorem 9.13 is fulfilled. Thus, $\{\psi_j^k\}$ is an orthonormal basis. ∎

In particular, if one chooses v as the Gaussian $e^{-\zeta(x-\frac{1}{4})^2}$, $\zeta>0$, the window w defined in (9.69) has exponential decay in time and frequency, as shown in [26].

Gaussian bases of Coifman and Meyer. The results in [26] motivated Coifman and Meyer [22] to investigate Wilson bases with the Gaussian

$$w_0(x)=w_{-1}(x-\tfrac{1}{2})=w(x):=e^{-\zeta(x-\frac{1}{4})^2},\quad \zeta>0,$$

as window function. For that purpose they took advantage of special properties of the Gaussian. In particular, they proved that the Gaussian bases are Riesz bases and they determined the corresponding dual bases. Using our approach we can derive the results of Coifman and Meyer in a simple way. As a new result we gain the Riesz bounds for the Gaussian bases.

Obviously, $w(\frac{1}{2}-x) = e^{-\zeta(\frac{1}{2}-x-\frac{1}{4})^2} = e^{-\zeta(x-\frac{1}{4})^2} = w(x)$, i.e., we have one real-valued, symmetric window function and we can apply the results from the example on p. 449. First, we determine the function $|Zw(x,\xi)|^2 + |Zw(-x,\xi)|^2$, which we need to determine the Riesz bounds and the dual window. Using the identity $(j+t)^2 + (\ell+t)^2 = \frac{1}{2}(2t+(j+\ell))^2 + \frac{1}{2}(j-\ell)^2$ we obtain

$$\begin{aligned} Zw(x,\xi)\overline{Zw(x,\xi)} &= \sum_{j\in\mathbb{Z}}\sum_{\ell\in\mathbb{Z}} e^{-\zeta\left((j+x-\frac{1}{4})^2+(\ell+x-\frac{1}{4}))^2\right)} e^{2\pi\imath(j-\ell)\xi} \\ &= \sum_{j\in\mathbb{Z}}\sum_{\ell\in\mathbb{Z}} e^{-\frac{\zeta}{2}(2x-\frac{1}{2}+j+\ell)^2} e^{-\frac{\zeta}{2}(j-\ell)^2} e^{2\pi i(j-\ell)\xi}. \end{aligned}$$

With the substitution $k = j+\ell$ this equality reads as

$$Zw(x,\xi)\overline{Zw(x,\xi)} = \sum_{k\in\mathbb{Z}} e^{-\frac{\zeta}{2}(2x-\frac{1}{2}+k)^2} \sum_{j\in\mathbb{Z}} e^{-\frac{\zeta}{2}(2j-k)^2} e^{2\pi i(2j-k)\xi}.$$

Analogously the substitution $k = 1-j-\ell$ leads to

$$\begin{aligned} Zw(-x,\xi)\overline{Zw(-x,\xi)} &= \sum_{j\in\mathbb{Z}}\sum_{l\in\mathbb{Z}} e^{-\frac{\zeta}{2}(-2x-\frac{1}{2}+j+\ell)^2} e^{-\frac{\zeta}{2}(j-\ell)^2} e^{2\pi i(j-\ell)\xi} \\ &= \sum_{k\in\mathbb{Z}} e^{-\frac{\zeta}{2}(2x-\frac{1}{2}+k)^2} \sum_{j\in\mathbb{Z}} e^{-\frac{\zeta}{2}(2j-1+k)^2} e^{2\pi\imath(2j-1+k)\xi}. \end{aligned}$$

Summing up, we obtain

$$|Zw(x,\xi)|^2 + |Zw(-x,\xi)|^2 = \Big(\sum_{k\in\mathbb{Z}} e^{-\frac{\zeta}{2}(2x-\frac{1}{2}+k)^2}\Big)\Big(\sum_{j\in\mathbb{Z}} e^{-\frac{\zeta}{2}j^2} e^{2\pi i j\xi}\Big),$$

i.e., a tensor product of two 1-periodic functions. Furthermore, by the Poisson summation formula we obtain

$$|Zw(x,\xi)|^2 + |Zw(-x,\xi)|^2 = \sqrt{\frac{\pi}{2\zeta}}\Big(\sum_{k\in\mathbb{Z}} e^{-\frac{\zeta}{2}(2x-\frac{1}{2}+k)^2}\Big)\Big(\sum_{j\in\mathbb{Z}} e^{-\frac{2\pi^2}{\zeta}(\xi+j)^2}\Big).$$

Applying Theorem 9.11 we can now determine the Riesz bounds of the Gaussian bases.

THEOREM 9.15

Let $\zeta > 0$ be given. The functions $\psi_j^k(x) := e^{-\zeta(x-\frac{1}{4}-j)^2}\, d_j^k(x)$, $j \in \mathbb{Z}$, $k \in \mathbb{N}_0$, constitute a Riesz basis of $L^2(\mathbb{R})$ with the best possible Riesz bounds

$$A_0 = \left(\sum_{k\in\mathbb{Z}} e^{-\frac{\zeta}{2}(k+\frac{1}{2})^2}\right)\left(\sum_{j\in\mathbb{Z}}(-1)^j\, e^{-\frac{\zeta}{2}j^2}\right) > 0, \tag{9.70}$$

$$B_0 = \left(\sum_{k\in\mathbb{Z}} e^{-\frac{\zeta}{2}k^2}\right)^2 < \infty. \tag{9.71}$$

PROOF We define $g_\zeta(x) := \sum_{k\in\mathbb{Z}} e^{-\zeta(x+k)^2}$ such that

$$|Zw(x,\xi)|^2 + |Zw(-x,\xi)|^2 = \sqrt{\tfrac{2\pi}{\zeta}}\; g_{\frac{\zeta}{2}}(2x - \tfrac{1}{2})\; g_{\frac{2\pi^2}{\zeta}}(\xi).$$

To prove the theorem we have to find lower and upper bounds for the 1-periodic function g_ζ.

From the Poisson summation formula it follows that

$$g_\zeta(x) = \sqrt{\tfrac{\pi}{\zeta}} \sum_{k\in\mathbb{Z}} e^{-\frac{\pi^2 k^2}{\zeta}}\, e^{2\pi i k x} = \sqrt{\tfrac{\pi}{\zeta}}\; \vartheta\left(2\pi x, e^{-\frac{\pi^2}{\zeta}}\right)$$

with the Theta function

$$\vartheta(x,q) := \sum_{k\in\mathbb{Z}} q^{k^2} e^{ikx}, \quad x \in \mathbb{R},\ q \in \mathbb{C},\ |q| < 1.$$

The Theta function enjoys the infinite product formula (see [46], Sect. 21.3)

$$\vartheta(x,q) = \prod_{\ell=0}^{\infty} \left(1 - q^{2\ell}\right)\left(1 + q^{2\ell+1}\,\cos(x) + q^{4\ell+2}\right).$$

Hence, the function $g_\zeta(x) = \sqrt{\pi/\zeta}\; \vartheta(2\pi x, e^{-\frac{\pi^2}{\zeta}})$ is decreasing for $x \in [0, \frac{1}{2}]$ (cf. [9], Proposition 2.3). Since g_ζ is even and 1-periodic, this monotonicity implies $\max_{x\in\mathbb{R}} g_\zeta(x) = g_\zeta(0)$ and $\min_{x\in\mathbb{R}} g_\zeta(x) = g_\zeta(\frac{1}{2})$. From (9.64), we obtain the best possible Riesz

$$A_0 = \sqrt{\tfrac{2\pi}{\zeta}}\, g_{\frac{\zeta}{2}}(\tfrac{1}{2})\, g_{\frac{2\pi^2}{\zeta}}(\tfrac{1}{2}) \qquad \text{and} \qquad B_0 = \sqrt{\tfrac{2\pi}{\zeta}}\, g_{\frac{\zeta}{2}}(0)\, g_{\frac{2\pi^2}{\zeta}}(0) = g_{\frac{\zeta}{2}}^2(0),$$

i.e., the identities (9.70) and (9.71) are proved. ∎

In Fig. 9.7 the Riesz bounds in dependence of ζ are shown. If ζ is close to 2π then the Gaussian basis is stable, while for very large or very small ζ the Riesz stability becomes worse. This can also be seen from the behavior of the dual window functions (cf. Figs. 9.8 and 9.9). The factorization of $|Zw(x,\xi)|^2 + |Zw(-x,\xi)|^2$ leads to a simple representation of the dual window $\tilde{w}$, which was also found by Coifman and Meyer [22]. Applying our results we can present a short proof of their statement.

THEOREM 9.16

Coifman and Meyer [22] *For the window function $w(x) := e^{-\zeta(x-\frac{1}{4})^2}$ the dual window is given by*

$$\tilde{w}(x) = \frac{\sum_{\ell\in\mathbb{Z}} \gamma_\ell \, e^{-\zeta(x-\frac{1}{4}+\ell)^2}}{\sum_{k\in\mathbb{Z}} e^{-\frac{\zeta}{2}(2x-\frac{1}{2}+k)^2}},$$

where

$$\gamma_\ell := \sqrt{\frac{\zeta}{2\pi}} \int_{-\frac{1}{2}}^{\frac{1}{2}} \frac{e^{2\pi i\ell\xi}}{\sum_{j\in\mathbb{Z}} e^{-\frac{2\pi^2}{\zeta}(\xi+j)^2}} \, d\xi.$$

PROOF By Fourier series expansion with respect to ξ one obtains

$$\frac{1}{|Zw(x,\xi)|^2 + |Zw(-x,\xi)|^2} = \frac{\sum_{\ell\in\mathbb{Z}} \gamma_\ell \, e^{-2\pi i\ell\xi}}{\sum_{k\in\mathbb{Z}} e^{-\frac{\zeta}{2}(2x-\frac{1}{2}+k)^2}}.$$

Now, the application of (9.65) leads to

$$\begin{aligned}
\tilde{w}(x) &= \int_{-\frac{1}{2}}^{\frac{1}{2}} \frac{Zw(x,\xi)}{|Zw(x,\xi)|^2 + |Zw(-x,\xi)|^2} \, d\xi \\
&= \frac{1}{\sum_{k\in\mathbb{Z}} e^{-\frac{\zeta}{2}(2x-\frac{1}{2}+k)^2}} \sum_{j,\ell\in\mathbb{Z}} \left(w(x+j) \, \gamma_\ell \int_{-\frac{1}{2}}^{\frac{1}{2}} e^{2\pi i(j-\ell)\xi} \, d\xi \right) \\
&= \frac{\sum_{\ell\in\mathbb{Z}} \gamma_\ell \, e^{-\zeta(x-\frac{1}{4}+\ell)^2}}{\sum_{k\in\mathbb{Z}} e^{-\frac{\zeta}{2}(2x-\frac{1}{2}+k)^2}}.
\end{aligned}$$

Note, that $1/\big(\sum_{j\in\mathbb{Z}} e^{-\frac{2\pi^2}{\zeta}(\xi+j)^2}\big)$ is analytic for each $\xi \in \mathbb{R}$ and therefore the Fourier coefficients γ_l have exponential decay. Hence, all sums in the proof converge absolutely. ∎

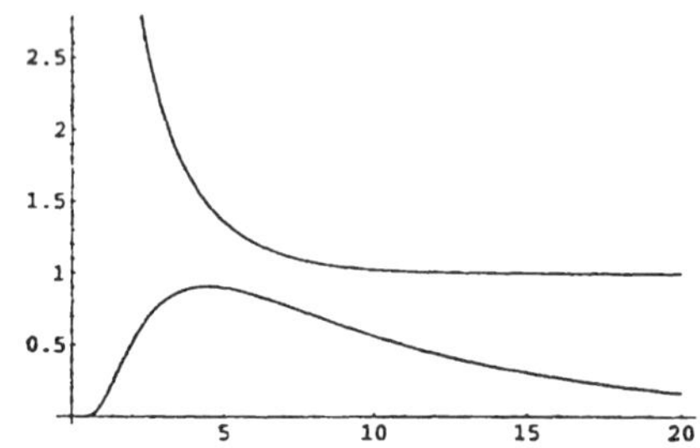

FIGURE 9.7
Upper and lower Riesz bound for Gaussian bases in dependence of ζ.

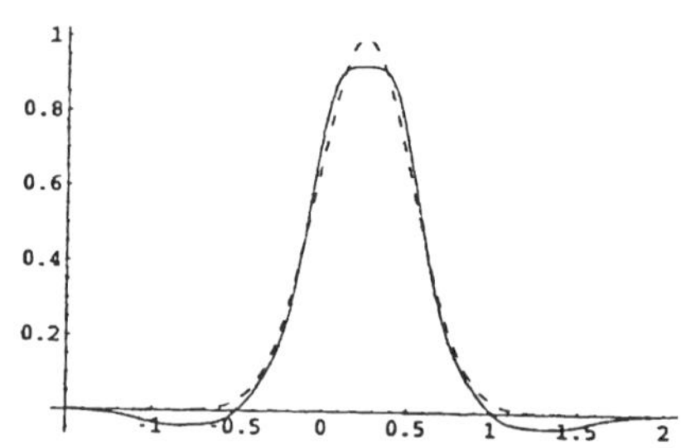

FIGURE 9.8
Dual window (solid line) for Gaussian bell (dashed) with $\zeta = 2\pi$.

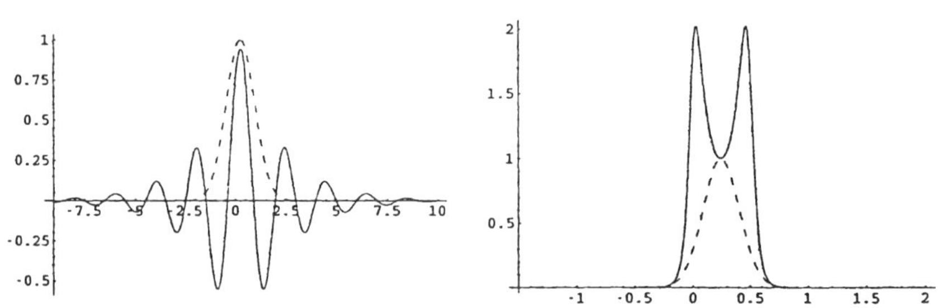

FIGURE 9.9
Dual window (solid line) for Gaussian bell (dashed) with $\zeta = 1$ (left) and $\zeta = 20$ (right).

Wilson bases with two-overlapping windows. For the implementation of effective algorithms it is often useful to consider basis functions with compact support. An example of such basis functions are the local trigonometric bases in the two-overlapping setting described in Sections 9.3–9.5. For the Wilson bases we obtain a two-overlapping setting if

$$\operatorname{supp} w_j \subset [j - \tfrac{1}{4}, j + \tfrac{3}{4}], \quad j \in \mathbb{Z}. \tag{9.72}$$

Note that the Wilson bases which are obtained with such window functions can be seen as a special case of the local trigonometric bases of Chui and Shi described in Section 9.5. Here, we will use our assertions on general Wilson bases to investigate these two-overlapping bases. All results for these examples can be obtained also immediately by applying the results of Chui and Shi.

For the investigation of Wilson bases the matrix $\boldsymbol{M}(x,\xi)$ plays an essential role. Therefore, we will investigate how the structure of $\boldsymbol{M}(x,\xi)$ is influenced by the two-overlapping setting. The entries of $\boldsymbol{M}(x,\xi)$ are function values of $Zw_0(x,\xi)$ and $Zw_{-1}(x,\xi)$. Since the support of w_j is contained in $[j - \frac{1}{4}, j + \frac{3}{4}]$ the Zak transform of the window functions is given by

$$Z_2 w_j(x,\xi) = w_j(x) e^{2\pi i (j-r)\xi}, \quad \text{if } x \in [\tfrac{r}{2} - \tfrac{1}{4}, \tfrac{r}{2} + \tfrac{3}{4}), \tag{9.73}$$

i.e., for a fixed x the entries of $\boldsymbol{M}(x,\xi)$ are trigonometric monomials in ξ. In particular, we obtain

$$\boldsymbol{M}(x,\xi) = \begin{cases} \begin{pmatrix} w_0(x) & -w_{-1}(x) \\ w_0(-x) & w_{-1}(-x) \end{pmatrix}, & \text{if } x \in (-\frac{1}{4}, \frac{1}{4}) \\ \begin{pmatrix} w_0(x) & -w_{-1}(x-1)e^{-2\pi i\xi} \\ w_0(1-x)e^{2\pi i\xi} & w_{-1}(-x) \end{pmatrix}, & \text{if } x \in (\frac{1}{4}, \frac{3}{4}). \end{cases}$$

For simplification we introduce the matrices

$$\boldsymbol{M}_0(x) \quad := \quad \begin{pmatrix} w_0(x) & -w_{-1}(x) \\ w_0(-x) & w_{-1}(-x) \end{pmatrix}, \tag{9.74}$$

$$\boldsymbol{M}_{-1}(x) \quad := \quad \begin{pmatrix} w_{-1}\left(-\frac{1}{2} + x\right) & w_0\left(\frac{1}{2} + x\right) \\ -w_{-1}\left(-\frac{1}{2} - x\right) & w_0\left(\frac{1}{2} - x\right) \end{pmatrix}. \tag{9.75}$$

These matrices do not depend on ξ but for $x \in (-\frac{1}{4}, \frac{1}{4})$ we have $\boldsymbol{M}_0(x) = \boldsymbol{M}(x,\xi)$ and just $\det \boldsymbol{M}_{-1}(x) = \det \boldsymbol{M}(\frac{1}{2} - x, \xi)$ as well as $\|\boldsymbol{M}_{-1}(x)\|_2 = \|\boldsymbol{M}(\frac{1}{2} - x, \xi)\|_2$. Applying Theorem 9.11 we obtain the following result on the Riesz stability of two-overlapping Wilson bases.

COROLLARY 9.4
Let the functions $\psi^k_{2j+s} := w_s(\cdot - 2j)\, d^k_{2j+s}$, $k \in \mathbb{N}_0$, $j \in \mathbb{Z}$, $s \in \{-1,0\}$, *be given, where* $\operatorname{supp} w_s \in [s - \frac{1}{4}, s + \frac{3}{4}]$ *and* $w_s \in L^\infty(\mathbb{R})$. *Furthermore, we set*

$$\Delta_s(x) := |w_0(-\tfrac{s}{2} + x)|^2 + |w_0(-\tfrac{s}{2} - x)|^2 + |w_{-1}(\tfrac{s}{2} + x)|^2 + |w_{-1}(\tfrac{s}{2} - x)|^2.$$

The Wilson system $\{\psi^k_j\}$ *is a Riesz basis with Riesz bounds* A *and* B *if and only if* $0 < A \le A_0 \le B_0 \le B < \infty$, *where*

$$A_0 := \min_{s\in\{-1,0\}} \left(\operatorname*{ess\,inf}_{x\in(0,1/4)} \frac{\Delta_s(x)}{2} - \sqrt{\frac{\Delta_s^2(x)}{4} - |\det \boldsymbol{M}_s(x)|^2} \right),$$

$$B_0 := \max_{s\in\{-1,0\}} \left(\operatorname*{ess\,sup}_{x\in(0,1/4)} \frac{\Delta_s(x)}{2} + \sqrt{\frac{\Delta_s^2(x)}{4} - |\det \boldsymbol{M}_s(x)|^2} \right).$$

The corollary follows immediately from Theorem 9.11 taking into account that for $x \in (0, \frac{1}{4})$ and $s \in \{0, 1\}$ it holds that $\det \boldsymbol{M}_{-s}(x) = \det \boldsymbol{M}(\frac{s}{2} - x, \xi)$ and $\Delta_{-s}(x) = \Delta(\frac{s}{2} - x, \xi)$. The corollary can also be derived immediately from Theorem 9.6 with $w^k_{2j+s} := w_s(\cdot - 2j)$, $j \in \mathbb{Z}$, $s \in \{-1, 0\}$.

Furthermore, the dual basis is a two-overlapping Wilson basis, too.

COROLLARY 9.5
Let window functions $w_0, w_{-1} \in L^\infty(\mathbb{R})$ *with* $\operatorname{supp} w_s \in [s - \frac{1}{4}, s + \frac{3}{4}]$ *be given such that* $\{\psi^k_j\} = \{w_s(\cdot - 2j)\, d^k_{2j+s}\}$ *is a Riesz basis. Then, the dual basis is constituted by the functions* $\tilde{\psi}^k_{2j+s} = \tilde{w}_s(\cdot - 2j)\, d^k_{2j+s}$ *with the dual window functions*

$$\tilde{w}_0(x) := \begin{cases} \dfrac{\overline{w_{-1}(-x)}}{\overline{\det \boldsymbol{M}_0(x)}}, & \text{if } x \in [-\frac{1}{4}, \frac{1}{4}), \\[2ex] \dfrac{\overline{w_{-1}(-x)}}{\overline{\det \boldsymbol{M}_{-1}\left(\frac{1}{2} - x\right)}}, & \text{if } x \in [\frac{1}{4}, \frac{3}{4}), \\[2ex] 0, & \text{otherwise}, \end{cases}$$

$$\tilde{w}_{-1}(x) := \begin{cases} \dfrac{\overline{w_0(-x)}}{\overline{\det \boldsymbol{M}_0(x)}}, & \text{if } x \in [-\frac{1}{4}, \frac{1}{4}), \\[2ex] \dfrac{\overline{w_0(-x)}}{\overline{\det \boldsymbol{M}_{-1}\left(\frac{1}{2} + x\right)}}, & \text{if } x \in [-\frac{3}{4}, -\frac{1}{4}), \\[2ex] 0, & \text{otherwise}. \end{cases}$$

PROOF Theorem 9.12 states that

$$\tilde{w}_0(x) = \int_{-\frac{1}{2}}^{\frac{1}{2}} \frac{\overline{Z_2 w_{-1}(-x,\xi)}}{\overline{\det \boldsymbol{M}(x,\xi)}}\, d\xi.$$

Using the quasi-periodicity (9.42) of the Zak transform we obtain for $x \in [-\frac{1}{4}, \frac{1}{4})$ and $j \in \mathbb{Z}$ that

$$\tilde{w}_0(x+j) = \int_{-\frac{1}{2}}^{\frac{1}{2}} \frac{\overline{Z_2 w_{-1}(-x,\xi)}\, e^{-2\pi i j\xi}}{\overline{\det \boldsymbol{M}(x,\xi)}}\, d\xi.$$

Applying (9.73) and $\det \boldsymbol{M}(x,\xi) = \det \boldsymbol{M}_0(x)$ we conclude

$$\tilde{w}_0(x+j) = \int_{-\frac{1}{2}}^{\frac{1}{2}} \frac{\overline{w_{-1}(-x)}}{\overline{\det \boldsymbol{M}_0(x)}}\, e^{-2\pi i j\xi}\, d\xi = \delta_{j,0}\, \frac{\overline{w_{-1}(-x)}}{\overline{\det \boldsymbol{M}_0(x)}}.$$

For $x \in [\frac{1}{4}, \frac{3}{4})$ we have $\det \boldsymbol{M}(x,\xi) = \det \boldsymbol{M}_{-1}(\frac{1}{2} - x)$. Now, as for $x \in [-\frac{1}{4}, \frac{1}{4})$ one shows

$$\tilde{w}_0(x+j) = \delta_{j,0}\, \frac{\overline{w_{-1}(-x)}}{\overline{\det \boldsymbol{M}_{-1}\left(\frac{1}{2} - x\right)}}, \qquad x \in [\tfrac{1}{4}, \tfrac{3}{4}),\ j \in \mathbb{Z}.$$

For $\tilde{w}_{-1}$ the proof is analogous. ∎

The corollary also follows from Theorem 9.7 of Chui and Shi.

References

[1] G. Aharony, A. Averbuch, R. R. Coifman, and M. Israeli. Local cosine transform – a method for the reduction of the blocking effect in JPEG. In A. F. Laine, editor, *Mathematical Imaging: Wavelet Applications in Signal and Image Processing*, SPIE Proceedings Series vol. 2034, pages 205–217, San Diego, CA, 1993.

[2] G. Aharony, A. Averbuch, R. R. Coifman, and M. Israeli. Local cosine transform – a method for the reduction of the blocking effect in JPEG. *J. Math. Imag. Vision*, 3:7–38, 1994.

[3] P. Auscher. Remarks on the local Fourier bases. In J. J. Benedetto and M. W. Frazier, editors, *Wavelets: Mathematics and Applications*, pages 203–218. CRC Press, Boca Raton, FL, 1994.

[4] P. Auscher, G. Weiss, and M. V. Wickerhauser. Local sine and cosine bases of Coifman and Meyer and the construction of smooth wavelets. In C. K. Chui, editor, *Wavelets – A Tutorial in Theory and Applications*, pages 237–256. Academic Press, Boston, 1992.

[5] R. Balian. Un principe d'incertitude fort en théorie du signal ou en mécanique quantique. *C. R. Acad. Sci. Paris*, 292:1357–1362, 1981.

[6] N. K. Barry. Sur les bases dans l'espace de Hilbert. *Dokl. Akad. Nauk. SSSR*, 54:379–382, 1946.

[7] N. K. Barry. Biorthogonal systems and bases in Hilbert spaces, *Uchen. Zap. Moskov. Gos. Univ.*, 148:69–107, 1951.

[8] G. Battle. Heisenberg proof of the Balian-Low theorem. *Lett. Math. Phys.*, 15:175–177, 1988.

[9] B. J. C. Baxter and N. Sivakumar. On shifted cardinal interpolation by Gaussians and multiquadrics. *J. Approx. Theory*, 87:36–59, 1996.

[10] J. J. Benedetto and D. F. Walnut. Gabor frames of L^2 and related spaces. In J. J. Benedetto and M. W. Frazier, editors, *Wavelets: Mathematics and Applications*, pages 97–162. CRC Press, Boca Raton, FL, 1994.

[11] K. Bittner. Bivariate local trigonometric bases on triangular partitions. In *Wavelets and Multiscale Methods, Proceedings of the International Wavelet Conference, Tanger*, pages 96–99. Institut National de Recherche en Informatique et en Automatique, 1998.

[12] K. Bittner. On the reproduction of linear functions by local trigonometric bases. In C. K. Chui and L. L. Schumaker, editors, *Approximation Theory IX*, volume II, *Computational Aspects of Innovations in Applied Mathematics*, pages 1–8. Vanderbilt University Press, Nashville, 1998.

[13] K. Bittner. Biorthogonal Wilson bases. In M. A. Unser, A. Aldroubi, and A. F. Laine, editors, *Wavelet Applications in Signal and Image Processing VII*. SPIE, 1999, submitted.

[14] K. Bittner. Error estimates and reproduction of polynomials for biorthogonal local trigonometric bases. *Appl. Comput. Harmon. Anal.*, 6:75–102, 1999.

[15] K. Bittner. *Verallgemeinerte Klappoperatoren und Biorthogonale Wilson-Basen.* Ph.D. thesis, Technische Universität München, 1999, submitted.

[16] K. Bittner, C. K. Chui, and J. Prestin. Multivariate cosine wavelets. In G. Nürnberger, J. W. Schmidt, and G. Walz, editors, *Multivariate Approximation and Splines*, ISNM 125, pages 29–44. Birkhäuser, Basel, 1998.

[17] C. K. Chui and X. Shi. Characterization of biorthogonal cosine wavelets. *J. Fourier Anal. Appl.*, 3:559–575, 1997.

[18] C. K. Chui and X. Shi. A study of biorthogonal sinusoidal wavelets. In C. Rabut, A. LeMehaute and L. L. Schumaker, editors, *Surface Fitting and Multiresolution Methods, Innovations in Applied Mathematics*, pages 51–66. Vanderbilt University Press, Nashville, 1997.

[19] C. K. Chui and X. Shi. Wavelets of Wilson type with arbitrary shapes. CAT Report 376, Texas A&M University, 1997.

[20] I. Cohen, S. Raz, and D. Malah. Orthonormal shift-invariant adaptive local trigonometric decomposition. *Signal Processing*, 57:43–64, 1997.

[21] R. R. Coifman and Y. Meyer. Remarques sur l'analyse de Fourier à fenêtre. *C. R. Acad. Sci. Paris*, 312:259–261, 1991.

[22] R. R. Coifman and Y. Meyer. Gaussian bases. *Appl. Comput. Harmon. Anal.*, 2:299–302, 1995.

[23] R. R. Coifman and M. V. Wickerhauser. Entropy-based algorithms for best basis selection. *IEEE Trans. Inform. Theory*, 38:713–718, 1992.

[24] I. Daubechies. The wavelet transform, time-frequency localization and signal analysis. *IEEE Trans. Inform. Theory*, 36:961–1005, 1990.

[25] I. Daubechies. *Ten Lectures on Wavelets.* CBMS-NSF Reg. Conf. Ser. Appl. Math. 61. SIAM, Philadelphia, PA, 1992.

[26] I. Daubechies, S. Jaffard, and J. L. Journé. A simple Wilson orthonormal basis with exponential decay. *SIAM J. Math. Anal.*, 22:554–573, 1991.

[27] I. Daubechies and A. J. E. M. Jensen. Two theorems on lattice expansions. *IEEE Trans. Inform. Theory*, 39:3–6, 1993.

[28] X. Fang and E. Séré. Adapted multiple folding local trigonometric transforms and wavelet packets. *Appl. Comput. Harmon. Anal.*,

1:169–179, 1994.

[29] H. G. Feichtinger, K. Gröchenig, and D. Walnut. Wilson bases and modulation spaces. *Math. Nachr.*, 155:7–17, 1992.

[30] H. G.Feichtinger and T. Strohmer, editors. *Gabor Analysis and Algorithms: Theory and Applications.* Birkhäuser, Boston, 1998.

[31] D. Gabor, Theory of communication. *J. IEE*, 93:429–457, 1946.

[32] I. M. Gelfand, Eigenfunction expansions for equations with periodic coefficients. *Dokl. Akad. Nauk SSSR*, 73:1117–1120, 1950.

[33] K. Gröchenig and S. Samarah. Non-linear approximation with local Fourier bases. Preprint, University of Connecticut, 1998.

[34] A. J. E. M. Jansen. The Zak transform: A signal transform for sampled time-continuous signals. *Philips J. Res.*, 43:23–69, 1988.

[35] B. Jawerth, Y. Liu, and W. Sweldens. Signal compression with smooth local trigonometric bases. *Opt. Eng.*, 33:2125–2135, 1994.

[36] B. Jawerth and W. Sweldens. Biorthogonal smooth local trigonometric bases. *J. Fourier Anal. Appl.*, 2:109–133, 1995.

[37] E. Laeng. Une base orthonormale de $L^2(\mathbb{R})$ dont les éléments sont bien localisés dans l'espace de phase et leurs supports adaptés à toute partition symétrique de l'espace des fréquences. *C. R. Acad. Sci. Paris*, 311:677–680, 1990.

[38] F. Low. Complete sets of wave packets. In C. DeTar, editor, *A Passion for Physics - Essays in Honor of Geoffrey Chew.* World Scientific, Singapore, 1985.

[39] H. Malvar, *Signal Processing with Lapped Transforms.* Artech House, Inc., Norwood, MA, 1992.

[40] H. S. Malvar. Lapped transforms for efficient transform/subband coding. *IEEE Trans. Acoustic, Speech, and Signal Processing*, 38:969–978, 1990.

[41] G. Matviyenko, Optimized local trigonometric bases. *Appl. Comput. Harmon. Anal.*, 3:301–323, 1996.

[42] Y. Meyer, *Wavelets - Algorithms & Applications.* SIAM, Philadelphia, PA, 1993.

[43] D. J. Sullivan, J. J. Rehr, J. W. Wilkins, and K. G. Wilson. Phase space Warnier functions in electronic structure calculations. Preprint, Cornell University, 1987.

[44] A. Weil. Sur certains groups d'opérateurs unitaires. *Acta Math.*,

111:143–211, 1964.

[45] E. Wesfreid and M. V. Wickerhauser. Adapted local trigonometric transforms and speech processing. *IEEE Trans. SP*, 41:3596–3600, 1993.

[46] E. T. Whittaker and G. N. Watson. *A Course of Modern Analysis.* Cambridge Univ. Press, Cambridge, 1902.

[47] M. V. Wickerhauser, Smooth localized orthonormal bases. *C. R. Acad. Sci. Paris*, 316:423–427, 1993.

[48] M. V. Wickerhauser, *Adapted Wavelet Analysis from Theory to Softwave.* A. K. Peters, Wellesley, MA, 1994.

[49] K. G. Wilson. Generalized Warnier functions. Preprint, Cornell University, 1987.

[50] X. G. Xia and B. W. Suter. A family of two-dimensional nonseparable Malvar wavelets. *Appl. Comput. Harmon. Anal.*, 2:243–256, 1995.

[51] X. G. Xia and B. W. Suter. Construction of Malvar wavelets on hexagons. *Appl. Comput. Harmon. Anal.*, 3:65–71, 1996.

[52] R. Young, *An Introduction to Nonharmonic Fourier Series.* Academic Press, New York, 1980.

[53] J. Zak. Finite translations in solid state physics. *Phys. Rev. Lett.*, 19:1385–1397, 1967.

Chapter 10

Pólya-Type Inequalities

Charles E. M. Pearce,* Josip Pečarić,* Sanja Varošanec**

**Applied Mathematics Department, The University of Adelaide, Adelaide, SA 5005, Australia, E-mail addresses: cpearce@maths.adelaide.edu.au, jpecaric@maths.adelaide.edu.au, pecaric@hazu.hr*

***Department of Mathematics, University of Zagreb, Bijenička 30, 10000 Zagreb, Croatia, E-mail address: varosans@math.hr*

Contact author: C. Pearce

Abstract In their volumes *Problems and Theorems in Analysis*, Pólya and Szegö [32] presented the following related results.

Theorem A Let $f : [0,1] \to \mathbf{R}$ be a nonnegative and *increasing* function. If a and b are nonnegative real numbers, then

$$\left(\int_0^1 x^{a+b} f(x)dx\right)^2 \geq \left[1 - \left(\frac{a-b}{a+b+1}\right)^2\right] \int_0^1 x^{2a} f(x)dx \int_0^1 x^{2b} f(x)dx.$$

Theorem B Let $f : [0,\infty) \to \mathbf{R}$ be a nonnegative and *decreasing* function. If a and b are nonnegative real numbers, then

$$\left[\int_0^\infty x^{a+b} f(x)dx\right]^2 \leq \left[1 - \left(\frac{a-b}{a+b+1}\right)^2\right] \int_0^\infty x^{2a} f(x)dx \int_0^\infty x^{2b} f(x)dx.$$

These inequalities have since been improved, generalized and applied in a number of articles. There are several distinct streams of generalization. In one the class of weighted function f is broadened, in another the interval of integration extended. There are results in which the power functions x^{a+b}, x^{2a} and x^{2b} are replaced by more general functions. One line of generalization leads to different kinds of means. Discrete Pólya-type inequalities have also been obtained. Beyond all of this, techniques for the real scalar case have been adapted to operator theory. As a consequence, a number of results for positive linear operators have been derived.

In this chapter we give an account of this fertile and rapidly growing field, including general ideas, characteristic proofs and applications to probability theory.

Contents

10.1 Pólya's Inequalities

Two similar but opposite inequalities have been presented by Pólya and Szegö in their treatise *Problems and Theorems in Analysis* I & II. One [32, Vol. I p. 83] involves integrals of power functions with a nondecreasing weight function on the interval $[0, 1]$, the other [32, Vol. II, p. 114] integrals of power functions with a nonincreasing weight function on $[0, \infty)$. We have the following.

THEOREM 10.1

Suppose $f : [0, \infty) \to \mathbf{R}$ *is nonnegative and decreasing. If* a *and* b *are nonnegative real numbers, then*

$$\left[\int_0^\infty x^{a+b} f(x)dx\right]^2 \leq \left[1 - \left[\frac{a-b}{a+b+1}\right]^2\right] \int_0^\infty x^{2a} f(x)dx \int_0^\infty x^{2b} f(x)dx. \tag{10.1}$$

Equality occurs if and only if $f = C\chi_{(0,\lambda)}$ *for some positive number* λ, *where* χ *represents the characteristic function.*

THEOREM 10.2

Suppose $f : [0,1] \to \mathbf{R}$ *is nonnegative and increasing. If* a *and* b *are nonnegative real numbers, then*

$$\left[\int_0^1 x^{a+b} f(x)dx\right]^2 \geq \left[1 - \left[\frac{a-b}{a+b+1}\right]^2\right] \int_0^1 x^{2a} f(x)dx \int_0^1 x^{2b} f(x)dx. \tag{10.2}$$

Equality occurs if and only if f *is constant.*

PROOF The proofs in [32] utilise the quadratic polynomial

$$K(t) := (2a+1)\left(\int_0^B x^{2a} f(x)\, dx\right) t^2 + 2(a+b+1)\left(\int_0^B x^{a+b} f(x)\, dx\right) t$$
$$+(2b+1)\int_0^B x^{2b} f(x)\, dx,$$

with $B = \infty$ in the first theorem and $B = 1$ in the second.

Let $B = \infty$. If $x^r f(x) \in L_\infty(0,\infty)$, then

$$(r+1)\int_0^\infty x^r f(x)\, dx = -\int_0^\infty x^{r+1}\, df(x)$$

and the polynomial K takes the form

$$K(t) = -t^2 \int_0^\infty x^{2a+1}\, df(x) - 2t\int_0^\infty x^{a+b+1}\, df(x) - \int_0^\infty x^{2b+1}\, df(x)$$
$$= -\int_0^\infty x(x^a t + x^b)^2\, df(x),$$

so K is a nonnegative quadratic function. Hence the discriminant is a nonpositive real number; that is,

$$\left[2(a+b+1)\int_0^\infty x^{a+b}f(x)\,dx\right]^2$$

$$-4(2a+1)(2b+1)\int_0^\infty x^{2a}f(x)\,dx\int_0^\infty x^{2b}f(x)\,dx \leq 0,$$

which is easily transformed into (10.1).

If $B=1$, integration by parts of the coefficients of K yields

$$K(t) = f(1)(t+1)^2 - \int_0^1 x(x^a t + x^b)^2\,df(x).$$

For $a, b \leq 0$ and $x \in [0,1]$ we have $x(x^a + x^b)^2 \leq 2^2$ and

$$\int_0^1 x(x^a + x^b)^2\,df(x) \leq \int_0^1 4\,df(x) \leq 4f(1),$$

so $K(1) \geq 0$.

If $t = -1$, then $K(-1) = -\int_0^1 x(x^b - x^a)^2\,df(x) \leq 0$, because f is nondecreasing. Thus K has at least one real root and so its discriminant is nonnegative and (10.2) is established. ∎

Over the years these two results have been treated separately. In the next two sections we give some results relating to Theorems A and B in turn.

10.2 Nondecreasing Weight Functions

For the case of a nondecreasing weight function f, Fink and Jodeit [8] (or see [15, p. 260]) generalized (10.2), extending the domain of the exponents a and b. They showed that (10.2) is valid not only for nonnegative a and b, but for a, b greater than $-1/2$.

They observed that (10.2) can be written as

$$(a+b+1)^2 \int_0^1 \int_0^1 f(x)f(y)x^{a+b}y^{a+b}\,dx\,dy$$

$$\geq \frac{(2a+1)(2b+1)}{2} \int_0^1 \int_0^1 f(x)f(y)\left(x^{2a}y^{2b}+x^{2b}y^{2a}\right)\,dx\,dy$$

or

$$\int_0^1 \int_0^1 K(x,y)f(x)f(y)\,dx\,dy \geq 0, \tag{10.3}$$

where

$$K(x,y) = (a+b+1)^2 x^{a+b}y^{a+b} - \frac{(2a+1)(2b+1)}{2}(x^{2a}y^{2b}+x^{2b}y^{2a}).$$

If $G(s,t) = \displaystyle\int_s^1 \int_t^1 K(x,y)\,dx\,dy$, then $G(s,t) \geq 0$ on $[0,1] \times [0,1]$ and $G(s,t) = 0$ only if $(s,t) = (0,0)$ or $s = 1$ or $t = 1$. To see this, set $\alpha = a + 1/2$, $\beta = b + 1/2$, $t_1 = t^\alpha$, $t_2 = t^\beta$, $s_1 = s^\alpha$ and $s_2 = s^\beta$. We get

$$2G = (s_1 - s_2)^2 + (t_1 - t_2)^2 - (s_1t_2 - s_2t_1)^2$$

$$= (s_1 - s_2)^2\left(1 - \left(\frac{t_1+t_2}{2}\right)^2\right) + (t_1 - t_2)^2\left(1 - \left(\frac{s_1+s_2}{2}\right)^2\right)$$

$$+2\left(\frac{t_1+t_2}{2}\right)\left(\frac{s_1+s_2}{2}\right)(s_1 - s_2)(t_1 - t_2).$$

Since $s_1 - s_2$ and $t_1 - t_2$ have the same sign, each of the three terms is nonnegative and if $(s,t) \neq (0,0)$ or $s \neq 1$ or $t \neq 1$, then one of them is positive.

Inequality (10.3) may be written as

$$\int_0^1 \int_0^1 K(x,y) \int_0^x \int_0^y d\sigma(t)d\sigma(s)\,dx\,dy \geq 0,$$

because $f(x) = \int_0^x d\sigma(t)$ for σ a suitable nonnegative measure with mass $f(0)$ at zero. By Fubini's theorem we get

$$\int_0^1 \int_0^1 G(t,s)\,d\sigma(t)\,d\sigma(s) \geq 0.$$

This proves (10.2) and gives the case of equality. If equality holds in (10.2), $d\sigma(t)d\sigma(s)$ must be zero except at $(0,0)$; that is, $f(t) \equiv f(0)$.

Pečarić ([27], [15, p. 261]) has remarked that the product $f(x) \cdot f(y)$ can be replaced by a function $f(x,y)$ satisfying the conditions that $f(0,0)$, $f_1(x,0)$, $f_2(0,y)$ and $f_{12}(x,y)$ are nonnegative, where f_1, f_2 and f_{12} are, respectively, the partial derivatives $\partial f/\partial x$, $\partial f/\partial y$ and $\partial^2 f/\partial x \partial y$.

The special case $f(x,y) = h(x)g(y)$ with nonnegative increasing functions h and g yields the following.

THEOREM 10.3
If $a, b > -\frac{1}{2}$ and $h, g : [0,1] \to \mathbb{R}$ are nonnegative and nondecreasing, then

$$(2a+1)(2b+1)\left[\int_0^1 x^{2a}h(x)dx \int_0^1 x^{2b}g(x)dx + \int_0^1 x^{2b}h(x)dx \int_0^1 x^{2a}g(x)dx\right]$$

$$< 2(a+b+1)^2 \int_0^1 x^{a+b}h(x)\,dx \int_0^1 x^{a+b}g(x)\,dx$$

unless $h(x) \equiv h(0)$ and $g(x) \equiv g(0)$, when equality occurs.

The most general result for two variables is the following theorem from [36]. In the case when $n = 2$ and $f(x,y) = \varphi(x)\varphi(y)$, this result reduces to (10.2).

THEOREM 10.4
Suppose n is an even positive integer, a and b positive numbers and $f : [0,1] \times [0,1] \to \mathbb{R}$ a function with continuous partial derivatives f_1, f_2, f_{12}. If $f_1(x,0) \geq 0, f_2(0,y) \geq 0$ and $f_{12}(x,y) \geq 0$ for $x, y \in [0,1]$, then

$$\sum_{k=0}^{n}(-1)^k \binom{n}{k} \int_0^1 \int_0^1 f\left(x^{\frac{1}{(n-k)a+kb}}, y^{\frac{1}{(n-k)b+ka}}\right) dx\, dy \leq 0\ .$$

PROOF Using substitution and integration by parts,

$$\int_0^1 \int_0^1 f(x^{1/p}, y^{1/q})\, dx\, dy$$

$$= pq \int_0^1 \int_0^1 f(x,y)x^{p-1}y^{q-1}\, dx\, dy$$

$$= q\int_0^1 y^{q-1}\left(f(1,y) - \int_0^1 f_1(x,y)x^p dx\right) dy$$

$$= f(1,1) - \int_0^1 f_1(x,1)x^p\,dx - \int_0^1 f_2(1,y)y^q\,dy$$

$$+ \int_0^1\int_0^1 f_{12}(x,y)x^p y^q\,dx\,dy.$$

Hence

$$\sum_{k=0}^{n}(-1)^k\binom{n}{k}\int_0^1\int_0^1 f\left(x^{\frac{1}{(n-k)a+kb}}, y^{\frac{1}{(n-k)b+ka}}\right) dx\,dy$$

$$= \sum_{k=0}^{n}(-1)^k\binom{n}{k}\left\{f(1,1) - \int_0^1 f_1(x,1)x^{(n-k)a+kb}\,dx\right.$$

$$- \int_0^1 f_2(1,y)y^{(n-k)b+ka}\,dy$$

$$\left.+ \int_0^1\int_0^1 f_{12}(x,y)x^{(n-k)a+kb}y^{(n-k)b+ka}\,dx\,dy\right\}$$

$$= -\int_0^1 f_1(x,1)\left(\sum_{k=0}^{n}(-1)^k\binom{n}{k}x^{(n-k)a}x^{kb}\right) dx$$

$$- \int_0^1 f_2(1,y)\left(\sum_{k=0}^{n}(-1)^k\binom{n}{k}y^{(n-k)b}y^{ka}\right) dy$$

$$+ \int_0^1\int_0^1 f_{12}(x,y)\left(\sum_{k=0}^{n}(-1)^k\binom{n}{k}x^{(n-k)a}y^{(n-k)b}x^{kb}y^{ka}\right) dx\,dy$$

$$= -\int_0^1 f_1(x,1)(x^a - x^b)^n\,dx - \int_0^1 f_2(1,y)(y^a - y^b)^n\,dy$$

$$+ \int_0^1\int_0^1 f_{12}(x,y)(x^a y^b - x^b y^a)^n\,dx\,dy.$$

Substituting

$$f_1(x,1) = \int_0^1 f_{12}(x,y)\,dy + f_1(x,0)$$

and

$$f_2(1,y) = \int_0^1 f_{12}(x,y)\,dx + f_2(0,y),$$

we get

$$\sum_{k=0}^{n}(-1)^k\binom{n}{k}\int_0^1\int_0^1 f\left(x^{\frac{1}{(n-k)a+kb}}, y^{\frac{1}{(n-k)b+ka}}\right)dx\,dy$$

$$= -\int_0^1 f_1(x,0)(x^a - x^b)^n\,dx - \int_0^1 f_2(0,y)(y^a - y^b)^n\,dy$$

$$-\int_0^1\int_0^1 f_{12}(x,y)\Big[(x^a - x^b)^n + (y^a - y^b)^n - (x^a y^b - x^b y^a)^n\Big]\,dx\,dy.$$

The desired result follows from the inequality

$$(a-b)^n + (c-d)^n - (ad-bc)^n \geq 0$$

for $1 \geq a \geq b \geq 0$ and $1 \geq c \geq d \geq 0$ with n even. ∎

In [8] Fink and Jodeit also obtained an inequality similar to (10.2) for the class of functions

$$M_p := \left\{\phi : \phi(x) = \int_0^x (x-t)^p d\mu(t), \mu \text{ a nonnegative Borel measure}\right\}.$$

The class M_0 is that of nondecreasing nonnegative functions. Also if $\phi \in M_p$, then $\phi(x)x^{-a} \in M_{p-a}$ for $0 \leq \alpha \leq p$.

THEOREM 10.5
If $\phi \in M_k$ and $p, q > -k-1$, then

$$\left(\frac{p+q}{2} + k + 1\right)^2 \left(\int_0^1 x^{\frac{p+q}{2}}\phi(x)\,dx\right)^2$$

$$> (p+k+1)(q+k+1)\int_0^1 x^p\phi(x)\,dx \int_0^1 x^q\phi(x)\,dx \quad (10.4)$$

unless $\phi(x) = cx^k$, in which case equality holds.

The truth of (10.4) follows from (10.2) on substituting $2a = p + k$, $2b = q+k$ and $f(x) = x^{-k}\phi(x)$. Inequality (10.2) arises as the special case $k = 0$.

10.3 Nonincreasing Weight Functions

In this section we give some results connected with (10.1). In fact, inequality (10.1) was not the first result of its kind. In 1914 Bernstein and Kraft [6] derived a similar result using $[0, x]$ instead of $[0, \infty)$ as the domain of integration. Their result was as follows.

THEOREM 10.6
Suppose f is positive nonincreasing. Then

$$(2m+n+1)\int_0^x t^{2m+n} f(t)\,dt \cdot (n+1)\int_0^x t^n f(t)\,dt$$

$$\geq (m+n+1)^2 \left(\int_0^x t^{m+n} f(t)\,dt\right)^2 .$$

In [40] Volkov gave a generalization of (10.1). He considered nonnegative functions g_1 and g_2 defined on $(0, \infty)$ and a function f with a representation

$$f(x) = \int_0^\infty K(x,t)dh(t),$$

where $K(x,t) \geq 0$ for $x, t \in (0, \infty)$ and h is nondecreasing. Volkov showed that

$$\int_0^\infty g_1(x)g_2(x)f(x)\,dx \leq C\left[\int_0^\infty g_1^p(x)f(x)\,dx\right]^{1/p}\left[\int_0^\infty g_2^q(x)f(x)\,dx\right]^{1/q} \tag{10.5}$$

for $p, q > 0$ such that $1/p + 1/q = 1$ and

$$C = \sup_t \frac{\int_0^\infty K(x,t)g_1(x)g_2(x)\,dx}{\left(\int_0^\infty K(x,t)g_1^p(x)\,dx\right)^{1/p}\left(\int_0^\infty K(x,t)g_2^q(x)\,dx\right)^{1/q}}.$$

If we set $g_1(x) = x^a$, $g_2(x) = x^b$ and

$$K(x,t) = \begin{cases} 1, & 0 \le x < t, \\ 0, & x > t, \end{cases}$$

we obtain

$$\int_0^\infty x^{a+b} f(x)\,dx \le C\left(\int_0^\infty x^{ap} f(x)\,dx\right)^{1/p}\left(\int_0^\infty x^{bq} f(x)\,dx\right)^{1/q}$$

for $p, q > 0$ such that $p^{-1} + q^{-1} = 1$ and

$$C = \frac{(ap+1)^{1/p}(bq+1)^{1/q}}{1+a+b}.$$

If the kernel K assumes the form

$$K(x,t) = \begin{cases} t-x, & 0 \le x < t, \\ 0, & x > t, \end{cases}$$

so that f is a convex nonincreasing function, then (10.5) holds with

$$C = \frac{((ap+1)(ap+2))^{1/p}((bq+1)(bq+2))^{1/q}}{(1+a+b)(2+a+b)}.$$

Related results are given in [17] and [34]. In [34] a result is given for $2n$ functions $g_1, g_2, \ldots, g_{2n}$, which we now present.

THEOREM 10.7
Suppose $g_i : \mathbb{R}^+ \to \mathbb{R}$ $(i = 1, 2, \ldots, 2n)$ are nonnegative functions and $f : \mathbb{R}^+ \to \mathbb{R}$ a function with the representation

$$f(x) = \int_0^\infty K(x,t)dh(t),$$

where $K(x,t) \ge 0$ for $x, t \in \mathbb{R}^+$ and h is nondecreasing. If p_i $(i = 1, 2, \ldots, n)$ are positive numbers such that $\sum_{i=1}^n p_i^{-1} = 1$, then whenever the integrals exist

$$\int_0^\infty \prod_{j=1}^n g_j(x) f(x)\,dx \le C \prod_{j=1}^n \left(\int_0^\infty g_{n+j}(x) f(x)\,dx\right)^{1/p_j},$$

where

$$C = \sup_t \frac{\int_0^\infty K(x,t) g_1(x) \dots g_n(x)\, dx}{\prod_{j=1}^n (\int_0^\infty K(x,t) g_{n+j}(x)\, dx)^{1/p_j}}.$$

PROOF For fixed $\lambda > 0, \lambda \neq 1$, we define

$$c_i := \frac{1}{\ln \lambda}\left(\ln \frac{\prod_{j=1}^n (\int_0^\infty g_{n+j}(x) f(x)\, dx)^{1/p_j}}{\int_0^\infty g_{n+i}(x) f(x)\, dx}\right), \quad i = 1, \dots, n.$$

These numbers have the property that

$$\begin{aligned}\sum_{i=1}^n \frac{c_i}{p_i} &= \frac{1}{\ln \lambda}\left[\sum_{i=1}^n \frac{1}{p_i} \ln \prod_{j=1}^n \left(\int_0^\infty g_{n+j}(x) f(x)\, dx\right)^{1/p_j}\right. \\ &\qquad \left. - \sum_{i=1}^n \frac{\ln \int_0^\infty g_{n+i}(x) f(x)\, dx}{p_i}\right] \\ &= \frac{1}{\ln \lambda}\left[\ln \prod_{j=1}^n \left(\int_0^\infty g_{n+j}(x) f(x)\, dx\right)^{1/p_j}\right. \\ &\qquad \left. - \sum_{i=1}^n \ln \left(\int_0^\infty g_{n+i}(x) f(x)\, dx\right)^{1/p_i}\right] \\ &= 0,\end{aligned}$$

whence we obtain

$$\prod_{i=1}^n (\lambda^{c_i})^{1/p_i} = 1. \tag{10.6}$$

The representation of f yields

$$\begin{aligned}\int_0^\infty g_1(x) \dots g_n(x) f(x) dx &= \int_0^\infty \int_0^\infty g_1(x) \dots g_n(x) K(x,t) dx dh(t) \\ &= \int_0^\infty dh(t) \left[\int_0^\infty g_1(x) \dots g_n(x) K(x,t) dx\right]\end{aligned}$$

$$\le C\int_0^\infty dh(t)\prod_{j=1}^n\left[\int_0^\infty g_{n+j}(x)K(x,t)dx\right]^{1/p_j}$$

$$= C\int_0^\infty dh(t)\prod_{j=1}^n\left[\lambda^{c_j}\int_0^\infty g_{n+j}(x)K(x,t)dx\right]^{\frac{1}{p_j}}$$

by (10.6). The well-known inequality between geometric and arithmetic means gives that the last expression is less than or equal to

$$C\int_0^\infty dh(t)\sum_{j=1}^n\frac{\lambda^{c_j}\int_0^\infty g_{n+j}(x)K(x,t)\,dx}{p_j}.$$

From the definition of c_i we have

$$\int_0^\infty g_1(x)\dots g_n(x)f(x)\,dx$$

$$\le C\sum_{j=1}^n\frac{\lambda^{c_j}}{p_j}\int_0^\infty g_{n+j}(x)f(x)\,dx$$

$$= C\sum_{j=1}^n\frac{1}{p_j}\frac{\prod_{i=1}^n(\int_0^\infty g_{n+i}(x)f(x)\,dx)^{1/p_i}}{\int_0^\infty g_{n+j}(x)f(x)\,dx}\cdot\int_0^\infty g_{n+j}(x)f(x)\,dx$$

$$= C\cdot\prod_{i=1}^n\left(\int_0^\infty g_{n+i}(x)f(x)\,dx\right)^{1/p_i}.$$

The desired result follows. ∎

For $n = 2$ and $g_1 = g_3$, $g_2 = g_4$ we get Volkov's result. Furthermore, if $g_j(x) = g_{n+j}(x) = x^{a_j}$, $a_j > -1/p_j$ $(j = 1,\dots,n)$ and

$$K(x,t) = \begin{cases}1, & 0\le x<t\\ 0, & x>t,\end{cases}$$

a simple calculation gives

$$\int_0^\infty x^{a_1+\dots+a_n} f(x)\,dx \geq \frac{\prod_{i=1}^n (a_i p_i + 1)^{1/p_i}}{\sum_{i=1}^n a_i + 1} \prod_{i=1}^n \left[\int_0^\infty x^{a_i p_i} f(x)\,dx \right]^{1/p_i}, \tag{10.7}$$

where f is a nonincreasing function. This is a generalization of (10.1) with an increased number of integrals. For $n = 2$ and $p_1 = p_2 = 1/2$ we have inequality (10.1), but now this holds not only for nonnegative numbers a_1, a_2, but for $a_1, a_2 > -1/2$.

10.4 Unified Treatment

The techniques of the previous two sections are not appropriate for proving or generalizing both of Pólya's inequalities at the same time. In [38] Varošanec and Pečarić have given a method for a unified treatment of Pólya's inequalities. This is achieved by combining the general Hölder inequality with its reverse version, the Popoviciu inequality (see [14, p. 118]).

THEOREM 10.8

Suppose $g : [a,b] \to \mathbb{R}$ is a nonnegative increasing differentiable function and $f : [a,b] \to \mathbb{R}$ a nonnegative function such that f/g' is nondecreasing. Let p_i $(i = 1,\dots,n)$ be positive real numbers such that $\sum_{i=1}^n 1/p_i = 1$. If a_i $(i = 1,\dots,n)$ are real numbers with $a_i > -1/p_i$, then

$$\int_a^b g(x)^{a_1+\dots+a_n} f(x)\,dx$$

$$\geq \frac{\prod_{i=1}^n (a_i p_i + 1)^{1/p_i}}{1 + \sum_{i=1}^n a_i} \prod_{i=1}^n \left(\int_a^b g(x)^{a_i p_i} f(x)\,dx \right)^{1/p_i}. \tag{10.8}$$

If $g(a) = 0$ and f/g' is nonincreasing, then the inequality is reversed.

PROOF Inequality (10.8) reduces to an equality for $f \equiv 0$ and thus we may, without loss of generality, assume that $f(b) > 0$. Denote f/g' by F.

First suppose that F is nondecreasing. Applying integration by parts, we conclude that

$$\left(1+\sum_{i=1}^{n} a_i\right)\int_a^b g(x)^{a_1+\ldots+a_n} f(x)\,dx$$

$$= F(b)g(b)^{a_1+\ldots+a_n+1} - F(a)g(a)^{a_1+\ldots+a_n+1}$$

$$-\int_a^b g(x)^{a_1+\ldots+a_n+1}\,dF(x)$$

$$= F(b)g(b)^{a_1+\ldots+a_n+1} - F(a)(g(a))^{a_1+\ldots+a_n+1}$$

$$-\int_a^b \prod_{i=1}^{n}(g(x)^{a_ip_i+1})^{1/p_i}\,dF(x)$$

$$\geq F(b)g(b)^{a_1+\ldots+a_n+1} - F(a)g(a)^{a_1+\ldots+a_n+1}$$

$$-\prod_{i=1}^{n}\left(\int_a^b g(x)^{a_ip_i+1}\,dF(x)\right)^{1/p_i}, \tag{10.9}$$

where for the last step we use Hölder's inequality.

We now invoke Popoviciu's inequality, which is a reversed version of Hölder's inequality that states the following.

Suppose $u_1 > 0,\ u_2,\ldots,u_r \leq 0$ *with* $b_{ij} \geq 0$ *for* $i = 1,\ldots,r$ *and* $j = 1,\ldots,s$. *Assume* $q_i > 0$ *are such that* $\sum_{i=1}^{s} 1/q_i = 1$ *and* $\sum_{i=1}^{r} u_i b_{ij}^{q_j} \geq 0$ *for* $j = 1,\ldots,s$. *Then*

$$\sum_{i=1}^{r} u_i b_{i1}\ldots b_{is} \geq \prod_{j=1}^{s}\left(\sum_{i=1}^{r} u_i b_{ij}^{q_j}\right)^{1/q_j}. \tag{10.10}$$

In this result we set $r = 3$, $s = n$, $q_i = p_i$ and

$$u_1 = F(b) > 0,\ \ u_2 = -F(a),\ \ u_3 = -1,\ \ b_{1i} = (g(b)^{a_ip_i+1})^{1/p_i},$$

$$b_{2i} = (g(a)^{a_ip_i+1})^{1/p_i},\ b_{3i} = \left(\int_a^b g(x)^{a_ip_i+1}\,dF(x)\right)^{1/p_i} \quad \text{for } i = 1,\ldots,n.$$

This yields that the first expression in (10.9) is greater than or equal to

$$\left\{ \prod_{i=1}^{n} \left(F(b)g(b)^{a_i p_i + 1} - F(a)g(a)^{a_i p_i + 1} - \int_a^b g(x)^{a_i p_i + 1}\, dF(x) \right)^{1/p_i} \right\}$$

$$\times \prod_{i=1}^{n} \left((a_i p_i + 1) \int_a^b g(x)^{a_i p_i} f(x)\, dx \right)^{1/p_i},$$

giving the first part of the theorem.

If $g(a) = 0$ and f/g' is nonincreasing, we can use the discrete version of Hölder's inequality instead of Popoviciu's to provide a similar proof for the second part. ∎

If $g(x) = x$, we have the following theorem.

THEOREM 10.9

Suppose $f : [a, b] \to \mathbb{R}$ is nonnegative and nondecreasing. Let p_i ($i = 1, \dots n$) be positive real numbers satisfying $\sum_{i=1}^{n} 1/p_i = 1$. If a_i ($i = 1, \dots, n$) are real numbers such that $a_i > -1/p_i$, then

$$\int_a^b x^{a_1 + \dots + a_n} f(x)\, dx \geq \frac{\prod_{i=1}^{n} (a_i p_i + 1)^{1/p_i}}{\sum_{i=1}^{n} a_i + 1} \prod_{i=1}^{n} \left(\int_a^b x^{a_i p_i} f(x)\, dx \right)^{1/p_i}.$$

If $a = 0$ and f is nonincreasing, the inequality is reversed.

Inequality (10.2) derives as the special case $n = 2$, $p_1 = p_2 = 2$, $a = 0$, $b = 1$, $a_1, a_2 > -1/2$. If $a = 0$ and $b \to \infty$ we obtain (10.7) and for $n = 2$, $p_1 = p_2 = 2$, $a = 0$, $b \to \infty$, $a_1, a_2 > -1/2$ we obtain (10.1).

10.5 Inequalities Involving Derivatives

Another direction for generalizing Pólya's inequality (10.2) was opened up by Alzer [3] starting from the observation that $(2a+1)x^{2a}$ is the derivative of x^{2a+1}.

THEOREM 10.10

Suppose $f : [a,b] \to \mathbb{R}$ is nonnegative and nondecreasing and let $g, h : [a,b] \to \mathbb{R}$ be nonnegative and nondecreasing functions with a continuous first derivative. If $g(a) = h(a)$ and $g(b) = h(b)$, then

$$\left(\int_a^b \left(\sqrt{g(x)h(x)}\right)' f(x)\,dx\right)^2 \geq \int_a^b g'(x)f(x)\,dx \int_a^b h'(x)f(x)\,dx. \tag{10.11}$$

PROOF Integration by parts and the inequality between arithmetic and geometric means give us that

$$\sqrt{\int_a^b g'(x)f(x)\,dx \int_a^b h'(x)f(x)\,dx}$$

$$\leq \frac{1}{2}\int_a^b (g'(x) + h'(x))\, f(x)\,dx$$

$$= \frac{1}{2}\left[(g(b) + h(b))\, f(b) - (g(a) + h(a))\, f(a) - \int_a^b (g(x) + h(x))\, df(x)\right]$$

$$\leq \frac{1}{2}\left[(g(b) + h(b))\, f(b) - (g(a) + h(a))\, f(a) - \int_a^b 2\sqrt{g(x)h(x)}\, df(x)\right]$$

$$= \frac{1}{2}\left[\left(g(b) + h(b) - 2\sqrt{g(b)h(b)}\right) f(b)\right.$$

$$\left. - \left(g(a) + h(a) - 2\sqrt{g(a)h(a)}\right) f(a) - \int_a^b 2\left(\sqrt{g(x)h(x)}\right)' df(x)\right]$$

$$= \int_a^b \left(\sqrt{g(x)h(x)}\right)' df(x),$$

giving the desired result. ∎

The basic inequality (10.2) arises as the special case $g(x) = x^{2a+1}$ and $h(x) = x^{2b+1}$, $a = 0$, $b = 1$.

Using the A–G inequality in the previous proof is the main obstruction to obtaining a reversed version of the inequality. Pečarić and Varošanec [30] generalized Alzer's result, expanding it to an inequality of Hölder type. This enabled them to establish a reversed version of the result. Also boundary conditions for the functions x_i at the edges of the interval of integration disappear when the weight function f is nondecreasing, while when f is nonincreasing we have assumptions only on $x_i(a)$.

THEOREM 10.11

Suppose $f : [a,b] \to \mathbb{R}$ *is nonnegative and let* $x_i : [a,b] \to \mathbb{R}$ $(i = 1, ..., n)$ *be nonnegative increasing functions with a continuous first derivative. Let* p_i $(i = 1, ..., n)$ *be positive real numbers such that* $\sum_{i=1}^{n} 1/p_i = 1$.

If f *is nondecreasing, then*

$$\int_a^b \left(\prod_{i=1}^{n} x_i(t)^{1/p_i} \right)' f(t)\,dt \geq \prod_{i=1}^{n} \left(\int_a^b x_i'(t) f(t)\,dt \right)^{1/p_i}. \tag{10.12}$$

If f *is nonincreasing and* $x_i(a) = 0$ *for all* $i = 1, \ldots, n$, *then the inequality is reversed.*

PROOF The inequality (10.12) reduces to an equality for $f \equiv 0$ and thus we may, without loss of generality, assume that $f(b) > 0$.

Let f be a nondecreasing function. Integration by parts yields

$$\int_a^b \left(\prod_{i=1}^{n} x_i^{1/p_i}(t) \right)' f(t)\,dt \tag{10.13}$$

$$= f(b) \prod_{i=1}^{n} x_i^{1/p_i}(b) - f(a) \prod_{i=1}^{n} x_i^{1/p_i}(a) - \int_a^b \prod_{i=1}^{n} x_i^{1/p_i}(t) df(t)$$

$$\geq f(b) \prod_{i=1}^{n} x_i^{\frac{1}{p_i}}(b) - f(a) \prod_{i=1}^{n} x_i^{\frac{1}{p_i}}(a) - \prod_{i=1}^{n} \left(\int_a^b x_i(t) df(t) \right)^{\frac{1}{p_i}}.$$

Hölder's inequality provides the last step.

Set $r = 3$, $s = n$, $u_1 = f(b) > 0$, $u_2 = -f(a)$, $u_3 = -1$, $q_i = p_i$, $b_{1i} = x_i^{1/p_i}(b)$, $b_{2i} = x_i^{1/p_i}(a)$ and $b_{3i} = \left(\int_a^b x_i(t) df(t) \right)^{1/p_i}$ $(i = 1, 2, \ldots, n)$

in Popoviciu's inequality. Then the last expression in (10.14) is seen to be greater than or equal to

$$\prod_{i=1}^{n}\left(f(b)x_i(b)-f(a)x_i(a)-\int_a^b x_i(t)df(t)\right)^{\frac{1}{p_i}}=\prod_{i=1}^{n}\left(\int_a^b x_i'(t)f(t)dt\right)^{\frac{1}{p_i}}$$

and (10.12) is established. The proof of the second part of the theorem is similar. ∎

In [37] one can find several integral results involving higher-order derivatives of the functions x_i. We mention some without proof.

THEOREM 10.12

Let $f, x_i : [a,b] \to \mathbb{R}$ $(i = 1,\dots,m)$ be nonnegative functions with a continuous n-th order derivative $(n \geq 2)$ and such that

1° $(-1)^n f^{(n)}(t) \geq 0$ *and* $x_i^{(n)}(t) \geq 0$ *for all* $t \in [a,b], i = 1,\dots,m;$

2° $(-1)^k f^{(k)}(b) \geq 0$ *for* $k = 0,1,\dots,n-1;$

3° $x_i^{(k)}(a) = 0$ *and* $x_i^{(k)}(b) \geq 0$ *for* $k = 0,1,\dots,n-1$ *and* $i = 1,\dots,m.$

If p_i $(i = 1,\dots,m)$ are positive numbers such that $\sum_{i=1}^{m} 1/p_i = 1$, then

$$\int_a^b \left(\prod_{i=1}^{m} x_i^{1/p_i}(t)\right)^{(n)} f(t)\,dt \leq \prod_{i=1}^{m}\left(\int_a^b x_i^{(n)}(t)f(t)\,dt\right)^{1/p_i} + \Delta, \quad (10.14)$$

where

$$\Delta = \sum_{k=0}^{n-2}(-1)^k f^{(k)}(t)\left[\left[\prod_{i=1}^{m} x_i^{1/p_i}(t)\right]^{(n-k-1)} - \prod_{i=1}^{m}\left[x_i^{(n-k-1)}(t)\right]^{1/p_i}\right]\Bigg|_{t=b}.$$

With the assumptions of the previous theorem and the additional requirements that

$$x_i^{(k)}(b) = x_j^{(k)}(b) \text{ for all } i,j \in \{1,\dots,m\} \text{ and } k \in \{0,1,\dots,n-2\},$$

we have

$$\int_a^b \left(\prod_{i=1}^m x_i^{1/p_i}(t)\right)^{(n)} f(t)\,dt \le \prod_{i=1}^m \left(\int_a^b x_i^{(n)}(t) f(t)\,dt\right)^{1/p_i}.$$

THEOREM 10.13
Suppose $f, x_i : [a,b] \to \mathbb{R}$ $(i = 1, \ldots, m)$ are nonnegative with continuous n-th order derivatives $(n \ge 2)$ and such that

1° $(-1)^n f^{(n)}(t) \le 0, x_i^{(n)}(t) \ge 0, f(b) > 0$ *for all* $t \in [a,b], i = 1, \ldots, m$;

2° $(-1)^k f^{(k)}(b) \le 0$ *for every* $k = 1, \ldots, n-1$;

3° $x_i^{(k)}(b) \ge 0$ *and* $x_i^{(k)}(a) = 0$ *for* $i = 1, \ldots, m$ *and* $k = 0, 1, \ldots, n-1$.

Then inequality (10.14) *is reversed.*

With the assumptions of Theorem 10.13 and the condition

$$x_i^{(k)}(b) = x_j^{(k)}(b) \text{ for all } i, j \in \{1, \ldots, m\} \text{ and } k \in \{0, 1, \ldots, n-1\},$$

we have

$$\int_a^b \left(\prod_{i=1}^m x_i^{1/p_i}(t)\right)^{(n)} f(t)\,dt \ge \prod_{i=1}^m \left(\int_a^b x_i^{(n)}(t) f(t)\,dt\right)^{1/p_i}. \tag{10.15}$$

THEOREM 10.14
Suppose that f, $x_i : [a,b] \to \mathbb{R}$ $(i = 1, \ldots, m)$ are nonnegative functions with continuous n-th order derivatives and are such that the functions $(-1)^{n-1} f^{(n)}$, $(\prod_{i=1}^m x_i^{1/p_i})^{(n)}$ and $x_i^{(n)}$ $(i = 1, \ldots, m)$ are nonnegative and continuous. Then

$$\int_a^b \left(\prod_{i=1}^m (x_i^{1/p_i}(t)\right)^{(n)} f(t)\,dt \ge \prod_{i=1}^m \left(\int_a^b x_i^{(n)}(t) f(t)\,dt\right)^{1/p_i} + \Delta_1,$$

where

$$\Delta_1 = \sum_{k=0}^{n-1} \left[(-1)^{n-k-1} f^{(n-k-1)}(t) \left[\sum_{i=1}^m p_i x_i^{(k)}(t) - \left(\prod_{i=1}^m x_i^{1/p_i}(t)\right)^{(k)}\right]\right]\Bigg|_a^b.$$

If the assumptions of Theorem 10.14 hold and

$$x_i^{(k)}(a) = x_j^{(k)}(a) \quad \text{and} \quad x_i^{(k)}(b) = x_j^{(k)}(b)$$

for all $i,\ j \in \{1, 2, \ldots, m\}$ and $k \in \{0, 1, \ldots, n-1\}$, then (10.15) holds.

All of the foregoing results in this section have a form similar to Hölder's inequality. The inequalities in the following two theorems are of Minkowski type. The first involves the first derivative and the second higher-order derivatives. Detailed proofs can be found in [35].

THEOREM 10.15

Suppose $f : [a, b] \to \mathbb{R}$ is a nonnegative and nondecreasing function and $x_i : [a, b] \to \mathbb{R}$ $(i = 1, \ldots, n)$ are nonnegative and nondecreasing functions with continuous first derivatives. If $p > 1$, then

$$\left(\int_b^a \left(\left(\sum_{i=1}^n x_i(t) \right)^p \right)' f(t)\, dt \right)^{1/p} \geq \sum_{i=1}^n \left(\int_a^b (x_i^p(t))' f(t)\, dt \right)^{1/p}.$$

If f is nonincreasing and $x_i(a) = 0$ for all $i = 1, \ldots, n$, then the inequality is reversed.

THEOREM 10.16

Suppose $p > 1$ and $f,\ x_i : [a, b] \to \mathbb{R}$ $(i = 1, \ldots, m)$ are nonnegative functions, with continuous n-th order derivatives, which satisfy

1° $(-1)^k f^{(k)}(b) \geq 0$ *for* $k = 1, \ldots, n-1$;

2° $x_i^{(k)}(a) = 0$ *for* $k = 0, \ldots, n-1$ *and* $i = 1, \ldots, m$ *and* $(\sum_{i=1}^m x_i^p(b))^{(k)} \geq 0$ *for* $k = 1, \ldots, n-1$;

3° $(x_i^p(t))^{(n)} \geq 0$ *for* $t \in [a, b]$ *and* $i = 1, \ldots, m$;

4° $(-1)^n f^{(n)}(t) \geq 0$ *for* $t \in [a, b]$.

If

$$x_i^{(k)}(b) = x_j^{(k)}(b) \text{ for all } i, j \in \{1, 2, \ldots, m\} \text{ and } k = 0, 1, \ldots, n-1, \tag{10.16}$$

then

$$\int_a^b \left(\left(\sum_{i=1}^m x_i(t) \right)^p \right)^{(n)} f(t)\, dt \le \left(\sum_{i=1}^m \left(\int_a^b (x_i(t)^p)^{(n)} f(t)\, dt \right)^{1/p} \right)^p . \tag{10.17}$$

If (10.16) *is not satisfied, then*

$$\int_a^b \left(\left(\sum_{i=1}^m x_i(t) \right)^p \right)^{(n)} f(t)\, dt \le \Delta + \left(\sum_{i=1}^m \left(\int_a^b (x_i^p(t))^{(n)} f(t)\, dt \right)^{1/p} \right)^p , \tag{10.18}$$

where

$$\Delta = \sum_{k=0}^{n-1} (-1)^k f^{(k)}(t) \left(\left(\left(\sum_{i=1}^m x_i(t) \right)^p \right)^{(n-k-1)} \right.$$

$$\left. - \left(\sum_{i=1}^m \left((x_i^p(t))^{(n-k-1)} \right)^{1/p} \right)^p \right) \Bigg|_{t=b} .$$

If the inequalities in conditions 1° *and* 4° *are reversed, then* (10.17) *and* (10.18) *hold with the inequalities reversed.*

Putting $n = 2$, $p = 2$, $x_1 = \sqrt{f}$, $x_2 = \sqrt{g}$, $f = \varphi$ with φ a nondecreasing function in Theorem 10.15, we get

$$\int_a^b \left((\sqrt{f(x)} + \sqrt{g(x)})^2 \right)' \varphi(x) dx$$

$$\ge \left(\sqrt{\int_a^b f'(x)\varphi(x)dx} + \sqrt{\int_a^b g'(x)\varphi(x)dx} \right)^2 . \tag{10.19}$$

If the squares are expanded and common terms cancelled we derive Alzer's inequality (10.11). However (10.19) gives us a new direction of generalization. The following theorem holds (see [24]).

THEOREM 10.17

Suppose $f, g, \varphi : [a,b] \to \mathbb{R}$ *are nonnegative with* φ *nondecreasing and possessing a continuous first derivative. Further, let* p, q *be real numbers satisfying* $p + q = 1$.

(a) If φ is nondecreasing and $p, q > 0$, then

$$\int_a^b [(f^p(x) \pm g^p(x))] [(f^q(x) \pm g^q(x))]' \varphi(x)\, dx \\ \gtreqless \left[\left(\int_a^b f'(x)\varphi(x)\, dx \right)^p \pm \left(\int_a^b g'(x)\varphi(x)\, dx \right)^p \right] \\ \times \left[\left(\int_a^b f'(x)\varphi(x)\, dx \right)^q \pm \left(\int_a^b g'(x)\varphi(x)\, dx \right)^q \right].$$

Here the convention is that the greater than or equal possibility is associated with taking the plus throughout and the less than or equal with the minus.
(b) If φ is a nonincreasing, $p, q > 0$ and $f(a) = g(a) = 0$, the inequality is reversed.
(c) If φ is nondecreasing and $pq < 0$, the inequality is reversed.

Results with higher-order derivatives are given in [24].

In [1] the authors considered a new way of generalizing Pólya's inequalities. In the same paper they give a sequence of results with different conditions on three functions W_1, W_2, W_3. A characteristic result is the following.

THEOREM 10.18

Suppose w_i $(i = 1, 2, 3)$ are nonnegative and integrable functions on $[a, b]$ and let W_i be defined by

$$W_i(x) = \frac{\int_a^x w_i(t)dt}{\int_a^b w_i(t)dt} \quad (i = 1, 2, 3).$$

Let p_1, p_2 be positive real numbers such that $1/p_1 + 1/p_2 = 1$.

a) If g is a nonnegative nonincreasing function on $[a, b]$ and

$$W_1^{1/p_1}(x) W_2^{1/p_2}(x) \geq W_3(x) \quad \text{for all} \quad x \in [a, b], \tag{10.20}$$

then

$$\frac{\int_a^b w_3(x)g(x)dx}{\int_a^b w_3(x)dx} \leq \prod_{i=1}^{2} \left(\frac{\int_a^b w_i(x)g(x)dx}{\int_a^b w_i(x)dx} \right)^{1/p_i}. \tag{10.21}$$

b) If g is a nonnegative nondecreasing function on $[a, b]$ and the inequality (10.20) is reversed, then the inequality (10.21) is reversed.

It is readily seen that an analogous result holds for $n+1$ functions W_i. Let x_i be as in Theorem 10.11. Setting $w_i = x_i'$ for $i = 1, 2, \ldots, n$ and $w_{n+1} = (\prod_{i=1}^n x_i^{1/p_i})'$, we get

$$W_i(x) = \frac{x_i(x)}{x_i(b)}, \quad i = 1, 2, \ldots, n, \quad W_{n+1}(x) = \frac{\prod_{i=1}^n x_i(x)^{1/p_i}}{\prod_{i=1}^n x_i(b)^{1/p_i}}$$

and

$$\prod_{i=1}^n W_i^{1/p_i} = W_{n+1}.$$

Thus Theorem 10.18 gives the result of Theorem 10.11 and so constitutes a new generalization of Pólya's inequality. The above-mentioned theorem is of Hölder type. In the same paper [1] there are similar results of Minkowski type which generalize Theorem 10.15.

10.6 Inequalities Involving Means

The Alzer inequality (10.11) can be written in the form

$$\int_a^b G(g(t), h(t))' f(t) dt \geq G\left(\int_a^b g'(t) f(t) dt, \int_a^b h'(t) f(t) dt\right),$$

where $G(x, y) := \sqrt{xy}$ is the geometric mean of positive numbers x and y. This suggests that it may be possible to obtain more general results with different weighted power means. We shall give a generalized result involving quasiarithmetic means (see [29]). A quasiarithmetic mean is set up as follows.

Suppose f is a monotone real function on $[a, b]$ with inverse f^{-1} and $\mathbf{p} = (p_1, \cdots, p_n) = (p_i)_i$, $\mathbf{a} = (a_1, \cdots, a_n) = (a_i)_i$ are real n-tuples. The quasiarithmetic mean of the n-tuple $\mathbf{a}$ with respect to f is defined by

$$M_f(\mathbf{a}; \mathbf{p}) := f^{-1}\left(\frac{1}{P_n} \sum_{i=1}^n p_i f(a_i)\right),$$

where $P_n = \sum_{i=1}^n p_i$.

If f is an elementary power function $f(x) = x^r$, the quasiarithmetic mean $M_f(\mathbf{a}; \mathbf{p})$ becomes the weighted mean of the n-tuple $\mathbf{a}$ of order r with weights $\mathbf{p}$. If $f(x) = \log x$, the quasiarithmetic mean becomes a weighted geometric mean.

A result which involves a quasiarithmetic mean is given in the following theorem. For $f(x) = g(x) = \log x$ and $n = 2$, $p_1 = p_2 = 1$, this reduces to Alzer's inequality (10.11).

THEOREM 10.19

Suppose $\mathbf{p}$ *is a positive* n*-tuple and* $x_i : [a,b] \to \mathbb{R}$ $(i = 1, \cdots, n)$ *nonnegative functions with continuous first derivatives such that* $x_i(a) = x_j(a)$, $x_i(b) = x_j(b)$ $(i,j = 1, \ldots, n)$.

a) If φ *is a nonnegative nondecreasing function on* $[a,b]$ *and* f *and* g *convex increasing or concave decreasing functions, then*

$$M_f\left(\left(\int_a^b x_i'(t)\varphi(t)dt\right)_i ; \mathbf{p}\right) \geq \int_a^b M_g'\left((x_i(t))_i; \mathbf{p}\right)\varphi(t)dt. \qquad (10.22)$$

If f *and* g *are concave increasing or convex decreasing, then the inequality* (10.22) *is reversed.*

b) If φ *is a nonnegative nonincreasing function on* $[a,b]$, f *convex increasing or concave decreasing and* g *concave increasing or convex decreasing, then* (10.22) *holds.*

If f *is concave increasing or convex decreasing and* g *convex increasing or concave decreasing, then the inequality* (10.22) *is reversed.*

PROOF Suppose that φ is nondecreasing and f and g are convex functions. Integration by parts and the Jensen inequality for convex functions yield

$$M_f\left(\left(\int_a^b x_i'(t)\varphi(t)dt\right)_i ; \mathbf{p}\right)$$

$$= f^{-1}\left(\frac{1}{P_n}\sum_{i=1}^n p_i f\left(\int_a^b x_i(t)\varphi(t)dt\right)\right)$$

$$\geq \frac{1}{P_n}\sum_{i=1}^n p_i \int_a^b x_i'(t)\varphi(t)dt$$

$$= \int_a^b \frac{1}{P_n}\left(\sum_{i=1}^n p_i x_i'(t)\right)\varphi(t)dt$$

$$= \frac{1}{P_n}\sum_{i=1}^{n} p_i x_i(t)\varphi(t)|_a^b - \int_a^b \frac{1}{P_n}\left(\sum_{i=1}^{n} p_i x_i(t)\right) d\varphi(t)$$

$$\geq \frac{1}{P_n}\sum_{i=1}^{n} p_i x_i(t)\varphi(t)|_a^b - \int_a^b g^{-1}\left(\frac{1}{P_n}\left(\sum_{i=1}^{n} p_i g(x_i(t))\right)\right) d\varphi(t)$$

$$= \frac{1}{P_n}\sum_{i=1}^{n} p_i x_i(t)\varphi(t)|_a^b - \int_a^b M_g\left(x_i(t))_i; \mathbf{p}\right) d\varphi(t)$$

$$= \frac{1}{P_n}\sum_{i=1}^{n} p_i x_i(t)\varphi(t)|_a^b - M_g\left((x_i(t))_i; \mathbf{p}\right)\varphi(t)|_a^b$$

$$+ \int_a^b M_g'\left((x_i(t))_i; \mathbf{p}\right)\varphi(t)dt$$

$$= \int_a^b M_g'\left((x_i(t))_i; \mathbf{p}\right)\varphi(t)dt,$$

giving the first case in (a). The proofs of the other parts are similar. ∎

These results can be extended using so-called generalized quasiarithmetic means.

DEFINITION 10.1 *Suppose $\phi = (\phi_1, \ldots, \phi_n) : (\mathbb{R}^+)^n \to \mathbb{R}^+$, and $M : \mathbb{R}^+ \to \mathbb{R}^+$ is strictly monotone. A generalized quasiarithmetic mean is given by*

$$M_n(\mathbf{a}, \phi) := M_n(a_1, \ldots, a_n; \phi) = M^{-1}\left(\frac{\sum_{=1}^{n} \phi_i(a_i)M(a_i)}{\sum_{=1}^{n} \phi_i(a_i)}\right).$$

We shall say that a generalized quasiarithmetic mean $M_n(\mathbf{a}; \chi)$ is subarithmetic if

$$M_n(\mathbf{a}, \chi) \leq A_n(\mathbf{a}, \mathbf{w}) := \frac{w_1 a_1 + \cdots + w_n a_n}{w_1 + \cdots + w_n}. \tag{10.23}$$

If the inequality (10.23) is reversed, $M_n(\mathbf{a}, \chi)$ is superarithmetic.

It can be shown that (10.23) is equivalent to the condition that for all $u, t \in I\!R_+$,

$$\frac{M(u) - M(t)}{M'(t)} \frac{\chi_1(u)}{\chi_n(t)} \leq \frac{w_i}{w_n}(u - t).$$

We have the following (see [25]).

THEOREM 10.20

Suppose $g_1, \ldots, g_n : [a, b] \to \mathbb{R}$ are nonnegative nondecreasing functions with continuous first derivatives and $g_1(a) = \cdots = g_n(a)$, $g_1(b) = \cdots = g_n(b)$.

a) Let f be nonnegative nondecreasing on $[a, b]$. If $M_n(\mathbf{a}, \chi)$ and $L_n(\mathbf{a}, \phi)$ are subarithmetic, then

$$M_n\left(\int_a^b g_1'(t)f(t)dt, \ldots \int_a^b g_n'(t)f(t)dt; \chi\right)$$

$$\leq \int_a^b (L_n(g_1(t), \ldots, g_n(t); \phi))' f(t)dt. \tag{10.24}$$

If $M_n(\mathbf{a}, \chi)$ and $L_n(\mathbf{a}, \phi)$ are superarithmetic, the inequality is reversed.

b) Let f be nonnegative and nonincreasing. If $M_n(\mathbf{a}, \chi)$ is subarithmetic and $L_n(\mathbf{a}, \phi)$ superarithmetic, then (10.24) holds. If $M_n(\mathbf{a}, \chi)$ is superarithmetic and $L_n(\mathbf{a}, \phi)$ subarithmetic, the inequality is reversed.

Similar results exist for a variety of other means such as the logarithmic, Gini, Stolarsky, symmetric and complete symmetric means.

10.7 Discrete Analogues

The following results are discrete analogues of the previous theorems. These and some related results can be found in [39], [23] and [25]. The next theorem is a discrete analogue of Theorem 10.11. We use the notation $\Delta a_i := a_{i+1} - a_i$ for finite differences throughout.

THEOREM 10.21

Let $\mathbf{w} = (w_1, \ldots, w_n)$, $\mathbf{a_1} = (a_{11}, \ldots, a_{1n}), \ldots, \mathbf{a_m} = (a_{m1}, \ldots, a_{mn})$ be nonnegative n-tuples such that sums $\sum_{i=1}^{n-1} w_i \Delta a_{ji}$ are nonnegative for all $j = 1, \ldots, m$.

a) If p_i $(i = 1, \ldots, m)$ are nonnegative real numbers such that $\sum_{i=1}^{m} p_i = 1$ and $\mathbf{w}$ is a nondecreasing n-tuple, then

$$\sum_{i=1}^{n-1} w_i \Delta(a_{1i}^{p_1} \ldots a_{mi}^{p_m}) \geq \prod_{j=1}^{m} \left(\sum_{i=1}^{n-1} w_i \Delta a_{ji} \right)^{p_i} . \tag{10.25}$$

If $\mathbf{w}$ is a nonincreasing n-tuple with $a_{j1} = 0$ for $j = 1, \ldots, m$, then the inequality is reversed.

b) If p_i $(i = 1, \ldots, m)$ are real numbers such that $p_1 > 0$, $p_2 < 0, \ldots, p_m < 0$, $\sum_{i=1}^{m} p_i = 1$ and $\mathbf{w}$ is a nondecreasing n-tuple, then the inequality (10.25) *is reversed. If $\mathbf{w}$ is a nonincreasing n-tuple with $a_{j1} = 0$ for $j = 1, \ldots, m$, then the inequality* (10.25) *holds.*

PROOF If $\mathbf{w}$ is nondecreasing, then $\Delta w_{i-1} \geq 0$ and

$$\sum_{i=1}^{n-1} w_i \Delta \left(\prod_{j=1}^{m} a_{ji}^{p_j} \right) = w_n \prod_{j=1}^{m} a_{jn}^{p_j} - w_1 \prod_{j=1}^{m} a_{j1}^{p_j} - \sum_{i=2}^{n} \Delta w_{i-1} \prod_{j=1}^{m} a_{ji}^{p_j} .$$

We now utilise Popoviciu's inequality with $r = n + 1$, $s = m$, $u_1 = w_n$, $u_2 = -w_1$, $u_i = -\Delta w_{i-2}$ for $2 < i \leq n + 1$, $b_{1j} = a_{jn}^{p_j}$, $b_{2j} = a_{j1}^{p_j}$, $b_{ij} = a_{ji}^{p_j}$ for $i > 2$ and $q_j = 1/p_j$. Popoviciu's condition that each $\sum_{i=1}^{s} u_i b_{ij}^{q_j} \geq 0$ becomes $\sum_{i=1}^{n-1} w_i \Delta a_{ji} \geq 0$, so that Popoviciu's inequality is applicable. We derive

$$\begin{aligned} \sum_{i=1}^{n-1} w_i \Delta \left(\prod_{j=1}^{m} a_{ji}^{p_j} \right) &\geq \prod_{j=1}^{m} \left(w_n a_{jn} - w_1 a_{j1} - \sum_{i=2}^{n} a_{ji} \Delta w_{i-1} \right)^{p_j} \\ &= \prod_{j=1}^{m} \left(\sum_{i=1}^{n-1} w_i \Delta a_{ji} \right)^{p_j} . \end{aligned}$$

If $\mathbf{w}$ is a nonincreasing n-tuple, then Hölder's inequality is used instead of Popoviciu's to give the second part of (a). The proof of (b) is similar. ∎

The following theorem is a discrete analogue of the integral Minkowski-type inequality given in Theorem 10.15 of [39]. The proof is similar to that of the integral version.

THEOREM 10.22

Let $\mathbf{w}, \mathbf{a_1}, \ldots, \mathbf{a_m}$ *be nonnegative* n*-tuples.*

a) If $\mathbf{w}$ *is a nondecreasing* n*-tuple and* $p > 1$ *or* $p < 0$ *such that the sums* $\sum_{i=1}^{n-1} w_i \Delta a_{ji}^p$ $(j = 1, \ldots, m)$ *are nonnegative, then*

$$\sum_{i=1}^{n-1} w_i \Delta (a_{1i} + \ldots + a_{mi})^p \geq \left(\sum_{j=1}^{m} \left(\sum_{i=1}^{n-1} w_i \Delta a_{ji}^p \right)^{1/p} \right)^p . \tag{10.26}$$

If $0 < p < 1$, *the inequality is reversed.*

b) If $\mathbf{w}$ *is a nonincreasing* n*–tuple,* $a_{j1} = 0$ *for all* $j = 1, \ldots, m$ *and* $p > 1$ *or* $p < 0$ *such that sums* $\sum_{i=1}^{n-1} w_i \Delta a_{ji}^p$ $(j = 1, \ldots, m)$ *are nonnegative, then the inequality* (10.26) *is reversed. If* $0 < p < 1$, *then* (10.26) *applies.*

There are also discrete versions of the results involving different means (see [39]). Here we mention a discrete version of Theorem 10.19 which deals with quasiarithmetic means.

THEOREM 10.23

Let $\mathbf{p}$ *be a positive* n*-tuple and* $\mathbf{x_i}$ $(i = 1, \ldots, n)$ *nonnegative* m*-tuples such that* $x_{i1} = x_{j1}$ *and* $x_{im} = x_{jm}$ *for all* $i, j = 1, \ldots, n$.

a) If $\mathbf{w}$ *is a nonnegative nondecreasing* m*-tuple and* f *and* g *convex increasing or concave decreasing functions, then*

$$M_f \left(\left(\sum_{k=1}^{m-1} w_k \Delta x_{ik} \right)_i ; p \right) \geq \sum_{k=1}^{m-1} w_k \Delta M_g \left((x_{ik})_i ; p \right) . \tag{10.27}$$

If f *and* g *are concave increasing or convex decreasing, then the reverse inequality holds.*

b) If $\mathbf{w}$ *is a nonnegative nonincreasing* m*-tuple,* f *convex increasing or concave decreasing and* g *concave increasing or convex decreasing, then* (10.27) *holds.*

If f *is concave increasing or convex decreasing and* g *convex increasing or concave decreasing, then the inequality is reversed.*

(We suppose the relevant quasiarithmetic means are all well-defined.)

10.8 Pólya-Type Inequalities in Operator Theory

The foregoing discrete results have analogues in operator theory. First, we give some definitions and known properties. For further details see [4], [9] and [11].

We consider bounded, linear and positive (that is, positive semi-definite) operators on infinite-dimensional Hilbert space. The identity operator is denoted by 1. The order relation $A \leq B$ means that operator $B - A$ is positive. That $A_1 \geq A_2 \geq \ldots$ and A_n converges strongly to A is denoted by $A_n \downarrow A$.

A binary operator σ on the class of positive operators is called a *connection* if the following requirements are fulfilled (see [11]):

1. $A \leq C$ and $B \leq D \Rightarrow A\sigma B \leq C\sigma D$;
2. $C(A\sigma B)C \leq (CAC)\sigma(CBC)$;
3. $A_n \downarrow A$ and $B_n \downarrow B$ imply $(A_n \sigma B_n) \downarrow A\sigma B$.

As noted in [11], an immediate consequence of the second requirement is that

$$C(A\sigma B)C = (CAC)\sigma(CBC)$$

for C invertible, and thus

$$\alpha(A\sigma B) = (\alpha A)\sigma(\alpha B) \tag{10.28}$$

for α a positive constant.

If the connection relation

4. $1\sigma 1 = 1$ holds, then the connection is called a mean.

If $A\sigma B = B\sigma A$, the connection σ is termed *symmetric*.

Every mean σ possesses the property $A\sigma A = A$ for every A. The simplest means are the arithmetic, geometric and harmonic means, defined, respectively, by

$$A\nabla_\lambda B := \lambda A + (1-\lambda)B,$$

$$A\sharp_\lambda B := A^{1/2}(A^{-1/2}BA^{-1/2})^{1-\lambda}A^{1/2},$$

$$A!_\lambda B := (\lambda A^{-1} + (1-\lambda)B^{-1})^{-1},$$

where A, B are invertible operators and $\lambda \in (0,1)$.

As in the scalar case we have

$$A!_\lambda B \leq A\sharp_\lambda B \leq A\nabla_\lambda B. \tag{10.29}$$

General inequalities such as Hölder's, Cauchy's, Minkowski's and their reverse versions the Popoviciu, Aczél and Bellman inequalities have analogues in operator theory (see [20]).

In particular, if A_i, B_i $(1 \leq i \leq n)$ are bounded, linear and positive operators and σ a connection, we have

$$\sum_{i=1}^{n} (A_i \sigma B_i) \leq \left(\sum_{i=1}^{n} A_i \right) \sigma \left(\sum_{i=1}^{n} B_i \right). \tag{10.30}$$

If further $A_1 - \sum_{i=2}^{n} A_i \geq 0$ and $B_1 - \sum_{i=2}^{n} B_i \geq 0$, then

$$A_1 \sigma B_1 - \sum_{i=2}^{n} A_i \sigma B_i \geq \left(A_1 - \sum_{i=2}^{n} A_i \right) \sigma \left(B_1 - \sum_{i=2}^{n} B_i \right). \tag{10.31}$$

Using these results, we give an operator version of the Pólya inequalities (see [20]).

THEOREM 10.24

Suppose $A_1 \leq \ldots \leq A_n$ (not all equal) and $B_1 \leq \ldots \leq B_n$ (not all equal) are bounded, linear, positive operators. Let $\mathbf{a} = (a_1, \ldots, a_n)$ be a nondecreasing positive n-tuple of real numbers and σ a connection. Then

$$\left(\sum_{i=1}^{n-1} a_i \Delta A_i \right) \sigma \left(\sum_{i=1}^{n-1} a_i \Delta B_i \right) \leq \sum_{i=1}^{n-1} a_i \Delta (A_i \sigma B_i).$$

If a is a nonincreasing positive n-tuple of real numbers and $A_1 = B_1 = 0$, then the inequality is reversed.

PROOF Let $\mathbf{a}$ be a nondecreasing positive n-tuple of real numbers. Then applying in turn (10.28) and (10.30) we derive

$$\sum_{i=1}^{n-1} a_i \Delta (A_i \sigma B_i)$$

$$= a_n (A_n \sigma B_n) - a_1 (A_1 \sigma B_1) - \sum_{i=2}^{n} (A_i \sigma B_i) \Delta a_{i-1}$$

$$= a_n(A_n\sigma B_n) - a_1(A_1\sigma B_1) - \sum_{i=2}^{n}\Big((A_i\Delta a_{i-1})\sigma(B_i\Delta a_{i-1}\Big)$$

$$\geq (a_nA_n)\sigma(a_nB_n) - (a_1A_1)\sigma(a_1B_1)$$

$$- \left(\sum_{i=2}^{n} A_i\Delta a_{i-1}\right)\sigma\left(\sum_{i=2}^{n} B_i\Delta a_{i-1}\right).$$

Now

$$a_nA_n - a_1A_1 - \sum_{i=2}^{n} A_i\Delta a_{i-1} = \sum_{i=1}^{n-1} a_i\Delta A_i \geq 0$$

and a similar result holds for the operators B_i. Hence by (10.31) (for three terms) we have

$$\sum_{i=1}^{n-1} a_i\Delta(A_i\sigma B_i) \geq \left(\sum_{i=1}^{n-1} a_i\Delta A_i\right)\sigma\left(\sum_{i=1}^{n-1} a_i\Delta B_i\right).$$

This establishes the first part.

If $\mathbf{a}$ is a nonincreasing positive n-tuple of real numbers, then $-\mathbf{a}$ is nondecreasing, so that an application of (10.31) provides

$$\sum_{i=1}^{n-1} a_i\Delta(A_i\sigma B_i) = a_n(A_n\sigma B_n) - a_1(A_1\sigma B_1) + \sum_{i=2}^{n}(A_i\sigma B_i)\Delta(-a_{i-1})$$

$$= (a_nA_n)\sigma(a_nB_n) - a_1(A_1\sigma B_1)$$

$$+ \sum_{i=2}^{n}\Big((A_i\Delta(-a_{i-1}))\sigma(B_i\Delta(-a_{i-1}))\Big)$$

$$\leq (a_nA_n)\sigma(a_nB_n) - (a_1A_1)\sigma(a_1B_1)$$

$$+ \left(\sum_{i=2}^{n} A_i\Delta(-a_{i-1})\right)\sigma\left(\sum_{i=2}^{n} B_i\Delta(-a_{i-1})\right)$$

$$\leq \left(\sum_{i=1}^{n-1} a_i\Delta A_i\right)\sigma\left(\sum_{i=1}^{n-1} a_i\Delta B_i\right),$$

completing the second part. ∎

In [20] a Pólya-type inequality is also given for abstract solidarities, with special reference to operator entropy.

In [19] a generalization of Pólya's inequalities is given with quasiarithmetic means of operators. Denote by $S(J)$ the set of all self-adjoint operators on a Hilbert space whose spectra are contained in an interval J. Let $X = (X_1, \ldots, X_m)$ be an m-tuple of operators from $S(0,\infty)$ and $A = (A_1, \ldots, A_m)$ an m-tuple of contractions such that $\sum_{j=1}^{m} A_j^* A_j = I$.

The quasiarithmetic mean of X with respect to a monotone function f is defined by

$$M_f(X, A) := f^{-1}\left(\sum_{j=1}^{n} A_j^* f(X_j) A_j\right).$$

For quasiarithmetic means the following result holds (see [19]).

THEOREM 10.25

Let $C_{j1} \leq \ldots \leq C_{jn}$ (not all equal), $j = 1, \ldots, m$, be operators from $S(0,\infty)$ such that $C_{11} = \ldots = C_{m1}$ and $C_{1n} = \ldots C_{mn}$. Suppose $A_j (j = 1, \ldots, m)$ are contractions with $\sum_{j=1}^{m} A_j^* A_j = I$.

i) Let $a_1 \leq \ldots \leq a_n$ be positive numbers. If either f and g are operator convex and f^{-1} and g^{-1} operator increasing, or f and g operator concave and f^{-1} and g^{-1} operator decreasing, then

$$M_f\left(\sum_{i=1}^{n-1} a_i \Delta C_{1i}, \ldots, \sum_{i=1}^{n-1} a_i \Delta C_{mi}; A\right) \geq \sum_{i=1}^{n-1} a_i \Delta M_g(C_{1i}, \ldots C_{mi}; A). \tag{10.32}$$

If either f and g are operator concave and f^{-1} and g^{-1} operator increasing, or f and g operator convex and f^{-1} and g^{-1} operator decreasing, the reverse inequality holds.

ii) Let $a_1 \leq \ldots \leq a_n$ be positive numbers. If either f is operator convex and f^{-1} operator increasing, or f operator concave and f^{-1} operator decreasing, and either g is operator concave and g^{-1} operator increasing, or g operator convex and g^{-1} operator decreasing, then (10.32) applies.

If either f is operator concave and f^{-1} operator increasing, or f operator convex and f^{-1} operator decreasing, and either g is operator convex and g^{-1} operator increasing, or g operator concave and g^{-1} operator decreasing, the inequality is reversed.

If in place of A_j we consider real numbers $w_1 > 0$, $w_i < 0$ $(i = 2, \ldots, m)$ with $w_1 + \ldots + w_m = 1$, then under the same assumptions as in the previous theorem we have

$$M_f\left(\sum_{i=1}^{n-1} a_i\Delta C_{1i},\ldots,\sum_{i=1}^{n-1} a_i\Delta C_{mi};w\right)\geq\sum_{i=1}^{n-1} a_i\Delta M_g(C_{1i},\ldots C_{mi};w).$$

If $f(t) = t^r$ $(r \neq 0)$ the quasiarithmetic mean M_f is called the *power mean* of X with weights A of order r and denoted by $M_m^{[r]}(X, A)$. The inequality

$$M_m^{[r]}(X,A)\leq M_m^{[s]}(X,A)$$

holds if either

a) $r \leq s$, $r \notin (-1,-1)$, $s \notin (-1,1)$; or
b) $s \geq 1 \geq r \geq \frac{1}{2}$ or
c) $r \leq -1 \leq s \leq -\frac{1}{2}$

(see [18]).

The following result is also given in [19].

THEOREM 10.26

Suppose the conditions on the operators C_{ij} and contractions A_j of the previous theorem hold and let $a_1 \leq \ldots \leq a_n$ (respectively, $a_1 \geq \ldots \geq a_n$) be positive numbers.

(a) If $r, s \in J_1 = (-\infty, -1] \cup [1/2, 1)$ (respectively, $r \in J_1$ and $s \in J_2 = (1, \infty)$), then

$$M_m^{[r]}\left(\sum_{i=1}^{n-1} a_i\Delta C_{1i},\ldots,\sum_{i=1}^{n-1} a_i\Delta C_{mi};A\right)\leq\sum_{i=1}^{n-1} a_i\Delta M_m^{[s]}(C_{1i},\ldots C_{mi};A).$$

(b) If $r, s \in J_2$ (respectively, $r \in J_2$ and $s \in J_1$), then the reverse inequality holds.

10.9 Applications

In probability theory the r-th absolute moment is defined by

$$\nu_r=\int_0^{\infty} x^r dQ(x),$$

where the distribution function $Q : [0, \infty) \to [0, 1]$ is a nondecreasing function such that $Q(0) = 0$ and $\lim_{x\to\infty} Q(x) = 1$. In that case, if Q' is positive, continuous and nonincreasing, we have from Theorem 10.9 that

$$\nu_{a_1+\ldots+a_m} \le \frac{\prod_{i=1}^m (a_i p_i + 1)^{1/p_i}}{\sum_{i=1}^m a_i + 1} \prod_{i=1}^m (\nu_{a_i p_i})^{1/p_i},$$

where $p_i, (i = 1, \ldots, m)$ are positive real numbers such that $\sum_{i=1}^m 1/p_i = 1$.

Furthermore, if $(-1)^{k-1}Q^{(k)}$ is positive, continuous and nonincreasing for $k = 1, 2, \ldots, n$, a proof similar to that of Theorem 10.9; that is, Theorem 10.8 and the relation

$$(r+1)\int_0^\infty x^r Q'(x)dx = -\int_0^\infty x^{r+1} dQ'(x),$$

give us that

$$\nu_{a_1+\ldots+a_m} \le \frac{\prod_{i=1}^m \binom{a_i p_i + n}{n}^{1/p_i}}{\binom{\sum_{i=1}^m a_i + n}{n}} \prod_{i=1}^m (\nu_{a_i p_i})^{1/p_i}.$$

Thus the function

$$f_n(r) = \binom{r+n}{n}\nu_r$$

is log-convex, leading to the following consequences.

THEOREM 10.27

a) If $p > q > r$, then

$$\left(\binom{q+n}{n}\nu_q\right)^{p-r} \le \left(\binom{r+n}{n}\nu_r\right)^{p-q}\left(\binom{p+n}{n}\nu_p\right)^{q-r}. \tag{10.33}$$

b) If $p \ge q$, $r \ge s$ i $p > r$, $q > s$, then

$$\left(\frac{\binom{p+n}{n}\nu_p}{\binom{r+n}{n}\nu_r}\right)^{\frac{1}{p-r}} \ge \left(\frac{\binom{q+n}{n}\nu_q}{\binom{s+n}{n}\nu_s}\right)^{\frac{1}{q-s}}. \tag{10.34}$$

c) If $r \ge 0$, $r_1, \ldots, r_m > 0$, then

$$\left(\binom{r+n}{n}\nu_r\right)^{m-1}\binom{r_1+\ldots+r_m+r+n}{n}\nu_{r_1+\ldots+r_m+r} \tag{10.35}$$

$$\geq \prod_{i=1}^{m} \binom{r_i + r + n}{n} \nu_{r_i + r}.$$

d) If $q > s > r > p$, $p \leq t \leq q$, *then*

$$\left[\binom{p+n}{n}\nu_p\right]^{\frac{q-t}{q-p}} \left[\binom{q+n}{n}\nu_q\right]^{\frac{t-p}{q-p}} \geq \left[\binom{r+n}{n}\nu_r\right]^{\frac{s-t}{s-r}} \left[\binom{s+n}{n}\nu_s\right]^{\frac{t-r}{s-r}}. \tag{10.36}$$

e) If $p > q > 0$, *then*

$$\left(\binom{p+n}{n}\nu_p\right)^{1/p} \leq \left(\binom{q+n}{n}\nu_q\right)^{1/q}. \tag{10.37}$$

PROOF a) This is a consequence of the inequality

$$\begin{vmatrix} H(p) & H(q) & H(r) \\ p & q & r \\ 1 & 1 & 1 \end{vmatrix} \geq 0 \quad \text{for } p > q > r$$

for a convex function H (see [14, p. 1]).

b) For any convex function H, the inequality

$$\frac{H(p) - H(r)}{p - r} \geq \frac{H(q) - H(s)}{q - s}$$

holds for $p \geq q$ and $r \geq s$ (see [14, p. 2]). Therefore (10.34) is a simple consequence of the previous inequality if we set $H = f_1$.

c) Setting $r = s$, $p = r_1 + \ldots + r_m + r$, $q = r_i + r$ in (10.34) provides

$$\left(\frac{\binom{r_1+\ldots+r_m+r+n}{n}\nu_{r_1+\ldots+r_m+r}}{\binom{r+n}{n}\nu_r}\right)^{\frac{r_i}{r_1+\ldots+r_m}} \geq \frac{\binom{r_i+r+n}{n}\nu_{r_i+r}}{\binom{r+n}{n}\nu_r}.$$

On multiplying together these inequalities for $i = 1, \ldots, m$, we obtain (10.36).

d) This is a consequence of Narumi's inequality

$$\frac{q-t}{q-p}H(p) + \frac{t-p}{q-p}H(q) \geq \frac{s-t}{s-r}H(r) + \frac{t-r}{s-r}H(s)$$

(see [22]), where H is a convex function and $q > s > r > p$, $p \leq t \leq q$.

e) Set $r = s = 0$ in (10.36) and use the fact that $\nu_0 = 1$. ∎

For $n = 1$, inequality (10.37) becomes

$$\left((p+1)\nu_p\right)^{1/p} \geq \left((q+1)\nu_q\right)^{1/q}, \qquad p > q > 0 \tag{10.38}$$

which is known as the Gauss–Winckler inequality. The special case

$$\nu_4 \geq \frac{9}{5}\nu_2^2$$

for $q = 2, p = 4$ was stated without proof by Gaüss [10]. Inequality (10.38) was obtained by Winckler [41] but by an invalid argument, whilst the first correct proof was given by Faber [7]. More about inequality (10.38) and some other inequalities for the absolute moments of a distribution can be found in [5], [16], [31].

The special case $Q(x) = 1 - e^{-x}$ $(x \geq 0)$ for which $(-1)^{k-1}Q^k(x)$ is a positive, continuous, nonincreasing function for an arbitrary $k > 0$ leads to inequalities involving the gamma function [16]. Since $\nu_r = \Gamma(r+1)$ for that choice of Q and because $\Gamma(r+1) = r\Gamma(r)$, the inequalities (10.33)–(10.37) reduce to

$$\Gamma(q+k+1)^{p-r} \leq \Gamma(r+k+1)^{p-q}\Gamma(p+k+1)^{q-r}, \quad p > q > r > 0,\ k = 1, 2, \ldots;$$

$$\left(\frac{\Gamma(p+k+1)}{\Gamma(r+k+1)}\right)^{\frac{1}{p-r}} \geq \left(\frac{\Gamma(q+k+1)}{\Gamma(s+k+1)}\right)^{\frac{1}{q-s}} \quad p \geq q,\ r \geq s,\ p > r,\ q > s;$$

$$\Gamma(r+k+1)^{n-1}\Gamma(r_1 + \ldots r_n + r + 2)$$

$$\geq (k!)^{n-1}\prod \Gamma(r_i + r + 2) \quad r \geq 0,\ r_1, \ldots r_n > 0.$$

If $q > s > r > p$, $p \leq t \leq q$, then

$$\Gamma(p+n+1)^{q-t/q-p}\Gamma(q+n+1)^{\frac{t-p}{q-p}} \geq \Gamma(r+n+1)^{s-t/s-r}\Gamma(s+n+1)^{\frac{t-r}{s-r}};$$

$$\left(\frac{\Gamma(p+k+1)}{k!}\right)^{1/p} \geq \left(\frac{\Gamma(q+k+1)}{k!}\right)^{1/q} \qquad p > q > 0,\ k = 1, 2, \ldots .$$

Theorem 10.14 leads us to consider the r-th moment of a distribution defined on the finite interval $[0, 1]$. We have the following result.

THEOREM 10.28

Let $Q : [0,1] \to [0,1]$ be a nondecreasing function with a continuous $(n+1)$-th order derivative such that $(-1)^{n-1}Q^{(n+1)}$ is nonnegative. Then

$$\nu_{a_1+\ldots+a_m} \geq \frac{\prod_{i=1}^{m}\binom{a_i p_i + n}{n}^{1/p_i}}{\binom{\sum_{i=1}^{m} a_i + n}{n}} \prod_{i=1}^{m} (\nu_{a_i p_i})^{1/p_i} + \Delta_3, \tag{10.39}$$

where

$$\Delta_3 = \begin{cases} 0, & \text{for} \quad n = 1, 2 \\ \sum_{k=2}^{n-1} (-1)^{n-k-1} Q^{(n-k)}(1) \left(\sum_{i=1}^{m} \frac{1}{p_i}(a_i p_i + n)^{[k]} \right. & \\ \qquad \left. -(\sum_{i=1}^{m} a_i + n)^{[k]}\right), & \text{for} \quad n \geq 3 \end{cases}$$

and $r^{[k]} = r(r-1)\ldots(r-k+1)$, $r^{[0]} = 1$.

If $(-1)^{n-k-1}Q^{(n-k)}(1) \geq 0$ for $k = 2, \ldots, n-1$ $(n \geq 3)$, then

$$\nu_{a_1+\ldots+a_m} \geq \frac{\prod_{i=1}^{m}\binom{a_i p_i + n}{n}^{1/p_i}}{\binom{\sum_{i=1}^{m} a_i + n}{n}} \prod_{i=1}^{m} (\nu_{a_i p_i})^{1/p_i}. \tag{10.40}$$

PROOF Inequality (10.39) is a consequence of Theorem 10.14 for $x_i(x) = x^{a_i p_i + n}$ and $a_i p_i > -1$ $(i = 1, \ldots, m)$. If we prove that $\Delta_3 \geq 0$, the validity of inequality (10.40) follows. For this we have

$$\sum_{i=1}^{m} \frac{1}{p_i}(a_i p_i + n)^{[k]} - \left(\sum_{i=1}^{m} a_i + n\right)^{[k]}$$

$$= \sum_{i=1}^{m} \frac{1}{p_i}\left[\sum_{j=1}^{k} N_j (a_i p_i)^j\right] - \sum_{j=1}^{k} N_j \left(\sum_{i=1}^{m} a_i\right)^j$$

$$= \sum_{j=1}^{k} N_j g_j(p_1, \dots, p_m),$$

where each N_j $(j = 1, \dots, k)$ is a positive number and

$$g_j(p_1, \dots, p_m) = \sum_{i=1}^{m} a_i^j p_i^{j-1} - \left(\sum_{i=1}^{m} a_i \right)^j .$$

It is easy to see that $g_1 = 0$ and $g_j(p_1, \dots, p_m) \geq 0$ for $j = 1, \dots, k$, so $\Delta_3 \geq 0$. ∎

Suppose Q in the definition of ν_r satisfies the assumptions of Theorem 10.28 and

$$f_n(r) := \binom{r+n}{n} \nu_r .$$

Then f_n is a log-concave function and the inequalities (10.33)–(10.37) are reversed.

Setting $n = 1$, $m = 2$ and $r = 0$ in the reverse of inequality (10.36) gives

$$(r_1 + r_2 + 1)\nu_{r_1 + r_2} \leq (r_1 + 1)(r_2 + 1)\nu_{r_1} \nu_{r_2} .$$

This inequality appeared in Stolarsky [33] and has had a number of generalizations and improvements (see [2], [12], [13], [26], [28]).

References

[1] S. Abramovich, J. Pečarić and S. Varošanec, New generalization of Gauss–Pólya's inequality, *Math. Ineq. and Applic.* **1** (1998) 331–342.

[2] S. Abramovich, J. Pečarić and S. Varošanec, Further extension of Stolarsky's inequality with general weights, to appear in *Math. Pannonica.*

[3] H. Alzer, An extension of an inequality of G. Pólya, *Buletinul Institutului Politehnic Din Iasi* Tomul **XXXVI(XL)**, Fasc. 1–4, Sectia I (1990), 17–18.

[4] W. N. Anderson, Jr. and R. J. Duffin, Series and parallel addition of matrices, *J. Math. Anal. and Applic.* **26** (1969), 576–594.

[5] P. R. Beesack, Inequalities for absolute moments of a distribution: from Laplace to Von Mises, *J. Math. Anal. and Applic.* **98** (1984), 435–457.

[6] F. Bernstein and M. Kraft, Integralungleichungen konvexer Funktionen, *Nachr. Gess. Wiss. Göttingen* (1914), 299–308.

[7] G. Faber, Bemerkungen zu Sätzen der Gausschen Theoria Combinationis Observationum, *Bayerische Akademie der Wissenschaften; Math. Phys. Klasse*, München (1923), 7–21.

[8] A. M. Fink and M. Jodeit Jr., Jensen inequalities for functions with higher monotonicities, *Aequat. Math.* **40** (1990), 26–43.

[9] J. I. Fujii, M. Fujii and Y. Seo, An extension of the Kubo–Ando theory: solidarities, *Math. Japon.* **35** (1990), 387–396.

[10] C. F. Gauss, Theoria combinationis observationum, 1821, German transl. in *Abhandlungen zur Methode der kleinsten Quadrate.* Neudruck, Würzburg (1964), pp. 9 and 12.

[11] F. Kubo and T. Ando, Means of positive linear operators, *Math. Ann.* **246** (1980), 205–224.

[12] D.H. Luecking, Without commensurability, *Am. Math. Monthly* **99** (1972), 668.

[13] L. Maligranda, J. Pečarić and L. E. Persson, Stolarsky's inequality with general weights, *Proc. Amer. Math. Soc.* **123** (7) (1995), 2113–2117.

[14] D. S. Mitrinović, J. E. Pečarić and A. M. Fink, *Classical and New Inequalities in Analysis*, Dordrecht, Kluwer Acad. Publishers (1993).

[15] D. S. Mitrinović, J. E. Pečarić and A. M. Fink, *Inequalities Involving Functions and Their Integrals and Derivatives*, Dordrecht, Kluwer Acad. Publishers (1993).

[16] D. Mitrinović and J. Pečarić, Note on the Gauss–Winckler inequality, *Österreichischen Akademie der Wissenschaften* **6** (1986), 89–92.

[17] D. Mitrinović and J. Pečarić, Two integral inequalities, *SEA Bull. Math.*, **15** (2) (1991), 153–155.

[18] B. Mond and J. Pečarić, On Jensen's inequality for operator-convex

functions, *Houston J. Math.* **21** (4) (1995), 739–754.

[19] B. Mond, J. Pečarić, J. Šunde and S. Varošanec, Inequalities of Pólya type for positive linear operators, *Houston J. Math.* **22** (4) (1996), 851–858.

[20] B. Mond, J. Pečarić, J. Šunde and S. Varošanec, Operator versions of some classical inequalities, *Lin. Alg. Appl.* **264** (1997), 117–126.

[21] B. Mond, J. Pečarić, J. Šunde and S. Varošanec, Pólya's inequality for positive linear operators, *Rad. HAZU* **472** (13) (1997), 101–105.

[22] S. Narumi, On a generalized Gauss–Faber's inequality for integral, *Jap. J. Math.* **4** (1927), 33–39.

[23] C. E. M. Pearce, J. Pečarić and J. Šunde, Generalization of Pólya's inequality to Stolarsky and Gini means, *Math. Inequal. Appl.* **1** (1998), 211–222.

[24] C. E. M. Pearce, J. Pečarić and S. Varošanec, Inequalities of Gauss–Minkowski type, *Intern. Ser. Num. Math.* **123** (1997), 27–37.

[25] C. E. M. Pearce and J. Pečarić, Some inequalities for generalized quasiarithmetic means, manuscript.

[26] J. Pečarić, A reverse Stolarsky's inequality, *Amer. Math. Monthly*, **6** (1994), 566–568.

[27] J. Pečarić, On two integral inequalities, *Buletinul Institutului Politehnic Din Iasi* Tomul XL (XLIV) Fasc. 1–4 (1994), 45–50.

[28] J. Pečarić, On Stolarsky's quotient, *Macedonian Acad. of Sci. and Arts* **14** (2) (1993), 55–60.

[29] S. Varošanec, J. Pečarić and J. Šunde, On Gauss-Pólya inequality, *Österreichische Akademie der Wissenschaften*, Abt. II **207** (1998), 1–12.

[30] J. Pečarić and S. Varošanec, A generalization of Pólya's inequalities, in *Inequalities and Applications*. World Scientific Publ., Singapore (1994), 501–504.

[31] J. Pečarić and S. Varošanec, Remarks on Gauss–Winckler's and Stolarsky's inequality, *Utilitas Math.* **48** (1995), 233–241.

[32] G. Pólya and G. Szegö, *Aufgaben und Lehrsätze aus der Analysis*, I, II, Berlin (1925).

[33] K. Stolarsky, From Wythoff's Nim to Chebyshev's inequality, *Amer.*

Math. Monthly **98** (1991), 889–900.

[34] S. Varošanec, A remark on Volkov's result, *Rad HAZU* **470** (12) (1995), 199–203.

[35] S. Varošanec, Inequalities of Minkowski's type, *Real Anal. Exchange* **20** (1) (1994–95), 250–255.

[36] S. Varošanec and J. Pečarić, On Pólya's inequality, *Proc. of the Postgrad. School and Conf. in Dubrovnik, Nov. 10–17, 1993*, Aarhus Univ. (1994), 275–278.

[37] S. Varošanec and J. Pečarić, Some integral inequalities with application to bounds for moments of a distribution, *J. Austral. Math. Soc., Ser. B* **38** (1997), 325–335.

[38] S. Varošanec and J. Pečarić, Gauss' and related inequalities, *Z. Anal. Anwendungen* **14** (1) (1995), 175–183.

[39] S. Varošanec, J. Pečarić and J. Šunde, Some discrete inequalities, *Z. Anal. Anwendungen* **15** (4) (1996), 1033–1044.

[40] V.N. Volkov, Utočnenie neravenstva Gel'dera, *Mosk. Gos. Ped. Inst. im V.I. Lenina. Učen. Zap.-Ncravenstva* **460** (1972), 110–115.

[41] A. Winckler, Allgemeine Sätze zur Theorie der unregelmäβigen Beobachtungsfehler, *Sitzungsber. Math.-Natur. Kl. K. Akad. Wiss. Wien, Zweite Abt.* **53** (1866), 6–41.

Chapter 11

Limit Theorems and Best Constants in Approximation Theory

Michael I. Ganzburg

Department of Mathematics, Hampton University, Hampton, VA 23668
E-mail: michael.ganzburg@hamptonu.edu

Abstract We review various limit relations between polynomial, harmonic, and spline approximations and discuss their applications in approximation theory. We also prove some new limit theorems in φ-spaces and μ-quasi-normed rearrangement-invariant spaces and apply these results to sharp constants of multidimensional approximation theory.

Contents

11.1 Introduction

The basic problem in the classical approximation theory, the development of which was strongly influenced by numerical and functional analysis, applied mathematics, Fourier analysis, and differential equations, is to find or estimate the distance $E(f, B_\alpha, F)$ (or $E(K, B_\alpha, F)$) between an individual element f from a space F (or a class K of elements from F) and an indexed set $B_\alpha \subseteq F$.

The problem of finding $E(f, B_\alpha, F)$ or $E(K, B_\alpha, F)$ has been solved for some functions f, functional spaces F, subspaces B_α and classes K, by Chebyshev, Bernstein, Akhiezer, Krein, Favard, Nikolskii, Dzyadyk, Korneichuk, DeVore, and others.

For many functions and classes of functions, only the asymptotics of the form

$$E(f, B_\alpha, F) = a_\alpha(1+o(1)),\ E(K, B_\alpha, F) = b_\alpha(1+o(1)), \quad \alpha \to \infty, \quad (11.1)$$

have been established.

For example, Bernstein [6] showed that the error of best polynomial approximation $E(f_\lambda, \mathcal{P}_{n,1}, C[-1,1])$ of $f_\lambda(x) = |x|^\lambda$ in $C[-1,1]$ asymptotically decays as $A_\lambda n^{-\lambda}$, $n \to \infty$, where A_λ is the error of best approximation of f_λ by entire functions of exponential type 1 in $L_\infty(\mathbb{R}^1)$.

In the 1940s, Bernstein [7, 8, 9, 10] came up with the idea of general limit relations between polynomial and harmonic approximations which effectively allow one to obtain the asymptotics like (11.1).

These limit theorems can be described in the following general form

$$\lim_{n\to\infty} E(f_n, B_n, F_n) = E(f, B, F), \quad (11.2)$$

$$\lim_{n\to\infty} E(K_n, B_n, F_n) = E(K, B, F), \quad (11.3)$$

where f_n, F_n, B_n and K_n, $n = 1, 2, \ldots$, are some sequences of elements, spaces, subspaces and classes, respectively.

Bernstein's results have been developed, refined and generalized in [1, 24, 30, 32, 33, 35, 47, 51, 53, 54].

Powerful analytic methods based on limit relations (11.2) and (11.3) have been applied to various areas of approximation theory, in particular, to sharp and asymptotically sharp constants, Jackson's theorems, and Bernstein's inequalities [1, 22, 23, 24, 25, 28, 29, 33, 35, 47, 51, 53].

In this chapter, we survey all known limit theorems and their applications in approximation theory (Section 11.2), present a more general approach to relations (11.2) and (11.3) in φ-spaces and μ-quasi-normed rearrangement-invariant spaces (Sections 11.3 and 11.4), and apply these results to asymptotically sharp constants of multidimensional approximation by trigonometric polynomials (Section 11.5).

The proofs of all new results in Sections 11.3, 11.4, and 11.5 are based on analytic methods and facts from approximation theory, functional and Fourier analysis, distribution theory and theory of functional spaces.

Notation

Let $\mathbb{R}^m$ be the m-dimensional Euclidean space; $x = (x_1, \ldots, x_m)$ and $y = (y_1, \ldots, y_m)$ points in $\mathbb{R}^m$; $(x, y) := \sum_{i=1}^m x_i y_i$; $|x| := (x, x)^{1/2}$; $\mathbb{C}^m$ the m-dimensional complex space; $z = (z_1, \ldots, z_m)$ a point in $\mathbb{C}^m$; $\mathbb{Z}^m$ the integral lattice points in $\mathbb{R}^m$; V a centrally symmetric (with respect to the origin) convex compactum in $\mathbb{R}^m$ with the nonempty interior; $V^* := \{y \in \mathbb{R}^m\colon \sup_{x \in V} \left|\sum_{i=1}^m x_i y_i\right| \leq 1\}$ the polar of V;

$$Q_a := \{x \in \mathbb{R}^m : |x_i| \leq a,\ i = 1, \ldots, m\}$$

the cube in $\mathbb{R}^m$; χ_Ω the characteristic function of $\Omega \subseteq \mathbb{R}^m$; $|\Omega|$ the m-dimensional Lebesgue measure of a measurable set $\Omega \subseteq \mathbb{R}^m$.

For every function f measurable on $\mathbb{R}^m$, we define its decreasing rearrangement $f^* : [0, \infty) \to [0, \infty]$ by $f^*(t) := \inf\{\tau \geq 0 : E_\tau \leq t\}$, where $E_\tau = |\{x \in \mathbb{R}^m : |f(x)| > \tau\}|$ (see [15, 16, 42]).

Let $L_p(\Omega)$, $0 < p \leq \infty$, be the space of measurable functions f defined on the measurable set $\Omega \subseteq \mathbb{R}^m$ with the finite norm (quasi-norm)

$$\|f\|_{L_p(\Omega)} := \left(\int_\Omega |f|^p \, dx\right)^{1/p};$$

$L_{p,a}$, $0 < p \leq \infty$, the space of measurable functions f, $2a$-periodic in each variable, with the finite norm (quasi-norm)

$$\|f\|_{L_{p,a}} := \|f\|_{L_p(Q_a)};$$

$C(\Omega)$ the space of functions f continuous on $\Omega \subseteq \mathbb{R}^m$ with the finite norm

$$\|f\|_{C(\Omega)} := \sup_{x \in \Omega} |f(x)|;$$

$M_{A,N}$, $A > 0$, $N \geq 0$, the class of functions f measurable on $\mathbb{R}^m$ and satisfying the inequality

$$|f(x)| \leq A(1 + |x|)^N, \quad x \in \mathbb{R}^m;$$

$H^\lambda(\Omega)$ the class of functions f satisfying Hölder's condition of order λ, $0 < \lambda \le 1$, on a convex set $\Omega \subseteq \mathbb{R}^m$, i.e.,

$$|f(x) - f(y)| \le |x - y|^\lambda, \quad x, y \in \Omega;$$

$\tilde{H}^\lambda(Q_\pi) := H^\lambda(\mathbb{R}^m) \cap L_{\infty,\pi}$ Hölder's class of functions 2π-periodic in each variable.

For a measurable function f defined on a convex $\Omega \subseteq \mathbb{R}^m$, we set $(0 < p \le \infty)$

$$\omega_r(f, t, L_p(\Omega)) := \sup_{|y| \le t} \left\| \sum_{s=0}^{r} (-1)^{r-s} \binom{r}{s} f(\cdot + sy) \right\|_{L_p(\Omega_y)},$$

where $\Omega_y = \{x : x - sy \in \Omega,\ s = 0, 1, \ldots, r\}$ is the domain of definition of the rth difference.

We denote by $\hat{f}$ the Fourier transform and by $\check{f}$ the inverse Fourier transform of a function or a distribution f. In particular, for $f \in L_1(\mathbb{R})$,

$$\hat{f}(y) = (2\pi)^{-m/2} \int_{\mathbb{R}^m} f(x) \exp(-(x, y))\, dx;$$

$$\check{f}(y) = (2\pi)^{-m/2} \int_{\mathbb{R}^m} f(x) \exp((x, y))\, dx.$$

Now we define some classes of entire functions of exponential type and algebraic and trigonometric polynomials.

DEFINITION 11.1 *We say that an entire function $g(z)$ is of exponential type σV, $\sigma > 0$, if for every $\varepsilon > 0$ there exists $A_\varepsilon > 0$ such that for every $z \in \mathbb{C}^m$ we have*

$$|g(z)| \le A_\varepsilon \exp\left((1 + \varepsilon) \sup_{x \in \sigma V} \left| \sum_{i=1}^{m} z_i x_i \right| \right).$$

We denote by $B_{\sigma V}$ the class of all entire functions of exponential type σV. A similar class of entire functions was considered in [49]. In the case when $m = 1$ and $V = [-1, 1]$, this class coincides with B_σ, the class of all entire functions of exponential type σ.

Let $\mathcal{T}_V$, $\mathcal{T}_{a,V}$ be the classes of trigonometric polynomials of the form

$$T(x) = \sum_{k \in \mathbb{Z}^m \cap V} c_k \exp(i(k, x)),$$

$$T(x) = \sum_{k \in \mathbb{Z}^m \cap (a/\pi)V} c_k \exp(i(\pi/a)(k, x)),$$

respectively. The embeddings $\mathcal{T}_V \subset B_V$ and $\mathcal{T}_{a,V} \subset B_V$ are obvious, and $\mathcal{T}_V = \mathcal{T}_{\pi,V}$. In the case when $V = [-n, n]$, we set $\mathcal{T}_n := \mathcal{T}_V$.

Let $\mathcal{P}_{n,m}$ be the class of algebraic polynomials in m variables and of degree n, i.e.,

$$\mathcal{P}_{n,m} := \left\{ P : P(x) = \sum_{k=0}^{n} \sum_{\alpha_1 + \cdots + \alpha_m = k} c_{\alpha_1, \ldots, \alpha_m} x_1^{\alpha_1} \cdots x_m^{\alpha_m} \right\}.$$

Finally, let H be a vector space, B a subspace of H, K a subset of H, and F a normed subspace of H with a norm $\|\cdot\|_F$. If $f \in H - F$, then set $\|f\|_F = \infty$. For $f \in H$ we denote

$$E(f, B, F) := \inf_{g \in B} \|f - g\|_F,$$

$$E(K, B, F) := \sup_{f \in K} E(f, B, F).$$

Throughout the chapter $C, C_1, C_2, \ldots$ denote positive constants independent of n, N, σ, x, t, f, g, P. The same symbol does not necessarily denote the same constant in different occurrences.

11.2 Limit Theorems of Approximation Theory and their Applications

The survey of the limit relations in approximation theory and their applications may be regarded as an extension of earlier surveys [33, 35].

11.2.1 Polynomial Approximation

We discuss the limit theorems of approximation by univariate and multivariate polynomials in L_p.

11.2.1.1 Approximation by Univariate Polynomials

The first limit theorem and one of the first results in approximation theory was established in 1885 by Weierstrass [55]: if $f \in C[a,b]$, then

$$\lim_{n\to\infty} E(f, \mathcal{P}_{n,1}, C[a,b]) = E(f, C[a,b], C[a,b]). \tag{11.4}$$

In other words, the set $\bigcup_{n=0}^{\infty} \mathcal{P}_{n,1}$ is dense in $C[a,b]$, and (11.4) can be reformulated in the following generalized form: if B_n, $n = 1, 2, \ldots$, and B are subspaces of a normed space F, and $\bigcup_{n=1}^{\infty} B_n$ is dense in B, then for every $f \in F$,

$$\lim_{n\to\infty} E(f, B_n, F) = E(f, B, F). \tag{11.5}$$

This statement easily follows from the closure condition $\overline{\bigcup_{n=1} B_n} = B$. Relations (11.4) and (11.5) play an important role in many areas of analysis, while for some reasons, they are practically useless in approximation theory. First, the closure condition does not hold for many subspaces B_n and B which are popular in approximation theory. For example, it is clear that $\bigcup_{n=0}^{\infty} \mathcal{P}_{n,1}$ is not dense in $B_\sigma \cap L_p(R^1)$, $p > 0$. Second, in many problems one rather needs to approximate $A_n f$ in a normed (or quasi-normed) space F_n by elements from B_n, where $A_n : F \to F_n$, $n = 1, 2, \ldots$, is a sequence of certain linear operators, and $f \in F$.

The first limit relation of a new type for the errors of polynomial approximation of a function $f_\lambda(x) := |x|^\lambda$, $\lambda > 0$, was obtained in 1938 by Bernstein [6] as follows:

$$\lim_{n\to\infty} E(f_\lambda, \mathcal{P}_{n,1}, C[-n/\sigma, n/\sigma]) = \sigma^{-\lambda} \lim_{n\to\infty} n^\lambda E(f_\lambda, \mathcal{P}_{n,1}, C[-1,1])$$

$$= E(f_\lambda, B_\sigma, C(\mathbb{R}^1)). \tag{11.6}$$

In this case, $F = C(\mathbb{R}^1)$, $F_n = C[-n/\sigma, n/\sigma]$, $B = B_\sigma$, $B_n = \mathcal{P}_{n,1}\big|_{[-n/\sigma,n/\sigma]}$, and $A_n f_\lambda = f_\lambda\big|_{[-n/\sigma,n/\sigma]}$, $n = 1, 2, \ldots$.

In 1946, (11.6) was extended [8, 51, p. 48] to a continuous function $f \in M_{A,N}$, defined on $\mathbb{R}^1$, in the form of the following limit inequalities:

$$E(f, B_\sigma, C(\mathbb{R}^1)) \le \liminf_{n\to\infty} E(f, \mathcal{P}_{n,1}, C[-n/\sigma, n/\sigma])$$

$$\leq \limsup_{n\to\infty} E(f, \mathcal{P}_{n,1}, C[-n/\sigma, n/\sigma])$$

$$\leq \inf_{\gamma\in(0,\sigma)} E(f, B_\gamma, C(\mathbb{R}^1)). \tag{11.7}$$

Since $E(f, B_\sigma, C(\mathbb{R}^1))$ is a non-increasing function in σ, (11.7) implies that for a.a. $\sigma > 0$,

$$\lim_{n\to\infty} E(f, \mathcal{P}_{n,1}, C[-n/\sigma, n/\sigma]) = E(f, B_\sigma, C(\mathbb{R}^1)). \tag{11.8}$$

It was shown [32] that (11.8) does not hold for $f(x) = \cos\sigma x$, and a version of (11.8) which is valid for all $\sigma > 0$ and every continuous function f from $M_{A,N}$ was obtained in [32]:

$$\lim_{n\to\infty} E\left(f, \mathcal{P}_{n,1}, C\left[-\frac{n-\lambda_n}{\sigma}, \frac{n-\lambda_n}{\sigma}\right]\right) = E(f, B_\sigma, C(\mathbb{R}^1)). \tag{11.9}$$

Here, $\{\lambda_n\}_{n=1}^\infty$ is a sequence of real numbers satisfying the conditions:

(a) $0 \leq \lambda_n \leq n$, $n = 1, 2, \ldots$,

(b) $\lim_{n\to\infty} \lambda_n/n = 0$;

(c) $\liminf_{n\to\infty} \lambda_n n^{-1/3} \ln^{-2/3} n > 1/2$.

Unlike the approximation in C, an L_p-analogue of (11.8) holds for all $\sigma > 0$ and every $f \in L_p(\mathbb{R}^1)$, $1 \leq p < \infty$,

$$\lim_{n\to\infty} E(f, \mathcal{P}_{n,1}, L_p[-n/\sigma, n/\sigma]) = E(f, B_\sigma, L_p(\mathbb{R}^1)). \tag{11.10}$$

This result was announced by Bernstein [10] and proved by Raitsin [47].

11.2.1.2 Approximation by Multivariate Polynomials

In the 1980s–1990s, nontrivial multidimensional generalizations of (11.7), (11.8), (11.9), and (11.10) were established in [21, 24, 30, 32]. In particular, the following limit relations hold for all $\sigma > 0$ [24, 32]

$$\lim_{n\to\infty} E(f, \mathcal{P}_{n,m}, L_p((n/\sigma)V^*)$$

$$= E(f, B_{\sigma V}, L_p(\mathbb{R}^m)), \; f \in L_p(\mathbb{R}^m), \; 1 \leq p < \infty, \tag{11.11}$$

$$\lim_{n\to\infty} E\left(f, \mathcal{P}_{n,m}, L_\infty\left(\frac{n-\lambda_n}{\sigma}V^*\right)\right)$$

$$= E(f, B_{\sigma V}, L_\infty(\mathbb{R}^m)), \ f \in M_{A,N}, \tag{11.12}$$

where $\{\lambda_n\}_{n=1}^\infty$ is a sequence of real numbers satisfying the conditions (a), (b), and (c′) $\liminf\limits_{n\to\infty} \lambda_n n^{-\delta} > 0$ for some $\delta \in (1/3, 1)$.

We remark that the necessity of condition (a) for validity of (11.9) and (11.12) is trivial. It was shown in [32] that condition (b) is necessary for the class of all uniformly continuous functions $f \in M_{A,N} - \bigcup_{N=0}^\infty \mathcal{P}_{N,m}$, and the exponent 1/3 in conditions (c) and (c′) cannot be replaced by any $\gamma \in [0, 1/3)$ for all $f \in M_{A,N}$. In particular, (11.12) does not hold for $\lambda_n = 0$, $n \geq 1$, and all $f \in M_{A,N}$.

A multidimensional generalization of (11.9) for polynomials with "a cubic spectrum" was obtained in [30]. Let $\mathcal{P}_{n,m}^*$ be the class of all polynomials of the form

$$P(x) := \sum_{\alpha_1=1}^{m} \cdots \sum_{\alpha_m=1}^{n} c_{\alpha_1,\dots,\alpha_m} x_1^{\alpha_1} \cdots x_m^{\alpha_m}.$$

Then for all $\sigma > 0$ and every $f \in L_\infty(\mathbb{R}^m)$,

$$\lim_{n\to\infty} E\left(f, \mathcal{P}_{n,m}^*, L_\infty\left(\frac{n-\lambda_n}{\sigma}Q_1\right)\right) = E(f, B_{Q_\sigma}, L_\infty(\mathbb{R}^m)),$$

where $\{\lambda_n\}_{n=1}^\infty$ is a sequence of numbers satisfying (a), (b), and (c′).

11.2.1.3 Weighted Approximation

Finally, we consider a limit theorem for weighted polynomial approximation obtained in [33].

Let us set $\gamma_n := ((n-2)/2)^{1/2}$, $W_n(x) := (1-(x/\gamma_n)^2)^{\gamma_n^2}$, $n = 1, 2, \dots$, $W(x) := \exp(-x^2)$. Then for every $f \in L_{p,W}(\mathbb{R}^1) := \{f : \|f\|_{L_{p,W}(\mathbb{R}^1)} := \left(\int_{\mathbb{R}^1} |f|^p W\, dx\right)^{1/p} < \infty\}$, $1 \leq p \leq \infty$,

$$\lim_{n\to\infty} E(f, \mathcal{P}_{k,1}, L_{p,W_n}[-\gamma_n, \gamma_n]) = E(f, \mathcal{P}_{k,1}, L_{p,W}(\mathbb{R}^1)).$$

11.2.1.4 Polynomial Approximation of an Entire Function of Exponential Type

One of the basic tools used in the proofs of limit relations (11.6)–(11.13) is an upper estimate of the error of polynomial approximation of a function from B_σ or $B_{\sigma V}$. In particular, the following result holds [32]: for every $g \in B_{\sigma V} \cap M_{A,N}$,

$$E\left(g, \mathcal{P}_{n,m}, C\left(\frac{n-\lambda_n}{\sigma} V^*\right)\right) \le C_1 n^{C_2} \exp\left(-C_3 n^{3\delta/2 - 1/2}\right), \qquad (11.13)$$

where δ is the constant from condition (c'). It immediately follows from (11.13) that for every $g \in B_{\sigma V} \cap M_{A,N}$, and any $q > 1$,

$$\lim_{n\to\infty} E(g, \mathcal{P}_{n,m}, C((n)q)V^*)) = 0. \qquad (11.14)$$

For $m = 1$, (11.14) was proved in [8], while for $m > 1$, in [24].

11.2.2 Approximation by Trigonometric Polynomials and by Splines

The following limit theorem is an analogue of (11.8) and (11.10) for approximation by trigonometric polynomials: for every $f \in M_{A,N} \cap L_p(\mathbb{R}^m))$, $1 \le p \le \infty$,

$$\lim_{n\to\infty} E\left(f, \mathcal{T}_{a_n,V} \cap M_{A,N}, L_p(Q_{L_n})\right) = E(f, B_V, L_p(\mathbb{R}^m)), \qquad (11.15)$$

where $\{a_n\}_{n=1}^\infty$ and $\{L_n\}_{n=1}^\infty$ are sequences of positive numbers satisfying the condition $\lim_{n\to\infty} L_n^{N+2+m/p} a_n^{-1/(2N+m+2)} = 0$.

Relation (11.15) was obtained for $m = 1$, while for $m > 1$, in [33].

In 1981, Velikin [53] found a striking limit relation between spline and trigonometric approximations of a 2π-periodic function $f \in L_p[0, 2\pi]$.

Let S_{2N+1}^{2n+1} be the class of 2π-periodic spline functions of degree $2n+1$ and of defect 1 with respect to the uniform partition $\{2j\pi/(2N+1)\}_{j=1}^{2N}$. Then the following relation holds

$$\lim_{n\to\infty} E\left(f, S_{2N+1}^{2n+1}, L_p[0, 2\pi]\right) = E(f, \mathcal{T}_N, L_p[0, 2\pi]). \qquad (11.16)$$

Some generalizations of (11.16) to approximation in Banach spaces were given in [54].

11.2.3 Sharp Constants of Approximation Theory

The second type of the limit theorems is connected with the establishment of an asymptotic of the suprema of the error of best approximation by algebraic or trigonometric polynomials on certain classes of functions.

In particular, Bernstein [7] obtained the following asymptotics for the classes of the functions that satisfy the Hölder condition of order λ, $0 < \lambda \leq 1$,

$$C_\lambda := \lim_{n\to\infty} n^\lambda E(H^\lambda[-1,1], \mathcal{P}_{n,1}, C[-1,1])$$

$$= \lim_{n\to\infty} n^\lambda E(\tilde{H}^\lambda[0,2\pi], \mathcal{T}_n, C[0,2\pi])$$

$$= E(H^\lambda(\mathbb{R}^1), B_1, C(\mathbb{R}^1)). \tag{11.17}$$

For $\lambda = 1$, the similar asymptotics were established by Nikolskii [44]. Korneichuk [39] completed this cycle of investigations by finding the exact value of the corresponding trigonometric constant $C_\lambda = (\pi/n)^\lambda$.

A multidimensional analogue of (11.17) was obtained in [24, 28, 29, 33]:

$$C_{\lambda,m} := \lim_{n\to\infty} n^\lambda E(H^\lambda(v), \mathcal{P}_{n,m}, C(V)) = \lim_{n\to\infty} n^\lambda E(\tilde{H}^\lambda(Q_\pi), \mathcal{T}_{nV}, C(Q_\pi))$$

$$= E(H^\lambda(\mathbb{R}^m), B_{V^*}, C(\mathbb{R}^m)). \tag{11.18}$$

In the multidimensional case, the value of the corresponding limit constant $C_{\lambda,m}$ is unknown even for $\lambda = 1$ and a set V that coincides with a cube or a ball.

The following relation

$$\lim_{n\to\infty} E\big(\mathcal{M}^r, S^{2n+1}_{2N+1}, L_p[0,2\pi]\big) = E(\mathcal{M}^r, \mathcal{T}_N, L_p[0,2\pi]) \tag{11.19}$$

was established by Velikin [53] for a subset $\mathcal{M}$ of the class of all 2π-periodic r-differentiable function on the real line.

11.2.4 Limit Relations in Normed Spaces

A general approach to the limit theorems in normed spaces was developed in [33]. It includes the limit relations for individual elements (like (11.1), (11.5), (11.6), (11.8), (11.9), (11.10), (11.11), (11.12), (11.15), (11.16)) as well as the limit relations for classes of elements (like (11.2), (11.17), (11.18), (11.19)). All the results in more general settings (in φ-spaces and in μ-quasi-normed rearrangement-invariant spaces) will be presented in the next two sections.

11.2.5 Applications of the Limit Theorems

The limit theorems of approximation theory can be treated as the special "bridges" between different methods of approximation. They make it easier to carry over certain inequalities or equalities for the error of best approximation to the corresponding one of a different nature. Here, we discuss applications to Jackson–Bernstein theorems and approximation of individual functions. Other applications of the limit theorems will be given in Section 11.5.

11.2.5.1 Jackson's Theorems

It is well-known [16, p. 219] that a Jackson-type estimate of the error of approximation $E(f, \mathcal{P}_{n,1}, L_p[-1,1])$ of $f \in L_p[-1,1]$, $1 \leq p < \infty$, by algebraic polynomials cannot be proved by the substitution $g(t) = f(\cos t)$. That is why this estimate had been unknown until the 1960s. A proof of the inequality

$$E(f, \mathcal{P}_{n,1}, L_p[-1,1]) \leq C_r \omega_r(f, n^{-1}, L_p[-1,1]), \quad n \geq r, \tag{11.20}$$

based on K-functionals, can be found in [16, p. 220].

To illustrate an application of the limit relations to Jackson's theorems, we give a different proof of (11.20) that is based on a version of (11.13).

Note first that for any function $h \in L_p(\mathbb{R}^1)$ and $\sigma > 0$, the following estimate is valid

$$E(h, B_\sigma, L_p(\mathbb{R}^1)) \leq C\omega_r(h, \sigma^{-1}, L_p(\mathbb{R}^1)). \tag{11.21}$$

A simple proof of (11.21) is given in [51, p. 260]. Next, for every $f \in L_p[-1,1]$, there exists a function $h \in L_p(\mathbb{R}^1)$ such that $h - f \in \mathcal{P}_{r-1,1}$ on $[-1,1]$ and

$$\omega_r(h, t, L_p(\mathbb{R}^1)) \leq C\omega_r(f, t, L_p[-1,1]), \quad t \in [0,2] \tag{11.22}$$

(see [3]).

To prove (11.20), we also need a univariate version of (11.13) for $V = [-1,1]$, $\delta = 1/2$, and $g \in B_\sigma \cap C(\mathbb{R}^1)$ (see [32]),

$$E\left(g, \mathcal{P}_{n,1}, C\left(-\frac{n-\sqrt{n}}{\sigma}, \frac{n-\sqrt{n}}{\sigma}\right)\right)$$
$$\leq C_1 n^{C_2} \exp(-C_3 n^{1/4}) \omega_r(g, 2, L_p(\mathbb{R}^1)). \tag{11.23}$$

If $f \in L_p[-1,1]$, then there exists an extension h of f to $\mathbb{R}^1$ satisfying (11.22). Let us set

$$h_n(x) = (n - \sqrt{n})^{-1/p} h((n - \sqrt{n})^{-1} x), \quad n > 0,$$

and let $g \in B_1 \cap L_p(\mathbb{R}^1)$ be a function such that

$$\|h_n - g\|_{L_p(\mathbb{R}^1)} = E(h_n, B_1, L_p(\mathbb{R}^1)) = E(h, B_{n-\sqrt{n}}, L_p(\mathbb{R}^1)). \tag{11.24}$$

Further, (11.24) implies

$$\omega_r(g, 2, L_p(\mathbb{R}^1)) \leq 2^{r+1} \|h_n - g\|_{L_p(\mathbb{R}^1)} + \omega_r(h_n, 2, L_p(\mathbb{R}^1))$$

$$\leq C\big(E(h, B_{n-\sqrt{n}}, L_p(\mathbb{R}^1)) + \omega_r(h, 2, L_p(\mathbb{R}^1))\big). \tag{11.25}$$

Now using (11.21), (11.22), (11.23), and (11.25), we obtain

$$E(f, \mathcal{P}_{n,1}, L_p[-1,1]) \leq \|h - (n - \sqrt{n})^{1/p} g(n\cdot)\|_{L_p(\mathbb{R}^1)}$$

$$+ E(g, \mathcal{P}_{n,1}, L_p[-n + \sqrt{n}, n - \sqrt{n}]) \leq E(h, B_{n-\sqrt{n}}, L_p(\mathbb{R}^1))$$

$$+ C_1 n^{C_2} \exp(-C_3 n^{1/4})\big(E(h, B_{n-\sqrt{n}}, L_p(\mathbb{R}^1)) + \omega_r(h, 2, L_p(\mathbb{R}^1))\big)$$

$$\leq C\omega_r(f, n^{-1}, L_p[-1,1])(1 + C_1 n^{C_2 + r} \exp(-C_3 n^{1/4}))$$

$$\leq C\omega_r(f, n^{-1}, L_p[-1,1]).$$

Thus (11.20) follows.

By a similar method, (11.20) can be extended to the multidimensional approximation. Moreover, by using (11.13), it is possible [22, 23] to obtain a multidimensional version of Jackson's theorem in which the constant does not depend on the dimension of the set V on which the approximation takes place. By a different method, the corresponding result for a cube or a sphere and modulus of continuity of first order was proved by Newman and Shapiro [43]. They posed the problem on the possibility of obtaining similar theorems for other sets and/or with regard for smoothness of higher order. Yudin [56] has extended these results to the case of functions defined on the m-dimensional torus by trigonometric polynomials with spectrum in V and modulus of smoothness of second order.

The following inequality was established in [23]:

$$E(f, \mathcal{P}_{n,m}, C(V)) \leq 64(1 + \varepsilon_n)\omega_1(f, \lambda(V^*)n^{-1}, C(V)), \tag{11.26}$$

where $\lim_{n\to\infty} \varepsilon_n = 0$ and $\lambda(V^*)$ is the first eigenvalue of the eigenvalue problem

$$\left(-\sum_{i=1}^{m} \frac{\partial^2}{\partial x_i^2}\right) u = \lambda u, \quad u = 0 \text{ on } \partial V^*.$$

Let $H_\omega(\Omega)$ denote the set of all functions $f \in C(\Omega)$ satisfying the condition $\omega_1(f, t, C(\Omega)) \leq \omega(t)$, $0 \leq t \leq \operatorname{diam} \Omega$, where ω is a modulus of continuity [16, p. 41]. The class of those functions from $H_\omega(\mathbb{R}^m)$ which are 2π-periodic in each variable is denoted by H_ω^*.

The following relations show [23] that (11.26) cannot be improved in a certain sense for $V := V_q := \left\{x \in \mathbb{R}^m : \left(\sum_{i=1}^m |x_i|^q\right)^{1/q} \leq 1\right\}$, $1 \leq q \leq \infty$,

$$\begin{aligned} C_1\omega(m^{1/2+1/q}\sigma^{-1}) &\leq \sup_{f\in H_\omega^*} E(f, \mathcal{T}_{\sigma V_q}, C(Q_\pi)) \\ &\leq \sup_{f\in H_\omega(\mathbb{R}^m)} E(f, B_{\sigma V_q}, C(\mathbb{R}^m)) \\ &\leq C_2\omega(m^{1/2+1/q}\sigma^{-1}), \quad \sigma > m^{1/2}+1; \\ C_3\omega(m^{3/2-1/q}n^{-1}) &\leq \sup_{f\in H_\omega(V_q)} E(f, \mathcal{P}_{n,m}, C(V_q)) \\ &\leq C_4(1+\varepsilon_n)\omega(m^{3/2-1/q}n^{-1}), \end{aligned}$$

where $\lim_{n\to\infty} \varepsilon_n = 0$ and C_1, C_2, C_3, C_4 are constants that depend only on q.

An application of (11.13) to monotone simultaneous approximation by multivariable polynomials was given in [3].

11.2.5.2 Bernstein-Type Inequalities

By using limit relations like (11.13), it is possible to reduce certain Bernstein-type inequalities to V.A. Markov-type inequalities for the coefficients of algebraic polynomials.

To illustrate this idea, we derive the classical Bernstein inequality [51, p. 208]

$$|g^{(k)}(x)| \leq \sigma^k \|g\|_{C(\mathbb{R}^1)}, \quad x \in \mathbb{R}^1, \tag{11.27}$$

where $g \in B_\sigma \cap C(\mathbb{R}^1)$, $\sigma > 0$, from the V.A. Markov estimates [51, p. 227]

$$|P_n^{(k)}(0)| \le (n/a)^k \|P_n\|_{C[-a,a]}, \quad k = 0, 1, \ldots, n, \tag{11.28}$$

where $P_n \in \mathcal{P}_{n,1}$ and $a > 0$.

Note first that it suffices to prove (11.27) for $x = 0$. Next, let $P_n \in \mathcal{P}_{n,1}$ satisfy the relation $\|g - P_n\|_{C[-a_n,a_n]} = E(g, \mathcal{P}_{n,1}, C[-a_n, a_n])$, where $a_n = (n - \sqrt{n})/\sigma$, $n = 1, 2, \ldots$. Then using the A.A. Markov inequality [51, p. 218] and relation (11.13) for $V = [-1, 1]$ and $\delta = 1/2$, we obtain

$$\begin{aligned} \|g^{(k)} - P_n^{(k)}\|_{C[-a_n,a_n]} &\le \sum_{s=n}^{\infty} \|P_{s+1}^{(k)} - P_s^{(k)}\|_{C[-a_n,a_n]} \\ &\le 2\sum_{s=n}^{\infty} (s+1)^{2k} a_s^{-k} E_s(g, \mathcal{P}_{s,1}, C[-a_s, a_s]) \\ &\le C_1 \sum_{s=n}^{\infty} s^{2k+C_2} \exp(-C_3 s^{1/4}). \end{aligned} \tag{11.29}$$

Finally, (11.28) and (11.29) imply

$$\begin{aligned} |g^{(k)}(0)| &\le \liminf_{n\to\infty} \left(|P_n^{(k)}(0)| + \|g^{(k)} - P_n^{(k)}\|_{C[-a_n,a_n]}\right) \\ &\le \sigma^k \|g\|_{C(\mathbb{R}^1)} + \sigma^k \lim_{n\to\infty} \|g - P_n\|_{C[-a_n,a_n]} \\ &\quad + \lim_{n\to\infty} \|g^{(k)} - P_n^{(k)}\|_{C[-a_n,a_n]} = \sigma^k \|g\|_{C(\mathbb{R}^1)}. \end{aligned}$$

Thus (11.27) follows.

The following general relations can be proved similarly [25]:

$$\begin{aligned} \sup_{f\in \mathcal{T}_V} \frac{\|Lf\|_{C(Q_\pi)}}{\|f\|_{C(Q_\pi)}} &\le \sup_{f \in B_V} \frac{\|Lf\|_{C(\mathbb{R}^m)}}{\|f\|_{C(\mathbb{R}^m)}} \\ &\le \sup_{n\ge 1} \sup_{P\in \mathcal{P}_{n,m}} \frac{|LP(0)|}{\|P\|_{C((n-\sqrt{n})V^*)}}, \end{aligned}$$

where L is a linear differential operator with coefficients from $C(\mathbb{R}^m)$.

This approach is also efficient for some nonlinear differential operators. In particular, the following sharp inequality [25] holds for $f \in B_{\sigma V_q}$, where $V_q = \{x \in R^m : (\sum_{i=1}^m |x_i|^q)^{1/q} \le 1\}$,

$$\left(\sum_{i=1}^m \left|\frac{\partial f(x)}{\partial x_i}\right|^p\right)^{1/p} \le m^{(1/p-1/q)_+}\sigma\|f\|_{C(\mathbb{R}^m)}, \quad 1 \le q \le \infty,\ 0 < p \le \infty.$$

11.2.5.3 Approximation of Individual Functions

The asymptotic relation

$$\lim_{n\to\infty} n^\lambda E(f_\lambda, \mathcal{P}_{n,1}, C[-1,1]) = A_\lambda < \infty, \tag{11.30}$$

where $f_\lambda(x) = |x|^\lambda$, $\lambda > 0$, is an immediate consequence of limit theorem (11.9). Bernstein [6] proved (11.30) by a different method and gave some estimates of A_λ. Varga and Carpenter [52] computed $A_1 = 0.2801\ldots$ and conjected that A_λ "might admit a closed-form expression in term of classical hypergeometric functions and/or known mathematical constants." However, the problem of finding A_λ is still open.

Nikolskii [46] established an L_1-analogue of (11.30):

$$\lim_{n\to\infty} n^{\lambda+1} E(f_\lambda, \mathcal{P}_{n,1}, L_1[-1,1]) = B_\lambda < \infty, \quad \lambda > -1, \tag{11.31}$$

where

$$B_\lambda = \frac{4|\sin(\pi\lambda/2)|}{\pi}\int_{-\infty}^{\infty} |\cos t| \int_0^\infty \frac{u^{\lambda+1}\,du}{(t^2+u^2)(e^u+e^{-u})}\,dt.$$

Bernstein [11] noticed that

$$B_\lambda = \frac{8}{\pi}|\sin(\pi\lambda/2)|\sum_{k=0}^\infty (-1)^k(2k+1)^{-\lambda-2}. \tag{11.32}$$

Below we give a shorter proof of (11.31), based on limit relation (11.10) and some ideas from [26].

First we show that $E(f_\lambda, B_1, L_1(\mathbb{R}^1)) < \infty$. Moreover, for every $\varepsilon > 0$ there exists $g_\varepsilon \in B_\varepsilon \cap M_{A,N}$ such that $\|f_\lambda - g_\varepsilon\|_{L_1(\mathbb{R}^1)} < \infty$. Indeed, it is known [37, 50, p. 51] that the Fourier transform of the tempered distribution f_λ is

$$\hat{f}_\lambda(y) = -(2/\pi)^{1/2}\sin(\lambda\pi/2)\Gamma(\lambda+1)|y|^{-\lambda-1}, \quad y \in \mathbb{R}^1,\ \lambda > -1. \tag{11.33}$$

Then there exists an even extension $h(y)$ of $\hat{f}_\lambda(y)$ from $\mathbb{R}^1 \setminus [-\varepsilon, \varepsilon]$ to $\mathbb{R}^1$ that satisfies the conditions

$$\sup_{0\le\ell\le2} \|h^{(\ell)}\|_{C(\mathbb{R}^1)} < \infty, \quad h \in L_1(\mathbb{R}^1). \tag{11.34}$$

Hence the function $H = \hat{f}_\lambda - h$ is a tempered distribution with the support in $[-\varepsilon, \varepsilon]$. Then $g_\varepsilon = \check{H} \in B_\varepsilon \cap M_{A,N}$ due to the generalized Paley–Wiener theorem [50, p. 114], and it follows from (11.34) that $f_\lambda - g_\varepsilon \in L_1(\mathbb{R}^1)$. Thus $E(f_\lambda, B_1, L_1(\mathbb{R}^1)) < \infty$. Next, we note that (11.13) yields the relation

$$\lim_{n\to\infty} E(g_\varepsilon, \mathcal{P}_{n,1}, L_1[-n,n]) = 0, \quad \varepsilon \in (0,1). \tag{11.35}$$

Now, applying (11.10) to $f_\lambda - g_\varepsilon$, $\varepsilon \in (0,1)$, and taking into account (11.35), we obtain

$$\begin{aligned} \lim_{n\to\infty} n^{\lambda+1} E(f_\lambda, \mathcal{P}_{n,1}, L_1[-1,1]) &= \lim_{n\to\infty} E(f_\lambda, \mathcal{P}_{n,1}, L_1[-n,n]) \\ &= \lim_{n\to\infty} E(f_\lambda - g_\varepsilon, \mathcal{P}_{n,1}, L_1[-n,n]) = E(f_\lambda - g_\varepsilon, B_1, L_1(\mathbb{R}^1)) \\ &= E(f_\lambda, B_1, L_1(\mathbb{R}^1)). \end{aligned}$$

Thus to prove (11.31) with B_λ, defined by (11.32), it remains to show that

$$E(f_\lambda, B_1, L_1(\mathbb{R}^1)) = E(f_\lambda - g_\varepsilon, B_1, L_1(\mathbb{R}^1)) = B_\lambda. \tag{11.36}$$

We apply to $K = a_\lambda(f_\lambda - g_\varepsilon)$, where $a_\lambda = -\operatorname{sgn}\sin(\lambda\pi/2)$ and $\varepsilon \in (0,1)$, the following Nagy theorem [1]: if K is an even function from $L_1(\mathbb{R}^1)$ and $\hat{K}$ is three times differentiable and for some $\sigma > 0$ and all $t \ge \sigma$,

$$\hat{K}(t) > 0, \quad \frac{d\hat{K}(t)}{dt} \le 0, \quad \frac{d^2\hat{K}(t)}{dt^2} \ge 0, \quad \frac{d^3\hat{K}(t)}{dt^3} \le 0,$$

then

$$E(K, B_\sigma, L_1(\mathbb{R}^1)) = 4(2/\pi)^{1/2} \sum_{k=0}^{\infty} (-1)^k \frac{\hat{K}((2k+1)\sigma)}{2k+1}. \tag{11.37}$$

It is easy to verify that $\hat{K}(t) = a_\lambda \hat{f}_\lambda(t)$ for $t \ge 1$, and $a_\lambda \hat{f}_\lambda(t)$, given by (11.33), satisfy the conditions of the Nagy theorem. Hence (11.36) follows from (11.33) and (11.37). This proves relation (11.31).

Other applications of the limit theorem to approximation of individual functions of a single variable in L_∞ and L_1 can be found in [1, 9, 51].

It was proved in [24], by using a limit relation like (11.12), that the limit $\lim_{n\to\infty} n^\lambda E(f_\lambda, \mathcal{P}_{n,m}, C(V))$ exists and is finite for every λ-homogeneous function f_λ in m variables if and only if f is "λ-smooth". In particular, the limit is infinite for $m > 1$ and $f_\lambda(x) = \prod_{i=1}^m |x_i|^{\lambda_i}$, $0 \le \lambda_1 \le \cdots \le \lambda_m < \lambda$, $\lambda_{i_0} \ne 0, 2, 4, \ldots, \sum_{i=1}^m \lambda_i = \lambda$.

Some applications of the limit theorems to approximation of individual functions of several variables in L_1 will be given in Section 11.5.

11.3 Limit Theorems in φ-Spaces

In this section we discuss the limit theorems in vector spaces.

11.3.1 φ-Spaces and Best Approximations

Let F be a vector space, and let $\varphi_F : F \to [0, \infty)$ be a nonnegative subadditive functional on F; that is, φ_F satisfies the triangle inequality

$$\varphi_F(f + h) \le \varphi_F(f) + \varphi_f(h) \tag{11.38}$$

for all $f \in F$ and $h \in F$.

DEFINITION 11.2 *A vector space F, equipped with the functional φ_F, is said to be the φ-space with the φ-norm φ_F.*

The examples of φ-spaces include all normed and semi-normed spaces with $\varphi_F = \|\cdot\|_F$, all μ-quasi-normed spaces (see the next section), and the space $L_\psi = L_\psi(\Omega, \mu)$ of all μ-measurable functions f on $\Omega \le \mathbb{R}^m$ with the finite φ-norm $\varphi_{L_\psi}(f) = \int_\Omega \psi(|f|)\, d\mu$, where μ is a positive measure on Ω, and $\psi : [0, \infty) \to [0, \infty)$ is a subadditive function.

If a φ-space F is a subspace of a vector space H, then we set $\varphi_F(f) = \infty$ for every $f \in H - F$. Let us define the error of best approximation of $f \in H$ by elements from $B \subseteq H$ in F by

$$E(f, B, F) = \inf_{g\in B} \varphi_F(f - g). \tag{11.39}$$

The following properties of the error of best approximation immediately follow from (11.38) and (11.39).

PROPERTY 11.1 *If B is a subspace of F, then*

$$E(f, B, F) \leq \varphi_F(f) \tag{11.40}$$

for every $f \in F$.

PROPERTY 11.2 *If $f \in H$, $h \in H$, and $B \subseteq H$, then*

$$E(f+h, B, F) \leq E(f, B, F) + E(h, B, F). \tag{11.41}$$

We define the ball in F with the radius A by

$$S_A(F) := \{f \in F : \varphi_F(f) \leq A\}.$$

The following property holds.

PROPERTY 11.3 *If φ_F is an even functional, that is, $\varphi_F(-f) = \varphi_F(f)$ for all $f \in F$, then for any $f \in F$,*

$$E(f, B, F) = \inf_{g \in B \cap S_{2A}(F)} \varphi_F(f-g),$$

where $A = \varphi_F(f)$.

PROOF For every $\varepsilon > 0$ there exists $g \in B$ such that

$$\varphi_F(f-g) \leq E(f, B, F) + \varepsilon.$$

Hence taking into account Property 1, we obtain

$$\varphi_F(g) \leq \varphi_F(g-f) + \varphi_F(f) = \varphi_F(f-g) + \varphi_F(f) \leq 2\varphi_F(f) + \varepsilon.$$

Then

$$E(f, B, F) \geq \inf_{g \in B \cap \bigcap_{\varepsilon > 0} S_{2A+\varepsilon}(F)} \varphi_F(f-g) - \varepsilon.$$

Consequently, to prove the property, it is sufficient to note that $\bigcap_{\varepsilon>0} S_{2A+\varepsilon}(F) = S_{2A}(F)$. ∎

For a set $K \subseteq H$, we denote

$$E(K, B, F) := \sup_{f \in K} E(f, B, F).$$

In addition, let

$$\overline{A}(\Omega) = \{\overline{A}f : f \in \Omega\}$$

be the set of images of elements from $\Omega \subseteq K$ at the mapping $\overline{A} : K \to F$.

In this section, we consider the limit theorems in φ-spaces for individual elements of the form

$$\lim_{n\to\infty} E(h_n, B_n, F_n) = E(f, B, F),$$

and for classes of elements of the form

$$\lim_{n\to\infty} E(K_n, B_n, F_n) = E(K, B, F).$$

The corresponding results in normed spaces were established in [33]. Our limit theorems follow the proofs of those in [33].

11.3.2 Formulation of Conditions

Let H be a vector space, B a subspace of H, B' a subset of B, K a subset of H, F a φ-space and a subspace of H, F_n a φ-space, B_n a subspace of F_n, K_n a subset of F_n, $n = 1, 2, \ldots$.

Furthermore, let $\overline{A}_n : H \to F_n$ be a linear operator and $A_n : F \to F_n$ its restriction to F, $n = 1, 2, \ldots$.

We define the following conditions:

1) $\lim\limits_{n\to\infty} E(\overline{A}_n g, B_n, F_n) = 0$ for every $g \in B'$;

2) $\limsup\limits_{n\to\infty} \sup\limits_{f\in F} \varphi_{F_n}(A_n f)/\varphi_F(f) \le 1$;

3) $E(f, B, F) = E(f, B', F)$ for every $f \in K$;

4) $\lim\limits_{n\to\infty} \sup\limits_{g\in B'} E(\overline{A}_n g, B_n, F_n) = 0$;

5) there exist mappings $D_n = K_n \to K$, $n = 1, 2, \ldots$, such that

$$\limsup_{n\to\infty} \sup_{h_n\in K_n} \varphi_{F_n}(h_n - \overline{A}_n D_n h_n) = 0;$$

6) there exists a family of linear operators $A_{nm} : F_n \to F_m$, $n > m \ge 1$, with the properties

   a) $\overline{A}_m = A_{nm}\overline{A}_n$,

   b) $\sup\limits_{1\le m<n<\infty} \sup\limits_{f\in F_n} \varphi_{F_m}(A_{nm}f)/\varphi_{F_n}(f) \le 1$;

7) for any $f \in K$ there exist sets $B'_n = B'_n(f) \subseteq B_n$ such that

$$E(\overline{A}_n f, B_n, F_n) = E(\overline{A}_n f, B'_n, F_n), \quad n = 1, 2, \ldots;$$

8) the condition of compactness: for any sequence $\{g_n\}_{n=1}^{\infty}$, $g_n \in B'_n$, $n = 1, 2, \ldots$, there exist a subsequence $\{g_{n_k}\}_{k=1}^{\infty}$ and an element $g \in B$ such that

$$\lim_{k\to\infty} \varphi_{F_{n_k}}(g_{n_k} - \overline{A}_{n_k} g) = 0;$$

9) the weak condition of compactness: for any sequence $\{g_n\}_{n=1}^{\infty}$, $g_n \in B'_n$, $n = 1, 2, \ldots$, and any $m = 1, 2, \ldots$, there exist a subsequence $\{g_{n_k}\}_{k=1}^{\infty}$ ($n_k = n_k(m)$, $k = 1, 2, \ldots$) and $g \in B$ such that

$$\lim_{k\to\infty} \varphi_{F_m}(A_{n_k m} g_{n_k} - \overline{A}_m g) = 0, \quad m = 1, 2, \ldots;$$

10) $\liminf_{n\to\infty} \varphi_{F_n}(A_n f) \geq \varphi_F(f)$ for every $f \in F$;

11) $\lim_{n\to\infty} \varphi_{F_n}(A_n f) = \varphi_F(f)$ for every $f \in F$;

12) for any $f \in K$ there exists $n_0 = n_0(f)$ such that $\overline{A}_n f \in K_n$ for all $n > n_0$.

REMARK 11.1 It follows from the relations

$$\varphi_{F_m}(\overline{A}_m g - A_{nm} g_n) = \varphi_{F_m}(A_{nm}(\overline{A}_n g - g_n)) \leq \varphi_{F_n}(\overline{A}_n g - g_n)$$

that condition 9) is a consequence of conditions 6) and 8). ∎

REMARK 11.2 Condition 12) is satisfied if $\{\overline{A}_n f : f \in K\} \subseteq K_n$, $n = 1, 2, \ldots$. ∎

11.3.3 Limit Theorems

THEOREM 11.1
(a) If conditions 1), 2), and 3) are satisfied, then for any $f \in K$,

$$\limsup_{n\to\infty} E(\overline{A}_n f, B_n, F_n) \leq E(f, B, F). \tag{11.42}$$

(b) If conditions 2), 3), 4), and 5) are satisfied, then

$$\limsup_{n\to\infty} E(K_n, B_n, F_n) \le E(K, B, F). \tag{11.43}$$

PROOF Let $f \in K$. Without loss of generality, we can assume $E(f, B, F) = E(f, B', F) < \infty$, by condition 3). For any $g \in B'$, satisfying $\varphi_F(f - g) < \infty$, we have, by (11.40) and (11.41),

$$E(\overline{A}_n f, B_n, F_n) \le E(\overline{A}_n f - \overline{A}_n g, B_n, F_n) + E(\overline{A}_n g, B_n, F_n)$$

$$\le \varphi_{F_n}(\overline{A}_n(f - g)) + E(\overline{A}_n g, B_n, F_n). \tag{11.44}$$

Next, condition 2) implies the inequality

$$\limsup_{n\to\infty} \varphi_{F_n}(\overline{A}_n(f - g)) \le \varphi_F(f - g). \tag{11.45}$$

Thus (11.44), (11.45), and conditions 1) and 3) yield (11.42).

To prove statement (b) of Theorem 11.1, we consider an element $h_n \in K_n$, $n = 1, 2, \ldots$. Then $f_n = D_n h_n \in K$, where D_n, $n = 1, 2, \ldots$, are the mappings from condition 5). Now using (11.40) and (11.41), we obtain

$$E(h_n, B_n, F_n) \le E(\overline{A}_n f_n, B_n, F_n) + E(h_n - \overline{A}_n f_n, B_n, F_n)$$

$$\le E(\overline{A}_n f_n, B_n, F_n) + \sup_{h_n \in K_n} \varphi_{F_n}(h_n - \overline{A}_n f_n). \tag{11.46}$$

Next, similar to (11.44),

$$E(\overline{A}_n f_n, B_n, F_n) \le \sup_{f\in F}(\varphi_{F_n}(A_n f)/\varphi_F(f)) \sup_{f\in K} E(f, B', F)$$

$$+ \sup_{g\in B'} E(\overline{A}_n g, B_n, F_n). \tag{11.47}$$

Taking into account condition 2), we deduce from (11.46) and (11.47)

$$\limsup_{n\to\infty} E(K_n, B_n, F_n) = \limsup_{n\to\infty} \sup_{h_n\in K_n} E(h_n, B_n, F_n)$$

$$\le \sup_{f\in K} E(f, B', F) + \limsup_{n\to\infty} \sup_{g\in B'} E(\overline{A}_n g, B_n, F_n)$$

$$+ \limsup_{n\to\infty} \sup_{h_n\in K_n} \varphi_{F_n}(h_n - \overline{A}_n f_n). \tag{11.48}$$

Thus (11.43) follows from (11.48) and conditions 3), 4), and 5). This completes the proof of the theorem. ∎

THEOREM 11.2

(a) If either conditions 7), 8), 10) or conditions 6), 7), 9), 10) are satisfied, then for any $f \in K$,

$$\liminf_{n\to\infty} E(\overline{A}_n f, B_n, F_n) \geq E(f, B, F). \tag{11.49}$$

(b) If either conditions 7), 8), 10), 12) or conditions 6), 7), 9), 10), 12) are satisfied, then

$$\liminf_{n\to\infty} E(K_n, B_n, F_n) \geq E(K, B, F). \tag{11.50}$$

PROOF Let $f \in K$. Without loss of generality, we can assume that the limit $\lim_{n\to\infty} E(\overline{A}_n f, B_n, F_n)$ exists and is finite. By condition 7), for every $\varepsilon > 0$ there exists $g_{n,\varepsilon} \in B'_n$ such that

$$E(\overline{A}_n f, B_n, F_n) \geq \varphi_{F_n}(\overline{A}_n f - g_{n,\varepsilon}) - \varepsilon, \quad n = 1, 2, \ldots . \tag{11.51}$$

If condition 8) is satisfied, then there exist a subsequence $\{g_{n_k,\varepsilon}\}_{k=1}^{\infty}$ and $g \in B$ such that

$$\lim_{k\to\infty} \varphi_{F_{n_k}}(g_{n_k,\varepsilon} - \overline{A}_{n_k} g) = 0.$$

Hence, using condition 10) and relations (11.41) and (11.51), we obtain

$$\lim_{k\to\infty} E(\overline{A}_{n_k} f, B_{n_k}, F_{n_k}) \geq \liminf_{k\to\infty} \varphi_{F_{n_k}}(\overline{A}_{n_k} f - \overline{A}_{n_k} g)$$

$$- \lim_{k\to\infty} \varphi_{F_{n_k}}(g_{n_k,\varepsilon} - \overline{A}_{n_k} g) - \varepsilon$$

$$\geq \varphi_F(f - g) - \varepsilon \geq E(f, B, F) - \varepsilon. \tag{11.52}$$

Thus (11.49) follows from (11.52).

If conditions 6) and 9) are satisfied, then there exists a subsequence $\{g_{n_k,\varepsilon}\}_{k=1}^{\infty}$ and $g \in B$ such that

$$\lim_{k\to\infty} \varphi_{F_m}(A_{n_k m} g_{n_k} - \overline{A}_m g) = 0, \quad m = 1, 2, \ldots . \tag{11.53}$$

Next, for any $n_k > m$, by (11.41) and condition 6),

$$\varphi_{F_{n_k}}(\overline{A}_{n_k} f - g_{n_k,\varepsilon}) \geq \varphi_{F_m}(A_{n_k m}(\overline{A}_{n_k} f - g_{n_k,\varepsilon}))$$

$$= \varphi_{F_m}(\overline{A}_m f - A_{n_k m} g_{n_k,\varepsilon}) \geq \varphi_{F_m}(\overline{A}_m f - \overline{A}_m g)$$

$$- \varphi_{F_m}(A_{n_k m} g_{n_k,\varepsilon} - \overline{A}_m g), \; m = 1, 2, \dots . \tag{11.54}$$

Finally, relations (11.51), (11.53), (11.54) and condition 10) yield

$$\lim_{k\to\infty} E(\overline{A}_{n_k} f, B_{n_k}, F_{n_k}) \geq \liminf_{k\to\infty} \varphi_{n_k}(\overline{A}_{n_k} f - g_{n_k,\varepsilon}) - \varepsilon$$

$$\geq \liminf_{m\to\infty} \varphi_{F_m}(\overline{A}_m f - \overline{A}_m g) - \varepsilon \geq E(f, B, F) - \varepsilon.$$

Thus (11.49) follows. This proves statement (a).

To prove statement (b), we note first that (11.49) holds true for all $f \in K$. Next by condition 12), $\overline{A}_n f \in K$, $n \geq n_0(f)$, if $f \in K$, and hence for every $f \in K$,

$$\liminf_{n\to\infty} E(K_n, B_n, F_n) \geq \liminf_{n\to\infty} E(\overline{A}_n f, B_n, F_n) \geq E(f, B, F).$$

This yields (11.50). ∎

11.3.4 Some Corollaries

COROLLARY 11.1

(a) If either conditions 1), 3), 7), 8), 11), or conditions 1), 3), 6), 7), 9), 11) are satisfied, then for any $f \in K$,

$$\lim_{n\to\infty} E(\overline{A}_n f, B_n, F_n) = E(f, B, F). \tag{11.55}$$

(b) If conditions 1), 3), 6), 7), 9), 11) are satisfied and there exists $g_n \in B_n$ *such that*

$$\varphi_{F_n}(\overline{A}_n f - g_n) = E(\overline{A}_n f, B_n, F_n), \quad n = 1, 2, \dots,$$

then there exists $g_0 \in B$ *such that*

$$\varphi_F(f - g_0) = E(f, B, F),$$

and for an appropriate sequence $\{n_k\}_{k=1}^{\infty}$,

$$\lim_{k\to\infty} \varphi_{F_m}(A_{n_k m} g_{n_k} - \overline{A}_m g_0) = 0, \quad m = 1, 2, \ldots. \tag{11.56}$$

(c) If conditions 1), 3), 6), 7), 9), 11) are satisfied, then for any $f \in K \subseteq F$,

$$\lim_{n\to\infty} E(A_n f, B_n, F_n) = E(f, B, F). \tag{11.57}$$

(d) If either conditions 3), 4), 5), 7), 8), 11), 12) or conditions 3), 4), 5), 6), 7), 9), 11), 12) are satisfied, then

$$\lim_{n\to\infty} E(K_n, B_n, F_n) = E(K, B, F). \tag{11.58}$$

PROOF Relations (3.18) and (3.21) follow from Theorems 11.1 and 11.2, while (11.57) is an immediate consequence of (11.55) for $f \in F$.

Next, arguing similarly to the proof of Theorem 11.2(a), we obtain that there exist a sequence $\{n_k\}_{k=1}^{\infty}$ and $g_0 \in B$ such that (11.56) holds, and by (11.55),

$$E(f, B, F) = \lim_{k\to\infty} \varphi_{F_{n_k}}(\overline{A}_{n_k} f - g_{n_k}) \geq \varphi_F(f - g_0) \geq E(f, B, F).$$

Hence g_0 is an element of best approximation to f in F. Thus statement (b) follows. ∎

COROLLARY 11.2

Let φ_F *be an even functional, and let* B_n, $n = 1, 2, \ldots$, *and* B *be* N-*dimensional subspaces of* F *with bases* $\{u_{in}\}_{i=1}^{N}$ *and* $\{u_i\}_{i=1}^{N}$, *respectively. If condition 8) is satisfied, and*

$$\lim_{n\to\infty} \varphi_F(u_{in} - u_i) = 0, \quad i = 1, \ldots, N, \tag{11.59}$$

then

$$\lim_{n\to\infty} E(f, B_n, F) = E(f, B, F), \quad f \in F; \tag{11.60}$$

$$\lim_{n\to\infty} E(K, B_n, F) = E(K, B, F), \quad K \subseteq S_A(F). \tag{11.61}$$

PROOF In the given case,

$$H = F_n = F, \quad B'_n = B_n \cap S_{2A^*}(F), \quad K \subseteq S_{A^*}(F),$$

$$B' = B \cap S_{2A^*}(F), \quad A^* = \varphi_F(f),$$

and $A_n = \overline{A}_n$ is the unit operator, $n = 1, 2, \ldots$. Then the validity of conditions 1) and 4) follows from (11.59), conditions 3) and 7) follow from Property 11.3 of the errors of best approximation, while condition 11) is obviously satisfied. Thus relation (11.60) is a consequence of Corollary 11.1(a).

Now, let $K = K_n \subseteq S_A(F)$, $n = 1, 2, \ldots$. Since conditions 5) and 12) are obviously satisfied, (11.61) follows from Corollary 11.1(d). ▮

COROLLARY 11.3

Let $K^0 \subseteq F$, $K_n^0 \subseteq F_n$, $n = 1, 2, \ldots$, and for every $\varepsilon > 0$ there exist sets $K^\varepsilon \subseteq F$, $K_n^\varepsilon \subseteq F_n$, $n = 1, 2, \ldots$ such that

$$E(K^0, K^\varepsilon, F) < \varepsilon, \quad E(K_n^\varepsilon, K_n^0, F_n) < \varepsilon, \quad n = 1, 2, \ldots. \tag{11.62}$$

If for every $\varepsilon > 0$,

$$\liminf_{n \to \infty} E(K_n^\varepsilon, B_n, F_n) \geq E(K^\varepsilon, B, F), \tag{11.63}$$

then

$$\liminf_{n \to \infty} E(K_n^0, B_n, F_n) \geq E(K^0, B, F). \tag{11.64}$$

PROOF First we prove that the following "triangle inequality"

$$E(K, B, F) \leq E(K, K_1, F) + E(K_1, B, F) \tag{11.65}$$

holds true for any sets $K \subseteq F$ and $K_1 \subseteq F$. Indeed, for any $f \in K$ and every $\varepsilon > 0$, we choose $f_\varepsilon \in K_1$ and $g_\varepsilon \in B$ so that

$$\varphi_F(f - f_\varepsilon) < E(f, K_1, F) + \varepsilon, \quad \varphi_F(f_\varepsilon - g_\varepsilon) < E(f_\varepsilon, B, F) + \varepsilon. \tag{11.66}$$

It follows from (11.38) and (11.66) that

$$E(f, B, F) \leq \varphi_F(f - g_\varepsilon) \leq \varphi_F(f - f_\varepsilon) + \varphi_F(f_\varepsilon - g_\varepsilon)$$

$$< E(f, K_1, F) + E(f_\varepsilon, B, F) + 2\varepsilon$$

$$\leq E(K, K_1, F) + E(K_1, B, F) + 2\varepsilon.$$

This yields (11.65).

Next, using (11.63) and (11.65), we obtain

$$\liminf_{n\to\infty} E(K_n^0, B_n, F_n) \ge \liminf_{n\to\infty} E(K_n^\varepsilon, B_n, F_n) - \varepsilon$$

$$\ge E(K^\varepsilon, B, F) - \varepsilon \ge E(K^0, B, F) - 2\varepsilon.$$

Hence (11.64) follows. ∎

COROLLARY 11.4

Let $K \subseteq F$, $K^1 \subseteq F$, and $K_n \subseteq F_n$, $n = 1, 2, \ldots$, and let for every $f \in K$ and every $\varepsilon > 0$, there exist $n_0 = n_0(f, \varepsilon)$, $f_\varepsilon \in K^1$, and sequences $f_{\varepsilon,n} \in F_n$, $f_{\varepsilon,n}^* \in K_n$, $n = 1, 2, \ldots$, such that for all $n > n_0$,

$$\varphi_F(f - f_\varepsilon) < \varepsilon, \quad \varphi_F(f_{\varepsilon,n} - f_{\varepsilon,n}^*) < \varepsilon. \tag{11.67}$$

If

$$\liminf_{n\to\infty} E(f_{\varepsilon,n}, B_n, F_n) \ge E(f_\varepsilon, B, F), \tag{11.68}$$

then

$$\liminf_{n\to\infty} E(K_n, B_n, F_n) \ge E(K, B, F). \tag{11.69}$$

PROOF Using (11.40), (11.41), (11.67), (11.68), we obtain

$$\liminf_{n\to\infty} E(K_n, B_n, F_n) \ge \liminf_{n\to\infty} E(f_{\varepsilon,n}^*, B_n, F_n)$$

$$\ge \liminf_{n\to\infty} (E(f_{\varepsilon,n}, B_n, F_n) - E(f_{\varepsilon,n} - f_{\varepsilon,n}^*, B_n, F_n)$$

$$\ge \liminf_{n\to\infty} E(f_{\varepsilon,n}, B_n, F_n) - \varepsilon \ge E(f_\varepsilon, B, F) - \varepsilon$$

$$\ge E(f, B, F) - 2\varepsilon.$$

Thus (11.69) follows. ∎

REMARK 11.3 Relations (11.55) and (11.58) hold true even in the case when $E(f, B, F) = \infty$ or $E(K, B, F) = \infty$. ∎

REMARK 11.4 All limit relations in Section 11.2 can be obtained as corollaries of Theorems 11.1 and 11.2.

In particular, the limit theorem for spline approximation (11.16) can be easily derived from Corollary 11.2, since condition 8) is satisfied, by the finite dimensionality of S_{2N+1}^{2n+1} and $\mathcal{T}_N$, and (11.59) in this case was proved in [48] for the corresponding interpolated bases when $F = L_p[0, 2\pi]$. ∎

11.4 Limit Theorems in Rearrangement-Invariant Spaces

In this section, we apply the limit theorems for φ-spaces to certain functional spaces and classes.

11.4.1 Functional Spaces and Classes

In the capacity of H, we consider the vector space $M(\mathbb{R}^m)$ of all measurable on $\mathbb{R}^m$ functions.

In the capacity of F and F_n, we consider some functional spaces described below.

DEFINITION 11.3 *We say that a linear space $F = F(\mathbb{R}^m)$ of measurable functions f defined on $\mathbb{R}^m$ is a μ-quasi-normed rearrangement-invariant space (μ-QRIS) with the μ-quasi-norm $\|\cdot\|_F$ if the following conditions hold:*

(i) $\|f\|_F = 0 \Leftrightarrow f = 0$ *a.e. on* $\mathbb{R}^m$;

(ii) $\|\alpha f\|_F = |\alpha| \, \|f\|_F$, $\alpha \in \mathbb{R}^1$;

(iii) *there exists* $\mu \in (0, 1]$ *such that* $\|\cdot\|_F^\mu$ *is a subadditive functional, that is*

$$\|f + h\|_F^\mu \leq \|f\|_F^\mu + \|h\|_F^\mu \tag{11.70}$$

for all $f \in F$ *and* $h \in F$.

(iv) *if* $f \in F$ *and for all* $t \in [0, \infty)$, $h^*(t) \leq f^*(t)$, *then* $h \in F$ *and* $\|h\|_F \leq \|f\|_F$.

Throughout this section, $F = F(\mathbb{R}^m)$ denotes a μ-QRIS, and $S_A(F) = \{f \in F : \|f\|_F \leq A\}$.

We consider first some properties of these spaces.

(a) *For all $f \in F$ and $h \in F$,*

$$\|f + h\|_F \leq D(\|f\|_F + \|h\|_F), \tag{11.71}$$

where $0 < D \leq 2^{1/\mu - 1}$.

(11.71) easily follows from (11.70). This property shows that every μ-QRIS is a quasi-normed space. On the other hand, it is known [16, p. 20] that for every quasi-normed space F there exists an equivalent μ-quasi-norm $\|\cdot\|_F$.

(b) *F is a φ-space with $\varphi_F = \|\cdot\|_F^\mu$.*

(c) *A normed rearrangement-invariant space (NRIS) is a 1-QRIS.*

For example, $L_p(\mathbb{R}^m)$, $1 \leq p \leq \infty$, Orlitz, Marcinkiewicz, and Lorentz spaces are NRISes. For all this, see [15, 16, 42].

(d) *$L_p(\mathbb{R}^m)$, $0 < p < 1$, is a p-QRIS.*

(e) *F contains the characteristic functions of all measurable sets of finite measure.*

PROOF F contains at least one function $f(x) \neq 0$. Then (iv) implies that there exists $E_0 \subset \mathbb{R}^m$, $0 < |E_0| = \lambda < \infty$, such that $\chi_{E_0} \in F$. Hence $\chi_E \in F$ for all $E \subset \mathbb{R}^m$, $|E| \leq \lambda$. Since any $\Omega \subset \mathbb{R}^m$, $0 < |\Omega| < \infty$, can be represented as $\Omega = \bigcup_{i=1}^N E_i$, where $|E_i| \leq \lambda$, $1 \leq i \leq N$, we have

$$\|\chi_\Omega\|_F^\mu \leq \sum_{i=1}^N \|\chi_{E_i}\|_F^\mu < \infty.$$

Thus property (e) follows. ∎

DEFINITION 11.4 *A function $\psi_F : [0, \infty) \to [0, \infty)$, defined by*

$$\psi_F(t) = \|\chi_E\|_F, \quad E \subset \mathbb{R}^m, \quad |E| = t \geq 0,$$

is called the fundamental function of F.

(f) *ψ_F has the following properties:*

(f1) *ψ_F is a non-decreasing function on $[0, \infty)$;*

(f2) *$\psi_F(2t) \leq 2^{1/\mu}\psi_F(t)$, $t > 0$;*

(f3) $\psi_F(t) \le Ct^{1/\mu}$, $t > 1$.

PROOF Note first that (f1) follows from (iv). Next, for any $E \subset \mathbb{R}^m$, $|E| = 2t$, we have $E = E_1 \cup E_2$, where $E_1 \cap E_2 = \emptyset$ and $|E_i| = t$, $i = 1, 2$. Then

$$\psi_F^\mu(2t) = \|\chi_E\|_F^\mu \le \|\chi_{E_1}\|_F^\mu + \|\chi_{E_2}\|_F^\mu = 2\psi_F^\mu(t).$$

This yields (f2). Finally, for $t \in [2^{k-1}, 2^k)$, $k \ge 1$, it follows from (f1) and (f2) that

$$\psi_F(t) \le \psi_F(2^k) \le 2^{k/\mu}\psi_F(1) \le Ct^{1/\mu}.$$

Thus property (f) follows. ∎

(g) *Every μ-QRIS $F = F(\mathbb{R}^m)$ generates a μ-QRIS $F(\Omega)$ of measurable functions defined on a measurable set $\Omega \subseteq \mathbb{R}^m$ by the relation $F(\Omega) := \{h = f\chi_\Omega : f \in F(\mathbb{R}^m)\}$ with a μ-quasi-norm $\|h\|_{F(\Omega)} = \|f\chi_\Omega\|_F$.*

(h) *For every $\Omega \subset \mathbb{R}^m$, $|\Omega| < \infty$, and any $f \in L_\infty(\Omega)$,*

$$\|f\|_{F(\Omega)} \le \psi_F(|\Omega|)\|f\|_{L_\infty(\Omega)}.$$

This easily follows from condition (iv) and the definition of ψ_F and $F(\Omega)$.

DEFINITION 11.5 *A μ-QRIS F is called a space with the uniformly extended μ-quasi-norm if for any increasing sequence of measurable sets $\{\Omega_n\}_{n=1}^\infty$, $\bigcup_{n=1}^\infty \Omega_n = \mathbb{R}^m$, and for a function $f \in F$, the following relation holds:*

$$\lim_{n\to\infty} \|f\|_{F(\Omega_n)} = \|f\|_{F(\mathbb{R}^m)}. \tag{11.72}$$

DEFINITION 11.6 *A μ-QRIS F is called regular if for any decreasing sequence of measurable sets $\{\Omega_n\}_{n=1}^\infty$, $\bigcap_{n=1}^\infty \Omega_n = \emptyset$, and for a function $f \in F$,*

$$\lim_{n\to\infty} \|f\|_{F(\Omega_n)} = 0. \tag{11.73}$$

DEFINITION 11.7 *A μ-quasi-norm is said to have the Fatou property if $\lim_{n\to\infty} f_n(x) = f(x)$ a.e., where $\sup_n \|f_n\|_F < \infty$, implies $f \in F$ and $\|f\|_F \le \liminf_{n\to\infty} \|f_n\|_F$.*

We remark that for NRIS the definitions of a regular space and a norm with the Fatou property were given in [42].

(j) *If F is regular or has a μ-quasi-norm with the Fatou property, then F is the space with the uniformly extended μ-quasi-norm.*

PROOF Let $\{\Omega_n\}_{n=1}^{\infty}$ be an increasing sequence of sets, $\bigcup_{n=1}^{\infty}\Omega_n = \mathbb{R}^m$. If F is regular, then for $f \in F$, (11.73) implies

$$\liminf_{n\to\infty} \|f\|_{F(\Omega_n)} \geq \lim_{n\to\infty} \left(\|f\|^{\mu}_{F(\mathbb{R}^m)} - \|f\|^{\mu}_{F(\mathbb{R}^m-\Omega_n)}\right)^{1/\mu} = \|f\|_{F(\mathbb{R}^m)}. \tag{11.74}$$

If the μ-quasi-norm has the Fatou property, then from the relation $\lim_{n\to\infty} f(x)\chi_{\Omega_n}(x) = f(x)$ a.e. on $\mathbb{R}^m$ we deduce

$$\liminf_{n\to\infty} \|f\|_{F(\Omega_n)} \geq \|f\|_{F(\mathbb{R}^m)}. \tag{11.75}$$

Now property (j) follows from (11.74), (11.75), and (iv). ∎

(k) *If $f \in F$, then there exists $C > 0$ such that $|E_C| := |\{x \in \mathbb{R}^m\colon |f(x)| > C| < \infty$.*

PROOF For any $C > 0$,

$$\|f\|_{F(\mathbb{R}^m)} \geq \|f\|_{F(E_C)} \geq C\psi_F(|E_C|). \tag{11.76}$$

Further, by property (f1), either $\lim_{t\to\infty}\psi_F(t) = \infty$, or $\lim_{t\to\infty}\psi_F(t) = A < \infty$. If $\lim_{t\to\infty}\psi_F(t) = \infty$, then $|E_C| < \infty$, by (11.76). If $\lim_{t\to\infty}\psi_F(t) = A$, then choosing $C = 2A^{-1}\|f\|_{F(\mathbb{R}^m)}$, we obtain from (11.76) the estimate $\psi_F(|E_C|) \leq A/2$. Hence $|E_C| < \infty$. ∎

(l) *For a function $f \in F$ and any $\alpha > 0$, $\alpha \neq 1$,*

$$\|f(\alpha\cdot)\|_F \leq ([\alpha^{-1}] + 1)^{1/\mu}\|f\|_F. \tag{11.77}$$

PROOF Note first that $(f(\alpha\cdot))^*(t) = f^*(\alpha t)$. Since $f^*(\alpha_1 t) \geq f^*(\alpha_2 t)$ for $0 \leq \alpha_1 \leq \alpha_2$, we have that $\|f(\alpha\cdot)\|_F$ is a non-increasing function of $\alpha \geq 0$. Hence

$$\|f(\alpha\cdot)\|_F \leq \|f\|_F, \quad \alpha \geq 1. \tag{11.78}$$

Thus (11.77) is valid for $\alpha > 1$.

To prove (11.77) for $\alpha \in (0,1)$, we first need to show that for a function $f \in F$ and any $\gamma > 1$ there exists a simple function f_γ (that is a measurable nonnegative function on $\mathbb{R}^m$ with a countable set of values) such that for a.a. $x \in \mathbb{R}^m$,

$$\gamma^{-1} f_\gamma(x) \leq |f(x)| < f_\gamma(x). \tag{11.79}$$

Note that $f(x)$ is finite a.e. on $\mathbb{R}^m$, by (11.76). Let

$$E_k = \{x \in \mathbb{R}^m : \gamma^k \leq |f(x)| < \gamma^{k+1}\}, \quad -\infty < k < \infty.$$

Then $\{E_k\}_{k=-\infty}^{\infty}$ is the pairwise disjoint family of sets and $\left|\mathbb{R}^m - \bigcup_{k=-\infty}^{\infty} E_k\right| = 0$. Set

$$f_\gamma(x) = \begin{cases} \gamma^{k+1}, & x \in E_k,\ -\infty < k < \infty, \\ 0, & x \in \mathbb{R}^m - \bigcup_{k=-\infty}^{\infty} E_k. \end{cases}$$

Thus f_γ is a simple function satisfying (11.79).

We now return to the proof of property (l) in the case $\alpha \in (0,1)$. We first prove (11.77) for a simple function f_γ and $\alpha = p^{-1}$, $p = 2, 3, \ldots$. Let $\{E_{ks}\}$ be a partition of E_k with sets E_{ks} of finite measure, $-\infty < k < \infty$. Then

$$f_\gamma(x) = \sum_{k=-\infty}^{\infty} \sum_s \gamma^{k+1} \chi_{E_{ks}}(x).$$

Further, we divide each E_{ks} into p pairwise disjoint subsets $E_{ks\ell}$, $1 \leq \ell \leq p$, and set

$$y_\ell(x) = \sum_{k=-\infty}^{\infty} \sum_s \gamma^{k+1} \chi_{E_{ks\ell}}(x).$$

Then $f_\gamma(x) = \sum_{\ell=1}^{p} y_\ell(x)$ and $y_\ell^*(t/p) = f_\gamma^*(t)$. Hence

$$\|f_\gamma(p^{-1}\cdot)\|_F \leq \left(\sum_{\ell=1}^{p} \|y_\ell(p^{-1}\cdot)\|_F^\mu \right)^{1/\mu} = p^{1/\mu} \|f_\gamma\|_F. \tag{11.80}$$

Next, taking into account the monotonicity of $\|f_\gamma(\alpha\cdot)\|_F$ and using (11.80), we obtain for $\alpha \in (0,1)$,

$$\|f_\gamma(\alpha\cdot)\|_F \leq ([\alpha^{-1} + 1)^{1/\mu} \|f_\gamma\|_F. \tag{11.81}$$

Finally, by (11.79) and (11.81),

$$\|f(\alpha\cdot)\|_F \leq \|f_\gamma(\alpha\cdot)\|_F \leq ([\alpha^{-1}] + 1)^{1/\mu} \|f_\gamma\|_F \leq \gamma([\alpha^{-1}] + 1)^{1/\mu} \|f\|_F.$$

Now, letting $\gamma \to 1^+$, we obtain (11.77) in the case $\alpha \in (0,1)$. Thus we have established (11.77) for all $\alpha > 0$, $\alpha \neq 1$. ∎

(m) *Every μ-QRIS F generates a μ-QRIS F_a of all measurable functions defined on $\mathbb{R}^m$ and $2a$-periodic in each variable such that $f\chi_{Q_a} \in F(Q_a)$, with the μ-quasi-norm $\|f\|_{F_a} = \|f\|_{F(Q_a)}$.*

For example, if $F = L_p(\mathbb{R}^m)$, $0 < p \leq \infty$, then $F_a = L_{p,a}$.

It is clear that F_a is isomorphic and "isometric" (with respect to the corresponding μ-quasi-norms) to $F(Q_a)$. That is why in further considerations, we shall replace F_a with $F(Q_a)$.

In the capacity of K and K_n, we consider the following classes of periodic and nonperiodic convolutions with the kernels $\tilde{Q} \in L_{1,a}$ and $Q \in L_1(\mathbb{R}^m)$.

$$W_p(Q) := \left\{ f : f(x) = (Q \star \varphi)(x) \right.$$

$$\left. := \int_{\mathbb{R}^m} Q(y)\varphi(x-y)\,dy, \|\varphi\|_{L_p(\mathbb{R}^m)} \leq 1 \right\},$$

$$\tilde{W}_p(\tilde{Q}) := \left\{ f : f(x) = \int_{Q_a} \tilde{Q}(y)\varphi(x-y)\,dy, \|\varphi\|_{L_{p,a}} \leq 1 \right\},$$

$$\tilde{W}_{p,a}(Q) := \left\{ f : f(x) = (Q \star \varphi)(x), \|\varphi\|_{L_{p,a}} \leq 1 \right\}.$$

Here, $1 \leq p \leq \infty$.

It is known [49] that if $Q \in L_1(\mathbb{R}^m)$, then the series

$$\tilde{Q}_{(a)}(x) := \sum_{K \in \mathbb{Z}^m} Q(x + 2ak) \tag{11.82}$$

converges in $L_{1,a}$. A connection between two periodic classes is given in the following lemma.

LEMMA 11.1

For $Q \in L_1(\mathbb{R}^m)$,

$$\tilde{W}_{p,a}(Q) = \tilde{W}_p(\tilde{Q}_{(a)}). \tag{11.83}$$

PROOF For every $\varphi \in L_{p,a}$, the sequence

$$H_N(x) = \sum_{K \in \mathbb{Z}^m \cap Q} \int_{Q_a} Q(y + 2ak)\varphi(x-y)\,dy - \int_{Q_a} \tilde{Q}_{(a)}(y)\varphi(x-y)\,dy$$

converges to 0 in $L_{p,a}$ for a.a. $x \in Q_a$. This implies

$$\int_{\mathbb{R}^m} Q(y)\varphi(x-y)\,dy = \int_{Q_a} \tilde{Q}_{(a)}(y)\varphi(x-y)\,dy \tag{11.84}$$

for a.a. $x \in Q_a$. Now, (11.83) follows from (11.84). ∎

11.4.2 Properties of Entire Functions of Exponential Type

We need some properties of entire functions from $B_{\sigma V}$ with the power growth on $\mathbb{R}^m$.

LEMMA 11.2
(See [38]) *Let $E \subset \mathbb{R}^m$ be a measurable set such that there exist numbers $L > 0$ and $\delta > 0$ satisfying the condition*

$$\inf_{y \in \mathbb{R}^m} |\{(Q_L + y) \cap E\}| \geq \delta.$$

If $g \in B_{\sigma V}$, then

$$\|g\|_{L_\infty(\mathbb{R}^m)} \leq C\|g\|_{L_\infty(E)}, \tag{11.85}$$

where C depends only on L and δ.

LEMMA 11.3
(See [33]) *If $g \in B_{\sigma V}$ satisfies the condition $|g(x)| \leq A(1 + |x|)^N$, $x \in E$, where $|\mathbb{R}^m - E| < \infty$, then $g \in M_{AC,N}$. Here, C is independent of g.*

LEMMA 11.4
If $f \in M_{A,N}$, $g \in B_{\sigma V}$, and $f - g \in F$, then $g \in M_{A_1,N}$, where A_1 depends on f, g, and A.

PROOF By property (k) of a μ-QRIS, there exists $C > 0$ such that $|E_C| = |\{x \in \mathbb{R}^m \colon |f(x) - g(x)| > C\}| < \infty$. Then for every $x \in E = \mathbb{R}^m - E_C$, we have

$$|g(x)| \leq (C + A)(1 + |x|)^N.$$

Now, using Lemma 11.3, we conclude that $g \in M_{A_1,N}$. ∎

The next result shows that a set of functions from $B_{\sigma V} \cap M_{A,N}$ is compact in a certain sense.

LEMMA 11.5
(See [23]) *Let $g_n \in B_{\sigma V} \cap M_{A,N}$, $n = 1, 2, \ldots$, be a sequence of entire functions of exponential type, where A and N are independent of n. Then*

there exist a subsequence $\{g_{n_s}\}_{s=1}^{\infty}$ *and* $g \in B_{\sigma V}$ *such that*

$$\lim_{s \to \infty} f_{n_s}(x) = f(x)$$

uniformly on every compactum of $\mathbb{R}^m$.

A version of the Nikolskii inequality for entire functions of exponential type from a μ-QRIS F is presented below.

LEMMA 11.6
If $g \in B_{\sigma V} \cap F$, *then*

$$\|g\|_{L_\infty(\mathbb{R}^m)} \leq C\|g\|_{F(\mathbb{R}^m)}, \tag{11.86}$$

where C *is independent of* g.

PROOF First we note that for a set

$$E_D = \{x \in \mathbb{R}^m : |g(x)| > D\|g\|_{F(\mathbb{R}^m)}\},$$

the following estimate holds:

$$\psi_F(|E_D|) \leq D^{-1} \tag{11.87}$$

for all $D > 0$.
Next, choosing

$$D = \begin{cases} 1, & \text{if } \lim_{t\to\infty} \psi_F(t) = \infty, \\ 2/A, & \text{if } \lim_{t\to\infty} \psi_F(t) = A, \end{cases}$$

we obtain from (11.87) $|E_D| \leq C_0 < \infty$, where C_0 depends only on F. Then using Lemma 11.3, we deduce

$$\|g\|_{L_\infty(\mathbb{R}^m)} \leq C\|g\|_{L_\infty(\mathbb{R}^m - E)} \leq CD\|g\|_{F(\mathbb{R}^m)}.$$

Hence (11.86) follows. ∎

The following three lemmas show that some "periodizations" of entire functions of exponential type coincide with trigonometric polynomials.

LEMMA 11.7
(See [23]) $B_V \cap L_{\infty,a} = \mathcal{T}_{a,V}$.

LEMMA 11.8
If $g \in B_V \cap L_1(\mathbb{R}^m)$, then the series $\sum_{k\in\mathbb{Z}^m} g(x+2ak)$, $a>0$, converges in $L_\infty(Q_a)$ to some trigonometric polynomial from $\mathcal{T}_{a,V}$.

PROOF We first need a Sobolev-type inequality

$$\|g\|_{L_\infty(Q)} \leq \sum_{k=0}^{m} (2a)^{k-m} \sum_{1\leq s_1<\cdots<s_k\leq m} \int_Q \left| \frac{\partial^k g(y)}{\partial y_{s_1}\cdots \partial y_{s_k}} \right| dy, \qquad (11.88)$$

where Q is a cube of the form $Q = Q_a + x$, $x \in \mathbb{R}^m$, $a > 0$, and $g \in W_1^m(Q)$. This estimate can be easily proved by induction in m.

Next, the Bernstein inequality in $L_1(\mathbb{R}^m)$ for multivariate entire functions of exponential type [51, p. 217] implies that for all $1 \leq s_1 < \cdots < s_k \leq m$, $\frac{\partial^k g}{\partial y_{s_1}\cdots\partial y_{s_k}} \in L_1(\mathbb{R}^m)$. Then the uniform convergence of the series on Q_a follows from (11.88).

Moreover, (11.88) implies that

$$\left\| \sum_{k\in Q_N\cap\mathbb{Z}^m} g(\cdot + 2ak) \right\|_{L_\infty(\mathbb{R}^m)}$$

$$\leq C \sum_{k=0}^{m} \sum_{1\leq s_1<\cdots<s_k\leq m} \int_{\mathbb{R}^m} \left| \frac{\partial^k g(y)}{\partial y_{s_1}\cdots \partial y_{s_k}} \right| dy.$$

Now, applying Lemma 11.5 to $g_N = \sum_{k\in Q_N\cap\mathbb{Z}^m} g(\cdot + 2ak)$, we obtain that the series $\sum_{k\in\mathbb{Z}^m} g(\cdot + 2ak)$ converges in $L_\infty(Q_a)$ to a function $g_0 \in B_V \cap L_{\infty,a}$. Finally, using Lemma 11.7, we conclude that $g_0 \in \mathcal{T}_{a,V}$. ∎

LEMMA 11.9
Let for $N = 1, 2, \ldots$,

$$(A_n^{(p)} g)(x) := (A_{N,m}^{(p)} g)(x) := (2N+1)^{-m(p-1)/p} \sum_{k\in Q_N\cap\mathbb{Z}^m} g(x+2ak), \qquad (11.89)$$

be a sequence of linear operators on $L_p(\mathbb{R}^m)$, $1 < p \leq \infty$. Then for any sequence $g_N \in B_V \cap L_p(\mathbb{R}^m)$, $N = 1, 2, \ldots$, satisfying $\sup_N \|g_N\|_{L_p(\mathbb{R}^m)} <$

∞, *there exist a subsequence* $\{N_s\}_{s=1}^{\infty}$ *and* $g_0 \in \mathcal{T}_{a,V}$ *such that for all* $L > 0$,

$$\lim_{s\to\infty} \|A_{N_s}^{(p)} g_{N_s} - g_0\|_{L_p(Q_L)} = 0. \tag{11.90}$$

PROOF Using Minkowski's and Hölder's inequalities, we obtain for all $f \in L_p(\mathbb{R}^m)$, $1 \le p < \infty$, and $N = 1, 2, \ldots$,

$$\|A_N^{(p)} f\|_{L_p(Q_a)} \le \left\| \sum_{k \in Q_N \cap \mathbb{Z}^m} |f(\cdot + 2ak)| \right\|_{L_p(Q_a)} \le \|f\|_{L_p(\mathbb{R}^m)}. \tag{11.91}$$

For $p = \infty$, (11.91) trivially holds.

Setting $h_N = A_N^{(p)} g_N$, we have $h_N \in B_V$ and $h_N(x + 2a\ell) = (A_N^{(p)} g_N(\cdot + 2a\ell))(x)$ for all $\ell \in \mathbb{Z}^m$, $N = 1, 2, \ldots$. It follows from (11.91) that

$$\sup_{\ell \in \mathbb{Z}^m} \sup_N \|h_N(\cdot + 2a\ell)\|_{L_p(Q_a)} \le C < \infty.$$

Hence for all $\ell \in \mathbb{Z}^m$ and $N = 1, 2, \ldots$,

$$|\{x \in Q_a + 2a\ell : |h_N(x)| \le C(|Q_a|/2)^{-1/p}\}| \ge |Q_a|/2.$$

Now, applying Lemma 11.2 for $L = 2a$, $\delta = |Q_a|/2$, to h_N, we obtain

$$\sup_N \|h_N\|_{L_\infty(\mathbb{R}^m)} < \infty.$$

Hence, by Lemma 11.5, there exist a subsequence $\{h_{N_s}\}_{s=1}^{\infty}$ (without loss of generality, we assume that it coincides with $\{h_N\}_{N=1}^{\infty}$) and $g_0 \in B_V \cap L_\infty(\mathbb{R}^m)$ such that $\lim_{N\to\infty} \|h_N - g_0\|_{L_\infty(Q_L)} = 0$ for every $L > 0$. This yields (11.90) for $p \in [1, \infty]$.

It remains to prove that for $p \in (1, \infty]$, $g_0 \in \mathcal{T}_{a,V}$. Let $e_i = (0, \ldots, 1, \ldots, 0)$ be the i-th vector of the natural basis in $\mathbb{R}^m$, $1 \le i \le m$. Then for all $\ell \in \mathbb{Z}^m$ and $p \in (1, \infty]$,

$$\max_{1\le i\le m} \|h_N(\cdot + 2a(e_i + \ell)) - h_N(\cdot + 2a\ell)\|_{L_p(Q_a)}$$

$$\le \max_{1\le i\le m} \Bigg((2N+1)^{-m(p-1)/p}$$

$$\times \left\| \sum_{k \in Q_N \cap \mathbb{Z}^{m-1}} g_N(\cdot + 2a(k + \ell + (N+1)e_i)) \right\|_{L_p(Q_a)}$$

$$+ (2N+1)^{-m(p-1)/p} \left\| \sum_{k \in Q_N \cap \mathbb{Z}^{m-1}} g_N(\cdot + 2a(k + \ell - Ne_i)) \right\|_{L_p(Q_a)} \Bigg)$$

$$= \max_{1 \le i \le m} (I_1(i) + I_2(i)). \tag{11.92}$$

To estimate $I_1(i)$ in (11.92), we use the representation

$$I_1(i) = (2N+1)^{-(p-1)/p} \Bigg(\int_{-a}^{a} \Bigg(\int_{Q_a^{m-1}} |A^{(p)}_{N,m-1} g_N(\cdot + 2a(\ell + (N+1)e_i))|^p$$

$$\times dx_1 \cdots dx_{i-1} dx_{i+1} \cdots dx_m \Bigg) dx_i \Bigg)^{1/p}. \tag{11.93}$$

Next, applying (11.91) to $A^{(p)}_{N,m-1}$, we obtain from (11.93)

$$\max_{1 \le i \le m} I_1(i) \le CN^{-(p-1)/p}. \tag{11.94}$$

Similarly,

$$\max_{1 \le i \le m} I_2(i) \le CN^{-(p-1)/p}. \tag{11.95}$$

Now, combining (11.92) with (11.94) and (11.95), we obtain

$$\max_{1 \le i \le m} \|h_N(\cdot + 2a(e_i + \ell) - h_N(\cdot + 2a\ell)\|_{L_p(Q_a)} \le CN^{-(p-1)/p}. \tag{11.96}$$

Letting $N \to \infty$ in (11.96), we have that $g_0(x + 2a(e_i + \ell)) = g_0(x + 2a\ell)$, $x \in Q_a$, $1 \le i \le m$. Since ℓ is an arbitrary vector from $\mathbb{Z}^m$, g_0 is $2a$-periodic in each variable. Finally, Lemma 11.7 implies $g_0 \in \mathcal{T}_{a,V}$. ∎

11.4.3 Limit Theorems in a μ-QRIS

Let $H = M(\mathbb{R}^m)$ be a set of all measurable functions on $\mathbb{R}^m$ and $F = F(\mathbb{R}^m)$ a μ-QRIS. Property (b) of a μ-QRIS shows that F is a φ-space with $\varphi_F = \|\cdot\|_F^\mu$.

We consider some special cases (we call them situations) of the limit theorems in φ-spaces given in Section 11.3.

Throughout this subsection, we shall use the constant $\lambda = 2^{1/\mu}$.

Situation 1. Let $K \subseteq M_{A,N}$; $B' = B_{\sigma V} \cap M_{A_1,N_1}$, where $A_1 \geq AC_1$, $N_1 \geq N$ (or $K \subseteq S_A(F)$, $B' = S_{\lambda A}(F)$); $F_n = F(\Omega_n)$, where $\{\Omega_n\}_{n=1}^{\infty}$ is an increasing sequence of bounded measurable sets in $\mathbb{R}^m$, $\bigcup_{n=1}^{\infty} \Omega_n = \mathbb{R}^m$; A_{nm} be the operator of the restriction of functions from $F(\Omega_n)$ to Ω_m, $1 \leq m \leq n-1$; $\overline{A}_n$ the operator of the restriction of functions from $M(\mathbb{R}^m)$ to Ω_n, $n = 1, 2, \ldots$.

In the given situation, conditions 2) and 6b) are satisfied due to the monotonicity of the μ-quasi-norm in F (condition (iv) of a μ-QRIS), condition 6a) is obvious, and condition 3) for $K \subset M_{A,N}$ and $B' = M_{A_1,N_1}$, follows from Lemma 11.4, while for $K \subseteq S_A(F)$ and $B' = B \cap S_{\lambda A}(F)$, it is an immediate consequence of Property 11.3 of φ-spaces. Conditions 10) and 11) are equivalent to the following condition:

Condition 1_1) For every $f \in F$,

$$\lim_{L \to \infty} \|f\|_{F(Q_L)} = \|f\|_{F(\mathbb{R}^m)}.$$

Property (j) of μ-QRIS shows that this condition is satisfied, in particular, for a regular μ-QRIS or a μ-QRIS that has the Fatou property.

Thus Theorems 11.1 and 11.2 and Corollary 11.1 yield the following corollary.

COROLLARY 11.5

In situation 1 condition 1) implies (11.42); conditions 4) and 5) imply (11.43); conditions 7), 9), and 1_1) imply (11.49); conditions 7), 9), 12), and 1_1) imply (11.50); while conditions 1), 7), 9), 1_1) imply (11.55); and conditions 4), 5), 7), 9), 12), 1_1) imply (11.58).

Situation 2. Let $B \subset M(\mathbb{R}^m)$; $K \subseteq S_A(F)$; $B' = B \cap S_{\lambda A}(F)$; $B_1 = B_2 = \cdots = B_n = \cdots = B_0 \subset F(Q_a)$; $F_1 = \cdots = F_n = \cdots = F(Q_a)$; $K_1 = \cdots = K_n = \cdots = K_0 \subseteq F(Q_a)$; and $A_n : F(\mathbb{R}^m) \to F(Q_a)$, $n = 1, 2, \ldots$, be a sequence of linear operators. Here, $A > 0$ and $a > 0$ are some fixed numbers.

In this situation, condition 3) is satisfied due to Property 11.3 of φ-spaces, since $\varphi_F = \|\cdot\|_F^{\mu}$ is an even functional.

Furthermore, let us consider the following conditions of "periodization".

Condition 2_1) For every $g_n \in B \cap S_{\lambda A}(F)$, $n = 1, 2, \ldots$, there exist a sequence $\{n_s\}_{s=1}^{\infty}$ and $g_0 \in B_0$ such that

$$\lim_{s \to \infty} \|A_{n_s} g_{n_s} - g_0\|_{F(Q_a)} = 0.$$

Condition 2_2) For every $g \in K_0$ there exist $\varphi_n \in K$, $n = 1, 2, \ldots$, such that

$$\lim_{n \to \infty} \|A_n \varphi_n - g\|_{F(Q_a)} = 0.$$

We first show that condition 2_1) implies conditions 1) and 4). It suffices to prove that $2_1) \Rightarrow 4)$, since condition 1) is an immediate consequence of condition 4).

For an arbitrary sequence $\{n_k\}_{k=1}^{\infty}$ and $\varepsilon > 0$, we choose functions $g_{n_k} \in B \cap S_{\lambda A}(F)$ satisfying the inequalities

$$\sup_{g \in B'} E(A_{n_k} g, B_0, F(Q_a)) < E(A_{n_k} g_{n_k}, B_0, F(Q_a)) + \varepsilon, \quad k = 1, 2, \ldots .$$

By condition 2_1), there exist a subsequence of $\{g_{n_k}\}_{k=1}^{\infty}$ (without loss of generality, we can assume that the subsequence coincides with $\{g_{n_k}\}_{k=1}^{\infty}$) and $g_0 \in B_0$ such that

$$0 = \lim_{k \to \infty} \|A_{n_k} g_{n_k} - g_0\|_{F(Q_a)} \geq \liminf_{k \to \infty} E(A_{n_k} g_{n_k}, B_0, F(Q_a))$$

$$\geq \liminf_{k \to \infty} \sup_{g \in B'} E(A_{n_k} g, B_0, F(Q_a)) - \varepsilon.$$

Hence for any sequence $\{n_k\}_{k=1}^{\infty}$,

$$\liminf_{k \to \infty} \sup_{g \in B'} E(A_{n_k} g, B_0, F(Q_a)) = 0. \tag{11.97}$$

Now condition 4) follows from (11.97) and the following elementary property of limits: if $a_n \geq 0$, $n = 1, 2, \ldots$ and for any sequence $\{n_k\}_{k=1}^{\infty}$, $\liminf_{k \to \infty} a_{n_k} = 0$, then $\lim_{n \to \infty} a_n = 0$.

Since condition 2_2) is a reformulation of condition 5) (for $D_n g = \varphi_n$, $n = 1, 2, \ldots$), by Theorem 11.1, we obtain

COROLLARY 11.6

In situation 2, conditions 2) and 2_1) imply (11.42), while conditions 2), 2_1), and 2_2) imply (11.43).

Now we consider a special case of situation 2.

Situation 3. Let $B = B_{\sigma V}$; $K \subset S_A(F)$; $F = L_p(\mathbb{R}^m)$; $B' = B_{\sigma V} \cap S_{\lambda A}(F)$; $B_1 = \cdots = B_n = \cdots = \mathcal{T}_{a,V}$; $F_1 = \cdots = F_n = \cdots = L_p(Q_a)$, $1 < p \leq \infty$; $K_1 = \cdots = K_n = \cdots = K_0 \subset L_p(Q_a)$; $A_n := A_n^{(p)}$, where $A_n^{(p)}$ is defined by (11.89), $n = 1, 2, \ldots$.

In this situation, condition 2_1) is satisfied by Lemma 11.9. Next, estimate (11.91) implies the relation

$$\sup_n \|A_n^{(p)} f\|_{L_p(Q_a)} \leq \|f\|_{L_p(\mathbb{R}^m)}.$$

Hence condition 2) is satisfied as well. Thus Corollary 11.6 yields

COROLLARY 11.7

In situation 3, relation (11.42) is valid, while condition 2_2) implies (11.43).

Next, we present two simplified versions of situation 2.

Situation 4. Let $K \subseteq F$, $B' = B \cap F$, $B_1 = \cdots = B_n = \cdots = B_0 \subset F(Q_a)$; $F_1 = \cdots = F_n = \cdots = F(Q_a)$; $K_1 = \cdots K_n = \cdots = K_0 \subset F(Q_a)$; and $A_1 = \cdots = A_n = \cdots = A_0$, where $\sup_{f \in F} \|A_0 f\|_{F(Q_a)} / \|f\|_F \leq 1$.

We consider the following conditions:

Condition 4_1) For every $g_0 \in B \cap F$, $A_0 g \in B_0$.

Condition 4_2) For every $g \in K_0$ there exists $\varphi \in K$ such that $A_0 \varphi = g$.

Then conditions 2) and 3) are trivially satisfied, while conditions 1) and 4) follow from condition 4_1), and condition 5) is a consequence of condition 4_2).

Thus Theorem 11.1 yields

COROLLARY 11.8

In situation 4, condition 4_1) implies the relation

$$E(A_0 f, B_0, F(Q_a)) \leq E(f, B, F), \quad f \in K, \tag{11.98}$$

while conditions 4_1) and 4_2) imply the inequality

$$E(K_0, B_0, F(Q_a)) \leq E(K, B, F). \tag{11.99}$$

Situation 5. Let $F := L_1(\mathbb{R}^m)$; $K \subseteq L_1(\mathbb{R}^m)$; $B := B_{\sigma V}$, $B' := B_{\sigma V} \cap L_1(\mathbb{R}^m)$; $B_1 = \cdots = B_n = \cdots = \mathcal{T}_{a,\sigma V}$; $F_1 = \cdots = F_n = \cdots = L_1(Q_a)$; $K_1 = \cdots = K_n = \cdots = K_0 \subset L_1(Q_a)$; and $A_1 f = \cdots = A_n f = \cdots = \tilde{f}_{(a)}$, $f \in L_1(\mathbb{R}^m)$, where $\tilde{f}_{(a)}$ is the operator, defined by (11.82).

This situation is a special case of situation 4 for $F = L_1(\mathbb{R}^m)$, $B = B_{\sigma V}$, $B_0 = \mathcal{T}_{a,\sigma V}$, and $A_0 f = \tilde{f}_{(a)}$.

It is easy to verify that $\sup_{f \in L_1(\mathbb{R}^m)} \|\tilde{f}_{(a)}\|_{L_1(\mathbb{R}^m)} / \|f\|_{L_1(\mathbb{R}^m)} \leq 1$. Next, by Lemma 11.8, condition 4_1) is satisfied. Thus the following corollary holds.

COROLLARY 11.9

In situation 5, inequality (11.98) holds, while condition 4_2) implies (11.99).

Situation 6. Let $B_n \subseteq F(Q_{a_n})$; $F_n = F(Q_{a_n})$; $K_n \subseteq S_A(F(Q_{a_n}))$; $\overline{A}_n$ be the operator of the restriction of functions from $M(\mathbb{R}^m)$ to Q_{a_n}; A_{nm} the operator of the restriction of functions from $F(Q_{a_n})$ to Q_{a_m}, $m = 1, \ldots, n-1$, $n = 1, 2, \ldots$, where $\{a_n\}_{n=1}^{\infty}$ is an increasing sequence of positive numbers, $\lim_{n \to \infty} a_n = \infty$. Let $K \subseteq F$ and F_1 be a μ-QRIS satisfying the condition of the continuous embedding $F_1(Q_{a_n}) \to F(Q_{a_n})$ (that is, $F_1(Q_{a_n}) \subseteq F(Q_{a_n})$ and for every $f \in F_1(Q_{a_n})$, $\|f\|_{F(Q_{a_n})} \leq C(n)\|f\|_{F_1(Q_{a_n})}$), $n = 1, 2, \ldots$.

In this situation, condition 6) is satisfied. Let us consider the following conditions.

Condition 6_1) $B_n \subseteq B$, $n = 1, 2, \ldots$.

Condition 6_2) the condition of compactness: if $h_n \in \overline{A}_n(B) \cap S_C(F_1)$, $n = 1, 2, \ldots$, where C is independent of n, then there exist a sequence $\{n_s\}_{s=1}^{\infty}$ and $h_0 \in B$ such that

$$\lim_{s \to \infty} \|h_{n_s} - h_0\|_{F_1(Q_{a_m})} = 0, \quad m = 1, 2, \ldots .$$

Condition 6_3) the Nikolskii-type condition: for every $T \in B_n$, $n = 1, 2, \ldots$,

$$\|T\|_{F_1(Q_a)} \leq C\|T\|_{F(Q_a)},$$

where C is independent of a and T.

Condition 6_4) the extension condition: for every $f \in K$ there exist $n_0 = n_0(f)$ and $f_n \in K_n$ such that $f_n(x) = f(x)$, $x \in Q_{a_n}$, $n > n_0$.

Furthermore, condition 12) follows from condition 6_4) since $\overline{A}_n f \in K_n$, $n = n_0 + 1, n_0 + 2, \ldots$, for every $f \in K$. Next, $B'_n = B_n \cap S_{\lambda A}(F(Q_{a_n}))$, $n \geq n_0$, for every $f \in K$. Then taking into account Property 11.3 of φ-spaces, we conclude that condition 7) is satisfied, since $\|\overline{A}_n f\|_{F(Q_{a_n})} \leq A$, $n = 1, 2, \ldots$. Further, for every $t_n \in B_n$, $n = 1, 2, \ldots,$, conditions 6_1) and 6_3) imply that $h_n = \overline{A}_n t_n \in \overline{A}_n(B) \cap S_{\lambda AC}(F_1(Q_{a_n}))$.

Then using condition 6_2) and the embedding $F_1(Q_{a_m}) \to F(Q_{a_m})$, $m = 1, 2, \ldots$, we obtain for an appropriate subsequence $\{n_s\}_{s=1}^{\infty}$ and some $h_0 \in B$

$$\lim_{s\to\infty} \|A_{n_s m} t_{n_s} - \overline{A}_m h_0\|_F \le \lim_{s\to\infty} \|h_{n_s} - h_0\|_{F(Q_{a_m})}$$

$$\le C \lim_{s\to\infty} \|h_{n_s} - h_0\|_{F_1(Q_{a_m})} = 0, \quad m = 1, 2, \ldots .$$

Hence condition 9) is satisfied. Thus Theorem 11.2 implies

COROLLARY 11.10

In situation 6, conditions 10), 6_1), 6_2), and 6_3) imply (11.49), while conditions 10), 6_1), 6_2), 6_3), and 6_4) imply (11.50).

REMARK 11.5 Some special cases of Corollaries 11.6 and 11.10 for $B = B_n = \{0\}$, F is a space of measurable functions with bounded supports, and A_n is the operator of $2a_n$-periodic extension, were obtained in [13]. ∎

The following situation is a special case of situation 6.

Situation 7. Let F be a NRIS, $B_n = \mathcal{T}_{a_n,\sigma V}$; $F_n = F(Q_{a_n})$; $K_n \subseteq S_A(F(Q_{a_n}))$; A_{nm} and $\overline{A}_n$ be the operators from situation 6, $m = 1, \ldots, n-1$, $n = 1, 2, \ldots$; and $F_1 = L_\infty$.

In this situation, the embedding condition holds, due to property (h) of a μ-QRIS. Next, condition 6_1) is trivially satisfied, while condition 6_2) follows from Lemma 11.5.

It follows from the Nikolskii-type inequality ($T \in \mathcal{T}_{a_n,\sigma V}$)

$$\max_{Q_{a_n}} |T| \le C(\psi_F((\pi/\sigma)^m))^{-1} \|T\|_{F(Q_{a_n})}, \quad n = 1, 2, \ldots,$$

established by Berkolaiko and Ovchinnikov [5], that condition 6_3) is satisfied. Thus Corollary 11.10 yields

COROLLARY 11.11

In situation 7, condition 1_1) implies (11.49), while conditions 1_1) and 6_4) imply (11.50).

11.4.4 Limit Theorems for Polynomial Approximation in a μ-QRIS

11.4.4.1 Statement of Main Results

Following is a generalization of relation (11.9) to a μ-QRIS.

THEOREM 11.3
Let F be a μ-QRIS with the uniformly extended μ-quasi-norm, and let $\{\lambda_n\}_{n=1}^{\infty}$ be a sequence of real numbers satisfying the conditions:

(a) $0 \leq \lambda_n \leq n$;

(b) $\lim_{n\to\infty} \lambda_n/n = 0$;

(c') *there exist* $\delta \in (1/3, 1)$ *such that* $\liminf_{n\to\infty} \lambda_n n^{-\delta} > 0$.

Then for any $f \in F$,

$$\lim_{n\to\infty} E\left(f, \mathcal{P}_{n,m}, F\left(\frac{n-\lambda_n}{\sigma}V^*\right)\right) = E(f, B_{\sigma V}, F(\mathbb{R}^m)). \qquad (11.100)$$

Moreover, if $f \in M_{A,N}$ and $E(f, B_{\sigma V}, F(\mathbb{R}^m)) < \infty$, then (11.100) holds.

It turns out that for a regular μ-QRIS, (11.100) holds true with $\lambda_n = 0$.

THEOREM 11.4
If F is a regular μ-QRIS, then for any $f \in F$,

$$\lim_{n\to\infty} E(f, \mathcal{P}_{n,m}, F((n/\sigma)V^*)) = E(f, B_{\sigma V}, F(\mathbb{R}^m)). \qquad (11.101)$$

REMARK 11.6 Note that, by property (j) of a μ-QRIS, every regular μ-QRIS is a μ-QRIS with the uniformly extended μ-quasi-norm. An example of $F = L_\infty$ shows that the converse statement is false. Moreover, there is an example of a function $f_0 \in C(\mathbb{R}^m)$ for which (1.101) does not hold (see [32]). ∎

To prove these theorems, we need to study some properties of $E(f, B_{\sigma V}, F(\mathbb{R}^m))$ and to obtain certain inequalities for multivariate polynomials.

11.4.4.2 Properties of $E(f, B_{\sigma V}, F(\mathbb{R}^m))$

LEMMA 11.10
Let F be a μ-QRIS with the uniformly extended μ-quasi-norm. If $E(f, B_{\sigma V}, F(\mathbb{R}^m)) < \infty$, then there exists $g_0 \in B_{\sigma V}$ such that

$$E(f, B_{\sigma V}, F(\mathbb{R}^m)) = \|f - g_0\|_{F(\mathbb{R}^m)}. \tag{11.102}$$

PROOF Without loss of generality, we assume that $f \in F$. Note first that by Property 11.3 of φ-spaces,

$$E(f, B_{\sigma V}, F(\mathbb{R}^m)) = \inf_{g \in B_{\sigma V} \cap S_{\lambda A}(F)} \|f - g\|_F,$$

where $\lambda = 2^{1/\mu}$ and $A = \|f\|_F$. Then there exists a sequence $g_k \in B_{\sigma V} \cap S_{\lambda A}(F)$, $k = 1, 2, \ldots$, such that

$$E(f, B_{\sigma V}, F(\mathbb{R}^m)) = \lim_{k \to \infty} \|f - g_k\|_{F(\mathbb{R}^m)}. \tag{11.103}$$

Taking into account Lemma 11.6, we obtain

$$\sup_k \|g_k\|_{L_\infty(\mathbb{R}^m)} \le C\lambda \|f\|_F. \tag{11.104}$$

It follows from (11.104) and Lemma 11.5 that there exist a subsequence $\{g_{k_s}\}_{s=1}^\infty$ and $g_0 \in B_{\sigma V}$ such that for every $L > 0$,

$$\lim_{s \to \infty} \|g_0 - g_{k_s}\|_{L_\infty(Q_L)} = 0. \tag{11.105}$$

Next, by (11.103), (11.105), and property (h) of a μ-QRIS, we obtain

$$\begin{aligned} E(f, B_{\sigma V}, F(\mathbb{R}^m)) &\ge \limsup_{s \to \infty} \|f - g_{k_s}\|_{F(Q_L)} \\ &\ge \limsup_{s \to \infty} \left(\|f - g_0\|_{F(Q_L)}^\mu - \|g_0 - g_{k_s}\|_{F(Q_L)}^\mu\right)^{1/\mu} \\ &\ge \limsup_{s \to \infty} \left(\|f - g_0\|_{F(Q_L)}^\mu - (\psi_F(|Q_L|)\|g_0 - g_{k_s}\|_{L_\infty(Q_L)})^\mu\right)^{1/\mu} \\ &= \|f - g_0\|_{F(Q_L)}. \end{aligned} \tag{11.106}$$

Letting now $L \to \infty$, we have from (11.106)

$$\|f - g_0\|_{F(\mathbb{R}^m)} \geq E(f, B_{\sigma V}, F(\mathbb{R}^m)) \geq \lim_{L\to\infty} \|f - g_0\|_{F(Q_L)}$$

$$= \|f - g_0\|_{F(\mathbb{R}^m)}.$$

This yields (11.102). ∎

LEMMA 11.11

Let F be a μ-QRIS with the uniformly extended μ-quasi-norm. If $E(f, B_{\sigma_0 V}, F(\mathbb{R}^m)) < \infty$, $\sigma_0 > 0$, then for $\sigma \geq \sigma_0$,

$$\lim_{\tau\to\sigma+0} E(f, B_{\tau V}, F(\mathbb{R}^m)) = E(f, B_{\sigma V}, F(\mathbb{R}^m)). \tag{11.107}$$

PROOF Note first that there exists $g \in B_{\sigma_0 V}$ such that $f - g \in F$ and $E(f - g, B_{\sigma V}, F(\mathbb{R}^m)) = E(f, B_{\sigma V}, F(\mathbb{R}^m))$ for all $\sigma \geq \sigma_0$. Hence without loss of generality, we can assume $f \in F$.

Next, by Lemma 11.10, for a sequence of positive numbers $\{\varepsilon_k\}_{k=1}^{\infty}$, $\lim_{k\to\infty} \varepsilon_k = 0$, there exists a sequence of functions $g_k \in B_{(\sigma+\varepsilon_k)V}$, satisfying the relations

$$\|f - g_k\|_F = E(f, B_{(\sigma+\varepsilon_k)V}, F(\mathbb{R}^m)), \quad k = 1, 2, \ldots. \tag{11.108}$$

Since $g_k \in S_{\lambda A}(F)$, $k = 1, 2, \ldots$, where $\lambda = 2^{1/\mu}$ and $A = \|f\|_F$, the following estimate holds, by Lemma 11.6, $\sup_k \|g_k\|_{L_\infty(\mathbb{R}^m)} \leq C\lambda\|f\|_F$.

Hence, by Lemma 11.5, we obtain that there exist a subsequence $\{g_{k_s}\}_{s=1}^{\infty}$ and g_0 such that

$$\lim_{s\to\infty} g_{k_s} = g_0 \tag{11.109}$$

uniformly on any cube Q_L, $L > 0$. It is clear that the limit function g_0 belongs to $B_{\sigma V}$. Then for every $L > 0$, (11.108), (11.109), and property (h) of a μ-QRIS imply

$$E^\mu(f, B_{\sigma V}, F(\mathbb{R}^m)) \leq \lim_{L\to\infty} \|f - g_0\|^\mu_{F(Q_L)}$$

$$\leq \lim_{L\to\infty} \left(\liminf_{s\to\infty} \|f - g_{k_s}\|^\mu_{F(Q_L)} + \lim_{s\to\infty} \|g_0 - g_{k_s}\|^\mu_{F(Q_L)} \right)$$

$$\leq \liminf_{k\to\infty} E^\mu(f, B_{(\sigma+\varepsilon_k)V}, F(\mathbb{R}^m)). \tag{11.110}$$

Now (11.107) follows from (11.110) and the inequality

$$\limsup_{k\to\infty} E(f, B_{(\sigma+\varepsilon_k)V}, F(\mathbb{R}^m)) \le E(f, B_{\sigma V}, F(\mathbb{R}^m)).$$

∎

LEMMA 11.12

Let F be a regular μ-QRIS. If $E(f, B_{\sigma_0 V}, F(\mathbb{R}^m)) < \infty$, $\sigma_0 > 0$, then for $\sigma > \sigma_0$,

$$\lim_{\tau\to\sigma-0} E(f, B_{\tau V}, F(\mathbb{R}^m)) = E(f, B_{\sigma V}, F(\mathbb{R}^m)). \tag{11.111}$$

PROOF Without loss of generality, we can assume $f \in F$. By Lemma 11.10, for every $\sigma > \sigma_0$, we can choose $g_0 \in B_{\sigma V}$ such that

$$\|f - g_0\|_F = E(f, B_{\sigma V}, F(\mathbb{R}^m)). \tag{11.112}$$

Given $\delta \in (0, \sigma/2)$ we consider a function $g_\delta(x) = g_0\left(\frac{\sigma-\delta}{\sigma}x\right) \in B_{(\sigma-\delta)V}$. Then (11.112) implies

$$\begin{aligned} E^\mu(f, B_{(\sigma-\delta)V}, F(\mathbb{R}^m)) &\le \|f - g_\delta\|^\mu_{F(\mathbb{R}^m)} \le \|f - g_0\|^\mu_{F(\mathbb{R}^m)} \\ &+ \|g_0 - g_\delta\|^\mu_{F(\mathbb{R}^m)} \le E^\mu(f, B_{\sigma V}, F(\mathbb{R}^m)) + \|g_0 - g_\delta\|^\mu_{F(Q_L)} \\ &+ \|g_0 - g_\delta\|^\mu_{F(\mathbb{R}^m - Q_L)}, \end{aligned} \tag{11.113}$$

where $L > 0$.

Next, by property (h) of a μ-QRIS,

$$\begin{aligned} \|g_0 - g_\delta\|_{F(Q_L)} &\le \psi_F(|Q_L|)\|g_0 - g_\delta\|_{L_\infty(Q_L)} \\ &\le \delta\sigma^{-1}\sqrt{4m}\, L\psi_F(|Q_L|) \max_{x\in Q_L} |\nabla g_0(x)|, \end{aligned} \tag{11.114}$$

where $\nabla f(x) := (\partial f(x)/\partial x_1, \ldots, \partial f(x)/\partial x_m)$ is the gradient of f.

Furthermore, using properties (g) and (l) of a μ-QRIS, we obtain

$$\|g_0 - g_\delta\|^\mu_{F(\mathbb{R}^m - Q_L)} \le \|g_0\|^\mu_{F(\mathbb{R}^m \setminus Q_L)} + \left\|g_0\left(\left(\frac{\sigma-\delta}{\sigma}\right)\cdot\right)\chi_{\mathbb{R}^m - Q_L}\right\|^\mu_{F(\mathbb{R}^m)}$$

$$\leq \|g_0\|^{\mu}_{F(\mathbb{R}^m - Q_L)} + 2\|g_0\|^{\mu}_{F(\mathbb{R}^m - Q_{L/2})}.$$

Now taking into account regularity of F, for every $\varepsilon > 0$, we can choose L large enough so that $\|g_0 - g_\delta\|^{\mu}_{F(\mathbb{R}^m - Q_L)} < \varepsilon^{\mu}$. Finally, it follows from (11.114) that $\|g_0 - g_\delta\|^{\mu}_{F(Q_L)} < \varepsilon^{\mu}$ for δ small enough. Thus (11.113) implies that for every $\varepsilon > 0$ there exists $\delta > 0$ such that

$$E^{\mu}(f, B_{(\sigma-\delta)V}, F(\mathbb{R}^m)) \leq E^{\mu}(f, B_{\sigma V}, F(\mathbb{R}^m)) + 2\varepsilon^{\mu}.$$

This yields (11.111). ∎

REMARK 11.7 Lemmas 11.10, 11.11, and 11.12 in the special case when $F = L_p(\mathbb{R}^1)$, $1 \leq p \leq \infty$, were proved in [1]. ∎

11.4.4.3 Properties of Multivariate Polynomials

LEMMA 11.13

(See [24]) *Let $\{P_n\}_{n=0}^{\infty}$ be a sequence of polynomials, $P_n \in \mathcal{P}_{n,m}$, $n \geq 0$, satisfying the condition*

$$\sup_n \|P_n\|_{C((n/\sigma)V^*)} < \infty.$$

Then there exist a subsequence $\{n_s\}_{s=1}^{\infty}$ and a function $g_0 \in B_{\sigma V}$ such that

$$\lim_{n\to\infty} P_n(x) = g_0(x)$$

uniformly on a compact set in $\mathbb{R}^m$.

LEMMA 11.14

(See [36]) *If $P \in \mathcal{P}_{n,m}$, then for every $a > 0$,*

$$|P(0)| \leq \frac{C}{\psi_F((a/(2n))^m)} \|P\|_{F(Q_a)}, \tag{11.115}$$

where C is an absolute constant.

LEMMA 11.15
If $P \in \mathcal{P}_{n,m}$, then for every $\varepsilon > 0$,

$$\|P\|_{C((n/\sigma)V^*)} \leq C\|P\|_{F((1+\varepsilon)n\sigma^{-1}V^*)}, \tag{11.116}$$

where C is independent of n and P.

PROOF Let the point $x_0 \in (n/\sigma)V^*$ satisfy the relation $\max_{x \in (n/\sigma)V^*} |P(x)| = |P(x_0)|$. Then the cube $Q_{n\delta}(x_0) := Q_{n\delta} + x_0$, where $\delta = \varepsilon M/2\sqrt{m}$ and M is the width of $(\sigma^{-1})V^*$, is a subset of $(1+\varepsilon)(n/\sigma)V^*$. Next, using (11.115), we obtain

$$|P(x_0)| \leq \frac{C}{\psi_F((\delta/2)^m)}\|P\|_{F(Q_{n\delta}(x_0))} \leq C\|P\|_{F((1+\varepsilon)(n/\sigma)V^*)}.$$

Thus (11.116) follows. ∎

11.4.4.4 Proofs of the Theorems

PROOF **(of Theorem 11.3)** We first prove that for a function f from a μ-QRIS F with the uniformly extended μ-quasi-norm, the relation

$$E(f, B_{(\sigma+\varepsilon)V}, F(\mathbb{R}^m)) \leq \liminf_{n\to\infty} E(f, \mathcal{P}_{n,m}, F((n/\sigma)V^*) \tag{11.117}$$

holds true for every $\varepsilon > 0$.

To prove (11.117), we consider the special case of situation 1 when

$$\Omega_n = (n/\sigma)V^*, \ \sigma > 0; \ \ B_n = \mathcal{P}_{n,m}\big|_{\Omega_n}; \ \ B = B_{(\sigma+\varepsilon)V};$$

$$B' = B_{(\sigma+\varepsilon)V} \cap F(\mathbb{R}^m); \ f \in K = S_A(F), \ A = \|f\|_F;$$

$$B'_n = S_{\lambda A}(F(n/\sigma)V^*) \cap \mathcal{P}_{n,m}, \ \lambda = 2^{1/\mu}, \ n = 1, 2, \ldots .$$

Then condition 7) is valid, due to Property 11.3 of φ-spaces, while condition 9) holds, by Lemmas 11.13 and 11.15. Finally, condition 1_1) is satisfied, by (11.72). Thus (11.117) follows from Corollary 11.5.

Next, the relation

$$E(f, B_{(\sigma+2\varepsilon)V}, F(\mathbb{R}^m)) \leq \liminf_{n\to\infty} E\left(f, \mathcal{P}_{n,m}, F\left(\frac{n-\lambda_n}{\sigma}V^*\right)\right) \tag{11.118}$$

is an immediate consequence of (11.117).

Furthermore, by Lemma 11.11, $E(f, B_{\sigma V}, F(\mathbb{R}^m))$ is continuous on the right in $\sigma > 0$. Thus (11.117) and (11.118) yield

$$E(f, B_{\sigma V}, F(\mathbb{R}^m)) \leq \liminf_{n\to\infty} E\left(f, \mathcal{P}_{n,m}, F\left(\frac{n-\lambda_n}{\sigma}V^*\right)\right)$$

$$\leq \liminf_{n\to\infty} E(f, \mathcal{P}_{n,m}, F((n/\sigma)V^*)). \qquad (11.119)$$

To obtain a lower bound, we consider the following version of situation 1:

$$\Omega_n = \frac{n-\lambda_n}{\sigma}V^*,\ \sigma > 0;\quad B_n = \mathcal{P}_{n,m}\big|_{\Omega_n};\quad B = B_{\sigma V};$$

$$f \in K = S_A(F),\ A = \|f\|_F;\quad B' = B_{\sigma V} \cap S_{\lambda A}(F),\ \lambda = 2^{1/\mu}.$$

Then using Lemma 11.6, estimate (11.13), and properties (f3) and (h) of a μ-QRIS, we obtain that for every $g \in B' = B_{\sigma V} \cap S_{\lambda A}(F)$, the function g belongs to $B_{\sigma V} \cap L_\infty(\mathbb{R}^m)$ and

$$E\left(g, \mathcal{P}_{n,m}, F\left(\frac{n-\lambda_n}{\sigma}V^*\right)\right)$$

$$\leq \psi_F\left(\frac{(n-\lambda_n)^m}{\sigma^m}|V^*|\right) E\left(g, \mathcal{P}_{n,m}, C\left[\frac{n-\lambda_n}{\sigma}V^*\right]\right)$$

$$\leq C_4 n^{C_5} \exp\left(-C_3 n^{3\delta/2-1/2}\right), \quad n \geq 1, \qquad (11.120)$$

where $\delta \in (1/3, 1)$ and C_3, C_4, C_5 are positive constants independent of n.

Now, it follows from (11.120) that condition 1) is satisfied in this situation. Hence Corollary 11.5 implies the inequality

$$\limsup_{n\to\infty} E\left(f, \mathcal{P}_{n,m}, F\left(\frac{n-\lambda_n}{\sigma}V^*\right)\right) \leq E(f, B_{\sigma V}, F(\mathbb{R}^m)). \qquad (11.121)$$

Thus relations (11.119) and (11.121) yield (11.100) in the case when $f \in F$.

If $f \in M_{A,N}$ and $E(f, B_{\sigma V}, F(\mathbb{R}^m)) < \infty$, then there exists $g \in B_{\sigma V}$ such that $\|f - g\|_F < \infty$. It follows from Lemma 11.4 that $g \in M_{A_1,N}$.

Hence (11.120) is valid, by (11.13), that is,

$$\lim_{n\to\infty} E\left(g, \mathcal{P}_{n,m}, F\left(\frac{n-\lambda_n}{\sigma}V^*\right)\right) = 0. \tag{11.122}$$

Now (11.41) and (11.122) yield the following relations

$$\begin{aligned}
\lim_{n\to\infty} E^\mu\left(f, \mathcal{P}_{n,m}, F\left(\frac{n-\lambda_n}{\sigma}V^*\right)\right) &= \lim_{n\to\infty} E^\mu\left(f, \mathcal{P}_{n,m}, F\left(\frac{n-\lambda_n}{\sigma}V^*\right)\right) \\
- \lim_{n\to\infty} E^\mu\left(g, \mathcal{P}_{n,m}, F\left(\frac{n-\lambda_n}{\sigma}V^*\right)\right) & \\
&\leq \lim_{n\to\infty} E^\mu\left(f-g, \mathcal{P}_{n,m}, F\left(\frac{n-\lambda_n}{\sigma}V^*\right)\right) \\
&\leq \lim_{n\to\infty} E^\mu\left(f, \mathcal{P}_{n,m}, F\left(\frac{n-\lambda_n}{\sigma}V^*\right)\right) \\
&\quad + \lim_{n\to\infty} E^\mu\left(g, \mathcal{P}_{n,m}, F\left(\frac{n-\lambda_n}{\sigma}V^*\right)\right) \\
&= \lim_{n\to\infty} E^\mu\left(f, \mathcal{P}_{n,m}, F\left(\frac{n-\lambda_n}{\sigma}V^*\right)\right).
\end{aligned} \tag{11.123}$$

Finally, using Theorem 11.3 for $f_1 = f - g \in F$ and taking into account (11.123), we obtain

$$\begin{aligned}
&\lim_{n\to\infty} E\left(f, \mathcal{P}_{n,m}, F\left(\frac{n-\lambda_n}{\sigma}V^*\right)\right) \\
&\quad = \lim_{n\to\infty} E\left(f_1, \mathcal{P}_{n,m}, F\left(\frac{n-n_\lambda}{\sigma}V^*\right)\right) \\
&\quad = E(f_1, B_{\sigma V}, F(\mathbb{R}^m)) = E(f, B_{\sigma V}, F(\mathbb{R}^m)).
\end{aligned}$$

This establishes (11.100) in the case when $f \in M_{A,N}$ and $E(f, B_{\sigma V}, F(\mathbb{R}^m)) < \infty$.

Thus Theorem 11.3 is proved. ∎

PROOF **(of Theorem 11.4)** Note first that (11.121) implies the relation

$$\limsup_{n\to\infty} E(f,\mathcal{P}_{n,m},F((n/\sigma)V^*)) \le E(f,B_{(\sigma-\varepsilon)V},F(\mathbb{R}^m)) \qquad (11.124)$$

for every $\varepsilon \in (0,\sigma)$. Next, F is a regular μ-QRIS. Consequently, by Lemma 11.12, $E(f,B_{\sigma V},F(\mathbb{R}^m))$ is continuous on the left in $\sigma > 0$ for every $f \in F$. Hence (11.124) holds for $\varepsilon = 0$. Now, (11.101) follows from (11.119) and (11.124). Thus Theorem 11.4 is established. ∎

11.4.4.5 Approximation by Trigonometric Polynomials

Limit relation (11.15) was proved in [33] by using a special case of situation 1 for normed spaces. This relation can be easily generalized to a μ-QRIS with the uniformly extended μ-quasi-norm as a special case of Corollary 11.5. In particular, (11.15) holds true for all $p \in (0,\infty]$.

11.5 Some Asymptotics for the Errors of Best Approximation on Convolution Classes

11.5.1 A Brief Review

Sharp constants of harmonic approximation on convolution classes $W_p(Q)$ and $\tilde{W}_p(\tilde{Q})$, defined in Section 11.4, play an important role in approximation theory.

In particular, Akhiezer and Krein [2], Favard [20] and Korneichuk [40] showed that

$$E(\tilde{W}_\infty(B_r),\mathcal{T}_n,C[0,2\pi]) = K_r(n+1)^{-r}, \qquad (11.125)$$

where r is a natural number, $B_r(x) = \sum_{k=1}^{\infty} k^{-r}\cos(kx-\pi r/2)$ the Bernoulli kernel and

$$K_r = (4/\pi)\sum_{k=0}^{\infty}(-1)^{k(r+1)}(2k+1)^{-r-1}$$

the Favard constant. A nonperiodic analogue of (11.125) was established by Akhiezer [1] and Krein [41]:

$$E(W_\infty(\phi_r),B_{(\sigma-0)},L_\infty(\mathbb{R}^1)) = K_r\sigma^{-r}, \qquad (11.126)$$

where $B_{(\sigma-0)} = \bigcup_{0<\tau<\sigma} B_\tau$ and $\phi_r(x) = \pi^{-1}\int_0^\infty t^{-r}p(t)\cos(tx-\pi r/2)$ is the nonperiodic Bernoulli kernel. Here, p is an infinitely differentiable even function on $\mathbb{R}^1$ satisfying the condition: $p(x) = 0$ on $[-\varepsilon/2, \varepsilon/2]$ and $p(x) = 1$ for $|x| \geq \varepsilon$, where ε is small enough.

It is known [1, 40] that $\tilde{W}_\infty(B_r)$ coincides with the class of all 2π-periodic functions f such that $\|f^{(r)}\|_{C[0,2\pi]} \leq 1$, and $W_\infty(\phi_r)$ coincides with the class of all r-differentiable functions f on $\mathbb{R}^1$ such that $\|f^{(r)}\|_{L_\infty(\mathbb{R}^1)} \leq 1$.

Generalizations of (11.125) and (11.126) to all $r > 0$ were given by Dzyadyk [17, 18] and Bushanskii [14], respectively.

Using the dual relations for the errors of best harmonic approximation, Korneichuk [40] and Nikolskii [45] showed that for every $\tilde{Q} \in L_{1,\pi}$ and $Q \in L_1(\mathbb{R}^1)$,

$$E(\tilde{W}_1(\tilde{Q}), \mathcal{T}_n, L_1[0, 2\pi]) = E(\tilde{Q}, \mathcal{T}_n, L_1[0, 2\pi]), \tag{11.127}$$

$$E(W_1(Q), B_\sigma, L_1(\mathbb{R}^1)) = E(Q, B_\sigma, L_1(\mathbb{R}^1)). \tag{11.128}$$

In particular, for a natural r,

$$E(\tilde{W}_1(B_r), \mathcal{T}_n, L_1[0, 2\pi]) = K_r(n+1)^{-r}, \tag{11.129}$$

$$E(W_1(\phi_r), B_\sigma, L_1(\mathbb{R}^1)) = K_r \sigma^{-r}. \tag{11.130}$$

It is clear that (11.125), (11.129) and (11.130) imply the relations

$$\begin{aligned} E(\tilde{W}_\infty(B_r), \mathcal{T}_n, C[0, 2\pi]) &= E(\tilde{W}_1(B_r), \mathcal{T}_n, L_1[0, 2\pi]) \\ &= E(W_1(\phi_r), B_{n+1}, L_1(\mathbb{R}^1)). \end{aligned} \tag{11.131}$$

A necessary and sufficient condition for

$$E(\tilde{W}_\infty(\tilde{Q}), \mathcal{T}_n, C[0, 2\pi]) = E(\tilde{W}_1(\tilde{Q}), \mathcal{T}_n, L_1[0, 2\pi])$$

to be true was found in [27]. Some applications of sharp constants of harmonic approximation on convolution classes in statistics were given in [19].

In the 1980s–1990s, the author [29, 31, 34] developed a "weighted" approach to relations like (11.126) and (11.130) and obtained generalizations of these equalities to some nonperiodic classes of multidimensional convolutions. However, multidimensional analogues of (11.125), (11.129), and (11.131) are unknown.

In the next section, we obtain an asymptotic L_p-version of the second relation in (11.131) for some multidimensional classes (Corollary 11.6), by using general limit theorems for multidimensional convolution classes in L_p (Theorem 11.5 and Corollary 11.12). Then we find an asymptotically sharp constant of best approximation by trigonometric polynomials in L_1 for a two-dimensional analogue of the class $\tilde{W}_1(B_r)$. We derive this result from a new limit theorem in L_1 for individual functions (Corollary 11.7) and the corresponding nonperiodic relation [29, 34].

11.5.2 A Limit Relation for Harmonic Approximations on Convolution Classes in L_p

THEOREM 11.5
Let $Q \in L_1(\mathbb{R}^m)$. Then for $p \in [1, \infty)$, the following relation holds

$$\lim_{n\to\infty} n^m E(\tilde{W}_p(\widetilde{(Q(n\cdot))}_{(\pi)}), \mathcal{T}_{nV}, L_p(Q_\pi)) = E(W_p(Q), B_V, L_p(\mathbb{R}^m)), \tag{11.132}$$

where $\tilde{Q}_{(a)}$ is defined by (11.82).

PROOF We first show that for every $g \in \tilde{W}_{p,\pi n}(Q)$, $1 < p \le \infty$, there exists a sequence $\varphi_N \in W_p(Q)$, $N = 3, 4\ldots$, such that

$$\lim_{N\to\infty} \|A_N^{(p)}\varphi_N - g\|_{L_p(Q_{\pi n})} = 0, \tag{11.133}$$

where $A_N^{(p)}$ is defined by (11.89).

Let $g = Q \star \varphi$, where $\|\varphi\|_{L_{p,\pi n}} \le 1$. Setting

$$\psi_N(x) = ((2N+1)n)^{-m/p}\varphi(x)\chi_{Q_{(2N+1)\pi n}}(x); \quad \varphi_N(x) = (Q \star \psi_N)(x),$$

we have

$$\|\psi_n\|_{L_p(\mathbb{R}^m)} = \|\varphi\|_{L_p,\pi n} \le 1.$$

Hence $\varphi_N \in W_p(Q)$, $N = 3, 4, \ldots$. Next, for a sequence $\{s_N\}_{N=3}^{\infty}$ of odd numbers satisfying the conditions $3 \le s_N \le N$, $\lim_{N\to\infty} s_N = \infty$, $\lim_{N\to\infty} s_N/N = 0$, we obtain

$$(A_N^{(p)}\varphi_N)(x) = \int_{Q_{\pi n s_N}} Q(y)(A_N^{(p)}\psi_N)(x-y)\,dy$$

$$+ \left(A_N^{(p)} \int_{\mathbb{R}^m - Q_{\pi n s_N}} Q(y)\psi_N(\cdot - y)\,dy\right)(x)$$

$$= I_1(x) + I_2(x). \tag{11.134}$$

Note that the following estimate of I_2 follows from (11.91)

$$\|I_2\|_{L_p(Q_{\pi n})} \le \|Q\|_{L_1(\mathbb{R}^m - Q_{\pi n s_N})}. \tag{11.135}$$

Further, the following formula for $D_N(x) = (A_N^{(p)}\psi_N)(x)$ holds true

$$D_N(x - y - 2k\pi n) = \gamma(x - y - 2k\pi n)\varphi(x - y)(2N+1)^{-m}, \tag{11.136}$$

where $x, y \in Q_{\pi n}$, $k \in \mathbb{Z}^m \cap Q_{s_N - 1/2}$, and $\gamma(u)$ is the multiplicity of covering of a point u by the family of cubes $\{2n\pi\ell + Q_{(2N+1)\pi n}\}_{\ell \in Q_N \cap \mathbb{Z}^m}$. Then, by (11.136),

$$1 \ge \delta_k(x-y) := \gamma(x - y - 2k\pi n)(2N+1)^{-m} \ge \prod_{j=1}^{m}(2N - |k_j|)/(2N+1)$$

$$\ge (1 - (s_N + 1)/(2N+1))^m \ge 1 - m s_N/N. \tag{11.137}$$

Now using relations (11.134), (11.136), and (11.137), we obtain

$$|g(x) - A_N^{(p)}\varphi_N(x)|$$

$$\le \left| \int_{Q_{\pi n}} \sum_{k \in \mathbb{Z}^m \cap Q_{s_N - 1/2}} Q(y + 2k\pi n)(1 - \delta_k(x-y))\varphi(x-y)\,dy \right|$$

$$+ \int_{Q_{\pi n}} \sum_{k \in \mathbb{Z}^m - Q_{s_N - 1/2}} |Q(y + 2k\pi n)|\,|\varphi(x-y)|\,dy + |I_2(x)|$$

$$= J_1(x) + J_2(x) + |I_2(x)|. \tag{11.138}$$

Next, by estimate (11.137),

$$\|J_1\|_{L_p(Q_{\pi n})} \le \|Q\|_{L_1(\mathbb{R}^m)} \|\varphi\|_{L_{q,\pi n}} \max_{k \in \mathbb{Z}^m \cap Q_{s_N - 1/2}} \max_{x \in Q_{\pi n}} (1 - \delta_k(x))$$

$$\le C s_N/N \|Q\|_{L_1(\mathbb{R}^m)}. \tag{11.139}$$

Furthermore, it is easy to see that

$$\|J_2\|_{L_p(Q_{\pi n})} \leq \|Q\|_{L_1(\mathbb{R}^m - Q_{s_N \pi n})}. \tag{11.140}$$

Thus inequalities (11.135), (11.139), (11.140) yield (11.133) for $1 < p \leq \infty$.

Note that (11.133) holds also for $p = 1$. But in this case, $A_N^{(p)}$ can be replaced by an operator that does not depend on N. Indeed, setting $\psi = \varphi\chi_{\pi n}$ and $\varphi_0 = Q \star \psi$ for $g = Q \star \varphi$, we obtain $\|\psi\|_{L_1(\mathbb{R}^m)} = \|\varphi\|_{L_{1,\pi n}} \leq 1$. It is easy to verify that

$$\sum_{k \in \mathbb{Z}^m} \varphi_0(x + 2k\pi n) = g(x) \tag{11.141}$$

in $L_1(Q_{\pi n})$.

Let us consider a special case of situation 3 for $K = W_p(Q)$; $A = \|Q\|_{L_1(\mathbb{R}^m)}$; $a = \pi n$; $K_0 = \tilde{W}_{p,\pi n}(Q)$, $1 < p \leq \infty$. Then relation (11.133) shows that condition 2_2) is satisfied in this situation. Thus Corollary 11.7 yields the inequality

$$E(\tilde{W}_{p,\pi n}(Q), \mathcal{T}_{\pi n, V}, L_{p,\pi n}) \leq E(W_p(Q), B_V, L_p(\mathbb{R}^m)), \quad n = 1, 2, \ldots. \tag{11.142}$$

Then it follows from a special case of situation 5 for $K = W_1(Q)$; $a = \pi n$; $K_0 = \tilde{W}_{1,\pi n}(Q)$, that condition 4_2) is satisfied, by (11.141). Thus Corollary 11.9 implies (11.142) for $p = 1$.

Further, let us put $h^{(s)}(x) = h(x)\chi_{Q_{\pi s/2}}(x)$, $x \in \mathbb{R}^m$. Note that the set of functions $K^1 = \{Q^{(s)} \star \varphi^{(s)} : \|\varphi\|_{L_p(\mathbb{R}^m)} \leq 1,\ s = 1, 2, \ldots\}$ is dense in $W_p(Q)$ in the metric of $L_p(\mathbb{R}^m)$, where $1 \leq p < \infty$. Indeed, for every $f = Q \star \varphi \in W_p(Q)$,

$$\begin{aligned}\|Q \star \varphi - Q^{(s)} \star \varphi^{(s)}\|_{L_p(\mathbb{R}^m)} &\leq \|Q \star (\varphi - \varphi^{(s)})\|_{L_p(\mathbb{R}^m)} \\ &\quad + \|(Q - Q^{(s)}) \star \varphi^{(s)}\|_{L_p(\mathbb{R}^m)} \\ &\leq \|Q\|_{L_1(\mathbb{R}^m)}\|\varphi - \varphi^{(s)}\|_{L_p(\mathbb{R}^m)} + \|Q - Q^{(s)}\|_{L_1(\mathbb{R}^m)}. \end{aligned} \tag{11.143}$$

Since $L_p(\mathbb{R}^m)$ is a regular NRIS, (11.143) implies

$$\lim_{s \to \infty} \|Q \star \varphi - Q^{(s)} \star \varphi^{(s)}\|_{L_p(\mathbb{R}^m)} = 0. \tag{11.144}$$

Then for any $n \geq s$, by (11.84),

$$\|(Q - Q^{(s)}) \star (\widetilde{\varphi^{(s)}})_{(\pi n)}\|_{L_p,\pi n} \leq \|Q - Q^{(s)}\|_{L_1(\mathbb{R}^m)}. \tag{11.145}$$

Now for every $f = Q \star \varphi \in W_p(Q)$ and $n \geq n_0$, we define

$$f_\varepsilon = Q^{(n_0)} \star \varphi^{(n_0)}; \quad f_{\varepsilon,n} = Q^{(n_0)} \star (\widetilde{\varphi^{(n_0)}})_{(\pi n)}; \quad f^*_{\varepsilon,n} = Q \star (\widetilde{\varphi^{(n_0)}})_{(\pi n)}.$$

Next, by (11.144) and (11.145), we can choose $n_0 = n_0(f, \varepsilon)$ so that

$$\|f - f_\varepsilon\|_{L_p(\mathbb{R}^m)} < \varepsilon, \quad \|f_{\varepsilon,n} - f^*_{\varepsilon,n}\|_{L_{p,\pi n}} < \varepsilon \tag{11.146}$$

for $n > n_0$. Note that $f_{\varepsilon,n} = f_\varepsilon$ on $Q_{\pi n}$. Now we consider a special case of situation 7 for $F = L_p(\mathbb{R}^m)$, $1 \leq p < \infty$, $a_n = \pi n$, $n \geq n_0$. Then condition 1_1) is satisfied due to the regularity of $L_p(\mathbb{R}^m)$, and $\overline{A}_n f_{\varepsilon,n} = f_\varepsilon$. Hence Corollary 11.11 yields

$$\liminf_{n\to\infty} E(f_{\varepsilon,n}, \mathcal{T}_{\pi n,V}, L_{p,\pi n}) \geq E(f_\varepsilon, B_V, L_p(\mathbb{R}^m)). \tag{11.147}$$

Finally, inequalities (11.146) and (11.147) show that all conditions of Corollary 11.4 are satisfied for $F = L_p(\mathbb{R}^m)$; $K = W_p(Q)$; $K_n = \tilde{W}_{p,\pi n}(Q)$; $F_n = L_{p,\pi n}$, $1 \leq p < \infty$ and, by (11.69), we obtain

$$\liminf_{n\to\infty} E(\tilde{W}_{p,\pi n}(Q), \mathcal{T}_{\pi n,V}, L_{p,\pi n}) \geq E(W_p(Q), B_V, L_p(\mathbb{R}^m)). \tag{11.148}$$

Inequalities (11.142) and (11.148) show that (11.132) immediately follows from the relation $(1 \leq p \leq \infty)$

$$E(\tilde{W}_{p,\pi n}(Q), \mathcal{T}_{\pi n,V}, L_{p,\pi n}) = n^m E(\tilde{W}_p((\widetilde{Q(n\cdot)}_{(\pi)}), \mathcal{T}_{nV}, L_p(Q_\pi)). \tag{11.149}$$

To prove (11.149), we first note that, by Lemma 11.1,

$$\tilde{W}_p((\widetilde{Q(n\cdot)})_{(\pi)}) = \tilde{W}_p(Q(n\cdot)). \tag{11.150}$$

Next, the following relation is valid:

$$W := \{f(n\cdot) : f \in \tilde{W}_{p,\pi n}(Q)\} = n^{m(p-1)/p}\tilde{W}_p(Q(n\cdot)). \tag{11.151}$$

Indeed, if $f \in \tilde{W}_{p,\pi n}(Q)$, then $f(nx) = n^m \int_{\mathbb{R}^m} Q(ny)\varphi(nx - ny)\,dy$, where $\|\varphi(n\cdot)\|_{L_p(Q_\pi)} \le n^{-m/p}$. Hence $W \subseteq n^{m(p-1)/p}\tilde{W}_p(Q(n\cdot))$. The inclusion $n^{m(p-1)/p}\tilde{W}_p(Q(n\cdot)) \subseteq W$ can be proved similarly. Hence (11.151).

Now making the substitution $x = ny$ in the left-hand side of (11.149), we conclude that (11.149) follows from (11.150) and (11.151).

Thus (11.132) is established. ∎

REMARK 11.8 In the case $p = \infty$, (11.142) implies the inequality

$$\limsup_{n\to\infty} n^m E(\tilde{W}_\infty((\tilde{Q}(n\cdot))_{(\pi)}, \mathcal{T}_{nV}, L_\infty(Q_\pi)) \le E(W_\infty(Q), B_V, L_\infty(\mathbb{R}^m)). \tag{11.152}$$

Equality in (11.152) holds for $m = 1$ when Q is a Benoulli-type kernel. Thus the conjecture that (11.132) is valid also for $p = \infty$ seems plausible. ∎

The following corollary is a consequence of Theorem 11.5 and the Poisson summation formula [49].

COROLLARY 11.12

If $Q \in L_1(\mathbb{R}^m)$ satisfies the estimate

$$\max\{|Q(x)|, |\hat{Q}(x)|\} \le C(1 + |x|)^{-m-\delta}$$

for all $x \in \mathbb{R}^m$ and some $\delta > 0$, then for $p \in [1, \infty)$,

$$\lim_{n\to\infty} E(\tilde{W}_p(Q_n), \mathcal{T}_{nV}, L_p(Q_\pi)) = E(W_q(Q), B_V, L_q(\mathbb{R}^m)), \tag{11.153}$$

where

$$Q_n(x) = \sum_{\ell \in \mathbb{Z}^m} \hat{Q}(\ell/n)\exp(-(\ell, x)).$$

In particular, for $\lambda > m$, we consider the kernels

$$H_\lambda(x) = \check{\gamma}(x); \quad h_\lambda(x) = \sum_{\ell \in \mathbb{Z}^m} |\ell|^{-\lambda}\exp(-i(\ell, x)),$$

where $\gamma(x) = \mu_\varepsilon(y)|y|^{-\lambda}$, $\varepsilon \in (0,1)$, and μ_ε is an infinitely differentiable function on $\mathbb{R}^m$ such that $\mu_\varepsilon(y) = 0$ if $|y| \le \varepsilon/2$, and $\mu_\varepsilon(y) = 1$ if $|y| \ge \varepsilon$. Since

$$\sum_{\ell \in \mathbb{Z}^m} \gamma(\ell/n)\exp(-(\ell, x)) = n^\lambda h_\lambda(x),$$

Corollary 11.12 yields the following statement.

COROLLARY 11.13
If $1 \le p < \infty$ and $\lambda > m$, then

$$\lim_{n\to\infty} n^{\lambda} E(\tilde{W}_p(h_\lambda), \mathcal{T}_{nV}, L_p(Q_\pi)) = E(W_p(H_\lambda), B_V, L_p(\mathbb{R}^m)). \quad (11.154)$$

REMARK 11.9 If $\lambda = 2N$, $N \ge 1$, is an even number, then class $\tilde{W}_p((2\pi)^m h_\lambda)$ coincides with the class of all 2π-periodic functions f in m variables satisfying the condition $\|\Delta^N f\|_{L_p(Q_\pi)} \le 1$, where $\Delta = \sum_{i=1}^m \partial^2/\partial x_i^2$ is the Laplace operator (see [4, 12]). Some upper estimates of $\sup_n n^{\lambda} E(\tilde{W}_p(h_\lambda), \mathcal{T}_{nv}, L_p(Q_\pi))$, $\lambda = 2N$, were obtained for $m = 2$ in [12], while for $m > 2$ in [4]. ▮

11.5.3 A Limit Relation for Individual Functions in $L_1(\mathbb{R}^m)$

COROLLARY 11.14
For $Q \in L_1(\mathbb{R}^m)$,

$$\lim_{n\to\infty} n^m E(\widetilde{(Q(n\cdot))}_{(\pi)}, \mathcal{T}_{nV}, L_1(Q_\pi)) = E(Q, B_V, L_1(\mathbb{R}^m)), \quad (11.155)$$

where $\tilde{Q}_{(a)}$ is defined by (11.82).

PROOF Note first that the following multidimensional analogues of Nikolskii's dual relations (11.127) and (11.128) hold:

$$E(\tilde{W}_1(\tilde{Q}), \mathcal{T}_{nV}, L_1(Q_\pi)) = E(\tilde{Q}, \mathcal{T}_{nV}, L_1(Q_\pi)), \tilde{Q} \in L_{1,\pi}, \quad (11.156)$$

$$E(W_1(Q), B_V, L_1(\mathbb{R}^m)) = E(Q, B_V, L_1(\mathbb{R}^m), Q \in L_1(\mathbb{R}^m). \quad (11.157)$$

The proofs of (11.156) and (11.157) are similar to those of (11.127) and (11.128) in [40, 45].

Now (11.155) follows from Theorem 11.5 for $p = 1$ and relations (11.156) and (11.157). ▮

COROLLARY 11.15
For $m = 2$ and $\lambda > 2$,

$$\lim_{n\to\infty} n^{\lambda} E(\tilde{W}_1(h_\lambda), \mathcal{T}_{nS_1^{(2)}}, L_1(Q_\pi)) = \frac{2^{-\lambda+3}\Gamma(\lambda)}{\Gamma^2(\lambda/2)} \sum_{s=0}^{\infty}(2s+1)^{-\lambda-1}, \tag{11.158}$$

where $S_r^{(m)} = \{x \in \mathbb{R}^m : |x| \le r\}$ is the ball in $\mathbb{R}^m$ of radius r.

PROOF Note first that the equality

$$E(H_\lambda, B_{S_1^{(2)}}, L_1(\mathbb{R}^2)) = \frac{2^{-\lambda+3}}{\Gamma^2(\lambda/2)} \sum_{s=0}^{\infty}(2s+1)^{-\lambda-1} \tag{11.159}$$

was established in [29, 34]. Then it follows from (11.155) that for $m = 2$,

$$\lim_{n\to\infty} n^{\lambda} E(h_\lambda, \mathcal{T}_{n,S_1^{(2)}}, L_1(Q_\pi)) = E(H_\lambda, B_{S_1^{(2)}}, L_1(\mathbb{R}^2)). \tag{11.160}$$

Next, (11.156) implies

$$E(\tilde{W}_1(h_\lambda), \mathcal{T}_{nS_1^{(2)}}, L_1(Q_\pi)) = E(h_\lambda, \mathcal{T}_{nS_1^{(2)}}, L_1(Q_\pi)). \tag{11.161}$$

Thus (11.159), (11.160), and (11.161) yield (11.158). ∎

REMARK 11.10 Theorem 11.5 and Corollaries 11.12, 11.13, and 11.14 are new even when $m = 1$. ∎

References

1. N.I. Akhiezer, *Lectures on Approximation Theory*, 2nd edition, Nauka, Moscow, 1965.

2. N.I. Akhiezer and M.G. Krein, On the best approximation of periodic functions, *Dokl. Akad. Nauk SSSR*, **15** (1937) 107–112.

3. G.A. Anastassiou and M.I. Ganzburg, L-positive approximation, *Revue Roumaine de Math. Pures et Appl.*, **43** (1998), 475–494.

4. E.S. Belinsky, An application of the Fourier transform to summation of Fourier series, *Sibirskii Mat. J.*, **43** (1977), 497–511.

5. M.Z. Berkolaiko and V.I. Ovchinnikov, Inequalities for entire functions of exponential type in symmetric spaces, *Proc. Steklov Inst. Math.*, **161** (1984), 1–17.

6. S.N. Bernstein, On the best approximation of $|x|^p$ by means of polynomials of extremely high degree, *Izv. Akad. Nauk SSSR*, **2** (1938), 169–180.

7. S.N. Bernstein, Generalization of a result of S.M. Nikolskii, *Dokl. Akad. Nauk SSSR*, **53** (1946), 587–589.

8. S.N. Bernstein, On the best approximation of continuous functions on the whole real axis by means of functions of given degree, *Dokl. Akad. Nauk SSSR*, **54** (1946), 103–108, 479–482.

9. S.N. Bernstein, New derivation and generalization of certain formulas of best approximation, *Dokl. Akad. Nauk SSSR*, **54** (1946), 667–668.

10. S.N. Bernstein, Limit laws of the theory of best approximation, *Dokl. Akad. Nauk SSSR*, **58** (1947), 525–528.

11. S.N. Bernstein, On the best approximation of $|x - c|^p$. *Collected Works*, vol. II, pp. 273–280, *Akad. Nauk SSSR*, Moscow, 1954.

12. J.S. Bugrov, Approximation by trigonometric polynomials of classes of functions defined by the polyharmonic operator, *Yspehi Mat. Nauk*, **13** (1958), 149–156.

13. V.I. Burenkov and M.L. Goldman, On intercommunication of norms of operators in periodic and nonperiodic functional spaces, *Proc. Steklov Inst. Math.*, **161**, (1984).

14. A.V. Bushanskii, An analogue of the Favard problem on the whole real axis. In N.P. Korneichuk, S.B. Stechkin, and S.A. Telyakovsky, eds., *Theory of Approximation of Functions* (Proc. Internat. Conf., Kiev, 1983), pp. 82–85, Nauka, Moscow, 1987.

15. K.M. Chong and N.M. Rice, *Equimeasurable Rearrangements of Functions*, Queen's University, Kingston, Ontario, Canada, 1971.

16. R.A. DeVore and G.G. Lorentz, *Constructive Approximation*, Springer-Verlag, 1993.

17. V.K. Dzyadyk, On the best approximation on a class of periodic functions having a bounded s-th derivative $(0 < s < 1)$, *Izv. Akad. Nauk SSSR*, **17** (1953), 135–162.

18. V.K. Dzyadyk, On best approximation on classes of periodic functions defined by the kernels that are integrals of absolutely monotone

functions, *Izv. Akad. Nauk SSSR*, **23** (1959), 933–950.

19. S. Efromovich and M. Ganzburg, Best Fourier approximation and application in efficient blurred signal reconstruction, *J. Comp. Anal. and Appl.*, **1** (1999), 43–62.

20. J. Favard, Sur les meilleures procédés d'approximation de certaines classes des fonctions par des polynômes trigonométriques, *Bull. Sci. Math.*, **61** (1937), 209–224, 243–256.

21. M.I. Ganzburg, A multidimensional limit theorem in the theory of best polynomial approximation, *Soviet Math. Dokl.*, **19** (1979), 1027–1030.

22. M.I. Ganzburg, The theorems of Jackson and Bernstein in $\mathbb{R}^m$, *Russian Math. Surveys*, **34** (1979), 221–222.

23. M.I. Ganzburg, Multidimensional Jackson theorems, *Siberian Math. J.*, **22** (1981), 223–231.

24. M.I. Ganzburg, Multidimensional limit theorems of the theory of best polynomial approximation, *Siberian Math. J.*, **23** (1983), 316–331.

25. M.I. Ganzburg, Multidimensional Bernstein-type inequalities, *Ukrainian Math. J.*, **34** (1983), 607–610.

26. M.I. Ganzburg, On criteria of best approximation of local integrable functions in $L(\mathbb{R})$. In V.P. Motornii, ed., *Studies in Current Problems of Summation and Approximation of Functions and their Applications*, pp. 11–16, Dniepropetrovsk State University, Ukraine, 1983.

27. M.I. Ganzburg, On the best harmonic approximation of convolutions in L_1 and L_∞, *Soviet Math. (Iz. VUZ)*, **29**(5) (1985), 22–27.

28. M.I. Ganzburg, A harmonic and polynomial approximation of functions of several variables. In N.P. Korneichuk, S.B. Stechkin, and S.A. Telyakovsky, eds., *Theory of Approximation of Functions* (Proc. Internat. Conf., Kiev, 1983), pp. 109–111, Nauka, Moscow, 1987.

29. M.I. Ganzburg, On sharp constants for best harmonic approximation of functions of several variables, *Soviet Math. Dokl.*, **37** (1988), 142–145.

30. M.I. Ganzburg, Best approximation of sums of elements and a theorem of Newman and Shapiro, *Ukrainian Math. J.*, **41** (1989), 1395–1401.

31. M.I. Ganzburg, A Markov-type theorem for approximation in in-

tegral metric with power weight, *Soviet Math.* (*Iz. VUZ*), **34**(1) (1990), 9–18.

32. M.I. Ganzburg, Limit theorems for the best polynomial approximations in the L_∞ metric, *Ukrainian Math. J.*, **43**(3) (1991), 299–305.

33. M.I. Ganzburg, Limit theorems in approximation theory, *Anal. Math.*, **18** (1992), 37–57.

34. M.I. Ganzburg, Markov's type theorems for approximation of radial functions in the integral metric, *Russian Math.* (*Iz. VUZ*), **36**(2) (1992), 7–16.

35. M.I. Ganzburg, Limit relations in approximation theory and their applications. In C.K. Chui and L.L. Schumaker, eds., *Approximation Theory VIII*, vol. 1: *Approximation and Interpolation*, pp. 223–232, World Scientific, Singapore, 1995.

36. M.I. Ganzburg, Polynomial inequalities on measurable sets and their applications, submitted.

37. I.M. Gelfand and G.E. Shilov, *Generalized Functions*, vol. I, Academic Press, 1964.

38. V.E. Kacnel'son, Equivalent norms in spaces of entire functions of exponential type, *Math. USSR Sbornik*, **21**(1) (1973), 33–55.

39. N.P. Korneichuk, The best uniform approximation on certain classes of continuous functions, *Soviet Math.*, **2** (1961), 1254–1257.

40. N.P. Korneichuk, *Exact Constants in Approximation Theory*, Cambridge Univ. Press, New York, 1991.

41. M.G. Krein, On the best approximation of continuous differentiable functions on the whole real axis, *Dokl. Akad. Nauk SSSR*, **18** (1938), 615–624.

42. S.G. Krein, Yu.I. Petunin, and E.M. Semenov, *Interpolation of Linear Operators*, American Mathematical Society, Providence, 1982.

43. D.J. Newman and M.S. Shapiro, Jackson's theorems in higher dimensions. In *International Series of Numerical Mathematics*, vol. 5, *On Approximation Theory* (Proc. Conf. Oberwalfach 1963), pp. 208–219, Birkhauser Verlag, Basel and Stuttgart, 1964.

44. S.M. Nikolskii, On the best approximation by polynomials of the functions that satisfy the Lipschitz condition, *Izv. Akad. Nauk SSSR*, **10** (1946), 295–318.

45. S.M. Nikolskii, Approximation of functions by trigonometric polynomials in the mean, *Izv. Akad. Nauk SSSR*, **10** (1946), 207–256.

46. S.M. Nikolskii, On the best mean approximation by polynomials of the functions $|a-x|^s$, *Izv. Akad. Nauk SSSR*, **11** (1947), 139–180.

47. R.A. Raitsin, S.N. Bernstein's limit theorem for the best approximation in the mean and some of its applications, *VUZ Mat.*, **12**(10) (1968), 81–86.

48. J.J. Schoenberg, Notes on spline functions, I. The limits of interpolating periodic spline functions as their degree tends to infinity, *Indag. Math.*, **34** (1972), 412–422.

49. E.M. Stein and G. Weiss, *Introduction to Harmonic Analysis on Euclidean Spaces*, Princeton Univ. Press, Princeton, 1971.

50. R.S. Strichartz, *A Guide to Distribution Theory and Fourier Transforms*, CRC Press, Boca Raton, 1994.

51. A.F. Timan, *Theory of Approximation of Functions of a Real Variable*, Pergamon Press, New York, 1963.

52. R.S. Varga and A.J. Carpenter, On the Bernstein conjecture in approximation theory, *Constr. Approx.*, **1** (1985), 333–348.

53. V.L. Velikin, On an asymptotic connection between the approximation of periodic functions by splines and by trigonometric polynomials, *Soviet Math. Dokl.*, **23** (1981), 540–544.

54. V.L. Velikin, Best approximation in a Banach space by a convergence sequence of subspaces of fixed dimension, *Soviet Math. Dokl.*, **35** (1987), 483–486.

55. K. Weierstrass, Über die analytische Darstellbarkeit sogennanter willkürlicher Funktionen reeler Argumente, *Sitzungsberichte der Acad. Berlin*, (1885), 633–639, 789–805.

56. V.A. Yudin, A multi-dimensional Jackson theorem, *Math. Notes*, **20** (1976), 801–804.

Chapter 12

Approximation Theory in Random Setting

Sorin G. Gal

Department of Mathematics, University of Oradea
Str. Armatei Române 5, 3700 Oradea, ROMANIA
E-mail: galso@math.uoradea.ro

Abstract In this survey main results in classical approximation theory such as Weierstrass-type results, Korovkin-type results, Jackson, Timan, Teljakovskii-type estimates, interpolating results, best approximation results, etc. to random setting are generalized. The results are mainly known but new results are also proved.

Contents

Motto:
"Although it may seem a paradox, all exact science is dominated by the idea of approximation."
(Bertrand Russell, 1872–1970)

12.1 Introduction

The classical approximation theory is an important tool in all areas of mathematics, in general, and in applied mathematics especially, supplying methods of investigation of the real (deterministic) world.

On the other hand, randomness is an essential aspect of the real world.

By combining now the concept of approximation with that of randomness we obtain a powerful tool in the investigation of real world: the random (stochastic) approximation.

In this survey main results in classical approximation theory such as Weierstrass-type results, Korovkin-type results, Jackson, Timan, Teljakovskii-type estimates, interpolation results, best approximation results, etc. to random setting are generalized.

The results are mainly known but new results are also proved. Where the authors are not specified, the results are new and belong to the present author.

12.2 Preliminaries

In this paragraph we consider the main concepts that will be used in those which follow. For details see any book of probability theory.

One of the first basic concepts in stochastic mathematics is that of the random variable.

DEFINITION 12.1 *Let S be a non-empty set, $\mathcal{B} \subset \mathcal{P}(S)$ a field of parts on S (i.e., $S, \phi \in \mathcal{B}$, $A \in \mathcal{B}$ implies $S \setminus A \in \mathcal{B}$ and $A_i \in \mathcal{B}$, $i \in \mathbf{N}$, implies $\bigcup_{i \in \mathbf{N}} A_i \in \mathcal{B}$), and $P : \mathcal{B} \to [0,1]$ a complete probability on $\mathcal{B}$ (i.e.,*

$$P(S) = 1,\ P\left(\bigcup_{i \in \mathbb{N}} A_i\right) = \sum_{i \in \mathbb{N}} P(A_i),\ \text{if } A_i \cap A_j = \emptyset,\ i \neq j).$$

Then $(S, \mathcal{B}, P)$ is called complete probability space.

A (real) random variable is a $\mathcal{B}$-measurable mapping $g : S \to \overline{\mathbb{R}}$ (i.e., for any open set in the usual topology on $\overline{\mathbb{R}} = \mathbb{R} \cup \{-\infty\} \cup \{+\infty\}$, $U \subset \overline{\mathbb{R}}$, we have $g^{-1}(U) = \{s \in S;\ g(s) \in U\} \in \mathcal{B}$).

The random variable g is called almost surely (a.s.) finite, if

$$P(\{s \in S;\ |g(s)| = +\infty\}) = 0.$$

The set of all a.s. finite real random variables is denoted by $L(S, \mathcal{B}, P)$. Also, for any $0 < q < +\infty$ let us denote

$$L^q(S, \mathcal{B}, P) = \left\{ g \in L(S, \mathcal{B}, P);\ \int_S |g(s)|^q dP(s) < +\infty \right\},$$

where $\int_S |g(s)|^q dP(s)$ means the abstract Lebesgue integral with respect to the measure (probability) P.

From real functions theory we have

DEFINITION 12.2 *Let $g_n, g \in L(S, \mathcal{B}, P)$, $n \in \mathbb{N}$. We say that $g_n \to g$, almost surely, if $P(\{s \in S;\ g_n(s) \to g(s),\ \text{when } n \to +\infty\}) = 0$ (we write $g_n \xrightarrow[n\to\infty]{\text{"}a.s.\text{"}} g$);*
$g_n \to g$, almost uniformly, if $\forall\ \delta > 0$, $\exists\ e \in \mathcal{B}$, $P(e) < \delta$, such that $g_n \xrightarrow{n\to\infty} g$, uniformly on $S \setminus e$ (we write $g_n \xrightarrow[n\to\infty]{\text{"}a.u.\text{"}} g$);
$g_n \to g$, in probability, if $\forall\ \varepsilon, \eta > 0$, $\exists\ n_{\varepsilon,\eta} \in \mathbb{N}$ such that $P(\{s \in S;\ |g_n(s) - g(s)| \geq \varepsilon\}) < \eta$, for all $n \geq n_{\varepsilon,\eta}$ (we write $g_n \xrightarrow[n\to\infty]{\text{"}P\text{"}} g$);
$g_n \to g$, in q-mean $(0 < q < +\infty)$, if $\lim_{n\to\infty} \int_S |g_n(s) - g(s)|^q dP(s) = 0$

(we write $g_n \xrightarrow[n\to\infty]{\text{"}L^q\text{"}} g$).

REMARK 12.1 If we define $d : L(S, \mathcal{B}, P) \times L(S, \mathcal{B}, P) \to [0, 1]$ by

$$d(g, h) = \int_S \frac{|g(s) - h(s)|}{1 + |g(s) - h(s)|} dP(s), \quad g, h \in L(S, \mathcal{B}, P),$$

then it is known (see, e.g., Dunford and Schwartz [37]) that d is a semi-metric on $L(S, \mathcal{B}, P)$ and $d(g_n, g) \overset{n\to\infty}{\longrightarrow} 0$ iff $g_n \xrightarrow[n\to\infty]{\text{"}P\text{"}} g$. ∎

REMARK 12.2 Let us introduce the equivalence relation on $L(S, \mathcal{B}, P)$

$$g \sim h \quad \text{iff} \quad P(\{s \in S;\ g(s) \neq h(s)\}) = 0.$$

Denote $\tilde{L}(S, \mathcal{B}, P) = L(S, \mathcal{B}, P)/_\sim$ = the set of all equivalent classes. If we define $\|\cdot\|_P : \tilde{L}(S, \mathcal{B}, P) \to \mathbb{R}_+$ by

$$\|g\|_P = \inf\{\alpha + P(\{s \in S;\ |g(s)| > \alpha\});\ \alpha > 0\}$$

and

$$\rho(g, h) = \|g - h\|_P, \quad g, h \in \tilde{L}(S, \mathcal{B}, P),$$

then it is known (see, e.g., Dunford and Schwartz [37]) that $(\tilde{L}(S, \mathcal{B}, P), \rho)$ is a complete metric space and $\rho(g_n, g) \overset{n\to\infty}{\longrightarrow} 0$ iff $g_n \overset{\text{"}P\text{"}}{\longrightarrow} g$. ∎

DEFINITION 12.3 *Let $(T, \mathcal{T})$ be a topological space. A mapping $f : T \to L(S, \mathcal{B}, P)$ is called a random function (or stochastic process) on T. We denote $f(t)(s) = f(t, s)$, $t \in T$, $s \in S$.*

The concepts in Definition 12.2 allow us to consider different concepts of continuity, differentiability and integrability for random functions, as follows (see, e.g., Wentzell [82] or any book on stochastic processes).

DEFINITION 12.4 *Let $(T, \mathcal{T})$ be a topological space, $f : T \to L(S, \mathcal{B}, P)$ be a random function and let us denote $\mathcal{C} = \{$"a.s.", "P", "L^q"$\}$, where the elements of $\mathcal{C}$ have the meaning in Definition 12.2. Let $c \in \mathcal{C}$.*

We say that f is c-continuous in $t_0 \in T$, if for any $t_n \overset{\mathcal{T}}{\longrightarrow} t_0$, with t_n, $t_0 \in T$, $n \in \mathbb{N}$, we have $f(t_n, s) \xrightarrow[n\to\infty]{c} f(t_0, s)$.

Let $T = [a, b] \subset \mathbb{R}$. We say that f is c-differentiable in $t_0 \in [a, b]$, if there exists $g \in L(S, \mathcal{B}, P)$ such that for any $t_n \in [a, b]$, $t_n \overset{n\to\infty}{\longrightarrow} t_0$, $t_n \neq t_0$, we have

$$\frac{f(t_n, s) - f(t_0, s)}{t_n - t_0} \xrightarrow[n \to \infty]{c} g(s).$$

We say that f is c-integrable on $[a, b]$, if for each sequence of divisions $(\mathcal{D}_n)_{n\in\mathbb{N}}$ of $[a, b]$, with $\Delta(\mathcal{D}_n) \overset{n\to\infty}{\longrightarrow} 0$, where $\Delta(\mathcal{D}_n)$ is the norm of $\mathcal{D}_n$, we have

$$\sum_{k=0}^{n-1} f(\xi_k^{(n)}, s)(t_{k+1}^{(n)} - t_k^{(n)}) \xrightarrow[n \to \infty]{c} I(s) \in L(S, \mathcal{B}, P),$$

and $I(s) \overset{def.}{=} (c) \int_a^b f(t, s)dt$, where $\mathcal{D}_n : a = t_0^{(n)} < \ldots < t_n^{(n)} = b$, $\xi_k^{(n)} \in [t_k^{(n)}, t_{k+1}^{(n)}]$, $k = \overline{0, n-1}$.

REMARK 12.3 The concepts in Definition 12.4 can easily be formulated in ε, δ language, too. ∎

REMARK 12.4 Any usual (non-random) real function $f : T \to \mathbb{R}$, can be identified with the degenerate random function $f(t, s) = f(t)$, for all $t \in R$, $s \in S$. ∎

REMARK 12.5 Sums and products of random functions are defined pointwise and the natural ordering is $f \leq g$ iff $f(t, s) \leq g(t, s)$, for all $t \in T$, a.s. $s \in S$ (passing to equivalence classes, random functions that coincide with probability one for each fixed $t \in T$, will not be distinguished). ∎

REMARK 12.6 The concepts in Definition 12.4 permit extension in three ways for random functions of many results in classical mathematical analysis. In a similar manner, if X and Y are two Banach spaces, a random operator from X to Y will be a map $\phi : X \times S \to L_Y(S, \mathcal{B}, P)$, where $(S, \mathcal{B}, P)$ is a complete probability space and

$$L_Y(S, \mathcal{B}, P) = \{g : S \to Y;\ g \text{ is Bochner measurable}\}.$$

By Definition 12.4, many classical results in functional analysis (as, e.g., the Banach-Steinhauss theorem, the closed graph theorem or the condensation of singularities principle), can be extended to random operators, too; see Vellasco and Villena [75], [76], Gal and Villena [43]. ∎

REMARK 12.7 A main idea in the field of stochastic mathematics is that of "randomization," i.e., if $(Y, +, \cdot, \mathcal{T}_Y)$ is a topological linear space, then a randomization of Y will be the linear space of all Y-valued Bochner random variables on $(S, \mathcal{B}, P)$. Here $g : S \to Y$ is called the Bochner random variable if there exists a sequence $g_n : S \to Y$, $n \in \mathbb{N}$, of step functions (i.e., $g_n(x) \in \{y_1, \ldots, y_r\}$, $\forall\, x \in S$ and $g_n^{-1}(y_i) \in \mathcal{B}$, $i = \overline{1, r}$, $n \in \mathbb{N}$), such that $g_n \overset{\text{"}a.s.\text{"}}{\longrightarrow} g$ (i.e., there is $S_0 \subset S$ with $P(S_0) = 0$, such that $g_n(s) \overset{\mathcal{T}_Y}{\longrightarrow} g(s)$, $\forall\, s \in S \setminus S_0$). ∎

DEFINITION 12.5 *An expression of the form* $\sum_{k=0}^{r} c_k(s)x^k$, $x \in \mathbb{R}$, $c_k \in L(S, \mathcal{B}, P)$, $k = \overline{0, r}$, *will be called an algebraic random polynomial. An expression of the form*

$$\sum_{k=0}^{r} [a_k(s) \cos kx + b_k(s) \sin kx], \quad x \in \mathbb{R}, \ a_k, b_k \in L(S, \mathcal{B}, P), \ k = \overline{0, r},$$

will be called a trigonometric random polynomial.

12.3 Weierstrass-Type Results

The theory of approximation of random functions is a natural extension of the classical approximation theory.

We shall call them "random approximation theory" and "deterministic approximation theory," respectively.

It is interesting to briefly trace some aspects in the development of random approximation theory. The scope of results obtained there is extremely wide. Some results due to Andrus and Brown [4] and Brown and Schreiber [16] generalize some results in "deterministic" theory, with the help of generalized random functions. Another approach helps us to handle real problems determining the classes of approximation functions. An early result in this direction is due to Ky Fan [51], which asserts that a random function can be approximated by the so-called random splines of order one. Later Dugue [36] proves that a random function can be approximated by random polynomials, representing in fact a generalization of the Weierstrass theorem. Thus the class of random polynomials becomes of special interest. The book of Bharucha, Reid and Sambandham [14] is concerned mainly with an algebraic point of view, but the approximation problem is also mentioned (see pp. 16, 17 there).

Recall the well-known Weierstrass result (algebraic case) in deterministic theory.

THEOREM 12.1

For any $f : [a,b] \to \mathbb{R}$ continuous function on $[a,b]$, there exists a sequence of algebraic polynomials $(P_n(x))_n$, such that $P_n \overset{n\to\infty}{\longrightarrow} f$, uniformly on $[a,b]$.

A very elegant proof can be made by using as $P_n(x)$ the Bernstein polynomials.

We present the following two random variants of Theorem 12.1.

Let $(S, \mathcal{B}, P)$ be a complete probability space.

THEOREM 12.2 (Onicescu and Istrăţescu [60], Cenuşă and Săcuiu [29], or Onicescu, Cenuşă and Săcuiu [59], pp. 57-59)

If the random function $f : [0,1] \to L(S, \mathcal{B}, P)$ is "P"-continuous on $[0,1]$ (i.e., in probability) and if there is $0 < M < +\infty$ such that

$$\sup\{|f(t,s)|;\ t \in [0,1]\} \le M, \quad a.s. \quad s \in S,$$

then the sequence of random polynomials of Bernstein-type

$$B_n(f)(x,s) = \sum_{k=0}^{n} \binom{n}{k} x^k (1-x)^{n-k} f\left(\frac{k}{n}, s\right), \quad x \in [0,1],\ s \in S,$$

satisfies $\rho(B_n(f)(x), f(x)) \overset{n\to\infty}{\longrightarrow}$ uniformly with respect to $x \in [0,1]$ (we can say that $B_n(f)(x,s)$ converges uniformly in probability to $f(x,s)$). Here ρ is defined in the previous paragraph.

THEOREM 12.3 (Onicescu and Istrăţescu [60], Cenuşă and Săcuiu [29], or Onicescu, Cenuşă and Săcuiu [59], p .60)

Let $f : [0,1] \to L^1(S, \mathcal{B}, P)$ be a random function having the properties:

1) f is L^1-continuous on $[0,1]$;

2) $\sup\left\{\int_S |f(x,s)|dP(s);\ x \in [0,1]\right\} \le M < +\infty.$

Then for $B_n(f)(x,s)$ defined as in Theorem 12.2, we have

$$B_n(f)(x,s) \xrightarrow[n\to\infty]{\text{“}L^1\text{”}} f(x,s),$$

uniformly with respect to $x \in [0,1]$.

DEFINITION 12.6 *Let $Q_n(x,s) = \sum_{k=0}^{n} a_{k,n}(s)x^k$, $n \in \mathbb{N}$, be a sequence of random polynomials. We say that*

1) $(Q_n(x,s))_n$ has the coefficients uniformly bounded in probability, if there is $0 < M < +\infty$ such that

$$P(\{s \in S;\ |a_{k,n}(s)| \geq M\}) = 0,\ \forall\ n \in \mathbb{N},\ \forall\ k \in \mathbb{N} \cup \{0\}.$$

2) $(Q_n((x,s))_n$ has the coefficients bounded in L^1 (in mean), if

$$\int_S |a_{k,n}(s)|dP(s) \leq M < +\infty,\ \forall\ n \in \mathbb{N},\ \forall\ k \in \mathbb{N} \cup \{0\}.$$

THEOREM 12.4 (Onicescu and Istrăţescu [61])
The necessary and sufficient condition for the existence of a sequence of random polynomials with coefficients uniformly bounded in probability, uniformly convergent in probability to the random function $f : [-1,1] \to L(S,\mathcal{B},P)$, “P”-continuous on $[-1,1]$, is that f admits the representation

$$f(x,s) = \sum_{k=0}^{\infty} a_k(s)x^k, \quad x \in [-1,1],\ s \in S,$$

where $(a_k(s))_{k\in\mathbb{N}\cup\{0\}}$ is uniformly bounded in probability.

THEOREM 12.5 (Săcuiu and Cenuşă [66])
The necessary and sufficient condition for the existence of a sequence of random polynomials with coefficients uniformly bounded in L^1, uniformly convergent in L^1 to the random function $f : [-1,1] \to L^1(S,\mathcal{B},P)$, “$L^1$”-continuous on $[-1,1]$, is that f admits the representation

$$f(x,s) = \sum_{k=0}^{\infty} a_k(s)x^k, \quad x \in [-1,1],\ s \in S,$$

where $(a_k(s))_{k\in\mathbb{N}\cup\{0\}}$ is uniformly bounded in L^1.

REMARK 12.8 For other generalizations of Theorems 12.2, 12.3, 12.4 and 12.5 (especially for the case of so-called vector random functions, when the real random variable is replaced with a m-dimensional complex random variable), see, e.g., [28]–[29], [59]–[62], [66]. ∎

In what follows we will present an application of the random Weierstrass-type results. Thus, in the deterministic theory, a function $f : \mathbb{R} \to \mathbb{R}$ is called almost periodic on $\mathbb{R}$ if for any $\varepsilon > 0$ there is a trigonometric polynomial P_ε, such that $|f(t) - P_\varepsilon(t)| < \varepsilon$, for all $t \in \mathbb{R}$ (see, e.g., Corduneanu [33]).

Taking into account Definition 12.2 and inspired from the deterministic case, the following can be introduced.

DEFINITION 12.7 *(Onicescu, Cenuşă and Săcuiu [59], Cenuşă and Săcuiu [29], Corduneanu [33]) Let $f : \mathbb{R} \to L(S, \mathcal{B}, P)$ be a random function, $\mathcal{C} = \{$"a.s.", "P", "L^q"$\}$ and $c \in \mathcal{C}$.*

We say that f is c-almost periodic if there exists a sequence $(P_n(x,s))_n$ of trigonometric random polynomials such that $P_n(x,s) \xrightarrow[n\to\infty]{c} f(x,s)$, uniformly with respect to $x \in \mathbb{R}$.

REMARK 12.9 Detailed studies of the above concepts in Definition 12.7 together with other new concepts of almost periodic random functions can be found in Cenuşă and Săcuiu [30]–[31], Cenuşă [19]–[27], Onicescu, Cenuşă and Săcuiu [59], Corduneanu [33], Cenuşă and Săcuiu [29]. ∎

12.4 Qualitative and Quantitative Korovkin-Type Results

According to different objectives and based on quite different assumptions, the approximation of stochastic processes has been studied under various aspects, e.g., convergence of empirical processes with cadlag sample paths (Billingsley [15], Gänssler and Stute [44], Pollard [64]), simulating functionals of continuous Gaussian processes (Eplett [38]–[40])

and approximation of weakly stationary random processes by convolution processes (Splettstösser [72]).

We do not present these aspects here (the interested reader can consult the mentioned papers), because our aim is to give a powerful general method of approximation based on Korovkin-type theorems, which permits treating the previous particular aspects, too.

The well-known Korovkin theorem in deterministic theory states that $\lim_{n\to+\infty} T_n(f) = f$, uniformly on $[a,b]$, for all $f \in C[a,b] = \{g : [a,b] \to \mathbb{R};\ g \text{ continuous on } [a,b]\}$, if the operators $T_n : C[a,b] \to C[a,b]$ are linear, positive (monotone) and if $\lim_{n\to+\infty} T_n(f_i) = f_i$, uniformly on $[a,b]$, $i = 1,2,3$, where $f_1(t) = 1,\ f_2(t) = t,\ f_3(r) = t^2,\ t \in [a,b]$.

While the above result is called a qualitative Korovkin-type result, many quantitative Korovkin-type results have also been proved (i.e., results giving estimates of $\|T_n(f) - f\|$ for the above operators, where $\|\cdot\|$ is the uniform norm on $C[a,b]$).

Here we present random variants of the qualitative and quantitative Korovkin-type results in classical theory.

Contrary to the nonrandom case, additional conditions are required in order to obtain these random variants.

First, we introduce some known notations and definitions.

Keeping the notations in the previous paragraphs, let $(S,\mathcal{B},P)$ be a complete probability space and $L^q(S,\mathcal{B},P)$, $q \geq 1$ be the set of all real-valued random variables with finite qth moments.

Let V be a real normed space and $\Gamma = \{\gamma_1,\ldots,\gamma_m\}$, $m \geq 1$ be a fixed set of bounded linear functionals on V with the following property of separation: Given $v,w \in V$ with $v \neq w$, one can find a functional $\gamma \in \Gamma$ satisfying $\gamma(v) \neq \gamma(w)$.

Let T be a compact convex subset of V. For a stochastic process $f : T \to L^q(S,\mathcal{B},P)$, let us define

$$\|f\|_{q,S} = \sup\left\{\left(\int_S |f(t,s)|^q dP(s)\right)^{1/q};\ t \in T\right\}$$

and

$$B_S(T)_q = \{f : T \to L^q(S,\mathcal{B},P);\ \|f\|_{q,S} < +\infty\}.$$

Then $(B_S(T)_q, \|\cdot\|_{q,S})$ becomes a normed vector lattice and

$$C_S(T)_q = \{f : T \to L^q(S,\mathcal{B},P);\ f \text{ is ``}L^q\text{''-continuous on } T\}$$

becomes a linear sublattice of $B_S(T)_q$ (for the order relation between two stochastic processes, see Remark 12.5 of Definition 12.4).

REMARK 12.10 Obviously, $C(T;R) = \{g : T \to \mathbb{R};\ g \text{ continuous on } T\}$ and $B(T;\mathbb{R}) = \{g : T \to \mathbb{R};\ \|g\| < +\infty\}$, where $\|\cdot\|$ is the usual uniform norm, are subsets of $C_S(T)_q$ and of $B_S(T)_q$, respectively ($f(t,s) \equiv f(t),\ \forall\, t \in T,\ s \in S$). ∎

Given $f \in B_S(T)_q$, we denote

$$\|f(t)\|_q = \left(\int_S |f(t,s)|^q dP(s)\right)^{1/q}, \quad t \in T$$

and the expectation of f by

$$E(f)(t) = \int_S f(t,s)dP(s), \quad t \in T.$$

Smoothness of $f \in B_S(T)_q$ with respect to the system Γ is expressed by the stochastic modulus of continuity

$$\Omega(f;\Gamma;\delta)_q = \sup\left\{ \|f(t_1) - f(t_2)\|_q;\ t_1, t_2 \in T,\right.$$

$$\left.\sum_{i=1}^{m} [\gamma_i(t_1) - \gamma_i(t_2)]^2 \le \delta^2 \right\}, \quad \delta \ge 0.$$

DEFINITION 12.8 (Weba [78]). *A bounded linear mapping $L : C_S(T)_q \to B_S(T)_q$ is called monotone if*

$$0 \le f \textit{ implies } 0 \le L(f), \textit{ for all } f \in C_S(T)_q.$$

Also, L is called E-commutative if

$$E(L(f)) \equiv L(E(f)), \textit{ for all } f \in C_S(T)_q.$$

For each $A \in \mathcal{B}$, the stochastic process $1_A \in C_S(T)_q$ defined by

$$1_A(x,s) = \begin{cases} 1, & \textit{if } s \in A,\ x \in T \\ 0, & \textit{if } s \notin A,\ x \in T, \end{cases}$$

is called simple.

A mapping $L : C_S(T)_q \to B_S(T)_q$ *is called stochastically simple if there exists a real-valued function* $a_L : T \to \mathbb{R}$ *such that*

$$L(1_A)(x,s) = a_L(x) \cdot 1_A(x,s), \quad x \in T, \quad s \in S.$$

REMARK 12.11 The operators of the form

$$L(f)(x,s) = \sum_{k=1}^{n} g_k(x) f(x_k, s), \quad x \in T, \quad s \in S,$$

with $x_1 < \ldots < x_n$ fixed in T and $g_k : T \to \mathbb{R}_+$ continuous on T, are monotone, E-commutative and stochastically simple on $C_S(T)_q$. ▮

The following two qualitative Korovkin-type results hold.

THEOREM 12.6 (Weba [77])
Let $K_S(T)$ *be a finite subset of* $C_S(T)_1$, *which contains the constant process* $f(x,s) \equiv 1$, $\forall\ x \in T$, $\forall\ s \in T$. *For each* $x_0 \in T$ *assume the existence of a linear combination* $f_{x_0} \in spanK_S(T)$, *such that* $E(f_{x_0})(x)$ *is nonnegative on* T *and vanishes if and only if* $x = x_0$.

Then $K_S(T)$ *is a Korovkin system in the following sense:*

If a sequence $L_n : C_S(T)_1 \to C_S(T)_1$, $n \in \mathbb{N}$, *of monotone, E-commutative and stochastically simple operators satisfies*

$$\lim_{n\to\infty} \|f - T_n(f)\|_{1,S} = 0, \ \forall\ f \in K_S(T),$$

then one can conclude

$$\lim_{n\to\infty} \|f - T_n(f)\|_{1,S} = 0, \ \forall\ f \in C_S(T)_q,$$

where $\|f\|_{q,S} = \sup\left\{\left(\int_S |f(x,s)|^q dP(s)\right)^{1/q} ; x \in T\right\}$ *is a norm on* $C_S(T)_q$, $q \geq 1$.

THEOREM 12.7 (Weba [77])
Let $T = [a,b]$, $0 \leq a < b$ *and* $N_\lambda(x,s)$ *be a Poisson process with parameter* $\lambda > 0$.

If $K_S([a,b])$ consists of the following three stochastic processes

$$f_1(x,s) \equiv 1, \quad f_2(x,s) \equiv N_\lambda(x,s), \quad f_3(x,s) = N_\lambda^2(x,s),$$

$$x \in [a,b], \quad s \in S,$$

then $K_S([a,b])$ is a Korovkin system in the sense of Theorem 12.6.

The first quantitative Korovkin-type result is

THEOREM 12.8 (Weba [78])
Assume $L : C_S(T)_q \to B_S(T)_q$ is monotone, E-commutative and stochastically simple. Given $\delta > 0$, for each $f \in C_S(T)_q$ we have

$$\|f - L(f)\|_{q,S} \leq \|f\|_{q,S}\|\gamma_0 - L(\gamma_0)\|_{q,S} +$$

$$+\Omega(f;\Gamma;\delta)_q \left(\|L(\gamma_0)\|_{q,S} + \delta^{-2} \sup_{x \in T} \left\{ \sum_{i=1}^{m} L(\gamma_{i,x})(x) \right\} \right),$$

where $\gamma_0(x) = 1$, $\forall\, x \in T$ and

$$\gamma_{i,x_0}(x) = [\gamma_i(x) - \gamma_i(x_0)]^2, \ \forall\, x \in T,$$

for each $x_0 \in T$ and $1 \leq i \leq m$.

COROLLARY 12.1
(Weba [78]) *Let $L_n : C_S(T)_q \to B_S(T)_q$, $n \in \mathbb{N}$, be a sequence of monotone, E-commutative and stochastically simple operators. If*

$$\lim_{n \to +\infty} \|\gamma_i - L_n(\gamma_i)\|_{q,S} = 0, \ \forall\, i = \overline{0,m}$$

and

$$\lim_{n \to +\infty} \|\gamma_i^2 - L_n(\gamma_i^2)\|_{q,S} = 0, \ \forall\, i = \overline{1,m},$$

then

$$\lim_{n \to +\infty} \|f - L_n(f)\|_{q,S} = 0, \ \forall\, f \in C_S(T)_q.$$

REMARK 12.12 In the particular case $q = 2$ and $T = [a,b] \subset \mathbb{R}$, we can take in Theorem 12.8 and Corollary 12.1, $m = 1$, $\Gamma = \{\gamma_1\}$, with $\gamma_1(x) = x$, $\forall\, x \in K$ (see Weba [81]). In this case, the quantitative

Korovkin-type results have been applied to the discretization of stationary times series with trend, to the simulated distributions of functionals and to the solving of linear stochastic differential equations. ∎

REMARK 12.13 Abstract generalizations of the results in Weba [77], [78], [81] were obtained in Weba [79]. ∎

In the case when the operators are not stochastically simple, quantitative Korovkin-type results were obtained in Anastassiou [1]. The method of proving is different from that in Weba's papers, being mainly probabilistic and the estimates are pointwise.

In what follows we present the followings.

THEOREM 12.9 (Anastassiou [1])

Let $L : C_S([a,b])_1 \to B_S([a,b])_1$ be monotone and E-commutative. Let $L(1)(x_0) = m_{x_0}$, $r > 0$, $[L((x-x_0)^2)(x_0)]^{1/2} = d_2(x_0) > 0$ and $L(x-x_0)(x_0) = 0$, for some $x_0 \in [a,b]$. Then for all $f \in C_S^1([a,b])_1 = \left\{f : [a,b] \to L^1(S,\mathcal{B},P);\ \exists \dfrac{\partial f}{\partial x}(x,s) \text{ continuous on } [a,b], \text{ for all } s \in S\right\}$, we have

$$|E(L(f))(x_0) - E(f)(x_0)| - |E(f)(x_0)| \cdot |m_{x_0} - 1| \leq$$

$$\begin{cases} \omega_1\left(E\left(\dfrac{\partial f}{\partial x}\right); rd_2(x_0)\right) \dfrac{d_2(x_0)}{8r}\left(2 + r\sqrt{m_{x_0}}\right)^2, & \text{when } 0 < r \leq 2/\sqrt{m_{x_0}}, \\ \sqrt{m_{x_0}} \cdot \omega_1\left(E\left(\dfrac{\partial f}{\partial x}\right); rd_2(x_0)\right) d_2(x_0), & \text{when } r > 2/\sqrt{m_{x_0}}, \end{cases}$$

where ω_1 is the usual modulus of continuity on $[a,b]$.

COROLLARY 12.2

[(Anastassiou [1])] If $m_{x_0} = 1$ and $r = \dfrac{1}{2}$, then

$$|E(L(f))(x_0) - E(f)(x_0)| \leq 1.5625\omega_1\left(E\left(\frac{\partial f}{\partial x}\right); \frac{1}{2}d_2(x_0)\right) d_2(x_0).$$

REMARK 12.14 For other details and for extensions to the multidimensional case see Anastassiou [1]. ∎

12.5 Approximation by Random Polynomials. Quantitative Estimates

As we have seen, the random Bernstein polynomial is a main tool of approximation of random functions. What is left to be determined is the rate of approximation. In this sense, based on the Definition 12.2 and on the Remarks 12.1 and 12.2, various kinds of moduli of smoothness can be introduced.

DEFINITION 12.9 (Ignatov, Mills and Tzankova [46], Gal [42]) *Let $(S, \mathcal{B}, P)$ be a complete probability space and $f : [a,b] \to L(S, \mathcal{B}, P)$ be a random function.*

The random modulus of smoothness of order $n \in \mathbb{N}$ is given by

$$W_n(f, \delta) = \sup\{\sup\{|\Delta_h^n f(t,s)|; t \in [a, b - nh]\}; 0 \le h \le \delta\},$$

where $\delta \in [0, b-a]$, $s \in S$, $\Delta_h^n f(t,s) = \sum_{k=1}^{n} (-1)^{n-k} \binom{n}{k} f(x + kh, s)$.

The L^q-modulus of smoothness of order $n \in \mathbb{N}$ is given by

$$\Omega_n(f;\delta)_q = \sup\left\{ \sup\left\{ \left(\int_S |\Delta_h^n f(t,s)|^q dP(s) \right)^{1/q}; \right.\right.$$

$$\left.\left. t \in [a, b - nh] \right\}; 0 \le h \le \delta \right\},$$

where $q \ge 1$, $\delta \in [0, b-a]$.

The d-modulus of continuity is given by

$$\omega_d(f;\delta) = \sup\{d[f(t_1,s), f(t_2,s))]; t_1, t_2 \in [a,b], |t_1 - t_2| \le \delta\},$$

and the ρ-modulus of continuity is given by

$$\omega_\rho(f;\delta) = \sup\{\rho[f(t_1,s), f(t_2,s)]; t_1, t_2 \in [a,b], |t_1 - t_2| \le \delta\},$$

where d and ρ are defined in the Remarks 12.1 and 12.2 after Definition 12.2.

Concerning the approximation by random Bernstein polynomials

$$B_n(f)(x,s) = \sum_{k=0}^{n} f\left(\frac{k}{n}, s\right)\binom{n}{k} x^k(1-x)^{n-k},$$

the following estimates are known.

THEOREM 12.10

Let $f : [0,1] \to L(S, \mathcal{B}, P)$ be "P"-continuous on $[0,1]$.

(i) (Ignatov, Mills and Tzankova [46]). We have

$$|f(t,s) - B_n(f)(t,s)| \leq [1 + t(1-t)]W_1\left(f; \frac{1}{\sqrt{n}}\right)(s),$$

$$t \in [0,1], \quad n \in \mathbb{N}, \quad s \in S;$$

there exists $n_0 \in \mathbb{N}$, such that

$$|f(t,s) - B_n(f)(t,s)| \leq 2W_1\left(f; \sqrt{\frac{t(1-t)}{n}}\right)(s),$$

$$n \geq n_0, \quad t \in [0,1], \quad s \in S,$$

and also

$$|f(t,s) - B_n(f)(t,s)| \leq \frac{5}{4}W_1\left(f; \frac{1}{\sqrt{n}}\right)(s),$$

$$n \in \mathbb{N}, \quad t \in [0,1], \quad s \in S.$$

(ii) (Ignatov, Mills and Tzankova [46], Kamolov [49])
Let $f : [0,1] \to L^2(S, \mathcal{B}, P)$ be "L^2"-continuous on $[0,1]$. Then

$$\|f(t,s) - B_n(f)(t,s)\|_{L^2} \leq [1 + 2t(1-t)/e]\Omega_1\left(f; \frac{1}{\sqrt{n}}\right)_2,$$

$$t \in [0,1], \quad n \in \mathbb{N};$$

there exists $n_0 \in \mathbb{N}$ such that

$$\|f(t,s) - B_n(f)(t,s)\|_{L^2} \leq \left(1 + \frac{2}{e}\right)\Omega_1\left(f; \sqrt{\frac{t(1-t)}{n}}\right)_2,$$

$$t \in [0,1], \; n \geq n_0,$$

and also

$$\|f(t,s)-B_n(f)(t,s)\|_{L^2} \le \left(1+\frac{1}{2e}\right)\Omega_1\left(f;\frac{1}{\sqrt{n}}\right)_2,$$

$$t\in[0,1], \quad n\in\mathbb{N},$$

where $\|f(t,s)\|_{L^q} = \left(\int\limits_S |f(t,s)|^q dP(s)\right)^{1/q}, \; q\ge 1.$

For $n\in\mathbb{N}$ and $q\ge 1$, let us consider that $L^q_{2\pi}(\mathbb{R})\times S) = \{f:\mathbb{R}\to L^q(S,\mathcal{B},P); f(t,s)$ is $\mathcal{L}(\mathbb{R})\times\mathcal{B}$ measurable and $\exists\, A\subset S$ with $P(A)=1$, such that $f(t+2\pi,s)=f(t,s)$, $\forall\, t\in\mathbb{R}$, $\forall\, s\in A$ and $(\mathcal{L})\int_{-\pi}^{\pi}|f(t,s)|^q dt < +\infty$, $\forall\, s\in A\}$, where $\mathcal{L}(\mathbb{R})$ denotes the Lebesgue measurable sets on $\mathbb{R}$ and $(\mathcal{L})\int_{-\pi}^{\pi}|f(t,s)|^q dt$ means the Lebesgue integral (with respect to t), for each fixed $s\in A$.

Also, for $p\in\mathbb{N}$, let us introduce the trigonometric random polynomials of degree n,

$$I_n(f)(x,s) = -\int_{-\pi}^{\pi} K_{n,r}(t)\sum_{k=1}^{p+1}\binom{p+1}{k} f(x+kt,s)dt, \; x\in\mathbb{R}, \; s\in S,$$

where the integral with respect to t is the Lebesgue integral, $K_{n,r}(t)$ is the kernel defined in Lorentz [56], p.57, relation (5), i.e.,

$$K_{n,r}(t) = L_{n',r}(t), \quad n' = \left[\frac{n}{r}\right]+1,$$

and

$$L_{n',r}(t) = \lambda_{n',r}^{-1}\left(\frac{\sin(nt/2)}{\sin(t/2)}\right)^{2r}, \quad \int_{-\pi}^{\pi} L_{n',r}(t)dt = 1.$$

Denote

$$CM^q(\mathbb{R}) = \{f\in L^q_{2\pi}(\mathbb{R}\times S); f(t,s) \text{ is "}L^q\text{"-continuous on } \mathbb{R}\}.$$

THEOREM 12.11 (Gal [41])

Let $p,n\in\mathbb{N}$, $q>1$ and r be the smallest integer such that $r\ge(p+1)q/2+1$. For each $f\in CM^q(\mathbb{R})$ we have

$$\|f(x,s)-I_n(f)(x,s)\|_{L^q} \le C_{p+1,q}\Omega_{p+1}\left(f;\frac{1}{n}\right)_q, \quad x\in\mathbb{R}, \quad n\in\mathbb{N},$$

where $C_{p+1,q} > 0$ is independent of f and n.

REMARK 12.15 The case $q = 1$ in Theorem 12.11 was proved in Gal [42]. ▮

REMARK 12.16 Theorem 12.11 is the random version of the well-known Jackson-type estimate in Lorentz [56], p.57-58. ▮

In what follows we will obtain the random version of the Timan-type result in classical approximation theory.

Let us consider $\mathcal{L}[a,b] = \{A \subset [a,b]; A$ is Lebesgue measurable$\}$, $L^q([a,b] \times S) = \{f : [a,b] \to L^q(S,\mathcal{B},P);\ f(t,s)$ is $\mathcal{L}[a,b] \times \mathcal{B}$ measurable and $(\mathcal{L}) \int_a^b |f(t,s)|^q dt < +\infty$, a.s. $s \in S\}$, $CM^q[a,b] = \{f : [a,b] \to L^q(S,\mathcal{B},P); f$ is "L^q"-continuous on $[a,b]\}$.

Also, let us introduce the following random algebraic polynomials (of degree $3n-3$)

$$T_n(f)(x,s) = \frac{1}{\pi} \int_{-\pi}^{\pi} f[\cos(\arccos t + v), s] K_{3n-3}(v) dv,$$

where K_{3n-3} is the kernel introduced in Lehnhoff [54], i.e.,

$$K_{3n-3}(v) = \{10/[n(11n^4 + 5n^2 + 4)]\}[\sin(nv/2)/\sin(v/2)]^6.$$

THEOREM 12.12 (Gal [41], [42])

Let $1 \leq q < +\infty$. If $f \in L^q([-1,1] \times S) \cap CM^q[-1,1]$, then the Timan-type estimate

$$\|f(x,s) - T_n(f)(x,s)\|_{L^q} \leq 4\Omega_1\left(f; \frac{\sqrt{1-t^2}}{n} + \frac{|t|}{n^2}\right)_q, \quad t \in [-1,1]$$

holds.

At the end of this section we will obtain a random version of the classical Teljakovskii's result. The proof needs some auxiliary results.

LEMMA 12.1

For all $v \in [-\pi,\pi]$, $t \in [0,1]$ with $\sqrt{1-t^2} \leq 1/n$, we have

$$\Omega_1(f; |\cos(\arccos t + v) - \cos v|)_q \leq (2 + n|v|)\Omega_1\left(f; \sqrt{1-t^2}/n\right)_q,$$

where $n \in \mathbb{N},\ t \in [-1,1],\ 1 \leq q < +\infty,\ f : [-1,1] \to L(S, \mathcal{B}, P)$.

PROOF Reasoning exactly as in Lehnhoff [55], p.179 and taking into account the properties of $\Omega_1(f;\delta)_q$ (see, e.g., Weba [78]), we get

$$\Omega_1(f; |\cos(\arccos t + v) - \cos v|) \leq$$

$$\leq \Omega_1\left(f; \sqrt{1-t^2}/n\right)_q [1 + n|\cos(\arccos t + v) - \cos v|/\sqrt{1-t^2} =$$

$$= \Omega_1\left(f; \sqrt{1-t^2}/n\right)_q [1 + |(1-t)\cos v + \sqrt{1-t^2}\sin v|/\sqrt{1-t^2}] \leq$$

$$\leq \Omega_1\left(f; \sqrt{1-t^2}/n\right)_q [1 + n(1-t)|\cos v|/\sqrt{1-t^2} + n|\sin v|] \leq$$

$$\leq \Omega_1\left(f; \sqrt{1-t^2}/n\right)_q [1 + 1/(1+t) + n|v|] \leq$$

$$\leq (2 + n|v|)\Omega_1\left(f; \sqrt{1-t^2}/n\right)_q,$$

which proves the lemma. ∎

Now, for an integer $r \geq 3$ and $g : [-1,1] \to \mathbb{R}$, continuous on $[-1,1]$, let us consider the Lehnhoff's operator in Lehnhoff [54],

$$L_{n,r}(g)(t) = \frac{1}{\pi}\int_{-\pi}^{\pi} g[\cos(\arccos t + v)]K_{r(n-1)}(v)dv,$$

where

$$K_{r(n-1)}(v) = \left\{\pi/\int_{-\pi}^{\pi}[\sin(nt/2)/\sin(t/2)]^{2r}dt\right\}[\sin(nv/2)/\sin(v/2)]^{2r}.$$

Obviously $L_{n,r}(g)(t)$ is an usual algebraic polynomial of degree $r(n-1)$ and satisfies $L_{n,r}(e_0) = e_0$, where $e_0(t) = 1,\ \forall\, t \in [-1,1]$.

LEMMA 12.2
There exists a constant $C_r > 0$ *such that for any* $g \in C[-1,1]$, *we have*

$$|g(t) - L_{n,r}(g)(t)| \leq C_r\omega_1\left(f; |t|/n^2 + \sqrt{1-t^2}/n\right),\ \forall\, t \in [-1,1],\ n \in \mathbb{N},$$

where

$$\omega_1(g;\delta) = \sup\{|g(t_1) - g(t_2)|; t_1, t_2 \in [-1,1], |t_1 - t_2| \leq \delta\}.$$

PROOF Keeping the notations by Lehnhoff [54] for all $t \in [-1,1]$, we have

$$L_{n,r}((u-t)^2)(t) = (1/2)(1-\rho_{1,r(n-1)})(1-t^2)+$$

$$+\left(3/2 - 2\rho_{1,r(n-1)} + \frac{1}{2}\rho_{2,r(n-1)}\right)t^2,$$

which together with $1-\rho_{1,r(n-1)} = \mathcal{O}(n^{-2})$ (see, e.g., Cao and Gonska [18], p.88) and with $3/2 - \rho_{1,r(n-1)} + (1/2)\rho_{2,r(n-1)} = \mathcal{O}(n^{-4})$ (see Cao and Gonska [18], p.88) implies

$$L_{n,r}((u-t)^2)(t) = \mathcal{O}((1-t^2)n^{-2} + t^2 n^{-4}).$$

The lemma is an immediate consequence of Lehnhoff [54], Lemma 2.2 and Theorem 2.3. ∎

Returning now to the random case, for $f \in L^q([-1,1]\times S) \cap CM^q[-1,1]$ and $r = [q]+2$, let us consider the random algebraic polynomial of degree $r(n-1) = ([q]+2)(n-1)$, defined by

$$R_{n,r}(f)(t,s) = (1/\pi)\int_{-\pi}^{\pi} f[\cos(\arccos t + v), s]K_{r(n-1)}(v)dv.$$

It is easy to see that $R_{n,r}$ satisfies the conditions in Gal [41], Theorem 2.1 and taking in Gal [41], relation (2)

$$\delta = R_{n,r}(|u-t|)(t) \leq [R_{n,r}(|u-t|^2)(t)]^{1/2} = [L_{n,r}((u-t)^2)(t)]^{1/2},$$

by the last relation in the proof of Lemma 12.2 we obtain the following.

COROLLARY 12.3

There exists a constant $C_q > 0$ such that for any $f \in L^q([-1,1]\times S) \cap CM^q[-1,1]$ we have

$$\|f(t,s) - R_{n,r}(f)(t,s)\|_{L^q} \leq C_q \Omega_1\left(f; \sqrt{1-t^2}/n + |t|/n^2\right)_q,$$

$$n \in \mathbb{N}, \quad t \in [-1,1].$$

Now we are in position to prove the announced Teljakowskii-type estimate.

THEOREM 12.13
For any $f \in L^q([-1,1] \times S) \cap CM^q[-1,1]$, $1 \le q < +\infty$, there exists a random algebraic polynomial of degree $r(n-1)$ with $r = [q] + 2$, $\overline{R}_{n,r}(f)(t,s)$, such that

$$\|f(t,s) - \overline{R}_{n,r}(f)(t,s)\|_{L^q} \le C_q \Omega_1 \left(f; \sqrt{1-t^2}/n\right)_q, \quad n \in \mathbb{N},\ t \in [-1,1],$$

where $C_q > 0$ is independent of f, t and n.

PROOF For $r = [q] + 2$, let us define

$$\overline{R}_{n,r}(f)(t,s) = R_{n,r}(f)(t,s) + [(t+1)/2][f(1,s) - R_{n,r}(f)(1,s)] +$$
$$+ [(1-t)/2][f(-1,s) - R_{n,r}(f)(-1,s)],$$

where $R_{n,r}(f)(t,s)$ is given by Corollary 12.3.

As in the proof of Theorem 1.3 in Lehnhoff [55], we have three cases:

a) $1/n \le \sqrt{1-t^2}$;
b) $\sqrt{1-t^2} \le 1/n,\ t \ge 0$;
c) $\sqrt{1-t^2} \le 1/n,\ t < 0$.

We will use Lemma 12.1 and Corollary 12.3, and the fact that for fixed t we have

$$\|f(t,s) - g(t,s)\|_{L^q} \le \|f(t,s)\|_{L^q} + \|g(t,s)\|_{L^q},$$

$$\|\lambda f(t,s)\|_{L^q} = |\lambda| \cdot \|f(t,s)\|_{L^q}$$

and the following properties of $\Omega_1(f;\cdot)_q$ (see, e.g., Weba [78]):

$$\Omega_1(f; n\delta)_q \le n\Omega_1(f;\delta)_q,$$

$$\Omega_1(f;\lambda\delta)_q \le (\lambda+1)\Omega_1(f;\delta)_q, \ \forall\ \delta \ge 0,\ n \in \mathbb{N},\ \lambda > 0.$$

Case a). For all $t \in [-1,1]$ we have

$$\|f(t,s) - \overline{R}_{n,r}(f)(t,s)\|_{L^q} \le \|f(t,s) - R_{n,r}(f)(t,s)\|_{L^q} +$$
$$+ [(t+1)/2]\|f(1,s) - R_{n,r}(f)(1,s)\|_{L^q} +$$
$$+ [(1-t)/2]\|f(-1,s) - R_{n,r}(f)(-1,s)\|_{L^q} \le$$
$$\le C_q \Omega_1 \left(f; \sqrt{1-t^2}/n + |t|/n^2\right)_q + 2C_q \Omega_1(f; 1/n^2)_q \le$$
$$\le C_q \Omega_1 \left(f; 2\sqrt{1-t^2}/n\right)_q + 2C_q \Omega_1 \left(f; \sqrt{1-t^2}/n\right)_q \le$$

$$\leq 4C_q\Omega_1\left(f;\sqrt{1-t^2}/n\right)_q.$$

Case b). Reasoning as in the proof of Theorem 1.3, case b, in Lehnhoff [55], for all $t\in[-1,1]$ we obtain

$$\|f(t,s)-\overline{R}_{n,r}(f)(t,s)\|_{L^q}\leq\|f(t,s)-f(1,s)\|_{L^q}+$$

$$+\|R_{n,r}(f)(t,s)-R_{n,r}(f)(1,s)\|_{L^q}+$$

$$+[(1-t)/2]\{\|f(1,s)-R_{n,r}(f)(1,s)\|_{L^q}+$$

$$+\|f(-1,s)-R_{n,r}(f)(-1,s)\|_{L^q}\}\leq$$

$$\leq\Omega_1(f;1-t)_q+C_q(1-t)\Omega_1\left(f;\frac{1}{n^2}\right)_q+$$

$$+\|R_{n,r}(t)(t,s)-R_{n,r}(f)(1,s)\|_{L^q}\leq$$

$$\leq\Omega_1(f;1-t^2)_q+C_q(1-t^2)\Omega_1\left(f;\frac{1}{n^2}\right)_q+$$

$$+\|R_{n,r}(f)(t,s)-R_{n,r}(f)(1,s)\|_{L^q}\leq\text{(see Lehnhoff [55])}\leq$$

$$\leq\Omega_1\left(f;\sqrt{1-t^2}/n\right)_q+$$

$$+C_q(1-t^2)\left[1+1/\left(n\sqrt{1-t^2}\right)\right]\Omega_1\left(f;\sqrt{1-t^2}/n\right)_q+$$

$$+\|R_{n,r}(f)(t,s)-R_{n,r}(f)(1,s)\|_{L^q}\leq$$

$$\leq\left[1+2C_q\sqrt{1-t^2}/n\right]\Omega_1\left(f;\sqrt{1-t^2}/n\right)_q+D\leq$$

$$\leq[1+2C_q/n^2]\Omega_1\left(f;\sqrt{1-t^2}/n\right)_q+D\leq$$

$$\leq(1+2C_q)\Omega_1\left(f;\sqrt{1-t^2}/n\right)_q+D,$$

for all $n\geq 1$, where $D:=\|R_{n,r}(f)(t,s)-R_{n,r}(f)(1,s)\|_{L^q}$.

Now, if we wish to estimate D, we distinguish two possibilities:

1) $q=1$;

2) $q>1$.

In the first case we get $r=3$ and

$$\|R_{n,r}(f)(t,s)-R_{n,r}(f)(1,s)\|_{L^1}=$$

$$(1/\pi)\int_S\left|\int_{-\pi}^{\pi}[f(\cos(\arccos t+v),s)-f(\cos v,s)]K_{3n-3}(v)dv\right|dP(s)\leq$$

$$(1/\pi)\int_S \left[\int_{-\pi}^{\pi} |f(\cos(\arccos t+v),s)-f(\cos v,s)|K_{3n-3}(v)dv\right] dP(s) =$$

$$(1/\pi)\int_{-\pi}^{\pi}\left[\int_S |f(\cos(\arccos t+v),s)-f(\cos v,s)|dP(s)\right] K_{n-3}(v)dv \leq$$

$$\leq (1/\pi)\int_{-\pi}^{\pi} \Omega_1(f;|\cos(\arccos t+v)-\cos v|)_1 K_{3n-3}(v)dv \leq$$

$$\leq \Omega_1\left(f;\sqrt{1-t^2}/n\right)_1 \left[1+n/\left(\pi\sqrt{1-t^2}\right)\cdot\right.$$

$$\left.\cdot\int_{-\pi}^{\pi} |\cos(\arccos t+v)-\cos v|K_{3n-3}(v)dv\right] \leq$$

$$\leq (\text{see Lehnhoff [55], pp.179}-180) \leq C_q'\Omega_1\left(f;\sqrt{1-t^2}/n\right)_q.$$

In the second case $q>1$, from $1/p+1/q=1$, by Hölder's inequality, a.s. $s\in S$, we get

$$I_{n,r}(t,s) := \int_{-\pi}^{\pi} |f(\cos(\arccos t+v),s)-$$

$$-f(\cos v,s)|K_{r(n-1)}^{1/q}(v)K_{r(n-1)}^{1/p}(v)dv \leq$$

$$\leq \left(\int_{-\pi}^{\pi} |f(\cos(\arccos t+v),s)-f(\cos v,s)|^q K_{r(n-1)}(v)dv\right)^{1/q}\cdot$$

$$\cdot\left(\int_{-\pi}^{\pi} K_{r(n-1)}(v)dv\right)^{1/p}.$$

Hence

$$\|R_{n,r}(f)(t,s)-R_{n,r}(f)(1,s)\|_{L^q} \leq$$

$$\leq (1/\pi)\left\{\int_S \left[\int_{-\pi}^{\pi} |f(\cos(\arccos t+v),s)-\right.\right.$$

$$\left.\left.-f(\cos v,s)|K_{r(n-1)}(v)dv\right]^q dP(s)\right\}^{1/q} =$$

$$= (1/\pi)\left\{\int_S [I_{n,r}(t,s)]^q dP(s)\right\}^{1/q} \leq$$

$$\leq (1/\pi)\left\{\int_S \left[\int_{-\pi}^{\pi} |f(\cos(\arccos t + v), s) - f(\cos v, s)|^q K_{r(n-1)}(v)dv\cdot\right.\right.$$

$$\left.\left.\cdot\int_{-\pi}^{\pi} K_{r(n-1)}(v)dv\right]^{q/p} dP(s)\right\}^{1/q} = \left[(1/\pi)\int_{-\pi}^{\pi} K_{r(n-1)}(v)dv\right]^{1/p}\cdot$$

$$\cdot(1/\pi)^{1/q}\left\{\int_{-\pi}^{\pi}\left[\int_S |f(\cos(\arccos t + v), s) - \right.\right.$$

$$\left.\left. - f(\cos v, s)|^q dP(s)\right]^{(1/q)q} K_{r(n-1)}(v)dv\right\}^{1/q} \leq$$

$$\leq (1/\pi)^{1/q}\left\{\int_{-\pi}^{\pi} \Omega_1^q(f; |\cos(\arccos t + v) - \cos v|)K_{r(n-1)}(v)dv\right\}^{1/q} \leq$$

$$\leq \text{(by Lemma 12.1)} \leq$$

$$\leq (1/\pi)^{1/q}\left\{\int_{-\pi}^{\pi}\left[(2 + n|v|)\Omega_1\left(f; \sqrt{1-t^2}/n\right)_q\right]^q K_{r(n-1)}(v)dv\right\}^{1/q} =$$

$$= (1/\pi)^{1/q}\Omega_1\left(f; \sqrt{1-t^2}/n\right)_q \left\{\int_{-\pi}^{\pi} (2 + n|v|)^q K_{r(n-1)}(v)dv\right\}^{1/q} =$$

$$= 2^{1/q}\Omega_1\left(f; \sqrt{1-t^2}/n\right)_q \left\{(1/\pi)\int_0^{\pi} (2 + nv)^q K_{r(n-1)}(v)dv\right\}^{1/q} \leq$$

$$\leq C_q\Omega_1\left(f; \sqrt{1-t^2}/n\right)_q,$$

since $[q] + 1 < 2r - 2 = 2([q] + 2) - 2 = 2[q] + 2$ and by Lorentz [56], p.57, relation (3), we have

$$(1/\pi)\int_0^{\pi} (2 + nv)^q K_{r(n-1)}(v)dv \leq C'_q.$$

Case c). Analogously with Lehnhoff [55], p.181, we get

$$\|f(t,s) - \overline{R}_{n,r}(f)(t,s)\|_{L^q} \leq \|f(t,s) - f(-1,s)\|_{L^q} +$$

$$+\|R_{n,r}(f)(-1,s) - R_{n,r}(f)(t,s)\|_{L^q}+$$

$$+[(1+t)/2]\{\|f(-1,s)-R_{n,r}(f)(-1,s)\|_{L^q}+\|R_{n,r}(f)(1,s)-f(1,s)\|_{L^q}\}.$$

Since $t < 0$ implies $1+t \leq 1-t^2$, reasoning as in case b) we arrive at

$$\|f(t,s) - \overline{R}_{n,r}(f)(t,s)\|_{L^q} \leq C_q\Omega_1\left(f; \sqrt{1-t^2}/n\right)_q,$$

which proves the theorem. ∎

The following two open questions are suggested to the reader.

Open question A. Let $0 < q < 1$. In this case it is known that $L^q(S, \mathcal{B}, P)$ endowed with the metric

$$\|g\|_q^* = \int_S |g(s)|^q dP(s), \quad g \in L^q(S, \mathcal{B}, P)$$

is a complete metric space and for $f : [a,b] \to L^q(S, \mathcal{B}, P)$, the L^q-modulus of smoothness of order $n \in \mathbb{N}$ is given by

$$\Omega_n^*(f;\delta)_q = \sup\left\{\sup\left\{\int_S |\Delta_h^n f(t,s)|^q dP(s);\right.\right.$$

$$\left.\left. t \in [a, b-nh]\right\}; 0 \leq h \leq \delta\right\}.$$

Prove similar results with Theorems 12.11, 12.12, 12.13 in this case, too.

Open question B. Find estimates in approximation by random Bernstein polynomials or of Jackson, Timan or Teljakowskii type, in terms of the d-modulus and ρ-modulus considered in Definition 12.9.

12.6 Fourier Series of Periodic Stochastic Processes and Applications

In this section it is more convenient to consider instead of real random variables, the more general case of complex random variables and correspondingly, the complex-valued stochastic processes. All the concepts remain the same (the absolute value $|\cdot|$ on $\mathbb{R}$ is replaced by the modulus on $\mathbb{C}$). Let $(S, \mathcal{B}, P)$ be a complete probability space.

We will denote by $L(S, \mathcal{B}, P)_{\mathbb{C}} = \{g : S \to \tilde{\mathbb{C}}; g$ is $\mathcal{B}$-measurable and a.s. finite on $S\}$. Then a complex-valued stochastic process is a mapping $f : [a, b] \to L(S, \mathcal{B}, P)_{\mathbb{C}}$.

For $1 \leq q < +\infty$, let us denote

$L^q_{2\pi}[-\pi, \pi] = \{f : \mathbb{R} \to L^q(S, \mathcal{B}, P)_{\mathbb{C}}; f(t, s)$ is $\mathcal{L}(\mathbb{R}) \times \mathcal{B}$ measurable and $\exists\, A \subset S$ with $P(A) = 1$ such that $\int_S |f(t+2\pi, s) - f(t, s)| dP(s) = 0$, for all $t \in \mathbb{R}$ and $(\mathcal{L}) \int_{-\pi}^{\pi} |f(t, s)|^q dt < +\infty, \ \forall\, s \in A\}$.

REMARK 12.17 Comparing with the set $L^q_{2\pi}(\mathbb{R} \times S)$ introduced after Theorem 12.10, we see that

$$L^q_{2\pi}(\mathbb{R} \times S) \subset L^q_{2\pi}[-\pi, \pi], \text{ strictly,}$$

(if $L^q_{2\pi}(\mathbb{R} \times S)$ is also defined for complex-valued stochastic processes). ∎

DEFINITION 12.10 *(Kawata [50]) Let $f \in L^q_{2\pi}[-\pi, \pi]$. The Fourier coefficients and the Fourier sum of order n are given by*

$$C_k(s) = \frac{1}{2\pi} (\mathcal{L}) \int_{-\pi}^{\pi} f(t, s) e^{-ikt} dt, \quad s \in S$$

and

$$F_n(t, s) = \sum_{k=-n}^{n} C_k(s) e^{ikt}, \quad t \in \mathbb{R}, \quad s \in S,$$

respectively.

The L^q-modulus of smoothness of order $n \in \mathbb{N}$ is given by

$$\Omega_n(f; \delta)_q =$$

$$= \sup \left\{ \sup \left\{ \left(\int_S |\Delta^n_h f(t, s)|^q dP(s) \right)^{1/q} ; t \in \mathbb{R} \right\} ; 0 \leq h \leq \delta \right\},$$

and the integrated L^q-modulus of smoothness of order $n \in \mathbb{N}$ is given by

$$\Omega^*_n(f; \delta)_q =$$

$$= \sup \left\{ \left(\frac{1}{2\pi} (\mathcal{L}) \int_{-\pi}^{\pi} \left[\int\limits_{S} |\Delta_h^n f(t,s)|^q dP(s) \right] dt \right)^{1/q} ; 0 \leq h \leq \delta \right\}.$$

Known arguments usually used in the theory of absolute convergence of ordinary Fourier series can be adapted to the random case and the obtained results are mostly the analogues of what are well known in classical theory. However, they are of particular interest, because some of them are directly applied to get some theorems on sample continuity and differentiability of stochastic processes, which seems to provide the simplest way of deriving them, at least for the periodic case.

DEFINITION 12.11 *(Kawata [50]) Let $f \in L_{2\pi}^q[-\pi, \pi]$, $1 \leq q < +\infty$. If*

$$\sup_D \sum_{j=1}^{n} \left(\int\limits_S |f(t_j, s) - f(t_{j-1}, s)|^q dP(s) \right)^{1/q} = V_q < +\infty,$$

where sup is taken for all divisions

$$D : -\pi \leq t_0 < t_1 < \ldots < t_n \leq \pi,$$

then we say that $f(t,s)$ is of bounded variation and we write $f \in BV_{2\pi}^q$.

In what follows we present some selected results (for other details and for applications to sample continuity and differentiability of stochastic processes, see Kawata [50]).

THEOREM 12.14 (Kawata [50])
If $f \in BV_{2\pi}^q$, $1 < q \leq 2$, and if

$$\Omega_n(f; \delta)_q = \mathcal{O}(\delta^\beta), \text{ for some } \beta > 0,$$

then

$$\sum_{k=-\infty}^{+\infty} |C_k(s)| < +\infty, \text{ a.s. } s \in S,$$

where $C_k(s)$ are the Fourier coefficients of f.

REMARK 12.18 This result is the random version of the well-known Zygmund theorem on absolute convergence of Fourier series. ∎

THEOREM 12.15 (Kawata [50])

$$(i)\ \inf\left\{\frac{1}{2\pi}(\mathcal{L})\int_{-\pi}^{\pi}\left[\int_{S}\left|f(t,s)-\sum_{k=-n}^{n}a_k(s)e^{ikt}\right|^2 dP(s)\right]dt;\right.$$

$$\left.a_k\in L(S,\mathcal{B},P)_{\mathbb{C}}\right\}=\frac{1}{2\pi}(\mathcal{L})\int_{-\pi}^{\pi}\left[\int_{S}\left|f(t,s)-\sum_{k=-n}^{n}C_k(s)e^{ikt}\right|^2 dP(s)\right]dt,$$

where $C_k(s)$ *are the Fourier coefficients of* f.

(ii) Let $1\le q<+\infty$. *Denoting*

$$e_n^{(q)}(f)=\inf\left\{\left(\frac{1}{2\pi}(\mathcal{L})\int_{-\pi}^{\pi}\left[\int_{S}\left|f(t,s)-\sum_{k=-n}^{n}a_k(s)e^{ikt}\right|^q\right]dt\right)^{1/q};\right.$$

$$\left.a_k\in L(S,\mathcal{B},P)_{\mathbb{C}}\right\},$$

we have

$$e_{2n-1}^{(q)}(f)\le C_q\Omega_2^*\left(f;\frac{2\pi}{n}\right)_q.$$

Moreover, if $f\in BV_{2\pi}^1$, *then* $e_n^{(1)}(f)=\mathcal{O}(1/n)$.

(iii) Denoting by

$$\sigma_n(f)(t,s)=\frac{1}{2\pi n}(\mathcal{L})\int_{-\pi}^{\pi}f(v+t,s)\left[\frac{\sin(nv/2)}{\sin(v/2)}\right]^2 dv$$

the Fejér random trigonometric polynomials, if $f\in L_{2\pi}^q[-\pi,\pi]$ *is "P"-continuous on* $\mathbb{R}$, *then*

$$\sigma_n(f)(t,s)\xrightarrow[n\to\infty]{"P"}f(t,s).$$

12.7 Quadratures of Stochastic Processes

Let $f:[a,b]\to L^q(S,\mathcal{B},P)$, $1\le q<+\infty$. Consider the problem of estimating for the integral (see, e.g., Definition 12.4)

$$(c)\int_a^b w(t)f(t,s)dt,\ \text{where } c\in\mathcal{C}=\{\text{"a.s."},\text{"}P\text{"},\text{"}L^q\text{"}\},$$

by means of a finite linear combination $\sum_{j=1}^{n} c_j f(t_j, s)$, where $f(t_j, s)$ are given observations at times $t_j \in [a, b]$ and w is a known non-random weight function. There are many ways to select appropriate sample weights $c_j \in \mathbb{R}$ and to determine nodes $t_j \in [a, b]$. In the literature, the estimation problem is treated as a problem of statistical design (or as a special regression problem), see, e.g., Cambanis [17] or Ylvisaker [83].

In general, knowledge of covariances is required and the theory is restricted to L^2-processes.

A different approach to the estimation problem is the following: Apply classical well-behaved quadrature formulae on the paths of $f(t, s)$ and describe the estimation error with the aid of an extension argument for bounded functionals. Convergence of these estimates may be proved for processes being sufficiently "smooth" but no additional assumptions are made. Covariances are not required and the results are valid with respect to arbitrary L^q-norms.

For details of this last approach see Weba [80].

12.8 Random Condensation of Singularities and Divergence Results

Divergence results for interpolation and quadrature formulas in classical theory, derived from the classical condensation of singularities principle, are well-known (see Muntean [57]-[58]).

In this section we will present a random version of the above principle, from which are derived some divergence results for interpolation and quadrature of stochastic processes.

First we need some preliminaries (see for details Velasco and Villena [75]).

Let $(S, \mathcal{B}, P)$ be a complete probability space. Given the Banach spaces $(X, \|\cdot\|_X)$ and $(Y, \|\cdot\|_Y)$, a map $\phi : X \times S \to Y$ is said to be a random operator from X to Y if, for each $x \in X$, $\phi(x, s)$ as a function of s lies in $L_Y(S, \mathcal{B}, P)$, the linear space of all Y-valued Bochner random variables on S (see Remark 12.7). The random operator ϕ is said to be linear if

$$P(\{s \in S; \phi(\alpha_1 x_1 + \alpha_2 x_2, s) = \alpha_1 \phi(x_1, s) + \alpha_2 \phi(x_2, s)\}) = 1,$$

for all $x_1, x_2 \in X$ and α_1, α_2 scalars. It should be noted that $L_Y(S, \mathcal{B}, P)$ with almost surely identification becomes a metrizable complete linear topological space for the convergence in probability ("P"). Accordingly, the continuity in probability of ϕ at $x_0 \in X$ means that

$$\lim_{x \to x_0} P(\{s \in S; \|\phi(x,s) - \phi(x_0,s)\|_Y > \varepsilon\}) = 0, \ \forall\, \varepsilon > 0.$$

A family $\mathcal{A}$ of random operators from X to Y is equi-continuous in probability if for every $\varepsilon > 0$ there exists $M_\varepsilon > 0$ such that

$$P(\{s \in S; \|\phi(x,s)\|_Y \leq M_\varepsilon \|x\|_X\}) \geq \varepsilon, \ \forall\, x \in X, \ \forall\, \phi \in \mathcal{A}.$$

We say that a linear random operator ϕ is probably continuous if there exists $0 < \delta < 1$ with the following property: For every $0 < \delta' < \delta$, there exists $M_{\delta'} > 0$ such that

$$P(\{s \in S; \|\phi(x,s)\|_Y \leq M_{\delta'} \|x\|_X\}) \geq \delta', \ \forall\, x \in X.$$

The random version of the condensation of singularities principle can be stated as follows.

THEOREM 12.16 (Gal and Villena [43])
Let $(X, \|\cdot\|_X)$ and $(Y, \|\cdot\|_Y)$ be Banach spaces and let $\mathcal{A}$ be a family of probably continuous linear random operators from X to Y. Let $\delta > 0$ be fixed and let us consider the set C_δ of those $x \in X$ for which there is $M_x > 0$ such that

$$P(\{s \in S; \|\phi(x,s)\|_Y \leq M_x\}) \geq \delta,$$

for all $\phi \in \mathcal{A}$. If C_δ is of second category in X, then there exists $M > 0$ such that

$$P(\{s \in S; \|\phi(x,s)\|_Y \leq M\|x\|_X\}) \geq (2\delta - 1) - (1 - \alpha(\phi)),$$

$\forall\, x \in X$, $\phi \in \mathcal{A}$, where $\alpha(\phi) = \lim_{\varepsilon \to 0} \liminf_{x \to 0} P(\{s \in S; \|\phi(x,s)\|_Y \leq \varepsilon\})$.

Accordingly, if there is no $M > 0$ for which the above inequality holds, then the set S_δ of those $x \in X$ for which for each $M > 0$ there is $\phi_M \in \mathcal{A}$ such that

$$P(\{s \in S; \|\phi_M(x,s)\|_Y \leq M\}) < \delta,$$

is residual and therefore dense in X.

REMARK 12.19 If $C_\delta = X$ then we get the random version of the Banach-Steinhaus theorem (see Velasco and Villena [75]). ∎

At the end of this section we assume $X = \{\phi : [a,b] \to L^q(S, \mathcal{B}, P);$ ϕ is "L^q"-continuous on $[a,b]\}$. It is worth pointing out that X, with the identification of equivalent processes, becomes a Banach space with the norm given by

$$\|\phi\|_X = \max\left\{\left(\int_S |\phi(t,s)|^q dP(s)\right)^{1/q}; t \in [a,b]\right\}.$$

The random Lagrange interpolation polynomial attached to a $\phi \in X$ on the fixed knots $a \le t_n^{(1)} < \ldots < t_n^{(n)} \le b$ is the random polynomial given by

$$L_n(\phi)(t,s) = \sum_{k=1}^{n} l_{k,n}(t)\phi(t_n^{(k)}, s),$$

where $l_{k,n}(t)$ are the classical fundamental Lagrange interpolation polynomials. Obviously L_n are random operators from X to $C[a,b]$.

As a consequence of Theorem 12.16 we obtain

THEOREM 12.17 (Gal and Villena [43])
The set of those random functions ϕ for which the set $\{\|L_n(\phi)\|^*; n \in \mathbb{N}\}$ is unbounded in probability is dense in X
(where $\|L_n(\phi)\|^* = \sup\{|L_n(\phi)(t,s)|; t \in [a,b]\}$).

Now, for every $\phi \in X$, the quadrature formulas give the value of the integral ("P") $\int_a^b \phi(t,s)dt$ (see Definition 12.4), approximately by the following expression

$$Q_n(\phi)(s) = \sum_{k=1}^{n} c_{k,n}\phi(t_n^{(k)}, s), \quad s \in S,$$

where the knots $a \le t_n^{(1)} < \ldots < t_n^{(n)} \le b$ and the nonrandom coefficients $c_{k,n}$ are independent of ϕ.

Another consequence of Theorem 12.16 is the following.

THEOREM 12.18 (Gal and Villena [43])
If

$$\sup\left\{\sum_{k=1}^{n}|c_n^{(k)}|; n \in \mathbb{N}\right\} = +\infty,$$

then the set of those $\phi \in X$ *for which the set* $\{Q_n(\phi)(s); n \in \mathbb{N}\}$ *is unbounded in probability is dense in* X.

12.9 Abstract Random Best Approximation

The abundance of results concerning best approximation in usual normed spaces, on one hand, and the connections of best approximation with fixed point theory, on the other hand, are well-known.

The aim of this section is to present random variants of these results. In the first part, we deal with random versions of some results of best approximation in usual normed spaces. The basic concept of random (probabilistic) normed space has been introduced in Serstnev [68]-[70], where some problems regarding the best approximation are also studied (see also Istrăţescu [48], p.155–165).

First, we need some preliminaries.

Let D be the set of all left continuous real-valued functions defined on $\mathbb{R}$, with the properties:

(i) $f(x) = 1$, for all $x \leq 0$;

(ii) $f(\infty) = 0$;

(iii) f is nonincreasing on $\mathbb{R}$.

On D is considered the natural ordering, i.e., if $f, g \in D$, then $f \leq g$ means $f(x) \leq g(x)$, $\forall\, x \in \mathbb{R}$ and $f < g$ means $f \leq g$ and there is an $x_0 \in \mathbb{R}$ with $f(x_0) < g(x_0)$.

Denote by $\Delta \in D$, the function having the property $\Delta(x) = 0$, for all $x > 0$. Obviously $\Delta \leq f$, $\forall\, f \in D$.

Let E be a subset of D containing the element Δ.

DEFINITION 12.12 *(Serstnev [68]) An application* $\mu : E \times E \to E$ *is called a triangular function if satisfies the properties:*

(i) $\mu(a, b) = \mu(b, a)$, $\mu(a, \mu(b, c)) = \mu(\mu(a, b), c)$, $\forall\, a, b, c \in E$;

(ii) $\mu(a, \Delta) = a$, $\forall\, a \in E$;

(iii) $\mu(a_1, a) \leq \mu(a_2, a)$, $\forall\, a, a_1, a_2 \in E$, $a_1 \leq a_2$;

(iv) $\mu(a,b) \le \mu_1(a,b)$, $\forall\ a,b \in E$; *where*

$$\mu_1(a,b)(x) = \inf_{t\in[0,1]} \min\{a(tx) + b[(1-t)x], 1\}.$$

A real linear space $\mathcal{L}$ is called random (probabilistic) normed space if there exist the mappings $f : \mathcal{L} \to D$ and $\mu : E \times E \to E$ ($f(\mathcal{L}) \subset E \subset D$) triangular function, such that the following properties hold:

(i) $f(\varphi) = \Delta$ *iff* $\varphi = 0_{\mathcal{L}} \in \mathcal{L}$;

(ii) $f(\alpha\varphi)(x) = f(\varphi)(x/\alpha)$, $\forall\ \varphi \in \mathcal{L}$, $\alpha \in \mathbb{R}$, $\alpha \neq 0$, $x \in \mathbb{R}$;

(iii) $f(\varphi + \psi)(x) \le \mu(f(\varphi), f(\psi))$, $\forall\ \varphi, \psi \in \mathcal{L}$.

For analogy with classical normed spaces we use the notation

$$f(\varphi)(x) = \|\varphi; x\|, \quad \varphi \in \mathcal{L}, \quad x \in \mathbb{R} \quad (or\ f(\varphi) = \|\varphi; \cdot\|).$$

DEFINITION 12.13 *Let M be a subspace of a random normed space $\mathcal{L}$. Then*

(i) M is called proximinal if for all $\phi \in \mathcal{L}$, there exists $\phi \in M$ such that

$$\|\phi - \psi;\| = \inf_{\eta \in M} \|\phi - \eta;\| := d(\phi, M).$$

Here ψ is called best approximation of ϕ by elements in M.

(ii) M is called Chebyshev, if for each $\phi \in \mathcal{L}$ there is a unique $\phi \in M$ such that $\|\phi - \psi;\| = d(\phi, M)$.

From the results of best approximation in random normed spaces we selected the following.

THEOREM 12.19 (Serstnev [70])

Let $\mathcal{L}$ be a random normed space and M be a finite-dimensional subspace of $\mathcal{L}$. Then M is proximinal.

Exactly as in the usual case, the uniqueness of the element of best approximation holds in some particular cases.

DEFINITION 12.14 *(Serstnev [70], Beg [5]) A random normed space $(\mathcal{L}, \|;\|, \mu)$ is called strictly convex, whenever $\varphi, \psi \in \mathcal{L}$ and*

$$\|\varphi + \psi; \cdot\| = \mu(\|\varphi; \cdot\|, \|\psi; \cdot\|),$$

then there exists $\lambda > 0$ such that $\varphi = \lambda\psi$.

A random normed space $\mathcal{L}$ is said to be uniformly convex if for any $\alpha > 0$, there exists $\beta > 0$ such that for any two-unit vector $\varphi, \psi \in \mathcal{L}$ (i.e., $\|\varphi; x\| = \|\psi; x\| = \Delta(x-1)$, for $x \in \mathbb{R}$)

$$\left\|\frac{\varphi + \psi}{2}; 1 - \beta\right\| = 1 \text{ implies } \|\varphi - \psi; \alpha\| = 0.$$

THEOREM 12.20 (Serstnev [70])
If $(\mathcal{L}, \|;\|, \mu)$ is a strictly convex random normed space and M is a finite-dimensional subspace of $\mathcal{L}$, then M is a Chebyshev space.

THEOREM 12.21 (Beg [5])
Let M be a closed convex subset of an uniformly convex random Banach space. Then M is Chebyshev space with $\mu(a, b) = \max\{a, b\}$.

In what follows we pass to the second part of this section. The starting point is the classical Ky Fan result that combines fixed point theory with best approximation (in the usual setting).

THEOREM 12.22 (Ky Fan [52])
Let M be a non-empty compact convex set in an usual normed space $(X, \|\cdot\|)$. For any continuous map $T : M \to X$, there exists a point $x \in M$ such that

$$\|x - T(x)\| = d(T(x), M),$$

where $d(y, M) = \inf\{\|y - z\|; z \in M\}$.

The aim of this second part is to present some random versions of Ky Fan's result and applications to random fixed point theory.

First we need some preliminaries. Let $(X, \|\cdot\|)$ be a normed space and $(S, \mathcal{B}, P)$ be a complete probability space.

Let $M \subset X$. A mapping $T : X \times M \to X$ is called a (abstract) random operator if for any fixed $x \in M$, $T(\cdot, x)$ is measurable (i.e., if we denote $g(s) = T(s, x)$, $s \in S$, $g : S \to X$, then for any open $U \subset X$ we have $g^{-1}(U) \in \mathcal{B}$).

A measurable mapping $\xi : S \to M$ is called a random fixed point of the random operator $T : S \times M \to X$, if $\xi(s) = T(s, \xi(s))$, for all $s \in S$. A random operator $T : S \times M \to X$ is called non-expansive if

$$\|T(s, x) - T(s, y)\| \leq \|x - y\|, \text{ for all } x, y \in M \text{ and all } s \in S.$$

A random operator $T : S \times M \to X$ is said to be almost affine if

$$\|T(s,x) - y\| \le \lambda\|T(s,x_1) - y\| + (1-\lambda)\|T(s,x_2) - y\|,$$

for all $s \in S,\ x_1, x_2 \in M,\ 0 < \lambda < 1,\ x = \lambda x_1 + (1-\lambda)x_2,\ y \in X$.
From a great variety we selected the following very recent results.

THEOREM 12.23 (Beg and Shahzad [8])
Let M be a compact and convex subset of a Banach space $(X, \|\cdot\|)$ and $g : S \times M \to M$ be a continuous almost affine onto random operator. For each continuous random operator $f : S \times M \to X$, there exists a measurable map $\xi : S \to M$ satisfying

$$\|f(s,\xi(s)) - g(s,\xi(s))\| = d(f(s,\xi(s)), M),\ \forall\ s \in S.$$

THEOREM 12.24 (Beg and Shahzad [9], Beg [6])
Let M be a non-empty separable closed convex subset of a Hilbert space H and $f : S \times M \to H,\ M \subset H$, a non-expansive random operator with $\overline{f(s,M)}$ compact, for any $s \in S$. Then there exists a measurable map $\xi : S \to M$ such that

$$\|\xi(s) - f(s,\xi(s))\| = d(f(s,\xi(s)), M),\ \forall\ s \in S.$$

An application of Theorem 12.23 to random fixed point theory is

THEOREM 12.25 (Beg and Shahzad [9])
Let M be a non-empty separable closed convex subset of a Hilbert space H and $f : S \times M \to H$ be a non-expansive random operator with $\overline{f(s,M)}$ compact for any $s \in S$, such that for each x on the boundary of M,

$$\|f(s,x) - \xi_1(s)\| \le \|x - \xi_1(s)\|, \quad s \in S,$$

for some measurable map $\xi_1 : S \to M$. Then f has a random fixed point.

A random version of the Singer-type results in classical approximation theory (see, e.g., Singer [71]) could be

THEOREM 12.26 (Beg [7])
Let M be a non-empty convex subset of a complex normed space $(X, \|\cdot\|)$, $T : S \times M \to X$ be a random operator and $\xi : S \to M$ be a

measurable map such that $T(s,\xi(s)) \notin \overline{M}$. *Then* ξ *satisfies*

$$\|\xi(s) - T(s,\xi(s))\| = d(T(s,\xi(s)), M),$$

if and only if there exists $f \in X'$ *(here* $(X', |||\cdot|||)$ *denotes the dual of* X*), with the following properties:*
(i) $|||f||| = 1$,
(ii) $f(T(s,\xi(s)) - \xi(s)) = \|T(s,\xi(s)) - \xi(s)\|$,
(iii) $Re\, f(x - \xi(s)) \leq 0$, *for all* $x \in M$.

REMARK 12.20 The above results have been extended to random multivalued operators, too (see, e.g., Beg and Shahzad [10], Papageorgiou [63], Beg and Shahzad [13] and their references). ▮

Details on the part two of this section can be also found in Lin [53], Sehgal and Singh [67], Tan and Yuan [74], Beg and Shahzad [11], [12] and in their references.

12.10 Stochastic Global Smoothness Preservation

In approximating a stochastic process $f = f(t,s)$ by means of approximation operators $\tilde{L}_n$, it is interesting to find out which properties of f are preserved by the approximants $\tilde{L}_n(f)$.

For instance, one can be interested in comparing global smoothness characteristics of f and $\tilde{L}_n(f)$. In this section we present the random version of the main theorem in Anastassiou, Cottin and Gonska [2].

Let $(S, \mathcal{B}, P)$ be a complete probability space and let us denote

$$B([a,b]; L^1(S,\mathcal{B},P)) =$$

$$= \left\{ f : [a,b] \to L^1(S,\mathcal{B},P);\ \sup_{t\in[a,b]} \int_S |f(t,s)| dP(s) < +\infty \right\},$$

$$CM^1[a,b] = \{f : [a,b] \to L^1(S,\mathcal{B},P); f \text{ is } (L^1) - \text{continuous on } [a,b]\}.$$

Obviously we have the inclusions

$$B[a,b] \subset B([a,b]; L^1(S,\mathcal{B},P)),$$

$$C[a,b] \subset CM^1[a,b] \subset B([a,b]; L^1(S,\mathcal{B},P)),$$

where $B[a,b] = \{g : [a,b] \to \mathbb{R}; g \text{ bounded on } [a,b]\}$ and $C[a,b] = \{g : [a,b] \to \mathbb{R}; g \text{ continuous on } [a,b]\}$.

THEOREM 12.27 (Anastassiou and Gonska [3])

Let V be a linear subspace of $CM^1[a,b]$ such that $C[a,b] \subset V \subset CM^1[a,b]$. Let $\tilde{L} : V \to CM^1[a,b]$ be linear and such that the following hold:

(i) $\tilde{L}$ is E-commutative on V (i.e., $E(\tilde{L}(f)) = \tilde{L}(E(f))$, where $E(f)(t) = \int_S f(t,s)dP(s)$ denotes the expectation operator);

(ii) The restriction $L := \tilde{L}|_{C[a,b]}$ is bounded, with norm $|||L||| \neq 0$;

(iii) $L : C^1[a,b] \to C^1[a,b]$ is such that $\|(L(g))'\|_\infty \le C\|g'\|_\infty$, for all $g \in C^1[a,b]$ (where $\|\cdot\|_\infty$ is the uniform norm).

Then for all $f \in V$ and $0 \le \delta \le b-a$ one has

$$\omega_1(E(\tilde{L}(f));\delta) \le (|||L||| + c)\omega_1(E(f);\delta).$$

REMARK 12.21 Applying Theorem 12.27 for Bernstein random polynomials

$$\tilde{B}_n(f)(t,s) = \sum_{k=0}^{n} \binom{n}{k} t^k (1-t)^{n-k} f\left(\frac{k}{n}, s\right),$$

it follows (see Anastassiou and Gonska [3])

$$\omega_1(E(\tilde{B}(f));\delta) \le 2\omega_1(E(f);\delta), \quad 0 \le \delta \le 1, \quad n \in \mathbb{N}.$$

∎

REMARK 12.22 Applications of Theorem 12.27 to various random operators can be found in Anastassiou and Gonska [3]. ∎

REMARK 12.23 An interesting open question would be to find the global smoothness preservation for $\tilde{B}_n(f)$ in terms of the moduli ω_d and ω_ρ introduced by Definition 12.9.

For example, the question is if there is a constant $c > 0$, such that for all $f : [0,1] \to L(S,\mathcal{B},P)$, (P)-continuous on $[a,b]$, we have

$$\omega_d(\tilde{B}_n(f);\delta) \le c\omega_d(f;\delta), \quad 0 \le \delta \le 1?$$

12.11 Approximation of Wiener Processes

The Wienner process is widely applied and studied in mathematics, physics, communication theory and other branches of science (for various properties of Wienner processes, see, e.g., Csörgö and Révész [35]). If we denote it by $W(t,s)$, $t \in \mathbb{R}$, $s \in S$, where $W(t,s) = 0$, for $t < 0$, it is known that with probability one, a path of a Wienner process (i.e., $W(\cdot, s) : t \to W(t,s)$, $s \in S$, where $(S, \mathcal{B}, P)$ is a complete probability space) is an everywhere continuous but nowhere differentiable function. So nobody can imagine a picture of such function. Usually one observes the path of a Wiener process only at discrete points or one considers a smoothness version of it. In general, instead of the Wiener process we can observe one of the following processes:

$$W_1^{(n)}(t,s) = W\left(\frac{i}{n}, s\right) + n\left(t - \frac{i}{n}\right)\left[W\left(\frac{i+1}{n}, s\right) - W\left(\frac{i}{n}, s\right)\right],$$

if $\frac{i}{n} \le t < \frac{i+1}{n}$, $i = 0, 1, \ldots, n-1$, $n \in \mathbb{N}$, $s \in S$,

$$W_2^{(n)}(t,s) = n\int_{-\infty}^{+\infty} K[n(t-y)]W(y,s)dy, \quad n \in \mathbb{N}, \quad s \in S,$$

where the kernel K satisfies $K \in L^1(\mathbb{R})$, $\int_{-\infty}^{+\infty} K(v)dv = 1$, and K has $[-1, 1]$ as support,

$$W_3^{(n)}(t,s) = y_0(s)t + \sqrt{2}\sum_{k=1}^{n} y_k(s)\frac{\sin k\pi t}{k\pi},$$

where $y_0, \ldots, y_n, \ldots$ is a sequence of independent, normal $(0,1)$ random variables and W itself is represented in the form

$$W(t,s) = y_0(s)t + \sqrt{2}\sum_{k=1}^{+\infty} y_k(s)\frac{\sin k\pi t}{k\pi}.$$

It is natural to ask how far $W_i^{(n)}$, $i = 1, 2, 3$, are from W and also how far different characteristics of W, say the local time, will be from the corresponding characteristics (local time) of $W_i^{(n)}$, $i = 1, 2, 3$, when measuring the distance in some suitable norms.

In this sense we present the following.

THEOREM 12.28

(i) (Révész [65])

$$\lim_{n\to\infty}\left[\left(\frac{n}{2\log n}\right)^{1/2} \sup_{0\le t\le 1}|W_1^{(n)}(t,s) - W(t,s)|\right] = 1, \ \ a.s. \ \ s \in S;$$

(ii) (Révész [65])

$$\limsup_{n\to\infty}\left[\left(\frac{n}{\log n}\right)^{1/2} \sup_{0\le t\le 1}|W_2^{(n)}(t,s) - W(t,s)|\right] \le \sqrt{2}, \ \ a.s. \ \ s \in S;$$

(iii) (Stadtmüller [73])

$$\lim_{n\to\infty}\left[\left(\frac{1}{2\log n}\right)^{1/2} \sup_{0\le t\le 1}|W_2^{(n)}(t,s) - W(t,s)|\right] = 1, \ \ a.s. \ \ s \in S;$$

(iv) (Hall [45])

$$\limsup_{n\to\infty}\left[\left(\frac{n}{\log n}\right)^{1/2} \sup_{0\le t\le 1}|W_3^{(n)}(t,s) - W(t,s)|\right] \le \frac{2}{\sqrt{\pi}}, \ \ a.s. \ \ s \in S.$$

REMARK 12.24 For other results concerning approximation of Wiener processes and their characteristics, see, e.g., Révész [65], Stadtmüller [73], Csörgö and Révész [34]–[35], Hall [45]. ∎

REMARK 12.25 In approximation of Wiener processes we suggest using, for example, the random version of the Szász-Mirakjan operators defined by

$$S_n(W)(t,s) = e^{-nt}\sum_{k=0}^{+\infty} W\left(\frac{k}{n}, s\right)\frac{(nt)^k}{k!}, \quad n \in \mathbb{N},$$

and obtaining estimates of $|S_n(W)(t,s) - W(t,s)|$ by using various moduli of continuity (see, e.g., Definition 12.9, adapted to the interval $[0, +\infty)$). ∎

References

[1] Anastassiou, G.A., Korovkin inequalities for stochastic processes, *J. Math. Anal. Appl.* 157(1991), 366-384.

[2] Anastassiou, G.A., Cottin, C. and Gonska, H.H., Global smoothness of approximating functions, *Analysis* 11(1991), 43-57.

[3] Anastassiou, G.A. and Gonska, H.H., On stochastic global smoothness, *Rev. Acad. Cienc. Zaragoza* 49(1994), 119-136.

[4] Andrus, G.F. and Brown, L., Uniform approximation by random functions, *Rocky Mountain J. Math* 14(1984), 373-381.

[5] Beg, I., Approximation in random spaces, *Indian J. Pure Appl. Math.* 28(10)(1997), 1435-1442.

[6] Beg, I., Random fixed points of non-self maps and random approximations, *J. Appl. Math. Stoch. Anal.* 10(1997), 127-130.

[7] Beg, I., A characterization of random approximations, *Intern. J. Math. and Math. Sci.* 22(1999), No.1, 209-211.

[8] Beg, I. and Shahzad, N., Some random approximation theorems with applications, *J. Nonlinear Anal.* 35(1999), 609-616.

[9] Beg, I. and Shahzad, N., Random approximation and random fixed point theorems, *J. Appl. Math. Stoch. Anal.* 7(2)(1994), 143-148.

[10] Beg, I. and Shahzad, N., On random approximation and coincidence point theorems for multivalued operators, *J. Nonlinear Anal.* 26(1996), 1035-1041.

[11] Beg, I. and Shahzad, N., Applications of the proximity map to random fixed point theorems in Hilbert spaces, *J. Math. Anal. Appl.* 196(1995), 606-613.

[12] Beg, I. and Shahzad, N., Random fixed points and approximations in random convex metric spaces, *J. Appl. Math. Stoch. Anal.* 6(1993), No.3, 237-246.

[13] Beg, I. and Shahzad, N., Measurable selections: in random approximations and fixed point theory, *Stoch. Anal. Appl.* 15(1)(1997), 19-29.

[14] Bharucha-Reid, A.T. and Sambandham, M., *Random Polynomials*, Academic Press, Orlando, 1986.

[15] Billingsley, P., *Convergence of Probability Measures*, Wiley, New York, 1968.

[16] Brown, L. and Schreiber, B.M., Approximation and extension of random functions, *Monatsh. Math.* 107(1989), 111-123.

[17] Cambanis, S., Sampling designs for time series, in *Handbook of Statistics* vol. 5, pp.337-362, Elsevier, Amsterdam, New York, 1985.

[18] Cao, J.-D. and Gonska, H.H., Approximation by Boolean sums of positive linear operators, II. Gopengauz-type estimates, *J. Approx. Theory* 57(1989), 77-89.

[19] Cenuşă, G., Sequences of random vector variables ρ-almost periodic in probability (in Romanian), *Econ. Comput. Econom. Cybernet. Stud. Res.* (Bucharest) 1(1980), 57-60.

[20] Cenuşă, G., Ergodic properties of random functions almost periodic in probability, *Stud. Cerc. Mat.* (Bucharest) 32(1980), 4, 371-375.

[21] Cenuşă, G., Properties of mean and of covariance of an almost periodic in probability stochastic process (in Romanian), *Stud. Cerc. Mat.* (Bucharest), 32(1980), 4, 377-383.

[22] Cenuşă, G., Random vector functions ρ-almost periodic, *Bull. Math. Soc. Sci. Math. Roumanie* 1(1981), 15-31.

[23] Cenuşă, G., Properties of the mean of a random function almost periodic in probability. The associated random Fourier series, *Bull. Math. Soc. Sci. Math. Roumanie* 4(1980), 339-346.

[24] Cenuşă, G., Random functions weak almost periodic in probability (in Romanian), *Stud. Cerc. Mat.* (Bucharest) 33(1981), 4, 409-417.

[25] Cenuşă, G., Random functions asymptotically weak almost periodic in probability (in Romanian), *Stud. Cerc. Mat.* (Bucharest) 33(1981), 4, 419-424.

[26] Cenuşă, G., On the mean of an almost periodic in probability random function, *Econ. Comput. Econom. Cybernet. Stud. Res.* (Bucharest) 3(1981).

[27] Cenuşă, G., On random functions almost periodic in Bohr sense, *Econ. Comput. Econ. Cybernet. Stud. Res.* (Bucharest), 1(1982).

[28] Cenuşă, G., A Weierstrass-type theorem for vector random functions, *Anal. Univ. Bucharest* XXIX(1980), 21-25.

[29] Cenuşă, G. and Săcuiu, I., On some stochastic approximation for random functions, *Rend. Mat.* (Roma), 12, ser.VI(1979), 143-156.

[30] Cenuşă, G. and Săcuiu, I., Sequences of random variables almost periodical in probability, *Econ. Comput. Econom. Cybernet. Stud. Res.* (Bucharest), 2(1979), 77-82.

[31] Cenuşă, G. and Săcuiu, I., Some properties of random functions almost periodic in probability, *Rev. Roum. Math. Pures Appl.* 25(1980), 1317-1325.

[32] Constantin, G. and Istrăţescu, I., *Elements of Probabilistic Analysis and Applications*, Academic Press, Bucharest, 1981.

[33] Corduneanu, C., *Almost Periodic Functions*, Second Edition, Chelsea, New York, 1989.

[34] Csörgö, M. and Révész, P., Three strong approximations of the local time of a Wiener process and their applications to invariance, in *Colloq. Math. Soc. "Ianos Bolyai", Limit Theorems in Probability and Statistics*, Veszprém, Hungary, 1982, 223-254.

[35] Csörgö, M. and Révész, P., *Strong Approximations in Probability and Statistics*, Academic Press, New York, 1981.

[36] Dugue, D., *Traité de Statistique Théorique et Appliquée*, Masson et Cie Editeurs, Paris, 1958.

[37] Dunford, N. and Schwartz, J., *Linear Operators*, vol.I,II, Interscience, New York, 1958.

[38] Eplett, W.J.R., Approximation theory for the simulation of continuous Gausssian processes, *Probab. Th. Rel. Fields* 73(1986), 159-181.

[39] Eplett, W.J.R., The inadmissibility of linear rank tests under Bahadur efficiency, *Ann. Stat.* 9(1981), 1079-1086.

[40] Eplett, W.J.R., Two Mann-Whitney type rank tests, *J. Roy. Stat. Soc.* B, 44(1982), 270-286.

[41] Gal, S.G., On Jackson's and Timan's estimates in the approximation of random functions by random polynomials, in *Proc. Int. Conf. Approx. Optim.*, Cluj-Napoca, July 29-August 1, 1996 (D.D. Stancu et al., eds.), vol. II, Transylvania Press, Cluj-Napoca, 1997, pp. 101-106.

[42] Gal, S.G., Jackson type estimates in the approximation of random functions by random polynomials, *Rend. Mat.* (Roma), ser. VIII, 14(1994), 543-556.

[43] Gal, S.G. and Villena, A.R., Random condensation of singularities and applications, *Random Oper. Stoch. Equ.* 5(1997), 3, 263-268.

[44] Gänssler, P. and Stute, W., *Wahrscheinlichkeitstheorie*, Springer-Verlag, Berlin, 1977.

[45] Hall, P., The order of the approximation to a Wiener process by its Fourier series, *Math. Proc. Cambridge Philos. Soc.* 92(1982), 547-562.

[46] Ignatov, Z.G., Mills, T.M. and Tzankova, I.P., On the rate of approximation of random functions, *Serdica, Bulg. Math. Publ.* 18(1992), 240-247.

[47] Istrăţescu, V.I., *Strict Convexity and Complex Strict Convexity*, Marcel Dekker Inc., New York, 1984.

[48] Istrăţescu, V.I., *Introduction in the Theory of Probabilistic Metric Spaces with Applications* (in Romanian), Ed. Tehnica, Bucharest, 1974.

[49] Kamolov, A.I., On exact estimates of approximation of random processes by Bernstein polynomials (in Russian), *Dokl. Akad. Nauk SSSR* 11(1986), 4-6.

[50] Kawata, T., Absolute convergence of Fourier series of periodic stochastic processes and its applications, *Tôhoku Math. J.* 35(1983), 459-474.

[51] Ky Fan, Sur l'approximation et l'integration des fonctions aleatoires, *Bull. Soc. Math. France* 72(1944).

[52] Ky Fan, Extensions of two fixed point theorems of F.E. Browder, *Math. Z.* 112(1969), 234-240.

[53] Lin, T.C., Random approximations and random fixed point theorems for non-self maps, *Proc. Amer. Math. Soc.* 103(1988), 1129-1135.

[54] Lehnhoff, H.G., A simple proof of A.F. Timan's theorem, *J. Approx. Theory* 38(1983), 172-176.

[55] Lehnhoff, H.G., A new proof of Teljakowskii's theorem, *J. Approx. Theory* 38(1983), 177-181.

[56] Lorentz, G.G., *Approximation of Functions*, Chelsea, New York, 1986.

[57] Muntean, I., The Lagrange interpolation operators are densely divergent, *Studia Univ. "Babeş-Bolyai"* (Cluj-Napoca), 21(1976), 28-30.

[58] Muntean, I., Unbounded divergence of simple quadrature formulas, *J. Approx. Theory* 67(1991), 303-310.

[59] Onicescu, O., Cenuşă, G. and Săcuiu, I., *Random Functions Almost Periodic in Probability* (in Romanian), Academic Press, Bucharest, 1983.

[60] Onicescu, O. and Istrăţescu, V.I., Approximation theorems for random functions, *Rend. Mat.* (Roma), (1), vol. 8, ser. VI(1975), 65-81.

[61] Onicescu, O. and Istrăţescu, V.I., Approximation theorems for random functions, II, *Rend. Mat.* (Roma), (4), vol. 11, ser. VI(1978), 585-589.

[62] Onicescu, O., Sâmboan, G. and Theodorescu, R., A remark on the Weierstrass theorem, *Anal. Univ. Bucharest* 29(1961), 255-257.

[63] Papageorgiou, N.S., Fixed points and best approximations for measurable multifunctions with stochastic domain, *Tamkang J. Math.* 23(1992), 197-203.

[64] Pollard, D., *Convergence of Stochastic Processes*, Springer-Verlag, New York, 1984.

[65] Révész, P., Approximation of the Wiener process and its local time, in *Proc. 4th Pannonian Symp. Math. Statist.*, Bad. Tatzmannsdorf, Austria, 1983 (F. Konecny, J. Mogyoródi, W. Wertz eds.), 57-65.

[66] Săcuiu, I. and Cenuşă, G., On some approximation theorems for random functions, *Rend. Mat.* (Roma), (3), ser. VII(1981), 399-409.

[67] Sehgal, V.M. and Singh, S.P., On random approximations and a random fixed point theorem for set-valued mappings, *Proc. Amer. Math. Soc.* 95(1985), 91-94.

[68] Serstnev, A.N., The notion of random normed space (in Russian), *Dokl. Akad. Nauk SSSR* 149(2), 280-283; *Soviet. Math.* 4(2)(1963), 388-391.

[69] Serstnev, A.N., Best approximation problems in random normed spaces (in Russian), *Dokl. Akad. Nauk SSSR* 149(3), 539-542; *Soviet. Math.* 4(2)(1963), 446-448.

[70] Serstnev, A.N., Some best approximation problems in random normed spaces (in Russian), *Rev. Roum. Math. Pures Appl.*, Nr.9(1964), 771-789.

[71] Singer, I., *Best Approximation in Normed Linear Spaces by Elements of Linear Subspaces* (in Romanian), Academic Press, Bucharest, 1967.

[72] Splettstösser, W., On the approximation of random processes by convolution processes, *ZAMM* 6(1981), 235-241.

[73] Stadtmüller, U., Kernel approximation of a Wiener process, *Period. Math. Hung.* vol.19(1)(1988), 79-90.

[74] Tan, K.K. and Yuan, X.Z., Random fixed point theorems and approximation in cones, *J. Math. Anal. Appl.* 185(1994), 378-390.

[75] Velasco, M.V. and Villena, A.R., A random Banach-Steinhauss theorem, *Proc. Amer. Math. Soc.* 123(1995), 2489-2497.

[76] Velasco, M.V. and Villena, A.R., A random closed graph theorem, *Bolletino U.M.I.* (7), 11-A(1997), 105-113.

[77] Weba, M., Korovkin systems of stochastic processes, *Math. Z.* 192(1986), 73-80.

[78] Weba, M., A quantitative Korovkin theorem for random functions with multivariate domains, *J. Approx. Theory* 61(1990), 74-87.

[79] Weba, M., Monotone approximation of random functions with multivariate domains with respect to lattice semi-norms, *Results Math.* 20(1991), 554-576.

[80] Weba, M., Quadrature of smooth stochastic processes, *Prob. Th. Rel. Fields* 87(1991), 333-347.

[81] Weba, M., Quantitative results on monotone approximation of stochastic processes, *Probab. Math. Statist.* 11(1)(1990), 109-120.

[82] Wentzell, A.D., *A Course in the Theory of Stochastic Processes*, McGraw-Hill, 1981.

[83] Ylvisaker, D., Prediction and design, *Ann. Statist.* 15(1987), 1-19.

Chapter 13

Approximation Theory in Fuzzy Setting

SORIN G. GAL

Department of Mathematics, University of Oradea
Str. Armatei Române 5, 3700 Oradea, ROMANIA
E-mail: galso@math.uoradea.ro

Abstract In this survey main results in classical approximation theory such as Weierstrass and Stone and Weierstrass-type results, quantitative estimates in approximation by polynomials, interpolation results, best approximation results, etc. to fuzzy setting are generalized. Besides known results, several new results are proved.

Contents

13.1 Introduction

Besides the randomness, another kind of uncertainty of the real world is that of fuzziness. Although randomness and the fuzziness are essentially different phenomena, sometimes there are even mixed phenomena (for a detailed interesting discussion concerning these aspects see, e.g., Wang, Lui, Zhang, Zhang and Xu [26]).

By now combining the concept of approximation with that of fuzziness, one obtains another powerful tool in the investigation of the real world: the fuzzy approximation theory.

In this survey, main results in classical approximation theory to fuzzy setting are generalized. Many results appear for the first time here. Where the authors are not specified, the results are new and belong to the present author.

In comparison with the results in random approximation theory, the results in fuzzy approximation theory are very recent, belonging to the last decade (see References).

For this reason, many results obtained in random approximation do not have a correspondent yet in fuzzy setting, as, for example, those of best approximation in random normed spaces (see Gal [15]), although various concepts of fuzzy normed spaces were studied.

We hope that this survey will stimulate future research in the field of fuzzy approximation theory.

Acknowledgement. I express my gratitude to my colleague A.I. Ban for his help with the References.

13.2 Preliminaries

In this section we present the basic concepts that will be used in the next sections.

The concept of fuzzy sets was introduced by Zadeh [30] and was suggested by the following simple remark: if $X \neq \emptyset$ is a "total" set, then $(\mathcal{P}(X), \cup, \cap, C)$ and $(\mathcal{C}h(X), \max, \min, ^{-})$ are isomorphic Boole algebras, where $\mathcal{P}(X) = \{A; A \subseteq X\}$, $\mathcal{C}h(X) = \{\chi : X \to \{0,1\}\}$, by the isomorphism $A \to \chi_A$ = characteristic function of the set $A \subset X$ (here $\overline{\chi} = 1 - \chi$).

Then a fuzzy subset of X will be a function $f : X \to [0,1]$.

The value $f(x)$ is called the degree of membership of x to f and $\mathrm{supp}(f) = \{x \in X; f(x) > 0\}$ is called the support of the fuzzy set f. If we denote by $\mathcal{F}(X)$ the class of all fuzzy subsets of X, it is known that $(\mathcal{F}(X), \max, \min, ^{-})$ is no longer a Boole algebra (see, e.g., Negoiţă and Ralescu [23], [24]).

An important concept in fuzzy mathematics is that of fuzzy real numbers. There are various definitions but in this paper we adopt the following.

DEFINITION 13.1 *(see, e.g., Wu Congxin and Ma Ming [27]–[28], Ma Ming [22]) A fuzzy real number is a function $u : \mathbb{R} \to [0,1]$ satisfying the following properties:*

(i) u is normal, i.e., $\exists\ x_0 \in \mathbb{R}$ with $u(x_0) = 1$;

(ii) u is a convex fuzzy set (i.e., $u(\lambda x + (1-\lambda)y) \geq \min\{u(x), u(y)\}$, $\forall\ x, y \in \mathbb{R}$, $\lambda \in [0,1]$); equivalently, all the level sets $\{x \in \mathbb{R}; u(x) \geq \alpha\}$, $\alpha \in [0,1]$, are convex;

(iii) u is upper semi-continuous on $\mathbb{R}$;

(iv) the set $\overline{\mathrm{supp}(u)}$ is compact, where $\overline{A}$ means the closure of A.

The set of all fuzzy real numbers is denoted by $\mathbb{R}_{\mathcal{F}}$.

For $0 \leq r \leq 1$, let us denote $[u]^r = \{x \in \mathbb{R}; u(x) \geq r\}$.

THEOREM 13.1 (Ma Ming [22])

If $u \in \mathbb{R}_{\mathcal{F}}$, then

(i) $[u]^r$ is a closed interval, $\forall\ r \in [0,1]$;

(ii) $0 \leq r_1 \leq r_2 \leq 1 \ \Rightarrow\ [u]^{r_2} \subset [u]_{r_1}$.

REMARK 13.1 It is known (see Wu Congxin and Gong Zengtai [29]) that for $u, v \in \mathbb{R}_{\mathcal{F}}$, $k \in \mathbb{R}$, the addition and the scalar multiplication are defined by the equations (which have always unique solutions!)

$$[u+v]^r = [u]^r + [v]^r, \quad [ku]^r = k[u]^r, \quad r \in [0,1].$$

The sum and the product with scalars will be denoted by $u \oplus v$ and by $k \odot u$, $(u, v \in \mathbb{R}_{\mathcal{F}})$, respectively. ∎

Also, if we introduce $D : \mathbb{R}_{\mathcal{F}} \times \mathbb{R}_{\mathcal{F}} \to \mathbb{R}_+ \cup \{0\}$ by

$$D(u, v) = \sup_{r \in [0,1]} \max\{|u_-^r - v_-^r|, |u_+^r - v_+^r|\} = \sup_{r \in [0,1]} \{d_H([u]^r, [v]^r)\},$$

where $[u]^r = [u_-^r, u_+^r]$, $[v]^r = [v_-^r, v_+^r] \subset \mathbb{R}$ and d_H is the Hausdorff distance, then the following is known.

THEOREM 13.2 (Wu Congxin and Gong Zengtai [29])
(i) $(\mathbb{R}_{\mathcal{F}}, D)$ *is a complete metric space;*
(ii) $D(u \oplus w, v \oplus w) = D(u, v)$, $\forall\ u, v, w \in \mathbb{R}_{\mathcal{F}}$;
(iii) $D(k \odot u, k \odot v) = |k| D(u, v)$, $\forall\ u, v \in \mathbb{R}_{\mathcal{F}}$, $k \in \mathbb{R}$;
(iv) $D(u \oplus v, w \oplus e) \leq D(u, w) + D(v, e)$, $\forall\ u, v, w, e \in \mathbb{R}_{\mathcal{F}}$.

REMARK 13.2 The above properties of D are very close to those of a metric derived from a norm, which gives the possibility of extending many concepts and results from the classical mathematics to the concepts based on the set $\mathbb{R}_{\mathcal{F}}$. ∎

REMARK 13.3 On $\mathbb{R}_{\mathcal{F}}$ a partial order by $u \leq v$ iff $[u]^r \leq [v]^r$, $r \in [0, 1]$ iff $u_-^r \leq v_-^r$, $u_+^r \leq v_+^r$, $\forall\ r \in [0, 1]$ can be also introduced. ∎

REMARK 13.4 A useful property used in the next paragraph is the following:

$$(a + b) \odot u = a \odot u \oplus b \odot u,$$

for all usual real numbers $a, b \geq 0$ and $u \in \mathbb{R}_{\mathcal{F}}$. Indeed, if we denote $[u]^r = [u_-^r, u_+^r]$ = closed interval, by Remark 13.1, we obtain

$$[(a + b)u]^r = (a + b)[u]^r = (a + b)[u_-^r, u_+^r] = [(a + b)u_-^r, (a + b)u_+^r] =$$

$$= [au_-^r + bu_-^r, au_+^r + bu_+^r] = [au_-^r, au_+^r] + [bu_-^r, bu_+^r] = a[u]^r + b[u]^r,$$

which means

$$(a + b) \odot u = a \odot u \oplus b \odot u$$

(here the usual sum between two closed intervals and the usual product between a real scalar and a closed interval are used, i.e.,

$$[a_1, b_1] + [a_2, b_2] = [a_1 + a_2, b_1 + b_2]$$

and

$$\alpha[a_1, b_1] = [\alpha a_1, \alpha b_1], \quad \alpha > 0).$$

■

In fuzzy mathematics there are various definitions for the concept of fuzzy function (see, e.g., Negoiţă and Ralescu [24]).

In this survey we will consider the following kinds of fuzzy functions.

DEFINITION 13.2 *A mapping $f : [a, b] \to \mathbb{R}_{\mathcal{F}}$ will be called fuzzy (real) function defined on $[a, b] \subset \mathbb{R}$. Correspondingly, a fuzzy algebraic polynomial (of degree n) is an expression of the form*

$$P_n(x) = \sum_{k=0}^{n}{}^{*} x^k \odot c_k,$$

where $c_k \in \mathbb{R}_{\mathcal{F}}$, $k = \overline{0, n}$ and $x \in \mathbb{R}$ (here the sum $\sum^*$ and the product with scalars are those in Remark 13.1). Also, a fuzzy trigonometric polynomial (of degree n) is an expression of the form

$$\sum_{k=0}^{n}{}^{*} \{[\cos kx] \odot a_k + [\sin kx] \odot b_k\},$$

where $x \in \mathbb{R}$ and $a_k, b_k \in \mathbb{R}_{\mathcal{F}}$, $k = \overline{0, n}$.

The distance between $f; g : [a, b] \to \mathbb{R}_{\mathcal{F}}$ will be the uniform distance given by

$$D_U(f, g) = \sup\{D(f(x), g(x)); x \in [a, b]\}.$$

Another different possible definition for the concept of fuzzy function is based on the idea that any usual mapping between two metric spaces, $g : (X, d) \to (Y, \rho)$, can be identified with its graph, i.e., $Graph(g) = \{(x, y); y = g(x)\}$, which is subset of the metric space $(X \times Y, \max\{d, \rho\})$.

DEFINITION 13.3 *(Gal [10], [11]) A fuzzy function from $[a, b]$ to $\mathbb{R}$ is the fuzzified graph of a usual function $f : [a, b] \to \mathbb{R}$, i.e., a pair (f, φ_f), where $f : [a, b] \to \mathbb{R}$ and $\varphi_f : [a, b] \times \mathbb{R} \to [0, 1]$ is the function which measures the degree of membership of a point $(x, y) \in [a, b] \times \mathbb{R}$, to $Graph(f) = \{(x, y) \in [a, b] \times \mathbb{R}; y = f(x)\}$. Obviously we can write*

$$\varphi_f(x, y) = \begin{cases} F(x, y), & (x, y) \in Graph(f) \\ 0, & otherwise \end{cases}$$

where $0 < F(x,y) \leq 1$, *for all* $(x,y) \in Graph(f)$.

Correspondingly, a fuzzy algebraic polynomial from $[a,b]$ *to* $\mathbb{R}$ *will be a pair* (P, φ_P), *where* P *is a usual algebraic polynomial and*

$$\varphi_P(x,y) = \begin{cases} Q(x,y), & (x,y) \in Graph(P), \\ 0, & otherwise, \end{cases}$$

where $Q(x,y)$ *is a bivariate algebraic polynomial in* x *and* y, *satisfying* $0 < Q(x,y) \leq 1$, *for all* $(x,y) \in Graph(P)$.

A fuzzy trigonometric polynomial from $\mathbb{R}$ *to* $\mathbb{R}$ *will be a pair* (T, φ_T), *with* T *a trigonometric polynomial and with the membership function* $\varphi_T : \mathbb{R} \times \mathbb{R} \to [0,1]$ *defined by*

$$\varphi_T(x,y) = \begin{cases} F(x,y), & (x,y) \in Graph(T), \\ 0, & otherwise, \end{cases}$$

where $0 < F(x,y) \leq 1$, *for all* $(x,y) \in Graph(T)$ *and* $F(x,y)$ *is a trigonometric polynomial with respect to* x *and an algebraic polynomial with respect to* y.

The distance between two fuzzy functions $(f, \varphi_f), (g, \varphi_g)$ *will be the usual Hausdorff distance,* d_H, *between the sets (called support graphs of* φ_f *and* φ_g*)*

$$G^*(\varphi_f) = \{(x,y,z) \in [a,b] \times \mathbb{R} \times (0,1]; y = f(x),$$

$$0 < z = F(x,y), (x,y) \in Graph(f)\},$$

$$G^*(\varphi_g) = \{(x,y,z) \in [a,b] \times \mathbb{R} \times (0,1]; y = g(x),$$

$$0 < z = G(x,y), (x,y) \in Graph(g)\},$$

where

$$\varphi_f(x,y) = \begin{cases} F(x,y), & (x,y) \in Graph(f) \\ 0, & otherwise, \end{cases}$$

$$\varphi_g(x,y) = \begin{cases} G(x,y), & (x,y) \in Graph(g) \\ 0, & otherwise. \end{cases}$$

That is, we can write

$$D_{\mathcal{F}}[(f, \varphi_f), (g, \varphi_g)] = d_H(G^*(\varphi_f), G^*(\varphi_g)) =$$

$$= \max \left\{ \sup_{a \in G^*(\varphi_f)} \inf_{b \in G^*(\varphi_g)} \rho(a,b), \sup_{b \in G^*(\varphi_g)} \inf_{a \in G^*(\varphi_f)} \rho(a,b) \right\}$$

where for $a = (a_1, a_2, a_3)$, $b = (b_1, b_2, b_3)$ *we have*

$$\rho(a, b) = \max\{|a_1 - b_1|, |a_2 - b_2|, |a_3 - b_3|\}.$$

REMARK 13.5 If $G^*(\varphi_f)$ and $G^*(\varphi_g)$ are compact sets, then $D_{\mathcal{F}}[(f, \varphi_f), (g, \varphi_g)]$ is a metric and in the next paragraphs we will consider subclasses of fuzzy functions for which this holds. ∎

REMARK 13.6 The concept of fuzzy functions from $[a, b]$ to $\mathbb{R}$ in Definition 13.3 can easily be generalized to fuzzy functions from (X, d_1) to (Y, d_2), where $(X, d_1), (Y, d_2)$ are metric spaces. Correspondingly, $G^*(\varphi_f), G^*(\varphi_g) \subset X \times Y \times [0, 1]$ and in the expression of $D_{\mathcal{F}}[(f, \varphi_f), (g, \varphi_g)]$,

$$\rho(a, b) = \max\{d_1(a_1, b_1), d_2(a_2, b_2), |a_3 - b_3|\}.$$

∎

REMARK 13.7 It is interesting to note that in the particular case when $Y \subseteq (0, 1]$, a fuzzy function from X to Y in Definition 13.3 can be interpreted as a "doubly fuzzy subset" of X. ∎

REMARK 13.8 If $X \neq \emptyset$ and $g : X \to [0, 1]$ is a fuzzy subset of X, then any pair $(x, g(x))$ with $g(x) > 0$ is called a fuzzy point of the fuzzy subset g. In Negoiţă and Ralescu [23], another definition for the concept of a fuzzy function from X to Y is that of a usual function from a fuzzy subset of X, denoted g_X, to a fuzzy subset of Y, denoted by g_Y, i.e., which maps any fuzzy point $(x, g_X(x))$ of g_X to a fuzzy point $(y, g_Y(y))$ of g_Y. ∎

The following definition take into account Remark 13.8, generalizing at the same time Definition 13.3.

DEFINITION 13.4 *A generalized fuzzy function from* X *to* Y *is a triplet* $(\varphi, f, \gamma_{\varphi,f})$ *where* $\varphi : X \to [0, 1]$, $f : X \to Y$ *and* $\gamma_{\varphi,f} : X \times Y \times [0, 1] \to [0, 1]$ *satisfy* $\gamma_{\varphi,f}(x, y, z) > 0$ *iff* $(x, y) \in Graph(f)$ *and* $(x, z) \in G^*(\varphi)$, *where* $G^*(\varphi) = \{(x, z); 0 < z = \varphi(x), x \in X\}$ *is the support graph of* φ.

Correspondingly, a generalized fuzzy algebraic polynomial from $[a, b]$ *to* $\mathbb{R}$ *will be a triplet* (P, Q, R), *where* $P : [a, b] \to [0, 1]$, $Q : [a, b] \to \mathbb{R}$

are algebraic polynomials on $[a,b]$ *and* $R(x,y,z)$ *is a trivariate algebraic polynomial with respect to* x, y, z*, satisfying*

$$R(x,y,z) \in (0,1], \text{ for } y = Q(x), \quad z = P(x), \quad x \in [a,b].$$

Also, the distance between $(\varphi, f, \gamma_{\varphi,f})$ *and* $(\phi, F, \Gamma_{\phi,F})$ *will be the usual Hausdorff distance between the sets*

$$G^*(\gamma_{\varphi,f}), G^*(\Gamma_{\phi,F}) \subset Z = X \times Y \times [0,1] \times [0,1],$$

i.e.,

$$D_{\mathcal{F}}[(\varphi, f, \gamma_{\varphi,f}), (\phi, F, \Gamma_{\phi,F})] = d_H[G^*(\gamma_{\varphi,f}), G^*(\Gamma_{\phi,F})],$$

where, e.g.,

$$G^*(\gamma_{\varphi,f}) = \{(x,y,z,u); (x,y) \in Graph(f),$$

$$(x,z) \in G^*(\varphi), 0 < u = \gamma_{\varphi,f}(x,y,z)\}$$

and if $(X,d), (Y,\rho)$ *are metric spaces, then* Z *is endowed with the metric*

$$d^*(a,b) = \max\{d(a_1,b_1), \rho(a_2,b_2), |a_3 - b_3|, |a_4 - b_4|\},$$

$$a = (a_1,a_2,a_3,a_4), \; b = (b_1,b_2,b_3,b_4) \in Z.$$

REMARK 13.9 If $\varphi(x) = 1, \; \forall \, x \in X$, obviously Definition 13.4 reduces to Definition 13.3. ▮

REMARK 13.10 The generalized fuzzy function $(\varphi, f, \gamma_{\varphi,f})$ in Definition 13.4 takes into account Remark 13.8 of Definition 13.3, because in fact any fuzzy point $(x, \varphi(x)), x \in G^*(\varphi)$ is mapped to the fuzzy point $(f(x), \gamma_{\varphi,f}(x, f(x), \varphi(x)))$. ▮

13.3 Approximation of Fuzzy Sets

Many applications of fuzzy sets restrict attention to the metric space $(\mathbb{R}_{\mathcal{F}}, D)$ of all fuzzy real numbers (see Definition 13.1), or more generally, to the metric space $(\mathbb{R}^p_{\mathcal{F}}, D_p)$, $(p \in \mathbb{N}$ fixed, $\mathbb{R}^1_{\mathcal{F}} \equiv \mathbb{R}_{\mathcal{F}}, \; D_1 \equiv D)$ defined

(similarly with Definition 13.1) as $\mathbb{R}^p_{\mathcal{F}} = \{u : \mathbb{R}^p \to [0,1];\ \exists\ x_0 \in \mathbb{R}^p,$ $u(x_0) = 1, u$ is convex fuzzy set, u is upper-semicontinuous on $\mathbb{R}^p, \overline{\text{supp}u}$ is compact in $\mathbb{R}^p\}$
and

$$D_p(u,v) = \sup_{r\in[0,1]} d_H([u]^r, [v]^r),$$

where $[u]^r = \{x \in \mathbb{R}^p; u(x) \geq r\}$ and d_H is the Hausdorff distance between usual sets in $\mathbb{R}^p$. However, these fuzzy sets can be complicated in form and may be unsuitable for computational purpose. Thus appears as natural the problem of approximation of a fuzzy set in $\mathbb{R}^p_{\mathcal{F}}$ (in the metric D_p) by simpler fuzzy sets, e.g., by those having continuous membership functions or by elementary piecewise linear fuzzy sets called spikes, triangles and broken planes.

In this sense, first we present

THEOREM 13.3 (Collings and Kloeden [5])
For each $u \in \mathbb{R}^p_{\mathcal{F}}$, $(p \in \mathbb{N})$, there exists a sequence $u_n \in \mathbb{R}^p_{\mathcal{F}}$, $n \in \mathbb{N}$, with $u_n : \mathbb{R}^p \to [0,1]$, continuous on $\mathbb{R}^p$ (here $\mathbb{R}^p$ is considered endowed with the usual norm $\|\cdot\|$), such that

$$D_p(u_n, u) \leq \frac{1}{n}, \text{ for all } n = 1, 2, \ldots$$

A consequence of Theorem 13.3 is

COROLLARY 13.1
(Collings and Kloeden [5]) Let $U : [0,T] \to (\mathbb{R}^p_{\mathcal{F}}, D_p)$ be continuous on $[0,T]$ (as an application between two metric spaces).

Then for each $n \in \mathbb{N}$, there exists $U_n : [0,T] \to (\mathbb{R}^p_{\mathcal{F}}, D_p)$ continuous on $[0,T]$ such that $U_n(t) : \mathbb{R}^p \to [0,1]$, $\forall\ t \in [0,T]$ are continuous on $\mathbb{R}^p$ and moreover

$$D_p(U_n(t), U(t)) \leq \frac{1}{n},\ \forall\ t \in [0,T],\ n \in \mathbb{N}.$$

In what follows, for $K \subset \mathbb{R}^p$, non-empty compact convex set in $\mathbb{R}^p$, let us denote

$$\mathbb{R}^p_{\mathcal{F}}(K) = \{u \in \mathbb{R}^p_{\mathcal{F}}; \text{supp}(u) \subseteq K\},$$

$S^p_{\mathcal{F}}(K) = \{u : \mathbb{R}^p \to [0,1]; u$ is convex fuzzy set, u is upper-semicontinuous on $\mathbb{R}^p$ and $\text{supp}(u) \subseteq K\}$.

Obviously $S^p_{\mathcal{F}}(K) \subset \mathbb{R}^p_{\mathcal{F}}(K) \subset \mathbb{R}^p_{\mathcal{F}}$.

DEFINITION 13.5 *(Diamond, Kloeden and Vladimirov [6]) A fuzzy set in $S^p_{\mathcal{F}}(K)$ of the form*

$$s_{y,\alpha}(x) = \begin{cases} \alpha, & \text{if } x = y \\ 0, & \text{elsewhere} \end{cases}$$

will be called a spike (here $x, y \in \mathbb{R}^p$, y is fixed, $\alpha \in [0,1]$ is fixed).

Obviously $s_{y,1}$ is exactly the singleton set $\{y\}$.

A fuzzy set in $S^p_{\mathcal{F}}(K)$ of the form

$$u(x) = \max\{0, \alpha - N\|y - x\|\}, \quad x, y \in \mathbb{R}^p, \quad y \text{ fixed}, \quad \alpha \in [0,1] \text{ fixed},$$

will be called N-triangle (denoted by $T_{y,\alpha,N}$).

The convex hull of the family $u_\lambda \in S^p_{\mathcal{F}}(K)$, $\lambda \in \Lambda$, is the minimal fuzzy set $u \in S^p_{\mathcal{F}}(K)$ which satisfies

$$u_\lambda(x) \leq u(x), \ \forall \ x \in \mathbb{R}^p, \ \forall \ \lambda \in \Lambda.$$

Let $u \in \mathbb{R}^p_{\mathcal{F}}(K)$ and $N > 0$. Let us denote by $u^+_N(x)$ the convex hull of all triangles $\{T_{y,u(y),N}\}_{y \in [u]^0}$.

THEOREM 13.4 (Diamond, Kloeden and Vladimirov [6])
We have $u^+_N(x) \nearrow u(x)$, for all $x \in \mathbb{R}^p$ (when $N \to +\infty$) and

$$\lim_{N \to +\infty} D_p(u^+_N, u) = 0.$$

Another approximation result is

THEOREM 13.5 (Diamond, Kloeden and Vladimirov [6])
For any $u \in \mathbb{R}^p_{\mathcal{F}}(K)$ and any $\varepsilon > 0$, there exist $N(\varepsilon)$ spikes, where $N(\varepsilon)$ is some integer satisfying

$$\lim_{\varepsilon \searrow 0} \frac{\log_2 N(\varepsilon)}{\log_2 \varepsilon} = -\frac{n+1}{2},$$

such that for the convex hull of these spikes, denoted by s, we have

$$D_p(u, s) < \varepsilon.$$

REMARK 13.11 For details and other results see Diamond, Kloeden and Vladimirov [6]. ▮

Another kind of approximation of a fuzzy set (more exactly by an usual set!) is suggested by the followings.

Let $(X, \|\cdot\|)$ be a normed space, $Y \subset X$, $x \in X \setminus Y$.

It is known that the best approximation of x by elements in Y is defined as

$$E(x, Y) = \inf\{\|x - y\|; y \in Y\}.$$

Now, if $x \in V \subset X$, then the best approximation of V by Y is naturally expressed by

$$E(V, Y) = \sup_{x \in V} E(x, Y)$$

Considerations in Asmuss and Šostak [2] show that it seems natural to consider V as a fuzzy set, rather than an usual set.

To increase the generality, let $L = (L, \leq, \wedge, \vee)$, be a complete distributive lattice with universal bounds 0 and 1.

(Two main examples are $L = \{0, 1\}$ and $L = [0, 1]$).

It is known that an L-fuzzy subset of the normed space $(X, \|\cdot\|)$ will be a function $u : X \to L$ (see, e.g., Negoiţă and Ralescu [24]).

Also, we consider the concept of L-fuzzy real number as in Asmuss and Šostak [2], defined as a function $z : \mathbb{R} \to L$ with the properties:

$$x_1 \leq x_2 \text{ implies } z(x_1) \leq z(x_2);$$

$$\sup\{z(x); x \in \mathbb{R}\} = 1, \quad \inf\{z(x); x \in \mathbb{R}\} = 0;$$

$$z \text{ is left semi-continuous, i.e., } \bigwedge_{t<x} z(t) = z(x).$$

The set of all L-fuzzy real numbers will be denoted by $\mathbb{R}(L)$.

DEFINITION 13.6 *(Asmuss and Šostak [2]) The best approximation of an L-fuzzy subset u (of $(X, \|\cdot\|)$ by the (usual) set $U \subset X$ is defined as the L-fuzzy real number $E_{\mathcal{F}}(u, U) : \mathbb{R} \to L$, given by*

$$E_{\mathcal{F}}(u, U) = SUP(A),$$

where $A : \mathbb{R} \to L$ is the L-fuzzy set defined by

$$A(r) = \sup\{u(x); E(x, U) = r\},$$

$r \in \mathbb{R}$ (if $\{u(x); E(x,U) = r\} = \emptyset$, then by definition $A(r) = 0$) and $SUP(A) : \mathbb{R} \to L$ is defined by

$$SUP(A)(x) = \inf_{t<x} \left\{ \sup_{u>t} A(u) \right\}$$

REMARK 13.12 $E_{\mathcal{F}}(u, U)$ can be interpreted as the deviation of the L-fuzzy subset u of $(X, \|\cdot\|)$, from the usual set $U \subset X$. ∎

THEOREM 13.6 (Asmuss and Šostak [2])
For each $\alpha \in L$, the real value

$$v_\alpha = \sup\{E_{\mathcal{F}}(u,U)^{-1}([\alpha, 1])\} = \sup\{x \in \mathbb{R}; \alpha \leq E_{\mathcal{F}}(u,U)(x) \leq 1\}$$

gives the best approximation of the set

$$M_\alpha = u^{-1}([\alpha, 1]) = \{x \in X; \alpha \leq u(x) \leq 1\}$$

by U, in the usual sense in the normed space $(X, \|\cdot\|)$, i.e.,

$$v_\alpha = \sup_{y \in M_\alpha} \left\{ \inf_{x \in U} \|x - y\| \right\}.$$

REMARK 13.13 In the same paper by Asmuss and Šostak [2], the concept of optimal linear approximation of an L-fuzzy set is introduced and a duality theorem in the particular case of

$$X = L_p^r[0,1] = \{f : [0,1] \to \mathbb{R}; f^{(r)} \in L_p[0,1]\}$$

is proved (i.e., the problem of estimation of the value $E_{\mathcal{F}}(u, U)$ is reduced to a more visible extremal problem in the conjugate space $L_q^r[0,1]$, $\frac{1}{p} + \frac{1}{q} = 1$). ∎

At the end of this section we consider the problem of discrete approximation of a fuzzy set and of an intuitionistic fuzzy set, in a general context.

Let X be non-empty and let $\mathcal{A} \subset \mathcal{P}(X)$ be a σ-algebra (i.e., $X \in \mathcal{A}$, $A, B \in \mathcal{A} \Rightarrow A \setminus B \in \mathcal{A}$, $A_n \in \mathcal{A}$, $n \in \mathbb{N} \Rightarrow \bigcup_{n \in \mathbb{N}} A_n \in \mathcal{A}$).

A fuzzy subset of X, $g \in \mathcal{F}(X)$, is called $\mathcal{A}$-measurable if for any open subset of $[0,1]$ (in the topology on $[0,1]$ induced by the usual topology of $\mathbb{R}$), $D \subset [0,1]$, we have $g^{-1}(D) \in \mathcal{A}$, where $g^{-1}(D) = \{x \in X; g(x) \in D\}$.

A fuzzy subset of X, $g \in \mathcal{F}(X)$, is called discrete if the set $\{g(x); x \in X\}$ is finite.

From a classical result in the real function theory, it follows that for each $g \in \mathcal{F}(X)$, $\mathcal{A}$-measurable, there exists a sequence $g_n \in \mathcal{F}(X)$, discrete and $\mathcal{A}$-measurable, $n \in \mathbb{N}$, such that $g_n \nearrow g$ uniformly on X (in real function theory, g_n are called step functions).

The fact that $g_n \nearrow g$, can be expressed by

$$g_n \subset_{\mathcal{F}} g_{n+1} \subset_{\mathcal{F}} g, \ \forall\, n \in \mathbb{N},$$

where for $f, h \in \mathcal{F}(X)$, $f \subset_{\mathcal{F}} h$ means $f(x) \leq h(x)$, for all $x \in X$ (for details concerning basic concepts in fuzzy sets theory, see, e.g., Negoiţă and Ralescu [23], [24], Dubois and Prade [7]).

An important generalization of the concept of fuzzy sets is that of "intuitionistic fuzzy set" introduced by K.T. Atanassov.

DEFINITION 13.7 *(Atanassov [3]) If X is a non-empty set, an intuitionistic fuzzy subset of X is an object of the form $\langle \mu, \nu \rangle$, where $\mu, \nu : X \to [0,1]$ satisfy $\mu(x) + \nu(x) \leq 1$, for all $x \in X$.*

Here $\mu(x)$ defines the degree of membership of x and $\nu(x)$ defines the degree of non-membership of x, to $\langle \mu, \nu \rangle$.

The usual operations of $\cap_{I\mathcal{F}}, \cup_{I\mathcal{F}}$ and $\subset_{I\mathcal{F}}$ are defined by

$$\langle \mu_1, \nu_1 \rangle \cap_{I\mathcal{F}} \langle \mu_2, \nu_2 \rangle = \langle \min\{\mu_1, \mu_2\}, \max\{\nu_1, \nu_2\} \rangle,$$

$$\langle \mu_1, \nu_1 \rangle \cup_{I\mathcal{F}} \langle \mu_2, \nu_2 \rangle = \langle \max\{\mu_1, \mu_2\}, \min\{\nu_1, \nu_2\} \rangle,$$

$$\langle \mu_1, \nu_1 \rangle \subset_{I\mathcal{F}} \langle \mu_2, \nu_2 \rangle$$

iff $\mu_1(x) \leq \mu_2(x)$ and $\nu_1(x) \geq \nu_2(x)$, $\forall\, x \in X$.

The class of all intuitionistic fuzzy subsets of X will be denoted by $IFS(X)$.

REMARK 13.14 Any $g \in \mathcal{F}(x)$ generates $\langle g, 1-g \rangle \in IFS(X)$, so we can consider $\mathcal{F}(X) \subset IFS(X)$. ∎

REMARK 13.15 As in the case of $\mathcal{F}(X)$, $\langle \mu, \nu \rangle \in IFS(X)$ will be called $\mathcal{A}$-measurable if μ, ν are $\mathcal{A}$-measurable.

Also, $\langle \mu, \nu \rangle \in IFS(X)$ will be called discrete, if the sets $\{\mu(x); x \in X\}, \{\nu(x); x \in X\}$ are finite. ∎

Now, a direct application of the classical result in approximation of measurable functions by step functions to μ and ν obviously yields a sequence $\langle \mu_n, \nu_n \rangle \in IFS(X)$, μ_n, ν_n - discrete, $n \in \mathbb{N}$, such that $\mu_n \nearrow \mu$, $\nu_n \nearrow \nu$, uniformly on X, but for which the monotony property (i.e., $\langle \mu_n, \nu_n \rangle \subset_{I\mathcal{F}} \langle \mu_{n+1}, \nu_{n+1} \rangle \subset_{I\mathcal{F}} \langle \mu, \nu \rangle$, $n \in \mathbb{N}$) fails to exist.

But as was shown in Ban and Gal [4], this monotony property is essential in the definition of the concept of decomposable measure over $IFS(X)$.

In analytical terms we have to solve the following non-trivial problem of approximation with restrictions:

if $f, g : X \to [0,1]$ are $\mathcal{A}$-measurable and satisfy $f(x) + g(x) \leq 1$, for all $x \in X$, construct two sequences of step functions $f_n, g_n : X \to [0,1]$ such that

$$f_n(x)+g_n(x) \leq 1, \quad f_n(x) \leq f_{n+1}(x) \leq f(x), \quad g(x) \leq g_{n+1}(x) \leq g_n(x),$$

$\forall\, x \in X$, $\forall\, n \in \mathbb{N}$ and $f_n \overset{n\to\infty}{\longrightarrow} f$, $g_n \overset{n\to\infty}{\longrightarrow} g$ uniformly on X.

THEOREM 13.7 (Ban and Gal [4])

If $\langle \mu, \nu \rangle \in IFS(X)$ is $\mathcal{A}$-measurable, then there exists a sequence $\langle \mu_n, \nu_n \rangle \in IFS(X)$, discrete and $\mathcal{A}$-measurable, $n \in \mathbb{N}$, such that

$$\langle \mu_n, \nu_n \rangle \subset_{I\mathcal{F}} \langle \mu_{n+1}, \nu_{n+1} \rangle \subset_{I\mathcal{F}} \langle \mu, \nu \rangle, \ \forall\, n \in \mathbb{N},$$

and $\mu_n \overset{n\to\infty}{\longrightarrow} \mu$, $\nu_n \overset{n\to\infty}{\longrightarrow} \nu$, uniformly on X.

PROOF By the classical approximation result (see, e.g., Precupanu [25], p. 124–126), there exist the sequences of step functions $F_n, h_n : X \to [0,1]$, $n \in \mathbb{N}$, such that $F_n \nearrow \mu + \nu$ and $h_n \nearrow \nu$, uniformly on X, and

$$0 \leq \nu - h_n(x) \leq \frac{1}{2^n}, \text{ for all } n \in \mathbb{N},\ x \in X.$$

Now, by using the idea in Gal [9], we obtain a monotonous decreasing sequence of step functions,

$$h_n^*(x) = h_n(x) + \frac{2}{2^{n-1}}, \quad x \in X, \quad n \in \mathbb{N},$$

such that $h_n^* \overset{n\to\infty}{\longrightarrow} \nu$, uniformly on X and

$$h_n^*(x) - h_{n+1}^*(x) = h_n(x) - h_{n+1}(x) + \frac{2}{2^n} > 0, \ \forall\, x \in X, \ n \in \mathbb{N},$$

taking into account that

$$h_n(x) - h_{n+1}(x) < |h_n(x) - h_{n+1}(x)| \le |h_n(x) - \nu(x)| + |\nu(x) - h_{n+1}(x)| \le$$

$$\le \frac{1}{2^n} + \frac{1}{2^{n+1}} \le \frac{2}{2^n}, \quad x \in X, \quad n \in \mathbb{N}.$$

Because $h_n^*(x)$ can take values greater than 1, let us define

$$g_n(x) = \begin{cases} h_n^*(x), & \text{if } h_n^*(x) \le 1 \\ 1, & \text{if } h_n^*(x) > 1. \end{cases}$$

Obviously $g_n(x) \in [0,1]$ and g_n is step function, $n \in \mathbb{N}$.

By $\nu(x) \le g_n(x) \le h_n^*(x)$, $x \in X$, $n \in \mathbb{N}$, it follows that $g_n \overset{n\to\infty}{\longrightarrow} \nu$, uniformly on X. Also, by simple reasonings we get $g_{n+1}(x) \le g_n(x)$, $\forall\, x \in X$, $n \in \mathbb{N}$.

As a conclusion, $g_n : X \to [0,1]$, $n \in \mathbb{N}$, is a sequence of step functions satisfying $g_n \searrow \nu$, uniformly on X.

Now, let us define

$$\mu_n(x) = (F_n - g_n)^+(x) \quad \text{(i.e. the positive part of } F_n - g_n\text{)},$$

and

$$\nu_n(x) = g_n(x),$$

for all $x \in X$, $n \in \mathbb{N}$.

Obviously each μ_n is a step function. Then $0 \le \mu_n(x)$ and if $0 < \mu_n(x)$, then

$$\mu_n(x) = F_n(x) - g_n(x) \le \mu(x) + \nu_A(x) - g_n(x) \le \mu(x) \le 1,$$

i.e., $\mu_n(x) \in [0,1]$, for all $x \in X$, $n \in \mathbb{N}$.

Also,

$$\mu_n(x) = (F_n - g_n)^+(x) = \max\{F_n(x) - g_n(x), 0\} =$$

$$= \frac{F_n(x) - g_n(x) + |F_n(x) - g_n(x)|}{2} \overset{n\to\infty}{\longrightarrow}$$

$$\frac{\mu(x) + \nu(x) - \nu(x) + |\mu(x) + \nu(x) - \nu(x)|}{2} =$$

$$= \frac{\mu(x) + |\mu(x)|}{2} = \mu^+(x) = \mu(x),$$

uniformly on X.

Moreover, by simple reasonings we get

$$\mu_n(x) \le \mu_{n+1}(x),\ \forall\, x \in X,\ n \in \mathbb{N}.$$

Finally,

$$0 \le \mu_n(x) + \nu_n(x) = (F_n - g_n)^+(x) + g_n(x) =$$

$$= \begin{cases} F_n(x) \le 1, \text{ if } F_n(x) - g_n(x) \ge 0 \\ g_n(x) \le 1, \text{ if } F_n(x) - g_n(x) < 0 \end{cases}$$

which proves the theorem. ∎

REMARK 13.16 It would be interesting to extend Theorem 13.3, Corollary 13.1, Theorem 13.4, Theorem 13.5 and Theorem 13.6 to intuitionistic fuzzy sets too. ∎

13.4 Weierstrass and Stone and Weierstrass-Type Results in Approximation of Fuzzy Functions

Let us denote $FC[a,b] = \{(f, \varphi_f)$; where f, φ_f, F, are given in Definition 13.3, such that f and F are continuous on $[a,b]$ and on $Graph(f)$, respectively$\}$.

For a fuzzy function $(f, \varphi_f) \in FC[a,b]$, since f is continuous on $[a,b]$, it follows that $Graph(f)$ is closed in $\mathbb{R}^2$. Then, by the continuity of F on $Graph(f)$ we get that $G^*(\varphi_f)$ (see Definition 13.3) is closed and because moreover it is bounded, it follows that $G^*(\varphi_f)$ is compact, too.

As a conclusion, the concept $D_{\mathcal{F}}$ (see Definition 13.3) is a metric on $FC[a,b]$.

Also, it is obvious that any fuzzy polynomial $(P, \varphi_P) \in FC[a,b]$.

The first fuzzy variant of the Weierstrass approximation theorem is the following.

THEOREM 13.8 (Gal [10])

For any fuzzy function $(f, \varphi_f) \in FC[a, b]$, there exists a sequence of fuzzy algebraic polynomials from $[a, b]$ to $\mathbb{R}$, $(P_n, \varphi_{P_n})_n$, such that

$$\lim_{n \to \infty} D_{\mathcal{F}}[(f, \varphi_f), (P_n, \varphi_{P_n})] = 0.$$

REMARK 13.17 The proof in Gal [10] is not of a constructive type. ∎

As was shown in Remark 13.6 after Definition 13.3, the interval $[a, b]$ can be replaced by (K, d) - a compact metric space. Then $FC(K) = \{(f, \varphi_f); f, \varphi_f, F$ are as in Definition 13.3, such that f and F are continuous on K and on $Graph(f)$, respectively$\}$, and in Gal [10] it is proved that $D_{\mathcal{F}}$ becomes a metric on $FC(K)$.

Let us consider $C(K, \mathbb{R}) = \{g : K \to \mathbb{R}; g$ continuous on $K\}$, where $A_1(K) \subset C(K; \mathbb{R})$, $A_2(\mathbb{R}) \subset C(\mathbb{R}, \mathbb{R})$ are subalgebras.

DEFINITION 13.8 *(Gal [12]) A pair (P, φ_P) will be called (A_1, A_2)-fuzzy function from K to $\mathbb{R}$, if $P \in A_1(K)$ and $\varphi_P : K \times \mathbb{R} \to [0, 1]$ is of the form*

$$\varphi_P(x, y) = \sum_{i=1}^{m} f_i(x) Q_i(y) \in (0, 1] \text{ if } (x, y) \in Graph(P),$$

$$\varphi_P(x, y) = 0, \text{ otherwise},$$

with $f_i \in A_1(K)$, $Q_i \in A_2(\mathbb{R})$, $\forall\, i = \overline{1, m}$.

A fuzzy variant of the Stone-Weierstrass theorem is the following.

THEOREM 13.9 (Gal [12])

Let (K, d) be a compact metric space, $A_1(K) \subset C(K; \mathbb{R})$, $A_2(\mathbb{R}) \subset C(\mathbb{R}, \mathbb{R})$ two subalgebras containing the constants on K and $\mathbb{R}$, respectively, and separating the points of K and of $\mathbb{R}$, respectively. For any fuzzy function $(f, \varphi_f) \in FC(K)$, there exists a sequence of (A_1, A_2)-fuzzy functions from K to $\mathbb{R}$, denoted by $(f_n, \varphi_{f_n})_n$, such that

$$\lim_{n \to \infty} D_{\mathcal{F}}[(f, \varphi_f), (f_n, \varphi_{f_n})] = 0.$$

REMARK 13.18 For $K = [a, b]$, Theorem 13.9 becomes Theorem 13.8. ∎

REMARK 13.19 If $(f, \varphi_f) \in FC(K)$ is such that $\varphi_f : K \times \mathbb{R} \to [0, 1]$, $\varphi_f(x, y) = 1$, $\forall\ (x, y) \in Graph(f)$ (i.e., (f, φ_f) can be identified with f), in Gal [12] it is proved that Theorem 13.9 reduces to the classical Stone-Weierstrass approximation theorem. ∎

REMARK 13.20 For fuzzy functions considered by Definition 13.2 and by Definition 13.4, in the next section we will give constructive proofs for the corresponding fuzzy variants of the Weierstrass theorem, together with quantitative estimates. Because of this, they are not considered in the present section. ∎

REMARK 13.21 Theorem 13.9 can be extended to a little more general type of fuzzy functions, as follows. ∎

DEFINITION 13.9 *(Gal [13]) A fuzzy set-valued function from $[a, b]$ to $\mathcal{P}(\mathbb{R})$ is a pair (f, φ_f) with $f : [a, b] \to \mathcal{P}(\mathbb{R})$ a usual set-valued function, characterized by the membership function $\varphi_f : [a, b] \times \mathbb{R} \to [0, 1]$, which satisfies $\varphi_f(x, y) > 0$ iff $(x, y) \in Graph(f) = \{(x, y) \in [a, b] \times \mathbb{R}; y \in f(x)\}$.*

The distance between two fuzzy set-valued functions from $[a, b]$ to $\mathcal{P}(\mathbb{R})$ is

$$D_{\mathcal{F}}[(f, \varphi_f), (g, \varphi_g)] = d_H(G^*(\varphi_f), G^*(\varphi_g)),$$

where d_H represents the usual Hausdorff distance in $\mathbb{R}^3$ between $G^*(\varphi_f)$ and $G^*(\varphi_g)$, and

$$G^*(\varphi_f) = \{(x, y, z) \in [a, b] \times \mathbb{R} \times [0, 1]; z = \varphi_f(x, y), (x, y) \in Graph(f)\},$$

($G^*(\varphi_g)$ is defined similarly).

REMARK 13.22 A fuzzy set-valued function from $[a, b]$ to $\mathcal{P}(\mathbb{R})$ can be considered as a fuzzy relation defined as in, e.g., [24]. When f is single valued, Definition 13.9 reduces to Definition 13.3. ∎

DEFINITION 13.10 *(Gal [13]) Let (f, φ_f) be a fuzzy set-valued function from $[a, b]$ to $\mathcal{P}(\mathbb{R})$ and $\varepsilon > 0$. We say that (F, φ_F), a fuzzy* *(single-valued) function from* $[a,b]$ *to* $\mathbb{R}$ *(see Definition 13.3), is an* ε*-approximate selection for* (f,φ_f)*, if*

$$D_{\mathcal{F}}[(f,\varphi_f),(F,\varphi_F)) < \varepsilon.$$

Let (S,φ_S) *be a fuzzy relation on* $[a,b]\times\mathbb{R}$ *(i.e.* $S\subset[a,b]\times\mathbb{R}$ *and* $\varphi_S:[a,b]\times\mathbb{R}\to[0,1]$ *satisfies* $\varphi(x,y)>0$ *iff* $(x,y)\in S$*). We say that the fuzzy (single-valued) function* (F,φ_F) *is an* ε*-approximate selection of* (S,φ_S) *if*

$$D_{\mathcal{F}}[(S,\varphi_S),(F,\varphi_F)] < \varepsilon.$$

(For $(S,\varphi_S), G^*(\varphi_S)=\{(x,y,z)\in[a,b]\times\mathbb{R}\times[0,1]; z=\varphi_S(x,y),$ $(x,y)\in Graph(S)\}$*, where* $Graph(S)=\{(x,y)\in[a,b]\times\mathbb{R};(x,y)\in S\}$*).*

If moreover F *and* φ_F *are continuous on* $[a,b]$ *and on* $Graph(F)$*, respectively, then* (F,φ_F) *will be called a continuous* ε*-approximate selection.*

THEOREM 13.10 (Gal [13])

(i) If (f,φ_f) *is a fuzzy set-valued function from* $[a,b]$ *to* $\mathcal{P}(\mathbb{R})$*, which for each* $\varepsilon>0$ *admits a continuous* ε*-approximate selection, then there exists a sequence of fuzzy algebraic polynomials (see Definition 13.3),* $(P_n,\varphi_{P_n})_n$*, such that*

$$\lim_{n\to+\infty} D_{\mathcal{F}}[(f,\varphi_f),(P_n,\varphi_{P_n})] = 0.$$

(ii) If (S,φ_S) *is a fuzzy relation on* $[a,b]\times\mathbb{R}$*, which for each* $\varepsilon>0$ *admits a continuous* ε*-approximate selection, then there exists a sequence of fuzzy algebraic polynomials,* $(P_n,\varphi_{P_n})_n$*, such that*

$$\lim_{n\to+\infty} D_{\mathcal{F}}[(S,\varphi_S),(P_n,\varphi_{P_n})] = 0.$$

REMARK 13.23 Examples of (f,φ_f) and of (S,φ_S) which for each $\varepsilon>0$ admits continuous ε-approximate selections are given by Gal [13], Theorems 3.1 and 3.2. For instance, if the fuzzy relation (S,φ_S) satisfies the conditions:

a) $Dom(S)=\{x\in[a,b];\ \exists\ y\in\mathbb{R}$ with $(x,y)\in S\}$ is dense in $[a,b]$;

b) For each isolated point $x\in Dom(S)$, the vertical section

$$S(x)=\{y\in\mathbb{R};(x,y)\in S\}$$

is a singleton;

c) $|\varphi_S(x,y)-\varphi_S(x',y')| \le M \max\{|x-x'|, |y-y'|, \forall (x,y),(x',y') \in S$;
d) S is closed in the metric space $([a,b] \times \mathbb{R}; d_0)$, where

$$d_0(u,u') = \max\{|x-x'|, |y-y'|\}, \ u=(x,y), \ u'=(x',y') \in [a,b] \times \mathbb{R};$$

then for each $\varepsilon > 0$, (S, φ_S) admits a continuous ε-approximate selection. ∎

REMARK 13.24 For (K,d) a compact metric space, let us denote $C_{\mathcal{F}}(K) = \{f : (K,d) \to (\mathbb{R}_{\mathcal{F}}, D); f$ continuous at each $x \in K$, as mapping between two metric spaces$\}$.

It would be interesting to prove a Stone and Weierstrass-type result in $C_{\mathcal{F}}(K)$. ∎

13.5 Quantitative Estimates in Approximation by Fuzzy Polynomials

In this section we present some constructive proofs for Weierstrass-type results in fuzzy setting, obtaining quantitative estimates of the approximation's error.

DEFINITION 13.11 *(Gal [11]) Let (f,φ_f) be a fuzzy function from $[0,1]$ to $\mathbb{R}$ (see Definition 13.3).*

The fuzzy algebraic polynomial $(B_n(f),\varphi_n)$ given by

$$B_n(f)(x) = \sum_{i=0}^{n} f\left(\frac{i}{n}\right)\binom{n}{i} x^i(1-x)^{n-i},$$

$$\varphi_n : [0,1] \times \mathbb{R} \to [0,1],$$

$$\varphi_n(x,y) = \begin{cases} Q_n(x,y), & \text{if } (x,y) \in Graph[B_n(f)], \\ 0, & \text{otherwise} \end{cases}$$

where

$$Q_n(x,y) = \sum_{i=0}^{n} \varphi_f\left(\frac{i}{n}, f\left(\frac{i}{n}\right)\right)\binom{n}{i} x^i(1-x)^{n-i} + y - B_n(f)(x),$$

is called of Bernstein-type, attached to (f,φ_f).

We say that (f, φ_f) *is* 2π*-periodic fuzzy function from* $\mathbb{R}$ *to* $\mathbb{R}$ *if* $f : \mathbb{R} \to \mathbb{R}$ *is* 2π*-periodic and if* $\varphi_f : \mathbb{R} \times \mathbb{R} \to [0,1]$ *satisfies the conditions* $0 < \varphi_f(x, y)$ *iff* $(x, y) \in Graph(f)$, $\varphi_f(x + 2\pi, y) = \varphi_f(x, y)$, $\forall\ (x, y) \in Graph(f)$. *Denote* $FC_{2\pi}(\mathbb{R}) = \{(f, \varphi_f); (f, \varphi_f)$ *is* 2π *periodic, with* $f(x)$ *and* $H(x) = \varphi_f(x, f(x))$ *continuous on* $\mathbb{R}\}$.

The trigonometric fuzzy polynomial $(J_n(f), \varphi_n)$, *where* $J_n(f)(x)$ *is the Jackson trigonometric operator given in Lorentz [20], p.56, relation (7),* $\varphi_n : \mathbb{R} \times \mathbb{R} \to [0,1]$ *given by*

$$\varphi_n(x, y) = \begin{cases} T_n(x, y), & \text{if } (x, y) \in Graph[J_n(f)], \\ 0, & \text{otherwise} \end{cases}$$

with

$$T_n(x, y) = \int_{-\pi}^{\pi} \varphi_f(t, f(t)) K_n(x - t) dt + y - J_n(f)(x) =$$

$$= \int_{-\pi}^{\pi} \varphi_f(x + t, f(x + t)) K_n(t) dt + y - J_n(f)(x),$$

is called of Jackson-type attached to the 2π*-periodic fuzzy function from* $\mathbb{R}$ *to* $\mathbb{R}$, (f, φ_f).

THEOREM 13.11 (Gal [11])

(i) For any $(f, \varphi_f) \in FC[0,1]$ *we have*

$$D_{\mathcal{F}}[(f, \varphi_f), (B_n(f), \varphi_n)] \le \frac{3}{2} \left[\omega_1 \left(f; \frac{1}{\sqrt{n}} \right) + \omega_1 \left(H; \frac{1}{\sqrt{n}} \right) \right], \quad n \in \mathbb{N},$$

where

$$H(x) = \varphi_f(x, f(x)), \quad x \in [0,1],$$

$$\omega_1(f; \delta) = \sup\{|f(x) - f(y)|; x, y \in [0,1], |x - y| \le \delta\},$$

$$\omega_1(H; \delta) = \sup\{|H(x) - H(y)|; x, y \in [0,1], |x - y| \le \delta\}.$$

(ii) For any $(f, \varphi_f) \in FC_{2\pi}(\mathbb{R})$ *we have*

$$D_{\mathcal{F}}[(f, \varphi_f), (J_n(f), \varphi_n)] \le M \left[\omega_2 \left(f; \frac{1}{n} \right) + \omega_2 \left(H; \frac{1}{n} \right) \right], \quad n \in \mathbb{N},$$

where

$$H(x) = \varphi_f(x, f(x)), \quad x \in \mathbb{R},$$

$$\omega_2(f; \delta) = \sup\{|f(x + h) - 2f(x) + f(x - h)|; x \in \mathbb{R}, |h| \le \delta\}$$

($\omega_2(H;\delta)$ is defined similarly) and $M > 0$ is independent of n, f, H and φ_f.

REMARK 13.25 If $\varphi_f(x,y) = 1$, $\forall\ (x,y) \in Graph(f)$ (i.e. (f,φ_f) can be identified with f), then $H(x) \equiv 1$, therefore $\omega_1(H;\delta) \equiv 0$, $\omega_2(H;\delta) \equiv 0$, and (i), (ii) become the well-known estimates in classical approximation theory. ∎

REMARK 13.26 Passing to limit with $n \to +\infty$, from (i), (ii) we obtain constructive proofs for the Weierstrass-type theorems in the previous paragraph. ∎

REMARK 13.27 In the case of 2π-periodic fuzzy functions, we have

$$D_{\mathcal{F}}[(f,\varphi_f),(g,\varphi_g)] = d_H(G^*(\varphi_f), G^*(\varphi_g)),$$

where

$$G^*(\varphi_f) = \{(x,y,z) \in \mathbb{R} \times \mathbb{R} \times (0,1]; y = f(x), z = \varphi_f(x,y),$$
$$(x,y) \in Graph(f)\}$$

($G^*(\varphi_g)$ is defined similarly). ∎

Theorem 13.11 can be extended to approximation of generalized fuzzy functions from $[a,b]$ to $\mathbb{R}$ in Definition 13.4.

Thus, let us denote

$$GFC[a,b] = \{(\varphi, f, \gamma_{\varphi,f}); \text{supp}(\varphi) = [a,b], \varphi \text{ and } f \text{ continuous on}$$
$$[a,b] \text{ and } \gamma_{\varphi,f}|_M \text{ continuous on } M\},$$

where

$$M = \{(x,y,z) \in [a,b] \times \mathbb{R} \times (0,1]; (x,y) \in Graph(f), (x,z) \in G^*(\varphi)\}.$$

REMARK 13.28 If we denote $H\colon [a,b] \to \mathbb{R}^2$ by $H(x) = (f(x), \varphi(x))$, then obviously $M = Graph(H)$, where because f and φ are continuous on $[a,b]$, it follows that H is a continuous vector function, $\mathbb{R}^2$ being endowed with the box metric

$$d_2(x,y) = \max\{|x_1 - y_1|, |x_2 - y_2|\},$$

$$x = (x_1, x_2),\ y = (y_1, y_2) \in \mathbb{R}^2.$$

Since $[a, b]$ is compact, it follows that M is a compact subset of $[a, b] \times \mathbb{R} \times (0, 1]$. Then, since $G^*(\gamma_{\varphi,f}) = Graph(\gamma_{\varphi,f}|_M)$, we obtain that $G^*(\gamma_{\varphi,f})$ is a compact subset of $[a, b] \times \mathbb{R} \times (0, 1] \times (0, 1]$.

As a conclusion, $D_{\mathcal{F}}$ introduced by Definition 13.4 is a metric on $GFC[a, b]$. ∎

DEFINITION 13.12 *Let $(\varphi, f, \gamma_{\varphi,f}) \in GFC[0, 1]$. The generalized fuzzy algebraic polynomial from $[0, 1]$ to $\mathbb{R}$, $(B_n(\varphi), B_n(f), Q_n)$, where*

$$B_n(\varphi)(x) = \sum_{i=0}^{n} \varphi\left(\frac{i}{n}\right)\binom{n}{i}x^i(1-x)^{n-i},$$

$$B_n(f)(x) = \sum_{i=0}^{n} f\left(\frac{i}{n}\right)\binom{n}{i}x^i(1-x)^{n-i}$$

and $Q_n : [0, 1] \times \mathbb{R} \times [0, 1] \to [0, 1]$ is given by

$$Q_n(x, y, z) = \begin{cases} R_n(x, y, z), & (x, y) \in Graph(B_n(f)), (x, z) \in G^*(B_n(\varphi)) \\ 0, & otherwise, \end{cases}$$

with

$$R_n(x, y, z) = \sum_{i=0}^{n} \gamma_{\varphi,f}\left(\frac{i}{n}, f\left(\frac{i}{n}\right), \varphi\left(\frac{i}{n}\right)\right)\binom{n}{i}x^i(1-x)^{n-i} +$$

$$+ y - B_n(f)(x) + z - B_n(\varphi)(x),$$

is called the Bernstein-type attached to $(\varphi, f, \gamma_{\varphi,f})$.

First, we need the following.

LEMMA 13.1
For all $(\varphi, f, \gamma_{\varphi,f}), (\phi, F, \Gamma_{\phi,F}) \in GFC[0, 1]$, we have

$$D_{\mathcal{F}}[(\varphi, f, \gamma_{\varphi,f}), (\phi, F, \Gamma_{\phi,F})] \le \|\varphi - \phi\| + \|f - F\| +$$

$$+ \sup\{|\gamma_{\varphi,f}(a_1, f(a_1), \varphi(a_1)) - \Gamma_{\phi,F}(a_1, F(a_1), \phi(a_1))|; a_1 \in [0, 1]\},$$

where $\|\cdot\|$ is the uniform norm.

PROOF Let $a = (a_1, a_2, a_3, a_4)$, $b = (b_1, b_2, b_3, b_4)$ be with $a_2 = f(a_1)$, $a_3 = \varphi(a_1)$, $a_4 = \gamma_{\varphi,f}(a_1, a_2, a_3)$, $b_2 = F(a_1)$, $b_3 = \phi(a_1)$, $b_4 = \Gamma_{\phi,F}(b_1, b_2, b_3)$ and $d^*(a,b) = \max\{|a_i - b_i|; i = \overline{1,4}\}$.

For fixed $a_1 \in [0,1]$ we get

$$\inf\{d^*(a,b); b \in G^*(\Gamma_{\phi,F})\} \leq \inf\{|a_1 - b_1| + |f(a_1) - F(b_1)| +$$

$$+|\varphi(a_1) - \phi(b_1)| + |\gamma_{\varphi,f}(a_1, f(a_1), \varphi(a_1)) -$$

$$-\Gamma_{\phi,F}(b_1, F(b_1), \phi(b_1))|; b \in G^*(\Gamma_{\phi,F})\} \leq$$

$$\leq |f(a_1) - F(a_1)| + |\varphi(a_1) - \phi(a_1)| + |\gamma_{\varphi,f}(a_1, f(a_1), \varphi(a_1)) -$$

$$-\Gamma_{\phi,F}(a_1, F(a_1), \phi(a_1))|.$$

Passing now to supremum with $a \in G^*(\gamma_{\varphi,f})$, we obtain

$$\sup_{a \in G^*(\gamma_{\varphi,f})} \inf_{b \in G^*(\Gamma_{\phi,F})} d^*(a,b) \leq \|f - F\| + \|\varphi - \phi\| +$$

$$+ \sup_{a_1 \in [0,1]} |\gamma_{\varphi,f}(a_1, f(a_1), \varphi(a_1)) - \Gamma_{\phi,F}(a_1, F(a_1), \phi(a_1))| = E.$$

Entirely analogous we get

$$\sup_{b \in G^*(\Gamma_{\phi,F})} \inf_{a \in G^*(\gamma_{\varphi,f})} d^*(a,b) \leq E,$$

which proves the lemma. ∎

THEOREM 13.12

For any $(\varphi, f, \gamma_{\varphi,f}) \in GFC[0,1]$ we have

$$D_{\mathcal{F}}[(\varphi, f, \gamma_{\varphi,f}), (B_n(\varphi), B_n(f), Q_n)] \leq$$

$$\leq \frac{3}{2}\left[\omega_1\left(f; \frac{1}{\sqrt{n}}\right) + \omega_1\left(\varphi; \frac{1}{\sqrt{n}}\right) + \omega_1\left(H; \frac{1}{\sqrt{n}}\right)\right], \quad n \in \mathbb{N},$$

where $H(x) = \gamma_{\varphi,f}(x, f(x), \varphi(x))$, $x \in [0,1]$, and ω_1 is the usual first order modulus of continuity.

PROOF By Lemma 13.1 we get

$$D_{\mathcal{F}}[(\varphi, f, \gamma_{\varphi,f}), (B_n(\varphi), B_n(f), Q_n)] \leq \|f - B_n(f)\| + \|\varphi - B_n(\varphi)\| +$$

$$+ \sup_{a_1 \in [0,1]} |\gamma_{\varphi,f}(a_1, f(a_1), \varphi(a_1)) - Q_n(a_1, B_n(f)(a_1), B_n(\varphi)(a_1)|.$$

Since a classical result in approximation theory gives

$$\|f - B_n(f)\| \leq \frac{3}{2}\omega_1\left(f; \frac{1}{\sqrt{n}}\right),$$

$$\|\varphi - B_n(\varphi)\| \leq \frac{3}{2}\omega_1\left(\varphi; \frac{1}{\sqrt{n}}\right),$$

we obtain

$$D_{\mathcal{F}}[(\varphi, f, \gamma_{\varphi,f}), (B_n(\varphi), B_n(f), Q_n)] \leq$$

$$\leq \frac{3}{2}\left[\omega_1\left(\varphi; \frac{1}{\sqrt{n}}\right) + \omega_1\left(f; \frac{1}{\sqrt{n}}\right)\right] +$$

$$+ \sup_{a_1 \in [0,1]} |H(a_1) - B_n(H)(a_1)| \leq$$

$$\leq \frac{3}{2}\left[\omega_1\left(\varphi; \frac{1}{\sqrt{n}}\right) + \omega_1\left(f; \frac{1}{\sqrt{n}}\right) + \omega_1\left(H; \frac{1}{\sqrt{n}}\right)\right],$$

which proves the theorem. ▮

REMARK 13.29 Similarly, an estimate in approximation by generalized fuzzy trigonometric polynomials of Jackson-type can be proved. ▮

In what follows we will prove quantitative estimates in approximation of fuzzy functions in Definition 13.2.

DEFINITION 13.13 *Let $f : [0,1] \to \mathbb{R}_{\mathcal{F}}$ be a fuzzy (real) function (see Definition 13.2). The fuzzy algebraic polynomial defined by*

$$B_n^{(\mathcal{F})}(f)(x) = \sum_{k=0}^{n}{}^{*} \binom{n}{k} x^k (1-x)^{n-k} \odot f\left(\frac{k}{n}\right)$$

will be called a Bernstein-type (here the sum $\sum^*$ and the product with scalars $\odot$, are those given in Remark 13.1).

Let us denote
$C_{\mathcal{F}}[a,b] = \{f : [a,b] \to \mathbb{R}_{\mathcal{F}}; f$ is continuous in each $x \in [a,b]$ as mapping between the metric spaces $([a,b], |\cdot|)$ and $(\mathbb{R}_{\mathcal{F}}, D)\}$.

THEOREM 13.13
If $f \in C_{\mathcal{F}}[0,1]$, then

$$D(B_n^{(\mathcal{F})}(f)(x), f(x)) \leq \frac{3}{2}\omega_1^{(\mathcal{F})}\left(f; \frac{1}{\sqrt{n}}\right), \quad n \in \mathbb{N}, \quad x \in [0,1],$$

i.e.,

$$\lim_{n\to+\infty} D(B_n^{(\mathcal{F})}(f)(x), f(x)) = 0,$$

uniformly with respect to $x \in [0,1]$.

(Here $\omega_1^{(\mathcal{F})}(f;\delta) = \sup\{D(f(x), f(y)); |x-y| \leq \delta, x, y \in [0,1]\}$, with D in Theorem 13.2.)

PROOF Denote $p_{n,k}(x) = \binom{n}{k} x^k (1-x)^{n-k}$.

By $\sum_{k=0}^{n} p_{n,k}(x) = 1$ and by

$$f(x) = \left[\sum_{k=0}^{n} p_{n,k}(x)\right] \odot f(x) = \sum_{k=0}^{n}{}^{*} [p_{n,k}(x) \odot f(x)]$$

(see Remark 13.4 after Theorem 13.2), we get

$$D(B_n(f)(x), f(x) =$$

$$= D\left(\sum_{k=0}^{n}{}^{*}\left[p_{n,k}(x) \odot f\left(\frac{k}{n}\right)\right], \left[\sum_{k=0}^{n} p_{n,k}(x)\right] \odot f(x)\right) =$$

$$= D\left(\sum_{k=0}^{n}{}^{*} p_{n,k}(x) \odot f\left(\frac{k}{n}\right), \sum_{k=0}^{n}{}^{*} p_{n,k}(x) \odot f(x)\right) \leq$$

$$\leq (\text{Theorem 13.2, (iv),(iii)}) \leq \sum_{k=0}^{n} p_{n,k}(x) D\left(f\left(\frac{k}{n}\right), f(x)\right) \leq$$

$$\leq \sum_{k=0}^{n} p_{n,k}(x)\omega_1^{(\mathcal{F})}\left(f; \left|\frac{k}{n} - x\right|\right) =$$

$$= \sum_{k=0}^{n} p_{n,k}(x)\omega_1^{(\mathcal{F})}\left(f; \delta\left|\frac{k}{n} - x\right|\frac{1}{\delta}\right) = \otimes.$$

But it is easy to check that $\omega_1^{(\mathcal{F})}$ has the following properties:

$$\omega_1^{(\mathcal{F})}(f;\delta_1+\delta_2) \leq \omega_1^{(\mathcal{F})}(f;\delta_1) + \omega_1^{(\mathcal{F})}(f;\delta_2), \quad \delta_1,\delta_2 > 0$$

$$\omega_1^{(\mathcal{F})}(f;n\delta) \leq n\omega_1^{(\mathcal{F})}(f;\delta), \quad \delta > 0, \quad n \in \mathbb{N},$$

$$\omega_1^{(\mathcal{F})}(f;\lambda\delta) \leq (\lambda+1)\omega_1^{(\mathcal{F})}(f;\delta), \quad \delta > 0, \quad \lambda > 0,$$

$$f \in C_{\mathcal{F}}[0,1] \text{ implies } \lim_{\delta \searrow 0} \omega_1^{(\mathcal{F})}(f;\delta) = 0.$$

For example, we will prove the first property. Let $|x''-x'| \leq \delta_1 + \delta_2$, $0 \leq x' \leq x \leq x'' \leq 1$, such that $x - x' = \delta_1$. We have

$$D(f(x''),f(x')) = (Theorem\, 13.2,(i)) = D(f(x'')\oplus f(x), f(x')\oplus f(x)) =$$

$$D(f(x'')\oplus f(x), f(x)\oplus f(x')) \leq (Theorem 13.2,(iv)) \leq$$

$$\leq D(f(x''),f(x)) + D(f(x),f(x')) \leq \omega_1^{(\mathcal{F})}(f;|x''-x|) + \omega_1^{(\mathcal{F})}(f;|x-x'|) \leq$$

$$\leq \omega_1^{(\mathcal{F})}(f;\delta_1) + \omega_1^{(\mathcal{F})}(f;\delta_2).$$

We get

$$\otimes \leq \omega_1^{(\mathcal{F})}(f;\delta) \sum_{k=0}^{n} \left(1 + \frac{\left|\frac{k}{n} - x\right|}{\delta} \right) p_{n,k}(x) =$$

$$= \omega_1^{(\mathcal{F})}(f;\delta) + \frac{\omega_1^{(\mathcal{F})}(f;\delta)}{\delta} \sum_{k=0}^{n} \left|\frac{k}{n} - x\right| p_{n,k}(x),$$

for all $\delta > 0$, $n \in \mathbb{N}$, $x \in [0,1]$.

But from the classical approximation theory it is well-known that

$$\sum_{k=0}^{n} \left|\frac{k}{n} - x\right| p_{n,k}(x) \leq \frac{1}{2\sqrt{n}}, \quad x \in [0,1], \; n \in \mathbb{N}.$$

Choosing above $\delta = \dfrac{1}{\sqrt{n}}$, we obtain

$$D(B_n(f)(x), f(x)) \leq \omega_1^{(\mathcal{F})}\left(f; \frac{1}{\sqrt{n}}\right) + \frac{1}{2}\omega_1^{(\mathcal{F})}\left(f; \frac{1}{\sqrt{n}}\right)$$

$$= \frac{3}{2}\omega_1^{(\mathcal{F})}\left(f; \frac{1}{\sqrt{n}}\right),$$

which proves the theorem. ∎

In what follows, we need a particular case of the Henstock integral introduced in Wu Congxin and Gong Zengtai [29], Definition 2.1.

DEFINITION 13.14 *Let $f : [a,b] \to \mathbb{R}_{\mathcal{F}}$. We say that f is Riemann integrable to $I \in \mathbb{R}_{\mathcal{F}}$ if for any $\varepsilon > 0$, there exists $\delta > 0$ such that for any division $P = \{[u,v];\xi\}$ of $[a,b]$ with the norm $\Delta(P) < \delta$, we have*

$$D\left(\sum_{P}^{*}(v-u)\odot f(\xi), I\right) < \varepsilon.$$

We write

$$I = (R)\int_a^b f(x)dx.$$

Reasoning exactly as in the proof of Theorem 3.7 in Wu Congxin and Gong Zengtai [29], we obtain

COROLLARY 13.2
If $f \in C_{\mathcal{F}}[a,b]$ then f is Riemann integrable on $[a,b]$.

Let $f : \mathbb{R} \to \mathbb{R}_{\mathcal{F}}$ be 2π-periodic, i.e., $f(x) = f(x+2\pi)$, for all $x \in \mathbb{R}$.
In order to obtain a Jackson-type estimate in approximation of 2π-periodic fuzzy functions, first we need the following.

LEMMA 13.2
(i) If $f : \mathbb{R} \to \mathbb{R}_{\mathcal{F}}$ is continuous and 2π-periodic, then

$$(R)\int_0^{2\pi} f(u)du = (R)\int_a^{a+2\pi} f(u)du, \text{ for all } a \in \mathbb{R}.$$

(ii) If $f,g : \mathbb{R} \to \mathbb{R}_{\mathcal{F}}$ are continuous and 2π-periodic, then the function $F : \mathbb{R} \to \mathbb{R}_+$ defined by $F(x) = D(f(x), g(x))$ is continuous, 2π-periodic and

$$D\left[(R)\int_{-\pi}^{\pi} f(u)du, (R)\int_{-\pi}^{\pi} g(u)du\right] \le \int_{-\pi}^{\pi} F(x)dx.$$

PROOF (i) The proof is similar to that in the case of usual real 2π-periodic functions, because it is easy to prove that we can construct a sequence of divisions of $[0, 2\pi]$, $P_n = \{[u_n, v_n]; \xi_n\}$, $n \in \mathbb{N}$, with $\Delta(P_n) \searrow 0$ (when $n \to +\infty$), such that for each P_n, there exists $Q_n = \{[y_n, w_n]; \eta_n\}$, division of $[a, a+2\pi]$, satisfying $\Delta(P_n) = \Delta(Q_n)$ and

$$\sum^*(P_n) := \sum_{P_n}^*(v_n - u_n) \odot f(\xi_n) = \sum_{Q_n}(w_n - y_n) \odot f(\eta_n) =: \sum^*(Q_n).$$

Then by

$$D\left[(R)\int_0^{2\pi} f(u)du, (R)\int_a^{a+2\pi} f(u)du\right] \leq$$

$$\leq D\left[(R)\int_0^{2\pi} f(u)du, \sum^*(P_n)\right] +$$

$$+D\left(\sum^*(P_n), \sum^*(Q_n)\right) + D\left(\sum^*(Q_n), (R)\int_a^{a+2\pi} f(u)du\right),$$

passing to limit with $n \to +\infty$, we get

$$D\left((R)\int_0^{2\pi} f(u)du, (R)\int_a^{a+2\pi} f(u)du\right) = 0,$$

i.e.,

$$(R)\int_0^{2\pi} f(u)du = (R)\int_a^{a+2\pi} f(u)du.$$

(ii) First, obviously F is 2π-periodic. Then, let $x_n \in \mathbb{R}$, $n \in \mathbb{N}$, be with $x_n \overset{n\to+\infty}{\longrightarrow} x_0$. We have

$$D(f(x_n), g(x_n)) \leq D(f(x_n), f(x_0)) + D(f(x_0), g(x_0)) + D(g(x_0), g(x_n))$$

and

$$D(f(x_0), g(x_0)) \leq D(f(x_0), f(x_n)) + D(f(x_n), g(x_n)) + D(g(x_n), g(x_0)).$$

Passing to the limit with $n \to +\infty$, from the continuity of f and g we get

$$\lim_{n\to\infty} D(f(x_n), g(x_n)) = D(f(x_0), g(x_0)),$$

that is F is continuous at each $x_0 \in \mathbb{R}$.

Now, let $P_n = \{[u_n, v_n]; \xi_n\}$, $n \in \mathbb{N}$, be a sequence of divisions of $[-\pi, \pi]$ with $\Delta(P_n) \searrow 0$, when $n \to +\infty$.

We have

$$D\left((R)\int_{-\pi}^{\pi} f(u)du, (R)\int_{-\pi}^{\pi} g(u)du\right) \leq$$

$$\leq D\left((R)\int_{-\pi}^{\pi} f(u)du, \sum_{P_n}^{*}(v_n - u_n)\odot f(\xi_n)\right) +$$

$$+D\left(\sum_{P_n}^{*}(v_n - u_n)\odot f(\xi_n), \sum_{p_n}^{*}(v_n - u_n)\odot g(\xi_n)\right) +$$

$$+D\left(\sum_{P_n}^{*}(v_n - u_n)\odot g(\xi_n), (R)\int_{-\pi}^{\pi} g(u)du\right) \leq$$

$$\leq D\left((R)\int_{-\pi}^{\pi} f(u)du, \sum_{P_n}^{*}(v_n - u_n)\odot f(\xi_n)\right) +$$

$$+\sum_{P_n} D(f(\xi_n), g(\xi_n))(v_n - u_n) +$$

$$+D\left(\sum_{P_n}^{*}(v_n - u_n)\odot g(\xi_n), (R)\int_{-\pi}^{\pi} g(u)du\right).$$

Passing to the limit with $n \to +\infty$, we easily get

$$D\left((R)\int_{-\pi}^{\pi} f(u)du, (R)\int_{-\pi}^{\pi} g(u)du\right) \leq \int_{-\pi}^{\pi} D(f(x), g(x))dx,$$

which proves the lemma. ∎

Let us denote

$$C_{2\pi}^{(\mathcal{F})}(\mathbb{R}) = \{f : \mathbb{R} \to \mathbb{R}_{\mathcal{F}}; f \text{ is continuous and } 2\pi - \text{periodic on } \mathbb{R}\}.$$

Suggested by Lorentz [20], p. 55-56, for $f \in C_{2\pi}^{(\mathcal{F})}(\mathbb{R})$ let us introduce the Jackson operator

$$J_n(f)(x) = (R)\int_{-\pi}^{\pi} K_n(t)\odot f(x+t)dt,$$

where

$$K_n(t) = L_{n'}(t), \quad n' = \left[\frac{n}{2}\right] + 1,$$

$$L_n(t) = \lambda_n^{-1}\left[\frac{\sin(nt/2)}{\sin(t/2)}\right]^4,$$

$$\int_{-\pi}^{\pi} L_n(t)dt = 1,$$

$K_n(t) \geq 0$ being an even trigonometric polynomial of the form

$$\left(\sum_{j=0}^{n'-1} A_j \cos jt\right)^2, \quad A_j \in \mathbb{R}, \quad j = \overline{0, n'-1}.$$

It is known that $K_n(t)$ is of degree n. Then by Lemma 13.2, (i), we get

$$J_n(f)(x) = (R)\int_{x-\pi}^{x+\pi} K_n(u-x)\odot f(u)du = (R)\int_{-\pi}^{\pi} K_n(u-x)\odot f(u)du,$$

i.e., $J_n(f)(x)$ is a fuzzy trigonometric polynomial as in Definition 13.2 (taking into account Remark 13.4 after Theorem 13.2).

THEOREM 13.14

There exists a constant $C > 0$ (independent of n and f), such that for all $f \in C_{2\pi}^{(\mathcal{F})}(\mathbb{R})$ we have

$$D(J_n(f)(x), f(x)) \leq C\omega_1^{(\mathcal{F})}\left(f; \frac{1}{n}\right), \quad n \in \mathbb{N}, \quad x \in \mathbb{R},$$

where

$$\omega_1^{(\mathcal{F})}(f;\delta) = \sup\{D(f(x), f(y)); |x-y| \leq \delta, x, y \in \mathbb{R}\}.$$

PROOF By $f(x) = (R)\int_{-\pi}^{\pi} K_n(t)\odot f(x)dt$ and by Lemma 13.2, (ii), we easily obtain

$$D(J_n(f)(x), f(x)) =$$

$$= D\left((R)\int_{-\pi}^{\pi} K_n(t)\odot f(x+t)dt, (R)\int_{-\pi}^{\pi} K_n(t)\odot f(x)dt\right) \leq$$

$$\leq \int_{-\pi}^{\pi} D(K_n(t)\odot f(x+t), K_n(t)\odot f(x))dt =$$

$$= \int_{-\pi}^{\pi} K_n(t) D(f(x+t), f(x)) dt \leq$$

$$\leq \int_{-\pi}^{\pi} K_n(t) \omega_1^{(\mathcal{F})}(f; |t|) dt = 2 \int_0^{\pi} K_n(t) \omega_1^{(\mathcal{F})}(f; t) dt =$$

$$= 2 \int_0^{\pi} K_n(t) \omega_1^{(\mathcal{F})} \left(f; \frac{1}{n}(tn) \right) dt \leq 2\omega_1^{(\mathcal{F})} \left(f; \frac{1}{n} \right) \int_{-\pi}^{\pi} (nt+1) K_n(t) dt \leq$$

$$\leq C\omega_1^{(\mathcal{F})} \left(f; \frac{1}{n} \right),$$

because $\int_{-\pi}^{\pi} K_n(t) dt = 1$ and by Lorentz [20], p.56, relation (6), we have

$$\int_{-\pi}^{\pi} t K_n(t) dt \leq \frac{c}{n}.$$

The theorem is proved. ∎

REMARK 13.30 For any $f \in C_{\mathcal{F}}[-1, 1]$ we can attach an algebraic polynomial with fuzzy-numbers coefficients by

$$T_n(f)(x) = \frac{1}{\pi}(R) \int_{-\pi}^{\pi} K_{3n-3}(v) \odot f[\cos(\arccos x + v)] dv,$$

where $K_{3n-3}(v)$ is the kernel introduced in Lehnhoff [18].

It is left to the reader to attempt to prove the following Timan-type estimate

$$D(f(x), T_n(f)(x)) \leq 4\omega_1^{(\mathcal{F})} \left(f; \frac{\sqrt{1-x^2}}{n} + \frac{|x|}{n^2} \right), \; n \in \mathbb{N}, \; x \in [-1, 1].$$

Also, we conjecture that by using the Lehnhoff-type polynomials in Lehnhoff [19], even a Teljakowskii-type estimate can be obtained (here $\omega_1^{(\mathcal{F})}$ is defined on $C_{\mathcal{F}}[-1, 1]$). ∎

REMARK 13.31 Let us denote by $\mathcal{P}_n^{(\mathcal{F})}$ = the class of all fuzzy algebraic polynomials of degree $\leq n$, introduced by Definition 13.2.

For $f \in C_{\mathcal{F}}[a, b]$, the best approximation of f by elements in $\mathcal{P}_n^{(\mathcal{F})}$ can be defined as

$$E_n^{(\mathcal{F})} = \inf\{D_U(f, P); P \in \mathcal{P}_n^{(\mathcal{F})}\}, \quad n \in \mathbb{N}.$$

It is an open question if there exists a polynomial of best approximation $P_n^* \in \mathcal{P}_n^{(\mathcal{F})}$, i.e., which satisfies

$$E_n^{(\mathcal{F})} = D_U(f, P_n^*)?$$

(D_U is given in Definition 13.2.)

The existence (and possibly the uniqueness) of P_n^*, would permit extending many other results in classical approximation theory to this case.

For example, the fundamental role of the Chebyshev's alternation theorem in the proof of the uniqueness in the classical case is well known. It would be interesting to see if (and how) this theorem holds in our case too. ∎

REMARK 13.32 It is known (see Wu Congxin and Gong Zengtai [29]) that if $u, v \in \mathbb{R}_{\mathcal{F}}$, then $D(u,v) = \sup\limits_{r\in[0,1]} d_H([u]^r, [v]^r)$, where $d_H([u]^r, [v]^r)$ is the classical Hausdorff distance (between the closed intervals $[u]^r$ and $[v]^r$).

A question is what happens if D is replaced, for example, by the following distance (denoted by D^*),

$$D^*(u,v) = d_H(\overline{\text{supp}(u)}, \overline{\text{supp}(v)})?$$

∎

REMARK 13.33 More general approximation results could be developed if the usual metric $D(u,v)$ between $u, v \in \mathbb{R}_{\mathcal{F}}$, was replaced by a fuzzy metric between $u, v \in \mathbb{R}_{\mathcal{F}}$.

For example, a fuzzy metric can be suggested as a mapping $d_{\mathcal{F}} : \mathbb{R}_{\mathcal{F}} \times \mathbb{R}_{\mathcal{F}} \to \mathbb{R}_{\mathcal{F}}$ having the properties:

(i) $d_{\mathcal{F}}(u,v) = 0_{\mathcal{F}}$ iff $u = v$;

(ii) $d_{\mathcal{F}}(u,v) = d_{\mathcal{F}}(v,u)$, $\forall\, u, v \in \mathbb{R}_{\mathcal{F}}$;

(iii) $d_{\mathcal{F}}(u,v) \le d_{\mathcal{F}}(u,w) + d_{\mathcal{F}}(w,v)$, $\forall\, u, v, w \in \mathbb{R}_{\mathcal{F}}$;

where $0_{\mathcal{F}} : \mathbb{R} \to [0,1]$ is defined by $0_{\mathcal{F}}(x) = 1$, if $x = 0$, $0_{\mathcal{F}}(x) = 0$, otherwise, and the order $\le$ in $\mathbb{R}_{\mathcal{F}}$ is that defined in Remark 13.3 of Theorem 13.2.

The answer to this question is left to the reader. ∎

REMARK 13.34 Let $f : \mathbb{R} \to \mathbb{R}_{\mathcal{F}}$ be a fuzzy function. By analogy with the classical case, we can say that f is almost periodic if, for any

$\varepsilon > 0$, there exists a fuzzy trigonometric polynomial T, as in Definition 13.2, such that

$$D(f(x), T(x)) < \varepsilon, \text{ for all } x \in \mathbb{R}.$$

It would be interesting to study the properties of almost periodic fuzzy functions. ▮

13.6 Interpolation in Fuzzy Setting

The following problem of interpolating "fuzzy values" was initially stated by Zadeh [30] and formulated by Lowen [21].

Let $x_0 < \ldots < x_n$ be $n+1$ points in $\mathbb{R}$ and let $\mu_i \in \mathbb{R}_{\mathcal{F}}$, $i = \overline{0, n}$, be $n+1$ fuzzy (real) numbers.

Construct a function $F : \mathbb{R} \to \mathbb{R}_{\mathcal{F}}$ fulfilling the properties:

(i) $F(x_i) = \mu_i$, $i = \overline{0, n}$;

(ii) F is continuous on $\mathbb{R}$, as a mapping between the metric spaces $(\mathbb{R}, d_{\mathbb{R}}), (\mathbb{R}_{\mathcal{F}}, E)$, where $d_{\mathbb{R}} = |x - y|$, $x, y \in \mathbb{R}$ and

$$E(u, v) = d_H[end(u), end(v)],$$

$$end(u) = \{(x, r) \in \mathbb{R} \times (0, 1]; r \leq u(x)\}, \quad u, v \in \mathbb{R}_{\mathcal{F}}$$

(see Kloeden [17]);

(iii) if for all $i = 0, \ldots, n$,

$$\mu_i(t) = \begin{cases} 1, & t = y_i \\ 0, & \text{otherwise} \end{cases}$$

where $y_i \in \mathbb{R}$ and if P_n is the unique usual algebraic polynomial of degree $\leq n$ which satisfies $P_n(x_i) = y_i$, $i = 0, \ldots, n$, then $F(x) = P_n(x)$, for all $x \in \mathbb{R}$.

Lowen [21] proved the existence and uniqueness of the interpolating function F, which will be called fuzzy interpolating polynomial.

Kaleva [16] obtained the following representation of F in terms of α-level sets (together with some numerical methods for calculating $\{y \in \mathbb{R}; F(x)(y) \geq \alpha\}$, $\forall\, x \in \mathbb{R}$):

$$\{y \in \mathbb{R}; F(x)(y) \geq \alpha\} =$$

$$= \{y \in \mathbb{R}; y = P_n(d_0, \ldots, d_n)(x), d_i \in \mu_i^\alpha,\ i = \overline{0,n}\},\ \forall\ \alpha \in (0,1],$$

where $\mu_i^\alpha = \{t \in \mathbb{R}; \mu_i(t) \geq \alpha\}$ and $P_n(d_0, \ldots, d_n)$ is the Lagrange interpolation polynomial of degree $\leq n$, which satisfies

$$P_n(d_0, \ldots, d_n)(x_i) = d_i, \quad i = \overline{0,n},$$

i.e.,

$$F(x) = \sum_{i=0}^{n}{}^{*}\, l_i(x) \odot \mu_i,$$

where $l_i(x)$ are the fundamental classical Lagrange polynomials given by

$$l_i(x) = \prod_{\substack{j=0 \\ j \neq i}}^{n} \frac{(x - x_j)}{(x_i - x_j)}, \quad i = \overline{0,n},$$

and $\sum^*, \odot$ are given by Remark 13.1.

REMARK 13.35 Obviously F is a polynomial with fuzzy (real) number coefficients (see Definition 13.2) and is continuous as a mapping between the metric spaces $(\mathbb{R}, d_{\mathbb{R}})$ and $(\mathbb{R}_{\mathcal{F}}, D)$. ▮

Similarly we can express the concept of fuzzy interpolating spline of order p (see Abbasbandy and Babolian [1], Floreno and Novelli [8], Kaleva [16]).

Formally, a fuzzy interpolating spline of order p, denoted $S : \mathbb{R} \to \mathbb{R}_{\mathcal{F}}$, can be written as

$$S(x) = \sum_{i=0}^{n}{}^{*}\, s_i(x) \odot \mu_i, \quad x \in \mathbb{R},$$

where $s_i(x)$ are the usual splines of order p that satisfy

$$s_i(x_j) = \begin{cases} 1, & j = i \\ 0, & j \neq i \end{cases}, \quad i, j \in \{0, \ldots, n\},$$

and $\mu_i \in \mathbb{R}_{\mathcal{F}}$.

The problem of interpolation can be considered in the contex of Definition 13.3, too. Keeping the notations there, let us consider the following.

Question A (Gal [14]) Let (f, φ_f) be a fuzzy function, where $f : [a,b] \to \mathbb{R}$ and $\varphi_f : [a,b] \times \mathbb{R} \to [0,1]$ (see Definition 13.3).

Given $x_1 < \ldots < x_n$ in $[a, b]$, find a fuzzy algebraic polynomial (P, φ_P) which satisfies

$$P(x_i) = f(x_i),$$

$$F(x_i, f(x_i)) = \varphi_f(x_i, f(x_i)) = \varphi_P(x_i, P(x_i)) = Q(x_i, P(x_i)),\ i = \overline{1, n}.$$

REMARK 13.36 Obviously, we want the polynomials P and Q above to be of minimum degree. ∎

REMARK 13.37 A natural way to define the above fuzzy interpolating polynomial (P, φ_P) would be the following:

$P(x) = L_{n-1}(f; x_1, \ldots, x_n)(x) =$ classical Lagrange interpolation polynomial of degree $\leq n - 1$ that satisfies $P(x_i) = f(x_i),\ i = \overline{1, n}$ and

$$Q(x, y) = L_{n-1}(H; x_1, \ldots, x_n)(x) + y - L_{n-1}(f; x_1, \ldots, x_n)(x),$$

where $H(x) = F(x, f(x)),\ x \in [a, b]$.

But in this case, unfortunately, $Q(x, y)$ does not satisfy the required condition $0 < Q(x, y) \leq 1$, for all $(x, y) \in Graph(L_{n-1}(f; x_1, \ldots, x_n))$, because although $0 < H(x) \leq 1$, for all $x \in [a, b]$.

However in general it does not hold that

$$0 < L_{n-1}(H; x_1, \ldots, x_n)(x) \leq 1,\ \forall\, x \in [a, b].$$

∎

Consequently, Remark 13.37 generates (in the context of classical approximation theory) the following open

Question B (Gal [14]) Given $H : [a, b] \to \mathbb{R},\ m \leq H(x) \leq M,\ \forall\, x \in [a, b]$, and $x_1 < x_2 < \ldots < x_n$ in $[a, b]$, find an usual algebraic polynomial P (of minimum degree if it is possible) such that

$$P(x_i) = H(x_i),\ i = \overline{1, n} \text{ and } m \leq P(x) \leq M,\ \forall\, x \in [a, b].$$

A partial answer to Question A is

THEOREM 13.15 (Gal [14])
Let (f, φ_f) be a fuzzy function from $[a, b]$ to $\mathbb{R}$ such that $0 < m \leq \varphi_f(x, f(x)) \leq 1,\ \forall\, x \in [a, b]$.

If $H(x) = \varphi_f(x, f(x))$, $x \in [a,b]$ is continuous on $[a,b]$ and if $x_1 < \ldots < x_n$ are in $[a,b]$ with the property $|H(x_i)| \leq \|H\|$, $i = \overline{1,n}$, then there exists a fuzzy algebraic polynomial (P, φ_P) which satisfies the conditions in Question A.

Finally, let us introduce the following.

DEFINITION 13.15 *(Gal [14]) Let (f, φ_f) be a fuzzy function from $[a,b]$ to $\mathbb{R}$ and $a = x_1 < x_2 < \ldots < x_{n+1} = b$.*

We say that $(L_n(f), \varphi_{L_n(f)})$ is a fuzzy interpolating polygonal line of (f, φ_f) at x_i, $i = \overline{1, n+1}$ if:

$L_n(f)$ is the polygonal line inscribed in the curve $y = H(x)$, satisfying $L_n(f)(x_i) = f(x_i)$, $i = \overline{1, n+1}$;

$L_n(H)$ is the polygonal line inscribed in the curve $y = H(x)$, that satisfies $L_n(H)(x_i) = H(x_i)$, $i = \overline{1, n+1}$;
and

$$\varphi_{L_n(f)}(x, y) = L_n(H)(x) + y - L_n(f)(x),$$

where $H(x) = \varphi_f(x, f(x))$, $x \in [a,b]$.

REMARK 13.38 By $0 < H(x) \leq 1$, $x \in [a,b]$, it is obvious that $0 < L_n(H)(x) \leq 1$, $x \in [a,b]$ and therefore $(L_n(f), \varphi_{L_n(f)})$ is indeed a fuzzy function (as in Definition 13.3). ∎

THEOREM 13.16 (Gal [14])
If $(f, \varphi_f) \in FC[a,b]$ and the points x_i, $i = \overline{1, n+1}$ divide $[a,b]$ into n equal parts, then

$$D_{\mathcal{F}}[(f, \varphi_f), (L_n(f), \varphi_{L_n(f)})] \leq 2\left[\omega_1\left(f; \frac{b-a}{n}\right) + \omega_1\left(H; \frac{b-a}{n}\right)\right],$$

$n = 1, 2, \ldots$

13.7 Fuzzy-Random and Random-Fuzzy Approximation

The random (stochastic) concepts can be mixed with fuzzy concepts in two non-commutative ways, obtaining fuzzy-random functions, fuzzy-random polynomials, fuzzy-random approximation and random-fuzzy functions, random-fuzzy polynomials, random-fuzzy approximation, respectively.

DEFINITION 13.16 *Let S be a non-empty set, $\mathcal{B} \subset \mathcal{P}(S)$ a field of parts on S and $P : \mathcal{B} \to [0,1]$ a complete probability on $\mathcal{B}$ (i.e., $(S, \mathcal{B}, P)$ is complete probability space).*

A fuzzy-random variable is a $\mathcal{B}$-measurable mapping $g : S \to \mathbb{R}_{\mathcal{F}}$ (i.e., for any open set U, in the topology on $\mathbb{R}_{\mathcal{F}}$ generated by the metric D, we have $g^{-1}(U) = \{s \in S; g(s) \in U\} \in B$).

The set of all fuzzy-random variables is denoted by $L_{\mathcal{F}}(S, \mathcal{B}, P)$.

Let $g_n, g \in L_{\mathcal{F}}(S, \mathcal{B}, P)$, $n \in \mathbb{N}$. We say that
$g_n(s) \xrightarrow[n\to\infty]{\text{"}a.s.\text{"}} g(s)$ *(almost surely) if*

$$P(\{s \in S; D(g_n(s), g(s)) \overset{n\to\infty}{\not\longrightarrow} 0\}) = 0;$$

$g_n(s) \xrightarrow[n\to\infty]{\text{"}P\text{"}} g(s)$ *(in probability) if*
$\forall\ \varepsilon, \eta > 0,\ \exists\ n_{\varepsilon,\eta} \in \mathbb{N}$ *such that*

$$P(\{s \in S; D(g_n(s), g(s)) \geq \varepsilon\}) < \eta,\ \forall\ n \geq n_{\varepsilon,\eta};$$

$g_n(s) \xrightarrow[n\to\infty]{\text{"}q-mean\text{"}} g(s)$ *(where $0 < q < +\infty$) if*

$$\lim_{n\to+\infty} \int_S [D(g_n(s), g(s))]^q dP(s) = 0.$$

REMARK 13.39 If $f, g \in L_{\mathcal{F}}(S, \mathcal{B}, P)$, let us denote $F : S \to \mathbb{R}_+ \cup \{0\}$ by $F(s) = D(f(s), g(s))$, $s \in S$. In this case, F is $\mathcal{B}$-measurable, because $F = G \circ H$, where $G : \mathbb{R}_{\mathcal{F}} \times \mathbb{R}_{\mathcal{F}} \to \mathbb{R}_+ \cup \{0\}$, $G(u,v) = D(u,v)$ is continuous in the product topology on $\mathbb{R}_{\mathcal{F}} \times \mathbb{R}_{\mathcal{F}}$ (D is a metric), and $H : S \to \mathbb{R}_{\mathcal{F}} \times \mathbb{R}_{\mathcal{F}}$, $H(s) = (f(s), g(s))$, $s \in S$, is $\mathcal{B}$-measurable. (The topology on $\mathbb{R}_{\mathcal{F}}$ is that induced by the metric D.) This shows that the above convergences in probability and in q-mean have sense. ∎

DEFINITION 13.17 *Let $(T, \mathcal{T})$ be a topological space. A mapping $f : T \to L_{\mathcal{F}}(S, \mathcal{B}, P)$ will be called fuzzy-random function (or fuzzy-stochastic process) on T. We denote $f(t)(s) = f(t,s)$, $t \in T$, $s \in S$.*

Let $\mathcal{C} = \{$"a.s.", "P", "q − mean"$\}$, where the elements in $\mathcal{C}$ have the meaning in Definition 13.16. Let $c \in \mathcal{C}$.

We say that f is c-continuous in $t_0 \in T$, if for any $t_n \xrightarrow{T} t_0$, where $t_n, t_0 \in T$, $n \in \mathbb{N}$, we have $f(t_n, s) \xrightarrow[n\to+\infty]{c} f(t_0, s)$.

Let $T = [a, b]$. We say that f is c-integrable on $[a, b]$, if for each sequence of divisions $(\mathcal{D}_n)_{n\in\mathbb{N}}$ of $[a, b]$, with $\Delta(\mathcal{D}_n) \overset{n\to+\infty}{\longrightarrow} 0$, where $\Delta(\mathcal{D}_n)$ is the norm of $\mathcal{D}_n$, we have

$$\sum_{k=0}^{n-1}{}^{*} (t_{k+1}^{(n)} - t_k^{(n)}) \odot f(\xi_k^{(n)}, s) \xrightarrow[n \to +\infty]{c} I(s) \in L_{\mathcal{F}}(S, \mathcal{B}, P),$$

where $\mathcal{D}_n : a = t_0^{(n)} < \ldots < t_n^{(n)} = b$.

Then $I(s)$ is denoted by $(c) \int_a^b f(t, s)dt$.

Here the sum $\sum^*$ and the product $\odot$ are those introduced by Remark 13.1.

REMARK 13.40 The concepts of convergences in Definition 13.16 and those in Definition 13.17 can be easily formulated in ε, δ language, too. ∎

REMARK 13.41 Any usual fuzzy (real) function $f : T \to \mathbb{R}_{\mathcal{F}}$ (see Definition 13.2) can be identified with the degenerate fuzzy-random function $f(t, s) = f(t)$, $\forall\, t \in T$, $s \in S$. ∎

REMARK 13.42 Fuzzy-random functions that coincide with probability one for each $t \in T$ will be considered equivalent. ∎

REMARK 13.43 Let $f, g : T \to L_{\mathcal{F}}(S, \mathcal{B}, P)$. Then $f \oplus g$ and $k \odot f$ are defined pointwise, i.e.,

$$(f \oplus g)(t, s) = f(t, s) \oplus g(t, s),$$

$$(k \odot f)(t, s) = k \odot f(t, s), \quad t \in T, \quad s \in S.$$

∎

DEFINITION 13.18 *An expression of the form $\sum_{k=0}^{r}{}^{*} x^k c_k(s)$, $x \in \mathbb{R}$, $c_k \in L_{\mathcal{F}}(S, \mathcal{B}, P)$, $k = \overline{0, r}$ will be called algebraic fuzzy-random* *polynomial. Similarly, an expression of the form*

$$\sum_{k=0}^{r}{}^{*}[(\cos kx) \odot a_k(s) \oplus (\sin kx) \odot b_k(s)], \quad x \in \mathbb{R},$$

$a_k, b_k \in L_{\mathcal{F}}(S, \mathcal{B}, P)$, $k = \overline{0, r}$, *will be called a trigonometric fuzzy-random polynomial.*

For $f : [0,1] \to L_{\mathcal{F}}(S, \mathcal{B}, P)$, *the fuzzy-random polynomials defined by*

$$B_n^{(\mathcal{F})}(f)(x,s) = \sum_{k=0}^{n}{}^{*} \binom{n}{k} x^k (1-x)^{n-k} \odot f\left(\frac{k}{n}, s\right), \; x \in [0,1], \; s \in S,$$

will be called a Bernstein-type.

For $f : [a,b] \to L_{\mathcal{F}}(S, \mathcal{B}, P)$, *let us define the modulus of continuity*

$$\Omega_1^{(\mathcal{F})}(f;\delta)_{L^q} = \sup\left\{\left(\int_S D^q(f(x,s), f(y,s))dP(s)\right)^{1/q}; \right.$$

$$\left. x, y \in [a,b], |x-y| \leq \delta \right\}, \; 0 < \delta, \; 1 \leq q < +\infty,$$

and for $f : \mathbb{R} \to L_{\mathcal{F}}(S, \mathcal{B}, P)$, *the modulus of continuity*

$$\Omega_1^{(\mathcal{F})}(f;\delta)_{L^q} = \sup\left\{\left(\int_S D^q(f(x,s), f(y,s))dP(s)\right)^{1/q}; \right.$$

$$\left. x, y \in \mathbb{R}, |x-y| \leq \delta \right\}, \; 0 < \delta, \; 1 \leq q < +\infty.$$

The results in Gal [15] can be extended in this case, too. For example, we have

THEOREM 13.17

For $f : [0,1] \to L_{\mathcal{F}}(S, \mathcal{B}, P)$ *we have the estimate*

$$\int_S D(B_n^{(\mathcal{F})}(f)(x,s), f(x,s))dP(s) \leq \frac{3}{2}\Omega_1^{(\mathcal{F})}\left(f; \frac{1}{\sqrt{n}}\right)_{L^1},$$

$\forall \, x \in [0,1], \; n \in \mathbb{N}.$

If moreover f satisfies the condition

$$\lim_{\delta \searrow 0} \Omega_1^{(\mathcal{F})}(f, \delta)_{L^1} = 0,$$

then $B_n^{(\mathcal{F})}(f)(x,s) \xrightarrow[n \to +\infty]{\text{"1-mean"}} f(x,s)$, *uniformly in respect with* $x \in [0,1]$.

PROOF Reasoning as in the proof of Theorem 13.13, we get

$$D(B_n^{(\mathcal{F})}(f)(x,s), f(x,s)) \leq \sum_{k=0}^{n} \binom{n}{k} x^k (1-x)^{n-1} D\left(f\left(\frac{k}{n}, s\right), f(x,s)\right).$$

Integrating with respect to s (and taking into account that we easily get $\Omega_1^{(\mathcal{F})}(f; \lambda\delta)_{L^1} \leq (\lambda+1)\Omega_1^{(\mathcal{F})}(f;\delta)_{L^1}$, $\lambda > 0$, $\delta > 0$), we get (denoting $p_{n,k}(x) = \binom{n}{k} x^k (1-x)^{n-k}$)

$$\int_S D(B_n^{(\mathcal{F})}(f)(x,s), f(x,s)) dP(s) \leq \sum_{k=0}^{n} p_{n,k}(x) \Omega_1^{(\mathcal{F})}\left(f; \left|\frac{k}{n} - x\right|\right)_{L^1}$$

$$\leq \Omega_1^{(\mathcal{F})}(f;\delta)_{L^1} \sum_{k=0}^{n} \left(1 + \frac{\left|\frac{k}{n} - x\right|}{\delta}\right) p_{n,k}(x), \ \delta > 0, \ x \in [0,1], \ n \in \mathbb{N}.$$

Choosing $\delta = \frac{1}{\sqrt{n}}$, we easily obtain the desired estimate.

The second statement in the theorem follows from the first estimate. ∎

REMARK 13.44 If $f : [0,1] \to L_{\mathcal{F}}(S, \mathcal{B}, P)$ is "1-mean"-continuous on $[0,1]$ (see Definition 13.17) then

$$\lim_{\delta \searrow 0} \Omega_1^{(\mathcal{F})}(f;\delta)_{L^1} = 0.$$

∎

Now, let us denote

$CM_{2\pi}^1(\mathbb{R} \times S) = \{f : \mathbb{R} \to L_{\mathcal{F}}(S, \mathcal{B}, P); f(x,s)$ is measurable $\mathcal{L}(\mathbb{R}) \times \mathcal{B}$ and there exists $A \subset S$ with $P(A) = 1$, such that $f(x + 2\pi, s) = f(x,s)$, for all $x \in \mathbb{R}$, $s \in A$ and $f(\cdot, s) \in C_{2\pi}^{(\mathcal{F})}(\mathbb{R})$, for all $s \in A\}$.

THEOREM 13.18
Let $f \in CM^1_{2\pi}(\mathbb{R}\times S)$. For the trigonometric fuzzy-random polynomials of Jackson-type defined by

$$J_n^{(\mathcal{F})}(f)(x,s) = (a.s.)\int_{-\pi}^{\pi} K_n(u-x)\odot f(u,s)du =$$

$$= (a.s.)\int_{-\pi}^{\pi} K_n(u)f(x+u)du, \quad s\in A, \quad n\in\mathbb{N}$$

(where the kernel $K_n(t)$ is given in Lorentz [20], p.55, relation (5), see also after Lemma 13.2), we have

$$\int_S D(J_n(f)(x,s), f(x,s))dP(s) \le C\Omega_1^{(\mathcal{F})}\left(f;\frac{1}{n}\right)_{L^1}, \ \forall\, x\in\mathbb{R},\ n\in\mathbb{N},$$

($C>0$ is an absolute constant, independent of f, n, x and s).

PROOF By $f(\cdot,s)\in C_{2\pi}^{(\mathcal{F})}(\mathbb{R})$ and by the proof of Theorem 13.14, we obtain

$$D(J_n(f)(x,s), f(x,s)) \le (a.s.)\int_{-\pi}^{\pi} K_n(t)D(f(x+t,s), f(x,s))dt,$$

$s\in A,\ x\in\mathbb{R},\ n\in\mathbb{N}$.

If we denote $F(t,s)=D(f(x+t,s),f(t,s)),\ t\in\mathbb{R},\ s\in S$ (here $x\in\mathbb{R}$ is fixed), reasoning exactly as in Remark 13.39 of Definition 13.16 we easily obtain that $F(t,s)$ is $\mathcal{L}(\mathbb{R})\times\mathcal{B}$ measurable.

Consequently, because $K_n(t)\cdot F(t,s)\ge 0,\ t\in\mathbb{R},\ s\in A,\ n\in\mathbb{N}$, from a classical well-known result in real function theory (i.e., a Fubini-type result for positive functions) it follows that the function $H: A\to\mathbb{R}_+$,

$$H(s) = (a.s.)\int_{\pi}^{\pi} K_n(t)D(f(x+t,s), f(x,s))dt$$

is $\mathcal{B}$-measurable and

$$\int_S H(s)dP(s) = (a.s.)\int_{-\pi}^{\pi}\left[K_n(t)\int_S D(f(x+t,s), f(x,s))dP(s)\right]dt.$$

As a conclusion,

$$\int_S D(J_n(f)(x,s), f(x,s))dP(s) \le$$

$$\leq (a.s.) \int_{-\pi}^{\pi} K_n(t) \left[\int_S D(f(x+t,s), f(x,s)) dP(s) \right] dt \leq$$

$$\leq \int_{-\pi}^{\pi} \Omega_1^{(\mathcal{F})}(f; |t|)_{L^1} \cdot K_n(t) dt \leq$$

$\leq$ (reasoning exactly as in the proof of Theorem 13.14) $\leq$

$$\leq C\Omega_1^{(\mathcal{F})} \left(f; \frac{1}{n} \right)_{L^1},$$

which proves the theorem. ∎

REMARK 13.45 If in the statement of Theorem 13.18, f is moreover "1-mean" continuous on $\mathbb{R}$, then $\lim_{\delta \searrow 0} \Omega_1^{(\mathcal{F})}(f; \delta)_{L^1} = 0$ and passing to limit with $n \to +\infty$ in Theorem 13.18 we obtain

$$\lim_{n \to +\infty} \int_S D(J_n(f)(x,s), f(x,s)) dP(s) = 0,$$

uniformly with respect to $x \in \mathbb{R}$. ∎

REMARK 13.46 Similarly, other results from Gal [15] (as, e.g., Timan-type results, Teljakowskii-type results) can be extended to fuzzy-random approximation. ∎

In what follows we will sketch the second way in which the randomness and fuzziness can be combined.

DEFINITION 13.19 *A random-fuzzy function on* $[a,b]$ *is a pair* (f, φ_f), *where* $f : [a,b] \to L^q(S, \mathcal{B}, P)$ *is an usual random function (see Gal [11]) and* $\varphi_f : [a,b] \times L^q(S, \mathcal{B}, P) \to [0,1]$ *is the membership function given by*

$$\varphi_f(x,y) = \begin{cases} F(x,y), & (x,y) \in Graph(f) \\ 0, & otherwise \end{cases},$$

$$Graph(f) = \{(x,y) \in [a,b] \times L^q(S, \mathcal{B}, P); y = f(x)\},$$

$0 < F(x,y) \leq 1$, *for all* $(x,y) \in Graph(f)$, *(here* $1 \leq q < +\infty$*). A random-fuzzy function is in fact a fuzzified graph of a usual random function.*

A random-fuzzy algebraic polynomial is a pair (R, φ_R), *with* R *a usual random algebraic polynomial and* $\varphi_R : [a,b] \times L^q(S, \mathcal{B}, P) \to [0,1]$ *is given by*

$$\varphi_R(x,y) = \begin{cases} Q(x,y), & (x,y) \in Graph(R) \\ 0, & otherwise \end{cases},$$

$Q(x, R(x,s))$ *is an algebraic polynomial with respect to* x *which satisfies* $Q(x,y) \in (0,1]$, *for all* $(x,y) \in Graph(R) = \{(x,y) \in [a,b] \times L^q(S,\mathcal{B},P); y = R(x)\}$.

REMARK 13.47 Similarly to the concepts in Gal [15] (Definitions 1.3, 1.5), the 2π-periodic case can be introduced in the random-fuzzy case, too. ∎

DEFINITION 13.20 *Let* (f, φ_f) *be a random-fuzzy function on* $[-1,1]$, *where* $\phi(t) = \varphi_f(t, f(t))$, $t \in [-1,1]$ *is supposed to be Lebesgue integrable on* $[-1,1]$.

Let us define $Q_n : [-1,1] \times L^q(S,\mathcal{B},P) \to \mathbb{R}$, $(1 \le q < +\infty)$, *by*

$$Q_n(x,y) = L_n(\phi)(x) + \left(\int_S |y(s) - T_n(f)(x,s)|^q dP(s) \right)^{1/q},$$

where for $g \in C[-1,1]$, $f : [-1,1] \to L^q(S,\mathcal{B},P)$,

$$L_n(g)(x) = \frac{1}{\pi} \int_{-\pi}^{\pi} g[\cos(\arccos x + v)] K_{3n-3}(v) dv$$

(see Lehnhoff [18]),

$$T_n(f)(x,s) = \frac{1}{\pi} \int_{-\pi}^{\pi} f[\cos(\arccos t + v), s] K_{3n-3}(v) dv$$

(see Gal [15]), are the Lehnhoff-type polynomials, with the kernel

$$K_{3n-3}(v) = \{10/[n(11n^4 + 5n^2 + 4)]\}[\sin(nv/2)/(\sin(v/2)]^6.$$

Then the random-fuzzy polynomial $(T_n(f), \varphi_n)$, *where* $\varphi_n : [-1,1] \times L^q(S,\mathcal{B},P) \to [0,1]$ *is given by*

$$\varphi_n(x,y) = \begin{cases} Q_n(x,y), & (x,y) \in Graph(T_n(f)) \\ 0, & otherwise, \end{cases}$$

is called a Lehnhoff-type.

REMARK 13.48 Since $\phi(t) = \varphi_f(t, f(t)) \in (0,1]$, $\forall\, t \in [-1,1]$, by the properties of $L_n(g)(x)$ in Lehnhoff [18], it easily follows that $Q_n(x,y) \in (0,1]$, $\forall\, (x,y) \in Graph[T_n(f)]$, i.e., $(T_n(f), \varphi_n)$ is indeed a random-fuzzy polynomial on $[-1,1]$. ∎

REMARK 13.49 Random-fuzzy functions on $[a,b]$ can be constructed in the following simple way. Let $f : [a,b] \to L^q(S,\mathcal{B},P)$ be an usual random function and let $F : [a,b] \times \mathbb{R} \to (0,1]$, $F = F(x,y)$, continuous with respect to y. Also, let us define $G(x,s) = F(x, f(x,s))$, $x \in [a,b]$, a.s. $s \in S$, and the membership function $\varphi_f : [a,b] \times L^q(S,\mathcal{B},P) \to [0,1]$ by

$$\varphi_f(x,y) = \begin{cases} \displaystyle\int_S G(x,s)dP(s), & \text{if } (x,y) \in Graph(f) \\ 0, & \text{otherwise.} \end{cases}$$

Then obviously (f, φ_f) is a random-fuzzy function on $[a,b]$. ∎

Now, let us consider the following subclass of random-fuzzy functions $FRC^q[a,b] = \{(f,\varphi_f); f(x,s)$ is $\mathcal{L}[a,b] \times \mathcal{B}$ measurable, $(\mathcal{L}) \int_a^b |f(t,s)|^q dt < +\infty$, a.s. $s \in S$, f is "L^q"-continuous on $[a,b]\}$ (for details regarding the random functions see, e.g., Gal [15]).

Also, recall the q-modulus of continuity $(1 \le q < +\infty)$

$$\eta_1(f;\delta)_q = \sup\left\{ \left(\int_S |f(t_1,s) - f(t_2,s)|^q dP(s) \right)^{1/q} ;\right.$$

$$\left. t_1, t_2 \in [a,b], |t_1 - t_2| \le \delta \right\}.$$

DEFINITION 13.21 *Let $(f,\varphi_f), (g,\varphi_g) \in FRC^q[a,b]$ and let us denote $F = \varphi_f|_{Graph(f)}$, $G = \varphi_g|_{Graph(g)}$.*

Let us define the support graphs of φ_f and of φ_g by

$$G^*(\varphi_f) = \{(x,y,z); (x,y) \in Graph(f), z = F(x,y)\}$$

$$(\subset [a,b] \times L^q(S,\mathcal{B},P) \times [0,1])$$

and

$$G^*(\varphi_g) = \{(x,y,z); (x,y) \in Graph(g), z = G(x,y)\},$$

respectively.

The distance between (f,φ_f) and (g,φ_g) denoted by $D_{\mathcal{F}}[(f,\varphi_f),(g,\varphi_g)]$ will be the Hausdorff distance between the sets $G^*(\varphi_f)$ and $G^*(\varphi_g)$, i.e.,

$$D_{\mathcal{F}}[(f,\varphi_f),(g,\varphi_g)] = d_H(G^*(\varphi_f), G^*(\varphi_g)) =$$

$$= \max\left\{ \sup_{A\in G^*(\varphi_f)} \inf_{B\in G^*(\varphi_g)} \rho(A,B), \sup_{B\in G^*(\varphi_g)} \inf_{A\in G^*(\varphi_f)} \rho(A,B) \right\},$$

where for $A = (x_1,y_1,z_1)$, $B = (x_2,y_2,z_2)$, we have

$$\rho(A,B) = \max\left\{ |x_1 - x_2|, \left(\int_S |y_1(s) - y_2(s)|^q dP(s) \right)^{1/q}, |z_1 - z_2| \right\}.$$

REMARK 13.50 It is easily seen that if $(f,\varphi_f) \in FRC^q[a,b]$, then $G^*(\varphi_f)$ is a compact subset of the metric space $([a,b] \times L^q(S,\mathcal{B},P) \times [0,1], \rho])$, i.e., $D_{\mathcal{F}}$ is even a metric on $FRC^q[a,b]$. ∎

Similar reasonings with those in, e.g., Lemma 13.1 easily give

LEMMA 13.3

For any $(f,\varphi_f),(g,\varphi_g) \in FRC^q[a,b]$ we have

$$D_{\mathcal{F}}[(f,\varphi_f),(g,\varphi_g)] \le \|f-g\|_q + \sup_{a_1\in[a,b]} |F(a_1,f(a_1)) - G(a_1,g(a_1))|,$$

where $F = \varphi_f|_{Graph(f)}$, $G = \varphi_g|_{Graph(g)}$ and

$$\|f-g\|_q = \sup_{t\in[a,b]} \left(\int_S |f(t,s) - g(t,s)|^q dP(s) \right)^{1/q}.$$

We are in position to prove the following

THEOREM 13.19

For any $(f,\varphi_f)\in FRC^q[a,b]$ *and for the random-fuzzy polynomial of Lehnhoff-type* $(T_n(f),\varphi_n)$ *(introduced in Definition 13.20) we have*

$$D_{\mathcal{F}}[(f,\varphi_f),(T_n(f),\varphi_n)]\le C_q\eta_1\left(f;\frac{1}{n}\right)_q+4\omega_1\left(\phi;\frac{1}{n}\right),\quad n\in\mathbb{N},$$

where $\phi(t)=\varphi_f(t,f(t)),\ t\in[-1,1]$ *and*

$$\omega_1(\phi;\delta)=\sup\{|\phi(x)-\phi(y)|;x,y\in[-1,1],|x-y|\le\delta\}.$$

PROOF By Lemma 13.3 we get

$$D_{\mathcal{F}}[(f,\varphi_f),(T_n(f),\varphi_n)]\le\|f-T_n(f)\|_q+$$
$$+\sup_{a_1\in[-1,1]}|\varphi_f(a_1,f(a_1))-Q_n(a_1,T_n(f)(a_1))|.$$

But by Gal [15], Theorem 4.4, we immediately obtain

$$\|f-T_n(f)\|_q\le C_q\eta_1\left(f;\frac{1}{n}\right)_q.$$

On the other hand,

$$\sup_{a_1\in[-1,1]}|\varphi_f(a_1,f(a_1))-Q_n(a_1,T_n(f)(a_1))|=$$
$$=\sup_{a_1\in[-1,1]}|\phi(a_1)-L_n[\phi(t)](a_1)|\le$$
$$\le\sup_{a_1\in[-1,1]}L_n[|\phi(a_1)-\phi(t)|](a_1)\le\sup_{a_1\in[-1,1]}L_n[\omega_1(\phi;|t-a_1|)](a_1)\le$$
$$\le\sup_{a_1\in[-1,1]}\left\{\omega_1(\phi;\delta)L_n\left[1+\frac{|t-a_1|}{\delta}\right](a_1)\right\}=$$
$$=\omega_1(f;\delta)\sup_{a_1\in[-1,1]}\left\{1+\frac{1}{\delta}L_n[|t-a_1|](a_1)\right\}\le$$
$$\le\text{(by Lehnhoff [18], relation (3.2))}\le$$
$$\le\omega_1(\phi;\delta)\sup\left\{1+\frac{1}{\delta}\left[2\sqrt{1-a_1^2}/n+|a_1|/n^2\right]\right\}\le$$
$$\le\omega_1(\phi;\delta)\left\{1+\frac{1}{\delta}\cdot\frac{3}{n}\right\}.$$

Choosing now $\delta=\dfrac{1}{n}$, we obtain the theorem. ∎

References

[1] Abbasbandy, S. and Babolian, E., Interpolation of fuzzy data by natural splines, *Korean J. Comput. Appl. Math.* 5(1998), No. 2, 457-463.

[2] Asmuss, S.V. and Šostak, A., Extremal problems of approximation theory in fuzzy context, *Fuzzy Sets and Systems* 105(1999), 249-257.

[3] Atanassov, K.T., Intuinitionistic fuzzy sets, *Fuzzy Sets and Systems* 20(1986), 87-96.

[4] Ban, A.I. and Gal, S.G., Decomposable measures and information measures for intuitionistic fuzzy sets, accepted by *Fuzzy Sets and Systems.*

[5] Collings, I.L. and Kloeden, P.E., Continuous approximation of fuzzy sets, *J. Fuzzy Math.* 3(1995), No. 2, 449-453.

[6] Diamond, P., Kloeden, P.E. and Vladimirov, A., Spikes, broken planes and the approximation of convex fuzzy sets, *Fuzzy Sets and Systems* 99(1998), 225-232.

[7] Dubois, D. and Prade, H., *Fuzzy Sets and Systems: Theory and Applications*, Academic Press, New York, 1980.

[8] Floreno, M.C. and Novelli, G., Implementing fuzzy polynomial interpolation (FPI) and fuzzy linear regression (LFR), *Le Matematiche* vol. LI(1996), fasc. I, 59-76.

[9] Gal, S.G., Sur les théorèmes d'approximation de Weierstrass, *Mathematica* (Cluj) 23(46)(1981), No. 1, 25-30.

[10] Gal, S.G., A fuzzy variant of the Weierstrass' approximation theorem, *J. Fuzzy Math* 1(4)(1993), 865-872.

[11] Gal, S.G., Degree of approximation of fuzzy mappings by fuzzy polynomials, *J. Fuzzy Math.* 2(4)(1994), 847-853.

[12] Gal, S.G., Fuzzy variant of the Stone-Weierstrass approximation theorem, *Mathematica* (Cluj) 37(60)(1995), No. 1-2, 103-108.

[13] Gal, S.G., Approximate selections for fuzzy-set valued mappings and applications, *J. Fuzzy Math* 3(4)(1995), 941-947.

[14] Gal, S.G., Interpolation of fuzzy mappings, *Mathematica* (Cluj) 38(61)(1996), No. 1-2, 61-65.

[15] Gal, S.G., Approximation theory in random setting, this volume.

[16] Kaleva, O., Interpolation of fuzzy data, *Fuzzy Sets and Systems* 61(1994), 63-70.

[17] Kloeden, P.E., Compact supported endographs and fuzzy sets, *Fuzzy Sets and Systems* 4(1980), 193-201.

[18] Lehnhoff, H.G., A simple proof of A.F. Timan's theorem, *J. Approx. Theory* 38(1983), 172-176.

[19] Lehnhoff, H.G., A new proof of Teljakowskii's theorem, *J. Approx. Theory* 38(1983), 177-181.

[20] Lorentz, G.G., *Approximation of Functions*, Chelsea, New York, 1986.

[21] Lowen, R., A fuzzy Lagrange interpolation theorem, *Fuzzy Sets and Systems* 34(1990), 33-38.

[22] Ma Ming, On embedding problem of fuzzy number space, Part 4, *Fuzzy Sets and Systems* 58(1993), 185-193.

[23] Negoiţă, C.V. and Ralescu, D.A., *Fuzzy Sets and their Applications* (in Romanian), Ed. Tehnica, Bucharest, 1974.

[24] Negoiţă, C.V. and Ralescu, D.A., *Applications of Fuzzy Sets to System Analysis*, John Wiley and Sons, New York, 1975.

[25] Precupanu, A., *Mathematical Analysis. Real Functions* (in Romanian), Ed. Did. Ped., Bucharest, 1976.

[26] Wang, P.Z., Lui, H.C., Zhang, X.H., Zhang, H.M. and Xu, W., Win-Win strategy for probability and fuzzy mathematics, *J. Fuzzy Math.* 1(1993), No. 1, 223-231.

[27] Wu Congxin and Ma Ming, On embedding problem of fuzzy number space, Part I, *Fuzzy Sets and Systems* 44(1991), 33-38.

[28] Wu Congxin and Ma Ming, On embedding problem of fuzzy number space, Part II, *Fuzzy Sets and Systems* 45(1992), 189-202.

[29] Wu Congxin and Gong Zengtai, On Henstock integral of fuzzy-number-valued functions, Part I, *Fuzzy Sets and Systems* to appear.

[30] Zadeh, L.A., Fuzzy sets, *Inform. and Control* 8(1965), 338-353.

Chapter 14

Global Smoothness Preservation by Multivariate Bernstein-Type Operators

Jesús de la Cal and Ana M. Valle

Departamento de Matemática Aplicada, Estadística e Investigación Operativa, Facultad de Ciencias, Universidad del País Vasco, Apartado 644, 48080 Bilbao (Spain)
E-mail: (de la Cal) mepcaagj@lg.ehu.es and (Valle) mepvamaa@lg.ehu.es
Contact author: J. de la Cal

Abstract For multivariate Bernstein-type operators, we consider preservation of the usual l_p-modulus of continuity, as well as preservation of classes of functions determined by moduli of continuity. Using a probabilistic approach based on representations in terms of stochastic processes, we establish general bounds and formulae for the involved best constants. Special attention is paid to tensor product operators and to simplicial operators. Some applications and illustrations of the general results to particular operators are also given.

Contents

14.1 Introduction: Best Constants

Let $I^{\langle k \rangle}$ be a (nonempty) convex subset of $\mathbb{R}^k$. A Bernstein-type operator $L^{\langle k \rangle}$ on $I^{\langle k \rangle}$ is a positive linear operator acting on real functions defined on $I^{\langle k \rangle}$ and having the form

$$L^{\langle k \rangle} f(\mathbf{x}) = \int_{I^{\langle k \rangle}} f(\mathbf{z})\, d\mu^{\mathbf{x}}(\mathbf{z}), \qquad \mathbf{x} \in I^{\langle k \rangle},\ f \in \mathcal{L}^{\langle k \rangle}, \tag{14.1}$$

where $\mu^{\mathbf{x}}$ is a Borel probability measure on $I^{\langle k \rangle}$, and $\mathcal{L}^{\langle k \rangle}$ stands for the domain of $L^{\langle k \rangle}$, i.e., the set of all real functions on $I^{\langle k \rangle}$ for which the right-hand side in (14.1) makes sense.

Many classical examples of this kind of operator usually considered in approximation theory are constructed by using well-known families of probability distributions. For illustration and reference purposes, twelve families of one-dimensional operators have been collected in the Appendix at the end of this work. The most paradigmatic family is that of Bernstein operators on the interval $[0, 1]$ ((I) in the Appendix), which are constructed from the binomial distributions, and give the name for the class. On the other hand, by the Riesz representation theorem, when $I^{\langle k \rangle}$ is compact, every positive linear operator $L^{\langle k \rangle}$ on $C(I^{\langle k \rangle})$ such that $L^{\langle k \rangle} 1 = 1$ has the form (14.1).

When dealing with a Bernstein-type operator $L^{\langle k \rangle}$, several topics are of interest. In particular, it is interesting to know which properties of $f \in \mathcal{L}^{\langle k \rangle}$ are preserved by $L^{\langle k \rangle}$ (i.e., retained by $L^{\langle k \rangle} f$). Here, we will be concerned with preservation of global smoothness as measured by usual moduli of continuity. This means finding sharp estimates (even better, the exact value) for the best constant $C_p^{\langle k\rangle}(\delta)$ (not depending on f) such that the inequality

$$\omega_p(L^{\langle k\rangle} f;\delta) \le C_p^{\langle k\rangle}(\delta)\,\omega_p(f;\delta), \qquad \delta \ge 0,$$

holds for every $f \in \mathcal{L}^{\langle k\rangle}$, where $\omega_p(f;\cdot)$ stands for the usual modulus of continuity of f with respect to the l_p-norm $\|\cdot\|_p$ on $\mathbb{R}^k$ ($p \in [1,\infty]$), i.e.,

$$\omega_p(f;\delta) := \sup\{|f(\mathbf{x}) - f(\mathbf{y})| : \mathbf{x},\, \mathbf{y} \in I^{\langle k\rangle},\ \|\mathbf{x}-\mathbf{y}\|_p \le \delta\}, \quad \delta \ge 0,$$

with

$$\|\mathbf{x}\|_p := \begin{cases} \left(\sum_{i=1}^k |x_i|^p\right)^{1/p} & \text{if } p \in [1,\infty) \\ \max_{1\le i\le k} |x_i| & \text{if } p = \infty, \end{cases} \qquad \mathbf{x} := (x_1,\ldots,x_k) \in \mathbb{R}^k.$$

We observe that $I^{\langle k\rangle}$ being convex, the modulus $\omega_p(f;\cdot)$ is subadditive, and, therefore,

$$\omega_p(f;a\delta) \le \lceil a\rceil\, \omega_p(f;\delta) \le (1+a)\,\omega_p(f;\delta), \qquad a,\, \delta \ge 0,$$

where $\lceil\cdot\rceil$ stands for the *ceiling* function

$$\lceil a\rceil := \text{the smallest integer not less than } a.$$

This fact together with the equivalence of the l_p-norms implies that the condition $\omega_1(f;1) < \infty$ is equivalent to $\omega_p(f;\delta) < \infty$, for all $\delta \ge 0$ and all $p \in [1,\infty]$. We shall denote by $M^{\langle k\rangle}$ the set of all real measurable functions f on $I^{\langle k\rangle}$ satisfying such conditions (if all the probability measures $\mu^{\mathbf{x}}$ in (14.1) are discrete, the measurability assumption on f can be dropped).

Let $e(\cdot) := \|\cdot\|_1$. If the operator $L^{\langle k\rangle}$ satisfies

$$L^{\langle k\rangle} e(\mathbf{x}) = \int_{I^{\langle k\rangle}} \|\mathbf{z}\|_1\, d\mu^{\mathbf{x}}(\mathbf{z}) < \infty, \qquad \mathbf{x} \in I^{\langle k\rangle}, \tag{14.2}$$

then $M^{\langle k\rangle} \subset \mathcal{L}^{\langle k\rangle}$, as it follows from the inequalities

$$\int_{I^{\langle k\rangle}} |f(\mathbf{z})|\, d\mu^{\mathbf{x}}(\mathbf{z}) \le \int_{I^{\langle k\rangle}} |f(\mathbf{z}_0)|\, d\mu^{\mathbf{x}}(\mathbf{z}) + \int_{I^{\langle k\rangle}} |f(\mathbf{z}) - f(\mathbf{z}_0)|\, d\mu^{\mathbf{x}}(\mathbf{z})$$

$$\le |f(\mathbf{z}_0)| + \omega_1(f;1) \int_{I^{\langle k\rangle}} (1 + \|\mathbf{z}-\mathbf{z}_0\|_1)\, d\mu^{\mathbf{x}}(\mathbf{z}),$$

where $\mathbf{z}_0$ is a fixed element of $I^{\langle k\rangle}$. Throughout the chapter, we shall only consider operators of the form (14.1) satisfying condition (14.2), which will be called *S-operators*. It is obvious that, in such a case, we can write

$$C_p^{\langle k\rangle}(\delta) := \sup_{\text{const.}\neq f\in M^{\langle k\rangle}} \frac{\omega_p(L^{\langle k\rangle}f;\delta)}{\omega_p(f;\delta)}, \qquad \delta > 0.$$

A related problem to be considered is that of preservation of classes of functions determined by moduli of continuity. A real, nondecreasing and subadditive function $\omega(\cdot)$ defined on $[0,\infty)$, such that $\omega(0) = 0$, and $\omega(\delta) > 0$ for $\delta > 0$, will be called a modulus of continuity. If $\omega(\cdot)$ is in addition concave, we say that it is a c.m.c. We shall denote this by

$$M_{p,\omega}^{\langle k\rangle} := \{f \in M^{\langle k\rangle} : \omega_p(f;\cdot) \leq \omega(\cdot)\}.$$

In particular, for the c.m.c. $\omega^{(\alpha)}(\delta) := \delta^\alpha$, $(\alpha \in (0,1])$, $M_{p,\omega^{(\alpha)}}^{\langle k\rangle}$ is nothing but the Lipschitz class of order α (and constant 1) for the l_p-norm.

Given a modulus $\omega(\cdot)$, we are interested in the best constant $D_p^{\langle k\rangle}(\omega)$ such that

$$\omega_p(L^{\langle k\rangle}f;\delta) \leq D_p^{\langle k\rangle}(\omega)\,\omega(\delta), \qquad \delta \geq 0,\ f \in M_{p,\omega}^{\langle k\rangle},$$

that is,

$$D_p^{\langle k\rangle}(\omega) := \sup_{\delta>0}\ \sup_{f\in M_{p,\omega}^{\langle k\rangle}} \frac{\omega_p(L^{\langle k\rangle}f;\delta)}{\omega(\delta)}.$$

Problems of the above kinds have been discussed by several authors in the recent past, as can be read about in the historical notes (Section 14.13 below). In the present work, we attempt to develop a little theory for the C-constants and the D-constants from a probabilistic point of view. The next section contains some basic definitions and the first general results.

14.2 Stochastic Cores and Probabilistic Bounds

The operator $L^{\langle k\rangle}$ in (14.1) is determined by the family of probability measures $\{\mu^{\mathbf{x}} : \mathbf{x} \in I^{\langle k\rangle}\}$, but, for our work, the representation of $L^{\langle k\rangle}$ by means of a stochastic process will play a paramount role.

DEFINITION 14.1 *Let $Z^{\langle k \rangle} := \{Z^{\langle k \rangle}(\mathbf{x}) : \mathbf{x} \in I^{\langle k \rangle}\}$ be an $I^{\langle k \rangle}$-valued stochastic process defined on an appropriate probability space. If $Z^{\langle k \rangle}(\mathbf{x})$ has the probability distribution $\mu^{\mathbf{x}}$ ($\mathbf{x} \in I^{\langle k \rangle}$), then $Z^{\langle k \rangle}$ will be called a stochastic core for $L^{\langle k \rangle}$.*

REMARK 14.1 If $Z^{\langle k \rangle}$ is a stochastic core for $L^{\langle k \rangle}$, we can write

$$L^{\langle k \rangle} f(\mathbf{x}) = Ef\left(Z^{\langle k \rangle}(\mathbf{x})\right), \qquad \mathbf{x} \in I^{\langle k \rangle},\ f \in \mathcal{L}^{\langle k \rangle}, \tag{14.3}$$

where (here and hereafter) E denotes mathematical expectation. It is worth noting that, for such a stochastic core, we can always claim existence but not uniqueness. ∎

The following theorem gives bounds for both the C-constants and the D-constants of $L^{\langle k \rangle}$ in terms of a stochastic core.

THEOREM 14.1

Let $Z^{\langle k \rangle}$ be a stochastic core for $L^{\langle k \rangle}$ and let $p \in [1, \infty]$. We have, for all $\delta > 0$

$$a_p^{\langle k \rangle}(\delta) \le C_p^{\langle k \rangle}(\delta) \le b_p^{\langle k \rangle}(\delta), \tag{14.4}$$

and, for every modulus of continuity $\omega(\cdot)$,

$$c_p^{\langle k \rangle}(\omega) \le D_p^{\langle k \rangle}(\omega) \le d_p^{\langle k \rangle}(\omega), \tag{14.5}$$

where

$$a_p^{\langle k \rangle}(\delta) := \sup_{\substack{\mathbf{x}, \mathbf{y} \in I^{\langle k \rangle} \\ \|\mathbf{x}-\mathbf{y}\|_p \le \delta}} \left| E\left\lceil \frac{\|Z^{\langle k \rangle}(\mathbf{x})\|_p}{\delta} \right\rceil - E\left\lceil \frac{\|Z^{\langle k \rangle}(\mathbf{y})\|_p}{\delta} \right\rceil \right|, \tag{14.6}$$

$$b_p^{\langle k \rangle}(\delta) := \sup_{\substack{\mathbf{x}, \mathbf{y} \in I^{\langle k \rangle} \\ \|\mathbf{x}-\mathbf{y}\|_p \le \delta}} E\left\lceil \frac{\|Z^{\langle k \rangle}(\mathbf{x}) - Z^{\langle k \rangle}(\mathbf{y})\|_p}{\delta} \right\rceil, \tag{14.7}$$

$$c_p^{\langle k \rangle}(\omega) := \sup_{\delta > 0} \sup_{\substack{\mathbf{x}, \mathbf{y} \in I^{\langle k \rangle} \\ \|\mathbf{x}-\mathbf{y}\|_p \le \delta}} \frac{|E\omega(\|Z^{\langle k \rangle}(\mathbf{x})\|_p) - E\omega(\|Z^{\langle k \rangle}(\mathbf{y})\|_p)|}{\omega(\delta)}, \tag{14.8}$$

$$d_p^{\langle k\rangle}(\omega) := \sup_{\delta>0} \sup_{\substack{\mathbf{x},\mathbf{y}\in I^{\langle k\rangle} \\ \|\mathbf{x}-\mathbf{y}\|_p\le\delta}} \frac{E\omega\left(\|Z^{\langle k\rangle}(\mathbf{x})-Z^{\langle k\rangle}(\mathbf{y})\|_p\right)}{\omega(\delta)}. \tag{14.9}$$

PROOF Let $\delta > 0$, $p \in [1,\infty]$, and $f \in M^{\langle k\rangle}$ be fixed, and let $\mathbf{x}$, $\mathbf{y} \in I^{\langle k\rangle}$, with $\|\mathbf{x}-\mathbf{y}\|_p \le \delta$. From (14.3), we have

$$|L^{\langle k\rangle}f(\mathbf{x}) - L^{\langle k\rangle}f(\mathbf{y})| \le E\left|f\left(Z^{\langle k\rangle}(\mathbf{x})\right) - f\left(Z^{\langle k\rangle}(\mathbf{y})\right)\right|$$

$$\le E\omega_p\left(f; \|Z^{\langle k\rangle}(\mathbf{x}) - Z^{\langle k\rangle}(\mathbf{y})\|_p\right). \tag{14.10}$$

By the subadditivity of $\omega_p(f;\cdot)$, we obtain

$$|L^{\langle k\rangle}f(\mathbf{x}) - L^{\langle k\rangle}f(\mathbf{y})| \le E\left\lceil \frac{\|Z^{\langle k\rangle}(\mathbf{x}) - Z^{\langle k\rangle}(\mathbf{y})\|_p}{\delta}\right\rceil \omega_p(f;\delta)$$

$$\le b_p^{\langle k\rangle}(\delta)\,\omega_p(f;\delta),$$

where $b_p^{\langle k\rangle}(\delta)$ is given in (14.7), and we conclude that the second inequality in (14.4) holds true. To show the first inequality in (14.4), we observe that the function $\phi_{p,\delta}^{\langle k\rangle}$ given by

$$\phi_{p,\delta}^{\langle k\rangle}(\mathbf{x}) := \left\lceil \frac{\|\mathbf{x}\|_p}{\delta} \right\rceil, \qquad \mathbf{x} \in I^{\langle k\rangle}, \tag{14.11}$$

belongs to $M^{\langle k\rangle}$ and fulfills $\omega_p(\phi_{p,\delta}^{\langle k\rangle};\delta) \le 1$. Therefore

$$C_p^{\langle k\rangle}(\delta) \ge \omega_p(L^{\langle k\rangle}\phi_{p,\delta}^{\langle k\rangle};\delta) = \sup_{\substack{\mathbf{x},\mathbf{y}\in I^{\langle k\rangle} \\ \|\mathbf{x}-\mathbf{y}\|_p\le\delta}} \left|L^{\langle k\rangle}\phi_{p,\delta}^{\langle k\rangle}(\mathbf{x}) - L^{\langle k\rangle}\phi_{p,\delta}^{\langle k\rangle}(\mathbf{y})\right|$$

$$= a_p^{\langle k\rangle}(\delta),$$

completing the proof of (14.4). On the other hand, let $\omega(\cdot)$ be a modulus of continuity, and assume that $f \in M_{p,\omega}^{\langle k\rangle}$. We have from (14.10)

$$\omega_p(L^{\langle k\rangle}f;\delta) \le \sup_{\substack{\mathbf{x},\mathbf{y}\in I^{\langle k\rangle} \\ \|\mathbf{x}-\mathbf{y}\|_p\le\delta}} E\omega_p(f; \|Z^{\langle k\rangle}(\mathbf{x}) - Z^{\langle k\rangle}(\mathbf{y})\|_p)$$

$$\leq \sup_{\substack{\mathbf{x},\mathbf{y}\in I^{\langle k\rangle} \\ \|\mathbf{x}-\mathbf{y}\|_p\leq\delta}} E\omega(\|Z^{\langle k\rangle}(\mathbf{x}) - Z^{\langle k\rangle}(\mathbf{y})\|_p)$$

$$\leq d_p^{\langle k\rangle}(\omega)\,\omega(\delta),$$

showing the second inequality in (14.5). Finally, since $\omega(\cdot)$ is subadditive and nondecreasing, we have that the function $\psi_{p,\omega}^{\langle k\rangle}$ given by

$$\psi_{p,\omega}^{\langle k\rangle}(\mathbf{x}) := \omega(\|\mathbf{x}\|_p), \qquad \mathbf{x}\in I^{\langle k\rangle}, \tag{14.12}$$

belongs to $M_{p,\omega}^{\langle k\rangle}$, and the proof of the first inequality in (14.5) is analogous to that of the first inequality in (14.4). This completes the proof of the theorem. ∎

REMARK 14.2 The lower bounds in Theorem 14.1 only depend upon the family $\{\mu^{\mathbf{x}} : \mathbf{x}\in I^{\langle k\rangle}\}$ (i.e., upon $L^{\langle k\rangle}$), whilst the upper bounds depend upon the distributions of the random vectors $Z^{\langle k\rangle}(\mathbf{x}) - Z^{\langle k\rangle}(\mathbf{y})$, and these may vary from one stochastic core to another. Thus, when considering particular operators, an important ingredient in the probabilistic approach is to find appropriate representations. ∎

In the Appendix, we provide useful stochastic cores for the one-dimensional S-operators collected there. They all have been constructed from a few types of well-known stochastic processes.

In the rest of the chapter, the bounds in Theorem 14.1 will be refined under additional assumptions on the operator $L^{\langle k\rangle}$. In Sections 14.3–14.10, $L^{\langle k\rangle}$ will be a tensor product operator, while, in Sections 14.11 and 14.12, it will be a simplicial operator.

14.3 Tensor Product Operators

Throughout Sections 14.3–14.10, unless otherwise specified, $L^{\langle k\rangle}$ $(k = 1, 2, \ldots)$ will be the tensor product operator

$$L^{\langle k\rangle} := L_1\otimes\cdots\otimes L_k,$$

where $L_1, L_2, \ldots$ is a fixed sequence of one-dimensional S-operators on the intervals $I_1, I_2, \ldots$, respectively. This means that $I^{\langle k\rangle} := I_1 \times \cdots \times I_k$, and

$$L^{\langle k\rangle} f(\mathbf{x}) := \int_{I^{\langle k\rangle}} f(z_1, \ldots, z_k)\, d\mu_1^{x_1}(z_1) \cdots d\mu_k^{x_k}(z_k),$$

where $\mathbf{x} := (x_1, \ldots, x_k) \in I^{\langle k\rangle}$, and, for $j \geq 1$, $\{\mu_j^{x_j} : x_j \in I_j\}$ is the family of probability measures associated to L_j. It is immediate that $L^{\langle k\rangle}$ is an S-operator on $I^{\langle k\rangle}$.

A stochastic core for $L^{\langle k\rangle}$ is constructed in the following way: Let $Z_1 := \{Z_1(x) : x \in I_1\}$, $Z_2 := \{Z_2(x) : x \in I_2\}, \ldots$, be stochastic cores for $L_1, L_2, \ldots$, respectively. We can always assume that these stochastic processes are independent and defined on the same probability space. Then, for $\mathbf{x} := (x_1, \ldots, x_k) \in I^{\langle k\rangle}$, the probability distribution of the random vector

$$Z^{\langle k\rangle}(\mathbf{x}) := (Z_1(x_1), \ldots, Z_k(x_k)), \tag{14.13}$$

is the product measure $\mu_1^{x_1} \otimes \cdots \otimes \mu_k^{x_k}$. In other words, the stochastic process $Z^{\langle k\rangle} := \{Z^{\langle k\rangle}(\mathbf{x}) : \mathbf{x} \in I^{\langle k\rangle}\}$ given by (14.13) is a stochastic core for $L^{\langle k\rangle}$.

For the preceding operators, the following can be claimed.

PROPOSITION 14.1

Let $p \in [1, \infty]$. We have, for all $\delta > 0$,

$$C_1^{\langle 1\rangle}(\delta) = C_p^{\langle 1\rangle}(\delta) \leq C_p^{\langle 2\rangle}(\delta) \leq C_p^{\langle 3\rangle}(\delta) \leq \cdots,$$

and, for every modulus of continuity $\omega(\cdot)$,

$$D_1^{\langle 1\rangle}(\omega) = D_p^{\langle 1\rangle}(\omega) \leq D_p^{\langle 2\rangle}(\omega) \leq D_p^{\langle 3\rangle}(\omega) \leq \cdots.$$

Also, if $\sigma_1, \ldots, \sigma_k$ is a permutation of $1, \ldots, k$, then the C-constants and the D-constants for $L_\sigma^{\langle k\rangle} := L_{\sigma_1} \otimes \cdots \otimes L_{\sigma_k}$ coincide with those for $L^{\langle k\rangle}$.

PROOF For each $k = 1, 2, \ldots$, let T_k be the operator assigning to each real function f on $I^{\langle k\rangle}$, the real function $T_k f$ on $I^{\langle k+1\rangle}$ given by $(T_k f)(x_1, \ldots, x_{k+1}) := f(x_1, \ldots, x_k)$. It is clear that T_k is injective. Moreover, we have

$$\omega_p(T_k f; \cdot) = \omega_p(f; \cdot), \quad T_k(M^{\langle k\rangle}) \subset M^{\langle k+1\rangle}, \quad T_k(M_{p,\omega}^{\langle k\rangle}) \subset M_{p,\omega}^{\langle k+1\rangle},$$

and

$$L^{\langle k+1\rangle}(T_k f) = T_k(L^{\langle k\rangle} f), \qquad f \in M^{\langle k\rangle}.$$

From these facts, it immediately follows that

$$C_p^{\langle k\rangle}(\delta) \leq C_p^{\langle k+1\rangle}(\delta), \qquad D_p^{\langle k\rangle}(\omega) \leq D_p^{\langle k+1\rangle}(\omega),$$

showing the first part of the proposition. The second part follows by an obvious argument. ∎

In the next section, we consider a more specific class of tensor product operators.

14.4 Tensor Products of S*-Operators

DEFINITION 14.2 *Let L be an S-operator on the interval I. We say that L is an S*-operator if I is either $[0,1]$ or $[0,\infty)$, and it has a stochastic core $Z := \{Z(x) : x \in I\}$ satisfying the following two conditions:*

(H_1) *$Z(x) \leq Z(y)$ a.s., for all $x, y \in I$ with $x < y$.*

(H_2) *$\{Z(x) : x \in I\}$ has stationary increments, i.e., for all $x, y \in I$ with $x < y$, the random variables $Z(y) - Z(x)$ and $Z(y-x) - Z(0)$ have the same probability distribution.*

DEFINITION 14.3 *If, in addition, Z satisfies the condition*

(H_3) *$Z(0) = 0$ a.s.,*

then we say that L is an S_0^*-operator.

All of the operators in the Appendix (except those of family (X)) are examples of S*-operators. The operators of families (I), (II), (IV), (VI), (VIII), (XI) and (XII) are also examples of S_0^*-operators.

REMARK 14.3 Condition (H_3) merely means that Lf interpolates f at 0, for each f in the domain of L. On the other hand, condition (H_1) obviously implies that L preserves monotonicity; in fact, the converse assertion (if L preserves monotonicity, then it has a stochastic core satisfying

(H_1)) is also true, as follows by extending in an obvious way the proof of [39; Theorem 1.A.1]. ∎

In the following theorem, the results in Theorem 14.1 are refined under the assumption that $L^{\langle k\rangle}$ is a tensor product of one-dimensional S*-operators. Furthermore, if all of the factors are S_0^*-operators, exact formulae for both the C-constants and the D-constants are obtained. We denote by $a \wedge b$ the minimum of a and b, and $\rho_p^{\langle k\rangle}$ is the diameter of $I^{\langle k\rangle}$ for the l_p-norm. Since each of the involved intervals is $[0,1]$ or $[0,\infty)$, we have that $\rho_p^{\langle k\rangle}$ equals $k^{1/p}$ or ∞ according to $I_1 = \cdots = I_k = [0,1]$, or $I_i = [0,\infty)$ for some $i \in \{1,\ldots,k\}$.

THEOREM 14.2

Let $k \geq 1$, and let $L^{\langle k\rangle}$ and $Z^{\langle k\rangle}$ be the same as in Section 14.3. For $p \in [1,\infty]$, $\delta > 0$, and any modulus of continuity $\omega(\cdot)$, we have:

(a) *If $L_1, L_2, \ldots,$ are S*-operators, then*

$$C_p^{\langle k\rangle}(\delta) \leq \sup_{\substack{\mathbf{x}\in I^{\langle k\rangle} \\ \|\mathbf{x}\|_p=\delta\wedge\rho_p^{\langle k\rangle}}} E\left\lceil \frac{\|Z^{\langle k\rangle}(\mathbf{x}) - Z^{\langle k\rangle}(\mathbf{0})\|_p}{\delta}\right\rceil,$$

$$D_p^{\langle k\rangle}(\omega) \leq \sup_{\delta>0} \sup_{\substack{\mathbf{x}\in I^{\langle k\rangle} \\ \|\mathbf{x}\|_p=\delta\wedge\rho_p^{\langle k\rangle}}} \frac{E\omega\left(\|Z^{\langle k\rangle}(\mathbf{x}) - Z^{\langle k\rangle}(\mathbf{0})\|_p\right)}{\omega(\delta)}.$$

(b) *If $L_1, L_2, \ldots,$ are S_0^*-operators, then*

$$C_p^{\langle k\rangle}(\delta) = \sup_{\substack{\mathbf{x}\in I^{\langle k\rangle} \\ \|\mathbf{x}\|_p=\delta\wedge\rho_p^{\langle k\rangle}}} E\left\lceil \frac{\|Z^{\langle k\rangle}(\mathbf{x})\|_p}{\delta}\right\rceil, \tag{14.14}$$

$$D_p^{\langle k\rangle}(\omega) = \sup_{\delta>0} \sup_{\substack{\mathbf{x}\in I^{\langle k\rangle} \\ \|\mathbf{x}\|_p=\delta\wedge\rho_p^{\langle k\rangle}}} \frac{E\omega\left(\|Z^{\langle k\rangle}(\mathbf{x})\|_p\right)}{\omega(\delta)}. \tag{14.15}$$

The proof of Theorem 14.2 is based upon the following lemma, where we use the following notation. For any k-dimensional vector $\mathbf{x} := (x_1,\ldots,x_k) \in \mathbb{R}^k$, we set $\mathbf{x}^* := (|x_1|,\ldots,|x_k|)$. It is apparent that $\|\mathbf{x}\|_p = \|\mathbf{x}^*\|_p$, for all $p \in [1,\infty]$.

LEMMA 14.1

Let k, p, δ, ω, $I^{\langle k \rangle}$, and $Z^{\langle k \rangle}(\cdot)$ be the same as in the statement of Theorem 14.2(a). Then:

(a) *For all $\mathbf{x}, \mathbf{y} \in I^{\langle k \rangle}$, we have $(\mathbf{x}-\mathbf{y})^* \in I^{\langle k \rangle}$, and the random variables $\|Z^{\langle k \rangle}(\mathbf{x}) - Z^{\langle k \rangle}(\mathbf{y})\|_p$ and $\|Z^{\langle k \rangle}((\mathbf{x}-\mathbf{y})^*) - Z^{\langle k \rangle}(\mathbf{0})\|_p$ have the same probability distribution.*

(b) *The functions*

$$\mathbf{x} := (x_1, \ldots, x_k) \in I^{\langle k \rangle} \quad \mapsto \quad E\left\lceil \frac{\|Z^{\langle k \rangle}(\mathbf{x}) - Z^{\langle k \rangle}(\mathbf{0})\|_p}{\delta} \right\rceil,$$

and

$$\mathbf{x} := (x_1, \ldots, x_k) \in I^{\langle k \rangle} \quad \mapsto \quad E\omega\left(\|Z^{\langle k \rangle}(\mathbf{x}) - Z^{\langle k \rangle}(\mathbf{0})\|_p\right),$$

are nondecreasing in each coordinate.

PROOF (of Lemma 14.1). (a) Let $\mathbf{x}, \mathbf{y} \in I^{\langle k \rangle}$, and set

$$\mathbf{x} := (x_1, \ldots, x_k), \qquad \mathbf{y} := (y_1, \ldots, y_k).$$

Since I_i $(i = 1, \ldots, k)$ is either the interval $[0, 1]$ or the interval $[0, \infty)$, it is obvious that $|x_i - y_i| \in I_i$, and, therefore, $(\mathbf{x}-\mathbf{y})^* \in I^{\langle k \rangle}$. On the other hand, for $i = 1, 2, \ldots, k$, the ith component of the random vector $\left(Z^{\langle k \rangle}(\mathbf{x}) - Z^{\langle k \rangle}(\mathbf{y})\right)^*$ is the random variable $|Z_i(x_i) - Z_i(y_i)|$ which, by assumptions (H_1) and (H_2), has the same probability distribution as $Z_i(|x_i - y_i|) - Z_i(0)$. From the independence of the components, we therefore have that the random vectors $\left(Z^{\langle k \rangle}(\mathbf{x}) - Z^{\langle k \rangle}(\mathbf{y})\right)^*$ and $Z^{\langle k \rangle}((\mathbf{x}-\mathbf{y})^*) - Z^{\langle k \rangle}(\mathbf{0})$ are identically distributed, and the conclusion in (a) follows. On the other hand, part (b) is an immediate consequence of assumption (H_1) together with the monotonicity of mathematical expectation and the fact that the functions $\phi_{p,\delta}^{\langle k \rangle}$ and $\psi_{p,\omega}^{\langle k \rangle}$ defined in (14.11) and (14.12), respectively, are nondecreasing in each coordinate. This finishes the proof of Lemma 14.1. ∎

PROOF (of Theorem 14.2). Using Lemma 14.1(a) and Lemma 14.1(b), we have

$$b_p^{\langle k \rangle}(\delta) = \sup_{\substack{\mathbf{x}, \mathbf{y} \in I^{\langle k \rangle} \\ \|(\mathbf{x}-\mathbf{y})^*\|_p \le \delta \wedge \rho_p^{\langle k \rangle}}} E\left\lceil \frac{\|Z^{\langle k \rangle}((\mathbf{x}-\mathbf{y})^* - Z^{\langle k \rangle}(\mathbf{0}))\|_p}{\delta} \right\rceil$$

$$= \sup_{\substack{\mathbf{x}\in I^{\langle k\rangle} \\ \|\mathbf{x}\|_p \le \delta\wedge\rho_p^{\langle k\rangle}}} E\left[\frac{\|Z^{\langle k\rangle}(\mathbf{x}) - Z^{\langle k\rangle}(\mathbf{0})\|_p}{\delta}\right]$$

$$= \sup_{\substack{\mathbf{x}\in I^{\langle k\rangle} \\ \|\mathbf{x}\|_p = \delta\wedge\rho_p^{\langle k\rangle}}} E\left[\frac{\|Z^{\langle k\rangle}(\mathbf{x}) - Z^{\langle k\rangle}(\mathbf{0})\|_p}{\delta}\right],$$

where $b_p^{\langle k\rangle}(\delta)$ is defined in (14.7). The same argument yields

$$d_p^{\langle k\rangle}(\omega) = \sup_{\delta>0} \sup_{\substack{\mathbf{x}\in I^{\langle k\rangle} \\ \|\mathbf{x}\|_p = \delta\wedge\rho_p^{\langle k\rangle}}} \frac{E\omega\left(\|Z^{\langle k\rangle}(\mathbf{x}) - Z^{\langle k\rangle}(\mathbf{0})\|_p\right)}{\omega(\delta)},$$

where $d_p^{\langle k\rangle}(\omega)$ is defined in (14.9). This shows part (a). On the other hand, if $L_1, L_2, \ldots$, are S_0^*-operators, we have by (H_3) that $Z^{\langle k\rangle}(\mathbf{0}) = \mathbf{0}$ a.s., and the conclusions in part (b) follow from those in part (a) and the first inequalities in (14.4) and (14.5) (on taking $\mathbf{y} = \mathbf{0}$ in (14.6) and (14.8)). The proof of Theorem 14.2 is complete. ∎

REMARK 14.4 By (14.3), formulae (14.14) and (14.15) can be rewritten in the following way:

$$C_p^{\langle k\rangle}(\delta) = \sup_{\substack{\mathbf{x}\in I^{\langle k\rangle} \\ \|\mathbf{x}\|_p = \delta\wedge\rho_p^{\langle k\rangle}}} L^{\langle k\rangle}\phi_{p,\delta}^{\langle k\rangle}(\mathbf{x}), \quad D_p^{\langle k\rangle}(\omega) = \sup_{\delta>0} \sup_{\substack{\mathbf{x}\in I^{\langle k\rangle} \\ \|\mathbf{x}\|_p = \delta\wedge\rho_p^{\langle k\rangle}}} \frac{L^{\langle k\rangle}\psi_{p,\omega}^{\langle k\rangle}(\mathbf{x})}{\omega(\delta)},$$

where $\phi_{p,\delta}^{\langle k\rangle}$ and $\psi_{p,\omega}^{\langle k\rangle}$ are the functions defined in (14.11) and (14.12), respectively. The case $p = \infty$ is of particular interest. ∎

14.5 The Case $p = \infty$

Assume that $L_1, \ldots, L_k$ are one-dimensional S_0^*-operators on the same interval I (=$[0,1]$ or $[0,\infty)$) (in particular, this takes place when $L_1, \ldots, L_k$ are copies of the same S_0^*-operator). Then, for all $\delta > 0$ such that $\delta =$

$\delta \wedge \rho_\infty^{\langle k \rangle}$, we have that

$$\mathbf{d} := (\delta, \ldots, \delta) \in I^{\langle k \rangle} = I^k. \tag{14.16}$$

Therefore, for $p = \infty$, the formulae in Theorem 14.2(b) become

$$C_\infty^{\langle k \rangle}(\delta) = E\left\lceil \frac{\|Z^{\langle k \rangle}(\mathbf{d})\|_\infty}{\delta} \right\rceil = L^{\langle k \rangle} \phi_{\infty,\delta}^{\langle k \rangle}(\mathbf{d}), \tag{14.17}$$

$$D_\infty^{\langle k \rangle}(\omega) = \sup_{\substack{\delta > 0 \\ \delta = \delta \wedge \rho_\infty^{\langle k \rangle}}} \frac{E\omega\left(\|Z^{\langle k \rangle}(\mathbf{d})\|_\infty\right)}{\omega(\delta)} = \sup_{\substack{\delta > 0 \\ \delta = \delta \wedge \rho_\infty^{\langle k \rangle}}} \frac{L^{\langle k \rangle} \psi_{\infty,\omega}^{\langle k \rangle}(\mathbf{d})}{\omega(\delta)}, \tag{14.18}$$

as it follows from Lemma 14.1(b).

When dealing with concrete operators, (14.17) and (14.18) readily lead to expressions appropriate to be worked by standard theoretical or computational methods, in order to obtain the exact values of the constants or further insights into the problem under consideration. We shall illustrate these ideas by a simple example.

Example 14.1

Denote by $C_n(\delta)$ the C-constant for the nth Bernstein operator B_n on $[0, 1]$. Using (14.17) and the representation in the Appendix, we can write for $\delta \in (0, 1]$

$$C_n(\delta) = E\left\lceil \frac{S_n(\delta)}{n\delta} \right\rceil = \sum_{j=0}^{n} \left\lceil \frac{j}{n\delta} \right\rceil \binom{n}{j} \delta^j (1-\delta)^{n-j},$$

where $S_n(\delta)$ is a random variable having the binomial distribution with parameters n, δ. We therefore have, on the one hand,

$$C_n(\delta) \le 1 + \frac{ES_n(\delta)}{n\delta} = 2,$$

and, on the other hand,

$$\lim_{\delta \uparrow 1} C_n(\delta) = \lim_{\delta \uparrow 1} \left\lceil \frac{1}{\delta} \right\rceil \delta^n = 2,$$

showing that $2 = \sup_{0 < \delta \le 1} C_n(\delta) = \lim_{\delta \uparrow 1} C_n(\delta)$ is the uniform C-constant for B_n. $\square$

REMARK 14.5 In the case $k = 1$, formulae (14.17) and (14.18) (with $\omega = \omega^{(\alpha)}$) were first established by Adell and Pérez-Palomares [9]. These authors also computed the C-constants and the D-constants for a number of one-dimensional operators. The preceding example has been taken from that paper (but it should be noticed that the fact $\sup_{0<\delta\leq 1} C_n(\delta) = 2$ was early obtained by Anastassiou *et al.* [15] by using a completely different approach). Additional examples can be found in [24, 37]. ∎

In Section 14.10, we have collected the graphs of $C_\infty^{\langle k\rangle}(\cdot)$ corresponding to several tensor products of Bernstein operators, in order to illustrate some theoretical facts.

14.6 Upper Numerical Bounds

In this section, we return to the general setting of tensor products of one-dimensional S-operators and obtain upper numerical bounds for the C-constants and the D-constants, under additional assumptions (on the factors L_i) different from those considered in Section 14.4.

DEFINITION 14.4 *We say that the one-dimensional S-operator L on the interval I is A-Lipschitz ($A > 0$), if Le_1 (where $e_1(x) := x$) is a Lipschitz function of constant A and order 1, i.e., if the following condition is fulfilled:*

(H_4) $|Le_1(x) - Le_1(y)| \leq A|x-y|$, *for all* $x, y \in I$.

REMARK 14.6 Each of the operators collected in the Appendix is A-Lipschitz for some $A > 0$. On the other hand, condition (H_4) can be expressed in the form

$$|E(Z(x) - Z(y))| \leq A|x-y|, \quad \text{for all} \quad x,\, y \in I,$$

where Z is *any* stochastic core for L. ∎

THEOREM 14.3
Let $p \in [1,\infty]$, let $L^{\langle k\rangle}$ and $Z^{\langle k\rangle}$ be the same as in Section 14.3, and assume that, for $i = 1,\ldots,k$, the S-operator L_i is A_i-Lipschitz and satisfies (H_1)

(with Z replaced by Z_i). We have, for all $\delta > 0$,

$$C_p^{\langle k\rangle}(\delta) \leq 1 + Ak^{1-1/p}, \tag{14.19}$$

and, for every c.m.c. $\omega(\cdot)$,

$$D_p^{\langle k\rangle}(\omega) \leq \sup_{\delta>0} \frac{\omega\left(Ak^{1-1/p}\delta\right)}{\omega(\delta)} \leq \left\lceil Ak^{1-1/p}\right\rceil, \tag{14.20}$$

where

$$A := \max_{1\leq i\leq k} A_i. \tag{14.21}$$

PROOF Let p, k, δ and $\omega(\cdot)$ be fixed. First, we observe that we have for all $\mathbf{x} := (x_1, \ldots, x_k)$, $\mathbf{y} := (y_1, \ldots, y_k) \in I^{\langle k\rangle}$,

$$E\|Z^{\langle k\rangle}(\mathbf{x}) - Z^{\langle k\rangle}(\mathbf{y})\|_1 = \sum_{i=1}^{k} E|Z_i(x_i) - Z_i(y_i)| = \sum_{i=1}^{k} |E(Z_i(x_i) - Z_i(y_i))|$$

$$< \sum_{i=1}^{k} A_i|x_i - y_i| \leq A\|\mathbf{x} - \mathbf{y}\|_1$$

(where A is given in (14.21)), the second equality because each Z_i satisfies (H_1), and the first inequality by the assumption that each L_i is A_i-Lipschitz. Combining this fact with the inequalities

$$\lceil\cdot\rceil \leq 1 + \cdot, \qquad \|\cdot\|_p \leq \|\cdot\|_1,$$

we have from Theorem 14.1

$$C_p^{\langle k\rangle}(\delta) \leq 1 + \sup_{\substack{\mathbf{x},\mathbf{y}\in I^{\langle k\rangle} \\ \|\mathbf{x}-\mathbf{y}\|_p\leq\delta}} \frac{E\|Z^{\langle k\rangle}(\mathbf{x}) - Z^{\langle k\rangle}(\mathbf{y})\|_p}{\delta}$$

$$\leq 1 + \sup_{\substack{\mathbf{x},\mathbf{y}\in I^{\langle k\rangle} \\ \|\mathbf{x}-\mathbf{y}\|_p\leq\delta}} \frac{E\|Z^{\langle k\rangle}(\mathbf{x}) - Z^{\langle k\rangle}(\mathbf{y})\|_1}{\delta}$$

$$\leq 1 + \sup_{\substack{\mathbf{x},\mathbf{y}\in I^{\langle k\rangle} \\ \|\mathbf{x}-\mathbf{y}\|_p\leq\delta}} \frac{A\|\mathbf{x} - \mathbf{y}\|_1}{\delta}$$

$$\leq 1 + \frac{A}{\delta} \sup_{\substack{\mathbf{x} \in \mathbb{R}^k \\ \|\mathbf{x}\|_p \leq \delta}} \|\mathbf{x}\|_1$$

$$= 1 + Ak^{1-1/p},$$

showing (14.19). Analogously, (14.20) follows from the second inequality in (14.5), taking into account that, $\omega(\cdot)$ being concave, we have by Jensen's inequality (cf. [29; p. 26])

$$E\omega\left(\|Z^{\langle k\rangle}(\mathbf{x}) - Z^{\langle k\rangle}(\mathbf{y})\|_p\right) \leq \omega\left(E\|Z^{\langle k\rangle}(\mathbf{x}) - Z^{\langle k\rangle}(\mathbf{y})\|_p\right),$$

for all $\mathbf{x}$, $\mathbf{y} \in I^{\langle k\rangle}$. The proof of the theorem is complete. ∎

14.7 Lower Numerical Bounds

Different assumptions from those used in Theorem 14.3 lead to lower numerical bounds for the C-constants and the D-constants. We introduce the following definitions, where L stands for a one-dimensional S-operator on the interval I, and Z is a stochastic core for L.

DEFINITION 14.5 *We say that L is A-linear ($A > 0$), if it satisfies the condition*

(H$_5$) $Le_1(x) = EZ(x) = Ax$, *for all* $x \in I$.

1-linear operators are called centered.

DEFINITION 14.6 *Under the assumption that $0 \in I \subset [0, \infty)$, L is said to be accumulating at 0, if the following condition is fulfilled:*

(H$_6$) $L1_{\{0\}}(x) \to 1$, *as* $x \downarrow 0$ *(i.e.,* $\lim_{x\downarrow 0} P(Z(x) = 0) = 1$*)*

(where 1_A stands for the indicator function of the set $A \subset \mathbb{R}$).

The operators of families (I), (II), (IV), (VI), (VIII), (X)-(XII) are centered, and those of families (I), (II), (IV), (VI), (VIII), (X) are accumulating at 0.

If, for some $j \in \{1, \ldots, k\}$, L_j is a centered S-operator, then L_j reproduces all the linear functions on I_j, and we have by Proposition 14.1 that

$C_p^{\langle k \rangle}(\delta) \geq 1$, and $D_p^{\langle k \rangle}(\omega^{(1)}) \geq 1$ (where $\omega^{(1)}(x) := x$). The following theorem gives other lower bounds.

THEOREM 14.4

Let $p \in [1, \infty]$, *let* $L^{\langle k \rangle}$ *and* $Z^{\langle k \rangle}$ *be the same as in Section 14.3, and assume that, for each* $i = 1, \ldots, k$, *we have* $0 \in I_i \subset [0, \infty)$ *and the S-operator* L_i *on* I_i *is* A_i*-linear and accumulating at 0. Then*

$$\liminf_{\delta \downarrow 0} C_p^{\langle k \rangle}(\delta) \geq k^{-1/p} \sum_{i=1}^{k} A_i,$$

and, for the c.m.c. $\omega^{(1)}(x) := x$,

$$D_p^{\langle k \rangle}(\omega^{(1)}) \geq k^{-1/p} \sum_{i=1}^{k} A_i. \tag{14.22}$$

In particular, if $A_1 = \cdots = A_k = A$ *and all of the* L_i *satisfy* (H_1), *we have*

$$D_p^{\langle k \rangle}(\omega^{(1)}) = A k^{1-1/p}. \tag{14.23}$$

PROOF Fix $p \in [1, \infty]$ and $k \geq 1$. Since $0 \in I_i \subset [0, \infty)$ and the S-operator L_i on I_i is A_i-linear ($i = 1, \ldots, k$), we have $Z_i(0) = 0$ a.s. ($i = 1, \ldots, k$), (i.e., each L_i satisfies condition (H_3)), and we conclude that $Z^{\langle k \rangle}(\mathbf{0}) = \mathbf{0}$ a.s.

Let $\delta > 0$ be small enough so that $k^{-1/p}\mathbf{d} \in I^{\langle k \rangle}$, where $\mathbf{d}$ is given by (14.16). Using the first inequality in (14.4) and the fact that

$$\|Z^{\langle k \rangle}(k^{-1/p}\mathbf{d})\|_p \geq \sum_{i=1}^{k} Z_i(k^{-1/p}\delta) \prod_{j \neq i} 1_{\{0\}} \left(Z_j(k^{-1/p}\delta) \right),$$

we have

$$\begin{aligned} C_p^{\langle k \rangle}(\delta) &\geq \frac{E\|Z^{\langle k \rangle}(k^{-1/p}\mathbf{d})\|_p}{\delta} \\ &\geq \frac{1}{\delta} \sum_{i=1}^{k} E \left(Z_i(k^{-1/p}\delta) \prod_{j \neq i} 1_{\{0\}} \left(Z_j(k^{-1/p}\delta) \right) \right) \end{aligned}$$

$$= \frac{1}{\delta}\sum_{i=1}^{k} EZ_i(k^{-1/p}\delta)\prod_{j\neq i} P\left(Z_j(k^{-1/p}\delta)=0\right)$$

$$= k^{-1/p}\sum_{i=1}^{k} A_i\prod_{j\neq i} P\left(Z_j(k^{-1/p}\delta)=0\right),$$

the first equality by the independence of $Z_1, \ldots, Z_k$, and the second equality by the hypothesis that each L_i is A_i-linear. From this and the assumption that each L_i is accumulating at 0, we conclude that

$$\liminf_{\delta\downarrow 0} C_p^{\langle k\rangle}(\delta) \geq \liminf_{\delta\downarrow 0} \frac{E\|Z^{\langle k\rangle}(k^{-1/p}\mathbf{d})\|_p}{\delta} \geq k^{-1/p}\sum_{i=1}^{k} A_i.$$

On the other hand, from the first inequality in (14.5), we have

$$D_p^{\langle k\rangle}(\omega^{(1)}) \geq \liminf_{\delta\downarrow 0} \frac{E\|Z^{\langle k\rangle}(k^{-1/p}\mathbf{d})\|_p}{\delta} \geq k^{-1/p}\sum_{i=1}^{k} A_i.$$

Finally, (14.23) obviously follows from (14.20) and (14.22). This completes the proof of the theorem. ∎

For tensor products of S_0^*-operators, we can also assert the following.

THEOREM 14.5
Let $p \in [1,\infty]$, let $L^{\langle k\rangle}$ and $Z^{\langle k\rangle}$ be the same as in Section 14.3, and assume that, for each $i = 1, 2, \ldots$, L_i is an A-linear and accumulating at 0 S_0^*-operator. Then

$$\lim_{\delta\downarrow 0} C_p^{\langle k\rangle}(\delta) = Ak^{1-1/p}.$$

PROOF Fix $p \in [1,\infty]$, $k = 1, 2, \ldots$ and $\delta \in (0,1)$. Using the inequalities $\|\cdot\|_p \leq \|\cdot\|_1 \leq k^{1-1/p}\|\cdot\|_p$, and the fact that

$$E\lceil U\rceil \leq P(U>0) + EU$$

(which holds true for every nonnegative random variable U), we have for all $\mathbf{x} := (x_1, \ldots, x_k) \in I^{\langle k\rangle}$ such that $\|\mathbf{x}\|_p \leq \delta$

$$E\left\lceil \frac{\|Z^{\langle k\rangle}(\mathbf{x})\|_p}{\delta}\right\rceil \leq P\left(\|Z^{\langle k\rangle}(\mathbf{x})\|_p > 0\right) + \frac{E\|Z^{\langle k\rangle}(\mathbf{x})\|_1}{\delta}$$

$$\leq \sum_{i=1}^{k} P\left(Z_i(x_i) > 0\right) + \frac{A\|\mathbf{x}\|_1}{\delta}$$

$$\leq \sum_{i=1}^{k} P\left(Z_i(\delta) > 0\right) + Ak^{1-1/p},$$

the second (resp. third) inequality by the fact that each Z_i satisfies (H_5) (resp. (H_1)). From Theorem 14.2(b) and the assumption that each L_i is accumulating at 0, we therefore conclude that

$$\limsup_{\delta \downarrow 0} C_p^{\langle k \rangle}(\delta) \leq Ak^{1-1/p} + \sum_{i=1}^{k} \lim_{\delta \downarrow 0} P\left(Z_i(\delta) > 0\right) = Ak^{1-1/p}.$$

On the other hand, the application of Theorem 14.4 to the case at hand yields

$$\liminf_{\delta \downarrow 0} C_p^{\langle k \rangle}(\delta) \geq Ak^{1-1/p},$$

and the conclusion follows. ∎

14.8 Some Applications

As said in Remark 14.5, we already know the actual values of the C-constants and the D-constants for many of the concrete operators collected in the Appendix. Such one-dimensional results combined with Proposition 14.1 and Theorem 14.3 above readily lead to results for tensor product operators and the l_1-norm. For instance, we can ennunciate the following.

COROLLARY 14.1

Let $L^{\langle k \rangle}$ be the same as in Section 14.3, where $L_1, \ldots, L_k$ are assumed to be 1-Lipschitz S-operators satisfying (H_1).

(a) *If, for some $i \in \{1, \ldots, k\}$, the operator L_i belongs to one of the families* (I), (II), (IV), (XI), *then*

$$\sup_{\delta > 0} C_1^{\langle k \rangle}(\delta) = 2,$$

and, for every c.m.c. $\omega(\cdot)$,

$$D_1^{\langle k \rangle}(\omega) = 1.$$

(b) *If, for some* $i \in \{1, \dots, k\}$, *the operator* L_i *is the operator* H_t *of family* (VIII), *or the operator* M_t *of family* (X), *or the operator* G_t *of family* (XII), *then*

$$\sup_{t,\delta>0} C_{t,1}^{\langle k \rangle}(\delta) = 2. \tag{14.24}$$

(c) *If, for some* $i \in \{1, \dots, k\}$, *the operator* L_i *is the operator* S_t *of family* (VI), *or the operator* H_t *of family* (VIII), *or the operator* G_t *of family* (XII), *then we have, for every c.m.c.* $\omega(\cdot)$,

$$\sup_{t>0} D_{t,1}^{\langle k \rangle}(\omega) = 1. \tag{14.25}$$

(In formulae (14.24) and (14.25), the subscript t obviously reflects the dependence of $L^{\langle k \rangle}$ on the parameter $t > 0$, in the cases under consideration.)

On the other hand, the following corollary is a direct application of Theorems 14.3–14.5.

COROLLARY 14.2

Let $p \in [1, \infty]$, *and let* $L^{\langle k \rangle}$ *be the same as in Section 14.3, where it is assumed that, for each* $i \in \{1, \dots, k\}$, *the operator* L_i *belongs to one of the families* (I), (II), (IV), (VI), (VIII) *(not necessarily the same family for all* i*). Then*

$$k^{1-1/p} = \lim_{\delta \downarrow 0} C_p^{\langle k \rangle}(\delta) \le \sup_{\delta>0} C_p^{\langle k \rangle}(\delta) \le 1 + k^{1-1/p}, \tag{14.26}$$

$$D_p^{\langle k \rangle}(\omega^{(\alpha)}) \le k^{\alpha(1-1/p)}, \qquad \alpha \in (0, 1],$$

(where $\omega^{(\alpha)}(x) := x^\alpha$*), and*

$$D_p^{\langle k \rangle}(\omega^{(1)}) = k^{1-1/p}.$$

REMARK 14.7 In special situations, (14.26) can be considerably strengthened. In a forthcoming paper [25], we actually show that, for $L^{\langle k \rangle} := B_n \otimes \cdots \otimes B_n$ (the tensor product of k copies of the nth Bernstein operator B_n), the following curious phenomenon takes place:

(a) If $k \geq 3$, we have (for all n)

$$\sup_{\delta>0} C_\infty^{\langle k \rangle}(\delta) = \lim_{\delta \downarrow 0} C_\infty^{\langle k \rangle}(\delta) = k. \tag{14.27}$$

(b) If $k = 2$, $\sup_{\delta>0} C_\infty^{\langle 2 \rangle}(\delta)$ *depends* upon n, and

$$2 = \lim_{\delta \downarrow 0} C_\infty^{\langle 2 \rangle}(\delta) \leq \sup_{\delta>0} C_\infty^{\langle 2 \rangle}(\delta) \leq 5/2.$$

In the same paper, we also show that similar results hold true for tensor products of identical copies of Szász and Baskakov operators. ∎

14.9 Derivatives of Bernstein Polynomials

In this section, we apply Theorem 14.3 above to draw some conclusions about the global smoothness of the derivatives of multivariate Bernstein polynomials.

We shall use the following notations. For a given $k = 1, 2, \ldots$, let $\mathbf{n} := (n_1, \ldots, n_k)$ be a k-tuple of positive integers, and let $\mathbf{m} := (m_1, \ldots, m_k)$ be a k-tuple of nonnegative integers. We denote by $D^{\mathbf{m}}$ the differential operator

$$D^{\mathbf{m}} := \frac{\partial^{m_1 + \cdots + m_k}}{\partial x_1^{m_1} \cdots \partial x_k^{m_k}},$$

and we set

$$B_{\mathbf{n},\mathbf{m}} := B_{n_1,m_1} \otimes \cdots \otimes B_{n_k,m_k},$$

where B_{n_i,m_i} is the Kantorovič operator K_{n_i,m_i} (family (V)) or the Bernstein operator B_{n_i} according to $m_i \geq 1$ or $m_i = 0$.

It is not hard to check that, for any real function f on $[0,1]^k$ having a continuous derivative $D^{\mathbf{m}} f$, we have the following formula generalizing (14.41)

$$D^{\mathbf{m}}\left(B_{\mathbf{n}+\mathbf{m},\mathbf{0}} f\right) = c_{\mathbf{n},\mathbf{m}} B_{\mathbf{n},\mathbf{m}}\left(D^{\mathbf{m}} f\right),$$

where

$$c_{\mathbf{n},\mathbf{m}} := \prod_{i=1}^{k} \prod_{j=1}^{m_i} \frac{n_i + j}{n_i + m_i}$$

($\prod_{j=1}^{0}$ being understood as 1). Since the operator B_{n_i,m_i} is A_i-Lipschitz (with $A_i := n_i/(n_i + m_i)$) and satisfies condition (H_1), we have by Theorem 14.3

$$\omega_p\left(D^{\mathbf{m}}\left(B_{\mathbf{n}+\mathbf{m},\mathbf{0}}f\right);\delta\right) \le c_{\mathbf{n},\mathbf{m}}\left(1+Ak^{1-1/p}\right)\omega_p\left(D^{\mathbf{m}}f;\delta\right)$$

(for all $p \in [1,\infty]$ and $\delta > 0$), where

$$A := \max_{1\le i\le k} \frac{n_i}{n_i+m_i}.$$

Also, if $\omega(\cdot)$ is a c.m.c., and $D^{\mathbf{m}}f \in M_{p,\omega}^{\langle k\rangle}$, then

$$D^{\mathbf{m}}\left(B_{\mathbf{n}+\mathbf{m},\mathbf{0}}f\right) \in M_{p,c\,\omega}^{\langle k\rangle},$$

where

$$c := c_{\mathbf{n},\mathbf{m}} \sup_{\delta>0} \frac{\omega\left(Ak^{1-1/p}\delta\right)}{\omega(\delta)}.$$

REMARK 14.8 The global smoothness of derivatives of multivariate Szász or Baskakov operators can be worked in much the same way, by using the Szász-Kantorovič operators (family (VII)) and the Baskakov-Kantorovič operators (family (IX)), respectively. We omit the details. ∎

14.10 Some Graphics

In this section, we show the graphs of the functions $C_\infty^{\langle k\rangle}(\cdot)$ corresponding to several tensor products of Bernstein operators, as they can be obtained from (14.17). They all illustrate (14.26). Figure (a) below corresponds to the univariate Bernstein operator B_2 and also illustrates the fact that $\sup_{0<\delta\le 1} C_n(\delta) = \lim_{\delta\uparrow 1} C_n(\delta) = 2$ (as showed in Section 14.5). Figure (b) is the graph for $B_1 \otimes B_1 \otimes B_1$ and illustrates (14.27). Finally, Figures (c) and (d) correspond to $B_2 \otimes B_2$ and $B_1 \otimes B_1 \otimes B_2$, respectively, and show that (14.27) fails to be true if $k = 2$, or when $k \ge 3$ but the factors are not identical.

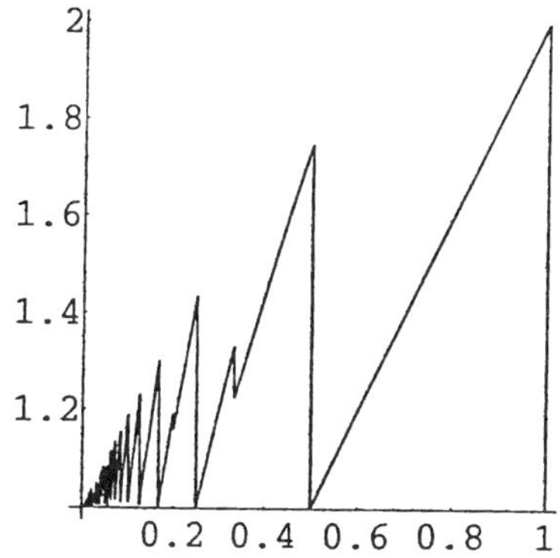

Figure (a)

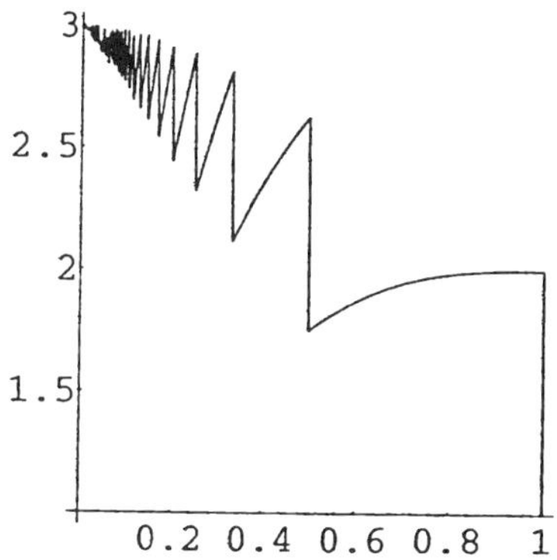

Figure (b)

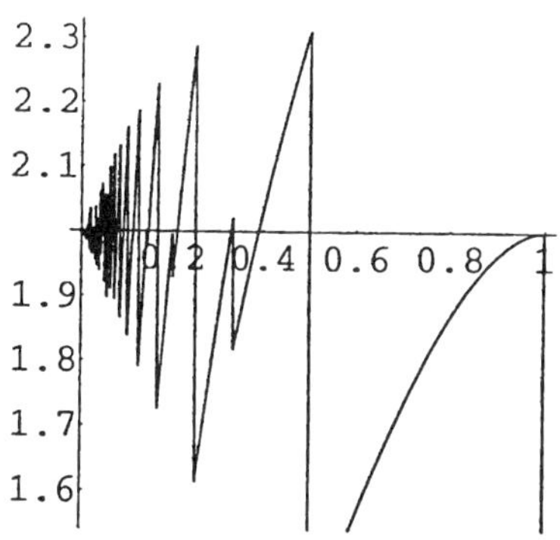

Figure (c)

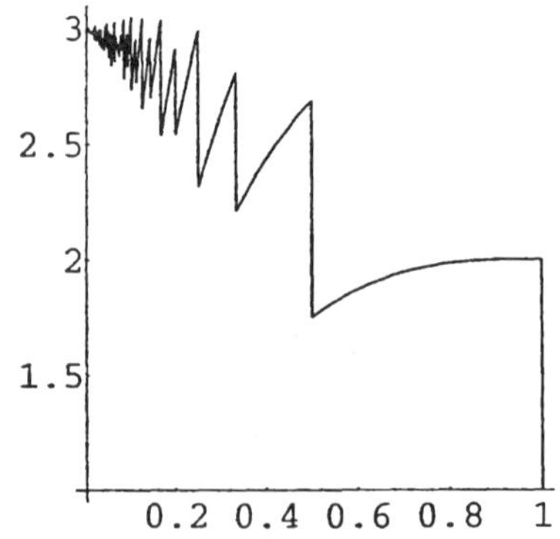

Figure (d)

14.11 Simplicial Operators

In what follows, I will be either the interval $[0,1]$ or the interval $[0,\infty)$, and, for $k = 1, 2, \ldots$, $I^{\langle k \rangle}$ will stand for the k-simplex generated by I, i.e.,

$$I^{\langle k \rangle} := \{\mathbf{x} = (x_1, \ldots, x_k) \in I^k : x_1 + \ldots + x_k \in I\}.$$

Obviously, $I^{\langle k \rangle}$ is either the standard k-simplex or $[0,\infty)^k$ according to $I = [0,1]$ or $I = [0,\infty)$.

In Sections 14.11 and 14.12, we will be concerned with simplicial operators which are generated by S_0^*-operators according to the following definition.

DEFINITION 14.7 *Let L be an S_0^*-operator on I, let $Z := \{Z(x) : x \in I\}$ be a fixed stochastic core for L satisfying conditions* (H_1)-(H_3) *above, and set, for $k = 1, 2, \ldots$,*

$$Z^{\langle k \rangle}(\mathbf{x}) := \left(Z_1^{\langle k \rangle}(\mathbf{x}), \ldots, Z_k^{\langle k \rangle}(\mathbf{x})\right), \qquad \mathbf{x} \in I^{\langle k \rangle}, \tag{14.28}$$

where

$$Z_j^{\langle k \rangle}(\mathbf{x}) := Z\left(\sum_{i=1}^{j} x_i\right) - Z\left(\sum_{i=1}^{j-1} x_i\right), \qquad j = 1, \ldots, k. \tag{14.29}$$

Then, the operator $L^{\langle k\rangle}$ on $I^{\langle k\rangle}$ given by

$$L^{\langle k\rangle}f(\mathbf{x}) := Ef(Z^{\langle k\rangle}(\mathbf{x})), \qquad \mathbf{x} \in I^{\langle k\rangle}, \tag{14.30}$$

will be called the kth L-simplicial operator (associated to Z).

REMARK 14.9 It is clear that $L^{\langle k\rangle}$ fulfills (14.2), and therefore is an S-operator. Moreover, for all $\mathbf{x} := (x_1, \ldots, x_k) \in I^{\langle k\rangle}$ and $f \in \mathcal{L}$, we have that

$$L^{\langle k\rangle}f_i(\mathbf{x}) = Lf(x_i), \qquad i = 1, \ldots, k,$$

and

$$L^{\langle k\rangle}f_*(\mathbf{x}) = Lf(x_1 + \cdots + x_k),$$

where

$$f_i(\mathbf{x}) := f(x_i), \qquad f_*(\mathbf{x}) := f(x_1 + \cdots + x_k).$$

In particular, $L^{\langle 1\rangle} = L$. We also observe that, when $I = [0, \infty)$ and the process Z has independent increments, $L^{\langle k\rangle}$ coincides with the tensor product $L \otimes \cdots \otimes L$ of k copies of L (this takes place, for instance, when L is the Szász operator). ▮

The preceding framework is general enough to include many simplicial operators usually considered in the literature on approximation theory.

Example 14.2

If L is the nth Bernstein operator B_n on $[0, 1]$, and Z is the stochastic core given in the Appendix, then the random vector $nZ^{\langle k\rangle}(\mathbf{x})$ has the multinomial distribution with parameters n, $\mathbf{x}$, and $B_n^{\langle k\rangle}$ becomes the classical nth Bernstein operator on the standard k-simplex, i.e.,

$$B_n^{\langle k\rangle}f(\mathbf{x}) := \sum_{\mathbf{u}\in U_n} f\left(\frac{\mathbf{u}}{n}\right) \frac{n!}{\prod_{i=0}^{k} u_i!} \prod_{i=0}^{k} x_i^{u_i},$$

where

$$x_0 := 1 - \sum_{i=1}^{k} x_i, \qquad u_0 := n - \sum_{i=1}^{k} u_i,$$

and U_n is the set of all k-tuples of nonnegative integers $\mathbf{u} := (u_1, \ldots, u_k)$ such that $\sum_{i=1}^{k} u_i \leq n$. Other examples can be found in [24]. □

PROPOSITION 14.2

Let L be a one-dimensional S_0^*-operator, and let $\{L^{\langle k\rangle} : k \geq 1\}$ be the sequence of the L-simplicial operators. Then, both the sequences $\{C_p^{\langle k\rangle}(\delta) : k \geq 1\}$ and $\{D_p^{\langle k\rangle}(\omega) : k \geq 1\}$ are nondecreasing.

PROOF The embedding argument used to show Proposition 14.1 also works in the present case. ∎

14.12 The Case of S_0^{**}-Operators

To achieve further results on the C-constants and the D-constants for L-simplicial operators, we need additional assumptions on L.

DEFINITION 14.8 *An S-operator L on the interval I is said to be an S_0^{**}-operator if it has a stochastic core $Z := \{Z(x) : x \in I\}$ fulfilling hypotheses* (H_1), (H_3), *and*

(H_2') *The process Z is superstationary, i.e., For each $k = 1, 2, \ldots,$ and each finite family $\mathcal{I} := \{(a_{ij}, b_{ij}] : 1 \leq i \leq k,\ 1 \leq j \leq r_i\}$ ($r_1, \ldots, r_k$ being arbitrary natural numbers) of pairwise disjoint subintervals of I, the k-dimensional random vector $(Z_1^*, \ldots, Z_k^*)$ given by $Z_i^* := \sum_{j=1}^{r_i}\big(Z(b_{ij}) - Z(a_{ij})\big)$ ($1 \leq i \leq k$) has the same probability distribution as the random vector $Z^{\langle k\rangle}(\mathbf{x})$ defined by (14.28)–(14.29), where $\mathbf{x} := (x_1, \ldots, x_k)$ is given by $x_i := \sum_{j=1}^{r_i}(b_{ij} - a_{ij})$. (In other words, the distribution of $(Z_1^*, \ldots, Z_k^*)$ depends on the family $\mathcal{I}$ only through the vector $\mathbf{x}$ above.)*

REMARK 14.10 The notion of a superstationary process was introduced in [24]. It is immediate that condition (H_2') implies (H_2), so that any S_0^{**}-operator is also an S_0^*-operator. All of the operators of families (I), (II), (IV), (VI), (VIII), (XI) and (XII) in the Appendix are examples of S_0^{**}-operators (see [24]). ∎

Our main result in this section is the following theorem. Observe that it gives formulae for the C-constants and the D-constants, when $I = [0, \infty)$.

THEOREM 14.6

Let $p \in [1, \infty]$, $k \geq 1$, let L be an S_0^{**}-operator having the stochastic core Z, and let $L^{\langle k\rangle}$ and $Z^{\langle k\rangle}$ be the same as in (14.28)–(14.30).

(a) *If L is A-Lipschitz, then, for all $\delta > 0$,*

$$C_p^{\langle k \rangle}(\delta) \le 1 + Ak^{1-1/p}, \tag{14.31}$$

and, for every c.m.c. $\omega(\cdot)$,

$$D_p^{\langle k \rangle}(\omega) \le \sup_{\delta>0} \frac{\omega\left(Ak^{1-1/p}\delta\right)}{\omega(\delta)}. \tag{14.32}$$

(b) *In case that $I = [0, \infty)$, we have for all $\delta > 0$*

$$C_p^{\langle k \rangle}(\delta) = \sup_{\substack{\mathbf{x} \in I^{\langle k \rangle} \\ \|\mathbf{x}\|_p \le \delta}} E\left\lceil \frac{\|Z^{\langle k \rangle}(\mathbf{x})\|_p}{\delta} \right\rceil, \tag{14.33}$$

and, for any modulus of continuity $\omega(\cdot)$,

$$D_p^{\langle k \rangle}(\omega) = \sup_{\delta>0} \sup_{\substack{\mathbf{x} \in I^{\langle k \rangle} \\ \|\mathbf{x}\|_p \le \delta}} \frac{E\omega\left(\|Z^{\langle k \rangle}(\mathbf{x})\|_p\right)}{\omega(\delta)}. \tag{14.34}$$

The key to the proof is the following lemma. In it, we use the notation $\mathbf{x}^* := (|x_1|, \ldots, |x_k|)$ introduced in Section 14.4 above, and $\mathbf{U} \simeq \mathbf{V}$ means that $\mathbf{U}$ and $\mathbf{V}$ are random vectors having the same probability distribution.

LEMMA 14.2

Let $Z^{\langle k \rangle}$ be as in the preceding theorem, and let $\mathbf{x}$, $\mathbf{y} \in I^{\langle k \rangle}$. Then:

(a) *There exists an $I^{\langle k \rangle} \times I^{\langle k \rangle}$-valued random vector $(\mathbf{X}, \mathbf{Y})$ such that $Z^{\langle k \rangle}(\mathbf{x}) \simeq \mathbf{X}$, $Z^{\langle k \rangle}(\mathbf{y}) \simeq \mathbf{Y}$, and, for each $i \in \{1, 2, \ldots, k\}$, $X_i \le Y_i$ a.s. or $Y_i \le X_i$ a.s. (X_i (resp. Y_i) being the ith component of $\mathbf{X}$ (resp. $\mathbf{Y}$)).*

(b) *Moreover, when $I = [0, \infty)$, $(\mathbf{X}, \mathbf{Y})$ can be chosen in such a way that $(\mathbf{X} - \mathbf{Y})^* \simeq Z^{\langle k \rangle}\left((\mathbf{x} - \mathbf{y})^*\right)$.*

PROOF (of Lemma 14.2). Set, for $i = 1, \ldots, k$,

$$u_i := \min(x_i, y_i),$$

$$v_i := \begin{cases} x_i - y_i & \text{if } x_i \ge y_i \\ 0 & \text{otherwise} \end{cases} \qquad w_i := \begin{cases} y_i - x_i & \text{if } x_i < y_i \\ 0 & \text{otherwise,} \end{cases}$$

where x_i (resp. y_i) is the ith coordinate of $\mathbf{x}$ (resp. $\mathbf{y}$). If $\mathbf{U}:= (U_1, \dots, U_k)$, $\mathbf{V}:= (V_1, \dots, V_k)$ and $\mathbf{W}:= (W_1, \dots, W_k)$ are the k-dimensional random vectors whose respective jth components ($j = 1, 2, \dots, k$) are given by

$$U_j := Z\left(\sum_{i=1}^{j} u_i\right) - Z\left(\sum_{i=1}^{j-1} u_i\right),$$

$$V_j := Z\left(\sum_{i=1}^{k} u_i + \sum_{i=1}^{j} v_i\right) - Z\left(\sum_{i=1}^{k} u_i + \sum_{i=1}^{j-1} v_i\right),$$

$$W_j := Z\left(\sum_{i=1}^{k} u_i + \sum_{i=1}^{j} w_i\right) - Z\left(\sum_{i=1}^{k} u_i + \sum_{i=1}^{j-1} w_i\right),$$

then we have that $\mathbf{U}+\mathbf{V} \simeq Z^{\langle k\rangle}(\mathbf{x})$, as it follows from assumption (H_2') on Z, the fact that the $2k$ subintervals of I

$$\left(\sum_{i=1}^{j-1} u_i, \sum_{i=1}^{j} u_i\right], \qquad j = 1, 2, \dots, k, \tag{14.35}$$

$$\left(\sum_{i=1}^{k} u_i + \sum_{i=1}^{j-1} v_i, \sum_{i=1}^{k} u_i + \sum_{i=1}^{j} v_i\right], \qquad j = 1, 2, \dots, k, \tag{14.36}$$

are pairwise disjoint, and the fact that $u_j + v_j = x_j$ ($j = 1, 2, \dots, k$). The analogous argument shows that $\mathbf{U}+\mathbf{W} \simeq Z^{\langle k\rangle}(\mathbf{y})$. Moreover, for each fixed $j = 1, 2, \dots, k$, we have that $U_j+V_j \le U_j+W_j$ (a.s.) or $U_j+V_j \ge U_j+W_j$ (a.s.), since, by assumption (H_1) on Z, the components of $\mathbf{U}$, $\mathbf{V}$ and $\mathbf{W}$ are (a.s.) nonnegative random variables and, we have either $V_j = 0$ or $W_j = 0$ according to $x_j < y_j$ or $x_j \ge y_j$. This completes the proof that the random vector ($\mathbf{U}+\mathbf{V}$,$\mathbf{U}+\mathbf{W}$) satisfies the requirements in Lemma 14.2(a). On the other hand, in the case that $I = [0, \infty)$, we can take the random vector $(\mathbf{X}, \mathbf{Y})$ given by

$$\mathbf{X} := \mathbf{U} + \mathbf{V}, \qquad \mathbf{Y} := \mathbf{U} + \mathbf{W}',$$

where $\mathbf{U}$ and $\mathbf{V}$ are the same as above, and $\mathbf{W}' := (W_1', \dots, W_k')$ is defined by

$$W_j' := Z\left(\sum_{i=1}^{k}(u_i + v_i) + \sum_{i=1}^{j} w_i\right) - Z\left(\sum_{i=1}^{k}(u_i + v_i) + \sum_{i=1}^{j-1} w_i\right),$$

for $j = 1, \ldots, k$. It is seen as before that $(\mathbf{X}, \mathbf{Y})$ fulfills the requirements in Lemma 14.2(a). Moreover, we have $(\mathbf{X} - \mathbf{Y})^* \simeq Z^{\langle k \rangle}((\mathbf{x} - \mathbf{y})^*)$, as it follows from the assumptions on Z, and the fact that the $3k$ subintervals of $[0, \infty)$ given in (14.35), (14.36) and

$$\left(\sum_{i=1}^{k} (u_i + v_i) + \sum_{i=1}^{j-1} w_i, \sum_{i=1}^{k} (u_i + v_i) + \sum_{i=1}^{j} w_i \right], \qquad j = 1, 2, \ldots, k,$$

are pairwise disjoint. The proof of Lemma 14.2 is complete. ∎

REMARK 14.11 The conclusion in Lemma 14.2(b) cannot be achieved when $I = [0, 1]$ and $(\mathbf{x} - \mathbf{y})^* \notin I^{\langle k \rangle}$, because $Z^{\langle k \rangle}((\mathbf{x} - \mathbf{y})^*)$ is not defined in such a case. In the same direction, observe that the construction of $\mathbf{W}'$, as done in the preceding proof, makes sense if and only if $\sum_{j=1}^{k}(u_j + v_j + w_j) = \sum_{j=1}^{k} \max(x_j, y_j) \in I$. ∎

PROOF (of Theorem 14.6). For each $(\mathbf{x}, \mathbf{y}) \in I^{\langle k \rangle} \times I^{\langle k \rangle}$, let $(\mathbf{X}, \mathbf{Y})$ be a fixed random vector satisfying the requirements in Lemma 14.2(a). Then, for all $f \in M^{\langle k \rangle}$, we can write

$$L^{\langle k \rangle} f(\mathbf{x}) = Ef(\mathbf{X}), \qquad L^{\langle k \rangle} f(\mathbf{y}) = Ef(\mathbf{Y}),$$

and the same arguments showing (the second inequalities in) (14.4) and (14.5) also yield

$$C_p^{\langle k \rangle}(\delta) \le \sup_{\substack{\mathbf{x}, \mathbf{y} \in I^{\langle k \rangle} \\ \|\mathbf{x} - \mathbf{y}\|_p \le \delta}} E\left\lceil \frac{\|\mathbf{X} - \mathbf{Y}\|_p}{\delta} \right\rceil, \tag{14.37}$$

for all $\delta > 0$, and

$$D_p^{\langle k \rangle}(\omega) \le \sup_{\delta > 0} \sup_{\substack{\mathbf{x}, \mathbf{y} \in I^{\langle k \rangle} \\ \|\mathbf{x} - \mathbf{y}\|_p \le \delta}} \frac{E\omega(\|\mathbf{X} - \mathbf{Y}\|_p)}{\omega(\delta)}, \tag{14.38}$$

for any modulus of continuity $\omega(\cdot)$. Moreover, since each component of $\mathbf{X} - \mathbf{Y}$ is (a.s.) nonnegative or nonpositive, we also have

$$E\|\mathbf{X} - \mathbf{Y}\|_1 = \sum_{i=1}^{k} E|X_i - Y_i| = \sum_{i=1}^{k} |E(X_i - Y_i)|$$

$$\leq \sum_{i=1}^{k} A|x_i - y_i| = A\|\mathbf{x} - \mathbf{y}\|_1, \tag{14.39}$$

the inequality by the fact that L is A-Lipschitz. Using (14.37)–(14.39), the proofs of inequalities (14.31) and (14.32) can be completed in the same way as those of (14.19) and (14.20), respectively. To show part (b), let $I = [0, \infty)$, and assume, by Lemma 14.2(b), that each of the random vectors $(\mathbf{X}, \mathbf{Y})$ above has been chosen in such a way that $(\mathbf{X} - \mathbf{Y})^* \simeq Z^{\langle k \rangle}((\mathbf{x} - \mathbf{y})^*)$. Then, inequalities (14.37) and (14.38) become, respectively,

$$C_p^{\langle k \rangle}(\delta) \leq \sup_{\substack{\mathbf{x}, \mathbf{y} \in I^{\langle k \rangle} \\ \|\mathbf{x}-\mathbf{y}\|_p \leq \delta}} E\left\lceil \frac{\|Z^{\langle k \rangle}((\mathbf{x} - \mathbf{y})^*)\|_p}{\delta} \right\rceil = \sup_{\substack{\mathbf{x} \in I^{\langle k \rangle} \\ \|\mathbf{x}\|_p \leq \delta}} E\left\lceil \frac{\|Z^{\langle k \rangle}(\mathbf{x})\|_p}{\delta} \right\rceil,$$

and

$$D_p^{\langle k \rangle}(\omega) \leq \sup_{\delta > 0} \sup_{\substack{\mathbf{x}, \mathbf{y} \in I^{\langle k \rangle} \\ \|\mathbf{x}-\mathbf{y}\|_p \leq \delta}} \frac{E\omega\left(\|Z^{\langle k \rangle}((\mathbf{x} - \mathbf{y})^*)\|_p\right)}{\omega(\delta)}$$

$$= \sup_{\delta > 0} \sup_{\substack{\mathbf{x} \in I^{\langle k \rangle} \\ \|\mathbf{x}\|_p \leq \delta}} \frac{E\omega\left(\|Z^{\langle k \rangle}(\mathbf{x})\|_p\right)}{\omega(\delta)}.$$

Since, by Theorem 14.1, the converse inequalities also hold true, the conclusions in Theorem 14.6(b) follow. ∎

For $I = [0, \infty)$ and $p = 1$, we have the following result.

COROLLARY 14.3

Let I, L, Z, δ and $\omega(\cdot)$ be the same as in Theorem 14.6(b). Then, for all $k \geq 1$, we have

$$C_1^{\langle k \rangle}(\delta) = C_1^{\langle 1 \rangle}(\delta), \qquad D_1^{\langle k \rangle}(\omega) = D_1^{\langle 1 \rangle}(\omega).$$

PROOF From the properties of the stochastic core Z, it is easily shown that $\|Z^{\langle k \rangle}(\mathbf{x})\|_1 \simeq Z(\|\mathbf{x}\|_1)$, for all $\mathbf{x} \in I^{\langle k \rangle}$. Combining such a fact with the formulae in Theorem 14.6(b), the conclusions in the corollary are readily obtained. ∎

In [24], the reader can find applications of Theorem 14.6(a) (with $p = 1$) and Corollary 14.3 to the simplicial operators associated to the operators of families (I), (II), (IV), (VI), (VIII), (XI) and (XII) in the Appendix. The following example provides further information about the simplicial Bernstein operator. Other examples can be worked in a similar way.

Example 14.3
Let L be the nth Bernstein operator B_n. We claim that, for all $p \in [1, \infty]$,

$$\liminf_{\delta \downarrow 0} C_p^{\langle k \rangle}(\delta) \geq k^{1-1/p}.$$

To show this claim, let $\mathbf{d} := (\delta, \ldots, \delta)$, with $\delta > 0$ small enough so that $k^{-1/p}\mathbf{d} \in I^{\langle k \rangle}$. Since $\|k^{-1/p}\mathbf{d}\|_p = \delta$, we have from the first inequality in (14.4)

$$C_p^{\langle k \rangle}(\delta) \geq k \sum_{j=1}^{n} \frac{j}{n\delta} \binom{n}{j} \left(k^{-1/p}\delta\right)^j \left(1 - k^{1-1/p}\delta\right)^{n-j}$$

$$= k^{1-1/p} \left(1 - (k-1)k^{-1/p}\delta\right)^{n-1},$$

and the claim follows by letting $\delta \downarrow 0$. On the other hand, in view of (14.32), the equality

$$D_p^{\langle k \rangle}(\omega^{(1)}) = k^{1-1/p},$$

can be showed in the same way. ▯

14.13 Notes

1. Lindvall [34], and Brown, Elliott and Paget [19], among others, proved that the one-dimensional Bernstein polynomials preserve Lipschitz classes. Della Vecchia [26], Rasa [38], and Khan and Peters [31] showed the same property for other operators. The probabilistic method introduced in [31] (based on the notion of a splitting probability distribution) has been also followed in other works (see, for instance, [7, 22]).

The classes of functions determined by moduli of continuity $M_{p,\omega}^{\langle k \rangle}$ were introduced by Nikol'skii [35, 36] (see also the book by Korneichuk [32]). The problem of the D-constants for such classes was first discussed by de la Cal and Valle [24].

Kratz and Stadtmüller [33] were the first to achieve preservation inequalities concerning the usual modulus of continuity; they treated certain discrete univariate operators by means of techniques depending heavily on the situation at hand. Anastassiou *et al.* [15, 16] considered positive linear operators from $C(X)$ into itself, X being a compact metric space, and proposed an approach based on K-functional techniques and the least concave majorant of a modulus of continuity. They obtained general results, and discussed applications to some univariate and multivariate operators over compact domains; among other things, they showed that 2 is the uniform C-constant for the simplicial Bernstein operator (and $p = 1$). The same approach was followed by Cottin and Gonska [20, 21]. On the other hand, Anastassiou [13, 14], and Anastassiou and Gonska [17, 18] have also considered a particular class of univariate and multivariate integral operators over noncompact domains.

2. The approach in the present chapter dates back to Lindvall [34]. This author succeeded in deriving several properties of one-dimensional Bernstein polynomials from the representation given in the Appendix below. The idea was later generalized and extensively applied by Adell and de la Cal [2-4]. It has been also followed in other works [6, 8, 9, 23, 24]. The first application to multivariate operators was given by us in [24]. It should be said that this approach has shown to be very useful to deal with other problems concerning Bernstein-type operators (see, for instance, [1, 5, 6, 10-12]).

3. The terminology in the present chapter (S-operator, stochastic core, etc.) has been introduced for the sake of brevity, and, therefore, it is not standard. Also, the main results are new, although some of them overlap in different ways with other results in the literature. As said in Section 14.5, formulae (14.17) and (14.18) (and, therefore, (14.14) and (14.15)) extend those given by Adell and Pérez-Palomares [9] for one-dimensional S_0^*-operators (without using this name). On the other hand, it is easy to see that the tensor product operators in Theorem 14.3 above have the *splitting property*, in the sense of [31]; therefore, inequality (14.20) can be derived by the method in that paper; by the same reason, inequality (14.19) can be derived from the results in [16], when all of the intervals $I_1, \ldots, I_k$ are compact. The results in Section 14.9 concerning derivatives of multivariate Bernstein polynomials should be compared with those obtained by Cottin and Gonska [21]. Finally, the basic ideas in Sections 14.11 and 14.12 come from our paper [24]; in such a paper we only considered the l_1-norm; the proof of Lemma 14.2 above is essentially the same as that of [24; Lemma 2], but the conclusion for the case $I = [0, \infty)$ is stronger; Theorem 14.6 above is much more general than Theorem 2 in [24]; observe also that formulae (14.33) and (14.34) are new partial generalizations of the one-dimensional formulae in [9].

4. Preservation inequalities concerning the usual modulus of second order have been obtained by Adell and de la Cal [3], Cottin and Gonska [21], Gonska and Kovacheva [27], Adell and Pérez-Palomares [8], Pérez-Palomares [37], and Zhou [40], among others. However, the determination of the best constant seems to be a very difficult problem, even in the case of one-dimensional Bernstein polynomials. Other moduli have been considered by Kratz and Stadtmüller [33] and by de la Cal and Cárcamo [23].

Appendix

In this appendix, we list several families of S-operators usually considered in the literature on approximation theory. For each example, we give both its analytical expression and its probabilistic representation in terms of an appropriate stochastic core; we also specify which hypotheses (H_i) are fulfilled. The list is not (cannot be) exhaustive, and we refer to [2, 4, 5] for additional examples. All of the stochastic cores will be constructed from the following four types of well-known stochastic processes:

(a) *The empirical process* $\{S_n(x) : x \in [0,1],\ n = 1, 2, \ldots\}$ given by

$$S_n(x) := \sum_{i=1}^{n} 1_{[0,x]}(X_i),$$

where $X_1, X_2, \ldots$ are independent and on the interval $[0,1]$ uniformly distributed random variables.

(b) *The process* $\{T_n : n = 1, 2, \ldots\}$ given by

$$T_n := \sum_{i=1}^{n} Y_i,$$

where $Y_1, Y_2, \ldots$ are independent and on the interval $[0,1]$ uniformly distributed random variables.

(c) *The standard Poisson process* $\{N(u) : u \geq 0\}$.

(d) *The standard gamma process* $\{U(t) : t \geq 0\}$, i.e., a stochastic process starting at 0, having stationary independent increments, and such that, for $t > 0$, $U(t)$ has the gamma distribution with density

$$g_t(u) := \Gamma(t)^{-1} u^{t-1} e^{-u} 1_{(0,\infty)}(u).$$

For the sake of brevity we use the following convention: If the construction of a stochastic core requires two or more of the preceding processes, they all are assumed to be independent and defined on the same probability space.

(I) *Bernstein Operators.* For $n = 1, 2, \ldots$, and $x \in [0, 1]$,

$$B_n f(x) := \sum_{k=0}^{n} f(k/n) p_{n,k}(x) = Ef(Z_n(x)),$$

where

$$p_{n,k}(x) := \binom{n}{k} x^k (1-x)^{n-k}, \qquad Z_n(x) := \frac{S_n(x)}{n}. \tag{14.40}$$

We have $EZ_n(x) = x$, and $P(Z_n(x) = 0) = p_{n,0}(x) = (1-x)^n$. All of the hypotheses (H_1)-(H_6) are fulfilled ((H_4) and (H_5) with $A = 1$). The superstationarity condition (H_2') is also satisfied.

(II) *Stancu Operators.* For $n = 1, 2, \ldots, a > 0$ and $x \in [0, 1]$,

$$L_{n,a} f(x) := \sum_{k=0}^{n} f(k/n) \binom{n}{k} \frac{x^{\langle k,a \rangle} (1-x)^{\langle n-k,a \rangle}}{1^{\langle n,a \rangle}} = Ef\left(Z_{n,a}(x)\right),$$

where

$$v^{\langle l,a \rangle} := \begin{cases} \prod_{j=0}^{l-1} (v + ja) & \text{if } l \geq 1 \\ 1 & \text{if } l = 0, \end{cases} \qquad Z_{n,a}(x) := \frac{S_n\left(U(xa^{-1}) U(a^{-1})^{-1}\right)}{n}.$$

We have $EZ_{n,a}(x) = x$ and $P\left(Z_{n,a}(x) = 0\right) = (1-x)^{\langle n,a \rangle} / 1^{\langle n,a \rangle}$. This operator satisfies the same hypotheses (H_i) as the Bernstein operator B_n.

(III) *Bernstein-Durrmeyer Operators.* For $n = 1, 2, \ldots$ and $x \in [0, 1]$,

$$D_n f(x) := (n+1) \sum_{k=0}^{n} p_{n,k}(x) \int_0^1 f(u) p_{n,k}(u) du = Ef\left(Z_n(x)\right),$$

where $p_{n,k}(x)$ is the same as in (14.40), and

$$Z_n(x) := \frac{U\left(S_n(x) + 1\right)}{U(n+2)}.$$

We have $EZ_n(x) = (nx+1)/(n+2)$, and $P(Z_n(x) = 0) = 0$. This operator satisfies (H_1), (H_2) and (H_4) (with $A := n/(n+2)$).

(IV) *Modified Bernstein-Durrmeyer Operators* (cf. [28]). For $n = 1, 2, \ldots$ and $x \in [0, 1]$,

$$\begin{aligned} D_n^* f(x) :&= f(0)p_{n,0}(x) \\ &\quad + (n-1)\sum_{k=1}^{n-1} p_{n,k}(x) \int_0^1 f(u)p_{n-2,k-1}(u)du \\ &\quad + f(1)p_{n,n}(x) \\ &= Ef\left(Z_n(x)\right), \end{aligned}$$

where

$$Z_n(x) := \frac{U\left(S_n(x)\right)}{U(n)}.$$

We have $EZ_n(x) = x$, and $P(Z_n(x) = 0) = p_{n,0}(x) = (1-x)^n$. This operator satisfies the same hypotheses (H_i) as the Bernstein operator B_n.

(V) *Kantorovič Operators.* For n, $m = 1, 2, \ldots$ and $x \in [0, 1]$,

$$\begin{aligned} K_{n,m}f(x) :&= \sum_{k=0}^{n} p_{n,k}(x) \int_0^1 \cdots \int_0^1 f\left(\frac{k + y_1 + \cdots + y_m}{n+m}\right) dy_1 \cdots dy_m \\ &= Ef(Z_{n,m}(x)), \end{aligned}$$

where

$$Z_{n,m}(x) := \frac{S_n(x) + T_m}{n+m}.$$

We have $EZ_{n,m}(x) = (nx + (m/2))/(n+m)$, and $P(Z_{n,m}(x) = 0) = 0$. This operator satisfies (H_1), (H_2) and (H_4) (with $A := n/(n+m)$). Moreover, we have for every $f \in C^m[0, 1]$

$$(B_{n+m}f)^{(m)} = \prod_{j=1}^{m} \frac{n+j}{n+m} K_{n,m}f^{(m)}, \tag{14.41}$$

where B_{n+m} is the $(n+m)$th Bernstein operator.

(VI) *Szász Operators.* For $t > 0$ and $x \geq 0$,

$$S_t f(x) := e^{-tx} \sum_{k=0}^{\infty} f(k/t) \frac{(tx)^k}{k!} = Ef\left(Z_t(x)\right),$$

where

$$Z_t(x) := \frac{N(tx)}{t}.$$

We have $EZ_t(x) = x$, and $P\left(Z_t(x) = 0\right) = e^{-tx}$. This operator satisfies the same hypotheses (H_i) as the Bernstein operator B_n.

(VII) *Szász-Kantorovič Operators.* For $t > 0$, $m = 1, 2, \ldots$, and $x \geq 0$,

$$S_{t,m} f(x) : = e^{-tx} \sum_{k=0}^{\infty} \frac{(tx)^k}{k!} \int_0^1 \cdots \int_0^1 f\left(\frac{k + y_1 + \cdots + y_m}{t}\right) dy_1 \cdots dy_m$$

$$= Ef\left(Z_{t,m}(x)\right),$$

where

$$Z_{t,m}(x) := \frac{N(tx) + T_m}{t}.$$

We have $EZ_{t,m}(x) = x + (m/2t)$, and $P\left(Z_{t,m}(x) = 0\right) = 0$. This operator satisfies (H_1), (H_2) and (H_4) (with $A := 1$).

(VIII) *Baskakov Operators.* For $t > 0$ and $x \geq 0$,

$$H_t f(x) := \sum_{k=0}^{\infty} f(k/t) b_{t,k}(x) = Ef\left(Z_t(x)\right),$$

where

$$b_{t,k}(x) := \binom{t + k - 1}{k} \frac{x^k}{(1 + x)^{t+k}}, \qquad Z_t(x) := \frac{N\left(U(t)x\right)}{t}. \tag{14.42}$$

We have $EZ_t(x) = x$, and $P\left(Z_t(x) = 0\right) = b_{t,0}(x) = (1 + x)^{-t}$. This operator satisfies the same hypotheses (H_i) as the Bernstein operator B_n.

(IX) *Baskakov-Kantorovič Operators.* For $t > 0$, $m = 1, 2, \ldots$, and $x \geq 0$,

$$H_{t,m} f(x) : = \sum_{k=0}^{\infty} b_{t,k}(x) \int_0^1 \cdots \int_0^1 f\left(\frac{k + y_1 + \cdots + y_m}{t}\right) dy_1 \cdots dy_m$$

$$= Ef\left(Z_{t,m}(x)\right),$$

where $b_{t,k}(x)$ is the same as in (14.42), and

$$Z_{t,m}(x) := \frac{N\left(U(t)x\right) + T_m}{t}.$$

We have $EZ_{t,m}(x) = x + (m/2t)$, and $P\left(Z_{t,m}(x) = 0\right) = 0$. This operator satisfies (H_1), (H_2) and (H_4) (with $A := 1$).

(X) *Modified Meyer-König and Zeller Operators.* For $t > 0$ and $x \in [0, 1)$,

$$M_t f(x) := (1-x)^{t+1} \sum_{k=0}^{\infty} f\left(\frac{k}{t+k}\right) \binom{t+k}{k} x^k = Ef\left(Z_t(x)\right),$$

where

$$Z_t(x) := \frac{N\left(U(t+1)q(x)\right)}{t + N\left(U(t+1)q(x)\right)}, \qquad q(x) := \frac{x}{1-x}.$$

We have $EZ_t(x) = x$, and $P\left(Z_t(x) = 0\right) = (1-x)^{t+1}$. This operator satisfies (H_1) and (H_3)-(H_6) ((H_4) and (H_5) with $A := 1$).

(XI) *Beta Operators.* For $t > 0$ and $x \in [0, 1]$,

$$\beta_t f(x) := \begin{cases} \displaystyle\int_0^1 f(u) \frac{u^{tx-1}(1-u)^{t(1-x)-1}}{B(tx, t(1-x))} du & \text{if } x \in (0,1) \\ f(x) & \text{if } x = 0 \text{ or } x = 1 \end{cases}$$

$$= Ef\left(Z_t(x)\right),$$

where $B(\cdot,\cdot)$ is the beta function, and $Z_t(x) := U(tx)U(t)^{-1}$. We have $EZ_t(x) = x$, and $P\left(Z_t(x) = 0\right) = 1$ or 0, according to $x = 0$ or $x \in (0, 1]$. This operator satisfies (H_1)-(H_5) ((H_4) and (H_5) with $A := 1$), and (H_2'). (For a different representation of β_t useful for other purposes, see [1].)

(XII) *Gamma Operators.* For $t > 0$ and $x \geq 0$,

$$G_t f(x) := \begin{cases} \displaystyle\frac{1}{\Gamma(t)} \int_0^{\infty} f\left(\frac{xu}{t}\right) u^{t-1} e^{-u} du & \text{if } x > 0 \\ f(0) & \text{if } x = 0, \end{cases}$$

$$= Ef\left(Z_t(x)\right),$$

where $Z_t(x) := xU(t)t^{-1}$. We have $EZ_t(x) = x$, and $P(Z_t(x) = 0) = 1$ or 0, according to $x = 0$ or $x > 0$. This operator satisfies the same hypotheses (H_i) as the preceding beta operator.

Acknowledgment

This work was supported by the University of the Basque Country, and by Grant PB95-0809 of the Spanish DGICYT.

References

1. J.A. Adell, F.G. Badía, J. de la Cal and F. Plo, On the property of monotonic convergence for beta operators, *J. Approx. Theory* **84** (1996), 61-73.

2. J.A. Adell and J. de la Cal, Using stochastic processes for studying Bernstein-type operators, *Rend. Circ. Mat. Palermo (2) Suppl.* **33** (1993), 125-141.

3. J.A. Adell and J. de la Cal, Preservation of moduli of continuity for Bernstein-type operators, in *Approximation, Probability and Related Fields* (G.A. Anastassiou and S.T. Rachev, Eds.), pp. 1-18, Plenum Press, New York, 1994.

4. J.A. Adell and J. de la Cal, Bernstein-Durrmeyer operators, *Comput. Math. Appl.* **30** (1995), 1-14.

5. J.A. Adell and J. de la Cal, Bernstein-type operators diminish the φ-variation, *Constr. Approx.* **12** (1996), 489-507.

6. J.A. Adell, J. de la Cal and A. Pérez-Palomares, On the Cheney and Sharma operator, *J. Math. Anal. Appl.* **200** (1996), 663-679.

7. J.A. Adell, J. de la Cal and M. San Miguel, Inverse beta and generalized Bleimann-Butzer-Hahn operators, *J. Approx. Theory* **76** (1994), 54-64.

8. J.A. Adell and A. Pérez-Palomares, Second modulus preservation inequalities for generalized Bernstein-Kantorovich operators, in *Approximation and Optimization. Vol. I* (D.D. Stancu, G. Coman, W.W. Breckner and P. Blaga, Eds.), pp. 147-156, Transylvania Press, Cluj-Napoca, 1997.

9. J.A. Adell and A. Pérez-Palomares, Best constants in preservation inequalities concerning the first modulus and Lipschitz classes for Bernstein-type operators, *J. Approx. Theory* **93** (1998), 128-139.

10. J.A. Adell and A. Pérez-Palomares, Stochastic orders in preservation properties by Bernstein-type operators, *Adv. in Appl. Probab.* **31** (1999), 1-17.

11. J.A. Adell and C. Sangüesa, Direct and converse inequalities for positive linear operators on the positive semi-axis, *J. Austral. Math. Soc. (Series A)* **66** (1999), 90-103.

12. J.A. Adell and C. Sangüesa, A strong converse inequality for Gamma type operators, *Constr. Approx.*, to appear.

13. G.A. Anastassiou, Differentiated shift-invariant multivariate integral operators, *Southwest J. Pure Appl. Math.* **1** (1995), 13-16.

14. G.A. Anastassiou, Differentiated shift-invariant integral operators, multivariate case, *Bull. Inst. Math. Acad. Sinica* **25** (1997), 243-275.

15. G.A. Anastassiou, C. Cottin and H.H. Gonska, Global smoothness of approximating functions, *Analysis* **11** (1991), 43-57.

16. G.A. Anastassiou, C. Cottin and H.H. Gonska, Global smoothness preservation by multivariate approximation operators, in *Israel Mathematical Conference Proceedings, Vol. IV* (S. Baron and D. Leviatan, Eds.), pp. 31-44, Bar-Ilan University, Ramat-Gan, 1991.

17. G.A. Anastassiou and H.H. Gonska, On some shift-invariant integral operators, multivariate case, in *Approximation, Probability and Related Fields* (G.A. Anastassiou and S.T. Rachev, Eds.), pp. 41-63, Plenum Press, New York, 1994.

18. G.A. Anastassiou and H.H. Gonska, On some shift-invariant integral operators, multivariate case, *Ann. Polon. Math.* **61** (1995), 225-243.

19. B.M. Brown, D. Elliott and D.F. Paget, Lipschitz constants for the Bernstein polynomials of a Lipschitz continuous function, *J. Approx. Theory* **49** (1987), 196-199.

20. C. Cottin and H.H. Gonska, A note on global smoothness preservation by Boolean sum operators, in *Constructive Theory of Functions, Varna 91* (K.G. Ivanov *et al.*, Eds.), pp. 85-91, Sofia, 1992.

21. C. Cottin and H.H. Gonska, Simultaneous approximation and global smoothness preservation, *Rend. Circ. Mat. Palermo (2) Suppl.* **33**

(1993), 259-279.

22. J. de la Cal, On Stancu-Mühlbach operators and some connected problems concerning probability distributions, *J. Aprox. Theory* **74** (1993), 59-68.

23. J. de la Cal and J. Cárcamo, Preservation of p-continuity by Bernstein type operators, *J. Approx. Theory*, in press.

24. J. de la Cal and A.M. Valle, Best constants in global smoothness preservation inequalities for some multivariate operators, *J. Approx. Theory* **97** (1999), 158-180.

25. J. de la Cal and A.M. Valle, Best constants for tensor products of Bernstein, Szász, and Baskakov operators, Preprint.

26. B. Della Vecchia, On the preservation of Lipschitz constants for some linear operators, *Boll. Un. Mat. Ital. B* **3** (1989), 125-136.

27. H.H. Gonska and R.K. Kovacheva, The second order modulus revisited: remarks, applications, problems, *Conferenze del Seminario di Matematica dell'Università di Bari* **257** (1994).

28. T.N.T. Goodman and A. Sharma, A Bernstein type operator on the simplex, *Math. Balkanica (N.S.)* **5** (1991), 129-145.

29. O. Kallenberg, *Foundations of Modern Probability*, Springer, New York, 1997.

30. M.K. Khan, Approximation properties of Beta operators, in *Progress in Approximation Theory* (P. Nevai and A. Pinkus, Eds.), pp. 483-495, Academic Press, New York, 1991.

31. M.K. Khan and M.A. Peters, Lipschitz constants for some approximation operators of a Lipschitz continuous function, *J. Approx. Theory* **59** (1989), 307-315.

32. N. Korneichuk, *Exact Constants in Approximation Theory*, Cambridge University Press, New York, 1991.

33. W. Kratz and U. Stadtmüller, On the uniform modulus of continuity of certain discrete approximation operators, *J. Approx. Theory* **54** (1988), 326-337.

34. T. Lindvall, Bernstein polynomials and the law of large numbers, *Math. Scientist* **7** (1982), 127-139.

35. S.M. Nikol'skii, Approximation of periodic functions by trigonometric polynomials, *Trudy Math. Inst. Stekhlov* **15** (1945), 1-76. [In Russian]

36. S.M. Nikol'skii, Fourier series of functions with a given modulus of continuity, *Dokl. Akad. Nauk SSSR* **52** (1946), 191-194. [In Russian]

37. A. Pérez-Palomares, Global smoothness preservation properties for generalized Szász-Kantorovich operators, Preprint.

38. I. Rasa, Altomare projections and Lototsky-Schnabl operators, *Rend. Circ. Mat. Palermo (2) Suppl.* **33** (1993), 439-451.

39. M. Shaked and J.G. Shantikhumar, *Stochastic Orders and Their Applications*, Academic Press, Boston, 1994.

40. Ding-xuan Zhou, On a problem of Gonska, *Results Math.* **28** (1995), 169-183.

Chapter 15

Summation Kernels for Orthogonal Polynomial Systems

Frank Filbir, Rupert Lasser, and Josef Obermaier

Institute of Biomathematics and Biometrics, GSF – National Research Center for Environment and Health, D-85764 Neuherberg, Germany
E-mail: filbir@gsf.de, lasser@gsf.de, josef.obermaier@gsf.de
Contact author: F. Filbir

Abstract The convergence of weighted Fourier expansions with respect to orthogonal polynomial systems $\{P_n : n \in \mathbb{N}_0\}$ is studied in certain Banach spaces $B \subseteq L^1(\pi)$, where the support of the orthogonality measure π is assumed to be infinite and compact. We focus on orthogonal polynomial systems which induce a hypergroup structure on $\mathbb{N}_0$ and a convolution structure on $supp\,\pi$. Especially the Dirichlet kernel, a Fejér-type kernel and the de la Vallée-Poussin kernel are studied, where we stress the analogy to the trigonometric case.

Contents

15.1 Introduction

Approximation methods of periodic functions by Fourier series have a long history and there is lot of fundamental work on this topic. Approximation can be pointwise or in norm and there exist many approximation procedures, the most familiar being the Fejér and the de la Vallée-Poussin approximation. In Fourier series the basic functions are the harmonic oscillations $e_n(t) = e^{int}$, $t \in [0, 2\pi[$, $n \in \mathbb{Z}$. In our contribution we will concentrate on orthogonal Fourier expansions on a compact subset D_S of the real line, where a system $\{P_n : n \in \mathbb{N}_0\}$ of orthogonal polynomials constitutes the basic functions. Our goal is to investigate in which way concepts such as Fejér and de la Vallée-Poussin approximations have their appropriate analogues for orthogonal Fourier expansions. An essential tool for classical Fourier series is the representation of the approximation procedure as a convolution operator, where the group structure of $\mathbb{T} \cong [0, 2\pi[$ is applied. In fact, the group structure $\mathbb{Z}$ is also applied in the analysis of Fourier series. In the present chapter we will essentially use certain convolution structures on D_S and on $\mathbb{N}_0$, too. They are related to the concept of hypergroups. Hence we start with a concise introduction to discrete hypergroups, in particular polynomial hypergroups on $\mathbb{N}_0$ and an investigation of a weak dual convolution structure on D_S. In the next section we recall basic functional analytic facts for approximation by weighted orthogonal expansions in Banach spaces. In the next section we collect results on the Dirichlet kernel especially in the case of Jacobi expansions. Afterward, a Fejér-type approximation is studied in detail. Next, very recent results on de la Vallée-Poussin kernels for orthogonal expansions are presented.

Without any claim to be exhaustive we give some references on summation kernels for orthogonal expansions. Usually the discussion of this subject is restricted to the special case of Jacobi polynomials. Many results can be found, of course, in Szegö's book [38] or in Freud's book [13]. All modern studies on mean convergence of Fourier-Jacobi expansions refer back to Pollard's work [33], [34], [35]. A very nice discussion of Pollard's achievements is contained in the survey of Lubinsky [28]. There are many further contributions on Jacobi expansions as, for example, by Newman and Rudin [32], by Askey and Hirschmann [1] and in particular by Bavinck [3],

[4]. Far-reaching estimates for Cesàro sums of Jacobi series are the principal results of the memoirs of Chanillo and Muckenhoupt [8]. Extensions of Pollard's result can be found in Badkov [2]. Tauberian theorems for Jacobi series are studied by Bingham [5]. Yadav studied various approximation processes in [43] and saturation problems in [42], [44] all in the context of Jacobi polynomials. Generalized Jacobi series are considered in Xu [40], [41]. Pointwise convergence of Jacobi series is investigated by Li [24] and Meaney [29]. Following the ideas of Muckenhoupt and Stein [31] Li also studied conjugate Jacobi series, see [25], [26]. For other topics such as the weak behaviour, the Littlewood-Paley theory or the $\Lambda(p)$-problem for Jacobi polynomials, we refer to [16], [10] and [18], respectively. Finally we mention two references of the authors [22] and [23], where the subsequent contributions have their origin.

15.2 Discrete Hypergroups

In this chapter we introduce discrete hypergroups. We restrict ourselves to main results which we will need for the study of polynomial hypergroups. For the proofs and a more extensive investigation of general hypergroups we refer to standard literature like [17] and [6].

Let K be a non-void set equipped with the discrete topology. A Borel measure m on K is completely determined by the values $m(x) = m(\{x\})$ for all $x \in K$. Hence one may take Borel measures on K as functions. Denote by $M(K)$ the Banach space of all complex Borel measures on K and by $M_c^1(K)$ the subset of probability measures with finite support.
We write ϵ_x, $x \in K$, for the dirac measure or function with $\epsilon_x(x) = 1$, and $\epsilon_x(y) = 0$ if $y \neq x$. For $\mu \in M(K)$ there is a unique representation

$$\mu = \sum_{x \in I} a_x \epsilon_x, \tag{15.1}$$

where $supp\, \mu = I$ is a countable set and $a_x \in \mathbb{C} \setminus \{0\}$ for all $x \in I$.
The norm of the measure is given by

$$\|\mu\| = \sum_{x \in I} |a_x|. \tag{15.2}$$

For a given map $* : K \times K \to M_c^1(K)$, $(x, y) \to x * y$, let us define an extension $* : M(K) \times M(K) \to M(K)$ by

$$\mu * \nu = \sum_{x\in I} \sum_{y \in J} a_x b_y \, x * y, \tag{15.3}$$

with $\mu = \sum_{x\in I} a_x \epsilon_x$ and $\nu = \sum_{y\in J} b_y \epsilon_y$. We have $\|\mu * \nu\| \le \|\mu\| \|\nu\|$.
Note that $\epsilon_x * \epsilon_y = x * y$.
Additionally, for a given map $\tilde{} : K \to K$, $x \to \tilde{x}$, let us define an extension $\tilde{} : M(K) \to M(K)$ by

$$\tilde{\mu} = \sum_{x\in I} a_x \epsilon_{\tilde{x}}. \tag{15.4}$$

Note that $\tilde{\epsilon_x} = \epsilon_{\tilde{x}}$.

DEFINITION 15.1 *$(K, *, \tilde{})$ is called a discrete hypergroup if the following conditions hold:*

(DH1) *$* : K \times K \to M_c^1(K)$ is a map and its extension (15.3) fulfills the associativity law $\epsilon_x * (\epsilon_y * \epsilon_z) = (\epsilon_x * \epsilon_y) * \epsilon_z$ for all $x, y, z \in K$.*

(DH2) *$\tilde{} : K \to K$ is a bijective map with $\tilde{\tilde{x}} = x$ and its extension (15.4) fulfills $(x * y)\tilde{} = \tilde{y} * \tilde{x}$ for all $x, y \in K$.*

(DH3) *There exists an element $e \in K$ with $x * e = \epsilon_x = e * x$ for all $x \in K$.*

(DH4) *$e \in \text{supp}\,(x * \tilde{y})$ if and only if $x = y$.*

We call $*$ convolution, $\tilde{}$ involution and e unit element. A hypergroup is called commutative if $x * y = y * x$ and hermitian if $\tilde{x} = x$ for all $x, y \in K$.

The involution $\tilde{}$ and the unit element e in Definition 15.1 are necessarily unique. With (DH2) a hermitian hypergroup is always commutative. We should mention that every discrete group is also a discrete hypergroup if we set $x * y = \epsilon_{xy}$ and $\tilde{x} = x^{-1}$.
If we define $\mu^* = (\tilde{\mu})^-$ for $\mu \in M(K)$, where $^-$ is the complex conjugation, then $(M(K), *, +)$ is a Banach *- algebra with unit ϵ_e.

For $f : K \to \mathbb{C}$ and $y \in K$ we define a left-translation $L_y f$ by

$$L_y f(x) = f(y * x) = \int_K f(z)\, d\epsilon_y * \epsilon_x(z) = \sum_{u \in I} a_u f(u) \tag{15.5}$$

for all $x \in K$, where $\epsilon_y * \epsilon_x = \sum_{u\in I} a_u \epsilon_u$.

To be able to describe the size of the support of functions and measures on K we introduce the following notation:

For $A, B \subseteq K$ let

$$A * B = \bigcup_{x \in A, y \in B} supp\, x * y \quad \text{and} \quad \tilde{A} = \{\tilde{x} : x \in A\}. \tag{15.6}$$

Now (DH1) implies

$$A * (B * C) = (A * B) * C, \tag{15.7}$$

and (DH2) and (DH4) give

$$(A * B) \cap C = \emptyset \quad \text{if and only if} \quad (\tilde{A} * C) \cap B = \emptyset. \tag{15.8}$$

Applying (15.8) we see that $supp\, L_y f \subseteq \{\tilde{y}\} * supp\, f$. In particular, if $supp\, f$ is finite then $supp\, L_y f$ is finite, too.

A discrete measure h on K with $0 \leq h(x) < \infty$ for all $x \in K$ is called left-invariant if for all f with finite support and all $y \in K$ the equation

$$\sum_{x \in K} f(y * x) h(x) = \sum_{x \in K} f(x) h(x) \tag{15.9}$$

is satisfied.

A left-invariant measure $h \neq 0$ is called a Haar measure. For a discrete hypergroup there always exists a Haar measure which is unique up to a positive multiplicative constant. It holds $h(x) \geq h(e)$, and if K is commutative, then we have $h(x) = h(\tilde{x})$ for all $x \in K$. Henceforth we will denote by h the normalized Haar measure with $h(e) = 1$. The Haar measure can be determined explicitly:

$$h(x) = (\epsilon_{\tilde{x}} * \epsilon_x(e))^{-1} \quad \text{for all} \quad x \in K. \tag{15.10}$$

We denote as usual by $l^p(h)$, $1 \leq p < \infty$, and $l^\infty(h) = l^\infty$ the Banach spaces of functions $f : K \to \mathbb{C}$ with p-norm $\|f\|_p = (\sum_{x \in K} |f(x)|^p h(x))^{1/p}$, respectively, with sup-norm $\|f\|_\infty = \sup_{x \in K} |f(x)|$.
If $f \in l^p(h)$, $1 \leq p \leq \infty$, then $L_y f \in l^p(h)$ and $\|L_y f\|_p \leq \|f\|_p$.
For $\mu \in M(K)$ and $f \in l^p(h)$, $1 \leq p \leq \infty$, we define a convolution $\mu * f$ by

$$\mu * f(x) = \sum_{z \in K} f(\tilde{z} * x) \mu(z) \quad \text{for all} \quad x \in K. \tag{15.11}$$

It is straightforward to prove $\mu * f \in l^p(h)$ and $\|\mu * f\|_p \leq \|\mu\| \|f\|_p$.
For $f, g \in l^1(h)$ we define a convolution $f * g$ by

$$f * g(x) = \sum_{z \in K} f(z) g(\tilde{z} * x) h(z) \quad \text{for all} \quad x \in K. \tag{15.12}$$

We have $f * g \in l^1(h)$ and $\|f * g\|_1 \leq \|f\|_1 \|g\|_1$.
If we define $f^* = (\tilde{f})^-$, where $\tilde{f}(x) = f(\tilde{x})$, then $(l^1(h), *, +)$ is a Banach *-algebra with unit ϵ_e. The map $f \to fh$ is an isometric isomorphism from the Banach space $l^1(h)$ onto $M(K)$ and also a *- algebra homomorphism. Of course, if $(K, *, \tilde{})$ is commutative, then the Banach algebras $l^1(h)$ and $M(K)$ are commutative, too.

Henceforth we assume $(K, *, \tilde{})$ to be commutative. Let us now introduce the dual object

$$\hat{K} = \{\alpha \in l^\infty : \alpha \neq 0,\, \alpha(x * y) = \alpha(x)\alpha(y),\, \alpha(\tilde{x}) = \overline{\alpha(x)} \text{ for all } x, y \in K\}. \tag{15.13}$$

The elements of $\hat{K}$ are called characters. It is obvious that the constant function 1 is in $\hat{K}$. Furthermore it holds $|\alpha(x)| \leq \alpha(e) = 1$ for all $\alpha \in \hat{K}$ and $x \in K$. If $\hat{K}$ is equipped with the topology of pointwise convergence, then $\hat{K}$ becomes a compact Hausdorff space.

Denote by $\Delta(l^1(h))$ the structure space of the Banach algebra $l^1(h)$, i.e., the space of all non-zero continuous functionals φ on $l^1(h)$ with $\varphi(f * g) = \varphi(f)\varphi(g)$ and $\varphi(f^*) = \overline{\varphi(f)}$ for all $f, g \in l^1(h)$. Then the mapping $\hat{K} \to \Delta(l^1(h))$, $\alpha \to \varphi_\alpha$, with

$$\varphi_\alpha(f) = \sum_{z \in K} \overline{\alpha(z)} f(z) h(z) \quad \text{for all} \quad f \in l^1(h), \tag{15.14}$$

is a homeomorphism, where the topology on $\Delta(l^1(h))$ is the weak-$*$-topology. Note that $\Delta(l^1(h))$ is a subset of the dual space of the Banach space $l^1(h)$, which is isometric isomorphic to $l^\infty(h)$.

For $f \in l^1(h)$ the Fourier transform $\hat{f}$ is defined on $\hat{K}$ by

$$\hat{f}(\alpha) = \varphi_\alpha(f) \quad \text{for all} \quad \alpha \in \hat{K}. \tag{15.15}$$

The following statements can be easily checked:
$\hat{f}$ is continuous with $\|\hat{f}\|_\infty \leq \|f\|_1$, $\widehat{f + g} = \hat{f} + \hat{g}$, $\widehat{\lambda f} = \lambda \hat{f}$, $\widehat{f * g} = \hat{f}\hat{g}$ and $\widehat{f^*} = (\hat{f})^-$ for all $f, g \in l^1(h)$ and $\lambda \in \mathbb{C}$.
One should note that the Fourier transform $\hat{f}$ is exactly the Gelfand transform of the function $f \in l^1(h)$.
Furthermore, by applying the Theorem of Stone-Weierstraß the space $\{\hat{f} : f \text{ has finite support}\}$ is a dense self-adjoint subalgebra of $C(\hat{K})$.

Now we define an important subset of $\hat{K}$. Let

$$S = \{\alpha \in \hat{K} : |\hat{g}(\alpha)| \leq \|U_g\| \quad \text{for all} \quad g \in l^1(h)\}, \tag{15.16}$$

where $\|U_g\|$ denotes the norm of the operator $U_g : l^2(h) \to l^2(h)$, $f \to (gh) * f$.
S is a nonvoid compact subset of $\hat{K}$. To prove this one can apply the Gelfand theory of commutative C^*-algebras. In fact S is the structure space of the C^*-algebra generated by the operators U_g in the space of all continuous linear operators on the Hilbert space $l^2(h)$.
If $f \in l^1(h)$ with $\hat{f}(\alpha) = 0$ for all $\alpha \in S$, then $f = 0$.
Therefore, the Fourier transformation is a injective linear mapping from $l^1(h)$ into $C(S)$. Hence if K is infinite, then S is infinite, too.

Applying again the Gelfand theory of commutative C^*-algebras one can show the existence of a unique regular positive bounded Borel measure π on $\hat{K}$ with

$$\sum_{z\in K} f(z)\overline{g(z)}h(z) = \int_{\hat{K}} \hat{f}(\alpha)\overline{\hat{g}(\alpha)}\, d\pi(\alpha) \quad \text{for all} \quad f, g \in l^1(h). \tag{15.17}$$

Moreover $supp\, \pi = S$.
The measure π is called Plancherel measure and it holds $\pi(\hat{K}) = 1$.
Especially we get

$$\sum_{z\in K} |f(z)|^2 h(z) = \int_{\hat{K}} |\hat{f}(\alpha)|^2\, d\pi(\alpha) \tag{15.18}$$

for all $f \in l^1(h)$. Note that $l^1(h) \subseteq l^2(h)$.

As one might expect from Fourier analysis one can define for $\varphi \in L^1(\hat{K}, \pi) = L^1(\pi)$ the inverse Fourier transform by

$$\check{\varphi}(x) = \int_{\hat{K}} \varphi(\alpha)\alpha(x)\, d\pi(\alpha) \quad \text{for all} \quad x \in K. \tag{15.19}$$

The dual uniqueness theorem states that $\check{\varphi} = 0$ implies $\varphi = 0$.

15.3 Polynomial Hypergroups

Now having quoted some of the most important parts of the general theory of discrete hypergroups in the preceding section we will introduce polynomial hypergroups on $\mathbb{N}_0$ in detail. We start by collecting some basic facts about orthogonal polynomials in a form suitable for us.

Let $(a_n)_{n\in\mathbb{N}}$, $(c_n)_{n\in\mathbb{N}}$ be sequences of positive real non-zero numbers, $(b_n)_{n\in\mathbb{N}}$ a sequence of positive real numbers with property

$$a_n + b_n + c_n = 1 \quad \text{for all} \quad n \in \mathbb{N}, \tag{15.20}$$

and $a_0, b_0 \in \mathbb{R}$ with $a_0 > 0$, $a_0 + b_0 = 1$.
We define a polynomial sequence $(P_n)_{n\in\mathbb{N}_0}$ by

$$P_0(x) = 1, \; P_1(x) = (x - b_0)/a_0, \quad \text{and}$$

$$P_1(x)P_n(x) = a_n P_{n+1}(x) + b_n P_n(x) + c_n P_{n-1}(x) \tag{15.21}$$

for all $n \in \mathbb{N}$, $x \in \mathbb{R}$. The Theorem of Favard states that there is a probability measure π on $\mathbb{R}$ so that

$$\int_{\mathbb{R}} P_n(x)P_m(x)\,d\pi(x) = \delta_{n,m}\mu_m \quad \text{for all} \quad n, m \in \mathbb{N}_0, \tag{15.22}$$

where $\mu_m > 0$ and $\delta_{n,m}$ denotes Kronecker's symbol.
So the polynomial sequence $(P_n)_{n\in\mathbb{N}_0}$ is an orthogonal polynomial sequence. We point out that in view of our assumption (15.20) we have

$$P_n(1) = 1 \quad \text{for all} \quad n \in \mathbb{N}_0. \tag{15.23}$$

Note that the polynomials $(P_n)_{n\in\mathbb{N}_0}$ are not orthonormal. One can read the recurrence relation (15.21) as a linearization of the product $P_1 P_n$. The next lemma extends this linearization formula a little bit.

LEMMA 15.1
Let $(P_n)_{n\in\mathbb{N}_0}$ be a real orthogonal polynomial sequence defined by a recurrence relation of the type (15.21). Then we have

$$P_n(x)P_m(x) = \sum_{k=|n-m|}^{n+m} g(n,m,k)P_k(x) \quad \textit{for all} \quad n, m \in \mathbb{N}_0, \tag{15.24}$$

where $g(n,m,k) \in \mathbb{R}$ for $|n-m| \le k \le n+m$.
Moreover we have $g(n,m,|n-m|) \neq 0$ and $g(n,m,n+m) \neq 0$.

PROOF Since the orthogonal polynomials $\{P_n : n \in \mathbb{N}_0\}$ defined by (15.21) form a basis of the vector space of all polynomials we obtain the representation

$$P_n(x)P_m(x) = \sum_{k=0}^{n+m} g(n,m,k)P_k(x) \quad \text{for all} \quad n,m \in \mathbb{N}_0.$$

Obviously we have $g(n,m,k) \in \mathbb{R}$ for $0 \leq k \leq n+m$ and because $\deg(P_nP_m) = n+m$ we have $g(n,m,n+m) \neq 0$.
It remains to show that $g(n,m,k) = 0$ for all $0 \leq k < |n-m|$ and $g(n,m,|n-m|) \neq 0$. Let us assume $m < n$. For $k < n-m$ we have $\deg(P_mP_k) < n$. Hence

$$0 = \int_{\mathbb{R}} P_m(x)P_k(x)P_n(x)\,d\pi(x)$$

$$= \sum_{j=0}^{n+m} g(n,m,j) \int_{\mathbb{R}} P_j(x)P_k(x)\,d\pi(x) = g(n,m,k)\mu_k.$$

Since $\mu_k > 0$ we obtain $g(n,m,k) = 0$ for all $k = 0,1,\ldots,|n-m|-1$.
Assuming $g(n,m,n-m) = 0$ we get

$$0 = g(n,m,n-m)\mu_{n-m} = \int_{\mathbb{R}} P_n(x)P_m(x)P_{n-m}(x)\,d\pi(x)$$

$$= \sum_{k=|n-2m|}^{n} g(m,n-m,k) \int_{\mathbb{R}} P_k(x)P_n(x)\,d\pi(x) = g(m,n-m,n)\mu_n,$$

which contradicts $\mu_n > 0$ and $g(m,n-m,n) \neq 0$. ∎

Now we compose a list of relations for the linearization coefficients $g(n,m,k)$. Most of these identities are easy to check so we skip the proofs.

PROPOSITION 15.1
Let $g(n,m,k)$ be a linearization coefficient given by (15.24) and $|n-m| \leq k \leq n+m$. Then the following identities hold:

(i) $g(n,m,k) = g(m,n,k)$ *for all* $n,m \in \mathbb{N}_0$.

(ii) $g(0,n,n) = g(n,0,n) = 1$ *for all* $n \in \mathbb{N}_0$.

(iii) $g(1,n,n+1) = a_n$, $g(1,n,n) = b_n$ *and* $g(1,n,n-1) = c_n$ *for all* $n \in \mathbb{N}$.

(iv) $\sum_{k=|n-m|}^{n+m} g(n,m,k) = 1$ *for all* $n, m \in \mathbb{N}_0$.

(v) $g(n,n,0) = \mu_n$ *for all* $n \in \mathbb{N}_0$.

(vi) $g(n,m,k)\mu_k = g(n,k,m)\mu_m$ *for all* $n, m \in \mathbb{N}_0$.

(vii)

$$g(n,m,n+m) = \frac{a_m a_{m+1} \cdots a_{m+n-1}}{a_1 a_2 \cdots a_{n-1}} \quad \textit{and}$$

$$g(n,m,m-n) = \frac{c_m c_{m-1} \cdots c_{m-n+1}}{a_1 a_2 \cdots a_{n-1}} \quad \textit{for} \quad 2 \leq n \leq m.$$

Those orthogonal polynomial sequences $(P_n)_{n\in\mathbb{N}_0}$ which have non-negative linearization coefficients are most important for us.
We say that property (P) holds if the linearization coefficients satisfy

$$g(n,m,k) \geq 0 \quad \text{for all} \quad m, n \in \mathbb{N}_0, \ |n-m| \leq k \leq n+m. \tag{15.25}$$

Every orthogonal polynomial system with property (P) induces a hypergroup structure on $\mathbb{N}_0$. More precisely we have the following statement.

THEOREM 15.1

Let $(P_n)_{n\in\mathbb{N}_0}$ *be an orthogonal polynomial sequence defined by a recurrence relation of the type (15.21) and assume that property (P) holds.*
Define a map $* : \mathbb{N}_0 \times \mathbb{N}_0 \to M_c^1(\mathbb{N}_0)$ *by*

$$n * m = \sum_{k=|n-m|}^{n+m} g(n,m,k)\epsilon_k \tag{15.26}$$

and a map $\tilde{}: \mathbb{N}_0 \to \mathbb{N}_0$ *by* $\tilde{n} = n$.
Then $(\mathbb{N}_0, *, \tilde{})$ *is a discrete commutative hypergroup with unit element* 0, *which we shall call a polynomial hypergroup (induced by* $(P_n)_{n\in\mathbb{N}_0}$*).*

PROOF We have to check the conditions (DH1) - (DH4) of Definition 15.1. The only non-trivial part of the proof is the associativity law. Let

$k \in \mathbb{N}_0$. Since $\epsilon_n * \epsilon_m(k) = g(n,m,k)$ we obtain

$$\epsilon_l * (\epsilon_n * \epsilon_m)(k) = \sum_{j=0}^{\infty} g(n,m,j)\epsilon_j * \epsilon_l(k) = \sum_{j=|n-m|}^{n+m} g(n,m,j)g(j,l,k)$$

$$= \sum_{j=|n-m|}^{n+m} g(n,m,j)\frac{1}{\mu_k}\int_{\mathbb{R}} P_j(x)P_l(x)P_k(x)\,\pi(x)$$

$$= \frac{1}{\mu_k}\int_{\mathbb{R}} P_l(x)P_n(x)P_m(x)P_k(x)\,\pi(x)$$

$$= \sum_{j=|l-m|}^{l+m} g(l,n,j)g(j,m,k) = (\epsilon_l * \epsilon_n) * \epsilon_m(k)$$

for all $l, n, m \in \mathbb{N}_0$. ∎

There is an abundance of orthogonal polynomial sequences $(P_n)_{n\in\mathbb{N}_0}$ defined by (15.21), which enjoy the crucial non-negativity condition of Theorem 15.1. But generally it is not easy to check property (P) for a concrete sequence $(P_n)_{n\in\mathbb{N}_0}$. For instance the Askey conditions [20, Propostion 5] $c_2a_1 \geq c_1$, $a_nc_{n+1} \geq a_{n-1}c_n$ and $b_n \geq b_{n-1}$ for all $n \geq 2$ imply the non-negativity of the linearization coefficients.

One should note that due to the theorem of Favard every commutative hypergroup on $\mathbb{N}_0$ with the identity map as involution, 0 as the unit element and

$$\{n-1, n+1\} \subseteq supp\,\epsilon_1 * \epsilon_n \subseteq \{n-1, n, n+1\} \quad \text{for all} \quad n \in \mathbb{N}, \tag{15.27}$$

is a polynomial hypergroup induced by a certain orthogonal polynomial sequence $(P_n)_{n\in\mathbb{N}_0}$.

For a polynomial hypergroup the translation operator (15.5) reads as

$$L_m f(n) = \sum_{k=|m-n|}^{m+n} g(m,n,k)f(k) \quad \text{for all} \quad m, n \in \mathbb{N}_0. \tag{15.28}$$

Using (15.10) we obtain for the Haar measure

$$h(n) = (\epsilon_n * \epsilon_n(0))^{-1} = g(n,n,0)^{-1} = \mu_n^{-1} \quad \text{for all} \quad n \in \mathbb{N}_0. \tag{15.29}$$

For the convenience of the reader we note the relation of the three-term recurrence of $(P_n)_{n\in\mathbb{N}_0}$ and that of the orthonormal polynomials $p_n(x) = \sqrt{\mu_n}P_n(x)$, $n \in \mathbb{N}_0$. With $p_0(x) = 1$ they fulfill the recurrence relation

$$xp_n(x) = \lambda_{n+1}p_{n+1}(x) + \beta_n p_n(x) + \lambda_n p_{n-1}(x) \tag{15.30}$$

for all $n \in \mathbb{N}_0$, $x \in \mathbb{R}$, where

$$\lambda_0 = 0,\ \lambda_1 = a_0\sqrt{c_1},\ \lambda_n = a_0\sqrt{c_n a_{n-1}} \quad \text{for} \quad n \geq 2,$$

and $\beta_0 = b_0$, $\beta_n = a_0 b_n + b_0$ for $n \geq 1$.

Now we have to determine the dual object $\hat{K}$, when $K = \mathbb{N}_0$ is a polynomial hypergroup. For $x \in \mathbb{R}$ define

$$\alpha_x : \mathbb{N}_0 \to \mathbb{R},\ \alpha_x(n) = P_n(x), \tag{15.31}$$

and let

$$D_S = \{x \in \mathbb{R} : (P_n(x))_{n\in\mathbb{N}_0} \text{ is bounded}\}. \tag{15.32}$$

PROPOSITION 15.2
Let $(P_n)_{n\in\mathbb{N}_0}$ induce a polynomial hypergroup on $\mathbb{N}_0$.
Then we have:

(i) $\sup_{n\in\mathbb{N}_0} |P_n(x)| = \|\alpha_x\|_\infty = 1 = P_n(1)$ *for all* $x \in D_S$.
In particular $D_S = \{x \in \mathbb{R} : |P_n(x)| \leq 1$ *for all* $n \in \mathbb{N}_0\}$.

(ii) $\widehat{\mathbb{N}_0} = \{\alpha_x : x \in D_S\}$.

(iii) *The mapping* $D_S \to \widehat{\mathbb{N}_0}$, $x \to \alpha_x$, *is a homeomorphism.*
The inverse mapping is given by $\alpha \to a_0\alpha(1) + b_0$.
In particular D_S *is a compact set and* $D_S \subseteq [1 - 2a_0, 1]$.

PROOF (i): The characters $\alpha \in \widehat{\mathbb{N}_0}$ are exactly the bounded non-zero sequences $\alpha \in l^\infty$ enjoying $\alpha(n)\alpha(m) = \alpha(n * m)$ and $\alpha(n) = \overline{\alpha(n)}$. Hence for every $x \in D_S$ by formula (15.24) we have $\alpha_x \in \widehat{\mathbb{N}_0}$. Since each character $\alpha \in \widehat{\mathbb{N}_0}$ fulfills $\|\alpha\|_\infty = 1$ the assertion follows.
(ii): We have already shown that $\alpha_x \in \widehat{\mathbb{N}_0}$ for all $x \in D_S$. Conversely let $\alpha \in \widehat{\mathbb{N}_0}$. Put $x_0 = a_0\alpha(1) + b_0$. Then $\alpha(1) = P_1(x_0)$ and we get by induction $\alpha(n) = P_n(x_0) = \alpha_{x_0}(n)$ for all $n \in \mathbb{N}_0$. Since $x_0 \in D_S$ the assertion follows.

(iii): The map $x \to \alpha_x$ is bijective and the inverse is continuous. Since $\widehat{\mathbb{N}_0}$ is compact, $x \to \alpha_x$ is a homeomorphism. Considering the polynomial $P_1(x)$ the inclusion $D_S \subseteq [1-2a_0, 1]$ follows by means of (i). ∎

In the following we identify $\widehat{\mathbb{N}_0}$ with the corresponding compact subset D_S of the real line. Hence the Fourier transform $\hat{f}$ is an orthogonal expansion

$$\hat{f}(x) = \sum_{k=0}^{\infty} P_k(x) f(k) h(k) = \sum_{k=0}^{\infty} p_k(x) f(k) \sqrt{h(k)} \quad \text{for all} \quad x \in D_S, \tag{15.33}$$

where $f \in l^1(h)$.

To determine $S \subseteq \widehat{\mathbb{N}_0}$ we have only to observe that $S = supp\,\pi$, where π is characterized by identity (15.17). Choosing in (15.17) $f = \epsilon_n$ and $g = \epsilon_m$ and switching from $\widehat{\mathbb{N}_0}$ onto D_S we see that π is the orthogonalization measure of the orthogonal polynomial sequence $(P_n)_{n\in\mathbb{N}_0}$. We will not distinguish between the measure π defined on $\widehat{\mathbb{N}_0}$ and that defined on D_S. In particular we see that

$$supp\,\pi \subseteq D_S. \tag{15.34}$$

So the orthogonalization measure π of $(P_n)_{n\in\mathbb{N}_0}$ is supported by a compact set, and thus unique [9, Theorem 5.6]. We also note that

$$h(n) = \left(\int_{D_S} P_n^2(x)\, d\pi(x) \right)^{-1} \quad \text{for all} \quad n \in \mathbb{N}_0. \tag{15.35}$$

Finally we explicitly collect the main facts of polynomial hypergroups in a theorem.

THEOREM 15.2

Let $(a_n)_{n\in\mathbb{N}}$, $(b_n)_{n\in\mathbb{N}}$ and $(c_n)_{n\in\mathbb{N}}$ be real sequences with $a_n > 0$, $b_n \geq 0$, $c_n > 0$ and $a_n + b_n + c_n = 1$ for all $n \in \mathbb{N}$. Also let $a_0, b_0 \in \mathbb{R}$ with $a_0 > 0$ and $a_0 + b_0 = 1$.
Define a sequence of polynomials by
$P_0 = 1$, $P_1(x) = (x - b_0)/a_0$, $x \in \mathbb{R}$, and

$$P_1 P_n = a_n P_{n+1} + b_n P_n + c_n P_{n-1} \quad \textit{for all} \quad n \in \mathbb{N}.$$

If the coefficients in

$$P_n P_m = \sum_{k=|n-m|}^{n+m} g(n,m,k) P_k \quad \textit{for all} \quad n, m \in \mathbb{N}_0,$$

are non-negative, i.e., property (P) holds, then the convolution

$$n * m(k) = g(n,m,k),$$

for all $n, m \in \mathbb{N}_0$ *and* $|n-m| \leq k \leq n+m$, *defines a polynomial hypergroup with*

(i) $\widehat{\mathbb{N}_0}$ *is homeomorphic to the compact set*

$$D_S = \{x \in \mathbb{R} : |P_n(x)| \leq 1 \quad \text{for all} \quad n \in \mathbb{N}_0\}$$

and $D_S \subseteq [1 - 2a_0, 1]$.

(ii) *The Plancherel measure* π *has infinite compact support* $\operatorname{supp} \pi \subseteq D_S$ *and is the orthogonality measure of the sequence* $(P_n)_{n \in \mathbb{N}_0}$.

(iii) *The Haar measure* h *is given by* $h(0) = 1$, $h(1) = 1/c_1$ *and* $h(n) = \prod_{k=1}^{n-1} a_k / \prod_{k=1}^{n} c_k$ *for* $n \geq 2$.

We present now some important examples of polynomial hypergroups:

Example 15.1

Hypergroups with respect to Jacobi polynomials.

Let $\alpha > -1$ and $\beta > -1$.

Define

$$a_0 = \frac{2(\alpha+1)}{\alpha+\beta+2}, \quad b_0 = \frac{\beta-\alpha}{\alpha+\beta+2},$$

$$a_n = \frac{(\alpha+\beta+2)(n+\alpha+\beta+1)(n+\alpha+1)}{(\alpha+1)(2n+\alpha+\beta+1)(2n+\alpha+\beta+2)},$$

$$b_n = \frac{\alpha-\beta}{2(\alpha+1)}\left(1 - \frac{(\alpha+\beta)(\alpha+\beta+2)}{(2n+\alpha+\beta)(2n+\alpha+\beta+2)}\right) \quad \text{and}$$

$$c_n = \frac{(\alpha+\beta+2)n(n+\beta)}{(\alpha+1)(2n+\alpha+\beta)(2n+\alpha+\beta+1)} \quad \text{for all} \quad n \in \mathbb{N}.$$

By recurrence relation (15.21) we get an orthogonal polynomial sequence $(J_n^{(\alpha,\beta)})_{n \in \mathbb{N}_0}$ of Jacobi polynomials.

Property (P) holds if and only if $(\alpha,\beta) \in V$, where $V = \{(\alpha,\beta) \in \mathbb{R}^2 : \alpha \geq \beta > -1 \text{ and } a(a+5)(a+3)^2 \geq (a^2-7a-24)b^2,\ a = \alpha+\beta+1,\ b = \alpha-\beta\}$, see [14, Theorem 1]. In particular $(\alpha,\beta) \in V$ if $\alpha \geq \beta > -1$ and $\alpha+\beta+1 \geq 0$.

The Haar measure is computed by

$$h(n) = \frac{(2n+\alpha+\beta+1)\Gamma(\beta+1)\Gamma(n+\alpha+1)\Gamma(n+\alpha+\beta+1)}{\Gamma(\alpha+1)\Gamma(\alpha+\beta+2)\Gamma(n+1)\Gamma(n+\beta+1)}$$

for all $n \in \mathbb{N}$.

It holds $D_S = [-1,1]$ and the Plancherel measure is given by $d\pi(x) = 2^{-(\alpha+\beta+1)} \frac{\Gamma(\alpha+\beta+2)}{\Gamma(\alpha+1)\Gamma(\beta+1)}(1-x)^{\alpha}(1+x)^{\beta}\,dx$ with $supp\,\pi = D_S$ [38].

□

Example 15.2

Hypergroups with respect to ultraspherical polynomials.

The ultraspherical polynomials are a special subset of the Jacobi polynomials, i.e., $\alpha = \beta$.

As stated in [20] the convolution $m * n$, $m, n \in \mathbb{N}_0$, $m \leq n$ is given in case $\alpha = \beta > -1/2$ by

$$m * n(k) = \frac{(k+\alpha+1/2)\Gamma(2\alpha+1)\Gamma(m+1)\Gamma(n+1)}{((m+n+k+1)/2+\alpha)\Gamma(\alpha+1/2)^2\Gamma(m+2\alpha+1)}$$

$$\times \frac{\Gamma((m+n-k+1)/2+\alpha)\Gamma((m-n+k+1)/2+\alpha)}{\Gamma(n+2\alpha+1)\Gamma((m+n-k)/2+1)\Gamma((m-n+k)/2+1)}$$

$$\times \frac{\Gamma((n-m+k+1)/2+\alpha)\Gamma((m+n+k)/2+2\alpha+1)}{\Gamma((n-m+k)/2+1)\Gamma((n+m+k+1)/2+\alpha)}$$

for all $k \in \{n-m, n-m+2, \ldots, n+m\}$, and

$$m * n(k) = 0 \quad \text{for all} \quad k \in \{n-m+1, n-m+3, \ldots, n+m-1\}.$$

For the **hypergroup with respect to the Chebychev polynomials of the first kind**, i.e., $\alpha = \beta = -1/2$, it holds that

$$m * n(k) = \frac{1}{2} \quad \text{for all} \quad m, n > 0 \quad \text{and} \quad k = |m-n| \text{ or } k = m+n.$$

□

Example 15.3
Hypergroups with respect to generalized Chebychev polynomials. Let $\alpha > -1$ and $\beta > -1$.

Define

$$a_0 = 1, \quad b_0 = 0,$$

$$a_n = \begin{cases} \dfrac{q+\alpha+\beta+1}{2q+\alpha+\beta+1}, & \text{if } n = 2q, \\ \dfrac{q+\alpha+1}{2q+\alpha+\beta+2}, & \text{if } n = 2q+1, \end{cases}$$

$$c_n = \begin{cases} \dfrac{q}{2q+\alpha+\beta+1}, & \text{if } n = 2q, \\ \dfrac{q+\beta+1}{2q+\alpha+\beta+2}, & \text{if } n = 2q+1, \end{cases}$$

and $b_n = 0$ for all $n \in \mathbb{N}$.
By recurrence relation (15.21) we get an orthogonal polynomial sequence $(C_n^{(\alpha,\beta})_{n\in\mathbb{N}_0}$ of generalized Chebychev polynomials.
Using the relationship to Jacobi polynomials

$$C_n^{(\alpha,\beta)}(x) = \begin{cases} J_q^{(\alpha,\beta)}(2x^2-1), & \text{if } n = 2q, \\ x J_q^{(\alpha,\beta+1)}(2x^2-1), & \text{if } n = 2q+1, \end{cases}$$

for all $n \in \mathbb{N}_0$, we are able to calculate the linearization coefficients of the generalized Chebychev polynomials in dependence of the linearization coefficients of the Jacobi polynomials.
If $\alpha \geq \beta + 1 > 0$ then property (P) holds, see [20].
For $n \in \mathbb{N}$ the Haar measure is computed by

$$h(n) = \frac{(2q+\alpha+\beta+1)\Gamma(\beta+1)\Gamma(q+\alpha+1)\Gamma(q+\alpha+\beta+1)}{\Gamma(\alpha+1)\Gamma(\alpha+\beta+2)\Gamma(q+1)\Gamma(q+\beta+1)},$$

if $n = 2q$, and

$$h(n) = \frac{(2q+\alpha+\beta+2)\Gamma(\beta+1)\Gamma(q+\alpha+1)\Gamma(q+\alpha+\beta+2)}{\Gamma(\alpha+1)\Gamma(\alpha+\beta+2)\Gamma(q+1)\Gamma(q+\beta+2)},$$

if $n = 2q+1$.
We have $D_S = [-1,1]$ and the Plancherel measure is given by $d\pi(x) = \frac{\Gamma(\alpha+\beta+2)}{\Gamma(\alpha+1)\Gamma(\beta+1)}(1-x^2)^{\alpha}|x|^{2\beta+1}\,dx$ with $supp\,\pi = D_S$ [20]. □

Example 15.4
Hypergroups with respect to polynomials connected by homogeneous trees.
Let $\alpha \geq 2$.
Define $a_0 = 1$, $b_0 = 0$,

$$a_n = \frac{(\alpha-1)}{\alpha}, \quad b_n = 0 \quad \text{and} \quad c_n = \frac{1}{\alpha} \quad \text{for all} \quad n \in \mathbb{N}.$$

By recurrence relation (15.21) we get an orthogonal polynomial sequence $(H_n^\alpha)_{n\in\mathbb{N}_0}$ of polynomials connected by homogeneous trees. In case $\alpha = 2$ we get the Chebychev polynomials of the first kind.
It is shown in [20] that property (P) holds for all $\alpha \geq 2$.
The convolution is defined by

$$m * n(k) = \begin{cases} \dfrac{1}{\alpha(\alpha-1)^{m-1}}, & \text{if } k = n - m, \\ \dfrac{\alpha-2}{\alpha(\alpha-1)^{(n+m-k)/2}}, & \text{if } \operatorname{mod}(n+m-k, 2) = 0, \\ \dfrac{\alpha-1}{\alpha}, & \text{if } k = n + m, \\ 0, & \text{otherwise}, \end{cases}$$

for all $m, n \in \mathbb{N}$ with $m \leq n$ and $n - m \leq k \leq n + m$.
The Haar measure is computed by $h(n) = \alpha(\alpha-1)^{n-1}$ for all $n \in \mathbb{N}$.
Once again $D_S = [-1, 1]$ and the Plancherel measure is given by $d\pi(x) = \dfrac{\alpha\sqrt{4(\alpha-1)/\alpha^2 - x^2}}{2\pi(1-x^2)}\,dx$ with $supp\,\pi = [-2\sqrt{\alpha-1}/\alpha, 2\sqrt{\alpha-1}/\alpha]$ [12]. Hence for $\alpha > 2$ we have $S \neq D_S$. $\square$

15.4 Dual Convolution

Next we have to investigate a dual convolution structure on $D_S \cong \widehat{\mathbb{N}_0}$. Let $M(D_S)$ denote the Banach space of all complex Borel measures on D_S with norm $\|\mu\| = |\mu|(D_S)$, where $|\mu|$ is the total variation of $\mu \in M(D_S)$, see [37]. Recall that $M(D_S)$ can be identified with $C(D_S)^*$, the dual Banach space of $C(D_S)$.

DEFINITION 15.2 *We say that for $(P_n)_{n\in\mathbb{N}_0}$ a uniformly bounded* *dual product formula holds if for all $x, y \in D_S$ there exists a complex Borel measure $\mu_{x,y} \in M(D_S)$ with*

$$\|\mu_{x,y}\| \leq M \quad \textit{for all} \quad x, y \in D_S, \quad \textit{and} \tag{15.36}$$

$$P_n(x)P_n(y) = \int_{D_S} P_n(z)\, d\mu_{x,y}(z) \quad \textit{for all} \quad n \in \mathbb{N}_0, \tag{15.37}$$

where $M > 0$ is independent from $x, y \in D_S$.

Note that (15.37) implies $\mu_{x,y}(D_S) = 1$ for all $x, y \in D_S$.

LEMMA 15.2
Assume that for $(P_n)_{n\in\mathbb{N}_0}$ a uniformly bounded dual product formula holds. The mapping $D_S \times D_S \to M(D_S)$, $(x, y) \to \mu_{x,y}$, is continuous, where $M(D_S)$ is equipped with the weak-$*$-topology.

PROOF Given some $\varphi \in C(D_S)$ and $\epsilon > 0$ Weierstraß's theorem yields a function $f \in l^1(h)$ with finite support, such that $|\varphi(x) - \hat{f}(x)| < \epsilon$ for every $x \in D_S$. Now for $x_0, y_0, x, y \in D_S$ we obtain

$$|\int_{D_S} \varphi(z)\, d\mu_{x,y}(z) - \int_{D_S} \varphi(z)\, d\mu_{x_0,y_0}(z)| \leq$$

$$2M\epsilon + |\int_{D_S} \hat{f}(z)\, d\mu_{x,y}(z) - \int_{D_S} \hat{f}(z)\, d\mu_{x_0,y_0}(z)| \leq$$

$$2M\epsilon + \sum_{k=0}^{\infty} |f(k)||P_k(x)P_k(y) - P_k(x_0)P_k(y_0)||h(k)|,$$

where the last series is only a finite sum. Choosing $x, y \in D_S$ from appropriate neighbourhoods of $x_0, y_0 \in D_S$, respectively, such that $|P_k(x)P_k(y) - P_k(x_0)P_k(y_0)|$ is small enough for $k \in \textit{supp}\, f$, the assertion follows. ∎

Given $\varphi \in C(D_S)$ define the translation $T_y\varphi$ of φ by $y \in D_S$ as

$$T_y\varphi(x) = \int_{D_S} \varphi(z)\, d\mu_{x,y}(z) \quad \text{for all} \quad x \in D_S. \tag{15.38}$$

By Lemma 15.2 we have $T_y\varphi \in C(D_S)$ and $\|T_y\varphi\|_\infty \leq M\|\varphi\|_\infty$.

It is easy to check that

$$T_y\hat{f}(x) = \sum_{k=0}^{\infty} f(k)P_k(y)P_k(x)h(k) \quad \text{for all} \quad f \in l^1(h). \tag{15.39}$$

Hence in turn it follows that

$$\int_{D_S} T_y\hat{f}(x)\,d\pi(x) = f(0) = \int_{D_S} \hat{f}(x)\,d\pi(x) \quad \text{for all} \quad f \in l^1(h), \tag{15.40}$$

and

$$\int_{D_S} T_y\varphi(x)\,d\pi(x) = \check{\varphi}(0) = \int_{D_S} \varphi(x)\,d\pi(x) \quad \text{for all} \quad \varphi \in C(D_S), \tag{15.41}$$

i.e., π is a translation invariant Borel measure on D_S.
Moreover from (15.39) we get a more general formula

$$\int_{D_S} T_y\hat{f}(x)\hat{g}(x)\,d\pi(x) = \sum_{k=0}^{\infty} f(k)g(k)P_k(y)h(k)$$

$$= \int_{D_S} \hat{f}(x)T_y\hat{g}(x)\,d\pi(x) \tag{15.42}$$

for all $f, g \in l^1(h)$, and therefore

$$\int_{D_S} T_y\varphi(x)\psi(x)\,d\pi(x) = \sum_{k=0}^{\infty} \check{\varphi}(k)\check{\psi}(k)P_k(y)h(k)$$

$$= \int_{D_S} \varphi(x)T_y\psi(x)\,d\pi(x) \tag{15.43}$$

for all $\varphi, \psi \in C(D_S)$.
Now we introduce a dual convolution on $C(D_S)$ by

$$\varphi * \psi(y) = \int_{D_S} \varphi(x)T_y\psi(x)\,d\pi(x) = \int_{D_S} T_y\varphi(x)\psi(x)\,d\pi(x) \tag{15.44}$$

for all $y \in D_S$.
This convolution could be extended to $L^1(D_S, \pi)$.

With this multiplication $L^1(D_S, \pi)$ becomes a Banach algebra. But we shall not continue this point of view.
The following properties we shall use later on.

PROPOSITION 15.3

Assume that for $(P_n)_{n\in\mathbb{N}_0}$ a uniformly bounded product formula holds. Then the following two statements are true.

(i) *The map $D_S \to C(D_S)$, $x \to T_x\varphi$, is continuous for all $\varphi \in C(D_S)$.*

(ii) *$\varphi * \psi \in C(D_S)$ and $\|\varphi * \psi\|_\infty \le M\|\varphi\|_1\|\psi\|_\infty$ for all $\varphi, \psi \in C(D_S)$.*

PROOF (i): Let $x_0 \in D_S$ and $\epsilon > 0$. For any $y \in D_S$ there exist a neighbourhood $U_y(x_0)$ of x_0 and a neighbourhood $V(y)$ of $y \in D_S$ such that $|T_x\varphi(z) - T_{x_0}\varphi(y)| < \epsilon$ for all $x \in U_y(x_0)$ and $z \in V(y)$. Since D_S is compact finitely many $V(y_1), V(y_2), \ldots, V(y_n)$ cover D_S. Let $U(x_0) = \bigcap_{i=1}^n U_{y_i}(x_0)$. For any $z \in D_S$ we find some y_i such that $z \in V(y_i)$. Then for $x \in U(x_0)$ it holds $|T_x\varphi(z) - T_{x_0}\varphi(z)| < 2\epsilon$.
(ii): We have $|\varphi * \psi(x) - \varphi * \psi(x_0)| \le \|T_x\psi - T_{x_0}\psi\|_\infty\|\varphi\|_1$ and therefore (i) implies $\varphi * \psi \in C(D_S)$. Likewise we obtain for every $x \in D_S$ the inequality $|\varphi * \psi(x)| \le \|T_x\psi\|_\infty\|\varphi\|_1 \le M\|\varphi\|_1\|\psi\|_\infty$. ∎

We finish this section by stating some examples for which a uniformly bounded product formula holds:

Example 15.5

Jacobi polynomials.
If $\alpha \ge \beta > -1$ and $\alpha + \beta \ge -1$, then for the corresponding sequence $(J_n^{(\alpha,\beta)})_{n\in\mathbb{N}_0}$ of Jacobi polynomials, see Example 15.1, a uniformly bounded product formula holds [15, Theorem]. □

Example 15.6

Generalized Chebychev polynomials.
If ($\alpha \ge \beta > -1$ and $\alpha+\beta > -1$) or $\alpha = \beta = -\frac{1}{2}$, then for the corresponding sequence $(C_n^{(\alpha,\beta)})_{n\in\mathbb{N}_0}$ of generalized Chebychev polynomials, see Example 15.3, a uniformly bounded product formula holds [19, Theorem 1], [39, Theorem 2]. □

15.5 General Convergence Theorems

Henceforth, B denotes one of the Banach spaces $C(D_S)$ or $L^p(D_S, \pi) = L^p(\pi)$, $1 \leq p < \infty$ and $\|.\|_B$ denotes the corresponding norm. Note that $B \subseteq L^1(\pi)$.
Let

$$(a_{n,k})_{n \in \mathbb{N}_0, 0 \leq k \leq n}$$

be a triangular scheme of complex numbers. Our goal is to investigate the convergence of weighted Fourier expansions

$$A_n \varphi = \sum_{k=0}^{n} a_{n,k} \check{\varphi}(k) P_k h(k), \quad \text{for all} \quad n \in \mathbb{N}_0, \tag{15.45}$$

with respect to the norm $\|.\|_B$.

THEOREM 15.3

We have $\lim_{n\to\infty} \|A_n\varphi - \varphi\|_B = 0$ *for all* $\varphi \in B$ *if and only if the following two conditions hold:*

(i) $\lim_{n\to\infty} a_{n,k} = 1$ *for all* $k \in \mathbb{N}_0$.

(ii) There exists a constant $C > 0$ *with* $\|A_n\varphi\|_B \leq C\|\varphi\|_B$ *for all* $\varphi \in B$ *and* $n \in \mathbb{N}_0$.

PROOF The map $A_n : B \to B, \varphi \to A_n\varphi$, is a continuous linear operator and the linear span $W =< \{P_n : n \in \mathbb{N}_0\} >$ is a dense subset of B.
It holds that $A_n P_k = a_{n,k} P_k$, if $k \leq n$, respectively, $A_n P_k = 0$, if $k > n$.
By (i) we get $\lim_{n\to\infty} A_n \mu = \mu$ for all $\mu \in W$.
Hence for $\epsilon > 0$ there exist $n_0 \in \mathbb{N}$ and $\mu_\epsilon \in W$ such that

$$\|A_n\varphi - \varphi\|_B \leq C\|\varphi - \mu_\epsilon\|_B + \|A_n\mu_\epsilon - \mu_\epsilon\|_B + \|\mu_\epsilon - \varphi\|_B < \epsilon.$$

for all $n \geq n_0$, where we have used (ii).
Conversely, assume $\lim_{n\to\infty} \|A_n\varphi - \varphi\|_B = 0$ for all $\varphi \in B$.
Again with $A_n P_k = a_{n,k} P_k$ for $k \leq n$ it follows (i). Finally, the Banach-Steinhaus theorem [7, Proposition 0.7.3] yields (ii). ∎

Additionally, let us now introduce the kernel function

$$K_n(x,y) = \sum_{k=0}^{n} a_{n,k} P_k(x) P_k(y) h(k) \quad \text{for all} \quad x, y \in D_S,\, n \in \mathbb{N}_0, \qquad (15.46)$$

and let us abbreviate

$$\|K_n(x,.)\|_1 = \int_{D_S} |K_n(x,y)| \, d\pi(y). \qquad (15.47)$$

Moreover, denote by $\|A_n\|^B$ the operator norm of A_n.

PROPOSITION 15.4
The operator norm fulfills the inequality

(i) $\|A_n\|^B \le \sup_{x \in D_S} \|K_n(x,.)\|_1$ *for all* $n \in \mathbb{N}_0$.

Moreover, in the special case $B = L^1(\pi)$ *and* $B = C(D_S)$ *we get the equality*

(ii) $\|A_n\|^{L^1(\pi)} = \|A_n\|^{C(D_S)} = \sup_{x \in D_S} \|K_n(x,.)\|_1$ *for all* $n \in \mathbb{N}_0$.

PROOF It holds that $A_n\varphi(x) = \int_{D_S} K_n(x,y)\varphi(y)\, d\pi(y)$.
(i): If $\varphi \in C(D_S)$ we get $\|A_n \varphi\|_\infty \le \sup_{x \in D_S} \|K_n(x,.)\|_1 \|\varphi\|_\infty$.
Now let $B = L^p(\pi), 1 \le p < \infty$. For a pair of conjugate exponents $1/q + 1/p = 1$ Hölder's inequality implies

$$\left(\int_{D_S} |\varphi(y)||K_n(x,y)| \, d\pi(y) \right)^p \le \|K_n(x,.)\|_1^{p/q} \int_{D_S} |\varphi(y)|^p |K_n(x,y)| d\pi(y).$$

Hence

$$\begin{aligned} \|A_n\varphi\|_p^p &\le \big(\sup_{x \in D_S} \|K_n(x,.)\|_1)^{p/q} \int_{D_S} |\varphi(y)|^p \int_{D_S} |K_n(x,y)| \, d\pi(x) \, d\pi(y) \\ &\le \big(\sup_{x \in D_S} \|K_n(x,.)\|_1)^p \|\varphi\|_p^p. \end{aligned}$$

Note that Fubini's theorem was used.

(ii): Using a well-known equality for the L^1-norm, we obtain

$$\begin{aligned}\|K_n(x,.)\|_1 &= \sup_{\|\varphi\|_\infty=1} \left|\int_{D_S} \varphi(y)K_n(x,y)\,d\pi(y)\right| \\ &= \sup_{\|\varphi\|_\infty=1} |A_n\varphi(x)| \\ &\leq \sup_{\|\varphi\|_\infty=1} \|A_n\varphi\|_\infty \\ &= \|A_n\|^{C(D_S)} \quad \text{for all} \quad x \in D_S.\end{aligned}$$

Applying $\int_{D_S} A_n\varphi(y)\mu(y)\,d\pi(y) = \int_{D_S} \varphi(y)A_n\mu(y)\,d\pi(y)$ we get

$$\begin{aligned}\|A_n\|^{L^1(\pi)} &= \sup_{\|\mu\|_1=1} \|A_n\mu\|_1 \\ &= \sup_{\|\mu\|_1=1} \sup_{\|\varphi\|_\infty=1} \left|\int_{D_S} \varphi(y)A_n\mu(y)\,d\pi(y)\right| \\ &= \sup_{\|\varphi\|_\infty=1} \sup_{\|\mu\|_1=1} \left|\int_{D_S} A_n\varphi(y)\mu(y)\,d\pi(y)\right| \\ &= \sup_{\|\varphi\|_\infty=1} \|A_n\varphi\|_\infty = \|A_n\|^{C(D_S)}.\end{aligned}$$

The equality follows by (i). ∎

The proof of the following theorem is now easy and is omitted.

THEOREM 15.4
If $\lim_{n\to\infty} a_{n,k} = 1$ *for all* $k \in \mathbb{N}_0$ *and* $\sup_{n\in\mathbb{N}_0} \sup_{x\in D_S} \|K_n(x,.)\|_1 < \infty$, *then* $\lim_{n\to\infty} \|A_n\varphi - \varphi\|_B = 0$ *for all* $\varphi \in B$.
In case $B = L^1(\pi)$ *or* $B = C(D_S)$ *the opposite direction is also true.*

Now we will consider those $(P_n)_{n\in\mathbb{N}_0}$ for which a uniformly bounded product formula holds. We get the following useful statement.

THEOREM 15.5
Let

$$A_n(x) = \sum_{k=0}^{n} a_{n,k} P_k(x) h(k) \quad \text{for all} \quad x \in D_S, n \in \mathbb{N}_0, \tag{15.48}$$

and assume that for $(P_n)_{n \in \mathbb{N}_0}$ a uniformly bounded product formula holds. Then with the convolution on $C(D_S)$ defined by (15.44) we have

$$A_n \varphi = A_n * \varphi \quad \text{for all} \quad \varphi \in C(D_S).$$

If $\lim_{n \to \infty} a_{n,k} = 1$ for all $k \in \mathbb{N}_0$, and there exists $R > 0$ with $\|A_n\|_1 \leq R$ for all $n \in \mathbb{N}_0$, then

$$\lim_{n \to \infty} \|A_n \varphi - \varphi\|_B = 0 \quad \text{for all} \quad \varphi \in B.$$

PROOF Using (15.45), (15.38) and (15.44) we get

$$\begin{aligned} A_n \varphi(x) &= \sum_{k=0}^{n} a_{n,k} \check{\varphi}(k) P_k(x) h(k) \\ &= \int_{D_S} \sum_{k=0}^{n} a_{n,k} \varphi(y) P_k(y) P_k(x) h(k) \, d\pi(y) \\ &= \int_{D_S} \int_{D_S} A_n(z) \varphi(y) \, d\mu_{x,y}(z) \, d\pi(y) \\ &= \int_{D_S} \varphi(y) T_y A_n(x) \, d\pi(y) = A_n * \varphi(x) \end{aligned}$$

for all $x \in D_S$ and $\varphi \in C(D_S)$.
Therefore, by Proposition 15.3 (ii) we obtain

$$\|A_n \varphi\|_\infty = \|A_n * \varphi\|_\infty \leq M \|A_n\|_1 \|\varphi\|_\infty \leq MR \, \|\varphi\|_\infty$$

for all $n \in \mathbb{N}_0$. Hence Theorem 15.3 implies

$$\lim_{n \to \infty} \|A_n \varphi - \varphi\|_\infty = 0 \quad \text{for all} \quad \varphi \in C(D_S).$$

Now by Theorem 15.4 we get $\sup_{n\in\mathbb{N}_0}\sup_{x\in D_S}\|K_n(x,.)\|_1<\infty$ and

$$\lim_{n\to\infty}\|A_n\varphi-\varphi\|_B=0 \quad \text{for all} \quad \varphi\in B,$$

and the proof is complete. ∎

In the proof of Theorem 15.5 we have used the uniformly bounded dual product formula to obtain a boundedness condition for the double sums $K_n(x,y)$. It is a remarkable fact that the validity of this product formula is equivalent to a boundedness condition for triple sums.
Given a triangular scheme of complex numbers $(a_{n,k})_{n\in\mathbb{N}_0,0\le k\le n}$ set

$$K_n(x,y,z)=\sum_{k=0}^{n}a_{n,k}P_k(x)P_k(y)P_k(z)h(k) \quad \text{for all} \quad x,y,z\in D_S, n\in\mathbb{N}_0, \tag{15.49}$$

and again for brevity

$$\|K_n(x,y,.)\|_1=\int_{D_S}|K_n(x,y,z)|\,d\pi(z). \tag{15.50}$$

THEOREM 15.6
If $\lim_{n\to\infty}a_{n,k}=1$ *for all* $k\in\mathbb{N}_0$ *and*

$$\sup_{n\in\mathbb{N}_0}\sup_{x,y\in D_S}\|K_n(x,y,.)\|_1=M<\infty,$$

then for all $x,y\in D_S$ *there exists a complex Borel measure* $\mu_{x,y}$ *such that* $\|\mu_{x,y}\|\le M$ *and*

$$P_n(x)P_n(y)=\int_{D_S}P_n(z)\,d\mu_{x,y}(z) \quad \textit{for all} \quad n\in\mathbb{N}_0,$$

i.e., for $(P_n)_{n\in\mathbb{N}_0}$ *a uniformly bounded product formula is valid.*

PROOF Let $\mathcal{P}(D_S)$ be the linear space of all polynomials on D_S. Set $u_k=P_k(x)P_k(y)$ for fixed $x,y\in D_S$ and define a linear functional G on $\mathcal{P}(D_S)$ by

$$G(Q)=\sum_{k=0}^{m}v_ku_kh(k),$$

when $Q(z) = \sum_{k=0}^{m} v_k P_k(z)h(k)$.
Since $\check{Q}(k) = v_k$ for $k = 0, 1, \ldots, m$, and $\check{Q}(k) = 0$ for $k \geq m+1$, we obtain

$$
\begin{aligned}
|G(Q)| &= |\sum_{k=0}^{\infty} \check{Q}(k)u_k h(k)| \\
&= \lim_{n\to\infty} |\sum_{k=0}^{n} a_{n,k}\check{Q}(k)u_k h(k)| \\
&= \lim_{n\to\infty} |\int_{D_S} Q(z) \sum_{k=0}^{n} a_{n,k}P_k(z)P_k(x)P_k(y)h(k)\, d\pi(z)| \\
&\leq M\|Q\|_\infty.
\end{aligned}
$$

Thus G is continuous with respect to $\|.\|_\infty$.
The space $\mathcal{P}(D_S)$ is dense in $C(D_S)$.
Hence by $G(\varphi) = \lim_{m\to\infty} G(Q_m)$ for $\lim_{m\to\infty} Q_m = \varphi$, a continuous linear functional is defined on $C(D_S)$, which fulfills $\|G\| \leq M$ and $G(P_k) = u_k = P_k(x)P_k(y)$.
Riesz's representation theorem, see [37], yields a complex Borel measure $\mu_{x,y}$ on D_S such that

$$G(\varphi) = \int_{D_S} \varphi(z)\, d\mu_{x,y}(z) \quad \text{for all} \quad \varphi \in C(D_S),$$

and $\|\mu_{x,y}\| \leq M$. ∎

Finally we mention that a converse result is valid, too.

PROPOSITION 15.5
Let $(a_{n,k})_{n\in\mathbb{N}_0, 0\leq k\leq n}$ be a triangular scheme such that

$$\sup_{n\in\mathbb{N}_0} \sup_{u\in D_S} \|K_n(u,.)\|_1 = M < \infty.$$

If for $(P_n)_{n\in\mathbb{N}_0}$ a uniformly bounded dual product formula holds, then

$$\sup_{n\in\mathbb{N}_0} \sup_{x,y\in D_S} \|K_n(x,y,.)\|_1 < \infty.$$

PROOF Given $x, y \in D_S$ and $n \in \mathbb{N}_0$ we get applying Fubini's theorem

$$\begin{aligned}
\|K_n(x,y,.)\|_1 &= \int_{D_S} |\sum_{k=0}^{n} a_{n,k} P_k(x) P_k(y) P_k(z) h(k)|\, d\pi(z) \\
&= \int_{D_S} |\sum_{k=0}^{n} a_{n,k} P_k(z) \int_{D_S} P_k(u)\, d\mu_{x,y}(u) h(k)|\, d\pi(z) \\
&\leq \int_{D_S} \int_{D_S} |\sum_{k=0}^{n} a_{n,k} P_k(u) P_k(z) h(k)|\, d\pi(z)\, \mu_{x,y}(u) \\
&= \int_{D_S} \|K_n(u,.)\|_1\, d\mu_{x,y}(u) \leq M,
\end{aligned}$$

which proves the stated boundedness. ∎

15.6 Dirichlet Kernel

In the case of classical Fourier series, see [21], the Dirichlet kernel $(D_n)_{n\in\mathbb{N}_0}$ is defined by

$$D_n(t) = \hat{\xi}_n(t) = \sum_{k=-n}^{n} e^{\imath kt} \quad \text{for all} \quad t \in [0, 2\pi[,\ n \in \mathbb{N}_0, \tag{15.51}$$

where ξ_n denotes the characteristic function of the set $\{-n, -n+1, \ldots, n\}$ and $\hat{}$ the ordinary Fourier transform.

Denote by χ_n the characteristic function of the set $\{0, 1, \ldots, n\}$ for all $n \in \mathbb{N}_0$. Likewise, in case of orthogonal polynomial systems we make the following definition.

DEFINITION 15.3 *Let $(\mathbb{N}_0, *)$ be a polynomial hypergroup induced by a set of polynomials $\{P_n : n \in \mathbb{N}_0\}$ according to Theorem 15.2.*
We define the Dirichlet kernel (D_n) by

$$D_n(x) = \hat{\chi}_n(x) = \sum_{k=0}^{n} P_k(x) h(k) \quad \textit{for all} \quad n \in \mathbb{N}_0,\ x \in D_S. \tag{15.52}$$

With respect to the preceding section the Dirichlet kernel corresponds to the case, where $a_{n,k} = 1$ for all $n \in \mathbb{N}_0, 0 \leq k \leq n$. In particular the convergence and divergence have been thoroughly studied for Jacobi expansions.

Example 15.7
Jacobi polynomials.
For clearness we sign those quantities which correspond with parameters α and β, see Example 15.1, by (α, β).
It is easy to derive that

$$D_n^{(\alpha,\beta)} = \chi_n * \chi_n^{(\alpha,\beta)}(0) J_n^{(\alpha+1,\beta)}$$

$$= \frac{\Gamma(n+\alpha+\beta+2)\Gamma(\beta+1)\Gamma(n+\alpha+2)}{\Gamma(\alpha+\beta+2)\Gamma(n+\beta+1)\Gamma(\alpha+2)\Gamma(n+1)} J_n^{(\alpha+1,\beta)}$$

for all $n \in \mathbb{N}_0$.
We define $K_n^{(\alpha,\beta)}(x,y)$ by (15.46).
Rau [36] and Lorch [27] have shown that

$$\|K_n^{(\alpha,\beta)}(1,.)\|_1 \sim \ln(n), \quad \text{if} \quad \alpha = -\frac{1}{2}, \quad \text{and}$$

$$\|K_n^{(\alpha,\beta)}(1,.)\|_1 \sim n^{(\alpha+\frac{1}{2})}, \quad \text{if} \quad \alpha > -\frac{1}{2},$$

where $\beta > -1$ is arbitrary.
Interchanging α and β and applying [38, (4.1.3),(4.3.4)] we obtain

$$\|K_n^{(\alpha,\beta)}(-1,.)\|_1 \sim \ln(n), \quad \text{if} \quad \beta = -\frac{1}{2}, \quad \text{and}$$

$$\|K_n^{(\alpha,\beta)}(-1,.)\|_1 \sim n^{(\beta+\frac{1}{2})}, \quad \text{if} \quad \beta > -\frac{1}{2},$$

where $\alpha > -1$ is arbitrary.
Therefore, Theorem 15.4 implies the existence of $\varphi \in B$ with $B = L^1(\pi^{(\alpha,\beta)})$ or $B = C([-1,1])$ such that $D_n^{(\alpha,\beta)}\varphi$ does not converge in the corresponding B-norm, provided we assume $\max(\alpha,\beta) \geq \frac{1}{2}$. For $-1 < \alpha, \beta < -\frac{1}{2}$ the convergence or non-convergence of $D_n^{(\alpha,\beta)}\varphi$ seems to be an open question.

The convergence problem of $D_n^{(\alpha,\beta)}\varphi$ for the spaces $L^p(\pi^{(\alpha,\beta)})$, $1 < p < \infty$, was completely solved by Muckenhoupt [30]. We are formulating here Muckenhoupt's result explicitly.
Given $\alpha, \beta > -1$ denote $\gamma = \max(-\frac{1}{2}, \alpha, \beta)$.
For $1 < p < \infty$ the following is true:
If

$$p \in]\frac{4(\gamma+1)}{2\gamma+3}, \frac{4(\gamma+1)}{2\gamma+1}[,$$

then $\lim_{n\to\infty} \|D_n^{(\alpha,\beta)}\varphi - \varphi\|_p = 0$ for all $\varphi \in L^p(\pi^{(\alpha,\beta)})$.

If $p \notin]\frac{4(\gamma+1)}{2\gamma+3}, \frac{4(\gamma+1)}{2\gamma+1}[$, then there exists a function $\varphi \in L^p(\pi^{(\alpha,\beta)})$ such that $D_n^{(\alpha,\beta)}\varphi$ does not converge with respect to the p-norm.
For $\alpha, \beta \geq -\frac{1}{2}$ the convergence result is due to Pollard [35] and the non-convergence result is contained in [35] and [11]. $\square$

15.7 Fejér-type Kernel

In the trigonometric case the Fejér kernel $(F_n)_{n\in\mathbb{N}_0}$ is defined as the (C,1)-series of the Dirichlet kernel $(D_n)_{n\in\mathbb{N}_0}$ by

$$F_n(t) = \frac{1}{n+1}\sum_{k=0}^{n} D_k(t) = \sum_{k=-n}^{n}(1 - \frac{|k|}{n+1})e^{ikt} \tag{15.53}$$

for all $t \in [0, 2\pi[$, $n \in \mathbb{N}_0$.
We have in the even case

$$F_{2n}(t) = \sum_{k=-2n}^{2n} \frac{\xi_n * \xi_n(k)}{\xi_n * \xi_n(0)} e^{ikt} \quad \text{for all} \quad t \in [0, 2\pi[, n \in \mathbb{N}_0, \tag{15.54}$$

where $*$ is the usual convolution in $l^1(\mathbb{Z})$. Therefore, we also have the representation

$$F_{2n}(t) = \frac{\widehat{\xi_n * \xi_n}(t)}{\xi_n * \xi_n(0)} = \frac{D_n(t)^2}{2n+1} = \frac{1}{2n+1}\left(\frac{\sin\frac{n+1}{2}t}{\sin\frac{t}{2}}\right)^2 \tag{15.55}$$

for all $t \in [0, 2\pi[, n \in \mathbb{N}_0$.
It is well-known that $F_n * \varphi \to \varphi$ for all $\varphi \in C([0, 2\pi[)$ or $\varphi \in L^p([0, 2\pi[)$, $1 \leq$

$p < \infty$ with respect to the corresponding norm, where $*$ is the usual convolution in $L^1([0, 2\pi[)$, see [21].

In case of orthogonal polynomial systems the convergence of (C,1)-series is not as universal as for the classical system $(e^{int})_{n\in\mathbb{Z}}$. Our goal is the construction of a Fejér-type kernel which guarantees such a universal convergence at least for some important classes of orthogonal polynomial systems.

DEFINITION 15.4 *Let $(\mathbb{N}_0, *)$ be a polynomial hypergroup induced by a set of polynomials $\{P_n : n \in \mathbb{N}_0\}$ according to Theorem 15.2.*
We define Fejér-type weights by

$$\varphi_{2n,k} = \frac{\chi_n * \chi_n(k)}{\chi_n * \chi_n(0)} \quad \text{for all} \quad n, k \in \mathbb{N}_0, \tag{15.56}$$

and a Fejér-type kernel $(F_{2n})_{n\in\mathbb{N}_0}$ by

$$F_{2n}(x) = \sum_{k=0}^{2n} \varphi_{2n,k} P_k(x) h(k) \quad \text{for all} \quad n \in \mathbb{N}_0,\ x \in D_S. \tag{15.57}$$

By $F_{2n}(x) = \widehat{\chi_n * \chi_n}(k)/\chi_n * \chi_n(0)$ we get

$$F_{2n}(x) = \frac{D_n(x)^2}{\chi_n * \chi_n(0)} \quad \text{for all} \quad n \in \mathbb{N}_0,\ x \in D_S, \tag{15.58}$$

where $(D_n)_{n\in\mathbb{N}_0}$ is the Dirichlet kernel. Hence, using a proper version of the Christoffel-Darboux identity [9, Theorem 4.5] it holds that $F_0(x) = 1$ and

$$F_{2n}(x) = \frac{1}{\chi_n * \chi_n(0)} \left(a_n h(n) \frac{P_{n+1}(x) - P_n(x)}{P_1(x) - P_0(x)} \right)^2 \tag{15.59}$$

for all $n \in \mathbb{N}$, $x \in D_S \setminus \{1\}$.

The Fejér-type weights depend on the hypergroup convolution and therefore vary with the orthogonal polynomial system.
In the case of Chebychev polynomials of the first kind $(T_n)_{n\in\mathbb{N}_0}$, see Example 15.2, we get

$$F_{2n}(x) = 1 + 2\sum_{k=1}^{2n}(1 - \frac{k}{2n+1})T_k(x) \quad \text{for all} \quad n \in \mathbb{N}_0,\ x \in [-1, 1] \tag{15.60}$$

which coincides with the classical Fejér kernel.

Depending on the linearization coefficients of the polynomials and on the Haar measure we have by convolution formula (15.12)

$$\chi_n * \chi_n(k) = \sum_{j=0}^{n} \sum_{i=|k-j|}^{\min(n,k+j)} g(k,j,i)h(j) \quad \text{for all} \quad n,k \in \mathbb{N}_0. \tag{15.61}$$

In the following lemma we summarize some properties of this convolution function. The proof is easy and is omitted.

LEMMA 15.3

For the sequence $\chi_n * \chi_n$ we have:

(i) $\chi_n * \chi_n(0) = \sum_{j=0}^{n} h(j)$.

(ii) $\chi_n * \chi_n(k) = \sum_{j=0}^{n} h(j) - \sum_{j=n+1-k}^{n} \sum_{i=n+1}^{k+j} g(k,j,i)h(j)$ *for* $k = 0, 1, \ldots, n$.

(iii) $\chi_n * \chi_n(k) = \sum_{j=k-n}^{n} \sum_{i=k-j}^{n} g(k,j,i)h(j)$ *for* $k = n, n+1, \ldots, 2n$. *Especially* $\chi_n * \chi_n(2n) = g(2n,n,n)h(n)$.

(iv) $\chi_n * \chi_n(k) = 0$ *for* $k > 2n$.

(v) $\chi_n * \chi_n(0) \geq \chi_n * \chi_n(k)$ *for all* $k \in \mathbb{N}_0$.

With regard to Theorem 15.4 we have to check whether $\lim_{n\to\infty} \varphi_{2n,k} = 1$ for all $k \in \mathbb{N}_0$. Therefore we make the following definition:
We say that property (H) holds if

$$\lim_{n\to\infty} \frac{h(n)}{\sum_{i=0}^{n} h(i)} = 0. \tag{15.62}$$

PROPOSITION 15.6

Let $(\mathbb{N}_0, *)$ be a polynomial hypergroup with property (H). Then we have $\lim_{n\to\infty} \varphi_{2n,k} = 1$ for all $k \in \mathbb{N}_0$.

PROOF By Lemma 15.3 we get

$$\varphi_{2n,k} = \frac{\chi_n * \chi_n(k)}{\chi_n * \chi_n(0)} = 1 - \frac{\sum_{j=n+1-k}^{n} \sum_{i=n+1}^{k+j} g(k,j,i)h(j)}{\sum_{i=0}^{n} h(i)}$$

for all $n \geq k > 0$. With

$$0 \leq \frac{\sum_{j=n+1-k}^{n} \sum_{i=n+1}^{k+j} g(k,j,i)h(j)}{\sum_{i=0}^{n} h(i)} \leq \sum_{j=n+1-k}^{n} \frac{h(j)}{\sum_{i=0}^{n} h(i)}$$

the assertion follows. ∎

A condition which implies property (H), and is easy to handle, is given in the next proposition.

PROPOSITION 15.7

Let

$$r_n = \frac{c_n}{a_{n-1}} = \frac{g(1,n,n-1)}{g(1,n-1,n)} \quad \textit{for all} \quad n \in \mathbb{N}. \tag{15.63}$$

If $\lim_{n\to\infty} r_n = 1$, *then property (H) holds.*

PROOF If $0 < \epsilon < 1$, then there exists $k \in \mathbb{N}$ such that $1 - \epsilon \leq r_{n+k} \leq 1 + \epsilon$ for all $n \in \mathbb{N}$. By Theorem 15.2 (iii) we get

$$0 < \frac{h(n+k)}{\sum_{j=0}^{n+k} h(j)} \leq \frac{h(n+k)}{\sum_{j=k}^{n+k} h(j)}$$

$$= \frac{h(n+k)}{h(n+k)(1 + r_{n+k} + r_{n+k}r_{n+k-1} + \ldots + r_{n+k}r_{n+k-1}\cdots r_{k+1})}$$

$$\leq \frac{\epsilon}{1-(1-\epsilon)^{n+1}},$$

and thus the assertion follows. ∎

In the following examples we check property (H) for a couple of orthogonal polynomial systems:

Example 15.8

Jacobi polynomials.

In the case of Jacobi polynomials, see Example 15.1, we derive

$$r_n = \frac{n(n+\beta)(2n+\alpha+\beta-1)}{(n+\alpha)(n+\alpha+\beta)(2n+\alpha+\beta+1)} \quad \text{for all} \quad n > 1.$$

Hence $\lim_{n\to\infty} r_n = 1$ and property (H) holds by Proposition 15.7. □

Example 15.9

Generalized Chebychev polynomials.

In the case of generalized Chebychev polynomials, see Example 15.3, we have

$$r_n = \begin{cases} \dfrac{(q+\beta+1)(2q+\alpha+\beta+1)}{(q+\alpha+\beta+1)(2q+\alpha+\beta+2)}, & \text{if } n = 2q+1, \\ \dfrac{q(2q+\alpha+\beta)}{(q+\alpha)(2q+\alpha+\beta+1)}, & \text{if } n = 2q, \end{cases}$$

for all $n \in \mathbb{N}$. Hence $\lim_{n\to\infty} r_n = 1$ and property (H) hold by Proposition 15.7. $\square$

Example 15.10

Polynomials connected with homogeneous trees.

In the case of polynomials connected by homogeneous trees, see Example 15.4, we get for $\alpha > 2$,

$$\lim_{n\to\infty} \frac{h(n)}{\sum_{j=0}^{n} h(j)} = \lim_{n\to\infty} \frac{\alpha(\alpha-1)^{n-1}(\alpha-2)}{\alpha(\alpha-1)^n - 2} = \frac{\alpha-2}{\alpha-1}.$$

Thus in case $\alpha > 2$ property (H) does not hold. $\square$

Assuming a convolution structure on D_S we get the following important statement.

THEOREM 15.7

Let $(P_n)_{n\in\mathbb{N}_0}$ define a polynomial hypergroup with property (H). Assume that for $(P_n)_{n\in\mathbb{N}_0}$ a uniformly bounded product formula holds.
Then

$$\lim_{m\to\infty} \|F_{2n}\varphi - \varphi\|_B = 0 \quad \textit{for all} \quad \varphi \in B.$$

PROOF By Proposition 15.6 and Theorem 15.5 it is sufficient to show that there exists $R > 0$ with $\|F_{2n}\|_1 \le R$ for all $n \in \mathbb{N}_0$.
By (15.58) we have $F_{2n}(x) \ge 0$ for all $x \in D_S$. Thus we easily derive

$$\|F_{2n}\|_1 = \int_{D_S} \sum_{k=0}^{2n} \varphi_{2n,k} P_k(x) h(k)\, d\pi(x) = \varphi_{2n,0} h(0) = 1$$

for all $n \in \mathbb{N}_0$, and the assertion follows. ∎

Finally, we take a look at our main examples:

Example 15.11

Jacobi polynomials.
For the Fejér kernel $(F_{2n}^{(\alpha,\beta)})_{n\in\mathbb{N}_0}$ we derive by (15.58)

$$F_{2n}^{(\alpha,\beta)} = \chi_n * \chi_n^{(\alpha,\beta)}(0)(J_n^{(\alpha+1,\beta)})^2$$

$$= \frac{\Gamma(n+\alpha+\beta+2)\Gamma(\beta+1)\Gamma(n+\alpha+2)}{\Gamma(\alpha+\beta+2)\Gamma(n+\beta+1)\Gamma(\alpha+2)\Gamma(n+1)}(J_n^{(\alpha+1,\beta)})^2$$

for all $n \in \mathbb{N}_0$.
With regard to Examples 15.5 and 15.8 we may apply Theorem 15.7 for Jacobi polynomials in case $\alpha \geq \beta > -1$ and $\alpha + \beta \geq -1$, that is

$$\lim_{n\to\infty} F_{2n}^{(\alpha,\beta)}\varphi = \varphi$$

with respect to the B-norm for all $\varphi \in B$.
We should note that for $-1 < \alpha < \beta$ and $\alpha + \beta \geq -1$ we achieve similar results by normalizing $Q_n^{(\alpha,\beta)}(x) = \dfrac{J_n^{(\alpha,\beta)}(x)}{J_n^{(\alpha,\beta)}(-1)}$ and applying the equality $Q_n^{(\alpha,\beta)}(x) = J_n^{(\beta,\alpha)}(-x)$. □

Example 15.12

Generalized Chebychev polynomials.
With regard to Examples 15.6 and 15.9 we may apply Theorem 15.7 for generalized Chebychev polynomials in case ($\alpha \geq \beta > -1$ and $\alpha + \beta > -1$) or $\alpha = \beta = -\frac{1}{2}$. □

Example 15.13

Polynomials connected to homogeneous trees.
In the case of polynomials connected to homogeneous trees, see Example 15.4, we get $\lim_{n\to\infty} F_{2n}H_2^{\alpha} = \frac{1}{\alpha-1}H_2^{\alpha}$. Hence for $\alpha > 2$ there is no general convergence with respect to the B-norm. □

15.8 de la Vallée-Poussin Kernel

A characteristic property of the de la Vallée-Poussin kernel in the trigonometric case is to reproduce polynomials up to a certain degree. Note that the Fejér kernel reproduces exactly only the constant functions. The classical de la Vallée-Poussin kernel is defined by

$$V_m^n(t) = \frac{1}{2m+1} \sum_{k=n-m}^{n+m} D_k(t) \quad \text{for all} \quad t \in [0, 2\pi[,\ m, n \in \mathbb{N}_0, \tag{15.64}$$

with $n \geq m$.
We have

$$V_m^n(t) = \sum_{k=-(n+m)}^{n+m} \frac{\xi_n * \xi_m(k)}{\xi_n * \xi_m(0)} e^{ikt} = \frac{\widehat{\xi_n * \xi_m}(t)}{\xi_n * \xi_m(0)} = \frac{D_n(t)D_m(t)}{2m+1} \tag{15.65}$$

for all $t \in [0, 2\pi[$, $m, n \in \mathbb{N}_0$, $n \geq m$.

Likewise, we make the following definition.

DEFINITION 15.5 *Let $(\mathbb{N}_0, *)$ be a polynomial hypergroup induced by a set of polynomials $\{P_n : n \in \mathbb{N}_0\}$ according to Theorem 15.2.*
We define de la Vallée-Poussin weights by

$$\delta_{m,k}^n = \frac{\chi_n * \chi_m(k)}{\chi_n * \chi_m(0)} \quad \textit{for all} \quad n, m, k \in \mathbb{N}_0, n \geq m, \tag{15.66}$$

and the de la Vallée-Poussin kernel $(V_m^n)_{n,m \in \mathbb{N}_0, n \geq m}$ by

$$V_m^n(x) = \frac{D_n(x)D_m(x)}{\chi_n * \chi_m(0)} = \sum_{k=0}^{n+m} \delta_{m,k}^n P_k(x) h(k) \quad \textit{for all} \quad x \in D_S. \tag{15.67}$$

At first we prove some useful properties of the sequence $\chi_n * \chi_m$.

LEMMA 15.4
For the sequence $\chi_n * \chi_m$ with $n \geq m$ we have:

(i) $\chi_n * \chi_m(k) = \sum_{j=0}^{m} h(j)$ *for* $k = 0, 1, \ldots, n-m$.

(ii) $\chi_n * \chi_m(k) = \sum_{j=0}^{m} \sum_{l=|k-j|}^{\min(n,k+j)} g(j,k,l)h(j)$
for $k = n - m + 1, n - m + 2, \ldots, n + m$.
Especially $\chi_n * \chi_m(n + m) = g(n + m, m, n)h(m)$.

(iii) $\chi_n * \chi_m(k) = 0$ *for* $k > n + m$.

PROOF (i),(ii): By definition of the convolution we obtain for any $k \in \mathbb{N}_0$:

$$\chi_n * \chi_m(k) = \sum_{j=0}^{m} \sum_{l=|k-j|}^{k+j} g(j,k,l)\chi_n(l)h(j).$$

If $k \in \{0, 1, \ldots, n - m\}$, then for $j \in \{0, 1, \ldots, m\}$ we get $0 \leq k + j \leq n$ and therefore it follows in view of Proposition 15.1 (iv)

$$\sum_{l=|k-j|}^{k+j} g(j,k,l)\chi_n(l) = \sum_{l=|k-j|}^{k+j} g(j,k,l) = 1.$$

(iii): If $k > n + m$ we get $k - j > n$ for $j \in \{0, 1, \ldots, m\}$ and hence

$$\sum_{l=|k-j|}^{k+j} g(j,k,l)\chi_n(l) = 0,$$

and this proves (iii). ∎

The next proposition states a reproducing property.

PROPOSITION 15.8
For the de la Vallée-Poussin mean we have $V_m^n q = q$ *for every polynomial* q *with* $\deg q \leq n - m$.

PROOF For $\varphi \in B$ we get by Lemma 15.4 (iii)

$$V_m^n \varphi(x) = \sum_{k=0}^{n+m} \delta_{m,k}^n \check{\varphi}(k) P_k(x) h(k).$$

Let $q(x) = \sum_{j=0}^{n-m} d_j P_j(x)$. Then we have $\check{q}(k) = \int_{D_S} q(x) P_k(x) d\pi(x) = d_k / h(k)$ for $k = 0, 1, \ldots, n - m$, and $\check{q}(k) = 0$ for $k > n - m$. By Lemma 15.4 (i) we get $\delta_{m,k}^n = 1$ for $k = 0, 1, \ldots, n - m$. Hence $V_m^n q = q$. ∎

The norm $\|V_m^n\|_1$ is not as easy to calculate as for the Fejér-type kernel. But we get the following result.

PROPOSITION 15.9
For the de la Vallée-Poussin kernel $(V_m^n)_{n,m\in\mathbb{N}_0,n\geq m}$ we have

$$\|V_m^n\|_1 \leq \frac{\sum_{j=0}^n h(j)}{\sum_{j=0}^m h(j)} \quad \textit{for all} \quad n,m\in\mathbb{N}_0.$$

PROOF We observe that

$$\|V_m^n\|_1 = \frac{1}{\chi_n * \chi_m(0)} \int_{D_S} |D_n(x)D_m(x)|\, d\pi(x).$$

Using $D_n(x)D_m(x) = (D_n^2(x) + D_m^2(x) - (D_n(x) - D_m(x))^2)/2$ we obtain

$$\begin{aligned} 2\int_{D_S} |D_n(x)D_m(x)| d\pi(x) &\leq \int_{D_S} D_n^2(x) d\pi(x) + \int_{D_S} D_m^2(x) d\pi(x) \\ &\quad + \int_{D_S} (D_n(x) - D_m(x))^2 d\pi(x) \\ &= \sum_{j=0}^n h(j) + \sum_{j=0}^m h(j) + \sum_{j=m+1}^n h(j) \\ &= 2\sum_{j=0}^n h(j). \end{aligned}$$

Hence by Lemma 15.4 (i) we get $\|V_m^n\|_1 \leq \sum_{j=0}^n h(j)/\sum_{j=0}^m h(j)$. ∎

In the classical trigonometric approximation the case $n = 2m$ is often considered. Our final result is restricted to the case $n = 2m$, too.

THEOREM 15.8
Let $(P_n)_{n\in\mathbb{N}_0}$ define a polynomial hypergroup and assume that for $(P_n)_{n\in\mathbb{N}_0}$ a uniformly bounded product formula holds. Furthermore assume that the growth condition $h(n) \sim n^\gamma$ for some $\gamma > 0$ is true.
Then

$$\lim_{m\to\infty} \|V_m^{2m}\varphi - \varphi\|_B = 0 \quad \textit{for all} \quad \varphi \in B.$$

PROOF Using Theorem 15.5 we have only to check that there exists some bound $R > 0$ such that $\|V_m^{2m}\|_1 \leq R$ for all $m \in \mathbb{N}$. Since

$$\frac{\sum_{j=0}^{2m} h(j)}{\sum_{j=0}^{m} h(j)} \sim \frac{(2m)^{\gamma+1}}{m^{\gamma+1}},$$

Proposition 15.9 implies the boundedness of $\|V_m^{2m}\|_1$. ∎

Now we take a last look at our examples:

Example 15.14
Jacobi polynomials.
For Jacobi polynomials we have $h(n) \sim n^{2\alpha+1}$ which is independent of the parameter β. Hence we may apply Theorem 15.8 in case $\alpha \geq \beta > -1$ and $\alpha + \beta \geq -1$. See also Examples 15.1 and 15.5. □

Example 15.15
Generalized Chebychev polynomials.
For generalized Chebychev polynomials we have $h(n) \sim n^{2\alpha+1}$ which is independent of the parameter β. Thus we may apply Theorem 15.8 in case ($\alpha \geq \beta > -1$ and $\alpha + \beta > -1$) or $\alpha = \beta = -\frac{1}{2}$. See also Examples 15.3 and 15.6. □

References

[1] R. Askey and I.I. Hirschmann, Jr. Mean summability for ultraspherical polynomials. *Math. Scand.*, 12:167–177, 1963.

[2] V.M. Badkov. Convergence in mean and almost everywhere of Fourier series in polynomials orthogonal on an interval. *Math. USSR Sbornik*, 24:223–256, 1974.

[3] H. Bavinck. A special class of Jacobi series and some applications. *J. Math. Anal. Appl.*, 37:767–797, 1972.

[4] H. Bavinck. Approximation processes for Fourier-Jacobi expansions. *Applicable Analysis*, 5:193–312, 1976.

[5] N.H. Bingham. Tauberian theorems for Jacobi series. *Proc. London Math. Soc.*, 36:285–309, 1978.

[6] W.R. Bloom and H. Heyer. *Harmonic Analysis of Probability Measures on Hypergroups.* de Gruyter, Berlin, New York, 1995.

[7] P.L. Butzer and R.J. Nessel, *Fourier Analysis and Approximation.* Birkhäuser, Basel, 1971.

[8] S. Chanillo and B. Muckenhoupt. Weak type estimates for Cesàro sums of Jacobi polynomial series. *Mem. Amer. Math. Soc.*, 487, 1993.

[9] T.S. Chihara. *An Introduction to Orthogonal Polynomials.* Gordon and Breach, New York, London, Paris, 1978.

[10] W.C. Connett and A.L. Schwartz. The Littlewood-Paley theory for Jacobi expansions. *Trans. Amer. Math. Soc.*, 251:219–234, 1979.

[11] B. Breseler and P.M. Soardi. A Cohen-type inequality for Jacobi expansions and divergence of Fourier series of compact symmetric spaces. *J. Approx. Theory*, 35:214–221, 1982.

[12] J.L. Dunau. *Etude d'une classe de marches aléatoires sur l'arbre homogène.* Publication du Laboratoire de Statistique Probabilities, Université Paul Sabatier, Toulouse, 1983.

[13] G. Freud. *Orthogonale Polynome.* Birkhäuser, Basel, 1969.

[14] G. Gasper. Linearization of the product of Jacobi polynomials II. *Can. J. Math.*, 22:582–593, 1970.

[15] G. Gasper. Banach algebras for Jacobi series and positivity of a kernel. *Ann. of Math.*, 95:261–280, 1972.

[16] J.J. Gaudalupe, M. Pérez, and J.L. Verona. Weak behaviour of Fourier-Jacobi series. *J. Approx. Theory*, 61:222–238, 1990.

[17] R.I. Jewett. Spaces with an abstract convolution of measures. *Ann. of Math.*, 18:1–101, 1975.

[18] N.J. Kalten and L. Tzafriri. The behaviour of Legendre and ultraspherical polynomials in L_p-spaces. *Can. J. Math.*, 50:1236–1252, 1998.

[19] T.P. Laine. The product formula and convolution structure for the generalized Chebyshev polynomials. *SIAM J. Math. Anal.*, 11:133–146, 1980.

[20] R. Lasser. Orthogonal polynomials and hypergroups. *Rend. Mat.*, 3:185–209, 1983.

[21] R. Lasser. *Introduction to Fourier Series.* Marcel Dekker, New York, 1996.

[22] R. Lasser and J. Obermaier. On Fejér means with respect to orthogonal polynomials: A hypergroup theoretic approach. In P. Nevai and A. Pinkus, editors, *Progress in Approximation Theory*, pages 551–565, Academic Press, Boston, 1991.

[23] R. Lasser and J. Obermaier. On the convergence of weighted Fourier expansions. *Acta Sci. Math.*, 61:345–355, 1995.

[24] Z. Li. Pointwise convergence of Fourier-Jacobi series. *Approx. Theory & its Appl.*, 11:58–77, 1995.

[25] Z. Li. Conjugate Jacobi series and conjugate functions. *J. Approx. Theory*, 86:179–196, 1996.

[26] Z. Li. On the Cesáro means of conjugate Jacobi series. *J. Approx. Theory*, 91:103–116, 1997.

[27] L. Lorch. The Lebesgue constants for Jacobi series I. *Proc. Amer. Math. Soc.*, 10:756–761, 1959.

[28] D.S. Lubinsky. A survey of mean convergence of orthogonal polynomial expansions. In K. Jarosz, editor, *Function Spaces. The second conference*, volume 172 of *Lecture Notes Pure and Appl. Math.*, pages 281–310, Marcel Dekker, New York, 1995.

[29] C. Meaney. Divergent Jacobi polynomial series. *Proc. Amer. Math. Soc.*, 87:459–462, 1983.

[30] B. Muckenhoupt. Mean convergence of Jacobi series. *Proc. Amer. Math. Soc.*, 23:306–310, 1969.

[31] B. Muckenhoupt and E.M. Stein. Classical expansions and their relation to conjugate harmonic functions. *Trans. Amer. Math. Soc.*, 118:17–92, 1965.

[32] J. Newman and W. Rudin. Mean convergence of orthogonal series. *Proc. Amer. Math. Soc.*, 3:219–222, 1952.

[33] H. Pollard. The mean convergence of orthogonal polynomial series I. *Trans. Amer. Math. Soc.*, 62:387–403, 1947.

[34] H. Pollard. The mean convergence of orthogonal polynomial series II. *Trans. Amer. Math. Soc.*, 63:355–367, 1948.

[35] H. Pollard. The mean convergence of orthogonal polynomial series III. *Duke Math. J.*, 16:189–191, 1949.

[36] H. Rau. Über die Lebegueschen Konstanten der Reihenentwicklung nach Jacobischen Polynomen. *J. Reine Angew. Math.*, 161:237–254, 1929.

[37] W. Rudin. *Real and Complex Analysis.* McGraw-Hill, New York, third edition, 1986.

[38] G. Szegö. *Orthogonal Polynomials.* Amer. Math. Soc., New York, 1959.

[39] R. Szwarc. Orthogonal polynomials and a discrete boundary value problem II. *SIAM J. Math. Anal.*, 23:965–969, 1992.

[40] Y. Xu. Mean convergence of generalized Jacobi series and interpolating polynomials I. *J. Approx. Theory*, 72:237–251, 1993.

[41] Y. Xu. Mean convergence of generalized Jacobi series and interpolating polynomials II. *J. Approx. Theory*, 76:77–92, 1994.

[42] S.P. Yadav. On the saturation order of approximation processes involving Jacobi polynomials. *J. Approx. Theory*, 58:36–49, 1989.

[43] S.P. Yadav. A note on matrix-transform of Fourier-Jacobi expansions. *Indian J. Math.*, 34:111–117, 1992.

[44] S.P. Yadav. Saturation orders of some approximation processes in certain Banach spaces. *Studia Sci. Math. Hungar.*, 28:1–18, 1993.

Chapter 16

Digitized PDE Method for Data Restoration

Stanley Osher and Jianhong Shen

Department of Mathematics, UCLA, Los Angeles, CA 90095
E-mail: {sjo, jhshen}@math.ucla.edu
Contact author: J. Shen

Abstract Towards finding the connections between wavelets and the PDE method in image processing, we bring the two subjects closer in this chapter by developing the digital version of the PDE method. Continuous image domains are digitized to graphs and the Euler-Lagrange PDE equations become algebraic ones. Algorithms of these equations connect the digital PDE method to iterations of local digital filters (linear or nonlinear). We develop a self-contained theory of existence and uniqueness of the solutions, which avoids the technical difficulties appearing in classical continuous models and thus makes it easier for general readers without PDE backgrounds to understand and apply it. The PDE method has various applications in image analysis. In this chapter, we focus on denoising and restoration.

Contents

16.1 Introduction

16.1.1 Wavelets and the PDE method in image analysis

Wavelets and the PDE method are currently two major mathematical tools for image analysis. Both have been developed independently into mature methodologies within the past two decades. They have remarkably distinct origins, contents, and applications (see Daubechies [6] for wavelets and Morel and Solimini [8] for the PDE method). These differences have separated the two subjects for a long time, and we intend to understand their relations.

The classical applied wavelet method can be traced to digital filters and filter banks, and to a large extend, is a linear method. Therefore, frequency analysis (or Fourier method), and its polished new version — multiresolution analysis, have been the two indispensable tools for wavelet analysis. The multiresolution decomposition makes wavelets one of the most efficient tools for data compression, denoising and restoration, edge detection, and indexing of image banks, etc. The P.D.E. method, on the other hand, originates in variational models (for denoising and restoration, edge enhancement, segmentation, and object detection, etc.). In addition, image restoration (which we shall address in this chapter) is typically connected to various types of diffusions — isotropic or anisotropic, free or constrained, and plain or enhanced. The method is usually nonlinear and dynamical. The mathematical tools are mainly the variational calculus and mathematical morphology. The existing PDE models are mostly the Euler-Lagrange equations for certain energy functionals, or their modified versions. These fundamental differences make it less surprising to see their quite "orthogonal" development in the past two decades.

So the question is: Is it possible to make certain (even incomplete) connections between the two? We just noticed that we are not the first ones asking the question. For instance, one can read the recent two papers by

Coifman and Sowa [4, 5]. In this chapter, we also try to answer it positively, but with a different, and more general viewpoint. To us, the basic idea is to "digitize" the PDE method. We see this as a first step towards a unified (if possible) study of these two major mathematical tools in the field of image processing and computer vision.

The classical wavelet method starts with the four structural digital filters (one or two if the wavelets are orthogonal). The PDE method, naturally, begins with partial differential equations. Hence, for image processing, wavelets seem to be more digital (or discrete), while the latter is more analog (or continuous). Thus, our first intention is to make some digital interpretation for the effective actions of the differential equations. This perhaps premature connection, in the long run, may serve as a stepping-stone for much deeper and wider understandings.

By "digitizing" the PDE method, we do not mean the discrete numerical implementation of the existing differential equations. Instead, we digitize the *methodology*! That is, we start directly with the discrete variational problem, from which algebraic equilibrium equations analogous to the PDEs are established. This methodology is thus self-contained. It is unnecessary to search for the meaning of these algebraic equations from their PDE counterparts. Remarkably, the underlying domain can be fairly irregular. A digital version for a continuous domain is modeled by a graph. Therefore, we shall study the digitized PDE method on graphs.

Finite digital filters appear when we apply linear or nonlinear Gauss-Jacobi, Gauss-Seidel, or infinitesimal steepest descent algorithms to solve these algebraic equations. Unlike in wavelet theory, where digital filters accomplish multiresolution analysis or synthesis, the filters here solve a global optimization problem. These filters can be both linear and nonlinear.

The PDE method has found its wide applications in image segmentation, restoration, and object detection (active contours), etc. In this chapter, the target application in mind is data denoising and restoration.

16.1.2 Why the digitized PDE method or the graph model

Variational model-based PDE denoising and restoration method has been one of the most significant tools in image analysis (see Morel and Solimini [8], Mumford and Shah [9], Rudin and Osher [10], and Rudin, Osher and Fatemi [11]). The classical methodology is displayed in Figure 16.1.

Images are assumed to be defined on a continuous domain in $\mathbb{R}^2$. Then a continuous variational functional is constructed, which typically involves a regularization term (or *internal energy*) and a constraint (or *external energy*). The Euler-Lagrange equation associated with the variational functional is usually of second order in spatial variables. The diffusion equation corresponds to the infinitesimal steepest descent marching for the constructed energy functional. Finally, the resulting differential equations (ei-

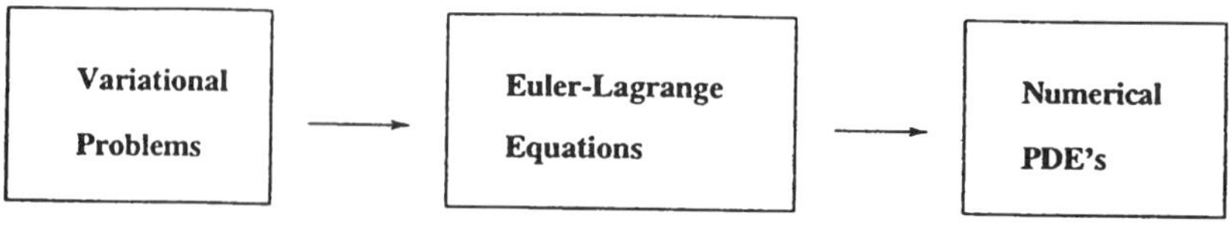

FIGURE 16.1
Classical methodology

ther the diffusion equation or the equilibrium equation) are discretized numerically on a rectangular grid. The advantage of this methodology is that one can easily adapt many existing numerical methods for PDEs to this problem and establish efficient algorithms.

This approach may seem difficult for the digital group to understand. It lacks a digital sense in both its formulation and interpretation of the results. To build a bridge between these two territories, it seems necessary to formulate a direct digital version of the classical restoration model. This means that we shall work on a general discrete domain—a graph. In the one-dimensional case, it means working on discrete signals, which are familiar objects in wavelet theory.

The importance of digitizing continuous models resides in many classical examples. For instance, although Fourier integrals, Fourier series, and discrete Fourier series are logically linked, without an independent point of view on the Fourier matrices, the Fast Fourier Transform is impossible. Another example is the design of digital filters. In the beginning of digital signal processing, digital filters were generally considered as a discretized version of analog filters. But later on we witnessed a complete independent rich theory of digital filters.

"Digitizing" the PDE method for graphs is not only led by the above digitization philosophy, but also motivated by applications. Diffusion-based denoising method has been historically stimulated by image analysis and constantly finds its applications in it. But it is not only images that need denoising or restoration. General measured, observed, or experimental data with certain "topological spatial structure" need it as well. For general data sets, the underlying topology may be quite irregular, unlike the rectangular grid for images. For instance, an oceanographer may measure the depth of an ocean floor at different spots, very reasonably unnecessary on some prescribed rectangular grid points (Figure 16.2).

For such an irregular topology, the direct numerical discretization of the continuous partial differential equations usually fail to work. This is because that partial derivatives like u_x, u_y and the gradient ∇u are not well-defined on a general graph.

Fortunately, as we shall see in the next section, quantities like $\|\nabla u\|$ and

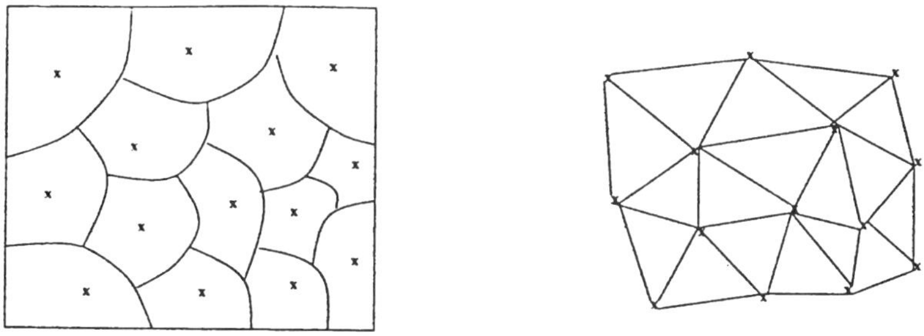

Irregular Measuring Spots **Graph Topology of Measuring Spots**

(An edge connects two neighbors)

FIGURE 16.2
Data are measured on irregular spatial spots. (Here we assume that each measuring spot (or station) has its influence region described in the left plot. Then two spots are connected by an edge if their influence regions share borders.

Δu (the Laplacian) do have their analogy on an arbitrary graph, which shall be denoted by $\|\nabla_g u\|$ and $\Delta_g u$. In fact, the existence of the Laplacian graph brings birth to the whole subject of *spectral graph theory* (see Chung [3]). The spectra of a Laplacian graph reveal many combinatorial secrets of the associated graph. Both $\|\nabla_g u\|$ and $\Delta_g u$ are directly related to certain variational problems. Therefore, the right approach for formulating a restoration scheme for data existing on a graph is to start from the variational problem in the first box of Figure 16.1, rather than attempting to discretize the Euler-Lagrange equations numerically.

In this chapter, we

1) construct the energy functions and the restoration equations (Section 2);
2) analyze properties of these equations (existence and uniqueness, etc.) (Section 2);
3) design algorithms leading to digital filters and show numerical results (Sections 3, 4).

16.2 Energy Functions and Restoration Equations

Let $[\Omega, G]$ denote a graph with a finite set Ω of nodes (or vertices) and a dictionary G of edges. The size D of the graph is the total number of vertices. We shall assume that the graph is connected and has no self-loops (i.e., no direct edge from a vertex to itself). General vertices are denoted by $\alpha, \beta, \cdots$. If α and β are linked by an edge, then we write $\alpha \sim \beta$, and

$$N_\alpha = \{\beta \in \Omega | \beta \sim \alpha\}$$

denotes all the neighbors of α.

Let u^0 be a given raw (noisy) data distribution on Ω. In applications, u^0 is frequently modeled as the measurement of a clean data u, contaminated at each spot by an independent noise n, i.e.,

$$u^0 = u + n, \quad \text{or} \quad u^0_\alpha = u_\alpha + n_\alpha, \qquad \text{for all } \alpha \in \Omega.$$

It is often assumed that the noise at each spot has mean zero and variance σ^2 (which usually models correctly the measurement or observation error), i.e.,

$$\mathrm{E}u_\alpha = \mathrm{E}u^0_\alpha \qquad \text{and} \qquad \mathrm{E}(u_\alpha - u^0_\alpha)^2 = \sigma^2,$$

where E is the expectation operator. In practice, a measurement or observation u^0 is only one particular realization of the assumed random distribution. Suppose that the size D is large enough, then the *Law of Large Numbers* makes it safe to assume that

$$\frac{1}{D}\sum_\alpha u_\alpha = \frac{1}{D}\sum_\alpha u^0_\alpha \quad \text{and} \quad \frac{1}{D}\sum_{\alpha\in\Omega}(u^0_\alpha - u_\alpha)^2 = \sigma^2. \tag{16.1}$$

Geometrically, in the space $\mathbb{R}^D$, the first constraint defines a hyperplane π that sweeps through u^0 and is perpendicular to $\mathbf{1} = (1, 1, \cdots, 1)$. The second constraint is a hypersphere centered at u^0 and with radius $\sqrt{D}\sigma$. Together, they define a hypercircle (of dimension $D - 2$) in the space of all possible distributions (Figure 16.3).

In applications, σ is easy to estimate. Thus, Eq. (16.1) provides a natural fitting constraint. Under this constraint, we try to find a smooth optimal restoration u. Define the strength function $\|\nabla_g u_\alpha\|$ by

$$\|\nabla_g u_\alpha\| := \left[\sum_{\beta\in N_\alpha} (u_\beta - u_\alpha)^2\right]^{\frac{1}{2}}.$$

Though $\nabla_g u$ misses its meaning for a general graph, the above notation is convenient since we do retain the two properties:

(i) $\|\nabla_g(bu)_\alpha\| = |b|\|\nabla_g u_\alpha\|$, for any scalar b.

(ii) $\|\nabla_g(u+v)_\alpha\| \le \|\nabla_g u_\alpha\| + \|\nabla_g v_\alpha\|$ for any two data u and v.

The L^2 *total energy* $E^{(2)}$ is defined as

$$E^{(2)}(u) = \sum_{\alpha\in\Omega} \frac{1}{2}\|\nabla_g u_\alpha\|^2 = \sum_{\alpha\sim\beta}(u_\beta - u_\alpha)^2,$$

where the second summation goes through all edges $\alpha \sim \beta$ ($\beta \sim \alpha$ is considered identical to $\alpha \sim \beta$). The *total variational* (TV) type of total energy is

$$E^{\mathrm{TV}}(u) = \sum_{\alpha\in\Omega} \|\nabla_g u_\alpha\|.$$

The constrained restoration model associated with these energy functions is

$$\min_{u \text{ subject to } (16.1)} E^{(2)}(u) \quad (\text{or} E^{\mathrm{TV}}(u)). \tag{16.2}$$

The existence of an optimal restoration is instant since both $E^{(2)}$ and E^{TV} are continuous functions and the constraint manifold (hypercircle) is compact. To ensure uniqueness (and non-triviality), we need another condition on the estimation of σ. Recall that we have assumed

$$u^0_\alpha = u_\alpha + n_\alpha,$$

and the noise n is independent of u. Therefore,

$$\mathrm{Var}(u^0_\alpha) = \mathrm{Var}(u_\alpha) + \mathrm{Var}(n_\alpha).$$

Especially,

$$\mathrm{Var}(u^0_\alpha) \ge \mathrm{Var}(n_\alpha) = \sigma^2,$$

which leads to the consistency condition of Chambolle and Lions [1]

$$\frac{1}{D}\sum_\alpha (u^0_\alpha - \overline{u}^0)^2 \ge \sigma^2, \tag{16.3}$$

where $\overline{u}^0$ is the empirical mean $1/D\sum_\alpha u^0_\alpha$.

The following theorem belongs to Chambolle and Lions [1] in the continuous case. We present a simpler proof in the digital case. However, it should be noted that our proofs are not of the usual (more difficult) numerical analysis type involving convergence to the continuous limit.

THEOREM 16.1 (Uniqueness)
Suppose that u^0 satisfies the Chambolle-Lions consistency condition (16.3). Then the constrained optimization problem (16.2) has a unique optimal solution.

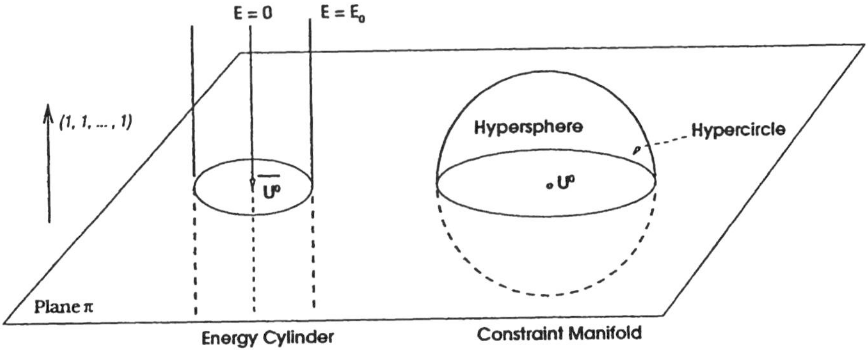

FIGURE 16.3
Hypersphere, hypercircle, and energy cylinder.

PROOF (Refer to Figure 16.3.) Let E denote either $E^{(2)}$ or E^{TV}. In $\mathbb{R}^D$, consider the E_0-equipotential hypersurface (i.e., the level set):

$$E = E_0.$$

It is a cylindrical surface in the direction of $\mathbf{1} = (1, 1, \cdots, 1)$, since

$$E(u + t) \equiv E(u)$$

for any constant distribution $t = (t, t, \cdots, t)$. Hence, $E \leq E_0$ is a *cylinder* in direction $\mathbf{1}$. The central axis of the cylinder is $E = 0$, a straight line along $\mathbf{1}$. The orthogonal projection of u^0 on this line is simply its empirical mean $\overline{u}^0 = (\overline{u}^0, \overline{u}^0, \cdots, \overline{u}^0)$. Hence the Chambolle-Lions consistency condition amounts to saying that $E = 0$ lies outside the hyperball, especially, outside the hypercircle defined by constraint (16.1) and the hyperdisk encircled by it.

Let E_c be the critical energy value such that the cylinder $E \leq E_c$ just contacts the hypercircle (or equivalently, the hyperdisk). Then

$$E_c = \min_{u \text{ subject to } (16.1)} E(u).$$

We now show that the contacting set $\mathcal{A}$ must contain only one "point," and restrict ourselves to the hyperplane π where the hyperdisk is on, which is of $D - 1$ dimension. $\mathcal{A}$ is also the contacting set for the hyperdisk and the section $\pi \bigcap \{E \leq E_0\}$. From the definition of the energy function E, it is easy to deduce that the latter is a one point set (if $E_0 = 0$) or a $D-1$-dimensional *convex* body. The hyperdisk is a $D - 1$-dimensional *strictly* convex body. By the meaning of E_c, it is now easy to conclude that $\mathcal{A}$ can contain no more than one point. This is the uniqueness of the restoration solution. ∎

Notice that the central axis of the energy cylinder is in the direction **1**, which is also the normal direction of the hyperdisk. Therefore, the critical energy cylinder contacts the hyperball, hyperdisk, and hypercircle at the same unique point. This means that the original constraints in (16.1) are equivalent to any of the following three:

$$\frac{1}{D} \sum_{\alpha \in \Omega} (u_\alpha^0 - u_\alpha)^2 = \sigma^2, \tag{16.4}$$

or

$$\frac{1}{D} \sum_{\alpha \in \Omega} (u_\alpha^0 - u_\alpha)^2 \leq \sigma^2,$$

or

$$\frac{1}{D} \sum_{\alpha} u_\alpha = \frac{1}{D} \sum_{\alpha} u_\alpha^0 \quad \text{and} \quad \frac{1}{D} \sum_{\alpha \in \Omega} (u_\alpha^0 - u_\alpha)^2 \leq \sigma^2.$$

The latter two constraints are convex constraints. In the continuous case, the discussion can also be found in Chambolle and Lions [1].

From the above analysis and Figure 16.3, we easily obtain the well-known result in the continuous case about the Lagrange multiplier.

PROPOSITION 16.1

Suppose that u^0 satisfies the Chambolle-Lions consistency condition. Then the Lagrange multiplier of the restoration problem associated with constraint (16.4) is non-negative.

This is because that the gradients of the energy function and the quadratic constraint function at the unique contacting point have opposite directions.

What is well-practiced and proven efficient in the literature of image analysis is the unconstrained formulation. Let λ be a positive constant (corresponding to the Lagrange multiplier. Then the *fitted* L^2 total energy is defined to be

$$E_\lambda^{(2)}(u) := E^{(2)}(u) + \frac{\lambda}{2}\|u - u^0\|_\Omega^2, \tag{16.5}$$

where, $\|u - u^0\|_\Omega$ is the Euclidean distance in $\mathbb{R}^D$. Similarly, the *fitted* TV type of energy is

$$E_\lambda^{\mathrm{TV}}(u) := E^{\mathrm{TV}}(u) + \frac{\lambda}{2}\|u - u^0\|_\Omega^2. \tag{16.6}$$

In practice, the constant λ is either chosen *a priori* (Marquina and Osher [7]), or estimated using the projected gradient method (Rudin and Osher [10] and Rudin, Osher and Fatemi [11]).

It is easy to show that

PROPOSITION 16.2

Both the fitted L^2 total energy and TV total energy are strictly convex functions of $u \in \mathbb{R}^D$.

In particular, since both energy functions tend to ∞ as $u \to \infty$ in $\mathbb{R}^D$, we have

THEOREM 16.2

Each of the restoration problems (16.5) and (16.6) has a unique solution.

To find the optimal restoration, we take derivatives:

$$\frac{\partial E_\lambda^{(2)}}{\partial u_\alpha} = -2\Delta_g u_\alpha + \lambda(u_\alpha - u_\alpha^0), \qquad \alpha \in \Omega.$$

Here the Laplacian graph $\Delta_g u$ (one simplest version; there are couple of other definitions out of different considerations; see Chung [3]) is defined by

$$\Delta_g u_\alpha = \sum_{\beta \in N_\alpha} (u_\beta - u_\alpha).$$

Hence, the restoration equation associated with $E^{(2)}$ is

$$-2\Delta_g u_\alpha + \lambda(u_\alpha - u_\alpha^0) = 0, \qquad \alpha \in \Omega.$$

This is a system of linear equations in u_α's.

Let d_α be the degree of a vertex α, i.e.,

$$d_\alpha = \sum_{\beta \in N_\alpha} 1.$$

Then the last linear system of equations becomes

$$(\lambda + 2d_\alpha)u_\alpha - 2\sum_{\beta \sim \alpha} u_\beta = \lambda u_\alpha^0, \tag{16.7}$$

for all $\alpha \in \Omega$. Since λ is positive, the linear system (16.7) is strictly *positive definite.* In fact, written in a matrix form, the associated matrix is diagonally dominant. In particular, we again have

PROPOSITION 16.3

The graph "variational problem" (16.5) has a unique solution, which depends on λ and u^0 algebraically, and is also the unique solution to Eq. (16.7).

Similarly, for the total variational model, the restoration equation is

$$0 = \sum_{\beta \in N_\alpha} (u_\alpha - u_\beta)\left(\frac{1}{\|\nabla_g u_\alpha\|} + \frac{1}{\|\nabla_g u_\beta\|}\right) + \lambda(u_\alpha - u_\alpha^0), \quad \alpha \in \Omega. \tag{16.8}$$

In practice, to avoid a zero denominator, we replace $\|\nabla_g u_\alpha\|$ by

$$\|\nabla_g u_\alpha\|_a := \sqrt{a^2 + \|\nabla_g u_\alpha\|^2},$$

for some fixed small (like 10^{-4}) regulization constant a. Then the above equation is fixed to

$$0 = \sum_{\beta \sim \alpha} (u_\alpha - u_\beta)\left(\frac{1}{\|\nabla_g u_\alpha\|_a} + \frac{1}{\|\nabla_g u_\beta\|_a}\right) + \lambda(u_\alpha - u_\alpha^0). \tag{16.9}$$

From the variational point of view, this corresponds to minimizing

$$\sum_{\alpha \in \Omega} \|\nabla_g u_\alpha\|_a + \frac{\lambda}{2}\|u - u^0\|_\Omega^2,$$

which is still strictly convex. Both Eqs. (16.8) and (16.9) have a unique solution since the associated energy functions are strictly convex.

Define

$$w_{\alpha\beta}(u) = \frac{1}{\|\nabla_g u_\alpha\|_a} + \frac{1}{\|\nabla_g u_\beta\|_a}.$$

Then $w_{\alpha\beta} = w_{\beta\alpha}$, and Eq. (16.9) becomes

$$\left(\lambda + \sum_{\beta \in N_\alpha} w_{\alpha\beta}(u)\right) u_\alpha - \sum_{\beta \in N_\alpha} w_{\alpha\beta}(u) u_\beta = \lambda u_\alpha^0, \tag{16.10}$$

for all $\alpha \in \Omega$. This is a system of nonlinear equations.

Some comments are in order now.

1. The main advantage of this self-contained "digital" version of the classical methodology illustrated in Figure 16.1 is that the restoration equations are automatically ready for computers. Tricky numerical discretization becomes unnecessary, although it is sometimes desirable [11].

2. As in the continuous case, the equations associated with the TV energy are nonlinear algebraic equations. Therefore, a complete discussion with real applications in mind must include efficient algorithms. This is the partial task of the next section.

3. The diffusion equations in the continuous case correspond to the steepest descent marching method here. For example, for the fitted L^2 total energy, the "digital" version of the diffusion equations is a system of O.D.E.s:

$$\frac{du_\alpha(t)}{dt} = 2\Delta_g u_\alpha - \lambda(u_\alpha - u_\alpha^0), \qquad \alpha \in \Omega.$$

This marching method does have its own computational advantages for a graph with large sized or complicated edge structures.

4. One can of course introduce other types of (convex) energy functions. The two discussed here correspond to the most widely used in the continuous case. Especially, the TV model and its regularized version also allow geometric explanations. The major well-known advantage of the TV model is that it can "intelligently" recognize real edges from noise. In contrast, the L^2 energy model usually smears edges and creates spurious oscillations at edges in the presence of noise (see the numerical results later).

16.3 Algorithms and Local Digital Filters

16.3.1 Restoration by the L^2 total energy

We first solve the linear Eq. (16.7)

$$(\lambda + 2d_\alpha)u_\alpha - 2\sum_{\beta\in N_\alpha} u_\beta = \lambda u_\alpha^0.$$

Direct methods such as Gauss-elimination are not feasible when the graph has a large size (e.g., for an image of 256 by 256, the matrix will be $D = 65536$ by D) or complex topology. Therefore, in applications, iterative methods are preferable.

Due to factors like the complexity of a graph and parallel implementation, usually we prefer "local" iterative methods to those "global" ones. By locality, we mean that at each spot α, the updating of u_α only depends on the current data existing close to the spot. The classical steepest descent method and conjugate gradient method are both global since at each iteration, global quantities like $\mathbf{x}^T A\mathbf{x}$ have to be worked out.

The simplest local iteration is the Gauss-Jacobi method

$$u_\alpha^{n+1} = \sum_{\beta\in N_\alpha} \frac{2}{\lambda + 2d_\alpha} u_\beta^n + \frac{\lambda}{\lambda + 2d_\alpha} u_\alpha^0, \tag{16.11}$$

where n denotes the iteration step. This process can be independently explained as a forced local lowpass digital filter. The updated u_α is a weighted average of the existing u_β on its direct neighbors and the raw data at α. The raw data serve as an attracting force preventing u from wandering far away.

The above digital filter interpretation brings more insights to our digital models. It can be generalized easily. Let $h_{\alpha\beta} \geq 0$ (for all $\alpha, \beta \in \Omega$ and $\alpha \sim \beta$ or $\alpha = \beta$) be any set of non-negative filter coefficients, such that

$$h_{\alpha\alpha} + \sum_{\beta\in N_\alpha} h_{\alpha\beta} = 1, \qquad \alpha \in \Omega.$$

This condition corresponds to the lowpass condition in signal processing. Then the forced lowpass filtering generated from h is defined by

$$u_\alpha^{n+1} = \sum_{\beta\in N_\alpha} h_{\alpha\beta} u_\beta^n + h_{\alpha\alpha} u_\alpha^0.$$

Assume $h_{\alpha\alpha} < 1$, and

$$m_{\alpha\beta} = \frac{h_{\alpha\beta}}{1 - h_{\alpha\alpha}}$$

is symmetric with respect to the node indices, i.e., $m_{\alpha\beta} = m_{\beta\alpha}$. Then the above h-filtering process marches to the equilibrium state of the energy function

$$E^{(h)}(u) := \frac{1}{2} \sum_{\alpha\sim\beta} m_{\alpha\beta}(u_\alpha - u_\beta)^2 + \frac{1}{2}\|u - u^0\|_\Omega^2.$$

Compared to $E^{(2)}$, the regularization term is now anisotropic.

Such a general h-filtering process still ignores the characteristics of the existing data since it depends only on the graph. It has serious drawbacks for data with intrinsic singularities (see Figure 16.4). Meanwhile, this digital filter explanation easily shows the horizon of successful modifications by requiring

$$h_{\alpha\beta} = h_{\alpha\beta}(u),$$

which is exactly the idea of *nonlinear* adaptive filtering. The question is how to define h properly from u. The TV model provides the right types of clues we need, as will be seen in the coming subsection.

Back to the original equation. Another possible scheme is Gauss-Seidel. We first label all vertices in a fixed linear order $...\gamma < \alpha < \beta....$ At each step $n|\alpha$, we compute

$$u_\alpha^{n+1} = \sum_{\beta\in N_\alpha \& \beta<\alpha} \frac{2}{\lambda + 2d_\alpha} u_\beta^{n+1} + \sum_{\beta\in N_\alpha \& \beta>\alpha} \frac{2}{\lambda + 2d_\alpha} u_\beta^n + \frac{\lambda}{\lambda + 2d_\alpha} u_\alpha^0, \tag{16.12}$$

and then move to the vertex right "greater" than α. Gauss-Seidel is a local but sequential iterative scheme.

Finally, though the classical steepest descent method is non-local, we can still consider the *infinitesimal* steepest descent method which is local. In the classical case, this corresponds to the diffusion-based restoration (see Morel and Solimini [8]). In our "digital" setting, we try to find the steady solution u to

$$\frac{du_\alpha(t)}{dt} = 2\Delta_g u_\alpha - \lambda(u_\alpha - u_\alpha^0), \qquad \alpha \in \Omega.$$

One can choose $u(0) = u^0$, for instance. To establish a local iteration scheme, we can try Euler method or Runge-Kutta 2 for the time derivative. For instance, for Runge-Kutta 2, we first compute and store the intermediate values.

$$K_1 = 2\Delta_g u_\alpha^n - \lambda(u_\alpha^n - u_\alpha^0), \qquad \overline{u}_\alpha^{n+1} = u_\alpha^n + hK_1;$$

and then

$$K_2 = 2\Delta_g \overline{u}_\alpha^{n+1} - \lambda(\overline{u}_\alpha^{n+1} - u_\alpha^0), \quad u_\alpha^{n+1} = u_\alpha^n + h\frac{K_1 + K_2}{2}.$$

Here h is the infinitesimal marching size. Since the scheme is of second order, the local computation involves the existing data on all neighbors of the neighbors of α. A filtering explanation can be established as before.

16.3.2 Restoration by the TV total energy

The TV restoration equation (16.10) is nonlinear

$$\left(\lambda + \sum_{\beta \in N_\alpha} w_{\alpha\beta}(u)\right) u_\alpha - \sum_{\beta \in N_\alpha} w_{\alpha\beta}(u) u_\beta = \lambda u_\alpha^0.$$

The corresponding linearized Gauss-Jacobi iteration is given by

$$u_\alpha^{n+1} = \sum_{\beta \in N_\alpha} \frac{w_{\alpha\beta}(u^n)}{\lambda + \sum_{\beta \in N_\alpha} w_{\alpha\beta}(u^n)} u_\beta^n + \frac{\lambda}{\lambda + \sum_{\beta \in N_\alpha} w_{\alpha\beta}(u^n)} u_\beta^0, \quad (16.13)$$

for all $\alpha \in \Omega$. Like before, this can be interpreted independently as a forced adaptive digital filter. The filter coefficients are adapted to local variations of the existing data. Most importantly, the form of the filter coefficients answers the question we raised in the previous subsection about the h-filtering process.

The adaptivity of this nonlinear filter is easy to understand qualitatively. The key is the competition between the Lagrange multiplier λ and the local weights $w_{\alpha\beta}(u)$. The local weights dominate λ if the current data u^n are very flat near a node α. Then the fitting term becomes less important and the filter acts like lowpass filtering purely on u^n, which makes the output u_α^{n+1} even flatter at the spot. On the other hand, if the current data u^n undergo an abrupt change of large amplitude at α, then the local weights $w_{\alpha\beta}$ are insignificant compared with λ. If this is the case, the filter intelligently sacrifices smoothness for faithfulness. This mechanism is obviously important for faithful restoration of edges (in image processing) or intrinsic singularities of general data (see Figures 16.4–16.6).

Because of the lowpass filtering property, the above schemes all satisfy the maximum principle:

PROPOSITION 16.4

Suppose that we start any of the above schemes with the given raw data

u^0. *Then at each step* n,

$$\min_\beta u_\beta^0 \leq u_\alpha^n \leq \max_\beta u_\beta^0,$$

for any node α.

Similarly, Gauss-Seidel can be established as long as we compute $w_{\alpha\beta}(u)$ by both u^n and u^{n+1}, depending on the linear ordering of vertices (for the case of image analysis, see also Chan and Shen [2]).

Finally for the time marching problem, nonlinearity causes no problem as long as we use explicit schemes like Euler or Runge-Kutta. Implicit schemes are non-local.

In order to get a *faster* time marching toward the steady restoration solution, as inspired by the recent work of Marquina and Osher [7], we can "precondition" the *direct* infinitesimal steepest descent equation (see Eq. (16.9))

$$\frac{du_\alpha}{dt} = \sum_{\beta\sim\alpha}(u_\beta - u_\alpha)\left(\frac{1}{\|\nabla_g u_\alpha\|_a} + \frac{1}{\|\nabla_g u_\beta\|_a}\right) - \lambda(u_\alpha - u_\alpha^0),$$

to

$$\frac{du_\alpha}{dt} = \sum_{\beta\sim\alpha}(u_\beta - u_\alpha)\left(1 + \frac{\|\nabla_g u_\alpha\|_a}{\|\nabla_g u_\beta\|_a}\right) - \lambda\|\nabla_g u_\alpha\|_a(u_\alpha - u_\alpha^0).$$

The marching speed is modified. Yet the steady optimal restoration remains the same since the factors $\|\nabla_g u_\alpha\|_a$ are positive. Numerically this modification improves speed and stability dramatically for graphs with large size.

16.4 Numerical Results and Concluding Remarks

16.4.1 Numerical results

We start with an application for the simplest graph — a linear graph or a chain

$$\Omega: \quad \alpha_1 < \alpha_2 < \cdots < \alpha_N,$$

with the natural edge dictionary

$$G = \{\alpha_1 \sim \alpha_2, \alpha_2 \sim \alpha_3, \cdots, \alpha_{N-1} \sim \alpha_N\}.$$

A chain graph can model the irregular measuring spots for a one-dimensional continuous landscape. In the second example, the digital model is applied to restore data noisily measured on a Sierpinski graph.

16.4.1.1 Linear graph

In Figure 16.4, the top one plots the real shape of a landscape with two cliffs. Plotted right below is the measured height field data at irregularly distributed measuring spots (α_k's). Due to the error of measuring tools and human factors, the set of measured data is noisy, which is modeled in computers by polluting the real height data with a random noise (we use uniform distribution on $[-.5, +.5]$ in this example).

The third one shows the restoration result using the L^2 total energy model (16.7) with an optimal $\lambda = .5$. The plot at the bottom shows the restoration by the TV model (16.10) with $\lambda = 6$ and $a = 0.001$.

From the comparison, it is clear that the TV model works better than the model with L^2 energy. This is generally true when the landscape has steep cliffs or sharp slopes. As in the classical literature, TV model is an adaptive filtering process and thus preserves cliffs faithfully. The L^2 energy inevitably smears intrinsic singularities.

16.4.1.2 Sierpinski graph

Our second example applies the TV restoration model (16.10) to data existing on a level 3 Sierpinski graph (Figure 16.5; see Yamaguti, et al. [12]). In order to clearly visualize the noisy data and the restoration result, we have labeled the 42 nodes.

The subplot on top in Figure 16.6 shows the noisy data u^0 according to the labeling in Figure 16.5. The TV restoration equation (16.10) with $\lambda = 4$ and $a = 0.0001$ is applied to u^0, and the restoration result is plotted at the bottom. Notice that here we apply the restoration equation to the original graph topology, not the labeled linear chain. As a result, we can easily interpret the overshooting at node 25, which is the "boundary" node between two groups of nodes.

16.4.2 Concluding remarks

The key to the digitized PDE method is the energy functions which measure the regularity of a given data on a graph. The L^2 total energy $E^{(2)}$ and TV total energy E^{TV} discussed in this chapter correspond to

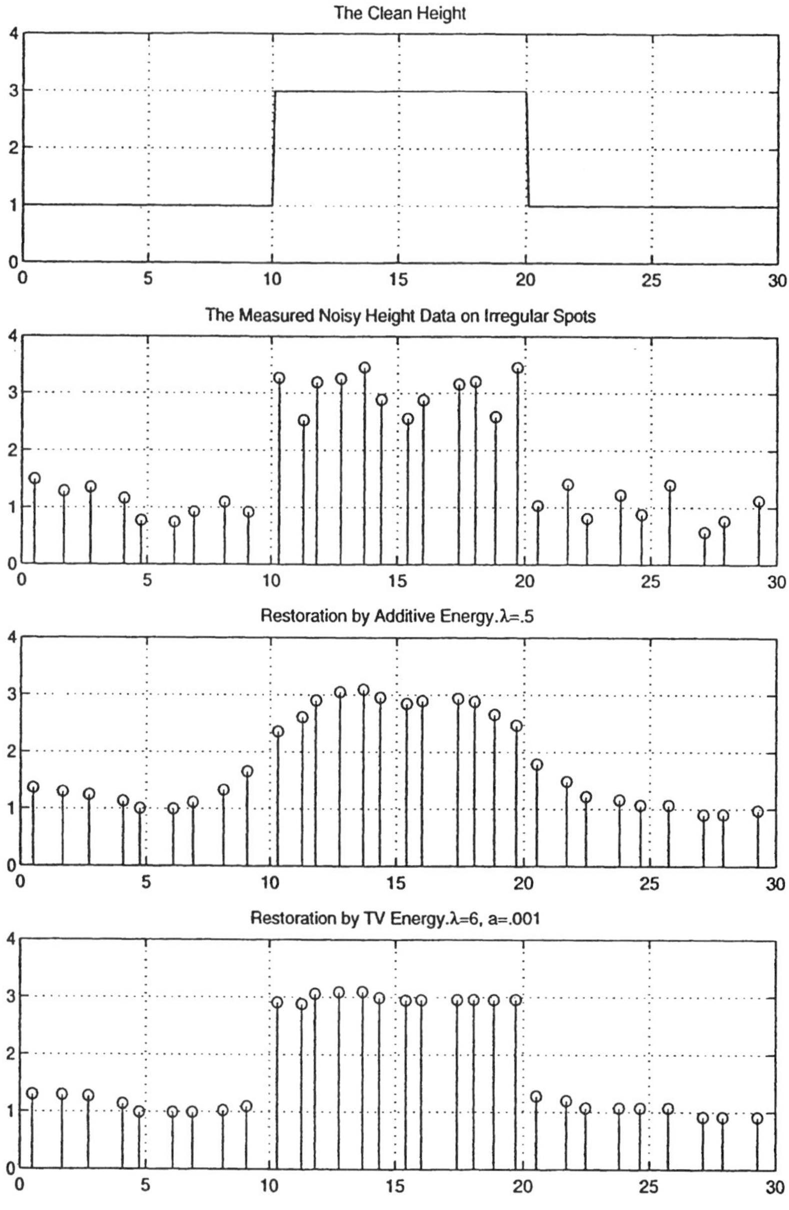

FIGURE 16.4
Application to a linear graph (chain): restoration of the height field data of a one-dimensional landscape.

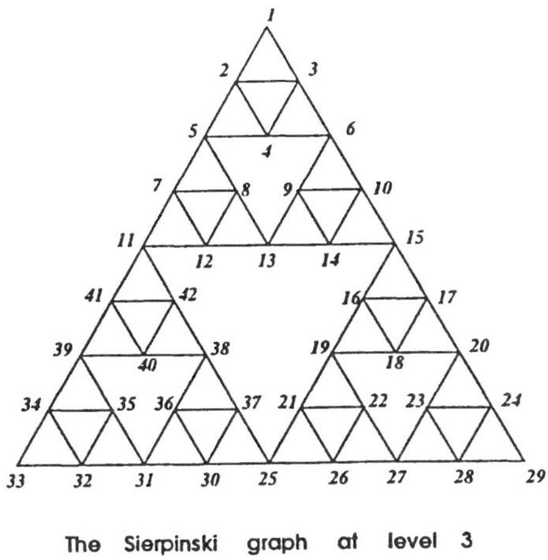

FIGURE 16.5
The 42 nodes of the Sierpinski graph at level 3.

the two most commonly used functionals in the classical literature. $E^{(2)}$ performs efficiently if the contaminated data are originally smooth, while the TV model can faithfully restore intrinsic singularities.

In some other applications, it is advisable to modify $E^{(2)}$ and E^{TV} according to the underlying graph topology. For a general *plain* graph or weight-free graph, the distance between two nodes is naturally measured by the length of the shortest linking path. Therefore, the direct neighbors in N_α are the equally nearest to α, for instance. This assumption is apparently unreasonable if the graph is weighted, especially when the weight on an edge does have something to do with the closeness of the two linked nodes. For instance, in Figure 16.2, the graph on the right is better weighted by the physical distance. For such type of weighted graphs, the strength function $\|\nabla_g u\|$ leading to global energy functions must be modified accordingly. Let $m_{\alpha,\beta}$ denote the weight on the edge $\alpha \sim \beta$. Then one possible fixing strategy is to modify $\|\nabla_g u\|$ to $\|\nabla_g^w u\|$

$$\|\nabla_g^w u\| = \left[\sum_{\beta \in N_\alpha} (u_\beta - u_\alpha)^2 W(m_{\alpha,\beta}) \right]^{\frac{1}{2}} .$$

Here $W(x)$ is a real positive function describing the actual influence of the weights. As long as the convexity properties are preserved, the analysis in this chapter still works effectively.

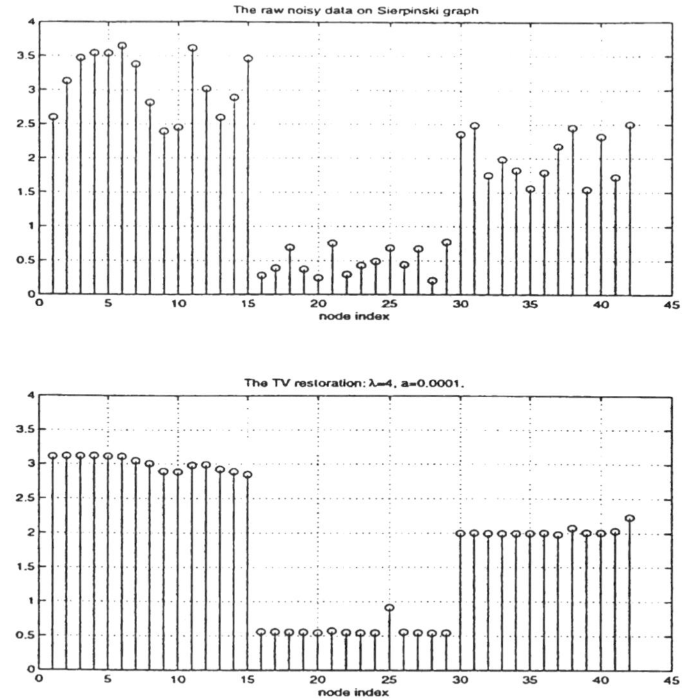

FIGURE 16.6
TV restoration for data on the Sierpinski graph. (See text explanation for the "spurious" node 25.)

Acknowledgements Stanley Osher's research was supported by NSF Grant DMS-9706827. Jianhong Shen's research was supported by NSF Grant DMS-9626755.

References

[1] A. Chambolle and P. L. Lions. Image recovery via total variational minimization and related problems. *Numer. Math.*, 76:167–188, 1997.

[2] T. Chan and J. Shen. Variational restoration of non-flat image features: models and algorithms. Submitted to *SIAM J. Appl. Math.*, 1999.

[3] F. R. K. Chung. *Spectral Graph Theory*, volume 92 of *Regional Con-*

ference Series in Mathematics. Amer. Math. Soc., 1997.

[4] R. R. Coifman and A. Sowa. Combining variational calculus and wavelets for image enhancement. To appear in *Appl. Comp. Harmon. Anal.*, 1999.

[5] R. R. Coifman and A. Sowa. New methods of controlled total variation reduction for digital functions. Preprint, 1999.

[6] I. Daubechies. *Ten Lectures in Wavelets*, volume 61 of *CBNS-NSF Regional Conference Series in Applied Mathematics*. SIAM, Philadelphia, 1992.

[7] A. Marquina and S. Osher. Explicit algorithms for a new time dependent model based on level set motion for nonlinear deblurring and noise removal. UCLA, CAM Report 99-5, 1999.

[8] J.-M. Morel and S. Solimini. *Variational Methods in Image Segmentation*, volume 14 of *Progress in Nonlinear Differential Equations and Their Applications*. Birkhäuser, Boston, 1995.

[9] D. Mumford and J. Shah. Optimal approximations by piecewise smooth functions and associated variational problems. *Comm. Pure Applied. Math.*, XLII:577–685, 1989.

[10] L. Rudin and S. Osher. Total variation based image restoration with free local constraints. *Proc. 1st IEEE ICIP*, 1:31–35, 1994.

[11] L. Rudin, S. Osher, and E. Fatemi. Nonlinear total variation based noise removal algorithms. *Physica D*, 60:259–268, 1992.

[12] M. Yamaguti, M. Hata, and J. Kigami. *Mathematics of Fractals*, volume 167 of *Translations of Mathematical Monographs*. AMS, Providence, 1993.

Chapter 17

Boundary Quadrature Formulas and Their Applications

Tian-Xiao He

Department of Mathematics and Computer Science, Illinois Wesleyan University, Bloomington, IL 61702-2900
E-mail: the@sun.iwu.edu

Abstract This chapter surveys the analytical approach for constructing multivariate numerical integration formulas that use only boundary points as evaluation points. The applications of boundary quadrature formulas to boundary value problems of partial differential equations are also discussed.

Contents

17.1 Introduction

Although numerical multivariate integration is an old subject, it has never been applied as widely as it is now. We can find its applications everywhere in math, science, and economics. A good example is the collateralized mortgage obligation (CMO), which is a bundle of loans that generates cash flows from interest payments and the repayment of principal. It is clear that fluctuations in interest rates alter the cash flows. Let us consider a CMO consisting of 15-year mortgages with monthly payments. There are 180 cash flows in all, depending on 180 interest rates, which are basically 180 variables. The goal is to compute the present expected value of the CMO, averaged over all possible fluctuations of the 180 interest rates. Parameterized by the probabilities with which fluctuations occur, this calculation can be formulated as a multivariate integral over the 180-dimensional unit cube ([4]).

This chapter discusses a technique for numerical integration by using dimensionality-reducing expansions (DRE) to reduce a higher dimensional integral to lower dimensional integrals with or without a remainder. Hence, a type of boundary quadratures can be constructed by using the expansions.

The DRE without remainder is also called an exact DRE. Obviously, a DRE can be used to reduce the computation load of many very high dimensional numerical integrations, such as the CMO problem mentioned above. Most DREs are based on Green's Theorem in a real or complex field. In 1963, using the theorem, Hsu [15] devised a way to construct a DRE with algebraic precision for multivariate integrations. From 1978 to 1986, Hsu, Zhou, and the author [16-19] developed a more general method to construct a DRE with algebraic precision and estimate its remainder. In 1972, with the aid of Green's Theorem and the Schwarz function, P.J. Davis [6] gave an exact DRE for a double integral over a complex field. In 1979, also by using Green's Theorem, Kratz [23] constructed an exact DRE for a function that satisfied a type of partial differential equations. Lastly, if we want this introduction to be complete, we must not overlook Burrows' DRE for measurable functions. His DRE can reduce a multivariate integration into a one-dimensional integral. Some important applications of DRE include the construction of boundary quadrature formulas and asymptotic formulas for oscillatory integrals.

A boundary quadrature formula is an approximate integration formula with all its evaluation points lying on the boundary of the domain of integration. Such a formula may be particularly useful for the case where the values of the integrand function and its derivatives inside the domain are not given or are not easily determined. Indeed, boundary quadrature formulas are not really new. From the viewpoint of numerical analysis, the classical Euler-Maclaurin summation formula and the Hermite two-end multiple nodes quadrature formulas may be regarded as one-dimensional boundary quadrature formulas since they make use of only the integrand function values and their derivatives at the limits of integration. The earliest example of a boundary quadrature formula with some algebraic precision for multivariate integration is possibly the formula of algebraic precision (or degree) 5 for a triple integral over a cube given by Sadowsky [27] in 1940. He used 42 points on the surface of a cube to construct the quadrature, which has been modified by the author with a quadrature with 26 points, the fewest possible boundary points [11,12]. Some 20 years later, Levin [24, 25], Federenko [8], and Ionescu [20] investigated individually certain optimal boundary quadrature formulas for double integration over a square using partial derivatives at some boundary points of the region. Despite these advances, however, both the general principle and the general technique for construction remained lacking for many years.

During 1978-87, Hsu, Zhou, and the author developed a general process for the construction of boundary quadrature formulas using some basic ideas of DRE of multivariate integration in [16-19]. In this chapter, we will discuss all of these results. We will also look at some recent developments on this topic, such as using boundary quadrature formulas to develop a boundary element scheme (please see Section 5), solving the problem of the fewest possible evaluation points of a boundary quadrature formula, and constructing boundary quadrature formulas with wavelet weight functions.

In Section 2, we will show that a DRE can be made an effective tool for constructing boundary quadrature formulas with preassigned algebraic precision. Some optimal DREs will also be discussed. Section 3 will discuss general Hermite formulas and a class of boundary quadrature formulas with the highest possible algebraic precision. A general process to construct quadratures for boundary integrals by using periodic wavelets will be given in Section 4. Section 5 will describe numerical solutions of boundary value problems of partial differential equations using boundary quadrature formulas.

17.2 Dimensionality-Reducing Expansions

Suppose that we are given a class of multiple integrals of the form

$$J := \int_{\Omega} F(X)\, dX, \ F(X) \in K,$$

where K is a certain class of real functions defined on a bounded domain $\Omega \subset R^n$, $X \equiv (x_1, \ldots, x_n)$ represents a point-vector, and $dX \equiv dx_1, \ldots, dx_n$. The $n-1$-dimensional boundary surface of Ω is denoted by $\partial\Omega$. Our goal is to obtain approximate integration formulas of the form

$$J \approx \sum_{i,j} A_{ij} \Lambda_i(F)(X_{ij}), \quad X_{ij} \in \partial\Omega,$$

where each $\Lambda_i(F)$ stands for a certain linear combination of F and some of its partial derivatives, X_{ij} are evaluation points (nodes) lying on the boundary of Ω, and A_{ij} are the corresponding quadrature coefficients.

One general process of construction consists of three main steps: (1) Replace J by a set of boundary surface integrals with an error term that can be reflected or well controlled. (2) Choose a suitable auxiliary function based on the question in order to minimize or diminish the error term. (3) Choose suitable numerical integration formulas that can be used to approximate those boundary surface integrals with a certain sense of accuracy. Note that the first step may easily be achieved by means of successive applications of the Green formula in higher dimensions.

In general, for a differential operator L of order m, if there exists a differential operator M of the same order such that

$$\int_\Omega (F \cdot L(G) - G \cdot M(F))\, dX = \int_{\partial\Omega} W\, dS,$$

where W is a certain linear combination of the products of F and G or their partial derivatives with their total order not exceeding $m-1$, then M is called the conjugate differential operator of L. If M is equal to L, then L is said to be a self-conjugate differential operator.

In this section we will discuss a differential operator L defined by

$$L = \sum_{m_1+\ldots+m_n \leq m} \lambda_{m_1,\ldots,m_n} D^{m_1,\ldots,m_n}, \tag{17.1}$$

where $\lambda_{m_1,\ldots,m_n}$ are constants and $D^{(m_1,\ldots,m_n)} = \partial^{m_1+\ldots+m_n} / \partial x_1^{m_1} \ldots \partial x_n^{m_n}$. From the following lemma, we know that the conjugate operator of L is $M = (-1)^m L$.

LEMMA 17.1

Let $\Omega \subset R^n$ be an n-dimensional bounded closed domain with its boundary $S = \partial\Omega$ being a piecewise smooth surface. In particular, if $n = 2$, $\partial\Omega$ is a *simple closed curve with finite length. Let $F(X), G(X) \in C^m(\Omega)$, where m is the order of the operator L. Then*

$$\int_\Omega F \cdot L(G)\, dX = \int_{\partial\Omega} W\, dS + (-1)^m \int_\Omega G \cdot L(F)\, dX, \tag{17.2}$$

where dS is the area element of $\partial\Omega$ and W is a certain linear combination of all the possible terms formed by the products of partial derivatives of F and of G with their total order not exceeding $m-1$.

Lemma 17.1 can be easily verified by repeatedly applying Green's Theorem. Thus, for instance, if $L = \Lambda = D^{(m_1,\ldots,m_n)}, m_1 + \ldots + m_n = m$, we find

$$W = \sum_{i=1}^{n} \sum_{k=0}^{m_i-1} (-1)^{m_1+\ldots+m_{i-1}+k} D^{(m_1,\ldots,m_{i-1},k,0,\ldots,0)} F(X)$$

$$= D^{(0,\ldots,0,m_i-k-1,m_{i+1},\ldots,m_n)} G(X) \frac{\partial x_i}{\partial \nu}, \tag{17.3}$$

where $\frac{\partial x_i}{\partial \nu}$ is the normal derivative of x_i on the surface $\partial\Omega$. In particular, if $\partial\Omega$ can be expressed as $Q(x_1, \ldots, x_n) = 0$, then $\frac{\partial x_i}{\partial \nu} = \frac{\partial Q}{\partial x_i} / [(\frac{\partial Q}{\partial x_1})^2 + \cdots + (\frac{\partial Q}{\partial x_n})^2]^{-\frac{1}{2}}$. For $i = 1$ and $i = n$, we take, respectively, $D^{(m_1,\ldots,m_{i-1},k,0,\ldots,0)} = D^{(k,0,\ldots,0)}$ and $D^{(0,\ldots,0,m_i-k-1,m_{i+1},\ldots,m_n)} = D^{(0,\ldots,0,m_n-k-1)}$ in (17.3). Obviously, the order of differentiation of each term of W is $m-1$.

By suitable choice of G we may make use of (17.2) to get an expansion of the integral J in terms of some boundary surface integrals with an error term $\int_\Omega G \cdot L(F)\, dX$. This is a DRE when the error term is dropped. Thus, the second step of the method is to minimize the error term by finding an appropriate G. This actually leads to a minimization problem of the functional $\int_\Omega G \cdot L(F)\, dX$ subject to certain required conditions for $G(X)$. Without loss of generality, in the following we assume $L = \Lambda = D^{(m_1,\ldots,m_n)}, m_1 + \ldots + m_n = m$. Thus (17.2) can be written as

$$\int_\Omega F\Lambda(G)\, dX = \int_{\partial\Omega} W\, dS + (-1)^m \int_\Omega G\Lambda(F)\, dX, \tag{17.4}$$

where W is defined by (17.3).

Obviously, $\Lambda(G)$ can be a constant for certain types of polynomials, and it follows that (17.4) is a DRE of the integral $\int_\Omega F\, dX$ with the error term $(-1)^m \int_\Omega G\Lambda(F)\, dX$. Hence, we will define G as the following three types of polynomials so that the error term will be minimized. Denote $i = i_1 + \ldots + i_n$ and $m = m_1 + \ldots + m_n$, and let $\sum'_{i_1,\ldots,i_n} a_{i_1\ldots i_n} x_1^{i_1} \ldots x_n^{i_n}$, where $0 \le i \le m$, the linear combination of all monomials $x_1^{i_1} \ldots x_n^{i_n}$, $0 \le i \le m$, except the term $x_1^{m_1} \ldots x_n^{m_n}$. Using these notations, we define $G_n^{(m_1,\ldots,m_n)} = \{P(x) = x_1^{m_1} \ldots x_n^{m_n} + \sum'_{i_1,\ldots,i_n} a_{i_1\ldots i_n} x_1^{i_1} \ldots x_n^{i_n}, 0 \le i \le m\}$, $K_n^{(m_1,\ldots,m_n)} = \{P(x) = x_1^{m_1} \ldots x_n^{m_n} + \sum_{i_1,\ldots,i_n} a_{i_1\ldots i_n} x_1^{i_1} \ldots x_n^{i_n}, 0 \le i_1, \ldots, i_n \le m-1, 0 \le i \le m-1\}$, and $H_n^{(m_1,\ldots,m_n)} = \{P(x) = x_1^{m_1} \ldots x_n^{m_n} + \sum_{i_1,\ldots,i_n} a_{i_1\ldots i_n} x_1^{i_1} \ldots x_n^{i_n}, 0 \le i_k \le m_k; k = 1, \ldots, n; 0 \le i \le m-1\}$. Obviously, $H_n^{(m_1,\ldots,m_n)} \subset K_n^{(m_1,\ldots,m_n)} \subset G_n^{(m_1,\ldots,m_n)} \subset C^m$. If $G \in H_n^{(m_1,\ldots,m_n)}$, we have the approximate DRE

$$\int_\Omega F(X)\, dX \approx \frac{1}{m_1! \ldots m_n!} \int_{\partial\Omega} W(X)\, dS, \tag{17.5}$$

where $W(X)$ is defined by (17.3) and the omitted error term or remainder is

$$\rho_m = \frac{(-1)^m}{m_1! \ldots m_n!} \int_\Omega G(X) \Lambda(F)(X)\, dX. \tag{17.6}$$

(17.5) holds exactly for any polynomial $F(X) = \sum'_{i_1,\ldots,i_n} a_{i_1\ldots i_n} x_1^{i_1} \ldots x_n^{i_n}$, $i \le m$. In particular, it is exactly true for all $F(X) \in \pi_{m-1}^n$, the set of all polynomials in R^n with total degree not more than $m - 1$.

In the following theorems, we will give the optimal approximate DRE (17.5) in the sense that the remainder ρ_m possesses the minimum estimation. Since $|\rho_m| \le \frac{1}{m_1! \ldots m_n!} ||\Lambda F||_{L_1} ||G||_C$, $\frac{1}{m_1! \ldots m_n!} ||\Lambda F||_C ||G||_{L_1}$, and $\frac{1}{m_1! \ldots m_n!} ||\Lambda F||_{L_2} ||G||_{L_2}$, we need to estimate $||G||_C$, $||G||_{L_2}$, and $||G||_{L_1}$, respectively.

THEOREM 17.1

Suppose $\Omega = V_n = \{(x_1, \ldots, x_n) : -1 \le x_i \le 1, i = 1, \ldots, n\}$ is an n-dimensional cube, $F(X) \in C^{(m_1,\ldots,m_n)}(\Omega)$, the set of all functions defined on Ω and have continuous partial derivatives $\frac{\partial^m F}{\partial x_1^{m_1} \ldots \partial x_n^{m_n}}$, and $G(X) \in K_n^{(m_1,\ldots,m_n)}$. Then the approximate DRE (17.5) is optimal if

$$G(X) = G^* = \prod_{i=1}^{n} \prod_{k=1}^{m_i} (x_i + \cos \frac{2k-1}{2m_i} \pi) \tag{17.7}$$

and the minimum estimation for ρ_m is

$$|\rho_m| \le \frac{1}{2^{m-n} m_1! \ldots m_n!} ||\Lambda F||_{L_1}. \tag{17.8}$$

PROOF It is sufficient to note that $G(X)$ in (17.7) is the best Chebyshev approximation polynomial or the minimax approximation polynomial, i.e., it satisfies $||G^*||_C = 2^{-m+n} = \min_{G \in K_n^{(m_1,\dots,m_n)}} ||G(X)||_C, X \in V_s$. ∎

Similarly, we have the following result.

THEOREM 17.2
Suppose $G(x,y) \in k_2^{(m_1,m_2)}$ and Ω are either the triangle region $\triangle(x \geq -1, y \geq -1, x+y \leq 0)$ or the unit disc R $(x^2+y^2 \leq 1)$. Then the DREs over $\triangle$ and R are optimal if

$$G(x,y) = B_{m_1 m_2}(x,y) = \prod_{i=1}^{m_1}(x + \cos \tfrac{2i-1}{2(m_1+m_2)}\pi) \prod_{j=1}^{m_2}(y + \cos \tfrac{2j-1}{2(m_1+m_2)}\pi)$$

and

$$G(x,y) = I_{m_1 m_2}(x,y) = \prod_{i=1}^{m_1}(x + \sin \tfrac{2i-m_1-1}{2(m_1+m_2)}\pi) \prod_{j=1}^{m_2}(y + \sin \tfrac{2j-m_2-1}{2(m_1+m_2)}\pi),$$

respectively. The optimal estimations for the above expansions are

$$|\rho_m| \leq ||D^{(m_1,m_2)}F||_{L_1(\triangle)}/[2^{m_1+m_2-1}m_1!m_2!]$$

and

$$|\rho_m| \leq ||D^{(m_1,m_2)}F||_{L_1(R)}/[2^{m_1+m_2-1}m_1!m_2!],$$

respectively.

From the well-known theory of orthogonal polynomials, we immediately have the following result.

THEOREM 17.3
There exists a unique $G^*(X) \in G_n^{(m_1,\dots,m_n)}$ with $\langle G^*, G^* \rangle = \min_{G \in K_n^{(m_1,\dots,m_n)}} \langle G, G \rangle$. Moreover, $G^*(X)$ is the orthogonal polynomial of degree m. The corresponding approximate DRE (17.5) is optimal for all $G \in G_s^{(m_1,\dots,m_s)}$ and the corresponding minimum estimation for the remainder is

$$|\rho_m| \leq \frac{1}{m_1! \dots m_n!} ||\Lambda F||_{L_2} ||G^*||_{L_2}. \tag{17.9}$$

As for $G(X) \in K_n^{(m_1,\dots,m_n)}$, it can be written as $G(X) = x_1^{m_1} \dots x_n^{m_n} - p(X)$, where $p(X) \in \pi_{m-1}^n$. Assume that an orthonormal basis of π_{m-1}^n

is $\{p_i(X), i = 1, \ldots, n\}$, then we have $n \leq \binom{m+n-2}{n-1}$ from Jackson Theorem. Therefore, $\langle G^*, G^* \rangle = \min_{G \in K_n^{(m_1,\ldots,m_n)}} \langle G, G \rangle$ if

$$G^*(X) = x_1^{m_1} \ldots x_n^{m_n} - \sum_{i=1}^{n} \langle x_1^{m_1} \ldots x_n^{m_n}, p_i \rangle p_i. \tag{17.10}$$

Hence, we have Theorem 17.4.

THEOREM 17.4
Suppose $F(X) \in C^{(m_1,\ldots,m_n)}$ and $G(X) \in K_n^{(m_1,\ldots,m_n)}$. Then the approximate DRE (17.5) is optimal if $G(X) = G^*(X)$ shown in (17.10). The corresponding remainder has the minimum estimation shown in (17.9).

THEOREM 17.5
In (17.5), suppose $\Omega = V_s = \{(x_1, \ldots, x_n) = -1 \leq x_i \leq 1, i = 1, \ldots, n\}$, an s-dimensional cube, $F(X) \in C^{(m_1,\ldots,m_n)}(\Omega)$, and $G(X) \in H_n^{(m_1,\ldots,m_n)}$. Then the approximate DRE (17.5) is optimal if $G(X) = \prod_{i=1}^{s} Q_{m_i}(x_i)$, where

$$Q_m(x) = \frac{\sin((m+1)\ \cos^{-1} x)}{2^m \sqrt{1-x^2}} \tag{17.11}$$

is the second kind of Chebyshev polynomial. The corresponding minimum estimation of the remainder is

$$|\rho_m| \leq \frac{1}{2^{m-n} m_1! \ldots m_n!} ||\Lambda F||_C. \tag{17.12}$$

In addition, let $A = \{G(X) : G(x) = \prod_{i=1}^{n} g_i(x_i),\ g_i(x_i) \in \pi_{m_i}^1,\ i = 1, 2, \ldots, n\}$. Also denoting $B_M = \{F(X) : F(X) \in C^{(m_1,\ldots,m_n)}(V_n)$ and $|\Lambda F| \leq M$ for some $M\}$, we have

$$\inf_{G \in A} \sup_{F \in B} |\rho_m(F, G)| = \frac{M}{2^{m-n} m_1! \ldots m_n!}. \tag{17.13}$$

PROOF Actually this is a consequence of the well-known Zolotarëff-Korkin theorem in a higher dimension which asserts that

$$\inf_{G \in A} \int_{V_n} |G(X)|\, dX = \inf_{G \in A} \prod_{i=1}^{n} \int_{-1}^{1} |Q_{m_i}(x_i)|\, dx_i = \prod_{i=1}^{n} \frac{1}{2^{m_i-1}} = \frac{1}{2^{m-n}}. \tag{17.14}$$

Moreover, it can be easily shown that

$$\sup_{F\in B} |\rho_m(F,G)| = \frac{M}{m_1!\dots m_n!}\int_{V_n} |G(X)|\, dX. \tag{17.15}$$

Hence, (17.13) follows from (17.14) and (17.15). ∎

Using expansion (17.5) and the Gaussian quadrature formula, we can construct an optimal boundary quadrature formula over a certain domain. Here the optimal boundary quadrature formula means a quadrature formula with the highest possible algebraic precision and the fewest possible nodes (evaluation points) on the boundary of the domain. More examples will be shown in the next section.

As an example, let us consider the boundary quadrature formula for the integral of $F : R_s \to R$, $F(X) = F(x_1, \dots, x_n)$, over $V_n = [-1,1]^n$. Assume $G(X) = \prod_{i=1}^n Q_{m_i}(x_i)$, where $Q_m(x)$ is the second kind of Chebyshev polynomial defined by (17.11). From (17.5) and (17.3) we have the expansion

$$\int_{V_n} F(X)\, dX \approx \frac{1}{m_1!\dots m_n!}\sum_{i=1}^{n}\int_{V_{n-1}} [W(x_1,\dots,x_n)]_{x_i=-1}^{x_i=1}\, dS$$

$$\approx \sum_{i=1}^{n}\int_{-1}^{1}\cdots\int_{-1}^{1}\frac{1}{m_1!\dots m_i!}\sum_{k=0}^{m_i-1}(-1)^{m_1+\dots+m_{i-1}+k}$$

$$[D^{(m_1,\dots,m_{i-1},k,0,\dots,0)}FQ_{m_1}(x_1)\dots Q_{m_{i-1}}(x_{i-1})$$

$$\frac{\partial^{m_i-k-1}}{\partial x_i^{m_i-k-1}}Q_{m_i}(x_i)]_{x_i=-1}^{x_i=1}\, dx_1\dots dx_{i-1}dx_{i+1}\dots dx_n. \tag{17.16}$$

The remainder of the expansion, $\rho_m = \frac{(-1)^m}{m_1!\dots m_n!}\int_{V_n} G\Lambda(F)\, dX$, can be estimated by (17.12).

Obviously, $\rho_m = 0$ and, therefore, expansion (17.16) holds exactly for a polynomial of degree of at most $m-1$. Hence, we can only expect to construct boundary quadrature formula with algebraic precision of at most $m-1$.

On the other hand, if $F \in \pi_{m-1}^n$, the integrand of the integrals over V_{n-1} in the summation of equation (17.16) are polynomials of degree $m-1$ in R^{n-1}. Therefore, each of these integrals can be expanded approximately to the summation of integrals over V_{n-2} by using (17.5), and these expansions are also exactly true for all $F \in \pi_{m-1}^n$.

By repeating this process of expansions successively, we can eventually and approximately expand the integral $\int_{V_n} F(X)\,dX$ as the summation of univariate integrals over $[-1,1]$; i.e., the integrals along $(x_1, \pm 1, \ldots, \pm 1)$, $(\pm 1, x_2, \ldots, \pm 1)$, $\ldots$, $(\pm 1, \pm 1, \ldots, x_n)$ with $-1 \le x_i \le 1$, $i = 1, 2, \ldots, n$, respectively. And the expansion is exact if $F \in \pi^n_{m-1}$. For each univariate integral, let us choose a quadrature formula of algebraic precision $m-1$ with the least number of evaluation points. For instance, if $m = 2r$, we choose

$$\int_{-1}^{1} f(t)\,dt \approx \sum_{j=0}^{r-1} C_j f(t_j) \quad (-1 \le t_j \le 1),$$

where $t_j, j = 0, 1, \ldots, r-1$, are the zeros of the nth degree Legendre polynomial. Hence, we obtain the following result.

THEOREM 17.6
There exists a kind of optimal boundary quadrature formulas for the integral $\int_{V_s} F(X)\,dX$ that possesses the highest possible algebraic precision $2r-1$ and uses $2^{n-1}nr$ evaluation points on the "edges" of the cubical domain $[-1,1]^n$ as follows: $(t_j, \pm 1, \ldots, \pm 1)$, $(\pm 1, t_j, \ldots, \pm 1)$, $\ldots$, $(\pm 1, \ldots, \pm 1, t_j)$ with $j = 0, 1, \ldots, r-1$ and t_j being the zeros of the nth degree Legendre polynomials.

17.3 Boundary Quadrature Formulas

In this section, we shall examine the 3-dimensional case, but all results can be extended to the n-dimensional case without essential difficulties.

Define $\Omega \subset R^3$ as

$$\Omega := \{(x, y, z) : \phi_1(x, y) \le z \le \phi_2(x, y), (x, y) \in E \subset R^2\}, \tag{17.17}$$

where ϕ_1 and ϕ_2 are bivariate smooth functions and E a convex region with a piecewise smooth closed curve, ∂E, as boundary. Denote $X \equiv (x, y, z)$ and $dX \equiv dxdydz$, and we assume that $F(X), G(X) \in C^m(\Omega)$, where m is a positive integer.

To construct a boundary quadrature formula with a simpler structure, we may choose $\Lambda = \frac{1}{m!}\left(\frac{\partial}{\partial z}\right)^m$ and assume that $G(X)$ takes the form

$$G(X) = z^m + h(x, y) z^{m-1} + \text{lower degree terms in z}, \tag{17.18}$$

where $h(x, y)$ and the coefficients of all the lower degree terms in z may be any smooth functions of (x, y). Then by using expression (17.2), we obtain

$$J \equiv \int_{\Omega} F(X)dX \simeq \frac{1}{m!}\sum_{k=0}^{m-1}(-1)^k \int_{\partial\Omega}\left(\frac{\partial}{\partial z}\right)^k F\left(\frac{\partial}{\partial z}\right)^{m-k-1} Gdxdy, \tag{17.19}$$

where the right-hand side is an expansion with the following remainder term being dropped.

$$P_m = \frac{(-1)^m}{m!}\int_{\Omega} G(X)\left(\frac{\partial}{\partial z}\right)^m FdX. \tag{17.20}$$

A direct estimate of $|P_m|$ was obtained in (17.9), namely,

$$|P_m| \le \frac{1}{m!}||\frac{\partial^m F}{\partial z^m}||_{L_2}||G||_{L_2}.$$

Theorem 17.3 minimized the L_2-norm of G, $||G||_{L_2}$, by using an orthogonal basis of π^2_{m-1} over Ω. However, for general Ω, we can only approximately minimize $||G||_{L_2}$ under appropriate conditions. Gröbner [10] has shown that Appell polynomials and Hermite-Didon polynomials are just the solutions for minimizing $||G||_{L_2}$ over simplex and spherical domains, respectively (see examples at the end of this section). Thus, as motivated by Gröbner's result concerning the above-mentioned polynomials, one may make an appropriate choice for $G(X)$ in the following form

$$G(X) = \frac{m!}{(2m)!}\left(\frac{\partial}{\partial z}\right)^m [(z-\phi_1(x,y))^m(z-\phi_2(x,y))^m]. \tag{17.21}$$

Obviously, $G(X)$ is of form (17.18) and implies the special polynomials mentioned previously as particular cases. Moreover, it can be shown that such a choice of $G(X)$ can even yield an extension of the Hermite formula.

THEOREM 17.7

Let $F(X) \in C^{2m}(\Omega)$ and denote $\phi_i \equiv \phi_i(x,y)$, $i = 1, 2$. Then we have the generalized Hermite formula

$$\int_{\Omega} F(X)dX = \sum_{v=1}^{m}\frac{1}{v!}\frac{\binom{m}{v}}{\binom{2m}{v}}\int_E H_v(x,y)dxdy + P_m, \tag{17.22}$$

where $H_v(x,y)$ *is defined by*

$$H_v(x,y) = (\phi_2 - \phi_1)^v [F_z^{(v-1)}(x,y,\phi_1) + (-1)^{v-1} F_z^{(v-1)}(x,y,\phi_2)], \quad (17.23)$$

$F_z^{(\alpha)} = (\partial/\partial z)^\alpha F$ *and the remainder term is given by*

$$P_m = \frac{1}{m!} \int_\Omega (z-\phi_1)^m (z-\phi_2)^m F_z^{(2m)} dX. \quad (17.24)$$

In addition, formula (17.22) is of the highest possible algebraic precision, $2m-1$.

PROOF Actually, (17.22) and (17.23) are only a special case of equation (17.5). It is sufficient to prove (17.24). The original form (17.20) of P_m can be written as

$$P_m = \frac{(-1)^m}{(2m)!} \int_\Omega \left(\frac{\partial}{\partial z}\right)^m F \left(\frac{\partial}{\partial z}\right)^m [(z-\phi_1)^m (z-\phi_2)^m] dX. \quad (17.25)$$

This can be transformed into (17.24) by using formula (17.2), in which $L = \Lambda = (\partial/\partial z)^m$ and $G = \bar{G} = (z-\phi_1)^m (z-\phi_2)^m$. Hence, for $X \in \partial\Omega$, $(\partial/\partial z)^j \bar{G}(X) = 0$, $j = 0, 1, \cdots, m-1$, so that the corresponding integral $\int_{\partial\Omega} W dS$ vanishes, and expression (17.24) is validated.

It is clear that $P_m = 0$ if F is a polynomial in x, y, and z, with degree $\leq 2m-1$. On the other hand, if F is a polynomial in z of degree $2m$, (17.24) gives $|P_m| > 0$. Hence, formula (17.22) has the highest possible algebraic precision $2m-1$.

By applying the Cauchy-Schwartz inequality to (17.24) and then using formula (17.2), noting that the corresponding boundary integrals $\int_{\partial\Omega} W dS$ vanishes, we have

$$|P_m| \leq \frac{1}{(2m)!} ||F_z^{(2m)}||_{L_2} \left(\int_\Omega (z-\phi_1)^{2m} (z-\phi_2)^{2m} dX \right)^{1/2}$$

$$= \frac{1}{(2m)!} ||F_z^{(2m)}||_{L_2} \left(\frac{(2m)!}{(4m)!} \int_\Omega (z-\phi_2)^{2m} \frac{\partial^{2m}}{\partial z^{2m}} (z-\phi_1)^{4m} dX \right)^{1/2}$$

$$= \frac{1}{(2m)!} ||F_z^{(2m)}||_{L_2} \left(\frac{(2m)!}{(4m)!} \int_\Omega (2m)!(z-\phi_1)^{4m} dX \right)^{1/2}$$

$$= \frac{1}{(4m)!}||F_z^{(2m)}||_{L_2}||(z-\phi_1)^{2m}||_{L_2}.$$

Here, the factor $||(z-\phi_1)^{2m}||_{L_2}$ may be replaced by $||(z-\phi_2)^{2m}||_{L_2}$. Hence, we have the following result. ∎

THEOREM 17.8
For the remainder P_m as given by (17.24) we have the estimate

$$|P_m| \leq \frac{1}{(4m)!}\mu||F^(2m)_z||_{L_2}, \tag{17.26}$$

where $\mu = \min\{||(z-\phi_1)^{2m}||, ||(z-\phi_2)^{2m}||\}$.

Let us now choose a numeral integration formula of the form

$$\int_E f(x,y)dxdy \simeq \sum_i a_i f(x_i, y_i), \tag{17.27}$$

where $\sum_i$ is a finite summation and a_i are the corresponding quadrature coefficients. Of course, the evaluation points to be used are assumed to be inside the region E of integration. Thus, applying (17.27) to each of the integrals contained in the summation of (17.22) without P_m, we obtain

$$J \equiv \int_\Omega F(X)dX \simeq \sum_{v=1}^{m} \frac{1}{v!}\sum_i a_i H_v(x_i, y_j), \tag{17.28}$$

where the values $H_v(x_i, y_i)$ are computed by (17.23) with $(x,y) = (x_i, y_i)$.

Clearly, if $\partial\Omega$ is formed by some algebraic surfaces, the boundary quadrature formula for $J(F)$ may possess some preassigned algebraic precision.

THEOREM 17.9
Suppose Ω is a bounded domain defined by (17.17) in which $\phi_1(x,y)$ and $\phi_2(x,y)$ are bivariate polynomials with degrees $deg\,\phi_1 = r$ and $deg\,\phi_2 = s$. Then in order to make formula (17.28) possess the highest possible algebraic precision, $2m-1$, the formula (17.27) to be employed should be of algebraic precision no less than t, which is given by

$$t = \begin{cases} 2m-1, & where r = s = 0; \\ 2m \cdot \max(r,s), & otherwise. \end{cases} \tag{17.29}$$

PROOF Since the expansion given by (17.22) without P_m is of the highest possible algebraic precision, $2m-1$, it suffices to consider the case when F is taken to be a polynomial in x, y, and z, with degree $deg\, F(x,y,z) = 2m-1$. Also, we shall use the notation $(x)_+ = \max\{0,x\}$. Ignoring the numerical coefficient, a typical term of the polynomial $F(x,y,z)$ may be written as $x^\alpha y^\beta z^\gamma \leq 2m-1$. Now it is easy to compute the maximum of degree $deg\, H_v(x,y)$, $v = 1, \cdots, m$, as follows.

$$\begin{aligned}\max_{1\leq\gamma\leq m}\{deg\, H_v(x,y)\} &= \max_{1\leq v\leq m}\{v\cdot \max(r,s)\\ &+ \max_{\alpha+\beta+\gamma\leq 2m-1}(\alpha+\beta+(\gamma-v+1)_+ \cdot \max(r,s))\}.\end{aligned}$$

Hence, when $r=s=0$, $\max_{1\leq\gamma\leq m}\{deg\, H_v(x,y)\} = 2m-1$; when $r+s \geq 1$, $\max_{1\leq\gamma\leq m}\{deg\, H_v(x,y)\} = 2m\cdot\max\{r,s\}$. Therefore, if (17.27) is of algebraic precision no less than t, then it will be exact for $f(x,y) = H_v(x,y)$, $v = 1,\cdots,m$, and consequently (17.28) is a formula of algebraic precision $2m-1$. Moreover, the algebraic precision of (17.27) should not be less than t, because the value $t := \max\{deg\, H_v(x,y)\}$ can be precisely reached. Hence, the theorem is true. ∎

In the following, we will give several examples of boundary quadrature formulas over simplex, cubical domain, and spherical surface.

Example 17.1

For the simplex $\Omega = \{(x,y,z) : x+y+z \leq a, x,y,z \geq 0\}$, we may express it in form (17.17) with $\phi_1 \equiv 0$, $\phi_2 \equiv a-x-y$, and $E \equiv \Delta := \{(x,y) : x+y \leq a, x,y \geq 0\}$. Accordingly, the required auxiliary function G takes the form

$$G(X) = \frac{m!}{(2m)!}\left(\frac{\partial}{\partial z}\right)^m [z^m(x+y+z-a)^m].$$

This is the well-known Appell polynomial in three variables. Notice that in the present case $r = deg\,\phi_1 = 0$ and $s = deg\,\phi_2 = 1$, so we should take $t = 2m$ from Theorem 9. Thus, if we choose any numerical integration formula (17.27) for the Δ-region with algebraic precision no less than $2m$, we can obtain a boundary quadrature formula of form (17.28) with

$$\begin{aligned}H_v(x_i,y_i) &= (a-x_i-y_i)^v(F_z^{(v-1)}(x_i,y_i,0)\\ &+(-1)^{v-1}F_z^{(v-1)}(x_i,y_i,a-x_i-y_i))\end{aligned}$$

and with the error term $|P_m| \leq \left(\frac{a^{4m+3}}{(4m+3)!}\right)^{1/2} ||F_z^{(2m)}||_{L_2}$. □

Example 17.2

For the cubical domain $\Omega = \{(x,y,z) : |x| \le 1, |y| \le 1, |z| \le 1\}$, we have $\phi_1(x,y) \equiv -1$, $\phi_2(x,y) \equiv 1$, and $E \equiv \{(x,y) : |x| \le 1, |y| \le 1\}$. Accordingly we take

$$G(X) = \frac{m!}{(2m)!}\left(\frac{\partial}{\partial z}\right)^m (z^2-1)^m.$$

This is just the m-th degree Legendre polynomial with leading term z^m. Since $r = s = 0$, we require $t = 2m-1$. Similar to the example shown at the end of Section 2, we apply the Gauss-Legendre m-point quadrature formula to the boundary integral in (17.27) and obtain a boundary quadrature formula for $\int_\Omega F(X)dX$ of precision $2m-1$:

$$\int_\Omega F(X)dX \simeq \sum_{v=1}^{m} \frac{2^v}{v!}\frac{\binom{m}{v}}{\binom{2m}{v}} \sum_{i=0}^{m-1}\sum_{j=0}^{m-1} c_i c_j H_v(x_i, y_j),$$

where $H_v(x_i,y_j) = F_z^{(v-1)}(x_i,y_j,-1) + (-1)^{v-1}F_z^{(v-1)}(x_i,y_j,1)$, c_i are Gaussian quadrature coefficients and x_i, y_j are the zeros of the m-th degree Legendre polynomial. Moreover, the corresponding error term P_m to be dropped has the estimate

$$|P_m| \le \left(\frac{2^{4m+3}}{(4m+1)!}\right)^{1/2} ||F_z^{(2m)}||_{L_2}.$$

Here, all those errors caused by using Gaussian quadrature formula are not considered. □

Example 17.3

For the spherical domain $\Omega = \{(x,y,z) : x^2+y^2+z^2 \le 1\}$, we have $\phi_1(x,y) \equiv -\sqrt{1-x^2-y^2}$, $\phi_2(x,y) \equiv \sqrt{1-x^2-y^2}$, and $E \equiv \{(x,y) : x^2+y^2 \le 1\}$. Accordingly, $G(X)$ takes the form

$$G(X) = \frac{m!}{(2m)!}\left(\frac{\partial}{\partial z}\right)^m (x^2+y^2+z^2-1)^m.$$

This is just the Hermite-Didon polynomial in 3 dimensions. But now ϕ_1 and ϕ_2 are not polynomial functions. Certainly one cannot expect to use the generalized Hermite formula to obtain boundary quadrature formulas of form (17.28) with algebraic precision. However, in this case one can directly make use of (17.19), thus obtaining

$$J \equiv \int_{\Omega} F(X)dX$$

$$\simeq \sum_{k=0}^{m-1} \frac{(-1)^k}{(2m)!} \int_{\partial\Omega} F_z^{(k)}[(x^2+y^2+z^2-1)^m]_z^{(2m-k-1)} \cdot zdS, (17.30)$$

where $\partial\Omega$ is the surface of the unit sphere Ω. $\square$

To see what precision can be attained, we may take F to be a polynomial in single variable z with $deg\,F \geq m-1$. Clearly, the degree of the integrand of each integral in sum of (17.30) is $(deg\,F-k)+k+2 = deg\,F+2$. Because of (17.20) and (17.24) we may assume that $deg\,F = m-1$ or $deg\,F = 2m-1$, so that the integrand mentioned above will get degree $m+1$ or $2m+1$. Thus, by applying fully symmetric integration formulas of algebraic precision $m+1$ or $2m+1$ for the spherical surface integrals (see Keast and Diaz [22]) to integral J in (17.30), we obtain a boundary quadrature formula with algebraic precision $m-1$ or $2m-1$, respectively.

17.4 Integrals with Wavelet Weight Functions

The approximate DRE (17.5) provides a general process for the dimensional reduction for a higher dimensional integral $\int_{\Omega} F\,dX$. However, we still need a general process to construct a numerical quadrature formula for the boundary integral $\int_{\partial\Omega} W\,dS$ over the boundary of an arbitrary domain Ω. Here, the integral $\int_{\partial\Omega} W\,dS$ can be considered as an integral of periodic function over $\partial\Omega$; and $\partial\Omega$ is a piecewise C^1 continuous surface or a piecewise C^1 continuous simple closed curve. In the following, we will give a general process for the construction of a quadrature formula for $\int_{\partial\Omega} W\,dS$ when $F: R^2 \to R$. This construction can be immediately generalized to functions in higher dimensions. Therefore, without loss of generality, we may assume $\partial\Omega$ is a piecewise C^1 continuous simple closed curve defined by $c(t):[0,1] \to R$,$c(t) = (x(t),y(t))$, $c(0) = c(1)$. In addition, $F(x,y) = F(x(t),y(t))$ and $G(x,y) = G(x(t),y(t))$ are functions in space C^{m-1} – periodic (the subspace of periodic functions in the space

C^{m-1}) with period 1. Hence the integral $\int_{\partial\Omega} W\, dS$ can be written as

$$\int_{\partial\Omega} W\, dS = \sum_{k=0}^{m_1-1} (-1)^k \int_0^1 D^{(k,0)} F(x(t), y(t)) D^{(m_1-k-1,m_2)}$$

$$G(x(t), y(t)) y'(t)\, dt + \sum_{k=0}^{m_2-1} (-1)^{m_1+k+1} \int_0^1 D^{(m_1,k)}$$

$$F(x(t), y(t)) D^{(0,m_2-k-1)} G(x(t), y(t)) x'(t)\, dt. \tag{17.31}$$

The integrals in (17.31) can be considered as integrals with periodic weight functions of period 1: $D^{(m_1-k-1,m_2)} \cdot G(x(t), y(t)) y'(t)$, $k = 0, 1, \ldots, m_1 - 1$, and $D^{(0,m_2-k-1)} G(x(t), y(t)) x'(t)$, $k = 0, 1, \ldots, m_2 - 1$, respectively. All these weight functions can be expanded as wavelet series in terms of periodic wavelets of period 1. Therefore all integrals in (17.31) can be decomposed to integrals of the form $\int_0^1 f(t)\psi^*(t)\, dt$; here $\psi^*(t)$ is a periodic wavelet of period 1 and $f(t)$ is a periodic function of period 1. Thus we focus on two problems: (a) periodic wavelet series approximating continuous functions, and (b) numerical quadrature formulas for the integral $\int_0^1 f(t)\psi^*(t)\, dt$.

To find periodic wavelet series approximating to continuous functions, we begin with the space S_r of rapidly decreasing C^r continuous functions defined on R, i.e., functions that satisfy

$$|g^{(k)}(t)| \le C_{pk}(1 + |t|)^{-p}, \quad k = 0, 1, \ldots, r, \quad p = 0, 1, \ldots, \quad t \in R. \tag{17.32}$$

Let $\phi \in S_r$ be a scaling function with associated multiresolution analysis and mother wavelet ψ. We define ϕ^*_{mn} and ψ^*_{mn} as the periodized version of $\phi_{mn}(t) = 2^{m/2}\phi(2^m t - n)$ and $\psi_{mn}(t) = 2^{m/2}\psi(2^m t - n)$,

$$\phi^*_{mn}(t) := \sum_k \phi_{mn}(t - k), \quad \psi^*_{mn} := \sum_k \psi_{mn}(t - k).$$

Both ϕ^*_{mn} and ψ^*_{mn} are periodic functions of period 1. Obviously $\phi^*_{00}(t) = 1$ and $\phi^*_{mn}(t) \in \text{span}\,\{\phi^*_{0k}(t)\}$ when $m < 0$. Thus, $\phi^*_{mn}(t)$ need only be considered for $m \ge 0$. We denote by V^*_m the space span $\{\phi^*_{mn}(t)\}_{n\in Z}$. V^*_m is clearly now a finite-dimensional space since $\phi^*_{mn} = \phi^*_{m,n+2^m}$. In fact, $\phi^*_{m,0}, \phi^*_{m,1}, \ldots, \phi^*_{m,2^m-1}$ are an orthogonal basis of V^*_m in the sense of $L^2(0,1)$ since $\int_0^1 \phi^*_{m0}\phi^*_{mn} = 0$, $0 < n < 2^m$. These V^*_m's satisfy nested property $V^*_0 \subset V^*_1 \subset \cdots \subset V^*_m \subset \cdots$ and $\overline{\bigcup_m V^*_m} = L^2(0,1)$. It also can be shown that the V_m^*'s are a reproducing kernel of Hilbert space with reproducing kernel $q_m^*(x,t) = \sum_{n=0}^{2^m-1} \phi_{mn}^*(x)\phi_{mn}^*(t)$, $x, t \in [0,1]$, and $\{q_m^*(x,t) = x_{[0,1)}(x)\, x_{[0,1)}(t)\}$ is a quasi-positive delta sequence. Hence we have the result:

PROPOSITION 17.1

[29] Let $g(t)$ be continuous and periodic, then the projection g_m^* of g onto V_m^* converges to g uniformly.

REMARK 17.1 The corresponding wavelets themselves are simpler in the periodic case as well. We denote $\psi_0(t) = \phi_{00}^*(t) = 1$ and $\psi_{2^m+k}(t) = \psi_{m,k}^*(t)$, $0 \leq m$, $k = 0, 1, \ldots, 2^m - 1$. The set $\{\psi_n\}$ is clearly an orthonormal basis of $L^2(0,1)$ consisting of periodic functions. In particular, the set $\{\psi_0, \psi_1, \ldots, \psi_{2^m-1}\}$ is also an orthonormal basis of V_m^* since there are 2^m of them and they are orthonormal. The orthogonal expansion of $g \in L^2(0,1)$ is given by $\sum_{n=0}^{\infty}\langle g, \psi_n\rangle\psi_n$. If g is continuous, from the proposition, we have $g = \sum_{n=0}^{\infty}\langle g, \psi_n\rangle\psi_n$. ▮

REMARK 17.2 The periodic wavelet series expansion can be extended to the space $H^{m-1}(0,1)$ – periodic, i.e., the subspace of the periodic functions of period 1 in Sobolev space H^{m-1}(see Jaffard and Laurençot [21] and Meyer [26]). The corresponding decomposition formulas can also be found in Jaffard and Laurençot [21] and Meyer [26]. ▮

REMARK 17.3 Obviously, the property of uniform convergence is not true for trigonometric Fourier series expansion. ▮

We now discuss the numerical quadrature formula for the integral $\int_0^1 f(t)\,\psi^*(t)dt$, here $f(t)$ and $\psi^*(t)$ are, respectively, periodic function and periodic wavelet of period 1. We first give a general quadrature for the integral $\int_{-\infty}^{\infty} f(t)\psi(t)\, dt$ with non-periodic wavelet weight function $\psi(t)$.

THEOREM 17.10

Let $\psi \in S_r$, with $\psi_{mn}(t) = 2^{m/2}\psi(2^m t - n)$ being an orthonormal system in $L^2(R)$, and $\int_{-\infty}^{\infty} t^{r+1}\psi(t)\, dt = A \neq 0$. Then for any set of distinct evaluation points $\{t_i, i = 0, 1, \ldots, r+1\}$ and $f \in C^{r+2}$, we have

$$\int_{-\infty}^{\infty} f(t)\psi(t)\, dt \approx \sum_{i=0}^{r+1} c_i f(t_i), \tag{17.33}$$

where

$$c_i = A \prod_{j=0, j\neq i}^{r+1} \frac{1}{t_i - t_j}. \tag{17.34}$$

In addition, if $\int_{-\infty}^{\infty} x^{r+2}\psi(t)\ dt = B$ *and* $|f^{(r+2)}(t)| \leq M$*, then the error in quadrature formula (17.19) is*

$$\left| \int_{-\infty}^{\infty} f(t)\psi(t)\ dt - \sum_{i=0}^{r+1} c_i f(t_i) \right| \leq \frac{M}{(r+2)!} [|B| + |A| \left| \sum_{j=0}^{r+1} t_j \right|]. \tag{17.35}$$

PROOF For the given set of distinct evaluation points $\{t_i : i = 0, 1, \ldots, r+1\}$, we have the Lagrange interpolation $f(t) \approx \sum_{i=0}^{r+1} f(t_i) l_i(t)$, where $l_i(t) = \prod_{j=0, j\neq i}^{r+1} \frac{t - t_j}{t_i - t_j}$. The error in the interpolation is $\frac{1}{(r+2)!} \prod_{j=0}^{r+1} (t - t_j) f^{(r+2)}(\zeta)$ for some ζ. Thus

$$\int_{-\infty}^{\infty} t^k \psi(t)\ dt \approx \sum_{i=0}^{r+1} f(t_i) \int_{-\infty}^{\infty} l_i \psi(t)\ dt.$$

Since $\psi(t) \in S_r$ and ψ_{mn} are an orthonormal system in $L^2(R)$, we have $\int_{-\infty}^{\infty} t^k \psi(t)\ dt = 0, \quad k = 0, 1, \ldots, r$. Therefore

$$\int_{-\infty}^{\infty} l_i(t)\psi(t)\ dt = \int_{-\infty}^{\infty} \left(\prod_{j=0}^{r+1} \frac{1}{t_i - t_j} \right) t^{r+1} \psi(t)\ dt = A \prod_{j=0, j\neq i}^{r+1} \frac{1}{t_i - t_j}.$$

∎

REMARK 17.4 If $\int_{-\infty}^{\infty} t^k \psi(t)\ dt = 0$ for $k = 0, 1, \ldots, r, \ldots, m$, but $\int_{-\infty}^{\infty} x^{m+1}\ \psi(x)\ dx \neq 0$, then we can increase the number of evaluation points to $m+2$ and construct the corresponding quadrature formula similarly. ∎

REMARK 17.5 If $\psi_{mn}(t)$ is not an orthonormal basis in $L^2(R)$, then we may use the regularity of the duals of $\psi_{mn}(t)$ to construct the numerical quadrature for $\int_{-\infty}^{\infty} f(t)\psi(t)\ dt$ (see Theorem 5.5.1 in Daubechies [5]). ∎

COROLLARY 17.1

Let $\psi \in S_r$*, with* $\psi_{mn}(t) = 2^{m/2}\psi(2^m t - n)$ *being an orthonormal system in* $L^2(R)$*, and* $\int_{-\infty}^{\infty} t^{r+1}\psi(t)\, dt = A \neq 0$*. Also let* $\psi^*(t) = \sum_k \psi(t-k)$

be the periodized version of ψ with period 1. Then for a C^{r+2} continuous periodic function f of period 1, we have the following quadrature formula:

$$\int_0^1 f(t)\psi^*(t)\,dt \approx \sum_{i=0}^{r+1} c_i f(t_i), \tag{17.36}$$

where

$$c_i = A \prod_{j=0,j\neq i}^{r+1} \frac{1}{t_i - t_j} \tag{17.37}$$

and $\{t_i : i = 0, 1, \ldots, r+1\}$ are $r+2$ distinct real numbers.

PROOF First, we note that, if $t_i \notin [0,1)$, we can always find $t_i' \in [0,1)$ such that $f(t_i) = f(t_i')$ because of the periodic property of f. Second, we have

$$\int_0^1 f(t)\psi^*(t)\,dt = \int_0^1 f(t)\sum_k \psi(t-k)\,dt$$

$$= \sum_k \int_k^{k+1} f(t+k)\psi(t)\,dt = \int_{-\infty}^{\infty} f(t)\psi(t)\,dt.$$

Hence, (17.36) and (17.37) hold because of (17.33) and (17.34), respectively. ∎

We go back to the integrals in (17.31). The weight function in each integral can be expanded in terms of sufficiently smooth periodic wavelet functions, such as the C^r continuous periodic wavelet $\psi^* = \sum_k \psi(t-k)$, $\psi \in S_r$. Then, for each integral with the form $\int_0^1 f(t)\psi^*_{mn}(t)\,dt$ in (17.31), we make use of quadrature (17.36), and the boundary quadrature formula for $\int_\Omega F\,dX$ is eventually obtained by replacing the integral $\int_{\partial\Omega} W\,dS$ in (17.5) with the quadrature constructed through the above process.

17.5 Applications

In this section, we will discuss an application of the boundary-type quadrature formula in the boundary element method, a method for solving partial differential equations numerically. We begin with the DRE related to the second order differential operator L, which is defined by

$$Lu = \sum_{i,j=1}^{n} a_{ij}(X)\frac{\partial^2 u}{\partial x_i \partial x_j} + \sum_{i=1}^{n} b_i(X)\frac{\partial u}{\partial x_i} + c(X)u, \tag{17.38}$$

where $a_{ij}(X) \in H_n^2(\Omega)$, $b_i(X)$ and $c(X) \in H_n^1(\Omega)$, and $H_n^\alpha(\Omega)(\alpha \geq 1)$ is the collection of all functions $f(X) = f(x_1, \ldots, x_n)$ that have continuous partial derivatives $D^{(i_1,\ldots,i_n)}f$, $0 \leq i_1 + \ldots + i_n \leq \alpha n$, $0 \leq i_k \leq \alpha$, $k = 1, 2, \ldots, n$. It is well-known that the conjugate operator of L can be defined by

$$Mv = \sum_{i,j=1}^{n} \frac{\partial^2(va_{ij}(X))}{\partial x_i \partial x_j} - \sum_{i=1}^{n} \frac{\partial(vb_i(X))}{\partial x_i} + c(X)v. \tag{17.39}$$

If we denote by $r_i(X)$ the following expression

$$r_i(X) = -v\sum_{j=1}^{n} a_{ij}\frac{\partial u}{\partial x_j} + u\sum_{j=1}^{n} a_{ij}\frac{\partial v}{\partial x_j} + uv\sum_{j=1}^{n} \frac{\partial a_{ij}}{\partial x_j} - b_i uv, \tag{17.40}$$

then we have

$$uMv - vLu = \sum_{i=1}^{n} \frac{\partial r_i}{\partial x_i}. \tag{17.41}$$

From this relation and Green's formula, we have the following result.

LEMMA 17.2

Let $\Omega \in R^n$ be an n-dimensional bounded closed domain with the boundary $\partial\Omega$ being a piecewise smooth surface. In particular, if $n = 2$, $\partial\Omega$ is a simple closed curve with finite length. Let $u = u(X)$ and $v = v(X)$ be functions in $C^2(\Omega)$, and let L and M be differential operators defined by (17.38) and (17.39), respectively. Then we have the identity

$$\int_\Omega (uMv - vLu)\, dX = \int_{\partial\Omega} \left(\sum_{i=1}^{n} r_i \frac{\partial x_i}{\partial \nu}\right) dS, \tag{17.42}$$

where $r_i(X)$ is given by (17.40). Using the notation

$$p_i = p_i(X) = \sum_{j=1}^{n} a_{ij}(X)\frac{\partial v}{\partial x_j} + \sum_{j=1}^{n} v\frac{\partial a_{ij}(X)}{\partial x_j} - b_i(X)v, \tag{17.43}$$

we can rewrite identity (17.42) as follows:

$$\int_\Omega (uMv - vLu)\, dX = \int_{\partial\Omega} \left[u \sum_{i=1}^{n} p_i \frac{\partial x_i}{\partial \nu} \right] dS -$$

$$\int_{\partial\Omega} \left[v \sum_{i=1}^{n} \left(\sum_{j=1}^{n} a_{ij}(X) \frac{\partial u}{\partial x_j} \right) \frac{\partial x_i}{\partial \nu} \right] dS. \qquad (17.44)$$

PROOF In Green's formula, $\int_\Omega \frac{\partial f(X)}{\partial x_i}\, dX = \int_{\partial\Omega} f(X) \frac{\partial x_i}{\partial \nu}\, dS$, we replace $f(X)$ with $r_i(X)$ and immediately obtain (17.42). Noting the expression of $r_i(X)$ given by (17.40), we also obtain identity (17.44).

By using identity (17.42) or (17.44) and taking a certain function v, we can construct some exact DREs for the integral $\int_\Omega u\, dX$.

For instance, if $v = v(X)$ is a solution to the differential equation $Mv = 1$, then for all $u = u(X)$ that satisfy the equation $Lu = g$, we have an approximate DRE

$$\int_\Omega u\, dX \doteq \int_{\partial\Omega} \left[u \sum_{i=1}^{n} p_i \frac{\partial x_i}{\partial \nu} \right] dS$$

$$- \int_{\partial\Omega} \left[v \sum_{i=1}^{n} \left(\sum_{j=1}^{n} a_{ij}(X) \frac{\partial u}{\partial x_j} \right) \frac{\partial x_i}{\partial \nu} \right] dS \qquad (17.45)$$

with the remainder $\int_\Omega vg\, dX$. If $g(X) = 0$ (i.e., u satisfies $Lu = 0$), then expansion (17.45) is exact.

In addition, if v satisfies the boundary condition $v(X) = 0$, $X \in \partial\Omega$ (i.e., v is the solution of the boundary problem $Mv = 1$ and $v = 0$ on $\partial\Omega$), then expansion (17.45) can be reduced to

$$\int_\Omega u\, dX \doteq \int_{\partial\Omega} \left[u \sum_{i=1}^{n} \left(\sum_{j=1}^{n} a_{ij}(X) \frac{\partial v}{\partial x_j} \right) \frac{\partial x_i}{\partial \nu} \right] dS \qquad (17.46)$$

with the remainder $\int_\Omega vg\, dX$. If $g(X) = 0$ (i.e., u satisfies $Lu = 0$), then expansion (17.46) is also exact. ∎

As an example, for all functions u that satisfy the elliptic equation

$$\sum_{i=1}^{n} \frac{\partial^2 u}{\partial x_i^2} + \sum_{i=1}^{n} b_i \frac{\partial u}{\partial x_i} + cu = g, \tag{17.47}$$

the integral of u over Ω can be written as

$$\int_{\Omega} u \, dX = \int_{\partial\Omega} \left[u \sum_{i=1}^{n} \left(\frac{\partial v}{\partial x_i} \cdot \frac{\partial x_i}{\partial \nu} \right) \right] dS + \int_{\Omega} vg \, dX, \tag{17.48}$$

where v is the solution to the following Dirichlet problem:

$$\begin{cases} \sum_{i=1}^{n} \frac{\partial^2 v}{\partial x_i^2} - \sum_{i=1}^{n} \frac{\partial (v b_i)}{\partial x_i} + cv = 1 & \text{on } \Omega, \\ v = 0 & \text{on } \partial\Omega. \end{cases}$$

If u satisfies the corresponding homogeneous equation (17.47) (i.e., equation (17.47) with $g = 0$), then the second integral, $\int_{\Omega} vg \, dX$, on the right-hand side of (17.48) vanishes.

Expansion (17.42) or (17.44) can also be used to solve the boundary value problem

$$\begin{cases} Lu(X) = g(X) & X \in \Omega, \\ u = \overline{u} & X \in \partial\Omega_1, \\ g = \frac{\partial u}{\partial n} = \overline{g} & X \in \partial\Omega_2, \end{cases}$$

where $\partial\Omega_1 \cup \partial\Omega_2 = \partial\Omega$. The process of solving this problem, as shown below, is analogous to the boundary element method.

Let v be the fundamental solution of $Mv = \delta(X - X^0)$. Here M is the conjugate differential operator of L, $\delta(X)$ is the Delta function, and X^0, called the source point, is an arbitrary point in Ω. Then, by using the limit process on (17.44), we have

$$\begin{aligned} -\alpha u(X^0) = & \int_{\Omega} vg \, dX + \int_{\partial\Omega} \left[u \sum_{i=1}^{n} p_i \frac{\partial x_i}{\partial \nu} \right] dS \\ & - \int_{\partial\Omega} \left[v \sum_{i=1}^{n} \left(\sum_{j=1}^{n} a_{ij}(X) \frac{\partial u}{\partial x_j} \right) \frac{\partial x_i}{\partial \nu} \right] dS, \end{aligned} \tag{17.49}$$

where $\alpha = 1$, if $X^0 \in \Omega \backslash \partial\Omega$, and α is a positive real number less than 1, if $X^0 \in \partial\Omega$ (in particular, $\alpha = \frac{1}{2}$ if X^0 is on the smooth boundary). Thus, we obtain an equation about $u(X^0)$ and boundary-type weighted integrals of u and $\frac{\partial u}{\partial n}$. Here $\frac{\partial u}{\partial n} = \sum_{i=1}^{n} \frac{\partial u}{\partial x_i}\frac{\partial x_i}{\partial \nu}$ is the outward normal derivative of u on $\partial\Omega$. Applying a quadrature formula (e.g., the formula given in Section 3) to these boundary integrals, we obtain an algebraic equation, called the basic algebraic equation, about $u(X^0)$ and values of u and $\frac{\partial u}{\partial n}$ at the nodes on $\partial\Omega$. Replacing the source point X^0 with each node of the boundary quadrature formula in the basic algebraic equation, we have a system of linear equations about the values of u and $\frac{\partial u}{\partial n}$ at nodes on $\partial\Omega$. Substituting in the given boundary conditions, which are the given values of u and $\frac{\partial u}{\partial n}$ at some nodes on $\partial\Omega$, we can solve for the remaining unknown values of u and $\frac{\partial u}{\partial n}$ at the other nodes on $\partial\Omega$. After finding all boundary values of u and $\frac{\partial u}{\partial n}$ at the nodes, the value of u at any interior point X^0 in Ω can be evaluated from the basic algebraic equation with the values of u and $\frac{\partial u}{\partial n}$ at the nodes on $\partial\Omega$.

As an example, we consider the following boundary value problem of the Helmholtz's equation

$$\begin{cases} (\nabla^2 + k)u = 0 \text{ in } \Omega, \\ u = \overline{u} \qquad \text{on } \partial\Omega_1, \\ q = \frac{\partial u}{\partial n} = \overline{q} \qquad \text{on } \partial\Omega_2, \end{cases} \tag{17.50}$$

where $\nabla^2 u = \frac{\partial^2 u}{\partial x^2} + \frac{\partial^2 u}{\partial y^2}$ and $\partial\Omega_1 \cup \partial\Omega_2 = \partial\Omega$. Obviously, $L = \nabla^2 + k$ is a self-conjugate operator. Therefore, $M = L = \nabla^2 + k$. The fundamental solution of $Mv = \delta(X - X^0)$ is $v = v(X, X^0) = -\frac{j}{4}H_0^{(2)}(kr)$; here $H_0^{(2)}$ is the zeroth order Hankel function of the second kind, and $r = |X - X^0|$ denotes the distance from a point $X = (x, y)$ to an arbitrarily fixed source point $X^0 = (x^0, y^0) \in \Omega$. From (17.49) and observing $g = 0$ and $a_{ij} = \delta_{ij}$, $1 \le i, j \le 2$, we have

$$\alpha u(x^0, y^0) = \int_{\partial\Omega} u\frac{\partial v}{\partial n}\, dS - \int_{\partial\Omega} v\frac{\partial u}{\partial n}\, dS, \tag{17.51}$$

where $\alpha = -1$ if $(x^0, y^0) \in \Omega\backslash\partial\Omega$; $\alpha = -\frac{1}{2}$ if $(x^0, y^0) \in \partial\Omega$ and $\partial\Omega$ is smooth. In (17.51), since $v = v(x, y; x^0, y^0) = -\frac{j}{4}H_0^{(2)}(kr)$, $r = |(x, y) - (x^0, y^0)|$, we have $\frac{\partial v}{\partial n} = \frac{\partial v}{\partial n}(x, y; x^0, y^0) = \frac{j}{4}kH_1^{(2)}(kr)\cos(\vec{r}, \vec{n})$, where $\vec{r} = (x, y)$, $\vec{n}$ is the outward normal vector at $(x, y) \in \partial\Omega$, and $H_1^{(2)}$ is the first order Hankel function of the second kind. Assume $\partial\Omega$ is defined by the parametric function $c(t) = (x(t), y(t))$, $0 \le t \le 1$. The solution of problem (17.50) for the circle $\partial\Omega$ by using the boundary element method was discussed by Chan, Chui, and He [2].

We denote $(x^j, y^j) = (x(t_j), y(t_j))$, $u_j = u(x^j, y^j) = u(x(t_j),\ y(t_j))$, $q_j = \frac{\partial u}{\partial n}(x^j, y^j) = \frac{\partial u}{\partial n}(x(t_j), y(t_j))$, and $t_j = \frac{j}{n}$, $j = 1, 2, \ldots, n$. Thus, u and

g can be expanded approximately as

$$u \doteq \sum_{j=1}^{n} u_j \phi_j^1(t) \quad \text{and} \quad q \doteq \sum_{j=1}^{n} q_j \phi_j^2(t),$$

where $\phi_j^k(t)$, $k = 1, 2$, are Lagrange interpolation basis functions. We can choose $\phi_j^k(t)$, $k = 1, 2$, either as the basis function $l_j(t) = \prod_{i=1, i\neq j}^{n} \frac{t-t_i}{t_j-t_i}$ shown in Section 3 or as some other basis functions.

For instance, we can assume that $\phi_j^k(t) = \phi^k(t - j/n)$, $j = 1, \ldots, n$, and $\{\phi^k(2nt - 2j)\}_{j\in Z}$, $k = 1, 2$, are the basis functions of the optimal spline Lagrange interpolation for the data at the even integers [13]. Here the optimal spline interpolation means the interpolation with a spline basis function that possesses the highest possible approximation order and the smallest possible compact support. Thus, we take

$$\phi^1(2nt) = -\frac{1}{8}N_3(t-1) + \frac{1}{8}N_3(t) + N_3(t+1)$$

$$+ N_3(t+2) + \frac{1}{8}N_3(t+3) - \frac{1}{8}N_3(t+4),$$

where $N_3(t)$ is the B-spline of order 3. Hence, the support of $\phi^1(2nt)$ is $[-4, 4]$. It is easy to understand that $\phi^1(i/n) = \delta_{i0}$ and that the corresponding interpolation on the interval [0,1] with basis $\{\phi_j^1(t)\}$ has the optimal approximation order of $O((1/2n)^3)$. As for $\phi^2(2nt)$, we assume it is $N_2(t+1)$, B-spline of order 2, which satisfies $\phi^2(i/n) = \delta_{i0}$.

Replacing (x^0, y^0) with $(x^i, y^i) = (x(\frac{i}{n}), y(\frac{i}{n}))$, $i = 1, \ldots, n$, from (17.51), we have

$$\frac{1}{2}u_i + \sum_{j=1}^{n} u_j \int_0^1 \phi_j^1(t)\frac{\partial v}{\partial n}(t, \frac{i}{n})\, ds(t) = \sum_{j=1}^{n} q_j \int_0^1 \phi_j^2(t) v(t, \frac{i}{n})\, ds(t), \tag{17.52}$$

where $v(t, \frac{i}{n}) = v(x(t), y(t); x(\frac{i}{n}), y(\frac{i}{n}))$ and $\frac{\partial v}{\partial n}(t, \frac{i}{n}) = \frac{\partial v}{\partial n}(x(t), y(t); x(\frac{i}{n}), y(\frac{i}{n}))$.

In order to evaluate the boundary integrals in (17.52), we expand v and $\frac{\partial v}{\partial n}$ in terms of $\psi_{mk}^*(t)$, the periodized version of $\psi_{mk}(t) = 2^{m/2}\psi(2^m - k)$. Here $\psi(t)$ is the wavelet associated with the scaling function $\phi(t) = \phi^1(nt)$ [13]; i.e., $\psi(t) = \sum_k \tilde{q}_k \phi(2t - k)$, where the coefficients $\{\tilde{q}_k\}$ are determined by their two-scale symbol $\tilde{Q}(z) = \frac{1}{2}\sum_k \tilde{q}_k z^k$, $z = e^{-iw/2}$. From the paper by the author [13], we have $\tilde{Q}(z) = \frac{c(-z)}{c(z^2)}Q(z)$, where $c(z) = -\frac{1}{8}z^{-1} + \frac{1}{8}z^0 +$

$z^1 + z^2 + \frac{1}{8}z^3 - \frac{1}{8}z^4$, $Q(z) = \frac{1}{2}\sum_k q_k z^k$, $z = e^{-\imath w/2}$, and

$$q_k = \begin{cases} \frac{(-1)^k}{4}\sum_{l=0}^{3}\binom{3}{l} N_6(k+1-l), 0 \le k \le 7, \\ 0, \qquad\qquad\qquad\qquad \text{otherwise.} \end{cases}$$

After finding all boundary integrals in (17.52), we denote $h_{ij} = \int_0^1 \phi_j^1(t) \frac{\partial v}{\partial n}(t, \frac{i}{n})\ ds(t) + \frac{\delta_{ij}}{2}$ and $g_{ij} = \int_0^1 \phi_j^2(t)\ v(t, \frac{i}{n})\ ds(t)$. Thus, (17.38) can be written as $Hu = Gg$, where $H = [h_{ij}]_{1\le i,j\le n}$, $G = [g_{ij}]_{1\le i,j\le n}$, $u = (u_1, \ldots, u_n)$, and $g = (g_1, \ldots, g_n)^{\mathrm{T}}$. Substituting the boundary conditions, u_j and g_j that are known on $\partial\Omega$, into the above linear system, we may solve it for the unknown u_j and g_j.

If $(x^0, y^0) \in \Omega\backslash\partial\Omega$, from (17.37), we obtain the value of u at the point (x^0, y^0) by using u_j and g_j on the boundary $\partial\Omega$:

$$u(x^0, y^0) = \sum_{j=1}^{n} q_j \int_0^1 \phi_j^2(t) v(t, t^0)\ ds(t)$$

$$- \sum_{j=1}^{n} u_j \int_0^1 \phi_j^1(t) \frac{\partial v}{\partial n}(t, t^0)\ ds(t).$$

Thus we obtain the numerical solution of u on the boundary $\partial\Omega$ and at any interior point of Ω.

References

[1] B.L. Burrows, *A new approach to numerical integration,* J. Inst. Math. Applics, 26(1980), 151-173.

[2] A.K. Chan, C.K. Chui, and T.X. He, *Application of generalized vertex splines to boundary element method for electromagnetic scattering,* Sixth Annual Review of Progress in Applied Computational Mathematics. Monterey, CA, March 19-22, 1990, 329-337.

[3] C.K. Chui, *An Introduction to Wavelets*, Academic Press, New York, 1992.

[4] B. Cipra, *What's Happening in the Mathematical Sciences*, American Mathematical Society, Providence, RI, 1996.

[5] I. Daubechies, *Ten Lectures on Wavelets*, SIAM, Philadelphia, 1992.

[6] P.J. Davis, *Double integrals expressed as single integrals or interpolatory functions,* J. Appro. Theory, 5(1972), 276-307.

[7] P.J. Davis and P. Rabinowitz, *Methods of Numerical Integration*, Academic Press, New York, 1975.

[8] J.D. Federenko, *A formula for the approximate evaluation of double integrals,* Dopovidi. Akad. Nauk Ukrain. RSR, (1964), 1000-1005.

[9] A. Ghizzetti and A. Ossicini, *Quadrature Formula*, Academic Press, New York, 1970.

[10] W. Gröbner, *Über die Konstruktion von Systemen orthogonaler Polynome in ein-und-zwei dimensionalen Bereiche,* Monatsh. Math., 52(1948), 48-54.

[11] T.X. He, *On the algebraic method for constructing the boundary-type quadrature formulas,* Numer. Math., (1985), 1-5.

[12] T.X. He, *Boundary-type quadrature formulas without derivative terms,* J. Math. Res. Expo. 2(1981), 93-102.

[13] T.X. He, *Spline interpolation and its wavelet analysis,* Proceedings of the Eighth International Conference on Approximation Theory, C.K. Chui and L.L. Schumaker (eds.), 1995, to appear.

[14] T.X. He, *Construction of boundary quadrature formulas using wavelets,* Wavelet Applications in Signal and Image Processing III, SPIE-The International Society for Optical Engineering, A.F. Laine and M.A. Unser (eds.) 1995, 825-836.

[15] L.C. Hsu, *On a method for expanding multiple integrals in terms of integrals in lower dimensions,* Acta. Math. Acad. Sci. Hung., 14(1963). 359-367.

[16] L.C. Hsu and T.X. He, *On the minimum estimation of the remainders in dimensionality lowering expansions with algebraic precision,* J. Math. (Wuhan), 2,3(1982), 247-255.

[17] L.C. Hsu and Y.S. Zhou, *Numerical integration in high dimensions,* Computational Methods Series. Science Press, Beijing, 1980.

[18] L.C. Hsu and Y.S. Zhou, *Two classes of boundary type cubature for-*

mulas with algebraic precision, Calcolo, 23(1986), 227-248.

[19] L.C. Hsu, Y.S. Zhou, and T.X. He, *Topics on Numerical Integration in High Dimensions,* Anhui Education Press, Hefei, 1985.

[20] D.V. Ionescu, *Generalization of a quadrature formula of N. Obreschkoff for double integrals* (Romanian), Stud. Cerc. Mat., 17(1965), 831-841.

[21] S. Jaffard and Ph. Laurençot, *Orthonormal wavelets, analysis of operators, and applications to numerical analysis*, Wavelet- A Tutorial in Theory and Applications, C.K. Chui (ed.), Academic Press, San Diego, 1992, 543-601.

[22] P. Keast and J.C. Diaz, *Fully symmetric integration formula for the surface of the sphere in S dimension,* SIAM J. Numer. Anal. 20(1983), 406-419.

[23] L.J. Kratz, *Replacing a double integral with a single integral,* J. Appro. Theory, 27(1979), 379-390.

[24] M. Levin, *On a method of evaluating double integrals,* Tartu Riikl. Ül. Toimetised, 102(1961), 338-341.

[25] M. Levin, *Extremal problems connected with a quadrature formula,* Eesti NSV Tead. Akad. Toimetised Füüs-Mat. Tehn. Seer., 12(1963), 44-56.

[26] Y. Meyer, *Ondelettes et Opóerateurs*, Herman, Paris, 1990.

[27] M. Sadowsky, *A formula for the approximate computation of a triple integral,* Amer. Math. Monthly, 47(1940), 539-543.

[28] A.H. Stroud, *Approximate Calculation of Multiple Integrals*, Prentice-Hall, Englewood Cliffs, N.H., 1971.

[29] G.G. Walter, *Wavelets and Other Orthogonal Systems with Applications*, CRC Press, Ann Arbor, 1994.

Chapter 18

Value at Risk: Recent Advances

Irina N. Khindanova and Svetlozar T. Rachev

Department of Economics, University of California at Santa Barbara, Santa Barbara, CA, 93106-3110. E-mail: khindan@econ.ucsb.edu.
Institute of Statistics and Mathematical Economics, School of Economics, University of Karlsruhe, Kollegium am Schloss, Bau II, 20.12, R210, Postfach 6980, D-76128, Karlsruhe, Germany. E-mail: rachev@lsoe.uni-karlsruhe.de
Contact author: I. N. Khindanova

Abstract The Value-at-Risk (VAR) measurements are widely applied to estimate exposure to market risks. The traditional approaches to VAR computations — the variance-covariance method, historical simulation, Monte Carlo simulation, and stress-testing — do not provide satisfactory evaluation of possible losses. In this chapter we review the recent advances in the VAR methodologies. The proposed improvements still lack a convincing unified technique capturing the observed phenomena in financial data such as heavy-tails, time-varying volatility, and short- and long-range dependence. We suggest using stable Paretian distributions in VAR modeling.

Contents

18.1 Introduction: VAR and the New Bank Capital Requirements for Market Risk

One of the most important tasks of financial institutions is evaluation of exposure to market risks, which arise from variations in prices of equities, commodities, exchange rates, and interest rates. The dependence on market risks can be measured by changes in the portfolio value, or profits and losses. A commonly used methodology for estimation of market risks is the *Value at Risk* (VAR).

Regulators and the financial industry advisory committees recommend VAR as a way of risk measuring. In July 1993, the Group of Thirty first advocated the VAR approaches in the study "Derivatives: Practices and Principles".[1] In 1993, the European Union instructed setting capital reserves to balance market risks in the Capital Adequacy Directive "EEC 6-93", effective from January 1996.[2] It was an improvement with respect to the 1988 Basle Capital Adequacy Accord of G10, which centered on credit risks and did not consider market risks in details.[3]

In 1994, the Bank for International Settlements in the Fisher report advised disclosure of VAR numbers.[4] In the April 1995 proposal "Supervisory Treatment of Market Risks", the Basle Committee on Banking Supervision suggested that banks can use their internal models of VAR estimations as the basis for calculation of capital requirements.[5] In January 1996, the Basle Committee amended the 1988 Basle Capital Accord.[6] The supplement suggested two approaches to calculate capital reserves for market risks: "standardized" and "internal models".[7] According to the in-house models approach, capital requirements are computed from multiplying the banks' VAR values by a factor between three and four. In August 1996,

[1] Kupiec [46]; Simons [76]; Fallon [23]; Liu [49].
[2] Liu [49].
[3] Jackson, Maude, and Perraudin [38] and [39].
[4] Hendricks and Hirtle [34].
[5] Kupiec [46]; Jorion [42]; Beder [5].
[6] Basle Committee on Banking Supervision [4].
[7] Simons [76]; Jackson, Maude, and Perraudin [39]; Hopper [37].

the US bank regulators endorsed the Basle Committee amendment.[8] The Federal Reserve Bank allowed the two-year period for its implementation. The proposal is effective from January 1998.[9] The US Securities and Exchange Commission suggested applying VAR for enhancing transparency in derivatives activity. Derivatives Policy Group has also recommended VAR techniques for quantifying market risks.[10]

The use of VAR models is rapidly expanding. Financial institutions with significant trading and investment volumes employ the VAR methodology in their risk management operations.[11] In October 1994, JP Morgan unveiled its VAR estimation system,[12] RiskMetrics™. Credit Swiss First of Boston developed proprietary Primerisk and PrimeClear (March 1997). Chase Manhattan's product is called Charisma. Bankers Trust introduced the RAROC in June 1996. Deutsche Bank uses the dbAnalyst 2.0 from January 1995. Corporations use VAR numbers for risk reporting to management, shareholders, and investors since VAR measures allow aggregating exposures to market risks into one number in money terms. It is possible to calculate VAR for different market segments and to identify the most risky positions. The VAR estimations can complement allocation of capital resources, setting position limits, and performance evaluation.[13] In many banks the evaluation and compensation of traders are derived from returns per unit VAR. Nonfinancial corporations employ the VAR technique to unveil their exposure to financial risks, to estimate riskiness of their cashflows, and to undertake hedging decisions. Primers of applying VAR analysis for estimating market risks by nonfinancial firms are two German conglomerates Veba and Siemens.[14] The Norwegian oil company Statoil implemented a system, which incorporates the VAR methodologies.[15] Corporations hedge positions to "buy insurance" against market risks. An appealing implication of VAR is as an instrument for corporate self-insurance.[16] VAR can be explained as the amount of uninsured loss that a corporation accepts. If the self-insurance losses are greater than the cost of insuring by hedging, the corporation should buy external insurance. Investment analysists employ VAR techniques in project valuations.[17] Institutional investors, for instance, pension funds, use VAR for quantifying market risks.

---

[8] Lopez [50].
[9] Hendricks and Hirtle [34].
[10] Kupiec [46].
[11] Heron and Irving [35]; The Economist [79].
[12] JP Morgan [45].
[13] Liu [49]; Jorion [42].
[14] Priest [67] and [68].
[15] Hiemstra [36].
[16] Shimko [74].
[17] Shimko [75].

The new market risk capital requirements became effective from January 1998. The US capital standards for market risks are imperative for banks with trading accounts (assets and liabilities) greater than $1 billion or 10 percent of total assets;[18] although, the regulators can apply these standards to banks with smaller trading accounts. The market risk capital requirements allow the calculation of capital reserves based either on "standardized" or "internal models" methods. The standardized method computes capital charges separately for each market (country) assigning percentage provisions for different exposures to equities, interest rate and currency risks.[19] The total capital charge equals the sum of the market capital requirements. The main drawback of the "standardized" approach is that it does not take into consideration global diversification effects.[20] The second approach determines capital reserves based on in-house VAR models. The VAR values should be computed with a 10-day time horizon at a 99 percent confidence level using at least one year of data.[21]

The new capital requirements classify market risk as general market risk and specific risk. *The general risk* is the risk from changes in the overall level of equity and commodity prices, exchange rates and interest rates. *Specific risk* is the risk from changes in prices of a security because of reasons associated with the security's issuer.

The capital requirement for general market risk is equal to the maximum of

(i) the *current VAR* (VAR_t) number and

(ii) the *average VAR* $\left(\frac{1}{60}\sum_{i=1}^{60}\text{VAR}_{t-i}\right)$ over the previous 60 days multiplied by a factor between three and four.

The capital charges for specific risk cover debt and equity positions. The specific risk estimates obtained from the VAR models should be multiplied by a factor of four. Thus, a market risk capital requirement at time t, C_t, is

$$C_t = A_t \, {}^*max\left\{\frac{1}{60}\sum_{i=1}^{60}\text{VAR}_{t-i}, \text{VAR}_t\right\} + S_t \ ,$$

where A_t is a multiplication factor between three and four, S_t is the capital charge for specific risk.

[18]Hendricks and Hirtle [34].

[19]Example: The required capital reserves for positions in the US market recognize hedging by the US instruments but do not consider hedging by the UK instruments.

[20]In other words, the "standardized" method ignores correlations across markets in different countries. See Jackson, Maude and Perraudin [39]; Liu [49].

[21]For the exact definition of VAR see (1) with $\tau = 10$ and $\alpha = .99$ later in this section.

The A_t values depend on the accuracy of the VAR models in previous periods.[22] Denote by K the number of times when daily actual losses exceeded the predicted VAR values over the last year, or the last 250 trading days.[23] Regulators split the range of values of K into three zones: the green zone ($K \leq 4$), the yellow zone ($5 \leq K \leq 9$), and the red zone ($K \geq 10$).[24] If K is within the green zone, then $A_t = 3$. If K is within the yellow zone, $3 < A_t < 4$, in the red zone, $A_t = 4$.

A *VAR measure* is the highest possible loss over a certain period of time at a given confidence level. Example: The daily VAR for a given portfolio of assets is reported to be \$2 million at the 95 percent confidence level. This value of VAR means that, without abrupt changes in the market conditions, one-day losses will exceed \$2 million 5 percent of the time.

Formally, a VAR = $\text{VAR}_{t,\tau}$ is defined as the upper limit of the one-sided confidence interval:

$$Pr[\Delta P(\tau) < -VAR] = 1 - \alpha \tag{18.1}$$

where α is the confidence level and $\Delta P(\tau) = \Delta P_t(\tau)$ is the *relative change* (*return*) in the portfolio value over the time horizon τ.

$$\Delta P_t(\tau) = P(t + \tau) - P(t),$$

where $P(t + \tau) = \log S(t + \tau)$ is the log-spot value at $t + \tau$, $P(t) = \log S(t)$, $S(t)$ is the portfolio value at t, the time period is $[t, T]$, with $T - t = \tau$, and t is the current time.

The time horizon, or the holding period, should be determined from the liquidity of assets and the trading activity. The confidence level should be chosen to provide a rarely exceeded VAR value.

The VAR measurements are widely used by financial entities, regulators, non-financial corporations, and institutional investors. Clearly, VAR is of importance for practitioners and academia alike. The aim of this chapter is to review the recent approaches to VAR and to outline directions for new empirical and theoretical studies.

In Section 18.2 we discuss traditional approaches to approximations of the distribution of ΔP and VAR computations. Section 18.3 analyzes components of VAR methodologies. Section 18.4 reports VAR strengths and

[22]The regulators recommend using the time horizon τ of 10 days (two weeks) in VAR estimations. For backtesting, the regulators use $\tau = 1$ day.

[23]For more detailed explanation of the time horizon and the window length see also Sections 3.3 and 3.4.

[24]Denote by $\hat{K}$ the fraction of days when the observed losses exceeded the VAR estimate. If $K = 10$, then $\hat{K}$ is $10/250 = 0.04$. However, the 99% confidence level implies probability of 0.01 for exceeding the VAR estimate of daily losses.

weaknesses. Section 18.5 presents recent VAR advances. Section 18.6 states conclusions.

18.2 Computation of VAR

From the definition of VAR $=$ VAR$_{t,\tau}$, (18.1), the VAR values are obtained from the probability distribution of portfolio value returns:

$$1-\alpha = F_{\Delta P}(-\mathrm{VAR}) = \int_{-\infty}^{-\mathrm{VAR}} f_{\Delta P}(x)dx \ ,$$

where $F_{\Delta P}(x) = Pr(\Delta P \leq x)$ is the cumulative distribution function (cdf) of portfolio returns in one period, and $f_{\Delta P}(x)$ is the probability density function (pdf) of ΔP.[25] The VAR methodologies mainly differ in ways of constructing $f_{\Delta P}(x)$.

The traditional techniques of approximating the distribution of ΔP are

- the *parametric method* (analytic or models-based),
- *historical simulation* (nonparametric or empirical-based),
- *Monte Carlo simulation* (stochastic simulation), and
- *the stress-testing* (scenario analysis).[26]

18.2.1 Parametric Method

If the changes in the portfolio value are characterized by a parametric distribution, VAR can be found as a function of distribution parameters. In this section we review:

- applications of two parametric distributions: normal and gamma,
- linear and quadratic approximations to price movements.

[25] If $f_{\Delta P}(x)$ does not exist, then VAR can be obtained from cdf $F_{\Delta P}$.

[26] JP Morgan [45]; Phelan [66]; Mahoney [51]; Jorion [42]; Simons [76]; Fallon [23]; Linsmeier and Pearson [48]; Hopper [37]; Dave and Stahl [18]; Gamrowski and Rachev [27]; Duffie and Pan [21]; Fong and Vasicek [25]; Pritsker [69].

18.2.1.1 VAR for a Single Asset

Assume that a portfolio consists of a single asset, which depends only on one risk factor. Traditionally, in this setting, the distribution of asset return is assumed to be *the univariate normal distribution*, identified by two parameters: the mean, μ, and the standard deviation, σ. The problem of calculating VAR is then reduced to finding the (1-α)th percentile of the standard normal distribution $z_{1-\alpha}$:

$$1-\alpha = \int_{-\infty}^{X^*} g(x)dx = \int_{-\infty}^{z_{1-\alpha}} \phi(z)dz = N(z_{1-\alpha}), \quad \text{with } X^* = z_{1-\alpha}\sigma + \mu \ ,$$

where $\phi(z)$ is the standard normal density function, N(z) is the cumulative normal distribution function, X is the portfolio return, $g(x)$ is the normal distribution function for returns with the mean μ and the standard deviation σ, and X^* is the lowest return at a given confidence level α.

Investors in many applications assume that the expected return μ equals 0. This assumption is based on the conjecture that the magnitude of μ is substantially smaller than the magnitude of the standard deviation σ and, therefore, can be ignored. Then, it can be assumed:

$$X^* = z_{1-\alpha}\sigma \ ,$$

and, therefore,

$$\text{VAR} = -Y_0 X^* = -Y_0 z_{1-\alpha}\sigma \ ,$$

where Y_0 is the initial portfolio value.

18.2.1.2 Portfolio VAR

If a portfolio consists of many assets, the computation of VAR is performed in several steps. Portfolio assets are decomposed into "building blocks", which depend on a finite number of risk factors. Exposures of the portfolio securities are combined into risk categories. The total portfolio risk is constructed, based on aggregated risk factors and their correlations. We denote:

- X_p is the portfolio return in one period,
- N is the number of assets in the portfolio,
- X_i is the i-th asset return in one period $(\tau = 1), X_i = \Delta P(1) = P_i(1) - P_i(0)$, where P_i is the log-spot price of asset $i, i = 1, ..., N$.

More generally, X_i can be the risk factor that enters linearly[27] in the portfolio return.

- w_i is the i-th asset's weight in the portfolio, $i = 1, \ldots, N$.

$$X_P = \sum_{i=1}^{N} w_i X_i \ .$$

In matrix notation,

$$X_P = w^T X \ ,$$

where $w = (w_1, w_2, \ldots, w_N)^T$, $X = (X_1, X_2, \ldots, X_N)^T$.
Then the portfolio variance is

$$V(X_p) = w^T \underline{\Sigma} w = \sum_{i=1}^{N} w_i^2 \sigma_{ii} + \sum_{i=1}^{N} \sum_{j=1, i \neq j}^{N} w_i w_j \rho_{ij} \sigma_i \sigma_j \ ,$$

where σ_{ii} is the variance of returns on the i-th asset, σ_i is the standard deviation of returns on the i-th asset, ρ_{ij} is the correlation between the returns on the i-th and the j-th assets, $\underline{\Sigma}$ is the covariance matrix, $\underline{\Sigma} = [\sigma_{ij}]$, $1 \leq i \leq N$, $1 \leq j \leq N$. If all portfolio returns are *jointly normally distributed*, the portfolio return, as a linear combination of normal variables, is also *normally distributed*. The portfolio VAR based on the normal distribution assumption is

$$\text{VAR} = -Y_0 z_{1-\alpha} \sigma(X_p) \ ,$$

where $\sigma(X_p)$ is the portfolio standard deviation (the *portfolio volatility*),

$$\sigma(X_p) = \sqrt{V(X_p)} \ .$$

Thus, risk can be represented by a combination of linear exposures to normally distributed factors.

In this class of parametric models, to estimate risk, it is sufficient to evaluate the covariance matrix of portfolio risk factors (in the simplest case, individual asset returns).

The estimation of the covariance matrix is based on the *historical data* or on *implied data* from securities pricing models.

[27]If the risk factor does not enter linearly (as in a case of an option), then a linear approximation is used.

If portfolios contain zero-coupon bonds, stocks, commodities, and currencies, VAR can be computed from correlations of these basic risk factors and the asset weights. If portfolios include more complex securities, then the securities are decomposed into building blocks.

The portfolio returns are often assumed to be normally distributed.[28] One of methods employing the normality assumption for returns is the *delta* method (the delta-normal or the variance-covariance method).

18.2.1.3 Delta Method

The *delta* method estimates changes in prices of securities using their "deltas" with respect to basic risk factors. The method involves a *linear* (also named as *delta* or *local*) *approximation* to (log) price movements:

$$P(X+U) \approx P(X) + P'(X)U,$$

or

$$\Delta P(X) = P(X+U) - P(X) \approx P'(X)U,$$

where X is the level of the basic risk factor (i.e., an equity, an exchange rate), U is the change in X, $P(X+U) = P(t+\tau, X+U)$, $P(X) = P(t, X)$,[29] $P(X)$ is the (log) price of the asset at the X level of the underlying risk factor, $P'(X) = \partial P/\partial X$ is the first derivative of $P(X)$, it is commonly called the *delta* ($\underline{\Delta} = \underline{\Delta}(X)$) of the asset.
Thus, the price movements of the securities approximately are

$$\Delta P(X) \approx P'(X)U = \underline{\Delta}U.$$

The *delta-normal* (the *variance-covariance*) method computes the portfolio VAR as

$$\text{VAR} = -Y_0 z_{1-\alpha}\sqrt{d^T \underline{\Sigma} d}\ ,$$

where $d = d(X) = (\underline{\Delta}_1(X), \underline{\Delta}_2(X), \ldots, \underline{\Delta}_n(X))^T$ is a vector of the delta positions, $\underline{\Delta}_j(X)$ is the security's delta with respect to the j-th risk factor, $\underline{\Delta}_j = \partial P/\partial X_j$.

18.2.1.4 VAR Based on the Gamma Distribution Assumption

Since the normal model for factor distributions is overly simplistic, Fong and Vasicek [25] suggest estimating the probability distribution of the portfolio value changes ΔP by another type of the parametric distributions -

[28] JP Morgan [45]; Phelan [66].

[29] Because the time horizon (τ) is fixed and t is the present time, we shall omit the time argument and shall write $P(X+U)$ instead of underlying $P(t+\tau, X+U)$ and $P(X)$ instead of $P(t, X)$. We shall consider the dependency of P on the risk factor X only.

the gamma distribution. They also assume that the basic risk factors X_i are jointly normally distributed with the zero mean and the covariance matrix Σ. However, Fong and Vasicek propose a *quadratic* (*gamma* or *delta-gamma*) *approximation* to the individual asset price changes:

$$\Delta P(X) = P(X_1 + U_1, \ldots, X_n + U_n) - P(X_1, \ldots, X_n)$$

$$\approx \sum_{j=1}^{n} \underline{\Delta}_j U_j + \frac{1}{2} \sum_{j=1}^{n} \Gamma_j U_j^2 \ ,$$

where ΔP is a security price change, n is the number of basic risk factors, U_j is the change in the value of the j-th risk factor, $\underline{\Delta}_j = \underline{\Delta}_j(X)$ is the security's delta at the level X with respect to the j-th risk factor, $\underline{\Delta}_j = \partial P/\partial X_j$, Γ_j is quadratic exposure (*the gamma*) at the level X to the j-th risk factor, $\Gamma_j = \Gamma_j(X) = \partial^2 P/\partial X_j^2$, $j = 1, \ldots, n$.

The delta-gamma approximation for the portfolio return in one period is defined by

$$\Delta P = \Delta P(X) = P(X + U) - P(X)$$

$$= \sum_{i=1}^{n} \underline{\Delta}_i w_i U_i + \frac{1}{2} \sum_{i=1}^{n} \sum_{j=1}^{n} \Gamma_{ij} w_i w_j U_i U_j \ , \tag{18.2}$$

where $X = (X_1, X_2, \ldots, X_n)^T$, X_i is i-th risk factor, U_i is the change in the risk factor X_i, w_i is the weight of the i-th risk factor, $\Gamma_{ij} = \Gamma_{ij}(X)$ is the *portfolio (i,j)-gamma*, $\Gamma_{ij}(X) = \partial^2 P(X)/\partial X_i \partial X_j$, $\Gamma_{jj} = \Gamma_j$, $i = 1, \ldots, n$, $j = 1, \ldots, n$. The variance of portfolio return can be estimated by

$$V(\Delta P(X)) = \sum_i \sum_j \underline{\Delta}_i \underline{\Delta}_j w_i w_j \mathrm{cov}(X_i, X_j)$$

$$+ \sum_i \sum_j \sum_k \underline{\Delta}_i \Gamma_{jk} w_i w_j w_k \mathrm{cov}(X_i, X_j X_k)$$

$$+ \frac{1}{4} \sum_i \sum_j \sum_k \sum_l \Gamma_{ij} \Gamma_{kl} w_i w_j w_k w_l \mathrm{cov}(X_i X_j, X_k X_l).$$

From (18.2), ΔP is a quadratic function of normal variates. This distribution of ΔP is, in general, non-symmetric. However, one can approximate the quantile by the skewness parameter and the standard deviation. In fact, Fong and Vasicek [25] used the approximation for the portfolio VAR value, based on a generalized "*gamma*" distribution:

$$\text{VAR} = -Y_0 k(\gamma, \alpha)\sigma(X_p) \ ,$$

where γ is the skewness of the distribution, $\gamma = \mu_3/\sigma^3, \mu_3$ is the third moment of ΔP, $k(\gamma, \alpha)$ is the ordinate obtained from the generalized gamma distribution for the skewness γ at the confidence level α. Fong and Vasicek [25] report the $k(\gamma, \alpha)$ values at $\alpha = 99\%$:

γ	$k(\gamma, \alpha)$	γ	$k(\gamma, \alpha)$
-2.83	3.99	0.50	1.96
-2.00	3.61	0.67	1.83
-1.00	3.03	1.00	1.59
-0.67	2.80	2.00	0.99
-0.50	2.69	2.83	0.71
0.0	2.33		

Source: Fong and Vasicek [25].

The gamma distribution takes into consideration the skewness of the ΔP distribution, whereas the normal distribution is symmetric and does not reflect the skewness.

18.2.2 Historical Simulation

The *historical simulation* approach constructs the distribution of the portfolio value changes ΔP from historical data without imposing distribution assumptions and estimating parameters. Hence, sometimes the historical simulation method is called a *nonparametric* method. The method assumes that trends of past price changes will continue in the future. Hypothetical future prices for time t+s are obtained by applying historical price movements to the current (log) prices:

$$P^*_{i,t+s} = P^*_{i,t+s-1} + \Delta P_{i,t+s-\kappa} \ ,$$

where t is the current time, $s = 1, 2, \ldots, \kappa$, κ is the horizon length of going back in time, $P^*_{i,t+s}$ is the hypothetical (log) price of the i-th asset at time $t + s$, $P^*_{i,t} = P_{i,t}$, $\Delta P_{i,t+s-\kappa} = P_{i,t+s-\kappa} - P_{i,t+s-1-\kappa}$, $P_{i,t}$ is the historical (log) price of the i-th asset at time t. Here we assumed that the time horizon $\tau = 1$.

A portfolio value $P^*_{p,t+s}$ is computed using the hypothetical (log) prices $P^*_{i,t+s}$ and the current portfolio composition. The portfolio return at time $t+s$ is defined as

$$R^*_{p,t+s} = P^*_{p,t+s} - P_{p,t} \ ,$$

where $P_{p,t}$ is the current portfolio (log) price.

The portfolio VAR is obtained from the density function of computed hypothetical returns. Formally, VAR $=$ VAR$_{t,\tau}$ is estimated by the negative of the $(1-\alpha)$th quantile, VAR*; namely, $F_{\kappa,\Delta P}(-\text{VAR}) = F_{\kappa,\Delta P}(\text{VAR}^*) = 1-\alpha$, where $F_{\kappa,\Delta P}(x)$ is the empirical density function

$$F_{\kappa,\Delta P}(x) = \frac{1}{\kappa}\sum_{s=1}^{\kappa} \mathbf{1}\left\{R^*_{p,t+s} \le x\right\}, \quad x \in \mathbf{R} \ .$$

18.2.3 Monte Carlo Simulation

The *Monte Carlo* method specifies statistical models for basic risk factors and underlying assets. The method simulates the behavior of risk factors and asset prices by generating random price paths. Monte Carlo simulations provide possible portfolio values on a given date T after the present time t, $T > t$. The VAR (VAR$_T$) value can be determined from the distribution of simulated portfolio values. The Monte Carlo approach is performed according to the following algorithm:

1. Specify stochastic processes and process parameters for financial variables and correlations.
2. Simulate the hypothetical price trajectories for all variables of interest. Hypothetical price changes are obtained by simulations, draws from the specified distribution.
3. Obtain asset prices at time T, $P_{i,T}$, from the simulated price trajectories. Compute the portfolio value $P_{p,T} = \sum w_{i,T} P_{i,T}$.
4. Repeat steps 2 and 3 many times to form the distribution of the portfolio value $P_{p,T}$.
5. Measure VAR$_T$ as the negative of the $(1-\alpha)$th percentile of the simulated distribution for $P_{p,T}$.

18.2.4 Stress Testing

The parametric, historical simulation, and Monte Carlo methods estimate the VAR (expected losses) depending on risk factors. The *stress testing* method examines the effects of large movements in key financial variables on the portfolio value. The price movements are simulated in line with the certain scenarios.[30] Portfolio assets are reevaluated under each scenario. The portfolio return is derived as

$$R_{p,s} = \sum w_{i,s} R_{i,s} \ ,$$

where $R_{i,s}$ is the hypothetical return on the i-th security under the new scenario s. $R_{p,s}$ is the hypothetical return on the portfolio under the new scenario s.

Estimating a probability for each scenario s allows construction of a distribution of portfolio returns, from which VAR can be derived.

18.3 Components of VAR Methodologies

Implementation of the VAR methodologies requires analysis of their components:

- distribution and correlation assumptions,
- volatility and covariance models,
- weighting schemes,
- the window length of data used for parameter estimations,
- the effect of the time horizon (holding period) on the VAR values, and
- incorporation of the mean of returns in the VAR analysis.

18.3.1 Distribution Assumptions

The parametric VAR methods assume that asset returns have parametric distributions. The parametric approaches are subject to "model risk": distribution assumptions might be incorrect. The frequent assumption is that asset returns have a multivariate normal distribution, although, many financial time-series violate the normality assumption. Empirical data exhibit asymmetric, leptokurtic or platokurtic distributions with heavy tails. Fong and Vasicek [25] suggest using the gamma distribution. The historical simulation technique does not place the distribution assumptions, thus, it

[30]Scenarios include possible movements of the yield curve, changes in exchange rates, etc. together with estimates of the underlying probabilities.

is free of model risk and "parameter estimation" risk. The Monte Carlo approach specifies the distributions of the underlying instruments.

18.3.2 Volatility and Covariance Models

The VAR methods apply diverse volatility and correlation models:[31]

- constant volatility (moving window),
- exponential weighting,
- GARCH,
- EGARCH (asymmetric volatility),
- cross-market GARCH,
- implied volatility,
- subjective views.[32]

18.3.2.1 Constant Volatility Models

In the *constant volatility* (*equally weighted*) models, variances and covariances do not change over time. They are approximated by sample variances and covariances over the estimation "window":

$$\hat{\sigma}_{t,T}^2 = \frac{1}{T-t} \sum_{i=t+1}^{T} (R_i - \hat{\mu}_{t,T})^2 \ ,$$

where $\hat{\sigma}_{t,T}$ is the estimated variance of returns R_i over the time window $[t, T]$, $\hat{\mu}_{t,T}$ is the estimated mean of returns over the time window $[t, T]$,

$$\hat{\mu}_{t,T} = \frac{1}{T-t} \sum_{i=t+1}^{T} R_i.$$

If the mean return is assumed to be sufficiently small,

$$\hat{\sigma}_{t,T}^2 = \frac{1}{T-t} \sum_{i=t+1}^{T} R_i^2 \ .$$

[31]Duffie and Pan [21]; Jackson, Maude, and Perraudin [39]; JP Morgan [45]; Phelan [66]; Hopper [37]; Mahoney [51]; Hendricks [34]).

[32]The method of "subjective views" means that analysts make predictions of volatility from their own views of market conditions. See Hopper [37].

18.3.2.2 Weighted Volatility Models

The empirical financial data do not exhibit constant volatility. The *exponential weighting* models take into account time-varying volatility and accentuate the recent observations:

$$\hat{\sigma}^2_{t,T} = \sum_{i=t+1}^{T} \Theta_i (R_i - \hat{\mu}_{t,T})^2 ,$$

where θ_i are the weighting values:

$$0 < \Theta_i < 1, \sum_{i=t+1}^{T} \Theta_i = 1 . \qquad (18.3)$$

The weighting schemes are divided on *uniform*[33] and *asset-specific*[34] schemes. The JP Morgan's RiskMetrics system adopted the *uniform weighting* approach:

$$\Theta_i = (1 - \lambda)\lambda^{T-i} c_{T-t} ,$$

where λ is the decay factor, $0 < \lambda < 1$, and $c_{T-t} > 0$ is chosen so that the constraints (18.3) are met. JP Morgan uses $\lambda = 0.94$ for a 1-day time horizon. Jackson, Maude and Perraudin [39] demonstrate that the weighting schemes with lower values of λ in parametric models lead to higher tail probabilities, proportions of actual observations exceeding the VAR predictions.[35] They point out a trade-off between the degree of approximating time-varying volatilities and the performance of the parametric methods. Hendricks and Hirtle [33] found that decreasing λ accompanies higher variability of the VAR measurements.

The CSFB's PrimeRisk employs *the asset-specific weighting schemes*. It develops specific volatility models (different weighting schemes) for different types of assets (i.e., equities, futures, OTC options).

18.3.2.3 ARCH Models

Popular models explaining time-varying volatility are *autoregressive conditional heteroscedasticity (ARCH)* models, introduced by Engle [22]. In the ARCH models the conditional variances follow autoregressive processes. The ARCH(q) model assumes the returns on the i-th asset $R_{i,1}$, $R_{i,2}$, ...

[33] JP Morgan [45]; Jackson, Maude, and Perraudin [39].

[34] Lawrence and Robinson [47].

[35] See Jackson, Maude, Perraudin [39, table 4, p. 179].

are explained by the process:

$$R_{i,t} = \mu_i + \sigma_{i,t} u_{i,t} \ ,$$

$$\sigma_{i,t}^2 = \alpha_i + \sum_{j=1}^{q} \beta_{ij} \left(R_{i,t-j} - \mu_i\right)^2 \ ,$$

where μ_i is the expectation of R_i, $\sigma_{i,t}^2$ is the conditional variance of R_i at time t, $u_{i,t+1}$ is a random shock with the mean of zero and the variance of 1 (a common assumption is $(U_{i,t})_{t \geq 1} \sim$ iid $\mathrm{N}(0,1)$), α_i and β_{ij} are constants, $\alpha_i > 0$, $\beta_{ij} \geq 0$, $j = 1, \dots, q$, $i = 1, \dots, n$.[36]

In the ARCH(1) model the conditional volatility at period t depends on the volatility at the previous period $t-1$. If volatility at time $t-1$ was large, the volatility at time t is expected to be large as well. Observations will exhibit clustered volatilities: one can distinguish periods with high volatilities and tranquil periods.

18.3.2.4 GARCH Models

Bollerslev [6] suggested the *generalized ARCH (GARCH)* model.[37] In the GARCH models the conditional variance contains both autoregressive and moving average components (it follows an ARMA process). In the GARCH(p,q) model, the return on the i-th asset has the representation

$$R_{i,t} = \mu_i + \sigma_{i,t} u_{i,t} \ ,$$

the conditional variance is assumed to follow

$$\sigma_{i,t}^2 = \alpha_i + \sum_{j=1}^{q} \beta_{ij} (R_{i,t-j} - \mu_i)^2 + \sum_{k=1}^{p} \gamma_{jk} \sigma_{i,t-k}^2 \ ,$$

where α_i, β_{ij}, γ_{ik} are constants, $\alpha_i > 0$, $\beta_{ij} \geq 0$, $\gamma_{ik} \geq 0$, $j = 1, \dots, q$, $k = 1, \dots, p$, $i = 1, \dots, n$.

The advantage of using the GARCH model ensues from the fact that an AR process of a high order might be represented by a more parsimonious ARMA process. Thus, the GARCH model will have less parameters to be estimated than the corresponding ARCH model.

[36] The dependence structure of the returns $R_t = (R_{1,t}, R_{2,t}, \dots, R_{n,t})$ needs additional specifications for each $t > 0$. See, for example, Section 18.3.2.6 further on.

[37] See also Bollerslev, Chou, and Kroner [7].

18.3.2.5 EGARCH Models

Nelson [62] introduced the *exponential GARCH (EGARCH)* model. In the general EGARCH(p, q) model, the conditional variance follows[38]

$$\log\ \sigma_t^2 = \alpha + \sum_{j=1}^{q} \beta_j \left\{ \left| \frac{R_{t-j} - \mu}{\sigma_{t-j}} \right| - E \left| \frac{R_{t-j} - \mu}{\sigma_{t-j}} \right| + \delta \frac{(R_{t-j} - \mu)}{\sigma_{t-j}} \right\}$$

$$+ \sum_{i=1}^{p} \gamma_i \log\ \sigma_{t-i}^2 .$$

The δ parameter helps explain asymmetric volatility. If $\beta_j > 0$ and $-1 < \delta < 0$, then negative deviations of R_t from the mean entail higher volatility than positive deviations do. If $\beta_j > 0$ and $\delta < -1$, then positive deviations lower volatility whereas negative deviations cause additional volatility.

The advantage of using the EGARCH model is that it does not impose positivity restrictions on coefficients, whereas the GARCH model requires coefficients to be positive.

18.3.2.6 Cross-market GARCH

The *cross-market GARCH* allows estimating volatility in one market from volatilities in other markets. Duffie and Pan [21] provide an example of a cross-market GARCH, which employs the *bivariate GARCH* model:

$$\begin{pmatrix} \sigma_{1,t}^2 \\ \sigma_{12,t} \\ \sigma_{2,t}^2 \end{pmatrix} = A + B \begin{pmatrix} R_{1,t}^2 \\ R_{1,t}\ R_{2,t} \\ R_{2,t}^2 \end{pmatrix} + \Gamma \begin{pmatrix} \sigma_{1,t-1}^2 \\ \sigma_{12,t-1} \\ \sigma_{2,t-1}^2 \end{pmatrix}$$

where $\sigma_{1,t-1}$ is the conditional standard deviation of $R_{1,t}$, $\sigma_{2,t-1}$ is the conditional standard deviation of $R_{2,t,}$, $\sigma_{12,t-1}$ is the conditional covariance between $R_{1,t}$ and $R_{2,t}$, $R_{1,t}$ is the return in the first market at time t, $R_{2,t}$ is the return in the second market at time t, A is a vector of three elements, B is a 3×3 matrix, Γ is a 3×3 matrix.

18.3.2.7 Implied Volatilities

Sometimes analysts use *implied volatilities* to estimate future volatilities. Implied volatilities are volatilities derived from pricing models. For

[38] Hamilton [32].

instance, implied volatilities can be obtained from the Black-Scholes option pricing model. Option prices calculated by the Black-Scholes formula $C_t = C(S_t, K, r, \sigma, \tau)$ are increasing in volatility σ. Hence, "inverting" the formula, one can obtain the implied volatility values $\sigma = \sigma(C_t, S_t, K, r, \tau)$. Here, C_t is the option price, S_t is the price of the underlying asset, K is the exercise price, r is the constant interest rate, and τ is the time to expiration.

The *implied tree* technique[39] assumes implied volatilities change over time and computes them relating the modeled and observed option prices.

18.3.2.8 Correlation Models

Besides the distribution assumptions and volatility models, the VAR computations also need specification of correlation assumptions on price changes and volatilities within and across markets.[40] Beder [5] illustrated the sensitivity of VAR results to correlation assumptions. She computed VAR using the Monte Carlo simulation method under different assumptions: (i) correlations across asset groups and (ii) correlations only within asset groups. The obtained VAR estimates were lower for the first type of correlation assumptions than for the second type.

18.3.3 Time Horizon

The *time horizon* (the *holding period*) in the VAR computations can take any time value. In practice, it varies from one day to two weeks (10 trading days) and depends on liquidity of assets and frequency of trading transactions. It is assumed that the portfolio composition remains the same over the holding period. This assumption constrains dynamic trading strategies. The Basle Committee recommends using the 10-day holding period. Users argue that the time horizon of 10 days is inadequate for frequently traded instruments and is restrictive for nonliquid assets. Long holding periods are usually recommended for portfolios with nonliquid instruments, though many model approximations are only valid within short periods of time.

Beder [5] analyzed the impact of the time horizon on VAR estimations. She calculated VAR for three hypothetical portfolios applying four different approaches for the time horizons of 1-day and 10-day. For all VAR calculations, with the exception of one case, Beder reported larger VAR estimates for longer time horizons.

[39] Derman and Kani [20]; Rubinstein [70]; Jackwerth and Rubinstein [40] and [41].
[40] Duffie and Pan [21]; Beder [5]; Liu [49].

18.3.4 Window Length

The *window length* is the length of the data subsample (the observation period) used for a VAR estimation. The window length choice is related to sampling issues and availability of databases. The regulators suggest using the 250-day (one-year) window length.

Jackson, Maude, and Perraudin [39] computed parametric and simulation VARs for the 1-day and 10-day time horizons using window lengths from three to 24 months. They concluded that VAR forecasts based on longer data windows are more reliable.[41] Beder [5] estimated VAR applying the historical simulation method for the 100-day and 250-day window lengths. Beder shows that the VAR values increase with the expanded observation intervals. Hendricks [33] calculated the VAR measures using the parametric approach with equally weighted volatility models and the historical simulation approach for window lengths of 50, 125, 250, 500, and 1250 days. He reports that the VAR measures become more stable for longer observation periods.

18.3.5 Incorporation of the Mean of Returns

In many cases the mean of returns is assumed to be zero. Jackson, Maude, and Perraudin [39] analyze the effects of (i) inclusion of the mean in calculations and (ii) setting the mean to zero on VAR results. Their analysis did not lead to certain conclusions.[42]

18.4 VAR Strengths and Weaknesses

The VAR methodologies are becoming necessary tools in risk management. It is important to be aware of *VAR strengths and weaknesses*.[43]

Institutions use the VAR measurements to estimate exposure to market risks and assess expected losses. Application of different VAR methods provides different VAR estimates. The choice of methods should mostly depend on the portfolio composition. If a portfolio contains instruments with linear dependence on basic risk factors, the *delta* method will be satisfactory. The strength of the delta approach is that computations of VAR are relatively easy. Drawbacks of the delta-normal method are (i) empirical observations

[41] Jackson, Maude, and Perraudin [39, table 5, p. 180].

[42] Jackson, Maude, and Perraudin [39, table 6, p. 181].

[43] Beder [5]; Mahoney [51]; Simons [76]; Jorion [42] and [43]; Hopper [37]; Shaw [73]; Schachter [71]; *Derivatives Strategy* [19].

on returns of financial instruments do not exhibit the normal distribution and, thus, the delta-normal technique does not fit well data with heavy tails; (ii) accuracy of VAR estimates diminishes with nonlinear instruments; in their presence, VAR estimates are understated. For portfolios with option instruments, *historical* and *Monte Carlo simulations* are more suitable. The *historical simulation* method is easy to implement having a sufficient database. The advantage of using the historical simulation is that it does not impose distributional assumptions. Models based on historical data assume that the past trends will continue in the future. However, the future might encounter extreme events. The historical simulation technique is limited in forecasting the range of portfolio value changes. The *stress-testing* method can be applied to investigate effects of large movements in financial variables. A weakness of stress testing is that it is subjective. The *Monte Carlo* method can incorporate nonlinear positions and non-normal distributions. It does not restrict the range of portfolio value changes. The Monte Carlo method can be used in conducting the sensitivity analysis. The main limitations in implementing the Monte Carlo methodology are (i) it is affected by model risk; (ii) computations and software are complex; (iii) it is time consuming.

VAR methodologies are subject to *implementation risk*: implementation of the same model by different users produces different VAR estimates. Marshall and Siegel [56] conducted an innovative study of implementation risk. They compared VAR results obtained by several risk management systems developers using one model, JP Morgan's RiskMetrics. Marshall and Siegel found that, indeed, different systems do not produce the same VAR estimates for the same model and identical portfolios. The varying estimates can be explained by the sensitivity of VAR models to users' assumptions. The degree of variation in VAR numbers was associated with the portfolio composition. Dependence of implementation risk on instrument complexity can be summarized in the following relative ascending ranking: foreign exchange forwards, money markets, forward rate agreements, government bonds, interest rate swaps, foreign exchange options, and interest rate options. Nonlinear securities entail larger discrepancies in VAR results than linear securities. In order to take into account implementation risk, it is advisable to accompany VAR computations for nonlinear portfolios with sensitivity analysis to underlying assumptions.

Other VAR weaknesses are

- Existing VAR models reflect observed risks and they *are not useful in transition periods* characterized by structural changes, additional risks, contracted liquidity of assets, and broken correlations across assets and across markets.
- The trading positions change over time. Therefore, *extrapolation*

of a VAR for a certain time horizon to longer time periods *might be problematic.* Duffie and Pan [21] point out that if intra-period position size is stochastic, then the VAR measure obtained under the assumption of constant position sizes should be multiplied by a certain factor.[44]

- The VAR methodologies assume that necessary database is available. For certain securities, *data over a sufficient time interval may not exist.*
- If historical information on financial instruments is not available, the instruments are mapped into known instruments; though *mapping reduces precision of VAR estimations.*
- *Model risks* can occur if the chosen stochastic underlying processes for valuing securities are incorrect.
- Since true parameters are not observable, estimates of parameters are obtained from sample data. The *measurement error* rises with the number of parameters in a model.

18.5 VAR Advances

In order to improve performance of VAR methodologies, researchers suggest numerous modifications of traditional techniques and new ways of VAR evaluations. This section presents modifications of

- delta,
- historical simulation,
- Monte Carlo,
- scenario analysis.

The section also describes new approaches to VAR estimations and interpretations.

[44]Duffie and Pan [21] provide an expression for the factor in the case of a single asset. If (i) the underlying asset returns have constant volatility σ, (ii) the position size is a martingale and follows lognormal process with volatility s, then a multiplication factor is approximately of $(\exp(at) - 1)/(at)$, where $a = 2s^2 + 4\rho s\sigma$, ρ is the correlation of the position size with the asset.

18.5.1 Modifications of Delta

Let U_t be the n-dimensional vector of changes in risk factors over one period, $U_t = \Delta X_t$. The standard *delta* and *delta-gamma* methods assume that changes in risk factors follow the normal distribution conditional on the current information:

$$(U_{t+1}|\Omega_t) \sim N(0, \Sigma_t) ,$$

where Ω_t is the information available until the current time t.[45]

Delta methods apply a linear approximation to the portfolio returns as a function of the underlying risk factors:

$$\Delta P_{t+1} \equiv P_{t+1}(X_t + U_t) - P_{t+1}(X_t) \approx \underline{\Delta}_t^T Y_{t+1} ,$$

where $\underline{\Delta}_t = (\underline{\Delta}_{1t}, \underline{\Delta}_{2t}, \ldots, \underline{\Delta}_{nt})^T$, $\underline{\Delta}_{it} = \underline{\Delta}_i(X_t) = \frac{\partial P_t(X_t)}{\partial X_{i,t}}$, X_t is the vector of risk factors, $X_t = (X_{1,t}, X_{2,t}, \ldots, X_{n,t})^T$, $P_t(X_t)$ is the portfolio log-price at time t, which depends on the current risk factors only, and $Y_{t+1} = (Y_{1,t+1}, Y_{2,t+1}, \ldots, Y_{n,t+1})^T$, where $Y_{i,t} = w_i\, U_{i,t}$, $i = 1, \ldots, n$.

Under the *delta* approach, $\Delta P_{t+1} \sim N(0, \underline{\Delta}_t^T \underline{\Sigma}_t^Y \underline{\Delta}_t)$, where $\underline{\Sigma}^Y$ is the covariance matrix of Y_{t+1}.

Delta-gamma methods use a quadratic approximation to the portfolio value changes:

$$\Delta P_{t+1} \approx \underline{\Delta}_t^T Y_{t+1} + \frac{1}{2} Y_{t+1}^T \Gamma_t\, Y_{t+1}$$

where

$$\Gamma_t = \Gamma(t, X_t) = \frac{\partial^2 P_t(X_t)}{\partial X_t \partial X_t^T} .$$

Under *delta-gamma methods*, the distribution of ΔP cannot be approximated by the normal distribution. Hence, a traditional technique of VAR derivation as a multiple of the $(1-\alpha)$-th percentile of the standard normal distribution cannot be employed.

We shall describe the following improvements of the delta-gamma method:

- Delta-gamma-Monte Carlo,
- Delta-gamma-delta,
- Delta-gamma-minimization,
- Delta-gamma-Johnson,

[45] Formally, $(\Omega_t)_{t\geq 0}$ is the filtration generated by the underlying market-shot-process (processes), typically white noise-type processes.

- Delta-gamma-Cornish-Fisher.

18.5.1.1 Delta-gamma-Monte Carlo

The *delta-gamma-Monte Carlo* method approximates the distribution of ΔP by the distribution of hypothetical portfolio value changes:[46]

(i) values of U_{t+1} are obtained by random draws from its distribution;

(ii) hypothetical values of ΔP are calculated at each draw using the delta-gamma approximation;

(iii) steps (i) and (ii) are repeated many times;

(iv) the distribution of ΔP is formed by ordering the ΔP values from step (iii).

A VAR estimate is derived as the negative of the $(1-\alpha)$-th percentile of the ΔP distribution constructed in step (iv).

18.5.1.2 Delta-gamma-delta

The *delta-gamma-delta* method also employs a delta-gamma approximation.[47] It assumes: (i) shocks to the portfolio value P are represented by U_{t+1} and elements of $U_{t+1}U'_{t+1}$, which correspond to $\underline{\Delta}$ and Γ terms in a delta-gamma approximation; (ii) the shocks are uncorrelated and normally distributed. For instance, if the portfolio value depends on a single factor, then shocks to P are assumed to come from jointly normally distributed U_{t+1} and U_{t+1}^2; though the assumptions of the normality U_{t+1}^2 and the joint normality are not correct. According to the delta-gamma-delta approach,

$$\Delta P_{t+1} \sim N(.5\Gamma_t\, \sigma_t^2,\ \underline{\Delta}_t^2 \sigma_t^2 + .5\Gamma_t^2\, \sigma_t^4),$$

where σ_t^2 is the variance of $Y_{t+1}, Y_{i,t+1} = w_i U_{i,t+1}$. Therefore, VAR can be calculated as

$$\text{VAR}_t = -\left(\frac{1}{2}\Gamma_t \sigma_t^2 + z_{1-\alpha}\sqrt{\underline{\Delta}_t^2\ \sigma_t^2 + \frac{1}{2}\Gamma_t^2\ \sigma_t^4}\right).$$

18.5.1.3 Delta-gamma-minimization

The *delta-gamma-minimization* method uses a delta-gamma approximation to ΔP and assumes that a vector of changes in risk factors U_{t+1} is

[46] Pritsker [69].

[47] Pritsker [69].

normally distributed.[48] Denote by Y_{t+1} a vector of shocks with the weights of risk factors, $Y_{i,t+1} = w_i U_{i,t+1}$. The delta-gamma-minimization technique determines VAR_t as the solution of the following minimization problem:

$$-\mathrm{VAR}_t = \min_{Y_{t+1}} \left\{ \underline{\Delta}_t{}^T Y_{t+1} + \frac{1}{2} Y_{t+1}{}^T \Gamma_t \, Y_{t+1} \right\}$$

subject to the constraint

$$Y_{t+1}^T \, \underline{\Sigma}^{-1} \, Y_{t+1} \leq \chi^2(\alpha, k) \ ,$$

where $\underline{\Sigma}$ is the covariance matrix of risk factors and $\chi^2(\alpha, k)$ is the $\alpha\%$ critical value of the central chi-squared distribution with k degrees of freedom. The method implicitly supposes that the ΔP values within the constraint set exceed the "external" ΔP values. In practice, this assumption might be violated. Hence, the fraction of the ΔP values, which are lower than $-\mathrm{VAR}$, can be less than $(1-\alpha)\%$ and the VAR estimate can be overstated. The strengths of the delta-gamma-minimization method are (i) it does not impose the assumption of joint normality as the delta-gamma-delta method does; (ii) it avoids Monte Carlo data generation.

18.5.1.4 Delta-gamma-Johnson

The *delta-gamma-Johnson* method[49] relies on the normal assumption for the distribution of $(U_t)_{t\geq 1}$. The method chooses a distribution function for ΔP and estimates its parameters matching the first four moments of the distribution and the delta-gamma approximation of ΔP. A VAR estimate is obtained from the cumulative density function of the chosen distribution. The strength of the method is that it is analytic. However, the delta-gamma-Johnson method uses information only up to the fourth moment and might be less precise than the delta-gamma-Monte-Carlo method, which uses all information on the delta-gamma Taylor expansion.

18.5.1.5 Delta-gamma-Cornish-Fisher

The *delta-gamma-Cornish-Fisher* method is based on a delta-gamma approximation of the ΔP distribution and the normality assumption for U_t. It uses a Cornish-Fisher expansion to estimate the $(1-\alpha)$-th percentile of

[48] Pritsker [69]; Fallon [23]; Wilson [80].

[49] Zangari [81]; Pritsker [69].

the standardized ΔP distribution ΔP^S:[50]

$$F_{\Delta P^S}(1-\alpha) = \Phi(\alpha) + \frac{1}{6}(\Phi(\alpha)^2 - 1)\, q_3 + \frac{1}{24}(\Phi(\alpha)^3 - 3\Phi(\alpha))q_4$$

$$- \frac{1}{36}(2\Phi(\alpha)^3 - 5\Phi(\alpha))q_3^2\ ,$$

where q_3 is the third cumulant of ΔP^S and q_4 is the fourth cumulant of ΔP^S. The cumulants q_i are determined from an expansion $\ln(G_{\Delta P^S}(t)) = \sum_{i=1}^{\infty} q_i \frac{t^i}{i!}$ where $G_{\Delta P^S}(t) = E[\exp(t(\Delta P^S))]$ is the moment generating function of ΔP^S.

The advantage of the delta-gamma-Cornish-Fisher approach is that it is analytic. The weakness of the method: it ignores part of information.

18.5.2 Modifications of Historical Simulation

The standard historical simulation technique forms the distribution of portfolio returns based on the historical data. It obtains future returns on assets by applying past changes to the current returns.

We consider here three modifications of the historical simulation method:

- bootstrapped historical simulation,
- combining kernel estimation with historical simulation, and
- hybrid approach - combining exponential smoothing with historical simulation.

18.5.2.1 Bootstrapped Historical Simulation

The *bootstrapped historical simulation* method generates returns of the risk factors by "bootstrapping" from historical observations.[51] Drawing data from the updated returns is suggested. Suppose that we have observations of returns on n-assets $[R_{it}]_{1\leq i\leq n, 1\leq t\leq T}$ with the covariance matrix $\underline{\Sigma}$. The past returns can be updated using new estimates of

(i) *volatilities*, and

(ii) *volatilities and correlations*.

[50] Fallon [23]; Pritsker [69]; Zangari [81].

[51] Duffie and Pan [21]; Jorion [42]; Pritsker [69]; Shaw [73].

In case (i), the updated returns are given by

$$R_{it}^U = R_{it}\frac{\sigma_i^U}{\sigma_i} ,$$

where σ_i is the historical standard deviation (volatility) of the i-th asset, σ_i^U is a new estimate of the standard deviation σ_i, R_{it}^U is an updated return for the i-th asset, $t = 1, \ldots, T$, $i = 1, \ldots, n$.

In case (ii), the updated return vector is determined by

$$R_t^U = (\underline{\Sigma}^U)^{\frac{1}{2}}(\underline{\Sigma})^{-\frac{1}{2}}R_t ,$$

where $R_t = (R_{1t,}, R_{2t}, \ldots, R_{nt})^T$ is the vector of returns at time t, $(\underline{\Sigma})^{-\frac{1}{2}}$ is the matrix square root of $\underline{\Sigma}^{-1}$, and $(\underline{\Sigma}^U)^{\frac{1}{2}}$ is the matrix square root of the updated covariance matrix $\underline{\Sigma}^U$.

One of future research questions would be to investigate impacts of updating approaches on the shape of the portfolio return distribution.

18.5.2.2 Combining Kernel Estimation with Historical Simulation

Butler and Schachter ([11] and [12]) suggest combining the historical simulation approach with kernel estimation. This modification allows estimating precision of VAR measures and the construction of confidence intervals around them. The combined approach is performed in three steps:

1. Approximation of the probability density function (pdf) $f_{\Delta P}(x)$ and the cumulative distribution function (cdf) $F_{\Delta P}(x)$ of portfolio returns,

2. Approximation of the distribution of the order statistic corresponding to the confidence level,[52] and

3. Estimation of VAR using moments of the pdf for the ith order statistic $R_{i,s}$, determined by $(1-\alpha)$th quantile. The first moment, the mean, $ER_{i,k}$, approximates the $-$VAR and the second moment, the variance, $\mathrm{Var}R_{i,k}$ reflects precision of the VAR estimate. The standard deviation can be used for constructing confidence intervals.

[52]Given a sample $R_1, \ldots, R_k$ of observations R_i, we rearrange the sample in increasing order $R_{1;k} \leq \cdots \leq R_{k;k}$ (the random sample), then the ith order statistic is given by $R_{i,k}$. The value of the q-quantile $(0 \leq q \leq 1)$ is the value of the (rounded) (qk)th term in the ranked sample. Using historical return data $R_1, \ldots, R_k$, VAR can be estimated by the negative of the $(1-\alpha)$th quantile.

Step 1. While the standard historical simulation technique derives the piecewise pdf, a kernel estimation forms a smooth pdf. Butler and Schachter [12] apply a normal kernel estimator:

$$\hat{f}_{\Delta P}(x) = \frac{1}{k(0.9\sigma k^{-0.2})} \sum_{i=1}^{k} \frac{1}{\sqrt{2\pi}} e^{-\frac{1}{2}\left(\frac{x-R_i}{0.9\sigma k^{-0.2}}\right)^2}, \tag{18.4}$$

where k is the sample size, R_i is the ith observation on portfolio return ΔP, $i = 1, \ldots, k, \sigma$ is the standard deviation of ΔP, x is a current point of estimation.

The cdf $F_{\Delta P}(x)$ can be approximated:

- from the estimated pdf $\hat{f}_{\Delta P}(x)$, given by (18.4), or
- by the empirical cumulative distribution function[53]

$$\hat{F}_{k,\Delta P}(x) = \frac{1}{k} \sum_{i=1}^{k} 1\{X_i \leq x\}. \tag{18.5}$$

Step 2. Let s be the ith order statistic, $h(s)$ the pdf of s, $H(s)$ the cdf of s. In order to assess the pdf $h(s)$ and the cdf $H(s)$, Butler and Schachter [12] employ the estimated pdf $\hat{f}_{\Delta P(x)}$ from (4) and the empirical cdf $\hat{F}_{\Delta P}(x)$:[54]

$$\hat{H}(s) = \sum_{j=i}^{k} \frac{k!}{j!(k-j)!} \hat{F}_{\Delta P}(s)^j (1 - \hat{F}_{\Delta P}(s))^{k-j} ,$$

$$\hat{h}(s) = \frac{k!}{i!(k-i)!} \hat{f}_{\Delta P}(x) \hat{F}_{\Delta P}(x)^{i-1} (1 - \hat{F}_{\Delta P}(x))^{k-i} .$$

Step 3. Moments of the pdf $\hat{h}(s)$ can be obtained using 12-point Gauss-Hermite integration:

$$E(s) = \int_{-\infty}^{\infty} sh(s)ds \approx \frac{1}{12} \sum_{j=1}^{12} a_j s_j e^{s_j^2} \hat{h}(s_j) = \hat{E}(s) ,$$

[53]Butler and Schachter [12] use the empirical cdf $\hat{F}_{\Delta P}(x)$ to approximate $F_{\Delta P}(x)$.

[54]Note that $\hat{h}(s)$ is not equal to the derivative of $\hat{H}(s)$, $\hat{h}(s)$ is only "close" to $\frac{\partial \hat{H}(s)}{\partial s}$.

$$\mathrm{Var}(s) = \int_{-\infty}^{\infty} s^2 h(s) ds - E(s)^2 \approx \frac{1}{12}\sum_{j=1}^{12} a_j s_j^2 e^{s_j^2} \hat{h}(s_j) - \hat{E}(s)^2 = \hat{\sigma}(s)^2,$$

where a_j is the j-th integration weight, s_j is the j-th point of integral approximation, $j = 1, \ldots, 12$.

By the combined historical simulation and kernel estimation method, the estimate of $-\mathrm{VAR}$, VAR^*, is

$$\mathrm{VAR}^* = \frac{1}{12}\sum_{j=1}^{12} a_j s_j e^{s_j^2} \hat{h}(s_j) \; . \tag{18.6}$$

The $\hat{\sigma}(s)$ can be used to form the confidence interval for VAR^* in large samples. If the sample is large enough, then the quantile is distributed normally. Therefore, for large samples,[55] the $\beta\%$ confidence interval can be constructed as $(\mathrm{VAR}^* - z_\beta\, \hat{\sigma}(s),\, \mathrm{VAR}^* + z_\beta\, \hat{\sigma}(s))$ where z_β is the β-th quantile of the standard normal distribution.

18.5.2.3 Hybrid Approach[56]

The *hybrid approach* combines *exponential smoothing* and the *historical simulation* method. The exponential smoothing accentuates the most recent observations and is aimed to take into account time-varying volatility. The historical simulation forms future returns using historical returns.

The algorithm of the hybrid approach:

1. Exponentially declining weights are attached to historical returns, starting from the current time and going back in time. Let R_{t-k+1}, $\cdots$, R_{t-1}, R_t be a sequence of k observed returns on a given asset, where t is the current time. The ith observation is assigned a weight $\theta_i = c\lambda^{t-i}$, where $0 < \lambda < 1$, $c = \frac{1-\lambda}{1-\lambda^k}$ is the constant, chosen such that $\sum_i \theta_i = 1$, $i = t-k+1, t-k+2, \ldots, t$.

2. Similarly to the historical simulation method, the hypothetical future returns are obtained from the past returns and sorted in increasing order. Let R_j^*, $j = t+1, \ldots, t+k$, be the sequence of ordered returns.

3. The VAR measure is computed from the empirical cumulative density function (cdf) F_{k,R^*}. If VAR is obtained at the α confidence

[55] $k > 50$ is typically sufficient, but exact estimates for the threshold k are not known because in practice the financial returns form non-stationary processes.

[56] Boudoukh, Richardson, and Whitelaw [8].

level, then VAR is the negative of the $(1-\alpha)$th quantile, r, of the cdf. The quantile r is determined by aggregating weights of ordered returns up to $1-\alpha$:[57]

$$F_{k,R^*}(r) = 1 - \alpha \approx \sum_{j=t+1}^{m} \Theta_j^* \,,$$

$$R_m^* \leq r \leq R_{m+1}^* \,, \quad m \leq t+k \,.$$

The value of r is calculated using linear interpolation between returns R_m^* and R_{m+1}^*, VAR $= -r$.

Example: Assume:

- VAR is estimated at the 95% confidence level.
- The hybrid approach is applied to a given series of size $K = 100$. The decay factor $\lambda = 0.98$.

The exponential weighting by the hybrid method is demonstrated in the following table.

Periods from the current time t	Hypothetical returns R*, %	Periods ago	Weight	Cumulative weight
1	-3.5	4	0.0217	0.0217
2	-3.1	2	0.0226	0.0443
3	-2.8	71	0.0056	0.0499
4	-2.6	6	0.0208	0.0707
5	-2.5	50	0.0086	0.0793
6	-2.3	20	0.0157	0.0950

The historical simulation method, in the case of $K = 100$, assigns to each observation a weight of 1%. Under the historical simulation, the 95% VAR

[57]In forming the empirical cdf, the historical simulation technique applies equal weights to returns. Example: Assume that a given sample contains 250 observed ordered returns $\{R_i^*\}$, $i = 1, \ldots, 250$. The empirical cdf is $F_{250,R}(r) = \frac{1}{250}\sum_{s=1}^{250} \mathbf{1}\{R^* \leq r\}$; $r \in \mathbf{R}$. Thus, observations are assigned equal weights of $1/250 = 0.004$. The q-th quantile is a value r such that $F(r) = q$. For example, the 1% quantile, $r_{0.01}$ will be represented by the third lowest observation.

measure is approximately 2.4% (the negative of the average between R_5^* and R_6^*).

In order to estimate VAR, the hybrid method accumulates weights up to 5%. The 5% quantile corresponds to R_3^*. Thus, by the hybrid approach, VAR is about 2.8%.

Boudoukh, Richardson, and Whitelaw [8] applied the hybrid approach to four different types of financial series and compared the performance of the hybrid, exponential smoothing, and historical simulation methods. In approximating the 1% VAR of the $S\&P500$, absolute errors[58] under the hybrid technique were from 30% to 40% smaller than by the exponential smoothing and from 14% to 28% smaller than by the historical simulation. The hybrid method was more accurate for exchange rates and heavy-tailed series: Brady bond index returns, oil prices. Boudoukh et al. [8] conclude that the hybrid approach is appropriate for VAR estimations of fat-tailed series.

18.5.3 Modifications of the Monte Carlo Method

The Monte Carlo method estimates the VAR measure from the simulated distribution of portfolio value changes $\Delta P_{p,T}$ on a certain date T. The simulations involve:

- specifying models for underlying factors $X_1, \ldots, X_s$,
- generating hypothetical paths for $X_1, \ldots X_s$.

Simulation of the $P_{p,T}$ values implies "filling" s-dimensional space of the changes in the underlying risk factors, $\epsilon_1, \ldots, \epsilon_s$, where $\epsilon_i = \{\epsilon_i(t) = X_i(t) - X_i(t-1),\ t \in [0,T]\}$, $i = 1, \ldots, s$. The traditional Monte Carlo (*Full Monte Carlo*) technique generates a set of vectors $\epsilon_t = (\epsilon_{1t}, \ldots, \epsilon_{st})$ by random draws from the specified distributions. Random draws are accomplished in two steps: (i) generating random variables U_i, uniformly distributed over the interval $[0,1]$, $i = 1, \ldots, s$; (ii) transformation of the U_i numbers into realizations of ϵ_i from the chosen distributions using inverse cumulative

[58] Denote by K_i the actual number of tails events over the 100-period window $[i, i+99]$ and by L_i the predicted number of tail events over the same 100-period interval $[i, i+99]$, $i = 1, \ldots, N-99$, N is the sample size. Boudoukh et al. [8] define the mean absolute error (MAE) as

$$\text{MAE} = \frac{1}{N-99} \sum_{i=1}^{N-99} |K_i - \bar{L}| ,$$

where $\bar{L}$ is the expected number of predicted tail events.

distribution functions.[59] Thus, "filling" s-dimensional space of underlying risk factors essentially means "filling" an s-dimensional cube $[0,1]^s$. We shall consider modifications of the Monte Carlo method, which apply more consistent (deterministic) approaches to filling the s-space (a cube $[0,1]^s$):

- the quasi-Monte Carlo,
- the grid Monte Carlo,
- the modified grid Monte Carlo.

18.5.3.1 Quasi-Monte Carlo[60]

The group of *shape quasi-Monte Carlo (QMC)* methods encompasses deterministic schemes with different levels of complexity, including *stratified sampling*, *Latin Hypercube sampling*, *Orthogonal Arrays*, (t,m,s)-*nets*, and (t,s)-*sequences*.

Suppose that N simulations are conducted for a single risk factor. The univariate simulations imply filling an interval $[0,1]$. Under the *stratified sampling*,[61] the interval $[0,1]$ is divided into M subintervals of the same length and N/M simulations are carried out in each subinterval. This simple stratified sampling is not suitable in the multivariate simulations. An alternate method, a *Latin Hypercube sampling*, can be applied. The *Latin Hypercube* sampling individually stratifies each factor. However, it entails sparse sampling when the number of the risk factors is large. Generalizations of Latin Hypercubes are called *Orthogonal Arrays* and (t,m,s)-*nets*. The Orthogonal Arrays are the samples composed in such way that stratification is done for each pair or triple of factors. The (t,m,s)-*nets*[62] are the samples of points, obtained under more general stratification of input factors. Denote by s the number of risk factors, $s \geq 1$. In order to give a formal definition of a (t,m,s)-*net*, we need to introduce an elementary interval. An *elementary interval* in base b is a hyperrectangular "cell" of the form $C = \prod_{i=1}^{s} \left[\frac{a_i}{b_i^{k_i}}, \frac{a_i+1}{b_i^{k_i}}\right)$ where k_i, a_i are integers, $k_i \geq 0$, $0 \leq a_i < b^{k_i}$. The *volume* of the cell C is equal to b^{-r}, where $r = \sum_{i=1}^{s} k_i$. A finite sequence $\{u_i \in [0,1)^s\}_{i=1}^{b^m}$ is a (t,m,s)-*net* in base b if any elementary interval in base b of volume b^{t-m} includes exactly b^t points of the sequence, where

[59] One of important problems of the transformation step is maintaining the correlation structure of the empirical data.

[60] Jorion [42]; Niederreiter [63]; Owen and Tavella [64]; Paskov and Traub [65]; Pritsker [69]; Sobol [78]; Shaw [73].

[61] Shaw [73].

[62] Owen and Tavella [64].

b, m, and t are the integers, $b \geq 2$, $m \geq t \geq 0$. An infinite sequence $\{u_i \in [0,1)^s\}_{i \geq 1}$ is a *(t,s)-sequence* in base b if for any $k \geq 0$ and $m \geq t$ the finite sequence $\{u_i \in [0,1)^s\}_{i=kb^m+1}^{(k+1)b^m}$ is a (t,m,s)-net in base b.

Owen and Tavella [64] compared performances of (t,s)-sequences and (t,m,s)-nets and the *randomized (t,s)-sequences* and *(t,m,s)-nets* (the scrambled nets). The randomized (t,s)-sequences and (t,m,s)-nets are formed from (t,s)-sequences and (t,m,s)-nets by random scrambling their digits. In two analyzed value-at-risk examples, the scrambled nets demonstrated: (i) the same level of accuracy as the deterministic QMC schemes and (ii) the convergence rate of the order one-fiftieth of the traditional Monte Carlo methods.

18.5.3.2 Grid Monte Carlo[63]

The *grid Monte Carlo* approach forms a grid of changes in risk factors ϵ_t. Next, it computes the portfolio values at each node of the grid. The possible realizations of ϵ_t are obtained by random draws from the chosen models. The portfolio values for new draws are approximated by interpolating portfolio values at adjacent grid points.

The disadvantage of the grid Monte Carlo method is that it is subject to the dimensionality problem.

18.5.3.3 Modified Grid Monte Carlo

Pritsker [69] proposed the *modified grid Monte Carlo*, which alleviates the dimensionality problem by considering lower dimension grids combined with a linear Taylor approximations.

The method assumes that the changes in the portfolio values are caused by two types of factors: linear and non-linear. Changes due to the non-linear factors are estimated using the grid points and changes due to the linear factors are evaluated applying linear approximations. Let P be the portfolio value. Assume that P depends on two risk factors: a linear factor X_1 and a non-linear factor X_2. Denote by ΔP changes in the portfolio value over a given time horizon. $\Delta P = \Delta P(\epsilon_1, \epsilon_2)$, where ϵ_i are changes in the factor X_i, $i = 1, 2$. The modified grid Monte Carlo divides variations in ΔP on two components: variations due to ϵ_2 and variations due to ϵ_1:

$$\Delta P(\epsilon_1, \epsilon_2) = \Delta P(0, \epsilon_2) + [\Delta P(\epsilon_1, \epsilon_2) - \Delta P(0, \epsilon_2)]$$

$$\approx \Delta P(0, \epsilon_2) + \Delta P_1(0, \epsilon_2)\epsilon_1$$

[63]Pritsker [69].

$$\approx \Delta P(0, \epsilon_2) + \Delta P_1(0, 0)\epsilon_1 ,$$

where

$$\Delta P_1(0, \epsilon_2) = \left. \frac{\partial \Delta P(\epsilon_1, \epsilon_2)}{\partial \epsilon_1} \right|_{\epsilon_1 = 0} .$$

The first component $\Delta P(0, \epsilon_2)$ is estimated by constructing a grid[64] for only the variable ϵ_2, the second component is approximated by the linear term of Taylor expansions for changes in P due to ϵ_1, $\Delta P_1(0, 0)\epsilon_1$.

Pritsker [69] compared the performance of several VAR methods. He found that the modified grid Monte Carlo entails the level of accuracy close to the delta-gamma-Monte Carlo method (see Section 18.5.1.1). However, in simulation exercises, the delta-gamma-Monte Carlo approach required less computational time by a factor of 8 than the modified grid Monte Carlo.

18.5.4 Modifications of Scenario Analysis

The *scenario analysis* (*stress-testing*) method estimates the VAR measure based on analysis of the impacts of variations in the financial variables on the portfolio value. The price variations are generated according to the chosen scenarios. We shall examine the following types of the scenario analysis:

- Worst Case Scenario analysis,
- Factor-based interest rate scenarios.

18.5.4.1 Worst Case Scenario Analysis

Recall from the definition of a VAR measure that it represents the highest possible loss over a certain period of time at a given confidence level. The estimated VAR measure allows describing the portfolio riskiness in the frequency terms.[65] Boudoukh, Richardson, and Whitelaw [9] suggest applying a complementary risk measure — *"worst-case scenario" risk* (*WCS*).

The WCS measure evaluates the magnitude of the worst losses:

$$\text{WCS} = \underline{R_p} = \min[R_{p1}, R_{p2}, \ldots, R_{pT}],$$

[64]The grid Monte Carlo method would estimate the ΔP values employing the grid for both variables ϵ_1 and ϵ_2.

[65]Example: The daily VAR is estimated to be \$ 2 million at the 95 percent confidence level. This value of VAR implies that one-day losses would exceed \$ 2 million 5 percent of the time.

where R_{pt} is t-th observation on the portfolio return R_p, $t = 1, \ldots, T$. WCS considers the distribution of the loss during the worst trading interval over a certain time horizon, $F(\underline{R_p})$. The distribution $F(\underline{R_p})$ can be simulated, for example, by random draws from the specified distribution.

Boudoukh et al. [9] demonstrate that the expected WCS exceeds the associated VAR. In the example of a one-year option on a 10-year zero-coupon bond, the 99% VAR was found to be 17.77%, whereas the expected WCS was 18.99% and the 1% quantile of the WCS distribution $F(\underline{R_p})$ was 27.24%.

WCS allows evaluation of the size of losses and can be employed in the risk management analysis in addition to the VAR measure.

18.5.4.2 Factor-based Interest Rate Scenarios

Frye [26] proposes applying to VAR estimations the *factor-based scenario method*. The method forms the distribution of the portfolio value changes in line with *interest rate scenarios*, which are based on a *principal component analysis* (PCA)[66] of the yield curve. The highest loss under these scenarios is accepted as a VAR measurement. The proposed method is suitable for portfolios with interest rate risk factors.

The factor-based scenario method develops interest rate scenarios assuming that changes in the yield curve depend on factors (principal components), which can be determined from PCA. Ranking of the principal components corresponds to their impacts on the yield curve. For example, the first principal component is the component which has the strongest influence on the data and the largest volatility.

We shall introduce formal definitions of the principal components. Denote by X a T × N matrix of the interst rate data, where N is the number of maturity buckets along the yield curve and T is the number of samplings over time. Let Σ be the variance-covariance matrix of X. The first principal component of X is the N × 1 vector c_1, which is obtained by solving the maximization problem $\max_{c_1}\{c_1^T X^T X c_1\}$ subject to the constraint $c_1^T c_1 = 1$. The second principal component of X is the Nx1 vector c_2, the solution of the maximization problem $\max_{c_2}\{c_2^T X^T X c_2\}$ subject to the constraint $c_2^T c_2 = 1$ and $c_2^T c_1 = 0$. The definitions for the remaining principal components are analogous. The principal components are equal to the eigenvectors of Σ. The variance-covariance matrix Σ can be written as

$$\Sigma = CDC^T ,$$

[66] For another application of PCA in VAR computations, see Singh [77].

where C is the N × T matrix with columns equal to the principal components c_j, $j = 1, \ldots, T$, D is a T × T diagonal matrix of the eigenvalues of Σ sorted in the descending order. Frye [26] names the columns of C as the principal component loadings and the columns of CD as the factors shift, twist, etc.

Interpretation of impacts of principal components on the yield curve can be described as follows:

- the first principal component is analogous to a level shift factor,
- the second principal component is comparable with a curve steepening factor,
- the third principal component controls intermediate-term yield shifts.

Frye [26] applied PCA for analyzing the shifts of the yield curve and observed that the first two principal components explain more than 93% of the movements in the data. From these two principal components he formed two factors, basic elements of interest rate scenarios: the shift factor and the twist factor. The shift factor is calculated as the product of the standard deviation and the loadings of the first principal component, the twist factor is equal to the product of the standard deviation and the loadings of the second principal component. Frye [26] constructed four interest rate scenarios for the shift and the twist factors: "UpUp", "UpDown", "DownUp", and "DownDown". The names of scenarios correspond to directions of changes in factors. For example, the "UpUp" scenario implies that both factors increase 2.33 times. Analysis of the four scenarios showed that the shift and the twist factors do not completely characterize variations of three-month and six-month interest rates. Adding two more factors,[67] "Bow-1" and "Bow-2", allowed formation of 16 scenarios. Implementation of the scenarios demonstrated that the scenarios, based on the first four factors, well characterize the yield curve movements.

The factor-based scenario method, applied for VAR calculations of non-option portfolios, has a tendency to entail overstated VAR estimates. In general, the method provides better VAR estimations if portfolios do not depend on many factors.

[67]The Bow-1 factor is computed as the product of the standard deviation and the loadings of the third principal component. Similarly, the Bow-2 factor is equal to the product of the standard deviation and the loadings of the fourth principal component.

18.5.5 A Generalized VAR Method Using Shortfall Risk Approach[68]

The VAR methodologies estimate risk exposures based on probabilities of highest losses. They concentrate on observations in the left tail of a distribution of portfolio value changes (portfolio returns). One of critiques of the VAR approach is that it does not analyze the overall shape of the distribution.[69] The notion of *shortfall risks* takes into account characteristics of the distribution and allows generalization of the VAR approach.

Shortfall (downside) risk implies risk of observing portfolio returns below a certain return. Common measures of shortfall risk are *lower partial moments (lpm).*[70]

An lpm of order n is computed as

$$lpm_n(R^*) = \int_{-\infty}^{R^*} (R^* - x)^n f_R(x)\, dx,$$

where R^* is the chosen threshold-return, $f_R(x)$ is the probability density function of portfolio returns R.

If the order of the lower partial moment n equals 0, then

$$lpm_0(-\text{VAR}) = \int_{-\infty}^{-\text{VAR}} f_R(x)\, dx = Pr(R < -\text{VAR}) = F_R(-\text{VAR}) = 1 - \alpha,$$

where $F_R(\text{x})$ is the cumulative distribution function of portfolio returns, α is the VAR confidence level. The lower partial moment of order 1 (lpm_1) is named *target shortfall*. It shows the average deviation to the left from the threshold-return. The lpm_2 is *target semivariance.*

Schröder [72] conjectured that, in cases of skewed distributions of portfolio returns, lower partial moments are superior indicators of risk compared to the standard deviation.

[68] Schröder [72].

[69] Some researchers claim that consideration of skewness of the distribution, in addition to analysis of tails, could improve characterization of the portfolio riskiness.

[70] Lower partial moments are statistical moments computed for returns under a chosen level. Recall that by definition of moments, the first moment is the mean, the second moment is the variance.

Under the generalized approach, VAR numbers, VAR_n, should be obtained correspondingly to the given values S_n:

$$S_n = lpm_n(-\text{VAR}_n) = \int_{-\infty}^{-\text{VAR}_n} (-\text{VAR}_n - x)^n f_R(x)\, dx.$$

S_0 can be directly associated with the confidence level of the traditional $\text{VAR} = \text{VAR}_0$:

$$S_0 = 1 - \alpha.$$

S_1 represents the expected reductions in returns, with respect to the VAR_1.

Schröder [72] observed that in an example of a portfolio with put options, with the skewed to the right distribution of returns, VAR_2 was lower than the VAR_2 for the normal distribution with the same VAR_0. He concluded the relatively defensive portfolios, which use hedging strategies, have lower VAR_1 or VAR_2. In these cases, the generalized VAR measurements VAR_1 and VAR_2 are preferable risk indicators.

18.5.6 A Semi-parametric Method for VAR Evaluation[71]

A *semi-parametric* method incorporates two techniques: *historical simulation*[72] and *parametric evaluation*.

The historical simulation method is employed to forecast the *interior part* of the portfolio return distributions and the parametric technique is applied to approximate tails.

Most financial data exhibit heavy-tailed distributions. Hence, analysis of the *tails* of the portfolio returns can be built on the theory of fat-tailed distributions.

Let R_i denote the observations of returns in descending order, $R_1 \geq R_2 \geq \cdots \geq R_M \geq \cdots \geq R_N$. Danielsson and de Vries [16] consider the extreme value method and suggest an estimator for tail probabilities:

$$\hat{F}(x) = p = \frac{M}{N}\left(\frac{R_{M+1}}{x}\right)^{\hat{\alpha}}, x > R_{M+1}\,, \qquad (18.7)$$

where p is the probability, M is a threshold index (the number of cases $R_i > R_{M+1}$), R_{M+1} is a high threshold observation, N is the sample size,

[71] Danielsson and de Vries [17].

[72] The historical simulation technique does not require estimation of the distribution parameters. It is a non-parametric method.

$\hat{\alpha}$ is an estimate of the tail index α.[73] An extreme quantile estimator can be derived from (18.7) as

$$\hat{x}_p = \hat{F}^{-1}(x) = R_{M+1}\left(\frac{M}{Np}\right)^{\frac{1}{\hat{\alpha}}}. \tag{18.8}$$

The semi-parametric method at first generates portfolio returns applying the historical simulation method. Then, the tails are approximated using formula (18.8). A VAR measurement is obtained from the estimated tails.

Danielsson and de Vries [17] calculated VAR for portfolios of 7 stocks using the semi-parametric method, historical simulation, and the variance-covariance method (the RiskMetrics version). The following table includes some empirical results.

Table: Average Number of Exceptions[74]

Confidence level, $\underline{\alpha}$	95%	97.5%	99%	99.5%
Expected number of exceptions	50	25	10	5
RiskMetrics	52.45 (7.39)	30.26 (4.41)	16.28 (3.13)	10.65 (2.73)
Historical Simulation	43.24 (10.75)	20.50 (7.22)	7.66 (3.90)	3.69 (2.39)
Extreme Value	43.24 (11.10)	20.84 (7.35)	8.19 (3.86)	4.23 (2.55)

Standard errors are in parenthesis. Source: Danielsson and de Vries [17].

Computations showed, at high values of $\underline{\alpha}$,[75] the VAR estimates obtained by the semi-parametric method are more accurate than those calculated by the historical simulation method or by the RiskMetrics approach.

18.5.7 An Economic VAR Measure

The traditional VAR measurements are obtained from distributions of portfolio value changes (returns), which, in turn, are derived using the statistical models of price movements of underlying risk factors. Aït-Sahalia

[73] One of the estimators of α is the Hill estimator:

$$\hat{a} = M\left(\sum_{i=M}^{N} \ln \frac{R_i}{R_{M+1}}\right)^{-1}.$$

[74] Exceptions are cases when actual portfolio returns were lower than VAR estimates.

[75] $\underline{\alpha} = 97.5\%$ or greater.

and Lo [1] point out that the statistical approach ignores any varying economic significance of VAR numbers in different market situations. They propose to incorporate *economic valuation* into VAR computations applying *state-price densities*. Authors name the VAR estimates, adjusted for economic evaluation, as *E-VAR*, and the VAR measures, based on the statistical modeling, as *S-VAR*.

The E-VAR methodology is built on the principles of the *Arrow-Debreu securities*, state-contingent claims with the \$1 payoff in a given state of nature and zero-payoffs in other states. Prices of the Arrow-Debreu securities are called the *Arrow-Debreu prices*. They are determined from the market equilibrium conditions and, thus, embody more economically relevant information than the prices obtained by statistical modeling do.

We shall examine a case of the *dynamic equilibrium models*, a dynamic exchange economy. The economy has a single consumption good, no endowed income, dynamically complete securities markets, one risky stock and one riskless bond. Denote by S_t the equilibrium price of the stock at time t, $\varphi(\underline{C}_T)$ payoff of the security at time T, $T = t + \tau$, τ the time to maturity, $\underline{C}_T$ the consumption. It can be shown that S_t can be derived as

$$S_t = E_t \left[\varphi \left(\underline{C}_t \right) M_{t,T} \right], \tag{18.9}$$

where $M_{t,T} \equiv \frac{U'(\underline{C}_T)}{U'(\underline{C}_t)}$ is marginal rate of substitution (MRS).
We can rewrite the asset price as a discounted expected payoff:

$$S_t = e^{-r_{t,\tau}\tau} E_t^* \left[\varphi \left(\underline{C}_T \right) \right] \ , \tag{18.10}$$

where r is the riskless interest rate. The conditional expectation in (18.10) is calculated with respect to the weighted probability density function f*. Weights of f* depend on marginal rates of substitution (MRS). The density function f* is named the *state-price density* (SPD), the continuous-state analog of Arrow-Debreu prices. The SPD reflects market equilibrium and agents' economic assessment of market conditions: degree of risk aversion, intertemporal preferences, consumption and investment decisions. An *economic VAR* (E-VAR)[76] estimate is derived from the SPD f*.

Under certain assumptions, the *no-arbitrage* or *dynamically-complete-markets models* also allow to derive an SPD.

The problem in realization of the E-VAR methodology is that the MRS are not directly observable. The existing approaches for estimation of the SPD can be classified on *parametric* and *nonparametric*. Parametric

[76] If, in the aggregate, investors are risk-neutral, then E-VAR equals S-VAR.

methods provide closed-form expressions for the SPD, while nonparametric methods involve numerical approximations. Some nonparametric techniques employ *kernel regressions*, *learning networks*, and *implied binomial trees*.

The parametric method, suggested by Banz and Miller [3] and Breeden and Litzenberger [10], produces the closed form for the SPD as the second derivative of a call option price with respect to the strike price:

$$\mathrm{f}_t^*(S_T) = e^{r_{t,\tau}\tau} \frac{\partial^2 C_t}{\partial X^2}, \tag{18.11}$$

where C_t is the call option price at the current time t, T is the maturity date, $T \equiv t + \tau$, τ is time to maturity, X is the strike price, and S_T is the price of the underlying asset at the maturity date T. Aït-Sahalia and Lo [1] demonstrate the method using the option price formula. Let $\delta_{t,\tau}$ be the dividend yield. By the Black-Scholes formula,

$$C(S_t, X, \tau, r_{t,\tau}, \delta_{t,\tau}, \sigma) = e^{-r_{t,\tau}\tau} \int_0^\infty \max\left[S_T - X, 0\right] \mathrm{f}_t^*(S_T) dS_T$$

$$= S_t \Phi(d_1) - X e^{-r_{t,\tau}\tau} \Phi(d_2), \tag{18.12}$$

where

$$d_1 = \frac{\ln(S_t/X) + (r_{t,\tau} - \delta_{t,\tau} + \sigma^2/2)\tau}{\sigma\sqrt{\tau}}, d_2 = d_1 - \sigma\sqrt{\tau}.$$

Taking the second derivative of C_t with respect to X, the SPD can be determined as

$$\mathrm{f}_t^*(S_T) = e^{r_{t,\tau}\tau} \left. \frac{\partial^2 C}{\partial X^2} \right|_{X=S_T}$$

$$= \frac{1}{S_T\sqrt{2\pi\sigma^2\tau}} \exp\left(- \frac{\left(\ln\left(S_T/S_t\right) - \left(r_{t,\tau} - \delta_{t,\tau} - \sigma^2/2\right)\tau\right)^2}{2\sigma^2\tau} \right).$$

Aït-Sahalia and Lo [1] offer a nonparametric method for approximation of the SPD. It is based on the formula (18.11) and implemented in two steps:

1. kernel estimation of the option price and

2. differentiation of the approximated option price with respect to the strike price X.

The method is illustrated for the case when the volatility factor σ in the formula (18.12) depends on $X/F_{t,\tau}$ and τ:

$$C(S_t, X, \tau, r_{t,\tau}, \delta_{t,\tau}, \sigma) = C\left(F_{t,\tau}, X, \tau, r_{t,\tau}, \delta_{t,\tau}; \sigma(X/F_{t,\tau}, \tau)\right),$$

where $F_{t,\tau}$ is the value of a futures contract on the same security with maturity τ.

Step 1. The factor σ is evaluated by the Nadaraya-Watson kernel estimator:

$$\hat{\sigma}(X/F_{t,\tau}, \tau) = \frac{\sum_{i=1}^{n} k_{X/F}\left(\frac{X/F_{t,\tau} - X_i/F_{t_i,\tau_i}}{h_{X/F}}\right) k_\tau\left(\frac{\tau-\tau_i}{h_\tau}\right)\sigma_i}{\sum_{i=1}^{n} k_{X/F}\left(\frac{X/F_{t,\tau} - X_i/F_{t_i,\tau_i}}{h_{X/F}}\right) k_\tau\left(\frac{\tau-\tau_i}{h_\tau}\right)},$$

where σ_i is the volatility associated with the option price C_i, $k_{X/F}$ and k_τ are the univariate kernel functions, $h_{X/F}$ and h_τ are the bandwidths. The option prices are derived as

$$\hat{C}(S_t, X, \tau, r_{t,\tau}, \delta_{t,\tau}) = C(F_{t,\tau}, X, \tau, r_{t,\tau}, \delta_{t,\tau}; \hat{\sigma}(X/F_{t,\tau}, \tau)).$$

Step 2. The SPD is estimated as follows:

$$\hat{f_t^*}(S_T) = e^{r_{t,\tau}\tau} \left.\frac{\partial^2 \hat{C}(S_t, X, \tau, r_{t,\tau}, \delta_{t,\tau})}{\partial X^2}\right|_{X=S_T}.$$

Aït-Sahalia and Lo [1] calculated S-VAR and E-VAR for daily S&P 500 option prices. These two VAR measures mainly differ in skewness of distributions. Empirical analysis demonstrated that S-VAR and E-VAR provide different estimates of risk. The authors' conclusion is that E-VAR allows taking into account components of risk assessment what S-VAR does not consider.

Generally, the nonparametric fit for the SPD will not provide adequate VAR measurements because the empirical state-price densities are very poor at the tails due to the lack of extreme observations. We propose to use stable laws in order to improve accuracy of the E-VAR estimates:

1. Estimate the parameters of the α-stable density from the empirical state-price density.

2. Derive the E-VAR at a given confidence level as a corresponding quantile of the fitted stable density.

18.6 Conclusions

The essence of VAR modeling is the prediction of the highest expected loss for a given portfolio. The VAR techniques estimate losses by approximating low quantiles in the portfolio return distribution. The delta methods are based on the normal assumption for the distribution of financial returns. However, financial data violate the normality assumption. The empirical observations exhibit "fat" tails and excess kurtosis. The historical method does not impose the distributional assumptions but it is not reliable in estimating low quantiles with a small number of observations in tails. The performance of the Monte Carlo method depends on the quality of distributional assumptions on underlying risk factors.

The existing methods do not provide satisfactory evaluation of VAR. The proposed improvements still lack a convincing unified technique capturing the observed phenomena in financial data such as heavy-tails, time-varying volatility, and short- and long-range dependence. The directions of future research would be to construct models that encompass these empirical features and to develop more precise VAR-estimation techniques based on the new models.

Adequate approximation of distributional forms of returns is a key condition for accurate VAR estimation. Given the leptokurtic nature (heavy tails and excess kurtosis) of empirical financial data, the stable Paretian distributions seem to be the most appropriate distributional models for returns.[77]

The conditional heteroscedastic models based on the α-stable hypotheses can be applied for describing both thick tails and time-varying volatility. The fractional-stable GARCH models can explain all observed phenomena: heavy-tails, time-varying volatility, and short- and long-range dependence. Hence, we suggest using stable distributions in VAR modeling.[78]

We demonstrate examples of stable and normal VAR modeling in Figures 1-16 with evaluation of the VAR measurements of four indices: DAX30,

[77]Cheng and Rachev [13] and [14]; Chobanov, Mateev, Mittnik, and Rachev [15]; Fama [24]; Gamrowski and Rachev [28], [29], and [30]; Mandelbrot [52], [53], [54], and [55]; McCulloch [57]; Mittnik and Rachev [58] and [59]; Mittnik, Rachev, and Chenyao [60]; Mittnik, Rachev, and Paolella [61].

[78]See also Gamrowski and Rachev [27].

S&P500, DJIA, and FTSE100.[79] The estimates of VAR* = −VAR at the confidence levels $\alpha = 99\%$ and $\alpha = 95\%$ are shown in Figures 4, 8, 12, and 16. The graphs illustrate that

(i) at the VAR confidence level $\alpha = 95\%$, the normal modeling is acceptable for VAR evaluation;

(ii) at α=99% the stable modeling generally provides conservative evaluation of VAR.[80] The 99% VAR is lower than the empirical 99% VAR only for the FTSE100 index;

(iii) normal modeling leads to overly optimistic forecasts of losses at $\alpha =$ 99%. For all considered data sets, the normal 99% VAR estimates are smaller than the empirical VAR.

We observe that stable VAR modeling outperforms the normal modeling at high values of the VAR confidence level. For legitimate conclusions on the stable VAR methodology, we need to test it on broader classes of risk factors and portfolios of assets of various complexity. A full analysis of performance of stable models in VAR computations will be given in a follow-up publication.

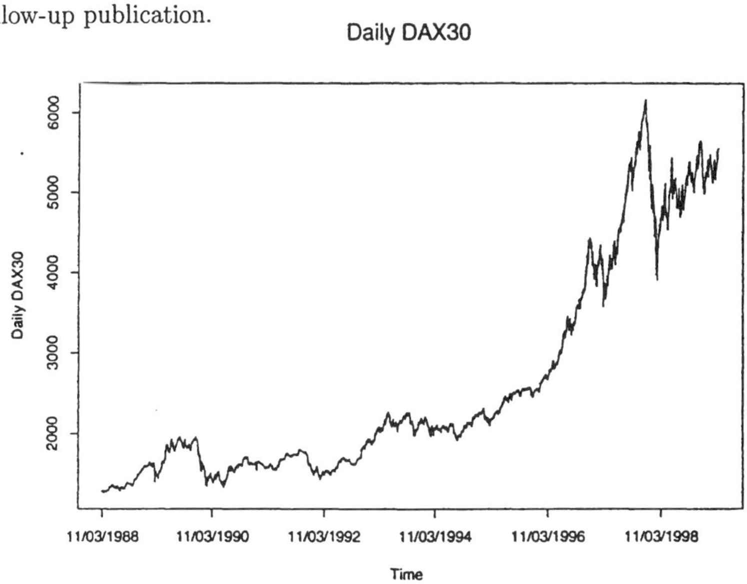

Figure 1. Daily DAX30 index

[79]The DAX30, S&P 500, DJIA, and FTSE100 series include daily observations from 11.3.1988 to 11.3.1999. The data were provided by Kristina Tetereva.

[80]The financial institutions and regulators prefer slightly overestimated VAR measurements rather than underestimated. If the models systematically underestimate VAR, then the higher multiplication factor is applied in calculation of market risk capital requirements. See Section 18.1 for discussion of capital charges.

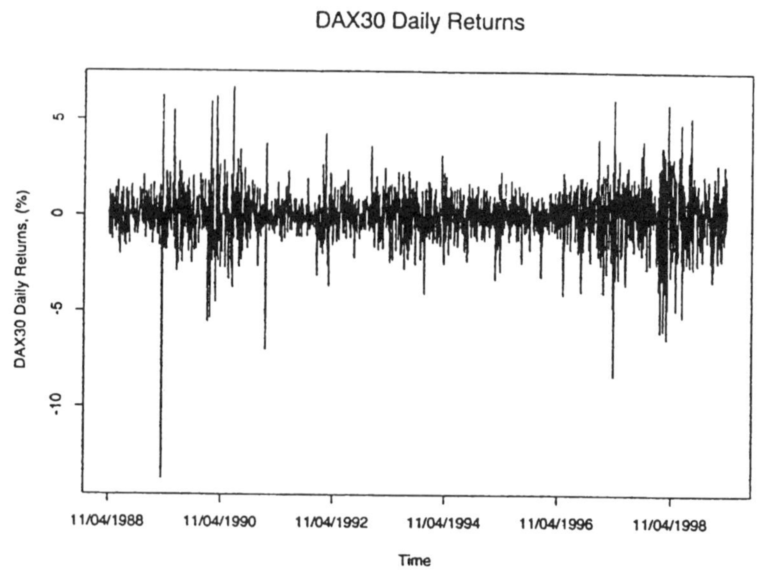

Figure 2. DAX30 daily returns

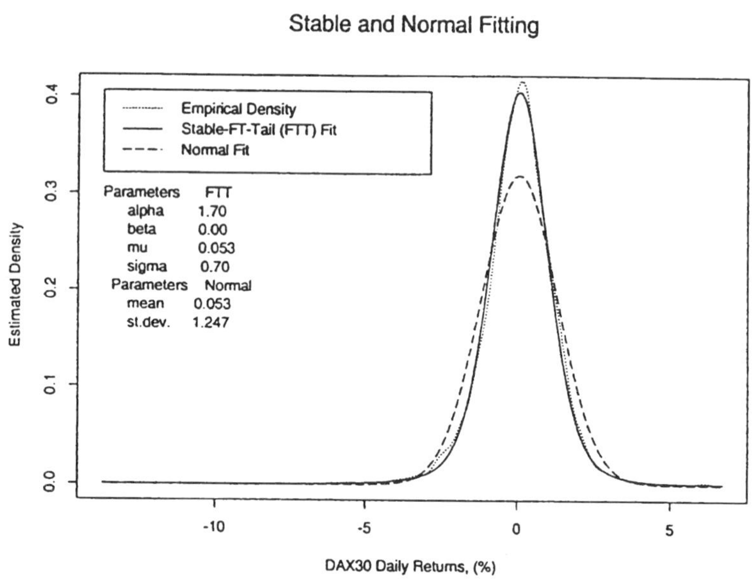

Figure 3. Stable and Normal Fitting of the DAX30 index

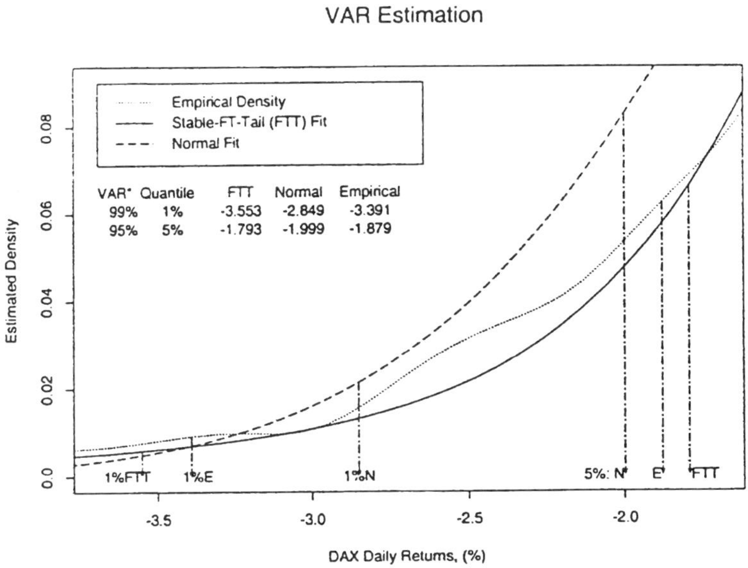

Figure 4. VAR estimation for the DAX30 index.

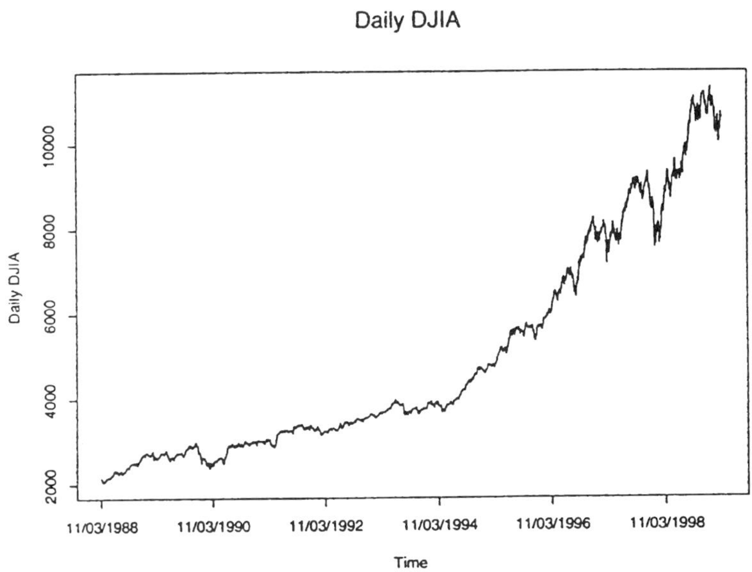

Figure 5. Daily DJIA

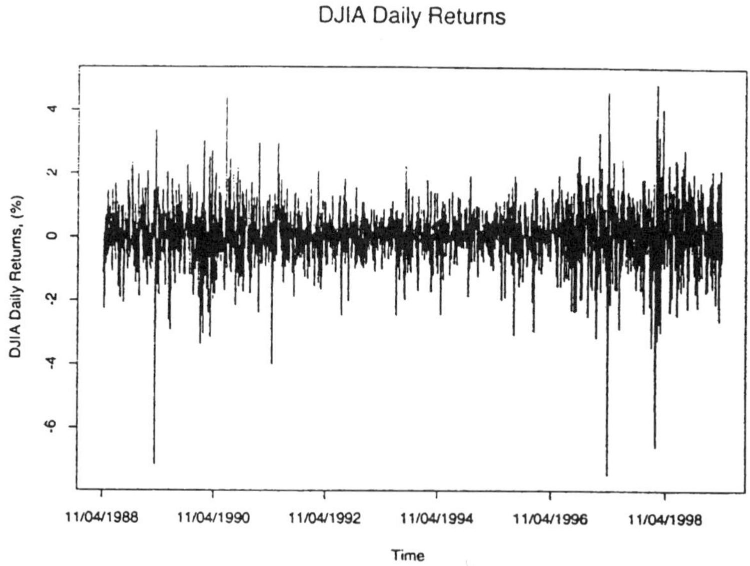

Figure 6. DJIA daily returns

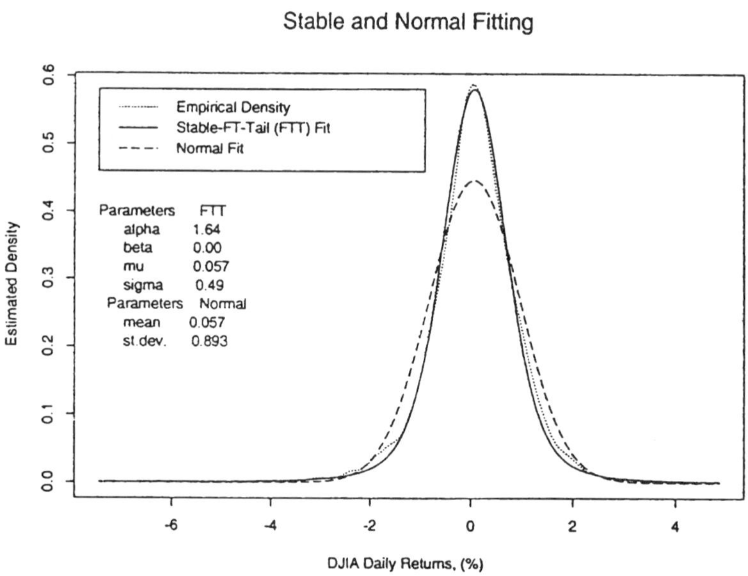

Figure 7. Stable and Normal Fitting of the DJIA index

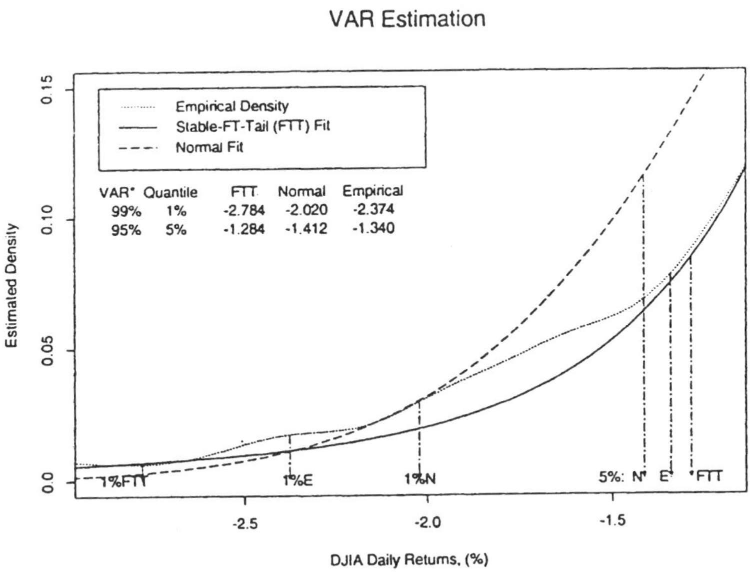

Figure 8. VAR estimation for the DJIA index.

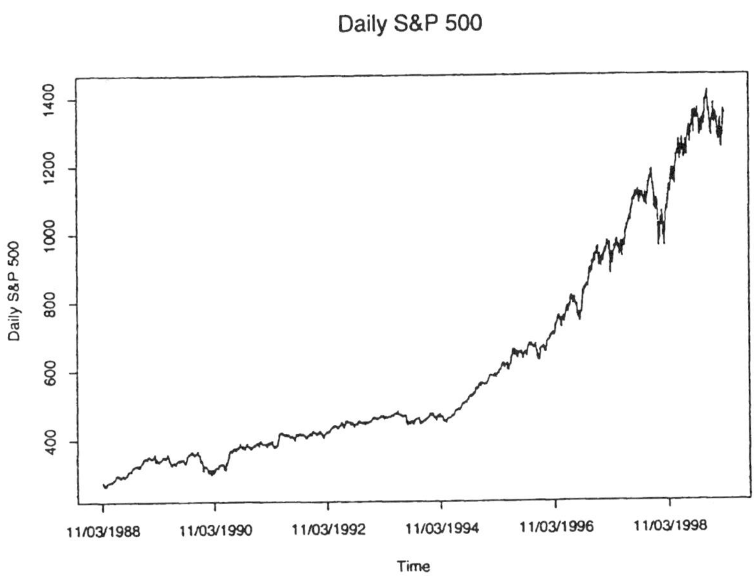

Figure 9. Daily S&P 500

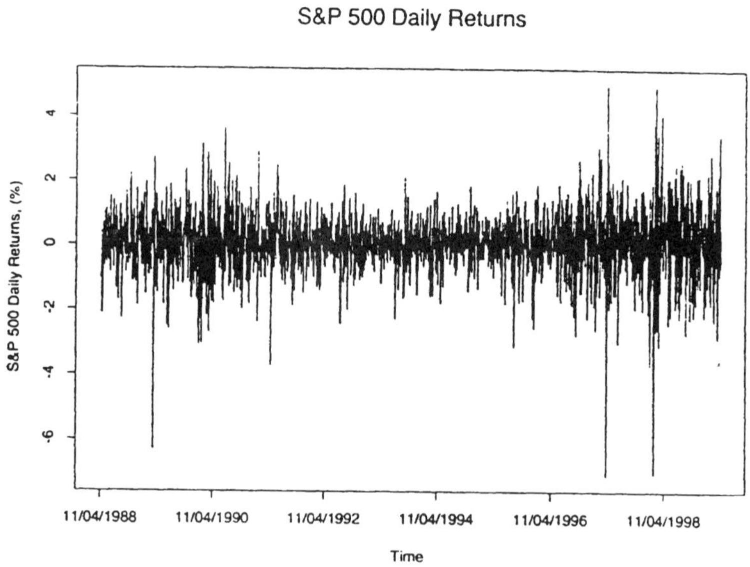

Figure 10. S&P 500 daily returns

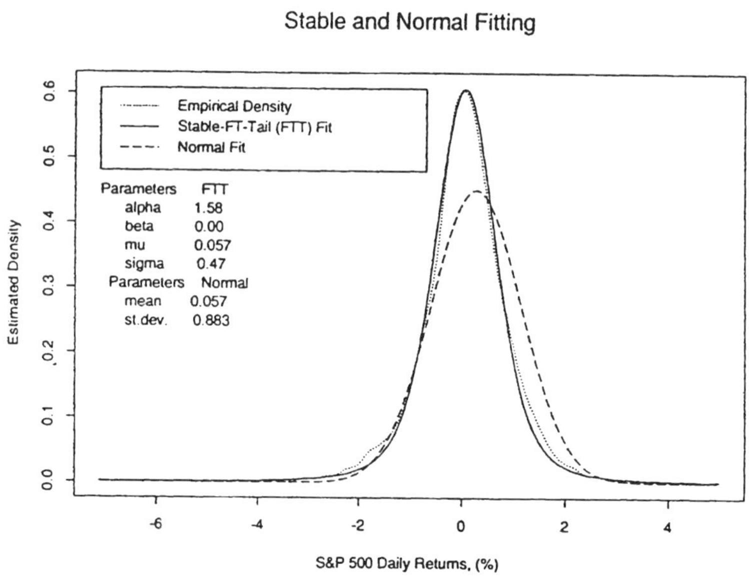

Figure 11. Stable and Normal Fitting of the S&P 500 index

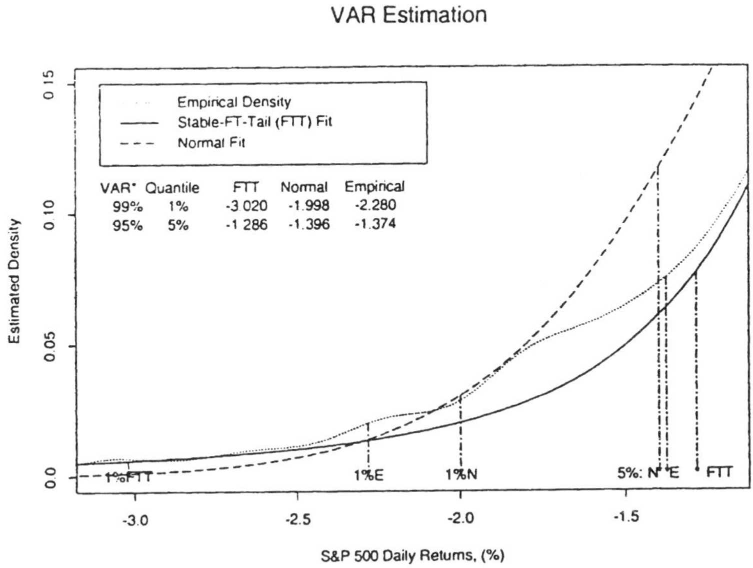

Figure 12. VAR estimation for the S&P 500 index.

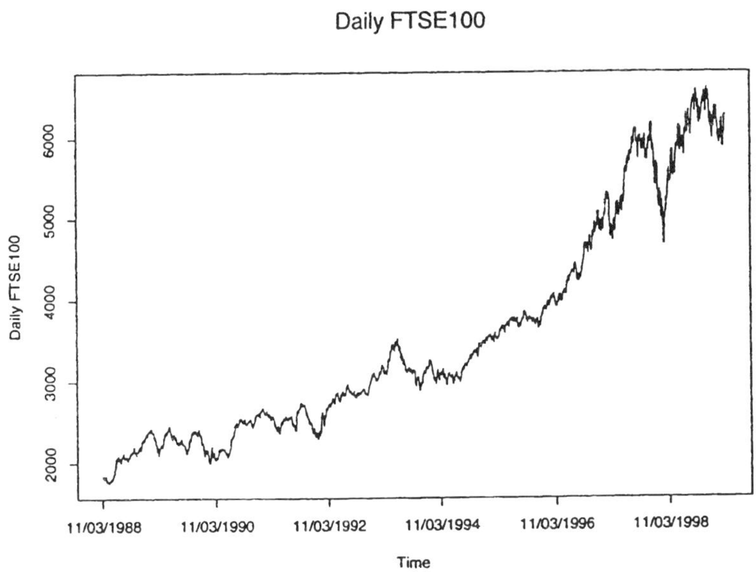

Figure 13. Daily FTSE100

FTSE100 Daily Returns

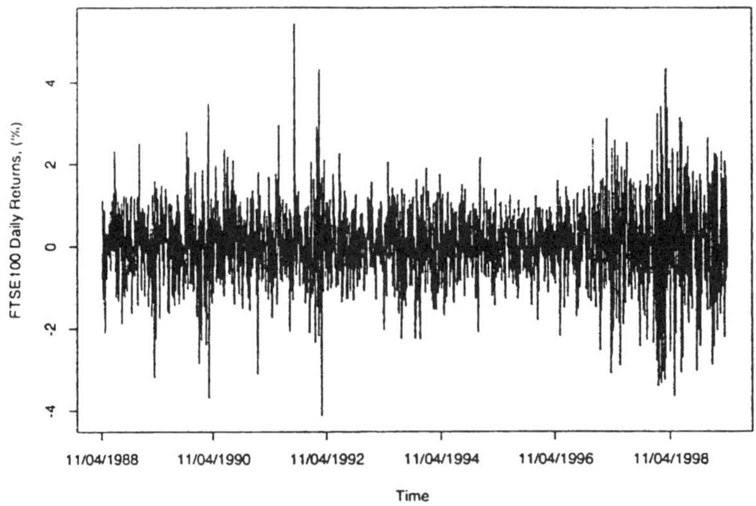

Figure 14.FTSE100 daily returns

Stable and Normal Fitting

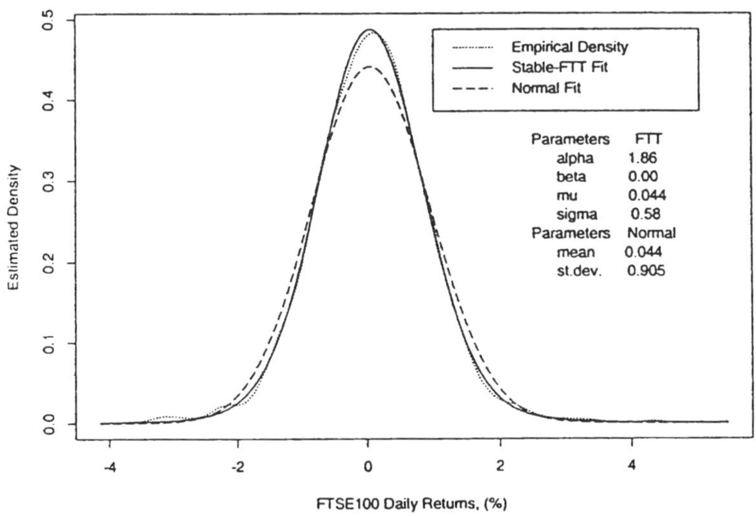

Figure 15. Stable and Normal Fitting of the FTSE100 index

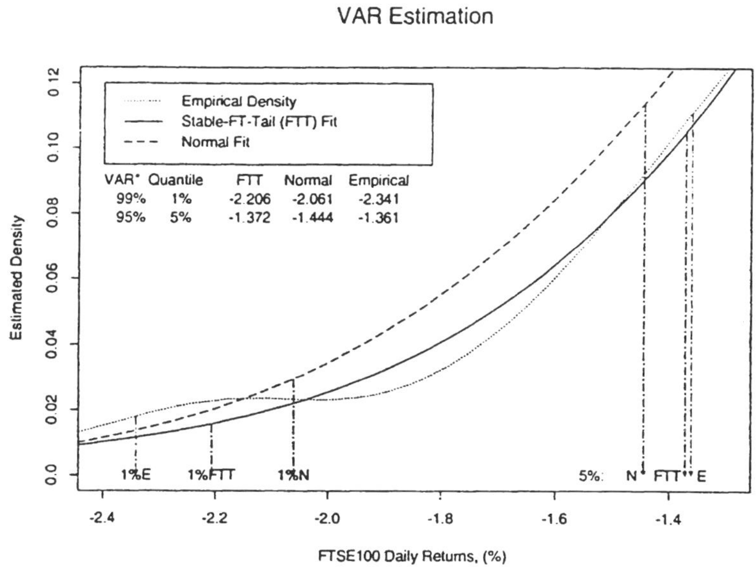

Figure 16. VAR estimation for the FTSE100 index.

Acknowledgements We thank the Market Risk Management Group of the Sanwa Bank, New York Branch, for valuable comments.

References

[1] Aït-Sahalia, Y. and A. Lo, 1997, *Nonparametric Risk Management and Implied Risk Aversion*, NBER Working Paper Series, Working Paper 6130 and 1998, "http://gsbwww.uchicago.edu/fac/~finance/papers/risk.pdf".

[2] Aït-Sahalia, Y. and A. Lo, 1995, *Nonparametric Estimation of State-Price Densities Implicit in Financial Asset Prices*, NBER Working Paper Series, Working Paper 5351, and 1998, *Journal of Finance*, 53, 499-548.

[3] Banz, R.W. and M.H. Miller, 1978, *Prices for State-contingent Claims: Some Estimates and Applications*, *Journal of Business*, 51(4), 653-672.

[4] Basle Committee on Banking Supervision, 1996, *Amendment to the Capital Accord to Incorporate Market Risks*.

[5] Beder, T., 1995, *VAR: Seductive But Dangerous*, *Financial Analysts Journal* (September/October),12-24, and 1997, in Grayling, S., editor, *VAR: Understanding and Applying Value at Risk*, London: Risk, 113-122.

[6] Bollerslev, T., 1986, *A Conditionally Heteroscedastic Time Series Model for Speculative Prices and Rates of Return*, *Review of Economics and Statistics*, 69, 542-547.

[7] Bollerslev, T., R. Chou, and K. Kroner, 1992, *ARCH Modeling in Finance: A Review of the Theory and Empirical Evidence*, *Journal of Econometrics*, 52, 5-59.

[8] Boudoukh, J., M. Richardson, and R. Whitelaw, 1998, *The Best of Both Worlds: A Hybrid Approach to Calculating Value at Risk*, Social Science Research Network Electronic Library, Financial Economics Network, "http://papers.ssrn.com/toptens/topten20360.html" and in *Risk* 11 (May), 64-67.

[9] Boudoukh, J., M. Richardson, and R. Whitelaw, 1997, *Expect the Worst*, in Grayling, Susan, editor, *VAR: Understanding and Applying Value at Risk*, London: Risk, 79-81.

[10] Breeden, D.T. and R.H. Litzenberger, 1978, *Prices of State-contingent Claims Implicit in Option Prices*, *Journal of Business*, 51(4), 621-651.

[11] Butler, J.S. and B. Schachter, 1998, *Estimating Value at Risk With a Precision Measure By Combining Kernel Estimation With Historical Simulation*, forthcoming, *Review of Derivatives Research*.

[12] Butler, J.S. and B. Schachter, 1996, *Improving Value-at-Risk Estimates by Combining Kernel Estimation with Historical Simulation*, Office of the Comptroller of the Currency, Economic & Policy Analysis, Working Paper 96-1.

[13] Cheng, B. and S.T. Rachev, 1995, *Multivariate Stable Futures Prices*, *Journal of Mathematical Finance*, 5, 133-153.

[14] Cheng, B. and S.T. Rachev, 1994, *The Stable Fit To Asset Returns*, University of California, Santa Barbara, Department of Statistics and Applied Probability, Technical Report 273.

[15] Chobanov, G., P. Mateev, S. Mittnik, and S.T. Rachev, 1996, *Modeling the Distribution of Highly Volatile Exchange-Rate Time Series*, in P. Robinson and M. Rosenblatt, eds, "Time Series", Springer Verlag, 130-144.

[16] Danielsson, J. and C.G. de Vries, 1997a, *Beyond the Sample: Extreme Quantile and Probability Estimation*, "http://www.hag.hi.is/~joind/research".

[17] Danielsson, J. and C.G. de Vries, 1997b, *Value-at-Risk and Extreme Returns*, "http://www.hag.hi.is/~joind/research".

[18] Dave, R.D. and G. Stahl, 1997, *On the Accuracy of VAR Estimates Based on the Variance-Covariance Approach*, Working Paper.

[19] *Derivatives Strategy*, 1998, *Roundtable. The Limits of VAR*, (April 1998), vol. 3, no. 4, 14-22.

[20] Derman, E. and I. Kani, 1994, *Riding on a Smile*, *Risk,* 7, 32-39.

[21] Duffie, D. and J. Pan, 1997, *An Overview of Value at Risk*, *The Journal of Derivatives* 4 (Spring), 7-49.

[22] Engle, R.F., 1982, *Autoregressive Conditional Heteroscedasticity with Estimates of the Variance of United Kingdom Inflation*, *Econometrica* 50 (July 1982), 987-1007.

[23] Fallon, W., 1996, *Calculating Value at Risk*, Wharton Financial Institutions Center Working Paper Series, Working Paper 96-49.

[24] Fama, E., 1965, *The Behavior of Stock Market Prices*, *Journal of Business* 38 (1965), 34-105.

[25] Fong, G. and O.A. Vasicek, 1997, *A Multidimensional Framework for Risk Analysis*, *Financial Analysts Journal*, July/August 1997, 51-58.

[26] Frye, J., 1997, *Principals of Risk: Finding Value-at-Risk Through Factor-Based Interest Rate Scenarios*, in Grayling, S., editor, *VaR: Understanding and Applying Value-at-Risk*, London: Risk, 275-287

[27] Gamrowski, B. and S.T. Rachev, 1996, *Testing the Validity of Value at Risk Measures*, in Heyde, Prohorov Pyke, and Rachev, eds., Athens Conference on Applied Probability and Time Series, Volume I: *Applied Probability*, Springer-Verlag, 307-320.

[28] Gamrowski, B. and S.T. Rachev, 1995a, *A Testable Version of the Pareto-stable CAPM*, University of California, Santa Barbara, Department of Statistics and Applied Probability, Technical Report 292.

[29] Gamrowski, B. and S.T. Rachev, 1995b, *Financial Models Using Stable Laws*, *Probability Theory and Its Applications, Surveys in Applied and Industrial Mathematics*, Yu. V. Prohorov (editor), 2, 556-604.

[30] Gamrowski, B. and S.T. Rachev, 1994, *Stable Models in Testable Asset Pricing, Approximation, Probability and Related Fields*, New York: Plenum Press, 223-236.

[31] Grayling, S., editor, 1997, *VaR: Understanding and Applying Value-at-Risk*, London: Risk.

[32] Hamilton, J.D., 1994, *Time Series Analysis*, Princeton, New Jersey: Princeton University Press.

[33] Hendricks, D., 1996, *Evaluation of Value-at-Risk Models Using Historical Data*, Federal Reserve Bank of New York Economic Policy Review 2 (April), 39-70, 1996, in Risk Measurement and Systemic Risk, *Proceedings of a Joint Central Bank Research Conference* Washington, DC: Board of Governors of the Federal Reserve System, 323-357, and 1997, in Grayling, S., editor, *VaR: Understanding and Applying Value-at-Risk*, London: Risk, 151-171.

[34] Hendricks, D. and B. Hirtle, 1997, *Bank Capital Requirements for Market Risk: The Internal Models Approach, Federal Reserve Bank of New York Economic Policy Review* 3 (December), 1-12.

[35] Heron, D. and R. Irving, 1997, *Banks Grasp the VAR Nettle*, in Grayling, S., editor, *VaR: Understanding and Applying Value-at-Risk*, London: Risk, 35-39.

[36] Hiemstra, M., 1997, *VAR with Muscles*, in Grayling, S., editor, *VaR: Understanding and Applying Value-at-Risk*, London: Risk, 359-361.

[37] Hopper, G., 1996, *Value at Risk: A New Methodology for Measuring Portfolio Risk, Federal Reserve Bank of Philadelphia Business Review* (July/August), 19-30, and 1997, in Grayling, S., editor, *VaR: Understanding and Applying Value-at-Risk*, London: Risk, 141-149.

[38] Jackson, P., D. Maude, and W. Perraudin, 1996, *Value-at-Risk Techniques: An Empirical Study*, in Risk Measurement and Systemic Risk, *Proceedings of a Joint Central Bank Research Conference*
Washington, DC: Board of Governors of the Federal Reserve System, 295-322.

[39] Jackson, P., D. Maude, and W. Perraudin, 1997, *Bank Capital and Value-at-Risk, Journal of Derivatives* 4 (Spring), 73-90, and 1997, in Grayling, S., editor, *VaR: Understanding and Applying Value-at-Risk*, London: Risk, 173-185.

[40] Jackwerth, J.C. and M. Rubinstein, 1995, *Implied Probability Distributions: Empirical Analysis*, University of California at Berkeley, Institute of Business and Economic Research, Research Program in

Finance Working Paper Series, Finance Working Paper 250.

[41] Jackwerth, J.C. and M. Rubinstein, 1996, *Recovering Probability Distributions from Option Prices*, *The Journal of Finance*, LI(5), 1611-1631.

[42] Jorion, P., 1996a, *Value at Risk: The New Benchmark for Controlling Market Risk*, Irwin Professional.

[43] Jorion, P., 1996b, *Risk^2: Measuring the Risk in Value at Risk*, *Financial Analysts Journal* 52 (November/December), and 1997, in Grayling, S., editor, *VaR: Understanding and Applying Value-at-Risk*, London: Risk, 187-193.

[44] Jorion, P., 1997, *In Defense of VaR*, *Derivatives Strategy* 2 (April), 20-23.

[45] JP Morgan, 1995, *RiskMetrics*, Third edition, JP Morgan.

[46] Kupiec, P., 1995, *Techniques for Verifying the Accuracy of Risk Measurement Models*, *Journal of Derivatives* 3 (Winter), 73-84, 1996, in Risk Measurement and Systemic Risk, *Proceedings of a Joint Central Bank Research* Conference, Washington, DC: Board of Governors of the Federal Reserve System, 381-401, and 1997, in Grayling, S., editor, *VaR: Understanding and Applying Value-at-Risk*, London: Risk, 195-204.

[47] Lawrence, C. and G. Robinson, 1995, *How Safe Is RiskMetrics? Risk* 8 (January), 26-29, and 1997, in Grayling, S., editor, *VaR: Understanding and Applying Value-at-Risk*, London: Risk, 65-69.

[48] Linsmeier, T. and N. Pearson, 1996, *Risk Measurement: An Introduction to Value at Risk*, University of Illinois at Urbana-Champaign, Department of Accountancy and Department of Finance.

[49] Liu, R., 1996, *VaR and VaR Derivatives*, *Capital Market Strategies* 5 (September), 23-33.

[50] Lopez, J.A., 1996, *Regulatory Evaluation of Value-at-Risk Models*, Wharton Financial Institutions Center Working Paper Series, Working Paper 96-51.

[51] Mahoney, J., 1996, *Empirical-Based Versus Model-Based Approaches to Value-at-Risk: An Examination of Foreign Exchange and Global Equity Portfolios*, in Risk Measurement and Systemic Risk, *Proceedings of a Joint Central Bank Research Conference* Washington, DC: Board of Governors of the Federal Reserve System, 199-217.

[52] Mandelbrot, B.B., 1967, *The Valuation of Some Other Speculative Prices*, *Journal of Business* 40 (1967), 393-413.

[53] Mandelbrot, B.B., 1963a, *New Methods in Statistical Economics, Journal of Political Economy* 71 (1963), 421-440.

[54] Mandelbrot, B.B., 1963b, *The Variation of Certain Speculative Prices, Journal of Business* 26 (1963), 394-419.

[55] Mandelbrot, B. B., 1962, *Sur Certain Prix Spéculatifs: Faits Empiriques et Modéle Basé sur les Processes Stables Additifs de Paul Lévy, Comptes Rendus* 254 (1962), 3968-3970.

[56] Marshall, C. and M. Siegel, 1997, *Value at Risk: Implementing A Risk Management Standard, Journal of Derivatives* 4 (Spring), 91-110, and 1997, in Grayling, Susan, editor, *VaR: Understanding and Applying Value-at-Risk*, London: Risk, 257-273.

[57] McCulloch, J. H., 1996, Financial Applications of Stable Distributions, in G. S. Maddala and C. A. Rao, editors, *Handbook of Statistics: Statistical Methods in Finance*, Amsterdam: Elsevier Science B. V., 14, 393-425.

[58] Mittnik, S. and S.T. Rachev, 1993, *Modeling Asset Returns with Alternative Stable Distributions, Econometric Reviews*, 12, 261-330.

[59] Mittnik, S. and S.T. Rachev, 1991, *Alternate Multivariate Stable Distributions and Their Applications to Financial Modeling*, in S. Cambanis et al., editors, *Stable Processes and Related Topics*, Boston: Birkhäuser, 107-119.

[60] Mittnik, S., S.T. Rachev, and D. Chenyao, 1996, *Distribution of Exchange Rates: A Geometric Summation-Stable Model, Proceedings of the Seminar on Data Analysis*, Sept, 12-17, 1996, Sozopol, Bulgaria.

[61] Mittnik, S., S.T. Rachev, and M.S. Paolella, 1997, *Stable Paretian Modeling in Finance: Some Empirical and Theoretical Aspects*, in R. Adler et al., eds, *A Practical Guide to Heavy Tails: Statistical Techniques for Analyzing Heavy Tailed Distributions*, Birkhäuser, Boston, in press.

[62] Nelson, D., 1991, *Conditional Heteroscedasticity in Asset Returns: A New Approach, Econometrica*, 59, 347-370.

[63] Niederreiter, H., 1992, *Random Number Generation and Quasi-Monte Carlo Methods*, Society for Industrial and Applied Mathematics, Philadelphia, Pennsylvania.

[64] Owen, A. and D. Tavella, 1997, *Scrambled Nets for Value-at-Risk Calculations*, in Grayling, S., editor, *VaR: Understanding and Applying Value-at-Risk*, London: Risk, 289-297.

[65] Paskov, S.H. and J.F. Traub, 1995, *Faster Valuation of Financial Derivatives*, *The Journal of Portfolio Management* (Fall 1995), 113-120.

[66] Phelan, M.J., 1995, *Probability and Statistics Applied to the Practice of Financial Risk Management: The Case of JP Morgan's Riskmetrics*™, Wharton Financial Institutions Center Working Paper Series, Working Paper 95-19.

[67] Priest, A., 1997a, *Veba's Way with VAR*, in Grayling, S., editor, *VaR: Understanding and Applying Value-at-Risk*, London: Risk, 355-357.

[68] Priest, A., 1997b, *Not So Simple for Siemens*, in Grayling, S., editor, *VaR: Understanding and Applying Value-at-Risk*, London: Risk, 363-365.

[69] Pritsker, M., 1996, *Evaluating Value at Risk Methodologies: Accuracy Versus Computational Time*, Wharton Financial Institutions Center Working Paper Series, Working Paper 96-48, and 1997, in Grayling, S., editor, *VaR: Understanding and Applying Value-at-Risk*, London: Risk, 233-255.

[70] Rubinstein, M., 1994, *Implied Binomial Trees*, *The Journal of Finance*, LXIX (3), 771-818.

[71] Schachter, B., 1997, *The Lay Person's Introduction to Value at Risk*, *Financial Engineering News* 1 (August).

[72] Schröder, M., 1997, *The Value-at-Risk Approach: Proposals on a Generalization*, in Grayling, S., editor, *VaR: Understanding and Applying Value-at-Risk*, London: Risk, 299-305.

[73] Shaw, J., 1997, *Beyond VAR and Stress Testing*, in Grayling, S., editor, *VaR: Understanding and Applying Value-at-Risk*, London: Risk, 211-223.

[74] Shimko, D., 1997a, *VAR for Corporates*, in Grayling, S., editor, *VaR: Understanding and Applying Value-at-Risk*, London: Risk, 345-347.

[75] Shimko, D., 1997b, *Investors' Return on VAR*, in Grayling, S., editor, *VaR: Understanding and Applying Value-at-Risk*, London: Risk, 349.

[76] Simons, K., 1996, *Value at Risk - New Approaches to Risk Management*, *Federal Reserve Bank of Boston New England Economic Review* (Sept/Oct), 3-13, and 1997, in Grayling, S., editor, *VaR: Understanding and Applying Value-at-Risk*, London: Risk, 133-139.

[77] Singh, M., 1997, *Value at Risk Using Principal Components Analysis*, *Journal of Portfolio Management* 24 (Fall), 101-110.

[78] Sobol, I.M., 1973, *Numerical Monte Carlo Methods* (in Russian), Moscow: Nauka.

[79] *The Economist*, 1998, *Model Behavior*, (February 28), p. 80.

[80] Wilson, T., 1994, *Plugging the GAP*, *Risk* 7,10 (October), 74-80.

[81] Zangari, P., 1996, *How Accurate is the Delta-Gamma Methodology?*, *RiskMetrics Monitor*, (Third Quarter 1996), 12-29.

Chapter 19

Asset and Liability Management: Recent Advances

Svetlozar Rachev and Yesim Tokat

Department of Economics, University of California at Santa Barbara, Santa Barbara, CA. E-mail: yesim@econ.ucsb.edu
Institute of Statistics and Mathematical Economics, School of Economics, University of Karlsruhe, Kollegium am Schloss, Bau II, 20.12, R210, Postfach 6980, D-76128, Karlsruhe, Germany. E-mail: rachev@lsoe.uni-karlsruhe.de
Contact author: Y. Tokat

Abstract Asset and liability management is the simultaneous consideration of assets and liabilities in strategic investment planning. The asset and liability management models in the literature are reviewed with an emphasis on the recently developed approaches. In most of the recent models, the uncertainty is described by a representative set of scenarios. The validity of the Gaussian assumption in scenario generation is questioned, and stable Paretian distribution is suggested as an alternative hypothesis.

Contents

PART I: Asset and Liability Management: Recent Advances and Critique

19.1 Introduction

Managing assets and liabilities is a concern for banks, pension funds and insurance companies. Before the deregulation of interest rates, the market value of liabilities changed very little from year to year. However, after interest rates were deregulated in 1979, they showed much more volatility. This lead the institutional investors mentioned above to consider assets and liabilities simultaneously during their strategic planning. Strategic investment planning is the allocation of portfolio across broad asset classes such as bonds, stocks, cash and real estate considering the legal and policy constraints facing the institution. Empirical evidence by Culp et al. [24] suggests that asset allocation is the most important factor in determining investment performance.

Most of the early models in this field are either myopic or represent deterministic formulations of multiperiod problems. Grauer and Hakansson [39] showed that solving a sequence of single period models optimizes investor's long-run wealth or the expected utility of wealth. They assumed absence of transaction costs, market impact costs, and liquidity considerations. However, these assumptions are not justifiable in many situations. Myopic models cannot capture long-term investment goals in the presence of transaction costs. Investment options with maturities exceeding a single period cannot be included. The effects of transaction costs are not accurately accounted for. Moreover, these models tend to produce high portfolio turnovers and opportunistic asset trades.

There has been a growing interest in the development of multiperiod stochastic models for asset and liability management (ALM). Kusy and Ziemba [54] developed a multiperiod stochastic linear programming model for Vancouver City Savings Credit Union for a 5-year planning period. Their work suggests that their stochastic ALM model is superior to 5-year deterministic models. Another successful application of multistage stochastic programming is the Russell-Yasuda Kasai model by Carino et al. [17]. The investment strategy suggested by the model resulted in extra income of $79 million during the first two years of its application (1991 and 1992). An ALM model designed by Mulvey [75] has been implemented by the Pacific Financial Asset Management Company. Boender [8] reported the success of a hybrid simulation/optimization scenario model for ALM of pension funds in the Netherlands. The application of the model to a particular pension fund lead to a reduction of the yearly expected contributions of $100 million.

Mathematical programming has been the dominant methodology in the literature of ALM modeling. Brennan et al. [14] suggest a stochastic optimal control model as an alternative to mathematical programming based models.

Traditional ALM was criticized in RiskMetrics [74] since it uses accrual accounting rather than marking all the trading items to market. However, modern models such as the Russell-Yasuda Kasai model consider both market and accounting values of assets. Another criticism is the reliability of long-term forecasts used in ALM. However, modern ALM models do not rely on a single forecast of the future uncertainty. Scenarios are used to represent the set of future possibilities to a sufficient degree. Due to alleged deficiencies, Riskmetrics suggests Value-at-Risk (VaR) as an evolving alternative to ALM. Generally VaR is only used to manage short-term (1-10 days) market risk. The use of VaR for longer horizons up to 2 years is suggested as an ALM alternative for pension funds in LongRun [48]. Different from VaR, ALM can be used to manage liquidity risk, credit risk, sector risk and residual risk as well as market risk. While managing relevant risks, ALM models consider legal and policy constraints in determining the asset allocation of the firm. As of now, the two approaches have different domains, the overlapping part being the management of market risk. ALM benefits from the experiences in successful implementations of VaR.

This chapter consists of two parts. In Part I, ALM models are reviewed with an emphasis on the ones that were recently developed. The advantages and drawbacks of the available approaches are also pointed out. In Section 19.2, the earlier proposed ALM models are briefly discussed. Sections 19.3, 19.4, 19.5, 19.6 analyze the stochastic programming, stochastic control, stochastic programming with decision rules, and capital growth approaches to ALM modeling, respectively. Section 19.7 presents the components of

ALM models. In Part II, the stable Paretian[1] distribution is introduced, and possible improvement in the representation of the future uncertainty is discussed in this framework. Section 19.8 questions the validity of the Gaussian assumption in scenario generation. In Section 19.9, we present the reasons for desirability of stable models. Section 19.10 describes the stable distribution, and Section 19.11 covers financial modeling and estimation. We conclude in Section 19.12 with a discussion of future research directions.

19.2 Earlier ALM Models

19.2.1 Immunization Model

The immunization approach is mainly used for fixed income portfolios. Immunization strategy matches interest rate risk of assets and liabilities in a portfolio. Traditional immunization models assume that the term structure of interest rate is flat and it only shifts in parallel. Extensions have been made to these models in order to control cash flow dollar convexity (shape risk and volatility risk), liquidity risk and credit risk.

Many different immunization models have been suggested. The most commonly used objective function is to maximize the asset portfolio yield. We present the mathematical formulation of a basic immunization model with this objective. Please refer to Zenios [104] for a more detailed review of the literature.

The following notation is used in the model:

$U = \{1, 2, 3, ..., I\}$ is the universe of securities,

$T = \{1, 2, 3, ...T_{\max}\}$ is a set of discrete points in time,

x_i is the nominal holdings of security $i \in U$,

r_i is the cash flow yield,

P_i is the present value of security i,

k_i is the dollar duration of security i.

The first order approximation to the true portfolio yield is the dollar duration weighted average yield of the individual securities in the portfolio. Maximizing is equivalent to maximizing $\sum_{i \in U} k_i r_i x_i$, which gives a linear objective function. The present value and dollar duration of liabilities are assumed as given. The present value and dollar duration of assets are set to be equal to those of liabilities. This linear programming problem is formulated as:

[1]We will call it stable distribution from now on.

$$
\begin{aligned}
\max_{s \in R^I} \quad & \sum_{i \in U} k_i r_i x_i \\
\text{subject to} \quad & \sum_{i \in U} P_i x_i = P_L \\
& \sum_{i \in U} k_i x_i = k_L \qquad\qquad x_i \geq 0.
\end{aligned}
$$

19.2.2 Dedication Model

The dedication model is used for the management of fixed income assets. This model establishes a portfolio with cash flows which always suffice to meet liability payments. If a sufficiently conservative reinvestment rate is chosen, this strategy eliminates all risk except for credit risk. The underlying assumption is that liabilities are known with certainty or good estimates of liabilities are available for the full time span. Extensions recognize the variability of the market value of liabilities and maximize the number of fully funded liabilities or aim to meet the first liabilities fully and later liabilities partially. A stochastic dedication model that explicitly recognizes stochasticity of interest rates is presented in Hiller and Eckstein [43].

We present an optimal dedication model that has the objective of minimizing the cost of bond purchases and initial cash holdings. Time is discretized by looking at the liability dates only. We first present the notation:

$U = \{1, 2, 3, ..., I\}$ is the universe of securities,
$T = \{1, 2, 3, ...T_{\max}\}$ is a set of discrete points in time,
x_i is the nominal holdings of security i,
P_i is the present value of security i,
S_τ is the surplus cash holding at time τ,
ρ is the reinvestment rate,
L_t is the liability payment at time τ,
C_{it} is the cash flow from bond i at time t,
$D_{i\tau} = \sum_{t \in [\tau-1, \tau]} C_{it}(1+\rho)^{\tau - t}$ is the reinvested value of bond cash flows between liability dates $\tau - 1$ and $\tau \in T_l$.

On each liability date, the revenue from reinvested bond cash flows and cash holdings is used to pay the liabilities. If any surplus is generated, it is carried forward to the next liability date at the reinvestment rate. This is enforced by the first constraint of the following linear program:

$$
\begin{aligned}
\min_{x \in R^I} \quad & \sum_{i \in U} P_i x_i + S_0 \\
\text{subject to} \quad & \sum_{i \in U} D_{i\tau} x_i + S_{\tau-1}(1+\rho)^{\Delta\tau} = L_\tau + S_\tau, \forall \tau \in T_l \\
& S_\tau \geq 0, \forall \tau \in T_l \\
& x_i \geq 0.
\end{aligned}
$$

The non-negativity constraints on the surplus ensure the precedence of cash inflows to liability outflows. See Zipkin [109] for further details of this formulation.

19.2.3 Mean-variance Model

Mean-variance analysis was suggested by Markowitz [66]. It is very widely used by practitioners. It is a systematic way for dealing with residual risk. However, the underlying assumptions are strict: either the returns have a multivariate normal distribution or the investor's preference can be represented by some utility function over the mean and variance of the portfolio return. Markowitz [67] shows that for a wide range of utility functions and historical distributions, knowing the mean and variance of the distribution almost gives the expected utility of the distribution.

The model generates the efficient frontier, rather than a single suggested action. This leaves the decision-maker with the problem of choosing a point on the efficient frontier on his own. Mean-variance models penalize both positive and negative deviations. In a lot of situations, the positive deviation is desirable. Mean absolute deviation optimization (Konno and Yamazaki [52]) and mean-semi-variance optimization (Markowitz et al. [68]) have been proposed as possible alternatives. Chopra and Ziemba [22] note that errors in the estimates of means, variances and covariances may have very significant effects on the optimal portfolio selected by using mean variance analysis. The appeal of the model is its simplicity and the availability of efficient algorithms for linearly constrained quadratic programs. However, it is not easily extended to multistage problems.

The following notation is used to present a sample mean-variance model:

$U = \{1, 2, 3, ..., I\}$ is the universe of securities,
q_{ij} is the covariance between securities $i, j \in U$,
μ_i is the expected return of security i,
x_i is the fraction of the portfolio that consists of security $i \in U$.

The model is

$$\min_{x \in R^I} \quad \sum_{i \in U} \sum_{j \in U} q_{ij} x_i x_j - \lambda \sum_{i \in U} \mu_i x_i$$
subject to $Ax = b$.

The parameter λ is used to trade-off the return against variance in order to trace out the efficient frontier. General linear constraints can be used to determine permissible combinations of x.

19.2.4 Expected Utility Model

An investor's risk bearing attitude under uncertainty is modeled as a von Neumann-Morgenstern [101] utility function. The user needs to determine the utility function and risk aversion attitude that represent his preferences best. The model also requires the specification of a probability distribution for the uncertain parameters. There is no specific distributional assumption on returns. Expected utility models have been extended to multiperiod models.

A very general one-period utility model can be formulated as the following:

$$\max_{x \in R^I} \quad E[u(r,x)]$$
$$\text{subject to} \quad Ax = b$$

where r is the vector of returns on securities $i \in U$,
x is the vector of nominal holdings of securities $i \in U$.

The general linear constraint imposes the desired limitations on the relationship among the security holdings. More detailed models are presented in the multiperiod context.

19.2.5 Multicriteria Decision Models

All of the models explained above have a single objective. In reality, however, there are a number of conflicting objectives taken into account by the management. For instance, the objectives of a bank may be related to expected profits, risk, liquidity, capital adequacy, growth, market share, etc. Multi-objective decision methodology models these objectives explicitly. However, these models are often very complex and the number of efficient solutions is very large. Hence, the effective analysis by the decision-maker is cumbersome.

Goal programming is one of the most widely used approaches to multicriteria decision-making. The major advantage of goal programming is its flexibility which enables the decision-maker to incorporate numerous variations of constraints and goals easily.

A goal programming model can be stated as follows:

$$\min_{x \in R^I} \quad f(d_g^+, d_g^-)$$

$$\text{subject to} \sum_{i \in U} c_{gi} x_i = e_g + d_g^+ - d_g^-, \quad \forall g \in G \tag{19.1}$$

$$Ax = b \tag{19.2}$$

where,

$U = \{1, 2, 3, ..., I\}$ is the universe of securities,
$G = \{1, 2, 3, ..., N\}$ is the set of goals,

x_i is the nominal holdings of security $i \in U$,
e_g is the target level for goal $g \in G$,
d_g^+, d_g^- are the positive and negative deviations from the target of goal $g \in G$,
c_{gi} is the coefficient related to the security i in goal g.

The objective function minimizes a function of the deviations from the targets set for the goals. The set of general linear constraints (19.2) can be used to determine permissible combinations of x. See Giokas and Vassiloglou [37] for an application of this model to a bank. Mulvey et al. [80] uses goal programing in a multiperiod context. See also Shing and Nagasawa [96].

19.3 Stochastic Programming ALM Models

This method provides a general-purpose modeling framework that conveniently addresses real world concerns such as transaction costs, taxes, legal and policy constraints. The number of decision variables becomes very large resulting in large scale optimization problems. The computational costs make it impractical to test the recommendations out of the sample.

We decribe various modeling approaches developed within this framework.

19.3.1 Chance-Constrained Model

Charnes and Kirby [21] developed a chance-constrained model that expresses future deposits and loan payments as jointly distributed random variables, and capital adequacy formula by chance-constraints on meeting withdrawal claims. A drawback of the model is that constraint violations are not penalized according to their magnitude.

Charnes et al. [20] applied this methodology to balance sheet management. Two papers by Li [55,56] used chance-constrained programming in portfolio analysis of insurance companies. Recently, Dert [29] developed a multistage chance-constrained ALM model for a defined benefit pension fund. As opposed to the original approach of Charnes and Kirby, Dert modeled the uncertainty using scenarios rather than making distributional assumptions.

His model minimizes the cost of funding while ensuring the stability of contributions and ability to make benefit payments timely with an acceptable level of insolvency risk. The solvency requirement is the asset level being at least equal to the product of required funding level with the value of the remaining liabilities (constraint (19.9)). The asset value falling below the required level is modeled as a probabilistic constraint. Since uncertainty is modeled through scenarios, binary variables are needed to formulate the chance constraint explicitly (constraints (19.10), (19.11), (19.12)). It is assumed that remedial contributions are made in case of underfunding (constraint (19.8)).

The ALM model is formulated as follows:

$$\min A_{01} + \sum_{t=1}^{T-1} \sum_{s=1}^{S_t} P(t,s)\gamma_{ts} Y_{ts} + \lambda \sum_{t=1}^{T} \sum_{s=1}^{S_t} P(t,s)\gamma_{ts} Z_{ts}$$

subject to

$$Y_{ts}^l \leq Y_{ts} \leq Y_{ts}^u \tag{19.3}$$

$$y_{ts}^l \leq \frac{Y_{ts}}{W_{ts}} \leq y_{ts}^u \tag{19.4}$$

$$\frac{Y_{ts}}{W_{ts}} - \frac{Y_{t-1,\hat{s}}}{W_{t-1,\hat{s}}} \leq \beta_t \tag{19.5}$$

$$A_{ts} + Y_{ts} - l_{ts} = \sum_{i=1}^{N} X_{its} \tag{19.6}$$

$$x_{its}^l (A_{ts} + Y_{ts} - l_{ts}) \leq X_{its} \leq x_{its}^u (A_{ts} + Y_{ts} - l_{ts})$$

$$t = 0, ..., T-1, s = 1, ..., S_t \tag{19.7}$$

$$A_{ts} = Z_{ts} + \sum_{i=1}^{N} e^{r_{its}} X_{i,t-1,\hat{s}} \tag{19.8}$$

$$A_{ts} \geq \alpha L_{ts} \tag{19.9}$$

$$Z_{ts} \geq f_{ts} M_{ts} \tag{19.10}$$

$$\sum_{s=1}^{S_t} P\left[(t,s) \mid (t-1,\hat{s})\right] f_{ts} \leq \Psi_{t-1,\hat{s}} \tag{19.11}$$

$$f_{ts} \in \{0,1\} \qquad t = 0,...,T-1, \qquad s = 1,...,S_t \tag{19.12}$$

where,

$t = 0,1,...,T$ is the time period,
$s = 1,2,...,S_t$ is the status of the world,
$i = 1,2,...,N$ is the asset class,
α is the demanded funding level,
β_t is the maximal raise in contribution per period as a fraction of the cost of wages at time t,
γ_{ts} is the discount factor for a cash flow at time t in state s,
l_{ts} is the benefit payments and costs to the fund at time t in state s,
L_{ts} is the actuarial reserve at time t in state s,
λ is the penalty parameter to penalize remedial contributions,
r_{its} is the continuous return on investment of each asset class i during period t in state s,
M_{ts} is the large constant at time t in state s,
W_{ts} is the cost of wages during period t in state s,
A_{ts} is the total asset value before receiving regular contributions and making benefit payments at time t in state s,
f_{ts} is the binary variable for remedial contributions at time t in state s,
Ψ_{ts} is the probability of underfunding at time $t+1$ given the world was in state s at time t,
X_{its} is the amount of money invested in asset class i at time t in state s,
x_{its} is the fraction of asset value invested in asset class i at time t in state s,
Y_{ts} is the regular contribution during period t in state s,
y_{ts} is the regular contribution as a fraction of the cost of wages during period t in state s,
Z_{ts} is the remedial contribution at time t in state s.

The first three constraints, namely, (19.3), (19.4), (19.5), limit the regular contribution amount, regular contribution as a fraction of wages and maximal raise in contribution as a fraction of cost of wages, respectively. After receiving regular contributions and making benefit payments, the assets are reallocated (19.6) considering the upper and lower bounds on the asset mix (19.7).

The price inflation, wage inflation, and asset return scenarios are generated using vector autoregressive model. The characteristics of participants are modeled using a Markov chain. A more detailed description of a similar model is given in Dert [28].

19.3.2 Dynamic Programming

The main idea behind dynamic programming is to solve the problem by dealing with one stage at a time. The procedure produces one solution per possible state in each stage. If there are many state variables or the objective function depends in an arbitrary way on the whole history up to the current period, this method is not very appropriate. It can handle a small number of financial instruments simultaneously. Therefore it is of limited use in practice.

Eppen and Fama [32] modeled a three-asset portfolio problem using this approach. At any point in time, they assume that the state of the system is described by two variables: m being the level of cash balance ($m \in N$), and b being the level of bond account ($b \in \{N - N^-\}$). Decisions concerning the state of the system are made at equally spaced discrete points in time. The stochastic changes in the cash balance between the periods are a sequence of independent identically distributed random variables with the discrete probability mass function $p(d)$. The function $p(d)$ is positive only on a finite state space, i.e., there is a finite K such that $p(d) = 0$ if $|d| > K$.

The notation is as follows:

$T(m, b; m', b')$ is the minimum transfer cost involved in changing the state from (m, b) to (m', b'),

c_h is the marginal opportunity cost of starting a period with an additional dollar of cash,

c_p is the loss of being a dollar short on cash which is incurred at the beginning of the period,

$L(m')$ is the penalty cost of carrying cash:

$$\begin{aligned} L(m') &= c_h m', m' \geq 0 \\ &= -c_p m', m' < 0, \end{aligned}$$

α is the discount factor,

$f_n(m, b)$ is the discounted expected cost for an n period problem whose state at the beginning of period n is (m, b).

The recursive relationship for $f_n(m, b)$ is given by:

$$f_n(m, b) = \min_{m', b'} [T(m, b; m', b') + G_n(m', b')],$$

where

$$G_n(m, b) = c_h^b.b + L(m) + \alpha \sum_{d=-K \text{ to } K} f_{n-1}(m + d, b).p(d).$$

$G_n(m, b)$ is the current expected holding penalty cost (the first two terms) plus the discounted expected cost if a decision is made to start period n in state (m, b) and an optimal policy is followed in period $n-1$ and all future periods.

19.3.3 Sequential Decision Analysis

This approach uses implicit enumeration to find an optimal solution. It results in extremely large equivalent linear programming problems since it enumerates all possible portfolio strategies for all scenarios in all periods of consideration. The method ensures feasibility of the first period for every possible scenario, this shrinks the feasible set and gives substantial importance to scenarios with low probabilities of occurrence.

The stochastic decision tree model by Bradley and Crane [11] overcomes the computational difficulties of the approach by using a decomposition algorithm. The objective is the maximization of expected terminal wealth of the firm. Constraint (19.13) guarantees that the firm cannot purchase more assets than it has funds available. The second set of constraints balances the inventory. The net realized capital losses in a period are controlled by some pre-specified upper bound using (19.15). Constraint (19.16) limits the holding of a particular asset.

Their linear programming formulation[2] is

$$\max \sum_{e_N \in E_N} p(e_N)\left\{\sum_{k=1}^{K}\sum_{m=0}^{N-1}[y_m^k(e_m) + u_{m,N}^k(e_N)]h_{m,N}^k(e_N) \right.$$

$$\left. + \left[y_N^k(e_N) + u_{N,N}^k(e_N)\right]b_N^k(e_N)\right\}$$

subject to

$$\sum_{k=1}^{K} b_n^k(e_n) - \sum_{k=1}^{K}\left[\sum_{m=0}^{n-2} y_m^k(e_m)h_{m,n-1}^k(e_{n-1}) + y_{n-1}^k(e_{n-1})b_{n-1}^k(e_{n-1})\right]$$

$$-\sum_{k=1}^{K}\sum_{m=0}^{n-1}[1 + g_{m,n}^k(e_n)]s_{m,n}^k(e_n) = f_n(e_n) \tag{19.13}$$

$$-h_{m,n-1}^k(e_n - 1) + s_{m,n}^k(e_n) + h_{m,n}^k(e_n) = 0, \ m = 0, 1, ..., n-2 \tag{19.14}$$

$$-b_{n-1}^k(e_{n-1}) + s_{n-1,n}^k(e_n) + h_{n-1,n}^k(e_n) = 0,$$

[2]The formulation is taken from Kusy and Ziemba [54]

$$h^k_{0,0}(e_0) = h^k_0$$

$$-\sum_{k=1}^{K}\sum_{m=0}^{n-1} g^k_{m,n}(e_n)s^k_{m,n}(e_n) \leq L_n(e_n) \tag{19.15}$$

$$\sum_{k\in K_i}\left[b^k_n(e_n) + \sum_{m=0}^{n-1} h^k_{m,n}(e_n)\right] \leq C^i_n(e_n),\ i = 1, 2, ..., I \tag{19.16}$$

$$y^k_{m,n}(e_n) \geq 0,$$

$$s^k_{m,n}(e_n) \geq 0,$$

$$h^k_{m,n}(e_n) \leq 0,\ m = 1, ..., n-1,$$

where

$e_n \in E_n$, $n = 1, 2, ..., N$,
$k = 1, 2, ..., K$,
e_n is an economic scenario from period 1 to n having probability $p(e_n)$,
E_n is the set of possible economic scenarios from period 1 to n,
K is the total number of assets,
K_i is the number of assets of type i,
N is the number of time periods,
$y^k_m(e_m)$ is the income yield per dollar of purchase price in period m of asset k conditional on scenario e_n,
$u^k_{m,N}(e_N)$ is the expected terminal value per dollar of purchase price in period m of asset k held at period N conditional on scenario e_n,
$b^k_n(e_n)$ is the dollar amount of asset k purchased in period n conditional on scenario e_n,
$h^k_{m,n}(e_n)$ is the dollar amount of asset k purchased in period m and held in period n conditional on scenario e_n,
$s^k_{m,n}(e_n)$ is the dollar amount of asset k purchased in period m and sold in period n conditional on scenario e_n
$g^k_{m,n}(e_n)$ is the capital gain or loss per dollar of purchase price of asset k purchased in period m and sold in period n conditional on scenario e_n,
$f_n(e_n)$ is the incremental increase or decrease of funds available for period n,
$L_n(e_n)$ is the dollar amount of maximum allowable net realized capital losses in period n,

$C_n^i(e_n)$ is the upper bound in dollars on the amount of funds invested in asset type i in period n.

They use a decomposition algorithm to break down the problem and use an efficient technique to solve the subproblems of the overall portfolio. However, the solution is still computationally intractable for real life problems.

19.3.4 Stochastic Linear Programming with Recourse (SLPR)

The basic formulation of the general T-stage SLPR model is

$$\min_{x_1}\left\{c_1'x_1 + E_{w_1}\left[\min_{x_2}\left(c_2(w_1)'x_2\right.\right.\right.$$
$$\left.\left.\left.+\ldots + E_{w_{T-1}}\left[\min_{x_T} c_T(w_{T-1})'x_T\right]\right)\right]\right\}$$

subject to

$$\begin{aligned}
&A_1x_1 &&= b_1\\
&B_2(w_1)x_1 + A_2(w_1)x_2 &&= b_2(w_1)\\
&\qquad B_3(w_2)x_2 + A_3(w_2)x_3 &&= b_3(w_2)\\
&\qquad\qquad B_T(w_{T-1})x_{T-1} + A_T(w_{T-1})x_T &&= b_T(w_{T-1})\\
&\qquad\qquad l_t \le x_t \le u_t, \quad t = 1, 2, \ldots, T,
\end{aligned}$$

where

w_t is the random vector that generates the coefficients b_t, c_t, A_t, and B_t of the decision problem at the tth stage, $t = 2, \ldots, T$,
l_t, u_t are the vector of deterministic bounds on x_t at stage t, $t = 2, \ldots, T$,
b_1, c_1, and A_1 are the deterministic first stage coefficient vectors or matrices, and
x_t is the vector decision variable.

The objective formalizes a sequence of optimization problems corresponding to different stages. At stage 1, the outcome completely depends on future realizations of the uncertainty. After the first period, decisions are allowed to be a function of the observed realization (x_{t-1}, w_t) only. One first decides on x_1, then observes w_1, then decides on x_2, then observes w_2, and so on. The recourse decisions depend on the current state of the system as determined by previous decisions and random events. The uncertainty is modeled by using finite scenarios which have pre-assigned probabilities.

In this case, the problem reduces to a large linear program of a special structure:

$$\min\left\{ c_1' x_1 + \sum_{k_2=2}^{K_2} p_{k_2} c_{k_2}' x_{k_2} + \sum_{k_3=K_2+1}^{K_3} p_{k_3} c_{k_3}' x_{k_3} + \ldots + \sum_{k_T=K_{T-1}+1}^{K_T} p_{k_T} c_{k_T}' x_{k_T} \right\}$$

subject to

$$\begin{aligned}
&A_1 x_1 && = b_1 \\
&B_{k_2} x_1 + A_{k_2} x_2 && = b_{k_2},\ \ k_2 = 2, \ldots, K_2 \\
&\quad B_{k_3} x_{a(k_3)} + A_{k_3} x_{k_3} && = b_{k_3},\ k_3 = K_2 + 1, \ldots, K_3 \\
&\qquad B_{k_T} x_{a(k_T)} + A_{k_T} x_{k_T} = b_{k_T}, && k_T = K_{T-1+1}, \ldots, K_T \\
&\qquad l_t \leq x_{k_t} \leq u_t, && k_t = K_{t-1} + 1, \ldots, K_t,\ t = 1, 2, \ldots, T.
\end{aligned}$$

The scenarios used determine the size, form and optimal solution of the linear program. There are finitely many sequences of possible realizations of the random coefficients $(c_{k_t}, A_{k_t}, B_{k_t}, b_{k_t})$ with path probabilities p_{k_t} of the subsequences of these realizations, $p_{k_t} > 0$, $\forall k_t$, $\sum_{k_t=K_{t-1}+1}^{K_t} p_{k_t} = 1$, $t = 2, \ldots, T$, that identify the discrete joint probability distribution of $w = \{w_1, \ldots, w_{T-1}\}$. In the program, $a(k_t)$ denotes the immediate predecessor of k_t, for example, $a(k_2) = k_1$.

An important application of stochastic linear programming with a simple recourse model is given by Kusy and Ziemba [54]. The model was developed for the Vancouver City Savings Credit Union for a 5-year planning period. The formulation has the following features: (1) Changing yield spreads across time, transaction costs associated with selling assets prior to maturity, and synchronization of cash flows across time are incorporated in a multiperiod context. (2) Assets and liabilities are considered simultaneously to satisfy basic accounting principles and match liquidities. (3) Transaction costs are included. (4) Uncertainty of withdrawal claims and deposits is reflected in uncertain cash flows. (5) Uncertainty of interest rates is explicitly recognized. (6) Legal and policy constraints are taken into account. Their two-stage model did not contain end effects. Three possible scenarios that are independent over time were considered to keep the computations tractable. Their results indicate that their model generates policies that are superior to stochastic decision analysis.

Another milestone after the Kusy and Ziemba model is the Russell-Yasuda Kasai model by Carino et al. [17]. The model builds on the previous research to design a large scale SLPR model with possibly dependent scenarios, end effects, and all the relevant institutional and policy constraints. We present their model next.

Decision variables are

V_t : total fund market value at time t,
X_{nt} : market value in asset n at time t,
w_{t+1} : income shortfall at time $t+1$, and
v_{t+1} : income surplus at time $t+1$.

Random coefficients are

RP_{nt+1} : price return of asset n from end of t to end of $t+1$,
RI_{nt+1} : income returns of asset n from end of t to end of $t+1$,
F_{t+1} : deposit inflow from end of t to end of $t+1$,
P_{t+1} : principal payout from end of t to end of $t+1$,
I_{t+1} : interest payout from end of t to end of $t+1$,
g_{t+1} : rate at which interest is credited to policies from end of t to end of $t+1$,
L_t : liability valuation at t.

The objective is to maximize the expected market value of the firm at the horizon net of penalties for the shortfalls. Expected amount by which goals are not achieved is a more tangible risk measure than variance. The penalty costs of shortfalls may be based on expected financial impact or psychological costs. The piecewise linear convex cost function for the shortfall is denoted by $c_t(w_t)$. (19.17) is the budget constraint. The return on assets and inflow of deposits net of principal and interest payout gives the total fund market value (19.18). Liability balances and cash flows are computed to model liability accumulation (19.20). If Yasuda does not achieve adequate income, recourse action must be taken at a cost. The income generation is modeled as a soft constraint (19.19), which permits surpluses or deficits.

$$\max E\left[[V_t - \sum_{t=1}^{T} c_t(w_t)\right]$$

subject to

$$\sum_n X_{nt} - V_t = 0 \tag{19.17}$$

$$V_{t+1} - \sum_n (1 + RP_{nt+1} + RI_{nt+1}) X_{nt} = F_{t+1} - P_{t+1} - I_{t+1} \tag{19.18}$$

$$\sum_n RI_{nt+1} X_{nt} + w_{t+1} - v_{t+1} = g_{t+1} L_t \tag{19.19}$$

$$L_{t+1} = (1 + g_{t+1})L_t + F_{t+1} - P_{t+1} - I_{t+1} \tag{19.20}$$

$$X_{nt} \geq 0, \; w_{t+1} \geq 0, \; v_{t+1} \geq 0.$$

The abbreviated formulation does not include some elements of the model. There are additional types of shortfalls, indirect investment types, regulatory restrictions, multiple accounts, loan assets, tax effects and end effects that are included in the original model. See Carino and Ziemba [19] for the details of the formulation. Carino, Myers and Ziemba [18] discuss the concepts, technical issues and uses of the model.

Korhonen [53] applied SLPSR to multicriteria decision making. Oguzsoy and Guven [88] used the SLPSR methodology for a bank ALM model in Turkey.

Some authors argue against linearizing the objective function. Bai et al. [1] demonstrate that nonlinear programs are not much more difficult than their linear counterparts. Zenios et al. [105] apply multistage stochastic nonlinear programming with recourse to fixed income portfolio management. Golub et al. [41] utilize this methodology in money management.

19.3.5 Dynamic Generalized Networks

Multistage stochastic nonlinear programs with recourse can be represented by generalized network formulations. This framework can be used to account for the dynamic aspects of ALM problems while considering uncertainty in all relevant parameters and accommodating random parameters by means of a moderate number of scenarios.

The network structure is exploited in the solution procedure. The problem is decomposed into its constituent scenario subproblems. Preserving the network structure of each subproblem is challenged by the existence of non-anticipativity constraints. These constraints dictate that scenarios that share common information history up to a specific period must yield the same decision up to that period, i.e., dependence on hindsight is avoided.

The desired decomposition is achieved by dualizing the non-anticipativity constraints. The algorithm by Rockafeller and Wets [94] operates on the split-variable form of the original problem. The problem is solved by progressively enforcing the non-anticipativity constraints. The algorithm is discussed briefly in a later section.

Mulvey [75] utilized this methodology in designing an asset allocation model for the Pacific Financial Asset Management Company. The single period portfolio model is formulated as a network model. In Figure 1,[3] the

[3] Source: Mulvey [75]

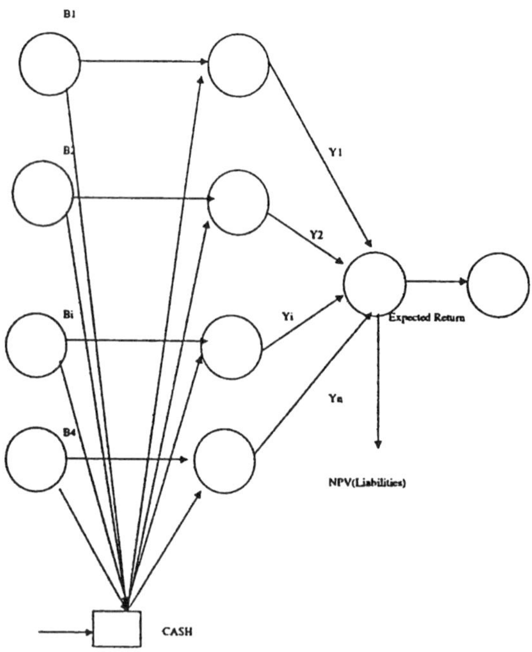

FIGURE 19.1
A sample network diagram.

supply nodes which are denoted as B represent the current asset mix. A cash node absorbs the transaction costs by means of arc multipliers and transforms the current asset mix. The revised asset mix, i.e., the Y arcs, are collected into a final node for the entire portfolio. The arcs can be constrained to impose legal or policy constraints. The liabilities are represented by the downward pointing arc. The objective function is the expected utility of surplus at the end of the planning horizon. The model was implemented in a PC environment with acceptable accuracy and efficiency.

Mulvey and Vladimirou [81,82] presented several aspects of stochastic generalized network models. See also the review of Mulvey and Ziemba [86] which discusses the model in a general context.

19.3.6 Scenario Optimization

According to the scenario optimization approach, one computes a solution to the deterministic problem under all scenarios then solves a coordinating model to find a single feasible policy. This approach can be compared to the scenario aggregation method suggested by Rockafeller and Wets [94]. It handles multistage stochastic programming problems, and allows for decisions to depend on future outcomes. On the other hand, scenario optimization is designed for two-period models only. It is assumed that scenario probabilities are functions of time, and estimates of the random parameters in the future stages are poor. Hence one only selects a policy for the immediate future.

Suppose the scenario subproblem is

$$v_s = \min \quad c_s^T x$$

subject to

$$A_s x = b_s \tag{19.21}$$

$$A_d x = b_d \tag{19.22}$$

$$x \geq 0,$$

where the objective function is a particular realization of the uncertainty under scenario s, (19.21) is also a particular realization of the uncertain constraints under scenario s, and (19.22) is the deterministic constraints.

A possible coordinating model could be

$$\begin{aligned} &\min \sum_s p_s \|c_s^T x - v_s\|^2 + \sum_s p_s \|A_s x - b_s\|^2 \\ &\text{subject to} \quad A_d x = b_d \\ &\qquad\qquad\quad\; x \geq 0. \end{aligned}$$

The coordinating model tracks the scenario solution as closely as possible while still maintaining feasibility. Alternative coordination models are discussed in Dembo ([26,27]. Illustrative applications in portfolio immunization and dedication are also presented therein.

19.3.7 Robust Optimization

Robust optimization approach integrates goal programming formulations with a scenario based description of the uncertainty in the data. The aim is to produce solutions that are relatively less sensitive to the realizations of different scenarios. The objective function, in its most general form, is composed of two terms: the first term trades off between mean value and the variability in the mean; the second term is a feasibility penalty function. Consider the following formulation.

$$\begin{aligned} &\min \sigma(x, y_1, ..., y_s) + w\rho(z_1, ..., z_s) \\ &\text{subject to} \\ &Ax = b \\ &B_s x + C_s y_s + z_s = e_s \quad \forall s \in \Omega \\ &x \geq 0, y_s \geq 0, \qquad\quad\; \forall s \in \Omega, \end{aligned}$$

where

x is the vector of decision variables whose value cannot be adjusted once a specific realization of the data is observed,

y is the vector of decision variables that are subject to adjustment once uncertain parameters are observed,

z is the vector of decision variables that measure infeasibility allowed,

$s \in \Omega = \{1, ..., S\}$ is the set of possible scenarios,

A, b, B_s, C_s, e_s are the coefficients related to the variables,

w is the goal programming weight that is used to derive a spectrum of answers that trade-off the two objectives.

The inclusion of higher order moments in the objective function reduces the variability of the solution. Hence, few adjustments become necessary as scenarios unfold. The model recognizes that it may not always be possible to find a feasible solution to the problem under all scenarios. The penalty function is used to detect the least amount of infeasibilities to be dealt with outside the model. See Mulvey et al. [80] for possible objective function choices and their applications. Bai et al. [1] argued that linear objective functions fail to identify robust solutions and concave utility functions produce much better results for risk averse decision-makers even when penalty terms are not used. Both papers compare robust optimization with the stochastic linear programming approach (SLP). Since SLP optimizes only the first moment of the distribution of the objective value, more adjustment is needed as scenarios are realized. However, there is no mechanism for choosing w, and the cost of the robust solution may be higher than that of SLP.

19.4 Stochastic Control ALM Models

Stochastic control approach captures uncertainty by allowing for a continuum of states which can be characterized by a small number of state variables that follow a joint Markov process. Due to the exponential growth of the problem with the number of state variables, a small number of states and securities can be used to describe the state of the world. It is difficult to specify general constraints on the process which hinders the practical applicability. The confidence limits are difficult to calculate and asset proportions are very sensitive to the input parameters.

A study by Brennan et al. [14] analyzed the portfolio problem of an investor who can invest in bonds, stock and cash. The time variation in expected asset returns is assumed to be driven by three state variables: the short-term interest rate (r), the rate on long-term bonds (l), and the dividend yield on a stock portfolio (δ). The formulation assumes that the investor has no liabilities. It is relatively straightforward to include liabilities whose expected rate of increase depends on the levels of the state variables.

The state variables follow a joint Markov process. The joint stochastic process for state variables is assumed to be of the form:

$$\begin{aligned} dr &= \mu_r dt + \sigma_r dz_r, \\ dl &= \mu_l dt + \sigma_l dz_l, \\ d\delta &= \mu_\delta dt + \sigma_\delta dz_\delta; \end{aligned}$$

the returns on the stock and the consol bond are given by

$$\frac{dS}{S} = \mu_s dt + \sigma_s dz_s,$$

$$\frac{dB}{B} + ldt = \left(l - \frac{\mu_l}{l} + \frac{\sigma_l^2}{l^2}\right) dt - \frac{\sigma_l}{l} dz_l,$$

where dS/S is the rate of return on the stock portfolio, dB/B is instantaneous total return on the consol bond. The parameters μ_i, σ_i $(i = r, l, \delta, S)$ are at most functions of the state variables r, l, δ, and dz_i are increments to the Wiener processes. The correlation coefficients between increments to the Wiener processes are denoted by ρ_{rl}, etc.

Defining W as the wealth, the utility is assumed to be of the isoelastic form, $V(r, l, \delta, W, \tau) = \frac{1}{\gamma} W^\gamma$, for $\gamma < 1$, when $\tau = 0$.

Defining x as the proportion invested in stocks, and y as the proportion invested in consol bond, the Bellman equation is

$$\max_{x,y} E[dV] = 0.$$

Solving the first order conditions, the optimal controls $x^*(r, l, \delta, \tau)$ and $y^*(r, l, \delta, \tau)$ are found. The processes are estimated from empirical data and the investor's optimal control problem is solved numerically for the resulting parameter values. The optimal portfolio proportions of an investor with a short horizon are found to be significantly different from those of an investor with a long horizon.

Brennan and Schwartz [12] extended that analysis by allowing the investor to take long and short positions in short-term interest rate futures as well as in stock, bonds and cash. They demonstrated that this new investment opportunity leads to significant improvement in expected utility.

Merton [69] used stochastic control theory to discuss optimal investment strategies for university endowment funds. Bielecki and Pliska [6] proposed a new stochastic control approach that incorporates underlying economic variables explicitly in the model along with a risk sensitive infinite horizon objective function.

19.5 Multistage Stochastic ALM Programming with Decision Rules

In this method, time is discretized into n-stages across the planning horizon, and investments are made using a decision rule, e.g., fixed mix, at the beginning of each time period. The decision rule can easily be tested with out-of-sample scenarios and confidence limits on the recommendations can be constructed. The use of this approach hinges on discovering policies that are intuitive and that will produce superior results. Decision rules may lead to non-convexities and highly nonlinear functions. Some decision rules used in the literature are fixed mix, no rebalancing, life cycle mix (Berger and Mulvey [3]), constant proportional portfolio insurance (Perold and Sharpe [90]), target wealth path tracking (Mulvey and Ziemba [87]).

Boender [8] and Boender et al. [9] describe an ALM model designed for Dutch pension funds. Their goal is to find efficient frontiers of initial asset allocations, which minimize the value of downside risk for given certain values of average contribution rates. The scenarios are generated across the time horizon of interest. The management selects a funding policy, an indexation policy of the earned pension rights, and an investment decision rule. These strategies are simulated against generated scenarios. Then, the objective function of the optimization problem is a completely specified simulation model except for the initial asset mix. The hybrid simulation/optimization model has the following three steps:

1) Randomly generate initial asset mixes, simulate them, and evaluate their contribution rates and downside risks.

2) Select the best performing initial asset mixes that are located at a minimal critical distance from each other.

3) Use a local search algorithm to identify the optimal initial asset mix.

Maranas et al. [65] adopted another approach to stochastic programming with decision rules. They determine the optimal parameters of the decision rule by means of a global optimization algorithm. They propose a dynamically balanced investment policy which is specified by the following parameters:

w_0: initial dollar wealth,

r_{it}^s: percentage return of asset $i \in \{1, 2, ..., I\}$ in time period $t \in \{1, 2, ..., T\}$ under scenario $s \in \{1, 2, ..., S\}$,

p^s: probability of occurrence of scenario s.

The decision variables are

w_t^s: dollar wealth at time t in scenario s,

λ_i: fraction of wealth invested in asset category i (note that it is constant over time).

The model is a multiperiod extension of mean-variance method. The multiperiod efficient frontier is obtained by varying $\beta (0 \leq \beta \leq 1)$. The formulation is as follows:

$$\max_{\lambda_i, w_t^s} \quad \beta \text{ mean}(w_T) - (1-\beta)\text{var}(w_T)$$

subject to

$$w_T^s = w_0 \prod_{t=1}^{T} \left[\sum_{i=1}^{I} (1 + r_{it}^s) \lambda_i \right], \quad s = 1, ..., S \tag{19.23}$$

$$\sum_{i=1}^{I} \lambda_i = 1 \tag{19.24}$$

$$0 \leq \lambda_i \leq 1, \quad i = 1, ..., I$$

The wealth accumulation is governed by (19.23). When (19.23) is substituted into the objective function, we get a nonconvex multivariable polynomial function in λ_i involving multiple local minima. A global optimization tool has been developed which obtains the above efficient frontier.

There are other models constructed using similar methodologies. Hiller and Eckstein [43] designed a stochastic dedication model to construct the optimal passive fixed income portfolio. Berger and Mulvey [3] describe Home Account Advisor^{TM} which assists individual investors in ALM using decision rules. Sweeney et al. [98] apply a simulation/optimization scenario approach to optimal insurance asset allocation in a multi-currency environment.

19.6 Capital Growth

This approach aims to maximize the long-run growth of assets by maximizing the expected log of asset wealth under the typical assumptions about capital markets: no transaction costs, no taxes, divisibility of assets which have limited liability, borrowing and lending allowed at the same rate, permissible short selling. Hakansson showed that the Kelly [46] or capital growth strategy is myopic, i.e., period by period optimization is optimal for general asset distributions. The strategy leads to risky and less diversified portfolios due to inherent low risk aversion. The wagers are typically very large and this yields high variance of wealth. The asset and liability generation processes of these models are kept simple. The policies tend to concentrate on a small number of superior assets and the asset proportions are very sensitive to the input parameters. We refer to MacLean and Ziemba [61] for important properties of the capital growth criterion.

MacLean and Ziemba [59], and MacLean et al. [62] developed a theory of growth versus security using a convex combination of Kelly strategy and cash, which they called fractional Kelly strategy. This strategy leads to a less volatile and safer sequence of wealth levels with less growth.

Along these lines, MacLean and Ziemba [61] designed a general investment model with three growth and three security measures (see Table 1). The measures are single valued aggregates of wealth over multiple periods, and constant proportion strategies can be evaluated using their explicit formulas. More general state and time dependent strategies can be assessed numerically by simulation.

The growth and security trade-offs are effective if a decrease in growth is offset by an increase in security. Analogous to static mean variance analysis, the trade-off is inefficient if another pair has either a higher growth rate and no lower security level or a higher security level and no lower growth rate. The effective path may be inside the efficient frontier. The effective path is efficient for lognormal assets, see Li et al. [57].

The optimal trade-off between security and growth can be formalized by using a utility function over these attributes. Another method of making this decision is suggested by MacLean and Ziemba [60]. They add probability constraints on the drawdown, then use scenario analysis to determine an optimal fractional Kelly strategy.

An alternative approach to growth versus security trade-off in the portfolio decision problem is proposed by Dohi et al. [30]. The criteria of optimality in the stopping rule profile and the wealth profile suggested by MacLean et al. [62] are reconsidered and modified by taking into account a finite horizon of investment and the financial agent's goal seeking behavior, respectively. Refer to Hakansson and Ziemba [42] for a review of the capital growth literature and various applications.

Table 1: Performance measures

Criterion	Characteristic	Measure
End of horizon wealth	Mean	$\mu_T(x) = E[W_T(x)]$ Mean accumulated wealth by time T
	Percentile	$\gamma_T(x) = \Pr(W_T(x) \geq c_T)$ Probability of achieving wealth c_T
Wealth trajectory	Mean	$\phi_T(x) = E\{\log[W_T(x)]^{1/T}\}$ Mean exponential growth rate over T periods
	Percentile	$\alpha_b(x) = \Pr[W_t(x) \geq b(t)W_{t-1}(x), t \geq 0]$ Probability that wealth drawdown is above $b(t)$
Wealth goals	Mean	$\eta_u(x) = E\{\tau[W(x) \geq U]\}$ Mean first passage time to achieving goal U
	Percentile	$\beta_{L,U}(x) = \Pr\{\tau[W(x) \geq U] \prec \tau[W(x) \geq L]\}$ Probability of reaching U before falling to L

Source: MacLean and Ziemba [61]

19.7 Components of ALM Models

19.7.1 Objective Function

There are numerous types of objective functions used in the literature. Some models employ mean-variance analysis which has been discussed in Section 1.2.3.

An alternative approach is the von Nuemann-Morgenstern expected utility of wealth which has been briefly analyzed in Section 1.2.4. This approach does not address the temporal aspects of the underlying preference structure over the planning horizon. A possible remedy is to consider discounted utility function over time and different states of the world:

$$\sum_s \sum_t p_s \frac{1}{(1+\rho)^t} u(w_s),$$

where p_s is the probability of state s, w_s is the wealth in state s, and ρ is the discount factor. Another approach is to use multi-attribute version of the Nuemann-Morgenstern utility which includes attributes such as the form of the wealth trajectory during the planning horizon.

Some researchers view utility function simply as a numerical value for determining the relative importance of some characteristics of a model's performance. For instance, Carino et al. [17] penalize shortfalls from wealth targets. Another example is goal programming where one selects a set of targets with corresponding importance levels which are penalized for any deviation from the target.

19.7.2 Scenario Generation

A scenario gives a single set of outcomes for the random variables in the model over the planning horizon. A representative set of scenarios describes the possible future environmental situations. Traditional quantitative forecasting methods extrapolate new ideas about future developments based on the knowledge of the past and present. However, the economic environment may change invalidating the past assumptions. Hence, subjective beliefs of the management have become an essential part of scenario building. See Bunn and Salo [15] for a review of qualitative scenario generation techniques.

The earlier ALM models used few number of independent scenarios to describe uncertainty. The recent models have become more sophisticated in scenario generation methods. The following are the widely accepted techniques in the literature.

19.7.2.1 Time Series Model

Multivariate Approach Using Vector Autoregression (VAR) in the analysis of economic systems became popular after an influential paper by Sims [97]. Following Sims, Dert [29] used this methodology to generate scenarios for a pension plan. He created future price inflation, wage inflation, stock returns, bond returns, cash return and real estate returns that are consistent with historical patterns in means, standard deviations, autocorrelations and cross correlations between state variables. A Markov model was used in determining future development of each individual participating in the pension plan. Carino et al. [17] also employed VAR in generating scenarios for the Yasuda Kasai model.

VAR may sometimes diverge from long-term equilibrium. Boender et al. [9] extended the VAR model to a Vector Error Correction Model (VECM) which additionally takes economic regime changes and long term equilibria into account. First, a submodel generates future economic scenarios. Then, a liability submodel determines the earned pension rights and payments corresponding to each economic scenario.

The economic scenario submodule uses time series analysis. The vector of the lognormal transformations of inflation, wage growth, bond return, cash return, equity return, real estate return and nominal GNP growth is

y_t. Diagnostic tests revealed that the order of the VAR process is 1.

$$y_t \sim N(\mu + \Omega * \{y_{t-1} - \mu\}, \Sigma)$$

where $N(\mu, \Sigma)$ denotes a Gaussian distribution with mean μ and covariance matrix Σ.

The extended VECM is given as

$$y_t \sim N(\Omega_1 y_{t-1} + \Omega_2 C^T (x_{t-1} - \mu_1 I_{\{T_1\}} - \mu_2 I_{\{T_2\}}), \Sigma)$$

where the Ω_1 corresponds to the short-term dynamics and the Ω_2 corresponds to the long-term correction. The index set T_1 specifies the period of an economic regime with growth vector μ_1, and T_2 gives the period of another economic regime with growth vector μ_2. The second term, $C^T(x_{t-1} - \mu_1 I_{\{T_1\}} - \mu_2 I_{\{T_2\}})$ generates the error correction to restore violations of the equilibria, while Ω_2 determines the speed of the response. They estimated the model by row wise ordinary least squares and seemingly unrelated regression methods. Then, scenarios are generated iteratively using the parameter estimates. They report that the VECM improves the explanatory power of the model. The VECM has a more clear economic interpretation which incorporates regime changes and long run equilibrium.

The liability submodule uses a push Markov model to determine the future status of each individual plan member depending on age, gender, and employee category. Given this information, the pull part of the model is used to determine additional promotions and new employees. Then, the pension rules are applied to compute the guaranteed pension payments and earned pension rights.

Cascade Approach Wilkie [102] suggested using a cascade structure rather than a multivariate model, in which each variable could affect each of the others. He considers inflation, ordinary shares and fixed interest securities as the main economic determinants of a stochastic investment model. The model includes the following variables: inflation, an index of share of dividends, the dividend yield (the dividend index divided by the corresponding price index) on these share indices, and the yield on consols (as a measure of the general level of fixed interest yields in the market).

Wilkie's investigations and actuarial experience led him to the conclusion that inflation is the driving force for the other investment variables. Figure 2,[4] where the arrows indicate the direction of influences, depicts the cascade structure of the model.

[4]Source: Wilkie [102]

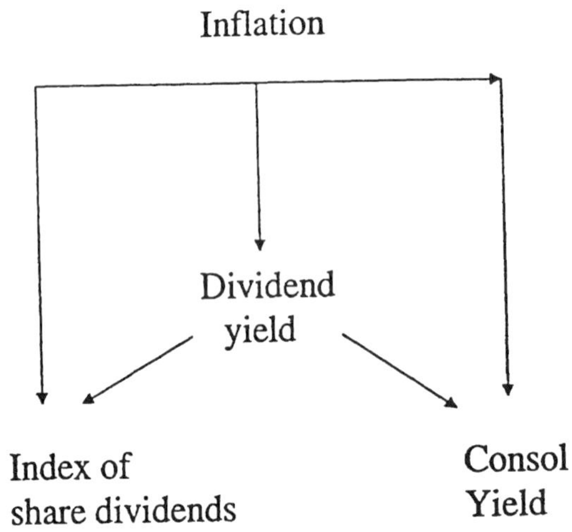

FIGURE 19.2
Wilkie's cascade structure.

The inflation is described first using a first order autoregressive model. The dividend yield depends on both the current level of inflation and the previous values of itself. The index of share dividends depends on inflation and the residual of the yield model. The consol yield also depends on inflation and the residual of the yield model along with the previous values of itself. Then, the estimated parameters are used to generate future economic scenarios. Wilkie [103] improves this basic model.

19.7.2.2 Stochastic Differential Equation

Mulvey [76] designed an economic projection model for Towers Perrin using stochastic differential equations. The model has a cascade structure as depicted in Figure 3.[5] First the Treasury yield curve, and then government bond returns, price and wage inflation, and large cap returns are generated. Lastly, returns on primary asset categories such as small cap stock and corporate bonds are projected.

It is assumed that short- and long-term interest rates (denoted by r_t and l_t, respectively) are linked through a correlated white noise term. The spread between the two is kept under control by using a stabilizing term. This variant of the two-factor Brennan and Schwartz [13] model is as follows:

$$dr_t = a(r_0 - r_t)dt + b\sqrt{r_t}dz_1,$$

[5]Resource: Mulvey (1996)

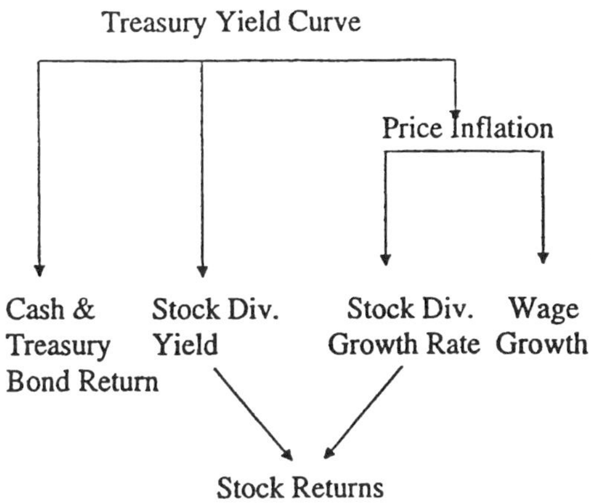

FIGURE 19.3
Mulvey's cascade structure.

$$dl_t = c(l_0 - l_t)dt + e\sqrt{l_t}dz_2,$$

where a and c are functions that depend on the spread between the long and short rates, b and e are constants, and dz_1 and dz_2 are correlated Weiner terms.

The price inflation rate is modeled as a diffusion process that depends on short-term interest rate:

$$dp_t = ndr_t + g(p_0 - p_t)dt + h(v_{pt})dz_3,$$

where p_t is the price inflation at time t, and v_{pt} is the stochastic volatility at time t. Since the volatility of inflation persists, it is represented using Autoregressive Conditional Heteroscedasticity (ARCH) model. The equation for the stochastic volatility is given by:

$$dv_{pt} = k(v_{p0} - v_{pt})dt + m\sqrt{v_{pt}}dz_4,$$

where g and k are functions that handle the independent movement of the underlying prices at time t for the price inflation and stochastic volatility, respectively, and h and m are constants.

Real yields are related to interest rates, current inflation, and expectations for future inflation. The diffusion equation for long-term yield is

$$dy_t = n(y_u, y_t, l_u, l_t, p_u, p_l)dl + q(y_u, y_t, l_u, l_t, p_u, p_l)dt + u(y_t)dz_5,$$

where y_u is the normative level of real yields, n and q are vector functions that depend upon various economic factors.

The wage inflation is connected to price inflation in a lagged and smoothed fashion. The stock returns are broken down into two components: dividends and capital appreciation, and they are estimated independently. Mulvey found that the decomposed structure provides more accurate linkages to the key economic factors such as inflation and interest rates.

The parameters of the model are calibrated by considering the overall market trends in the light of historical evidence and subjective beliefs of the management. This model has been in use at Towers Perrin since 1992. Mulvey and Thorlacius [79] extended the model to a global environment that links the economies of individual countries within a common framework.

Modeling term structure of interest rates is a very essential part of scenario generation. The use of binomial lattice models in the valuation of interest rate contingencies is prevalent. However, the number of scenarios grows very large if the valuation is to be precise. There are some sampling methods to reduce the size of the event tree such as Monte Carlo simulation, antithetic sampling and stratified sampling. However, Klaasen [49] points out that even if the underlying description is arbitrage free, a subset of it may include arbitrage opportunities that may lead to spurious profits. Instead of sampling paths, Klaasen [50] suggests an aggregation method that can be used to reduce the size of the event tree preserving the arbitrage-free description of uncertainty. In Klaasen [51], he presents a solution algorithm which iteratively disaggregates the condensed representation towards a more precise description of uncertainty.

19.7.3 Solution Methods and Computational Issues

Each modeling approach lends itself to different computational methods that exploit the problem's special structure. Hence, we have four sections corresponding to the underlying ALM framework.

19.7.3.1 Stochastic Programming

The number of decision variables in the deterministic equivalent of the stochastic program grow exponentially in the number of time periods and parameters. The solution of large scale multi-stage stochastic programming problems still remains a computational challenge. The primary algorithms fall into three categories:

Direct Solvers Simplex and interior point methods are two direct solvers for linear programming problems. The simplex method proceeds to an optimal solution by moving from vertex to vertex of the simplex. It is recognized as very efficient for the solution of relatively small dense problems.

Interior point method was first proposed by Karmarkar in 1984. The primal-dual interior point method is based on approximation of primal dual trajectories in the interior of the feasible region. The primal and dual problems are solved simultaneously with a progressive reduction of the duality gap at each iteration. The convergence for this algorithm is in general seen to be independent of the problem size. Lustig et al. [58] demonstrated that unlike simplex based methods, the interior point algorithms can significantly benefit from partial variable splitting formulations that yield a staircase structure. See also Berger et al. [4] and Gausmann [36] on this method.

Bender's Decomposition Based Approaches Bender's decomposition approach as applied to stochastic programming was developed by Van Slyke and Wets [99]. The multi-stage problem is split into a series of two-stage relations which are connected overall by a nesting scheme. It is a special kind of cutting plane algorithm which is updated as more information becomes available from the dual problems at the descendants of the current nodes. Refer to Dantzig and Infanger [25], Infanger [45], Fan and Carino [35] for more information on this approach.

Progressive Hedging Algorithm This algorithm was proposed by Rockafellar and Wets [94]. The problem is decomposed into its constituent subproblems. The solutions and the first stage decisions of the individual scenario subproblems do not necessarily coincide with those of the overall problem. An iterative procedure progressively imposes the non-anticipativity requirement by means of augmented Lagrangian functionals. Since all scenario subproblems are solved at each iteration of the algorithm, any special structure can be exploited to increase the efficiency of the method.

A state-of-the-art survey of the stochastic programming methods is presented in Birge [7]. There have been studies in the literature that compare the performance of these methods. It is generally accepted that simplex method and progressive hedging are adequate for small to medium problems and Bender's decomposition outperforms the others for large problems.

Mulvey and Vladimirou [82] applied the progressive hedging methodology to two-stage network problems. They compared the algorithm with a simplex based procedure and a primal-dual interior point method. They found that progressive hedging algorithm outperforms the simplex based procedure and is outperformed by the interior point method when the staircase form is used for linear programs. When nonlinear problems are considered, it fares well against the interior point method and still outperforms the simplex procedure. See also Mulvey and Ruszczynski [78].

Carino et al. [18] solved the Russell-Yasuda Kasai model using progressive hedging algorithm, OSL interior point code, OSL simplex code, and Bender's decomposition (parallel code running on a single processor). They concluded that, for small problems, progressive hedging is comparable to direct solve using the interior point method plus the simplex method to finish the optimization. The direct solve interior plus the simplex was the fastest algorithm for medium sized problems. The Bender's decomposition method was much faster than other methods in solving large problems.

Consigli and Dempster [23] discuss how simplex, interior point and Bender's decomposition algorithms can be improved. They propose methods and give computational results.

19.7.3.2 Stochastic Control

Optimal stochastic control algorithms are practical for relatively small state space sizes. The continuous problem is approximated by standard approaches such as finite differencing or dynamic programming. Refer to Brennan et al. [17] for an example.

19.7.3.3 Stochastic Programming with Decision Rules

This method may lead to non-convex optimization problems. There are mainly three approaches to the solution. A standard approach is using a non-linear programming algorithm and restarting the algorithm at widely separated points. Tabu search (Glover [38]) is an alternative method which prevents certain moves to often visited neighbors and moves away from the current local optimum. A third approach available in the literature is due to Maranas et al. [65]. This algorithm is based on a convex lower bounding of the original objective function and the successive refinement of converging lower and upper bounds by means of a standard branch and bound procedure.

19.7.3.4 Capital Growth

In many cases, the optimization problem resulting from this approach involves a concave program over a convex set. One can solve these problems using standard non-linear programming algorithms. See, for example, Grauer and Hakansson [40].

Part II: Stable Paretian Approach

19.8 Validity of Gaussian Assumption in Scenario Generation

The ALM models that have gained applicability are based on stochastic programming with or without decision rules. In these models, the future uncertainty is modeled using discrete scenarios. A representative set of scenarios describes the possible future environmental situations facing the institution. We have discussed two scenario generation techniques: time series analysis and stochastic differential equations. We would like to concentrate on the validity of the assumptions of the time series analysis. Following Sims [97], Dert [29] used VAR and Boender et al. [9] used VECM to generate economic scenarios for pension plans. Boender et al. created future price inflation, wage growth, bond return, cash return, equity return, real estate return and nominal GNP growth scenarios. The innovations were assumed to follow normal distribution.

However, there is contrary evidence in the literature. Fama [34] and Mandelbrot [63,64] found excess kurtosis in their investigation of the returns on financial assets, which led them to reject the normal assumption and propose stable distribution as an alternative. Balke and Fomby [2] show that most macroeconomic time series exhibit non-Gaussian behavior. They analyze fifteen post World War II US macroeconomic time series which include Consumer Price Index (a measure of price inflation), nominal compensation per hour in manufacturing (a measure of wage growth), yields on AAA bonds, Standard and Poor's 500 stock price index, and GNP deflator. They report that residuals from autoregressive models indicate that these series show significant evidence of excess kurtosis and skewness. They conclude that even after estimating GARCH models, significant excess kurtosis and/or skewness still remains.

The empirical density function, the normal fit, and the stable fit for monthly annualized GNP growth, monthly wage inflation, weekly change in 30-year government bond yield, weekly change in 3-month Treasury bill rate, and weekly change in S&P 500 stock price index are available in Figures 4, 5, 6, 7, and 8, respectively.[6] The graphs show the improvement of the stable fit over the normal fit.

Using stable distributions in portfolio optimization is not a new idea. For instance, Ziemba [107] utilized symmetric stable distribution in an ap-

[6]We would like to thank Carlo Marinelli for computational assistance.

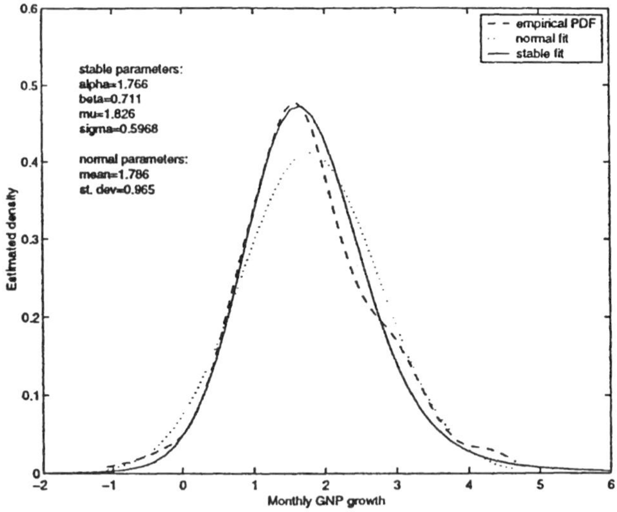

FIGURE 19.4
The normal and stable fit for monthly GNP growth.

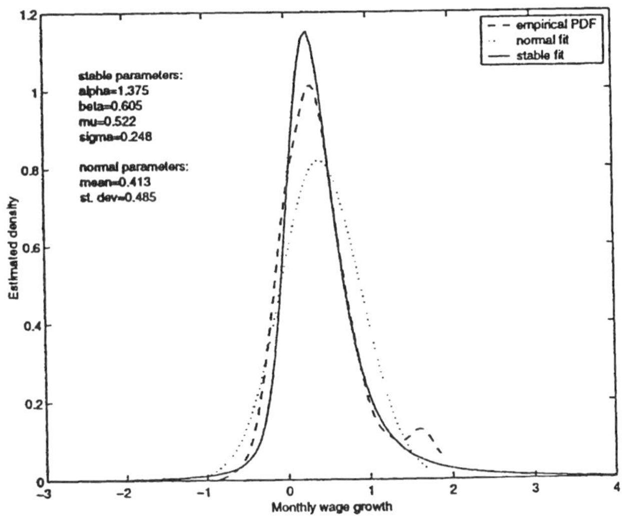

FIGURE 19.5
The normal and stable fit for monthly wage growth.

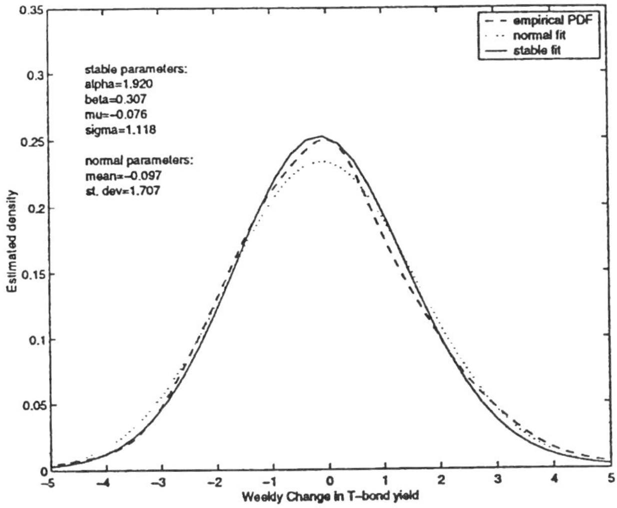

FIGURE 19.6
The normal and stable fit for weekly change in T-bond yield.

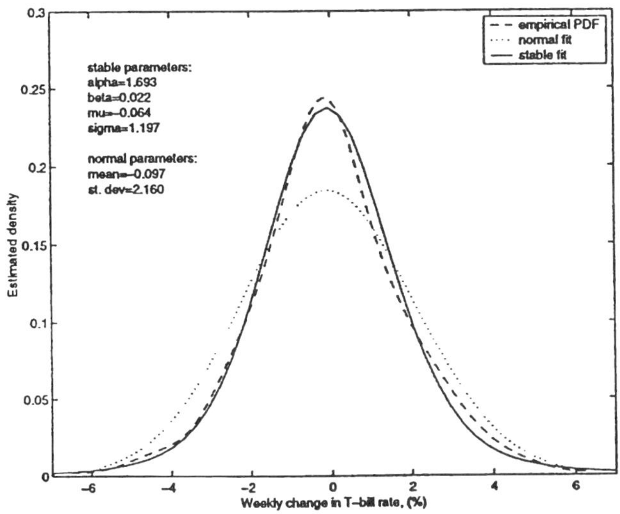

FIGURE 19.7
The normal and stable fit for weekly change in T-bill rate.

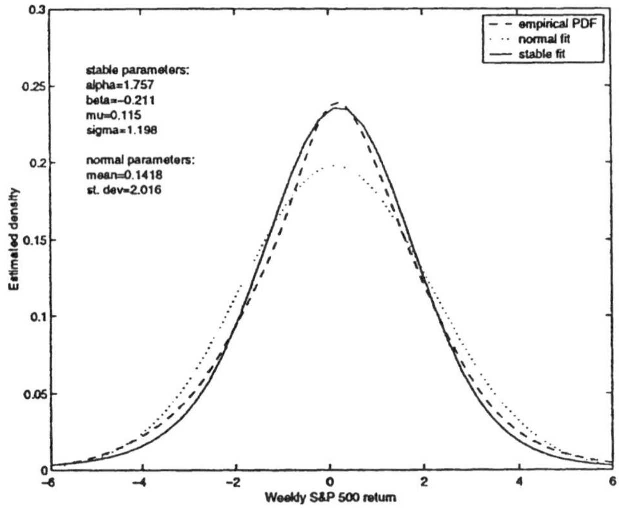

FIGURE 19.8
The normal and stable fit for weekly change in S&P 500.

proximate portfolio optimization problem. The interest in this approached faded due to theoretical and computational difficulties which outweighed its advantages. The recent theoretical contributions and increase in the computational power facilitates another look at the potential benefits. A recent paper by Rachev and Han [91] generalizes the mean-variance approach by using multivariate stable distribution rather than normal distribution. They do not report any empirical results on how this affects the resulting allocation.

The phenomenon of heavy tailedness and asymmetry in macroeconomic variables cannot be captured by linear, constant parameter models driven by Gaussian innovations. In order to accomodate such features while preserving the Gaussian hypothesis, econometricians have resorted to schemes such as outlier contamination, structural breaks or nonlinear dynamics. Kim [47] explores linear and asymmetric non-Gaussian data generating processes as a possible alternative explanation of business cycle asymmetry.

19.9 Reasons for Desirability of Stable Models

There are several important reasons for modeling financial variables using stable distributions. Stable distributions are leptokurtotic. When compared to normal distribution, they typically have fatter tails and a higher peak around the center. Asymmetry is allowed. Hence, they fit the empirical distribution of the financial data better.

Any distribution in the domain of attraction of a specified stable distribution will have properties which are close to the ones of stable distribution. Even if the observed data do not exactly follow the ideal distribution specified by the modeler, in principle, the resulting decision is not affected. The wide use of normal distribution is mainly due to the fact that it is the only distribution with finite variance that is stable with respect to the summation scheme.

Each stable distribution has an index of stability which remains the same regardless of the sampling interval adopted. The index of stability can be regarded as an overall parameter that can be employed in inference and decision making. However, we should note that for some financial data empirical analysis shows that the index of stability increases as the sampling interval increases.

It is possible to check whether a distribution is in the domain of attraction of a stable distribution or not by examining the tails of the distribution. The tails dictate the properties of the distribution.

19.10 Description of Stable Distributions

If the sums of linear functions of independent identically distributed (iid) random variables belong to the same family of distributions, the family is called stable. Formally, a random variable r has stable distribution if for any $a > 0$ and $b > 0$ there exist constants $c > 0$ and $d \in R$ such that

$$ar_1 + br_2 \stackrel{d}{=} cr + d,$$

where r_1 and r_2 are independent copies of r, and $\stackrel{d}{=}$ denotes equality in distribution. The distribution is described by the following parameters: $\alpha \in (0, 2]$ (index of stability), $\beta \in [-1, 1]$ (skewness parameter), $\mu \in R$ (location parameter), and $\sigma \in [0, \infty)$ (scale parameter). The variable is then represented as $r \sim S_{\alpha,\beta}(\mu, \sigma)$. Gaussian distribution is actually a special case of stable distribution when $\alpha = 2$, $\beta = 0$. The smaller the stability index is, the stronger the leptokurtic nature of the distribution becomes, i.e., with higher peak and fatter tails. If the skewness parameter is equal to zero, as in the case of Gaussian distribution, the distribution is symmetric. When $\beta > 0$ ($\beta < 0$), the distribution is skewed to the right (left). If $\beta = 0$ and $\mu = 0$, then the stable random variable is called symmetric

α-stable ($S\alpha S$). The scale parameter generalizes the definition of standard deviation. The stable analog of variance is variation, v_α, which is given by σ^α.

Stable distributions generally do not have closed form expressions for density and distribution functions. They are more conveniently described by characteristic functions. The characteristic function of random variable r, $\Phi_r(\theta) = E[\exp(ir\theta)]$, is given by

$$\Phi_r(\theta) = \exp\left\{-\sigma^\alpha |\theta|^\alpha \left(1 - i\beta \mathrm{sign}(\theta) \tan\frac{\pi\alpha}{2}\right) + i\mu\theta\right\}, \quad \text{if } \alpha \neq 1,$$

$$= \exp\left\{-\sigma |\theta| \left(1 - i\beta\frac{2}{\pi}\mathrm{sign}(\theta) \ln\theta\right) + i\mu\theta\right\}, \quad \text{if } \alpha = 1.$$

The pth absolute moment of r, $E|X|^p = \int_0^\infty P(|X|^p > y)dy$, is finite if $0 < p < \alpha$, and infinite otherwise. Hence, when $\alpha < 1$ the first moment is infinite, and when $\alpha < 2$ the second moment is infinite. The only stable distribution that has finite first and second moments is the Gaussian distribution.

In models that use financial data, it is generally assumed that $\alpha \in (1, 2]$. There are several reasons for this:

1) When $\alpha > 1$, the first moment of the distribution is finite. It is convenient to be able to speak of expected returns.

2) Empirical studies support this parametrization.

3) Although the empirical distributions of the financial financial data sometimes depart from normality, the deviation is not "too much".

In scenario generation, one may need to use multivariate stable distributions. The extension to the multivariate case is nontrivial. Although most of the literature concentrates on the univariate case, recently some new results have become available. See, for example, Samorodnitsky and Taqqu [95], Mittnik and Rachev [71].

If R is a stable d-dimensional stable vector, then any linear combination of the components of R is also a stable random variable. However, the converse is true under certain conditions (Samorodnitsky and Taqqu [95]). The characteristic function of R is given by:

$$\Phi_Y(\theta) = \exp\left\{-\int_{S_d} |\theta^T s| \left(1 - i\,\mathrm{sign}(\theta^T s) \tan\frac{\pi\alpha}{2}\right) \Gamma(ds) + i\theta^T\mu\right\},$$

$$\text{if } \alpha \neq 1,$$

$$= \exp\left\{-\int_{S_d} |\theta^T s| \left(1 + i\frac{2}{\pi}\text{sign}(\theta^T s)\ln|\theta^T s|\right)\Gamma(ds) + i\theta^T\mu\right\},$$

$$\text{if } \alpha = 1,$$

where Γ is the spectral measure which replaces the scale and skewness parameters that enter the desciption of the univariate stable distribution. It is a bounded nonnegative measure on the unit sphere S_d, and $s \in S_d$ is the integrand unit vector. The index of stability is again α, and μ is the vector of locations.

Stable distributions have infinite variances. The stable equivalent of covariance for $S\alpha S$ variables is covariation:

$$[R_1, R_2]_\alpha = \int_{S_d} s_1 s_2^{<\alpha-1>}\Gamma(ds),$$

where (R_1, R_2) is a $S\alpha S$ vector ($\alpha \in (1,2)$), and $x^{<\alpha-1>} = |x|^k\text{sign}(x)$. The matrix of covariations determines the dependence structure among the individual variables.

19.11 Financial Modeling and Estimation

Financial modeling frequently involves information on past market movements. Examples include technical analysis to derive investment decisions, or researchers assessing the efficiency of financial markets. In such cases, it is not the unconditional return distribution which is of interest, but the conditional distribution, which is conditioned on information contained in past return data, or a more general information set. The class of autoregressive moving average (ARMA) models is a natural candidate for conditioning on the past of a return series. These models have the property that the conditional distribution is homoskedastic. In view of the fact that financial markets frequently exhibit volatility clusters, the homoskedasticity assumption may be too restrictive. As a consequence, conditional heteroskedastic models, such as Engle's [33] autoregressive conditional heteroskedastic (ARCH) models and the generalization (GARCH) of Bollerslev [10], possibly in combination with an ARMA model, referred to as an ARMA–GARCH model, are now common in empirical finance. It turns out that ARCH-type models driven by normally distributed innovations imply unconditional distributions which themselves possess heavier tails. Thus, in this respect, ARCH models and stable distributions can be viewed as competing hypotheses.

Mittnik et al. [73] present empirical evidence favoring stable hypothesis over the normal assumption as a model for unconditional, homoskedastic conditional, and heteroskedastic conditional distributions of several asset return series.

19.11.1 Maximum Likelihood Estimation

We will describe an approximate conditional maximum-likelihood (ML) estimation procedure suggested by Mittnik et al. [72]. The unconditional ML estimate of $\theta = (\alpha, \beta, \mu, \sigma)$ is obtained by maximizing the logarithm of the likelihood function

$$L(\theta) = \prod_{t=1}^{T} S_{\alpha,\beta}\left(\frac{r_t - \mu}{\sigma}\right)\sigma^{-1}.$$

One needs to use conditional ML to estimate ARMA and ARMA-GARCH models. The ML estimation is conditional, in the sense that, when estimating, for example, an ARMA(p, q) model, one conditions on the first p realizations of the sample, r_p, $r_{p-1}, \ldots, r_1$, and, when $\alpha > 1$ holds, sets innovations ε_p, $\varepsilon_{p-1}, \ldots, \varepsilon_{p-q+1}$ to their unconditional mean $\mathbf{E}(\varepsilon_t) = 0$. The estimation of all stable models is approximate in the sense that the stable density function, $S_{\alpha,\beta}(\mu, \sigma)$, is approximated via fast Fourier transformation (FFT) of the stable characteristic function,

$$\int_{-\infty}^{\infty} e^{itx} dH(x) = \begin{cases} \exp\{-\sigma^\alpha |t|^\alpha [1 - i\beta \text{sign}(t) \tan \frac{\pi\alpha}{2}] + i\mu t\}, & \text{if } \alpha \neq 1, \\ \exp\{-\sigma |t| [1 + i\beta \frac{2}{\pi} \text{sign}(t) \ln |t|] + i\mu t\}, & \text{if } \alpha = 1, \end{cases}$$

where H is the distribution function corresponding to $S_{\alpha,\beta}(\mu, \sigma)$.

This ML estimation method essentially follows that of DuMouchel [31], but differs in that the stable density is approximated numerically by an FFT of the characteristic function rather than some series expansion. As DuMouchel shows, the resulting estimates are consistent and asymptotically normal with the asymptotic covariance matrix of $T^{1/2}(\hat{\theta} - \theta_0)$ being given by the inverse of the Fisher information matrix. The standard errors of the estimates are obtained by evaluating the Fisher information matrix at the ML point estimates. For details on stable ML estimation, see Mittnik et al. [72], Mittnik and Rachev [71], and Paulauskas and Rachev [89].

19.12 Conclusions

We have described four approaches to ALM. Each approach has its advantages and drawbacks. The stochastic control method does not require generation of scenarios but allows for a limited number of state variables. The capital growth criterion leads to risky and less diversified portfolios. These approaches have not gained much practical applicability. The ALM models that have been adopted by the industry for practical use utilize stochastic programming with and without decision rules. These models are harder to solve, but they admit more realistic assumptions. They represent the future uncertainty by discrete scenarios. A line of research is on developing more efficient solution algorithms. Designing models that combine the strengths of the existing models seems to be another direction of active research.

Generating scenarios that realistically represent the future uncertainty is important for the validity of the results of stochastic programming based ALM models. The assumption underlying the above described scenario generation methods is the normal distribution. The validity of normal distribution has been questioned in the finance and macroeconomics literature. The financial data exhibit heavy tails, time varying volatility, and long-range dependence. The macroeconomic variables exhibit significant kurtosis and asymmetry. The leptokurtic (heavy tailed and peaked) and asymmetric nature of the economic variables can be better captured by using stable distribution as opposed to normal distribution. The conditional heteroskedastic models (ARMA-GARCH) utilizing stable distributions can be used to describe the time varying volatility along with the asymmetric and leptokurtic behavior. In addition to these, the long-range dependence can also be modeled if fractional-stable GARCH models are employed.

References

1. Bai, D., Carpenter, T., and Mulvey, J. M. (1997), "Making a case for robust optimization models", Management Science, 43 (7), 895-907.

2. Balke, N. S. and Fomby, T. B. (1994), "Large shocks, small shocks, and economic fluctuations: outliers in macroeconomic time series", Journal of Applied Econometrics, 9, 181-200.

3. Berger, A. J. and Mulvey, J. M. (1998), "The Home Account Advi-

sor: asset and liability management for individual investors", in W. T. Ziemba and J. M. Mulvey (eds.), Worldwide Asset and Liability Modeling, Cambridge University Press, Cambridge, UK, 634-665.

4. Berger, A. J., Mulvey, J. M., Rothberg, E. and Vanderbei, R. (1995), "Solving multi-stage stochastic programs using tree dissection", Princeton University, Stochastic and Operations Research Report 95-7.

5. Bessler, W. and Booth, G G. (1994) "An interest rate risk management model for commercial banks", European Journal of Operational Research, 74(2), 243-256.

6. Bielecki, T. R. and Pliska, S. R. (1999), "Risk sensitive dynamic asset management", Applied Mathematics and Optimization, 39, 337-360.

7. Birge, J. R. (1997), "Stochastic programming computation and applications", Informs Journal on Computing, 9, 111-133.

8. Boender, G. C. E. (1997), "A hybrid simulation/optimization scenario model for asset/liability management", European Journal of Operational Research, 99, 126-135.

9. Boender, G. C. E., van Aalst, P., and Heemskerk, F. (1998), "Modeling and management of assets and liabilities of pension plans in the Netherlands", in W. T. Ziemba and J. M. Mulvey (eds.), Worldwide Asset and Liability Modeling, Cambridge University Press, Cambridge, UK, 561-580.

10. Bollerslev, T. (1986), "Generalized autoregressive conditional homoskedasticity", Journal of Econometrics, 52, 5-59.

11. Bradley, S. P. and Crane, D. B. (1972), "A dynamic model for bond portfolio management", Management Science, 19, 139-151.

12. Brennan, M. J. and Schwartz, E. S (1998), "The use of Treasury bill futures in strategic asset allocation programs", in W. T. Ziemba and J. M. Mulvey (eds.), Worldwide Asset and Liability Modeling, Cambridge University Press, Cambridge, UK, 205-228.

13. Brennan, M. J. and Schwartz, E. S (1982), "An equilibrium model of bond pricing and a test of market efficiency", Journal of Financial and Quantitative Analysis, 17(3), 301-329.

14. Brennan, M. J., Schwartz, E. S. and Lagnado, R. (1997), "Strategic asset allocation", Journal of Economic Dynamics and Control, 21, 1377-1403.

15. Bunn, D. W. and Salo, A. A. (1993), "Forecasting with scenarios",

European Journal of Operational Research, 68(3), 291-303.

16. Carpenter, T., Lustig, I. and Mulvey, J. M. (1991), Formulating stochastic programs for interior point methods", Operations Research, 39, 757-770.

17. Carino, D. R., Kent, T., Myers, D. H., Stacy, C., Sylvanus, M., Turner, A. L., Watanabe, K., and Ziemba, W. T. (1994), "The Russell-Yasuda Kasai Model: An asset/liability Model for a Japanese insurance company using multistage stochastic programming", Interfaces, 24, 29-49.

18. Carino, D. R., Myers, D. H., and Ziemba, W. T. (1998), "Concepts, technical issues, and uses of the Russel-Yasuda-Kasai financial planning model", Operations Research, 46(4), 450-462.

19. Carino, D. R. and Ziemba, W. T. (1998), "Formulation of the Russell-Yasuda Kasai financial planning model", Operations Research, 46(4), 433-449.

20. Charnes, A., Gallegos, C., and Yao, S. (1982), "A chance constrained approach to bank dynamic balance sheet management", Center for Cybernetic Studies CCS 428, University of Texas, Austin.

21. Charnes, A. and Kirby, M. J. L. (1966), "Optimal decision rules for the E-model of chance-constrained programming", Cahiers du Centre d'Etudes de Recherche Operationelle, 8, 5-44.

22. Chopra, V.K. and Ziemba, W. T. (1993), "The effect of errors in means, variances, and covariances on optimal portfolio choice", Journal of Portfolio Management, Winter, 6-11.

23. Consigli, G. and Dempster, M. A. H. (1998), "Dynamic stochastic programming for asset-liability management", Annals of Operations Research, 81, 131-161.

24. Culp, C., Tanner, K., and Mensink, R. (1997), "Risk, returns and retirement", RISK, 10 (10), 63-69.

25. Dantzig, G. B. and Infanger, G. (1993), "Multistage stochastic linear programs for portfolio optimization", Annals of Operations Research, 45, 59-76.

26. Dembo, R. S. (1991), "Scenario optimization", Annals of Operations Research, 30, 63-80.

27. Dembo, R. S. (1993), "Scenario immunization", in S. T. Zenios (ed.), Financial Optimization, Cambridge University Press, Cambridge, UK, 290-308.

28. Dert, C. L. (1995), "Asset and liability management for pension funds, a multistage chance constrained programming approach", Ph.D. thesis, Erasmus University, Rotterdam, The Netherlands.

29. Dert, C. L. (1998), "A dynamic model for asset liability management for defined benefit pension funds", W. T. Ziemba and J. M. Mulvey (eds.), Worldwide Asset and Liability Modeling, Cambridge University Press, Cambridge, UK, 501-536.

30. Dohi, T., Tanaka, H., Kaio, N., and Osaki, S. (1995), "Alternative growth versus security in continuous dynamic trading", European Journal of Operational Rèsearch, 84(2), 265-278.

31. DuMouchel, W. H. (1973), "On the asymptotic normality of the maximum likelihood estimate when sampling from a stable distribution", Annals of Statistics, 1, 948-957.

32. Eppen, L. and Fama, E. (1971), "Three asset cash balance and dynamic portfolio problems", Management Science, 17(5), 311-319.

33. Engle, R. (1982), "Autoregressive conditional heteroskedasticity with estimates of the variance of United Kingdom inflation", Econometrica, 50, 987-1007.

34. Fama, E. (1965), "The bahavior of stock market prices", Journal of Business, 38, 34-105.

35. Fan, Y. and Carino, D. R. (1994), "Nested versus non-nested decomposition for asset allocation problems", 15th International Symposium on Mathematical Programming, Ann Arbor, August.

36. Gausmann, H. I. (1990), "MSLiP: a computer code for multistage stochastic linear programming problem", Mathematical Programming, 47, 407-423.

37. Giokas, D. and Vassiloglou, M. (1991), "A goal programming model for bank assets and liability management", European Journal of Operational Research, 50, 48-60.

38. Glover, F. (1989), "Tabu search-part I", ORSA Journal on Computing, 1, 190-206.

39. Grauer, R. R. and Hakansson, N. H. (1982), "Higher return, lower risk: historical returns on long run, actively managed portfolios of stocks, bonds and bills", Financial Analysts Journal, March-April, 1936-1978.

40. Grauer, R. R. and Hakansson, N. H. (1998), "On timing the market: the empirical probability assesment approach with an inflation adapter", in W. T. Ziemba and J. M. Mulvey (eds.), Worldwide

Asset and Liability Modeling, Cambridge University Press, Cambridge, UK.

41. Golub, B., Holmer, M., McKendall, R., Pohlman, L., and Zenios, S. A. (1995), "A stochastic programming model for money management", European Journal of Operational Research, 85, 282-296.

42. Hakansson, N. H. and Ziemba, W. T. (1995), "Capital growth theory", in R. Jarrow, V. Maksimovic and W. T. Ziemba (eds.), Finance, North-Holland, 65-86.

43. Hiller, R. S. and Eckstein, J. (1993), "Stochastic dedication: designing fixed income portfolios using massively parallel Bender's decomposition", Management Science, 39 (11), 1422-1438.

44. Hiller, R. S. and Schaack, C. (1990), "A Classification of Structured Bond Portfolio Modeling Techniques", Journal of Portfolio Management, 17(1), 37-48.

45. Infanger, G. (1994), "Planning Under Uncertainty: Solving Large Scale Stochastic Linear Programs", The Scientific Press, Danvers, Massachusetts.

46. Kelly, J. (1956), "A new interpretation of information rate", Bell System Technology Journal, 35, 917-926.

47. Kim, J.-R. (1999), "Another look on the asymmetric behavior of unemployment rates over the business cycle", working paper, Technische Universitat, Dresden.

48. Kim, J., Malz, A. M., and Mina, J. (1999), "LongRun Technical Document", First edition, New York: RiskMetrics Group.

49. Klaasen, P. (1997), "Discretized reality and spurious profits in stochastic programming models for asset/liability management", European Journal of Operational Research, 101, 374-392.

50. Klaasen, P. (1998), "Financial asset-pricing theory and stochastic programming models for asset/liability management: a synthesis", Management Science, 44(1), 31-48.

51. Klaasen, P. (1998), "Solving stochastic programming models for asset/liability management using iterative disaggregation", in W. T. Ziemba and J. M. Mulvey (eds.), Worldwide Asset and Liability Modeling, Cambridge University Press, Cambridge, UK, 427-463.

52. Konno, H. and Yamazaki, H. (1991), "Mean-absolute deviation portfolio optimization model and its applications to Tokyo Stock Market", Management Science, 37 (5), 519-531.

53. Korhonen, A. (1987), "A dynamic bank portfolio planning model with multiple scenarios, multiple goals and changing priorities", European Journal of Operational Research, 30, 13-23.

54. Kusy, M. I. and Ziemba, W. T. (1986), "A bank asset and liability management model", Operations Research, 35, 356-376.

55. Li, S. X. (1995a), "An insurance and investment portfolio model using chance constrained programming", Omega, 23(5), 577-585.

56. Li, S. X. (1995b), "A satisfying chance constrained model in the portfolio selection of insurance lines and investments", Journal of the Operational Research Society, 46(9), 1111-1120.

57. Li, Y., MacLean, L. C. and Ziemba, W. T. (1996), "Security aspects of optimal growth models with minimum expected time criteria", Mimeo, University of British Columbia.

58. Lustig, J. L., Mulvey, J. M. and Carpenter, T. (1991), "Formulating two-stage stochastic programs for interior point methods", Operations Research, 39 (5), 757-770.

59. MacLean, L. C. and Ziemba, W. T. (1991), "Growth-security profiles in capital accumulation under risk", Annals of Operations Research, 31, 501-510.

60. MacLean, L. C. and Ziemba, W. T. (1998), "Capital growth with security", Mimeo, University of British Columbia.

61. MacLean, L. C. and Ziemba, W. T. (1999), "Growth versus security tradeoffs in dynamic investment analysis", Annals of Operations Research, 85, 193-225.

62. MacLean, L. C., Ziemba, W. T. and Blazenko, G. (1992), "Growth versus security in dynamic investment analysis", Management Science, 38, 1562-1585.

63. Mandelbrot, B. B. (1963), "The variation of certain speculative prices", Journal of Business, 26, 394-419.

64. Mandelbrot, B. B. (1967), "The variation of some other speculative prices", Journal of Business, 40, 393-413.

65. Maranas, C. D., Androulakis, I. P., Floudas, C. A., Berger, A. J., and Mulvey, J. M. (1997), "Solving long-term financial planning problems via global optimization", Journal of Economic Dynamics and Control, 21, 1405-1425.

66. Markowitz, H. M. (1952), "Portfolio selection", Journal of Finance, 7 (1), 77-91.

67. Markowitz, H. M. (1987), "Mean-Variance Analysis in Portfolio Choice and Capital Markets", Basil Blackwell, New York.

68. Markowitz, H. M., Todd, P., Xu, G., and Yamane, Y. (1993), "Computation of mean-semivariance efficient sets by the critical line algorithm", Annals of Operations Research, 45, 309-318.

69. Merton, R. C. (1993), "Optimal investment strategies for university endowment funds", in C. T. Clotfelter and M. Rothschild (eds.), Studies of Supply and Demand in Higher Education, University of Chicago Press, Chicago.

70. Mittnik, S. and Rachev, S. T. (1993), "Modeling asset returns with alternative stable distribution", Econometric Review, 12(3), 261-330.

71. Mittnik, S. and Rachev, S. T. (1999), "Stable Models in Finance", Wiley, New York.

72. Mittnik, S., Rachev, S. T., Doganoglu, T., and Chenyao, D. (1996), "Maximum likelihood estimation of stable Paretian models", unpublished manuscript, Institute of Statistics and Econometrics, Christian Albrechts University at Kiel.

73. Mittnik, S., Rachev, S. T., and Paolella, M. S. (1997), "Stable Paretian modeling in finance: some empirical and theoretical aspects", unpublished manuscript, Institute of Statistics and Econometrics, Christian Albrechts University at Kiel.

74. Morgan Guaranty Trust Company (1996), "RiskMetrics Technical Document", Fourth edition, JP Morgan.

75. Mulvey, J. M. (1994), "An asset-liability investment system", Interfaces, 24, 22-33.

76. Mulvey, J. M. (1996), "Generating scenarios for the Towers Perrin investment system", Interfaces, 26, 1-15.

77. Mulveyt, J. M. (1998), "Generating scenarios for global financial planning systems", International Journal of Forecasting, 14, 291-298.

78. Mulvey, J. M. and Ruszczynski, A. (1995), "A new scenario decomposition method for large scale stochastic optimization", Operations Research, 43, 477-490.

79. Mulvey, J. M. and Thorlacius, A. E. (1998), "The Tower Perrin global capital market scenario generation system", in W. T. Ziemba and J. M. Mulvey (eds.), Worldwide Asset and Liability Modeling, Cambridge University Press, Cambridge, UK, 286-312.

80. Mulvey, J. M., Vanderbrei, R. J., and Zenios, S. A. (1995), "Robust Optimization of Large-Scale Systems", Operations Research, 43 (2), 264-281.

81. Mulvey, J. M. and Vladimirou, H. (1989), "Stochastic network optimization models for investment planning", Annals of Operations Research, 20, 187-217.

82. Mulvey, J. M. and Vladimirou, H. (1991), "Applying the progressive hedging algorithm to stochastic generalized networks", Annals of Operations Research, 31, 399-424.

83. Mulvey, J. M. and Vladimirou, H. (1992), "Stochastic network programming for financial planning problems", Management Science, 38 (11), 1642-1664.

84. Mulvey, J. M. and Zenios, S. A. (1994), "Capturing correlations of fixed income instruments", Management Science, 40(10), 1329-1342.

85. Mulvey, J. M. and Zenios, S. A. (1994), "Dynamic diversification of fixed income portfolios", Financial Analysts Journal, Jan-Feb, 30-38.

86. Mulvey, J. M. and Ziemba, W. T. (1995), "Asset and liability allocation in a global environment", in R. Jarrow et al. (eds.), Handbooks in OR and MS, 9,435-463.

87. Mulvey, J. M. and Ziemba, W. T. (1998), "Asset and liability management systems for long-term investors: discussion of the issues", in W. T. Ziemba and J. M. Mulvey (eds.), Worldwide Asset and Liability Modeling, Cambridge University Press, Cambridge, UK, 3-38.

88. Oguzsoy, C. B. and Guven, S. (1997), "Bank asset and liability management under uncertainty", European Journal of Operational Research, 102, 575-600.

89. Paulauskas, V. and Rachev, S. T. (1999), "Maximum likelihood estimators in regression models with infinite variance innovation", working paper, Vilnius University, Lithuania.

90. Perold, J. M. and Sharpe, W. F. (1988), "Dynamic strategies for asset allocation", Financial Analysts Journal, January, 16-27.

91. Rachev, S. T. and Han, S. (1999), "Optimization problems in mathematical finance", working paper, University of Karlsruhe, Germany.

92. Rachev, S. T., Kim, J.-R. and Mittnik, S. (1999), "Stable paretian models in econometrics: Part I", working paper, University of Karlsruhe, Germany.

93. Rachev, S. T., Kim, J.-R. and Mittnik, S. (1999), "Stable paretian models in econometrics: Part I", working paper, University of Karlsruhe, Germany.

94. Rockafeller, R. T. and Wets, R. J.-B. (1991), "Scenarios and policy aggregation in optimization under uncertainty", Mathematics of Operations Research, 16, 119-147.

95. Samorodnitsky, G. and Taqqu, M. S. (1994), "Stable Non-Gaussian Random Variables", Chapman and Hall, New York.

96. Shing, C. and Nagasawa, S. (1999), "Interactive decision system in stochastic multiobjective portfolio selection", International Journal of Production Economics, 60-61, 187-193.

97. Sims, Christopher (1980), "Macroeconomics and reality", Econometrica, 48(1),1-48.

98. Sweeney, J. C., Sonlin, S. M., Correnti, S., and Williams, A. P. (1998), "Optimal insurance asset allocation in a multi-currency environment", in W. T. Ziemba and J. M. Mulvey (eds.), Worldwide Asset and Liability Modeling, Cambridge University Press, Cambridge, UK, 341-368.

99. Van Slyke, R. and Wets, R. J. -B. (1969), "L-shaped linear programs with applications to optimal control and stochastic programming", SIAM Journal of Applied Mathematics, 17, 638-663.

100. Vassoadou-Zeniou, C. and Zenios, S. A. (1996), "Robust optimization models for managing callable bond portfolios", European Journal of Operational Research, 91, 264-273.

101. von Neumann, J. and Morgenstern, O. (1947), "Theory of Games and Economic Behavior", Princeton University, Princeton, NJ

102. Wilkie, A. D. (1986), "A stochastic investment model for actuarial use", Transactions of the Faculty of Actuaries, 39, 391-403.

103. Wilkie, A. D. (1995), "More on a stochastic asset model for actuarial use", presented to the Institute Actuaries and Faculty of Actuaries, London.

104. Zenios, S. A. (1993), "Financial Optimization", Cambridge University Press, Cambridge, UK.

105. Zenios, S. A., Holmer, M. R., McKendall, R., and Vassiadou-Zeniou, C. (1998), "Dynamic models for fixed income portfolio management under uncertainty", Journal of Economic Dynamics and Control, 22, 1517-1541.

106. Zenios, S. A. and Shtilman, M. S. (1993), "Constructing optimal samples from binomial lattice", Journal of Information and Optimization Sciences, 14(2), 125-147.

107. Ziemba, W. T. (1974), "Choosing investment portfolios when the returns have stable distributions", in P. L. Hammer and G. Zoutendijk (eds.), North-Holland, Amsterdam, 443-482.

108. Ziemba, W. T. and Mulvey, J. M. (1998), "Worldwide Asset and Liability Modeling", Cambridge University Press, Cambridge.

109. Zipkin, P. (1992), "The Structure of Structured Bond Portfolio Models", Operations Research, Jan/Feb, S157-S169.

Chapter 20

Optimization Problems in Fair Division Theory

Marco Dall'Aglio

Dipartimento di Scienze, Università "G. d'Annunzio", Viale Pindaro, 42 65127 - Pescara, Italy, E-mail: maglio@sci.unich.it

Abstract The problem of dividing an object among two or more parties, known as the *fair division problem* has captured the attention of several notable mathematicians, such as Steinhaus, Banach, Knaster, Dubins and Spanier. The latter two authors, in particular, have defined two different optimization problems that single out satisfactory partitions of the object in question.

In this work we review the available results on optimization in fair division and we take a closer look at some open questions. In particular we add constraints to the first problem, in order to improve the quality of the optimal solution. We recur to functional Lagrangian duality in order to solve the constrained problem. Furthermore, we describe a constructive procedure (otherwise unavailable in the existing literature) to find a solution to the second problem, based on a geometrical setting. We apply the algorithms to several examples.

Contents

20.1 Introduction

The problem of dividing a non-homogeneous object, say a cake, among a given number of persons has been studied by several mathematicians. Steinhaus in [18] (in collaboration with Banach and Knaster) was the first to give a mathematical framework to the widely used method in which only two persons are involved and one person cuts the cake, while the other chooses one of the two parts. Provided the cake is "completely divisible" (this concept will be formulated below in more rigorous terms), the method yields a fair division of the cake. More importantly, the same framework allowed Steinhaus to succesfully generalize the procedure for an arbitrary number of persons.

Several years later, Dubins and Spanier in [6] employed probability measures to describe each person's preferences over the different parts of the cake. By doing so, these authors could apply a theorem of Lyapunov [13] on the ranges of vector measures, to show when fair divisions of the cake exist. In the same work they also defined two optimization problems to help choosing a satisfactory partition of the cake. In the first problem, which we denote as (DS1), we look at a division of the cake that maximizes a linear combination of the different participants' satisfactions. In the second problem, denoted (DS2), the lexicographic order of the persons' satisfactions is considered, and we look for a division of the cake which maximizes the ordered values, starting from the smallest one.

The work by Dubins and Spanier has also revived interest in fair division problems. Many new results have appeared since the publication of [6]. For a review of the most important results in the theory of fair division we refer to the review paper by Hill [11]. Also, algorithmic procedures specifically designed for fair division have been implemented which extend in various directions the methods described by Steinhaus and by Dubins and Spanier. A review of these methods appears in a recent book by Brams and Taylor [4].

Steinhaus also gave several applications of his methods to practical situations such as dividing an inheritance or assigning jobs to different teams. Since then, fair division methods have been applied to a wide range of situations where any sort of "goods" must be partitioned among two or more parties. Those applications range from legal settlements, to international treaty negotiations, to voting rules, and an account of this variety can be found in [4]. In addition — as pointed out by Hill in [11] and by Hill and Tong in [10], there is a close formal correspondence between fair division theory and multi-hypotheses testing procedures in statistical theory. This link allows transfer of results from either of the two fields into the other.

The goal of the present chapter is to take a closer look at the two optimization problems defined by Dubins and Spanier. We give a review of the available results on these problems (many of which by Dubins and Spanier themselves). While for (DS1), Dubins and Spanier show how to achieve an optimal division, for the second problem their result is purely existential. In fact, this result is based on the already mentioned theorem by Lyapunov which is non-constructive — and hence does not give any help in finding the optimal solution. Yet, an optimal solution for (DS2) satisfies more favorable properties than the corresponding solution for (DS1).

We suggest two directions to overcome the existing problems. Problem (DS1) can be modified through constraints, in order for the optimal solution to satisfy more criteria. This task is achieved with the help of general results on functional Lagrangian by Wagner and Stone [19]. A dual procedure is therefore illustrated to solve these constrained problems. It must be noted that a particular case has already been investigated in statistics by J.A. Anderson [1] with different computational methods, thus confirming the strong ties between the two fields of research.

We also analyze the structure of problem (DS2) from a geometric viewpoint. This framework enables us to design a dual optimization problem with a structure simpler than the original one. Consequently, we provide a constructive procedure to find the optimal solution for (DS2). Finally we provide examples for both (DS2) and the constrained version of (DS1) to verify the feasibility of the shown procedures.

After describing the basic mathematical framework for fair division in Section 20.2, we review the main criteria that, whenever fulfilled, identify a "good" partition (Section 20.3). In Section 20.4 we illustrate problems (DS1) and (DS2). We also examine the properties of the corresponding optimal solutions. In the same section we define a geometrical setting to help locate the optimal solution of both (DS1) and (DS2). In Section 20.5 we describe a constructive procedure to find the optimal solution for (DS2) based on the geometric dual structure. Section 20.6 deals with optimization problems in which we add constraints to the objective functions of (DS1) in order to improve the quality of the optimal solution. Section 20.7 reviews some available results on functional Lagrangians and their duality. These results are applied to fair division in Section 20.8. Finally, Section 20.9 shows an example where the constrained versions of (DS1) are successfully solved with the aid of Lagrangian duality.

Section 20.5 and Proposition 20.1 are taken from [5], where they are proved in a less general form. Sections 20.6 through 20.9 contain new material.

20.2 A Mathematical Framework for Fair Division

The mathematical framework for fair division theory (f.d.t.) requires the following basic elements. Following the description by Hill in [11], we consider a set of n probability measures $\mu_1, \ldots, \mu_n$, defined on the same probability space $(\Omega, \mathcal{G})$. Π denotes the class of all $\mathcal{G}$-measurable partitions of Ω, composed exactly of n sets, i.e., $P \in \Pi$ if $P = (C_1, \ldots, C_n)$, $\bigcup_{i=1}^n C_i = \Omega$, $C_i \cap C_j = \emptyset$ for $i \neq j$ and $C_i \in \mathcal{G}$ for all $i = 1, \ldots, n$. The goal is to choose one or more partitions in Π that are "good" in some sense.

Traditionally, a cake-cutting interpretation is given to this mathematical setting. A cake Ω has to be divided among n persons (or players). The cake is non-homogeneous and each player has subjective tastes which are expressed through a probability measure. Thus, if player i receives slice $C \subset \Omega$, his level of satisfaction will be expressed by the value $\mu_i(C)$. Here Π describes all possible divisions of the cake. The goal, therefore, is to make n slices so as to make all participants "as happy as possible" simultaneously. The use of measures to evaluate each participant's satisfaction implies that specific assumptions are made on the players' tastes. First of all, every player attributes the same value 1 to the whole cake. Moreover, their preferences are (countably) additive. Namely, for each person, the value of two disjoint pieces of the cake equals the sum of the values of each piece, and the same is true if an infinite number of pieces is considered. Both assumptions seem reasonable in most cases.

Following Dvoretzky et al. [7], we consider here a slight generalization of partitions. Namely, the class Φ of all non-negative, measurable vector functions $\phi(x) = (\phi_1(x), \ldots, \phi_n(x))$ with $\sum_{i=1}^n \phi_i(x) = 1$ for every $x \in \Omega$. The basic idea underlying these definitions is that each point x is assigned to some player according to the probability distribution $\phi_1(x), \ldots, \phi_n(x)$. This corresponds to the randomization technique used in statistical hypothesis testing. Clearly, the class Φ includes all partitions whenever ϕ is such that $\phi_i(x) \in \{0, 1\}$ for all $x \in \Omega$ and $i = 1, \ldots, n$. With a slight abuse of notation, let Π denote this smaller class of allocations. The use of allocations in place of partitions makes it easier to find "good" ways to cut the cake. Their use may face the same criticism reserved for randomization techniques in statistics. Still, they represent a very useful tool for f.d.t. and there is a common situation where allocations and partitions are equivalent (see Theorem 20.1). Also, in the long history of f.d.t. it has been pointed out by some authors (in particular see [6]), that we may consider densities in place of the original measures. In fact, we can always find a measure ν such that all measures μ_i are absolutely continuous w.r.t. ν. We may always choose $\nu = \left(\sum_i \mu_i\right)/n$, but often we can choose ν to be the Lebesgue measure on $(\Re, \text{Borels})$. Let f_i, $i = 1, \ldots, n$ be the density functions of μ_i w.r.t. ν. On intuitive bases, f_i yields a pointwise description of the i-th player's preferences.

With the simultaneous use of density and allocation functions, the quantities of interest in f.d.t. can be summarized by the n^2 values

$$a_{ij}(\phi) = \int_\Omega \phi_j \, f_i \, d\nu \qquad i, j = 1, \ldots, n \,. \tag{20.1}$$

Each of these numbers describes how player i evaluates the share of the cake received by person j. In particular, $a_{ii}(\phi)$ measures how player i considers the slice he himself receives, while $a_{ij}(\phi)$ with $i \neq j$ tells us how he evaluates the part of the cake given to another participant j. It is convenient, for our purposes, to group these values as an $n \times n$ matrix

$$A(\phi) = \left(a_{ij}(\phi)\right)_{i,j=1,\ldots,n} \tag{20.2}$$

Primarily we are intersted in allocations that yield high values on the principal diagonal of $A(\phi)$ and low values off it. There is no single method to obtain this, as will be illustrated in the next section.

In order to (partially) justify the use of allocations in place of partitions, we need to recall some important results of fair division theory. First of all we consider $\mathcal{A}$, the set of all values spanned by the matrix $A(\phi)$ as ϕ ranges over Φ.

$$\mathcal{A} = \{A(\phi) : \phi \in \Phi\} \tag{20.3}$$

$\mathcal{A}$ is a subset of the space of $n \times n$ real-valued matrices and will be called the *matrix allocation range*. A measurable set C is said to be an *atom* for μ_i if $\mu_i(C) \neq 0$ and, for every measurable set H, either $\mu_i(C \cap H) = 0$ or $\mu_i(C \setminus H) = 0$. The measure μ_i is called *atomless* if there exists no atom for it.

The following statement summarizes several well-known and important results (Theorems 1, 2 and 4) in [7].

THEOREM 20.1

(i) *The matrix allocation range $\mathcal{A}$ is a compact and convex set in the space of $n \times n$ real-valued matrices.*

(ii) *If $\mu_1, \ldots, \mu_d$ are atomless, then, for each $M \in \mathcal{A}$, there exists a partition $\phi^p \in \Pi$ such that $A(\phi^p) = M$.*

Part (i) is a generalization of a result first stated by Lyapunov in [13], and will be referred to as the Lyapunov Theorem. When all the probability measures are atomless, part (ii) states that partitions and allocations are equivalent in terms of the evaluations made by all players. Accordingly, for all the results that follow, we assume that $\mu_1, \ldots, \mu_n$ are always atomless.

Fair division theory is also important because the same framework can be employed in statistical theory. We refer in particular to discriminant analysis and multi-hypotheses testing theory. This second interpretation is fully explained in Hill [11]: a random variable X taking values in Ω is observed and it must be decided which of n known different distributions $\mu_1, \ldots, \mu_n$ it originated. Once X is observed, our aim is to guess the originating distribution, also called hypothesis. This is accomplished by choosing an allocation $\phi \in \Phi$. If $X = x$, we attribute the observation to one of the n hypotheses, according to the probability distribution $\phi_1(x), \ldots, \phi_n(x)$. In this context we have $a_{ij}(\phi) = P(\text{choose hypothesis } j | \mu_i \text{ is true})$ and we aim at minimizing errors which arise from a wrong classification by looking for high values in the principal diagonal of $A(\phi)$.

20.3 Some Criteria for a Division

We have not been rigorous in specifying what is a proper definition for a "good" or "optimal" partition. In fact there is no single criterion for it. Instead, it is common in f.d.t. to define several properties which an allocation should have. Consequently, an allocation should be judged according to the optimal properties it satisfies. Some definitions are stronger than others and we are going to review some known relationships among them. The criteria that have been considered in the long history of f.d.t. seek to maximize the "global" performance of the allocation, instead of taking a given participant's perspective.

Also, it is usually assumed that all players have the same importance and are treated symmetrically. Without much effort, however, we can consider the more general situation where each player has a different weight $s_i > 0$.

This measures the relative importance of the player in the division. Without loss of generality, we impose a normalizing constraint $\sum_{i=1}^{n} s_i = 1$ on the weights and we let $\mathbf{s} = (s_1, \ldots, s_n)$. This setting includes the more usual symmetrical situation where $\mathbf{s} = (1/n, \ldots, 1/n)$. In some situations another systems of weights $\boldsymbol{\xi}$ will be also considered, with the same normalizing constraints.

The most basic objective of allocations is to make sure each person receives a satisfaction from his slice at least equal to his given degree of importance s_i. Therefore an allocation is called *fair* (or proportional) w.r.t. $\mathbf{s}$ if

$$a_{ii}(\phi) \geq s_i \qquad \text{for all } i = 1, \ldots, n\,. \tag{20.4}$$

Let $\mathcal{F}(\mathbf{s})$ denote the class of fair allocations w.r.t. $\mathbf{s}$. If the inequalities in (20.4) are all strict, then the allocation is *super-fair.* Fair allocations always exist. Moreover, if we exclude the singular case where all measures are identical, we know that super-fair allocations also exist. In fact, we have the following from [6] (Theorem 1.2):

THEOREM 20.2

If for some $C \in \mathcal{G}$ and $i \neq j$, $\mu_i(C) \neq \mu_j(C)$, then there exists a super-fair allocation.

The proof of this result is based on the convexity of $\mathcal{A}$ stated in Theorem 20.1. In case of atomless measures, Theorem 20.2 extends automatically to partitions, as a consequence of part (ii) of Theorem 20.1.

Each element of $A(\phi)$ outside the principal diagonal describes how each player evaluates someone else's share. This value ought to be no greater than the value actually received by the player in question, in order to avoid envious feelings which may arise if he discovers that the slice he received is not the highest in value, among all slices, according to his own specified preferences. In a more general situation players may have different degrees of envy, which are measured once again by the weight in $\mathbf{s}$. Thus, an allocation ϕ will be called *envy-free* (e.f.) w.r.t. $\mathbf{s}$ whenever

$$\frac{a_{ii}(\phi)}{s_i} \geq \frac{a_{ij}(\phi)}{s_j} \qquad \text{for all } i \neq j \tag{20.5}$$

Let $\mathcal{E}(\mathbf{s})$ denote the class of envy-free allocations w.r.t. $\mathbf{s}$. A stronger type of envy-freeness bounds each evaluation of the other players' share, independently of the share actually received, in the following manner.

$$a_{ij}(\phi) \leq s_j \qquad \text{for all } i \neq j\,. \tag{20.6}$$

These allocations are called *strongly envy-free* (st.e.f.) w.r.t. $\mathbf{s}$ and they form the class $\tilde{\mathcal{E}}(\mathbf{s})$. Clearly, the allocation $\phi = (s_1, \ldots, s_n)$ is strongly envy-free. We may thus consider a smaller class of envy-free allocations that does not include this trivial case and introduce the notion of *super envy-free* allocation given by Barbanel, in which all the inequalities in (20.6) are strict. The same author in [3] gives sufficient conditions for an allocation with this property to exist (see Theorem 20.6 below).

Since strongly envy-free allocations satisfy $a_{ii}(\phi) \geq s_i$ for all i, they are also envy-free in the weaker sense. Moreover, since $\sum_{i=1}^n \phi_i(x) = 1$ for all $x \in \Omega$, an envy-free allocation is such that

$$a_{ii}(\phi) = 1 - \sum_{j \neq i} a_{ij}(\phi) \geq 1 - \frac{1 - s_i}{s_i} a_{ii}(\phi)$$

so $a_{ii}(\phi) \geq s_i$. Hence an envy-free allocation is fair, as long as the same system of weights is used. Thus, for a given $\mathbf{s}$, we conclude

$$\tilde{\mathcal{E}}(\mathbf{s}) \subset \mathcal{E}(\mathbf{s}) \subset \mathcal{F}(\mathbf{s}). \tag{20.7}$$

In order to verify fairness and envy-freeness of an allocation w.r.t. a system of weights $\mathbf{s}$, it is convenient to consider the weighted allocation matrix W in which each column i of A is divided by s_i. A fair allocation has all terms in the principal diagonal greater than or equal to 1. An envy-free allocation has the larger term of each row on the principal diagonal. A strongly envy-free allocation has all the off-diagonal terms less than or equal to 1.

The two definitions of envy-freeness arise from different applications. Usually envy-free partitions (in the weakest sense) are required in cake-cutting problems, where there is an ongoing competition among players and we want to ban any discontent arising from the desire to swap pieces with others in the given allocation. Strong envy-freeness is more suitable in a statistical context. In particular, according to J.A. Anderson [1], an allocation must verify strong envy-freeness when

> the sample points [$x \in \Omega$] represent objects whose individual fate is of interest [such as when diagnosing differences between diseases in a particular specialty], [since] then safeguards for the individuals are required to ensure that appropriate probabilities of misclassification are not too high.

Therefore a bound is put over each source of error.

In the same spirit of the envy-free requirements we may require that everybody receives the same value of the cake, so the allocation will be *equitable* w.r.t. s when, for some constant γ, we have

$$\frac{a_{ii}(\phi)}{s_i} \equiv \gamma \qquad \text{for all } i = 1, \ldots, n\,.$$

To make this allocation interesting, we require that $\gamma \geq 1$. Another desirable property is that, starting from the allocation ϕ we are considering, no overall improvement is possible. In other words, there is no other allocation $\tilde{\phi}$ which is strictly better in terms of the values received by the players. More formally,

$$a_{ii}(\tilde{\phi}) \geq a_{ii}(\phi) \quad \text{for all } i$$

implies

$$a_{ii}(\tilde{\phi}) = a_{ii}(\phi) \quad \text{for all } i$$

and ϕ is *Pareto-optimal* (or efficient). This definition is independent of the system of weights actually used. There is no implication between Pareto-optimality and fairness (resp. envy-freeness).

20.4 Optimization Problems in Fair Division Theory

The criteria we have reviewed up to this point — however desirable — usually define a rather broad class of allocations. More restrictive criteria are set by Dubins and Spanier in their very important work [6]. They define optimization problems in the f.d.t. context that pick out a restricted subclass of optimal allocations — usually consisting of only one allocation (or, more precisely, by the allocations that share the same values in $A(\phi)$).

In the first problem, a partition that maximizes the weighted sum of the values received by all participants is sought:

$$\text{Find } \phi \text{ that maximizes } \sum_{i=1}^{n} \xi_i \, a_{ii}(\phi) \tag{DS1}$$

with $\xi_i > 0$ for all i and $\sum_i \xi_i = 1$. The same authors give a general solution for (DS1) (Theorem 2, [6]).

THEOREM 20.3

A solution for (DS1) always exists.

PROOF For any $x \in \Omega$, define $\psi(x) = \max_i \xi_i f_i(x)$. Then, we will show

$$\sum_{i=1}^{n} a_{ii}(\phi) \leq \int_{\Omega} \psi(x)\,\nu(dx) \tag{20.8}$$

and that there exists a ϕ^* which attains the upper bound.

To see (20.8), simply observe that, for all $x \in \Omega$

$$\sum_{i=1}^{n} \xi_i\,\phi_i(x)\,f_i(x) \leq \left(\sum_{i=1}^{n} \phi_i(x)\right)\psi(x) = \psi(x) \tag{20.9}$$

and the same holds when we integrate the first and last term over Ω. To show that the bound is attained, let

$$K(x) = \{i : \xi_i\,f_i(x) = \psi(x)\} \tag{20.10}$$

and define ϕ^* as

$$\sum_{i \in K(x)} \phi_i^*(x) = 1 \qquad \text{for all } x \in \Omega. \tag{20.11}$$

Substituting ϕ^* into (20.8) shows that the upper bound is attained. ∎

As a consequence of Theorem 20.3, the application of the rule given in (20.11) always yields the optimal solution for (DS1). This solution is also Pareto-optimal. To verify this, suppose that there is a better allocation $\tilde{\phi}$ with

$$a_{ii}(\phi^*) \leq a_{ii}(\tilde{\phi}) \qquad \text{for all } i = 1, \ldots, n$$

with strict inequality for at least one index. Then

$$\sum_{i=1}^{n} \xi_i\,a_{ii}(\phi^*) < \sum_{i=1}^{n} \xi_i\,a_{ii}(\tilde{\phi})$$

thus contradicting the optimality of ϕ^*.

A DS1-solution, however, may not be fair, as the following simple example shows.

Example 20.1

Let $\Omega = [0,1]$, $\mathcal{G}$ = Borels and let S_1 and S_2 be the intervals $[0,1/2]$ and $(1/2,1]$ respectively. Consider two measures, μ_1 and μ_2 defined by the following density functions w.r.t. the Lebesgue measure

$$f_1(x) = 0.4\, I_{S_1}(x) + 1.6\, I_{S_2}(x) \quad \text{and} \quad f_2(x) = 0.8\, I_{S_1}(x) + 1.2\, I_{S_2}(x)\ ,$$

where $I_C(x)$ is the indicator function of the set $C \subset \Omega$. Moreover, let $\xi_1 = \xi_2 = 1/2$. Then,

$$\phi^*(x) = (0,1)\, I_{S_1}(x) + (1,0)\, I_{S_2}(x)\ ,$$

so $a_{11}(\phi^*) = 0.8$ and $a_{22}(\phi^*) = 0.4$. Thus ϕ^* is not a fair allocation w.r.t. the weights ξ_1 and ξ_2. A fortiori, envy-freeness may also fail for this solution. In fact, we have $0.8 = a_{22}(\phi^*)/s_2 < a_{21}(\phi^*)/s_1 = 1.2$. □

In order to acquire a better understanding of problem (DS1), we return to Theorem 20.1 and some of its direct consequences. First of all, we look at the diagonal of $A(\phi)$, i.e.,

$$\mathbf{d}_u(\phi) = (a_{11}(\phi), \ldots, a_{nn}(\phi)) \in \Re^n \tag{20.12}$$

and consider its range

$$\mathcal{D}_u = \{\mathbf{d}_u(\phi) : \phi \in \Phi\}\ . \tag{20.13}$$

Then the same conclusions of Theorem 20.1 hold for $\mathcal{D}_u$ as well. In particular, $\mathcal{D}_u$ is a compact and convex subset of $\Re^n$.

We now give a geometrical description of (DS1). More specifically, we describe how to plot the DS1-optimal value on $\mathcal{D}_u$. To this aim, consider the class of hyperplanes in $\Re^n$ with fixed coefficients given by $\boldsymbol{\xi}$:

$$\mathcal{H}(b) = \left\{ \mathbf{x} \in \Re^n : \sum_{i=1}^{n} \xi_i\, x_i = b \right\}\ .$$

We then choose the hyperplane with the highest value for b, say b^*, that intersects the partition range:

$$\text{Find } b^* = \sup\left\{ b : \text{ such that } \mathcal{H}(b) \bigcap \mathcal{D}_u \neq \emptyset \right\}\ .$$

Moreover, an allocation satisfies (20.11) and, therefore, is DS1-optimal, if and only if its value is on $\mathcal{H}(b^*) \cap \mathcal{D}_u$.

The second optimization problem defined by Dubins and Spanier requires a few preliminary notions. For each allocation $\phi \in \Phi$, arrange the values

$$\frac{a_{ii}(\phi)}{s_i} \qquad \text{for } i = 1, \ldots, n$$

in non-decreasing order and denote the resulting sequence as

$$\left\{\alpha_{(1)}(\phi), \ldots, \alpha_{(n)}(\phi)\right\} . \tag{20.14}$$

The allocation ϕ is DS2-*optimal* if, for any other allocation $\phi' \in \Phi$, either $\alpha_{(i)}(\phi) = \alpha_{(i)}(\phi')$ for all i, or, if j is the smallest i such that $\alpha_{(i)}(\phi) \neq \alpha_{(i)}(\phi')$, then $\alpha_{(j)}(\phi') < \alpha_{(j)}(\phi)$.

The optimal partition can be found through an iterative procedure. First of all we solve

$$\alpha^*_{(1)} = \sup\{\alpha_{(1)}(\phi) : \phi \in \Phi\} \tag{DS2$'$}$$

then, for given $\alpha^*_{(1)}, \ldots, \alpha^*_{(t)}$, $1 \leq t \leq n-1$ we look for

$$\alpha^*_{(t+1)} = \sup\{\alpha_{(t+1)}(\phi) : \phi \in \Phi, \alpha_{(j)}(\phi) = \alpha^*_{(j)}, j = 1, \ldots, t\} . \tag{DS2$''$}$$

We will show that each step has some allocation achieving the supremum. It is easy to verify that any allocation attaining the optimal value at step n is DS2-optimal.

Dubins and Spanier show that optimal solutions for (DS2) always exist. Unlike the result for (DS1), however, the existence here is based on the non-constructive Theorem 20.1. Therefore, no clue is given on how to find the optimal solution for (DS2).

A DS2-solution enjoys more desirable properties than the DS1 counterpart. In fact, not only is it always Pareto-optimal, but it also fair. To show fairness, simply observe that, if $\mu_i \neq \mu_j$ for some $i \neq j$, there is a super-fair allocation ϕ. So, for every $i = 1, \ldots, n$

$$\alpha^*_{(i)} \geq \min_j \frac{a_{jj}(\phi)}{s_j} > 1$$

and the DS2-solution must be super-fair.

The proof of Pareto-optimality comes from Weller [20]. Suppose that some allocation $\tilde{\phi}$ dominates the DS2-solution ϕ^*, i.e., for all $i = 1, \ldots, n$

$$a_{ii}(\tilde{\phi}) \geq a_{ii}(\phi^*) .$$

Now order the weighted values of $\tilde{\phi}$ as $\{\tilde{\alpha}_{(1)}, \ldots, \tilde{\alpha}_{(n)}\}$ and let t identify the smallest index, i.e., $a_{tt}(\phi)/s_t = \tilde{\alpha}_{(1)}$. By the following inequalities

$$\tilde{\alpha}_{(1)} = \frac{a_{tt}(\tilde{\phi})}{s_t} \geq \frac{a_{tt}(\phi^*)}{s_t} \geq \alpha^*_{(1)}$$

and the fact that $\tilde{\alpha}_{(1)} \geq \alpha^*_{(1)}$, it follows that $a_{tt}(\tilde{\phi}) = a_{tt}(\phi^*)$. The argument can be repeated iteratively to show that $\tilde{\phi}$ and ϕ^* have the same values.

The objective functions of (DS1) and (DS2) fulfill different needs. (DS1) is more appropriate in a statistical context, especially when the researcher has an *a priori* guess of the probability of each possible hypothesis, quantified by the weights ξ_i. In fact, this problem is considered in many reference texts for statistical inference, such as [2] Section 6.7 and [16] Section 7d. On the other hand, (DS2) is more suitable in the cake-cutting interpretation. Not only does it yield a fair solution, but it also makes sure that everyone's weighted evaluation is no less than $\alpha^*_{(1)}$. Moreover, since there could be many allocations with this property, (DS2) goes further and defines a unique allocation by applying the maximin criterion iteratively over each ordered value of the allocation.

As for (DS1), we can describe this optimization problem geometrically. The task is slightly more complicated than the previous case, but as a result we get better insight into the problem and also a constructive procedure. This time we consider the principal diagonal of $W(\phi)$:

$$\mathbf{d}(\phi) = \left(\frac{a_{11}(\phi)}{s_1}, \ldots, \frac{a_{nn}(\phi)}{s_n}\right) \in \Re^n \tag{20.15}$$

and its range

$$\mathcal{D} = \{\mathbf{d}(\phi) : \phi \in \Phi\} \ . \tag{20.16}$$

The set $\mathcal{D}$ is compact and convex by Theorem 20.1. Compactness here allowed Dubins and Spanier to conclude that a DS2-solution always exists. In fact, since $(x_1, \ldots, x_n) \mapsto \min_i x_i$ is a continuous mapping from the compact set $\mathcal{D}$ to $\Re$, then (DS2′) has a solution and, for the same reason, also (DS2″) admits a solution for each step $t \leq n$.

Some preliminary notions are required in order to locate the optimal value for the second problem. First of all, if we denote as $\mathbf{p}_i$ the vector with $1/s_i$ on the i-th place and 0 elsewhere, then

$$\text{conv}\ (\mathbf{p}_1, \ldots, \mathbf{p}_n) \subset \mathcal{D} \tag{20.17}$$

where "conv" denotes the convex hull. Also, in the following construction, we make repeated use of a famous theorem about convex sets, usually denoted as the *supporting hyperplane theorem* (see for instance [15], p. 32), where ∂C denotes the boundary of a set C.

THEOREM 20.4

If $C \subset \Re^n$ is a non-empty convex set with $\mathbf{z} \in \partial C$, then there exists a hyperplane

$$\mathcal{H} = \left\{ \mathbf{x} \in \Re^n : \sum_{j=1}^{n} w_j \, x_j = c \right\}$$

that contains $\mathbf{z}$ and such that C is contained in one of the half-spaces determined by $\mathcal{H}$, i.e.,

$$\sum_{j=1}^{n} w_j \, x_j \leq \sum_{j=1}^{n} w_j \, z_j = c \qquad \text{for all } \mathbf{x} \in C.$$

To plot a DS2-optimal solution on $\mathcal{D}$, we define an iterative scheme. First of all, let $J_1 = \{1, 2, \ldots, n\}$ and, for any $I \subset J_1$, define $\Delta(I)$ to be the "marginal" simplex on I, i.e., the set of non-negative values, indexed by I, which sum to 1

$$\Delta(I) = \left\{ (v_j)_{j \in I} : v_j \geq 0 \; \forall j \in I, \sum_{j \in I} v_j = 1 \right\} . \tag{20.18}$$

Now consider the equitable allocation on $\mathcal{D}$ with highest value

$$y_1 = \sup \{ y : (y, \ldots, y) \in \mathcal{D} \}$$

The above set is non-empty since $(1, \ldots, 1) \in \mathcal{D}$ by (20.17). Defining $\mathbf{y}_1^*$ as $(y_1, \ldots, y_1)$, we have

$$\mathbf{y}_1^* \in \partial \mathcal{D} \subset \mathcal{D}$$

since $\mathcal{D}$ is a closed set, as a result of Theorem 20.1, part (i). By Theorem 20.4, there exists a hyperplane

$$\mathcal{H}_1 = \left\{ \mathbf{x} : \sum_{j \in J_1} u_{1j} \, x_j = b_1 \right\} \tag{20.19}$$

which supports $\mathcal{D}$ in $\mathbf{y}_1^*$. A simple geometrical argument shows that we can choose the coefficients so that $\mathbf{u}_1 = (u_{1j})_{j \in J_1} \in \Delta(J_1)$ for all j and $\sum_j u_{1j} = 1$ (consequently $b_1 = y_1 > 0$). Now, if $u_{1j} > 0$ for all $j \in J_1$, we claim that $\mathbf{y}_1^*$ is the DS2-optimal solution. Otherwise, we group all the null component indices for u_{1j} into

$$J_2 = \{j \in J_1 : u_{1j} = 0\}.$$

Next, we consider the subspace $\mathcal{S}_2$, with dimension $|J_2|$, in which the components with indices corresponding to $J_1 \setminus J_2$ are set equal to y_1, while the other components are unrestricted

$$\mathcal{S}_2 = \{\mathbf{x} \in \Re^n : x_j = y_1 \text{ for all } j \in J_1 \setminus J_2\}$$

and consider the set

$$\mathcal{D}_2 = \mathcal{D} \bigcap \mathcal{S}_2 .$$

We now look for the allocation in $\mathcal{D}_2$ with the highest equitable value for the unrestricted coordinates

$$y_2 = \sup\{y : \mathbf{x} \in \mathcal{D}_2, \, x_j = y \text{ for all } j \in J_2\} .$$

Since $\mathbf{y}_1^*$, then $y_2 \geq y_1$. Now define $\mathbf{y}_2^*$ with components

$$y_{2j}^* = \begin{cases} y_1 & \text{if } j \in J_1 \setminus J_2 \\ y_2 & \text{if } j \in J_2 . \end{cases}$$

Also, $\mathbf{y}_2^* \in \partial\mathcal{D}_2 \subset \mathcal{S}_2$ and we can define a hyperplane, this time in $\mathcal{S}_2$, as

$$\mathcal{H}_2 = \left\{ \mathbf{x} \in \mathcal{S}_2 : \sum_{j \in J_2} u_{2j} \, x_j = b_2 \right\} . \tag{20.20}$$

Without loss of generality, we assume $\mathbf{u}_2 = (u_{2j})_{j \in J_2} \in \Delta(J_2)$ and $b_2 = y_2 > 0$. In case $u_{2j} > 0$ for all $j \in J_2$, we show next that $\mathbf{y}_2^*$ is the DS2-optimal solution. Otherwise, we continue iteratively, by defining r subsets of indices $J_1 \supset J_2 \supset \cdots \supset J_r$, and two sequences of r vectors $\mathbf{u}_1, \ldots, \mathbf{u}_r$ and $\mathbf{y}_1^*, \ldots, \mathbf{y}_r^*$. For any $t = 2, \ldots, r$, $\mathbf{u}_{t+1}$, J_{t+1} and $\mathbf{y}_{t+1}$ can be derived from $\mathbf{u}_t$, J_t and $\mathbf{y}_t^*$ in the following manner. The null component indices of $\mathbf{u}_t$ are grouped in J_{t+1} and the following subspace of $\Re^n$ with dimension $|J_{t+1}|$ is considered

$$\mathcal{S}_{t+1} = \left\{\mathbf{x} \in \Re^n : x_j = y_{tj}^* \text{ for all } j \in J_1 \setminus J_{t+1}\right\}$$

and $\mathcal{D}_{t+1} = \mathcal{D} \cap \mathcal{S}_{t+1}$. We then pick the highest possible allocation which is equitable in the unrestricted coordinates, i.e., we choose

$$y_{t+1} = \sup\{y : \mathbf{x} \in \mathcal{D}_{t+1}, x_j = y \text{ for all } j \in J_{t+1}\}$$

and define $\mathbf{y}^*_{t+1}$ with components

$$y^*_{t+1,j} = \begin{cases} y^*_{tj} & \text{if } j \in J_1 \setminus J_{t+1} \\ y_{t+1} & \text{otherwise.} \end{cases}$$

Now consider the hyperplane supporting $\mathcal{D}_{t+1}$ in $\mathbf{y}^*_{t+1}$ in the space $\mathcal{S}_{t+1}$. As in the previous steps, this hyperplane has linear coefficients $\mathbf{u}_{t+1} = (u_{t+1,j})_{j \in J_{t+1}} \in \Delta(J_{t+1})$. If all coefficients are positive, then the search for the solution stops since $\mathbf{y}^*_{t+1}$ is the optimal value. Otherwise, we must continue for at least one more step.

At the end of this iterative process, we will have a vector $\mathbf{y}^*_r$ and a sequence of indices $J_r \subset \cdots \subset J_1$ that fully characterize the solution for (DS2).

PROPOSITION 20.1

$\mathbf{y}^*_r$ is the DS2-optimal value. Moreover, if

$$|J_t \setminus J_{t+1}| = q_t \qquad t = 1, \ldots, r$$

with $J_{r+1} = \emptyset$ and $\sum_t q_t = n$, then

$$\alpha^*_{(1)} = \cdots = \alpha^*_{(q_1)} = y_1 \tag{20.21}$$

and, for each h, $2 \leq h \leq r$

$$\alpha^*_{(q_1+\cdots+q_{h-1}+1)} = \cdots = \alpha^*_{(q_1+\cdots+q_h)} = y_h. \tag{20.22}$$

PROOF Let $\boldsymbol{\alpha}^* = (\alpha^*_1, \ldots, \alpha^*_n) \in \mathcal{D}$ be the DS2-optimal value. By definition, $\boldsymbol{\alpha}^*$ dominates $\mathbf{y}^*_1$, i.e.,

$$\alpha^*_j \geq y_{1j} \qquad \forall j \in J_1. \tag{20.23}$$

Now the hyperplane $\mathcal{H}_1$ supports $\mathcal{D}$ in $\mathbf{y}^*_1$, so

$$\sum_{j \in J_1} u_{1j}\, \alpha^*_j \leq y_1 . \tag{20.24}$$

From (20.23), (20.24) and the definition of $\mathbf{y}_1^*$ we conclude,

$$\begin{array}{ll} \alpha_j^* = y_1 & \forall j \in J_1 \setminus J_2 \\ \alpha_j^* \geq y_1 & \forall j \in J_2 \; . \end{array} \tag{20.25}$$

Therefore, if $|J_1 \setminus J_2| = q_1$, equation (20.21) is satisfied.

Now continue inductively and suppose that (20.22) holds for every $h \leq t$. These equations yield two straightforward consequences

$$\alpha_j^* \geq y_{t+1,j}^* \qquad \forall j \in J_{t+1} \tag{20.26}$$

$$\boldsymbol{\alpha}^* \in \mathcal{D}_{t+1}. \tag{20.27}$$

Since $|J_{t+1} \setminus J_{t+2}| = q_{t+1}$, it follows that

$$\alpha^*_{(q_1+\cdots+q_t+1)} = \cdots = \alpha^*_{(q_1+\cdots+q_{t+1})} = y_{t+1} \; .$$

∎

20.5 A Procedure for the Second Dubins-Spanier Problem

The geometric framework of the previous section provides us with a constructive procedure that returns a DS2-optimal allocation. A more detailed account of this method can be found in [5] where the symmetric case $s_1 = \cdots = s_n = 1/n$ is considered. All the arguments used in [5] can be adapted to the more general situation with arbitrary weights described here, by an appropriate scaling of the densities, i.e., by considering

$$h_j = \frac{f_j}{s_j} \qquad \text{for all } j \in J_1 \tag{20.28}$$

in place of f_j throughout the procedure described in [5].

Much of the notation is the same as the previous section. Also, the procedure executes the same number (r) of steps as the geometrical construction. Beginning with step 1, we first identify the hyperplane $\mathcal{H}_1$, defined in (20.19), and then we characterize all allocations belonging to $\mathcal{H}_1$.

First of all, we let $F_1 = \Omega$. Now, for any $\mathbf{v}_1 = (v_{1j})_{j \in J_1} \in \Delta(J_1)$ and $x \in F_1$, we consider

$$\psi_1(x; \mathbf{v}_1) = \max_{j \in J_1} v_{1j} \, h_j(x)$$

and define $g_1 : \Delta(J_1) \to \Re_+$ as

$$g_1(\mathbf{v}_1) = \int_{F_1} \psi_1(x; \mathbf{v}_1)\, \nu(dx)\,. \tag{20.29}$$

It is easy to see that g_1 is convex and admits at least one minimizing argument (see Propositions 6.1 and 6.2 in [5]). Also, if $\mathbf{u}_1$ is the vector of coefficients characterizing $\mathcal{H}_1$, then

$$\mathbf{u}_1 \in \operatorname{argmin} \{g_1(\mathbf{v}) : \mathbf{v} \in \Delta(J_1)\}\,. \tag{20.30}$$

To verify (20.30), it can be shown by an argument similar to Theorem 20.3 that, for each $\mathbf{v} = (v_j) \in \Delta(J_1)$, the hyperplane

$$\sum_{j=1}^{n} v_j\, x_j = g_1(\mathbf{v}) \tag{20.31}$$

supports the set $\mathcal{D}$. This hyperplane contains the following point with equal coordinates

$$(g_1(\mathbf{v}), \ldots, g_1(\mathbf{v}))\ .$$

Since the hyperplane defined by (20.31) supports $\mathcal{D}$, while $\mathbf{y}_1^* \in \mathcal{D}$, it follows that

$$y_1^* \leq g_1(\mathbf{v}) \qquad \text{for all } \mathbf{v} \in \Delta(J_1)\,.$$

Equality here is attained only by those simplex vectors that identify an hyperplane supporting $\mathcal{D}$ in $\mathbf{y}_1^*$. Since the hyperplane $\mathcal{H}_1$ has these properties, we must conclude that (20.30) holds.

Once $\mathcal{H}_1$ is determined by minimizing the function g_1, we proceed by describing all the allocations that touch $\mathcal{H}_1$. Let

$$G_1 = \{x \in F_1 : \psi_1(x; \mathbf{u}_1) > 0\}$$

and for any $x \in G_1$ set

$$K_1(x) = \{j \in J_1 : u_j\, h_j(x) = \psi_1(x; \mathbf{u}_1)\}\,.$$

Once again, by an argument similar to Theorem 20.3, it can be shown that any allocation satisfying

$$\sum_{j \in K_1(x)} \phi_j(x) = 1 \qquad \text{for all } x \in G_1 \tag{20.32}$$

is such that $\mathbf{d}(\phi) \in \mathcal{H}_1 \cap \mathcal{D}$. Exactly as in the geometrical description of the last section, we now have to check whether $\mathbf{u}_1$ contains null elements, i.e., if J_2 is empty. Since $\mathbf{y}_1^* \in \mathcal{H}_1 \cap \mathcal{D}$, there exists one allocation satisfying (20.32) for which $\mathbf{d}(\phi)$ has components in $J_1 \setminus J_2$ with equitable values,

$$\frac{a_{ii}(\phi)}{s_i} = y_1^* \qquad \text{for all } i \in J_1 \setminus J_2 \; . \tag{20.33}$$

In case $J_2 = \emptyset$, by Proposition 20.1, an allocation satisfying (20.32) and (20.33) is DS2 optimal with value $\mathbf{y}_1^*$. Otherwise, we proceed with step 2.

Any allocation that satisfies (20.32) and (20.33) is not determined for the points outside G_1, and all the players in J_2 receive no part of G_1. Therefore, in this step we allocate the points outside G_1 to the participants in J_2. As anticipated by the geometric construction, however, the complete allocation of the "cake" Ω may require further additional steps.

To illustrate step 2 in detail, let

$$F_2 = G_1^c$$

and, for any $\mathbf{v}_2 = (v_{2j})_{j \in J_2} \in \Delta(J_2)$ and any $x \in F_2$, let

$$\psi_2(x; \mathbf{v}_2) = \max_{j \in J_2} v_{2j}\, h_j(x) \; .$$

Define now $g_2 : \Delta(J_2) \to \Re_+$ as

$$g_2(\mathbf{v}_2) = \int_{F_2} \psi_2(x; \mathbf{v}_2)\, \nu(dx) \; .$$

Once again, g_2 is convex, and its minimum is achieved. So, with the same arguments as step 1, we assert that the hyperplane $\mathcal{H}_2$, defined in (20.20), is characterized by the vector $\mathbf{u}_2$ verifying

$$\mathbf{u}_2 \in \text{argmin } \{g_2(\mathbf{v}) : \mathbf{v} \in \Delta(J_2)\} \; .$$

We next identify the allocations with values on $\mathcal{H}_2$. Let

$$G_2 = \{x \in F_2 : \psi_2(x; \mathbf{u}_2) > 0\}$$

and, for any $x \in G_2$, let

$$K_2(x) = \{J \in J_2 : u_{2j}\, h_j(x) = \psi_2(x; \mathbf{u}_2)\} \; .$$

Any allocation satisfying (20.32), (20.33) and such that

$$\sum_{j \in K_2(x)} \phi_j(x) = 1 \qquad \text{for all } x \in G_2 \tag{20.34}$$

has value on $\mathcal{H}_2$. Among these allocations, we pick the one which has equitable values for the components on $J_2 \setminus J_3$

$$\frac{a_{ii}(\boldsymbol{\phi})}{s_i} = y_2^* \qquad \text{for all } i \in J_2 \setminus J_3 \,. \tag{20.35}$$

In case $J_3 = \emptyset$, any allocation satisfying (20.32), (20.33), (20.34) and (20.35) is the desired DS2-optimal solution. It should be now clear how to proceed in case more steps are needed. We briefly describe how to move from step t to step $t+1$, provided $t < r$, with the purpose of describing the process in a form which is closer to an algorithmic development.

Suppose that step t has yielded two sets, F_t and G_t and an allocation $\boldsymbol{\phi}^*$ such that $d(\boldsymbol{\phi}^*) = \mathbf{y}_t^*$, but $J_{t+1} \neq \emptyset$. Then we proceed as follows:

- Define $F_{t+1} = F_t \setminus G_t = \bigcap_{i=1}^t G_i^c$
- For any $\mathbf{v} \in \Delta(J_{t+1})$
  - define $\psi_{t+1}(x; \mathbf{v}) = \max_{j \in J_{t+1}} u_{t+1,j}\, h_j(x)$
  - define $g_{t+1}(\mathbf{v}) = \int_{F_{t+1}} \psi_{t+1}(x; \mathbf{v})\, \nu(dx)$
- Choose $\mathbf{u}_{t+1} \in \operatorname{argmin} \{g_{t+1}(\mathbf{v}) : \mathbf{v} \in \Delta(J_{t+1})\}$
- Define
  - the set $G_{t+1} = \{x \in F_{t+1} : \psi_{t+1}(x; \mathbf{u}_{t+1}) > 0\}$
  - and $K_{t+1}(x) = \{j \in J_{t+1} : u_{t+1,j}\, h_j(x) = \psi_{t+1}(x; \mathbf{u}_{t+1})\}$
- Pick an allocation $\boldsymbol{\phi}^*$ such that

$$\sum_{j \in K_{t+1}(x)} \phi_j^*(x) = 1 \qquad \text{for all } x \in G_{t+1} \tag{20.36}$$

  and it is equitable in $J_{t+1} \setminus J_{t+2}$
- If $J_{t+2} \neq \emptyset$, repeat these substeps; otherwise, pick an allocation that satisfies (20.36) for each step i, $i = 1, \ldots, t+1$, and it is equitable within each step.

The above procedure makes use of a geometrical duality structure which has a long history in decision theory and has already been used by some authors in fair division. In particular, Elton, Hill and Kertz in [8] and, more explicitly, Legut and Wilczyński in [12] have applied duality theory to find the minimax partition, i.e., a partition that achieves $\alpha^*_{(1)}$ defined in (DS2′). This is, in essence, step 1 of the present algorithm.

Example 20.2

Suppose $\Omega = [0,2]$, $\mathcal{G}$ = Borels and 5 participants express their preferences on Ω through the following probability measures: μ_1 and μ_2 are two Beta distributions with parameters given, respectively, by $(2,5)$ and $(8,3)$. μ_3 is a uniform distribution on $[0.9, 1.1]$ while μ_4 and μ_5 are other two Beta distributions, shifted on $[1,2]$, with parameters $(2,7)$ and $(4,2)$, respectively. Suppose also that the vector of weights is given by:

$$\mathrm{s} = (0.175, 0.125, 0.25, 0.25, 0.2).$$

According to the algorithm now illustrated, we set $J_1 = \{1,2,3,4,5\}$, $F_1 = [0,2]$ and we look for the vector in $\Delta(J_1)$ which minimizes g_1. We obtain the following solution:

$$\mathrm{u}_1 = (0, 0, 0.3253, 0.6200, 0.0547)$$

which implies $G_1 = [0.9, 2]$ and $J_2 = \{1,2\} \neq \emptyset$. Therefore, in step 1 we distribute the set G_1 to participants 3, 4 and 5, according to the rule defined in (20.32). All such allocations take the following form:

$$\phi_3(x) = I_{C_3}(x) \qquad \phi_4(x) = I_{C_4}(x) \qquad \phi_5(x) = I_{C_5}(x) \tag{20.37}$$

with

$$C_3 = [0.9, 1.0746) \qquad C_4 = [1.0746, 1.5790) \qquad C_5 = [1.5790, 2]\,. \tag{20.38}$$

These allocations also satisfy (20.33).

In step 2, the remaining set, $F_2 = F_1 \setminus G_1 = [0, 0.9)$, must be assigned to persons 1 and 2. The function g_2 is minimized by the following simplex vector

$$\mathbf{u}_2 = (0.8949, 0.1051)$$

leading to $G_2 = F_2$ and $J_3 = \emptyset$. So there are no further steps of the procedure, and the remaining part of the cake is assigned the first two persons according to the rule (20.34), which yields:

$$\phi_1(x) = I_{C_1}(x) \qquad \phi_2(x) = I_{C_2}(x) \tag{20.39}$$

with

$$C_1 = [0, 0.6370) \qquad C_2 = [0.6370, 0.9). \tag{20.40}$$

The optimal allocation (actually a partition) defined by (20.37), (20.38), (20.39) and (20.40) yields the following allocation matrix:

$$A(\phi) = \begin{pmatrix} 0.9736 & 0.0263 & 0.0001 & 0 & 0 \\ 0.2344 & 0.6954 & 0.0702 & 0 & 0 \\ 0 & 0 & 0.8729 & 0.1271 & 0 \\ 0 & 0 & 0.1153 & 0.8729 & 0.0118 \\ 0 & 0 & 0.0001 & 0.3016 & 0.6983 \end{pmatrix}.$$

The weighted allocation matrix is now given by

$$W(\phi) = \begin{pmatrix} 5.5635 & 0.2107 & 0.0002 & 0 & 0 \\ 1.3392 & 5.5635 & 0.2808 & 0 & 0 \\ 0 & 0 & 3.4915 & 0.5084 & 0 \\ 0 & 0 & 0.4611 & 3.4915 & 0.0592 \\ 0 & 0 & 0.0005 & 1.2062 & 3.4915 \end{pmatrix}.$$

Two remarks can be made on the weighted allocation matrix. The optimal solution (on the principal diagonal of W) is not equitable, but its values are grouped in clusters of equitable values, as pointed out by Proposition 20.1. Also, if we look at the off-diagonal elements of W we note that the allocation is envy-free, but not strongly envy-free. An open question is to establish whether envy-freeness is a property always shared by all the optimal solutions of (DS2). $\square$

20.6 Optimization Problems with Constraints

The criteria set forth by Dubins and Spanier do not take into explicit account envy-free demands. This is especially important for (DS1) which may yield a solution that is neither fair nor envy-free, as the Example 20.1 shows. We now consider three optimization problems which redefine (DS1) by imposing constraints that force the solution to be fair, envy-free or strongly envy-free, respectively. Unlike the previous definitions, we consider two different systems of weights. The vector $\mathbf{s}$ defines the class of fair (resp. envy-free, strongly envy-free) allocations we restrict our search upon, while vector $\boldsymbol{\xi} = (\xi_1, \ldots, \xi_m)$, with the usual assumptions $\xi_i > 0$ and $\sum_i \xi_i = 1$, measures the importance of each player in the linear objective function.

Hence, we aim at solving the following type of optimization problems:

$$\text{Find } \phi \in \mathcal{I} \text{ that maximizes } \sum_{i=1}^{m} \xi_i \, a_{ii}(\phi) \tag{20.41}$$

where $\mathcal{I}$ can be any of the sets $\mathcal{F}(\mathbf{s})$, $\mathcal{E}(\mathbf{s})$ or $\tilde{\mathcal{E}}(\mathbf{s})$, respectively. In order to develop a solution method for those constrained problems it is better to write down the objective function, as well as the constraints, explicitly in terms of the density functions. For all three problems we are examining the objective function is the same as that for (DS1), i.e.,

$$\sum_{i=1}^{n} \xi_i \int_{\Omega} \phi_i(x) \, f_i(x) \, \nu(dx) \, . \tag{20.42}$$

In case we want to find an optimal fair allocation, we need the allocation to satisfy

$$\int_{\Omega} \phi_i(x) \, f_i(x) \, \nu(dx) \geq s_i \text{ for all } i \, . \tag{20.43}$$

Let (FA) denote the problem defined by the objective function (20.42) and the constraints (20.43).

Similarly, if we are looking for an optimal envy-free allocation, we set the constraints to

$$\int_{\Omega} \left(\frac{\phi_j(x)}{s_j} - \frac{\phi_i(x)}{s_i}\right) f_i(x) \, \nu(dx) \leq 0 \qquad \text{for all } i \neq j \tag{20.44}$$

and refer to the optimization problem as (EF).

Finally, in case we are considering strongly envy-free allocations only, we define the constraints as

$$\int_{\Omega} \phi_i(x) \, f_j(x) \, \nu(dx) \leq s_i \qquad \text{for all } i \neq j \, . \tag{20.45}$$

We denote this problem as (ST). Let w_{FA} (resp. w_{EF}, w_{ST}) denote the optimal value for the optimization problem (FA) (resp. (EF), (ST)). Let also w_{DS1} denote the optimal value for (20.42), i.e., the first Dubins-Spanier problem without constraints. In case the problems share the same system of weights, we have

$$w_{DS1} \geq w_{FA} \geq w_{EF} \geq w_{ST}.$$

As outlined in Section 20.3, (ST) is particularly sensible in the statistical interpretation of fair division theory. In fact, problem (ST) was first tackled in the work [1] by J.A. Anderson on discriminant analysis. Its statement, however, can be traced back to the work of Neyman and Pearson [14]. Here the same problem, together with (FA) and (EF), is examined in the light of the general duality schemes introduced by Wagner and Stone in [19]. As a result of this process, Lagrangian dual problems for all three constrained optimization problems are made available. This also yields a procedure to compute an optimal solution in all three situations.

20.7 Some Available Results on Functional Lagrangians

In this section, some results on Lagrangian duality theory contained in [19] are stated and then applied to the problem under investigation. Let X be a measurable space, and for each $x \in X$, let $Y(x)$ be a non-empty set. Define

$$\Upsilon = \{(x, y) : x \in X \text{ and } y \in Y(x)\} .$$

Fix d real-valued functions $c_1, \ldots, c_d$ and e on Υ, and fix a measure τ over X. τ has *finite substance* if, for each measurable $P \subset X$ for which $\tau(P) > 0$, there exists a measurable $P' \subset P$ such that $0 < \tau(P') < +\infty$. A σ-finite measure has finite substance.

For $x \in X$, we denote by $\mathbf{c}(x, \cdot)$ the vector function mapping y into $(c_1(x, y), \ldots, c_d(x, y))$ for $y \in Y(x)$. If $q(x) \in Y(x)$ for $x \in X$, $\mathbf{c}(\cdot, q(\cdot))$ is the function mapping x into $\mathbf{c}(x, q(x))$ for $x \in X$ and, similarly, for $e(\cdot, q(\cdot))$. We define

$$\mathcal{Q} = \{q : q(x) \in Y(x) \; \forall x \in X, \; \mathbf{c}(\cdot, q(\cdot)) \text{ and } e(\cdot, q(\cdot)) \text{ are integrable}\} ,$$

$$C_i(q) = \int_X c_i(x, q(x)) \, d\tau(x) \text{ for } i = 1, \ldots, d \quad q \in \mathcal{Q} ,$$

$$E(q) = \int_X e(x, q(x)) \, d\tau(x) \text{ for } q \in \mathcal{Q} .$$

Suppose $\boldsymbol{\lambda} \in \Re^d_+$ and $\mathbf{C} = (C_1, \ldots, C_d)$. We define the *pointwise Lagrangian*, ℓ_λ, as

$$\ell_\lambda(x, y) = e(x, y) - \boldsymbol{\lambda} \cdot \mathbf{c}(x, y) \qquad \text{for } (x, y) \in \Upsilon$$

and the *functional Lagrangian*, L_λ, by

$$L_\lambda(q) = E(q) - \boldsymbol{\lambda} \cdot \mathbf{C}(q) \qquad \text{for } q \in \mathcal{Q} .$$

We say that the functionals $\mathbf{C}$ and $-E$ are *convex* if

$$\hat{\mathcal{Q}} = \mathcal{Q} \cap \{q : C_i(q) < +\infty \ \forall i \text{ and } E(q) > -\infty\}$$

is a convex subset of a real vector space and, whenever $0 \leq \beta \leq 1$ and $q, p \in \hat{\mathcal{Q}}$, we have

$$\mathbf{C}(\beta\, q + (1-\beta)\, p) \leq \beta\, \mathbf{C}(q) + (1-\beta)\mathbf{C}(p)$$

$$E(\beta\, q + (1-\beta)\, p) \geq \beta\, E(q) + (1-\beta)E(p) .$$

The *range* of a function is denoted as $\operatorname{rng}(f)$, while $\operatorname{int} A$ denotes the interior of a set A. We now list the statements of the most relevant results about Lagrangian duality which can be found in [19] (Theorem 3.2).

THEOREM 20.5
Suppose τ has finite substance, $\mathbf{v} \in \Re^d_+$, $\mathbf{u} \in \operatorname{rng} \mathbf{C}$, $\mathbf{u} < \mathbf{v}$, $q^* \in \mathcal{Q}$, $E(q^*)$ and $\mathbf{C}(q^*)$ are finite and $\mathbf{C}(q^*) \leq \mathbf{v}$. Suppose also that either (a) τ is atomless or (b) $\mathbf{C}$ and $-E$ are convex. Then the following are equivalent:

(i) $E(q^*) = \max\{E(q) : \mathbf{C}(q) \leq \mathbf{v}\}$

(ii) *there exists $\boldsymbol{\lambda} \in \Re^d_+$ such that, whenever $q \in \mathcal{Q}$ and $E(q)$ and $\mathcal{C}(q)$ are finite, the following holds:*
For $\boldsymbol{\eta} \in \Re^d_+$

$$L_\lambda(q) + \boldsymbol{\lambda} \cdot \mathbf{v} \leq L_\lambda(q^*) + \boldsymbol{\lambda} \cdot \mathbf{v} \leq L_\eta(q^*) + \boldsymbol{\eta} \cdot \mathbf{v} \tag{20.46}$$

$$\boldsymbol{\lambda} \cdot [\mathbf{v} - \mathbf{C}(q^*)] = 0. \tag{20.47}$$

Theorem 20.5 suggests a practical method that helps us find the optimal $\boldsymbol{\lambda}$ and q^* characterizing (20.46) and (20.47).

PROPOSITION 20.2
Suppose that (20.46) and (20.47) of Theorem 20.5 hold, then

$$E(q^*) = L_\lambda(q^*) + \boldsymbol{\lambda} \cdot \mathbf{v} = \inf_{\boldsymbol{\eta} \in \Re^d_+} \sup_{q \in \mathcal{Q}} L_\eta(q) + \boldsymbol{\eta} \cdot \mathbf{v}. \tag{20.48}$$

Moreover

$$\mathbf{C}(q^*) \leq \mathbf{v} \,. \tag{20.49}$$

PROOF The first equality in (20.48) is obtained by recalling the definition of functional Lagrangian and (20.47).

To complete the proof of (20.48), we apply the first inequality in (20.46) which shows the existence of some $\boldsymbol{\lambda}$ for which

$$\sup_{q \in \mathcal{Q}} L_\lambda(q) + \boldsymbol{\lambda} \cdot \mathbf{v} \leq L_\lambda(q^*) + \boldsymbol{\lambda} \cdot \mathbf{v}$$

and therefore

$$\inf_{\boldsymbol{\eta} \in \Re^d_+} \left[\sup_{q \in \mathcal{Q}} L_\eta(q) + \boldsymbol{\eta} \cdot \mathbf{v} \right] \leq L_\lambda(q^*) + \boldsymbol{\lambda} \cdot \mathbf{v}. \tag{20.50}$$

On the other hand, by the second inequality in (20.46), we have

$$\sup_{q \in \mathcal{Q}} L_\eta(q) + \boldsymbol{\eta} \cdot \mathbf{v} \geq L_\eta(q^*) + \boldsymbol{\eta} \cdot \mathbf{v} \geq L_\lambda(q^*) + \boldsymbol{\lambda} \cdot \mathbf{v} \qquad \forall \boldsymbol{\eta} \in \Re^d_+$$

and therefore

$$\inf_{\boldsymbol{\eta} \in \Re^d_+} \sup_{q \in \mathcal{Q}} L_\eta(q) + \boldsymbol{\eta} \cdot \mathbf{v} \geq L_\lambda(q^*) + \boldsymbol{\lambda} \cdot \mathbf{v}. \tag{20.51}$$

Together (20.50) and (20.51) imply the second equality in (20.48). Furthermore, by (20.47) we know that if $\lambda_i > 0$ then $C_i(q) - v_i = 0$, so (20.49) is satisfied for these indices. Suppose instead that $\lambda_j = 0$ and $C_j(q) > v_j$ hold simultaneously for some j. If we consider $\boldsymbol{\eta}$ as the vector $\boldsymbol{\lambda}$ with the j-th component replaced by some $\eta_j > 0$, then the following inequality contradicts (20.46)

$$L_\eta(q^*) + \boldsymbol{\eta} \cdot \mathbf{v} = E(q^*) - \eta_j \, (C_j(q^*) - v_j) < E(q^*) = L_\lambda(q^*) + \boldsymbol{\lambda} \cdot \mathbf{v}$$

which establishes (20.49). ∎

20.8 Lagrangian Duality in Fair Division

The results of the last section can be applied to the fair division problems (FA), (EF) and (ST) when we consider $\tau = \nu$, $\mathcal{Q} = \Phi$, $X = \Omega$, $y = (t_1, \ldots, t_n) \in Y(x)$ with $t_j \geq 0$ and $\sum_j t_j = 1$.

First of all we focus on the objective function of all three problems, which is obtained by setting

$$e(x, y) = \sum_{i=1}^{n} \xi_i \, t_i \, f_i(x) \,. \tag{20.52}$$

Consequently, for a given (measurable) $\phi(x) = (\phi_1(x), \ldots, \phi_n(x)) \in Y(x)$, we have

$$E(\phi) = \int_\Omega \sum_{i=1}^{n} \xi_i \, \phi_i(x) \, f_i(x) \, \nu(dx) \,. \tag{20.53}$$

We now consider problem (FA) by setting

$$c_i^{FA}(x, y) = -t_i \, f_i(x) \qquad i = 1, \ldots, n$$

which, integrated over Ω, leads to

$$C_i^{FA}(\phi) = -\int_\Omega \phi_i(x) \, f_i(x) \, \nu(dx) \qquad i = 1, \ldots, n \,.$$

Problem (FA) then takes the following more concise form

$$\sup_{\phi \in \Phi} E(\phi)$$

$$\text{with } \phi \quad \text{such that} \quad \mathbf{C}^{FA}(\phi) \leq -\mathrm{s} \,,$$

where $\mathbf{C}^{FA} = (C_1^{FA}, \ldots, C_n^{FA})$. For fixed $\phi \in \Phi$ and $\boldsymbol{\lambda} \in \Re_+^n$, the functional Lagrangian is

$$L_\lambda^{FA}(\phi) = \sum_{i=1}^{n} \xi_i \int_\Omega \phi_i(x) \, f_i(x) \, \nu(dx) + \sum_{i=1}^{n} \lambda_i \int_\Omega \phi_i(x) \, f_i(x) \, \nu(dx) =$$

$$\int_\Omega \sum_{i=1}^{n} \phi_i(x) \, f_i(x) \, (\xi_i + \lambda_i) \, \nu(dx) \,.$$

Now $\hat{\Phi} = \Phi$, while $\mathbf{C}^{FA}$ and E are linear functions of ϕ. Therefore $\mathbf{C}^{FA}$ and $-E$ are convex (one of the sufficient conditions of Theorem 20.5). Moreover we know, by Theorem 20.2, that when all measures are not identical, then there exists a super-fair allocation. Therefore, we can apply the general results of the previous section for the problem (FA) in fair division.

PROPOSITION 20.3
If there exists C and two measures for which $\mu_i(C) \neq \mu_j(C)$, then the following statements are equivalent:

(i) $E(\phi^*) = \max\left\{E(\phi) : \phi \in \Phi \text{ and } \mathbf{C}^{FA}(\phi) \leq -\mathbf{s}\right\}$
with $\mathbf{C}^{FA}(\phi^*) \leq -\mathbf{s}$

(ii) *there exists* $\boldsymbol{\lambda} \in \Re^n$ *such that, for* $\phi \in \Phi$, *the following holds:*
For all $\boldsymbol{\eta} \in \Re^n$

$$L_{\lambda}^{FA}(\phi) - \boldsymbol{\lambda} \cdot \mathbf{s} \leq L_{\lambda}^{FA}(\phi^*) - \boldsymbol{\lambda} \cdot \mathbf{s} \leq L_{\eta}^{FA}(\phi^*) - \boldsymbol{\eta} \cdot \mathbf{s} \tag{20.54}$$

$$\boldsymbol{\lambda} \cdot \left(\mathbf{C}^{FA}(\phi^*) + \mathbf{s}\right) = 0. \tag{20.55}$$

PROOF This is a straightforward application of Theorem 20.5. The statement of this result is simpler than the original theorem since, by the Lyapunov Theorem, there always exists an allocation ϕ^* that satisfies (i). In more detail, since $\mathcal{A}$ is compact by part (i) of Theorem 20.1, the set

$$B = \{\mathbf{C}(\phi) : \phi \in \Phi \text{ and } \mathbf{C}(\phi) \leq -\mathbf{s}\} \subset \Re^n$$

also shares this property.

Now, since $(x_1, \dots, x_n) \mapsto \sum_i \xi_i x_i$ is a continuous mapping from $\Re^n$ to $\Re$, the supremum of $E(\cdot)$ is achieved on B.

Furthermore, by Theorem 20.2, if all measures are not identical, then there exists a super-fair allocation, or equivalently, in terms of the Lagrangian theory, a $\mathbf{u} \in \operatorname{rng} \mathbf{C}^{FA}$ with $\mathbf{u} < -\mathbf{s}$. ∎

We can also apply Proposition 20.2 which yields the following result:

COROLLARY 20.1
Under the hypotheses of Theorem 20.3,

$$w_{FA} = \inf_{\boldsymbol{\lambda} \in \Re_+^n} \sup_{\phi \in \Phi} L_{\lambda}^{FA}(\phi) - \boldsymbol{\lambda} \cdot \mathbf{s} \tag{20.56}$$

and any optimal allocation ϕ^{FA} reaching the optimal value w_{FA} in (20.56), also attains optimality in (FA).

To describe (20.56) more explicitly, let

$$r_i^{FA}(x) = f_i(x)\,(\xi_i + \lambda_i) \qquad i = 1, \ldots, n\,.$$

The dual problem can be written as:

$$\text{Find} \quad w_{FA} = \inf_{\boldsymbol{\lambda} \in \Re_+^n} h^{FA}(\boldsymbol{\lambda}) - \boldsymbol{\lambda} \cdot \mathrm{s} \tag{20.57}$$

with

$$h^{FA}(\boldsymbol{\lambda}) = \sum_{i=1}^{n} \int_\Omega \phi_i^{FA}(x)\, r_i^{FA}(x)\, \nu(dx)$$

and the allocation ϕ^{FA} defined by

$$\sum_{i \in K^{FA}(x)} \phi_i^{FA}(x) = 1 \qquad \text{a.s.}(\nu)$$

with

$$K^{FA}(x) = \left\{ i : r_i^{FA}(x) = \max_j r_j^{FA}(x) \right\}.$$

We can now turn our attention to problem (EF). We set $e(x, y)$ and $E(\phi)$ exactly as in (20.52) and (20.53). With regard to the constraints, we set

$$c_{ij}^{EF}(x, y) = \left(\frac{t_j}{s_j} - \frac{t_i}{s_i}\right) f_i(x) \qquad \text{for all } i \neq j\,. \tag{20.58}$$

Consequently

$$C_{ij}^{EF}(\phi) = \int_\Omega \left(\frac{\phi_j(x)}{s_j} - \frac{\phi_i(x)}{s_i} \right) f_i(x)\, \nu(dx) \qquad \text{for all } i \neq j\,. \tag{20.59}$$

Now arrange all the constraints in a single vector

$$\mathbf{C}^{EF} = (C_{12}^{EF}, C_{13}^{EF}, \ldots, C_{1,n}^{EF}, \ldots, C_{ij}^{EF}\ (i \neq j), \ldots, C_{n,1}^{EF}, \ldots, C_{n,n-1}^{EF}).$$

Problem (EF) assumes the following short form:

$$\text{Find} \ \sup_{\phi \in \Phi} E(\phi)$$

$$\text{with } \boldsymbol{\phi} \quad \text{such that} \quad \mathbf{C}^{EF}(\boldsymbol{\phi}) \leq \mathbf{0}$$

with $\mathbf{0} \in \Re^{(n-1)\times(n-1)}$. For fixed $\boldsymbol{\lambda} \in \Re^{(n-1)\times(n-1)}$ and $\boldsymbol{\phi} \in \Phi$, the functional Lagrangian is

$$L_\lambda^{EF}(\boldsymbol{\phi}) = \sum_i \xi_i \int_\Omega \phi_i(x)\, f_i(x)\, \nu(dx) -$$

$$\sum_{\forall i \neq j} \lambda_{ij} \int_\Omega \left(\frac{\phi_j(x)}{s_j} - \frac{\phi_i(x)}{s_i} \right) f_i(x) \nu(dx) =$$

$$\sum_i \int_\Omega \frac{\phi_i(x)}{s_i} \left[f_i(x) \left(s_i\, \xi_i + \sum_{j \neq i} \lambda_{ij} \right) - \sum_{h \neq i} \lambda_{hi}\, f_h(x) \right] \nu(dx)$$

Also in this case $\mathbf{C}^{EF}$ and $-E$ are convex, therefore, we can apply Proposition 20.5 and Proposition 20.2.

PROPOSITION 20.4

If there exists an envy-free allocation that satisfies all the defining inequalities (20.5) with strict inequality, then the following statements are equivalent:

(i) $E(\boldsymbol{\phi}^*) = \max\left\{E(\boldsymbol{\phi}) : \boldsymbol{\phi} \in \Phi \text{ and } \mathbf{C}^{EF}(\boldsymbol{\phi}) \leq \mathbf{0}\right\}$ *with* $\mathbf{C}^{EF}(\boldsymbol{\phi}^*) \leq \mathbf{0}$

(ii) *there exists* $\boldsymbol{\lambda} \in \Re_+^{(n-1)\times(n-1)}$ *such that, for* $\boldsymbol{\phi} \in \Phi$*, the following holds:*
For all $\boldsymbol{\eta} \in \Re_+^{(n-1)\times(n-1)}$

$$L_\lambda^{EF}(\boldsymbol{\phi}) \leq L_\lambda^{EF}(\boldsymbol{\phi}^*) \leq L_\eta^{EF}(\boldsymbol{\phi}^*) \tag{20.60}$$

$$\boldsymbol{\lambda} \cdot \mathbf{C}^{EF}(\boldsymbol{\phi}^*) = 0\,. \tag{20.61}$$

Moreover

$$w_{EF} = \inf_{\boldsymbol{\lambda} \in \Re_+^{(n-1)\times(n-1)}} \sup_{\boldsymbol{\phi} \in \Phi} L_\lambda^{EF}(\boldsymbol{\phi}) \tag{20.62}$$

and an optimal allocation for (20.62) is also optimal for (EF).

To derive an explicit expression for (20.62) let

$$r_i^{EF}(x) = \frac{1}{s_i} \left[f_i(x) \left(s_i\, \xi_i + \sum_{j \neq i} \lambda_{ij} \right) - \sum_{h \neq i} \lambda_{hi}\, f_h(x) \right]\,.$$

We then write the dual problem (20.62) as

$$\inf_{\boldsymbol{\lambda}} h^{EF}(\boldsymbol{\lambda}) \tag{20.63}$$

$$\text{with} \qquad h^{EF}(\boldsymbol{\lambda}) = \sum_i \int_\Omega \phi_i^{EF}(x)\, r_i^{EF}(x)\, \nu(dx)$$

and

$$\sum_{i \in K^{EF}(x)} \phi_i^{EF}(x) = 1 \quad \text{a.s.}(\nu)$$

$$\text{with} \quad K^{EF}(x) = \{i : r_i^{EF}(x) = \max_j r_j^{EF}(x)\} \,.$$

With a similar reasoning, we can find the Lagrangian dual problem for (ST). We consider the constraints

$$c_{ij}^{ST}(x, y) = t_j\, f_i(x) \qquad \text{for all } i \neq j \,. \tag{20.64}$$

Consequently, to a given $\boldsymbol{\phi}(x) = (\phi_1(x), \ldots, \phi_n(x)) \in Y(x)$, we associate

$$c_{ij}^{ST}(\boldsymbol{\phi}) = \int_\Omega \phi_j(x)\, f_i(x)\, \nu(dx) \qquad \text{for all } i \neq j \,. \tag{20.65}$$

Thus, problem (ST) can be transformed as

$$\text{Find} \ \sup_{\boldsymbol{\phi} \in \Phi} E(\boldsymbol{\phi})$$

$$\text{with } \boldsymbol{\phi} \in \Phi \quad \text{such that} \quad \mathbf{C}^{ST}(\boldsymbol{\phi}) \leq \tilde{\mathbf{s}} \,.$$

(Here $\mathbf{C}^{ST} = (C_{12}^{ST}, \ldots, C_{1,n}^{ST}, \ldots, C_{ij}^{ST}(i \neq j), \ldots, C_{n,1}^{ST}, \ldots, C_{n,n-1}^{ST})$ and $\tilde{\mathbf{s}} = (s_2, s_3, \ldots, s_n, s_1, s_3, \ldots, s_n, \ldots, s_2, \ldots, s_n))$.

The functional Lagrangian associated with (ST) can be written as

$$L^{ST}(\boldsymbol{\lambda}; \boldsymbol{\phi}) = \sum_i \int_\Omega \xi_i\, \phi_i(x)\, f_i(x)\, d\nu - \sum_{\forall\, i \neq j} \lambda_{ij} \left(\int_\Omega \phi_j(x)\, f_i(x)\, d\nu - s_j \right) =$$

$$\sum_{\forall\, i \neq j} s_j\, \lambda_{ij} + \sum_i \int_\Omega \phi_i(x) \left(\xi_i\, f_i(x) - \sum_{j \neq i} \lambda_{ji}\, f_j(x) \right) d\nu$$

with $\lambda_{ij} \geq 0\ \forall\, i \neq j$, $\boldsymbol{\lambda} = (\lambda_{1,2}, \ldots, \lambda_{1,n}, \ldots, \lambda_{n,1}, \ldots, \lambda_{n,n-1})$ and $\boldsymbol{\phi} \in \Phi$.

As for (FA) and (EF), Lagrangian duality can be used to solve this new optimization problem. In fact, we have:

PROPOSITION 20.5
Suppose there exists a super envy-free allocation, then the following statements are equivalent:

(i) $\sum_i \xi_i \, a_{ii}(\boldsymbol{\phi}^*) = \max\left\{\sum_i \xi_i \, a_{ii}(\boldsymbol{\phi}) : \boldsymbol{\phi} \in \tilde{\mathcal{E}}(\mathrm{s})\right\}$

(ii) *there exists* $\boldsymbol{\lambda} \in \Re_+^{(n-1)\times(n-1)}$ *such that, for* $\boldsymbol{\phi} \in \Phi$*, the following holds:*
For all $\boldsymbol{\eta} \in \Re_+^{(n-1)\times(n-1)}$

$$L_\lambda^{ST}(\boldsymbol{\phi}) + \tilde{\mathrm{s}} \cdot \boldsymbol{\lambda} \le L_\lambda^{ST}(\boldsymbol{\phi}^*) + \tilde{\mathrm{s}} \cdot \boldsymbol{\lambda} \le L_\eta^{ST}(\boldsymbol{\phi}^*) + \tilde{\mathrm{s}} \cdot \boldsymbol{\eta} \tag{20.66}$$

$$\boldsymbol{\lambda} \cdot \left(\tilde{\mathrm{s}} - \mathbf{C}^{ST}(\boldsymbol{\phi}^*)\right) = 0 \,. \tag{20.67}$$

Consequently,

$$w_{ST} = \inf_{\boldsymbol{\lambda} \in \Re_+^{(n-1)\times(n-1)}} \sup_{\boldsymbol{\phi} \in \Phi} \left(L_\lambda^{ST}(\boldsymbol{\phi}) + \tilde{\mathrm{s}} \cdot \boldsymbol{\lambda}\right)$$

Define now

$$r_i^{ST}(x) = \xi_i \, f_i(x) - \sum_{j \ne i} \lambda_{ij} \, f_j(x) \qquad i = 1, \ldots, n \,.$$

We re-write the dual problem as

$$\inf_{\boldsymbol{\lambda} \in \Re_+^{(n-1)\times(n-1)}} h^{ST}(\boldsymbol{\lambda}) + \boldsymbol{\lambda} \cdot \tilde{\mathrm{s}} \tag{20.68}$$

with

$$h^{ST}(\boldsymbol{\lambda}) = \sum_{i=1}^{n} \int_\Omega \phi_i^{ST}(x) \, r_i^{ST}(x) \, \nu(dx)$$

$$\sum_{i \in K^{ST}(x)} \phi_i^{ST}(x) = 1 \quad \text{a.s.}(\nu)$$

$$\text{and} \quad K^{ST}(x) = \left\{ i : r_i^{ST}(x) = \max_j r_j^{ST}(x) \right\} \,.$$

J.A. Anderson in [1] defines the same framework of Lagrangians for (ST) and states a set of conditions for the solution to be optimal in the case of equality constraints. The same author does the necessary computations for an example with 2 participants, but concludes that, for $n \geq 3$, his "analytic approach is usually too difficult, even for the multivariate normal distribution".

While for the first problem (FA) there is a very simple condition that guarantees the existence of super-fair allocations given by Theorem 20.2, we have not specified any practical sufficient condition that checks for the existence of an allocation ϕ which satisfies the constraints with strict inequality. The following result by Barbanel ([3], Corollary 2) can be employed in both problems, since it identifies conditions under which super envy-free allocations exist (which are also envy-free in the weaker sense). We consider the usual definition for linear independence of the densities: $f_1, \ldots, f_n$ are linearly independent if for any set of constants $d_1, \ldots, d_n$ such that $\sum_i d_i f_i \equiv 0$, we have $d_1 = \cdots = d_n = 0$.

THEOREM 20.6

Suppose $f_1, \ldots, f_n$ are linearly independent. Then there exists a super envy-free allocation.

We refer to the same paper to handle the more general case where linear dependencies among densities are allowed.

20.9 Constrained Optimization in Fair Division: An Example

Example 20.3

The same data are used throughout the following example to compare the behavior of the different constrained problems. All computations have been performed using the *Mathematica* programming language v.3.

Let $\Omega = [0, 1]$, $\mathcal{G}$ = Borels and let μ_1, μ_2 and μ_3 be three Beta distributions characterized by the following parameters, respectively, $\mu_1 = \text{Be}(4, 6)$, $\mu_2 = \text{Be}(7, 6)$ and $\mu_3 = \text{Be}(8, 4)$. We will be working with the density functions of these measures w.r.t. the Lebesgue measure. The system of weights for the linear objective function is

$$\xi_1 = 0.3 \qquad \xi_2 = 0.2 \qquad \xi_3 = 0.5$$

while the weights associated to the constraints are, respectively,

$$s_1 = 0.3 \qquad s_2 = 0.5 \qquad s_3 = 0.2 \, .$$

First of all, we look at the DS1-optimal allocation as described by Theorem 20.3. This is in fact a partition, and the same will be true for all the solutions that follow.

$$\phi^{DS1}(x) = \left(I_{B_1^d}(x), 0, I_{B_3^d}(x)\right)$$

with

$$B_1^d = [0, 0.4927) \qquad B_3^d = [0.4927, 1]$$

and optimal value $w_{DS1} = 0.6674$.

The corresponding allocation matrix is

$$A(\phi^{DS1}) = \begin{pmatrix} 0.7316 & 0. & 0.2684 \\ 0.3677 & 0. & 0.6323 \\ 0.1042 & 0. & 0.8958 \end{pmatrix} .$$

Person 2 receives no slice, so the allocation cannot be fair (nor envy-free in any sense). To force the solution to be fair, we compute a solution for (FA). The function (20.56) is minimized by the parameters

$$\lambda_2 = 0.2340 \qquad \lambda_1 = \lambda_3 = 0$$

and the corresponding optimal partition is

$$\phi^{FA}(x) = \left(I_{B_1^f}(x), I_{B_2^f}(x), I_{B_3^f}(x)\right)$$

with

$$B_1^f = [0, 0.3976) \qquad B_2^f = [0.3976, 0.5958) \qquad B_3^f = [0.5958, 1] \, .$$

The optimal value is now decreased to $w_{FA} = 0.6102$. The optimal partition yields the following allocation matrix (resp. weighted allocation matrix)

$$A(\phi^{FA}) = \begin{pmatrix} 0.5113 & 0.3846 & 0.1041 \\ 0.1539 & 0.5 & 0.3461 \\ 0.0282 & 0.2582 & 0.7136 \end{pmatrix}$$

$$W(\phi^{FA}) = \begin{pmatrix} 1.7042 & 0.7692 & 0.5208 \\ 0.5131 & 1. & 1.7304 \\ 0.0939 & 0.5164 & 3.5682 \end{pmatrix} .$$

Since $\lambda_2 > 0$, the constraint corresponding to person 2 is active, i.e., it attains the bound $s_2 = 0.5$. Now person 2 receives some part of the cake. This is enough to make him "happy" about the slice received, but not enough to avoid envy feelings, because in the weighted matrix, the second row is maximized by the third element. We therefore proceed by finding an optimal envy-free allocation as described by (EF). The function (20.63) is minimized by the following parameters:

$$\lambda_{23} = 0.0602 \qquad \lambda_{12} = \lambda_{13} = \lambda_{21} = \lambda_{31} = \lambda_{32} = 0,$$

which yield the following partition

$$\phi^{EF}(x) = \left(I_{B_1^e}(x), I_{B_2^e}(x), I_{B_3^e}(x)\right)$$

with

$$B_1^e = [0, 0.4399) \qquad B_2^e = [0.4399, 0.6477) \qquad B_3^e = [0.6477, 1]$$

and optimal value $w_{EF} = 0.5837$. The corresponding allocation matrices with or without weights are, respectively,

$$A(\phi^{EF}) = \begin{pmatrix} 0.6150 & 0.3297 & 0.0553 \\ 0.2378 & 0.5444 & 0.2178 \\ 0.0531 & 0.3661 & 0.5808 \end{pmatrix}$$

$$W(\phi^{EF}) = \begin{pmatrix} 2.0498 & 0.6595 & 0.2763 \\ 0.7927 & 1.0889 & 1.0889 \\ 0.1770 & 0.7323 & 2.9038 \end{pmatrix} .$$

As expected, since $\lambda_{23} > 0$, the constraint which refers to the second person judging the third slice is active. Therefore, in the weighted matrix this value is equal to the value actually received by person 2.

Finally, we consider the constrained problem (ST) with strongly envy-free allocations. The function (20.68) is minimized by:

$$\lambda_{21} = 0.0688 \qquad \lambda_{23} = 0.4594 \qquad \lambda_{12} = \lambda_{13} = \lambda_{31} = \lambda_{32} = 0 .$$

This solution corresponds to the following partition:

$$\phi^{ST}(x) = \left(I_{B_1^s}(x), I_{B_2^s}(x), I_{B_3^s}(x)\right)$$

with optimal value $w_{ST} = 0.5817$ and

$$B_1^s = [0, 0.4664) \qquad B_2^s = [0.4664, 0.6559) \qquad B_3^s = [0.6559, 1]\ .$$

The corresponding allocation matrix (without weights) is

$$A(\phi^{ST}) = \begin{pmatrix} 0.6758 & 0.2748 & 0.0494 \\ 0.3 & 0.5 & 0.2 \\ 0.0756 & 0.3664 & 0.5580 \end{pmatrix}.$$

Here the values of the second row are completely determined by the fact that both λ_{21} and λ_{23} are strictly positive. $\square$

Acknowledgments. The author wishes to thank professor Fabio Tardella for the kind advice on functional Lagrangian theory. The author would also like to thank Professor Theodore P. Hill for helpful advice on fair division theory and for giving him the opportunity to visit the School of Mathematics at the Georgia Institute of Technology, where part of the work was done. Grateful acknowledgment is made for hospitality.

This work is partially supported by a grant No. 203.10.39 from Consiglio Nazionale delle Ricerche — Italy.

References

[1] J.A. Anderson, Constrained discrimination between k populations, *J. Roy. Stat. Soc.*, Ser.B, **31** (1969), 123–139.

[2] T.W. Anderson, *An Introduction to Multivariate Statistical Analysis*, 2nd edition, (Wiley, 1984).

[3] J.B. Barbanel, On the possibilities for partitioning a cake, *Proceedings of the American Mathematical Society*, **124** (1996), 3443–3451.

[4] S.J. Brams and A. Taylor, *Fair Division* (Cambridge University Press, 1996).

[5] M. Dall'Aglio, The Dubins-Spanier optimization problem in fair division theory, preprint.

[6] L.E. Dubins and E.H. Spanier, How to cut a cake fairly, *American Mathematical Monthly*, **68** (1961), 1–17.

[7] A. Dvoretzky, A. Wald and J. Wolfowitz, Relations among certain ranges of vector measures, *Pacific Journal of Mathematics*, **1** (1951), 59–74.

[8] J. Elton, T.P. Hill and R.P. Kertz, Optimal-partitioning inequalities for nonatomic probability measures, *Transactions of the American Mathematical Society*, **296** (1986), 703–725.

[9] T.P. Hill, A sharp partitioning inequality for atomless probability measures based on the mass of the infimum of the measures, *Probability Theory and Related Fields*, **75** (1987), 143–147.

[10] T.P. Hill and Y.L. Tong, Optimal-partitioning inequalities in classification and multi-hypotheses testing, *The Annals of Statistics*, **17** (1989), 1325–1334.

[11] T.P. Hill, Partitioning inequalities in probability and statistics, in: M. Shaked and Y.L. Tong, eds., *Stochastic Inequalities, IMS Lecture Notes, Monograph Series*, Vol. 22, (1993) 116–132.

[12] J. Legut and M. Wilczyński, Optimal partitioning of a measurable space, *Proceedings of the American Mathematical Society*, **104** (1988), 262–264.

[13] A. Lyapunov, Sur les fonctions-vecteurs complétement additives, *Bulletin of the Academy of Sciences (URSS)*, **4** (1940), 465–478.

[14] J. Neyman and K. Pearson, Contributions to the theory of testing statistical hypotheses, I, *Stat. Res. Mem.*, **1** (1936), 1–37.

[15] M.J. Panik, *Classical Optimization: Foundations and Extensions*, North Holland, Amsterdam, 1976.

[16] C.R. Rao, *Linear Statistical Inference and its Applications*, 2nd edition, (Wiley, 1973).

[17] H. Steinhaus, The problem of fair division, *Econometrika*, **16** (1948), 101–104.

[18] H. Steinhaus, Sur la division pragmatique, *Econometrika (Supplement)*, **17** (1949), 315–319.

[19] D.H. Wagner and L.D. Stone, Necessity and existence results on constrained optimization of separable functionals by a multiplier rule, *SIAM J. Control*, **12** (1974), 356–372.

[20] D. Weller, Fair division of a measurable space, *Journal of Mathematical Economics*, **14** (1985), 5–17.

Chapter 21

Binomial-Type Coefficients and Classical Approximation Processes

Michele Campiti

Department of Mathematics, Polytechnic of Bari, Universita Degli Studi di Bari, via E Orabona, 4, 70125 Bari, Italy
E-Mail: campiti@dm.uniba.it

Abstract This chapter describes some connections between approximation processes and evolution problems and, in particular, the possibility of approximating the solutions of more general diffusion models using a generalization of the binomial coefficients in some classical approximation processes. This enlarged class of evolution problems includes some diffusion models of particular interest as gene frequency models in population genetics, when selection, migration, mutation and other factors occur. In the interval [0,1], the approximation processes taken into consideration are of Bernstein-type for continuous functions and of Bernstein-Kantorovitch-type for L^p-integrable functions; on the half-line, we consider Bernstein-Chlodovski-type operators and Baskakov-type operators; finally, we shall be also concerned with Bernstein-type operators on the hypercube and on the standard simplex of $\mathbf{R}^d$.

Contents

21.1 Introduction

In this chapter we discuss a generalization of the binomial coefficients in some classical sequences of operators, which preserves the recursive formulas but is associated to arbitrary sequences of real numbers.

The main motivation of this generalization is the connection between the approximation processes which arise in this manner and the solutions of suitable evolution problems which can be associated with them (by means of a Voronovskaja-type formula and the semigroup theory). Indeed, in many cases it is possible to apply the Trotter's theorem [50, Theorem 5.3] which allows the representation of the semigroup in terms of iterates of the approximation process.

This kind of connection was recently pointed out by Altomare [4] for Bernstein-Schnabl operators in an infinite dimensional context. In the sequel, many authors considered these problems either using different classical approximation processes or introducing new sequences of positive operators satisfying a prescribed Voronovskaja formula.

We refer to [8, Chapter 6] for a rather complete treatment of the subject and to [4, 5, 6, 7, 9, 10, 11, 12, 13, 14, 15, 16, 17, 24, 26, 27] and the papers cited therein for further developments.

In this context, we discuss a particular class of approximation processes which can be obtained from classical Bernstein-type operators when the binomial coefficients are substituted with more general ones satisfying a similar recursive formula. In this manner we obtain a general Voronovskaja-type formula where the coefficients of the first order derivative(s) do not necessarily vanish (as for the classical Bernstein operators) but depend on some sequences of real numbers and can be controlled by them. The introduction of these first order terms in the Voronovskaja-type formula is very useful since the corresponding evolution problems are diffusion models of particular interest in population genetics.

Section 21.2 is based on the papers by Campiti and Metafune [21, 22], where the general binomial-type coefficients are introduced and applied to the classical sequence of Bernstein operators on $[0,1]$. Parts of the results in [21, 22] are exposed in [3] and [39].

Section 21.3, based on [18, 19], we discuss further extensions of the recursive formula and their applications. Subsequently, again in the context of the interval [0,1], we consider in Section 21.4 the Bernstein-Kantorovitch operators on the space $L^p(0,1)$ which have been treated by Campiti and Metafune in [23] (the Voronovskaja's formula for these operators was obtained by I. Raşa).

We consider also Bernstein-Chlodovski and Baskakov-type operators on the half-line in Section 21.5, referring to the papers of Agratini [1, 2].

Finally, we describe some generalizations to the multi-dimensional case considering Bernstein-type operators on the hypercube and the standard simplex in $\mathbf{R}^d$, discussed in detail in [20].

21.2 Bernstein-type Operators on the Interval [0,1]

The generalization of the binomial coefficents introduced by Campiti and Metafune in [21, 22] is based on the following representation of the Pascal's triangle

$$\begin{array}{ccccccccccc} & & & & 1 & & 1 & & & & \\ & & & 1 & & 2 & & 1 & & & \\ & & 1 & & 3 & & 3 & & 1 & & \\ & 1 & & 4 & & 6 & & 4 & & 1 & \\ 1 & & 5 & & 10 & & 10 & & 5 & & 1 \\ \dots & \dots & \dots & \dots & \dots & \dots & \dots & & & & \end{array},$$

where the construction is made using two constant sequences of constant value 1 at the two sides and a recursive formula for the internal values. So, it is very natural to ask what it happens if one uses the same recursive formula with two different sequences of real numbers at the two sides.

This lead us to consider two sequences $(\lambda_m)_{m\geq 1}$ and $(\rho_m)_{m\geq 1}$ of real numbers and to define the binomial-type coefficients $\alpha_{n,k}$, $n \geq 1$, $k = 0, \ldots, n$ according to the following formulas

$$\alpha_{n,0} = \lambda_n, \qquad \alpha_{n,n} = \rho_n, \tag{21.1}$$

$$\alpha_{n+1,k} = \alpha_{n,k} + \alpha_{n,k-1}, \qquad k = 1, \ldots, n. \tag{21.2}$$

It is clear that if $\lambda_m = \rho_m = 1$ for every $m \geq 1$ then $\alpha_{n,k} = \binom{n}{k}$ for every $n \geq 1$ and $k = 0, \ldots, n$.

REMARK 21.1 The binomial-type coefficients $\alpha_{n,k}$ depend linearly and monotically on the sequences $(\lambda_m)_{m\geq 1}$ and $(\rho_m)_{m\geq 1}$. Indeed, if $(\lambda_m^{(i)})_{m\geq 1}$ and $(\rho_m^{(i)})_{m\geq 1}$, $i = 1, 2$, are sequences of real numbers and if we denote by $\alpha_{n,k}^{(i)}$ the corresponding coefficients defined according to (21.1) and (21.2), then the binomial coefficients $\alpha_{n,k}$ associated with the sequences $(\lambda_m^{(1)} + \lambda_m^{(2)})_{m\geq 1}$ and $(\rho_m^{(1)} + \rho_m^{(2)})_{m\geq 1}$ are given by $\alpha_{n,k} = \alpha_{n,k}^{(1)} + \alpha_{n,k}^{(2)}$.

Moreover, if $\lambda_m^{(1)} \leq \lambda_m^{(2)}$ and $\rho_m^{(1)} \leq \rho_m^{(2)}$ for every $m \geq 1$ then $\alpha_{n,k}^{(1)} \leq \alpha_{n,k}^{(2)}$ for every $n \geq 1$ and $k = 0, \ldots, n$. ∎

In this section, we shall apply this generalization of the binomial coefficients to the sequence of the classical Bernstein operators on the interval $[0, 1]$. Therefore, according to [21, (1.1)], we fix two sequences $(\lambda_m)_{m\geq 1}$ and $(\rho_m)_{m\geq 1}$ of real numbers and for every $n \geq 1$, $f \in C[0, 1]$ and $x \in [0, 1]$, we define the operators

$$A_n f(x) = \sum_{k=0}^{n} \alpha_{n,k} x^k (1 - x)^{n-k} f\left(\frac{k}{n}\right). \tag{21.3}$$

Obviously, the operators associated with the constant sequences of constant value 1 are just the classical Bernstein operators

$$B_n f(x) = \sum_{k=0}^{n} \binom{n}{k} x^k (1 - x)^{n-k} f\left(\frac{k}{n}\right).$$

By the preceding Remark 21.1, the operators A_n depend linearly and monotonically on the sequences $(\lambda_m)_{m\geq 1}$ and $(\rho_m)_{m\geq 1}$. Consequently, if $\lambda_m \leq M$ and $\rho_m \leq M$ for every $m \geq 1$ we also have $A_n \leq M \cdot B_n$.

The convergence of the operators A_n can be studied in a simple way introducing, for every $m \geq 1$, the so-called *m-left* (*right*, respectively) *elementary operators* $L_{m,n}$ ($R_{m,n}$, respectively) associated with the sequences $\lambda = (\delta_n^m)_{n\geq 1}$ and $\rho = 0$ ($\lambda = 0$ and $\rho = (\delta_n^m)_{n\geq 1}$, respectively). The corresponding operators defined according to (21.3) are given by

$$L_{m,n}f(x) = \sum_{k=1}^{n-m} \binom{n-m-1}{k-1} x^k (1-x)^{n-k} f\left(\frac{k}{n}\right),$$

$$R_{m,n}f(x) = \sum_{k=m}^{n-1} \binom{n-m-1}{k-m} x^k (1-x)^{n-k} f\left(\frac{k}{n}\right)$$

if $m < n$, while

$$L_{m,m}f(x) = (1-x)^m f(0), \qquad R_{m,m}f(x) = x^m f(1);$$

if $m > n$ the operators $L_{m,n}$ and $R_{m,n}$ vanish.

Using the elementary operators it is possible to decompose the general operators A_n in the following manner:

$$A_n = \sum_{m=1}^{n} (\lambda_m L_{m,n} + \rho_m R_{m,n}) \tag{21.4}$$

and, in particular, $B_n = \sum_{m=1}^{n} (L_{m,n} + R_{m,n})$.

If we evaluate the operator A_n at the constant function $\mathbf{1}$ we get

$$A_n\mathbf{1} = \sum_{m=1}^{n-1} (\lambda_m x(1-x)^m + \rho_m x^m(1-x)) + \lambda_n(1-x)^n + \rho_n x^n\ . \tag{21.5}$$

21.2.1 Convergence of the Bernstein-type operators

Let $(\lambda_m)_{m\geq 1}$ and $(\rho_m)_{m\geq 1}$ be two sequences of real numbers and consider the operators A_n associated with them by (21.3). For every $f \in C[0,1]$, we have $A_n f(0) = \lambda_n f(0)$ and $A_n f(1) = \rho_n f(1)$ and therefore the pointwise convergence of $(A_n)_{n\geq 1}$ implies that of the sequences $(\lambda_m)_{m\geq 1}$ and $(\rho_m)_{m\geq 1}$. For this reason, in the sequel we shall always assume that $(\lambda_m)_{m\geq 1}$ and $(\rho_m)_{m\geq 1}$ converge and we put

$$\lambda_\infty = \lim_{m\to+\infty} \lambda_m, \qquad \rho_\infty = \lim_{m\to+\infty} \rho_m.$$

Moreover, equation (21.5) suggests we consider the function $\ell : [0,1] \to \mathbf{R}$ defined by

$$\ell(t) = \begin{cases} \sum_{m=1}^{+\infty} \lambda_m \, t\,(1-t)^m, & t \neq 0, \\ \lambda_\infty, & t = 0, \end{cases}$$

where $(\lambda_m)_{m\geq 1}$ is a sequence of real numbers converging to λ_∞ (if necessary, we shall emphasize the dependence on $(\lambda_m)_{m\geq 1}$ using the notation $\ell(\lambda_m; \cdot)$).

The following lemma is taken from [20, Lemma 3.1].

LEMMA 21.1

The function ℓ is continuous at 0.

PROOF Let $\varepsilon > 0$ and put $M = \sup_{m\geq 1} |\lambda_m - \lambda_\infty|$; consider $N \geq 1$ such that $|\lambda_m - \lambda_\infty| \leq \varepsilon/3$ for every $m > N$ and $\delta > 0$ such that, for every $t \in]0,\delta]$, $NMt/(1-t) \leq \varepsilon/3$ and $|\lambda_\infty|\, t \leq \varepsilon/3$; for every $t \in]0,\delta]$, we have $\frac{t}{1-t} \sum_{m=1}^{+\infty} (1-t)^m = 1$ and consequently

$$\begin{aligned}
\left| \sum_{m=1}^{+\infty} \lambda_m \, t\,(1-t)^m - \lambda_\infty \right| &= \left| \frac{t}{1-t} \sum_{m=1}^{+\infty} ((1-t)\,\lambda_m - \lambda_\infty)\,(1-t)^m \right| \\
&= \left| \frac{t}{1-t} \sum_{m=1}^{+\infty} (\lambda_m - \lambda_\infty)(1-t)^m - \frac{t^2}{1-t} \lambda_\infty \sum_{m=1}^{+\infty} (1-t)^m \right| \\
&\leq \frac{t}{1-t} \sum_{m=1}^{N} |\lambda_m - \lambda_\infty|(1-t)^m + \frac{t}{1-t} \sum_{m=N+1}^{+\infty} |\lambda_m - \lambda_\infty|(1-t)^m \\
&\quad + |\lambda_\infty|\, t \leq \frac{t}{1-t} N\,M + \frac{\varepsilon}{3} + |\lambda_\infty|\, t \leq \frac{\varepsilon}{3} + \frac{\varepsilon}{3} + \frac{\varepsilon}{3} = \varepsilon.
\end{aligned}$$

∎

At this points, we can define the function $w : [0,1] \to \mathbf{R}$ setting

$$\begin{aligned}
w(x) &= \ell(\lambda_m; x) + \ell(\rho_m; 1-x) \\
&= \begin{cases} \sum_{m=1}^{+\infty} \left(\lambda_m x (1-x)^m + \rho_m x^m (1-x) \right), & 0 < x < 1, \\ \lambda_\infty, & x = 0, \\ \rho_\infty, & x = 1. \end{cases}
\end{aligned} \tag{21.6}$$

From Lemma 21.1, the function w is continuous on $[0,1]$. Moreover, if we introduce the functions $b_n : [0,1] \to \mathbf{R}$ taking

$$b_n\left(\frac{k}{n}\right) = \alpha_{n,k}\binom{n}{k}^{-1}, \qquad k = 0\ldots,n,$$

and affine on every segment $[k/n,(k+1)/n]$, $k = 0\ldots,n-1$, we have $A_n f = B_n(b_n \cdot f)$, $f \in C[0,1]$, and we can state the following properties of w (for a detailed proof we refer to [21, Theorem 2.4 and Proposition 2.5]).

PROPOSITION 21.1

1) *The sequence $(b_n)_{n\geq 1}$ converges uniformly to w.*
2) *If the sequences $(\lambda_m)_{m\geq 1}$ and $(\rho_m)_{m\geq 1}$ are increasing (respectively, decreasing), then every b_n and w are convex (respectively, concave).*
3) *If the sequence $(\lambda_m)_{m\geq 1}$ is increasing (respectively, decreasing) and $(\rho_m)_{m\geq 1}$ is decreasing (respectively, increasing) and if $\lambda_1 \geq \rho_1$ (respectively, $\lambda_1 \leq \rho_1$), then every b_n and w are decreasing (respectively, increasing).*
4) *If $\lambda_m = \rho_m$ for every $m \geq 1$, then $b_n(x) = b_n(1-x)$ and $w(x) = w(1-x)$ for every $n \geq 1$ and $x \in [0,1]$.*

In the following result we state the convergence properties of the sequence $(A_n)_{n\geq 1}$ in terms of the classical modulus of continuity defined by

$$\omega(f,\delta) = \sup\{|f(x) - f(y)| \mid x,y \in [0,1],\ |x-y| \leq \delta\}$$

for every $f \in C[0,1]$ and $\delta > 0$.

The quantitative estimate depends on the sequences $(\lambda_m)_{m\geq 1}$ and $(\rho_m)_{m\geq 1}$ by means of the quantities

$$r(n) = \sup_{m\geq n} \max\{|\lambda_m - \lambda_n|, |\rho_m - \rho_n|\},\ s(n) = \max_{m\leq n}\{|\lambda_m|, |\rho_m|\}. \quad (21.7)$$

THEOREM 21.1 [21, Theorems 2.1, 2.2]
The sequence $(A_n)_{n\geq 1}$ converges strongly on $C[0,1]$ if and only if the sequences $(\lambda_m)_{m\geq 1}$ and $(\rho_m)_{m\geq 1}$ converge and, in this case, we have

$$\lim_{n\to+\infty} A_n(f) = w \cdot f$$

uniformly on $[0,1]$ *for every* $f \in C[0,1]$.

Moreover, for every $x \in [0,1]$,

$$|A_n f(x) - w(x) \cdot f(x)| \le (1 + x(1-x))s(n)\omega\left(f, \frac{1}{\sqrt{n}}\right) + ((1-x)^n + x^n)\, r(n)|f(x)|$$

and consequently

$$\|A_n f - w \cdot f\| \le \frac{5}{4}\ s(n)\ \omega\left(f, \frac{1}{\sqrt{n}}\right) + r(n)\ \|f\|.$$

In [21, Section 4], the convergence of the derivatives of the operators A_n has also been studied. The main result states that if $f \in C[0,1]$ admits a derivative of order k at a point $x \in]0,1[$, then

$$\lim_{n\to+\infty} (A_n(f))^{(k)}(x) = \frac{d^k(w\cdot f)}{dx^k}(x).$$

Moreover, if $f \in C^k[0,1]$, then the convergence is uniform on every compact subinterval of $]0,1[$.

We conclude this section stating some examples of particular interest.

Example 21.1 [21, (2.25)–(2.26)]
As regards to the elementary operators $L_{m,n}$ and $R_{m,n}$, by Theorem 21.1, we have

$$\lim_{n\to+\infty} L_{m,n}f(x) = x(1-x)^m f(x), \qquad \lim_{n\to+\infty} R_{m,n}f(x) = x^m(1-x)f(x)$$

uniformly on $[0,1]$ for every $f \in C[0,1]$. A quantitative estimate of the convergence is given in [21, Proposition 2.3]. □

Example 21.2 [21, Example 2]
For every $m \ge 1$, define $\lambda_m = 1$ and $\rho_m = 0$ (respectively, $\lambda_m = 0$ and $\rho_m = 1$) and consider the operators B_n^ℓ (respectively, B_n^r) associated with the sequences $(\lambda_m)_{m\ge1}$ and $(\rho_m)_{m\ge1}$.

Then, for every $f \in C[0,1]$,

$$B_n^\ell f(x) = \sum_{k=0}^{n-1} \binom{n-1}{k} x^k(1-x)^{n-k} f\left(\frac{k}{n}\right),$$

$$B_n^r f(x) = \sum_{k=1}^{n} \binom{n-1}{k-1} x^k(1-x)^{n-k} f\left(\frac{k}{n}\right)$$

and further, from Theorem 21.1,

$$\lim_{n\to+\infty} B_n^\ell f(x) = (1-x)f(x), \qquad \lim_{n\to+\infty} B_n^r f(x) = xf(x)$$

uniformly on $[0,1]$.

In this manner, the classical Bernstein operator B_n can be decomposed into the sum of a *left operator* B_n^ℓ and a *right operator* B_n^r. □

Example 21.3 [21, Example 4]
Assume that the sequences $(\lambda_m)_{m\geq 1}$ and $(\rho_m)_{m\geq 1}$ are definitively constant and let $p \geq 1$ such that $\lambda_m = \lambda$ and $\rho_m = \rho$ for every $m \geq p$. In this case, the operators associated with the sequences $(\lambda_m)_{m\geq 1}$ and $(\rho_m)_{m\geq 1}$ are linear combinations of left Bernstein operators, right Bernstein operators (see the preceding example) and a finite number of elementary operators. The function w in (21.6) becomes

$$w(x) = \lambda(1-x) + \rho x + \sum_{m=1}^{p-1}((\lambda_m - \lambda)x(1-x)^m + (\rho_m - \rho)(1-x)x^m)$$

and is a polynomial of degree at most p.

Conversely, it is easy to recognize that if w is a polynomial then the sequences $(\lambda_m)_{m\geq 1}$ and $(\rho_m)_{m\geq 1}$ must be definitively constant. □

21.2.2 Voronoskaja-type formula and associated evolution problems

As explained in the introduction, one of the main motivation of the introduction of the operator A_n is the validity of a general Voronovskaja-type formula which gives the connections between these operators and the solutions of suitable second-order parabolic problems.

From now on, we always assume that $(\lambda_m)_{m\geq 1}$ and $(\rho_m)_{m\geq 1}$ are positive sequences of real numbers such that

$$\lim_{m\to+\infty} \lambda_m = \lambda_\infty > 0, \qquad \lim_{m\to+\infty} \rho_m = \rho_\infty > 0. \tag{21.8}$$

For sufficiently large $n \geq 1$, the above condition ensures that the function $A_n\mathbf{1}$ is strictly positive and hence it is possible to define the following positive linear operators on $C[0,1]$

$$L_n = \frac{A_n}{A_n\mathbf{1}}.$$

Since the sequence $(A_n)_{n\geq 1}$ converges uniformly to the function w, from Theorem 21.1, we infer that $(L_n)_{n\geq 1}$ is an approximation process on $C[0,1]$, i.e.,

$$\lim_{n\to+\infty} L_n = I \qquad \text{strongly on } C[0,1],$$

where I denotes the identity operator on $C[0,1]$.

The Voronovskaja-type formula for the operators L_n can be stated as follows (see [21, 22]).

THEOREM 21.2

For every $f \in C^2[0,1]$, we have

$$\lim_{n\to+\infty} n\,(L_n f(x) - f(x)) = \begin{cases} \dfrac{x(1-x)}{2} f''(x) + x(1-x)\dfrac{w'(x)}{w(x)} f'(x), & 0 < x < 1, \\ 0, & x = 0, 1 \end{cases}$$

uniformly on $[0,1]$.

The proof of the preceding formula is given in [21, Theorem 3.4]; it is also a particular case of the Voronovskaja-type formula stated in [20, Theorem 4.3 and Remark 4.4].

The preceding Voronovskaja-type formula connects the operators L_n (and hence A_n) to the solution of the following degenerate second-order parabolic problem

$$\begin{cases} \dfrac{\partial u}{\partial t}(x,t) = \alpha(x)\dfrac{\partial^2 u}{\partial x^2}(x,t) + \beta(x)\dfrac{\partial u}{\partial x}(x,t), & 0 < x < 1, \\ & t > 0, \\ \displaystyle\lim_{x\to 0^+,1^-} \alpha(x)\dfrac{\partial^2 u}{\partial x^2}(x,t) + \beta(x)\dfrac{\partial u}{\partial x}(x,t) = 0, & t > 0, \\ u(x,0) = u_0(x), & x \in [0,1], \end{cases} \tag{21.9}$$

where $u_0 \in C[0,1]$ satisfies the boundary conditions

$$\lim_{x\to 0^+,1^-} \alpha(x)u_0''(x) + \beta(x)u_0'(x) = 0, \tag{21.10}$$

$\alpha(x) = x(1-x)/2$ as in the case of the classical Bernstein operators and $\beta(x) = x(1-x)w'(x)/w(x)$ depends on the sequences $(\lambda_m)_{m\geq 1}$ and $(\rho_m)_{m\geq 1}$.

In order to study the preceding problem we introduce the following operator

$$Au(x) = \frac{x(1-x)}{2}u''(x) + x(1-x)\frac{w'(x)}{w(x)}u'(x), \qquad 0 < x < 1, \quad (21.11)$$

on the domain

$$D(A) = \left\{ u \in C[0,1] \cap C^2]0,1[\mid \lim_{x \to 0^+, 1^-} Au(x) = 0 \right\}.$$

By applying a well-known result by Clément and Timmermans [28], it easily follows that $(A, D(A))$ generates a C_0-semigroup

$$(T(t))_{t \geq 0}$$

of positive contractions on $C[0,1])$. Moreover, the space $C^2[0,1]$ is a core for $(A, D(A))$. We do not give more details on this property since we shall prove a more general result in Theorem 21.6.

THEOREM 21.3 [22, Theorem 2.2]
For every $t \geq 0$ and for every sequence $(k(n))_{n \geq 1}$ of positive integers such that $\lim_{n \to +\infty} k(n)/n = t$, we have

$$\lim_{n \to +\infty} L_n^{k(n)} = T(t) \qquad \textit{strongly on } C[0,1]$$

(L_n^m denotes the m-iterate of L_n).

PROOF Since $C^2[0,1]$ is a core for A, for sufficiently large $\lambda > 0$, the range $(\lambda I - A)(C^2[0,1])$ is dense in $C[0,1]$. Now, consider the operator $L : D(L) \to C[0,1])$ defined by $Lu = \lim_{n \to +\infty} n(L_n u - u)$ on the domain

$$D(L) = \left\{ u \in C[0,1] \mid \text{the limit } \lim_{n \to +\infty} n(L_n u - u) \text{ exists in } C[0,1] \right\}.$$

From the Voronovskaja formula stated in Theorem 21.2, we deduce that $A = L$ on $C^2[0,1]$ and hence $(\lambda I - L)(C^2[0,1])$ is dense in $C[0,1]$. Since each L_n is a positive contraction, we can apply Trotter's theorem [50, Theorem 5.3] (see also [38, Theorem 6.7, p. 96]) and obtain that the closure $\overline{L}$ of L generates a C_0-semigroup $(S(t))_{t \geq 0}$ of positive contractions on $C[0,1]$ such that, for every $t>0$ and for every sequence $(k(n))_{n\geq 1}$ of positive integers satisfying $\lim_{n\to+\infty} k(n)/n = t$, we have

$$\lim_{n\to+\infty} L_n^{k(n)} = S(t) \qquad \text{strongly on } C[0,1].$$

At this point, we observe that $C^2[0,1]$ is a core for $\overline{L}$ and hence $D(A) = D(\overline{L})$ and $A = \overline{L}$; consequently, the semigroups $(T(t))_{t\geq 0}$ and $(S(t))_{t\geq 0}$ coincide (see, e.g., [38, Theorem 1.2.6, p. 6]) and this completes the proof. ∎

As a consequence of Theorem 21.3, for every $u_0 \in D(A)$, the unique solution of the evolution problem (21.9) and (21.10) is given by

$$u(x,t) = (T(t))u_0(x) = \lim_{n\to+\infty} L_n^{k(n)} u_0(x), \tag{21.12}$$

where $(k(n))_{n\geq 1}$ is any sequence of positive integers satisfying the condition $\lim_{n\to+\infty} k(n)/n = t$ (for example, $k(n) = [nt]$) and the convergence is uniform on every compact subset of $[0,1] \times [0,+\infty[$.

Furthermore, (21.12) holds for every $u_0 \in C[0,1]$ since the semigroup $(T(t))_{t\geq 0}$ is also analytic and hence differentiable (see [25]).

We have seen that assigned two sequences $(\lambda_m)_{m\geq 1}$ and $(\rho_m)_{m\geq 1}$ of positive real numbers converging to $\lambda_\infty > 0$ and $\rho_\infty > 0$, the corresponding operators A_n and the differential operator A (and hence the evolution problem (21.9)) are connected by (21.12).

Conversely, assigned the evolution problem (21.9) and hence the operator A defined by (21.11), the problem arises of finding the sequences $(\lambda_m)_{m\geq 1}$ and $(\rho_m)_{m\geq 1}$ of positive real numbers converging to $\lambda_\infty > 0$ and $\rho_\infty > 0$ for which the function given in (21.6) coincides with w.

We have already observed in Example 21.3 that every polynomial w is always associated to definitively constant sequences, and this is already quite satisfactory from the point of view of approximating the solutions of problem (21.9)-(21.10).

Indeed, if in general $w \in C^1[0,1]$ is strictly positive, we can consider a sequence $(w_n)_{n\geq 1}$ of polynomials satisfying $\lim_{n\to+\infty} w_n = w$ and $\lim_{n\to+\infty} w_n' = w'$ uniformly on $[0,1]$. Now, let A_n^* be the differential operator defined as in (21.11) with w_n at the place of w. We have

$$\lim_{n\to+\infty} A_n^*(f) = A(f) \qquad \text{for every } f \in C^2[0,1]$$

and further the range $(I-A)(C^2[0,1])$ is dense in $C[0,1]$. Thus, applying the Trotter-Kato theorem (see, e.g., [38, Theorem 4.5, p. 89]), the semigroups $(T(t))_{t\geq 0}$ generated by A and $(T_n(t))_{t\geq 0}$ generated by A_n^* satisfy

$$\lim_{n\to+\infty} T_n(t) = T(t) \qquad \text{strongly on } C[0,1]$$

and the convergence is uniform with respect to t in bounded intervals.

As a consequence, if $u_0 \in D(A)(= D(A_n^*))$ and if u and u_n denote the solutions of problem (21.9)-(21.10) corresponding to A and A_n^*, respectively, we have

$$u(x,t) = \lim_{n\to+\infty} u_n(x,t)$$

uniformly on compact subsets of $[0,1] \times [0,+\infty[$.

We refer to [22, Theorem 3.3] for further developments of this result.

21.3 Further Generalizations of the Binomial Coefficients

In this section, we continue the analysis of more general binomial-type coefficients in the sequence of Bernstein operators introducing further generalizations of them. These generalizations do not affect the sequences at the two sides of the Pascal's triangle, but the recursive formula (21.2). We introduce a weight function p which may also depend on $n \geq 1$ in the second subsection.

In the first subsection, based on [18], we shall see the introduction of a weight p in the recursive formula produces a better generality of the coefficient of the second-order derivative in the Voronovskaja-type formula concerning the operators A_n defined in (21.3).

In the second subsection we shall deal with some approximation properties of the operators A_n defined as in (21.3), which are contained in [19] and concern the case where the weight function p introduced in the first subsection depends also on $n \geq 1$.

21.3.1 A weighted recursive formula

In this subsection, we consider the 'binomial-type' coefficients $\alpha_{n,k}(x)$ obtained as in Section 21.2, but the contributions of the left addends are weighted by a function $p(x)$ and those of the right addends by a function $q(x)$. Therefore, we shall continue to fix two sequences $(\lambda_m)_{m\geq 1}$ and

$(\rho_m)_{m\geq 1}$ of real numbers and, for every $x \in [0,1]$, we define

$$\alpha_{n,0} = \lambda_n \cdot q^n, \qquad \alpha_{n,n} = \rho_n \cdot p^n,$$

$$\alpha_{n+1,k} = \alpha_{n,k-1} \cdot p + \alpha_{n,k} \cdot q, \qquad k = 1, \dots, n.$$

We denote by C_n the corresponding operators defined as in (21.3); hence, for every $n \geq 1$, $C_n : C[0,1] \to C[0,1]$ is defined by

$$C_n f(x) = \sum_{k=0}^{n} \alpha_{n,k}(x) x^k (1-x)^{n-k} f\left(\frac{k}{n}\right) \tag{21.13}$$

for every $f \in C[0,1]$ and $x \in [0,1]$.

The study of the convergence of the sequence $(C_n)_{n\geq 1}$ is based on the following lemma established in [19, Lemma 1.1]. The proof will be omitted since we shall state a more general result in Lemma 21.3 of the next subsection.

LEMMA 21.2

For every $n \geq 2$, $k = 1, \dots, n-1$ and $x \in [0,1]$, we have

$$\alpha_{n,k}(x) = \left(\sum_{i=1}^{n-k} \binom{n-1-i}{k-1} \lambda_i + \sum_{j=1}^{k} \binom{n-1-j}{n-1-k} \rho_j \right) p(x)^k q(x)^{n-k}. \tag{21.14}$$

In the case of constant sequences of constant value 1, we have

$$\alpha_{n,k}(x) = \binom{n}{k} p(x)^k \, q(x)^{n-k}.$$

In the sequel, it will be useful to put, for every $x \in [0,1]$,

$$\varphi(x) = xp(x), \quad \psi(x) = (1-x)q(x), \quad \chi(x) = \varphi(x) + \psi(x).$$

From (21.14) and (21.13), we obtain

$$C_n f(x) = \lambda_n \, \psi(x)^n \, f(0) + \rho_n \, \varphi(x)^n \, f(1) + \sum_{k=1}^{n-1} \left(\sum_{i=1}^{n-k} \binom{n-1-i}{k-1} \lambda_i \right.$$

$$\left. + \sum_{j=1}^{k} \binom{n-1-j}{n-1-k} \rho_j \right) \varphi(x)^k \psi(x)^{n-k} f\left(\frac{k}{n}\right)$$

$$
\begin{aligned}
&= \lambda_n \, \psi(x)^n f(0) + \rho_n \, \varphi(x)^n f(1) \\
&\quad + \sum_{i=1}^{n-1} \lambda_i \, \varphi(x) \, \psi(x)^i \sum_{k=0}^{n-1-i} \binom{n-1-i}{k} \times \\
&\quad \times \varphi(x)^k \psi(x)^{n-1-i-k} f\left(\frac{k+1}{n}\right) \\
&\quad + \sum_{j=1}^{n-1} \rho_j \, \varphi(x)^j \, \psi(x) \sum_{k=0}^{n-1-j} \binom{n-1-j}{k} \times \\
&\quad \times \varphi(x)^k \psi(x)^{n-1-j-k} f\left(\frac{k+j}{n}\right)
\end{aligned} \tag{21.15}
$$

($f \in C[0,1]$, $x \in [0,1]$) and, in particular

$$C_n \mathbf{1} = \lambda_n \psi^n + \rho_n \varphi^n + \varphi \sum_{i=1}^{n-1} \lambda_i \, \psi^i \, \chi^{n-1-i} + \psi \sum_{j=1}^{n-1} \rho_j \, \varphi^j \, \chi^{n-1-j}. \tag{21.16}$$

The preceding expressions simplify the study of the convergence of the sequence $(C_n)_{n \geq 1}$. Since in the case $p = q = \mathbf{1}$ we just obtain the operators A_n considered in the preceding section, it is natural to require that the sequences $(\lambda_m)_{m \geq 1}$ and $(\rho_m)_{m \geq 1}$ converge.

Moreover, from the relation (see (21.16))

$$C_n \mathbf{1}(0) = \lambda_n \, q(0)^n, \qquad C_n \mathbf{1}(1) = \rho_n \, p(1)^n$$

the limit of $(C_n \mathbf{1})_{n \geq 1}$ may be a strictly positive function if the limits of $(\lambda_m)_{m \geq 1}$ and $(\rho_m)_{m \geq 1}$ are strictly positive. Consequently, the two last sums in (21.16) converge if and only if $\chi(x) \leq 1$ for every $x \in [0,1]$. To get a strictly positive limit of $(C_n \mathbf{1})_{n \geq 1}$, we impose

$$\chi(x) = 1 \qquad \text{for every } x \in [0,1]; \tag{21.17}$$

in particular $q(0) = p(1) = 1$ and, for every $x \in]0,1[$,

$$0 < x \, p(x) < 1, \qquad 0 < (1-x) \, q(x) < 1.$$

In the sequel we always assume that conditions (21.8) and (21.17) hold.

Finally, we define the functions

$$P(x) = \begin{cases} \lambda_\infty, & x = 0, \\ \varphi(x) \displaystyle\sum_{i=1}^{+\infty} \lambda_i \, \psi(x)^i, & 0 < x \leq 1, \end{cases} \tag{21.18}$$

$$Q(x) = \begin{cases} \rho_\infty, & x = 1, \\ \psi(x) \sum_{j=1}^{+\infty} \rho_j \, \varphi(x)^j, & 0 \le x < 1, \end{cases} \tag{21.19}$$

and we put

$$\gamma = P + Q. \tag{21.20}$$

Comparing (21.20), (21.18) and (21.19) with (21.6), we obtain $\gamma = w \circ \varphi$; hence, from Lemma 21.1, the function γ and consequently P and Q are continuous in [0,1]. Further, from Proposition 21.1, we can state analogous properties of the function γ.

The convergence properties of the operators C_n stated in the next results can be obtained from the corresponding properties of the operators A_n, by virtue of the following relation

$$C_n f(x) = A_n f(\varphi(x)), \qquad f \in C[0,1], \quad x \in [0,1],$$

which can be obtained comparing (21.14) and (21.15) with (21.3).

THEOREM 21.4 [18, Theorem 2.1]
Assume that conditions (21.8) and (21.17) hold. Then, for every $f \in C[0,1]$, we have

$$\lim_{n \to +\infty} C_n f = \gamma \cdot (f \circ \varphi)$$

uniformly on $[0,1]$. More precisely, for every $n \ge 1$ and $x \in [0,1]$,

$$\begin{aligned} |C_n f(x) - \gamma(x)\, f(x\, p(x))| &\le s(n)\,(1 + x\, p(x)(1 - x\, p(x)))\, \omega(f, \frac{1}{\sqrt{n}}) \\ &\quad + r(n)|f(x\, p(x)|((1 - x\, p(x))^n + (x\, p(x))^n) \end{aligned}$$

(see (21.7)) and, in particular,

$$\|C_n f - \gamma \cdot (f \circ \varphi)\| \le \frac{5}{4}\, s(n)\, \omega(f, \frac{1}{\sqrt{n}}) + r(n)\, \|f\|.$$

THEOREM 21.5 [18, Theorem 2.4]
If $f \in C[0,1]$ and p are two times differentiable at the point $x\,p(x) \in [0,1]$ and $\varphi'(x) \ne 0$, then

$$\lim_{n \to +\infty} n\,(C_n f(x) - f(x\, p(x))\, C_n \mathbf{1}(x))$$

$$= \begin{cases} \dfrac{x\,p(x)(1-x\,p(x))}{2}\,\gamma(x)\,f''(x\,p(x)) \\ \qquad +x\,p(x)(1-x\,p(x))\,\dfrac{\gamma'(x)}{\varphi'(x)}\,f'(x\,p(x)), & 0 < x\,p(x) < 1, \\ 0, & x\,p(x) = 0, 1. \end{cases}$$

If $f \in C^2[0,1]$, $p \in C^2[0,1]$ and $\varphi'(x) \neq 0$ for every $x \in [0,1]$, the convergence is uniform on $[0,1]$.

At this point, we investigate some applications of the operators C_n to the approximation of the solutions of suitable parabolic problems.

Besides (21.8) and (21.17), we also require that $p \in C^2[0,1]$ and $\varphi'(x) \neq 0$ for every $x \in [0,1]$; since $\varphi(0) = 0$ and $\varphi(1) = 1$, this ensures that φ is strictly increasing and therefore invertible.

Moreover, evaluating $D(f \circ \varphi^{-1})$ and $D^2(f \circ \varphi^{-1})$, we can write the preceding Voronovskaja-type formula as follows:

$$\begin{aligned} \lim_{n\to+\infty} n &\left(\frac{C_n(f\circ\varphi^{-1})(x)}{C_n\mathbf{1}(x)} - f(x) \right) \\ &= \frac{1}{2}\,\frac{x\,p(x)\,(1-x\,p(x))}{(p(x)+x\,p'(x))^2}\,f''(x) + \frac{x\,p(x)(1-x\,p(x))}{p(x)+x\,p'(x)} \times \\ &\quad \times \left(\frac{\gamma'(x)}{\gamma(x)} - \frac{2p'(x)+x\,p''(x)}{2(p(x)+x\,p'(x))^4} \right) f'(x) \end{aligned} \tag{21.21}$$

uniformly in $[0,1]$ for every $f \in C^2[0,1]$.

Finally, for every $n \geq 1$, we define the operator $D_n : C[0,1] \to C[0,1]$ by putting

$$D_n f = \frac{C_n(f\circ\varphi^{-1})}{C_n\mathbf{1}}, \qquad f \in C[0,1].$$

The sequence $(D_n)_{n\geq 1}$ converges strongly to the identity operator and from (21.21) it is strictly connected to the differential operator

$$\begin{aligned} Au(x) &= \frac{1}{2}\,\frac{\varphi(x)\,(1-\varphi(x))}{\varphi'(x)^2}\,u''(x) \\ &\quad + \frac{\varphi(x)\,(1-\varphi(x))}{\varphi'(x)} \left(\frac{\gamma'(x)}{\gamma(x)} - \frac{\varphi''(x)}{2\varphi'(x)^4} \right) u'(x), \end{aligned} \tag{21.22}$$

$0 < x < 1$, on the domain

$$D(A) = \left\{ u \in C[0,1] \cap C^2]0,1[\;\middle|\; \lim_{x\to 0^+,1^-} Au(x) = 0 \right\}.$$

Now, we can state the main result of this subsection.

THEOREM 21.6 [19, Theorem 3.1])
Assume that conditions (21.8) and (21.17) hold and that φ *is invertible,* $p \in C^2[0,1]$ *and* $\varphi'(x) \neq 0$ *for every* $x \in [0,1]$.

Then the operator $A : D(A) \to C[0,1]$ *defined by (21.22) is the generator of a strongly continuous semigroup* $(T(t))_{t\geq 0}$ *of positive contractions on* $C[0,1]$.

Moreover $C^2[0,1]$ *is a core for* A *and for every* $t > 0$ *and every sequence* $(k(n))_{n\geq 1}$ *of positive integers satisfying* $\lim_{n\to+\infty} k(n)/n = t$, *we have*

$$\lim_{n\to+\infty} D_n^{k(n)} = T(t)$$

strongly on $C[0,1]$.

PROOF For every $x \in [0,1]$ put, for brevity,

$$\alpha(x) = \frac{1}{2}\,\frac{\varphi(x)\,(1-\varphi(x))}{\varphi'(x)^2},$$

$$\beta(x) = \frac{\varphi(x)\,(1-\varphi(x))}{\varphi'(x)}\left(\frac{\gamma'(x)}{\gamma(x)} - \frac{\varphi''(x)}{2\varphi'(x)^4}\right)$$

and consider the function

$$W(x) = \exp\left(-\int_{1/2}^{x} \frac{\beta(t)}{\alpha(t)}\,dt\right), \qquad x \in [0,1];$$

since γ and φ' are strictly positive, the function W is integrable over the intervals $[0,1/2]$ and $[1/2,1]$ and therefore, by a well-known result of Clément and Timmermans [28], the operator A is the generator of a strongly continuous semigroup $(T(t))_{t\geq 0}$ of positive contractions.

Now, we show that the space $C^2[0,1]$ is a core for A using an argument similar to [23, Lemma 1.2]. For simplicity we argue only on the interval $[0,1/2]$. Since $\varphi'(0) > 0$ it follows also $p(0) > 0$ and therefore $x\,p(x) \sim x$ at the point 0.

If $u \in D(A)$, we can write $Au = \alpha W \dfrac{d}{dx}\left(\dfrac{1}{W}\dfrac{d}{dx}u\right)$ and consequently, if we consider $v = u'/W$ we obtain $v' = o(1/(\alpha W))$ ($\sim o(1/\alpha)$) at 0; thus $v'(x) = o(1/x)$ at 0 and therefore $v(x) = o(\log x)$ which yields $u'(x) = o(\log x)$ at 0. Finally, by $v' = u''/W + (1/W)'u'$ we get $u''(x) = o(1/x)$ at 0.

Now, for every $n \geq 1$, define

$$u_n(x) = \begin{cases} u(x), & \dfrac{1}{n} \leq x \leq 1, \\ \dfrac{1}{2}\left(x - \dfrac{1}{n}\right)^2 u''\left(\dfrac{1}{n}\right) + \left(x - \dfrac{1}{n}\right) u'\left(\dfrac{1}{n}\right) + u\left(\dfrac{1}{n}\right), & \\ & 0 \leq x < \dfrac{1}{n}. \end{cases}$$

Since

$$\begin{aligned} \|u_n - u\| &= \sup_{0 \leq x \leq 1/n} |u_n(x) - u(x)| \\ &\leq \frac{1}{n^2}\left|u''\left(\frac{1}{n}\right)\right| + \frac{1}{n}\left|u'\left(\frac{1}{n}\right)\right| + \omega\left(u, \frac{1}{n}\right), \end{aligned}$$

we have $\lim_{n \to +\infty} \|u_n - u\| = 0$. Moreover,

$$\begin{aligned} \|Au_n - Au\| &= \sup_{0 < x \leq 1/n} |Au_n(x) - Au(x)| \\ &\leq \sup_{0 < x \leq 1/n} |Au_n(x)| + \sup_{0 < x \leq 1/n} |Au(x)|; \end{aligned}$$

the second addend in the last inequality tends to 0 since $u \in D(A)$ while for the first addend we have

$$\begin{aligned} &\sup_{0 < x \leq 1/n} |Au_n(x)| \\ &\quad \leq \sup_{0 < x \leq 1/n} \alpha(x)\left|u''\left(\frac{1}{n}\right)\right| + |\beta(x)|\left|\left(x - \frac{1}{n}\right) u''\left(\frac{1}{n}\right) + u'\left(\frac{1}{n}\right)\right| \\ &\quad \leq \sup_{0 < x \leq 1/n} \left(\alpha(x) + \left|\beta(x)\left(x - \frac{1}{n}\right)\right|\right)\left|u''\left(\frac{1}{n}\right)\right| + \left|\beta(x)\, u'\left(\frac{1}{n}\right)\right| \\ &\quad \leq \frac{C}{n}\left(\left|u''\left(\frac{1}{n}\right)\right| + \left|u'\left(\frac{1}{n}\right)\right|\right) \end{aligned}$$

(C suitable constant) and therefore tends to 0 as $n \to +\infty$. Hence $\lim_{n \to +\infty} \|Au_n - Au\| = 0$ and this shows that the sequence $(u_n)_{n \geq 1}$ of elements of $C^2[0, 1/2]$ tends to u in the graph norm of A. Arguing in a similar way on the interval $[1/2, 1]$, we conclude that $C^2[0, 1]$ is a core for A. At this point, taking into account (21.21), the last part of the theorem directly follows by applying Trotter's theorem [50, Theorem 5.3] and thus the proof is complete. ∎

The hypothesis φ' strictly positive yields $p(0) > 0$ and therefore the operator A has a degeneration of the first order at 0 and 1 with regard to the coefficient of the second derivative. In the terminology of Feller [33], the endpoints are exit boundary points for A.

Then, if the function u_0 satisfies the boundary conditions

$$\lim_{x\to 0^+,1^-} \frac{1}{2}\frac{\varphi(x)\,(1-\varphi(x))}{\varphi'(x)^2}\frac{\partial^2 u}{\partial x^2}(x,t) + \frac{\varphi(x)\,(1-\varphi(x))}{\varphi'(x)}\left(\frac{\gamma'(x)}{\gamma(x)} - \frac{\varphi''(x)}{2\varphi'(x)^4}\right)\frac{\partial u}{\partial x}(x,t) = 0,$$

the unique solution of the evolution problem

$$\begin{cases} \dfrac{\partial u}{\partial t}(x,t) = \dfrac{1}{2}\dfrac{\varphi(x)\,(1-\varphi(x))}{\varphi'(x)^2}\dfrac{\partial^2 u}{\partial x^2}(x,t) \\ \qquad + \dfrac{\varphi(x)\,(1-\varphi(x))}{\varphi'(x)}\left(\dfrac{\gamma'(x)}{\gamma(x)} - \dfrac{\varphi''(x)}{2\varphi'(x)^4}\right)\dfrac{\partial u}{\partial x}(x,t), & 0 < x < 1,\quad t > 0, \\ \lim\limits_{x\to 0^+,1^-} \dfrac{1}{2}\dfrac{\varphi(x)\,(1-\varphi(x))}{\varphi'(x)^2}\dfrac{\partial^2 u}{\partial x^2}(x,t) \\ \qquad + \dfrac{\varphi(x)\,(1-\varphi(x))}{\varphi'(x)}\left(\dfrac{\gamma'(x)}{\gamma(x)} - \dfrac{\varphi''(x)}{2\varphi'(x)^4}\right)\dfrac{\partial u}{\partial x}(x,t) = 0, & t > 0, \\ u(x,0) = u_0(x), & x \in [0,1], \end{cases}$$

is given by

$$u(x,t) = \lim_{n\to+\infty} D_n^{k(n)} u_0(x), \qquad 0 \le x \le 1,\ t > 0$$

where $(k(n))_{n\ge 1}$ is any fixed sequence of positive integers such that $(k(n)/n)_{n\ge 1}$ converges to t.

At this point, the problem of studying the coefficients α and β (α strictly positive on $]0,1[$ with $\alpha(0) = \alpha(1) = 0$) for which the differential operator $Au = \alpha u'' + \beta u'$ can be put in the form (21.22) arises. So, we need a function $p \in C[0,1]$ such that the corresponding function φ is in $C^1[0,1]$ with $\varphi'(x) \ne 0$ for every $x \in [0,1]$ and further

$$\alpha(x) = \frac{\varphi(x)\,(1-\varphi(x))}{2\,\varphi'(x)^2}; \tag{21.23}$$

with regard to the coefficient β we have to find two sequences $(\lambda_m)_{m\geq 1}$ and $(\rho_m)_{m\geq 1}$ of positive real numbers such that

$$\beta(x) = \frac{\varphi(x)\,(1-\varphi(x))}{\varphi'(x)} \left(\frac{\gamma'(x)}{\gamma(x)} - \frac{\varphi''(x)}{2\varphi'(x)^4} \right) \tag{21.24}$$

(in this case we also have to require $\varphi \in C^2[0,1]$).

With regard to the coefficient α, since φ must be increasing and α positive, (21.23) is equivalent to

$$\int_0^x \frac{\varphi'(t)}{\sqrt{\varphi(t)\,(1-\varphi(t))}}\, dt = \int_0^x \frac{1}{\sqrt{2\,\alpha(t)}}\, dt$$

and therefore, with a straightforward argument (take into account that $\varphi(0) = 0$), we obtain

$$\varphi(x) = \sin^2 \left(\frac{1}{2} \int_0^x \frac{1}{\sqrt{2\alpha(t)}}\, dt \right), \qquad 0 < x < 1. \tag{21.25}$$

Moreover, we have also to impose $\lim_{x\to 1} \varphi(x) = 1$ and hence

$$\int_0^1 \frac{1}{\sqrt{2\alpha(t)}}\, dt = \sqrt{2}\pi \tag{21.26}$$

(observe that this last condition is satisfied if $\alpha(x) = x(1-x)/2$). Finally, φ must have a first-order degeneration at 0 and therefore, since

$$\lim_{x\to 0} \frac{\sin^2 \left(\frac{1}{2} \int_0^x \frac{1}{\sqrt{2\alpha(t)}}\, dt \right)}{x} = \lim_{x\to 0} \frac{\left(\int_0^x \frac{1}{\sqrt{\alpha(t)}}\, dt \right)^2}{8x},$$

$$\lim_{x\to 0} \frac{\int_0^x \frac{1}{\sqrt{\alpha(t)}}\, dt}{\sqrt{x}} = \lim_{x\to 0} 2\sqrt{\frac{x}{\alpha(x)}}$$

by evaluating $\lim_{x\to 0} \varphi(x)/x$, we find that also the function α must have a first-order degeneration at 0. Similarly, it can be also shown that the function α must have a first-order degeneration also at the point 1.

Thus, we have the following result.

THEOREM 21.7 [18, Theorem 3.2]
Let $\alpha \in C[0,1]$ be strictly positive on $]0,1[$ and with a first-order degeneration at the points 0 and 1. Then, there exists a function $p \in C[0,1]$

satisfying (21.23) if and only if condition (21.26) holds. In this case, the function φ (and consequently p) is given by (21.25).

Once that the function φ is assigned according to the preceding theorem, we may observe that problem (21.24) can be solved using the same arguments developed in [22, Section 3] (see the preceding Section 21.2). For brevity, we omit the discussion relative to the coefficient β.

21.3.2 A more general varying weighted recursive formula

Now we consider a further generalization of the recursive formula where the weight functions p and q introduced in the preceding subsection may depend on $n \geq 1$.

Namely, we fix two positive sequences $(\lambda_m)_{m\geq 1}$ and $(\rho_m)_{m\geq 1}$ of real numbers and two positive sequences $(p_n)_{n\geq 1}$ and $(q_n)_{n\geq 1}$ of continuous real functions on $[0,1]$.

For every $n \geq 1$, we define the coefficients $\alpha_{n,k}$ by

$$\alpha_{n,0} = \lambda_n \cdot \prod_{j=1}^{n} q_j, \quad \alpha_{n,n} = \rho_n \cdot \prod_{i=1}^{n} p_i, \tag{21.27}$$

and

$$\alpha_{n+1,k} = p_k \alpha_{n,k-1} + q_{n+1-k}\alpha_{n,k}, \qquad k = 1, \ldots, n. \tag{21.28}$$

The corresponding operators are obtained by setting, for every $n \geq 1$, $f \in C[0,1]$ and $x \in [0,1]$,

$$C_n f(x) = \sum_{k=0}^{n} \alpha_{n,k}(x)\, x^k (1-x)^{n-k} f\left(\frac{k}{n}\right). \tag{21.29}$$

The first step is a generalization of Lemma 21.2.

LEMMA 21.3

For every $n \geq 2$ and $k = 1, \ldots, n-1$, we have

$$\alpha_{n,k} = \left(\sum_{i=1}^{n-k} \binom{n-1-i}{k-1} \lambda_i + \sum_{j=1}^{k} \binom{n-1-j}{n-1-k} \rho_j \right) \prod_{i=1}^{k} p_i \prod_{j=1}^{n-k} q_j. \tag{21.30}$$

PROOF If $n = 2$ and $k = 1$, by (21.28) and (21.27),

$$\alpha_{2,1} = \alpha_{1,0}p_1 + \alpha_{1,1}q_1 = (\rho_1 + \lambda_1)p_1q_1$$

and therefore (21.30) holds.

If (21.30) holds for $n \geq 2$, then from (21.27) and (21.28) we obtain (21.30) for $\alpha_{n+1,k}$ when $k = 1$ and $k = n$; further, if $k = 2, \ldots, n-1$,

$$\begin{aligned}\alpha_{n+1,k} &= \left(\sum_{i=1}^{n-k}\left(\binom{n-1-i}{k-2} + \binom{n-1-i}{k-1}\right)\lambda_i + \lambda_{n+1-k}\right.\\ &\quad \left.+\sum_{j=1}^{k-1}\left(\binom{n-1-j}{n-k} + \binom{n-1-j}{n-1-k}\right)\rho_j + \rho_k\right)\prod_{i=1}^{k} p_i \prod_{j=1}^{n+1-k} q_j\\ &= \left(\sum_{i=1}^{n+1-k}\binom{n-i}{k-1}\lambda_i + \sum_{j=1}^{k}\binom{n-j}{n-k}\rho_j\right)\prod_{i=1}^{k} p_i \prod_{j=1}^{n+1-k} q_j\end{aligned}$$

and the proof is complete. ∎

As a justification of the assumption (21.27), we observe that if we use the convention $\prod_{i=1}^{0} a_i = 1$, Lemma 21.3 holds also for $k = 0$ and $k = n$.

In the preceding subsection, in the particular case $p_n = q_n = 1$, we have studied the convergence of the sequence $(C_n)_{n\geq 1}$ under the assumption (21.8), so it is natural to require also in this case that the sequences $(\lambda_m)_{m\geq 1}$ and $(\rho_m)_{m\geq 1}$ satisfy the same condition.

We need to recall the following consequence of the Toeplitz's theorem (see, e.g., [34, Theorem 6, p. 75]).

PROPOSITION 21.2

Let $(a_{n,k})_{n\geq 0, k=0,\ldots,n}$ be a system of positive real numbers satisfying the following conditions:

a) For every $k \geq 0$, we have $\lim_{n\to+\infty} a_{n,k} = 0$ and $\lim_{n\to+\infty} a_{n,n-k} = 0$;

b) $\lim_{n\to+\infty} \sum_{k=0}^{n} a_{n,k} = 1$.

Then, if $(u_n)_{n\geq 0}$ and $(v_n)_{n\geq 0}$ are sequences of real numbers converging to u and, respectively, v, we have

$$\lim_{n\to+\infty} \sum_{k=0}^{n} a_{n,k}\, u_k\, v_{n-k} = u \cdot v.$$

At this point, we assume that the infinite products $\prod_{n=1}^{+\infty} p_n$ and $\prod_{n=1}^{+\infty} q_n$ converge and consider the functions

$$p = \prod_{n=1}^{+\infty} p_n, \qquad q = \prod_{n=1}^{+\infty} q_n. \tag{21.31}$$

The following assumption ensures the convergence at 0 and 1:

$$p(0) = 1, \qquad q(1) = 1.$$

From Lemma 21.3, the operators C_n can be represented as follows:

$$C_n f(x) = \sum_{k=0}^{n} \beta_{n,k} \prod_{i=1}^{k} (x\, p_i(x)) \prod_{j=1}^{n-k} ((1-x)\, q_j(x))\, f\left(\frac{k}{n}\right). \tag{21.32}$$

At this point we can study the convergence of the sequence $(C_n)_{n\geq 1}$; for simplicity, a quantitative estimate will be obtained in terms of the quantities

$$\ell_n(x) = |C_n \mathbf{1}(x) - w(x)\, p(x)\, q(x)|, \qquad x \in [0,1] \tag{21.33}$$

and the partial products

$$r_n(x) = \max_{k=1,\dots,n} \left\{ \prod_{i=1}^{k} p_i(x),\ \prod_{j=1}^{k} q_j(x) \right\}, \qquad x \in [0,1] \tag{21.34}$$

of the infinite products (21.31). We have the following main result.

THEOREM 21.8 [19, Theorem 3]
Assume that $(\lambda_m)_{m\geq 1}$ and $(\rho_m)_{m\geq 1}$ satisfy (21.8) and that the infinite products $\prod_{n=1}^{+\infty} p_n$ and $\prod_{n=1}^{+\infty} q_n$ converge uniformly to the functions p and q, respectively. Then, for every $f \in C[0,1]$ and $x \in [0,1]$,

$$|C_n f(x) - w(x)\, p(x)\, q(x)\, f(x)| \leq \frac{5}{4}\, r_n(x)\, s(n)\, \omega(f, \frac{1}{\sqrt{n}}) + \ell_n(x)\, |f(x)|, \tag{21.35}$$

where $r_n(x)$ and $s(n)$ are defined by (21.34) and (21.7) and further

$$\lim_{n\to+\infty} C_n f = w\, p\, q\, f \tag{21.36}$$

uniformly on $[0,1]$.

PROOF Let $f \in C[0,1]$ and $x \in [0,1]$. We have

$$|C_n f(x) - w(x)\,p(x)\,q(x)\,f(x)| \le |C_n f(x) - f(x)\,C_n\mathbf{1}(x)| + |f(x)|\,|C_n\mathbf{1}(x) - w(x)\,p(x)\,q(x)|. \quad (21.37)$$

Since $\beta_{n,k} \le s(n)\binom{n}{k}$ (see the preceding subsection), from (21.32) and (21.34) the first term can be estimated as follows:

$$\begin{aligned}
&|C_n f(x) - f(x)\,C_n\mathbf{1}(x)| \\
&= \sum_{k=0}^{n} \beta_{n,k} \prod_{i=1}^{k}(x\,p_i(x)) \prod_{j=1}^{n-k}((1-x)\,q_j(x)) \left|f\left(\frac{k}{n}\right) - f(x)\right| \\
&\le r_n(x)\,s(n) \sum_{k=0}^{n} \binom{n}{k} x^k(1-x)^{n-k} \left|f\left(\frac{k}{n}\right) - f(x)\right| \\
&\le r_n(x)\,s(n) \sum_{k=0}^{n} \binom{n}{k} x^k(1-x)^{n-k} \left(1 + \frac{1}{\delta^2}\left(\frac{k}{n} - x\right)^2\right) \omega(f,\delta) \\
&= r_n(x)\,s(n) \left(1 + \frac{x(1-x)}{n\delta^2}\right) \omega(f,\delta)
\end{aligned}$$

for every $\delta > 0$, and taking $\delta = 1/\sqrt{n}$, we get

$$|C_n f(x) - f(x)\,C_n\mathbf{1}(x)| \le \frac{5}{4}\, r_n(x)\,s(n)\,\omega\left(f, \frac{1}{\sqrt{n}}\right).$$

Taking (21.33) into account, this completes the proof of (21.35).

Finally, by (21.37) we have only to show that the sequence $(C_n\mathbf{1})_{n\ge1}$ converges uniformly to $w\,p\,q$. Indeed, we can write $\beta_{n,k} = \binom{n}{k} b_{n,k}$ and the continuous piecewise affine functions b_n satisfying $b_n(k/n) = b_{n,k}$ for $k = 0, \ldots, n$ converge uniformly to w (see Proposition 21.1, 1)); by (21.32) we have

$$\begin{aligned}
&|C_n\mathbf{1}(x) - w(x)\,p(x)\,q(x)| \\
&\le \left|\sum_{k=0}^{n} \binom{n}{k} x^k(1-x)^{n-k} \prod_{i=1}^{k} p_i(x) \prod_{j=1}^{n-k} q_j(x)\,b_{n,k} - w(x)\,p(x)\,q(x)\right| \\
&\le r_n(x) \sum_{k=0}^{n} \binom{n}{k} x^k(1-x)^{n-k}\,|b_{n,k} - b_n(x)| \\
&\quad + \sum_{k=0}^{n} \binom{n}{k} x^k(1-x)^{n-k} \prod_{i=1}^{k} p_i(x) \prod_{j=1}^{n-k} q_j(x)\,|b_n(x) - w(x)|
\end{aligned}$$

$$+|w(x)|\left(\sum_{k=0}^{n}\binom{n}{k}\,x^k(1-x)^{n-k}\prod_{i=1}^{k}p_i(x)\prod_{j=1}^{n-k}q_j(x)\;-p(x)\,q(x)\right).$$

The first addend in the last sum tends uniformly to 0 since it coincides with $r_n(x)\,|B_nb_n(x)-b_n(x)|$ and $(b_n)_{n\geq 1}$ is equicontinuous. Since $(b_n)_{n\geq 1}$ tends uniformly to w also the second addend tends uniformly to 0 and finally by Proposition 21.2 the same holds for the third addend. Thus, (21.36) is also proved. ∎

The case considered in the preceding subsection corresponds to constant sequences $(p_n)_{n\geq 1}$ and $(q_n)_{n\geq 1}$, where the infinite products (21.31) converge if and only if $p_n = q_n = 1$. Theorem 21.4 considers the more general condition $xp(x)+(1-x)q(x) = 1$ (or also $xp(x)+(1-x)q(x) \leq 1$) which can be generalized as follows (see [19, Corollary 4]).

COROLLARY 21.1

Assume that the sequences $(\lambda_m)_{m\geq 1}$ and $(\rho_m)_{m\geq 1}$ converge and that there exist continuous functions p, q, u and v such that

a) $$\lim_{n\to+\infty}\frac{\prod_{i=1}^{n}p_i(x)}{u(x)^n}=p(x),\qquad \lim_{n\to+\infty}\frac{\prod_{j=1}^{n}q_j(x)}{v(x)^n}=q(x);$$

b) The function $x\mapsto x\,u(x)$ is strictly increasing and, for every $x\in[0,1]$, $x\,u(x)+(1-x)\,v(x)=1$.

Then, for every $f\in C[0,1]$,

$$\lim_{n\to+\infty}C_nf(x)=w(x\,u(x))\,p(x)\,q(x)\,f(x\,u(x))$$

uniformly in $x\in C[0,1]$.

PROOF For every $f\in C[0,1]$ and $x\in[0,1]$, we can write

$$C_nf(x)=\sum_{k=0}^{n}\beta_{n,k}(x\,u(x))^k((1-x)\,v(x))^{n-k}\,\frac{\prod_{i=1}^{k}p_i(x)}{u(x)^k}\times$$
$$\times\frac{\prod_{j=1}^{n-k}q_j(x)}{v(x)^{n-k}}\,f\left(\frac{k}{n}\right).$$

By assumption b), we have $(1-x)\,v(x) = 1-x\,u(x)$ and therefore the operator C_n can be considered as the operator arising by (21.29) corresponding to the sequences

$$\left(\frac{p_n(\varphi(x))}{u(\varphi(x))}\right)_{n\geq 1}, \qquad \left(\frac{q_n(\varphi(x))}{v(\varphi(x))}\right)_{n\geq 1}$$

and evaluated at the point $x\,u(x)$, where φ denotes the inverse of $x \mapsto x\,u(x)$.

At this point, by the assumption a) we may apply Theorem 21.8 and conclude the proof. ∎

A quantitative estimate of the convergence can be directly obtained from (21.35).

The assumption b) in Corollary 21.1 ensures that $C_n\mathbf{1}(x)$ does not tend to 0; without this restriction, we may assume the weaker condition $x\,u(x) + (1-x)\,v(x) \leq 1$, $0 \leq x \leq 1$.

Also in this context, different examples may be obtained restating those in Section 21.2 using particular sequences $(p_n)_{n\geq 1}$ and $(q_n)_{n\geq 1}$.

21.4 Bernstein-Kantorovitch-type Operators

In this section, we introduce and study the Kantorovitch-type operators obtained by considering the general binomial-type coefficients introduced in Section 21.2 in place of the classical one. The results in this section are taken from [23], while the Voronovskaja-type formula has been obtained by I. Raşa.

So, we fix two sequences $(\lambda_m)_{m\geq 1}$ and $(\rho_m)_{m\geq 1}$ of real numbers and define the coefficients $\alpha_{n,k}$ according to (21.1) and (21.2).

Let $1 \leq p < +\infty$; for every $n \geq 1$, we define the operator $K_n : L^p(0,1) \to L^p(0,1)$, by setting, for every $f \in L^p(0,1)$ and $x \in [0,1]$,

$$K_n f(x) = \sum_{k=0}^{n} \alpha_{n,k} x^k (1-x)^{n-k} (n+1) \int_{k/(n+1)}^{(k+1)/(n+1)} f(t)\,dt.$$

If we take $\lambda_m = \rho_m = 1$ for every $m \geq 1$, the operators K_n coincide with the classical Kantorovitch operators $U_n : L^p(0,1) \to L^p(0,1)$ defined by

$$U_n f(x) = \sum_{k=0}^{n} \binom{n}{k} x^k (1-x)^{n-k} (n+1) \int_{k/(n+1)}^{(k+1)/(n+1)} f(t)\,dt.$$

We recall that (see, e.g., (see, e.g., [35, p. 31] and [8, Section 5.3.7])

$$U_n(\mathbf{1}) = \mathbf{1}, \qquad U_n(\mathrm{id})(x) = \frac{n}{n+1}x + \frac{1}{2(n+1)},$$

$$U_n(\mathrm{id}^2)(x) = \frac{n(n-1)}{(n+1)^2}x^2 + \frac{2n}{(n+1)^2}x + \frac{1}{3(n+1)^2},$$

where $\mathrm{id}(x) = x$ for every $x \in [0,1]$. From the preceding expression, it is easy to estimate

$$U_n((\mathrm{id} - x\cdot\mathbf{1})^2)(x) = \frac{n-1}{(n+1)^2}x(1-x) + \frac{1}{3(n+1)^2} \le \frac{1}{4(n+1)}$$

and it is also possible to evaluate the general operator K_n at the function $\mathbf{1}$, for every $x \in [0,1]$ (see also (21.5))

$$\begin{aligned} K_n\mathbf{1}(x) &= \sum_{m=1}^{n-1} (\lambda_m x(1-x)^m + \rho_m x^m(1-x)) + \lambda_n(1-x)^n + \rho_n x^n \\ &= A_n\mathbf{1}(x)\,. \end{aligned} \tag{21.38}$$

Since $\sum_{k=0}^{n} \binom{n}{k} x^k(1-x)^{n-k} = 1$ and $x^k(1-x)^{n-k} \ge 0$ for every $x \in [0,1]$, by the convexity of the function $t \mapsto t^p$ and Jensen's inequality applied to the measure $(n+1)\,dt$, we get, for every $f \in L^p(0,1)$ (see also (21.7))

$$|K_n f(x)|^p \le s(n)^p \sum_{k=0}^{n} \binom{n}{k} x^k(1-x)^{n-k}(n+1)\int_{k/(n+1)}^{(k+1)/(n+1)} f(t)\,dt.$$

Consequently, the equality

$$\int_0^1 x^k(1-x)^{n-k}\,dx = \frac{1}{n+1}\binom{n}{k}^{-1}, \qquad k = 0,\dots,n$$

yields $\|K_n(f)\|_p \le s(n)\|f\|_p$ and therefore

$$\|K_n\|_p \le s(n). \tag{21.39}$$

Conversely, if we evaluate $\|K_n(f)\|_p$ taking $f = \mathrm{sign}(\lambda_n)\cdot\chi_{[0,1/(n+1)]}$ ($\chi_{[0,1/(n+1)]}$ denotes the characteristic function of the interval $[0,1/(n+1)]$), we get $\|f\|_p = 1/(n+1)^{1/p}$ and consequently

$$\|K_n\|_p \ge \frac{\|K_n(f)\|_p}{\|f\|_p} = |\lambda_n|\left(\frac{n+1}{np+1}\right)^{1/p} \ge p^{-1/p}|\lambda_n|;$$

thus $|\lambda_n| \le p^{1/p} |||K_n||_p$ and analogously $|\rho_n| \le p^{1/p} |||K_n||_p$.

From (21.39), we get the following result (see [23, Proposition 1]).

PROPOSITION 21.3

The sequence $(\|K_n\|_p)_{n\ge1}$ *is bounded if and only if both the sequences* $(\lambda_m)_{m\ge1}$ *and* $(\rho_m)_{m\ge1}$ *are bounded and, in this case,*

$$\sup_{n\ge1} \|K_n\|_p \le M,$$

where $M = \sup_{n\ge1}\{|\lambda_n|, |\rho_n|\}$.

In the sequel, we shall assume that both the sequences $(\lambda_m)_{m\ge1}$ and $(\rho_m)_{m\ge1}$ are bounded. In this case, the function w defined in (21.6) also satisfies $\|w\| \le M$. The function w is not necessarily continuous (since the sequences $(\lambda_m)_{m\ge1}$ and $(\rho_m)_{m\ge1}$ do not necessarily converge); indeed, for positive sequences $(\lambda_m)_{m\ge1}$ and $(\rho_m)_{m\ge1}$, the existence of the limit $\lim_{x\to0^+} w(x)$ (respectively, $\lim_{x\to1^-} w(x)$) is equivalent to that of the limit $\lim_{m\to+\infty}(\lambda_1 + \ldots + \lambda_m)/m$ (respectively, $\lim_{m\to+\infty}(\rho_1 + \ldots + \rho_m)/m$) and, in this case, the two limits coincide (see, e.g., [49, Chapter 7, Section 5, p. 226–229]).

We can now state a characterization of the convergence of the sequence $(K_n)_{n\ge1}$ in $L^p(0,1)$ toward the operator $K : L^p(0,1) \to L^p(0,1)$ defined by $K(f) = f \cdot w$ for every $f \in L^p(0,1)$; observe that K is continuous in the L^p-norm and satisfies $\|K\|_p = \|w\|$.

THEOREM 21.9 [23, Theorem 3]

For every $f \in L^p(0,1)$, *the sequence* $(K_n(f))_{n\ge1}$ *converges in the space* $L^p(0,1)$ *if and only if the sequences* $(\lambda_m)_{m\ge1}$ *and* $(\rho_m)_{m\ge1}$ *are bounded.*

In this case, we have

$$\lim_{n\to+\infty} \|K_n(f) - w \cdot f\|_p = 0 \tag{21.40}$$

for every $f \in L^p(0,1)$.

PROOF From the Banach-Steinhaus theorem and Proposition 21.3, we have only to show that if $(\lambda_m)_{m\ge1}$ and $(\rho_m)_{m\ge1}$ are bounded, then (21.40) holds. Since by Proposition 21.3, the sequence $(K_n)_{n\ge1}$ is equibounded in $L^p(0,1)$, it is enough to show (21.40) for every $f \in C[0,1]$.

But if $f \in C[0,1]$, we have

$$\|K_n(f) - w \cdot f\|_p \le \|K_n(f) - f \cdot K_n(\mathbf{1})\| + \|f\| \cdot \|K_n(\mathbf{1}) - w\|_p.$$

With regard to the first addend, with the same straighforward techniques used in Theorem 21.1, it is easy to obtain

$$\|K_n(f) - f \cdot K_n(\mathbf{1})\| \le \frac{5}{4} s(n)\omega\left(f, \frac{1}{\sqrt{n+1}}\right).$$

The second addend may be estimated using (21.38) and (21.6) as follows:

$$\begin{aligned}|K_n\mathbf{1}(x) - w(x)| &= \left|(1-x)^n \sum_{m=0}^{+\infty} (\lambda_n - \lambda_{n+m}) x (1-x)^m \right.\\ &\qquad \left. + x^n \sum_{m=0}^{+\infty} (\rho_n - \rho_{n+m}) x^m (1-x) \right| \\ &\le 2s(n)\left((1-x)^n + x^n\right).\end{aligned}$$

It follows that

$$\begin{aligned}\|f \cdot K_n(\mathbf{1}) - w\|_p &\le 2M \left(\int_0^1 ((1-x)^n + x^n)^p \, dx\right)^{1/p} \|f\| \\ &\le 4M \left(\int_0^1 x^{np} \, dx\right)^{1/p} \|f\| = 4M \left(\frac{1}{np+1}\right)^{1/p} \|f\|\end{aligned}$$

and this completes the proof. ∎

A quantitative estimate of the convergence in Theorem 21.9 may be stated using the so-called averaged modulus of smoothness (see [43]) defined by

$$\tau(f, \delta)_p = \left(\int_0^1 \omega(f, x, \delta)^p \, dx\right)^{1/p}$$

for every $f \in L^p(0,1)$, $\delta > 0$ and where

$$\omega(f, x, \delta) = \sup\left\{ |f(t+h) - f(t)| \mid t, t+h \in \left[x - \frac{\delta}{2}, x + \frac{\delta}{2}\right] \cap [0,1] \right\}.$$

As stated in [23, Theorem 5], for every $f \in L^p(0,1)$, we have

$$\|K_n(f) - w \cdot f\|_p \le 1683 M \tau\left(f, \frac{1}{\sqrt{n+1}}\right)_p + 4M\left(\frac{1}{np+1}\right)^{1/p} \|f\|.$$

With regard to the pointwise convergence of $(K_n(f)))_{n\geq 1}$ for a fixed $f \in L^p(0,1)$, we have the following result, whose proof is omitted for brevity.

THEOREM 21.10 [23, Proposition 4]
Assume that $(\lambda_m)_{m\geq 1}$ and $(\rho_m)_{m\geq 1}$ are bounded sequences of real numbers. If $f \in L^p(0,1)$ and if $x \in]0,1[$ is a Lebesgue point for f, i.e.,

$$\lim_{\delta\to 0}\frac{1}{\delta}\int_0^\delta |f(x+t)-f(x)|\,dt = 0$$

then

$$\lim_{n\to+\infty} K_n f(x) = w(x)\cdot f(x).$$

Hence $\lim_{n\to+\infty} K_n f = w\cdot f$ *almost everywhere.*

The analogousness of Theorem 21.10 for the classical Kantorovitch operator can be found in [35, p. 30].

In order to state a Voronovskaja-type formula for the operators K_n, we define

$$M_n = \frac{K_n}{K_n\mathbf{1}}.$$

We assume that the sequences $(\lambda_m)_{m\geq 1}$ and $(\rho_m)_{m\geq 1}$ are strictly positive and converge to $\lambda_\infty > 0$ and $\rho_\infty > 0$, respectively. This will ensure the validity of the Voronovskaja's formula (21.2) for the operators $(L_n)_{n\geq 1}$.

THEOREM 21.11 [Raşa]
For every $f \in C^2[0,1]$, we have

$$\lim_{n\to+\infty} n\left(M_n f(x) - f(x)\right) = \frac{x(1-x)}{2}f''(x) + \left(\frac{1-2x}{2} + x(1-x)\frac{w'(x)}{w(x)}\right)f'(x) \qquad (21.41)$$

uniformly in $x \in [0,1]$.

PROOF Fix $f \in C^2[0,1]$ and $n \geq 1$. We denote by f_n the continuous function which takes the value

$$f_n\left(\frac{k}{n}\right) = (n+1)\int_{k/(n+1)}^{(k+1)/(n+1)} f(t)\,dt$$

at every point k/n, $k = 0, \ldots, n$, and is extended linearly on each interval $[k/n, (k+1)/n]$, $k = 0, \ldots, n-1$. We observe that

$$M_n f = \frac{K_n f}{K_n \mathbf{1}} = \frac{A_n f_n}{A_n \mathbf{1}} = L_n f_n. \tag{21.42}$$

Moreover, we define the function $\varphi(t) = (1/2 - t) f'(t)$, $t \in [0,1]$.

Consider $x \in [0,1]$ and let $k = 0, \ldots, n-1$ be such that $k/n \le x \le (k+1)/n$; then

$$\begin{aligned}
n(f_n(x) - f(x)) - \varphi(x) &= n\Big[(n+1)(nx-k) \int_{(k+1)/(n+1)}^{(k+2)/(n+1)} f(t)\,dt \\
&\quad +(n+1)(k+1-nx) \int_{k/(n+1)}^{(k+1)/(n+1)} f(t)\,dt - f(x)\Big] \\
&\quad -\Big(\frac{1}{2} - x\Big) f'(x) \\
&= n\Big[(n+1)(nx-k) \int_{(k+1)/(n+1)}^{(k+2)/(n+1)} \big(f(x) + (t-x)f'(x) + R(t)\big)\,dt \\
&\quad +(n+1)(k+1-nx) \int_{k/(n+1)}^{(k+1)/(n+1)} \big(f(x) + (t-x)f'(x) + R(t)\big)\,dt \\
&\quad -f(x)\Big] - \Big(\frac{1}{2} - x\Big) f'(x) \\
&= -\frac{1}{n+1}\Big(\frac{1}{2} - x\Big) f'(x) + n(n+1)\Big[(nx-k) \int_{(k+1)/(n+1)}^{(k+2)/(n+1)} R(t)\,dt \\
&\quad +(k+1-nx) \int_{k/(n+1)}^{(k+1)/(n+1)} R(t)\,dt\Big],
\end{aligned}$$

where $|R(t)| \le (t-x)^2 \|f''\|/2$ and $0 \le nx - k \le 1$, $0 \le k+1-nx \le 1$. Thus, we have obtained

$$\begin{aligned}
|n(f_n(x) - f(x)) - \varphi(x)| &\le \frac{1}{2(n+1)} \|f'\| \\
&\quad + n(n+1) \int_{k/(n+1)}^{(k+2)/(n+1)} (t-x)^2 \frac{\|f''\|}{2}\,dt.
\end{aligned}$$

Since $k/n \le x \le (k+1)/n$ we have also $k/(n+1) \le x \le (k+2)/(n+1)$ and consequently $(t-x)^2 \le 4/(n+1)^2$ for every $t \in [k/(n+1), (k+2)/(n+1)]$. It follows that

$$\|n(f_n - f) - \varphi\| \le \frac{1}{2(n+1)} \|f'\| + \frac{4n}{(n+1)^2} \|f''\|$$

and this proves that

$$\lim_{n\to+\infty} n(f_n - f) = \varphi \qquad \text{uniformly on } [0,1]. \tag{21.43}$$

At this point we recall that $\|L_n\| = 1$ and $\lim_{n\to+\infty} L_n\varphi = \varphi$ uniformly on $[0,1]$; taking into account (21.42) and (21.43), we obtain

$$\begin{aligned}
\lim_{n\to+\infty} & n(M_n f(x) - f(x)) \\
&= \lim_{n\to+\infty} \big(L_n\big(n(f_n(x) - f(x)) - \varphi(x)\big) \\
&\qquad + L_n\varphi(x) + n(L_n f(x) - f(x))\big) \\
&= \varphi(x) + \frac{x(1-x)}{2} f''(x) + x(1-x)\frac{w'(x)}{w(x)} f'(x) \\
&= \left(\frac{x(1-x)}{2} f'(x)\right)' + x(1-x)\frac{w'(x)}{w(x)} f'(x)
\end{aligned}$$

uniformly in $x \in [0,1]$. ∎

In the case of the classical Kantorovitch operators on $L^p(0,1)$, we have $U_n\mathbf{1} = \mathbf{1}$ and consequently $M_n = U_n$; thus, formula (21.41) generalizes the classical Voronovskaja's formula for Kantorovitch operators.

21.5 Baskakov-type Operators on the Half-line

In this section, we consider some extensions of some results in Section 21.2 to the unbounded real interval $[0,+\infty[$.

The most natural way of extending them might seem the introduction of the general binomial-type coefficients defined in (21.1)-(21.2) in the sequence of Bernstein-Chlodowski operators (see, e.g., [8, Section 5.3.12, p. 347–348]), considering the operators $C_n^* : C[0,+\infty[\to C[0,+\infty[$ defined by

$$C_n^* f(x) = \begin{cases} \displaystyle\sum_{k=0}^{n} \alpha_{n,k} \left(\frac{x}{b_n}\right)^k \left(1 - \frac{x}{b_n}\right)^{n-k} f\left(\frac{b_n k}{n}\right), & 0 \le x \le b_n, \\ f(x), & x > b_n, \end{cases}$$

where $(b_n)_{n\geq 1}$ is a sequence of real numbers such that

$$\lim_{n\to+\infty} b_n = +\infty, \qquad \lim_{n\to+\infty} \frac{b_n}{n} = 0$$

and the coefficients $\alpha_{n,k}$ depend on two fixed sequences $(\lambda_m)_{m\geq 1}$ and $(\rho_m)_{m\geq 1}$ of real numbers satisfying (21.8).

Unfortunately, it is easy to observe that $C_n^*\mathbf{1}(x) = A_n\mathbf{1}(x/b_n)$ and therefore, the sequence $(C_n(\mathbf{1}))_{n\geq 1}$ converges to $\lambda_\infty \cdot \mathbf{1}$ (see Theorem 21.1); consequently, it is possible to show that the sequence $(C_n(f))_{n\geq 1}$ converges to $\lambda_\infty \cdot f$ for every $f \in C[0,+\infty[$ having a finite limit at $+\infty$. In the case of the operators A_n, it has been possible to consider a non-zero first-order perturbation term in the Voronovskaja-type formula since the derivative of the function w does not necessarily vanish (see Theorem 21.2); as a consequence, in the case of the Bernstein-Chlodowski-type operators defined above, we cannot obtain a more general Voronovskaja-type formula.

For this reason, we consider now a different extension based on the classical Baskakov operators on the unbounded interval $[0,+\infty[$. This extension has been performed by Agratini [1, 2] as follows.

Let $(\lambda_m)_{m\geq 1}$ and $(\rho_m)_{m\geq 1}$ be two sequences of real numbers and define the coefficients $\alpha_{n,k}$ as in (21.1)-(21.2) and, for every $n \geq 1$, define the operator $R_n : C[0,+\infty[\to C[0,+\infty[$ by putting

$$R_n f(x) = \sum_{k=0}^{+\infty} \alpha_{n+k-1,k} \frac{x^k}{(1+x)^{n+k}} f\left(\frac{k}{n}\right). \tag{21.44}$$

Also in this case, if $\lambda_m = \rho_m = 1$ for every $m \geq 1$, we obtain the classical Baskakov operators $V_n : C[0,+\infty[\to C[0,+\infty[$ by putting

$$V_n f(x) = \sum_{k=0}^{+\infty} \binom{n+k-1}{k} \frac{x^k}{(1+x)^{n+k}} f\left(\frac{k}{n}\right).$$

The definition of the operators R_n given in [1] is slightly different; in that paper two sequences $(a_m)_{m\geq 1}$ and $(b_k)_{k\geq 0}$ of real numbers are considered satisfying $a_1 = b_0$ and the coefficients $c_{n,k}$ are defined by setting

$$c_{n,0} = a_n, \quad n \geq 1, \qquad c_{1,k} = b_k, \quad k \geq 0$$
$$c_{n+1,k} = c_{n,k} + c_{n+1,k-1}, \qquad n \geq 1,\ k \geq 1.$$

Then the corresponding operators are defined as in (21.44) with $c_{n,k}$ in place of $\alpha_{n+k-1,k}$. Imposing $c_{n,k} = \alpha_{n+k-1,k}$, we have $\lambda_m = \alpha_{m,0} =$

$c_{m+1,0} = a_{m+1}$ and $\rho_m = \alpha_{m,m} = c_{1,m} = b_m$; therefore the two definitions are equivalent.

Using a decomposition similar to (21.4), it is possible to prove the following convergence result.

THEOREM 21.12 [1, Theorem 3]
Assume that the sequences $(\lambda_m)_{m\geq 1}$ and $(\rho_m)_{m\geq 1}$ are bounded and let

$$M = \max\left\{\sup_{m\geq 1}|\lambda_m|, \sup_{m\geq 1}|\rho_m|\right\}.$$

If $f \in C[0,+\infty[$ is bounded, we have, for every $x \geq 0$ and $n \geq 3$,

$$|R_n f(x) - \varphi(x)f(x)| \leq M\left((1+x+x^2)\omega\left(f, \frac{1}{\sqrt{n}}\right) + \frac{2|f(x)|}{(1+x)^{n-1}}\right),$$

where

$$\varphi(x) = \sum_{m=1}^{+\infty}\frac{\lambda_m x}{(1+x)^{m+1}} + \sum_{m=1}^{+\infty}\frac{\rho_m x^m}{(1+x)^{m+1}}.$$

Consequently,

$$\lim_{n\to+\infty} R_n f = \varphi\cdot f$$

uniformly on every compact subset K of $]0,+\infty[$.

For the operators R_n, an interesting Voronovskaja-type formula can be stated as follows.

THEOREM 21.13 [2, Theorems 3 and 4]
Let $f \in C[0,+\infty[$ and assume that $f(x)/(1+x^2)$ is bounded and that f is twice differentiable at a point $x > 0$. If the sequences $(\lambda_m)_{m\geq 1}$ and $(\rho_m)_{m\geq 1}$ are bounded, then

$$\lim_{n\to+\infty} n(R_n f(x) - f(x)\,R_n\mathbf{1}(x)) = \frac{x(1+x)}{2}\left(\varphi f''(x) + 2\varphi'(x)f'(x)\right).$$

21.6 Generalizations to the Multi-dimensional Case

In this last section, we briefly describe some recent developments concerning the generalization of the binomial coefficients in the sequence of Bernstein operators on the hypercube and the standard simplex of $\mathbf{R}^d$.

21.6.1 Bernstein-type operators on the hypercube

Let $(\lambda_m^{(i)})_{m\geq 1}$and $(\rho_m^{(i)})_{m\geq 1}$, $i = 1,\ldots,d$, be $2d$ sequences of real numbers. According to (21.1)-(21.2), for every $i = 1,\ldots,d$, we define the general recursive coefficients $\alpha_{n,k}^{(i)}$, $n \geq 1$, $k = 0,\ldots,n$ as follows:

1) $\alpha_{n,0}^{(i)} = \lambda_n^{(i)}$ and $\alpha_{n,n}^{(i)} = \rho_n^{(i)}$;

2) $\alpha_{n,k}^{(i)} = \alpha_{n-1,k-1}^{(i)} + \alpha_{n-1,k}^{(i)}$ for every $k = 1,\ldots,n-1$

and, for every $n \geq 1$, the operator $A_n : C([0,1]^d) \to C([0,1]^d)$ by

$$A_n f(x_1,\ldots,x_d) = \sum_{k_1,\ldots,k_d=0}^{n} \left(\prod_{i=1}^{d} \alpha_{n,k_i}^{(i)} x_i^{k_i}(1-x_i)^{n-k_i} \right) f\left(\frac{k_1}{n},\ldots,\frac{k_d}{n}\right)$$

$(f \in C([0,1]^d)$, $(x_1,\ldots,x_d) \in [0,1]^d)$.

Again, if we take the constant sequences $\lambda_m^{(i)} = 1$ for every $m \geq 1$ and $i = 1,\ldots,d$, we get the classical Bernstein operators $B_n : C([0,1]^d) \to C([0,1]^d)$ defined by

$$B_n f(x_1,\ldots,x_d) = \sum_{k_1,\ldots,k_d=0}^{n} \left(\prod_{i=1}^{d} \binom{n}{k_i} x_i^{k_i}(1-x_i)^{n-k_i} \right) f\left(\frac{k_1}{n},\ldots,\frac{k_d}{n}\right).$$

We can study the convergence of the sequence $(A_n)_{n\geq 1}$ following the same line of [21] and outlined in Section 21.2; hence, we assume that all the sequences $(\lambda_m^{(i)})_{m\geq 1}$ and $(\rho_m^{(i)})_{m\geq 1}$ converge and put

$$\lambda_\infty^{(i)} = \lim_{m\to+\infty} \lambda_m^{(i)}, \qquad \rho_\infty^{(i)} = \lim_{m\to+\infty} \rho_m^{(i)}$$

for every $i = 1,\ldots,d$. Moreover, for every $i = 1,\ldots,d$, we can define the function $w^{(i)} : [0,1] \to \mathbf{R}$ as in [21, (2.5)] (see also (21.6))

$$w^{(i)}(x) = \begin{cases} \lambda_\infty^{(i)}, & \text{if } x = 0, \\ \rho_\infty^{(i)}, & \text{if } x = 1, \\ \displaystyle\sum_{m=1}^{+\infty} \lambda_m^{(i)} x(1-x)^m + \sum_{m=1}^{+\infty} \rho_m^{(i)} (1-x)x^m, & \text{if } 0 < x < 1 \end{cases}$$

and finally the function $w : [0,1]^d \to \mathbf{R}$ defined by setting, for every $x = (x_1, \ldots, x_d) \in [0,1]^d$,

$$w(x) = \prod_{i=1}^{d} w^{(i)}(x_i).$$

If we consider the operator $A_n^{(i)} : C([0,1]) \to C([0,1])$, $i = 1, \ldots, d$, defined as in (21.3) and corresponding to the sequences $(\lambda_m^{(i)})_{m\geq 1}$ and $(\rho_m^{(i)})_{m\geq 1}$, i.e.,

$$A_n^{(i)} f(x) = \sum_{k=0}^{n} \alpha_{n,k}^{(i)} x^k (1-x)^{n-k} f\left(\frac{k}{n}\right),$$

$(f \in C([0,1]), x \in [0,1])$ by [21, Theorem 2.2], we have $\lim_{n\to+\infty} A_n^{(i)} f = w^{(i)} \cdot f$ uniformly on [0,1] for every $f \in C[0,1]$; in particular, since for every $x = (x_1, \ldots, x_d) \in [0,1]^d$, $A_n \mathbf{1}(x) = \prod_{i=1}^{d} A_n^{(i)} \mathbf{1}(x_i)$, we get $\lim_{n\to+\infty} A_n \mathbf{1} = w$.

Further, from Proposition 21.1, we can state analogous properties for the function w. The main convergence result can be stated as follows.

THEOREM 21.14
If the sequences $(\lambda_m^{(i)})_{m\geq 1}$ and $(\rho_m^{(i)})_{m\geq 1}$ converge for every $i = 1, \ldots, d$, then $(A_n)_{n\geq 1}$ converges strongly on $C([0,1]^d)$ and

$$\lim_{n\to+\infty} A_n f = w \cdot f \qquad \text{uniformly on } [0,1]^d$$

for every $f \in C([0,1]^d)$. More precisely, if we define

$$r(n) = \sup_{m\geq n} \max_{i=1,\ldots,d} \{|\lambda_m^{(i)} - \lambda_n^{(i)}|, |\rho_m^{(i)} - \rho_n^{(i)}|\},$$

we have, for every $n \geq 1$, $f \in C([0,1]^d)$ and $x = (x_1, \ldots, x_d) \in [0,1]^d$,

$$|A_n f(x) - w(x) f(x)| \leq s(n) \left(1 + \sum_{i=1}^{d} x_i(1-x_i)\right) \omega\left(f, \frac{1}{\sqrt{n}}\right)$$
$$+ r(n) \prod_{i=1}^{d} ((1-x_i)^n + x_i^n)\, |f(x)|,$$

where, as usual, $\omega(f,\delta) = \sup\{|f(x) - f(y)| \mid x, y \in [0,1]^d, \|x-y\| \leq \delta\}$.

In particular,

$$\|A_n f - w \cdot f\| \leq \left(1 + \frac{d}{4}\right) s(n)\, \omega\left(f, \frac{1}{\sqrt{n}}\right) + r(n)\, \|f\|.$$

Finally, we state the following Voronovskaja-type formula; as usual, we denote by $C^2([0,1]^d)$ the space of all functions in $C([0,1]^d)$ having a C^2-extension on a neighborhood of $[0,1]^d$.

THEOREM 21.15
Assume that the sequences $(\lambda_m^{(i)})_{m\geq 1}$ and $(\rho_m^{(i)})_{m\geq 1}$, $i = 1, \ldots, d$, converge. Then, for every $f \in C^2([0,1]^d)$, we have

$$\lim_{n\to+\infty} n(A_n f(x) - A_n \mathbf{1}(x)\, f(x)) = w(x) \sum_{i=1}^{d} \frac{x_i(1-x_i)}{2} \frac{\partial^2 f}{\partial x_i^2}(x) + \sum_{i=1}^{d} x_i(1-x_i) \frac{\partial w}{\partial x_i}(x) \frac{\partial f}{\partial x_i}(x)$$

uniformly in $x = (x_1, \ldots, x_d) \in [0,1]^d$.

The existence of a semigroup represented in terms of iterates of the operators A_n can be discussed in a straightforward way using the methods described in [8, Section 6.3.10, p. 470–476] and therefore, for the sake of brevity, we omit the details.

21.6.2 Bernstein-type operators on a simplex

Let

$$K_d = \left\{(x_1, \ldots, x_d) \in \mathbf{R}^d \mid x_1, \ldots, x_d \geq 0, \sum_{i=1}^{d} x_i \leq 1\right\}$$

be the standard simplex in $\mathbf{R}^d$, $d > 1$. The vertices of K_d are denoted by $v_i = (\delta_{ij})_{j=1,\ldots,d}$ for every $i = 0, \ldots, d$. Moreover, for every $n \geq 1$, $(x_1, \ldots, x_d) \in K_d$ and $h_1, \ldots, h_d \geq 0$ such that $h_1 + \ldots + h_d \leq n$, it is convenient to put

$$x_0 = 1 - x_1 - \ldots - x_d, \qquad h_0 = n - h_1 - \ldots - h_d$$

and to consider the set

$$H_n^0 = \left\{(h_0, h_1, \ldots, h_d) \in \mathbf{N}^{d+1} \mid h_1 + \ldots + h_d \leq n\right\}.$$

Now, consider $d+1$ sequences $(\lambda_m^{(i)})_{m\geq 1}$, $i=0,\ldots,d$, of real numbers and define the binomial-type coefficients $\alpha_{n,h}$ for every integer $n\geq 1$ and $h=(h_0,h_1,\ldots,h_d)\in H_n^0$ according to the following formulas:

1) If there exists $i=0,\ldots,d$ such that $h_i=n$, then $\alpha_{n,h}=\lambda_n^{(i)}$;

2) If $h_i<n$ for every $i=0,\ldots,d$, we put $\alpha_{n,h}=\sum_{i\in S(h)}\alpha_{n-1,k^{(i)}}$, where $S(h)=\{i=0,\ldots,d \mid h_i>0\}$ and, for every $i\in S(h)$, $k^{(i)}=(k_0^{(i)},\ldots,k_d^{(i)})$ is given by

$$k_j^{(i)}=\begin{cases} h_j, & \text{if } j\neq i; \\ h_i-1, & \text{if } j=i. \end{cases}$$

At this point, for every $n\geq 1$, the operator $A_n : C(K_d)\to C(K_d)$ is defined as the classical Bernstein operators, but considering the more general coefficients $\alpha_{n,h}$ in place of the binomial ones; thus, we put

$$A_n f(x_1,\ldots,x_d)=\sum_{h=(h_0,\ldots,h_d)\in H_n^0}\alpha_{n,h}\,x_0^{h_0}x_1^{h_1}\ldots x_d^{h_d}f\left(\frac{h_1}{n},\ldots,\frac{h_d}{n}\right)$$

for every $f\in C(K_d)$ and $(x_1,\ldots,x_d)\in K_d$.

We observe that, if we take $\lambda_m^{(i)}=1$ for every $m\geq 1$ and $i=0,\ldots,d$, then $\alpha_{n,h}=n!/(h_1!\ldots h_d!(n-h_1-\ldots-h_d)!)$ and consequently the operators A_n coincide with the classical Bernstein operators on K_d

$$B_n f(x_1,\ldots,x_d)$$
$$=\sum_{h=(h_0,\ldots,h_d)\in H_n^0}\frac{n!}{h_0!\ldots h_d!}x_0^{h_0}x_1^{h_1}\ldots x_d^{h_d}f\left(\frac{h_1}{n},\ldots,\frac{h_d}{n}\right).$$

As in the one-dimensional case, it is possible to decompose the operator A_n as $A_n=\sum_{m=1}^{n}\sum_{j=0}^{d}\lambda_m^{(j)}L_{m,n}^{(j)}$, where, for every $j=0,\ldots,d$ and $m\geq 1$, $L_{m,n}^{(j)}$ is the elementary operator associated with the sequences $(\delta_i^j\delta_n^m)_{n\geq 1}$, $i=0,\ldots,d$.

The study of the convergence of the sequence $(A_n)_{n\geq 1}$ can be carried out by writing explicit expressions for the operators $L_{m,n}^{(j)}$ and using the above decomposition. For every $i=0,\ldots,d$, we have $A_n f(v_i)=\lambda_n^{(i)}f(v_i)$ and therefore the convergence of the sequences $(\lambda_m^{(i)})_{m\geq 1}$, $i=0,\ldots,d$, is necessary for that of $(A_n)_{n\geq 1}$.

For this reason, from now on we assume that

$$\lim_{m\to+\infty}\lambda_m^{(i)}=\lambda_\infty^{(i)}\in\mathbf{R},\qquad i=0,\ldots,d.$$

This assumption allows us to define the function $w : K_d \to \mathbf{R}$ by setting, for every $x = (x_1, \ldots, x_d) \in K_d$,

$$w(x) = \begin{cases} \lambda_\infty^{(i)}, & \text{if } x = v_i \text{ for some } \\ & i = 0, \ldots, d, \\ \sum_{m=1}^{+\infty} \sum_{i=0}^{d} \lambda_m^{(i)} (1 - x_i) x_i^m, & \text{otherwise.} \end{cases} \tag{21.45}$$

The convergence properties of the sequence $(A_n)_{n \geq 1}$ are collected in the following theorem. We denote as usual by $C^2(K_d)$ the space of all functions in $C(K_d)$ with a C^2-extension on a neighborhood of K_d.

THEOREM 21.16 [20, Theorem 5.1]

1) (Properties of w) *The function w is continuous. If the sequences $(\lambda_m^{(i)})_{m \geq 1}$, $i = 0, \ldots, d$, are increasing (respectively, decreasing), then w is convex (respectively, concave).*

2) (Convergence properties) *The sequence $(A_n)_{n \geq 1}$ converges strongly on $C(K_d)$ if and only if the sequences $(\lambda_m^{(i)})_{m \geq 1}$, $i = 0, \ldots, d$, converge. In this case, we have*

$$\lim_{n \to +\infty} A_n f = w \cdot f \qquad \textit{uniformly on } K_d.$$

More precisely, for every $n \geq 1$ and $f \in C(K_d)$,

$$\|A_n f - w \cdot f\| \leq r_n \|f\| + 2 s_n \, \omega\left(f, 1/\sqrt{n}\right)$$

where

$$r_n = \sup_{m \geq n} \max_{i=0,\ldots,d} |\lambda_m^{(i)} - \lambda_n^{(i)}|, \qquad s_n = \max_{m \leq n} \max_{i=0,\ldots,d} |\lambda_m^{(i)}|,$$

$$(\omega(f, \delta) = \sup\{|f(x) - f(y)| \mid x, y \in K_d, \ \|x - y\| \leq \delta\}).$$

3) (Voronovskaja formula) *Assume that the sequences $(\lambda_m^{(i)})_{m \geq 1}$, $i = 0, \ldots, d$, converge. Then, for every $f \in C^2(K_d)$, we have*

$$\lim_{n \to +\infty} n(A_n f(x) - f(x) A_n \mathbf{1}(x)) = w(x) \times$$

$$\times \sum_{i,j=1}^{d} \frac{x_i(\delta_{ij} - x_j)}{2} \frac{\partial^2 f}{\partial x_i \, \partial x_j}(x) + \sum_{i,j=1}^{d} x_i(\delta_{ij} - x_j) \frac{\partial w}{\partial x_j}(x) \frac{\partial f}{\partial x_i}(x)$$

uniformly in $x = (x_1, \ldots, x_d) \in K_d$.

For simplicity, we consider some examples only in the two-dimensional case.

Example 21.4
If $m \geq 1$ and $j = 0, 1, 2$, consider the n-th elementary operator $L^{(j)}_{m,n}$ associated with the sequences $(\delta_i^j \cdot \delta_n^m)_{n\geq 1}$, $i = 0, 1, 2$. We have $L^{(j)}_{m,n} = 0$ if $n < m$ while, for every $f \in C(K_2)$ and $(x, y) \in K_2$,

$$L^{(0)}_{m,m} f(x,y) = (1 - x - y)^m\, f(0,0),$$

$$L^{(1)}_{m,m} f(x,y) = x^m\, f(1,0), \qquad L^{(2)}_{m,m} f(x,y) = y^m\, f(0,1)$$

and

$$L^{(0)}_{m,n} f(x,y) = \sum_{1\leq h+k\leq n-m} \binom{n-m-1}{h+k-1} \binom{h+k}{h} x^h\, y^k\, (1-x-y)^{n-h-k}\, f\left(\frac{h}{n}, \frac{k}{n}\right),$$

$$L^{(1)}_{m,n} f(x,y) = \sum_{\substack{h=m,\dots,n-1\\ k=0,\dots,n-h}} \binom{n-m-1}{n-h-1} \binom{n-h}{k} x^h\, y^k\, (1-x-y)^{n-h-k}\, f\left(\frac{h}{n}, \frac{k}{n}\right),$$

$$L^{(2)}_{m,n} f(x,y) = \sum_{\substack{k=m,\dots,n-1\\ h=0,\dots,n-k}} \binom{n-m-1}{n-k-1} \binom{n-k}{h} x^h\, y^k\, (1-x-y)^{n-h-k}\, f\left(\frac{h}{n}, \frac{k}{n}\right)$$

if $n > m$. Moreover, we have

$$\lim_{n\to+\infty} L^{(i)}_{m,n} f = (\mathbf{1} - \mathrm{pr_i})\, \mathrm{pr_i}^m\, f,$$

$(\mathrm{pr_i}(x) = x_i)$, $i = 0, 1, 2$, uniformly on K_2 for every $f \in C(K_2)$. ☐

Example 21.5
For every $j = 0, 1, 2$, denote by $B^{(j)}_n$ the n-th operator associated with the sequences $(\lambda^{(i)}_m)_{m\geq 1} = (\delta_{ij})_{m\geq 1}$ and let $w^{(j)}$ be the corresponding function defined in (21.45).

Then, for every $f \in C(K_2)$ and $(x,y) \in K_2$, we have

$$B_n^{(0)} f(x,y) = \sum_{1 \le h+k \le n-1} \binom{n-1}{h+k} \binom{h+k}{h} x^h\, y^k\, (1-x-y)^{n-h-k} f\left(\frac{h}{n}, \frac{k}{n}\right),$$

$$B_n^{(1)} f(x,y) = \sum_{h=1}^{n} \sum_{k=0}^{n-h} \binom{n-1}{n-h} \binom{n-h}{k} x^h\, y^k\, (1-x-y)^{n-h-k} f\left(\frac{h}{n}, \frac{k}{n}\right),$$

$$B_n^{(2)} f(x,y) = \sum_{k=1}^{n} \sum_{h=0}^{n-k} \binom{n-1}{n-k} \binom{n-k}{h} x^h\, y^k\, (1-x-y)^{n-h-k} f\left(\frac{h}{n}, \frac{k}{n}\right)$$

and further $w^{(i)} = \mathrm{pr}_i$, $i = 0,1,2$. Consequently, for every $f \in C(K_2)$ and $i = 0,1,2$, we have $\lim_{n \to +\infty} B_n^{(i)} f = \mathrm{pr}_i\, f$ uniformly on K_2 and the classical n-th Bernstein operator can obviously be decomposed as $B_n = B_n^{(0)} + B_n^{(1)} + B_n^{(2)}$. □

Example 21.6

For every $n \ge 1$ and $p \ge 1$, we can truncate the n-th Bernstein operator B_n by considering the operator $B_{p,n}$ corresponding to the sequences $(\lambda_m^{(i)})_{m \ge 1}$ defined by

$$\lambda_m^{(i)} = \begin{cases} 1, & m \le p, \\ 0, & m > p, \end{cases} \qquad i = 0,1,2, \quad m \ge 1.$$

In this case $B_{p,n} = \sum_{i=0}^{2} \sum_{m=1}^{p} L_{m,n}^{(i)}$ and the corresponding function $w_{p,n}$ is given by $w_{p,n}(x,y) = \sum_{i=0}^{2} \sum_{m=1}^{p} \mathrm{pr}_i{}^m$. □

Example 21.7

Finally, it is possible to consider linear combinations of Bernstein operators and a finite number of elementary operators taking the sequences $(\lambda_m^{(i)})_{m \ge 1}$, $i = 0,1,2$, definitively constant; in this case, the corresponding function w is a polynomial.

Conversely, it is easy to show that every function which is the sum of polynomials depending only on x, y and $x+y$, respectively, coincides with the function w associated with suitable definitively constant sequences $(\lambda_m^{(i)})_{m\geq 1}$, $i = 0, 1, 2$. □

We observe that if all the sequences $(\lambda_m^{(i)})_{m\geq 1}$, $i = 0, \ldots, d$, are positive and $\lambda_\infty^{(i)} > 0$, then the function w is strictly positive on K_d and the Voronovskaja formula can be written in the following form:

$$\lim_{n\to+\infty} n\left(\frac{A_n f(x)}{A_n \mathbf{1}(x)} - f(x)\right) = \sum_{i,j=1}^{d} \frac{x_i(\delta_{ij} - x_j)}{2} \frac{\partial^2 f}{\partial x_i\, \partial x_j}(x) + \sum_{i,j=1}^{d} \frac{x_i(\delta_{ij} - x_j)}{w(x)} \frac{\partial w}{\partial x_j}(x) \frac{\partial f}{\partial x_i}(x) \tag{21.46}$$

for every $f \in C^2(K_d)$ and uniformly in $x = (x_1, \ldots, x_d) \in K_d$.

This suggests considering in more detail the differential operator

$$Au(x) = \sum_{i,j=1}^{d} \frac{x_i(\delta_{ij} - x_j)}{2} \frac{\partial^2 u}{\partial x_i\, \partial x_j}(x) + \sum_{i=1}^{d} b_i(x) \frac{\partial u}{\partial x_i}(x) \tag{21.47}$$

$(u \in C^2(K_d),\ x = (x_1, \ldots, x_d) \in K_d)$, where

$$b_i(x) = \sum_{j=1}^{d} \frac{x_i(\delta_{ij} - x_j)}{w(x)} \frac{\partial w}{\partial x_j}(x), \qquad i = 1, \ldots, d. \tag{21.48}$$

This operator arises from the Voronovskaja formula (21.46) and is associated with the following second-order parabolic evolution problem:

$$\begin{cases} \dfrac{\partial u}{\partial t}(x,t) = \displaystyle\sum_{i,j=1}^{d} \frac{x_i(\delta_{ij} - x_j)}{2} \frac{\partial^2 f}{\partial x_i\, \partial x_j}(x) \\ \qquad\qquad + \displaystyle\sum_{i,j=1}^{d} \frac{x_i(\delta_{ij} - x_j)}{w(x)} \frac{\partial w}{\partial x_j}(x) \frac{\partial f}{\partial x_i}(x) \\ u(x,0) = u_0(x), \end{cases} \tag{21.49}$$

where, for every $t \geq 0$, $u(\cdot, t) \in C^2(K_d)$ and $u_0 \in C^2(K_d)$.

Feller [32] pointed out the importance of the operator A as a typical example of stochastic process; indeed, the evolution problem associated with this operator is a diffusion approximation of gene frequency models in population genetics.

The L^2-theory (see [45, 44]) is not appropriate for the study of this process since the movements of the process on the boundary are one of its main characteristics. For this reason, Shimakura and other authors studied this process in $C(K_d)$ (see [40, 41]); a survey of the main results and other references may also be found in [36, 42] and in [47, p. 221–224].

It is readily seen that A is a second-order elliptic operator degenerate at the boundary of K_d. Furthermore, the boundary of K_d is not smooth due to the presence of sides and corners.

In the case where $w = \mathbf{1}$ it is associated with the classical Bernstein operators on K_d and it is well-known that the closure $\overline{A}$ of A generates a C_0-semigroup in $C(K_d)$ and further $C^2(K_d)$ is a core for $\overline{A}$ (see, e.g., [8, Chapter 6]).

In general, if the population is affected by mutation, migration, selection or other factors, we have $w \neq \mathbf{1}$. The uniqueness of the solution of the martingale problem of the diffusion process associated with the coefficients of (21.47) has been studied by different authors. For the two-dimensional simplex, some formulas for the diffusion of the model can be found in the paper by Crow and Kimura [29], while for the d-dimensional simplex we refer to the papers of Fackerel and Littler [31], Shimakura [45] and Sato [40]. The uniqueness of the solution of the martingale problem has been proved in general in the paper of Ethier [30]; in the same paper the following result on the existence of a C_0-semigroup generated by an extension of A has been shown.

THEOREM 21.17 (Ethier [30])

Consider the differential operator A on $C^2(K_d)$ defined as in (21.47) and assume that $b \in C^{n+2}(K_d)$ satisfies

$$\langle b, \nabla \det(x_i(\delta_{ij} - x_j))\rangle \geq 0 \qquad \text{on } \partial K_d.$$

Then, there exists a strongly continuous positive contraction semigroup $(T(t))_{t\geq 0}$ on $C(K_d)$, generated by an extension of A, such that $T(t)(C^m(K_d)) \subset C^m(K_d)$ for every $m = 1, \ldots, n$ and $t \geq 0$; further $\|T(t)f\|_{C^m(K^d)} \leq e^{\lambda_m t}\|f\|_{C^m(K^d)}$ for every $f \in C^m(K^d)$, where λ_m is a suitable constant.

Moreover, for $m = 1, \ldots, n$, the restriction of the semigroup $(T(t))_{t\geq 0}$ to $C^m(K^d)$ is strongly continuous in the norm of $C^m(K^d)$.

If $n \geq 2$, then $(T(t))_{t\geq 0}$ is generated by the closure of A.

In concrete examples the functions b_i (and consequently the function w) are assigned. We assume that

$$w \in C^4(K_d) \tag{21.50}$$

and hence, with regard to the operators A_n, the sequences $(\lambda_m^{(i)})_{m\geq 1}$, $i = 0, \ldots, d$, have to be chosen in such a way that the corresponding function w satisfies (21.50). This is true, for example, in the case of definitively constant sequences, since the corresponding function w is a polynomial; this case is of particular interest in the applications since the coefficients of the differential operators related to diffusion models in population genetics are of the polynomial type and therefore we can always associate definitively constant sequences to them.

In order to apply Ethier's result on the existence of a C_0-semigroup generated by the closure of A, we also point out that the determinant of the matrix $(x_i(\delta_{ij} - x_j))_{i,j=1,\ldots,d}$ is given by

$$\det \left(x_i(\delta_{ij} - x_j)\right)_{i,j=1,\ldots,d} = x_1 \cdot \ldots \cdot x_d \,(1 - x_1 - \ldots - x_d)$$

and since every function b_i vanishes for $x_i = 0$ we have

$$\sum_{k=1}^{d} b_k(x) \frac{\partial}{\partial x_k} \left(\det \left(\frac{x_i(\delta_{ij} - x_j)}{2} \right)_{i,j=1,\ldots,d} \right) = 0$$

for every $x = (x_1, \ldots, x_d) \in \partial K_d$. Together with (21.50), this ensures that the hypotheses of Ethier's result are satisfied and hence we can state the following main result.

THEOREM 21.18

Let $(\lambda_m^{(i)})_{m\geq 1}$, $i = 0, \ldots, d$, be positive sequences of real numbers such that $\lambda_\infty^{(i)} > 0$ and such that the corresponding function w is in $C^4(K_d)$. Consider the sequence $(A_n)_{n\geq 1}$ of positive operators associated with the sequences $(\lambda_m^{(i)})_{m\geq 1}$, $i = 0, \ldots, d$, and the differential operator A defined by (21.47)-(21.48). Then, the closure $\overline{A}$ of A generates a C_0-semigroup $(T(t))_{t\geq 0}$ of positive contractions on $C(K_d)$, $C^2(K_d)$ is a core for $\overline{A}$ and further, for every $t \geq 0$ and for every sequence $(k(n))_{n\geq 1}$ of positive integers satisfying $\lim_{n\to+\infty} k(n)/n = t$, we have

$$\lim_{n\to+\infty} \left(\frac{A_n}{A_n \mathbf{1}} \right)^{k(n)} = T(t) \qquad \textit{strongly on } C(K_d).$$

By the preceding theorem, the unique solution of the evolution problem (21.49) is given by

$$u(x,t) = \lim_{n\to+\infty} \left(\frac{A_n}{A_n 1}\right)^{[nt]} u_0(x), \qquad t \geq 0, \quad x \in K_d,$$

where $[nt]$ denotes the integer part of nt.

In concrete application, the differential operator A is assigned and the coefficients b_i, $i = 1,\ldots,d$, have the form

$$b_i(x) = \sum_{j=1}^{d} x_i(\delta_{ij} - x_j)\,\varphi_i(x), \quad x = (x_1,\ldots,x_d) \in K_d;$$

if we can take a strictly positive function w such that

$$\varphi_i = \frac{1}{w}\frac{\partial w}{\partial x_i} = \frac{\partial}{\partial x_i}(\log w), \qquad i = 1,\ldots,d,$$

we can consider the problem of finding $d+1$ converging sequences $(\lambda_m^{(i)})_{m\geq 1}$, $i = 0,\ldots,d$, of positive real numbers for which the function w is defined by (21.45). This is always possible if w is the sum of $d+1$ polynomials, each depending only on one of the variables $x_0,\ldots,x_d$; in this case the sequences $(\lambda_m^{(i)})_{m\geq 1}$, $i = 0,\ldots,d$, satisfying our assumptions are definitively constant.

This case is of particular interest since we are able to approximate the solutions of the evolution problem (21.49) in the case where $w \in C^5(K_d)$ is the sum of strictly positive functions $w_0,\ldots,w_d$ such that w_i depends only on the variable x_i, $i = 0,\ldots,d$.

Indeed, for every $i = 0,\ldots,d$, we can consider a sequence $(w_n^{(i)})_{n\geq 1}$ of polynomials of one real variable satisfying $\lim_{n\to+\infty} w_n^{(i)} = w_i$ and $\lim_{n\to+\infty}(w_n^{(i)})' = w_i'$ uniformly on [0,1] (as functions of one real variable). For every $n \geq 1$, we set $w_n = w_n^{(0)} + \ldots + w_n^{(d)}$ and we denote by L_n the differential operator defined as in (21.47) with w_n at the place of w. We have $\lim_{n\to+\infty} L_n(u) = A(u)$ for every $u \in C^2(K_d)$; moreover $(I - A)(C^2(K_d))$ is dense in $C(K_d)$ since the coefficients b_i are of class $C^4(K_d)$ and hence the closure of A is the unique extension of A which generates a C_0-semigroup $(T(t))_{t\geq 0}$ (this is ensured by Ethier's result Theorem 21.17); therefore $C^2(K_d)$ must be also a core for $\overline{A}$. In turn, again by Ethier's result, the closure of every L_n generates a C_0-semigroup $(T_n(t))_{t\geq 0}$. As a consequence of the Trotter-Kato theorem, we have

$$\lim_{n\to+\infty} T_n(t) = T(t) \qquad \text{strongly on } C(K_d)$$

and the convergence is uniform with respect to t in bounded intervals. Thus, if we indicate by u_n and u, respectively, the solutions of the evolution problem corresponding to the differential operators L_n and A, we also have

$$u(x,t) = \lim_{n\to+\infty} u_n(x,t) \qquad \text{uniformly in } K_d \times [0,t], \quad t > 0.$$

References

[1] Agratini, O., Properties of a new class of recursively defined Baskakov-type operators, *Archivum Mathematicum (BRNO)* 34 (1998), 353–359.

[2] Agratini, O., Asymptotic formulae for recursively defined Baskakov-type operators, *preprint* 1999.

[3] Agratini, O., *Operatori de Aproximare*, Cluj-Napoca, 1998.

[4] Altomare, F., Limit semigroups of Bernstein-Schnabl operators associated with positive projections, *Ann. Scuola Norm. Sup. Pisa, Cl. Sci. (4)* 19 (1989), no. 2, 259–279.

[5] Altomare, F., Positive projections, approximation processes and degenerate diffusion equations, *Conf. Sem. Mat. Univ. Bari* 241 (1991), 43–68.

[6] Altomare, F., Lototsky-Schnabl operators on compact convex sets and their associated limit semigroup, *Mh. Math.* 114, (1992), 1–13.

[7] Altomare, F., Attalienti, A., Forward diffusion equations and positive semigroups, *Math. Z.* 225 (1997), 211–229.

[8] Altomare, F., Campiti, M., *Korovkin-type Approximation Theory and its Applications*, De Gruyter Studies in Mathematics 17, W. De Gruyter, Berlin-New York, 1994.

[9] Altomare, F., Carbone, I., On a new sequence of positive linear operators on unbounded intervals, *Rend. Circ. Mat. Palermo (2)*, Suppl. no. 40 (1996), 23–46.

[10] Altomare, F., Carbone, I., On some degenerate differential operators on weighted function spaces, *J. Math. Anal. Appl.*, 213 (1997), 308–333.

[11] Altomare, F., Mangino, E., On a generalization of Baskakov operators, *preprint* 1999.

[12] Altomare, F., Mangino, E., On a class of elliptic-parabolic equations on unbounded intervals, *preprint* 1999.

[13] Altomare, F., Raşa, I., Towards a characterization of a class of differential operators associated with positive projection, *Atti Sem. Mat. Fis. Univ. Modena*, Suppl. Vol. XLVI, (1998) 3–38.

[14] Attalienti, A., Generalized Bernstein-Durrmeyer operators and the associated limit semigroup, *J. Approx. Theory* 99 (1999), no. 3, 289–309.

[15] Attalienti, A., Campiti, M., Degenerate evolution problems and Beta-type operators, *preprint* 1999.

[16] Campiti, M., A generalization of Stancu-Mühlbach operators, *Constr. Approx.* 7 (1991), 1–18.

[17] Campiti, M., Limit semigroups of Stancu-Mühlbach operators associated with positive projections, *Ann. Scuola Norm. Sup. Pisa*, Cl. Sci. (4) 19 (1992), no. 1, 51–67.

[18] Campiti, M., General binomial coefficients in the sequence of Bernstein operators, *Approx. Theory & its Appl.* 15 (1999), no. 1, 92–102.

[19] Campiti, M., On a weighted recursive rule, in *"Approximation Theory IX, I: Theoretical Aspects"* (C. K. Chui, L. Schumaker, eds.), *Proc. Internat. Conf. AT98, Nashville, January 2-6, 1998*, Vanderbilt University Press, Nashville-London, 59–66.

[20] Campiti, M., Recursive Bernstein operators and degenerate diffusion processes, *preprint* 1999.

[21] Campiti, M., Metafune, G., Approximation properties of recursively defined Bernstein-type operators, *J. Approx. Theory* 87 (1996), no. 3, 243–269.

[22] Campiti, M., Metafune, G., Evolution equations associated with recursively defined Bernstein-type operators, *J. Approx. Theory* 87 (1996), no. 3, 270–290.

[23] Campiti, M., Metafune, G., L^p-Convergence of Bernstein-Kantorovitch-type operators, *Ann. Pol. Math.* 63 (1996), no. 3, 273–280.

[24] Campiti, M., Metafune, G., Approximation of solutions of some degenerate parabolic problems, *Numer. Funct. Anal. Optim.* 17 (1996), no. 1&2, 23–35.

[25] Campiti, M., Metafune, G., Ventcel's boundary conditions and analytic semigroups, *Arch. Math.* 70 (1998), 377–390.

[26] Campiti, M., Metafune, G., Pallara, G., General Voronovskaja formula and solutions of second-order degenerate differential equations, *preprint* 1999.

[27] Campiti, M., Raşa, I., On a best extension property with respect to a linear operator, *Applicable Analysis* 64 (1997), 189–202.

[28] Clément, Ph., Timmermans, C. A., On C_0-semigroups generated by differential operators satisfying Ventcel's boundary conditions, *Indag. Math. New Ser.* 89 (1986), 379–387.

[29] Crow, J., Kimura, M., Some genetic problems in natural populations, *Proc.* 3nd *Berkeley Sympos. Math. Statist. and Probab.* (1956), Vol. IV, 1–22.

[30] Ethier, S. N., A class of degenerate diffusion processes occurring in population genetics, *Comm. Pure Appl. Math.* 29 (1976), 483–493.

[31] Fackerel, E., Littler, R., Transition densities for neutral multi-allele diffusion models, *Biometrics* 31 (1975), 117–123.

[32] Feller, W., Diffusion processes in genetics, *Proc.* 2nd *Berkeley Sympos. Math. Statist. and Probab.* (1951), 227–246.

[33] Feller, W., The parabolic differential equations and the associated semi-groups of transformations, *Ann. of Math.* 55 (1952), 468–519.

[34] Knoop, K., *Theory and Application of Infinite Series*, Hafner Publishing Company, New York, 1971.

[35] Lorentz, G. G., *Bernstein Polynomials*, 2nd edition, Chelsea, New York, 1986.

[36] Maruyama, T., *Stochastic Problems in Population Genetics,* Lecture Notes in Biomath. 17, Springer, New York, 1977.

[37] Nagel R. (ed.), *One-parameter Semigroup of Positive Operators*, Lecture Notes in Math. 1184, Springer-Verlag, Berlin-Heidelberg-New York-Tokyo, 1986.

[38] Pazy, A., *Semigroups of Linear Operators and Applications to Partial Differential Equations*, Applied Mathematical Sciences 44, Springer-Verlag, New York-Berlin-Heidelberg-Tokyo, 1983.

[39] Raşa, I., Vladislav, T., *Analiză Numerică*, Editura Tehnică, Bucureşti, 1998.

[40] Sato, K., Diffusion processes and a class of Markov chains related to population genetics, *Osaka J. Math.* 13 (1976), 631–659.

[41] Sato, K., Convergence to a diffusion of a multi-allelic model in population genetics, *Adv. Appl. Prob.* 10 (1978), 538–562.

[42] Sato, K., *Diffusion Operators in Population Genetics and Convergence of Markov Chains* in Proc. Measure Theory Applications to Stochastic Analysis, Oberwolfach, 1977 (G. Kallianpur and D. Kölzow eds.), Lecture Notes in Math. 695 (1978), Springer, New York, 127–137.

[43] Sendov, B., Popov, V. A., *The Averaged Moduli of Smoothness,* Pure Appl. Math., Wiley, 1988.

[44] Shimakura, N., Equations différentielles provenant de la génétique des populations, *Tôhoku Math. Journ.* 29 (1977), 287–318.

[45] Shimakura, N., Formulas for diffusion approximations of some gene frequency models, *J. Mat. Kyoto Univ.* 21 (1981), no. 1, 19–45.

[46] Shimakura, N., Existence and uniqueness of solutions for a diffusion model of intergroup selection, *J. Mat. Kyoto Univ.* 25 (1985), no. 4, 775–788.

[47] Shimakura, N., *Partial Differential Operators of Elliptic Type, Transl. Math. Monogr.* 99, Amer. Math. Soc., Providence, RI, 1992.

[48] Stancu, D. D., On some polynomials of Bernstein type, *Acad. R. P. Romîne Fil. Iasşi. Stud. Cerç. Şti. Mat.* 11 (1960), 221–233.

[49] Titchmarsh, E. C., *The Theory of Functions*, Oxford University Press, Oxford, 1939.

[50] Trotter, H. F., Approximation of semi-groups of operators, *Pacific J. Math.* 8 (1958), 887–919.

Chapter 22

On Scattered Data Representations Using Bivariate Splines

Don Hong

Department of Mathematics, East Tennessee State University, Johnson City, TN 37614-0663, E-mail: hong@etsu.edu, www.etsu.edu/math/hong

Abstract The objective of this chapter is to present a study of scattered data representation using bivariate splines. First, we open a discussion with emphasis on the optimal order of approximation. When the polynomial degree is allowed to be sufficiently large as compared to the order of smoothness, it is shown that the spline elements can be used to represent scattered data with the optimal order of approximation over arbitrary triangulations. In real applications, where the polynomial degree is often required to be lower, it is necessary to find a so-called optimal triangulation so that the spline space can achieve the optimal approximation order. We present an algorithm to transform an arbitrary triangulation of the sample points into an optimal triangulation for representation of the scattered data using C^1 quartic splines. Then, we consider the possibilities of finding optimal triangulations for even lower degree spline spaces such as C^1 cubic and C^1 quadratic spaces. Some interpolation schemes and a stable local basis construction are also presented. Finally, we mention some recent results on representing scattered data using other spline elements, such as splines on spheres and natural splines.

Contents

22.1 Introduction

In many applications, it is desirable to approximate a given surface with a high degree of accuracy. Scattered data on the surface may be collected by recording the distance from sample points in a fixed plane to the surface. Once the scattered data have been collected, it is necessary to determine simple functions to interpolate, or best fit, the data. An ideal choice for these simple functions is splines, also called piecewise polynomial (*pp*) functions.

Since a bivariate spline is piecewise-defined over its planar domain, it is necessary to create a partition of the sample points in the plane. One of the most applicable partitions in this case is triangulation.

DEFINITION 22.1 *A triangulation of a finite set V of n sample points $\mathbf{v}_i = (x_i, y_i)$, $i = 1, \cdots, n$ in a plane $\mathbb{R}^2$ is defined as a collection Δ of triangles τ satisfying (i) the vertices of the triangles are precisely the sample points $\mathbf{v}_i$; (ii) the union of the triangles in Δ is a connected set; and (iii) the intersection of any two adjacent triangles in Δ is either a common vertex or a common edge.*

Note that V is also the vertex set of the triangulation Δ. Sometimes, we will denote V as $V(\Delta)$. In general, for a given set V of data sites, there are many different triangulations with vertex set V. On a triangulation Δ of a polygonal domain $\Omega \subset \mathbb{R}^2$ with vertex set V, one of most important problems in applications is to represent scattered data defined on V by C^r smooth spline functions. Of course, one usually wants to find an optimal triangulation of the given sample sites. Though the notion of optimality depends on the desirable properties in the approximation or modeling problems, here, we are concerned with optimal order of approximation with respect to the given order r of smoothness and degree k of the polynomial pieces of the smooth spline functions. In the study of spline functions on a triangulation Δ, the notation $S_k^r(\Delta)$ is used to denote the subspace of $C^r(\Omega)$ of all *pp* functions with total degree $\leq k$ and with grid lines given by the edges of Δ.

In scattered data representation using splines in the space $S_k^r(\Delta)$, it is critical to answer the question of how well the splines can approximate classes of smooth data. We give the definition of the approximation order of a function space S as follows.

DEFINITION 22.2 *The approximation order of a space S of functions on $\mathbb{R}^2$ is defined to be the largest real number ρ for which*

$$\operatorname{dist}(f, S) \leq \text{Const}\|D^{k+1}f\|\ |\Delta|^{\rho} \tag{22.1}$$

for any sufficiently smooth function f, with the distance measured in the maximum norm $\|\ \|$, and with the mesh size $|\Delta| := \sup_{\tau\in\Delta} \operatorname{diam}\tau$.

It is clear that the full order of approximation from the spline space $S_k^r(\Delta)$ cannot be better than $k+1$ regardless of r and is trivially $k+1$ in the case $r=0$.

We use the term *optimal triangulation* of a given set V of data sites to mean that (i) the set V of sample sites is the same as the set of vertices of the triangulation, and (ii) the space of *pp* functions with degree k and smoothness order r on this triangulation achieves the full order of approximation. More precisely, we make the following.

DEFINITION 22.3 *For a given set V of data sites, the degree k and the smoothness order r, any triangulation Δ with vertex set V is called optimal (of type (k, r)) if the spline space $S_k^r(\Delta)$ admits full approximation order $k+1$.*

In addition to the huge volume of research published on representing scattered data using multivariate splines, there are several survey articles that are related to this area (cf. Schumaker [61] and [62], Barnhill [7], Franke [32], Alfeld [1], Dahmen and Micchelli [25], Böhm et al. [8], and Hong [42]). Currently, box splines, thin-plate splines, and radial-basis functions are among the most commonly used tools for scattered data interpolation. However, from the computational point of view, a simple and efficient multivariate spline interpolation scheme for scattered data is still not available.

This gives rise to the problem of studying lower degree spline spaces and finding locally supported basis elements for the optimal order of approximation. This problem, however, is extremely complicated, and a general approach does not seem to be feasible. Here, in this chapter, we first consider finding optimal triangulations for C^1 quartic spline spaces. We present an algorithm to transform an arbitrary triangulation of the sample points into an optimal triangulation for representation of the scattered data using C^1 quartic splines. Then, we also discuss the possibility of even lower degree spline spaces such as C^1 cubic and C^1 quadratic spaces. Some interpolation schemes and a stable local basis construction are also presented. Finally, we present some recent results on representing scattered data using other spline elements such as splines on sphere and natural splines.

The outline of this chapter is as follows. Results on approximation order of spline spaces over arbitrary triangulations will be discussed first. Optimal triangulations and the algorithm to create optimal triangulations based on arbitrarily given data sites for C^1 quartic spline spaces is introduced in Section 22.3. The discussion of optimal triangulations for C^1 cubic and quadratic spline spaces and some other spline spaces is in Section 22.4. Section 22.5 is devoted to the study of existence of local basis and the local linear independence of basis functions. The approaches to represent scattered data using splines on sphere and natural splines are mentioned in the last section.

22.2 Approximation Order of Spline Spaces over Triangulations

As usual, let $\mathbb{R}$ be the set of all real numbers and $\mathbb{Z}_+$ the set of nonnegative integers. Thus $\mathbb{R}^3$ denotes the 3-dimensional Euclidean space and $\mathbb{Z}_+^3$ can be used as a multi-index set, while $\pi_k(\mathbb{R}^2)$ is the space of all polynomials of (total) degree $\leq k$ in two variables. Let $\tau = [\mathbf{v}_0, \mathbf{v}_1, \mathbf{v}_2]$ be a proper triangle with vertices $\mathbf{v}_0$, $\mathbf{v}_1$, $\mathbf{v}_2 \in \mathbb{R}^2$. Then for any $\mathbf{x} \in \mathbb{R}^2$, we have

$$\mathbf{x} = \xi_0\mathbf{v}_0 + \xi_1\mathbf{v}_1 + \xi_2\mathbf{v}_2 \quad \text{with} \quad \xi_0 + \xi_1 + \xi_2 = 1.$$

The 3-tuple $\xi = (\xi_0, \xi_1, \xi_2)$ is called the barycentric coordinate of $\mathbf{x}$ with respect to the triangle τ. For $\alpha = (\alpha_0, \alpha_1, \alpha_2) \in \mathbb{Z}_+^3$, the length of α is defined by $|\alpha| = \alpha_0 + \alpha_1 + \alpha_2$, and the factorial $\alpha!$ is defined as $\alpha_0!\alpha_1!\alpha_2!$. We define the Bernstein Polynomial $B_{\alpha,\delta}$ as

$$B_{\alpha,\tau}(\mathbf{x}) = \binom{|\alpha|}{\alpha}\xi^\alpha,$$

where $\xi^\alpha = \xi_0^{\alpha_0}\xi_1^{\alpha_1}\xi_2^{\alpha_2}$ and

$$\binom{|\alpha|}{\alpha} = \frac{|\alpha|!}{\alpha_0!\alpha_1!\alpha_s!}.$$

Moreover, we define the (domain) points

$$\mathbf{x}_{\alpha,\tau} := \frac{\alpha_0\mathbf{v}_0 + \alpha_1\mathbf{v}_1 + \alpha_2\mathbf{v}_2}{|\alpha|}, \quad |\alpha| = k. \tag{22.2}$$

It is well-known that any polynomial $p \in \pi_k$ can be written in a unique way as

$$p = \sum_{|\alpha|=k} b_{\alpha,\tau} B_{\alpha,\tau},$$

where $b_{\alpha,\tau}$ is called the B-net ordinate of p with respect to triangle τ. This gives rise to a mapping $b : \mathbf{x}_{\alpha,\tau} \to b_{\alpha,\tau}$, $|\alpha| = k$. Such a mapping b is called the B-net representation of p with respect to triangle τ.

Now, let us discuss the B-net representation of bivariate splines. Let Δ be a triangulation of a polygonal domain in $\mathbb{R}^2$ and $S_k^0(\Delta)$ the space of all continuous splines of degree k on Δ. Assume $s \in S_k^0(\Delta)$. On each triangle $\tau \in \Delta$, s agrees with some polynomial $p \in \pi_k$. Thus, we have

$$s|_\tau = \sum_{|\alpha|=k} b_{\alpha,\tau} B_{\alpha,\tau}.$$

Let X denote the set of all (domain) points $\mathbf{x}_{\alpha,\tau}$ as defined in (22.2). Then a mapping can be defined as follows:

$$b_s : \mathbf{x}_{\alpha,\tau} \mapsto b_{\alpha,\tau}, \quad |\alpha| = k, \ \tau \in \Delta. \tag{22.3}$$

Such a mapping b_f is called the B-net representation of the spline s.

It is well-known that to each triangle $\tau \in \Delta$, the matrix

$$(B_{\alpha,\tau}(\mathbf{x}_{\beta,\tau}))_{|\alpha|=k,|\beta|=k}$$

is invertible. Thus, the linear system

$$\sum_{|\gamma|=k} c_{\alpha,\gamma} B_{\beta,\tau}(\mathbf{x}_{\gamma,\tau}) = \delta_{\alpha,\beta} := \begin{cases} 1, & \alpha = \beta \\ 0, & \alpha \neq \beta \end{cases}$$

has a unique solution.

Since this linear system depends only on the barycentric coordinates of $\mathbf{x}_{\alpha,\tau}$, the solution $\{c_{\alpha,\beta}\}$ is independent of τ. Let $[\,\cdot\,]$ denote the point-evaluation functional, namely,

$$[\mathbf{x}_{\alpha,\tau}]f := f(\mathbf{x}_{\alpha,\tau}).$$

Then it is easy to see that the functionals

$$L_{\alpha,\tau} := \sum_{|\gamma|=k} c_{\alpha,\gamma}[\mathbf{x}_{\gamma,\tau}], \quad \alpha \in \mathbb{Z}_+^3, \quad |\alpha| = k,$$

form a dual basis of $\{B_{\alpha,\tau}; |\alpha| = k\}$ in the sense of

$$L_{\alpha,\tau}B_{\beta,\tau} = \delta_{\alpha,\beta}, \quad |\alpha| = |\beta| = k.$$

Furthermore, there is a positive constant C_k, depending only on the degree k, such that

$$||L_{\alpha,\tau}|| := \sup_{||f||_\infty=1} ||L_{\alpha,\tau}f||_\infty = \max_{|\beta|=k}|c_{\alpha,\beta}| \leq C_k, \tag{22.4}$$

for $\alpha \in \mathbb{Z}_+^3$, $|\alpha| = k$. From (22.4) and the fact that $b_s(\mathbf{x}_{\alpha,\tau}) = L_{\alpha,\tau}\, s$, we have the following.

LEMMA 22.1
If $s \in S_k^0(\Delta)$ and $b_s \in \mathbb{R}^X$ is the B-net representation of s, then

$$||s||_\infty \leq ||b_s||_\infty \leq C_k||s||_\infty. \tag{22.5}$$

Now, let $\tau = [\mathbf{v}_0, \mathbf{v}_1, \mathbf{v}_2]$ and $\tilde{\tau} = [\mathbf{v}_0, \mathbf{v}_1, \tilde{\mathbf{v}}_2]$ be two triangles in Δ with common edge $e = [\mathbf{v}_0, \mathbf{v}_1]$. Let S, S_0, S_1, and S_2 denote the oriented areas of the triangles τ, $[\tilde{\mathbf{v}}_2, \mathbf{v}_1, \mathbf{v}_2]$, $[\mathbf{v}_0, \tilde{\mathbf{v}}_2, \mathbf{v}_2]$, and $\tilde{\tau}$, respectively. The following result, which describes C^r-smoothness conditions on a spline function s in terms of its B-net representation, can be found in [44] (see also [10], [16], etc.).

THEOREM 22.1
Suppose that a bivariate spline function s is defined on the union of two triangles $\tau \cup \tilde{\tau}$ by

$$s|_\tau = \sum_{|\alpha|=k} b(\mathbf{x}_{\alpha,\tau})B_{\alpha,\tau};$$

$$s|_{\tilde{\tau}} = \sum_{|\alpha|=k} b(\mathbf{x}_{\alpha,\tilde{\tau}}) B_{\alpha,\tilde{\tau}}.$$

Then $s \in C^r(\tau \cup \tilde{\tau})$ *if and only if for all positive integers* $\ell \leq r$ *and* $\gamma = (\gamma_u, \gamma_v, 0) \in \mathbb{Z}_+^3$ *with* $|\gamma| = k - \ell$,

$$b(\mathbf{x}_{\gamma+\ell\mathbf{e}^3,\tilde{\tau}}) = \sum_{|\beta|=\ell} \binom{\ell}{\beta} b(\mathbf{x}_{\gamma+\beta,\tau}) \left(\frac{S_0}{S}\right)^{\beta_0} \left(\frac{S_1}{S}\right)^{\beta_1} \left(\frac{S_2}{S}\right)^{\beta_2}, \tag{22.6}$$

where $\beta = (\beta_0, \beta_1, \beta_2) \in \mathbb{Z}_+^3$ *and* $\mathbf{e}^3 = (0,0,1)$.

Let E_I be the set of interior edges of Δ. For $e \in E_I$ and two triangles $\tau = [\mathbf{v}_0, \mathbf{v}_1, \mathbf{v}_2]$ and $\tilde{\tau} = [\mathbf{v}_0, \mathbf{v}_2, \tilde{\mathbf{v}}_2]$ sharing the common edge $e = [\mathbf{v}_0, \mathbf{v}_1]$, and $\alpha = (\alpha_0, \alpha_1, \alpha_2) \in \mathbb{Z}_+^3$ with $1 \leq \alpha_2 \leq r$, we define the functionals $f_{e,\alpha}$ on $\mathbb{R}^X$, the space of all real functions, by

$$\lambda_{e,\alpha} b = b(x_{\alpha,\tilde{\tau}}) - \sum_{|\beta|=\alpha_{\tilde{w}}} \binom{\alpha_2}{\beta} b(x_{(\alpha_1,\alpha_2,0)+\beta,\tau}) \left(\frac{S_0^{\beta_0} S_1^{\beta_1} S_2^{\beta_2}}{S^{\alpha_2}}\right). \tag{22.7}$$

It is clear that the support of the functional $\lambda_{e,\alpha}$ is included in a diamond domain with vertices $\frac{(\alpha_0+\alpha_2)\mathbf{v}_0+\alpha_1\mathbf{v}_1}{k}$, $\frac{\alpha_0\mathbf{v}_0+(\alpha_1+\alpha_2)\mathbf{v}_1}{k}$, $\mathbf{x}_{\alpha,\tilde{\tau}}$ and $\mathbf{x}_{\alpha,\tau}$.

As an example, we can see that the C^1-smoothness conditions across the edge e for $s \in S_4^1(\Delta)$ are determined by the relation

$$b_{\alpha+\mathbf{e}^3,\tilde{\tau}} = c_1 b_{\alpha+\mathbf{e}^1,\tau} + c_2 b_{\alpha+\mathbf{e}^2,\tau} + c_3 b_{\alpha+\mathbf{e}^3,\tau}, \tag{22.8}$$

where $\alpha = (\alpha_0, \alpha_1, 0) \in \mathbb{Z}_+^3$ with $\alpha_0 + \alpha_1 = 3$, $\mathbf{e}^1$, $\mathbf{e}^2$, and $\mathbf{e}^3$ are the standard unit vectors in $\mathbb{R}^3$, $b_{\alpha,\tau} = b_s(\mathbf{x}_{\alpha,\tau})$ is the B-net representation of s, and c_i, $i = 1,2,3$, are the barycentric coordinates of $\tilde{\mathbf{v}}_2$ with respect to τ.

Recall that $\mathbb{R}^X$ is the space of all real functions. Let $|X|$ denote the cardinality of the set X of domain points. Then the spaces $S_k^0(\Delta)$ and $\mathbb{R}^X$ are isomorphic. Therefore, we have

$$\dim(S_k^0(\Delta)) = |X|.$$

Let $\Lambda = \Lambda_k^r$ be the set of all such functionals defined by (22.7) and

$$\Lambda^\perp = \{b \in R^X;\ \lambda b = 0, \forall \lambda \in \Lambda\}.$$

Then we see that the spaces $S_k^r(\Delta)$ and $(\Lambda_k^r)^\perp$ are isomorphic. Therefore, we have the following.

THEOREM 22.2

$$\dim(S_k^r(\Delta)) = |X| - \dim(\Lambda_k^r).$$

It is certainly not trivial to determine $\dim(\Lambda_k^r)$. However, this theorem is helpful to determine dimensions of spline spaces in the case of $k \geq 3r+2$ (see [41]). For some recent progress in the study of dimension problems of spline spaces, please see [42].

B-net representation of splines can also be applied to study the approximation order of $S_k^r(\Delta)$. For this purpose, we define the L_∞ norm on linear space $S_k^r(\Delta)$ and the ℓ_∞ norm on linear space $\mathbb{R}^X$. Then these two norms are equivalent by (22.5). Since $\mathbb{R}^X$ is finitely dimensional, the dual space of $\mathbb{R}^X$ has norm ℓ_1. For $g \in S_k^0(\Delta)$, using the dual theorem in functional analysis, we have

$$\operatorname{dist}(g, S_k^r(\Delta)) = \sup_{\lambda\in\Lambda} \frac{|\lambda g|}{\|\lambda\|}. \tag{22.9}$$

If f is a continuous function, then there is a unique function $g \in S_k^0(\Delta)$ such that f and g have the same values at points in X. Then $f \mapsto g$ gives a projection operator P from C to $S_k^0(\Delta)$. From (22.9), we have

THEOREM 22.3

Let f be a continuous function. Then

$$|\operatorname{dist}(f, S_k^r(\Delta)) - \sup_{\lambda\in\Lambda} \frac{|\lambda Pf|}{\|\lambda\|}| \leq \|f - Pf\|.$$

In the above theorem, $\|f - Pf\|$ turns out to be a local approximation problem and it is not so difficult to determine its approximation order. Thus, the key point is to determine

$$\sup_{\lambda\in\Lambda} \frac{|\lambda Pf|}{\|\lambda\|}.$$

In general, it is well-known that the approximation order of $S_k^r(\Delta)$ not only depends on k and r, but also on the geometric structure of the partition Δ. According to the results of finite elements analysis in [68] and [9], it was believed in the past that the full approximation order of $\rho = k+1$ can be obtained from the spline space $S_k^r(\Delta)$ only when the degree of the polynomial k is at least $4r+1$. de Boor and Höllig [14] applied this theorem and proved the following (see also [21] for a constructive proof).

THEOREM 22.4

For $k \geq 3r+2$ and sufficiently smooth function f, there is a constant Const *which depends only on the smallest angle of the partition Δ such that*

$$\mathrm{dist}(f, S_k^r(\Delta)) \leq \mathrm{Const}|\Delta|^{k+1}\|f\|_{k+1,\infty}.$$

Chui, Hong, and Jia in [21] provided a constructive scheme to achieve this optimal approximation order based on a stable basis of $S_k^r(\Delta)$ for the case of $k \geq 3r+2$. Therefore, any triangulation Δ is optimal for the spline space $S_k^r(\Delta)$ as long as $k \geq 3r+2$. It is natural to ask whether the condition $k \geq 3r+2$ is sharp. In other words, is there any pair of integers k and r with $k \leq 3r+1$ such that some triangulation is optimal for $S_k^r(\Delta)$? The first result in this direction was obtained by de Boor and Höllig (see [12]). They proved that $S_3^1(\Delta^{(1)})$ has approximation order 3 instead of 4. Here $\Delta^{(1)}$ stands for a three-directional mesh (also called a type-1 triangulation) which is formed by a uniform rectangular partition plus all northeast diagonals. Later, de Boor and Jia in [15] considered the approximation order of spline spaces over the three-direction mesh $\Delta^{(1)}$ for general smoothness order r. They obtained the following.

THEOREM 22.5

For a three-directional mesh $\Delta^{(1)}$, the approximation order of the space $S_k^r(\Delta^{(1)})$ is at most k provided that $k \leq 3r+1$.

Theorem 22.4 shows that if k is sufficiently large compared to r, the spline space $S_k^r(\Delta)$ provides the full accuracy expected of piecewise polynomials of degree k. If $k \leq 3r+1$, generally speaking, the space $S_k^r(\Delta)$ does not have full order of approximation as shown in Theorem 22.5 when Δ is the three-direction mesh $\Delta^{(1)}$. But, it is not known what its generic approximation order is. de Boor conjectured that bivariate C^1 cubic spline space has approximation order 0 generically (see [11], sect. 7.).

Recall that a triangulation formed from a uniform rectangular partition by drawing both northeast diagonals and northwest diagonals is called a type-2 or four-direction mesh, and it is denoted by $\Delta^{(2)}$. Notice that the intersections of diagonals are also in the vertex set V.

For a type-2 triangulation $\Delta^{(2)}$, Dahmen and Micchelli [26] proved that the space $S_4^1(\Delta^{(2)})$ arrives at the optimal approximation order 5. More generally, for a type-2 triangulation $\Delta^{(2)}$, Jia [46] proved the following general result by considering the local approximation order provided by the box splines in $S_k^r(\Delta^{(2)})$.

THEOREM 22.6
The approximation order of $S_k^r(\Delta^{(2)})$ for a four-direction mesh $\Delta^{(2)}$ is $k+1$ if $r \leq 1$ and $k \geq r+1$.

Theorem 22.6 showed that the space $S_k^1(\Delta^{(2)})$ has full order of approximation if $k \geq 2$. It already included a later result published in [48] on $S_3^1(\Delta^{(2)})$. This gives rise to a question: Is there any optimal triangulation for the spline space $S_k^r(\Delta)$ even though $k \leq 3r+1$?

It will be very interesting to find some triangulations Δ, which are somewhat specific but more general than the type-2 triangulation, such that the spline space $S_k^r(\Delta)$ still has full approximation order $k+1$. The question was well-answered for C^1 quartic spline spaces and we will discuss this in some detail in the next section.

Similar to the structure of type-2 triangulation, a quadrangulation of a connected polygonal domain Ω in $\mathbb{R}^2$ can be defined as follows.

DEFINITION 22.4 *A collection $\Diamond$ of quadrilaterals $q_i, i = 1, \cdots, N$, is called a quadrangulation diamond of a Ω if (i) $\Omega = \sum_{i=1}^N q_i$; (ii) the intersection of any two quadrilaterals is either empty, a single vertex, or a common edge; (iii) for any two quadrilaterals q_1, q_n, there is a sequence of quadrilaterals $q_1, \cdots, q_n$ in $\Diamond$ such that each pair q_i, q_{i+1} shares exactly one edge with each other. $\Diamond$ is called convex if all quadrilaterals are convex, and $\Diamond$ is said to be nondegenerate if none of the quadrilaterals is a triangle.*

For a nondegenerate convex quadrangulation of a polygonal domain Ω in $\mathbb{R}^2$, we use $\oplus$ to denote the triangulation obtained by inserting the diagonals of each quadrilateral of $\Diamond$. Spline spaces defined on triangulated quadrangulations have been studied both in the field of finite elements and spline theory. Finite elements in $S_3^1(\oplus)$ were constructed in [33] and [60]. The approximation properties of $S_3^1(\oplus)$ were investigated in [23] for L_2 norm and in [49] for L_∞ norm. Finite elements spanning a certain subspace of $S_{3r}^r(\oplus)$ for odd integer r and $S_{3r_1}^r(\oplus)$ for even integer r were constructed in [47] recently. The approximation properties under L_∞ norm were also studied there. In [52], the following result is obtained.

THEOREM 22.7
For integers $r \geq 1$ and $0 \leq m \leq 3r$. There exists a linear quasi-interpolation operator $Q_m : L_1(\Omega) \mapsto S_{3r}^r(\oplus)$ such that

$$\|D_x^\alpha D_y^\beta (f - Q_m f)\|_p \leq \text{Const}|\oplus|^{m+1-\alpha-\beta}|f|_{m+1,p},$$

for $1 \leq p \leq \infty$, $0 \leq \alpha + \beta \leq m$, and f in the Sobolev space $W_p^{m+1}(\Omega)$.

Here $|\diamondplus|$ *is the mesh size of* $\diamondplus$.

Therefore, the triangulation $\diamondplus$ is optimal for the spline space $S_k^r(\diamondplus)$ if $k \geq 3r$. Noticing the similar structures between the triangulation $\diamondplus$ and a four-directional mesh $\Delta^{(2)}$, we would like to make the following conjecture.

Conjecture. The triangulation $\diamondplus$ is optimal for the spline space $S_k^r(\diamondplus)$ if $k \geq 2r + 1$.

22.3 Optimal Triangulations for Lower-Degree Bivariate Spline Spaces

Since lower degree spline spaces are preferable for application purposes, it is beneficial to determine optimal triangulations for $S_k^r(\Delta)$ when $k \leq 3r+1$. This section will specifically focus on the spline space of C^1 quartic splines, $S_4^1(\Delta)$.

Recall that de Boor and Jia proved in 1993 in [15] that the bivariate spline space $S_k^r(\Delta^{(1)})$ attains an approximation order of at most k for $k \leq 3r+1$. So $\Delta^{(1)}$ is not an optimal triangulation for the spline space $S_k^r(\Delta)$ when $k \leq 3r+1$. In particular, $S_4^1(\Delta^{(1)})$ attains an approximation order of at most 4, but not the optimal approximation order of 5. So $S_4^1(\Delta^{(1)})$ is not optimal for C^1 quartic splines.

A couple of techniques have been implemented in recent years to determine optimal triangulations for C^1 quartic splines. In 1996, Chui and Hong developed in [19] a scheme known as a *Local Clough-Tocher Refinement Scheme* to transform an arbitrary triangulation of data points into an optimal triangulation for C^1 quartic splines. In that scheme, some triangles are refined into three subtriangles to become a Clough-Tocher cell. Here, locality means that the Clough-Tocher triangle is applied only to some isolated triangles in Δ, and as usual, a triangle is called a *Clough-Tocher triangle*, if it is subdivided, by using an interior point (such as the centroid of the triangle), into three subtriangles. A interpolation scheme was also constructed there by using certain locally supported Hermite elements, which are called star-vertex splines, to achieve this optimal approximation order.

Generation of an optimal mesh is one of the most important facets in finite element modeling. The method of local Clough-Tocher refinement of triangulations can be undertaken without any element distortion, and the local interpolation schemes will help in drastically decreasing the computational complexity as compared with the standard (global) Clough-Tocher scheme.

However, the disadvantage of this scheme is that it requires the inclusion of additional data points and often in applications no scattered data are available for additional data sites. To avoid introducing some new data points in addition to the vertex set of the triangulation Δ, as in the local Clough-Tocher refinement, Hong and Mohapatra later developed in 1997 in [43] a *mixed three-directional mesh* which is an optimal triangulation for C^1 quartic splines on the existing data points. A rectangle with northeast diagonal is called a *NE-rectangle.* Similarly, a rectangle with northwest diagonal is called a *NW-rectangle.* For a triangulation Δ which consists of NE- and NW-rectangles we may call Δ a mixed three-direction mesh and denote it by $\Delta^{(3)}$. In [43], we obtain the following.

THEOREM 22.8

For a mixed three-direction mesh $\Delta^{(3)}$ there is a linear interpolating operator $T: \ f \in C^1(\Delta^{(3)}) \mapsto s \in S_4^1(\Delta^{(3)})$ such that $Tp = p$ for any polynomial $p \in \pi_4$ and such that T achieves the optimal order of approximation; that is,

$$\|Tg - g\| \leq C\|g^{(5)}\||\Delta^{(3)}|^5, \text{ for } g \in C^5(\Delta^{(3)}), \tag{22.10}$$

where $|\Delta^{(3)}|$ is the mesh size of $\Delta^{(3)}$.

Therefore, the mixed three-directional mesh $\Delta^{(3)}$ is optimal for C^1 quartic spline space $S_4^1(\Delta^{(3)})$. And so, the mixed three-directional mesh $\Delta^{(3)}$ is better than the three-directional mesh in the sense that the corresponding spline space has a higher order of approximation. Also the mixed three-directional mesh $\Delta^{(3)}$ is better than local refinements in the sense that the C^1 quartic spline space achieves the optimal approximation order by using a smaller number of data sites in the interpolation. In comparison, the mixed three-directional mesh using the data only at the intersections of rectangle lines and with the optimal-order 5 can also be achieved by the space $S_4^1(\Delta^{(3)})$. Therefore, the mixed three-directional mesh is also better than the four-directional mesh on this point.

The uniform partition certainly restricts the application of the mixed three-directional elements to the arbitrarily given data points. In [20], we considered C^1 quartic spline space over arbitrary triangulations and provided an efficient method, called *Edge Swapping Algorithm*, for triangulating any finite arbitrarily scattered sample sites, such that for any discrete data given at these sample sites, there is a C^1 quartic polynomial spline on this triangulation that interpolates the given data with the optimal order of approximation. The MatLab complementation of the algorithm and numerical examples were given in [30]. To explain the idea of the Edge Swapping Algorithm here, we need to recall some notation from graph theory. The degree of any vertex $\mathbf{v} \in V$, which we will denote by $\deg(\mathbf{v})$, is the number of edges emanating from v. If deg($\mathbf{v}$) is an even integer, then we say that $\mathbf{v}$ is an even-degree vertex; otherwise, $\mathbf{v}$ is called an odd-degree vertex. In addition, an interior vertex $\mathbf{v}$ is called a singular vertex if (i) its degree is 4 and (ii) it is the intersection of two straight line segments. If e_{j-1}, e_j, e_{j+1} are three consecutive edges with a common vertex $\mathbf{v}$, then the edge e_j is called degenerate with respect to $\mathbf{v}$, provided that the two edges e_{j-1} and e_{j+1} are collinear. Now, we are ready to introduce the notion of type-O triangulation.

We denote by V_b the set of boundary vertices. Let $V_I = V \setminus V_b$. A vertex $\mathbf{u}$ will be called a *type-O vertex* of a triangulation Δ if $\mathbf{u}$ satisfies at least one of the following.

(a) $\mathbf{u}$ is a boundary vertex of Δ.
(b) $\mathbf{u} \in V_I$ with $deg(\mathbf{u}) = 4$.
(c) $\mathbf{u} \in V_I$ and $deg(\mathbf{u})$ is an odd integer.
(d) $\mathbf{u} \in V_I$ and there exists a vertex $\mathbf{v}$ of Δ that satisfies either (i) $\mathbf{v} \in V_I$ and $deg(\mathbf{v}) = 4$ or $deg(\mathbf{v}) =$ an odd integer, or (ii) $\mathbf{v} \in V_b$, such that $[\mathbf{u}, \mathbf{v}]$ is a nondegenerate edge of Δ with respect to $\mathbf{u}$.

We will use V_O to denote the collection of all type-O vertices in V.

DEFINITION 22.5 *A triangulation of V with only type-O vertices (i.e., $V = V_O$) is called a type-O triangulation.*

The reason for introducing the notion of type-O triangulations is the following (see [20]).

THEOREM 22.9
Any type-O triangulation Δ admits the optimal (5th) order of approximation from $S_4^1(\Delta)$.

As a consequence of the above theorem, we have

COROLLARY 22.1
If a triangulation Δ consists only of odd-degree interior vertices, then the spline space $S_4^1(\Delta)$ yields the optimal order of approximation.

To convert any triangulation to be a type-O triangulation, we introduce a so-called edge swapping algorithm. Every interior edge e of a triangulation Δ is the diagonal of a quadrilateral Q_e which is the union of two triangles of Δ with common edge e. Following [63], we say that e is a swappable edge if Q_e is convex and no three of its vertices are collinear. If an edge e of a triangulation Δ is swappable, then we can create a new triangulation by swapping the edge. That is, if $\mathbf{v}_1, \cdots, \mathbf{v}_4$ are the vertices of Q_e ordered

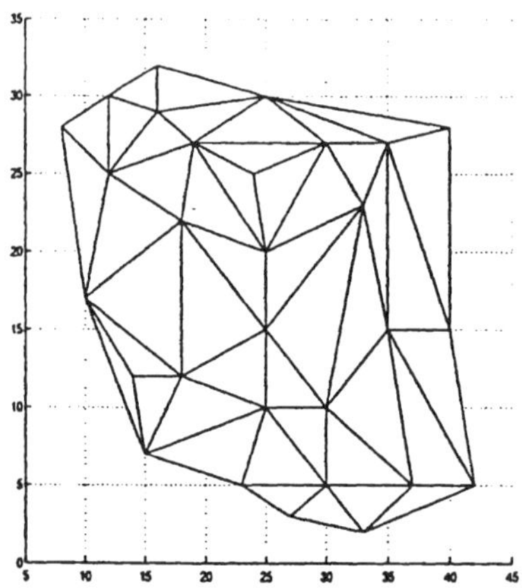

FIGURE 22.1
A triangulation Δ.

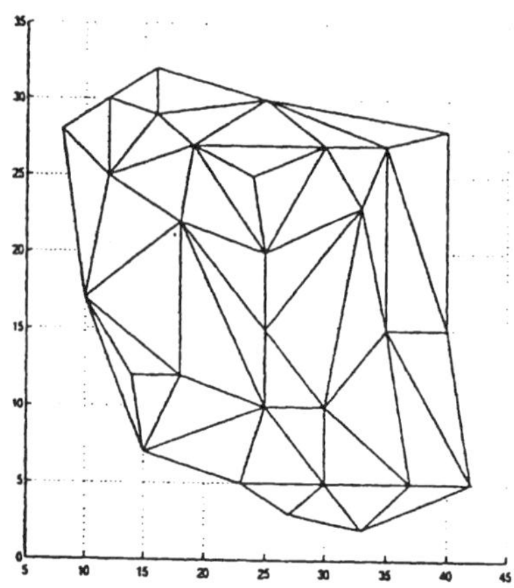

FIGURE 22.2
An optimal triangulation from Δ.

in the counterclockwise direction, and if e has endpoints $\mathbf{v}_1$ and $\mathbf{v}_3$, then the swapped edge has endpoints $\mathbf{v}_2$ and $\mathbf{v}_4$. Two vertices in Δ will be called *neighbors* of each other if they are the endpoints of the same edge in Δ. Hence, while $\mathbf{v}_1$ and $\mathbf{v}_3$ are neighbors in the original triangulation Δ, $\mathbf{v}_2$ and $\mathbf{v}_4$ become neighbors in the new triangulation after the edge e is swapped.

For any given set of sample sites, it is clear that, with the exception of those that are collinear, there is a triangulation with these sample sites as its only vertices. Let Δ be a triangulation associated with the given set V, and let V_O be the set of all type-O vertices in Δ. Set

$$\tilde{V} = V \setminus V_O.$$

If $\mathbf{u} \in \tilde{V}$, then $\mathbf{u}$ and all its neighbors with nondegenerate edges with respect to $\mathbf{u}$ must be even-degree vertices with $deg(\mathbf{u}) \geq 6$. We can see that, for every interior vertex $\mathbf{u}$ with $n := deg(\mathbf{u}) \geq 5$, there is a swappable edge $e \in E_{\mathbf{u}}$. Hence, there is at least one vertex $\mathbf{u}_i$ such that both $\angle \mathbf{u}_{i-1}\mathbf{u}\mathbf{u}_{i+1}$ and $\angle \mathbf{u}_{i-1}\mathbf{u}_i\mathbf{u}_{i+1}$ are less than π. Therefore, the quadrilateral $Q := [\mathbf{u}_{i-1}, \mathbf{u}_i, \mathbf{u}_{i+1}, \mathbf{u}]$ is convex; and hence, the edge $[\mathbf{v}, \mathbf{v}_i]$ is swappable.

Now we are ready to describe our *Edge Swapping Algorithm* for constructing a type-O triangulation $\widehat{\Delta}$, starting with any triangulation Δ.

Swapping Algorithm

Do while $(\tilde{V} \neq \emptyset)$
Pick any vertex $\mathbf{u}$ in $\tilde{V}$ and consider its neighbors.
Pick any neighbor v of $\mathbf{u}$ so that the edge $[\mathbf{u}, \mathbf{v}]$ is swappable.
Swap $[\mathbf{u}, \mathbf{v}]$, yielding a new edge $[\mathbf{u}', \mathbf{v}']$.
Form a subset of $\tilde{V}$ by deleting from $\tilde{V}$ all the neighbors $\mathbf{w}$ of $\mathbf{w}' := \mathbf{u}$, $\mathbf{v}$, $\mathbf{u}'$, or $\mathbf{v}'$, with $[\mathbf{w}, \mathbf{w}']$ being a nondegenerate edge with respect to $\mathbf{w}$.
Call this subset $\tilde{V}$.
Enddo

The new triangulation obtained by applying this Edge Swapping Algorithm is denoted by $\widehat{\Delta}$. It is clear that the triangulations Δ and $\widehat{\Delta}$ have the same number of triangles, singular vertices, interior and boundary vertices, and edges. Hence, it follows that

$$\dim S_4^1(\widehat{\Delta}) = \dim S_4^1(\Delta).$$

$\widehat{\Delta}$ is an optimal triangulation for C^1 quartic spline space. Combining the Edge Swapping Algorithm with Theorem 22.9, we have the following.

FIGURE 22.3
Type-1 triangulation $\Delta^{(1)}$.

THEOREM 22.10

Every finite set V of sample sites admits an optimal triangulation Δ, such that the C^1 quartic spline space $S_4^1(\Delta)$ has the optimal (fifth) order of approximation.

A MATLAB package which applies the Edge Swapping Algorithm to any triangulation on a finite set of vertices to construct a type-O triangulation of the sample points is described in [30]. The package includes a main function `swap.m` as well as subfunctions `consecv.m`, `delrow.m`, `findrow.m`, `findtri.m`, `nbors.m` and `trimesh2.m`, a modification of the MATLAB 5.0 function `trimesh.m`.

The `swap` program may be used to effectively implement the Edge Swapping Algorithm on any initial triangulation of sample points for which a triangulation admitting an optimal approximation with C^1 quartic splines is desired. Figure 22.1 shows a triangulation of some scattered sample points which has been defined in MATLAB using the x and y vectors and the `tri` matrix. This triangulation was transformed by `swap` to the type-O triangulation in Figure 22.2 with a single edge swap. The first non-type-O vertex encountered by `swap` was located at (25, 15). As the neighbors of this vertex were considered, the neighbor at (18, 22) was the first one found to form a swappable edge. The resulting edge swap was sufficient to create the type-O triangulation in the latter figure.

Recall that the three-directional mesh $\Delta^{(1)}$ is not optimal for C^1 quartic splines. Figure 22.4 depicts a type-O triangulation resulting from an application of `swap` to the sample $\Delta^{(1)}$ in Figure 22.3. Since `swap` considers the vertices of the initial triangulation in sequential order, the type-O

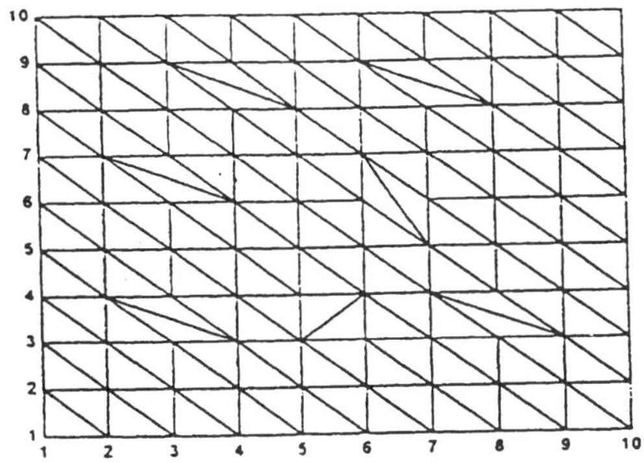

FIGURE 22.4
An optimal triangulation from $\Delta^{(1)}$.

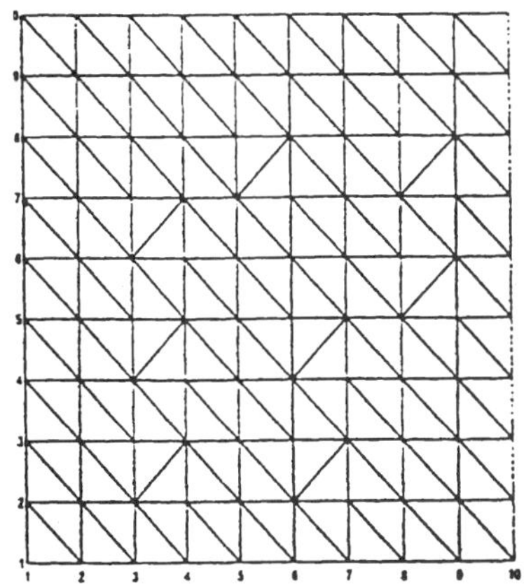

FIGURE 22.5
An optimal triangulation from $\Delta^{(1)}$.

triangulation returned by swap may be dependent on the order in which the vertices are defined in the x and y vectors. The result in Figure 22.5 was achieved by ordering the vertices from the bottom to the top of each column, beginning with the leftmost column. Figure 22.6 depicts a quite different type-O triangulation of this vertex set, where only the direction of the diagonals in the initial triangulation was changed. This illustrates how the output of swap on a particular vertex set may be changed, when desirable, by reordering the vertices or altering the initial triangulation.

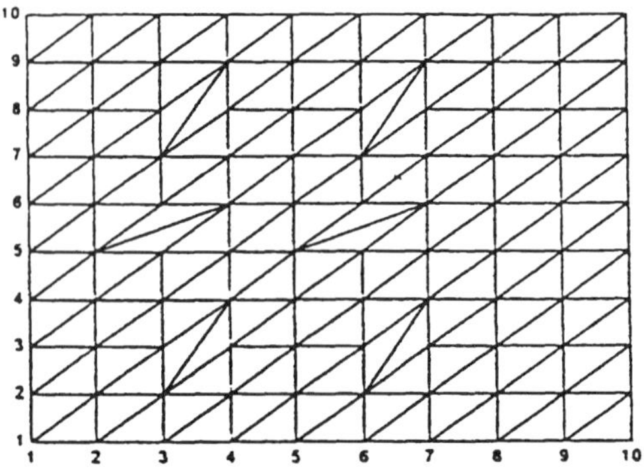

FIGURE 22.6
An optimal triangulation from $\Delta^{(1)}$.

22.4 Interpolation and Approximation using C^1 Cubic and C^1 Quadratic Splines

In this section, we would like to discuss some possibilities for finding the optimal triangulation for C^1 cubic or C^1 quadratic spline functions, and also some results on interpolation using cubic or quadratic spline elements.

As we can see from the previous sections, it was de Boor and Höllig who first applied B-net technique in [12] to study approximation order of C^1 cubic spline space. There they proved that $\Delta^{(1)}$ is not optimal for C^1 cubic splines. We've also seen that type-2 triangulations [46] and the triangulated quadrangulations [50] are optimal triangulations for C^1 cubic splines. From Theorem 22.6 we know that the type-2 triangulation is also optimal for C^1 quadratic splines. In general, the question of how to find an optimal triangulation for the bivariate spline space $S_k^r(\Delta)$ of C^1 cubic or C^1 quadratic is still open and will be a very challenging problem.

It is natural to consider the use of finite elements as interpolants, particularly since there is a large and sophisticated machinery available for handling them. For a piecewise polynomial interpolant to be differentiable globally, its polynomial degree must be at least 5 according to [68]. A well-known technique of reducing that degree is to subdivide the triangle into three subtriangles. Splitting a triangle about its centroid into three subtriangles and letting the polynomial degree be 3 gives rise to the widely used Clough-Tocher Scheme. Clough-Tocher splits were introduced in [24]. For a given triangulation Δ of a set Ω, we use Δ_{CT} to denote the Clough-Tocher (refinement) triangulation of Δ which is formed by connecting the centroid $\mathbf{v}_\tau$ of each triangle τ in Δ to the three vertices of τ. We can prove the following.

THEOREM 22.11
The Clough-Tocher triangulation Δ_{CT} is optimal for C^1 cubic spline space over Δ_{CT}.

Based on a similar consideration as in [19], we open the following.

Problem 1. Is there any local Clough-Tocher triangulation Δ_{LCT} so that it is optimal for C^1 cubic spline space over Δ_{LCD}?

Recall that a mixed three-directional mesh is optimal for C^1 quartic splines. It is certainly interesting to obtain a similar result for C^1 cubic splines.

Problem 2. Is there any mixed three-directional mesh which is optimal for C^1 cubic splines?

An alternative refinement approach is called the Powell-Sabin split, which splits each triangle into six subtriangles. More precisely, we give the following.

DEFINITION 22.6 *Given a triangulation Δ, the Powell-Sabin (refinement) triangulation Δ_{PS} is formed by connecting incenters of triangles of Δ and also by connecting incenters to central points of boundary edges for boundary triangles.*

It is well-known that Powell-Sabin triangulations can be used to study C^1 quadratic splines (see [58]). In [18], some computational schemes and optimization algorithms are introduced for interpolating discrete gridded data by C^1 quadratic spline surfaces that preserve the shape characteristics of the data. Also, some energy functionals are presented there for the characterization of optimal interpolants satisfying the required shape-preservation criteria.

A non-uniform type-1 (three-directional) mesh, we denote it by $\Delta_{MN}^{(1)}$, is constructed from a rectangular grid of a rectangular region $R = [a, b] \times [c, d]$, where

$$a = x_0 < \cdots < x_M = b, \ c = y_0 < \cdots < y_N = d$$

by drawing in all northeast diagonals. Similarly, $\Delta_{MN}^{(2)}$ will denote the non-uniform type-2 triangulation. R-representing scattered data using splines over triangulations $\Delta_{MN}^{(1)}$ and $\Delta_{MN}^{(2)}$ have attracted some attention. In [65], the following result was obtained.

THEOREM 22.12
For given data $f(\mathbf{v}_{i,j}), f'_x(\mathbf{v}_{i,j})$, $i = 0, 1, \cdots, n$, $j = 0, 1, \cdots, m$; $f'_y(\mathbf{v}_{i,n})$, $f''_{xy}(\mathbf{v}_{i,n})$, $i = 2, \cdots, m$; $f'_y(\mathbf{v}_{0,j})$, $f''_{xy}(\mathbf{v}_{1,j})$, $j = 1, \cdots, n$; and $f'_y(\mathbf{v}_{m,0})$,

$f'_y(\mathbf{v}_{m,1})$, $f'_y(\mathbf{v}_{0,0})$, then there is a unique $s \in S_3^1(\Delta_{MN}^{(1)})$ which satisfies the following interpolation conditions:

$$(s(\mathbf{v}_{i,j}), s'_x(\mathbf{v}_{i,j})) = (f(\mathbf{v}_{i,j}), f'_x(\mathbf{v}_{i,j})), i = 0, 1, \cdots, n, j = 0, 1, \cdots, m;$$

$$(s'_y(\mathbf{v}_{i,n}), s''_{xy}(\mathbf{v}_{i,n})) = (f'_x(\mathbf{v}_{i,n}), f''_{xy}(\mathbf{v}_{i,n})), i = 2, \cdots, m;$$

$$(s'_y(\mathbf{v}_{0,j}), s''_{xy}(\mathbf{v}_{1,j})) = (f'_y(\mathbf{v}_{0,j}), f''_{xy}(\mathbf{v}_{1,j})), j = 1, \cdots, m;$$

$$(s'_y(\mathbf{v}_{m,0}), s'_y(\mathbf{v}_{m,1}), s'_y(\mathbf{v}_{0,0})) = (f'_y(\mathbf{v}_{m,0}), f'_y(\mathbf{v}_{m,1}), f'_y(\mathbf{v}_{0,0})),$$

and also

$$\|f - s\|_\infty \leq \text{Const}|\Delta|^2[\omega(D^4 f, |\Delta|) + \|D^4 f\||\Delta|],$$

where $\|D^4 f\| = \|f\|_{4,\infty}$, and

$$\omega(D^4 f, |\Delta|) = \max_{0 \leq i \leq 4}\{\omega(\frac{\partial^4 f}{\partial x^i \partial y^{4-i}}, |\Delta|\}.$$

Many authors have also considered scattered data interpolation using bivariate splines over $\Delta_{MN}^{(2)}$. The following result is given by Ye [67] (see also [66]).

THEOREM 22.13

Let $f \in C^3([a, b] \times [c, d])$. Then there is a quadratic spline function $s \in S_2^1(\Delta_{MN}^{(2)})$ satisfying

$$\|f - s\| \leq 5\|D^3 f\||\Delta|^3 + \frac{|\Delta|^2}{16}[\omega_x(f_x^{(3)}, |\Delta|) + \omega_y(f_y^{(3)}, |\Delta|)].$$

Clearly, these interpolation results are not yet ideal since the order of approximation is not optimal due to either the triangulation structure or the interpolation scheme itself. It is interesting to seek both optimal triangulation construction and improved interpolation schemes for optimal order of approximation.

22.5 Stable Local Basis and Local Linear Independent Basis

For applications, one is required to construct an efficient scheme to achieve the full order $k+1$ of approximation. For this purpose, explicit bases for the spaces $S_k^r(\Delta)$ have been set up when $k \geq 3r+2$ and an approximation scheme using such bases to achieve the optimal order of approximation was discussed in [39]. However, the bases presented there, as well as in [22], are not stable. For spline space $S_k^r(\Delta)$, a basis $\{B_i\}_i^N$ of $S_k^r(\Delta)$ is said to be a local stable basis if each B_i is locally supported and there exist two positive constants C_1 and C_2, depending only on k and the smallest angle θ of the triangulation Δ, such that

$$C_1 \sup_\ell |a_\ell| \leq \left\| \sum_\ell a_\ell B_\ell \right\|_\infty \leq C_2 \sup_\ell |a_\ell|.$$

The construction of a local stable basis for a super spline space $S_k^{r,\mu}(\Delta)$ was presented in [21] (see also [40]). The subspace $S_k^{r,\mu}(\Delta)$ of super splines of smoothness r and degree $\leq k$ with enhanced smoothness order $\mu \geq r$ is defined as

$$S_k^{r,\mu}(\Delta) = \{s \in S_k^r(\Delta) : \; s \in C^\mu \text{ at each vertex of } \Delta\}.$$

In [21], the following result is obtained.

THEOREM 22.14

If $k \geq 3r+2$, then there is stable basis $\{B_\ell, \ell = 1, \cdots, N\}$ with $N = \dim(S_k^{r,\mu}(\Delta))$ for $S_k^{r,\mu}(\Delta)$. This basis is also local in the sense that, for any ℓ, there exists a vertex $\mathbf{u}$ such that the support of B_ℓ satisfies

$$\text{supp } B_\ell \subseteq \overline{St}^{\lfloor r/2 \rfloor + 1}(\mathbf{u}),$$

where the closed star of a vertex $\mathbf{v}$, denoted by $\overline{St}(\mathbf{v}) =: \overline{St}^1(\mathbf{v})$, is the union of all the triangles attached to $\mathbf{v}$, and the m-star of $\mathbf{v}$, denoted by $\overline{St}^m(\mathbf{v})$, is the union of all triangles that intersect with $\overline{St}^{m-1}(\mathbf{v})$, $m > 1$.

The star of a vertex is the set of triangles sharing that vertex. We call splines supported only on the star of a vertex *star-supported* splines.

Recently, Alfeld and Schumaker in [6] proved the following.

THEOREM 22.15
Suppose $r \geq 1$ and $k \leq 3r+1$. Then there are triangulations Δ for which $S_k^r(\Delta)$ does not have a star-supported basis.

The proof of this theorem is based on an analysis of spline spaces over a three-directional mesh.

Here, we can make a very easy argument to show the above theorem, based on the fact that a spline space over $\Delta^{(1)}$ will have full order of approximation provided that it has a locally supported basis.

By Theorem 22.5, the spline space $S_k^r(\Delta^{(1)})$ cannot have full order of approximation if $k \leq 3r+1$. Therefore, it cannot have a locally supported basis and, of course, there is no star-supported basis.

Some conjectures were made in [11] concerning approximation order of $S_k^r(\Delta)$, locally supported elements in $S_k^r(\Delta)$, and a local partition of unity in $S_k^r(\Delta)$. Here, local partition of unity means that a basis $\{B_i\}$ satisfies $\sum_i c_i B_i = 1$ with B_i nonnegative or $\{B_i\}$ is a local supported basis.

DEFINITION 22.7 *A basis $\{B_i\}_{i=1}^N$ of $S_k^r(\Delta)$ is said to be locally linearly independent (LLI) if for every $\tau \in \Delta$, the basis splines $\{B_i\}$ are linearly independent on τ.*

Local linear independence was first studied for the integer shift of a box spline (cf. [13], [27], and [45]). Usually, stability and local linear independence cannot hold simultaneously. An LLI basis for $S_k^r(\Delta)$ when $k \geq 3r+2$ was constructed recently in [29]. Using an LLI basis, a Hermite type interpolation scheme was presented in [28] for $S_k^r(\Delta)$, $k \geq 3r+2$, that possesses optimal approximation order in the same sense as in [21]. That is, the approximation constant does not depend on the geometric structure of Δ. The LLI basis construction is different from the B-net approach. The technique applied in [28] is based on nodal functionals, a common method in finite-element fields, and a so-called "weak interpolation" idea introduced in [57]. For comparison, we also mention that an alternative proof for the result of [21] can be found in [51].

22.6 Splines on Spheres and Natural Splines

A special but important and widely encountered problem arises when the data sites lie on a 2-dimensional surface embedded in $\mathbb{R}^3$. The most important instance of such a surface is a sphere. The problems of fitting data on a sphere arise in many areas, including, for example, geophysics and meteorology where the sphere is taken as a model of the earth. It is often unsatisfactory to project the surface into the plane. Instead, special methods have to be designed. Lawson [54], Renka [59], and Nielson and Ramaraj [56] independently propose schemes based on a triangulation of the surface of a sphere.

Very recently, the spaces of splines defined on triangulations lying on a sphere or on sphere-like surfaces have been discussed in [2] – [5]. These spaces arose out of a new kind of Bernstein-Beźier theory on such surfaces. A constructive theory for such spline spaces analogous to the well-known theory of polynomial splines on planar triangulations has been developed. Formulae for the dimension of such spline spaces, and locally supported bases for them, are given in [3]. Some applications of such spline spaces to fit scattered data on sphere-like surfaces are discussed in [5].

For many years people in the Computer Aided Geometric Design (CAGD) community believed that it was not possible to define barycentric coordinates on a spherical triangle. However, it was recognized in [4] that in fact there is a very natural way to define barycentric coordinates with respect to spherical triangles. It was later discovered that the same coordinates had been introduced and studied by Möbius more that 100 years ago.

The spherical spline space $S_k^r(\Delta)$ is the analog of the space of splines defined over a planar triangulation. As in the planar case, it is possible to identify the dimension of $S_k^r(\Delta)$, and construct locally supported bases for them, for all values of $k \geq 3r + 2$, see [3]. As shown in [5], the basic interpolation problem can solved as follows.

THEOREM 22.16
For given real values $\{f_i\}_{i=1}^n$ at the scattered points $\{\mathbf{v}_i\}_{i=1}^n$ on the unit sphere S, there is a spline function s defined on S such that

$$s(\mathbf{v}_i) = f_i, \ i = 1, \cdots, n.$$

The spherical triangulation Δ has vertices at the given points and the interpolant is local in the sense that the restriction of s to a triangle τ depends only on the data in that triangle.

Here, we need to mention a survey paper [31] which describes approaches to the interpolation and approximation data on the surface of a sphere. The authors discussed methods based on spherical harmonics, tensor-product spaces on a rectangular map of the sphere, functions defined over spherical triangulations, spherical splines, etc.

A generalized biharmonic spline interpolation scheme for data over a sphere was given in [34].

Before we discuss scattered data interpolation using bivariate natural splines, let's first recall the definition of the natural spline in the univariate setting. For points $a = x_0 < x_1 < \cdots < x_n < x_{n+1} = b$ and integer $m \geq 1$, we define

$$S_m(x_1, \cdots, x_n) = \{s \in C^{m-1}[a,b]; s|_{[x_i,x_{i+1}]} \in \pi_m, i = 0, 1, \cdots, n\}$$

the space of polynomial splines of degree m with n fixed knots $x_1, \cdots, x_n$.

DEFINITION 22.8 *A function $s \in S_{2r+1}(x_1, \cdots, x_n)$, where $r \geq 1$, is called a natural spline of degree $2r+1$ with knots $x_1, \cdots, x_n$, if*

$$s^{(j)}(a) = s^{(j)}(b) = 0, \ j = r+1, \cdots, 2r.$$

Clearly, a natural spline $s \in S_{2r+1}(x_1, \cdots, x_n)$ satisfies that s is a polynomial of degree $2r+1$ over each subinterval (x_i, x_{i+1}) for $i = 1, \cdots, n-1$, a polynomial of degree r over subintervals $[a, x_1)$ and $(x_n, b]$, and that $s \in C^{2r}[a,b]$.

Given a function $f \in C[a,b]$, the natural spline interpolation problem is to determine a natural spline $s \in S_{2r+1}(x_1, \cdots, x_n)$, $r \geq 1$, such that

$$s(x_i) = f(x_i), \ i = 1, \cdots, n.$$

It is well-known that if $n \geq r-1$, then the natural spline interpolation problem has a unique solution in $S_{2r+1}(x_1, \cdots, x_n)$. Furthermore, we have the following optimality properties for the natural interpolating spline function.

THEOREM 22.17

Let $n \geq r-1$, $f \in C^{r+1}[a,b]$ and suppose $s \in S_{2r+1}(x_1, \cdots, x_n)$ is the unique solution of the corresponding natural spline interpolation problem. Then for any natural spline $\widetilde{s} \in S_{2r+1}(x_1, \cdots, x_n)$, we have

(a)

$$\|f^{(r+1)} - s^{(r+1)}\|_2 \leq \|f^{(r+1)} - \widetilde{s}^{(r+1)}\|_2.$$

The equality holds if and only if $s - \widetilde{s} \in \pi_r$.

(b) for any $\mu \in I_f := \{u \in C^{r+1}[a,b]; \ u(x_i) = f(x_i), \ i = 1, \cdots, n\}$,

$$\|s^{(r+1)} - \widetilde{s}^{(r+1)}\|_2 \leq \|\mu^{(r+1)} - \widetilde{s}^{(r+1)}\|_2.$$

Let $Q = [a,b]$ and define

$$X := H^{r+1}(Q) = \{u(x); \ \frac{d^{r+1}u}{dx^{r+1}} \in L^2(Q), \ \frac{d^\alpha u}{dx^\alpha} \in AC(Q), \ \alpha = 0, \cdots, r\},$$

where $AC(Q)$ is the space of absolutely continuous functions over Q. From the property (b) in Theorem 22.17, if we choose $\widetilde{s} = 0$, we obtain

$$\|s^{(r+1)}\|_2 \leq \|\mu^{(r+1)}\|_2$$

for any $\mu \in I_f$. Therefore, a natural polynomial spline is the solution of the following problem: Given a function $f \in C[a, b]$, find a function $s(x) \in X$ satisfying the interpolation conditions:

$$s(x_i) = f(x_i), \ \ i = 1, \cdots, n$$

and

$$\int_a^b (s^{(r+1)}(x))^2 dx = \min_{u \in I_f} \int_a^b (u^{(r+1)}(x))^2 dx.$$

Let $Y := L^2(Q)$ and $t : X \mapsto Y$ be a linear operator defined by

$$t(u) = u^{(r+1)}(x) = \frac{d^{r+1}u(x)}{dx^{r+1}}.$$

Let $Z := \mathbb{R}^N$ be the N dimensional Euclidean space and $A : X \mapsto Z$ be an interpolation operator defined by

$$Au = (u(x_1), \cdots, u(x_N)).$$

Laurent [53] considered the following spline interpolation problem in the Hilbert space $H^{r+1}(Q)$: For a given N scattered data values $\{(x_i, f(x_i)), i = 1, \cdots, N\}$, find a function $s(x) \in X$ such that

$$\|t(s)\|_2^2 = \min_{u \in I_y} \|t(u)\|_2^2,$$

where $y = (f(x_1), \cdots, f(x_n))$ and $I_y = \{u \in X; \ Au = y\}$. The solution to this problem is also called a natural spline.

This natural spline interpolation problem in Hilbert space can be extended to the higher dimensional settings. For $R = [a, b] \times [c, d]$, let $X := H^{r,s}(R)$ denote the space

$$\{u(x, y); \ \frac{\partial^{r+s}u}{\partial x^r \partial y^s} \in L^2(R),$$

$$\frac{\partial^{\alpha+\beta}u}{\partial x^\alpha \partial y^\beta} \in AC(R), \alpha = 0, \cdots, r-1, \ \beta = 0, \cdots, s-1\},$$

where $AC(R)$ is the space of absolutely continuous functions over R. Let

$$Y = L^2(R) \times \prod_{\nu=0}^{s-1} L^2[a,b] \times \prod_{\mu=0}^{r-1} L^2[c,d]$$

and $T : X \mapsto Y$ be a linear operator defined by

$$T = t_0 \times \prod_{\nu=0}^{s-1} t_1^{(\nu)} \times \prod_{\mu=0}^{r-1} t_2^{(\mu)},$$

where

$$\begin{aligned} t_0(u) &= u^{(r,s)}(x,y) = \frac{\partial^{r+s} u(x,y)}{\partial x^r \partial y^s}; \\ t_1^{(\nu)}(u) &= u^{(r,\nu)}(x,c) = \frac{\partial^{r+\nu} u(x,y)}{\partial x^s \partial y^\nu}|_{y=c}, \ \nu = 0, ..., s-1; \\ t_2^{(\mu)}(u) &= u^{(s,\mu)}(a,y) = \frac{\partial^{s+\mu} u(x,y)}{\partial x^\mu \partial y^s}|_{x=a}, \ \mu = 0, ..., r-1. \end{aligned}$$

Let $Z = \mathbb{R}^N$ and $A : X \mapsto Z$ be an interpolation operator defined by

$$Au = (u(x_1, y_1), ..., u(x_N, y_N)).$$

Li and Guan [55] studied such a natural polynomial splines interpolation problem: Given N scattered data points and values $\{(x_i, y_i, z_i), i = 1, \cdots, N\}$, find a function $\sigma(x,y) \in X$ satisfying

$$\|T\sigma\|^2 = \min\{\|Tu\|^2, u \in X, Au = z\},$$

where, $z = (z_1, \cdots, z_N)$ and

$$\begin{aligned} \|Tu\|^2 = & \int\int_R (u^{(r,s)}(x,y))^2 dx dy + \sum_{\nu=0}^{n-1} \int_a^b (u^{(r,\nu)}(x,c))^2 dx, \\ & + \sum_{\mu=0}^{r-1} \int_c^d (u^{(\mu,s)}(a,y))^2 dy. \end{aligned}$$

The solution of this bivariate polynomial natural spline interpolation problem in Hilbert spaces $H^{r,s}(R)$ is called a bivariate polynomial natural spline.

The following results were obtained by Li and Guan in [55] (see also [17]) to characterize the null space of the operator T and to give a closed form of the bivariate natural spline function.

THEOREM 22.18
The null subspace of the operator T is

$$N(T) = P\langle r, s\rangle = \{u; u(x,y) = \sum_{i=0}^{r-1}\sum_{j=0}^{s-1} c_{ij}x^i y^j, c_{ij} \in \mathbb{R}\}.$$

THEOREM 22.19
The bivariate natural polynomial spline $\sigma(x,y)$ has the following explicit and closed-form expression:

$$\sigma(x,y) = \sum_{i=1}^{N} \lambda_i g_i(x,y) + \sum_{i=0}^{r-1}\sum_{j=0}^{s-1} c_{ij}x^i y^j$$

where $g_i(x,y) = G(x_i, y_i; x, y), i = 1, \cdots, N$, are $(2r-1, 2s-1)$ natural spline basis functions and

$$\begin{aligned} G(t,\tau;x,y) &= (-1)^{r+s}\frac{(t-x)_+^{2r-1}(\tau-y)_+^{2s-1}}{(2r-1)!(2s-1)!} + \\ &\sum_{\nu=0}^{s-1}(-1)^s \frac{(t-x)_+^{2r-1}(y-c)^\nu}{(2r-1)!\nu!}\left[\frac{(\tau-c)^\nu}{\nu!} - (-1)^{s-\nu}\frac{(\tau-c)^{2s-\nu-1}}{(2s-\nu-1)!}\right] + \\ &\sum_{\mu=0}^{r-1}(-1)^r \frac{(\tau-y)_+^{2s-1}(x-a)^\mu}{(2s-1)!\mu!}\left[\frac{(t-a)^\mu}{\mu!} - (-1)^{r-\mu}\frac{(t-a)^{2r-\mu-1}}{(2r-\mu-1)!}\right]. \end{aligned}$$

They also extended the results to the general k-dimension setting [17]. Guan [36] considered bivariate natural polynomial splines for smoothing or generalized interpolating of scattered data. In [37], a locally supported basis of bivariate natural polynomial splines was constructed. Recently, Guan and Hong in [38] constructed a locally supported basis of bivariate natural polynomial splines for scattered data on some lines (or for refinement grid points) to address a problem mentioned in [64]. For scattered data in a triangle, there were some similar discussions in [35] about the spline interpolations over triangulations.

Acknowledgment. This research was supported in part by a Research Development Grant #00-007/m from East Tennessee State University. This chapter has benefited from a thoughtful reading by Janice Huang and helpful comments by Lütai Guan.

References

[1] P. Alfeld, Scattered data interpolation in three or more variables, in: "*Mathematical Methods in Computer Aided Geometric Design*", T. Lyche and L.L. Schumaker (Eds.), pp. 1–33, Academic Press, Boston, 1989.

[2] P. Alfeld, M. Neamtu, and L. L. Schumaker, Circular Bernstein-B'ezier polynomials, in "*Mathematical Methods for Curves and Surfaces*", M. æhlen, T. Lyche, and L.L. Schumaker (Eds.), pp. 11–20, Vanderbilt University Press, 1995.

[3] P. Alfeld, M. Neamtu, and L. L. Schumaker, Dimension and local bases of homogeneous spline spaces, *SIAM J. Math. Anal.* **27** (1996), 1482–1501.

[4] P. Alfeld, M. Neamtu, and L. L. Schumaker, Bernstein-B'ezier polynomials on spheres and sphere-like surfaces, *Compt. Aided Geom. Design* **13** (1996), 333–349.

[5] P. Alfeld, M. Neamtu, and L. L. Schumaker, Fitting scattered data on sphere-like surfaces using spherical splines, *J. Comp. Appl. Math.* **73** (1996) 5–43.

[6] P. Alfeld and L. L. Schumaker, Non-existence of star-supported spline bases, *manuscript*, 1998.

[7] R.E. Barnhill, Representation and approximation of surfaces, In: *Mathematical Software III*, J.R. Rice ed., pp. 69–120, Academic Press, New York, 1977.

[8] W. Böhm, G. Farin, and J. Kehmann, A survey of curve and surface methods in CAGD. *Comput. Aided Geom. Design* **1** (1984), 1–60.

[9] J.H. Bramble and M. Zlamal, Triangular elements in the finite element method, *Math. Comp.* **24** (1970), 809–820.

[10] C. de Boor, B-form Basis, in: "*Geometric Modeling*", (G. Farin, Ed.), pp. 21–28, SIAM, Philadelphia, 1987.

[11] C. de Boor, Quasi-interpolants and approximation power of multivariate splines, in: "*Computations of Curves and Surfaces*", (Dahmen, Gasca, and Micchelli, Eds.), pp. 313–345, Kluwer (Dordrecht, Netherlands), 1990.

[12] C. de Boor and K. Höllig, Approximation order from bivariate C^1-cubics: A counterexample, *Proc. Amer. Math. Soc.* **87** (1983), 649–655.

[13] C. de Boor and K. Höllig, Bivariate box splines and smooth *pp* functions on a three-direction mesh, *J. Comput. Appl. Math.* **9** (1983), 13–28.

[14] C. de Boor and K. Höllig, Approximation power of smooth bivariate *pp* functions, *Math. Z.* **197** (1988), 343–363.

[15] C. de Boor and R. Q. Jia, A sharp upper bound on the approximation order of smooth bivariate *pp* functions, *J. Approx. Theory* **72** (1993), 24–33.

[16] C. K. Chui, "*Multivariate Splines*", CBMS Series in Applied Mathematics, no. **54**, SIAM, Philadelphia, 1988.

[17] C.K. Chui and L.T. Guan, Multivariate polynomial natural splines for interpolation of scattered data and other applications, In: "*Workshop on Computational Geometry*", A. Conte et al. eds., pp. 77–95, World Scientific Pub. Co., Singapore, 1993.

[18] C.K. Chui and T.X. He, Shape-preserving interpolation by bivariate C^1 quadratic splines, In: "*Workshop on Computational Geometry*", A. Conte, V. Demichelis, F. Fontanella, and I. Galligani (Eds.), pp. 21–75, World Scientific Pub. Co., Singapore, 1993.

[19] C. K. Chui and D. Hong, Construction of local C^1 quartic spline elements for optimal-order approximation, *Math. Comp.* **65** (1996), 85–98. MR **96d:**65023.

[20] C.K. Chui and D. Hong, Swapping edges of arbitrary triangulations to achieve the optimal order of approximation. *SIAM J. Numer. Anal.* **34** (1997), 1472–1482. MR **98h:**41036.

[21] C.K. Chui, D. Hong, and R.Q. Jia, Stability of optimal-order approximation by splines over arbitrary triangulations, *Trans. of Amer. Math. Soc.* **374** (1995), 3301–3318. MR **96d:**41012.

[22] C. K. Chui and M. J. Lai, On bivariate super vertex splines, *Constr. Approx.* **6** (1990), 399–419.

[23] J.F. Ciavaldini and J.C. Nedelec, Sur l'él'ement de Fraeijs de Veubeke et Sander, Rev. Francaise Automat. Infomat. Rech. Oper., *Anal. Numer.* **R2** (1974), 29–45.

[24] R. Clough and J. Tocher, Finite element stiffness matrices for analysis of plates in bending, In *Proc. of Conference on Matrix Methods in Structural Analysis*, Wright-Patterson Air Force Bases, 1965.

[25] W. Dahmen and C.A. Micchelli, Recent progress in multivariate splines, in "*Approximation Theory IV*", (C.K. Chui, L.L. Schumaker, and J.D. Ward, Eds.), pp. 27–121, Academic Press, New York, 1983.

[26] W. Dahmen and C.A. Micchelli, On the optimal approximation rates from criss-cross finite element spaces, *J. Comp. Appl. Math.* **10** (1984), 255–273.

[27] W. Dahmen and C.A. Micchelli, On the local linear independence of translates of a box spline, *Studia Math.* **82** (1985), 243–263.

[28] O. Davydov, G. Nürnberger, and F. Zeifelder, Bivariate spline interpolation with optimal order, *manuscript*, 1998.

[29] O. Davydov and L.L. Schumaker, Locally linearly independent bases for bivariate polynomial spline spaces, *manuscript*, 1999.

[30] B. Dyer and D.Hong, An algorithm for optimal triangulations on C^1 quartic spline approximation and MatLab implementation, *manuscript*, 1999.

[31] Greg Fasshauer and L.L. Schumaker, Scattered Data Fitting on the Sphere, in: *Mathematical Methods for Curves and Surfaces II*, (M. Daehlen, T. Lyche, and L. L. Schumaker eds.), 117–166, Vanderbilt Univ. Press, 1998.

[32] R. Franke, Scattered data interpolation: Tests of some methods, *Math. Comp.* **138** (1982), 181–200.

[33] B. Fraeijs de Veubeke, A conforming finite element for plate bending, *J. Solids Structures* **4** (1968), 95–108.

[34] L.T. Guan, Generalized biharmonic spline interpolations in a circular domain, *J. Engin. Math.* (Chinese), **1**(1987), 33–40.

[35] L.T. Guan, Multivariate spline interpolation to scattered data with continuous boundary conditions in a triangle, *J. of Numer. Math. & Appl.* (Chinese), **11:2** (1989), 11–23.

[36] L.T. Guan, Bivariate polynomial natural splines for smoothing or

generalized interpolating of scattered data. *J. of Numer. Math. & Appl.*(Chinese), **16:1** (1994), 1–12.

[37] L.T. Guan, Bivariate polynomial natural spline interpolation algorithms with local basis for scattered data. *Special issue on Wavelets and Approximation Theory*, D. Hong and M. Prophet (eds.), AMS conference, Austin, October 1999, *J. Comp. Anal. and Appl.*, submitted.

[38] L.T. Guan and D. Hong, On bivariate natural spline interpolation for scattered data over refined grid points, manuscript, 1999.

[39] D. Hong, "On bivariate spline spaces over arbitrary triangulations", *Master's Thesis*, Zhejiang University, Hangzhou, Zhejiang, China, 1987.

[40] D. Hong, "Construction of stable local spline bases over arbitrary triangulations for optimal order approximation", *Ph.D. Dissertation*, Texas A&M University, College Station, TX, 1993.

[41] D. Hong, Spaces of bivariate spline functions over triangulations, *Approx. Theory and Appl.* **7** (1991), 56–75. MR **92f:**65016.

[42] D. Hong, Recent progress on multivariate splines, in: "*Approximation Theory: In Memory of A.K. Varma*" (N.K. Govil, etc. Ed.), pp. 265–291, Marcel Dekker, Inc., New York, NY, 1998. (MR **99c**: 41022).

[43] D. Hong and R. N. Mohapatra, Optimal-order approximation by mixed three-directional spline elements, *Computers and Mathematics with Applications* (2000).

[44] R. Q. Jia, B-net Representation of Multivariate Splines, *Ke Xue Tong Bao* (*A Monthly Journal of Science*) **11** (1987), 804–807.

[45] R. Q. Jia, Local linear independence of the translates of a box spline, *Constr. Approx.* **1** (1985), 175–182.

[46] R.Q. Jia, *Lecture Notes on Multivariate Splines*, Department of Mathematics, University of Alberta, Edmonton, Canada. 1990.

[47] M. Laghchim-Lahlou and P. Sablonniere, Quadrilateral finite elements of FVS type and class C^r, *Numer. Math.* **70** (1995), 229–243.

[48] M.J. Lai, Approximation order from bivariate C^1-cubic on a four-directional mesh is full, *Comput. Aided Geometric Design* **11** (1994), 215-223.

[49] M.J. Lai, Scattered data interpolation and approximation by using bi-

variate C^1 piecewise cubic polynomials, *Comput. Aided Geom. Design* **13** (1996), 81–88.

[50] M.J. Lai and L.L. Schumaker, Scattered data interpolation using C^2 supersplines of degree six, *SIAM Numer. Anal.* **34** (1997), 905-921.

[51] M.J. Lai and L.L. Schumaker, On the approximation power of bivariate splines, *Advances in Comp. Math.* **9**(1998), 251–279.

[52] M.J. Lai and L.L. Schumaker, On the approximation power of splines on triangulated quadrangulations, *SIAM J. Numer. Anal.* **36** (1999), 143–159.

[53] P.J. Laurent, *Approximation et Optimization*, Hermann, Paris, 1972.

[54] C.L. Lawson, C^1 surface interpolation for scattered data on the surface of a sphere, *Rocky Mountain J. Math.* **14** (1984), 177–202.

[55] Y.S. Li and L.T. Guan, Bivariate polynomial natural spline interpolation to scattered data, *J. Comp. Math.* (Chinese), 1(1990), 135-146.

[56] G.M. Nielson and R. Ramaraj, Interpolation over a sphere based upon a minimum norm network, *Comput. Aided Geom. Design* **4** (1987), 41–57.

[57] G. Nürnberger, Approximation order of bivariate spline interpolation, *J. Approx. Theory* **78** (1996), 117–136.

[58] M.J.D. Powell and M.A. Sabin, Piecewise quadratic approximations on triangles, *ACM Trans. Math. Software* **3** (1977), 316–325.

[59] R.L. Renka, Interpolation of data on the surface of a sphere, *AMC Trans. Math. Software* **10** (1984), 417–436.

[60] G. Sander, Bornes superieures et inferieures dans l'analyse matricielle des plaques en flexion-torsion, *Bull. Soc. Royale Sciences Liege* **33** (1964), 456–494.

[61] L. L. Schumaker, Fitting surfaces to scattered data, in: *Approximation Theory II*, (G.G. Lorentz, C.K. Chui, and L.L. Schumaker eds.), pp. 203–268, Academic Press, New York, 1976.

[62] L.L. Schumaker, Recent progress on multivariate splines, in "*Mathematics of Finite Elements VII*", (J. Whiteman, Ed.), pp. 535–562, Academic Press, London, 1991.

[63] L.L. Schumaker, Computing optimal triangulations using simulated annealing, *Computer Aided Geometric Design* **10** (1993), 329–345.

[64] L.L. Schumaker, Alain Le Mèhauteé, and Leonardo Traversoni, Multivariate scattered data fitting, *J. Comp. Appl. Math.* **73** (1996), 1–4.

[65] Z. Sha, On interpolation by $S_3^1(\Delta_{MN}^{(1)})$, *J. Approx. Th. Appl.* **1:4** (1985), 1–18.

[66] Z. Sha, On interpolation by $S_2^1(\Delta_{MN}^{(2)})$, *J. Approx. Th. Appl.* **1:2** (1985), 71–82.

[67] M.D. Ye, Interpolations from $S_2^1(\Delta_{MN}^{(2)})$, *J. Comput. Math.* (Chinese), **4** (1986), 364–376.

[68] A. Ženišek, Interpolation polynomials on the triangle, *Numer. Math.* **15** (1970), 283-296.

Index